Engineering

ENCYCLOPEDIA/HANDBOOK

Materials,
Parts
AND
Finishes

Henry R. Clauser, Editor

MATERIALS CONSULTANT
FORMERLY EDITOR AND PUBLISHER, "MATERIALS ENGINEERING"

TECHNOMIC
PUBLISHING CO., INC.

Second Printing, 1984
Third Printing, 1987

Copyright ©Technomic Publishing Co., Inc., 1976
851 New Holland Ave., Box 3535
Lancaster, Pennsylvania 17604

Library of Congress Catalog Card No. 75-43010
ISBN 0-87762-189-6

Preface

It is the principal aim of this encyclopedic handbook to bring together in one volume concise information on the many thousands of materials, materials forms and parts, and finishes that are used in modern industry. The articles on materials and finishes provide information on their chemical or metallurgical nature, their engineering and service properties and their processing characteristics. The articles on materials parts and forms provide information on how they are produced, their size and shape characteristics and design capabilities.

The presentation and format of the book follow generally accepted encyclopedia style. The articles are arranged alphabetically and a comprehensive index is included at the end of the volume.

The identification of materials always presents some difficulties in a work of this size and nature. By and large, generic names are used wherever possible. However, in some instances, trade names are used for reasons of clarity or common usage. Whenever trade names have been used, the initial letter(s) is capitalized. Not only is this intended to inform the reader that the designation is a trade name, but also to indicate to all concerned that the word(s) has the legal status of a trade name.

Many of the articles in this book were published originally under the title, "Encyclopedia of Engineering Materials and Processes." A number of them have been modified by the editor in one way or another for this new volume, while others remain essentially unchanged. New articles have been added to fill gaps or to cover recently developed materials.

A list of original contributors to the encyclopedia is given at the end of the book (page 548). Because in a number of cases, the editor has made deletions or additions to the original contributions, the responsibility for the content of this work is his.

H. R. Clauser

Editor

A

ABS PLASTICS

ABS plastics are a family of opaque thermoplastic resins formed by copolymerizing acrylonitrile, butadiene and styrene (ABS) monomers. They are primarily notable for especially high impact strengths coupled with high rigidity or modulus. This differentiates them from the modified styrene plastics, in which increased impact strength is accompanied by decreased modulus.

At the present time two general types of ABS materials are commercially available: resin-rubber blends, and graft polymers wherein styrene and acrylonitrile monomer mixtures are polymerized in the presence of polybutadiene. In this latter process some of the resinous copolymer molecules become attached chemically (grafted) at active sites along the polybutadiene molecular spine, while a certain amount of free resinous copolymer is also formed. A broad range of mechanical properties is possible in the ABS plastics by varying the rubber-polymer ratio as well as styrene-acrylonitrile weight ratio.

Properties

The unique combinations of excellent impact strength with high mechanical strength and rigidity plus good long-term, load-carrying ability or creep resistance are characteristic of the ABS plastics family. In addition, all types of ABS plastics exhibit outstanding dimensional stability, good chemical and heat resistance, surface hardness, and light weight (low specific gravity). This available broad spectrum of useful properties coupled with the capability of maximizing one or more properties at will permits the production of plastic materials especially designed for the end application.

The individual commercially available ABS polymers span a wide range of mechanical properties. Most suppliers differentiate types on the basis of impact strength and fabrication method (extrusion or molding). Some compounds feature one particularly exceptional property, such as, high heat deflection temperature, abrasion resistance or dimensional stability. The natural color or pigmented materials are sold in the form of small dice with no further compounding required. A wide range of colored compounds is possible but the opaque nature of ABS materials does not permit the manufacture of a clear crystal type. Also available are undensified powders for those fabricators desiring to do their own compounding.

Property classification. For convenience, ABS plastics can be generally classified into five major types:

1. Medium impact: a hard, rigid, tough material used for appearance parts which must have high strength, good fatigue resistance, surface hardness and gloss.

2. High impact: used for similar applications, but where additional impact strength is required at some sacrifice in rigidity and hardness.

3. Extra-high impact: highest room temperature impact resistance, with a further decrease in rigidity, strength and hardness.

4. Low-temperature impact: tailored for high-impact strength at temperatures down to −40 F; strength, rigidity and heat resistance again suffer somewhat.

5. Heat resistant, high strength: provides maximum heat resistance (continuous use temperature of about 200 F; 264-psi heat distortion temperature of about 210-230 F, depending on molding conditions, etc.) with impact strength about comparable to that of high impact grades, but higher tensile and flexural strengths, modulus and hardness.

TYPICAL PROPERTIES OF ABS PLASTICS

Type	Medium Impact	Extra High Impact	Low Temp. Impact	Heat Resistant
Specific Gravity	1.05–1.07	1.01–1.06	1.02–1.04	1.06–1.08
Mod. Elast. in Tens., 10^5	3.3–4.0	2.0–3.1	2.0–3.1	3.5–4.2
Tens. Str., 1000 Psi	6.3–8.0	4.5–6.0	4–6	7–8
Elong., %	5–20	20–50	30–200	20
Hardness, Rockwell, R	108–115	85–105	75–95	107–116
Imp. Str., ft-lb/ in. notch	2–4	5–7	6–10	2–4
Max. Svc. Temp., F	180	180	180	220

1

Fabrication and forms

ABS plastics are readily formed by the various methods of fabricating thermoplastic materials—extrusion, injection molding, blow molding, calendering and vacuum forming. Molded products may be machined, riveted, punched, sheared, cemented, laminated, embossed or painted. Although the ABS plastics process easily and exhibit excellent moldability, they are generally more difficult flowing than the modified styrenes and higher processing temperatures are used. The surface appearance of molded articles are excellent and buffing may not be necessary.

Pipe. The ABS plastics as a whole are popular for extrusion and they offer a great deal for this type of forming. The outstanding contribution is their ability to be formed easily and to hold dimension and shape. In addition, very good extrusion rates are obtainable. Because ABS materials are processed at stock temperatures of 400 to 500 F, it is generally necessary to preheat and dry the material prior to extrusion.

The largest single ABS end product is plastic pipe, where the advantages of high long-term mechanical strength, toughness, wide-service temperature range, chemical resistance and ease of joining by solvent welding are used. ABS pipe has been used successfully for natural gas transmission and distribution, potable water distribution, irrigation and sprinkling systems, and chemical processing applications. Other commercial products made by extrusion include electrical conduit and ducts, textile bobbins, tubes, sleeves and numerous profiles.

Sheet. ABS sheet is manufactured by calendering or extrusion and molded articles are subsequently vacuum-formed. The hot strength of the ABS materials coupled with the ability to be drawn excessively without forming thin spots or losing embossing have made them popular with fabricators. The excellent mechanical strengths, formability and chemical resistance, particularly to fluorocarbons, are largely responsible for the rapid increase in the use of ABS for refrigerator and freezer door and box liners. In addition, ABS sheet is used to manufacture luggage, carrying cases, machine housings, automotive crash pads and headliners.

Moldings. The need for impact resistance and high mechanical properties in injection-molded parts has created a large use for ABS materials.

Advances in resin technology coupled with improved machinery and molding techniques have opened the door to ABS resins. Large complex shapes can be readily molded in ABS today. Some of the products that use ABS include: telephone handsets, vacuum cleaner housings and accessories, ladies' shoe heels, small appliance housings, office-machine housings, automobile armrests, automobile instrument clusters and dashboards, pipe fittings and valves, safety helmets and a wide variety of small component parts.

ACETAL PLASTICS

Acetals are highly crystalline plastics that were specifically developed to compete with zinc and aluminum die castings. The natural acetal resin is translucent white and can be readily colored. There are two basic types—a homopolymer (Delrin) and a copolymer (Celcon). In general, the homopolymers are harder, more rigid, and have higher tensile flexural and fatigue strength, but lower elongation. The copolymers are more stable in long-term high temperature service and have better resistance to hot water. Special types of acetals are glass filled, providing higher strengths and stiffness, and TFE filled, providing exceptional frictional and wear properties.

Acetals are among the strongest and stiffest of the thermoplastics. Their tensile strength ranges from 8,000 to about 13,600 psi, tensile modulus is about 500,000 psi, and fatigue strength at room temperature is about 5,000 psi. Acetals are also among the best in creep resistance. This combined with low moisture absorption (less than 0.4%) gives them excellent dimensional stability. They are useful for continuous service up to about 220 F.

Injection-molding powders and extrusion powders are the most frequently used forms of the material. Sheets, rods, tubes, and pipe are also available. Natural acetal resin is a translucent white. Colorability is excellent, and colored products have a high sparkle and brilliance.

The range of desirable design properties and processing techniques provides outstanding design freedom in the areas of (1) style (color, shape, surface texture and decoration), (2) weight reduction, (3) assembly techniques and (4) one-piece multifunctional parts (e.g. combined gear, cam, bearing and shaft).

TABLE 1–TYPICAL PROPERTIES OF ACETAL

Type	Homopolymer		Copolymer	
	Std.	20% Glass	Std.	25% Glass
Specific Gravity	1.43	1.56	1.41	1.61
Mod Elast. in Tens, 10^5	5.2	6–10	4.1	12.5
Tens. Str., 1000 psi	12	8.5	8.8	18.5
Elong. %	25	7	65	3
Hardness, Rockwell	M94	M90	M80	M79
Imp. Str., ft-lb in. notch	1.4	0.8	1.3	1.8
Max. Svc. Temp. F	195	195	220	220
Vol. Res., ohm-cm	10^{15}	5×10^{14}	10^{14}	1.2×10^{17}
Dielect. Str., v/mil	500	500	500	500

Properties

Typical properties of acetal resin are given in Table 1. Values are for molding grade resin. Extrusion grade resin has higher elongation and impact strength. Aside from these differences, the two grades have the same properties.

Strength and stiffness. Acetal resin is one of the strongest of all thermoplastics. Its load-bearing qualities in both tension and compression are outstanding when compared with other thermoplastic materials. At room temperature, tensile strength is 10,000 psi, and elongation is 16%, with no true yield point. At 212 F, there is a well-defined yield point at 6100 psi and considerable elongation, 350 to 500%. At 250 F, yield strength is still about 4000 psi.

Acetal resin is stiff as well as tough. Its high modulus of elasticity of 410,000 psi at room temperature is an important design property. More important is the fact that it retains much of its high modulus at elevated temperatures (90,000 psi at 250 F) and when submerged continuously in water.

The strength and stiffness of acetal resin are demonstrated by its use in pipe and in automotive, appliance, plumbing, hardware and machinery components—applications where the ability to withstand high loads with a minimum of deflection is a major design requirement.

Resilience. Inherent stiffness coupled with high strength makes acetal resin a leading choice for uses requiring the ability to resist loads with minimum deflection. It recovers well when the deflection-producing load is removed. In this respect acetal resin is somewhat like a steel spring: to deform it at all requires considerable force and if the deflection is not too great, re-covery is quick and complete. This recovery from loading suggests its use for snap-fit assemblies and light-duty, leaf type springs.

Impact and fatigue. Resistance to repeated impact characterizes the toughness of acetal resin. Toughness is retained at low temperatures due to the inherent chemical structure of the material rather than plasticizers. It has a notched Izod impact strength value of 1.2 at −40 F, just slightly less than its impact resistance of 1.4 ft-lb per in. notch at room temperature. The difference between notched and unnotched impact strength (20.5 ft-lb per in.) emphasizes the importance of eliminating all sharp corners (stress concentrators) in part design.

Acetal resin has the highest fatigue endurance limit of any commercial thermoplastic. It will withstand repeated cyclic stressing at 5,000 psi at room temperature almost indefinitely; the same is true with a 3,000 psi stress at 150 F and 100% relative humidity. Moisture or lubricating oils have little or no effect on fatigue strength.

This combination of good fatigue, impact and tensile properties (together with good frictional and wear characteristics) makes acetal resin suitable for gears, cams, bearings, bushings, pipe and many other structural uses. The material is being used for gear trains in motors, counters, instruments, photographic and communications equipment, and business machinery where certain parts must operate with only initial or no lubrication.

Friction. Acetal resin has a smooth, hard, glossy surface, somewhat slippery to the touch. Its unlubricated coefficient of friction is very low against steel (0.1 to 0.3); lubricated the value is even lower (0.05 to 0.1). An unusual property is the fact that its static and dynamic coefficients of friction against steel are the same. This reduces the possibility of stick-slip performance. Also, for any given lubrication conditions, the coefficient of friction against steel remains virtually unchanged over a wide range of temperatures (73 to 250 F), bearing loads (0.2 to 2,500 psi) and relative surface speeds (8 to 365 rpm).

Under light loads and low speeds, unlubricated acetal-to-acetal sliding friction is low. As loads and speeds increase, friction usually increases rapidly. Lubrication of acetal-to-acetal sliding surfaces will reduce or eliminate the problem.

Acetal resin is a desirable sliding-friction bearing material, because bearings can be used with little or no lubrication; fully lubricated, they can in some cases perform like metal bearings. Acetal also can be used for antifriction applications in which a portion of the molded object is designed to serve as its own bearing surface. Such parts include tabletop conveyor links, conveyor wheels, model train axles, idler gears, automotive brake pulleys, and a variety of hinged parts.

Abrasion resistance. Good abrasion resistance is provided by the hardness and good frictional properties of acetal resin. In general, where abrasion is a problem, parts perform as well as or better than parts made of most other engineering materials. Actual or simulated service testing is the best way to establish the suitability of a material where abrasion and wear are design requirements.

Practical evidence of wear resistance is the successful use of acetal resin in nonlubricated gears and bearings, table-top conveyor plates, pump impellers, toilet ballcock valves, spray nozzles and seed distribution plates of agricultural planters. In all of these uses there is some relative motion between the acetal and another material—either a solid, a liquid or a liquid carrying abrasive solids.

Although acetal resin is rather hard for a thermoplastic, it is soft compared to steel; hence, it will cut readily if forced across the sharp edge of a harder material. In cases where parts mate with metal parts, a very rough surface finish on the metal part may increase wear. A metal surface of 32 μin. rms, is a satisfactory finish; surfaces less than 16 μin. do not appear to provide appreciably increased benefits. For further improvements in abrasion resistance, parts should be lubricated or annealed in a suitable oil.

Electrical properties. Acetal resin is a good dielectric. Its dissipation factor and dielectric constant are low (0.0048 and 3.7 respectively) over a wide range of frequencies and temperatures. Volume resistivity is high (6 x 10^{14} ohm-cm) and does not change appreciably with moisture absorption. It has a dielectric strength of 500 volts per mil; further, it is nontracking.

These electrical properties, in conjunction with good mechanical properties, make acetal suitable for electrical applications where high strength, resistance to abrasion, dimensional stability and good high-temperature performance are also required and where self-extinguishing characteristics are not required. Arc chutes in high-voltage load break units and various switch components make use of this combination of electrical and mechanical properties.

Dimensional stability. Creep, or deformation with time under a continuously applied load, occurs with all plastics. Acetal resin, however, is particularly resistant to creep when compared with most other thermoplastics, especially at temperatures of 150 F and higher. High strength and creep resistance combine to make it suitable for applications where permanent holding power is required, as with self-tapping screws or interference fits. Other typical applications requiring these properties are submersible pump stages, bearings, aerosol containers, threaded plugs, and automobile instrument housings.

The coefficient of linear thermal expansion for acetal resin is 4.5×10^{-5} per F, roughly equal to that of other thermoplastics and about five times that of carbon steel.

In going from the dry, to the as-molded state, to equilibrium water content (0.22% absorption at 77 F and 50% relative humidity), dimensions increase by only 0.1%. When completely saturated with water at 77 F (0.9% absorption), the increase is only 0.4%. Water absorption rate is governed not only by the temperature and moisture environment, but by the thickness of the part as well. Thicker pieces require longer periods of time to reach equilibrium.

Allowance for dimensional changes due to creep, heat and moisture can be made in the part design by using more detailed data available from the material supplier.

Chemical resistance. Acetal resin has unusual resistance to organic solvents; it has no common solvents at temperatures under 160 F. This property, together with low moisture absorption, results in resistance to staining at room temperature by foods such as tea, catsup, margarine, and beet juice, as well as to industrial oils, greases, typewriter ribbon inks, and Hectograph duplicating masters. Exposure in the high temperature range may cause slight discoloration.

The polymer is not recommended for contact with strong acids, strong bases, or strong oxidizing agents. For applications in which weak acids, weak bases, or other questionable environments are involved, actual or simulated service tests should be made to determine the suitability of the material for the particular environment.

Tests have shown that acetal resin does not lose its properties when placed underground— nor is it attacked by fungi, rodents, or insects. These properties and solvent resistance contribute greatly to the utility of acetal resin pipe.

The permeability of acetal is exceptionally low to aliphatic, aromatic and halogenated hydrocarbons, alcohols and esters. Its permeability to some small polar molecules such as water, methyl alcohol and acetone is relatively high. Permeability characteristics make it a suitable material for containers, particularly the aerosol type, and pipe.

Heat resistance. Continuous service in air should usually be limited to about 180 F, intermittent service in air, to 250 F. This means it can be considered for under-the-hood automotive uses such as carburetors, fuel pumps and windshield-wiper drives. Continuous exposure to fresh water should be limited to 150 F. Intermittent exposure to higher temperatures is possible depending upon the service life requirements of the part. Since any physical deterioration of the polymer from heat is a function of the total exposure time, applications requiring heat resistance should be tested in actual or simulated service for the required life of the part to prove its suitability.

Weatherability. Acetal resin is affected by ultraviolet light. After prolonged exposure to intense ultraviolet radiation, its tensile and impact properties drop off significantly. Less severe and intermittent exposure usually results in chalking of the surface with little, if any, loss in mechanical properties. Weatherable compositions containing carbon black or ultraviolet filters have significantly improved retention of mechanical properties.

Processing

Acetal resin can be molded in standard injection molding equipment at conventional production rates. The processing temperature is around 400 F. Satisfactory performance has been demonstrated in full-automatic injection machines using multicavity molds. Successful commercial moldings point up the material's ability to be molded to form large-area parts with thin sections, heavy parts with thick sections, parts requiring glossy surfaces or different surface textures, parts requiring close tolerances, parts with undercuts for snap fits, parts requiring metal inserts, and parts requiring no flash.

It can also be extruded as rod, tubing, sheeting, jacketing, wire coating, or shapes on standard commercial equipment. Extrusion temperatures are in the range of 390 to 400 F.

Generally the same equipment and techniques for blow molding other thermoplastics work with acetal resin. Both thin-walled and thick-walled containers (aerosol type) can be produced in many shapes and surface textures.

Various sheet-forming techniques hold promise for use with acetal resin. Vacuum, pressure, and matched-mold forming have been accomplished experimentally.

Fabricating

Acetal resin is easy to machine (equal to or better than free-cutting brass) on standard production machine-shop equipment. It can be sawed, drilled, turned, milled, shaped, reamed, threaded and tapped, blanked and punched, filed, sanded and polished.

The material is easy to join, and offers wide latitude in the choice of fast, economical methods of assembly. Integral bonds of acetal-to-acetal can be formed by welding with a heated metal surface, hot gas, hot wire, or spin-welding techniques. High-strength joints result from standard mechanical joining methods such as snap fits, interference or press fits, rivets, nailing, heading, threads or self-tapping screws (thread cutting types ASA F or T). Where low joint strengths are acceptable, several commercial adhesives can be used for bonding acetal to itself and other substrates.

Acetal resin can be painted successfully with certain commercial paints and lacquers, using ordinary spraying equipment and a special surface treatment or followed by a baked top coat. Successful first-surface metallizing has been accomplished with conventional equipment and standard techniques for application of such coatings. Direct printing, process printing, and roll-leaf stamping (hot stamping) can be used for printing on acetal resin. Baking at elevated temperatures is required for good adhesion of the ink in direct and screen-process printing. In hot stamping, the heated die provides the elevated temperature. Printing produced by these processes resists abrasion and lifting by cellophane adhesive tape.

ACRYLIC PLASTICS

The most widely used acrylic plastics are based on polymers of methyl methacrylate. This primary constituent may be modified by copolymerizing or blending with other acrylic monomers or modifiers to obtain a variety of properties. Although acrylic polymers based on monomers other than methyl methacrylate have been investigated, they are not important as commercial plastics and are generally confined to uses in fibers, rubbers, motor oil additives, and other special products.

Standard acrylics

Poly(methyl methacrylate), the polymerized methyl ester of methacrylic acid, is thermoplastic. The method of polymerization may be varied to achieve specific physical properties, or the monomer may be combined with other components. Sheet materials may be prepared by casting the monomer in bulk. Suspension polymerization of the monomeric ester may be used to prepare molding powders.

Conventional poly(methyl methacrylate) is amorphous; however, reports have been published of methyl methacrylate polymers of regular configuration, which are susceptible to crystallization. Both the amorphous and crystalline forms of such crystallization-susceptible polymers possess physical properties which are different from those of the conventional polymer, and suggest new applications.

Service properties. Acrylic plastic is characterized by crystal clarity. good impact strength, formability, and excellent resistance to sunlight, weather and most chemicals. Maximum service temperature of heat resistant grades is about 200 F. Standard grades are rated as slow burning, but a special self-extinguishing grade of sheet is available. Although acrylic plastic weighs less than $\frac{1}{2}$ as much as glass, it has many times greater impact resistance. As a thermal insulator, it is approximately 20% better than glass. It is tasteless and odorless.

When poly(methyl methacrylate) is manufactured with scrupulous care, excellent optical properties are obtained. Light transmission is 92%; ultraviolet light transmission in the 360 millimicron range may be varied from 0 to 90% in special formulations; infrared light transmission is 92% at wavelengths up to 1100 millimicrons, falling irregularly to 0% at 2200 millimicrons; scattering effect is practically nil; refractive index is 1.49-1.50; critical angle is 42 deg; dispersion 0.008. Because of its excellent transparency and favorable index of refraction, acrylic plastic is often used in the manufacture of optical lenses. Superior dimensional stability makes it practicable to produce precision lenses by injection molding techniques.

In chemical resistance, poly(methyl methacrylate) is virtually unaffected by water, alkalies, weak acids, most inorganic solutions, mineral and animal oils, and low concentrations of alcohol. Oxidizing acids affect the material only in high concentrations. It is also virtually unaffected by paraffinic and olefinic hydrocarbons, amines, alkyl monohalides, and esters containing more than ten carbon atoms. Lower esters, aromatic hydrocarbons, phenols, aryl halides, aliphatic acids, and alkyl polyhalides usually have a solvent action.

TABLE 1 – TYPICAL PROPERTIES OF ACRYLICS

Type	Cast Sheet & Rods	General Purpose Moldings	High Impact Moldings
Specific Gravity	1.17–1.20	1.18–1.19	1.12–1.16
Mod. Elast in Tens., 10^5 psi	3.5–5.0	3.5–5.0	2.3–3.3
Tens. Str., 1000 psi	6–10	9.5–10.5	5.5–8.0
Elong., %	2–7	3–5	> 25
Hardness, Rockwell	M80–100	M80–103	R100–120
Imp. Str., Ft-lb/ in. notch	0.4	0.2–0.4	0.8–3
Max. Svc. Temp., F	150–225 ·	165–225	170–200
Vol. Res, ohm-cm	$> 10^{15}$	$> 10^{14}$	2×10^{16}
Dielect. Str., v/mil	450–525	400	400–500

Fabricating characteristics. When heated to a pliable state, acrylic sheet can be formed to almost any shape. The forming operation is usually carried out at about 290 to 340 F. Aircraft canopies, for instance, are usually made by differential air pressure, either with or without molds. Such canopies have been made from (1) monolithic sheet stock, (2) laminates of two layers of acrylic, bonded by a layer of polyvinyl butyral, and (3) stretched monolithic sheet. Irregular shapes, such as sign faces, lighting fixtures, or boxes, can be made by positive pressure forming, using molds.

Residual strains caused by forming are minimized by annealing, which also brings cemented joints to full strength. Cementing can be readily accomplished by using either solvent or polymerizable cements.

Acrylic plastic can be sawed, drilled and machined like wood or soft metals. Saws should be hollow ground or have set teeth. Slow feed and coolant will prevent overheating. Drilling can be done with conventional metal-cutting drills. Routing requires high speed cutters to prevent chipping. Finished parts can be sanded, and sanded surfaces can be polished with a high-speed buffing wheel. Cleaning should be by soap or detergent and water, not by solvent type cleaners.

Acrylic molding powder may be used for injection, extrusion, or compression molding. The material is available in several grades, having a varying balance of flow characteristics and heat resistance. Acrylics give molded parts of excellent dimensional stability. Precise contours and sharp angles, important in such applications as lenses, are achieved without difficulty, and this accuracy of molding can be maintained throughout large production runs.

Since dirt, lint and dust detract from the excellent clarity of acrylics, careful handling and storage of the molding powder are extremely important.

Acrylic plastics are available in a wide variety of clear, translucent, and opaque colors, as well as the conventional water-white colorless form. The sheet is furnished in sizes up to 120 by 144 inches. Some sheet sizes are supplied in thicknesses up to 4 in. Rods, tubes, and heavy castings are also manufactured.

Acrylic molding powders are available in ASTM (American Society for Testing and Materials) Grades 5, 6 and 8 (D-788 48T). They also are furnished in a wide variety of colors.

Applications. In merchandising, acrylic sheet has become the major sign material for internally lighted faces and letters, particularly for outdoor use where resistance to sunlight and weathering is important. In addition, acrylics are used for counter dividers, display fixtures and cases, transparent demonstration models of household appliances and industrial machines, and vending machine cases.

The ability of acrylics to resist breakage and corrosion, and to transmit and diffuse light efficiently has led to many industrial and architectural applications. Industrial window glazing, safety shields, inspection windows, machine covers, and pump components are some of the uses commonly found in plants and factories. Acrylics are employed to good advantage as the diffusing medium in lighting fixtures and large luminous ceiling areas. Dome skylights formed from acrylic sheet are an increasingly popular means of admitting daylight to industrial, commercial, and public buildings and even to private homes. Shower enclosures and decorated partitions are other typical applications. A large volume of the material is used for curved and flat windshields on pleasure boats, both inboard and outboard types.

Acrylic sheet is the standard transparent material for aircraft canopies, windows, instrument panels, and searchlight and landing light covers. To meet the increasingly severe service requirements of pressurized jet aircraft, new grades of acrylic have been developed that have improved resistance to heat and crazing. The stretching technique has made possible enhanced resistance to both crazing and shattering. Large sheets, edge-lighted, are used as radar plotting boards in shipboard and ground control stations.

In molded form, acrylics are used extensively for automotive parts, such as tail-light and stop-light lenses, medallions, dials, instrument panels and signal lights. The beauty and durability of molded acrylic products have led to their wide use for nameplates, control knobs, dials, and handles on all types of home appliances. Acrylic molding powder is also used for the manufacture of pen and pencil barrels, hairbrush backs, watch and jewelry cases, and other accessories. Large-section moldings, such as covers for fluorescent street lights, coin-operated phonograph panels and fruit juice dispenser bowls, are being molded from acrylic powder. The extrusion of acrylic sheet from molding powder is particularly effective in the production of thin sheeting for use in such applications as signs, lighting, glazing and partitions.

The transparency, strength, light weight, and edge-lighting characteristics of acrylics have led to applications in the fields of hospital equipment, medical examination instruments, and orthopedic devices. The use of acrylic polymers in the preparation of dentures is an established practice. Contact lenses are also made of acrylics. The embedment of normal and pathological tissues in acrylic for preservation and instructional use is an accepted technique. This has been extended to include embedment of industrial machine parts, as sales aids, and the preparation of various types of home decorative articles.

High-impact acrylics

High-impact acrylic molding powder has recently come into large volume, general use. It is used where toughness greater than that found in the standard acrylics is desired. Other advantages include resistance to staining, high surface gloss, dimensional stability, chemical resistance and stiffness.

High-impact acrylic is off-white and nearly opaque in its natural state and can be produced in a wide range of opaque colors. Several grades are available to meet requirements for different combinations of properties. Various members of the family have Izod impact strengths from about 0.5 to as high as 4 ft-lb per in. notch. Other mechanical properties are similar to those of conventional acrylics.

High-impact acrylics are used for hard service applications such as women's thin-style shoe heels, housings—ranging from electric razors to outboard motors, piano and organ keys, beverage vending machine housings and canisters—in short, applications where toughness, chemical resistance, dimensional stability, stiffness, resistance to staining, lack of unpleasant odor or taste, and high surface gloss are required.

ADHESIVES

Innumerable adhesives and adhesive formulations are available today. The selection of the proper type for a specific application can only be made after a complete evaluation of the design, the service requirements, production feasibility and cost considerations. Usually such selection is best left up to the adhesive supplier. Once he has been given the complete details of the application he is in the best position to select both the type and specific adhesive formulation.

Classification

Adhesives can be grouped into five classifications based on chemical composition. These are summarized in Table 1.

Natural. These include vegetable and animal-base adhesives and natural gums. They are inexpensive, easy to apply and have a long shelf life. They develop tack quickly, but provide only low strength joints. Most are water soluble. They are supplied as liquids or as dry powders to be mixed with water.

Casein-latex type is an exception. It consists of combinations of casein with either natural or synthetic rubber latex. It is used to bond metal to wood for panel construction and to join laminated plastics and linoleum to wood and metal. Except for this type, most natural adhesives are used for bonding paper, cardboard, foil and light wood.

Thermoplastic. These can be softened or melted by heating, and hardened by cooling. They are based on thermoplastic resins (including asphalt and oleoresin adhesives) dissolved in solvent or emulsified in water. Most of them become brittle at subzero temperatures and may not be used under stress at temperatures much above 150 F.

Being relatively soft materials, they have poor creep strength. Although lower in strength than all but natural adhesives and suitable only for noncritical service, they are also lower in cost than most adhesives. They are also odorless and tasteless and can be made fungus resistant.

Elastomeric. Based on natural and synthetic rubbers, elastomeric adhesives are available as solvent dispersions, latexes or water dispersions. They are primarily used as compounds which have been modified with resins to form some of the adhesive "alloys" discussed below. They are similar to thermoplastics in that they soften with heat, but never melt completely. They generally provide high flexibility and low strength, and without resin modifiers, are used to bond paper and similar materials.

Thermosetting. Based on thermosetting resins, they soften with heat only long enough for the cure to initiate. Once cured, they become relatively infusible up to their decomposition temperature. Although most such adhesives do not decompose at temperatures below 500 F, some are useful only to about 150 F. Different chemical types have different curing requirements. Some are supplied as two-part adhesives and mixed before use at room temperature; some require heat and/or pressure to bond.

As a group, these adhesives provide stronger

TABLE 1—ADHESIVES CLASSIFIED BY CHEMICAL COMPOSITION

Group →	Natural	Thermoplastic	Thermosetting	Elastomeric	Alloys [a]
Types Within Group	Casein, blood albumin, hide, bone, fish, starch (plain and modified); rosin, shellac, asphalt; inorganic (sodium silicate, litharge-glycerin)	Polyvinyl acetate, polyvinyl alcohol, acrylic, cellulose nitrate, asphalt, oleoresin	Phenolic, resorcinol, phenol-resorcinol, epoxy, epoxy-phenolic, urea, melamine, alkyd	Natural rubber, reclaim rubber, butadiene-styrene (GR-S), neoprene, acrylonitrile - butadiene (Buna-N), silicone	Phenolic-polyvinyl butyral, phenolic-polyvinyl formal, phenolic-neoprene rubber, phenolic-nitrile rubber, modified epoxy
Most Used Form	Liquid, powder	Liquid, some dry film	Liquid, but all forms common	Liquid, some film	Liquid, paste, film
Common Further Classifications	By vehicle (water emulsion is most common but many types are solvent dispersions)	By vehicle (most are solvent dispersions or water emulsions)	By cure requirements (heat and/or pressure most common but some are catalyst types)	By cure requirements (all are common); also by vehicle (most are solvent dispersions or water emulsions)	By cure requirements (usually heat and pressure except some epoxy types); by vehicle (most are solvent dispersions or 100% solids); and by type of adherends or end-service conditions
Bond Characteristics	Wide range, but generally low strength; good res to heat, chemicals; generally poor moisture res	Good to 150–200 F; poor creep strength; fair peel strength	Good to 200–500 F; good creep strength; fair peel strength	Good to 150–400 F; never melt completely; low strength; high flexibility	Balanced combination of properties of other chemical groups depending on formulation; generally higher strength over wider temp range
Major Type of Use [b]	Household, general purpose, quick set, long shelf life	Unstressed joints; designs with caps, overlaps, stiffeners	Stressed joints at slightly elevated temp	Unstressed joints on lightweight materials; joints in flexure	Where highest and strictest end-service conditions must be met; sometimes regardless of cost, as military uses
Materials Most Commonly Bonded	Wood (furniture), paper, cork, liners, packaging (food), textiles, some metals and plastics. Industrial uses giving way to other groups	Formulation range covers all materials, but emphasis on non-metallics—esp wood, leather, cork, paper, etc.	Epoxy-phenolics for structural uses of most materials; others mainly for wood; alkyds for laminations; most epoxies are modified (alloys)	Few used "straight" for rubber, fabric, foil, paper, leather, plastics films; also as tapes. Most modified with synthetic resins	Metals, ceramics, glass, thermosetting plastics; nature of adherends often not as vital as design or end-service conditions (i.e., high strength, temp)

[a] "Alloy," as used here, refers to formulations containing resins from two or more *different* chemical groups. There are also formulations which benefit from compounding two resin types from the same chemical group (e.g., epoxy-phenolic).
[b] Although some uses of the "nonalloyed" adhesives absorb a large percentage of the quantity of adhesives sold, the uses are narrow in scope; from the standpoint of diversified applications, by far the most important use of any group is the forming of adhesive alloys.

bonds than the other three groups discussed. Creep strength is good and peel strength is fair. Generally, bonds are brittle and have little resilience and low impact strength.

"Alloys." This term refers to adhesives compounded from resins of two or more different chemical families, e.g., thermosetting and thermoplastic, or thermosetting and elastomeric. In such adhesives the performance benefits of two or more types of resins can be combined. For example, thermosetting resins are plasticized by a second resin resulting in improved toughness, flexibility and impact resistance.

Structural adhesives

The term "structural adhesives" has come to mean those used to bond metals to other metals, to wood, or to rigid plastics, where bond strength is a critical requirement. Design concepts of structural adhesive bonding are covered elsewhere. But use of such adhesives has become so prevalent that a brief discussion of some of the most common structural adhesives is probably warranted.

Structural adhesives are generally of the "alloy"

or thermosetting type. Three of the most commonly used are the modified epoxies, neoprene-phenolics and vinyl formal-phenolics.

Modified epoxies. These types may be of either the room-temperature curing type which cure on addition of a chemical activator, or heat-curing type (either two-part, or one-part consisting of the resin and latent catalyst).

A primary advantage of the epoxies is that they are 100% solids and there is no problem of solvent evaporation after joining impervious surfaces. Other advantages include high shear strengths, rigidity, excellent self-filleting characteristics, and excellent wetting of metal and glass surfaces. Disadvantages include low peel strength, lack of flexibility and inability to withstand high impact. They are useful at temperatures up to about 350 F and slightly higher (heat-cured types).

Neoprene-phenolics. This "alloy" is characterized by excellent peel strength; shear strength is intermediate between the thermoplastic adhesives and the modified epoxies. They offer good flexibility and vibration absorption, and have good adhesion to most metals and plastics.

Neoprene-phenolics are usually moderately priced. Most are solvent types, but special two-part chemically curing types are sometimes used to obtain specific properties.

Vinyl formal-phenolic. The properties of this type fall between those of the modified epoxies and the thermoset-elastomer types. Vinyl formal-phenolics have good shear, peel, fatigue and creep strengths and good resistance to heat, though they soften somewhat at elevated temperatures.

They are supplied as solvent dispersions in solution or in film form. In the film form the adhesive is coated on both sides of a reinforcing fabric. Sometimes it is prepared by mixing a liquid phenolic resin with vinyl formal powder just prior to use.

Adhesive selection

In addition to chemical type, the form in which the adhesive is supplied is highly important. Today, adhesives are supplied as powders, liquids, light and medium pastes, heavy trowelable mastics, and in dry film form (both supported and unsupported). Each form provides specific benefits in terms of methods of application.

TABLE 1—TYPICAL PROPERTIES OF MOLDED ALKYDS

	Granular [a]	Putty [b]	Reinforced [c]
ELECTRICAL PROPERTIES			
Dielectric Constant			
60 cps	6.0–7.2	5.8–6.5	5.2–7
1 megacycle	4.2–5.7	4.4–5.0	4.0–5.9
Dissipation Factor			
60 cps	0.035–0.070	0.035–0.050	0.02–0.04
1 megacycle	0.014–0.040	0.014–0.035	0.015–0.022
Dielectric Strength, v/mil			
Short Time	350–400	350–40	350–400
Step-by-Step	300–350	300–350	300–350
Arc Resistance, sec	180	180	130–180
PHYSICAL PROPERTIES			
Specific Gravity	1.95–2.24	1.81–2.15	2.0–2.3
Compr Str, 1000 psi	16–24	18–24	24–31
Ten Str, 1000 psi	3–4.5	3–4	5.2–10
Flex Str, 1000 psi	7–10	8–10	12–17
Impact Str, ft-lb/in. notch	—	—	2–12
Water Absorption (24 hr), %	0.04–0.6	0.05–0.15	0.07–0.22
Max Service Temp, °F			
Long-Term	250–300	250	300
Short-Term	300–375	325	400

[a] Values cover range for General Purpose, Lower Water Absorption, and Lower Specific Gravity grades.
[b] Values cover range for Lower Specific Gravity and Lower Water Absorption grades.
[c] Values cover range for High Impact and Medium Impact grades.

ALKYD MOLDING COMPOUNDS

Alkyd molding compounds are thermosetting plastic materials characterized by excellent dimensional stability, electrical characteristics and moldability. They represent the introduction to the thermosetting plastics industry of the concept of low-pressure, high-speed molding. Typical properties are shown in Table 1.

As with other thermosetting materials, alkyd compounds are formulated with a resin (unsaturated polyesters, usually formulated with a diallyl phthalate crosslinking monomer), in conjunction with various inorganic mineral fillers, to produce the final molding compound. Depending upon the properties desired in the finished compound, the fillers used are clay, asbestos, fibrous glass, or combinations of these materials. The resulting alkyd compounds are characterized in their molding behavior by the following significant features: (1) no liberation of volatiles during the cure, (2) extremely soft flow, and (3) fast cure at molding temperatures.

While the general characteristics of fast cure

and low-pressure requirements are common to all alkyd compounds, they may be divided into three different groups which are easily discernible by the physical form in which they are manufactured. These groups are:

1. Granular types, which have mineral or modified mineral fillers, providing superior dielectric properties and heat resistance.

2. Putty types, which are quite soft and particularly well suited for low-pressure molding.

3. Glass fiber-reinforced types, which have superior mechanical strengths.

Each of these types has been developed to meet the requirements of certain fields of application. To accomplish this more specifically, within each of these three groups, several distinct types are available. Following is a more detailed description of the three general types.

Granular types

The physical form of materials in this group is that of a free-flowing powder. Thus, these materials readily lend themselves to conventional molding practices such as volumetric loading, preforming and high-speed automatic operations. The outstanding properties of parts molded from this group of compounds are high dielectric strength at elevated temperatures, high arc resistance, excellent dimensional stability and high heat resistance. Compounds are available within this group which are self-extinguishing and certain recently developed types display exceptional retention of insulating properties under high humidity conditions.

These materials have found extensive use as high-grade electrical insulation, especially in the electronics field. One of the major electronic applications for alkyd compounds is in the construction of vacuum tube bases, where the high dry insulation resistance of the material is particularly useful in keeping the electrical leakage between pins to a minimum. In the television industry, tuner segments are frequently molded from granular alkyd compound since electrical and dimensional stability are necessary to prevent calibration shift in the tuner circuits. Also, the granular alkyds have received considerable usage in automotive ignition systems where retention of good dielectric characteristics at elevated temperatures is vitally important.

Putty types

This group contains materials which are furnished in soft, puttylike sheets. They are characterized by very low pressure molding requirements (less than 800 psi), and are used in molding around delicate inserts and in solving special loading problems. Molders customarily extrude these materials into a ribbon of a specific size which is then cut into preforms before molding. Whereas granular alkyds are rather diversified in their various applications, putty has found widespread use in one major application: molded encapsulation of small electronic components, such as mica, polyester film, and paper capacitors; deposited carbon resistors; small coils; and transformers.

The purpose here is to insulate the components electrically, as well as to seal out moisture. Use of alkyds has become especially popular because of their excellent electrical and thermal properties which result in high functional efficiency of the unit in a minimum space, coupled with low-pressure molding requirements which prevent distortion of the subassembly during molding.

Glass fiber-reinforced types

This type of alkyd molding compound is used in a large number of applications requiring high mechanical strength as well as electrical insulating properties. Glass fiber-reinforced alkyds can be either compression or plunger molded permitting a wide variety of types of applications, ranging from large circuit breaker housings to extremely delicate electronic components.

Molding characteristics

Although full realization of the advantages of molding alkyds is best attained through the use of high-speed, lightweight equipment, nearly all modern compression presses are suitable for use with these materials. Since these compounds are quite fast curing, the press utilized in molding them should be capable of applying full pressure within approximately 6 to 8 sec after the mold has been charged. In selecting a press to operate a specific mold for alkyds, the following rule should prove useful: for average draws, the press should furnish about 1500 psi over the projected area of the cavity and lands for molding granular alkyds; about 800 psi for alkyd putty; and about 2000 psi for glass-reinforced alkyd.

Alkyd parts are in successful production in positive, semipositive, and flash molds. In general, the positive and semipositive types are recommended to obtain uniformly dense parts with lowest shrinkage. However, flash molds are frequently used with alkyd putty because of its low bulk factor. In any case, hardened, chromium-plated steel molds are recommended.

The resin characteristics of alkyd molding compounds are such that the material goes through a very low viscosity phase momentarily when heat and pressure are applied. This low viscosity phase makes possible the complete filling of the mold at pressures much lower than those required for other thermosets. Under ordinary conditions, alkyd materials have good release characteristics, and no lubrication is necessary to insure ejection from the mold.

ALLOY CAST IRONS

Alloyed cast irons are those to which alloying additions are intentionally added to impart or improve particular properties. The most common alloys used to increase tensile strength are as follows: copper (up to 1.5%), chromium (up to 1.0%), manganese (up to 1.0%), molybdenum (up to 1.0%), nickel (up to 3.0%), and vanadium (up to 0.35%).

These alloys are used in various combinations and operate by different mechanisms. Some dissolve in and strengthen ferrite. Others promote and/or refine pearlite. At the levels given above, these would be classified as "low-alloyed" irons.

Table 1 illustrates some of the strengths obtainable by alloying:

TABLE 1

	Iron 1	Iron 2	Iron 3
Alloy Content, %			
Nickel	1.7	3.2	—
Copper	—	—	1.06
Molybdenum	0.3	0.7	1.01
Chromium	—	0.2	—
Ten Str, 1000 psi	50	80	80
Mod of Elast, 10^6 psi	18	24	24
Compr Str, 1000 psi	145	200	200
Endurance Limit, 1000 psi	23	35	35
Bhn	260	350	350

Corrosion-resistant. There are three general types. The high-silicon type is widely used to

handle organic and inorganic acids. However, it has low resistance to shock and is very difficult to machine. High-chromium irons are

TABLE 2

	Typical Corrosion-Resistant Alloyed Irons		
	High-Silicon Type	High-Chromium Type	High-Nickel Type
Typical Composition, %			
Carbon	0.4–1.0	1.2–2.5	1.8–3.0
Silicon	14–17	0.5–2.5	1.0–2.7
Manganese	0.4–1.0	0.3–1.0	0.4–1.5
Nickel	—	0–5	14–30
Chromium	—	20–35	0.5–5.5
Copper	—	—	0–7
Molybdenum	0–3.5	—	0–1
Bhn	450–500	290–400	100–230
Ten Str, 1000 psi	13–18	30–90	25–45

tougher and more machinable, and reliable in many corrosive environments. High-nickel austenitic cast irons are tough and very serviceable. Typical examples of each are shown in Table 2.

Heat-resistant. These are useful for many applications such as furnace parts, stoker links, exhaust manifolds and other applications up to about 1800 F. The major alloys used to impart heat resistance are chromium, silicon, aluminum and chromium plus nickel. Typical examples of some of these are as follows:

TABLE 3

	Typical Heat-Resistant Alloyed Irons			
	High-Silicon Type	High-Chromium Type	High-Nickel Type	High-Aluminum Type
Typical Composition, %				
Carbon	1.6–2.5	1.8–3.0	1.8–3.0	1.3–1.7
Silicon	4.0–6.0	0.5–2.5	1.0–2.7	1.3–6.0
Manganese	0.4–0.8	0.3–1.5	0.4–1.5	0.4–1.0
Nickel	—	0–5	14–30	—
Chromium	—	15–35	1.7–5.5	—
Copper	—	—	0–7	—
Molybdenum	—	—	0–1	—
Aluminum	—	—	—	18–25
Bhn	170–250	250–500	130–250	180–350
Ten Str, 1000 psi	25–45	30–90	25–45	34–90
Max. Service Temp, °F	1650	1800	1740	1800

Other types. In addition to alloying for strength, resistance to corrosion, and resistance to heat, cast irons can be alloyed to adjust or control other properties such as thermal expansion, thermal conductivity, magnetic characteristics, electrical resistance and response to heat treatment.

ALLYLICS (DIALLYL PHTHALATE PLASTICS)

Allylics are thermosetting materials developed since World War II. Most important of these are diallyl phthalate (DAP) and diallyl isophthalate (DAIP), which are currently available in the form of monomers and prepolymers (resins).

Diallyl phthalate resin is the first all-allylic polymer commercially available as a dry, free-flowing white powder.

Chemically, it is a relatively linear partially polymerized resin which softens and flows under heat and pressure (as in molding and laminating), and crosslinks to a three-dimensional insoluble thermoset resin during curing. Unlike other thermosetting resins, diallyl phthalate undergoes an additional type of polymerization in the presence of a peroxide catalyst. No water or other volatile fragments are released during the curing reaction.

In preparing the resin, diallyl phthalate is polymerized to a point where almost all the change in specific gravity has taken place. Final cure, therefore, produces very little additional shrinkage.

Allylic resins enjoy certain specific advantages over other plastics, which make them of interest in various special applications. Allylics exhibit superior electrical properties under severe temperature and humidity conditions. These good electrical properties (insulation resistance, low loss factor, arc resistance, etc.) are retained despite repeated exposure to high heat and humidity. Diallyl phthalate resin is resistant to 300 to 350 F temperatures, while the diallyl isophthalate resin is good for continuous exposures up to 400 to 450 F temperatures. Allylic resins exhibit excellent post-mold dimensional stability, low moisture absorption, good resistance to solvents, acids, alkalis, weathering, and wet and dry abrasion. They are chemically stable, have good surface finish, mold well around metal inserts and can be formulated in pastel colors with excellent color retention at high temperatures.

Diallyl phthalate resin currently finds major use in (a) molding and (b) industrial and decorative laminates. Both applications utilize the desirable combination of low shrinkage, absence of volatiles and superior electrical and physical properties common to diallyl phthalate.

Molding compounds

A number of compounders currently market and sell molding powders compounded of diallyl phthalate resin, diallyl phthalate monomer and various fillers. These fillers include asbestos, "Orlon," "Dacron," cellulose, glass and other fibers. Inert fillers used include ground quartz and clays, calcium carbonate and talc.

Current uses of such molding compounds are largely in electrical and electronic parts, connectors, resistors, panels, switches, and insulators. Potential applications for molding compounds include appliance handles, control knobs, dinnerware and cooking equipment.

Laminates

Decorative laminates containing diallyl phthalate resin can be made from glass cloth (or other woven and nonwoven materials), glass mat or paper. Such laminates may be bonded directly to a variety of rigid surfaces at lower pressures (50 to 300 psi) than generally required for other plastic laminates. A short hot-to-hot cycle is employed, and pre s platens are always held at curing temperatures.

Diallyl phthalate laminates can, therefore, be used to give a permanent finish to high grade wood veneers (with a clear overlay sheet) or to upgrade low cost core materials (by means of a patterned sheet). It is no longer necessary to use high pressure equipment or cores of high internal strength to make commercial quality laminates.

In the area of industrial laminates, both diallyl phthalate and diallyl isophthalate resins are finding applications with woven and nonwoven fabrics of glass and synthetic fibers in making tubing, ducting, radomes, junction boxes, and aircraft and missile parts. Allylic resin-glass prepreg materials are now available from a number of suppliers. These can be stored for extended periods without special handling precautions and can be formulated to flow at either bag molding or matched metal molding pressures without altering the properties of the cured product. Only one reactive material and no diluents are present, thus simplifying processing.

Properties

Uncured diallyl phthalate resins are generally soluble in benzene, acetone, methyl ethyl ketone, dioxane, diallyl phthalate, many esters, chloroform and benzyl alcohol. They are partially soluble in methyl isobutyl ketone, toluene, xylene and acetic acid. They are insoluble in water, methanol, ethylene glycol, diethyl ether, carbon tetrachloride and aliphatic hydrocarbons.

TABLE 1–TYPICAL PROPERTIES OF ALLYLICS

Type	Diallyl Phthalate (DAP)	Diallyl Isophthalate (DAIP)	Allyl Diglycol Carbonate, Cast
Specific Gravity	1.27	1.26	1.32
Mod. Elast. in Tens., 10^5 psi	—	—	3.0
Tens. Str., 1000 psi	4	4.3	5–6
Elong., %	—	—	—
Hardness, Rockwell M	114–116	119–121	95–100
Imp. Str., ft-lb/ in. notch	0.2–0.3	0.2–0.3	0.2–0.4
Max. Svc. Temp., F	350	450	212
Vol. Res., ohm-cm	1.7×10^{16}	3.9×10^{17}	4×10^{14}
Dielect. Str., v/mil	450	420	290

Bulk density is 14-16 lbs./cu ft; specific gravity at 25 C, 1.267; shrinkage on cure, 0-1%; iodine No., 57; and softening range, 85-115 C.

Cured diallyl phthalate resins generally display the following properties: Rockwell M hardness, 114-116; Barcol hardness, 42-43; specific gravity at 25 C, 1.27; Izod impact strength, 0.2-0.3 ft-lb/in. of notch; heat distortion (264 psi stress), 155 C; compressive strength, 22,000-24,000 psi; flexural strength (ultimate), 7,000-9,000 psi; refractive index at 25 C, 1.571; water absorption (24 hrs at 25 C), 0.0-0.2%; modulus of elasticity in flexure, 0.5×10^6 psi; dielectric constant at 25 C, 3.6 (60 cps), 3.6 (1000 cps), 3.4 (1 mc); volume resistivity at 25 C, 1.8×10^{16} ohm-cm; surface resistivity at 25 C, 6×10^{15} ohms; dielectric strength at 25 C (step-by-step), 450 volts/mil; dissipation factor at 25 C, 1.0% (60 cps), 0.9% (1000 cps), 1.1% (1 mc); arc re-

sistance, 118 seconds; tensile strength, ultimate, 4,000 psi.

ALUMINIDES

True metals include the alkali and alkaline earth metals, beryllium, magnesium, copper, silver, gold and the transition elements. These metals exhibit those characteristics generally associated with the metallic state.

The B subgroups comprise the remaining metallic elements. These elements exhibit complex structures and significant departures from typically metallic properties. Aluminum, though considered under the B subgroup metals, is somewhat anomalous in that it exhibits many characteristics of a true metal.

The alloys of a true metal and a B subgroup element are very complex, since their components differ electrochemically. This difference gives rise to a stronger tendency toward definite chemical combination than to solid solution. Discrete geometrically ordered structures usually result. Such alloys are also termed electron compounds. The aluminides are phases in such alloys or compounds. A substantial number of beta, gamma and epsilon phases have been observed in electron compounds, but few have been isolated and evaluated.

Some early interest in the problem of providing molybdenum with a high temperature corrosion (oxidation primarily) resistant coating led to the use of aluminum as a diffusion layer coating. An investigation was directed toward the coating alloy itself.

The beta and gamma phases are found to exist in the molybdenum-aluminum alloy system, and though there is some disagreement concerning the precise composition of these phases, they are generally considered to correspond to the compositions $MoAl_5$ and $MoAl_2$ respectively.

Powder metallurgy techniques have proved feasible for the production of alloys of molybdenum and aluminum, provided care is taken to employ raw materials of high purity (99% +). Coarse granular HF-cleaned aluminum is preferred since this introduces a minimum of oxide into the resulting alloy. As the temperature of the compact is raised, a strong exothermic reaction occurs at about 640 C causing a rapid rise in temperature to above 960 C in a matter of seconds. Bloating occurs transforming the compact into a porous mass. Complete alloying, however, is accomplished. This porous, friable mass can be subsequently finely comminuted, repressed and sintered (or hot-pressed) to form a useful body quite uniform in composition. Vacuum sintering at 1300 C for one hour at 0.04 micron produces clean, oxide free metal throughout. Wet comminution prevents caking of the powder, and a pyrophoric powder can be produced by prolonged milling.

Hot pressing is a highly successful means of forming bodies of molybdenum and aluminum previously reacted as mentioned above. Graphite dies are employed to which resistance heating techniques are applicable. A parting compound is required since aluminum is highly reactive with carbon causing sticking to the die walls.

Oxidation resistance behavior is anomalous, improvement being observed at increasing temperatures.

Hot-pressed small bars exhibit modulus of rupture strengths ranging from 40,000 to 50,000 psi at room temperature decreasing to 38,000 to 40,000 psi at 1040 C.

Room temperature resistance to fuming nitric acids is excellent. A negative thermal coefficient of electrical conductivity obtains. To date only small test shapes have been fabricated.

Continued interest in aluminides disclosed other high melting point phases in metallic systems with aluminum. One system having a high potential to produce useful materials is the nickel-aluminum system. In the range of 14% to 34% aluminum by weight occur the two intermetallic phases Ni_3Al and $NiAl$. The alloys are prepared in a manner similar to that discussed for the molybdenum-aluminum system and by casting techniques. Compacts of $NiAl$ + 5% Ni, produced by powder techniques, exhibit room temperature modulus of rupture values of 144,000 psi, and heat shock resistance is considered excellent. Good resistance to red and white fuming nitric acids is obtained.

Additions of $MoAl_2$, TiC, Si, Ni + Cr_2O_3, B, Al_2O_3, TiB_2, W + Co, Ni + Mo_2C to $NiAl$ cause no improvement in properties, however, when titanium is added either as TiH_2 or sponge titanium, the high temperature strength properties are increased appreciably. This occurs at about 4% Ti.

The cast alloys of nickel and aluminum exhibit an increasing exothermic character with increasing aluminum content. Alloys with the exception of the 34% Al alloy and $NiAl$ (31.5% Al) produce sound castings free from excessive porosity. A room temperature tensile strength of approximately 49,000 psi is exhibited by the Ni_3Al compound with about 5% elongation. The 25% aluminum ($NiAl$) alloy has a room temperature tensile strength of 24,000 psi.

The 17.5% aluminum alloy which contained a mixture of the phases Ni_3Al and $NiAl$ exhibits the best strength properties. Room temperature tensile strength is approximately 80,000 psi with about 2% elongation, while at 815 C the tensile strength is about 50,000 psi. This alloy can be rolled at 1315 C, and possesses good thermal shock and oxidation resistance. Impact resistance is fair. The 100-hr stress-to-rupture strength at 734 C is 14,000 psi, but, it should be noted, creep rates are high.

Certain classes of permanent magnets contain CoAl, FeAl and NiAl together with certain binary and ternary combinations, and NiAl has shown some promise as a diffusion coating for certain alloys exhibiting poor oxidation resistance.

There has been some effort devoted to the use of intermetallic compounds as solid state devices, and it is in this area that the aluminides should receive attention, since their use as high-temperature structural materials leaves much to be desired.

ALUMINUM AND ALLOYS

Aluminum is the most abundant metallic element in the earth's crust. It is present as aluminum oxide (alumina) in the ore, known as bauxite, usually in amounts of from 40% to 60%. Aluminum is extracted from the ore by means of an electrolytic process in which the alumina is first dissolved in a molten electrolyte, and then an electric current is passed through it, causing the metallic aluminum to be deposited on the cathode.

Characteristics and properties

Aluminum and its alloys, like most ductile metals, have a face-centered cubic crystal structure. They offer an attractive combination of properties—light weight, excellent atmospheric corrosion resistance, high electrical and thermal conductivity, and ease of fabrication. Because their density is about one-third that of steel, some aluminum alloys have strength-to-weight ratios exceeding those of many steels and most nonferrous materials. Of the common conductor metals, aluminum is second only to copper on an equal volume basis, but twice that of copper on an equal weight basis. Due to the natural formation of a tough oxide film on the surface, aluminum materials resist corrosion in many atmospheric and chemical environments. The relatively good ductility and low hardness make aluminum easy to form and finish. In addition, aluminum is noted for its non-toxicity, non-magnetic and non-sparking properties.

As temperatures decrease below room temperature and down to −320 F, aluminum alloys increase in strength and toughness, thus making them attractive for cryogenic applications. However, all aluminum alloys lose strength rapidly as temperatures increase. Except for a few alloys, they are not used in products that are continuously exposed to temperatures above about 600 F.

Another attractive feature of aluminum is its workability. Its face-centered cubic crystal structure results in many slip systems, which make the metal very ductile and easily shaped. Consequently, aluminum is available in a great variety of cold rolled sheets and strip and in extruded forms in many intricate shapes. Foil can be obtained as thin as 0.00025 in. Hot working, however, is somewhat critical because the great decrease in strength at the elevated working temperatures can damage the metal.

Because aluminum metal has low strength, about 13,000 psi ultimate, alloying and/or cold working is required in order to develop useful properties for structural applications. The addition of alloying elements results in strength increases through two hardening mechanisms—solid solution strengthening and precipitation (age) hardening.

Types and conditions

Aluminum materials are classified in three major ways:

1. By composition they are divided into commercially pure aluminum and aluminum alloys.

2. By treatment they are divided into non-heat treatable and heat treatable alloys.

3. By production method they are divided into wrought and cast shapes or forms.

All wrought aluminum materials are identified by means of a four number system developed by the Aluminum Association. The first digit designates the alloy group, the second denotes alloy modifications or impurity limits, and the last two digits indicate the alloying elements or aluminum purity.

The numeral one identifies commercially pure aluminum grades. The last two digits give the aluminum content over 99% in hundredths of one percent. The second digit designates the degree of control over impurity limits.

The first digit designations for the various aluminum alloy groups are:

Copper	2
Manganese	3
Silicon	4
Magnesium	5
Magnesium and Silicon	6
Zinc	7
Other elements	8
Unused series	9

The last two of the four digits identify different alloys in the particular alloy group. The second digit indicates modifications of the original alloy. A zero denotes the original alloy. Integers 1 to 9 are assigned consecutively to alloy modifications. Experimental alloys carry the standard four digit number preceded by X until the alloy becomes standard.

Alloys in the 1, 3 and 5 series, which owe their strength to the hardening effects of manganese and magnesium, respectively, are the non-heat treatable classes of aluminum alloys. However, they can be strengthened by cold work. Alloys in the 2, 6 and 7 are the heat-treatable classes. Although some of the alloys in the 4 series are heat treatable, most of them are used only for brazing sheet and welding wire.

A standard system of letters and numbers is used to indicate the processed condition of aluminum alloys. The designation, called temper, follows the alloy identification number. The word temper, here, is used in a broader sense than in its connection with the heat treatment of steel. In relation to aluminum alloys the word temper covers the heat treated and/or production or fabricated condition of the alloy. The letter *H*, for example, indicates cold work (strain hardening), and the letter *T* indicates the heat treated condition. In addition, a number following the letter specifies the degree and/or combination of cold work and heat treatment. Thus, the temper designation also gives a rough indication of an alloy's mechanical properties.

Table 1 provides a summary of the aluminum alloy temper designation system.

TABLE 1—TEMPER DESIGNATIONS FOR ALUMINUM ALLOYS

Temper Desig-nation	Condition	Characteristics
F	As fabricated	Normal wrought or cast production operations. No guaranteed properties
O	Annealed, re-crystallized	Softest temper
H	Cold worked	Cold worked (strain hardened)
H1	Cold worked	Cold worked only
H12	Cold worked	Quarter-hard
H14	Cold worked	Half-hard
H18	Cold worked	Full-hard
H2	Cold worked and partially annealed	Work hardened and annealed to desired strength
H24	Cold worked and partially annealed	Half-hard
H28	Cold worked and partially annealed	Full-hard
W	Heat treated, unstable	Natural aging alloys (at room temperature after heat treatment)
T	Heat treated	Heat treated with or without supplementary work hardening.
T2	Annealed (castings only)	
T3	Annealed, cold worked, naturally aged	
T4	Solution treated, naturally aged	
T5	Artificially aged (no sol. treatment)	
T6	Solution treated, artificially aged	
T8	Solution treated, cold worked, artificially aged	
T9	Solution treated, artificially aged, cold worked	

The finish of as-supplied wrought aluminum materials is also designated by a standard system of letters and numerals. Finishes are classified into three major groups: mechanical finishes, chemical finishes, and coatings. Each of these groups is designated by a letter and spe-cific finishes in each group are identified by a two digit number. Furthermore, the sequence of operations leading to the final finish can be indicated by using more than designation. Table 2 gives a summary of the major finish designations.

TABLE 2—FINISH DESIGNATIONS FOR ALUMINUM ALLOYS

Mechanical Finishes

M1 – As fabricated
M2 – Buffed
M3 – Directional textured
M4 – Nondirectional textured

Chemical Finishes

C1 – Nonetched cleaned
C2 – Etched
C3 – Brightened
C4 – Chemical converstion coating

Coatings

A1 – Anodic
A2 – Protective and decorative (less than 0.4 mil)
A3 – Architectural class II (0.4–0.7 mil)
A4 – Architectural class I (0.7 mil and thicker)
R1 – Resinous and other organic coatings
V1 – Vitreous coatings–porcelain and ceramic types
E1 – Electroplated and other metal coatings
L1 – Laminated coatings

Non-heat treatable wrought alloys

The initial mechanical properties of non-heat treatable aluminum materials depend on the hardening and strengthening effect of elements such as manganese, silicon, iron and magnesium, which are present alone or in various combinations. These alloys can be further hardened by cold work. There are three major non-heat treatable groups as covered below (Table 3):

TABLE 3—PROPERTIES OF TYPICAL NON-HEAT TREATABLE WROUGHT ALUMINUM ALLOYS

Type & Cond.	Tens. Str., 1000 psi	Elong., % in 2 in.	Mod. Elast., 10^6 psi	Hard., Bhn	Elect. Res., microhm-cm
EC					
Annealed (O)	12	23	10	–	2.8
Hard (H19)	27	1.5	10	–	
1100					
Annealed (O)	13	40	10	23	2.92
Hard (H18)	24	10	10	44	3.02
3003					
Annealed (O)	16	35	10	28	3.45
Hard (H18)	30	7	10	55	4.31
5056					
Annealed (O)	42	35	10.3	65	5.90
Hard (H38)	60	15	10.3	100	6.40

1000 Series—Commercially pure aluminum, has a minimum aluminum content of 99%. The major difference between the grades in the series is the level of two impurities—iron and silicon. The electrical conductivity grade, designated EC, with a 99.45% aluminum content, is used widely as an electrical conductor in the form of wire and bus bars. The 1000 series alloys are especially noted for high electrical and thermal conductivity and excellent corrosion resistance. In the annealed condition they are relatively soft and weak. However, an annealed strength of 13,000 psi can be doubled by cold work, with a sacrifice in ductility. Typical uses of commercially pure aluminum grades are sheet metal work, foil, spun ware, chemical equipment and railroad tank cars.

3000 Series—Manganese, up to about 1.2%, is the major alloying element in this series. It provides a moderate improvement over the 1000 series in mechanical properties without significant loss of corrosion resistance and workability. One of the alloys, 3004, also contains magnesium, resulting in an additional strength improvement. Typical applications of this series: housings, cooking utensils, sheet metal work and storage tanks.

5000 Series—Magnesium is the main alloying element in this series of non-heat treatable alloys and normally ranges from less than 1% up to about 5%. The strongest of the non-heat treatable alloys, their strength increases with magnesium content. However, high magnesium grades are more difficult to hot work and are susceptible to stress corrosion about 150 F. Because of the presence of lower density magnesium, the density of these alloys is less than that of pure aluminum. Also, they are particularly resistant to marine atmospheres and various types of alkaline solutions. Typical applications are marine hardware, building hardware, appliances, welded structures and vessels, and cryogenic equipment.

Heat treatable wrought alloys

Heat treatable aluminum alloys develop their final mechanical properties through the solid solution hardening effects of alloying elements as well as by second-phase precipitation. Cold working is also sometimes employed to obtain optimum properties.

Two treatments are generally involved in the hardening process—solution heat treatment and age or precipitation hardening. The solution heat treatment cycle consists of: 1) heating the alloy up to between 800 F and 1000 F and holding it at temperature to allow the alloying element to go into solid solution; 2) quenching the alloy rapidly to hold the alloying element in

solution. By means of this treatment, the hardening element is dispersed uniformly throughout the material. Age hardening, performed after solution heat treatment, is done either at room temperature over an extended period (natural aging) or at a temperature somewhere between 240 F and 450 F over a shorter period (artificial aging). This treatment further increases strength and hardness by the precipitation of hard, second phase particles. The second phase, which is an intermetallic compound, appears as a fine network in the grain boundaries. Although loss of ductility is not great, the second phase penetrates the surface aluminum oxide layer and consequently lowers corrosion resistance.

2000 Series—This is the oldest and probably the most widely used aluminum alloy series. The principal alloying element, copper, combines with the base metal, aluminum, during heat treatment to form the hardening intermetallic compound, $CuAl_2$. Copper content ranges from about 2% to around 6%.

TABLE 4—PROPERTIES OF TYPICAL HEAT-TREATABLE WROUGHT ALUMINUM ALLOYS

	Tens. Str., 1000 psi	Elong., % in 2 in.	Mod. Elast., 10^6 psi	Hard., Bhn	Elect. Res., microhm-cm
2014					
Anneal. (O)	27	18	10.6	45	3.45
Heat Treat. (T6)	70	13	10.6	135	4.31
2024					
Anneal. (O)	27	20	10.6	47	3.45
Heat Treat. (T3)	70	18	10.6	120	5.75
6061					
Anneal. (O)	18	27	10	30	3.8
Heat Treat. (T6)	45	12	10	95	4.31
6262					
Anneal. (O)	–	–	10	–	3.9
Heat Treat. (T9)	58	10	10	120	–
7075					
Anneal. (O)	33	17	10.4	60	–
Heat Treat. (T6)	83	11	10.4	150	5.7

Until the introduction of the 7000 series, copper-aluminum alloys were the highest strength aluminum alloys. They possess relatively good ductility but are more susceptible to corrosion than other aluminum alloys, particularly in the aged condition. Also, except for 2014 and 2219, they have limited weldability. Because of the corrosion problem, the sheet forms are often clad with commercially pure aluminum or special aluminum alloys.

Because they can be heat treated to strengths up to 75,000 psi, these aluminum-copper alloys are used in structural applications. Alloys 2014 and 2024, the best known of the series, are widely used in the aircraft industry. Alloy 2014 is primarily a forging grade.

Alloy 2024 was the first high strength light weight alloy.

6000 Series—Alloys in this series contain silicon and magnesium in approximately equal amounts—up to about 1.3%. Small amounts of other metals, such as copper or chromium, or lead are also present in some of the alloys to provide improved corrosion resistance in the aged condition, or to increase strength or electrical conductivity. As a group, alloys in this series have the lowest strengths of the heat treatables, but possess good resistance to industrial and marine atmospheres. Typical uses include screw machine parts, moderate strength structural parts, furniture, bridge railing, and high strength bus bars.

7000 Series—This series, the newest standard group, has the highest strength of any aluminum alloy group. High strength is obtained by the addition of zinc from 1% to 7.5%, and magnesium in amounts of 2.5% to 3.3%. Chromium and copper also contribute added strength, but tend to lower weldability and corrosion resistance. Alloy 7075, the most widely used grade, and with a heat treated (T6) tensile strength of about 80,000 psi, has many aircraft structural applications. Other lesser used grades, such as alloy 7178 (T6), have tensile strengths from about 88,000 to over 90,000 psi.

Cast alloys

Although there are a great many cast alloys available, less than fifty are in common use. Some are used for all three major types of casting—sand, permanent mold and die casting—while others were specifically developed for one of the casting processes. Both heat treatable and non-heat treatable alloys are available. Those alloy compositions that do not respond to heat treatment are identified by an *F* following the alloy designation number or by ommission of a suffix. The heat treatable alloys, which can be solution heat treated and aged similar to wrought heat treatable grades, carry the temper designations T2, T4, T5, T6 or T7. Die castings are seldom solution heat treated because of the danger of blistering.

As is true of most cast metals, the mechanical properties of aluminum castings are considerably lower than those of the wrought forms. With one or two exceptions, tensile strengths in the heat treated condition do not exceed about

50,000 psi. Ductility and hardness are also lower.

Aluminum-Silicon—These are the most widely used aluminum casting alloys, primarily because of their excellent castability. They find considerable application in marine equipment and hardware because of high resistance to salt water and saline atmospheres. They are also used for decorative parts because of their resistance to natural environments and ability to reproduce detail.

TABLE 5—PROPERTIES OF TYPICAL ALUMINUM CASTING ALLOYS

Type & Condition	Tens. Str., 1000 psi	Elong., % in 2 in.	Hard., Bhn	Elect. Res., microhm-cm
443.0 (Si-Al) As-cast	20	9	42	4.66
A413.0 (Si-Al) As-cast	35	3.5	80	5.56
A380 (Si-Cu-Al) As-cast	47	4	80	7.50
319 (Cu-Si-Al) As-cast	30	2.2	75	6.40
Sol. Treat. & Aged	38	2.5	88	6.40
355 (Cu-Si-My-Al) Sol. Treat. & Aged	39	3.5	85	4.80

Aluminum-Copper-Silicon—Aluminum-copper compositions, the earliest aluminum castings alloys, have been largely replaced by those containing additions of silicon. Copper increases strength hardness, and machinability, and the silicon provides excellent casting properties. These alloys are especially suited to the production of castings of intricate design having large differences in section thickness or requiring pressure tightness.

Aluminum-Magnesium-Silicon—Aluminum-magnesium alloys are the most corrosion resistant of the casting alloys. Unfortunately, they are difficult to cast, and controlled melting and pouring practices are required. Castability is improved by the addition of silicon. Heat treatment produces mechanical properties that make them attractive for such applications as automotive and aircraft parts.

Other Alloys—Aluminum-zinc-magnesium alloys age at room temperature to provide relatively high tensile strength. They have good machinability and corrosion resistance, but are not recommended for elevated temperatures. Aluminum-tin alloys, developed primarily as bearing alloys, have high load carrying capacity and fatigue strength. Cast in sand or permanent molds, they are used for connecting rods and crankcase bearings.

Aluminum powder

Roughly a half dozen different aluminum alloy powders are being used in the production of powder metallurgy P/M parts. The major alloying elements used include copper up to 4%, magnesium up to 1%, zinc and silicon from 0.10 to 1%. One alloy contains 5.6% zinc. Aluminum P/M parts offer natural corrosion resistance, light weight, good electrical and thermal conductivity. Strengths range from about 15,000 to 50,000 psi depending on composition, density and heat treatment. In general, average fatigue limits are about half those of the wrought alloys. This is directly related to the lower density of P/M parts. Corrosion resistance, however, is not markedly affected by the porosity. Still relatively new in production use, perhaps the largest use of aluminum powder is for oil impregnated sleeve and spherical bearings.

ANODIC COATINGS

Anodic oxidation or anodizing is the common commercial term used to designate the electrolytic treatment of metals in which stable oxide films or coatings are formed on surfaces. Aluminum and magnesium are anodized to the greatest extent on a commercial basis. Some other metals such as zinc, beryllium, titanium, zirconium and thorium can also be anodized to form films of varying thicknesses but they are not used to any large extent commercially.

Aluminum

It is a well-known fact that a thin oxide film forms on aluminum when it is exposed to the atmosphere. This thin, tenacious film provides excellent resistance to corrosion. The ability of aluminum to form an adherent oxide film led to the development of electrochemical processes to produce thicker and more effective protective and decorative coatings.

Anodic coatings can be formed on aluminum in a wide variety of electrolytes utilizing either a.c. or d.c. current or combinations of both. Electrolytes of sulfuric, oxalic and chromic acids are considered to be the most important commercially. Other electrolytes such as borates, citrates, chromates, bisulfates, phosphates and carbonates can also be used for specialized applications.

Anodic coatings produced in a sulfuric acid type of electrolyte are generally translucent, can be produced with a wide range of properties by varying operating techniques, and have a greater number of pores which are smaller in diameter than other anodic coatings.

Coatings obtained by the chromic acid anodic process are much thinner than those produced by the sulfuric acid process based upon the same time of anodic oxidation. They require higher formation voltages than sulfuric acid anodic coatings and have a different structure. In general, there are fewer pores but they are larger in diameter than those in coatings formed in a sulfuric acid electrolyte. Although the coatings are thin, they provide high resistance to corrosion because of the presence of chromium compounds in combination with their relatively thick barrier layer. Because the conventional anodic coatings produced in a chromic acid electrolyte are thin, they have low resistance to abrasion but a high degree of flexibility. The anodic coatings produced in a chromic acid electrolyte have an opaque, slightly iridescent appearance and may be colored with organic dyestuffs such as the anodic coatings produced in sulfuric acid electrolytes.

The chromic acid anodic process is critical with respect to alloy composition. In general, the process is not recommended for aluminum alloys containing more than 7% silicon or 4.5% copper. Alloys containing over 7% total alloying elements are also not recommended. It is difficult to form films on these alloys, particularly on casting alloys. Anodic solution (surface attack) will generally occur unless the processing conditions are controlled very carefully.

On most aluminum alloys, anodic coatings of this type do not require sealing. However, sealing in a boiling dichromate or a dilute chromic acid solution produces coatings with the best resistance to salt spray corrosion on 2014-T6 and 7075-T6 alloys.

Anodic oxidation in dilute oxalic acid solutions produces coatings that are essentially transparent. Their color varies from a light yellow to bronze. Such coatings are dense and have little absorptive capacity but possess high resistance to abrasion. Combination electrolytes of sulfuric and oxalic acids are also sometimes used. They produce effects somewhat similar to lowering the temperature of the sulfuric acid electrolyte, i.e., a denser coating with greater resistance to abrasion.

Hard coatings. Procedures have been developed recently that produce finishes of greater thickness and density than the conventional anodic coatings. These procedures require high current densities and low-temperature electrolytes and are known commercially as "Alumilite" hard coatings, "Martin" hard coatings, "Sanford" or "Hardas." They have high resistance to abrasion, erosion and corrosion. The coatings have thicknesses in the range of 1 to 5 mils, but the thickness of conventional Alumilite finishes is in the approximate range of 0.1 to 1 mil, depending upon the application.

Hard coatings are rapidly becoming popular for applications requiring light weight in combination with high resistance to wear, erosion and corrosion. These applications include helicopter rotor blade surfaces, pistons, pinions, gears, cams, cylinders, impellers, turbines and many others. Also, due to their attractive gray color, hard anodic coatings are now being used for architectural applications.

The processing conditions for obtaining hard coatings are such that thick coatings with maximum density can be obtained on most aluminum alloys. The selection of alloy is of utmost importance.

Variations of the anode oxidation process that produce conventional hard anodic coatings have recently been developed which form thick, dense, colored anodic coatings for architectural applications. Attractive bronze, gray or black colored coatings are obtained by utilizing certain organic acids as electrolytes.

Other coatings. The "Ematal" process utilizes oxalic acid in combination with titanium, zirconium and thorium salts for the electrolyte. Oxide coatings produced by this anodic oxidation treatment are dense and have an opaque, light-gray appearance. Since the composition of the electrolyte is quite complex, difficulties may be encountered on a production basis.

Anodic oxidation of aluminum in sulfamic acid produces coatings that are denser than those produced by the sulfuric acid process. This anodic process is expensive owing to the high cost of sulfamic acid.

Anodic coatings produced in a phosphoric acid electrolyte are not considered commercially important. At one time they were used for the preparation of aluminum surfaces prior to electroplating. These coatings have large cells and a large pore diameter.

Alternating current may be used to form anodic coatings on aluminum and alloys with all of the electrolytes previously mentioned, but the aluminum surface is anodic only half of the cycle so that an oxide coating is formed only at half the rate of coatings formed with direct current.

Superimposed a.c.-d.c. is also used for anodizing. It has been most successful for the production of hard, thick, anodic coatings by the Hardas Process.

Double anodizing treatments utilize conventional anodizing in sulfuric acid electrolyte, with either d.c. or a.c. voltage, and after rinsing, the anodizing process is continued in other electrolytes which require much higher voltage. Such electrolytes are chromic, boric or tartaric acids or their salts. The higher voltages produce much thicker barrier layers. This condition results in anodic coatings with high resistance to corrosion.

Alloy selection. The previous discussion on the anodic oxidation characteristics of various electrolytes was based upon the use of relatively pure aluminum such as 1100 alloy. Alloy selection is important. Even aluminum of 99.3% minimum purity, such as the 1100 alloy, is in a sense an aluminum alloy from the standpoint of anodic oxidation, since the other elements present have an effect on the characteristics of the coating.

The response of the different constituents of aluminum alloys varies considerably and is an important factor. Some constituents will be dissolved by the anodic reaction, whereas others may be unaffected. For alloys where the constituents are dissolved during the anodic treatment, the coatings will have voids which decrease the density of the coating and also lower its resistance to corrosive action and abrasion. Aluminum-copper alloys are an example of this type.

Silicon in aluminum-silicon alloys is an example of a constituent that is unaffected by conventional anodic oxidation. The silicon particles remain unchanged in the coating in their original position.

Some constituents of aluminum alloys will themselves oxidize under anodic oxidation and the oxidation products will color the coating. For example, constituents such as manganese will produce a brownish opaque appearance due to the presence of manganese dioxide. Also, chromium as a constituent will give a yellowish tint to the coating from oxidation products of chromium.

The lower the concentration of constituents present or the purer the aluminum, the more continuous and transparent will be the oxide coating. The so-called "super purity" (99.99%) aluminum produces the most transparent oxide coating.

It should be appreciated that care must be used in selection of alloys for assemblies if it is desirable to obtain a uniform over-all appearance match; for example, castings and wrought products in the same assembly are undesirable if a uniform appearance is required. However, with careful selection of alloy, a surface pretreatment can minimize differences in appearance. Also, a surface that has been textured either mechanically or by etching will in many cases solve the problem of over-all appearance match.

Properties and applications

The oxide coatings produced by anodizing have many properties that make them commercially important. Anodic coatings are essentially aluminum oxide, which is a very hard substance. Because the aluminum oxide is integral with the surface it will not chip or peel from the surface; this outstanding characteristic is useful for architectural applications. The combination of high wear resistance and attractive satin sheen of the finish makes it a logical choice for aluminum hardware, handrails, moldings and numerous other architectural components. Because the anodic finish reproduces the texture of the surface from which it is formed, a wide variety of attractive effects are possible by variation in the surface preparatory procedures. Many commercial architectural applications can use a natural aluminum finish, but for applications where it is desirable to preserve initial appearance, the anodic finish will require less maintenance.

Because the oxide coating is brittle compared to the aluminum underneath, it will crack if the coated article is bent. It is possible, however, to produce oxide coatings that are relatively flexible. In general, the thicker coatings formed in sulfuric and oxalic acid electrolytes will crack or craze to a much greater extent than oxide coatings formed in a chromic acid electrolyte. For many applications this cracking may not be objectionable, because it is usually difficult to detect by visual observation. These fine cracks have an adverse effect on the bending properties of the metal, however, and may sometimes cause fracture if the bends are severe. For this reason, it is generally recommended that the finish be applied after forming.

If fatigue is a critical factor then proper allowance must be made for the reduction in endurance limits produced by relatively thick oxide coatings. Fatigue tests indicate that a coating 0.1 mil thick on smooth surfaces will have little effect on fatigue strength. Thicker coatings in the range of 0.3 to 0.5 mil on smooth surfaces have a slight detrimental effect at high stresses.

Anodic coatings also provide substantial protection against corrosion. There are many factors which must be considered in this connection such as the continuity of the coating and the choice of alloy. Since continuity is dependent upon constituents present in the alloy, anodic coatings on high-purity aluminum are the most resistant to corrosion. On the other hand, anodic coatings formed from aluminum-copper alloys have much lower resistance to corrosion.

Sealing of anodic coatings in dichromate solutions results in an appreciable improvement in resistance to corrosion, particularly by chlorides. The results of atmospheric exposure tests indicate that anodic coatings with a thickness of 0.4 mil or greater will provide greatly increased resistance to weathering.

The ability of anodic coatings to absorb coloring substances such as dyestuffs and pigments makes it possible to obtain finishes in a complete range of colors including black. The colors are unique because the luster of the underlying metals gives them a metallic sheen which is particularly attractive for applications which simulate metals such as gold, copper, bronze and brass. Colorants are available which, when used to color anodic coatings on the proper alloys, will reproduce the natural colors of the metals listed above. Colored finishes can be used in a wide range of applications for nameplates, panels, appliance trim, optical goods, cameras, fishing reels, instruments, giftware and jewelry.

Anodic coatings have good electrical insulating properties. Anodic films produced in boric acid electrolytes are used commercially on aluminum foil for electrical capacitors. The voltage necessary to break down the anodic coatings is generally proportional to the thickness. Anodic coatings 0.6 mil thick produced in H_2SO_4 electrolyte on 1100 alloy have an approximate breakdown voltage of 600. Breakdown voltages vary considerably even on the same thickness of coating on the same alloy. This variation is caused by weak spots in the coating. The coatings have considerable electrical insulation value, particularly at low voltages, and have the advantage of good resistance to temperature. Hard anodic coatings on certain alloys have been measured at values over 3,000 v.

Anodic oxidation in oxalic acid and phosphoric acid electrolytes produces coatings which have been successfully used as preparatory treatments for electroplating. Copper, nickel, cadmium, silver and iron have been successfull deposited over oxide coatings. Plating solutions which are highly alkaline should be avoided since they attack the coating and destroy the porous structure necessary for the best adhesion.

The "Krome-Alume" process utilizes an oxalic acid electrolyte to form the oxide coating and subsequent modification of the coating with hydrofluoric acid produces a structure satisfactory for electroplating. Furthermore, the anodic oxidation process utilizing a phosphoric acid electrolyte produces a structure which requires no further modification to condition it for electroplating. The alloy has an important effect on coating structure and, in general, the phosphoric anodic process is not recommended for preparing high purity aluminum, aluminum-magnesium wrought alloys and most die-casting alloys for electroplating.

Production methods. Anodic coatings are applied to aluminum and its alloys by a variety of methods which include batch, bulk, continuous conveyor and continuous strip.

The batch method of applying anodic coatings is similar to that used in electroplating except that the parts are anodic instead of cathodic. The continuous conveyor method is also similar to the conventional plating method.

The bulk method for applying anodic coatings to small parts such as rivets, washers and screws is radically different from bulk electroplating methods. The barrel-finishing method is not suit-able for applying anodic coatings because the initial flow of current forms an anodic coating on the parts and even if they contact each other during the rotation of the barrel, no current will flow because the coating is an insulator. The bulk methods employed for anodic coatings utilize special perforated nonmetallic cylindrical containers. Pressure, applied to the parts in the container through a threaded center contact post, maintains the initial contact between the surface of the parts.

The continuous strip process is being used commercially to apply anodic coatings to aluminum sheet which is subsequently roll-formed into weather strip. In Europe, this same process is used to apply anodic coatings to aluminum sheet which is formed into food containers such as sardine cans.

Magnesium

Although all of the magnesium alloys in commercial use today have good resistance to corrosion, many parts are provided with maximum resistance to corrosion and abrasion through electrolytic treatments based on anodic treatments. The anodic treatments produce relatively thick and dense coatings with excellent adhesion and high resistance to corrosion and abrasion. As in the case of aluminum, anodic coatings on magnesium alloys are an excellent base for lacquers and enamels.

It is a well-known fact that magnesium alloys are attacked by most inorganic and organic acids. However, because magnesium alloys are resistant to alkalies, fluorides, borates, and chromates, the electrolytes for anodizing are generally based upon these chemicals.

The simplest electrolyte for anodizing magnesium alloys is a 5% caustic soda solution (A.M.C. Treatment R). This electrolyte is used in a temperature range of 140 to 158 F. A current density of 18 amp per sq ft at a d.c. voltage of 5 to 6 is satisfactory; anodic oxidation time is generally 30 min. All magnesium alloys will respond to this treatment. The coatings produced are approximately 0.3 mil thick and are essentially crystalline magnesium hydroxide. They have relatively high resistance to abrasion and are gray or tan in color in the as-formed condition. The coatings may be colored for decorative applications by immersion in water-soluble dyestuffs in much the same way as the coloring procedures used for anodic coatings on aluminum alloys. If maximum resistance to corrosion is required, immersion (sealing) in a 5% sodium chromate solution at 170 to 180 F is recommended.

Another variation of the caustic soda electrolyte consists of approximately 23% sodium hydroxide with additions of 0.55 pints per gal of ethylene or diethylene glycol and 0.33 oz per gal of sodium oxalate (Dow #12).

The recommended procedure for anodizing in this electrolyte is 10 to 20 amp per sq ft at 6 to 24 v a.c. or 6 v d.c. Anodizing time is generally 15 to 25 min. Longer anodizing times produce thicker coatings which have higher resistance to abrasion. Since the magnesium-manganese alloys anodize more rapidly than the other magnesium alloys, the low current density range should be used.

After being anodized, the parts should be subjected to a stabilizing treatment. This is accomplished by leaving them in the electrolyte after the current is off for approximately 2 min.

If the parts are to be waxed, thorough rinsing of the surface with cold water followed by hot water is sufficient. If the parts are to be painted, a secondary treatment is required to neutralize the free alkali from the electrolyte from the pores of the anodic coating. The following solution and conditions are recommended:

Sodium Dichromate	6⅔ oz/gal
Sodium Acid Fluoride	6⅔ oz/gal
Temperature	70 to 90 F
Time	5 min

This treatment is also referred to as a sealing treatment because the coating adsorbs some dichromate; this adsorption imparts a yellowish tint to the coating. Anodic coatings produced in the caustic soda-ethylene glycol-sodium oxalate electrolyte may also be colored by immersion in a suitable dye solution.

Another anodic process (Dow #14) for magnesium alloys utilizes a well-agitated electrolyte with the following composition and operating conditions:

Sodium Metaborate	32 oz/gal
Sodium Metasilicate	9 oz/gal
Phenol	1 oz/gal
Sodium Hydroxide	⅓ oz/gal

Temperature Range	70 to 85 F
Current Density	10 to 15 amp/sq ft
Time	5 to 20 min

The anodic coating produced under these conditions varies from white to gray depending upon the alloy. The coating has high resistance to abrasion but waxing is recommended to reduce smudging. It is also an excellent base for paint provided that it is subjected to the sodium dichromate-sodium acid fluoride neutralization treatment.

Another variation of the alkaline electrolyte used for anodizing magnesium alloys has the following composition and conditions of operation:

Caustic Soda	40 oz/gal
Sodium Silicate	57 oz/gal
Phenol	½ oz/gal
Temperature Range	170 to 185 F
Current Density	10 to 20 amp/sq ft d.c.
	20 to 30 amp/sq ft a.c.
Time	5 to 25 min

This anodic treatment, known as the "Manodyz" Process, may be used on all types of magnesium alloys including the cerium and zirconium alloys. The voltage requirements are from 4 to 6 for both a.c. and d.c. The coatings produced by the a.c. process are gray in color and approximately 0.4 mil thick, whereas those produced by the d.c. process are light green and approximately 0.8 mil thick. The d.c. process is recommended for castings and heavier gauge sheet alloys. In order to obtain a 0.4 mil thick coating using a.c., a current density of 25 amp per sq ft for 25 min is required. Current densities as high as 50 amp per sq ft may also be used. In this case, a 0.4 mil coating is obtained in 5 to 10 min. An 0.8 mil coating is obtained with d.c. in 25 min at a current density of 15 amp per sq ft.

As in the case of all anodizing treatments for magnesium alloys, the surfaces must be thoroughly cleaned by either solvent or vapor degreasing and followed by an acid etch. After being anodized, the parts are rinsed in clean rinsing water, neutralized in an 0.08 oz per gal chromic acid solution maintained in a pH range of 2.1 to 2.5 for 2 to 5 min in a temperature range of 180 to 200 F. After neutralization, the parts may be painted or dyed. The Moh hardness of anodic coatings produced by the Manodyz Process is 6 to 7. The a.c. voltage treatment produces the hardest coating.

Another anodizing treatment for magnesium alloys capable of producing a coating with many desirable properties is the "H.A.E." Process. This process is also based upon an alkaline electrolyte. All magnesium alloys, both wrought and cast, respond to this process to give coatings with excellent resistance to corrosion, high dielectric strength and high hardness. The electrolyte used for this anodizing process (U.S. Patent 2,880,148) is also alkaline and consists of the following:

Potassium Chromate	20 g/l
Potassium Fluoride	35 g/l
Tripotassium Phosphate	35 g/l
Potassium Hydroxide	65 g/l
Potassium Aluminate	40 g/l
Temperature Range	10 to 65 C
Current Density	5 to 500 amps/sq ft
Time	15 to 25 min

The usual degreasing and etching of the surface is necessary to obtain uniform coatings. A.c. or d.c. voltages may be used but d.c. is preferred. The process also adapts to combinations of a.c. followed by d.c. or a.c. superimposed upon d.c. Current densities of 10 to 20 amp per sq ft are usually used. The voltages required vary greatly depending upon the alloy and the current densities employed; voltages up to 100 are sometimes used. The conditions employed for the hard coatings result in a thickness of approximately 1 mil. The Moh hardness of the H.A.E. anodic coating is 7 to 8. The softer coatings range from 0.3 to 0.4 mil in thickness and vary in color from olive drab to tan.

After the anodizing treatment, the parts are rinsed and immersed for approximately 1 min in the following solution:

Sodium or Potassium Dichromate	20 g/l
Ammonium Bifluoride	100 g/l
Temperature	Room

After drying, they are subjected to an aging treatment at a relative humidity of 95% for 7 to 16 hrs. This treatment gives an appreciable increase in salt-spray corrosion resistance.

Low-voltage work is immersed in a 2 g/l sodium or potassium dichromate solution at a temperature of 90 C for approximately 2 min. Parts should be dried without rinsing.

This process possesses great versatility and coatings can be produced with a wide variety of characteristics.

Another anodizing process (CR-22) recently developed for the application of hard corrosion-resistant coatings is based upon the following electrolyte:

Chromic Acid	25 g/l
Hydrofluoric Acid (52%)	25 cc/l
Phosphoric Acid (85%)	50 cc/l
Ammonium Hydroxide	160 to 180 cc/l
Temperature	25 to 95 C
Current Density	15 amp/sq ft
Time	12 min

Anodic coatings approximately 1 mil thick are formed in this electrolyte at 320 v a.c. The coatings are very durable and provide an excellent base for paint.

Another process (Dow #17) for anodizing magnesium alloys is based upon the following electrolyte:

Ammonium Acid Fluoride	32 to 40 oz/gal
Sodium Dichromate	13.3 oz/gal
Phosphoric Acid (85%)	11.5 fl oz/gal
Temperature	160 to 180 F
Current Density	5 to 10 amp/sq ft

This process employs voltages up to 110, either d.c. or a.c. The coating ranges in thickness from 0.2 to 1 mil and has excellent corrosion resistance. It has a green color whose depth of shade varies depending upon thickness. The deep green-colored coatings are approximately 1 mil thick.

Since the electrolyte is acidic, only degreasing is required to prepare the surface for anodizing. Also, no neutralizing treatment is required as in the case of the anodic coatings produced in alkaline electrolytes.

The hard thick coatings are recommended only for castings and extrusions because of their tendency to spall from the surface if the part is bent. The increase in surface dimension is approximately two-thirds of the coating thickness.

Zinc

Much of the work on anodic coatings for zinc has been conducted in alkaline electrolytes consisting of a two-stage process. The electrolyte for the first oxidation treatment is conducted in an alkali-carbonate solution followed by anodizing in a silicate solution.

Further work has centered on a single anodic treatment conducted in an alkaline electrolyte composed of sodium silicate, sodium borate and sodium tungstate. Both d.c. and a.c. voltages in the range of 100 v appear to have possibilities. The anodic treatment is carried out at approximately room temperature. Cooling of the electrolyte is required and power requirements when using d.c. voltage are very high (approximately ten times that required for anodizing aluminum alloys). A considerable reduction in power required results when alternating current is used; however, coatings formed with a.c. are powdery.

Other metals

Thin oxide films may be formed on beryllium by anodic oxidation in an electrolyte composed of 10% nitric acid containing 200 g/l of chromic acid. The anodic oxidation is carried out at approximately 25 amps per sq ft. Such films retard high temperature oxidation and corrosion.

Thin oxide films can also be formed on zirconium, titanium and thorium by anodizing in an electrolyte composed of 70% glacial acetic and 30% nitric acid at a current density of approximately 60 amps per sq ft. The corrosion resistance of titanium is substantially improved by anodizing for 10 to 15 min in a 15 to 22% (by weight) sulfuric acid solution (room temperature) at approximately 18 v. d.c. Anodic coatings on titanium are also used as a base for lubricants. In general the anodic oxidation of these metals is not practiced on a commercial scale comparable to that of aluminum and magnesium.

ANTIMONY

Antimony belongs to Group VA of the periodic table, which includes the elements nitrogen, phosphorus, arsenic and bismuth.

Chemically, antimony is classified as a non-metal or metalloid, although it has metallic characteristics in the trivalent state.

In the elemental state, antimony is a silvery-white, brittle crystalline solid which exhibits poor electrical and heat conductivity. The element can exist in two additional allotropic forms, which are the black amorphous modification and the yellow modification. Explosive antimony is sometimes referred to as a third modification.

It is produced by the electrolysis of antimony from a chloride electrolyte, under special conditions. The deposit is believed to be in a strained amorphous state which, when scratched or bent, will explode mildly, converting the antimony to the crystalline form.

Antimony is one of the few elements which exhibits the unique property of expanding on solidification.

Antimony is obtained commercially from the reduction of the mineral stibnite (Sb_2S_3).

An important nonmineral source of antimony is sodium antimonate, a by-product of the lead-softening operation. The metal is recovered from this compound by reducing it with pulverized coke fines.

Metallic antimony is used primarily for hardening lead and its alloys, with antimonial lead, bearing metal, and type metal accounting for about 90% of the consumption. In the form of compounds, it is used in flameproofing, ceramics, pigments and glass. Domestic consumption for these applications in 1961 was 12,342 short tons.

Physical constants

Symbol	Sb
Atomic No.	51
Atomic Weight	121.76
Melt Pt, °C	630.5
Boiling Pt, °C	1640
Density, 20 C, gm/cu cm	6.618
Latent Heat of Fusion, cal/gm	38.3
Latent Heat of Vaporization, cal/gm	161
Spec Heat, 0 C, cal/gm/°C	0.0494
Therm Cond, 20 C, cal/sec/sq cm/cm/°C	0.045
Lin Coef of Therm Exp, 20 C, μ in/°C	8–11
Elec Resis, 0 C, μ-ohm-cm	39.1
Magnetic Susceptibility, cgs	-0.87×10^{-6}
Crystal System	Hexagonal (Rhombohedral)
Lattice Constants, 26 C	a = 4.307 c = 11.273
Hardness, Mohs Scale	3.0

Chemical properties

Antimony is ordinarily stable and not readily attacked by air or moisture. Under controlled conditions it will react with oxygen to form the oxides Sb_2O_3, Sb_2O_4, and Sb_2O_5. Antimony tetroxide may be considered as a mixture of Sb_2O_3 and Sb_2O_5.

Antimony will react vigorously with chlorine to form the tri- and pentachlorides. Sulfur combines with antimony in all proportions and forms the compounds Sb_2S_3 and Sb_2S_5. The compound

Sb_2S_5 decomposes easily to Sb_2S_3 and S, and is formed most readily by chemical means. There is no direct combination between antimony and hydrogen. However, antimony hydride (stibine) SbH_3, which is extremely poisonous, may be formed by the action of hydrochloric acid on a compound such as ZnSb. Stibine may also be liberated from acidic solutions of antimony salts treated with Zn, Al, Mg, etc.

Antimony will react with nitric acid to form H_3SbO_4 and with hot concentrated sulfuric acid, to form the sulfate. In the absence of air, antimony is not attacked by HCl. It will be attacked by a warm solution of aqua regia. Antimony is also attacked by bases. In chemical reactions, antimony can exhibit a valence of -3, $+3$ or $+5$.

When antimony is alloyed, it will readily form antimonides with other metals.

Hygienic precautions

Generally, the toxicity of antimony compounds is classified as moderate. In the elemental form antimony is not considered harmful, but gaseous stibine is extremely toxic. Cases of dermatitis have been observed in the manufacture of antimony trioxide. It is therefore recommended that when handling antimony and its compounds good personal sanitation be practiced to avoid contamination or ingestion.

Antimony alloys and applications

Since elemental antimony is not easily fabricated due to its brittle nature, its important metallurgical uses are as an alloying agent. Several important applications of antimony-containing alloys are listed, with typical range of compositions . . .

Use	% Sb	% Sn	% Pb	% Other
Cable Sheathing	1.0	—	bal	—
Battery Grids	6–9	—	bal	—
Bearing Metal	10–15	75–3	bal	1–3 As
Sheet and Pipe	2–6	—	bal	—
Plumbers Solder	0–2	38–42	bal	—
Collapsible Tubes	2–3	—	bal	—
Type Metal	2.5–28	2–20	bal	1 Ag
Babbitt	4–8.5	bal	—	4–6.5 Cu
Pewter	0–7	bal	0–20	0–4 Cu
Britannia Metal	10–20	bal	—	—

Antimony is sometimes added to copper to increase its recrystallization temperature. It is also finding some uses in semiconductor applica-tions. It is used as an important alloying ingredient in bismuth telluride thermoelectric materials. In addition, it forms the III-V compounds AlSb, GaSb, and InSb. The latter compound has found some importance in infrared detecting devices and Hall effect devices.

There are many applications for antimony in compound form. Some of the more important compounds are the trioxide, trichloride and trisulfide. Antimony trioxide is an important ingredient in flameproofing fabrics and plastics, in porcelain enamels, in glass, and in the preparation of tartar emetic. Antimony trichloride is used in the preparation of medicine, in making mordants for dyes, and for staining dull finishes on iron and copper hardware. Antimony trisulfide finds applications in safety matches, in red rubber, in coloring ceramics, in tinting glass, and in camouflaging paint.

Forms available

Metallic antimony is usually sold in cakes weighing about 40 to 50 lb each, and available in commercial purities varying from 99% to 99.9%.

The term "Regulus" is often used to describe commercial antimony. A generally accepted, but unsound practice, is to judge the purity of antimony on the appearance of the fernlike or starlike as-cast surface structure of the cake. This practice is unreliable, since the surface of highly contaminated antimony can also be starred, using controlled conditions.

Extremely pure antimony, with purity higher than 99.999%, is available in fragmented form for electronics applications. This material is presently sold on a gram basis, although it can be made available in significant quantities should the demand develop.

ARSENIC

Arsenic is classified chemically as nonmetal or metalloid, and like antimony, belongs to Group V of the periodic table.

In the elemental state arsenic is a steel gray, brittle, crystalline material, which has low electrical and heat conductivity. It can also exist as a black, amorphous solid, i.e., arsenic mirror. Several other allotropic forms—yellow, gray to black, and pale reddish brown to dark brown, have been reported; however, there is very little evidence to support the existence of all these allotropes.

The metal does not exhibit a melting point under normal conditions, but sublimes on heating.

The largest demand for arsenic is in the form of arsenic trioxide (As_2O_3) which, in the refined form, is referred to as white arsenic (black arsenic is crude As_2O_3), arsenious acid anhydride, arsenious oxide, and also by the generally accepted consumer's misnomer, arsenic. The trioxide is primarily used for the manufacture of herbicides and insecticides.

Physical constants

Symbol	As
Atomic No.	33
Atomic Weight	74.91
Melt Pt, 28 atm, C	817
Boiling Pt, vapor pressure 1 atm, C	613
Density, 20 C, gm/cu cm	5.72
Latent Heat of Fusion, cal/gm	88.5
Latent Heat of Sublimation, cal/gm	102
Specific Heat, 20 C, cal/gm/°C	0.082
Lin Coef of Therm Exp, 20 C, μ-in./°C	4.7
Elec Resis, 20 C, μ-cm	33.3
Magnetic Susceptibility, 18 C, cgs	-0.31×10^{-6}
Crystal System	Hexagonal (Rhombohedral)
Lattice Constants, 26 C	a = 3.760 c = 10.548
Hardness, Mohs Scale	3.5

Chemical properties

Metallic arsenic is stable in dry air. When exposed to humid or moistened air the surface first becomes coated with a superficial golden bronze tarnish, which on further exposure turns black.

On heating in air arsenic will vaporize and burn to As_2O_3. A persistent garliclike odor accompanies this oxidation. When heated in the presence of halides it will combine to form definite compounds. It will react with molten sulfur to form the compounds As_2S_3, AsS, As_2S_5, and mixtures of various proportions.

Arsenic vapor will not combine directly with hydrogen to form hydrides. Arsine, (AsH_3) a highly poisonous gas, will however, be formed if an intermetallic compound such as AlAs is treated with hydrochloric acid or left in contact with moisture for prolonged periods of time. The compound can be thermally decomposed into its constituent elements by heating it to about 250 C.

Metallic arsenic is not readily dissolved by water, alkaline solutions, or nonoxidizing acids. It will react with nitric acid to form arsenic acid H_3AsO_4 and liberate nitrous oxide. Hydrochloric acid will attack arsenic only in the presence of an oxidant.

In chemical reactions, arsenic can exhibit a valence of -3, $+3$, or $+5$.

When alloyed, arsenic will form arsenides with many metals.

Hygienic precautions

Arsenic in the elemental form is not considered poisonous, although its compounds can range in toxicity from very low—in the case of arsenic trisulfide—to extremely high—in the case of gaseous arsine. The possibility of forming arsine should be considered and guarded against.

Dermatitis can result from handling arsenical compounds; hence it is desirable to use impervious gloves or protective creams. The best preventive for dermatitis is strict personal hygiene.

In areas where arsenical dusts and fumes are present, effective exhaust ventilation is necessary, or when impractical, a respirator should be used.

The most effective and specific antidote for arsenic poisoning is BAL (British Anti-Lewisite, 2,3 dimercapto-l-propanol). Freshly precipitated ferric hydroxide is also an effective antidote.

Uses

Arsenic lacks the properties of a ductile material, and for this reason it is used primarily as an alloying ingredient.

It is alloyed with lead in quantities ranging from $\frac{1}{2}$ to 2% for the manufacture of shot, since it improves the sphericity of the shot. The addition of up to 3% arsenic to lead-base bearing alloys improves their mechanical and especially their elevated temperature properties. A small percentage of arsenic is also added to lead-base battery grids and cable sheathing, in order to improve hardness of these materials.

Minor additions of arsenic to copper will improve its corrosion resistance and raises the recrystallization temperature. Copper containing 0.15 to 0.5% arsenic is recommended for elevated temperature applications such as locomotive staybolts, firebox straps and plates. A small amount of arsenic (0.02 to 0.05%) is a useful addition to brass for minimizing or preventing dezincification.

In the field of electronics, high-purity arsenic

is finding an important use as a constituent of the III-V semiconductor compounds. The compounds of most interest are indium arsenide, which may be used for Hall effect devices, and gallium arsenide, which may be used for making diodes, transistors, and solar cells.

Because of the high toxicity of many of the arsenical compounds, they are used in large quantities as insecticides and herbicides. Lead arsenate is used as an insect spray for fruit trees and plants. Calcium arsenate is used for boll weevil control on cotton plants. A double salt of copper orthoarsenite and copper acetate (Paris green) is also used for insect control. Sodium arsenite is used as a herbicide, particularly for crab grass control, and also for debarking trees which are to be used for wood pulp. Orthoarsenic acid ((H_3AsO_4)) is being used successfully as a defoliant for cotton plants. Sodium arsenate is useful as a wood preservative.

There are a myriad of minor uses for other arsenical compounds. Arsenic trisulfide is used in fireworks, leathermaking, and as a pigment. Arsenic trioxide is used, in addition to producing herbicides and pesticides, for decolorization of glass, in taxidermy for the preservation of hides, in the preparation of poison gases for warfare, and in some medicinal preparations.

Forms available

The commercial metal is normally sold in lump form (2 to 3 in.) with a minimum purity of 99%, and is packaged in drums of 200 lb each. The arsenic used for electronics applications has a purity in excess of 99.999+%, and is available in random size fragment 0.1 in. thick by about ¾ in. square. This material is sealed in evacuated glass ampules to prevent oxidation, and is packaged in units of 25, 50 an 100 gm.

ASBESTOS

Asbestos is a general term used to describe six fibrous minerals. The most important of the six is chrysotile which accounts for 93% of the world's production. The other types of commercial importance are amosite, crocidolite and anthophyllite. Those of least commercial importance are actinolite and tremolite.

Chrysotile is a hydrated magnesium silicate from the serpentine mineral group. Amosite and crocidolite are iron silicates from the amphibole

mineral group. Anthophyllite is a magnesium silicate of the amphibole group with the Mg isomorphically replaced by varying amounts of iron and aluminum.

The important characteristics of the asbestos minerals that make them unique are their fibrous form, high strength and surface area, resistance to heat, acids, moisture and weathering, and good bonding characteristics with most binders such as resins and cement. Table 1 lists the main properties for easy reference and comparison.

Grading

A dry screening operation known as the Quebec Standard Screen Test is the standard for grading and pricing. This test consists of mechanically shaking 16 oz of fiber through sieve boxes covered with ½ in., 4 mesh and 10 mesh screen. The relative proportions remaining on each screen define the grade and price. The higher the amount retained on the large opening screens, the longer the fiber and the higher the value.

Additional tests used to further measure the properties of the fiber are: Ro-Tap dry screening, dry and wet density, wet screening for length and dust by Bauer-McNett and Clark machines, dry length measurement by Suter-Webb Comb, water filtration, degree of openness or surface area, magnetite content, viscosity, color and soluble salts.

Considerable information is available on the basic chemical and physical properties of asbestos, such as thermal properties, magnetic rating, basicity, surface properties, morphology, and structure.

Chrysotile

Fifty per cent of the world's high grade chrysotile comes from Canada, mainly from the Province of Quebec. Russia is the second largest producer, with most of the fiber coming from the Ural Mountain area. Russian fiber is generally of fair quality but lacks uniformity. Southern Rhodesia is also a supplier of high grade chrysotile.

Chrysotile has very high surface area and tensile strength, excellent flexibility, heat, electrical and weathering resistance. It has poor resistance to acids but good resistance to alkalies.

The main uses of chrysotile are in the spinning of cloth for use in heat resistant and friction applications, as a reinforcing fiber in asbestos cement sheets and pipes, in brake linings, and in plastics. Chrysotile is superior to glass fiber in laminates subjected to high temperature. After an 8-hr, one-side exposure to 1500 F a silicone-chrysotile felt laminate retained 56% of original strength compared to 15.5% for glass. For short periods, chrysotile-plastic combinations successfully withstand 10,000 F in missile applications.

It is also used extensively to make paper and millboard for heat and electrical insulation; high temperature packings and gaskets; and in plastic coatings, moldings and paints.

Chrysotile is classified into two groups, crude fiber in vein form, and milled fiber. The milled fiber is further divided into various groups and subgroups according to length and general use. Each country has its own classification system.

Crocidolite

Ninety per cent of the crocidolite comes from Cape Province, Union of South Africa, with a small but growing amount from Australia. This fiber has very high tensile strength and surface area, good resistance to acids and alkalies and relatively poor heat resistance.

It is used as one of the reinforcing fibers in high-pressure asbestos cement pipes and in acid resistant packings and gaskets. It is spun with difficulty and is woven in Europe only. It is

TABLE 1—PROPERTIES OF ASBESTOS FIBERS [a]

	Chrysotile	Amosite	Anthophyllite	Crocidolite
Theoretical Formula	$3MgO \cdot 2SiO_2H_2O$	$(Fe \cdot MgO)SiO_3$ $1-5\%\ H_2O$	$(Mg \cdot Fe)_7Si_8O_{22}-$ $(OH)_2$	$Na \cdot Fe(SiO_3)_2-$ $FeSiO_3 \times H_2O$
Specific Gravity	2.4–2.6	3.1–3.3	2.9–3.1	3.2–3.3
Specific Heat, Btu/lb/°F	0.266	0.193	0.210	0.201
Tensile Strength, psi	80,000 100,000	16,000 90,000	4,000 & less	100,000 300,000
Filtration Properties	Slow	Fast	Medium	Fast
Electric Charge	Pos.	Neg.	Neg.	Neg.
Fusion Point, F	2,770	2,550	2,675	2,180
Resistance to Acids and Alkalies	Poor	Good	Very good	Good
Magnetite Content	0–5.2	0	0	3.0–5.9
Mineral Impurities Present	Iron, chrome, nickel, lime	Iron	Iron	Iron
Ionizable salts, micromhos [b]	1.82	1.34	0.58	0.84
Color	Green, grey to white	Yellowish-brown	Yellow-brown to white	Blue
% Wt. Loss after 2 hr at:				
400 F	0.30	0.23	0.05	0.08
800 F	2.17	0.98	0.38	0.73
1000 F	3.99	1.16	0.44	0.86
1100 F	10.38	1.36	0.52	1.0
1400 F	13.43	1.43	0.54	1.03
1800 F	13.77	1.53	2.39	0.77
% Wt Loss after 2 hr Reflux with:				
25% HCl	55.69	12.84	2.66	4.38
25% H_2SO_4	55.75	11.35	2.73	3.69
25% NaOH	0.99	6.97	1.22	1.35
Surface Area sq cm/gr	130,000 to 220,000	—	—	—
Fiber Dia, in.	.000000706 to .00000118	—	—	—

[a] Badollet, M. S. CIM Trans. **54**, 151-160 (1951).

[b] Relative electrical conductance.

available in long and short grades in 100-lb bags. Additional processing is usually required before use.

Amosite

Amosite of commercial grade comes exclusively from the Transvaal in the Union of South Africa. It has fair tensile strength and poor flexibility but has good resistance to heat, acids, and alkalies.

It is used mainly as a reinforcement in rigid 85% magnesia and calcium silicate thermal in-

sulations and in felted blanket form as heat insulation. It is available in long and short grades in 100-lb bags. Additional processing is usually necessary before use.

Anthophyllite

Commercial deposits of anthophyllite are found in North Carolina and Georgia in the United States, and in Finland. It is a weak short fiber which cannot be spun and is not used in asbestos cement. It has excellent acid resistance and has limited use as a filter in acid, beer and wine processing.

B

BERYLLIUM AND ALLOYS

Chemical properties

Beryllium is one of the Group II elements which are known as alkaline earth metals. These elements exhibit an oxidation state of +2, and are characterized by their good metallic properties and strength *vs* reducing agents. Although beryllium is relatively stable at room temperature, its reactions with various materials increase as temperature increases.

Beryllium metal exposed to air will retain its metallic luster indefinitely at room temperature. Despite the high oxygen affinity of beryllium it must be heated to 600 C in air before noticeable oxidation appears. Oxidation increases very rapidly as the temperature approaches the melting point of the metal. Beryllium oxide forms a protective coating. Its reaction with nitrogen is similiar to that of oxygen and the powder reacts with carbon at temperatures as low as 900 C. The reaction with carbon dioxide commences at 600 C and goes to completion at 950 C forming Be_2C. There is no appreciable reaction with hydrogen. Fluorine reacts at room temperature and chlorine, bromine and iodine at elevated temperatures. Beryllium's reaction with water is greatly dependent on the water purity and pH. The halogen acids react in all concentrations at room temperature. It reacts with dilute sulfuric, concentrated sulfuric slowly with heat, dilute nitric and dilute acetic, but does not react with glacial acetic or concentrated nitric acid at room temperature. Beryllium is slowly attacked by phosphoric acid and is rapidly dissolved by ammonium bifluoride. Alkali hydroxides in solution attack beryllium vigorously and molten alkali is said to react explosively. The salts of a large number of metals will be reduced except the alkali metal halides which are more stable. Since beryllium has a low atomic volume, it can be expected to have a high diffusion rate in many metals. Corrosion by liquid metals is dependent on the amount of oxygen present. Molten beryllium reacts with and reduces most oxides, nitrides, sulfides, and carbides including those of magnesium, calcium, aluminum, titanium and zirconium.

Melting and casting

Vacuum-cast beryllium offers a distinct advantage over hot-pressed powder because a higher purity product can be obtained. To date, however, most fabricated metal parts are made from hot-pressed powder. Methods have been studied to improve the cast metal and controlled solidification now produces sound ingots, up to 7 in. in diameter, which may be rolled, extruded or forged.

Beryllium is melted under vacuum, or in some cases, in an inert atmosphere such as argon. In most cases induction heating is used, but other methods such as consumable-arc melting or electron-beam melting have been used.

Molds for casting beryllium ingots are usually made of graphite. The graphite is dried to remove moisture and in some applications a beryllium oxide coating is applied to the mold. Metallic molds such as steel, cast iron or copper have been used with moderate success.

Powder metallurgy

Beryllium metal powder produced in the United States is prepared from vacuum cast ingots which are machined to produce chips. The chips are reduced to powder in a dry nitrogen atmosphere attritioning mill. The powder is sieved and oversize fractions returned to the mill. This method produces beryllium powder of 99% purity or better.

Beryllium products are produced from powder by hot pressing, warm pressing and cold pressing sintering techniques. Table 1 presents mechanical properties of beryllium after preparation by powder metallurgy techniques. Data are also included for forged and rolled beryllium.

Hot pressing. Hot-pressing techniques are used in the preparation of most of the beryllium products in this country. Simple shapes may be hot-pressed directly, but more complex shapes require subsequent machining. The desired pressing is formed by simultaneously heating and molding the powder in steel or graphite dies at temperatures around 1050 C and pressures of the order of 1000 psi.

TABLE 1—NOMINAL MECHANICAL PROPERTIES OF BERYLLIUM

	Vacuum Hot Pressed	Hot Extruded and Annealed	Hot Rolled	Hot Forged
Ten Str, 1000 psi	45	80	70	80
Yld Str, 0.2% offset, 1000 psi	35	45	50	50
Mod of Elast, 10^6 psi	42	42	42	42
Elongation, %	1	8	10	5

Warm pressing. Warm pressing requires the use of special dies fabricated from hot work tool steels or carbides because of the higher pressing pressure needed at lower temperature. This may be carried out in air with temperatures ranging from 400 to 650 C and pressures from 25 tsi to 100 tsi. It has the advantage of producing high-density compacts in a relatively short period of time. The technique is little used.

Cold pressing-sintering. Beryllium powder compacts can be formed at room temperature using a pressure of more than 10 tsi but there is no advantage in using pressures above 50 tsi.

Sintering of loose powder without the use of mechanical pressure may be carried out by tap or vibratory packing of unpressed powder in a graphite mold previously outgassed at 1300 C. A variety of shapes have been produced by this method including cladding for fuel elements. Somewhat lower densities are obtained compared to the hot-pressing process.

Fabrication

Fabrication procedures for beryllium metal generally follow accepted metallurgical and powder metallurgical techniques.

Contamination can be avoided by jacketing or cladding the beryllium with an impervious expendable material such as mild steel, or performing the operation under a vacuum or inert atmosphere. Proper mechanical properties may be achieved, with special attention to ductility, by controlling the crystallographic orientation or texture. Crystallographic texture is primarily controlled by a choice of fabrication techniques and direction of metal flow.

Forging. Forging is attractive as a method for fabricating beryllium parts from the mechanical and economic standpoint. Parts can potentially be produced close to final shape with an accompanying reduction in scrap losses and the rapidity of the forging operation minimizes the reaction of beryllium with oxygen and nitrogen in the atmosphere.

Hot-pressed powder billets have been used most frequently in beryllium forgings, with a reasonable degree of success. Forgings using loose powder in a steel jacket are also of interest because they eliminate the standard hot-pressing cycle normally used to produce the starting billets, and so reduce the fabricating costs.

Both press-forging and hammer-forging methods have been used to fabricate beryllium shapes. Since beryllium, with but a few exceptions, is sensitive to the rate of deformation, single-stroke press forging is the preferred method. Where a series of blows have been used, lateral displacement is appreciable and produces poorer forgings. Reductions in height ranging from 25 to 75% have been accomplished in unsupported beryllium by using press-forging techniques. Steel is normally used as jacketing or cladding material during the forging of hot-pressed beryllium powder billets. The most satisfactory temperature range for beryllium forging is reported to be 705 to 1065 C. Forgings at 750 C produced the best tensile properties. Pressures applied to beryllium forgings range from 5,000 to 60,000 psi.

Rolling. A high-purity, hot-pressed, slab block of beryllium is normally used for rolling by the "picture-frame rolling" method. The picture frame is prepared by removing the central portion of a steel plate equal in thickness to the beryllium block. The beryllium block is then placed in the frame and a top and bottom plate are welded to the picture frame to complete the assembly. The assembly is rolled and cross-rolled on a rolling mill at temperatures of approximately 700 to 800 C to final dimensions. The cladding is either sheared off or pickled off in nitric acid. Pass reduction ranges from 5 to 15%. Recent development work in the rolling of cast ingot has shown promise and fine grained structures have been produced. Properties achieved by rolling hot pressed block approach 75,000 psi tensile strength, 55,000 psi yield strength, and over 5% elongation.

Extrusion. Commercial beryllium is extruded either from a hot-pressed powder billet or from cold-pressed billet, both jacketed, because of finer grain size and better mechanical properties. This jacket acts as a lubricant which prevents beryllium's seizing and galling of tools. It also prevents oxidation during heating and extruding operations. Extrusion ratios of greater than 8:1 are normally used.

Machining

Beryllium can be machined on common machine tools although a great deal of skill is required. With proper training of the machinist, however, it is possible to machine intricate contours to a good surface finish. In some respects it machines like cast iron, but is more abrasive and brittle. Safety precautions are necessary since beryllium in finely divided form is toxic. High-speed exhaust equipment is used to collect all dust and chips around the cutting tool. This also offers an added economic advantage since the collected material can be recycled and reprocessed. Special precautions are also taken to prevent contamination of the chips by cutting fluids or pieces of broken tools.

Tool design is of utmost importance. High-speed steel can be used but tool life is limited. Tungsten-carbide tools are far superior in spite of their added cost, and give excellent tool life and a good surface finish. In addition, it is necessary to provide rigid tool support no matter what tool material is used.

In machining beryllium it is essential to prevent any excessive tensile stress. Thin sections require adequate support and thin-walled cylindrical sections should be supported internally. A back-up block must be used in drilling through thin sections to prevent cracking and to prevent spalling upon exit of the tool.

Sheet formability

Beryllium sheet has limited formability at room temperature, but hot forming is possible at temperatures of 370 to 815 C. In this range it has been possible to form such parts as small hemispheres, shallow cups and similar shapes. The exact forming temperature to be used is dependent on the forming operation to be performed, the properties required of the product, and available equipment.

It is recommended that a bend radius of four times the thickness of the sheet (under 0.125 in. thick) be used as a guide for forming between 370 and 540 C. Forming at higher temperatures or multiple-forming operations are necessary to accomplish bends of 2 to 3t. Multiple-forming operations also require reheating to develop a consistent process. It is preferable to carry out hot-forming and cutting operations at relatively slow speeds and to anneal at 730 to 760 C afterwards or between steps in extensive deformation operations.

Joining techniques

Shrink fits. Fitting of beryllium parts by expanding the outer member through heat to increase its diameter so that the inner member can be put in place is the commonest forming of joining beryllium. The outer member contracts, and holds the two pieces together.

Fusion welding. Most of the work on fusion welding, using beryllium metal as the fillet material, has employed tungsten arc and inert gas shield, usually argon or helium. Fusion welding does not yield satisfactory joints because of large grain size and poor mechanical properties of the as-cast metal.

Diffusion welding or self welding. Pieces of beryllium metal are held in close contact from one to several hours from 900 to 1100 C will join. These welds can be made in vacuum or argon atmosphere. Better dispersion of inclusions and better grain coalescence are obtained in vacuum.

Braze welding. Aluminum brazing alloy containing a small amount of silicon or silver-base alloys, are used as separate fillers.

Furnace brazing. Aluminum, silver, aluminum-silver, and silver-copper alloys are best materials for joining beryllium to itself or to other metals such as nickel, stainless steel and copper.

Applications

Nuclear. The greatest interest in beryllium metal was developed in research and development projects pertaining to the development of nuclear energy. This interest was created by its low atomic weight (9.02), low neutron-absorption cross section (0.0090 ± 0.005 barns/atom), high neutron scatter cross section (7.0 barns/atom), high melting point (1283 C), and slowing down length (9.9 cm) (fission energy to thermal energy). These properties make it useful as a moderator because it slows down neutrons and as a reflector because it reduces neutron losses, thereby increasing the efficiency of the nuclear reaction. Beryllium is in use in nuclear reactors in the United States and Europe. Many other physical properties of beryllium are given in Table 2.

Space and aircraft. Beryllium's possibilities as a structural material for space and aircraft applications have been of great interest for a period of years. Its low density (1.85 gm/cu cm) and high modulus of elasticity ($36\text{-}42 \times 10^6$ psi) would permit greater weight reduction with a corresponding savings in fuel consumption. In addition, beryllium has excellent thermal properties such as its unusually high heat-absorbing properties due to its high heat capacity, thermal conductivity, and melting point. Thermal coefficient is one-half and heat capacity is two and one-half times that of aluminum. These properties make it a leading contender for instrumentation cases and air brakes and as a heat sink material. Another space-age use for beryllium would be as a high-energy fuel since it has the second highest heat yield rating (29,000

TABLE 2—PHYSICAL PROPERTIES OF BERYLLIUM

Isotopes (natural occurring 100%)	Be^9
Atomic Number	4
Electronic Structure	4 electrons = $1S^2, 2S^2$
Valency	2
Atomic Dia, Å	2.221
Orbital Radii, Å	
1s electrons	0.143
2s electrons	1.19
Ionization Pot., eV	
First	9.320
Second (of a singly ionized atom)	18.20
Spin and Parity	3/2
Magnetic Dipole Moment	1.1774 nuclear magnetrons
Elec Quadropole Moment	0.02×10^{-24} sq cm
Binding Energy of Last Neutron, MeV	1.664
Thermal Neutron Absorption Cross Section, barns/atom	0.0090 ± 0.0005
At. Wt	9.02
Chemical Scale	9.013 ± 0.0004
Mass Spectrographic Scale	9.0126
Physical Scale	9.015043
Crystal Structure	
25 C	hexagonal close packed
	$a_0 = 2.2863$ Å
	$c_0 = 3.5830$ Å
	$c/a = 1.5671$
1262 C	body-centered cubic
Density, 25 C, gm/cu cm	1.85
Melting Point	1285 C
Boiling Point	2790 C
Vapor Pressure, atm	
Solid (905 C)	10^{-8}
Solid (1235 C)	10^{-5}
Log. p (atm) = 6.186 + 1.454×10^{-4} T $- 16,700/$T	
Latent Heat, cal/gm	
Fusion	259
Vaporization	5930
Sublimation (between 900 and 1280°C)	8490
Sp Ht, 20–100 C, cal/gm/°C	$0.43 - 0.52$
Electronic Sp Ht, cal/gm atom/°	5.4×10^{-5}
Entropy, 25 C, cal/mol/°K	
Condensed	2.28 ± 0.02
Vapor	32.56 ± 0.01
Enthalpy, 25 C, cal/mol	
Condensed	465
Vapor	1481
Heat Capacity, 25 C, cal/mol/°K	
Condensed	4.26
Vapor	4.97
Free Energy Function, 25 C, cal/mol/°K	
Condensed	0.72
Vapor	27.60
Lin Coef of Exp, in/in/°C $\times 10^{-6}$	
20–200 C	13.3
20–700 C	17.8
Ther Cond, 100 C, cal/sq cm/ cm/sec/°C	0.385
Ther Exp, 20–650 C, 10^{-6}/°C	
(100)	16.8
(001)	13.2
Elec Cond, % IACS	40
Elec Resistivity, μ ohm/cu cm	$3.9 - 4.3$
Electrode Potential, V	-1.69
Electrochemical Equivalent (valency 2), mg/C	0.04674
Magnetic Susceptibility, gauss/oersted (cgs)	0.79
Velocity of Sound, ft/sec	41,300

Reflectivity, (white light), %   50–55
Photoelectric Work Function, eV   3.92

Btu/lb.), among the elements surpassed only by hydrogen (52,000 Btu/lb.).

Electrical. The electrical conductivity of beryllium, is approximately 40% IACS, and remains relatively high at increased temperatures. Beryllium can be expected to minimize temperature surges when used as an instrument container box. Present and potential uses include high-temperature conductor wire and power leads, memory discs for computer applications (low inertia and high-temperature stability), gyroscope rotors, low inertia armatures, servo mechanism transfer components, actuators for hydraulic pumps and other equipment, rectangular wave guides, and antennas.

Other. Other applications and uses include X-ray cameras, high-speed aircraft cameras, aircraft brake discs, underwater instruments, high-velocity test equipment, aircraft instrument parts, accelerator targets, high-frequency electronic parts, aircraft fasteners, gas turbine engine parts, rocket engine components, hypersonic leading edges on aircraft wings, and rotor shafts in refrigeration equipment.

Beryllium-copper and beryllium oxide. In addition to its use in nuclear applications, the principal outlet for beryllium at present is in beryllium-copper because of its ability to precipitation-harden copper. Beryllium-copper alloys containing up to 2% beryllium have remarkable mechanical properties, such as high strength, wear and fatigue resistance, and electrical conductivity. Beryllium is also used as an additive to aluminum, titanium, steel, zinc and other alloy melts as an oxidizer. In magnesium it inhibits oxidizing tendencies and inflammability. Small quantities are used to refine grain size and increase fluidity of aluminum.

Present uses of beryllium oxide as a ceramic in the electrical industry and as a probable moderator for high-temperature ceramic reactors have increased the potential uses of beryllium. Beryllium oxide is a refractory of high melting point, excellent thermal conductivity, good thermal shock resistance, high electrical resistance, and chemical stability.

BISMUTH

Bismuth is a brittle, crystalline metal having a high metallic luster with a distinctive pinkish tinge. The metal is easily cast but not readily formed by working. Within a narrow range of temperature, around 225 C, it can be extruded. Its crystal structure is rhombohedral.

The metal is available commercially in purities of 99.99% (25 lbs bars, shot). World production of the metal is about 5 million lb per year of which a large proportion originates in the Western Hemisphere. The quoted price for the last few years has been $2.25 per lb (in 1 ton lots).

Physical constants

Some of the physical properties are as follows:

Atomic No.	83
Atomic Weight	209.00
Melting Point, C	271.3
Boiling Point, C	1560
Density, 20 C, gm/cu cm	9.8
Latent Heat of Fusion, cal/gm	12.5
Latent Heat of Vaporization, cal/gm .	204.3
Sp Ht, 20 C, cal/gm/°C	0.0294
Therm Cond, cal/sq cm/cm/°C/sec ...	0.020
Lin Coef of Therm Exp, μ in./°C	13.3
Elec Resis, 0 C, μ-ohm-cm	106.8

There are several unique physical properties of bismuth that are important to its uses.

Bismuth is one of the few metals (antimony and gallium the others) which increases in volume on solidification. The increase amounts to 3.3%.

Bismuth is the most diamagnetic of all metals (mass susceptibility of -1.35×10^6). It shows the greatest Hall effect (increase in resistance under influence of a magnetic field). It also has a low capture cross section for thermal neutrons (0.034 barn).

Alloys and applications

Bismuth combined with a number of metallic elements forms a group of interesting and useful low melting alloys. Some of the lowest melting of these are as follows:

	MP
49.5 Bi, 10.1 Cd, 13.1 Sn, 27.3 Pb	70 C (158 F)
49.0 Bi, 12.0 Sn, 18 Pb, 21 In	57 C (136 F)
44.7 Bi, 5.3 Cd, 8.3 Sn, 22.6 Pb,	
19.1 In	47 C (117 F)

The fusible alloys (trade names, "Asarcolo" and "Cerro" alloys) are used in many ingenious ways, e.g., sprinkler-system triggering devices, bending pipes, anchoring tools during machining, accurate die patterns, etc.

Bismuth metal (0.1%) is also added to cast iron and steel to improve machinability and mechanical properties. An alloy of 50% Bi and 50% Pb is added to aluminum for screw machine stock, to increase machinability.

A permanent magnet (bismanol) having excellent resistance to demagnetization is produced from manganese and bismuth.

The development of refrigerating systems depending on the Peltier effect for cooling uses a bismuth-tellurium or selenium alloy for thermocouples.

Bismuth is playing an important part in nuclear research. Its high density gives it excellent shielding properties for gamma rays while its low thermal neutron capture cross section allows the neutrons to pass through. For investigations in which it is desired to irradiate objects, i.e., animals, with neutrons but protect them from gamma rays, castings of bismuth are used as neutron windows in nuclear reactors.

Bismuth has been proposed as a solvent-coolant system for nuclear power reactors. The bismuth dissolves sufficient uranium so that when the solvent and solute are pumped through a moderator (graphite), criticality is reached and fission takes place. The heat generated from the fission reaction raises the temperature of the bismuth. The heated bismuth is then pumped to conventional heat exchangers producing the steam power required for eventual conversion to electricity.

The advantages of such a reactor are that (1) it has potential for producing low-cost power, (2) it has an integrated fuel processing system, and (3) it converts thorium to fissionable uranium.

Bismuth, like the other members of Group V_B, i.e., arsenic and antimony, forms two sets of compounds in which it has valences of three and five. The trivalent compounds are the more common.

Bismuth forms with oxygen a trioxide (B_2O_3), tetroxide (Bi_2O_4) and a pentoxide (Bi_2O_5). These compounds have basic properties. The trioxide is the most important oxide of bismuth. The salts of bismuth and the halogens hydrolyze in water to insoluble basic salts.

$$BiCl_3 + 2\ H_2O \longrightarrow Bi(OH)_2Cl + 2\ HCl$$

Dilution of the solution will precipitate the oxy salt and the addition of HC1 will dissolve the precipitate.

Bismuth compounds such as the subvarbonate, subnitrates and subgallate are used as medicines.

BITUMINOUS COATINGS

Bitumens have been defined as mixtures of hydrocarbons of natural or pyrogenous origin or combinations of both (frequently accompanied by their nonmetallic derivatives) which can be gaseous, liquid, semisolid, or solid and which are completely soluble in carbon disulfide.

Bitumens used in the manufacture of coatings are of the semisolid and solid variety and are derived from three sources:

1. Asphalt produced by the distillation of petroleum.

2. Naturally occuring asphalts and asphaltites.

3. Coal tar produced by the destructive distillation of coal.

It is customary to classify bituminous coatings by their application characteristics as well as by their generic composition. All of the coatings can be divided into two classes depending on whether or not they require heating prior to application.

Hot-applied coatings (Table 1) are either 100% bitumen or bitumen blended with selected fillers. A common loading for coatings employing fillers is 10 to 20% filler. Hot-applied coatings are brought to the desired application viscosity by heating. The majority of buried pipelines are coated with this type of bituminous coating.

Cold applied coatings (Table 2) employ both solvents and water to attain the desired application viscosity. A wide range of solvents is used and the choice depends mainly on the drying characteristics desired and the solvent power required to dissolve the particular bitumen being used. Various fillers are also used in cold-applied coatings to obtain specific applications and end-use properties.

Bituminous coatings can be formulated from many combinations of bitumens, solvents or dispersing agents and fillers. This makes possible a great variety of end products to meet application and service requirements. The coatings that can be produced range from thin-film (3 mil) coatings to protect machined parts in storage, up to thick (100 mil), tough coatings to protect buried pipelines.

As with all other coatings, the conditions of the surface to which a bituminous coating is applied is an important, life-determining factor. A good sandblast is preferred; especially if the surface is badly corroded and the exposure is severe. In any case the surface should be free of moisture, grease, dust, salts, loose rust and poorly adherent scale. A thin, penetrating bituminous primer can be beneficial on rusty surfaces that can only be cleaned by wire brushing and scraping.

End-use requirements

The application will dictate which performance properties of a coating should be given greatest consideration.

Service temperature. Many applications of bituminous coating require a moderate service temperature range, often no greater than that caused by weather changes. However, other applications, such as coatings for chemical processing vessels, may require much wider service temperature ranges. In any event, it is possible

to obtain good performance with bituminous coatings over a range of −100 to 325 F.

Thermal and electrical insulation. Bitumens themselves are relatively poor thermal insulators. However, by using low density fillers, coatings with good insulation properties can be formulated. These coatings both protect from corrosion and provide thermal insulation. An added advantage of the coatings is that the insulating material (low density filler) is completely surrounded by bitumen and is permanently protected from moisture. Thus, they are not subject to a loss in efficiency (as are conventional insulating systems) from damage or failure of the protecting vapor barrier.

Bitumens are naturally good electrical insulators. This is an important consideration in systems using cathodic protection as a complementary corrosion-prevention device.

Abrasion resistance. Many bituminous coatings need to have high abrason resistance. Automotive undercoatings, for example, need high abrasion resistance because they are continually being buffeted by gravel and debris thrown up by the wheels of the vehicle. These coatings can withstand 20 passes under a sandblast nozzle being operated with an air pressure of 90 to 100 psi. Abrasion-resistant coatings are also used

TABLE 1—PROPERTIES OF HOT-APPLIED PIPE COATINGS

	Asphalt-Base		Coal-Tar Base	
	Culvert Pipe Coating	Buried Pipe Coating	Buried Pipe Coating (Narrow Range)[a]	Buried Pipe Coating (Wide Range)[b]
Flash Point (Cleveland Open Cup), °F	575	600	465	—
Mineral Filler (by wt.), %	Nil	19	22	23
Softening Point (ASTM-D36), °F	200	240	195	225
Penetration (ASTM-D5)				
At 32 F, 200 g, 60 sec, mm/10	33	5	—	—
At 77 F, 100 g, 5 sec, mm/10	48	8	1	10
At 115 F, 50 g, 5 sec, mm/10	70	15	4	21
Max Temp without Sagging, F	—	160	120	160
Min Temp without Cracking, F	—	−20	+30	−20
Break Test at 77 F, 1 in. Mandrel	Pass	Pass	Fail	Pass
Resistivity, ohm-cm	4.5×10^{15}	6.5×10^{15}	1.8×10^{15}	6.2×10^{12}
Usual Film Thickness, mils	30	95	95	95
Coverage, sq ft/ton/mil	374,000	321,000	268,000	263,000

[a] 30 to 120 F service temperature range.
[b] −20 to 160 F service temperature range.

to protect interior surfaces of railroad cars or other vessels handling chemical solids or abrasive slurries.

Weathering resistance. Asphalt-base bituminous coatings generally weather better than coal-tar base coatings. Also, there is quite a wide difference in the performance of asphalt coatings derived from different petroleum crudes. By critically selecting the crude and the processing method, asphalt coatings can be formulated that will weather for many years. Numerous installations over ten years old are still to be found in good condition.

In industrial areas corrosive solids, solutions and vapors will affect weathering performance. In general, the resistance of bituminous coatings to corrosive media is equal to that of the best organic coatings. Bituminous coatings have good resistance to dilute hydrochloric, sulfuric and phosphoric acids as well as to sodium hydroxide. They also have good resistance to solutions of ammonium nitrate and ammonium sulfate. However, they have poor resistance to dilute nitric acid, and most coatings are not resistant to oils, greases and petroleum solvents.

Mechanical impact and thermal shock. Bituminous coatings generally have good adhesion when subjected to mechanical impact.

TABLE 2—PROPERTIES OF COLD-APPLIED COATINGS

	Asphalt-Base							Coal-Tar Base			
	Thin Film	Industrial Mastic	Vapor Barrier (breathing type)	Railroad Car Cement	Automotive Under-Coating	Sound Deadener	Insulating Coating	Thin Film	Mineral Filled	Industrial Mastic	Epoxy Modified
PHYSICAL PROPERTIES											
Flash Point (Pensky-Martens), °F	105	105	105	105	105	80	105	100	100	100	100
Ther Cond Btu/hr/sq ft/°F/in.	—	1.7	1.5	—	—	—	0.6	—	—	—	—
Water-Vapor Permeability (wet cup) grains/sq ft/hr/in. Hg/in.	—	0.0018	0.016	—	—	—	0.007	—	—	—	—
Sound-Deadening Ability (at 0.5 lb/sq ft), db decay											
At 0 F	—	—	—	—	13.6	10.4	—	—	—	—	—
At 50 F	—	—	—	—	8.2	15	—	—	—	—	—
At 70 F	—	—	—	—	5.9	12.2	—	—	—	—	—
At 100 F	—	—	—	—	3.8	6.8	—	—	—	—	—
MECHANICAL PROPERTIES											
Abrasion Resistance	—	Good	Good	—	Good	—	Good	—	—	—	—
Adhesion at Low Temp.	Good	Good	Good	Good	Good	Good	Good	—	—	—	Good
Service Temp. Range, F	—60 to 250	—60 to 325	—40 to 275	—60 to 325	—60 to 325	—40 to 325	—40 to 275	250 max	160 max	250 max	—30 to 400
CHEMICAL AND CORROSION RESISTANCE											
Dilute Hyd, Sulf and Phosphoric Acids	Good	Good	Good	Good	Good	—	Good	—	Good	—	Good
Nitric Acid	Poor	Poor	Poor	Poor	Poor	—	Poor	—	Poor	—	—
Sodium Hydroxide	Good	Good	Good	Good	Good	—	Fair	—	Good	—	Good
Amm Nitrate or Sulfate Sol'ns	Good	Good	Good	Good	Good	—	Good	—	Good	—	Good
Oils, Greases, Pet Solv	Poor	Poor	Poor	Poor	Poor	—	Poor	—	Good	—	Good
Salt Spray, min hr	360	500	—	—	500	—	—	—	—	—	—
Outdoor Weathering	Exc	Exc	Exc	Exc	Exc	—	Exc	Poor	Fair	Fair	Fair
APPLICATION											
Solids Content (by wt.), %	60	70	58	64	65	80	61	65	70	72	70
Mineral Filler (by wt.), %	Nil	22	25	18	28	60	—	Nil	25	25	20
Fibrous Filler Present	No	Yes	Yes	Yes	Yes	Yes	Yes	No	No	Yes	No
Application Methods [a]	B,D,S	B,S,T	B,S,T	B,S	B,S,T	S,T	S,T	B,D,S	B,S	B,S,T	B,S
Usual Film Thick, mils	3	63	63	6.5	63	27	250	3	15	72	15
Coverage, sq ft/gal/mil	905	1050	930	950	990	1010	1130	820	1040	960	970

[a] B = Brush, D = Dip, S = Spray, T = Trowel.

However, where severe mechanical impact is expected the coatings should first be field-tested or tested in the laboratory under simulated service conditions. A common laboratory test used to evaluate automotive coatings consists of a slamming device in which a coated panel is mounted on a hinged frame and allowed to swing freely against a stop. Properly formulated coatings can withstand this type of abuse even when the test is conducted at low temperatures representing the most severe operating conditions.

Adhesion after impact can also be determined by using a free falling weight on a coated panel. In a standard test used on hot-applied pipe coatings, a 2-in. steel ball is dropped from 1 ft on a coated specimen maintained at 32 F. A properly formulated coating will withstand this test without cracking or shattering.

Resistance to thermal shock is an important consideration in coatings used on some types of processing equipment. Laboratory tests in which a coated panel is transferred back and forth between hot and cold chambers can be used to predict field behavior. Bituminous coatings are available which will withstand thermal shock over a temperature range of −60 to 140 F.

Types of coatings

Because bitumens are very dark in color, heavy loadings of pigments are required to produce colors other than black. However, if color is an important consideration, certain colored paints, including whites, can be used as topcoats. Granules can also be blown into the coating before it has completely cured to produce a wide range of decorative effects.

Thin film coatings. Coatings less than 6 mils thick are arbitrarily included in this group. For the most part, these are solvent cutback coatings. They are black except in the case of aluminum paints employing bituminous bases. The coatings are inexpensive and can be used to give good protection from corrosion when color is not important.

Asphalt coatings of this type are used extensively to protect machined parts in storage. Because the coatings retain their solubility in low-cost petroleum solvents even after long weathering, kerosene or similar solvents can be used to remove a coating from the protected part just prior to placing it in service. Coal-tar base coatings are much more difficult to dissolve and cannot be used for this purpose. However, this property makes coal tar useful for protecting crude-oil tank bottoms and in other applications requiring resistance to petroleum fractions.

Industrial coatings. Heavy-bodied industrial coatings incorporate low density and fibrous mineral fillers. Coatings can be formulated that will not slump or flow on vertical surfaces when applied as thick as 250 mils. However, they are usually used in thicknesses from 6 to 120 mils.

The coatings are used extensively in industrial plants to protect tanks and structural steel from such corrosive environments as acids, alkalis, salt solutions, ammonia, sulfur dioxide, and hydrogen sulfide gases.

Industrial coatings are also used in large volume by the railroad industry. Complete exteriors of tank cars carrying corrosive liquids are often coated and provide good protection of the saddle area where spillage is likely to occur. Coatings on the exteriors and interiors of hopper cars in dry chemical service also provide protection from both corrosive action and abrasive wear.

Railroad car bituminous cements are used to seal sills and joints in box cars. An application of this material followed by an overcoating of granules makes an excellent roofing system for railroad cars.

Coatings for use over insulation. Practically all industrial insulating materials must be protected from the weather and moisture, otherwise they would lose their efficiency. Bituminous coatings formulated to have low rates of moisture vapor transmission give best results on installations operating at low (−100 F) to moderate (180 F) temperatures. Coatings that allow a higher rate of moisture transmission (breathing type) are need to protect the insulation on systems operating at 180 F and above. This is necessary so that moisture trapped beneath the coating can escape when the unit is brought to operating temperatures.

Thermal insulating coatings. Low-density fillers can be employed in bituminous mastics to produce coatings with relatively good insulating values, a k value of 0.6 Btu/sq ft/hr/°F being typical. Insulating coatings are usually applied somewhat thicker than conventional mastics in order to obtain the insulating value desired. They are commonly used in thicknesses of 250 to 375 mils and because of their thickness and resiliency they have excellent resistance to mechanical damage.

Automotive underbody coatings. These are mastic type coatings containing fibrous and other fillers. They are used to coat the undersides of floor panels, fenders, gasoline tanks and frames to protect against corrosion and provide sound deadening and joint sealing.

The coatings have high resistance to deicing salts, moisture and water. They also have sound-deadening properties which noticeably reduce the noise level inside an automobile. This provides for a more pleasant and less fatiguing ride. The sealing and bridging action of the coatings is also especially effective in reducing drafts and dust infiltration.

Sound-deadening coatings. High efficiency, sound-deadening coatings can be formulated from selected resinous bases and high-density fillers. They have better sound-deadening properties than automotive underbody coatings and are used on the wall, roof and door panels of automotive equipment where sound deadening is the primary need, rather than abrasion resistance of protection from corrosion. They are also used on railroad passenger cars, house trailers, stamped bath tubs, kitchen sinks, air-conditioning cabinets and ventilation ducts.

Pipe coatings. Industrial coatings are excellent for protection of pipe above ground. However, the environment of underground exposure and the complementary use of cathodic protection make it necessary to use specially designed coatings. The stresses created by shrinking and expanding soil require that the coating be very tough. Rocks and other sharp objects can be expected to cause high localized pressures on the coating surface. A coating must have good cold-flow properties to resist penetration by objects which can cause localized pressures as high as 100 psi.

Cathodic protection (an impressed negative electrical potential) is widely used to prevent the corrosion processes from occurring at flaws in the pipe coatings. Both asphalts and coal tars are good electrical insulators and make excellent coatings for cathodic protection applications.

Coatings for pipe are usually of the hot application type. Application can be made at the mill, at a special pipe-coating yard, or over the ditch, depending on the terrain, size of pipe and other factors. The coating may be given added strength by embedding it with a glass fabric while it is still hot. Outer wrappings of rag, asbestos and glass felts are sometimes used to give added resistance to damage by soil stresses.

Application

Bituminous coatings that are cut back with solvents or emulsified with water can be produced to consistencies suitable for application by dipping, brushing, spraying or troweling at ambient temperatures.

Dipping is usually used to coat small parts. As a rule, coating viscosity is adjusted to produce a thickness of 1 to 6 mils.

Brushing is used on areas that cannot be reached by spraying and on jobs that do not warrant setting up spray equipment. Coating thicknesses can range from 1 to 65 mils.

Spraying is the most popular method for applying cold coatings. The thickness required in one application determines the consistency of the formulation, and thicknesses of 1 to 250 mils are obtainable by spraying. Conventional paint-spray equipment can be used for coatings up to 6 mils thick. Heavier coatings require the use of mastic spray guns fed from pressure pots or heavy-duty pumps. Heated vessels and feed lines can also be used to decrease viscosity and permit faster application and the buildup of thicker films in one application.

Troweling is usually used in inaccessible areas and/or where it is necessary to produce a very heavy coating in one application. Trowel coats are usually applied in thicknesses above 250 mils.

Bituminous coatings can also be applied hot without the need for any diluents. Such coatings are widely used on piping in thicknesses of about 95 mils. They are heated to 350 to 550 F and then pumped into a special apparatus which surrounds and travels along the pipe.

BLOW MOLDINGS

Essentially, blow molding involves trapping a hollow tube of thermoplastic material in a mold. Air pressure applied to the inside of the heated tube blows the tube out to take the shape of the mold. There are many variations on the basic technique.

In short, the process is an economical, high-speed, high production rate method of forming thermoplastic parts of hollow shape, or parts that can be simply made from a hollow shape.

At present, its primary use is in the container and toy field, where bottles and toys of many different shapes are formed in large quantities at low cost. The most commonly used material is polyethylene. At this time, widespread use of the process for production of industrial parts using materials other than polyethylene has just begun.

Materials

Although any thermoplastic resin can be considered a candidate for blow molding, polyethylene is the most widely used. In fact, blow molding got its start with low density polyethylene for blow-molded squeeze bottles. Now, low, intermediate and high-density polyethylene resins are used, as well as special ethylene copolymers designed to provide greatly improved stress cracking resistance compared with polyethylene homopolymers, needed for detergent containers.

One of the main criteria of selection of a polyethylene resin for blow molding is the proper balance of physical properties required for the specific use. The squeeze bottle and normal container market use the softer, more resilient low density materials; the carrier container and riding toy end uses demand the higher rigidity, higher density resins. Regardless of resin density, blow molding grades of polyethylenes should not have a melt index higher than 2.0.

Other thermoplastics. With the extension of blow molding into broader use in industrial products, the need for engineering properties other than those of polyethylene have stimulated interest in other thermoplastics. The main plastics available for blow molding, other than polyethylene, include cellulosics, polyamides (nylons), polyacetals, polycarbonates, polypropylene, and vinyls.

The cellulosic family of plastics includes acetate, butyrate, propionate, and ethyl cellulose. For blow molding, the cellulosics offer strength, stiffness, transparency, and high surface gloss. They have unlimited color possibilities. Chemical resistance and availability of nontoxic resins make them potentially suitable for medicine and food packaging. Their strength, stiffness and transparency make them suitable for industrial

parts, toys, and numerous decorative and novelty items.

Polyamides or nylons, although relatively high cost materials, offer potential benefits in industrial parts and special containers, such as aerosols. Special developments have resulted in formulations tailored to special viscosity requirements for blow molding.

Polyacetal resins for blow molding offer toughness, rigidity, abrasion resistance, high heat distortion temperatures, and excellent resistance to organic solvents. Also, they are resistant to aliphatic and aromatic hydrocarbons, alcohols, ketones, strong detergents, weak organic acids, and to some weak inorganic bases. Aerosol containers appear to be one promising use for blow-molded acetal resins.

Polycarbonates, as yet relatively high in cost, should find their place in blow-molded industrial parts. Primarily they offer high toughness, strength, and heat resistance. They are transparent with almost unlimited colorability and are self-extinguishing.

Polypropylenes, somewhat similar to higher density polyethylenes, but with lower specific gravity, higher rigidity, strength and heat resistance, and lower permeability, offer interesting properties at low cost. Because of their lower permeability they may be used in containers where polyethylene is unsuited. They also have excellent stress-crack resistance.

Although few data are available on the use of PVC (polyvinyl chloride) for blow-molded parts, the materials should offer benefits in terms of variability of engineering properties. PVC's are available with properties ranging all the way from high rigidity in the unplasticized grades to highly flexible plasticized PVC's. The variability in performance resulting from the many possible formulations means that the engineer must consult with the materials supplier in attempting to obtain a formulation with the proper performance and processing characteristics to meet his needs.

Part design. Remember that the use of blow molding for parts other than symmetrical containers is still relatively new. Consequently, much development of design data is still to come. But there are several generalizations that may prove useful.

Design details. Uniform wall thickness is difficult to obtain. Reducing the blow-up ratio (maximum part dia/dia of parison) will improve uniformity. A maximum ratio of 5:1 should be observed, and where possible a ratio of 3:1 should be maintained. Excessive blow-up ratio will result in longer cycles.

Sharp angles and sharp edges must be avoided. The use of a generous radius will prevent excessive thinning of the plastic and prevent molded-in stresses.

(It should be pointed out that injection-blow molding permits greater wall thickness variations, and incorporation of greater amount of detail.)

Many techniques can be used by the blow molder to increase flexibility of design. These include: (1) Elliptical parisons to provide final rectangular shapes. (2) Parison temperature control to yield hot and cold sections in desired areas to provide controlled, nonuniform stretching of the tube. (3) Trapped objects, such as bells in blown toys, can be accomplished when blowing over a mandrel. (4) Inserts of a wide variety of types can be incorporated into the part.

Design possibilities. The scope of blow-molded design has already broadened beyond that of round hollow objects. Production of such parts as housings by blowing a unit and sawing the item along the parting line to produce two housings is already a reality. As many as ten cavities have been incorporated into a mold, using a wide tube and allowing the air to pass through a hollow sprue or runner system between parts.

The future design possibilities of blow molding appear bright. The constant improvements in equipment design continue to add flexibility to the blow-molding process.

Secondary operations. The equipment used will determine to a great extent the amount of finishing required on the parts. Parts must usually be trimmed and often decorated. Trimming may be by hand, or may involve highly automated trimming, reaming, and cutting.

Decorating of parts depends on the shape of the part and the type of material used. But generally, a variety of techniques is available, including labeling, hot stamping, silk screening, and offset printing.

Applications

As mentioned before, at present, by far the largest market for blow molding is in the container field, ranging from containers for food, drugs, and cosmetics to household and industrial chemicals. Toys, one of the first products to be blow-molded, represent a sizable market. The housewares market has been one of the slower ones to develop, because parts must compete on a quality basis with well-developed injection-molded products.

The industrial product area represents one of the biggest potential uses for the process. Here progress has as yet been slower than, for example, the container industry, because of the diversity of requirements in terms of both design shape and engineering performance. But among present products of this type are remote controls for television sets, rollers for lawn mowers, oil dispensers, toilet floats, molds for epoxy potting, and auto ducting for air conditioning.

BUTADIENE-STYRENE THERMOSETTING RESINS

Butadiene-styrene resins discussed here are low molecular weight, all-hydrocarbon, thermosetting copolymers designed primarily for surface coatings. They form tough, inert films by four different mechanisms: oxidation, polymerization, solvent evaporation, and through use of cross-linking agents.

Proper selection of curing conditions and/or modifiers provides conversions ranging from seconds to hours at temperatures from 70 to 1100 F.

In addition to coatings, the basic butadiene-styrene resin is an interesting chemical intermediate for reactions such as maleic anhydride addition, epoxidation and halogenation.

Four forms available

Butadiene-styrene resins, produced only by Enjay Chemical Co. and designated "Buton," are prepared in four forms: Resins 100, 150, 200 and 300.

Resin 100, the basic resin, is a viscous, clear material with a high degree of unsaturation. It is supplied in a solvent-free state for use in special applications such as can linings and thin film metal-strip coatings. It is readily soluble in hydrocarbon thinners and chlorinated hydrocarbons.

Resin 150 differs from Resin 100 in that it is

based on butadiene alone. It is similar to Resin 100 in molecular weight and physical properties, but possesses certain metal wetting and fabrication characteristics that make it superior for such uses as tinplate coatings.

Resins 200 and 300 are produced by modifying the base resin to introduce such polar groups as hydroxyls, carbonyls and carboxyls. As a result, they have a much more active chemical nature and are more compatible with other coating resins than either of the other two. Resin 300 differs from 200 in that it has higher polar modification.

Supplied in solution, these two resins provide superior pigment and metal-wetting characteristics, and have the ability to produce hard films at thicknesses greater than Resin 100 and 150. As coating vehicles, they can be used either alone or in conjunction with modifying agents.

All four resins are low-density materials with densities ranging from 7.65 to 9.2 lb per solid gal. Because of their high bulking value, plus a low initial cost, these resins offer coating manufacturers opportunities for economy when used alone or in blends with other resins.

Film properties

A wide range of film properties can be obtained by varying cure conditions or by modification with other resins and cross-linking agents. Basically, films have excellent hardness and flexibility as well as abrasion resistance equal or superior to other high-quality industrial coating resins.

Adhesion. Films provide excellent adhesion to steel, galvanized, tinplate, alminum, brass, and die cast zinc. A wide variety of plastics, wood, glass, untreated concrete, and most other surfaces can also be coated successfully.

Chemical resistance. Butadiene-styrene films have excellent resistance to corrosive atmospheres, particularly salt-spray and detergent solutions. In chemical resistance, the films are generally equivalent or superior to other chemically resistant coating systems. Typical properties of baked butadiene-styrene films include extended resistance to water, salt water, acids, ketones, alcohols, aromatics and other solvents.

Compatibility with other resins

Resins 200 and 300 are frequently combined with other resins to obtain specific coating prop-

erties. They are used as modifiers for such leading conventional coating resins as alkyd, vinyl, acrylic and epoxy. They can also be modified with other resins, primarily ureas, melamine and nitrocellulose.

In general, the compatibility of Resins 100 and 150 is limited; however, Resins 200 and 300 have a broad compatibility spectrum because of their polar nature.

For a variety of alkyd resins, Resins 200 and 300 upgrade the corrosion and chemical resistance of films at little or no increase in total resin costs. Alkyds modified with the polymer also have high gloss, flexibility, hardness and impact resistance.

Epoxy films modified with hydrocarbon-derived resins generally possess similar or better properties than epoxy-ester or all-epoxy systems, at lower material cost. Resin 300 is more compatible with lower molecular weight liquid epoxy resins than those of high molecular weight.

To vinyl formulations, Resins 200 and 300 contribute improved adhesion, chemical resistance and economy. In pigmented vinyl coatings, wetting and gloss characteristics are improved, and mill time is reduced by using these resins as the grinding vehicle.

Material cost savings and quicker, easier grinds are also obtained in acrylic systems by blending Resins 200 and 300. Satisfactory coatings with good mar resistance are produced at baking temperatures as low as 200 F.

Butadiene-styrene resins are modified by such resins as urea, melamine, and nitrocellulose mainly to provide lower temperature cure schedules. Low-bake films with corrosion and chemical resistance and high-build, chemically resistant lacquers are typical of the coatings which result from such modification.

Curing

Butadiene-styrene resins can be cured by several systems—air-dry, low-temperature bake, high-temperature bake, and "instant curing," a new approach to the curing of coatings.

With the addition of metallic driers, butadiene-styrene resins will air-dry in 8 to 24 hours. Curing time is rapid, however, when the resin is modified with nitrocellulose. This rapid, lacquer-type drying composition is particularly suitable for wood furniture and hardboard finishes.

In low-temperature bake systems, urea-formaldehyde or melamine-formaldehyde resins are used effectively with Resins 200 and 300 over a temperature range of 200 to 300 F. Ethyl acid phosphate or phenyl acid phosphate are recommended for quick conversion in the lower temperature range.

Unmodified butadiene-styrene resins are readily adaptable to baking at high temperatures (300 to 400 F). Films have excellent hardness, gloss and chemical resistance. Certain modifying resins or crosslinking agents can be used to optimize specific properties.

Instant curing is a brand new concept to the coatings industry and particularly applicable to the butadiene-styrene resins. High-quality metal coatings have been produced in production-line operations in as little as two seconds at 800 F.

Four means of instant cure have been found suitable: flame curing, high-density infrared heating, flame spraying and induction heating. The principle is simply to supply a great deal of heat to the coating as rapidly as possible.

Applications

Resin 100 is best suited for use in special thin-film coatings of up to 1.2 mils in such applications as can linings and flat steel primers. It may also be used as a chemical intermediate. Resin 150 displays improvement in metal-wetting and reduced tinplate sensitivity for can linings. Its uses include beverage-can base coats, as well as sanitary enamels.

The corrosion resistance, chemical resistance and enamel hold-out properties of Resins 200 and 300 make them suitable primers for a variety of metal-coating applications. Automobiles, appliances, metal wall partitions and pre-primed steel sheeting are among the uses in the primer field.

In the lacquer field, Resin 300 modified with nitrocellulose produces nonpenetrating, high-build, high-solids, chemically resistant finishes for such uses as wood furniture, metal cabinets and plastics. This resin system is also suitable as the primer-surfacer for hardboard and wall paneling.

Other uses of butadiene-styrene resins include metal strip coatings and pipe coatings produced

by instant cure methods and coatings for drum and tank liners.

BUTYL RUBBER

Butyl is a general-purpose synthetic rubber made by copolymerizing isobutylene with 1 to 3% isoprene at temperatures below −140 F in the presence of aluminum chloride catalyst. Both isobutylene and isoprene are available in substantial quantities from petroleum refinery operations.

Raw butyl rubber is used by numerous compounders and fabricators who manufacture and sell finished butyl products ranging from tires and tubes to label adhesives and stoppers for medicine bottles.

Formulating

Like other rubbers, butyl is seldom used in the pure gum state. It is usually compounded with fillers, vulcanizing or curing agents, stabilizers, processing aids, etc.; the plastic mass is then shaped or formed, and vulcanized (cured, crosslinked) to give a strong, dimensionally stable, elastic product. The properties of finished butyl articles can be varied over a wide range by skilled choice of grade of raw polymer, compounding ingredients and proportions, and curing conditions. Colored butyl articles can be made by incorporating red, blue, green, yellow, white and other pigments, though black is recommended where maximum strength and resistance to sunlight are important.

Butyl articles and shapes can be fabricated by the same techniques used with other rubbers: by molding, extruding, calendering or rolling, spread or solution coating, etc. Butyl is also available as a latex for use in coatings, adhesives, dips for textiles, etc. Vulcanization is usually accomplished by application of heat, though some systems, used in cements and sealants, cure at room temperatures.

Finished butyl articles generally cost about the same as similar products made with styrene-butadiene (SBR) synthetic rubbers, somewhat less than natural rubber products, and appreciably less than products made of other synthetic rubbers such as polychloroprenes, nitriles, polysulfides, etc.

Properties

Some of the properties of butyl vulcanizates

are summarized in Table 1. Butyl is superior to natural rubber and SBR in resistance to aging and weathering, sunlight, heat, air, ozone and oxygen, chemicals, flexing and cut-growth, and in impermeability to gases and moisture. Because of their resistance to deterioration, butyl vulcanizates retain their original properties in actual use longer than either natural rubber or SBR vulcanizates. On heating in an oxidative atmosphere, natural rubber and SBR tend to become hard and brittle, whereas butyl remains unchanged or becomes slightly softer. Butyl can be compounded to have abrasion resistance equal to that of natural rubber and SBR. Butyl also has excellent dielectric properties. In many of its performance properties butyl corresponds to the more expensive synthetics such as the polychloroprenes (neoprene), the polysulfides ("Thiokol"), and the various nitrile copolymers, though like natural rubber and SBR, butyl alone is neither oil-and-grease nor flame resistant. In common with other synthetics, butyl has advantages of uniformity and freedom from foreign matter, as compared with natural rubber.

Butyl has relatively high hysteresis loss, and thus unique dynamic properties which make it the preferred rubber in applications requiring shock, vibration, and sound absorption.

TABLE 1—PROPERTIES OF BUTYL VULCANIZATES

Specific Gravity of Polymer	0.92
Approximate Tensile Strength, psi	
Pure Gum	1500–2000
Black Reinforced	1500–3000
Colored	1000–2000
Practical Hardness Range, Shore A	
Durometer	35–80
Service Temperature Range, F	−70 to +400
Abrasion Resistance	G-E [a]
Acid, Alkali and Salt Resistance	E
Aging Resistance	E
Compression-Set Resistance	F-G
Cost	G
Dielectric Properties	
Dry	E
After Immersion in Water	E
Flex-Cracking Resistance	G
Flexibility at Low Temperatures	F-G
Heat Resistance	E
Impermeability to Gases	E
Moisture Resistance	E
Oxidation Resistance	E
Ozone Resistance	E
Shock Absorption	E
Solvent Resistance	
Hydrocarbons	P
Oxygenated Solvents	G
Animal and Vegetable Oils	E
Tear Resistance	G
Vibration Damping	E

[a] Qualitative ratings relative to other commonly used general purpose rubbers.
E = Excellent; G = Good; F = Fair; P = Poor.

Applications

Butyl has found a great many applications in the automotive transport, mechanical goods, electrical, chemical, proofed-fabric, building, and consumer-goods fields. Only a few applications can be specifically mentioned here, and in each case the characteristics which make butyl particularly suited to the application will be pointed out.

The original and still the largest-volume use for butyl is in pneumatic inner tubes. In the United States, butyl tubes virtually displaced all others years ago, and the trend continues around the world as butyl becomes increasingly available from foreign plants. Pertinent properties of butyl are its excellent impermeability to air (ten times better than natural rubber), its resistance to oxidation, its retention of strength, tear resistance, flex resistance, low-temperature flexibility, and dimensional stability even after years of use, as well as its availability in large quantities at low cost.

Butyl passenger-car tires have been successful because butyl's unique dynamic properties give a remarkably soft, cushioned ride, no squeal on braking or rounding corners, and higher coefficients of friction between tires and road which result in shorter stopping distances. The abrasion resistance of butyl tires is comparable to that of other commercial tires.

Butyl's resistance to weathering and to cutting and chipping has been demonstrated by experimental off-the-road-machinery and farm-tractor tires. One such tractor tire has been in daily service for almost ten years; experimental butyl tires in quarry service that wore out ordinary tires in six weeks were still in service after one and a half years.

Butyl's resistance to weathering, sunlight, and ozone has opened many applications in automotive weatherstrips, windshield gaskets, curtain-wall gaskets and sealers, caulking compounds and the like. Butyl-coated fabrics are extensively used as convertible tops, tarpaulins, outdoor furniture covers, etc. Such fabrics have been used as liners for irrigation ditches, adjacent widths of fabric being cemented together with butyl cement at the site to make linings wide enough for large ditches. One such installation in Utah has been in continuous use for more than ten years.

Resistance to heat is essential in many of butyl's applications, such as high-pressure steam hose. Butyl has become the preferred material for curing bags and bladders used in the vulcanization of tires. Butyl bags and bladders remain airtight through hundreds of successive 20-minute cycles at temperatures up to 340 F. Butyl is also used as the inner liner in many tubeless tires, particularly truck tires, because butyl liners remain impermeable even after exposure to the heat of several successive recapping operations.

Butyl's outstanding electrical properties, coupled with its age, ozone, and moisture resistance, have made it useful in many electrical applications. Its resistance to corona and tracking make it a preferred insulation material for power cable, and because of butyl's heat resistance, butyl-insulated cable can be used to carry more current than cable of equal diameter insulated with SBR or other rubbers. Butyl is also used in electrical encapsulation compounds, as busway, factory-wire, and communications-wire insulation, and in other miscellaneous electrical applications.

Many automotive and mechanical applications take advantage of butyl's ability to absorb shock, vibration and noise. These include axle and body bumpers, truck load-cushions, boat-dock bumpers, bowling alley bumpers, etc. Motor and machinery mounts, bridge pads, drive-shaft insulators, and gasketing applications also benefit from butyl's good resistance to permanent set from prolonged compression and vibration. Though not usually considered oil-resistant, butyl has given years of satisfactory under-the-hood service in spark-plug nipples, motor mounts, grommets, radiator hose, and other parts occasionally exposed to oil and grease.

Chemical and heat resistance requirements are satisfied by butyl in applications such as chemical tank linings, gaskets, hose, diaphragms, etc. Butyl is the preferred gasketing material for use in contact with ester type hydraulic fluids of low flammability. Its acid resistance is used to advantage in dairy hose, and in large rubber containers for the bulk transportation of foodstuffs. Chemical, heat and abrasion resistance are essential in butyl conveyor belts used to transport hot granular solids in chemical plants. Resistance to chemicals and compression set, and impermeability to gases, are important in butyl food-jar seals and medicine bottle stoppers.

C

CADMIUM

Cadmium is a silver-white, malleable metal which crystallizes in hexagonal pyramids.

The refining process produces a grade of cadmium 99.98% pure and special processes make available a high-purity grade exceeding 99.999%. (American Smelting and Refining Company is the sole U. S. producer of the high-purity metal). The commercial grades are sold as slabs (110 lb), ingots (25 lb), balls (1¼lb), sticks (¼ lb), and feathered.

Physical properties

Some physical properties are as follows:

Atomic No.	48
Atomic Weight	112.41
Melt Pt, C	321
Boil Pt, C	767
Latent Ht of Fusion, cal/gm	13.2
Latent Ht of Vaporization, cal/gm	286.4
Sp Ht, cal/gm	0.055
Therm Cond, cal/sq cm/cm/°C/sec	0.216
Electrolytic Soln Potential, v	0.4

A comparison of the electrical resistivity of some common metal coatings (μohm-cm) is as follows: Cd, 6.86; Zn, 5.92; Sn, 11.5; Ni 6.84; Pb, 20.65; Cr, 13.

Cadmium is a Group IIB, Period V, element which is divalent in practically all its stable compounds. A number of compounds are sold for commercial application, the chief ones being the oxide and sulfide.

Applications

The major use of cadmium is for protective coatings on base metals. Approximately 60% of the total consumption is in these applications.

Cadmium is usually electroplated on to ferrous metals for the purpose of protecting them against corrosion. Although the cadmium may be applied by vacuum deposition, dipping or spraying, electrodeposition is by far the major method of application. The properties of cadmium metal which make it an excellent coating are:

1. Ease and high rate of deposition (high throwing power, i.e., ability to deposit uniformly on intricate objects).
2. Good corrosion resistance to alkali and salt water.
3. Excellent ductility (parts plated can be stamped or otherwise formed).
4. Good solderability.
5. High retention of silver-white luster for extended periods.

Because it is relatively expensive, cadmium plating is commonly employed for more sheltered applications where a thin coating is sufficient for protection and its bright appearance is an advantage.

Cadmium compounds are widely used as pigments. The chief compounds are the sulfide and sulfoselenide which are often extended with barium sulfate (cadmium lithopones). They are used in automobile enamels, artificial leather, fabrics, plastics, glass, printing inks, ceramic glasses, etc. The compounds are known for their high color retention. The color of cadmium sulfide ranges from yellow to orange and cadmium sulfoselenide, from orange to deep maroon.

Cadmium oxide is used extensively in plastics. In conjunction with barium it forms a compound used to stabilize the color of finished plastics. It is also a major constituent of phosphors used in television tubes.

Cadmium is mutually soluble in a number of other metallic elements. It is combined with several of these elements to form a number of commercial alloys with special properties.

The alloys of cadmium with lead, tin, bismuth, and indium are unique because of their low melting points. For example:

Alloy	Melt Pt, C
33% Cd—67% Sn	176
18% Cd—51.2% Sn—30.6% Pb	145
40% Cd—60 Bi	144

48

20.2% Cd—25.9% Sn—53.9% Bi	103
10.1% Cd—13.1% Sn—49.5% Bi, 27.3% Pb	70
5.3% Cd—12.6% Sn—47.5% Bi, 25.4% Pb, 19.1 In	47

These fusible alloys are used in applications ranging from fire-detection apparatus to accurate proof casting and are sold under the trade names "Asarcolo" or "Cerro" alloy.

Cadmium has limited use in soft solders combined with tin, lead and zinc but its major application in the field of joining is for joints requiring higher temperature strength than can be obtained with the soft solders. Cadmium combined with silver, copper and zinc forms several brazing alloys. A typical alloy contains 35% Ag, 26% Cu, 21% Zn, 18% Cd and has a melting point of 608 C. Cadmium (0.7 to 1%) is added to copper to make a strong ductile metal which has a high annealing temperature but no serious loss of electrical conductivity. (Trolley wire is an example of its use.)

The importance of cadmium in the nuclear field depends on its high thermal neutron capture cross section. By absorbing neutrons, cadmium is employed to control the rate of fission in a nuclear reaction.

Cadmium sulfide exhibits both photosensitivity and electroluminescence, i.e., it can convert light to electricity and electricity to light.

The nickel-cadmiuum storage battery has been successfully used for supplying electric power for several decades. The battery is an alkaline type.

Toxicity

Cadmium compounds are toxic. The fumes of the cadmium as well as most of its compounds are poisonous. Therefore, use of the metal or its compounds must be accompanied by the proper precautions. Protective masks should be worn if there is any possibility of cadmium fumes.

CALENDERED SHEET

Calendering is the process of forming a continuous sheet of controlled size by squeezing a softened thermoplastic material between two or more horizontal rolls. Along with extrusion and casting, calendering is one of the major techniques used to process thermoplastics into film and sheeting. It also is used in the manufacture of flooring and to apply a plastic coating on paper, textiles and other supporting materials.

The calendering process was developed in the early days of the rubber industry. Today, the plastics, rubber, linoleum, paper and metals industries all use various roll-forming operations in the manufacture of sheeted materials. This article is concerned only with calendering of thermoplastic materials (plastics and elastomers the latter of which are calendered in a thermoplastic state prior to vulcanization).

The process consists of five steps: preblending, fluxing, calendering, cooling and wind-up. Blending of the resin powder with plasticizers, stabilizers, lubricants, colorants and fillers is usually done in large ribbon blenders. The preblend is discharged either batchwise into a Banbury mixer, or continuously into an extruder where, in either case, the compound is fluxed, i.e. heated and worked until it reaches a molten or dough-like consistency.

When a Banbury is used, the molten material from it is discharged to a 2-roll horizontal mill and thence to the calender either in a continuous strip or in batches. When an extruder is used for fluxing, the extrudate is fed directly to the calender. Alternatively, the preblend can in some cases be fed directly to the calender.

After passage through the calender, the continuous sheet of hot plastic is stripped off the last calender roll with a small, higher-speed stripping roll. The hot sheet is cooled as it travels over a series of cooling drums. The film or sheeting is finally automatically cut into individual sheets or wound up in a continuous roll.

Calender operation

On the calender itself, the plastic or rubber film or sheet from the thermoplastic melt is formed by squeezing through the clearance between one or more counterrotating pairs of rolls. A rotating bead of molten material, a so-called "bank," is usually maintained in front of each roll bite to homogenize and equalize the temperature of the melt. The last roll bite forms the melt into final film or sheet thickness. Proper tracking through the calender, i.e., transfer of the hot film or sheet from one roll to the following roll, is accomplished by differences in the temperature, speed and surface finish of the rolls.

Due to the pressure exerted on the rolls by the plastic passing between them, the rolls will deflect. For given roll dimensions, this deflection varies directly with melt viscosity and inversely with roll clearance. In the last roll nip, where the final forming of the film is done, one or both of the rolls must therefore have a "crown," i.e. a slight convexity, to offset this deflection. A given roll crown will produce a flat film within a narrow thickness at a given viscosity range. In order to extend the possible viscosity range, two methods are used: roll bending and roll axis crossing ("skewing").

Roll bending consists of the application of bending forces on a pair of auxiliary bearings located at an extension of the roll necks beyond the regular bearings. By changing the direction of the applied bending moments, the effective crown can be increased or decreased.

Roll crossing is a deliberate misalignment of the horizontal axis of the rolls which results in an effect similar to an actual increase in the roll crown. The crown can only be increased, not reduced by this technique.

The desired surface finish on the calendered film or sheeting is imparted by the last pair of calender rolls and may range from a high gloss to a heavy matte finish. Extra-high gloss or special engraved patterns can be made either by having a polished or engraved roll impinge on the last calender roll (contact embossing) or by passing the hot sheeting directly from the calender between an engraved metal roll and a rubber back-up roll (in-line embossing). Many attractive patterns are made by these techniques.

Materials and applications

The major groups of thermoplastics and elastomers which today are calendered into film or sheeting are polyvinyl chloride (plasticized and unplasticized) and natural and synthetic rubbers. ABS polymers (acrylonitrile-butadiene-styrene), polyolefins and silicones are also calendered, but in much smaller quantities.

Calendered vinyl is used extensively for floor tile and continuous flooring, rainwear, shower curtains, table covers, pressure-sensitive tape, automotive and furniture upholstery, wall coverings, luminous ceilings, signs and displays, credit cards, etc.

In contrast, calendered rubber—except for some fabric coating—is mainly an intermediate product used in the manufacture of a multitude of articles such as automobile tires, footwear, molded mechanical goods, etc.

In applications which involve the use of calenders for paper and cloth coating, the paper or fabric is fed into the last calender nip so that the plastic or rubber coating is formed on top of the material (calender coating). Frictioning is a variation of this technique by which a thin layer of rubber is squeezed or rubbed into the fabric itself, a technique widely used in the rubber industry for the manufacture of friction tapes and cord fabric for tires.

Calenders vary in size from laboratory calenders with 6×13 in. rolls to large production calenders with roll dimensions up to 36×90 in. or more. Production speeds vary greatly with materials, from 10 fpm up to 300 fpm. The effective take-off speed can in some instances be increased by stretching the hot film as it is stripped from the last calender roll. Vinyl film and sheeting are calendered in gauges from about two to 30 mils, with tolerances usually better than $\pm 10\%$ of the nominal thickness.

Calenders today are heavy, high-precision machines which are made in a variety of designs and with elaborate control equipment which automatically monitors film variations and readjusts the machine to produce the desired thickness.

CARBIDES

Of the several classes of carbides, only two types need be considered for engineering applications: the covalent carbides of silicon and boron, and the interstitial carbides of the transition metals (titanium, zirconium, hafnium, vanadium, niobium [columbium], tantalum, chromium, molybdenum, and tungsten). All of these may be characterized as hard, refractory materials of extreme chemical inertness. Other carbides such as those of aluminum, iron and manganese are too reactive to be considered for engineering applications.

Structure

Chemically, all the inert carbides, with the exception of silicon carbide, are unusual in that they exist in their typical form over a range of composition. In this respect they are more similar to alloys than true chemical compounds.

Structurally, interstitial carbides may be described as metal lattices into which the small carbon atoms have been inserted. In the ideal form this leads to a face-centered cubic structure for all of the MC types except tungsten carbide, which like the M_2C types has a simple hexagonal structure. Boron carbide is rhombohedral, and consists of a distorted boron lattice in which part of the boron atoms have been replaced by carbon atoms. Silicon carbide exists in a number of crystalline polytypes, both cubic (diamond or zinc blende structures) and hexagonal wurtzite type). All of these structures may be described on the basis of every silicon atom being surrounded tetrahedrally by four carbon atoms and every carbon atom being surrounded tetrahedrally by four silicon atoms.

Properties

Properties of carbides which make them unique are their extreme hardness, exceptional corrosion resistance, extreme refractoriness, high Young's modulus of elasticity, and high temperature strength.

Table 1 compares the properties of a few carbides with those of several metals and superalloys. The interstitial carbides exhibit many properties typical of metals, such as metallic luster and exceptionally high electrical conductivity. In fact this class of carbides is frequently referred to as the *hard metals* due to this interesting combination of properties.

All of these carbides have one serious limitation common to ceramic materials in general. They are quite brittle and have poor resistance to mechanical and thermal shock.

It has been found, however, that a dispersion of a soft material such as graphite in a hard carbide matrix yields a body which has greater resistance to thermal shock. Also, if it does fracture, it does not fail catastrophically as do homogeneous carbide bodies.

TABLE 1—PROPERTIES OF SOME TYPICAL CARBIDES VS. METALS

Melting Temperature, F	
Tantalum Carbide (TaC)	7035
"Inconel X"	2540–2600
Flexural Strength, 1000 psi	
Tantalum-Zirconium Carbide (8 TaC·ZrC)	
Room Temp	40
At 3975 F	18
Molybdenum	
Room Temp	350
At 2190 F	30
Young's Modulus, 10^6 psi	
Tungsten Carbide (WC)	102.5
Stainless Steel (typical)	30
Microhardness, kg/sq mm	
Tungsten Carbide (W_2C)	3000
Knife Steel	600

Fabrication

A variety of reactions may be used to prepare carbides. However, only two need be considered for quantity production of carbides. The simplest is a direct combination of the metal with carbon, used exclusively in the preparation of carbides where purity is more important than cost considerations. Most carbides are, however, prepared by reduction of metal oxides with carbon.

With the exception of silicon carbide, all carbide bodies are fabricated using powder metallurgical techniques, i.e., sintering and hot pressing. Many carbides are sintered in the presence of metals which increase sintering rates and yield denser structures. These metals may be used in very small amounts ($<0.5\%$) and removed by vacuum heating, or they may be used in comparatively large amounts to produce two phase metal-ceramic composite bodies (cermets).

Silicon carbide bodies are formed by infiltrating a preformed carbonaceous body with elemental silicon, forming silicon carbide in the shape desired.

Powder metallurgy techniques generally restrict the sizes available to the order of a foot in the longest dimension. Where the need exists, however, and cost is of minor importance, considerably larger bodies have been fabricated. Silicon carbide may be made in large pieces, particularly as extruded rods and tubes. All materials may be machined in the "green" (unfired) or partially sintered state with hardened steel tools. When densified, final grinding must be done with diamond, boron carbide, silicon carbide, or tungsten carbide tools.

To date more silicon carbide has been used in various engineering applications than any other single carbide. Following are several applications which demonstrate the diverse uses where these materials excell.

Complete pump assemblies have been built to transfer molten metals (corrosion resistance); pipes for heat exchangers (high thermal conductivity); cyclone separators and sand-blast nozzles (abrasion resistance); rocket nozzles (refractoriness and erosion resistance); electric-light sources (high melting temperature); and suction box covers for paper-making machines (ability to retain smooth finish without wear). All of these have been tested successfully.

CARBON AND GRAPHITE

Carbon and graphite have been used in industry for many years primarily as electrodes, arc carbons, brush carbons, and bearings. In the last decade or so, development of new types and emergence of graphite fibers as a promising reinforcement for high performance composites, has significantly increased the versatility of this family of materials.

Composition and structure

The raw materials for industrial carbon are usually petroleum or anthracite cokes, lamp-black, or carbon blacks. Combined with carbonaceous binders, such as tars, pitches and resins, the carbon is compacted by molding or extrusion, and baked at between 1500–3000 F to produce what is known as industrial carbon or baked carbon. Conventional industrial graphite is made by mixing mined, natural graphite with carbon to produce in effect a carbon-graphite composite; or baked carbon can be heat treated at about 5400 F, at which temperature the carbon graphitizes.

Manufactured or artificial carbon has a two-phase structure consisting of carbon particles (or grains) in a matrix of binder carbon. Both phases consist essentially of disordered, or uncrystallized, carbon surrounding embryonic carbon crystallites. The extent of crystallite development depends on raw materials and temperature used in manufacturing. The disordered structure is characteristic of carbons and accounts for their low electrical conductivity and for their abrasiveness as compared with graphite.

Graphites, except for the pyrolytic types, have a two-phase structure similar to that of carbon, but as the result of high-temperature processing contain well-developed graphite crystallites in both phases. These multicrystalline graphites exhibit many of the properties of single-crystal graphite, such as high electrical conductivity, lubricity, and anisotropy. Compared to carbon, graphite has higher electrical and heat conductivity, better lubricity and is easier to machine. Because of their more favorable properties, graphites have broader application as engineering materials than do carbons. Therefore, most of the discussion here will be related to graphites.

Types of graphite

Industrial and Composite Carbon and Graphite—Conventional industrial carbons and graphites are available in a large number of different grades, sizes and shapes. For many uses the composition is tailored to the application. Because conventionally produced materials are porous, they are often impregnated with synthetic resins, frequently phenolics. For bearing seal and wear applications, carbon and graphite can be blended with various metal powders or plastics depending on operating conditions. Forms of industrial carbons and graphites range in size from lamp filament dimensions to solid cylinders that are 63 in. in diameter. Custom shapes are readily machined from graphite. However, machinery of carbon requires special alloy or diamond-tipped tools.

Pyrolitic graphite, made by a gas or vapor plating process (thermal decomposition type), is a highly oriented graphite in which density (2.25 specific gravity) approaches theoretical and composition can be closely controlled. Essentially the process consists of depositing carbon from a heated hydrocarbon gas as it passes over a substrate of mandrel in a vacuum furnace. The resulting deposit of highly oriented graphite can either remain as a coating on the substrate or be stripped from the mandrel to form a self-supporting part. The orientation of the majority of graphite crystals provides a high degree of anisotropy in such properties as strength, and electrical and thermal conductivities.

Recrystallized graphite is produced by a proprietary hot-working process, which yields recrystallized or "densified" graphite with specific gravities in the 1.85 to 2.15 range, as compared with 1.4 to 1.7 for conventional graphites. Major attributes of the material are high degree of quality reproducibility, improved resistance to creep, a grain orientation that can be controlled from highly anisotropic

Mechanical properties

to relatively isotropic, lower permeability than usual, absence of structural macroflaw, and ability to take a fine surface finish.

Graphite fibers are produced from organic fibers such as rayon. The fiber or textile form (e.g., fabric, yarn, or felt), is graphitized at temperatures up to 5400 F. The resulting fibers are high purity (99.9%) graphite, with extremely high individual fiber strengths.

PT graphites consist essentially of graphite fiber impregnated or bonded with an organic resin (such as furfural) and then carbonized. The result is a graphite-reinforced carbonaceous material with a high degree of thermal stability. The composite has a low density (0.93 to 1.2 specific gravity), resulting in what are reported to be the highest strength-to-weight ratios of any known material at temperatures in the 4000 to 5000 F range.

Colloidal graphite consists of natural or artificial graphite in very fine particle form, coated with a protective colloid, and dispersed in a liquid. The selection of the liquid—water, oils or synthetics—is made on the basis of intended use of the product. Significant characteristics of colloidal graphite dispersions are that the graphite particles remain in suspension indefinitely, and the particles "wick"—that is, they are carried by the liquid to most places penetrated by the liquid.

Diamond—Diamond is the cubic crystalline form of carbon. When pure, diamond is water clear, but impurities add shades of opaqueness including black. It is the hardest natural material with a hardness on the Knoop scale ranging from 5500 to 7000. It will scratch and be scratched by the hardest man-made material, Borozon. It has a specific gravity of 3.5. Diamond has a melting point of around 7000 F, at which point it will graphitize and then vaporize. Diamonds are generally electrical insulators and nonmagnetic. Synthetic diamonds are produced from graphite at extremely high pressures (800,000 to 1,800,000 psi) and temperatures from 2200 to 4400 F. They are up to 0.01 carat in size and are comparable to the quality of industrial diamonds. In powder form they are used in cutting wheels. Of all diamonds mined, about 80% by weight are used in industry. Roughly 45% of the total industrial use is in grinding wheels. Tests have shown that under many conditions synthetic diamonds are better than mined diamonds in this application.

Strength—Compared to metals and most polymers, room temperature tensile strengths of conventional carbon and graphites are low, ranging from 1000 to 2000 psi. Compressive strengths range from 3000 to 8500 psi. These are generally with-grain values (i.e., specimen length is parallel to the grain). The degree of anisotropy in graphites varies, but cross-grain strengths are usually substantially lower. Tensile strengths of the newer engineering graphites are substantially higher, with that of pyrolytic graphite reaching around 14,000 psi.

The outstanding property of all graphites is their high-temperature strength. Unlike most materials, which rapidly lose strength at elevated temperatures, the strength of graphite increases continuously from room temperature to a maximum at about 4530 F. For graphites in general, strengths at 4500 F are about double those at room temperature (if protected from oxidation); and at 4500 F, graphite's strength exceeds that of any known material at that temperature. Pyrolytic graphite has a fracture strength in excess of 40,000 psi at 5000 F. Of great significance in some cases is strength-to-weight ratio of graphites. Because of low density and its increase in strength with temperature, graphite is unexcelled among materials at temperatures in the 4500 F range.

Stiffness—The modulus of elasticity in tension for conventionally manufactured graphites is about 0.5 to 1.75 million psi. The newer graphites have modulii up to around 2.5 million psi for recrystallized graphite, and as high as 5.5 to 8 million psi for pyrolytic along the grain. Higher modulii would be expected in the newer, higher density materials. Both conventional and recrystallized graphites increase in modulus with increasing temperature. On the other hand, modulus of pyrolytic graphite decreases with increasing temperature (with grain).

Creep—Graphite behaves like a viscoelastic material at high temperatures. Degree of creep depends on many variables, including density, particle-binder nature, and size, type and orientation of the crystallites and particles. Even at room temperature graphite is not perfectly elastic and yields plastically to some extent. The greater plasticity of graphite at temperatures above 2900 F greatly increases the rate of creep. Significant improvements in creep resistance are found in both the recrystallized graphites and the pyroelectric materials.

TABLE 1—CARBONS AND GRAPHITES

Typical Properties [a] (Values in bold are with grain, others across grain)

Material		Bulk Density, gm/cu cm (lb/cu in.)	Ten Str, 1000 psi (ASTM C190)	Spec Str, 1000 in.[h]	Flex Str, 1000 psi (ASTM C78)	Compr Str, 1000 psi (ASTM C109)	Young's Mod (sonic) 10^6 psi	Elec Res 10^{-4} ohm-cm	Coef of Ther Exp (70 F), 10^{-6}/F	Ther Cond, Btu/hr/sq ft/ °F/ft
Carbon-Graphites [b]		**1.6–2.5** (0.058–0.09)	**0.7–8.8**	**7.8–152**	**3.5–16**	**4.5–50**	**0.6–4.0**	**9–30**	**1.0–9.4**	**6.6–18**
Industrial Graphites [c]	General Purpose	**1.54–1.78** (0.055–0.064)	**0.4–1.4** 0.4–1.3	**0.6–24**	**0.75–3.1** 0.98–2.4	**1.9–5.9** 2.0–5.9	**0.5–1.8** 0.5–1.2	**7.99–11.3** 10.1–15	**0.7–1.3** 1.1–2.1	**65–94** 15–97
	Premium	**1.4–1.73** (0.05–0.063)	**1.1–1.7** 0.9–1.3	**18–34**	**2.2–3.3** 1.7–3.3	**3.9–8.4** 4.4–8.5	**0.8–1.7** 0.7–1.2	**7.7–50.3** 9.8–16.4	**0.6–2.2** —	**65–97** 46–77
Recrystallized Graphite [d]		**1.95** (0.07)	**4** 1.2	**57**	**2.4–5.4** 1.8–2.4	**5.5–7.2** 6.5–12.0	**1.5–2.7** 0.8–0.85	— —	**0.4–1.0** 3.1–4.5	**90–140** 39–80
Pyrolytic Graphite [e]		**1.90–2.26** (0.068–0.081)	**7–14** 0.5	**86–205**	— —	**15** 60	**4.0** —	**5.0** 2000–6000	**1.1–2.2** —	**215** 1.08
Graphite Textiles	Cloth	— —	**27** [g] 24	— —	— —	— —	— —	**0.47** [j] 0.51	**0.5–2** —	**22** —
	Felts	**0.08–0.16** (0.003–0.006)	**0.6–10.0** [g] 0.3–11.0	— —	— —	— —	— —	**0.3–0.6** [j] 0.3–0.9	— —	**0.1–0.4** —
PT Graphites [f]	Reinforced with Graphite Cloth	**0.91–1.06** (0.033–0.038)	— —	**31.6–203** [i]	**1.2–6.7**	— —	**0.3–0.9** ~ —	— —	— —	— —
	Reinforced with Graphite Felt	**0.81–1.09** (0.029–0.039)	— —	**82–289**	**3.2–8.4**	**13.7**	**1.1–1.7**	— —	— —	— —
	Reinforced with Carbon Felt	**0.94–1.11** (0.034–0.040)	— —	**57.5–335** [i]	**2.3–11.4**	**3.3–24.3**	**0.9–1.2**	— —	— —	— —

[a] General comparative properties at room temperature. As no standard grades exist, the grouping here is somewhat arbitrary. Properties represent general averages obtainable within the groups.

[b] General ranges obtainable in combined carbon-graphite materials, both impregnated and nonimpregnated. Properties should be considered only indicative. Since little orientation occurs, materials are essentially isotropic and thus only one set of values is given.

[c] Designations such as General Purpose and Premium are not standardized in the industry; they are used here arbitrarily.

[d] Value ranges cover three grades of "Z" series, developed by National Carbon Co., Div. of Union Carbide Corp. Note particularly the greater degree of anisotropy than is found in conventional and premium grades.

[e] Data primarily on materials produced by High Temperature Materials, Inc.

[f] Preliminary data. Materials are formed by impregnating indicated reinforcement with a carbon-forming material (e.g., furfural resin) and carbonizing the composite. Each range of values covers two or three grades reported.

[g] Textile properties measured and expressed in different fashion (ASTM D39), i.e., tensile strength is in lb per in., with bold face values lengthwise, light face values transverse.

[h] With-grain tensile strength (psi) divided by density (lb per cu in.); breadth of range is due to obtaining minimum by dividing lowest strength by highest density and maximum by dividing highest strength by lowest density.

[i] Flexural strength (psi) divided by density (lb per cu in.).

[j] In ohms per square.

Thermal properties

Carbon and graphite do not have a true melting point. They sublime at 7590 F. Conventional graphites have exceptionally high thermal conductivity at room temperature while carbon has only fair conductivity. Conductivity with the grain in graphite is comparable to that of aluminum; across the grain it is about the same as brass. Conductivity increases with temperature up to about 32 F; it then remains relatively high, but decreases slowly over a broad temperature range, before it drops sharply. In pyrolitic graphite, thermal conductivity with the grain approaches that of copper; across grain, it serves as a thermal insulator and is comparable to that of ceramics.

Thermal expansion of carbon and graphite is quite low (1 to 1.5×10^{-6} per °F)—less than one third those of many metals. Expansion of graphite across grain increases with increasing density, while along the grain it decreases with increasing density. But expansion increases in both directions with increasing temperatures.

An outstanding feature of graphites is their excellent shock resistance. No standard test exists to evaluate this property. However, on a comparative basis, the shock resistance of graphite far surpasses that of many ceramics and metals.

Chemical and oxidation resistance

Graphite is one of the most chemically inert materials, being subject to only three types of attack: oxidation, formation of lamellar compounds, and reaction with carbide-forming metals at very high temperatures.

Chemical Resistance—In resistance to acids and alkalies, carbon and graphite are highly inert to most, except to strong oxidizing media. Hydrofluoric acid does not attack carbon unless it is serving as the positive electrode in the electrolysis of aqueous hydrofluoric acid. The reaction with hydrochloric acid is quite similar. Carbon is slightly attacked by boiling sulfuric acid, the extent depending on the type of carbon. Carbon and graphite are both attacked by concentrated nitric acid, with the formation of mellitic acid, hydrocyanic acid, or carbon dioxide and nitrous oxide, depending on the conditions. Alkali hydroxides do not react with carbon or graphite in solution. When fused at elevated temperatures they react to form hydrogen.

Oxidation resistance—One serious shortcoming of graphite is that it begins to oxidize in air at about 800 F. At low temperatures, however, graphite is actually less reactive to oxygen than many metals. The problem is that, unlike many metal oxides, the oxide is volatile and does not form a protective film on the graphite. The oxidation of carbons and graphites varies widely, depending on such variables as the nature of the carbon, impurities present, degree of graphitization and particle size. No specific data on oxidation rate vs temperature can be given meaningfully.

Much of the work has been aimed at developing suitable high-temperature coatings to protect graphite from oxidation. One of the most successful of these has been silicon carbide or siliconized silicon carbide. Such coatings provide reasonable protection for a few hours at temperatures as high as 3000 F. For such uses, special grades of graphite are available with coefficients of thermal expansion matching that of silicon carbide.

Electrical properties

The electrical characteristics associated with carbon and graphite's use as electrodes or anodes are relatively well-known. Carbon and graphite are actually semiconductors, their electrical resistivity, or conductivity, falling between those of common metals and common semiconductors. At temperatures approaching absolute zero, carbon and graphite have few conducting electrons, the number increasing with increasing temperature. Thus, electrical resistivity decreases with increasing temperature. On the other hand, though increasing electron density tends to reduce resistivity as temperature rises, scattering effects may become dominant at certain temperatures in the range of 1800 F and thus modify or even reverse this trend. Pyrolytic graphite with its higher density has improved electrical conductivity (along the grain). Further, its high degree of anisotropy results in a high degree of electrical resistivity across the grain.

Other properties

Surface—Graphite has excellent lubricity and relatively low surface hardness; carbon has fair lubricity and relatively high surface hardness. Further, certain types of carbon graphitize relatively easily, others do not. Consequently, a wide variety of carbon, carbon-graphite and graphite materials are available, each designed to provide specific types of surface characteristics for such uses as bearings and seals. Grades are also available impregnated with a wide variety of substances, from synthetic resin or oil to a bearing metal.

Nuclear—The nuclear grades of carbon and graphite are of exceptionally high purity. As a moderator and reflector in nuclear reactors, they have no equal because of their low thermal neutron absorption cross section and high scattering cross section coupled with high strength at elevated temperatures and thermal stability in nonoxidizing environments. In general, the properties of carbon and graphite are improved by exposure to nuclear radiation. Hardness and strength increase while thermal and electrical conductivity decrease.

CARBON (PLAIN) STEELS, WROUGHT

By definition, plain carbon steels are those that contain up to about 1% carbon, not more than 1.65% manganese, 0.60% silicon and 0.60% copper, and only residual amounts of other elements, such as sulfur (0.05% max.) and phosphorus (0.04% max.). They are identified by means of a four digit numerical system established by the American Iron and Steel Institute (AISI). The first digit is the number one for all carbon steels. A zero after the one indicates non-resulfurized grades; a one for the second digit indicates resulfurized grades; and the number two for the second digit indicates resulphurized and rephosphorized grades. The last two digits give the nominal (middle of the range) carbon content in hundredths of a percent. For example, for grade 1040, the 40 represents a carbon range of 0.37 to 0.44%. If no prefix letter is included in the designation, this specifies that the steel was made by basic open hearth, basic oxygen or electric furnace process. The prefix, B, stands for the acid Bessmer process. The letter, L, between second and third digits identifies leaded steels, and the suffix, H, indicates that the steel was produced to hardenability limits.

Properties

For all plain carbon steels, carbon is the principal determinant of many performance properties. Carbon has a strengthening and hardening effect. At the same time it lowers ductility as evidenced by a decrease in elongation and reduction of area. In addition, a rise in carbon content lowers machinability and decreases weldability. The amount of carbon present also affects physical properties and corrosion resistance. With an increase in carbon content, thermal and electrical conductivity decline, magnetic permeability decreases drastically and corrosion resistance is lowered.

TABLE 1—MECHANICAL PROPERTIES OF TYPICAL PLAIN CARBON STEELS

AISI Type and Condition	Tens. Str., 1000 psi	Elong., % in 2 in.	Hardness, Bhn	Imp. Str., Izod, ft-lb
1010, As-rolled	40–65	25–50	110–140	–
1020, As-rolled	60–70	35–40	125–150	60–80
1030, Quench & Temp.	75–120	17–35	180–490	–
1050, Quench & Temp.	95–150	10–30	190–320	16–50
1080, Quench & Temp.	116–190	10–25	220–390	10–12
1095, Quench & Temp.	120–190	10–26	230–400	5–6

Note: Range of values represents differences in heat treatment

Carbon steel grades

Plain carbon steels are commonly divided into three groups, according to carbon content:

Low carbon   — up to 0.30%
Medium carbon   — 0.31% to 0.55%
High carbon   — 0.56% to 1%

Low carbon steels are the AISI grades 1005 to 1030. Sometimes referred to as mild steels, they are characterized by low strength and high ductility, and they are nonhardenable by heat treatment except by surface hardening processes. Because of their good ductility, low carbon steels are readily formed into intricate shapes. Cold work increases strength and decreases ductility. Where necessary annealing is used to improve ductility after cold working. These steels are also readily welded without danger of hardening and embrittlement in the weld zone. Although low carbon steels can not be through hardened they are frequently surface hardened by various methods which diffuse carbon into the surface. Upon quenching, a hard, abrasion resistant surface is obtained. The most common methods used are carburizing, carbonitriding and cyaniding.

Low Temperature carbon steels have been developed chiefly for use in low temperature equipment and especially for welded pressure vessels. They are low carbon (0.20 to 0.30%) high manganese (0.70–1.60%), silicon (0.15–0.60%) steels, which have a fine grain structure with uniform carbide dispersion. They feature moderate strength with toughness down to −50 F.

Medium carbon steels are the grades 1030 to 1055. They usually are produced as killed, semi-killed or capped steels and are hardenable by heat treatment. However, hardenability is limited to thin sections or to the thin outer layer on thick parts. Medium carbon steels in the quenched and tempered condition provide a good balance of strength and ductility. Strength can be further increased by cold work. The highest hardness practical for medium carbon steels is about 550 Bhn (Rockwell C55). Because of the good combination of properties, they are the most widely used steels for structural applications, where moderate mechanical properties are required.

High carbon steels are the grades 1060 to 1095. They are, of course, hardenable with a maximum surface hardness of about 710 Bhn (Rockwell C64) achieved in the 1095 grade. This high hardness suits these steels for wear resistant parts. So called spring steels are high carbon steels that are available in annealed and pretempered strip and wire. Besides their spring applications, these steels are used for such items as piano wire and saw blades.

Free machining steels are low and medium carbon grades with additions of sulfur (0.08 to 0.13%), sulfur-phosphorus combinations, and/or lead to improve machinability. They are grades 1108–1151 for sulfur grades and 1211–1215 for phosphorus and sulfur grades. The presence of relatively large amounts of sulfur and phosphorus cause some reduction in cold formability, weldability and forgeability as well as a lowering of ductility, toughness, and fatigue strength.

Production types

Steelmaking processes and methods used to produce mill products, such as plate, sheet and bars, have an important effect on a steel's properties and characteristics.

Deoxidation Practice—Steels are often identified as to the degree of deoxidation resulting during steel production. Killed steels, because they are strongly deoxidized, are characterized by high composition and property uniformity.

They are used for forging, carburizing and heat treating applications. Semi-killed steels have variable degrees of uniformity, intermediate between those of killed and rimmed steels. They are used for plate, structural sections, and galvanized sheets and strip. Rimmed steels are deoxidized only slightly during solidification. Carbon is highest at the center of the ingot. Because the outer layer of the ingot is relatively ductile, these steels are ideal for rolling. Sheet and strip made from rimmed steels have excellent surface quality and cold forming characteristics. Capped steels have a thin low-carbon rim which gives them surface qualities similar to rimmed steels. Their cross section uniformity approaches that of semi-killed steels.

Melting Practice—Steels are also classified as air melted, vacuum melted or vacuum degassed. Air melted steels are produced by conventional melting methods, such as open hearth, basic oxygen and electric furnace. Vacuum melted steels encompass those produced by induction vacuum melting and consumable electrode vacuum melting. Vacuum degassed steels are air melted steels that are vacuum processed before solidification. Compared to air melted steels, those produced by vacuum melting processes have lower gas content, fewer nonmetallic inclusions, and less center porosity and segregation. They are more costly, but have better mechanical properties, such as ductility and impact and fatigue strengths.

Rolling Practice—Steel mill products are produced from various primary forms such as heated blooms, billets and slabs. These primary forms are first reduced to finished or semifinished shape by hot working operations. If the final shape is produced by hot working processes, the steel is known as hot rolled. If it is finally shaped cold, the steel is known as cold finished, or more specifically as cold rolled or cold drawn. Hot rolled mill products are usually limited to low and medium, non-heat treated, carbon steel grades. They are the most economical steels, have good formability and weldability and are used widely for large structural shapes. Cold finished shapes, compared to hot rolled products have higher strength and hardness and better surface finish, but are lower in ductility.

CAST IRONS

Cast iron is a generic term for a group of

metals that basically are alloys of carbon and silicon with iron. Relative to steel, cast irons are high in carbon and silicon contents, as illustrated below:

Metal	Typical Composition, %				
	C	Si	Mn	P	S
Cast Steel	0.5–0.9	0.2–0.7	0.5–1.0	0.05	0.05
White Iron	1.8–3.6	0.5–2.0	0.2–0.8	0.18	0.10
Malleable Iron	2.0–3.0	0.6–1.3	0.2–0.6	0.15	0.10
Gray Iron	2.5–3.8	1.1–2.8	0.4–1.0	0.15	0.10
Ductile Iron	3.2–4.2	1.1–3.5	0.3–0.8	0.08	0.02

In addition to the major elements given above, all of these metals may contain other alloys added to modify their properties.

Size varies widely. Iron castings are produced in an exceptionally wide range of sizes and weights. Among the smallest are piston rings a fraction of an inch in diameter and weighing less than 1 oz. Among the largest are steam-turbine bases 20 ft long and weighing 180,000 lb.

Most cast iron is manufactured by melting a mixture of steel scrap, cast-iron scrap, pig iron, and alloys in a cupola using coke as a fuel. A small percentage is melted in electric furnaces. Most cast iron is poured into molds of silica sand bonded with bentonite, fireclay, and water. A small percentage is cast into metal molds or into molds of baked or fired ceramics. Internal cavities are formed by hard but collapsible cores of sand bonded with drying oils or synthetic resins. Small molds and cores usually are made by machine, using patterns of wood or metal. Very large molds are made mostly by hand using patterns made of wood.

CAST PLASTICS

Plastics casting materials can be generally classified in two groups: (1) those resins, usually thermosetting, which are cast as liquids and cured by chemical crosslinking either at room temperature or elevated temperatures, and (2) thermoplastics which are supplied essentially in suspension or monomeric form and fused or polymerized at elevated temperatures.

Materials

Each casting resin has a unique combination of properties, such as heat resistance, strength, electrical properties, chemical resistance, cost

and shrinkage, which dictate their use for specific applications.

The most commonly used casting resins are phenolics, polyesters and epoxies. Others are vinyls, acrylics, and urethane elastomers. Following is a brief discussion of each major type.

Phenolic resins. Phenolic resins are used for low-cost parts requiring good electrical insulating properties, heat resistance, or chemical resistance. The average shelf life of this resin is about one month at 70 F. This can be extended by storing it in a refrigerator at 35 to 50 F. Varying the catalyst (according to the thickness of the cast) and raising the cure temperature to 200 F will alter the cure time from as long as eight hours to as short as fifteen minutes.

Some shrinkage occurs in the finished casting (0.003 to 0.015 in. per in.), depending on the quantity of filler, amount of catalyst and the rate of cure. Faster cure cycles produce a higher rate of shrinkage. Since the cure cycle can be accelerated, phenolics are used in short-run casting operations.

Cast phenolic parts are easily removed from the mold if the parting agents recommended by the supplier are used. Postcuring improves the basic properties of the finished casting.

Polyester resins. Polyester resins are primarily used in large castings such as those required in the motion picture industry or for large sculptures for museums, parks and display purposes. Since polyester shrinkage is about 0.006 to 0.008 in. per in., castings are usually reinforced with glass cloth or mat and are generally cast in a flexible mold. Catalysts used to initiate the cure are peroxides or hydroperoxides and activators are cobalt naphthanate, alkyl mercaptans or dialkyl aromatic amines. Recently, isophthalic polyesters have been introduced containing isophthalic acid which provides improved heat, chemical and impact resistance.

Clear polyester castings can be made using diallyl or triallyl cyanurate type polyesters. Triallyl cyanurate polyesters are used in casting clear sheets since they have excellent scratch and heat resistance.

Acrylic resins. The process involved in casting acrylic resins is complex and forms a specialized field. The methyl-methacrylate monomer contains inhibitors which must be removed before adding the catalyst. The resin must be cured under very accurately controlled conditions.

The primary use of cast acrylics is in optically clear sheet, rod, or tube stock and in the embedment of specimens for museums and display of industrial parts; also, for the embedment of decorative motifs in the jewelry industry.

Vinyls. Plasticized vinyls (polyvinyl chloride or PVC) are used in industry in a variety of plastisol processing techniques, e.g., slush casting and rotational casting. Electroformed molds are commonly used for this purpose. Since the conversion of the semiliquid vinyl plastisol to a solid consists of fusing the suspended vinyl particles to each other, it is only necessary to raise the temperature of the mass to 180 to 350 F according to the formulation. Cast-vinyl prototype parts can be produced which are comparable to molded parts.

Epoxy resins. Most cast epoxy resins, other than those used in plastic tooling, are used in encapsulating electrical components and in casting prototype parts. The variations in properties possible, makes them very versatile materials. They can be made to have almost infinite shelf life, can be varied from a liquid to a thixotropic gel, and can be highly flexibilized by the addition of polysulfides and polyamides. They can be formulated to provide heat resistance up to 500 F.

Cure cycles may range from one hour to sixteen hours, at which time the casting is removed from the mold. Epoxy resins for casting are available in transparent water-white, semitransparent and opaque formulations. Room-temperature cures are effected when aliphatic amines are added to the resin in exact amounts. Heat resistance of such systems is about 180 F. Cure is relatively rapid, therefore exothermic reaction produces relatively high temperatures; thus, casting must generally be limited in thickness.

Proprietary amine hardeners and epoxy resin systems are available whose heat resistance is about 250 to 350 F. The use of liquid anhydrides yields castings with heat resistance above 400 F. Such systems permit casting in large masses since the exothermic reaction and the curing temperatures are low (about 250 F). The pot life is several days, since elevated temperatures are required to initiate the reaction.

Several facts are basic in the use of these systems. Since it is necessary to use acidic catalysts in order to achieve heat resistance,

castings tend to be more brittle; also more difficulty is encountered in releasing the cast from the mold. The use of the proper release agent with a given resin system is necessary in order to overcome this tendency.

CAST STEELS

The general nature and characteristics of cast steels are, in most respects, closely comparable to wrought steels. Cast and wrought steels of equivalent composition respond similarly to heat treatment and have fairly similar properties. A major difference between them is that cast steel is more isotropic in structure. Therefore, properties tend to be more uniform in all directions in contrast to wrought steel, whose properties generally vary, depending on the direction of hot or cold working.

Plain Carbon—Cast plain carbon steels can be divided into three groups—low, medium and high carbon. However, cast steel is usually specified by mechanical properties, primarily tensile strength, rather than composition. Standard classes are 60,000, 70,000, 85,000 and 100,000. Low carbon grades, used mainly in the annealed or normalized conditions have tensile strength ranging from 55,000 to 65,000 psi. Medium carbon grades, annealed and normalized, range from 70,000 to 100,000 psi. When quenched and tempered, strength exceeds 100,000 psi.

TABLE 1—MECHANICAL PROPERTIES OF TYPICAL CAST STEELS

Class & Condition	Tens. Str., 1000 psi	Elong., % in 2 in.	Hardness, Bhn	Imp. Str., Charpy ft-lb
Carbon Grades:				
60,000, Annealed	63	30	131	12
80,000, Norm. & Temp.	82	23	163	35
100,000, Quench & Temp.	105	19	212	40
Alloy Grades:				
65,000, Norm. & Temp.	68	32	137	60
80,000, Norm. & Temp.	86	24	170	48
105,000, Norm. & Temp.	110	21	217	58

Ductility and impact properties of cast steels are comparable, on average, to those of wrought carbon steel. However, the longitudinal properties of rolled and forged steels are higher than those of cast steel. Endurance limit strength ranges between 40 and 50% of ultimate tensile strength.

Low alloy steel castings are considered to be in the low alloy category if their total alloy content is less than about 8%. Although many alloying elements are used, the most common ones are manganese, chromium, nickel, molybdenum and vanadium. Small quantities of titanium and aluminum are also used for grain refinement. Carbon content is generally under 0.40%. The standard categories of low alloy cast steels for specification purposes, in terms of tensile strength are 65,000, 80,000, 105,000, 150,000 and 175,000. For service at elevated temperatures, however, chemical compositions as well as minimum mechanical properties are often specified.

Applications

The wide range of available compositions has led to a wide variety of applications which make use of one or more of the following properties: strength; impact, corrosion and abrasion resistance; weldability; machinability; ductility; high elastic limit; good fatigue properties; and fabricability.

Here are a few of the industries and some of the specific products which are being made from cast steel: automotive (frames, wheels, gears); electrical manufacturing (rotors, bases, housings, frames, shafts); transportation (couplings, draw bars, brake shoes, wheel truck frames); marine (rudders, stems, anchor chain, ornamental fittings, capstans); off-the-road equipment (crawler side frames, levers, shafts, tread links, turntables, buckets, dipper teeth); municipal (fire hydrants, catch basins, manhole frames and covers); miscellaneous (ingot and pig molds, rolling mill rolls, blast-furnace ingot buggies, engine housings, cylinder blocks and heads, crankshafts, flanges and valves).

CELLULOSE PLASTICS

Cellulose is a natural polymer present in all plant forms. For plastics, pure cellulose from wood pulp or cotton linters (pieces too short for textile use) is reacted with acids or alkalies and alkyl halides to produce a basic flake. Depending upon the reactants, any one of four esters of cellulose (acetate, propionate, acetate butyrate, or nitrate) or a cellulose ether (ethyl cellulose) may result. The basic flake is used for producing both solvent cast films and molding powders.

For molding powders, the flake is then compounded with plasticizers, pigments and sometimes other additives. At this stage of manufacture, the plastics producer is able to adjust hardness, toughness, flow, and other processing characteristics and properties. In general, these qualities are spoken of together as flow grades. The flow of a cellulose plastic is determined by the temperature at which a specific amount of the material will flow through a standard orifice under a specified pressure. Manufacturers offer cellulosic molding materials in a large number of standard flow grades, and, for an application requiring a nonstandard combination of properties, are often able to tailor a compound to fit.

Cellulosics provided both the first thermoplastic (cellulose nitrate, 1868) and the first injection molding material (cellulose acetate, 1932). All except cellulose nitrate are processed on conventional molding and extrusion equipment; special handling required for nitrate is noted below. Cellophane film, the largest volume and perhaps the best known cellulose material, is not a thermoplastic, but simply regenerated cellulose. Cellophane films with widely varied properties achieved by treating and coating are popular in packaging and other special uses such as windings on wire and cable.

General characteristics

Parts molded of cellulosic materials usually have a fine, lustrous surface. The cellulosics are available in a complete range of colors, including transparents and clear. Often as desirable as the toughness for which cellulosics are usually chosen is their relative freedom from static charges.

Cellulosic molding powders mold easily on conventional injection and extrusion equipment. Because cellulosics absorb moisture from the air, they must be dried before molding. Moisture absorption also affects the dimensions of a molded part, and should be considered in part design.

Special lacquers are available for painting cellulosic parts. Use of recommended lacquers will assure compatibility of lacquer and plastic. Occasionally, a cellulosic part may be employed directly in contact with another lacquered part —say, of wood or metal. To minimize possible softening of the adjacent lacquer by plasticizer migration, the cellulosic part in such instances should be molded in the hardest flow (lowest plasticizer level) possible.

Vacuum metallizing is successfully done on all of the cellulosics. When this type of surface treatment is contemplated, the end-user or molder should advise his plastics supplier in order that, should several formulations be suitable for the part involved, he may be assured of getting the best formulation for the metallizing process.

Cellulosic molding materials are soluble in numerous common solvents, including most low-boiling ketones and esters. When these solvents are used as adhesives, a fabricator is able to form strong bonds quickly between component cellulosic parts.

Included on the properties chart is a listing of forms of the cellulosics available as raw materials. Ranges are given for most properties in this table and it should be understood that specific materials are limited to possible combinations of properties; for example, Izod impact is highest on the more highly plasticized (softer) flow grades, and therefore the hardest flow cannot also possess the highest Izod impact value.

Among current large volume products in cellulosics are the following: optical frames, telephone handsets, automotive parts (steering wheels, armrests, etc.) toys, photographic film base, clear packaging films, and sheets for vacuum forming.

Cellulose acetate

Cellulose acetate molding material is made from cellulose acetate flake combined with plasticizers and stabilizers with or without addition of coloring agents. As the first injection-molding material, cellulose acetate has benefited from a variety of technological improvements in this process.

In practical use, cellulose acetate moldings exhibit toughness superior to most other general-purpose plastics. Flame-resistant formulations are currently specified for small appliance housings and for other uses requiring this property. Uses for cellulose acetate molding materials include toys, buttons, knobs and other parts where the combination of toughness and clear transparency is a requirement.

Extruded film and sheet of cellulose acetate provide packaging materials which maintain their properties over long periods. Here also the toughness of the material is advantageously used in blister packages, skin packs, window boxes, and overwraps. It is a breathing wrap, solvent and heat-sealable.

Cast cellulose acetate film and sheet provide gel-free transparency, and excellent optical clarity. Clear fabricated boxes of this material, less common than other package types, continue as a prime specialty.

Extruded cellulose acetate film and sheet, which are lower priced than cast film or sheet, do have slight optical defects. For packaging applications where the film is near the merchandise, these film irregularities are unnoticeable.

Large end uses for these films and sheets include photographic film base, protective cover sheets for notebook pages and documents, index tabs, sound recording tape, as well as the laminating of book covers. The grease resistance of cellulose acetate sheet allows its use in packaging industrial parts with enclosed oil for protection.

For eyeglass frames, cellulose acetate is the material in widest current use. Because fashion requires varied and sometimes novel effects, sheets of clear, pearlescent, and colored cellulose acetate are laminated to make special sheets from which optical frames are fabricated.

Cellulosic films are often laminated with other plastics, metallic foils and papers. By combining materials, desirable properties of one material may compensate for a property lacking in another and laminations can thus do jobs impossible with any one material alone.

The electrical properties of cellulosic films combined with their easy bonding, good aging, and available flame resistance bring about their specification for a broad range of electrical applications. Among these are: as insulations for capacitors; communications cable; oil windings; in miniaturized components (where circuits may be vacuum metallized); and as fuse windows.

Cellulose triacetate is widely used as a solvent cast film of excellent physical properties and good dimensional stability. Used as photographic film base and for other critical dimensional work such as graphic arts, cellulose triacetate is not moldable.

Cellulose propionate

Cellulose propionate, commonly called "CP" or propionate, is made by the same general method as cellulose acetate, but propionic acid is used in the reaction. Propionate offers several advantages over cellulose acetate for many applications. Because it is "internally" plasticized by the longer chain propionate radical, it requires less plasticizer than required for cellulose acetate of equivalent toughness.

Cellulose propionate absorbs much less moisture from the air and is thus more dimensionally stable than cellulose acetate. Because of better dimensional stability, cellulose propionate is often selected where metal inserts and close tolerances are specified.

Largest volume uses for cellulose propionate are as industrial parts (automotive steering wheels, armrests, and knobs, etc.), telephones, toys, findings, ladies' shoe heels, pen and pencil barrels, and toothbrushes.

Cellulose acetate butyrate

"CAB" or butyrate results from the reaction between cellulose and a mixture of acetic and butyric anhydrides. Butyrate has excellent toughness plus good dimensional stability.

As with propionate, butyrate requires less plasticization than acetate or nitrate with the result that use of inserts is less critical and closer tolerances may be held.

Special formulations of butyrate are recommended for outdoor use. Extruded monofilaments of CAB take advantage of this weatherability in garden furniture webbing. Clear sheets of cellulose acetate butyrate are finding use in vacuum forming applications. It is one of the few *clear* sheet materials with good outdoor aging properties. Large volume applications for butyrate include steering wheels, pen and pencil barrels, and rigid pipe. Films of cellulose acetate butyrate are also finding electrical uses.

CAB is somewhat tougher, and has lower moisture absorption and a higher softening point (190F) than cellulose acetate. Special formulations with good weathering characteristics plus transparency are used for outdoor applications such as signs, light globes and lawn sprinklers. Other typical uses include transparent dial covers, TV screen shields, tool handles and typewriter keys. Extruded pipe is used for electrical conduits, pneumatic tubing and low-pressure waste lines.

Cellulose nitrate

Cellulose nitrate is produced from the reaction of nitric and sulfuric acids on pure cellulose. Oldest of the thermoplastics, celluloid (as it was known), was originally developed as a material for billiard balls. Among thermoplastics, it is remarkable for toughness. For many applications today, however, cellulose nitrate is not practical because of serious property shortcomings: heat sensitivity, poor outdoor aging, and very rapid burning.

Cellulose nitrate cannot be injection molded or extruded by the nonsolvent process because it is unable to withstand the temperatures these processes require. It is sold as films, sheets, rods, or tubes, from which end products may then be fabricated.

Cellulose nitrate yellows with age; if continuously exposed to direct sunlight, it yellows faster and the surface cracks. Its rapid burning must be considered for each potential application to avoid unnecessary hazard.

The outstanding toughness properties of cellulose nitrate lead to its continuing use in such applications as optical frames, shoe eyelets, ping pong balls and pen barrels.

Ethyl cellulose

A cellulose ether produced by the reaction of ethyl chloride on alkali cellulose, ethyl cellulose has the toughness characteristics of all the cellulosics with the added advantage of maintaining serviceable impact resistance to temperatures as low as -40 F. Ethyl cellulose has less dimensional change with moisture variation than any other cellulosic plastic, and it is the lightest. Ethyl cellulose is somewhat limited in color range; though less abrasion and scratch resistant than the other cellulosics, it is often chosen, because of its toughness even at low temperatures, for articles subject to rough usage.

Moldable in all conventional ways, ethyl cellulose also requires predrying. Typical current ethyl cellulose applications include football helmets, equipment housings, refrigerator parts and luggage.

Fibers

Cellulose fibers, beginning with rayons, were the first of the man-made fibers. Acetates (di- and tri-) are later fiber developments. The technology of synthetic fibers has advanced to the point where now characteristics desired in a woven fabric can be foreseen and planned before the fiber is spun. Cellulosic fibers as a class can be as varied as are all the natural fibers taken as a group.

Tensile strength in fibers is seldom critical because the loading is small, but tied in with tensile strength are the properties of elongation (considered wet and dry for fibers) and elastic recovery (usually given as % recovery/% strain), which are important in the matters of wrinkle resistance and durability. Another property always considered in textile fibers is per cent moisture regain after thorough drying, in which cellulosic fibers cover a range greater than that between cotton and wool. In the manufacture of cellulose fibers, the temperatures, rates, solution types, orientation and method of spinning all combine to yield a given combination of properties for a given fiber or fabric. Blending with other man-made and natural fibers further broadens the range of possible textile property combinations.

CELLULOSIC PLASTICS PROPERTY TABLE

	ASTM Method	Cellulose Acetate	Cellulose Propionate	Cellulose Acetate Butyrate	Ethyl-Cellulose	Cellulose Nitrate
PHYSICAL PROPERTIES						
Spec Gr	D792	1.24–1.34	1.17–1.22	1.15–1.22	1.09–1.17	1.35–1.40
Ten Str, psi	D638	1900–8500	2000–7000	2600–6900	2000–8000	7000–8000
Izod Impact Str, ft-lb/in. notch	D256	0.4–5.2	0.6–11.0	0.8–6.3	2 0–8.0	5.0–7.0
Rockwell Hardness	D785	R35–125	R20–115	R30–115	R50–115	R95–115
Water Absorption (24 hr), %	D570	1.9–6.5	1.2–2.0	1.1–2.2	0.8–2.0	1.0–2.0
Refr Index	D542	1.48	1.48	1.48	1.47	1.48
Dielec Const (at 10^3 cycles)	D150	3.5–7.0	3.6–3.9	3.3–6.3	3.0–4.1	7.0
CHEMICAL PROPERTIES						
Effect of Weak Acids	D543	Slight	Slight	Slight	Slight	Slight
Effect of Strong Acids	D543	Decomposes	Decomposes	Decomposes	Decomposes	Decomposes
Effect of Weak Alkalies	D543	Slight	Slight	Slight	None	Slight
Effect of Strong Alkalies	D543	Decomposes	Decomposes	Decomposes	Slight	Decomposes
Unaffected or only Slightly Affected By:		Alcohols, hydrocarbons, most esters	Alcohols, hydrocarbons	Alcohols, hydrocarbons	Glycerine, n-hexane	Hydrocarbons, weak mineral acids @ Rm. Temp.
Soluble in: (Work as Adhesives)		Ketones and cyclic ethers	Ketones, esters	Ketones, esters	Ketones, esters	Ketones, esters
THERMAL PROPERTIES						
Burning Rate	D635	Slow to self-extinguishing	Slow	Slow	Slow	Very rapid
Heat Dist Temp, (66 psi), F	D648	110–205	110–200	115–200	115–190	140–160
Continuous Res to Heat, F		140–220	155–220	140–220	115–185	140
Injection-Molding Temp, F		335–490	335–515	335–480	350–500	— (Not Moldable)
AGING RESISTANCE						
Indoor		Excellent	Excellent	Excellent	Excellent	Very good
Outdoor		Poor	Good	Good	Fair	Very poor
Forms Available		Molding, powder, films, sheets rods tubes flakes	Molding powder, flake	Molding powder, film, sheet, flake	Molding powder, flake	Films sheets rod tube flake
Outstanding Property		Toughness, scratch resistance, easy molding	Toughness, dimensional stability, impact resistance over wide temp.	Toughness, dimensional stability, impact resistance over wide temp.	Toughness at very low temp (—40 F) impact resistance	Toughness, scratch resistance
Disadvantage		Less dimensional stability	Premium price	Some odor, premium price	Limited color range, scratches	Inflammability

CENTRIFUGAL CASTINGS

Centrifugal castings can be produced economically and with excellent soundness. They are used in the automotive, aviation, chemical and process industries for a variety of parts having a hollow, cylindrical form or for sections or segments obtainable from such a form.

There are three modifications of centrifugal casting: (1) true centrifugal casting, (2) semicentrifugal casting, and (3) centrifuging. This article is limited to a discussion of true centrifugal casting.

True centrifugal casting is used for the production of cylindrical parts. The mold is rotated, usually in a horizontal plane, and the molten metal is held against the wall by centrifugal force until it solidifies.

Semicentrifugal casting is used for disk- and wheel-shaped parts. The mold is spun on a vertical axis, the metal is introduced at the center of the mold and centrifugal force throws the metal to the periphery.

Centrifuging is used to produce irregular-shaped pieces. The method differs from static casting only in that the mold is rotated. Mold cavities are fastened at the periphery of a revolving turntable, the metal is introduced at the center and thrown into the molds through radial ingates.

The nature of the centrifugal casting process assures a dense, homogeneous cast structure free from porosity. Because the metal solidifies in a spinning mold under centrifugal force, it tends to be forced against the mold wall while impurities, such as sand, slag and gases, are forced toward the inside of the tube. Another advantage of centrifugal casting is that recovery can run as high as 90% of the metal poured.

Ferrous castings

Centrifugal castings can be made of many of the ferrous metals—cast irons, carbon and low alloy steels and duplex metals.

Mechanical properties. Regardless of alloy content, the tensile properties of irons cast centrifugally are reported to be higher than those of static castings produced from the same heat. Hydrostatic tests of cylinder liners produced by both methods show that centrifugally cast liners withstand about 20% more pressure than statically cast liners.

Freedom from directionality is one of the advantages that centrifugal castings have over forgings. Properties of longitudinal and tangential specimens of several stainless grades are substantially equal.

Shapes, sizes, tolerances. The external contours of centrifugal castings are not limited to circular forms. The contours can be elliptical, hexagonal or fluted, for example. However, the nature of the true centrifugal casting process limits the bore to a circular cross section.

Iron and steel centrifugally cast tubes and cylinders are produced commercially with diameters ranging from $1\frac{1}{8}$ to 50 in., wall thickness of $\frac{1}{4}$ to 4 in., and in lengths up to 50 ft. Generally it is impractical to produce castings with the o.d./i.d. ratio greater than about 4 to 1. The upper limit in size is governed by the cost of the massive equipment required to produce heavy castings.

As-cast tolerances for centrifugal castings are about the same as those for static castings. For example, tolerances on the outside diameter of centrifugally cast gray iron pipe range from ±0.06 in. for 3 in. dia to ±0.12 in. for 48 in. dia. Inside-diameter tolerances are greater, because they depend not only on the mold diameter, but also on the quantity of metal cast; the latter varies from one casting to another These tolerances are generally about 50% greater than those on outside diameters. Casting tolerances depend to some extent also on the shrinkage allowance for the metal being cast.

The figures given above apply to castings to be used in the unmachined state. For castings requiring machining, it is customary to allow $3/32$ to $\frac{1}{8}$ in. on small castings and up to $\frac{1}{4}$ in. on larger castings. If the end use requires a sliding fit, broader tolerances are generally specified to permit additional machining on the inside surface.

Cast irons. Large tonnages of gray iron are cast centrifugally. The relatively low pouring temperatures and good fluidity of the common grades make them readily adaptable to the process. Various alloy grades that yield pearlitic, acicular and chill irons are also used. In addition, specialty iron alloys such as "Ni-Hard" and "Ni-Resist" have been cast successfully.

Carbon and low alloy steels. Centrifugal castings are produced from carbon steels having carbon contents ranging from 0.05 to 0.90%. Practically all of the AISI standard low-alloy grades have also been cast.

Small diameter centrifugally cast tubing in the usual carbon steel grades is not competitive in price with mechanical tubing having normal wall thicknesses. However, centrifugally cast tubing is less expensive than statically cast material.

High alloy steels. Most of the AISI stainless and heat-resisting grades can be cast centrifugally. A particular advantage of the process is its use in producing tubes and cylinders from alloy compositions that are difficult to pierce and to forge or roll.

The excellent ductility resulting in the stainless alloys from centrifugal casting makes it possible to reduce the rough cast tubes to smaller diameter tubing by hot- or cold-working methods. For example, billets of 18-8 stainless steel, $4\frac{1}{2}$ in. o.d. by $\frac{5}{8}$ in. wall, have been reduced to 27-gauge capillary tubing without difficulty.

Duplex metals. Centrifugal castings with one metal on the outside and another on the inside are also in commercial production. Combinations of hard and soft cast iron, carbon steel and stainless steel have been produced successfully.

Duplex metal parts have been centrifugally cast by two methods. In one, the internal member of the pair is cast within a shell of the other. This method has been used to produce aircraft brake drums by centrifugally casting an iron liner into a steel shell.

In the second method, both sections of the casting are produced centrifugally; the metal that is to form the outer portion of the combination is poured into the mold and solidified and the second metal is introduced before the first has cooled. The major limitation of this method is that the solidification temperature of the second metal poured must be the same or lower than that of the first. This method is said to form a strongly bonded duplex casting.

The possibilities of this duplex method for producing tubing for corrosion-resistant applications and chemical pressure service are being investigated.

Nonferrous castings

Nonferrous centrifugal castings are produced from copper alloys, nickel alloys, and tin and lead-base bearing metals. Only limited application of the process is made to light metals because it is questionable whether any property improvement is achieved; for example, differences in density between aluminum and its normal impurities are smaller than in the heavy metals and consequently separation of the oxides, a major advantage of the process, is not so successful.

Shapes, sizes, tolerances. As with ferrous alloys, the external shapes of nonferrous centrifugal castings can be elliptical, hexagonal or fluted, as well as round. However, the greatest over-all tonnage of nonferrous castings is produced in plain or semiplain cylinders. The inside diameter of the casting is limited to a straight bore or one that can be machined to the required contour with minimum machining cost.

Nonferrous castings are produced commercially in o.d.'s ranging from about 1 in. to 6 ft and in lengths up to 27 ft. Weights of individual castings range from 8 oz to 60,000 lb.

Although tolerances on as-cast parts are about the same as those for sand castings, most centrifugal castings are finished by machining. An advantage of centrifugal casting is that normally only a small machining allowance is required; this allowance varies from as little as $\frac{1}{16}$ in. on small castings to $\frac{1}{4}$ in. on the o.d. of large diameter castings. A slightly larger machining allowance is required on the bore to permit removal of dross and other impurities that segregate in this area.

Copper alloys. A wide range of copper casting alloys is used in the production of centrifugal castings. The alloys include the plain brasses, leaded brasses and bronzes, tin bronzes, aluminum bronzes, silicon bronzes, manganese bronzes, nickel silvers and beryllium copper. The ASTM lists 32 copper alloys for centrifugal casting; in addition, there are a number of proprietary compositions that are regularly produced by centrifugal casting.

Most of these alloys can be cast without difficulty. Some trouble with segregation has been reported in casting the high leaded (over 10% lead) alloys. However, alloys containing up to 20% lead are being cast by some foundries, the requirements being (1) rapid chilling to prevent excessive lead segregation, and (2) close control of speed.

The mechanical properties of centrifugally cast

copper alloys vary with the composition and are affected by the mold material used. Centrifugal castings produced in chill molds have higher mechanical properties than those obtained by casting in sand molds. However, centrifugal castings made in sand molds have properties about 10% higher than those obtained on equivalent sections of castings produced in static sand molds. (Castings produced in centrifugal chill molds have properties 20 to 40% higher than those produced in static sand molds.)

Nickel alloys. Centrifugal castings of nickel 210, 213 and 305; "Monel" alloys 410, 505 and 506; and "Inconel" alloys 610 and 705, are commercially available in cylindrical tubes. Centrifugal castings are also produced from the heat-resisting alloys 60 nickel-12% chromium and 66 nickel-17% chromium. These alloys should behave like other materials and show improved density with accompanying improvement in mechanical properties. The nickel alloys are employed for service under severe corrosion, abrasion and galling conditions.

Bearing metals. Centrifugal casting is a standard method of producing lined bearings. Steel cylinders, after being cleaned, pickled and tinned, are rotated while tin or lead-base bearing alloys are cast into them. The composite cylinder is then cut lengthwise, machined and finished into split bearings.

CERAMIC FIBERS

Alumina-silica fibers, frequently referred to as ceramic fibers, are formed by subjecting a molten stream to a fiberizing force. Such force may be developed by high-velocity gas jets or rotors or intricate combinations of these. The molten stream is produced by melting high-purity aluminum oxide and silica, plus suitable fluxing agents, and then pouring this melt through an orifice. The jet or rotor atomizes the molten stream and attenuates the small particles into fine fibers as supercooling occurs.

The resulting fibrous material is a versatile high-temperature insulation for continuous service in the 1000 to 2300 F range. It thus bridges the gap between conventional inorganic fiber insulating materials (e.g., asbestos, mineral wool and glass) and insulating refractories.

Alumina-silica fibers have a maximum continuous use temperature of 2000 to 2300 F, and a melting point of over 3200 F. If the fiber is exposed to temperature in excess of 2000 F for extended periods of time a phenomenon called devitrification occurs. This is a change in the orientation of the molecular structure of the material from the amorphous state (random orientation) to the crystalline state (definitely arranged pattern). Insulating properties are not affected by this phase change but the material becomes more brittle.

The coefficient of thermal conductivity (K) ranges from 0.7 to 1.2 Btu/hr/sq ft/°F/in. for various forms of ceramic fibers at a mean temperature of 1000 F. The mean temperature is the average of the hot and cold face temperatures under thermal equilibrium conditions.

Most ceramic fibers have an alumina content from 40 to 60%, and a silica content from 40 to 60%. Also contained in the fibers are from $1\frac{1}{2}$ to 7% oxides of: sodium, boron, magnesium, calcium, titanium, zirconium and iron.

Fibers as formed resemble a cottonlike mass with individual fiber length varying from shorts to 10 in., and diameters from less than one micron to 10 microns. Larger diameter fibers are produced for specific applications. In all processes, some unfiberized particles are formed that have diameters up to 40 microns.

Low density, excellent thermal shock resistance, and very low thermal conductivity are the properties of alumina-silica fibers that make them an excellent high-temperature insulating material. Available in a variety of forms, ceramic fiber is in ever-increasing demand due to higher and higher temperatures now found in industrial and research processes.

A description of the available forms of alumina-silica fibers follows in Table 1.

Applications

Ceramic fibers were originally developed for application in insulating jet engines. Now, this is only one of numerous uses for this material. It can be found in aircraft and missile applications where a high-temperature insulating medium is necessary to withstand the searing heat developed by rockets and supersonic aircraft. Employed as a thermal-balance and pressure-distribution material, ceramic fiber in the form of paper has made possible the efficient brazing of metallic honeycomb-sandwich structures.

Successful trials have been conducted in aluminum processing where this versatile product in

TABLE 1—ALUMINA-SILICA FIBERS

Type	Description	Size & Density
Bulk Fibers	Manufactured at 2 to 3 lb/cu ft density as outlined.	Bulk quantities used at 6-12 lb/cu ft
Bulk Fiber Products:		
Washed	Fibers with unfiberized particles removed. Used primarily for further processing.	Bulk quantities
Chopped	Bulk fiber milled to shorten lengths and remove part of unfiberized particles.	Bulk quantities
Blanket or Batt	Fiber processed into blanket or batt form available in completely inorganic form or with an organic binder to improve handling strength.	¼ to 3 in. thick, up to 42 in. wide; 3 to 23 lb per cu ft density
Paper	Fibers processed into uniform sheets available in completely inorganic form or with organic binder to improve handling strength.	0.020, 0.040, 0.080 in. thickness, up to 60 in. wide; 9.8 lb/cu ft density
Block	Fibers mixed with inorganic binders and fillers to form low-density block with low conductivity and immunity to thermal shock. Very low heat capacity.	Thicknesses from 1 to 4 in.; 12 x 36 in.; density 17 lb/cu ft and 20 lb/cu ft
Board	Fibers mixed with inorganic binders to form hard, rigid board while retaining excellent insulating properties.	¼ x 12 x 36 in.; Density: 25 lb/cu ft
Fabricated Shapes	Fibers mixed with inorganic binders fabricated to intricate shapes and close tolerances for specific applications.	Wide variety of sizes and shapes; 10 to 100 lb/cu ft density
Tamping Mix	Fibers mixed with inorganic binders ready to cast into low-density shapes. No firing necessary.	No size limitations; 24 to 30 lb/cu ft density
Coating-Cement	Fibers mixed with inorganic binders and fillers. Forms a dense hard coating with excellent thermal shock resistance.	100 lb/cu ft density; minimum coating thickness 0.015 in.
Long Staple Fibers	Fibers of controlled diameters are especially processed for use in textile forms and as air filtration materials. Available in fine and medium diameters.	Average fine diameter: 7 microns, medium diameter 14 microns; bulk quantities used at 6 to 12 lb/cu ft
Long Staple Fiber Products:		
Blanket	Processed from fine or medium diameter fibers primarily for air filtration at elevated temperatures.	½ in. thick up to 36 in. wide; 6 lb/cu ft density
Roving	Fibers blended with organic carrier fibers condensed to a single strand without twist forming a bulky resilient material.	500 to 1000 yards/lb
Rope	Twisted three-stand rope manufactured from roving.	¼ to 1 in. diameter; 118 to 9 ft per lb
Yarn	Twisted strand of fibers, with wire or glass insert for additional tensile strength at elevated temperature.	450 to 1000 ft per lb; 1 to 1.5 lb per tube
Tape and Cloth	Broad-woven cloth, and tape feature resiliency, bulk, and very low thermal and electrical conductivity. Retain tensile strength at elevated temperature.	Tape 1 to 6 in. wide; 0.065 to 0.086 in. thick; cloth 36 in. wide

paper or molded form has been used to transport molten metal with very little heat loss. Such fibrous bodies are particularly useful in these applications since they are not readily wet by molten aluminum.

Industrial furnace manufacturers utilize lightweight ceramic fiber insulation between firebrick and the furnace shell. It is also used for "hot topping," heating element cushions, and as expansion joint packing to reduce heat loss and maintain uniform furnace temperatures.

Use of this new fiber as combustion chamber liners in oil-fired home heating units has materially improved heat-transfer efficiencies. The low heat capacity and light weight, compared to previously used firebrick, improve furnace performance and offers both customer and manufacturer many benefits. The end result is more heat for less money from an Underwriters' Laboratory approved design.

CERAMIC MATERIALS

Ceramics, one of the three major materials families, are crystalline materials composed of compounds of metallic and nonmetallic elements. The ceramic family is large and varied and includes such materials as refractories, glass, brick, cement and plaster, abrasives, sanitaryware, dinnerware, artware, porcelain enamel, ferroelectrics, ferrites and dielectric insulators. Here we will be concerned chiefly with industrial ceramics. There are other materials, which, strictly speaking, are not ceramics, but which nevertheless are often included in this family. These are carbon and graphite, mica, and asbestos. Also, intermetallic compounds, such as aluminides and beryllides, which are classified as metals, and cermets, which are mixtures of metals and ceramics, are usually thought of as ceramic materials because of physical characteristics that are similar to those of certain ceramics.

Technical and industrial ceramics

A wide variety of ceramic materials have been developed over the years for industrial and technical applications—particularly for chemical resistant, refractory, electrical and abrasion resistant service. Many of the commercial high volume use types are complex bodies composed of high-melting oxides or a combination of oxides of such elements as silicon, aluminum, magnesium, calcium and zirconium that are

similar in their structure to clay products. Other technical ceramics, developed in recent years for service at the very high temperatures encountered in gas turbines, jet engines, nuclear reactors and high temperature processes, are relatively simple crystalline bodies composed of very pure metallic oxides, borides, carbides, nitrides, sulfides and silicides. The major difference between the common and high grade technical ceramics is that the high grade types do not have a glassy matrix. Instead, in the sintering process, the fine particles of the ceramic material are bonded together by solid surface reactions between the individual particles to produce a crystalline bond.

Stoneware and porcelain

Porcelains and stoneware are highly vitrified ceramics that are widely used in chemical and electrical products. Electrical porcelains, which are basically classical clay type ceramics, are conventionally divided into low-voltage and high-tension types. The high-tension grades are suitable for voltages of 500 and higher and are capable of withstanding extremes of climatic conditions.

Chemical porcelains and stoneware are produced from blends of clay, quartz, feldspar, kaolin and certain other minerals. Porcelain is more vitrified than stoneware and is white in color. A hard glaze is generally applied to chemical porcelain. Stonewares can be classified into two types: a dense, vitrified body for use with corrosive liquids, and a less dense body for use in contact with corrosive fumes.

Both chemical porcelains and stoneware resist all acids except hydrofluoric. Strong, hot, caustic alkalies mildly attack the surface. They generally show low thermal shock resistance and tensile strength. Their universal chemical resistance explains the wide use of these ceramics in the chemical and process industries for tanks, reactor chambers, condensers, pipes, cooling coils, fittings, pumps, ducts, blenders, filters, etc.

Common refractories

Common refractory ceramics are produced from clays, and the final product is a glassy matrix binding together the crystalline constituents. The manufacturing methods used are designed to produce bodies in which the main ingredient is the glassy matrix, or ceramic bond.

The most widely used common refractories are the alumina-silica (aluminum oxide and silicon dioxide) types. The compositions range from nearly pure silica, through a wide range of alumina-silicas to nearly pure alumina. They also contain some impurities, such as basic oxides of iron and magnesium, and smaller amounts of alkaline metal oxides. Refractoriness increases with alumina content. Other common commercial refractories are silica, forsterite, magnesite, dolomite, silicon carbide and zircon.

Composition and structure

A broad range of metallic and nonmetallic elements are the primary ingredients in ceramic materials. Some of the common metals are aluminum, silicon, magnesium, beryllium, titanium and boron. Nonmetallic elements with which they are commonly combined are oxygen, carbon or nitrogen. Ceramics can be either simple, one-phase materials composed of one compound or multi-phase, consisting of a combination of two or more compounds. Two of the most common ceramic compounds are single oxides such as alumina (Al_2O_3) and magnesia (MgO), and mixed oxides such as cordierite (magnesia-alumina-silica) and forsterite (magnesia-silica). Other newer ceramic compounds include borides, nitrides, carbides and silicides.

Unlike the relatively simple crystals found in most metals, those of ceramics are quite complex. Because ceramic crystalline phases are compounds, the units cells often contain three or four different atoms. Also, there are relatively few free electrons in ceramic structures, as compared to metals. Instead, electrons of adjacent atoms are either shared to produce covalent bonds or they are transferred from one atom to another to produce ionic bonds. These strong bonding mechanisms are what account for many of ceramics' properties, such as high hardness, stiffness, and good high temperature and chemical resistance.

At the macrostructural level of ceramic materials there can be one, two or three major constituents or components. In classical ceramics, the component termed the body is an aggregate of the crystalline constituents. The other component is a vitreous, or glassy, matrix (or phase) that serves as a bonding agent to cement together the crystalline particles. It is often referred to as the ceramic bond. This glassy phase is the weaker of the two components. In most newer refractory or technical ceramics it is eliminated and replaced by what is termed crystalline bonding, in which the individual particles of the powder raw material are sintered together in the solid state. That is, the particles are heated to just short of complete melting, but to a temperature hot enough to cause a fusing together of the particles.

A third component often present is a surface glaze. This is a thin, glassy ceramic coating fired on a ceramic body to make it impervious to moisture or provide special surface properties. Macrostructurally, then, there are essentially three types of ceramics. 1. crystalline bodies with a glassy matrix, 2. crystalline bodies, sometimes referred to as holocrystalline, and 3. glasses.

Processing

The basic steps in producing ceramic products are: 1. preparation of the ingredients for forming, 2. shaping or forming the part, 3. drying, and 4. firing or sintering. Depending on type of ceramic and application, the drying step is sometimes not required.

The raw materials are usually in the form of particles or powder. In the preparation step, the ingredients are weighed, mixed and blended either wet or dry. In the case of dry processing, sometimes the mixture is heated in order to cause preliminary chemical reactions. In wet processing the required plasticity for shaping is obtained by grinding and blending plastic clays with finely pulverized nonplastic ingredients and addition of alkalies, acids, and salts.

Forming methods

Ceramics can be formed by a large number of methods either in a dry, semi-liquid or liquid state and either in a cold or hot condition.

Slip Casting consists of suspending powdered raw materials in liquid to form a slurry or slip that is poured into porous molds, usually made of gypsum. The mold absorbs the liquid, leaving a layer of solid material on the mold surface. If the part is to be hollow, excess slip is removed after the desired shell thickness has been built up. For solid parts the slip remains, and more is added as shrinkage occurs to produce the final solid shape. With slip casting, large and intricate shapes can be produced. The process is especially economical for short production runs. In general, because slip casting produces parts with green densities up to only about 70% of theoretical, a large amount of shrinkage occurs in the firing step.

There are several variations of the basic slip casting process. In pressure and vacuum casting the slip is shaped in the mold under pressure or vacuum. In centrifugal casting the porous mold is rotated, and in thixotropic casting, chemical agents are added to promote curing and to reduce the amount of water required.

Jiggering, limited to circular and oval shaped pieces, is extensively used for producing dinnerware and procelain electric insulators. In this process, which resembles the potter's wheel technique, a liquid or semiliquid ceramic body is placed on a porous gypsum mold and rotated while a profiling tool forms the surface of the part and cuts away excess material. Once a hand operation, jiggering is now done on automatic machines that are capable of turning out up to and over 1000 pieces per hour.

Pressing can be done with dry, plastic or wet raw materials. In dry pressing, ceramic mixtures with liquid levels up to 5% by weight are pressed under high pressure into shape in a metal die. Widely used for manufacturing non-clay refractories, electrical insulators and electronic ceramic parts, the method produces small uniform parts to close tolerances. Semidry and wet pressing, in which water content is from 5 to 15% and 15 to 20% respectively, uses lower pressures and less expensive dies.

In isostatic pressing, dry ceramic powder in a sealed rubber bag, or mold, of the approximate size and shape of the finished part is placed in chamber of hydraulic fluid. The part is formed by hydraulic pressure applied to the rubber bag. Small amounts of binder are used. Sintered densities range from about 80 to 95%. The process produces complex, accurate shapes, is widely used with high grade oxide ceramics, and is the common method of producing spark plug insulators.

Hot Pressing, which is comparatively new and produces ceramic parts of high density and improved mechanical properties, combines pressing and firing operations. Isostatic and uniaxial techniques are employed. In the uniaxial method, a plunger compresses the ceramic in either powder or precompacted form in a die that is heated to near the sintering temperature.

Extrusion—Simple cross sections and hollow shapes can be produced by extruding plastic ceramic material through a forming die under pressure, then cutting to length. Hydraulic extrusion machines are used for producing technical ceramics. Most clay ceramic products, such as bricks and drain tile, are made with auger extruders. In ram pressing, or plastic pressing, an extruded slug of ceramic is placed between two porous dies, which when moved to the closed position, dewater the slug and form the part.

Molding of ceramic parts is done in a similar way to the injection molding of plastics. The ceramic is mixed with a thermoplastic resin and heated sufficiently to provide the fluidity needed for the mixture to flow into the die cavity. The resin is later burned off in ovens prior to firing. Parts as thin as 0.02 in. and as thick as ¼ in. can be injection molded. Bulk molding is sometimes used for forming large or irregular shapes or when only a limited number of parts is wanted.

Several coating type processes can be used to build up shapes over a mandrel or a mold. In the chemical vapor deposition process, molecular or atomic particles are deposited on a heated substrate which can be any shape desired. In melt spraying ceramic particles are simultaneously melted and sprayed onto the mandrel. And in electrophoretic forming, charged ceramic particles in a suitable low dielectric fluid are deposited on an electrode mandrel.

Drying and firing

The purpose of drying is to remove any water present from the formed plastic ceramic body before the firing operation. Because excessive drying shrinkage may cause cracking or warping, drying must be performed very carefully. Low cost ceramic wares are usually dried in the atmosphere under a roof. Quality and technical ceramics often are processed from dry powders and, therefore, the amount of moisture present is not an important problem.

The function of firing or sintering is to convert the shaped, dry ceramic part into a permanent product. The firing process and temperatures depend on the ceramic composition and desired properties. The top temperature to which the ceramic is fired is termed the maturing temperature. In the case of traditional ceramic whiteware, maturing temperatures can range from 1700 to 2600 F. In the first stage of firing any moisture still present after drying is removed. In the next stage chemical reactions cause the clay to lose its plasticity. In the last state, vitrification of the ceramic begins and continues up to the maturing temperature. During vitrification, a liquid phase forms and fills the pore spaces. Upon cooling, the liquid solidifies to form a vitreous or glass matrix that bonds together the inert unmelted particles. Refractory and electronic ceramics are often fired at higher temperatures, sometimes above 3000 F, to obtain the desired vitrification, or ceramic bond. In the case of high grade refractories, the firing operation produces a crystalline bond, instead of the glassy phase bond resulting from vitrification.

Mechanical properties

As a class, ceramics are low tensile strength, relatively brittle materials. A few have strengths above 25,000 psi, but most are below this figure. Ceramics are notable for the wide difference between their tensile and compressive strengths. They are normally much stronger under compressive loading than in tension. It is not unusual for a compressive strength to be five to ten times that of the tensile strength. Tensile strength varies considerably depending on composition and porosity. The stress condition of the outer layers also greatly influences strength. For example, in glazed ceramic parts the flexural and tensile strength can be increased or decreased as much as 40 or 50% depending on the stress condition in the glaze.

One of the major distinguishing characteristics of ceramics, as compared to metals is their almost total absence of ductility. Being strictly elastic, ceramics exhibit very little yield or plastic flow under applied loads. This means that when a load is removed, the ceramic returns immediately to its original dimensions. Also, ceramics fail in a brittle fashion. That is, they will stretch or deform only slightly without fracturing. Lack of ductility is also reflected in low impact strength, although it depends to a large extent on the shape of the part. Parts with thin or sharp edges or curves and with notches have considerably lower impact resistance than those with thick edges and gentler curving contours.

As a class, ceramics are the most rigid of all materials. A majority of them are stiffer than most metals, and the modulus of elasticity in tension of a number of types runs as high as 50-65 million psi compared to 29 million psi for steel.

Ceramic materials, in general, are considerably harder than most other materials, making them especially useful as wear resistant parts and for abrasives and cutting tools.

Thermal properties

Materials with the highest known melting points are found among ceramics. Hafmium carbide and tantalum carbide, for example, have melting points slightly above 7000 F, compared to 6200 F for tungsten. The more conventional ceramic types, such as alumina melt at temperatures above 3500 F, which is considerably higher than the melting point of all commonly used metals.

In general, and as a class, thermal conductivities of ceramic materials fall between those of metals and polymers. However, thermal conductivity varies widely among ceramics. A two order magnitude variation is possible between different types, or even between different grades of the same ceramic. The thermal conductivity of refractories depends upon their composition, crystal structure, and texture. Simple crystalline structures usually have higher thermal conductivities, as for example silicon carbide. Thermal conductivity versus temperature depends on whether the amorphous (glassy phase) or crystalline constituent predominates. For example, fireclay bricks show an increase with rising temperature, whereas the more crystalline forsterite and some high aluminas show a decrease with rising temperature.

Compared to metals and plastics, the thermal expansion of ceramics is relatively low, although as in the case of thermal conductivity, it varies widely between different types and grades.

Thermal shock resistance is closely related to thermal conductivity and expansion in brittle materials such as ceramics and glasses. High thermal conductivity and low thermal expansion favors good shock resistance. Also, small differences between tensile and compressive strength lead to good shock resistance. Because the compressive strengths of ceramic materials are 5 to 10 times greater than tensile strength, and because of relatively low heat conductivity, ceramics as a class have fairly low thermal shock resistance. However, in a number of ceramics the low thermal expansion coefficient succeeds in counteracting to a considerable degree the effects of thermal conductivity and tensile-compressive strength differences. This is true in the case of special porcelains, cordierite, lithium ceramics and for such glasses as fused silica, Pyrex and other special compositions.

Chemical properties

Practically all ceramic materials have excellent chemical resistance, being relatively inert to all chemicals except hydrofluoric acid and, to some extent, hot caustic solutions. Organic solvents do not affect them. The high surface hardness of ceramics tends to prevent breakdown by abrasion, thereby retarding chemical attacks. All technical ceramics will withstand prolonged heating at a minimum of 1830 F. Therefore, atmospheres, gases and chemicals cannot penetrate the material surface and produce internal reactions which normally are accelerated by heat.

Electrical properties

Unlike metals, ceramics have relatively few free electrons and, therefore, are essentially nonconductive and considered to be dielectric. Most porcelains, aluminas, quartz, mica, and glass have volume electrical resistivity values greater than 10^{15} microhm-cm and dielectric constants up to 12. In general, dielectric strengths, which range between 200 and 350 v/mil, are lower than those of plastics. Electrical resistivity of many ceramics decreases rather than increases with an increase in porosity. It also decreases with an increase in impurities and is markedly affected by temperature.

The diverse types of ceramics for electrical use include everything from low loss, high frequency electrical insulation to conductors, semiconductors, ferroelectrics and ferromagnetics.

CERAMIC PARTS AND FORMS

Modern ceramic components are compounded, formed and fired to meet specific sets of chemical, mechanical, thermal, electrical, magnetic or nuclear requirements. In order to meet such requirements, both the purity and particle size of the inorganic raw materials must be carefully controlled. Most ceramic forming processes depend on the use of temporary lubricants and binders; such additives reduce the pressure required to compact the powders into a given shape. Unless actually formed at a high temperature, ceramic parts must be sintered or "fired" after forming, to a temperature high enough to bring about the densification, crystal growth and phase changes which determine the properties of the finished ceramic.

The forming method itself has a decided effect on the temperatures required to bring about these changes.

Extrusion

Extrusion is accomplished by forcing a plasticized ceramic mix through a hardened metal or ceramic orifice. Pressure is supplied by augers or hydraulically operated pistons. If the part has a continuous cross section normal to its axis, as in the case of a rod, a tube, a brick or a drain tile, extrusion is probably the most economical method of forming it.

Chief advantages are high production rates and low die cost. The only additional operations required are cutting to length, drying and firing.

Difficulties most often arise in controlling the deflocculation, moisture content, or viscosity of the mix. This in turn causes variations in part-to-part shrinkage and quality. Final physical strength and homogeneity of the structure is lower by this process than by those which use less moisture, less plasticizer and higher forming pressures. Air entrainment is a common cause of losses, as is lamination and warpage. For example, if very high quality were required it might be better to press a rod or tube isostatically, and then turn the outside diameter.

The high moisture content required leads to high shrinkages and an increase in the dimensional tolerances required. Tolerances must normally be ±2%.

Cross-sectional size is limited only by the equipment available and length only by the size of the kiln and the straightness required. The wall thickness of a tube will determine the roundness obtained since the extrusion is usually pliable until nearly dry.

Materials commonly extruded include graphite, silicon carbide, alumina, barium titanate, titanium dioxide and beryllia.

Dry pressing. Unidirectional dry pressing is easily the fastest and most economical method of forming simple, flat, ceramic shapes. Excellent automatic equipment has been developed.

Rotary tablet presses with multiple dies are capable of producing discs, washers, or plates at rates of up to 2000 per min. Some newer presses provide 2 or 3 directions of travel, permitting parts with variations in elevation to be pressed uniformly.

The greatest problem with this type of pressing is in filling the die rapidly and uniformly. Modern spray dryers give the best possible granulation for this purpose. Another problem is variation in unfired or green density, both from part to part and within one part. Such variations in density from part to part cause differences in fired part size. A variation in density within one piece will cause warpage and cracking due to differential shrinkage. In dry pressing normally only 1 to 4 per cent moisture will be used.

Nearly all ceramics can be pressed, if properly plasticized and granulated. If the body contains no talc, clay, or other naturally plastic material, water, waxes, oils or other binder-lubricants must be added to give the necessary flow characteristics. The greater the flow required, the more volatiles must be added to the ceramic, increasing shrinkage, likelihood of structural defects and difficulty of firing. Dry pressing at pressures of 8000 to 10,000 psi gives a product second in quality only to hot pressing. Dimensional tolerances of $\pm 1\%$ are easily met.

Wet pressing. In the heavier or more complex cross sections where greater flow of material within the die is required, wet or semidry pressing is used. This type of pressing requires moisture contents of up to 15%. The additional moisture increases the plasticity of the material and reduces friction both within the material and within the die, effecting a reduction in the pressures required, and increasing die life.

At the same time, this additional moisture is responsible for decreased production rates and a need for careful handling. The wetter material tends to stick together, usually making it necessary to fill the die manually. Slow, steady pressing rates are needed to take full advantage of the flowability of the mix. Until the parts are dry, they must be handled and stacked carefully to avoid deformation or slumping. One type of wet pressing depends on the use of hard plaster molds to remove some of the water and strengthen the pressed piece.

The high moisture contents used reduce the "green" or "as-formed" densities, thus increasing shrinkage and decreasing physical and electrical properties of the fired ceramic. Tolerances must be increased to $\pm 2\%$.

Isostatic pressing. Another form of pressing depends on the application of pressure from all directions to compact a fairly dry powder. The powder is contained in a flexible mold and the mold is immersed in a hydraulic fluid. Spark plug insulators and grinding balls have been formed by this method for many years, at high production rates. Equipment is commercially available which will produce several parts simultaneously at rates of 200 parts per hr.

Commonly called isostatic or hydrostatic molding, this process is identical to dry pressing in the quality of the ware produced and the close tolerances possible. The number of materials which can be isostatically pressed is unlimited. As in dry pressing, less work is needed to develop a suitable binder-lubricant mix than would be needed to develop a casting or extrusion mix.

Among its other advantages are the precise and smooth internal surfaces which are possible and the relatively low tooling costs. With the use of metal core pins, it is possible to produce configurations which could not be extruded, notably threads and counterbores. Through the use of thick rubber molds and spray-dried material, the outside configuration can be formed, although not to a close tolerance. Due to the high pressures and low moisture contents used, firing shrinkage is low to begin with, and since die wall friction is eliminated, much less cracking, warping or differential shrinkage is experienced.

The machining which usually must be done on the outside of the part and the difficulty of filling the mold make up the chief disadvantages of this method.

Hot pressing. Normally, hot pressing refers to the sintering of refractory powders at temperatures up to 2500 C and pressures of a few thousand psi. Graphite is most often employed as the die material, although metals or carbides are also employed, depending on temperature and pressure requirements.

Of all the forming processes known for high-purity oxides, carbides and borides, hot pressing results in the best combination of low fired porosity and small crystal size. This means that higher densities as well as improved strengths and other physical properties are obtained. At the same time, this is the only way by which some ceramic materials can be formed.

Chief among its advantages is the elimination of the sintering operation and the subsequent savings in manufacturing time. The fact that no temporary binders or plasticizers are needed is also an advantage.

Close tolerances can be achieved only by subsequent grinding and finishing, and the process is limited to relatively small, simple shapes which can be pressed uniaxially. The cost and short life of the dies together with the lack of automatic equipment make the process expensive.

Casting

A ceramic "slip" is a suspension of ceramic particles in a liquid. Such a slip may be poured into an absorbent mold, usually plaster of Paris, and allowed to "cast" or build up on the surface of the mold cavity until it forms either a solid or hollow piece of a desired wall thickness. Typical products range from large refractory shapes to lavatories and figurines.

The chief advantages of this process are the size and complexity of the parts which may be formed. It is common practice with cast shapes containing clay to stick several cast components together while they are still fairly wet. Plaster of Paris molds are easily fabricated, the material is inexpensive, and they require less lead time than the metal molds used in dry pressing or injection molding.

One disadvantage is that a large number of molds and a lot of mold storage space is required, primarily because the capillary action of the mold may require from several minutes to several hours to draw sufficient moisture from the cast part. The high fluid content also results in high drying shrinkages. This can mean high losses due to plastic deformation, cracking, and warpage; poor dimensional tolerances; and high manufacturing costs.

Particle size and shape, particle size distribution, absorbed ions, and the water used, all have such pronounced effects on the properties of ceramic slips that, even today, the process remains largely an art. This makes its success dependent on the craftsmanship of highly experienced men.

Research in recent years has taught ceramic engineers how to make good slips out of most of the oxides and newer ceramic materials. Such developments as centrifugal casting, pressure casting, and casting followed by hydrostatic pressing may bring out new applications for this process.

Cementitious casting. There are a large number of refractory concretes, castable insulations and hydraulically bonded ceramics. These materials are poured, vibrated, gunned or tamped in place or into a mold. They develop their own cementitious bonds and are not usually fired before they are put in use. Ceramics of this type are sold as prepared mixes and their chief uses are based on their thermal properties; acid, wear, or abrasion resistance; and economy.

Large shapes are more economical in these materials because they are cast by the consumer, using methods fairly well known to his labor force. The cost of equipment or molds is low enough to make this more attractive than buying a fabricated shape. Another advantage is that cast shapes may be patched and repaired.

Chief disadvantages are that: (1) properties of the final shape are limited by the cement used to something less than the optimum properties of the sintered aggregate material by itself; (2) the consumer seldom gains enough experience in the proper mixing, forming and curing methods to obtain the optimum properties; and (3) large cast shapes properly take a rather long time to cure.

Tape casting. Some ceramic materials, such as alumina, zircon, titanium dioxide, zirconia and barium titanate have been cast as thin tapes with continuous tape casting equipment. Thinner pieces are possible by this process than by dry pressing, and the unfired tapes, made flexible by the use of rubber or plasticized binders, can be punched or cut to the shape desired on high-speed machines.

Due to the large amounts of binder and solvent required, shrinkage is high and microstructure poorer than that obtained by dry pressing. Thus, warpage and cracking become serious problems and the mechanical and electrical properties of the ceramic are reduced.

Machining

Ceramic thread guides have for many years been produced by pressing or extruding a blank and machining to the shape desired. This process is ideal for short production runs or prototypes because it eliminates the need for a special mold and the time required to make such a mold. It also makes possible shapes which would be impossible or too expensive by other processes.

The disadvantages are: (1) high tool wear, (2) dust problems, (3) high losses from breakage, and (4) tool marks on the finished piece.

Injection molding

By intimately mixing together a ground ceramic powder, thermoplastic and oil and then applying heat and pressure, enough plasticity can be induced in a ceramic to make it flow through an orifice 0.020 in. square or less. Thus the ceramic can be forced through an opening of tiny size into a metal mold and made to conform to every detail of that mold; then as the mold cools the thermoplastic ingredient the part will become tough and rigid. Subsequent firing burns off organic constituents and consolidates the ceramic.

This process makes it possible to form thin, curved, or twisted shapes long considered impossible or too expensive by other methods. The part-to-part reproducibility and surface smoothness obtained is excellent, much superior to pressing followed by machining. The use of high-speed automatic machines and multiple-cavity molds together with the "quick set" of the thermoplastics makes it a very economical process for long production runs. The mixtures used are less sensitive to moisture and humidity than traditional ceramic mixes. This allows the manufacturer to hold dimensional tolerances of ±1%, as close as those held in dry pressing.

Alumina, forsterite, titania, zircon, barium titanate, spinel and other ceramics have been molded. It should be possible to mold any ceramic, remove the plasticizers, fire it, and achieve properties equal to or better than those obtained by extrusion.

The chief disadvantage is the cost of the molds. Because they must incorporate all the dimensions of an intricate part in hardened tool steel, and allow for shrinkage, these molds are costly. Secondly, the mold must have an opening into the cavity through which it is filled and it must open so that the part may be removed. Unless the mold can be designed so that these marks are on noncritical surfaces they must be removed before the part is fired. The plasticizers used are more expensive than most binders used for ceramics and the quantities used make their removal before firing a critical and time-consuming operation.

In general the process should not be considered for parts whose thickness exceeds 0.250 in. The mold must draw enough heat from the injected mix to harden the plastic, and heavy sections make for long molding cycles, as well as difficult bake out of the binders. At the same time, the size of the parts is limited by the mold clamping capacity of the machines used and their thinness by the rate at which the material can be injected.

Nucleated crystallization

One of the newest forming methods for polycrystalline ceramics is referred to as nucleated crystallization. Although only a few ceramics are now being formed in this manner, this method produces excellent microstructures. Shapes are produced by simple glass-forming techniques followed by treatment of the part to develop a crystalline structure. Raw material purity, concentration of the nucleating agent and good control of the crystallizing process are essential.

The advantages of the process are first that it uses simple glass-forming techniques which are highly developed and automated. It makes possible the production of difficult shapes such as large, thin-walled containers. Finally, it produces improved product properties such as higher strengths, negligible porosities and uniform crystal size.

Fusion casting

Some refractories, in particular those used in glass tanks, are fusion-cast in sand molds from liquid oxides such as alumina. Since crystal growth is not controlled, these refractories tend to be weak and subject to thermal shock. The process is capable of nothing better than crude shapes but the density of such refractories makes them extremely resistant to slag attack and the process is important for this reason.

Grinding

The hardness of fired ceramic materials dictates that once they are fired, they must be ground with silicon carbide or diamond wheels. Grinding can be accomplished only on relatively simple shapes, at slow speeds, and with expensive equipment. It can cause a reduction in the strength of a ceramic part by removing the "as-fired" surface.

For close dimensional tolerances, flatness, parallelism, concentricity, or extremely smooth surfaces the cost of grinding is justified. Tolerances of ± 0.0005 in. are easily achieved.

Drying

The drying process might be considered a part of the forming cycle for wet pressed, extruded, or cast ware. A large part of the shrinkage of parts formed by these processes occurs during drying, and this shrinkage or compaction is a significant part of the forming process.

If volatilized at too high a rate, the fluid in a piece could cause it to literally explode. At an uncontrolled rate it can cause warpage and cracking. Thus, the drying process becomes a disadvantage in that it is time consuming and increases the percentage of rejects.

Firing

With the exceptions of fusion cast shapes, castables with cementitious bonds, and hot-pressed parts, ceramic ware must be heat-treated. The firing operation may be performed at temperatures as high as 2000 C. Typical cycles range from several hours to several days. Depending on the ceramic and the properties desired in the finished product this firing may be done in oxidizing atmospheres, reducing atmospheres, vacuum, or an atmosphere saturated with lead, cobalt, or other vapors.

The parts must be stacked in the kiln in a fashion which will guarantee uniform heating, enough movement of air to carry away the volatiles generated, and as little drag as possible while the parts are shrinking.

Ceramics are fired somewhere below the melting point of the most refractory of the inorganic raw materials. In most compositions, however, enough liquid phase and plastic flow is created to cause slumping of a fragile part if unsupported. Some cast, extruded, or injection-molded parts are fired on a supporting ceramic piece made of the same batch of material as the part. This setter shrinks at the same rate as the part; eliminates the drag experienced on a refractory setter; supports the part during the period in which it might slump; and eliminates the possibility of reaction between part and setter.

CERMETS

In the broadest sense, cermets are a class of materials containing both ceramics and metals and hence combining the properties of the two. The ceramics include oxides, carbides and nitrides as well as the more conventional ceramic materials; the metal may be any pure metal or any alloy. Further, cermets may contain one or more of either or both types of constituents and the composition may vary from predominately ceramic to predominately metallic.

Important types

Although research and development work has been done on an infinite variety of combinations, only a few have practical importance. These include the TiC (titanium carbide) base cermets, Al_2O_3 (aluminum oxide) base cermets, and specially developed uranium dioxide cermets developed for nuclear reactors. Another material which appears to fit the classification is a proprietary friction material developed for and used in aircraft brakes and sold under the trade name, "Cerametalix."

By definition, the well-known WC (tungsten carbide) cutting tool compositions fit in the cermet category and hence they are included in this discussion for comparison purposes. They are very closely related to the TiC base cermets and in limited areas do have high temperature capabilities, namely, where their high density and limited oxidation resistance are not factors.

Properties

The development of cermets was undertaken to provide materials for service at higher temperatures than those which could be withstood by high temperature nickel and cobalt base alloys by combining (1) the ability of ceramics to retain high strength at high temperature with (2) the ductility and toughness of metals. This is largely accomplished in commercial cermets and although their resistance to impact damage has proved inferior to that of high temperature alloys and insufficient to permit their use in such applications as aircraft gas turbine engines (by present performance and design criteria) they have proved to be a class of materials with unique properties and a large field of utility.

Table 1 gives the range of physical and mechanical properties of various commercially important types of cermets, together with some information as to their composition (compared with properties of a superalloy). Of particular importance are the low density, high molulus of elasticity (rigidity) and high strength at high temperature of the TiC base cermets. These are the properties which, combined with their excellent resistance to oxidation at 2200 F and above, are important in most applications. In the newest application for TiC cermets, namely,

finishing of metals where light high-speed finishing cuts are used, the high hardness is also important.

The Al_2O_3 type cermets have perhaps slightly better oxidation resistance than the titanium carbide cermets; further, they are resistant to the corrosive action of molten metals. The Cr_2C_3 (chromium carbide) cermets are of particular interest because of their resistance to chemical corrosion in many media.

The stress-to-rupture properties of the TiC base cermets are outstanding when compared with those of the nickel and cobalt base high-temperature alloys as shown in the accompany-

TABLE 1—PHYSICAL AND MECHANICAL PROPERTIES OF CERMETS

Type	WC [a]	TiC [b]	Cr_2C_3 [c]	Al_2O_3 [d]	Superalloy [e] (Udimet 700)
Specific Gravity	11.1–15.2	5.4–7.3	7.0	4.6–8.8	7.7
Rockwell Hardness	A85.0–93.6	A73–93	A88	A70–90 [f]	A67–71
Modulus of Elasticity, 10^6 psi	65–94	36–55	—	37.5–52.3	32
Compressive Strength, 1000 psi	500–800	264–550	450	110–380	—
Conductivity Thermal, Btu/sq ft/hr/ °F/ft	16.5–50.8	17.4–20.6	—	5.3–5.6	—
Electrical, % Cu Std.	3.3–10.1	1.3–6.0	2	2	—
Thermal Expansion, 10^{-6} in/in/°F	2.5–4.1	4.2–7.5	6.0	4.6–5.0	—
Modulus of Rupture, 1000 psi	175–385	122–212	—	—	—
Tensile Strength, 1000 psi					
At 70 F	—	75–155	36	21–39	204
At 1600 F	—	59–95	—	20–21.5	100
At 2000 F	—	30–44.5	—	11.7	15
At 2400 F	—	3	—	3.4	—

[a] Predominately WC with 3-25% Co. Some other carbides such as TiC, TaC, etc. in some compositions.
[b] TiC: 30 to 90%, Ni and Ni alloys and usually CbC or Cr_2C_3 for oxidation resistance.
[c] Cr_2C_3: 80 to 90%, Ni or Ni alloys the balance.
[d] Al_2O_3: 30 to 70%, Cr or Cr alloys the balance.
[e] 60% Ni, 17% Co, 15% Cr, balance of Al, Mo, Ti, Fe, B and C.
[f] Estimated.

ing curves for representative materials (Fig 1).

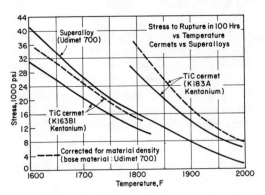

FIG. 1. Stress-rupture *vs* temperature for TiC-based cermets compared with that of a superalloy.

Fabrication

Cermets, by their very nature, require the use of powder-metallurgy techniques for their fabrication and this places some limitations on the size and complexity of components which can be provided. The high melting points of the ceramic constituents make melting impractical if not impossible, but even more, powder-metallurgy methods provide the means of retaining fine-grained uniform structures without the possibility of decomposition of the ceramic phase.

All methods of powder-metallurgy fabrication are used including cold press and sinter, cold press and sinter followed by infiltration, hot press, extrusion and slip casting. Auxiliary to several of these methods are secondary operations performed usually in the presintered condition, such as turning, drilling and other machining operations. Dimensional control is generally limited to 1% of the dimension in question although closer tolerances can be maintained with special precautions. Where greater accuracy or finer finishes are required, diamond or other forms of grinding are used and the finest surface finishes may be produced by lapping and polishing.

Various types of cermets present different degrees of difficulty in attachment to other materials. The TiC type involves the fewest problems since they usually can be brazed with silver solders and some high-temperature brazing alloys. In many instances they may be welded to the same or other TiC compositions by self-welding. Where the cermet has a high metal content and the difference in expansion is not too great, they may be welded to steels. When a large difference in expansion is involved, a gradation of the cermet composition can be used to compensate for the difference. With the aluminum oxide type cermets, mechanical attachment is generally required and where feasible, it is desirable with any of the cermets.

Cermets are produced as tubes, rods, rounds, flats, plates, rings, balls, as well as very complex shapes such as integrally bladed and shafted turbine wheels, multiple-diameter turbine shafts and both solid and segmented seats. Almost any shape can be produced provided there is economic justification. In view of the notch sensitivity of cermets, radii should be held above 0.2 in. whenever possible and in any case, above 0.1 in. Where no cross-sectional dimension exceeds 2 in., lengths up to 50 in. are available and discs and rings up to 15 in. dia can be made. Size limitations vary with other shapes.

Applications

Aside from the high-temperature applications in turbine buckets, nozzle vanes and impellers for auxiliary power turbines, there is a wide variety of applications for cermets based on various other properties. One of the most successful applications for the TiC base cermets is in elements of temperature sensing and controlling thermostats where their oxidation resistance together with their low coefficient of thermal expansion as compared with nickel base alloys are the important properties. Their ability to be welded directly to the alloys is also important.

The TiC base cermets are also used for bearings and thrust runners in liquid metal pumps, hot flash trimming and hot spinning tools, hot rod mill guides, antifriction and sleeve type bearings, hot glass pinch jaws, rotary seals for hot gases, oil well valve balls, etc. They have been found to be superior in some low temperature applications such as rotary seals operating at −300 F. The Al_2O_3 type cermets have been successful as thermocouple protection tubes for molten metal applications.

CHEMICAL MILLED PARTS

Chemical milling is the process of producing metal parts to predetermined dimensions by chemical removal of metal from the surface.

Acid or alkaline, pickling, or etching baths have been formulated to remove metal uniformly from surfaces without excessive etching, roughening or grain boundary attack. Simple immersion of a metal part will result in uniform removal from all surfaces exposed to the chemical solution. Selective milling is accomplished by use of a mask to protect the areas where no metal is to be removed. By such means optimum strength per unit of construction weight is achieved. Nonuniform milling can be done by the protective masking procedure or by programmed withdrawal of the part from the milling bath. Complex milling is done by multiple masking and milling or withdrawal steps.

Versatility offered

The aircraft industry, as an example, utilizes production chemical milling for weight reduction of large parts by means of precise etching. The process is the most economical means of removal of metal from large areas, nonplaner surfaces or complex shapes. A further advantage is that metal is just as easily removed from fully hardened as from annealed parts. The advantages of chemical milling result from the fact that metal removal takes place on all surfaces contacted by the etching solution. The solution will easily mill inside and reentrant surfaces as well as thin metal parts or parts that are multiple racked. The method does not require elaborate fixturing or precision setups and parts are just as easily milled after forming as in the flat. Job lots and salvage are treated, as well as production runs.

Maximum weight reduction is possible through a process of masking, milling, measuring and remilling with steps repeated as necessary. Planned processing is the key to production of integrally stiffened structures milled so that optimum support of stresses is attained without the use of stiffening by attachment, welding or riveting.

A level of ability comparable to that required for electroplating is necessary to produce chemically milled parts. Planned processing, solution control, and developed skill in masking and handling of the work are requisite to success. Periods to train personnel, however, are relatively short as compared to training for other precision metal-removal processes.

Tooling requirements are simple. Chemicals, tanks, racks, templates, a hoist, hangers and a few special hand and measuring tools are required.

Although chemical milling skill can be acquired without extensive training it is not feasible to expect to produce the extremes of complexity and precision without an accumulation of considerable experience. However, a number of organizations are available that will produce engineering quality parts on a job shop basis. The processes are well established, commercially, either in or out of plant.

Specific etchants needed

It is anticipated that any metal or alloy can be chemically milled. On the other hand, it does take time to develop a specific process and only those metals can be milled for which an etchant has been developed, tested and made available. Aluminum alloys have been milled for many years. Steel, stainless steel, nickel alloys, titanium alloys and superalloys have been milled commercially and a great number of other metals and alloys have been milled experimentally or on a small commercial scale.

It is advantageous to be able to mill a metal without changing the heat-treated condition or temper, as can be done chemically. Defective or nonhomogenous metal, however, can respond unfavorably. Porous castings will develop holes during milling and mechanically or thermally stressed parts will change in shape as stressed metal is removed. Good quality metal and controlled heat treating, tempering and stress relieving are essential to uniformity and reproducibility.

Process characteristics

Almost any metal size or shape can be milled, limitations being imposed only by extreme conditions such as complex shapes with inverted pockets that will trap gases released during milling, or very thin metal foil that is too flimsy to handle. Shapes can be milled that are completely impractical to machine. For example, the inside of a bent pipe could easily be reduced in section by chemical removal of metal. This possibility is used to advantage to reduce weight on many difficult-to-machine areas such as webs of forgings or walls of tubing. Thin sections are produced by milling when alternate machining methods are excessively costly and the optimum in design demands thin metal shapes that are beyond commercial casting, drawing or extruding capabilities.

Surface roughness is often reduced during milling from a rough-machined, cast or forged surface to a semimatte finish. The milled finish may vary from about 30 to 250 μ in., depending on the original finish, the alloy and the etchant. In some instances the production of an attractive finish reduces finishing steps and is a cost advantage. So-called "etching" that takes place during milling often causes a brightened finish and etchants have been developed that do not result in a loss of mechanical properties.

Complements machining

Chemical milling has flourished in the aircraft industry where paring away of every ounce of weight is important. It has spread to instrument industries where weight or balance of working parts is important to the forces required to initiate and sustain motion. It has also become a factor in the design of modern weight-limited portable equipment.

A realistic appraisal of the limitations and advantages of the process is essential to optimum designs. The best designs result from complementing mechanical, thermal and chemical processes. Chemical milling is not a substitute for mechanical methods but rather, is more likely as an alternate where machining is difficult or economically unfeasible. It does not compete with low-cost, mass production mechanical methods but rather, is successful where other methods are limited due to the configuration of the part.

Tolerances

It is good design practice to allow a complex shape to be manufactured by the most economical combinations of mechanical and chemical means. In order to allow this, print tolerances must reflect allowances that are necessary to apply chemical milling. Chemical milling will produce less well-defined cuts, radii and surface finishes. The tolerance of a milled cut will vary with the depth of the cut. For 1/10-in. cut a tolerance in depth of cut of $\pm$ .004 in. is commercial. This must be allowed in addition to the original sheet tolerance. Line definition (de-

viation from a straight line) is usually $\pm$ 1/32 in. Unmilled lands between two milled areas should be 1/10 in. minimum. Greater precision can be had at a premium price.

In general, milling rates are about .001 in. per min. and depth of cut is controlled by the immersion time. Cuts up to $\frac{1}{2}$ in. are not unrealistic although costs should be investigated before designs are made that are dependent on deep cuts.

Limitations

There are limitations to the process. Deep cuts on opposite sides of a part should not be taken simultaneously. One side can be milled at a time but it is less costly to design for one cut rather than two. Complex parts can be made by step milling or by programmed removal of parts to produce tapers. In general step milling is less expensive and more reliable. Chemical milling engineers should be consulted relative to the feasibility and cost of complex design. Very close tolerance parts can be produced by milling, checking, masking and remilling but such a multiple-step process could be more costly than

CHLORINATED POLYETHER

Chlorinated polyether is a thermoplastic resin used in the manufacture of process equipment. Chemically, it is a chlorinated polyether of high molecular weight, crystalline in character, and extremely resistant to thermal degradation at molding and extrusion temperatures. It possesses a unique combination of mechanical, electrical, and chemical-resistant properties, and can be molded in conventional injection and extrusion equipment.

Properties

Chlorinated polyether provides a balance of properties to meet severe operating requirements. It is second only to the fluorocarbons in chemical and heat resistance. Table 1 points up some typical property ranges that make it suitable for high-temperature corrosion service.

TABLE 1

Crystalline Melting Point, F	358
Specific Gravity	1.4
Mold Shrinkage, in./in.	0.004–0.006
Water Absorption, %	0.01
Heat-Distortion Temperature (66 psi), F	300
Deformation under 2000-lb Load for 6 hr at 122 F, %	1.4
Hardness, Rockwell R	100
Impact (Izod) ft-lb/in.	
Notched	0.4
Unnotched	>33
Tensile Strength (73 F), psi	6000
Tensile Strength (212 F), psi	3500
Ultimate Elongation (73 F), %	60–160
Dielectric Constant (60 cps)	3.1
Power Factor	0.016

Mechanical properties. A major difference between chlorinated polyether and other thermoplastics is its ability to maintain its mechanical strength properties at elevated temperatures. Heat distortion temperatures are above those usually found in thermoplastics and dimensional stability is exceptional even under the adverse conditions found in chemical plant operations. Resistance to creep is significantly high and in sharp contrast with the lower values of other corrosion-resistant thermoplastics. Water absorption is negligible, assuring no change in molded shapes between wet and dry environments.

Chemical properties. Chlorinated polyether offers resistance to more than 300 chemicals and chemical reagents, at temperatures up to 250 F and higher, depending on environmental conditions. It has a spectrum of corrosion resistance second only to certain of the fluorocarbons. Penton-steel constructions are being used in chemical processing equipment where chlorinated polyether liners or coatings on steel substrates provide the combination of protection against corrosion plus structural strength of metal.

Electrical properties. Along with the mechanical capabilities and chemical resistance, chlorinated polyether has good dielectric properties. Loss factors are somewhat higher than those of polystyrenes, fluorocarbons, and polyethylenes, but are lower than many other thermoplastics. Dielectric strength is high and electrical values show a high degree of consistency over a range of frequencies and temperatures.

Fabrication

The material is available as a molding powder for injection-molding and extrusion applications. It can also be obtained in stock shapes such as sheet, rods, tubes or pipe, and blocks for use in lining tanks and other equipment, and for machining gears, plugs, etc. In the form of a finely divided powder it is used in a variety of different coating processes.

The material can be injection-molded by conventional procedures and equipment. Molding cycles are comparable to those of other thermoplastics. Rods, sheet, tubes, pipe, blocks and wire coatings can be readily extruded on conventional equiment and by normal production techniques. Parts can be machined from blocks, rods, and tubes on conventional metal-working equipment.

Sheet can be used to convert carbon steel tanks into vessels capable of handling highly corrosive liquids at elevated temperatures. Using a conventional adhesive system and hot gas welding, sheet can be adhered to sandblasted metal surfaces.

Coatings of finely divided powder can be applied by several coating processes and offer chemical processors an effective and economical means for corrosion control. Using the fluidized bed process, pretreated, preheated metal parts are dipped in an air-suspended bed of finely divided powder to produce coatings, which after baking are tough, pinhole free and highly resistant to abrasion and chemical attack. Parts clad by this process are protected against corrosion both internally and externally.

Water-suspension coatings can be achieved by the Pfaudler process, using conventional spray equipment. Sandblasted shapes are first coated with resin suspensions, then, after the water has been driven off, oven-fused to form a tough continuous film.

The new Engelhard process for internally coating pipe provides an economical means of achieving coatings on aluminum or steel pipe in lengths up to 40 ft. The pipe is preheated in a large oven, then a charge of finely divided powder is applied to the interior surface to achieve a uniform film with excellent adhesion and great strength. Organic dispersion and flock-coating techniques are also recommended for use with chlorinated polyether.

Uses

Complete anticorrosive systems are available with chlorinated polyether, and lined or coated components, including pipe and fittings, tanks and processing vessels, valves, pumps and meters.

Rigid uniform pipe extruded from solid material is available in sizes ranging from ½ to 2 in. in either Schedule 40 or 80, and in lengths up to 20 ft. This pipe can be used with injection-molded fittings with socket or threaded connections.

Steel pipe with a heavy liner of extruded Penton can be obtained from a number of sources. This pipe, in standard lengths up to 10 feet, can be cut and fitted in the field for economical installation.

Clad pipe with both interior and exterior surfaces protected by resin, and pipe coated by the Engelhard process are other constructions available to the chemical and processing industries for economical corrosion control.

Lined tanks and vessels are useful in obtaining maximum corrosion and abrasion resistance in a broad range of chemical exposure conditions. Storage tanks, as well as processing vessels protected with this impervious barrier, offer a reasonably priced solution to many processing requirements.

A number of valve constructions can be readily obtained from leading valve manufacturers. Solid injection-molded ball valves, coated diaphragm and plug valves are among the variety available. Also available are diaphragm valves with solid chlorinated polyether bodies.

CHLOROSULFONATED POLYETHYLENE RUBBER

Chlorosulfonated polyethylene (produced by Du Pont as Hypalon synthetic rubber) was introduced commercially in 1952. It is used primarily where resistance to weather, abrasion, chemical or ozone attack, and heat are important in an elastomeric part. In addition, it can be permanently colored, eliminating one of the serious deficiencies of most other weather-resistant elastomers.

The material is made by reacting polyethylene with chlorine and sulfur dioxide to yield chlorosulfonated polyethylene. The reaction changes the thermoplastic polyethylene into a synthetic elastomer which can be compounded and vulcanized. The basic polyethylene contributes chemical inertness, resistance to damage by moisture, and good dielectric strength. Inclusion of chlorine in the polymer increases its resistance to flame (makes it self-extinguishing) and contributes to its oil and weather resistance.

In addition, the molecule is saturated, so there are no sites for ozone and oxygen to attack. Thus, properly compounded Hypalon is unexcelled among elastomers in its resistance to ozone degradation and weathering.

Properties

Products can be made to meet a wide range of specifications. Proper compounding is essential to realize the optimum properties of the polymer in vulcanizates. In the discussion of properties that follows, proper compounding is assumed in each case. In certain cases, where extreme performance is desired in a specific area, this can be obtained only through sacrifice of properties in other areas not critical to the application.

Hardness can be varied over the range of 50 to 95 durometer A. Tensile strength can range from 500 to 3000 psi. Compounds remain flexible at 0 F, and do not become brittle at −40 F or lower. At high temperatures, the material will perform satisfactorily after short-term exposure at 275 to 300 F, and up to 350 F if compounded for maximum heat resistance.

Flammability. Hypalon will not support combustion. It will burn slowly in the presence of flame, but is self-extinguishing when the flame source is removed. In cable sheathing, for instance, it has passed both the Underwriters' Laboratories vertical flame test and the U.S. Bureau of Mines burning test for mining cable. Even a thin coating will impart a high measure of flame resistance to a flammable material. A 15-mil coating on redwood, for example, lowers the wood's flame-spread characteristics from an index of 94 to an index of 8, as measured by the National Bureau of Standards' radiant panel test.

Ozone resistance. Vulcanizates are unaffected by concentrations of ozone as high as

10,000 parts per million parts of air, while most other elastomers are severely attacked by ozone in quantities of less than one part per million. Ozone resistance is so complete that the material has been used for years for gaskets in commercial ozone generators with no known failure to date. It is equally resistant to corona cutting. In one example, a #14 AWG copper wire with 4/64 in. sheathing was wrapped around an 8x mandrel and grounded against a steel plate. When subjected to 14,000 volts (225 volts per mil), the insulation lasted for over 3670 hours.

Electrical properties. Electrical properties of the elastomer make it appropriate for insulation in low-voltage applications up to 600 volts, and for jacketing for any type of wire and cable. Long-term weathering, direct burial, and water immersion appear to have little effect on original electrical characteristics. Typical properties as determined for RHH/RHW building wire jacketed with Hypalon, and tested at 60 F, are:

Insulation Resistance, megohms/	400–800
1000 ft	5×10^{13}
DC Resistivity, ohm-cm	
Specific Inductive Capacity	8–10
(1000 cps)	500
Dielectric Strength, volts/mil	1×10^{13}
Surface Resistivity, ohms	1200–2400
Insulation Resistance Constant "K"	

In addition, recent tests have demonstrated the elastomer's resistance to deterioration by radiation. Satisfactory performance was demonstrated under exposure to gamma radiation at a level of 5.5×10^7 Roentgens in air at 77 F.

When desired, chlorosulfonated polyethylene can be used as the base for conductive compounds. For instance, it has been loaded with magnetic iron oxide particles for application as the magnetic medium in heavy-duty announcing equipment.

Dynamic characteristics. Abrasion resistance surpasses that of most other elastomers. In shoe soles, for instance, ratings of 200% on the U.S. Bureau of Standards abrasion test are routine; equivalent to three times the wear life of other conventional shoe sole materials.

When subjected to heavy loading or constant deflection for long periods, vulcanizates show good compression recovery characteristics. Compressed 25% of their original thickness, typical vulcanizates exhibit the following performance under standard ASTM test conditions:

Time and Temperature	Compression Set, %
22 hours at 158 F	20–25
70 hours at 212 F	50–55
70 hours at 250 F	60–70

In dynamic applications, vulcanizates display good resistance to fatigue, cracking, and cut growth from continued flexing. Because of their resilience and abrasion resistance, they resist impact, crushing, cutting and other forms of mechanical abuse.

Weather and chemicals. Weatherability of properly compounded vulcanizates is excellent. They are immune to attack by atmospheric ozone, and highly resistant to deterioration by ultraviolet rays and oxidation. An added advantage is that they can be lastingly colored in a wide range of shades that will retain their original appearance indoors or out, without sacrifice in general properties.

The elastomer is highly resistant to chemicals, oils, grease and fluids, particularly to oxidizing chemicals such as concentrated sulfuric acid and hypochlorite solutions, to strong and weak alkalies, and to animal, vegetable and mineral oils. Exposure to aromatic solvents, aldehydes and chlorinated hydrocarbons should be avoided. Because chemical concentration, temperature, duration of exposure, and area exposed are all factors which must be taken into account, actual or simulated service tests are particularly recommended when considering the material for fluid exposure.

Like other chlorine-containing polymers, the elastomer is not attacked by microorganisms and will not propagate the growth of mold, fungus, or bacteria. It appears unaffected by soil chemicals, moisture, or other deteriorating factors associated with burial in earth.

Immersion in water for long periods appears to have little effect. A sample immersed for 28 days in water at 158 F, for instance, displayed a volume increase of only 2.5%. Four years' immersion in water at 122 F caused only moderate change in over all physical properties.

Selection and use

The elastomer is produced in various types, with generally similar properties. The design engineer can best rely on the rubber formulator to select the appropriate type for a given application, based on the nature of the part, the properties required, the exposure, and the performance necessary for successful use.

In combination with properly selected compounding ingredients, the polymer can be extruded, molded, or calendered. In addition, it can be dissolved to form solutions suitable for protective or decorative coatings.

Initially used in pump and tank linings, tubing, and comparable applications where chemical resistance was of prime importance, this synthetic rubber is now finding many uses where its weatherability, colorability, heat, ozone, and abrasion resistance and electrical properties are of importance. Included are jacketing and insulation for utility distribution cable, control cable for atomic reactors, automotive primary and ignition wire, and linemen's blankets. Among heavy-duty applications are conveyor belts for high-temperature use and industrial rolls exposed to heat, chemicals, or abrasion.

Interior, exterior, and underhood parts for cars and commercial vehicles are an increasingly important area of use. Representative automotive applications are headliners, window seals, spark-plug boots, and tractor seat coverings.

Chlorosulfonated polyethylene is used in a variety of mechanical goods, such as V-belts, motor mounts, O-rings, seals, and gaskets, as well as in consumer products like shoe soles and garden hose. It is also used in white sidewalls on automobile tires. In solution, it is used for fluid-applied roofing systems, for masonry coatings, and various protective-coating applications. It can also be extruded as a protective and decorative veneer for such products as sealing and glazing strips.

CHROMATE CONVERSION COATINGS

Chromate conversion coatings are formed by the chemical reaction that takes place when certain metals are brought in contact with acidified aqueous solutions containing basically water-soluble chromium compounds in addition to other active radicals. Although the majority of the coatings are formed by simple immersion, a similar type of coating can be formed by an electrolytic method.

Protective chromate conversion coatings are available for zinc and zinc alloys, cadmium, aluminum and aluminum alloys, copper and copper alloys, silver, magnesium and magnesium alloys. The appearance and protective value of the coatings depends on the base metal and on the treatment used.

Chromate conversion coatings both protect metals against corrosion and provide decorative appeal. They also have the characteristics of low electrical resistance, excellent bonding characteristics with organic finishes, and can be applied easily and economically. For these reasons the coatings have developed rapidly and they are now one of the most commonly used finishing systems. They are particularly applicable where metal is subjected to storage environments such as high humidity, salt and marine conditions.

The greatest majority of chromate conversion coatings are supplied as proprietary materials and processes. These are available usually as liquid concentrates or powdered compounds which are mixed with water. In the case of the powdered compounds, they are often adjusted with additions of acid for normal operation.

Coating composition and properties

Chromate conversion coatings are formed by immersing the metal in an aqueous acidified chromate solution consisting substantially of chromic acid or water-soluble salts of chromic acid together with various catalysts or activators. The chromate solutions, which contain either organic or inorganic active radicals or both, must be acid and must be operated within a prescribed pH range. When the metal is immersed in the bath, a small amount of the surface metal is dissolved which raises the pH of the solution at the metal solution interface to a point where a complex chrome gel is precipitated. Relatively little is known about the composition of the film although it has been established that its principal constituents are soluble hexavalent chromium, insoluble trivalent chromium, plus salts of the metal coated.

Solutions operated in the higher pH range tend to produce heavy coatings while solutions operated in the lower pH range often have a chemical polishing ability, particularly on zinc and cadmium, because continuous forming and redissolving of the deposited coating produces a brightening action. Although clear coatings can be obtained directly by this method, an iridescent yellow film is often deposited which can be removed by an alkaline leach following the chromate treatment.

Depending upon the metal treated and the process used, the amount of metal dissolved is usually 0.01 to 0.05 mil. A small amount of dissolved metal in a working solution is desirable and, in some cases, this can be obtained by introducing small amounts of metal into the bath prior to processing.

The nature of the coating is governed by the pH of the solution, the catalyst radical or radicals, and the ratio of their concentrations to that of the chromium compound or compounds used. Most chromate conversion coatings are soft while still wet and can be readily wiped from the metal surface, although, under some circumstances, hard coatings are developed in the wet state. When dried, however, all coatings form a relatively hard film which withstands normal handling or wear, but which does not add to the abrasion resistance of the coated surface.

Chromate conversion solutions are generally operated at room temperature between 70 and 90 F. Immersion times range from 5 to 20 sec when using chemically polishing solutions, and 15 sec to 3 min when heavier coatings are being formed.

The solutions do not fume or gas and need not be vented. Gentle agitation of the work or the solution usually improves over-all finish.

Chromate conversion coatings appear to become dehydrated when exposed to direct sunlight. (There is a question whether this is due to ultraviolet light or from heat from the sun.) Long exposure to sun will detract from their protective ability. Although the protective value of chromate conversion coatings is generally proportional to film thickness, thin, iridescent yellow coatings often provide corrosion protection equal or superior to heavy brown coatings, particularly under accelerated tests such as salt spray. Heavy coatings are usually quite soluble in aqueous solutions and the hexavalent chromium constituent readily leaches out. Lighter coatings, containing less soluble hexavalent chromium and probably a higher concentration of insoluble trivalent chromium, appear to be less readily affected by leaching.

Maximum corrosion protection is obtained by using olive drab or dark bronze coatings on zinc and cadmium surfaces, and yellow to brown-colored coatings on the other metals. Lighter iridescent yellow type coatings generally provide medium protection, and the clear-bright type coatings, produced either in one dip or by leaching, provide the least protection.

Chromate conversion coatings provide maximum corrosion protection in salt spray or marine types of environment, and in high humidity such as encountered in storage, particularly where stale air with entrapped water may be present. They also provide excellent protection against tarnishing, staining, and finger marking, or other conditions that normally produce surface oxidation.

Olive-drab type coatings are widely used on military equipment because of their high degree of corrosion protection coupled with a nonreflective surface. Iridescent yellow coatings are widely used for corrosion protection where appearance is not a deciding factor. The clear-bright chemically polishing type coatings for zinc and cadmium have been widely used to simulate nickel and chromium electroplate and are primarily used for decorative appeal rather than corrosion protection. Where additional corrosion protection or abrasion resistance is desired, these clear coatings act as an excellent base for a subsequent clear organic finish.

Heavy olive drab and yellow coatings for zinc, cadmium, and aluminum can be dyed various colors. Generally speaking, the dyed colors are used for identification purposes only since they are not lightfast and will fade upon exposure to direct sunlight or other sources of ultraviolet light. Within recent years, however, considerably more use has been made of the clear, bright treatments which have been dyed pastel shades and then coated with a clear lacquer. These treatments have proven very satisfactory for indoor use.

Although chromate conversion coatings have rather low abrasion resistance, coated metals can be cold-formed with smooth dies without appreciable damage to the film. The coating adheres strongly to the metal and makes a good base for organic coatings.

Because of their low electrical resistance, chromate conversion coatings are widely used for electronics equipment. Surface resistance depends on the type and thickness of the film deposited, the pressure exerted at the contact, and the nature of the contact. Low resistance coatings are particularly important on aluminum, silver, magnesium and copper surfaces.

Chromate conversion coatings can also be soldered and welded. A chromate coating on aluminum, for example, facilitates heliarc welding. Due to the slight increase in electrical resistance, an adjustment in current (depending upon the thickness of the coating) must be made in order to satisfactorily spot-weld. Soldering, using rosin fluxes, can be performed on cadmium-plated surfaces which have been treated with clear bright chromate conversion coatings. Clear, bright coatings on zinc-plate surfaces and colored coatings on both zinc and cadmium necessitate the use of an acid flux or removal of the film by an increase in soldering iron temperature, which burns through the coating, or by mechanical abrasion, which removes the film and provides a clean metal surface for the soldered joint.

Application and processing

Most chromate conversion treatments are applied by simple immersion in an acidified chromate solution. Because no electrical contacts need be made during immersion, the coatings can be applied by rack, bulk, or strip line operation. Under special situations, swabbing or brush coating can be used where small areas must be coated, as in a touch-up operation. Spray application is usually limited to continuous strip application although it can be used on racked parts. The spray should be a flow of liquid that covers the surface, rather than a fine mist spray such as is used when applying paints. Excess sprayed solutions should be rinsed off immediately after the normal coating has been formed in order to eliminate the possibility of powdering caused by drying of nonreactive conversion solution on the metal surface.

Chromate conversion coatings can also be applied by an electrolytic method in which the electrolyte is composed essentially of water-soluble chromium compounds and other radicals operated at neutral or slightly alkaline pH. This type of application is limited primarily to rack type operation.

In general, processing can be placed in two categories: (1) over freshly electroplated surfaces; and (2) over electroplated surfaces that have been aged or oxidized, or other metal surfaces such as zinc die castings, wrought metals, or hot-dipped surfaces. In the first category the chromate conversion process immediately follows the electroplating operation after thorough rinsing. In the second category it is necessary to preclean in alkaline solution to remove any oils, greases, or organic materials and to use a deoxidizing or neutralizing step (usually a mild acid dip) to remove any oxide contamination. Following is a typical sequence of operations for processing:

1. Electroplate, and/or clean and deoxidize.
2. Rinse in cold water.
3. Chromate conversion treatment.
4. Rinse in cold water.
5. Rinse in warm water; temperature should be kept within 100 to 160 F to facilitate drying.
6. Dry; temperature not to exceed 160 F in order to maintain maximum corrosion protection.

Where clear chromate conversion coatings obtained from chemically polishing iridescent yellow treatments are to be processed, an alkaline bleach type dip is placed after the rinse in step 4 and should be followed by an additional cold rinse prior to the warm rinse and dry.

Rinses should be kept clear of contaminants by the use of a good flow of running water. Immersion time in rinses after chromating should be as short as possible. The cold rinses are basically used to remove any adhering chromate solution and the final warm rinse to facilitate drying. Because the hexavalent chromium held in the coating is quite soluble before the film has dried, use of minimum immersion time in the rinses will prevent this corrosion-protective constituent from being lost and will assure maximum benefit from the film.

Chromate conversion films contain a certain quantity of water of hydration when they are dried at temperatures up to 150 F. Under these conditions, the hexavalent chromium trapped in the film is retained in a soluble state and can be leached out so as to increase the protective ability of the coating. In fact, it will provide a self-healing characteristic under humid conditions and protect scratched or abraded areas. Drying of the chromate conversion film at temperatures above 150 F drives off this normal water of hydration and changes the structure of the film from amorphous to crystalline. Because the hexavalent chromium in this dehydrated coating is no longer in a soluble state, it can greatly reduce the protective value of the coating.

If a hydrogen relief anneal is necessary on electroplated coatings, the baking step should be made after plating and prior to chromating. If the plated surface has become oxidized to a point where chromating is difficult, then the surface can be reactivated in a mild acid bath such as $\frac{1}{2}\%$ nitric acid.

Chromate solutions must be held in acid-resistant equipment. Tanks can be recommended by the materials supplier. An 18-8 type stainless steel is completely satisfactory except where chlorides are present. Plastic tanks, or tank linings, of such materials as tygon, koroseal and polyethylene are completely satisfactory except under high-temperature operation. Stoneware or ceramics can also be used except where solutions contain fluorides.

Under no circumstances should iron be brought in contact with the chromate solution because iron will quickly reduce the active hexavalent chromium in the operating bath and deplete the chromate solution.

Metals that can be coated

A wide variety of metals can be given a chromate conversion coating.

Zinc and cadmium. Where electroplated zinc and cadmium surfaces are to be treated, it is recommended that the plate thickness be a minimum of 0.2 mil (normal for commercial work). However, it has been noted that the corrosion-protective value of the chromate conversion coating will increase with an increase in plate thickness up to 0.7 mil. An increase of metal thickness beyond this point does not appear to improve the corrosion-protective value of the coating itself; however, it naturally increases the over-all life of the finish system.

For best results, electrodeposits should be fine grained and nonporous. Use of organic brighteners in the plating bath will provide initial smoothness and brightness of the electroplate deposited. This is particularly advantageous where clear, chemically polishing, bright chromate treatments are applied. Brightener addition agents for zinc and cadmium electroplating baths containing heavy metals or insoluble oils should be avoided. These tend to codeposit with the electroplate or form a thin film on the surface and can result in discoloration and/or loss of

corrosion protection. Inclusion of copper in the electrodeposit may adversely affect the protective value of the coatings and result in the formation of a black coating (e.g., on zinc).

Wherever possible it is recommended that a neutralizing dip of $\frac{1}{2}\%$ nitric acid be used after plating and prior to chromating in order to neutralize retained cyanide and alkali on the plated surface and activate the metal surface for the application of the chromate conversion coating.

The physical structure of the surface will also greatly affect the color and texture of the coating. Highly polished surfaces produce iridescent bright films; rough surfaces produce dull, lusterless finishes.

Electroplated finishes that have been stored prior to chromating, should be reactivated by using the $\frac{1}{2}\%$ nitric acid dip prior to treatment. This will remove any thin layer of oxide which may have developed. If organic substances such as oils or greases have contaminated the surface, they should be removed by using a mild alkaline cleaner, preferably followed by activation in the $\frac{1}{2}\%$ nitric acid solution.

Zinc-base die castings are usually cleaned in a mild alkaline cleaner, followed by an acid neutralizing dip of 1 to 2% sulfuric acid, prior to chromating.

Hot-dip galvanized surfaces (other than those coated by strip line operation) should be precleaned in a solution of caustic soda using a concentration of 3 to 5 oz per gal at a temperature of 150 to 160 F, with an immersion time of 30 to 60 sec. An acid cleaner containing phosphoric acid (2 to 5%), butyl cellosolve (2 to 5%) and a wetting agent (5 ml per gal) can also be used to replace the alkaline cleaner.

Strip line galvanized sheets can be processed immediately out of the galvanize kettle through the cooling tower into the chromate conversion bath. Operating temperature of the bath is usually between 140 and 180 F. Immersion times for this type of application range from 2 sec up to approximately 5 sec at standard line speeds. Excess solution is removed from the sheet by squeegee roll and treatment is usually followed by spray rinsing, additional squeegee rolls and drying. Coatings are available that will not change the appearance or coloration of the galvanized sheet and will provide a high degree of protection against the white rust that is usually formed during storage.

Dip type chromate conversion treatments for zinc and cadmium can be separated into two simple groups on the basis of their acidity (see table).

Coatings formed by solutions of Type I provide zinc and cadmium electroplated surfaces with a clear and bright finish, generally simulating either nickel or chromium plate in appearance. Minimum plate thicknesses of 0.2 to 0.3 mil are recommended. The appearance of finishes obtained from this type of process is greatly enhanced on smooth base metal and fine-grained electrodeposits.

Two types of treatments are available. The first is a dilute one-dip type of process which produces an extremely lustrous finish with a bluish cast and requires no bleach step. This finish can be dyed in a variety of beautiful pastel colors and is often used for decorative purposes.

The second process uses a solution of much greater concentration and produces a lustrous, iridescent yellow finish. This iridescent yellow color can be removed by immersion in a leach solution which leaches out the soluble hexavalent chromium causing the yellow color, and produces a clear, bright finish.

Several leaches are available, although the one most commonly used is composed of a solution

CHROMATE CONVERSION SOLUTIONS AND COATINGS FOR ZINC AND CADMIUM [a]

Number	Operation	Color	Surfaces Applied	Operating Temp, F	Immersion Time, sec	Time in Salt Spray to Produce White Salts on Zinc Plate, hr
Type I—Simple-dip chemical polishing solutions operating at a pH range of 0.0 or less to 1.5 producing lustrous clear to light iridescent yellow coatings of medium corrosion resistance.						
1	1 dip	Clear bright with a blue tint	Zinc and cadmium plate	70–100	5–20 to 60	24–100
2	1 dip	Bright, iridescent yellow	Zinc and cadmium plate; galv., zinc castings	70–100	5–20	100–200
2A	No. 2 using acid or alk bleach dip	a—clear bright b—blue tint c—iridescent	Zinc and cadmium plate; some galv. or zinc castings	70–100	5–20	24–100
Type II—One-dip, nonpolishing solutions operating at a pH of 1.0 to 3.5 producing medium to heavy films ranging in color from iridescent yellow to bronze, olive drab and black generally providing maximum corrosion protection.						
1	1 dip	Iridescent yellow to bronze	Zinc and cadmium plate; galv., zinc castings	60–90	15–45	100–200
2	1 dip	Olive drab	Zinc and cadmium plate; galv., zinc castings	70–90	15–45	100–200
3	1 dip	Black	Zinc plate	60–90	60–300	24–100

[a] Type I No. 1 and Type II No. 2 films can be dyed with certain alizarine and diazo dyes in various colors while still wet during processing. Clear coatings produce pastel shades, while olive drab coatings result in dark, opaque colors.

containing 3 oz per gal of caustic soda operated at room temperature. Care should be taken to prevent prolonged immersion or all of the protective value of the conversion coating may be lost.

Where increased corrosion protection is desired, and some slight iridescence without a yellow coloration can be tolerated, a leach dip composed of sodium carbonate at a concentration of 2 to 3 oz per gal, operating at a temperature of 120 to 130 F, with an immersion time of 15 to 60 sec, will provide excellent results. Should the slight iridescence be objectionable, it can be obscured by coating with a clear lacquer or other organic finish which will also increase the abrasion resistance and protective value.

In normal operation, the bleach type clear, bright chromate conversion film provides a higher degree of corrosion protection than can be obtained by the one-dip process. This type of finish has found wide application for electronic equipment, home utilities, automotive parts, tubular furniture, and hardware. Clear chromate conversion coatings for cadmium reduce oxidation during storage and are widely used in the electronic industry where ease of solderability after storage is required.

Clear, bright chromate conversion coatings are also available for zinc die castings and often eliminate the necessity of buffing (at the same time providing a high degree of corrosion protection). The conversion film can also be removed by a normal leach operation and the lustrous clear bright surface used as a base for subsequent copper-nickel-chromium plating.

Coatings of Type II provide colored films and maximum corrosion protection. They are usually used where corrosion protection, rather than decorative value, is of primary importance. Except for iridescent yellow coatings over cadmium, they also provide an excellent paint base.

Heavy olive drab and bronze coatings can be dyed various colors. Because dyed colors are organic in nature and are not lightfast they will fade on prolonged exposure to sunlight. However, applications have been found on small parts for identification and they are used on household utility, automotive, military, and other parts.

Aluminum. Because of their high degree of corrosion protection, excellent paint bonding

characteristics, and simplicity of operation, chromate conversion coatings can replace electrolytic anodizing of aluminum except under conditions where abrasion resistance is important.

Coatings range in color from clear through iridescent yellow to brown, and may be produced from a single bath by varying the treatment time. The degree of coloration varies considerably with alloy composition. Yellow to brown coatings can be dyed various colors and provide the same color and film characteristics as dyed coatings on zinc and cadmium surfaces.

The corrosion protection provided is in proportion to film thickness and also varies with the alloy; it is usually inversely proportional to the copper, silicon and iron content. The coatings are extremely important in the electronics industry due to their low electrical resistance, which is low enough for grounding applications as well as for use on shields and waveguides.

Proper cleaning of aluminum is extremely important and the following steps are recommended:

1. For wrought alloy extrusions and all alloys containing less than 1% silicon: etch type alkaline clean; rinse; deoxidize; rinse and chromate process.

2. For aluminum sand, die and permanent mold castings and all alloys with silicon over 1%: etch type alkaline clean; rinse; hydrofluoric nitric pickle; rinse and chromate.

3. For polished aluminum surfaces, and aircraft parts and similar applications where etching is undesirable (recommended for use on all alloys): nonetch alkaline clean; rinse; deoxidize; rinse and chromate.

Cleaned aluminum surfaces can be coated by spraying. Low pressure, slow type sprays rather than high pressure, mist type equipment are recommended. Treated surfaces should be thoroughly rinsed as quickly as possible after spraying before the chromate solution has dried. For this reason, treatment of small areas followed by rinsing may be advisable.

Chromate conversion coatings can also be applied to aluminum surfaces by swab or brush application. This is often useful in touching up or protecting abraded anodized films in limited areas and other bare areas on chromate-treated aluminum. The normal chromate treating solution is applied to the bare area and allowed to remain on the surface until a visible coating can be observed; it should then be thoroughly rinsed with clear water and dried.

Magnesium. A wide variety of chromate conversion treatments are universally used on magnesium surfaces to eliminate corrosion during shipment and storage prior to fabrication. Within the last few years promising proprietary processes have also been made available which appear to have entirely different operating characteristics than treatments originally available.

Chromate conversion coatings generally provide an excellent base for subsequent organic finishing of magnesium surfaces. Coatings with extremely low electrical resistance are also available.

Copper and copper alloys. Chromate treatments for copper and its alloys are usally available as proprietary liquid concentrates or powdered compounds. They will produce chemically polished, clear, bright surfaces which can be used as a final finish or, where the chromate conversion film has been completely removed, as a base for subsequent nickel-chromium plating. Maximum brightness is obtained on fine-grained homogeneous electrodeposits. Alloys having heterogeneous structure will result in a satin rather than a lustrous finish.

Two types of nonpolishing processes are available. One provides a heavy yellow-to-brown coating for increasing corrosion protection. The second is a light, colorless coating which is often used to eliminate spotting out from electroplating and as a base for subsequent painting. Clear colorless chromate conversion coatings also can be readily soldered.

Silver. Chromate conversion coatings can be applied to silver surfaces to eliminate tarnishing and corrosion. Finishes vary from clear through yellow and the degree of tarnish resistance varies in proportion to the film thickness. As with other chromate conversion coatings, only an extremely small increase in electrical resistance can be noted.

Thorough rinsing and neutralizing of any cyanide contamination on the silver electroplated surface is particularly important. A neutralizing dip of $\frac{1}{2}\%$ nitric acid is recommended after plating and prior to chromating. Surfaces to be chromated should be free of water-break prior to processing. Tarnish resistance is about the same as a good coating water-dip lacquer.

CHROMIUM AND ALLOYS

Chromium has an atomic number of 24 and is one of the transition elements of the first long period of the periodic table.

The atomic weight of the metal is 52.01, and it consists of four isotopes with mass numbers of 50 (4.31%), 52 (83.76%), 53 (9.55%), and 54 (2.38%).

The stable form has a body-centered cubic lattice with a value for the cube side of 2.884 Å (2.8790 KX). Although never found in massive metal, three other structures have been reported to occur in electrodeposited chromium. They are: (a) a hexagonal modification with a cell dimension of $a = 2.717$ KX and $c/a = 1.626$, (b) a cubic structure isomorphous with α-manganese with a lattice constant of 8.717 KX, and (c) a face-centered type with a lattice parameter of 3.8605 Å. All these modifications transform to the body-centered type on standing at room temperature for any length of time, and there is speculation that they may be due to the presence of hydrogen in electrolytic metal and may not be true allotropic forms.

Calculations of the density from the lattice parameter lead to a value of 7.194 g./cm.^{3} based on a value of 2.8848 Å for the cube side of the body-centered unit cell.

Thermal properties

Due to the variations in the purity of the metal used, and in particular the oxygen and nitrogen contents, there has been confusion as to the melting point of chromium. Recent work conducted at the National Bureau of Standards leads to a freezing point for high-purity metal of 1875 C $\pm$ 5°.

As calculated from the melting points of chromium-palladium alloys, the heat of fusion of chromium found to be 3.2–3.5 kcal/mol.

The heat contents of solid and liquid chromium as represented by H_T-H_{298} are in cal/mol.:

400 K	595
1000 K	4,640
2000 K	14,220
3000 K	29,090

and for gaseous chromium

400 K	505
1000 K	3,480
2000 K	8,765
3000 K	15,400
4000 K	23,335
6000 K	43,435
8000 K	69,600

The latent heat of vaporization of the metal is 94.85 kcal/mol. which, like that of the other transition metals with unfilled d shells, is high. The vapor pressure at various temperatures may be calculated from the expression

$$\log P_{atm.} = -(20{,}473)/T + 7.467$$

and the boiling point is 2665 C. Its entropy is 5.58 cal/degree/gm atomic wt. in the solid state at room temperature.

Chromium has a specific heat of 0.11 cal/gm at 20 C, and its heat capacity may be expressed by $C_p = 4.84 + 2.95 \times 10^{-3}T$ (0 $-$ 1600 C). Its absolute thermoelectric power is $16.2\mu V/°C$ at 20 C.

Thermal expansion measurements gave a temperature coefficient of 6.2×10^{-6} at 20 C, and for a temperature range of -100 to $+700$ C the length L_t at a temperature t is given by

$$L_t = L_o \ [1 + (5.88 + 0.0074t^2 - 0.00000388t^3) \ 10^6].$$

Mechanical properties

The type of metal used plays an important part in determining the mechanical properties of chromium. Chromium may be prepared by aqueous or fused salt electrolysis, by the aluminum, silicon, or carbon reduction of the oxide, and by the iodide process.

The first two types are too impure to fabricate and the iodide process metal is the purest, although on consumable electrode arc melting it picks up a considerable amount of oxygen. Arc-melted or sintered electrolytic chromium was used to determine most of the mechanical properties.

Brinell hardness of forged arc-cast metal decreases steadily from 125 to 70 as the temperature is raised from room to 700 C. Cast arc-melted metal shows a minimum value of 50 Brinell at 300 C.

The ultimate strength of electrolytic chromium plate has been reported to be between 15,000 and 80,000 psi. Young's modulus values were between 12 and 33×10^6 psi. The tensile strength of sintered powdered compacts drops from 13,500 psi at 20 C to 10,200 psi at 900 C, with a maximum elongation of 5% at 900 C. Tensile tests made at the Bureau of Mines on swaged arc-cast electrolytic metal which was recrystallized by annealing at 1200 C in hydrogen showed the following results at an average strain rate of 0.017 in./in./min:

Temp, C	Ult Str, 1000 psi	Yld Str, 0.2% Offset 1000 psi	Pro-port Limit, 1000 psi	Young's Modu-lus, 10⁶ psi	Elong., in 1 in., %	Red. in Area, %
Room	12	—	—	36	0	0
300	22.4	20.6	17	42	3	4
350	28.6	20.6	15.3	24.4	6	8
400	32.6	—	19.1	32.9	51	89
600	35.1	27.3	10	29.3	42	81
800	26	24.3	14	37	47	92

Tests made on wire at room temperature gave an ultimate strength of 127,000 psi with zero elongation. Room-temperature tensile tests on swaged iodide chromium showed at a crosshead speed of 0.005 in./min:

Condition	Yld Str, 0.2% Offset, 1000 psi	Ult Str, 1000 psi	Elong., in 0.75 in., %	Red. in Area, %
Wrought	52.5	60	44	78
Recryst	—	41	0	0

A creep rupture value for vacuum-cast electrolytic chromium of one minute at 20,000 psi at 871 C has been reported, and no creep was noted over a period of one week at 10,000 psi at 500 C. When the load was raised to 15,000 psi, the specimen elongated 0.01 in. in 1.53 in. in 1000 hr.

The impact properties of chromium at room temperature are low, not exceeding 1.5 ft-lb, on an unnotched Charpy specimen of as-cast electrolytic metal. At 400 C, the impact value increased to about 118 ft-lb.

A majority of the work done in the past to measure the ductility of chromium was by bend tests. Summarizing this work on sintered electrolytic chromium metal shows that for $2.5 \times 0.5 \times 0.5$ in. specimens the ductile-to-brittle transition temperature for the purest metal was between 25 and 50 C. Impurities, degree of working, and heat treatment change this temperature.

A value for the compressibility of chromium of 99% purity of 7×10^{-7} kg/sqcm between 100 and 600 megabars is given. The relationship between compressibility and pressure (p) is as follows:

at 86 F: $\delta V/V_0 = -5.187 \times 10^{-7}p + 2.19 \times 10^{-12}p^2$
at 167 F: $\delta V/V_0 = -5.310 \times 10^{-7}p + 2.19 \times 10^{-12}p^2$

Electrical and magnetic properties

The electrical resistivity of the metal at 20 C is 12.9 microhm-cm, and the temperature coefficient of resistivity from 20 to 100 C is 3×10^{-3}. Both values exhibit anomalies.

The magnetic susceptibility of chromium is 3.6×10^{-6} emu at 20 C and there are indications that the metal is other than paramagnetic. It has been postulated that the chromium atoms in high-purity chromium metal have a moment of 0.4 Bohr magnetrons and are coupled in an antiferromagnetic arrangement with a Curie (Néel) temperature of 200 C.

The values for the electronic work function are $\phi = .58 \pm 0.02$ ev and the constant in Richardson's equation is $A = 60 \pm 2A$ sq cm/deg sq.

Miscellaneous properties

The thermal conductivity of chromium at 20 C is 0.16 cal/cu cm/sec/°C.

The spectral emissivity at 6690 Å for solid chromium is 0.334, and for liquid chromium it is 0.39. The total emissivity in a nonoxidizing atmosphere at 100 C is given as 0.08.

The reflecting power at various wavelengths is

λ Å	3,000	5,000	10,000	40,000
R %	67	70	63	88

The refractive index (μ) and the absorption coefficient (ν) at various wavelengths are:

Å	2750	3250	3610	4440	5020	6680
μ	1.641	1.259	1.530	2.363	2.928	3.281
ν	3.69	2.91	3.21	4.44	4.55	4.30

Chromium becomes superconducting at a temperature of 0.08 K.

A value of 6.7 v was found for the first ionization potential and 16.6 v for the second ionization potential; a value of 2.89 v was found for the first resonance potential.

The Einstein characteristic frequency of chromium is between 8.3 and 8.43×10^{12}.

The electrochemical equivalent is 0.08983 mg/coulomb for hexavalent chromium and 0.17965 mg/coulomb for trivalent metal. The electrolytic solution potential for valence 3 against a hydrogen electrode is 0.5 v, and the hydrogen overvoltage is 0.38 v.

Chemical properties

The resistance of chromium to oxidation at temperatures from 700 to 900 C showed, for a specimen of 0.04% carbon content, that at 700 C the thickness of the oxide film formed was 1500 Å in 2 hr, while at 900 C it was 24,000 Å in 1 hr. The tests were carried out at a pressure of 760 mm of purified oxygen.

Chromium is resistant to the following acids: acetic, benzoic, butyric, carbonic, citric, fatty, hydrobromic, hydroiotic, lactic, nitric, oleic, oxalic, palmitic, phosphoric, picric, salicylic, stearic and tartaric. It is not resistant to hydrochloric and sulfuric acids. It is also resistant to the following reagents: acetone, air, methyl and ethyl alcohols, aluminum chloride, aluminum sulfate, ammonia, ammonium chloride, barium chloride, beer, benzyl chloride, calcium chloride, carbon dioxide, carbon disulfide, carbon tetrachloride, dry chlorine, chlorobenzene, chloroform, copper sulfate, ferric and ferrous chlorides, foodstuffs, formaldehyde, fruit products, glue, hydrogen sulfide, magnesium chloride, milk, mineral oils, motor fuels, crude petroleum products, phenols, photographic solutions, printing ink, sodium carbonate, chloride, hydroxide and sulfate, sugar, sulfur, sulfur dioxide, chlorinated, distilled and rain water, zinc chloride and sulfate. Ten per cent solutions at 13 C were used in the corrosion tests with the chemicals given above.

Fabrication

Chromium may be consolidated by powder-metallurgy methods or by arc melting in an inert atmosphere.

For powder-metal consolidation, the electrolytic chromium is first ground in a chromium-plated ball mill using plated balls. The resulting powder is then pressed at 20 to 30 tsi to form a compact. It is sometimes necessary to add a temporary bonding compound which is removed by heating to 300 C prior to sintering. The sintering operation takes place at 1450 to 1500 C in an atmosphere of purified hydrogen, helium, or argon. Porosity in the resulting compact is rarely less than 6%.

The best method for fusion consolidation is consumable-electrode arc melting using a water-cooled copper crucible under a partial pressure of argon or helium-argon.

Powder compacts sheathed in iron can be forged at 800 C, and the plate thus obtained could be rolled at 800 C with reductions of 10%. Compacts can also be swaged to wire. Sheathed arc-cast ingots can be forged and swaged at 1000 C. Hot extrusion at 1100 C has also proved successful. As the thickness is reduced by rolling, the working temperature can be lowered so that the final passes can be made at 400 C. After electropolishing, forged material, large-diameter rods, thin sheets, and wire can be bent at room temperature.

The machinability of pure chromium resembles that of cast steel, and machining is best done using cobalt-base high-speed tool steels at fast speeds and slow feed. Chromium can be readily ground, but milling and drilling are difficult.

Uses

Although pure chromium has been produced in the form of ingots, rod, sheet, and wire, the quantities have been small and, due to the low ductility at ordinary temperatures, its use as a constructional material is very limited. Most pure chromium is used for alloying purposes such as the production of nickel-chromium or other nonferrous alloys where the use of the cheaper ferrochrome grades of metal is not pos-sible. In metallurgical operations such as the production of low alloy and stainless steels, the chromium is added in the form of ferrochrome, an electric-arc furnace product which is the form in which most chromium is consumed.

CLAD METALS

The growth of steels clad with other metals has been one of the significant technical developments in the last quarter century of steel progress, yet their widespread acceptance in metalworking is relatively new.

Nor is the growth limited to steel. Many other metals in the nonferrous field, including such exotics as gold and platinum, are being clad in a variety of combinations for specific industrial assignments.

In the case of steel, *cladding* means the strong, permanent bonding of a high alloy and usually expensive material with a plain carbon or low alloy backing steel. The product of this bonding gives service equivalent to that of solid metals, but at about one-third less cost. It takes over the role of coated or plated metals when the environment is especially severe and the coating must be extra heavy. The cladding is normally 5 to 20% of total thickness, but can range much thicker.

Clad steel plate was first developed to supplement the use of solid nickel, and to permit the development of large, corrosion-resisting equipment which otherwise would have been economically impractical to build.

Most of the tonnage today in clad metals is accounted for by carbon or alloy steel clad with stainless steel, although steel is also clad with nickel, "Monel," "Inconel," copper, aluminum and titanium.

Cladding processes

In the first plate cladding process, an assembly was made from a slab of carbon steel and a plate of nickel. This was then heated and rolled to plate and at the same time bonded together.

Other cladding techniques, including a vacuum brazing process, have since been developed. The pack rolling process is still most widely used, however. In this system, a sandwich is built up consisting, for example, of a backing slab of carbon steel, a plate of stainless steel, then a thin "spread" of parting compound, another slab of stainless, and a final plate of carbon steel.

The pack is welded around the edges to keep it together and prevent contamination between the various plates. Next it is heated to the proper temperature and then rolled to the desired gauge. When this desired thickness is reached, the welded edges are trimmed off, and the result is two clad steel plates of uniform thickness.

Alloys. Generally speaking, the choice of alloys used in cladding is dictated by end use requirements such as corrosion, abrasion or strength.

The economy results from high alloy performance at lower cost. By using the most economical backing steel in the appropriate temperature range, and basing calculations on the total gauge of the clad steel plate, less any corrosion allowance desired, maximum savings in material cost are approached. Other advantages: long life, easy modification, easy cleaning and maintenance, and full design freedom.

Welding two types of metal, carbon steel backing and nickel cladding, for example, requires techniques similar to those used in conventional plate fabrication. Similarly, flame cutting of clad steel uses equipment normally used for cutting carbon steel.

Strength. The tensile strength of the clad plate depends on the tensile strengths of the components—which are not altered by the cladding process—and their ratio in the cross section. The bond strength of stainless-clad and nickel-clad steels usually exceeds the minimum ASME requirements of 20,000 psi (ASTM A263-4-5) and is generally in the range of 35,000 to 45,000 psi. Clad steels with a quenched and tempered yield strength of 100,000 psi in the backing steel have recently been developed.

Nonferrous clads

Although steel bulks heaviest in the clad metals picture, there are combinations available in the nonferrous field. Some examples:

Silver clad to copper for use in high-temperature coils, high-frequency conductors and radar cable braiding. It provides both corrosion resistance and thermal conductivity.

Silver clad to nickel for added wear resistance

in electrical contacts, and to beryllium-copper for good spring qualities.

Aluminum clad to copper for a less bulky magnet wire.

Gold clad to nickel, and platinum clad to copper, for chemical process equipment.

Nearly 100 other clad metal combinations have been introduced in the manufacture of semiconductors. The new combinations find their way into silicon or germanium base tabs, bases, lead wires and enclosures. Double-clad base materials and striped clad metals also are finding a growing market.

Availability

Plate gauges range from 3/16 in. to $8\frac{5}{8}$ in. and, on a 206-in. rolling mill, can be produced in sizes up to 178 in. wide or 480 in. long. Clad steel heads produced on a 276-in. spinning machine, can be formed in one piece up to 252 in. o.d. and 7 in. in thickness. Weight can range up to 48,000 lb.

Applications

Today clad metals are being applied in nuclear power reactor pressure vessels in submarines as well as in civilian power plants. Plates for the reactor vessel for the Shippingport atomic power station, for example, are believed to be the heaviest ever rolled in stainless-clad ($8\frac{5}{8}$ in. thick) and weighed as much as 21 tons before fabrication.

Other applications where use of clad is increasing are in such fields as fertilizer, chemicals, mining, food processing, and even sea-going wine tankers.

Stainless-clad is gaining more attention in critical structural applications, such as dam and bridge construction. It is also finding favor in markets where eye appeal is the factor. In architecture, for instance, stainless-clad has been used for the exterior of a 138-ft-high water tower at an automotive technical center and for the underside of a stairway for a new chemical company building. Producers see a market for clad metal curtain wall building panels, and even stainless-clad bus and automobile bumpers.

COBALT AND ALLOYS

Cobalt, the 27th element in the atomic series, stands between iron and nickel in the chart of elements. It is a silvery-white metal with a faint bluish tinge, closely resembling nickel in appearance and mechanical properties. Its chemical properties resemble, in part, those of both nickel and iron. Cobalt is the metal with the highest Curie temperature (2050F) and the lowest allotropic transformation temperature (790 F). Below 790 F, cobalt is close-packed hexagonal; above, it is face-centered cubic.

More than 800 cobalt-containing alloys have been patented; therefore, only the alloy types are given in Table 1. Engineering characteristics are indicated, together with percentage of cobalt contained and effective alloying elements present.

Properties

The outstanding engineering or service properties of the cobalt-containing alloys are listed as subtitles in Column 1 of Table 1.

The desirable high-temperature properties of the first group of alloys—high stress-rupture, creep, thermal shock resistance, and resistance to carburization—may be the result of the allotropic change of cobalt from a close-packed hexagonal at room temperature to a face-centered cubic lattice at high temperatures. All these alloys contain approximately 20% chromium, which results in good oxidation resistance. Below 1200 F, oxidation is negligible. Up to 1600 F, the oxidation is comparable with that of the best chromium-containing stainless steels; above 1600 F, good oxidation resistance coupled with strength make these cobalt-base alloys very attractive. In addition, the presence of tungsten and molybdenum as alloying elements helps strengthen the matrix.

The "Stellite" alloys are immune to all ordinary corroding media and are highly resistant to many corrosive acids and chemicals. Hydrochloric acid will slowly attack the Stellites, but fruit juices, salt water, nitric acid, and many other corrosive chemicals have no effect. Some Stellites combine high reflectivity, permanence of finish, and resistance to abrasion, probably to an extent unknown in any other material.

Cobalt's high Curie temperature imparts high damping characteristics useful for alloys subjected to vibration, such as in high-pressure, high-temperature steam turbines.

The interesting engineering properties of cobalt-containing permanent, soft, and constant-permeability magnets are a result of the electronic configuration of the metal and its high Curie temperature. In addition, cobalt in "Alnico" alloys decreases the grain size and increases the coercive force and residual magnetism. A 50% Co-Fe alloy has the highest permeability in medium fields and also higher electrical conductivity than that of either component. Additions of Co to Ni-Fe alloys result in constant permeability in low fields ("Perminvar"). Co-Pt alloys have magnetic energies per unit volume vastly superior to those of any material now available.

The reason for the good nongalling characteristics of cobalt-containing alloys, which makes them suitable, for example, as liners for machine gun barrels, has not been established. This property is important in alloys used for cutting (also known as "Stellites"), hard facing, and bearings (superalloys). In cutting alloys and high-speed steels, cobalt seems to contribute to low coefficients of friction and to maintaining "red hardness." Cobalt in high-speed steels (5 to 12%) allows tool use at high speeds on hard materials. In cemented carbides, the cobalt (3 to 25%) serves as a binder for the hard tungsten-carbide grains.

The Co-Cr-base alloys for dental and surgical applications are not attacked by the body fluids, and hence do not set up an electromotive force in the body to cause irritation of the tissue; thus the Vitallium alloys are also used as bone replacements. They are ductile enough to permit anchoring of dentures on neighboring teeth.

Cobalt is one of two elements that reduce the hardenability of steel. When dissolved in ferrite, cobalt provides resistance to softening at elevated temperatures. In a recently developed material, 1% cobalt was added to a high-silicon SAE 4130 steel to help improve its biaxial strength. At room temperature, the steel has a yield strength in excess of 200,000 psi.

Principal fabricating characteristics

The metal and alloy parts are both semi-finished or finished either by casting the molten metal or by using the powder-metallurgy technique. The finishing of these (or ingots) is accomplished by the multitude of techniques available to the materials engineer. Table 1 indicates whether the alloys are cast, wrought, or made by powder metallurgy.

TABLE 1—COBALT-CONTAINING ALLOYS

Application	Trade Names of Common Alloys [a]	Effective Alloying Elements, %
High-Temperature Alloys		
Superalloys		
Cobalt-base	HS-25 (w); HS-31 (c); S-816 (c,w); J-1570 (w); J-1650 (w); WI-52 (c)	Cr, 19–27; Ni, 0–32; W, 0–15; Mo, 0–5.5
Nickel-base	René 41 (w); Inco 717C (c); Udimet 700 (w); Nicrotung (c)	Co, 7–28; Mo or W, 0–10; Al, 1–6; Ti, 0.7–4
Iron-base	N-155 (c,w)	Co, 10–30; Cr, 13–21; Ni, 13–43; Mo, 0–10; W, 0–5; Ti; Al
Spring Alloys	HS-25 (51 Co); Elgiloy (w)(40); Dynavar (w)(42)	——
Bearing Alloys	HS-25; N-155 (20 Co)	——
Damping (high) Alloys	Nivco 10 (75 Co) (c)	Ni, 22.5; Ti, 1.8; Zr, 1.1
Hard-Facing Alloys	Stellites (c)	Cr, 26–33; W, 5–14; C, 1–2.5
Wear-Resistant Alloys	UMCo 50 'c), Stellites (c)	Cr, 30–32; Mo, 1.3–3.5; C, 0.05–2
Magnetic Materials		
Permanent Magnets		
Steels (high carbon)		Co, 2–35
Irons (low carbon)	Alnicos (c,p)	Co, 5–35; Al, 6–12; Ni, 14–28; Cu, 3–6; Ti, 0.5–8
Fe-W-Co, Fe-Mo-Co	Remalloy (c); Comol (c)	Co, 5–55
Fe-Co-V	Vicalloy (w)	Co, 52; V, 9.5–14
Nonferrous (Co-Ni-Cu)	Cunife (w), Cunico (w)	Co, 2.5–41
Ordered Structure		Pt, 77; Co, 23
Powder Magnets	Ferrites (p)	Co, 12–17.5
Fine-Particle	Fe-Co (p)	
High Permeability	Permendur, Supermendur	Co, 49–50; V, 0–2; Fe, bal
	Hiperco	Co, 35; Cr, 1; Fe, bal
Constant Permeability	Perminvar	Ni, 45–70; Co, 7.5–25; Mo, 7.5; Fe, bal
Cutting Materials		
Cobalt-Base Alloys	Stellites; Rexalloy 33	
Cemented Carbides	More than 250 trade names (p)	Co, 3–25
High-Speed Steels	More than 170 trade names (w)	Co, up to 12
Miscellaneous		
Glass-Metal Seals	Kovar (w); Fernico (w)	Ni, 28–29; Co, 17–18; Fe, bal
Low Expansion	Stainless Invar (54 Co)(w); Superinvar (5)(w)	——
Constant Modulus	Co-Elinvar (57–60 Co)(w); Velinvar (56–63 Co) (w)	——
Steels		
Hot work	More than 80 trade names (w)	Co, 1–5
Others	SAE 4130 + Co; etc.	Co, 1+
Dental Prosthesis and Osteosynthesis	Vitallium (c); Nobilium (c); Ticonium (c)	Co, 64; Cr, 30; Mo, 5
Tube Filament, Electronic	Cobanic (45 Co)(w); Konel (17)(w); Hilo (18)(w)	——
Electrical Resistance	Kanthal (2 Co)(w)	——
Bearing Alloys	Clevite 300 (15 Co)	——
Magnetron	——	Co, 90; Fe, 10
Thermocouple Wire	——	Co, 7; Au, bal
Magnetostrictive	——	Co-Ni, Co-Fe
Anodes	——	Co, Co-Ni
Spring Alloys	——	Cu-Be-Co, 1–2

[a] c = cast; w = wrought; p = powder metallurgy.

Close dimensional accuracy and exceptional high finish are attained by the precision casting process, and cobalt imparts good casting qualities to high-temperature alloys. Vacuum melting results in improved mechanical properties at room and high temperatures, particularly of the nickel-base superalloys containing titanium and aluminum.

The alloys can be forged, extruded, rolled, swaged and drawn, depending on their composition. They may be joined by brazing or welding. Miscellaneous heat-treating operations, such as annealing, solution heat treating, and precipitation hardening are common. Cold working of cobalt-base alloys before aging has caused improvements in tensile properties.

Some alloys, like the permanent magnets, are made "as cast" or shaped by powder metallurgy. The cemented carbides, however, are made only by powder metallurgy. The "Alnico" alloys are now cast and heat-treated while in a magnetic field to obtain optimum properties.

Available forms, sizes and shapes

The forms available are outlined in Table 2. As an example of the possible range of forms, HS-25 is available as sheet (36 in. wide and 8 ft long) and as tubing (0.052 in. o.d. and 0.029 in. i.d.). Forged blades of superalloys have been made weighing 6 lb each, while precision-cast supercharger wheel blades can weigh less than $\frac{1}{2}$ oz.

Applications

The major uses of cobalt are in cobalt-base and cobalt-containing materials for high-temperature alloys, permanent magnets and steels. One use of unalloyed cobalt is as Co^{60}, the radioisotope which is used in industrial radiography and as a tracer in metallurgical research and development.

The general fields of present application are indicated in Table 1. In addition, cemented carbides, which are considered cutting tools, are made into balls for ball-point pens and high-temperature ball bearings.

The hard-facing alloys are useful because of their resistance to corrosion, abrasion and oxidation at high temperatures for plowshares, oil bits, crushing equipment, tractor treads, rolling mill guides, knives, punches, shears, billet scrapers, valves for high-pressure steam, oil re-

fineries, and diesel and auto engines, to list only a few of the more than 300 applications.

The superalloys have found use as searchlight reflectors, although they were developed for use as components of jet aircraft engines (gas turbines, superchargers, turbojet nozzles and vanes, and ramjet tail cones). These cobalt-base alloys are also useful under the severe operating conditions of high-temperature nuclear reactors. Their superior elevated-temperature properties compensate, to some extent, for the high thermal-neutron absorption for cobalt. They are used in reactors in certain wear-resistant components as guides for control rods. In the nuclear submarines, these alloys are used where sereve wear in contact with sea water is encountered.

While the main application of Alnico permanent magnets is in motors, generators, regulating devices, instruments, radar, and loudspeakers, some are used in games, novelties and door latches.

TABLE 2—FORMS OF COMMERCIALLY AVAILABLE COBALT AND COBALT ALLOYS

Metal

Powder	Shot	Ingot	Wire	Foil
Rondel	Electrolytic	Bar	Plate	Tubing
Granule	Plate	Rod	Sheet	Anode

Alloys

Ingot	Wire	Sheet	Anode	Powder
Bar	Tubing	Foil	Casting	Electro-
			(Precision,	plate
Rod	Plate	Forging	Centrifugal,	Electro-
			etc.)	less
				plate

Cobalt-containing tool and high-speed steels are used for dies; the low-alloy steels are being developed for use in rocket and motor cases for the space age.

In summary, cobalt-containing materials are found in applications underground, on the surface of the earth (in the home and in industry), and in space.

COLD-MOLDED PLASTICS

Cold-molded plastics are one of the oldest of the so-called "plastic" materials; they were introduced in this country in 1908. For the first

time they provided the electrical engineer with
materials that could be molded into more complicated shapes than porcelain or hard rubber,
providing better heat resistance than hard rubber, and better impact strength than porcelain.
They could also incorporate metal inserts.

General nature and properties

So-called cold-molded plastics are formulated
and mixed by the molder (usually in a proprietary formulation). The materials fall into two
general categories: inorganic or refractory materials, and organic or nonrefractory materials.

Inorganic materials. Inorganic cold-molded
plastics consist of asbestos fiber filler and either
a silica-lime cement or Portland cement binder.
Clay is sometimes added to improve plasticity.
The silica-lime materials are easier to mold
although they are lower in strength than the
Portland cement types.

Mixing procedures are generally as follows:
(1) Asbestos, clay and dry binder materials
(either silica and calcium hydrate or Portland
cement) are mixed in proper proportion with
water to give a damp mix. (2) The mixture is
weighed or measured to provide the exact
amount required to fill the mold. (3) The mold
is loaded and closed, applying pressures in the
range of 4000 to 8000 psi. (4) After molding,
the press is opened and the part ejected by
means of a lower movable plunger. (5) Silica-
lime parts are placed on a tray and cured in an
autoclave under steam pressure; Portland cement parts first are sprayed with water, immersed in water tanks, or placed in high
humidity chambers, then cured by air drying,
steaming and oven baking.

Properties of inorganic cold-molded plastics
generally range as follows:

Dielectric Strength, v/mil	60
Arc Resistance, sec	480
Charpy Impact Strength, ft-lb/in. width	0.37
Flexural Strength, psi	5000
Compressive Strength, psi	18,000
Rockwell Hardness	M80–90
Specific Gravity	2.10
Water Absorption (48-hr immersion), %	10–14 ᵃ
Maximum Continuous Heat Resistance, F	750
Tensile Strength, psi	1550

Color	Generally gray or black

ᵃ Can be reduced to <1% by impregnation.

In general, advantages of such materials include: High arc resistance, heat resistance, good
dielectric properties, comparatively low cost,
rapid molding cycles, high production with
single cavity molds (thus low tool cost), and no
need for heating of mold.

On the other hand, the materials are relatively
heavy, cannot be produced to highly accurate
dimensions, are limited in color, and can be
produced only with a relatively dull finish.

They have been used generally for arc chutes,
arc barriers, supports for heating coils, underground fuse shells and similar applications.

Organic materials. Organic cold-molded
plastics consist of asbestos fiber filler materials
bound with bituminous (asphalt, pitches and
oils), phenolic or melamine binders. The binder
materials are mixed with solvents to obtain
proper viscosities, then thoroughly mixed with
the asbestos, ground and screened to form molding compounds. The bituminous-bound compounds are lowest in cost, and can be molded
more rapidly than the inorganic compounds; the
phenolic and melamine-bound compounds have
better mechanical and electrical properties than
the bituminous compounds, and have better
surfaces as well as being lighter in color. Like
the inorganic compounds, organic compounds
are cold-molded, followed by oven curing.

Following is a general comparison of properties of bituminous and phenolic-bound compounds:

	Bituminous	Phenolic
Dielectric Strength, v/mil	60	60
Arc Resistance, sec	175	450
Charpy Impact Str, ft-lb/in. of width	0.35	0.43
Flexural Strength, psi	4500	5500
Compressive Strength, psi	16,000	18,000
Rockwell Hardness	M75–80	M93
Specific Gravity	2.0	2.0
Water Absorption (48-hr immers.), %	3.0	0.9
Continuous Heat Resistance, °F	390	345
Tensile Strength, psi	1200	1700
Color	Black	Brown

Compounds with melamine binders are similar

to the phenolics, except that melamines have greater arc resistance, 360 sec; lower water absorption, 0.3–0.7% (24-hr immersion); are nontracking; and have higher dielectric strength, 80 to 100 v/mil.

Major disadvantages of these materials, again, are relatively high specific gravity, limited colors and inability to be molded to accurate dimensions. Also, they can be produced only with a relatively dull finish.

Compounds with bituminous binders are used for switch bases, wiring devices, connector plugs, handles, knobs and fuse cores. Phenolic and melamine compounds are used for similar applications where better strength and electrical properties are required.

Processing

An important benefit of cold-molded plastics is the relatively low tooling cost usually involved for short-run production. Most molding is done in single-cavity molds, in conventional compression-molding presses equipped for manual, semiautomatic, or fully automatic operation.

Because the parts are not hardened or cured in the molds, and thus may be easily broken or distorted, large ejecting areas must be provided to remove parts from the mold. In most cold molds, the entire bottom of the mold is formed by a movable plunger or pad which is used to eject the soft part.

Because of the abrasive nature of·the compounds, and the relatively high molding pressures used, molds should be sturdily built of a good grade of tool steel, chromium plated where high production runs are involved.

Where moisture absorption may be critical in service, parts can be impregnated with special waxes or asphalts. Water absorption can be reduced by such impregnation to less than 1%, although heat resistance is lowered to about 170 F for wax impregnation, or about 450 F for asphalt impregnation.

Design considerations

Cross sections are generally heavier than hot molded materials to provide durability in handling. Taper is not usually necessary on the part, except on projecting barriers or bosses, as well as on sides of recesses or depressions. Generous fillets should always be provided. Undercuts and reentrant angles should be avoided as

they will increase mold cost and reduce production rate.

In molding, a variation of ±0.015 in. must be allowed in thickness of part. Also, because parts are cured out of the mold, dimensional tolerances cannot be held very closely. Following are generally permissible tolerances:

For dimensions up to:

1 in. : ±0.010 in.
6 in. : ±0.025 in.
10 in. : ±0.040 in.

Dimensions over 10 in. : $\frac{1}{16}$ in. variation.

Lettering, figures and simple designs can be molded on surfaces; marking is usually of the raised type and placed on recessed surfaces to prevent rubbing off.

COLUMBIUM AND ALLOYS

Columbium's general characteristics include a high melting point, low vapor pressure, moderate density, excellent fabricability in the pure state, low cross section, and poor oxidation resistance, as well as rather low modulus of elasticity, and a susceptibility to embrittlement by impurities such as carbon, oxygen and nitrogen. Alloying has improved the oxidation resistance of columbium metal more than fifty fold and at the same time increased the high-temperature tensile strength by a factor of 4 or 5. These improvements are gained at some expense in the ease of fabrication.

TABLE 1—PHYSICAL PROPERTIES OF COLUMBIUM

Crystallographic	
Crystal Structure	b.c.c.–A$_2$
Lattice Constant, Å	3.2941
Co-ordination Number	8
Goldschmidt Radius (CN. 8) Å	1.426
Hermann-Mauguin Space Group	I $\frac{4}{m}\ \frac{2}{3}\ \frac{2}{m}$
Atomic and Nuclear	
Density, 20 C, gm/cu cm	8.66
Atomic Volume, cu cm/gm atom	10.83
Isotopes (natural)	93
Cross Section, Thermal Neutrons, barns/atom	1.1
Thermal	
Melting Point, °C	2468 ± 10
Boiling Point, °C	5127
Specific Heat, cal/gm (0 C)	0.06430
Heat Capacity	
cal/mol (0 C)	6.012
cal/mol (25 C)	5.95

Entropy, cal/mol (25 C)	8.73
Latent Heat of Fusion, cal/mol	6400
Latent Heat of Vaporization, kcal/gm-atom	166.5
Heat of Combustion, cal/gm	2379
Electrical and Magnetic	
Volume Conductivity, 18 C, % IACS	13.2
Electrochemical Equivalent, mg/coulomb	0.19256
Standard Electrode Potential, E°, volt	Cb/Cb^{+5}, 0.96
Magnetic Susceptibility at 25 C, cgs	2.28×10^{-6}
Optical, Thermionic, and Electronic	
Spectral Emissivity at $\lambda = 6500$ Å	0.37
Total Radiation, watts/sq cm	
1880 C	22
1980 C	30
Refractive Index	1.80
Ionization Potential, volts	6.67
Electron Emission (A value) amp/sq cm (K.)	37, 57
Work Function, eV	4.01
Positive-ion Emission, eV	5.5
Electron Configuration	$4d^4 5s^1$

TABLE 2—COMPOSITION AND DENSITY OF SOME COLUMBIUM-BASE ALLOYS

Alloy	Zr	V	Ti	Hf	Mo	W	Ta	Density lb/cu in.
1% Zr	0.6–1.2	—	—	—	—	—	—	0.31
D31	—	—	10	—	10	—	—	0.292
D41	—	—	10	—	6	20	—	0.31
FS-82	1	—	—	—	—	—	33	0.368
F48	1	—	—	—	5	15	—	0.34
F50	1	—	5	—	5	15	—	0.33
Cb-65	1	—	8	—	—	—	—	—
Cb-752	5	—	—	—	—	10	—	0.32
B33	—	4	—	—	—	—	—	0.306
B66	1	5	—	—	5	—	—	0.305
B77	1	5	—	—	—	10	—	0.319
FS-85	0.5	—	—	—	—	12	27	0.39
D14	5	—	—	—	—	—	—	0.31
D36	5	—	10	—	—	—	—	0.252
C103	—	—	1	10	—	—	—	0.32
C120	1	—	—	—	—	5	15	—
SCb-291	—	—	—	—	—	10	10	0.362

Mechanical properties

The tensile properties of columbium and its alloys as a function of temperature of test are shown in Table 3. On the basis of strength-weight ratio against temperature, columbium alloys are superior to the best currently available nickel- and cobalt-base alloys at temper-

TABLE 3—MECHANICAL PROPERTIES OF COLUMBIUM AND SOME COLUMBIUM-BASE ALLOYS

Alloy	Condn	Temp, C	Yld Str, 1000 psi	Ten Str, 1000 psi	Elong, %
Cb	Recryst	100	35	50	40
		800	10	17	63
		1100	—	10	75
Cb-1Zr	—	21	—	47	36
		1093	—	23	23
		1204	—	19	17
FS-82	—	21	—	55	—
		1093	—	30	—
		1316	—	13	19
D31	Cold Rolled	21	98	100	15
		1316	9.8	10.1	15
		1427	7	6.9	13
F48	Cold Rolled	21	85	1.25	25
		1093	42	64	18
		1204	30	50	22
Cb-752	Recryst	21	70	88	26
		1204	33	39	25
B33	—	−100	74.2	91.7	31
		24	59.5	81	32
		1093	30.3	33	34
		1316	13	13.7	55
B66	—	24	91	115	14
		1204	35	40	46
		1316	27	31	71
B77	—	24	106.5	132	18
		1316	27	30	34
FS-85	Cold Rolled	21	95	105	14
		1093	41	46	13
		1427	15	16	71
		1538	12	13	71
D14	Cold Rolled	21	62	78	15
		1093	27	35	35
		1316	14	16	112
D36	Recryst	21	71	80	21
		1093	22	23	50
C103	Cold Rolled	21	88	93.5	9
		1093	18.2	26.4	63
		1371	10.5	11.6	>75
		1482	7.3	8.1	>73

atures above 900 C and comparable to the best molybdenum-base alloys to temperatures on the order of 1400 C.

The stress-rupture data for columbium-base alloys are typified in Table 4 which shows the life to failure under various sustained loads at elevated temperatures.

TABLE 4—STRESS-RUPTURE PROPERTIES OF
SOME COLUMBIUM-BASE ALLOYS

Alloy	Temp, C	Time, hr	Stress, 1000 psi
Cb-1Zr	1093	100	19
		500	11
FS-82	1093	10	25
		100	18
D31	982	230	15
F48	1204	10	24
		100	17
Cb-752	1204	2.3	25
B33	1093	0.5	25
B66	1204	0.9	27.5
B77	1204	0.55	30
		1.9	27.5
FS-85	1204	10	19
	1316	10	13
D14	1204	100	5
SCb-291	1499	2.5	7.8
	1632	2.4	5
	1760	2.1	3.8

Like most body-centered-cubic metals, columbium exhibits a well-defined brittle-ductile transition temperature, which for recrystallized columbium tested in simple tension is reported to be from −125 to −215 C, the latter being for more highly purified metal. Alloying tends to embrittle columbium, i.e., raise the transition temperature, but the transition temperature in simple tension for most alloys is still below room temperature. Notch bar impact data are relatively sparse but both CB-65 and the 1% Zr alloy are notch tough at room temperature even after recrystallization which increases the transition temperature of these and most alloys.

Oxidation resistance

Columbium metal reacts readily with air at elevated temperatures but its alloys have superior oxidation resistance when compared with the other available refractory metals and alloys. Unlike many other refractory metals, colum-

bium alloys do not form liquid or volatile oxides to at least 1400 C and the scale formed is relatively adherent. In addition to the formation of oxide, however, the surface of the unconsumed columbium alloy adjacent to the oxide becomes contaminated by the progressive solution of oxygen and nitrogen which increases with temperature and time of exposure, resulting in a surface embrittlement.

Test results suggest that columbium-base alloys are serviceable for limited periods of time at temperatures on the order of 1200 C, but for longer times or for higher temperatures a protective coating will be required.

Coatings capable of protecting columbium or its alloys during extended high-temperature service are under active development.

Corrosion resistance

The wet corrosion behavior of columbium at room temperature is excellent.

The resistance of columbium to liquid metals is a factor which favors its use in nuclear reactors. Unalloyed columbium is resistant to the following metals to the indicated temperatures: gallium (400 C), bismuth (560 C), mercury (600 C), lead (980 C), lithium (1000 C), and solium-potassium alloys (1000 C). Data for alloys are forthcoming.

Unalloyed columbium corrodes at the rate of 0.20 mg/sq cm/hr in 360 C water while the alloys show rates on the order of 0.002.

Secondary fabrication

Pure columbium is considered one of the most workable of the refractory metals, and commercially fabricated columbium can be forged, rolled, swaged, drawn and stamped by existing commercial techniques. In the primary or mill fabrication, an ingot is hot-worked by forging or extruding, following which the surface is conditioned to remove the contaminated layer, annealed in vacuum or an inert atmosphere to obtain a recrystallized structure, and then cold-worked (with intermediate anneals, if required) by any desired technique to final shape and size. Columbium metal containing less than a total of 0.12% combined oxygen, nitrogen, and carbon can be given a cold reduction of over 90% in cross-sectional area.

Secondary fabrication is done cold to avoid

oxygen contamination and lubricants are used to minimize galling or seizing on the working tools. Rather straightforward procedures are used for making tubes, cups, flanged or flared sleeves, cones and other forms. Machining practice is most satisfactory with high-speed steel tools, rapid rates, and a strong flow of water-soluble oil coolant. The same tool rakes and angles as are used with soft copper will usually give satisfactory results. Extremely light finishing cuts should be avoided; it is better to use sharp tools and light feeds and finish the work in one cut rather than rough- and finish-cut, although the latter can be done. Milling cutters should be of the staggered-tooth type, using plenty of back and side relief. In drilling, the point of the drill should be relieved so that it does not rub the work. In threading large diameters, it is preferable to cut threads on a lathe rather than with a threading die. Grinding is difficult and should be avoided if possible. Vapor degreasing is an effective way to remove oils or grease from columbium parts. Immersion in various hot acids can be used for surface cleaning.

Columbium-base alloys, because of their greater strength at high temperatures, are more difficult to work than the unalloyed metal. Procedures have been and are still being developed for primary and secondary fabrication. Breakdown of the initial ingot requires higher temperatures and finish cold rolling involves more frequent annealing or, in some cases, hot rolling.

Availability

Various columbium mill products are available such as bar, rod, wire, strip, sheet, foil and tubes. Unalloyed columbium has been produced in sheet 36 in. wide by about 200 in. long.

Mill products of the alloys are less readily available since many of the techniques are still being developed. Several of the alloys are now being produced in sheet form in widths up to 24 in. and thicknesses down to 0.010 in. The 1% Zr alloy can also be obtained as foil, wire and tubing.

COMPRESSION AND TRANSFER MOLDED PLASTICS

Compression- and transfer-molding techniques are the most commonly used methods of molding thermosetting molding compounds as well as

rubber parts. They may also be used for forming thermoplastic materials (e.g., compression molding of vinyl phonograph records), but usually other methods are more economical for molding thermoplastics.

The two processes are somewhat similar in terms of sizes and shapes produced. The major difference lies in the greater control of material flow permitted in transfer molding, allowing use of more delicate inserts, and production of somewhat more complicated shapes.

Compression molding

This method involves forming a part by placing the material into an open heated mold, shaping the part by closing the mold and subsequently curing or hardening the part in the closed mold under pressure. (See Fig 1 and 2).

The materials to be molded are generally softened by preheating in conventional ovens, or, more frequently, in a dielectric preheating unit prior to placement in the mold.

Compression-molding techniques are used most extensively for the manufacture of products made from thermosetting plastic materials and rubbers. These materials require the relatively high pressures and temperatures afforded by the compression molding process. Such ma-

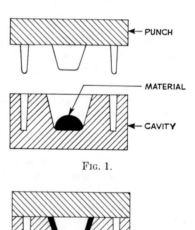

Fig. 1.

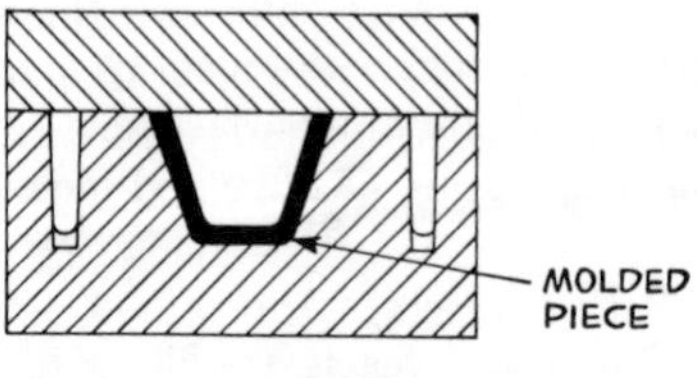

Fig. 2.

Compression molding.

terials include the phenolics, the melamines, the

ureas and the polyester resins. Under special circumstances, thermoplastics are compression-molded, but the injection-molding process is usually more economical for the production of thermoplastic parts.

The process is ideal for the production of such items as thermosetting radio cabinets, television cabinets, trays and other products which require resistance to heat. The size of the compression-molded articles is generally limited only by the platen size and tonnage capacity of the presses used. Through the use of multicavity molds, small electrical components such as switch plates and terminal blocks may be produced economically. Compression molding is also used to produce extremely large parts such as fiberglas boat hulls and the complete fuselage for radio-controlled target aircraft.

Compression molding is not practical for many intricate products where complicated molds are required. Thermosetting materials are extremely stiff or viscous plastic masses during the mold-closing period. The internal pressures developed by these materials tend to distort or break delicate core pins and other small mold components. The process also may be unsatisfactory for the production of extremely close tolerance articles, particularly where critical dimensions are influenced by the mold parting line. Flash thickness produced at the mold parting line tends to vary from cycle to cycle thus changing the dimensions in the direction of the stroke of the press.

Molding pressures range from as low as 50 psi for certain polyester compounds to as high as one and a half tons per sq in. for stiff high impact phenolic materials. Process temperatures range from 250 to 350 F depending upon the material used.

Thickness of the molded part influences production rates. A general rule of thumb allows a minute of mold close-time for each one-eighth inch of wall thickness. Standard dimensional tolerances are usually figured at ±0.002 in. per linear in., although closer tolerances can be held under special circumstances. Where phenolic and urea materials are to be molded, hardened steel is used for the mold construction. Such molds are usually carburized and hardened to about 50 Rockwell C. For use with polyester materials, where pressure requirements are not high, pre-

hardened steel at about 38 Rockwell C is usually satisfactory.

Transfer molding

Transfer molding is best described as a closed-mold technique wherein the material is injected or transferred into a closed mold through a gate or runner system. (See Fig 3 and 4.) Essentially, transfer molding is a one-shot injection molding technique. The process is particularly adaptable again to thermosetting materials, since these materials retain their plastic condition after preheating for only a short length of time. The technique is also used extensively for the molding of unplasticized polyvinylchloride, a material which tends to degrade when held at plasticizing temperatures for any length of time.

Generally, the comments made regarding compression molding apply to this process as well. The distinct advantage in the transfer-molding process lies in the fact that the mold is completely closed and under clamping pressure before the material is injected into the mold cavity. This results in little or no flash and accurate control of dimensions.

A preweighed material charge is plasticized generally in a dielectric preheating unit. The charge is then placed in a pot which usually is positioned above the closed mold. A ram enters the pot and forces the material through an orifice into the closed mold. The transfer plunger and mold are kept under pressure for a predetermined time to allow the chemical/heat hardening process to proceed. When the mold is opened, the small amount of material remaining in the transfer pot and that which has filled the orifice is removed as a cull and discarded.

Transfer molding generally operates at faster cycles than compression molding. Because of the highly plastic condition of the material, complex part designs involving cores, undercuts, and moving-die parts are best adapted to this process. Molded-in inserts can be held in position more easily in the transfer-molding operation. Since the mold is closed during the entire molding operation, control of dimensions is more satisfactory.

With a properly designed mold, transfer-molded articles require fewer finishing operations with a resultant lower net cost.

The process results in higher material costs due to the loss of material in the transfer pot and runner system. High-impact phenolic materials have generally lower physical properties when transfer-molded compared to those obtained by compression molding.

Transfer-molding techniques are used in the manufacture of a wide range of product shapes and sizes. Small, complex electrical components with molded-in terminals are made by this process. Radio and television cabinets weighing up to four pounds have been transfer-molded from phenolic materials.

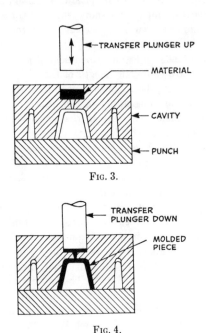

FIG. 3.

FIG. 4.

Transfer molding.

CONTINUOUS CASTINGS

Producing ingots with lengths many times those of the molds in which they are formed is known as continuous, semicontinuous or direct casting. The process now occupies an important place in the metallurgical industry in spite of the fact that the major developments were made within the last twenty years. The process is used in the copper industry but adoption in the aluminum industry has been most widespread. The process is also being used in the steel industry, although not on a wide scale.

Very simply, continuous casting consists of pouring metal into a mold which rapidly chills the metal to the solidification point and permits

it to be withdrawn from the mold. (The metal is poured continuously into the mold.)

Three processes

Reciprocating process. A reciprocating water-cooled copper mold is used, with the discharge rate of the solidified slab synchronized with the down stroke. Metal is poured at the rate of 15,000 to 20,000 lb per hr. As the slab is withdrawn from the mold a flying saw cuts it to the desired length.

Asarco process. Metal is gravity-fed into a mold which is integral with the furnace; the metal is continuously solidified and withdrawn by rolls located below the furnace. This process is used for continuously cast copper alloys. To start a casting cycle a rod of the same shape as the bar to be cast is inserted into the mold. The starting rod is tipped with an alloy of the same composition as that being cast so that when the tipped end melts, a perfect joint is formed between it and the metal being cast. The process produces various shaped rods and bars.

Direct chill process. Molten aluminum is poured from a holding furnace into stationary molds which rest on elevators. When the metal at the bottom of the mold is solidified, the elevator descends. The ingots are sprayed with water as they are lowered, completing the solidification process. Length and width of the ingots are set to produce slab sizes convenient for rolling. Most of the aluminum starting slabs in the United States are produced by this process.

COPPER AND ALLOYS

Copper and copper-base alloys are unique in their desirable combination of physical and mechanical properties. The following properties are found or may be developed in these materials:

1. Moderate to high strength and hardness.
2. Excellent corrosion resistance.
3. Ease of workability both in primary and secondary fabricating operations.
4. Pleasing color and wide color range.
5. High electrical and thermal conductivity.
6. Nonmagnetic properties.
7. Superior properties at subnormal temperatures.
8. Ease of finishing by polishing, plating.
9. Good-to-excellent machinability.

10. Excellent resistance to fatigue, abrasion and wear.

11. Relative ease of joining by soldering, brazing and welding.

12. Moderate cost.

13. Availability in a wide variety of forms and tempers.

The importance of copper and copper-base alloys is in no small part due to the fact that many other metals, whether singly or in combination, may be alloyed with copper. There are many binary alloys, and ternary and quaternary alloy systems are quite common. The copper content of alloys in commercial use ranges from about 58 to about 100%. Alloys of 63% to about 100% copper are of the alpha (single-phase) solid solution type, have excellent ductility, and are ideally suited for cold-working processes such as rolling, drawing and press working. Copper alloys in the lower range, 62% and less, are of the alpha plus beta (two-phase) type and are particularly suited to hot-working processes such as hot rolling, hot extrusion, and hot forging.

The chief elements alloyed with copper are zinc, tin, lead, nickel, silicon, and aluminum, and to a lesser extent, beryllium, phosphorus, cobalt, boron, arsenic, zirconium, antimony, cadmium, iron, manganese, chromium, and very recently, mercury. New alloys and modifications of existing alloys are continually being introduced to meet the increasingly exacting demands of industry for materials with more desirable or different combinations of properties, and lower-cost materials resulting from improved fabricating processes.

Copper is normally used for its combination of excellent resistance to atmospheric and liquid corrosion, its high electrical and thermal conductivity, and its excellent hot- and cold-working properties. Its strength, hardness, and other mechanical properties (creep strength, resistance to softening, fatigue strength, elastic limit, yield strength, strength at elevated temperatures, machinability, etc.,) are relatively poor, and it may be said that all commercial copper-base alloys result from the necessity for improving one or more of the basic requirements without materially affecting the others. In addition, the added cost of obtaining such improvements, or the possibility of lowering the cost in obtaining them, are always important factors.

Cast products

Foundry alloys are used in the form in which they are originally molded. In general, cast alloys are not easily workable and subsequent operations are limited to such treatment as machining, electroplating, soldering and brazing. The use of alloys in cast form permits the production of intricate and irregular shapes impractical or impossible to make by other means. The use of cores permits the making of hollow shapes. The alloying metals are generally greater in amount than for wrought alloys. Types of castings in commercial use include sand castings, die castings, permanent mold, plaster mold, shell mold, centrifugal and investment castings. The alloy involved, the properties desired, and the quantity to be made govern the choice of casting process. Where regularity of shape permits, a number of foundry alloys are available in the continuous-cast condition and may be either solid or hollow shapes. The classification of copper-base foundry alloys generally follows that used for wrought alloys. As in the wrought-alloy field, the number of cast alloys in commercial use runs well into the hundreds, many being modifications of standard alloys. The alloys listed in Table 1 are regarded as the more important and account for about 85% of the tonnage produced. In Table 1 and the tables which follow, only alloying elements are listed. The remainder is copper in all cases.

Wrought products

Wrought products are those originally cast into starting shapes such as billets, cakes (slabs) and wire bar, which are further fabricated into useful forms such as sheet, strip, plate, rod, bar, seamless tubes, extruded shapes, and forgings—by hot rolling, cold rolling, cold drawing, hot piercing, hot extrusion, or combinations of these primary fabricating processes. Secondary fabricating processes include such operations as machining, bending, spinning, cold heading, hot press forging, soldering, brazing, electroplating, cold extrusion, and operations normally associated with pressworking (blanking, deep drawing, coining, staking, embossing, drifting).

The various types of copper and the copper-base alloys are not fabricated with equal facility into usable forms or by secondary fabricating processes. This fact, together with the wide

TABLE 1—CAST COPPER BASE ALLOYS

Tin Bronze ("G" Bronze):
(10% Sn, 2% Zn)
General-utility structural bronze for heavy-duty service, pressure castings, bearings, bushings.

Tin Bronze (modified "G" bronze):
(8% Sn, 4% Zn)
General-utility structural bronze for heavy-duty service, pressure castings, bearings, bushings.

Tin Bronze (modified "G" bronze):
(11% Sn)
General-utility structural bronze for heavy-duty service, pressure castings, bearings, bushings.

Leaded Tin Bronze (Steam or valve bronze)
(Navy "M"):
(6% Sn, 1.5% Pb, 4.5% Zn)
High-grade valve bronze for service to 550 F, steam pressure castings, gears, pumps, backing for babbitt-lined bearings, bushings, ornamental work, electrical castings.

Leaded Tin Bronze (Navy P-c):
(8.5% Sn, 0.5% Pb, 4% Zn)
High-grade valve bronze for service to 550 F, steam pressure castings, gears, pumps, backing for babbitt-lined bearings, bushings, ornamental work, electrical castings.

High-Leaded Tin Bronze:
(10% Sn, 10% Pb)
Heavy-duty bearings operating under high pressure, high speeds, shock and vibration; corrosion resistance to mild acids.

High-Leaded Tin Bronze:
(7% Sn, 7% Pb, 3% Zn)
General-utility bearing alloy.

High-Leaded Tin Bronze:
(5% Sn, 9% Pb, 1% Zn)
General-utility bearing alloy.

High-Leaded Tin Bronze:
(7% Sn, 15% Pb)
Corrosion-resisting pumps, railway journal bearings.

Leaded Red Brass:
(5% Sn, 5% Pb, 5% Zn)
High-grade red brass for general use, applications requiring moderate strength, good machinability, good corrosion resistance; pipe and tube fittings, plumbing goods, small gears, pump impellers, electrical hardware.

Leaded Red Brass (commercial red brass):
(4% Sn, 6% Pb, 7% Zn)
Hydraulic pressure castings, builders' hardware, electrical hardware, plumbing goods, pipe and tube fittings, general-purpose free machining red brass.

Leaded Semi-Red Brass (valve composition):
(3% Sn, 7% Pb, 9% Zn)
Ornamental castings, low-pressure valves and fittings, builders' hardware.

Leaded Semi-Red Brass:
(2.5% Sn, 6.5% Pb, 15% Zn)
Plumbing fixtures, general hardware, air and gas fittings.

Leaded Yellow Brass (high-copper yellow brass):
(1% Sn, 3% Pb, 24% Zn)
Yellow brass alloy for general purposes.

Leaded Yellow Brass (commercial yellow brass):
(1% Sn, 3% Pb, 29% Zn)
Valves and fittings, lighting fixtures, hardware, ornamental castings, general-purpose yellow casting alloy.

Leaded Yellow Brass (naval brass):
(1% Sn, 1% Pb, 36.7% Zn, 0.3% Al)
Marine trim and fittings, hardware, ornamental castings.

Leaded High-Strength Yellow Brass (leaded manganese bronze):
(0.75% Sn, 0.75% Pb, 37% Zn, 1.25% Fe, 0.75% Al, 0.5% Mn)
Marine castings and fittings, gears, valve stems.

High-Strength Yellow Brass (manganese bronze):
(38.5% Zn, 1.25% Fe, 1.25% Al, 1% Mn)
Propellers and other marine parts and fittings requiring resistance to salt-water corrosion, structural parts in machinery construction, valve stems.

High-Strength Yellow Brass (high-strength manganese bronze):
(25.5% Zn, 3% Fe, 5% Al, 2.5% Mn)
High-strength structural parts, severe-wear gears, marine propellers, exceptional corrosion resistance.

Aluminum Bronze:
(3% Fe, 9% Al)
Valve seats, guides, stems; castings requiring strength at elevated temperatures, acid resisting pump and other castings, bearings, bushings.

Aluminum Bronze:
(1% Fe, 10% Al)
General-purpose, heat-treatable high-strength alloy, heavy-duty parts, marine equipment, gears, structural parts requiring high strength, excellent corrosion resistance, maintenance of strength at elevated temperatures, resistance to oxidation.

Aluminum Bronze:
(4% Fe, 11% Al)
Heat-treatable, gears, bushings, valve seats.

Aluminum Bronze:
(4% Ni, 4% Fe, 11% Al)
Heat-treatable, severe service.

TABLE 1—CAST COPPER BASE ALLOYS (*Continued*)

Leaded Nickel Brass (leaded nickel silver):
(2% Sn, 9% Pb, 20% Zn, 12% Ni)
Free-machining nickel silver, hardware, plumbing fixtures, valve trimmings, ornamental castings.

Leaded Nickel Bronze (leaded nickel silver):
(4% Sn, 4% Pb, 8% Zn, 20% Ni)
Free-machining leaded nickel bronze, marine castings, valves, hardware, ornamental castings.

Leaded Nickel Bronze (leaded nickel silver):
(5% Sn, 1.5% Pb, 2% Zn, 25% Ni)
For permanent white color, valves and valve parts for elevated temperature service, dairy equipment, hardware, musical instruments, ornamental castings.

Silicon bronze:
(0.5% Sn, 3% Zn, 1.5% Fe, 1% Al, 1% Mn, 3% Si)
High-strength, general-purpose structural alloy, good corrosion resistance, good weldability, chemical-plant equipment, gears.

Silicon Brass:
(0.5% Pb, 14% Zn, 3.5% Si)
Die castings, small gears, structural parts.

Silicon Brass:
(14% Zn, 4% Si)
Die castings, small gears, structural parts.

Beryllium Copper:
(2% Be, 0.6% Co)
Heat-treatable, very high strength.

Cobalt Beryllium Copper:
(0.6% Be, 2.4% Co)
Heat-treatable, moderate strength, high electrical conductivity.

Chromium Copper:
(0.7% Cr with or without small percentage of Si, Ag, Cd, Zr).
Heat-treatable alloy, high-strength structural parts, high electrical and thermal conductivity, structural parts in electrical equipment.

Copper:
(deoxidized with B, P, Li)
High conductivity, low strength, electrical equipment.

range and combination of physical and mechanical properties desirable for end use, accounts for the large number of commercial wrought copper-base alloys. Design considerations usually dictate a compromise between service properties desired, secondary fabricating properties, forms and sizes available and cost.

All alloys are not made or used in all forms, sizes or conditions.

"Temper" in copper-base alloys denotes final or finished condition, whether annealed or cold-worked. Annealed (soft) tempers are designated by nominal average grain size in millimeters. A number of nominal grain sizes are available in the single-phase alloys; the larger the grain size, the softer the material. Certain alloys, not single phase in composition, are commonly designated in the annealed condition as light annealed (roughly equivalent to 0.025 mm average grain size) or soft annealed (corresponding to 0.070 mm average).

Tempers of cold-rolled flat (sheet, strip) products and cold-drawn round wire are precise and are designated as shown in Table 2, which shows nominal percent reductions for the standard tempers.

TABLE 2
ROLLED TEMPERS (SHEET, STRIP)

Nominal Temper	% Red. of Thickness by Rolling
Eighth Hard	6
Quarter Hard	11
Half Hard	21
Three-Quarter Hard	29
Hard	37
Extra Hard	50
Spring	60
Extra Spring	69

DRAWN TEMPERS (WIRE)

Nominal Temper	% Red. of Area by Drawing
Eighth Hard	11
Quarter Hard	21
Half Hard	37
Hard	60
Extra Hard	75
Spring	84
Extra Spring	90
Special Spring	94
Rivet	5–21
Screw	About 15

For rod (rounds and hexagons) and bar (rectangles and squares) and shapes, the tempers produced by cold drawing are designated by quarter hard, half hard, hard, and extra hard, and the percentage reductions are arbitrary and vary with alloy and size. The hard and half-hard tempers are most commonly used.

For tubes, three drawn tempers are recognized, i.e., drawn general purpose, hard, and light drawn, with arbitrary reductions depending on alloy and size of finished product.

Tempers of mill products used, whether annealed or cold finished, are governed by the secondary fabricating operation to be performed.

When annealing is necessary during fabrication operations, the temperatures necessary will vary considerably with the alloy. Copper can be annealed as low as about 600 F, and the highest temperature generally necessary for any copper-base alloy will be approximately 1400 F. The actual temperature used for any given alloy will vary with the previous thermal and mechanical treatment and the degree of softening desired.

The alloys

In the copper industry, and in the brass mill section of that industry in particular, the alloys of copper have been designated by a sort of "word-terminology" which is often confusing and misleading. Names assigned to various alloys are not always descriptive or indicative of the composition. Several names may be in common use for the same composition. Numerous proprietary names are also used and, in some cases, several names in this category are in general use for essentially the same composition or type of alloy.

Since the more important alloying elements are used effectively in varying amounts, a convenient and logical classification is made possible by separating the alloys into well-recognized groups according to composition. Literally hundreds of wrought copper alloys are used. Those listed are standard alloys which account for the large majority of tonnage used. Alloys not listed are, in general, modifications of standard alloys often used for very special purposes.

Certain terms require clarification. Brass is an alloy of copper with zinc as the principal alloying element, used with or without additional elements. High brasses are copper-zinc alloys containing more than 20% zinc, more properly called yellow brasses. Low brasses are alloys containing 20% zinc or less and low brass is a specific alloy, 80% copper and 20% zinc.

Rich low brass is a term used for brass containing nominally 85% copper and 15% zinc, more properly called red brass. Straight brasses are those alloys containing copper and zinc only.

The term "bronze" in modern usage properly requires a modifying adjective. Early bronzes were alloys of copper with tin as the only, or principal, alloying metal, and thus the term "bronze" has been traditionally associated with alloys of copper and tin. The widespread use of numerous other alloy systems in recent times requires that a bronze be defined as an alloy of copper with some element, or elements, other than zinc as the chief alloying metal. Thus, "tin-bronze" is the proper name for alloys of copper and tin, silicon-bronze for alloys of copper and silicon, aluminum-silicon-bronze for an alloy of copper-aluminum-silicon. Tin-bronzes are also commonly called phosphor-bronzes because of the long established practice of deoxidizing these alloys with phosphorus.

The "coppers" include pure copper and its modified forms, and also certain materials which might more properly be classed as alloys but which are commonly called "coppers" where the alloying element is less than about 1%, e.g., "tellurium copper"—0.5% tellurium, the balance, copper.

Refractory copper-base alloys are those which, because of their hardness or abrasiveness, require greater dimensional tolerances than those established for nonrefractory alloys. Examples of refractory alloys are the phosphor bronzes, silicon bronzes, aluminum bronzes, nickel silvers.

In the tables which follow each section, in all cases, the available wrought forms are indicated as follows: S, sheet, strip, plate; W, wire; R, rod, bar; T, tubes; E, extruded shapes; F, forgings.

Coppers. Many types and variations of commercial coppers are available, the differences being slight or important depending upon the method of production or the intentional minor additions of other metals. Two types refer to the method or refining. Fire-refined copper is that finished by furnace refining. Electrolytically refined copper is that finished by electrolytic deposition where the original ore requires such refining because of certain undesirable impurities which can be satisfactorily removed only by this method, or where the original ore contains sufficient quantities of silver and gold, recoverable in electrolytic refining, to make the electrolytic process profitable. Over 85% of all copper is electrolytically refined. Three major types of commercial copper with respect to composition and method of casting are tough pitch, oxygen-free, and phosphorus deoxidized. Tough pitch copper contains a small but controlled amount of oxygen (about 0.04%) necessary to obtain a level set, i.e., the correct pitch, on the refinery casting (wire bar, cake, billet), the level set being necessary to provide adequate ductility for hot or cold primary fabricating operations, i.e., having sufficient "toughness." Electrolytic tough pitch copper is the standard copper of industry for electrical and many other purposes. Its low electrical resistance (0.15328 ohm/gm m sq at 68 F) equivalent to 100% IACS conductivity is the standard with which other metals are compared. Copper has the highest conductivity of any base metal and is exceeded only by that of silver.

Tough pitch copper has one distinct disadvantage. When heated above about 750 F in atmospheres containing reducing gases, particularly hydrogen, the cuprous oxide is reduced to copper and water vapor, the water changing to steam at the temperatures involved, thus "gassing" or embrittling the copper and destroying its ductility. Hence, the term hydrogen embrittlement. Under circumstances where hydrogen embrittlement is likely to occur (brazing, welding, annealing), oxygen-free copper may be used without sacrifice of conductivity. In cases where some conductivity loss may be tolerated, various types of deoxidized copper may be used. Tough pitch copper, while ductile enough for most fabricating work, is somewhat lacking in the ductility required for difficult forming operations (severe drawing, edgewise bending)and oxygen-free copper is often used in these instances. Also, tough pitch copper is not as free machining as the varieties made for that purpose, i.e., leaded copper, tellurium copper, sulful copper. Silver is added to copper (without sacrifice of conductivity) for the pur-

pose of increasing the recrystallization temperature, thus permitting soldering operations without reduction of strength and hardness. The amount of silver added varies with the application. Lake copper is a general term for silver-bearing copper, having varying but controlled amounts of silver up to about 30 oz/ton, the name deriving from the native silver-bearing copper deposits of the Lake Michigan region.

No.	
102	Oxygen-Free Copper—(S W R T)
104, 105	Silver Bearing Oxygen-Free Copper—Ag; 8, 10, 15 or 25 troy ounces/ton (S W R)
110	Electrolytic Tough Pitch—0.04% O_2 as cuprous oxide (S W R T E F)
113, 114, 116	Silver-Bearing Tough Pitch—Ag; 8, 10, 15, or 25 troy ounces/ton (one troy ounce = 0.0034%) (S W R T E F)
120	Low Phosphorus Deoxidized—About 0.01% P (R T)
122	High Phosphorus Deoxidized—About 0.02% P (S W R T E F)
—	Boron Deoxidized Copper—0.02% B (S W R)
—	Leaded Copper—1.0% Pb (R)
142	Phosphorized Arsenical Copper—0.3% As, 0.02% P (T)
145	Tellurium Copper—0.5% Te; 0.01% P (R)
147	Sulfur Copper—0.4% S (R)
	Other available types of copper with limited use: arsenical copper (0.35% As), phosphorized oxygen-free copper, phosphorized silver-bearing copper, selenium copper, silver-bearing arsenical tough pitch copper.

PROPERTIES OF SOME WROUGHT COPPERS

Type & Condition	Tens. Str., 1000 psi	Elong., % in 2 in.	Mod. Elast., 10^6 psi	Hard., Rockwell F	Elect. Res., microhm-cm
Oxygen Free					
Annealed	32–35	45–55	17	40–45	1.71
Hard	45–55	6–20	17	85–95	
Electrolytic Tough Pitch					
Annealed	32–35	45–55	17	40–45	1.71
Hard	50–55	6–20	17	85–95	
Phosphorous Deoxidized					
Annealed	32–35	45	17	40–45	2.03
Hard	55	8	17	95	
Silver Bearing					
Annealed	40	34–40	17	70	1.92
Hard	62	–	17	97	

Phosphorized copper, having excellent hot and cold workability, is the second largest type in use with respect to tonnage, being widely used for pipe and tube for domestic water service and refrigeration equipment.

Oxygen-free copper is made either by melting and casting copper in the presence of carbon or carbonaceous gases or by extruding compacted, specially prepared cathode copper under a protective atmosphere, so that no oxygen is absorbed. No deoxidizing agent is required; therefore, no residual deoxidants are present, and optimum electrical conductivity is maintained. Oxygen-free copper is used extensively in the electrical and electronics industries, e.g., metal-to-glass seals.

The brasses. The straight brasses are without question the most useful and widely used of all copper-base alloys. Zinc is used in substantial portions in five of the alloy groups listed. Most of the brasses are within the alpha field of the phase diagrams, thus maintaining good cold-working properties. With a higher zinc content of about 40%, the beta phase facilitates hot working. The addition of zinc decreases the melting point, density, electrical and thermal conductivities and modulus of elasticity. It increases the coefficient of expansion, strength and hardness. Work hardening increases with zinc content but 70 to 30 brass has the best combination of strength and ductility.

The low zinc brasses (20% zinc or less) have excellent corrosion resistance, are highly resistant to stress-corrosion cracking, and cannot dezincify. All of the low brasses are extensively used where these properties are required, along with good forming characteristics, and moderate strength. Commercial bronze is the standard alloy for domestic screen cloth, rotating bands on shells and weatherstripping. Red brass is highly corrosion resistant, more so than copper in many cases, and is used for condenser tubes and process piping. The brasses have a pleasing color range varying from the red of copper through bronze and gold colors to the yellow of the high zinc brasses. The alloy 87.5% copper, 12.5% zinc very closely matches the color of 14-carat gold, and the low brasses in particular are used in inexpensive jewelry, for closures, and in other decorative items.

The high brasses, cartridge brass and yellow brass, have excellent ductility and high strength and have innumerable applications for structural and decorative parts which are to be fabricated by drawing, stamping, cold heading, spinning and etching. Cartridge brass is able to withstand severe cold working in practically all fabricating operations, and derives its name from the deep-drawing operations necessary on small-arms ammunition cases, for which it is ideally suited. Muntz metal is primarily a hot-working alloy and is cold-formed with difficulty. It is used in applications where cold spinning, drawing or upsetting are not necessary. A typical use of Muntz metal is for condenser tube plates.

No.	
210	Gilding Metal—5.0% Zn (S W R)
220	Commercial Bronze—10.0% Zn (S W R T)
226	Jewelry Bronze—12.5% Zn (S W)
230	Red Brass—15.0% Zn (S W R T)
240	Low Brass—20.0% Zn (S W R T)
260	Cartridge Brass—30.0% Zn (S W R T)
268	Yellow Brass—34.0% Zn (S W T)
270	Yellow Brass—35.0% Zn (S W R)
274	Yellow Brass—37.0% Zn (W R T)
280	Muntz Metal—40.0% Zn (S R T E)

Silicon, aluminum and manganese brass.
Silicon red brass has the corrosion resistance of the low brasses, with higher electrical resistance than that normally inherent in those brasses, and is especially suited to applications involving resistance welding. Aluminum brass is a moderate-cost alloy primarily made in tube form for use in condensers and heat exchangers, where its improved resistance to the corrosive and erosive action of high-velocity sea water is desired. Manganese brass serves the same purpose as silicon brass.

No.	
667	Manganese Brass—29.0% Zn; 1.0% Mn (S T)
687	Aluminum Brass—20.95% Zn; 2% Al; 0.05% As (T)
694	Silicon Red Brass—17.0% Zn; 1.0 Si (S W)

PROPERTIES OF TYPICAL BRASSES

Type & Condition	Tens. Str., 1000 psi	Elong., % in 2 in.	Mod. Elast., 10^6 psi	Hard., Rockwell F	Elect. Res., microhm-cm
Commercial Bronze (220)					
Annealed	37	45	17	53R (F)	3.92
Hard	61	5	17	70R (B)	–
Red Brass (230)					
Annealed	39	48	17	56R (F)	4.66
Hard	70	5	17	77R (B)	–
Yellow Brass (268, 270)					
Annealed	46	65	15	58R (F)	6.39
Hard	74	8	15	80 R(B)	–
Muntz Metal (280)					
Annealed	54	45	15	80R (F)	6.16
Half Hard	70	13	15	–	–
Medium Leaded Brass (340)					
Annealed	51	53	15	72R (F)	6.63
Hard	74	7	15	80R (B)	–
Tin Brass (405)					
Annealed	42	47	18	47R (B)	4.3
Hard	62	10	18	59R (B)	–
Spring	72	4.5	18	68R (B)	–
Gilding Brass (210)					
Annealed	34	45	17	46R (F)	3.08
Hard	56	5	17	64R (B)	–
Low Brass (240)					
Annealed	43	52	16	57R (F)	5.39
Hard	75	6	16	82R (B)	–
Naval Brass (464)					
Annealed	57	47	15	55R (B)	6.63
Hard	75	20	15	82R (B)	–

Leaded brasses. The primary purpose in adding lead to any copper alloy is to improve machinability and related operations such as blanking or shearing. Lead also improves anti-friction properties and is sometimes useful in that respect. Lead has an adverse effect in that both hot- and cold-working properties are hindered by reduced ductility. Optimum machinability is reached in free-cutting brass rod with 3.25% lead, ideally suited for automatic screw-machine work. Other leaded brasses in rod and other forms, with lead content ranging from 0.5% to over 2%, and with varying copper content, are available to permit varying combinations of corrosion resistance, strength, and hardness, and equally important, secondary fabricating processes (flaring, bending, drawing, stamping, thread rolling). Applications of all of these alloys are found throughout industry.

No.	
310	Leaded Commercial Bronze—9.5% Zn; 0.5% Pb (S)
314	Leaded Commercial Bronze—9.0% Zn; 2.0% Pb (R)
316	Nickel-Leaded Commercial Bronze—6.90% Zn; 1.75% Pb; 1.0% Ni; 0.10% P (R T)
330	Low-Leaded Tube Brass—33.5% Zn; 0.5% Pb (T)
332	High-Leaded Tube Brass—31.9% Zn; 1.6% Pb (T)
340	Medium-Leaded Brass—34.5% Zn; 1.0% Pb (S W R)
350	Medium-Leaded Brass—37.0 Zn; 1.0% Pb (S R)
—	Leaded Flanging Brass—35.6% Zn; 1.9% Pb (R)
353	High-Leaded Brass—34.0% Zn; 2.0% Pb (S W R)
360	Free-Cutting Brass—35.25% Zn; 3.25% Pb (W R E)
370	Free-Cutting Muntz Metal—38.75% Zn; 3.25% Pb (R E)
377	Forging Brass—38.0% Zn; 2.0% Pb (R E F)
385	Architectural Bronze—38.25% Zn; 3.25% Pb (R E)
—	Hardware Bronze—13.25% Zn; 1.75% Pb (R)

Tin brasses. Tin improves the corrosion resistance and strength of copper-zinc alloys. Pleasing colors are obtainable when tin is alloyed with the low brasses. The tin brasses in sheet and strip form and with copper content of 80% or greater comprise a group of materials used as low-cost spring materials, for hardware, or for color effect alone, as in inexpensive jewelry, closures, and decorative items. Admiralty brass is the standard alloy for heat exchanger and condenser tubes, having good corrosion resistance to both sea-water and domestic-water supplies. The addition of a small amount of antimony, phosphorus, or arsenic inhibits dezincification, a type of corrosion to which the high-zinc alloys are susceptible. Naval brass and manganese bronze are widely used for applications requiring good corrosion resistance and high strength, particularly in marine equipment for shafting and fastenings.

No.	
—	Tin Brass—9.5% Zn; 0.5% Sn (S)
—	Tin Brass—4.0% Zn; 1.0 Sn; 0.03% P (S W)
—	Tin Brass—7.0% Zn; 1.0% Sn (S T)
422	Tin Brass—11.0% Zn; 1.0% Sn (S W T)
425	Tin Brass—10.0% Zn; 2.0% Sn (T)
430	Tin Brass—11.0% Zn; 2.0% Sn (S W R)
435	Tin Brass—18.0% Zn; 1.0% Sn (T)
443, 444, 445	Inhibited Admiralty Brass—27.95% Zn; 1% Sn; 0.05% As, Sb or P (T)
464	Naval Brass—39.25% Zn; 0.75% Sn (S W R T E F)
485	Leaded Naval Brass—36.75% Zn; 0.75% Sn; 2.0% Pb (R E F)
675	Manganese Bronze—39.25% Zn; 1.0% Sn; 1.0% Fe; 0.25% Mn (S R E F)

Phosphor bronzes. As a group these range in tin content from 1.25% to 10% nominal tin content. They have excellent cold-working characteristics, high strength, hardness, and endurance properties, low coefficient of friction and excellent general corrosion resistance, and hence find wide use as springs, diaphragms, bearing plates, Fourdrinier wire, bellows and fastenings.

No.	
502	Phosphor Bronze (Grade E)—1.25% Sn; 0.25% P (S)
510	Phosphor Bronze 5% (Grade A)—5.0% Sn; 0.25% P (S W R T)
521	Phosphor Bronze 8% (Grade C)—8.0% Sn; 0.25% P (S W R T)
524	Phosphor Bronze 10% (Grade D)—9.85% Sn; 0.15% P (S W R T)
534	Leaded Phosphor Bronze (Grade B)—4.75% Sn; 0.25% P; 1.0% Pb (R)
544	Free-Cutting Phosphor Bronze—4.0% Sn; 4.0% Zn; 4.0% Pb (S W R)

PROPERTIES OF TYPICAL BRONZES

Type & Condition	Tens. Str., 1000 psi	Elong., % in 2 in.	Mod. Elast., 10^6 psi	Hard., Rockwell B	Elect. Res., microhm-cm
Phosphor Bronze E (505)					
Annealed	40	48	17	–	–
Hard	75	8	17	75	3.59
Phosphor Bronze D (524)					
Annealed	66	68	16	55	15.7
Hard	100	13	16	97	–
Aluminum Bronze D (614)					
Annealed	72	35	17	–	12.0
Hard	82	30	17	90	–
Aluminum-Silicon-Bronze (638)					
Annealed	82	36	16.7	86	17.4
Hard	120	7	16.7	98	–
Low-Silicon Bronze (651)					
Annealed	40	50	17	55	14.5
Hard	70	15	17	80	–
High-Silicon Bronze (655)					
Annealed	57	58	15	55	25
Hard	92	22	15	90	–
Manganese Bronze (675)					
Annealed	65	33	15	65	7.2
Hard	82	25	15	–	–

Aluminum bronzes. The alpha aluminum bronzes (containing less than about 8% aluminum have good hot- and cold-working properties, while the alpha-beta alloys (8-12% aluminum, with nickel, iron, silicon, and manganese) are readily hot-worked. All have excellent corrosion resistance, particularly to acids, and high strength and wear resistance. Certain alloys in this category may be hardened by a heat-treatment process similar to that used for steels. Certain alloys in this group are used for spark-resistant tools.

No.	
606	Aluminum Bronze—5.0% Al (S W R T)
612	Aluminum Bronze—8.0% Al (S W R)
628	Aluminum Bronze—9.5% Al; 1.0% Mn; 5.0% Ni; 2.5% Fe (S R F)

Cupronickels. These are straight copper-nickel alloys with nickel content ranging from 2.5 to 30% nickel. The 10% and 30% nickel

alloys are more commonly used. The corrosion resistance of these alloys increases with increasing nickel content. Cupronickels are markedly superior in their resistance to the corrosive and erosive effects of high-velocity sea water. They are moderately hard but quite tough and ductile, so that they are particularly suited to the manufacture of tubes for condenser and heat-exchanger use, and have the flaring, rolling and bending characteristics necessary for installation.

No.	
706	Cupronickel 10%—10.0% Ni; 1.25% Fe; 0.40% Mn (S R T)
715	Cupronickel 30%—30.0% Ni; 0.60% Mn; 0.50% Fe (S W R T)

Cadmium bronzes. Alloys of copper and cadmium, with or without small additions of tin, are primarily used for electrical conductors where high electrical conductivity and improved strength and wear resistance over that of copper are required.

No.	
162	Cadmium Bronze—1.0% Cd (S W R)
164	Cadmium Bronze—0.8% Cd (W)
165	Cadmium Bronze—0.6% Sn; 0.8% Cd (S W R)

Nickel silvers. Actually nickel brasses, these alloys contain varying amounts of copper, zinc and nickel. They have a color ranging from ivory white in the lower nickel alloys to silver white in the higher nickel alloys. Their pleasing silver color, coupled with excellent corrosion resistance and cold formability, makes them ideally suited as a base for silver-plated flatware and hollow ware, and structural and decorative parts in optical goods, cameras and costume jewelry. The low-copper 18% nickel alloy is used for springs and other applications where high fatigue strength is desired.

No.	
745	Nickel Silver 10%—24.75% Zn; 10.0% Ni; 0.25% Mn (S W R T)
752	Nickel Silver 18%—17.25% Zn; 18.0% Ni; 0.25% Mn (S W R T)
754	Nickel Silver 15%—23.0% Zn; 15.0% Ni (S)
757	Nickel Silver 12%—24.0% Zn; 12.0% Ni (S W)

No.	
764	Nickel Silver 18%—20.25% Zn; 18.0% Ni; 0.25% Mn (T)
766	Nickel Silver 12%—31.25% Zn; 12.0% Ni; 0.25% Mn (S)
770	Nickel Silver 18%—26.75% Zn; 18.0% Ni; 0.25% Mn (S W R)
774	Leaded-Nickel Silver 10%—42.0% Zn; 10.0% Ni; 2.0% Mn; 1.0% Pb (R E F)

Silicon bronzes. The silicon bronzes are an extremely versatile series of alloys, having high strength, exceptional corrosion resistance, and excellent weldability, coupled with excellent hot and cold workability. The low silicon bronzes (1.5% Si) are widely used for electrical hardware. The high silicon bronzes are used for structural parts throughout industry. Aluminum silicon bronze is an important modification with exceptional strength and corrosion resistance, particularly suited to hot working and with better machinability than the regular silicon bronzes.

No.	
639	Aluminum Silicon Bronze—7.25% Al; 2.0% Si (R F)
651	Low Silicon Bronze—1.5% Si; 0.25% Mn or Zn (S W R T F)
655	High Silicon Bronze—3.0% Si; 1.0% Mn or Zn (S W R T E F)

PROPERTIES OF NICKEL-SILVER, CUPRONICKEL AND BERYLLIUM-COPPER

Type & Condition	Tens. Str., 1000 psi	Elong., % in 2 in.	Mod. Elast., 10^6 psi	Hard., Rockwell B	Elect. Res., microhm-cm
Nickel Silver, 65–18 (752)					
Annealed	56	35	18	40	29
Hard	85	3	18	87	–
Cupronickel 30 (715)					
Annealed	55	45	22	45	37.5
Hard	75	15	22	85	–
Beryllium Copper (824)					
Annealed	70	40	18.5	–	–
Hard	115	2	18.5	–	–
Heat Treated	120	2	18.5	38 R (C)	9.6

Precipitation hardenable alloys. This group of copper alloys, introduced within the last 30 years and increasingly and extensively used today, comprises the precipitation-hardening (age-hardening) alloys. Certain alloy systems are capable of being hardened and strengthened by heat treatment. The alloying element or elements are retained in solid solution by a high-temperature heat treatment (solution anneal) followed by a rapid water quench. In this condition, the alloy is relatively soft, thus facilitating both primary and secondary fabricating processes. After final forming, the alloy may be hardened and strengthened by an aging or precipitation-hardening heat treatment at somewhat lower temperature. The aging treatment precipitates a second phase consisting of the hardening constituent. Temperatures involved vary with the alloy system used.

The advantage of the precipitation-hardening alloys lies in their ability to retain strength and hardness at elevated temperatures which may be encountered in service applications. Also, brazing under properly controlled conditions may be employed in joining, when retention of strength and hardness is required. In addition, and equally important, fabricated parts are possible which develop full strength and hardness, whereas desired properties of parts made of cold-working type alloys are in many cases limited by tempers compatible with the necessary secondary fabricating operations. Further, the precipitation-hardening alloys are important in applications involving electrical conductivity, since electrical conductivity in the alloys is generally high, and is increased by the aging treat-

ment. Cold work between solution treatment and aging enhances the final properties after aging, increasing with increasing amounts of cold work. The alloys in this group have a wide range of possible combinations of hardness, strength, and electrical and thermal conductivity.

Copper Alloy Powders

Besides copper, there are a rather large range of compositions of copper alloy powders available, including the brasses, bronzes and copper-nickel. Brass powders are the most widely used for P/M structural parts. Conventional grades are available with zinc contents from around 10 to 30%. Sintered brass parts have tensile strengths up to around 35,000 and 40,000 psi and elongations of from 15 to around 40% depending on composition, design and processing. In machinability they are comparable to cost and wrought brass stock of the same composition. Brass P/M parts are well suited for applications requiring good corrosion resistance, and where free machining properties are desirable.

Copper-nickel, or nickel-silver, powders contain 10 or 18% nickel. Their mechanical properties are rather similar to the brasses, with slightly higher hardness and corrosion resistance. Because they are easily polished, they find considerable use in decorative applications.

Copper and bronze powders are used for filters, bearings, and electrical and friction products. Bronze powders are relatively hard to press to densities to give satisfactory strength for structural parts. Probably the most commonly used bronze contains 10% tin. The strength properties are considerably lower than iron-base and brass powders, being usually below 20,000 psi.

No.	
172	Beryllium Copper—2.0% Be; 0.30% Co (S W R)
—	Cobalt-Beryllium Copper—2.5% Co; 0.5% Be (S W R E F)
185	Chromium Copper—0.8% Cr; 0.1% Si, Ag, Cd, or Zr (S W R T F)
190	Copper-Nickel-Phosphorus Alloy—1.10% Ni; 0.25% P (R W)
191	Copper-Nickel-Tellurium-Phosphorus Alloy—1.1% Ni; 0.5% Te; 0.25% P (R)
—	Zirconium Copper—0.15% Zr (S W R)
—	Copper-Nickel-Silicon Alloy—2.0% Ni; 0.6% Si (S W R E)

CORK

Cork is the outer bark of a type of oak tree which grows in the countries around the Mediterranean Sea, particularly Spain, Portugal and Africa. Unlike the barks of most other trees which are fibrous, the bark from the cork oak tree is cellular in structure. The bark is made up of millions of microscopic 14-sided cells filled with trapped air and tightly bound together. At least 50% of each cell is air.

Cork is generally used in composition form. The cork is ground up and then molded into blocks under heat and pressure. Depending on the end use, the particles may be bound together by cork's own natural resins, or with other binders to provide additional desirable qualities. A cork tree cannot be stripped until it is about 25 years old. The first stripping is too coarse for anything but products made of ground cork. The second stripping nine or ten years later is of better quality but still coarse. The high-quality, even-grained cork needed for such applications as stoppers, floats, etc., comes with the third stripping about ten years later.

General properties

Cork is composed largely of celluloselike materials, together with fatty acids, waxy materials, lignin and tannins. The moisture and ash content vary between a few tenths of one per cent to several per cent, depending upon the source and quality of the specimen.

Although a number of plastic products have now been developed to take the place of cork, none of them can duplicate all of the engineering qualities of the natural product. The most important physical properties are compressibility, resilience, resistance to moisture and liquid penetration, ability to absorb vibration, frictional quality, low thermal conductivity, stability, and light weight.

A one-inch cube of cork has been compressed under pressures as great as 14,000 psi without breaking and with only a minimum of lateral flow. In other words, it compresses within itself, unlike rubber which squeezes sideways under pressure. The air entrapped within the cells compresses to take up the pressure. When released from this great pressure, the one-inch block of cork returns to 90% of its original height and shows no appreciable changes in its length and breadth dimensions.

The cellular structure of cork makes it highly resistant to penetration by most common liquids and it will resist boiling water for a considerable period.

Cork has a high coefficient of friction in contact with practically all surfaces. This high frictional characteristic is retained even in the presence of water and of various oils and greases. The air-filled cells give cork great vibration absorption qualities. It acts as a cushion to absorb both vibration and noise. Cork also has very low thermal conductivity because of its

minutely divided dead air spaces. Cork also has a high degree of stability even under changing humidity conditions. Since it is 50% air, it is one of the lightest solids, having a specific gravity of approximately 0.25.

Applications

As mentioned previously, cork is generally used in composition form. The cork is ground and baked into a block, bound together with its own natural resins or with other binders. Ground cork is mixed with a number of other binders to give it added qualities for specific applications. Among these are proteins, such as glue and gelatin, phenolic and other synthetic resins and the various synthetic and natural rubbers. The blocks can be used as they are or sliced into thin sheets. These cork compositions also can be molded into various forms. Cork also is used in the natural state, trimmed to the desired shape. As sheets it is easily cut or stamped into various shapes.

The general fields of application are almost infinite. In its natural form, it is used as stoppers, floats for the fishing and marine industries, as handles for fishing rods and many other such specialty items. Major use is in composition form.

One of the largest volume uses is as board insulation. It is formed in blocks of the desired dimensions and baked under pressure with its own natural resins. It is generally used as a low temperature insulation for cold storage rooms, refrigerator cars, freezer cabinets, roof insulations and duct insulations.

Another major use is in the form of gaskets and friction materials for automotive, airplane, and tractor applications, pipelines, compressors, clutch facings, and machinery of all types. Gaskets containing cork are used almost universally in engines where a seal is needed against water or oil and temperatures are not too high.

The textile industry also makes wide use of cork on cots, clutches and rolls.

Large blocks of corkboard which can be made of varying densities are also used to isolate vibration of heavy machinery. Acting as a cushion, the corkboard soaks up the vibration, gives longer life to the equipment and reduces noise. Since cork does not take a permanent set, isolation cork board will remain effective over long periods.

Cork is used in the shoe industry as cold filler and cushioning. It also is used in finely ground form as filler for linoleum, as cork floor tile, and as acoustical ceiling tile especially in high humidity areas.

Another major use is in liners for metal crowns and molded plastic closures. It gives a tight seal and does not impart any unwanted taste or affect on the contents.

Other uses are infinite such as nonskid bases for desk telephones, for bulletin boards, sporting equipment grips, etc.

D

DECORATIVE THERMOSETTING LAMINATES

Decorative thermosetting laminates consist of a combination of papers impregnated with thermosetting resins and consolidated under heat and high pressure. The top layer of the assembly has a decorative color or printed design.

Types

There are four types of laminated thermosetting decorative sheets: (1) General-purpose type, (2) cigarette-proof type, (3) postforming type, and (4) hardboard-core type.

The general-purpose type consists of a core of heavy kraft paper impregnated with a phenolic resin and a decorative surface treated with a melamine resin.

The cigarette-proof type is similar to the general-purpose type but has a heat-conducting layer to dissipate heat from a concentrated source.

The postforming type laminate is similar to the general-purpose type except that it can be formed under controlled temperature and pressure in accordance with the laminator's recommendations. The material will form to a minimum radius of $\frac{3}{4}$ in. for printed patterns and 1 in. for solid colors with the surface in tension and $\frac{1}{4}$ in. for printed patterns and $\frac{3}{8}$ in. for solid colors with the laminate surface in compression.

The hardboard-core type laminate has a surface similar to the general-purpose type laminated to a hardboard core which is generally self-supporting.

Properties

The outstanding properties of decorative laminates are durability, resistance to stains and cigarette burns, chemical inertness and resistance to discoloration. The basic physical properties are covered by the NEMA (National Electrical Mfrs' Assn.) standards for laminated thermosetting decorative sheets, as follows:

1. Thickness and Tolerance:

Normal Thickness, inches	Tolerance, inches (plus or minus)
$\frac{1}{16}$ (0.062)	0.005
$\frac{3}{32}$ (0.094)	0.007
$\frac{1}{8}$ (0.125)	0.008
$\frac{5}{32}$ (0.156)	0.009
$\frac{3}{16}$ (0.188)	0.010
$\frac{7}{32}$ (0.219)	0.011
$\frac{1}{4}$ (0.250)	0.012

2. Resistance of Surface to Wear:

When tested in accordance with standard procedure using a Taber abraser the rate of wear shall not exceed 0.08 gm per 100 cycles and the final wear value shall be 400 cycles minimum for laminates designated as Class I.

3. Resistance of Surface to Heat:

A) There shall be no blistering or other discernible surface disturbances when subjected to boiling water.

B) There shall be no blistering or other discernible surface disturbance with a satin-finish laminate when subjected to oil heated to 180 C. For furniture and glass-finish surfaces there may be slight impairment of the surface finish.

C) The surface shall not blister in less than 110 sec when subjected to a concentrated heat source of between 289 to 294 C.

4. Resistance of Surface to Stains:

The surface should be unaffected when contacted by normal household materials such as coffee, mustard, citric acid, alcohol, fly spray, acetone, detergent, etc. (For complete list see NEMA specifications.) The following material should not be allowed to remain in contact with the decorative material: hypochlorite bleach, hydrogen peroxide, mineral acids, lye, sodium bisulphate, potassium permanganate, berry juices, silver nitrate, gentian violet, and mild silver protein.

5. Color Fastness:

There shall not be more than a slight color change and no crazing during testing in a Fadeometer.

6. Dimensional Stability:

There shall be no crazing, chalking or delamination and 1/16 in. laminate shall show a gross dimensional change of not more than 0.5% with the grain and 0.9% across the grain when subjected to high and low humidity.

7. Flexual Strength, Modulus of Elasticity and Deflection at Rupture:

A) Average flexural strength shall be not less than 18,000 psi in either the lengthwise or crosswise direction with the decorative face in compression and not less than 12,000 psi in either the lengthwise or crosswise direction with the decorative face in tension.

B) Average modulus of elasticity shall not be less than 800,000 psi.

C) Minimum deflection at rupture for 1/16-in. laminate shall be 0.02 in. with the face in tension and 0.03 in. with the face in compression.

The back of the sheets may be sanded to permit bonding with adhesives to a suitable base material, such as plywood, for mechanical support. Information on adhesives and methods of installation can be obtained from the laminator. The postforming type under controlled heat and pressure can be used to produce a radiused edge.

Decorative sheets are available in a wide variety of color, decorative design and surface finishes. The material is used for good appearance and functional performance under hard service such as counter and table tops, both domestic and commercial, sink tops and kitchen

work surfaces, baseboards, wainscoting and wall paneling.

The material is available in flat sheets in the following sizes:

Lengths, in.	Widths, in.
145¼	60⅞, 48¾, 36⅝, 30⅜
120⅞	60⅞, 48¾, 36⅝, 30⅜
96⅝	60⅞, 48¾, 36⅝, 30¼
	24¼
84⅝	36⅝, 30¼, 24¼
72½	48¾, 36¼, 30¼, 24¼

DIAMOND

Most reference articles on diamond claim for that material a unique position based on three statements: Diamond is (1) the hardest substance known, (2) the most imperishable, and (3) the most brilliant. During the past few years technological progress has made each of these three statements false. A laboratory development, cubic boron nitride, or "Borazon," appears equal in hardness to diamond since it will scratch and be scratched by diamond. Diamond burns in air at 1100 C; Borazon slowly oxidizes at 1700 C. Brilliance is not an exactly defined term, but if we may take refractive index and transparency as a measure of brilliance, then laboratory-modified rutile outshines diamond.

However, diamond combines in one substance a sufficient number of desirable qualities so that it is an important industrial material. Of all diamonds mined 80 weight % are used in industry. U. S. industrial diamond imports vary considerably from year to year; 88 weight % of the world production was imported in 1956 but only 45 %, or 10,000,000 carats, in 1958 (one carat equals 200 mg).

Diamond is the cubic crystalline form of carbon. The eight atoms per unit cell are arranged on two interpenetrating, face-centered cubic latices. Unit cell edge length is 3.56Å. Each carbon atom is surrounded by four tetrahedrally disposed neighbors, each 1.54Å distant. When pure, diamond is water-clear, but small quantities of impurities or strain centers may add color to cover a wide range of shades and densities including opaque black.

Properties

Several characteristics of diamond are summarized in Table 1:

TABLE 1

Density, gms/cm³	3.52
Hardness, Mohs scale	10
Friction Coefficient (dry metal) [a]	0.05
Modulus of Elasticity, kg/cm² [b]	9×10^6
Bulk Modulus kg/cm² [c]	5.92×10^6
Individual Elastic Constants, dynes/cm²	
C_{11} [c]	9.5×10^{12}
C_{12} [c]	3.9×10^{12}
C_{44} [c]	4.3×10^{12}
C_{11} [d]	10.76×10^{12}
C_{12} [d]	1.25×10^{12}
C_{44} [d]	5.72×10^{12}
Elastic Wave Velocity [110], m/sec	18,100
Heat Capacity, cal/°C-gm mole [e]	1.46×20 C
Thermal Conductivity,[e] cal/sec-cm-°K	0.1818 at 4.2 K
	8.12 at 76 K
	1.57 at 273 K
Linear Thermal Expansion,[e] cm/cm-°C	1.2×10^{-6} (0–78 C)
	2.8×10^{-6} (0–400 C)
Thermal Diffusivity, cm²/sec	3.62 at 20 C
Heat of Combustion, cal/gm-mole [f]	−94,500
Melting Point	>4000 C and >200,000 atm press.
Refractive Index [g]	2.407 (red)
Dispersion [g]	0.058
Dielectric Const.[e]	5.5

[a] "The Friction of Solids at Very High Speeds," F. P. Bowden and E. H. Freitag, Proc. Roy. Soc., A 248, 350-367 (1958).
[b] "Diamond Tools," P. Grodzinski, Anton Smit. New York, N.Y., 1944, p. 52.
[c] "Elastic Constants of Diamond," S. Bhagavantam and J. Bhimasenachar, Proc. Roy. Soc., A 187, 381-384 (1946).
[d] "Elastic Moduli of Diamond," H. J. McScimin and W. L. Bond, Bell Lab. Record, July 1959.
[e] "American Institute of Physics Handbook," pp. 4-42, McGraw-Hill, New York, 1957.
[f] "Heat and Free Energy of Formation of Carbon Dioxide and the Transition Between Graphite and Diamond," F. D. Rossini and R. S. Jessup, Nat'l. Bur. Std. Res. Paper RP1141, 1948.
[g] "Gems and Gem Materials," E. H. Kraus and C. B. Slawson, p. 152, McGraw-Hill, New York, 1941.

In addition, diamonds are generally electrical insulators and nonmagnetic. They are quite transparent to X-rays, and thus make useful pressure cell windows for X-ray measurements. Nearly all diamonds show residual stresses in polarized light.

Mohs' scale of 10 for hardness quoted in the table does not give a true picture of diamond hardness. This figure indicates only that diamond will scratch all other materials including corundum which is at 9 on the Mohs' scale. Indentation tests indicate that diamond is 3 to 5 times as hard as corundum.

In this table the melting point of diamond is given in terms of temperature and pressure since the phase diagram of carbon is such that unless pressure is maintained on diamond when at a temperature greater than 4000 C, the diamond will graphitize and then vaporize.

Three values in this table are unusually high in comparison with corresponding values for other materials: elastic moduli, compressional wave transmission velocity and thermal diffusivity.

Applications

Several types of grinding wheels (usually distinguished by the bond material) are made which carry diamonds as the abrasive material. Wheel bonds may be resinoid, vitrified, metal, etc. Diamond grit sizes in these wheels may vary from 0.001 to 0.030 in. In general the diamond-carrying material is placed in a rather thin layer on the working surface of the wheel. Grinding wheels are made in several diamond concentrations, normally 25, 50, 75 and 100. Actually the volume percentage of diamonds in the working portion of a wheel is low since a 100 concentration indicates 72 carats of diamond per in.³ or 0.25 in.³ of diamond per in.³ of wheel abrasive material or 25%.

Wheel surface speeds of 4500 to 6000 fpm are recommended. Most grinding is done wet. By far the greatest use for diamond wheels is in shaping and sharpening tungsten carbide tools. In addition, diamond wheels are used for edging and beveling glass and for lens grinding. Also, artificial sapphire is sliced and formed with diamond grinding wheels.

Diamonds suitable for use in grinding wheels are now man-made and on the market at a price competitive with the natural material. Comparative tests have shown under many conditions an advantage in favor of man-made diamonds in the ratio of material removed to wheel wear. This is possible since the characteristics of the man-made diamonds may be altered somewhat during the growing process to suit grinding requirements. Diamond wheels remove 10 to 150 cu in. of carbide per cu in. of wheel wear depending on the type of carbide, wheel and grinding.

Diamond drills range from small diamond-tipped dental drills up to oil-well core bits containing 1000 carats. A 4-in. core bit may be set with 150 to 300 diamonds varying in size from 1/10 to 1/30 carat. 100 feet of limestone can be drilled before the diamonds require resetting in the hard metal matrix.

Diamond grinding wheel dressers are used to true grinding wheels or to shape wheels and maintain that shape accurately for form grinding. For the normal truing operations a single diamond is set in a steel shank. This tool is then fed across the working face of the rotating wheel, thereby truing the wheel. Stones weighing 1/4 to 4 carats are used for this purpose. Some users prefer natural octahedra since 6 natural points are provided. When one point becomes worn the diamond may be reset.

Some dressers for formed wheels are made from a metal or carbide matrix containing diamond. In certain cases the grinding wheel is run over the formed dresser at the beginning of each work pass to maintain an exact wheel shape.

Diamond wire drawing dies are required for tungsten filament wire and preferred for many other wire materials. The low friction coefficient noted in the table may contribute to the success of diamond in this application. Stones for wire drawing are large (up to 8 carats for 0.10 in. copper wire) and free from flaws and inclusions.

Single-point diamond lathe tools are made by setting a stone in the end of a steel shank suitable for mounting in a machine tool such as a lathe. The stone is mounted with consideration for its shape and crystallographic axis. After mounting, the tool is ground to the required shape on diamond grit wheels and polished on a scaife.

Scaifes have been used for many years by diamond gem polishers. A rotatable cast iron disk is charged with diamond grit in oil. Diamond held against this wheel will be polished provided the crystal orientation is correct for a "soft" grinding direction.

Shaping diamonds accurately is a slow process. Scaife polishing removes about 0.003 carats per min. Diamond grinding wheels will remove material about ten times faster but with an increase in surface roughness. Cleaving and chipping are more rapid methods of shaping but are rough and inexact.

Single-point diamond tools are designed with angles between faces of 80° or greater to avoid the breakage which would occur if slender points were used. Diamond tools are not suitable for heavy cuts particularly on ferrous material. Accurate, fine finish cuts may be made on nonferrous metals and plastics. Tool wear is slow, so accuracy is maintained for many work pieces. A diamond tool maintains its cutting edge about 25 times as long as a high-speed steel tool and 3 to 4 times as long as a carbide tool.

DIE CASTINGS

Die castings are made by forcing molten metal under high pressure into a steel die containing an impression of the part to be made, which is called the die cavity. The process is most useful for the high production manufacture of small to medium-size castings. Economies of the process are developed through:

1. High speed of production. The output from a die is many times more than for other casting processes.
2. Strategic use of metal. Through casting thinner walls, by coring holes and passages, and by contouring the die halves carefully, material can be saved and the casting made lighter.
3. Dimensional accuracy. The reproducibility is very good and dimensional tolerances can be held close enough to eliminate many machining operations.

Production

Die castings are made in a special machine containing a clamping mechanism to open and close the die and to hold the die halves together, and an injection mechanism to force the metal into the die. The injection mechanisms can be of two types. The immersed plunger type is

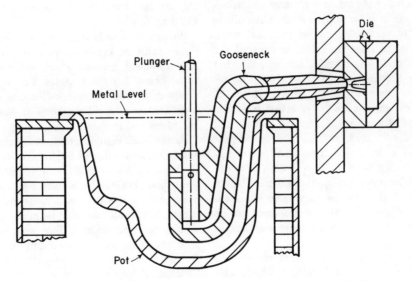

FIG. 1. Immersed plunger type injection system is commonly used where liquid metal is not likely to attack parts of the die casting mechanism.

schematically shown in Fig 1. This system is suitable for alloys such as tin, lead, zinc and sometimes magnesium where attack of the molten metal on the parts of the mechanism is insignificant. The die containing the cavity is closed and held tight in the clamping mechanism while the metal is injected by the downward motion of the plunger which is operated by a hydraulic cylinder. The filling of the die is accomplished in less than 1/10 sec. Dies can be opened in a few sec-

onds, the casting removed by ejector pins and the cycle repeated. Casting rates vary from 50 per hr for heavy parts to 15,000 per hr for small parts made on automatic machines.

Metals such as aluminum, brass and magnesium, which because of either their solvent action or high melting temperatures are not adaptable to the immersed plunger system, can be cast by the cold chamber system (Fig 2) where the metal is melted in a separate unit and

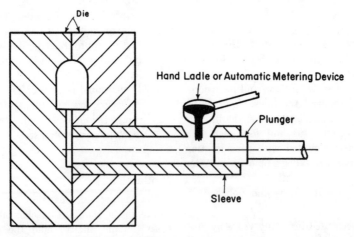

FIG. 2. Cold chamber injection system is used with metals which, because of their solvent action or high melting temperatures, cannot be used with immersed plunger system.

is hand-ladled or metered into a shot cylinder securely fastened to the die. The injection force is supplied by a plunger which in turn is operated by a hydraulic cylinder. Higher pressures (200 to 1000 atmospheres) are used in a cold chamber machine permitting improved density and mechanical properties of the die casting.

Casting machines are rated by their locking tonnage. The common commercial machines vary between 50 and 800 tons. The largest machines available are approximately 2000 tons. Castings weighing up to 75 lb in aluminum and zinc have been produced, and capacities are in existence for producing castings up to 100 lb in aluminum and 200 lb in zinc.

Design

The design of a die casting usually starts with the selection of the parting plane so as to keep the die cost at a minimum. A second step involves the removal of undercuts by redesign or through the use of slides, cores and angular ejectors. A third step is the selection of the optimum wall thickness to provide the best strength and finish with the least weight. Minimum wall thicknesses will vary with the type of metal being cast. The final design step involves elimination of the redundant metal through the use of metal savers or cores. This is done to effect greatest material economy. Careful consideration must be given in die design to "locating points" for subsequent machining operations to retain optimum accuracy of casting.

The die is designed so that the casting will stay in the ejector die when it is opened. It is removed from the ejector die by movable ejector pins which will leave marks on the casting surface and must be taken into consideration in the design of the part so that they do not interfere with subsequent finishing.

Characteristics

Die castings have smooth, dense surfaces which require a minimum amount of finishing. Zinc die castings can be electroplated by a color-buffing step followed by electrodeposition of copper, nickel and chromium, as with automotive bright work and hardware.

A major advantage of die casting is the close tolerances which can be held. These substantially reduce machining operations and are largely responsible for the economy of the process. Linear dimensions in zinc can be held as close as ± 0.001 in./in. and with aluminum as close as ± 0.0015 in./in.

Intricate casting shapes can be readily made through the use of cores and slides, ribs and bosses. Special properties can be obtained by casting in inserts such as bronze sleeves, laminated pole shoes, magnets and hardened steel plates.

One limitation of the process is the expense of the dies which must be amortized over the expected production. However, most automotive, appliance or business machine production is large enough so that the tool and die costs become a small portion of the total casting cost and they are completely offset by manufacturing economies due to elimination of machining of parts after processing.

The process is limited to certain types of alloys which are most adaptable to the process. It is limited in application only by the fact that undercuts or internal closed passages cannot be made. The process is also somewhat limited by the size of the parts which can be made economically. Aluminum die castings in excess of 10 lb are considered large. Average castings will run approximately 1 lb. Zinc die castings over 20 lb are considered large; average size is close to 2 lb. Magnesium castings are approximately the same size as aluminum but 1/3 lighter due to the difference in density.

Average wall thickness for medium-size parts in aluminum and magnesium is 0.08 in. and for large parts is 0.1 to 0.15 in. Wall thickness for zinc will run approximately 75% of aluminum and magnesium due to the ease of casting of zinc in comparison to the light metals. Minimum wall thicknesses vary with the size of the part—0.04 in. is generally accepted for zinc, 0.05 in. for aluminum and magnesium.

Applications

The largest single consumer of die castings is the automotive industry. Zinc is used for ornamentation and hardware such as radio grilles, instrument clusters and housings, as well as for functional parts such as carburetors, fuel pumps and speedometer housings. Zinc is used in appliances such as washing machines, dryers, electric ranges and refrigerators for decorative and functional parts. It is also used for camera cases and projector housings and for toys. Military uses include precision fuze parts.

Because aluminum and magnesium can be used almost interchangeably, relative cost is often the governing selection factor. Automotive uses of aluminum die castings include transmission cases, torque converter housings and extensions, valve bodies and carburetors. In the business machine field, typewriters are the largest application for aluminum die castings. The optical industry also uses aluminum die castings for binoculars, camera cases and projector frames and other parts. Military uses are varied and include ordnance parts such as fuzes, windshields, aircraft engine accessory parts and airframe parts, as well as some small missile airframe parts. The electrical industry uses aluminum and magnesium for instrument housings and frames and for brush holders.

Materials

Dies are generally made of alloy steels and are relatively expensive. Dies for zinc can be made from prehardened steels of the SAE 4140 type. Dies for aluminum and magnesium are made from the chrome-molybdenum H-11 or H-13 steels heat-treated to C46 to C48 Rockwell.

Zinc castings are made from high purity zinc-aluminum alloys containing 4% aluminum with or without copper and containing approximately 0.03% magnesium. Impurities such as tin, lead and cadmium must be kept at very low values, on the order of 0.005% or less, to avoid deterioration of the alloy in humid atmospheres. Zinc alloys have high tensile strength (43,000 psi) and high impact strength (40 ft lb). These properties change with decreasing temperature so that the useful temperature range for zinc is only recommended between 32 and 200 F.

Aluminum die-casting alloys are made from the aluminum-silicon, aluminum-silicon-copper or aluminum-magnesium series. These alloys have the best castability and physical properties for the die-casting process. Mechanical properties will vary between 40,000 to 48,000 psi tensile strength, 20,000 to 25,000 psi yield strength, 3 to 5% elongation. Die castings of aluminum have high fatigue strength (20,000 to 21,000 psi) which makes them particularly adaptable to structural uses where cyclic stresses are involved.

Magnesium alloys are of the aluminum-zinc type having tensile strength of 34,000 psi, yield strength of 21,000 psi and an elongation of 3%.

Copper alloys used are low melting silicon brasses with tensile strength of 50,000 to 85,000 psi, yield strength of 35,000 to 50,000 psi and an elongation of 15 to 25%.

Production factors

The largest expense in die casting is construction of the casting die which can be made as a single-cavity, multiple-cavity, combination of several parts, or interchangeable where several different sizes can be made with common parts of the cavities. Anticipated quantities must be high enough to amortize the die cost and still show production economies. Generally the saving in machining and finishing cost will offset the additional cost of the die for quantities in excess of 5000 parts. Die casting is seldom used for production of less than this amount. Production rates vary with the type of machine. High-speed, single-cavity dies are operated for extremely small parts at rates as high as 15,000 cycles per hour. Conventional 300-ton die casting machines will operate in zinc at rates up to 600 cycles per hr, or in aluminum at rates up to 250 cycles per hr. An aluminum die to produce an automotive rocker arm bracket in a 12-cavity die will produce 2400 pieces per hr with one-man operation. Parts can be multiple-trimmed or trimmed in single dies. Machining is usually confined to drilling, tapping and facing operations to remove gates or overflow marks. Surface preparation for finishing usually consists of a chemical dip.

DIFFUSION COATINGS

A large number of elements can be diffused into the surface of metals to improve their hardness and resistance to wear, corrosion, and oxidation. Diffusion coatings (sometimes called cementation coatings) are applied by heating the base metal in an atmosphere of the coating material, which diffuses into the metal.

Calorized

Aluminum (calorized) coatings are applied by diffusion to carbon and alloy steels to improve their resistance to high-temperature oxidation. They can be applied by treating the metal in a powdered aluminum compound or in aluminum chloride vapor, or by spraying the aluminum on and subsequently heat-treating it. The alloy coating formed (about 25% aluminum) protects the metal by sealing it from the surrounding air. The coatings range in depth from 5 to 40 mils and permit parts to remain serviceable for many years at temperatures up to 1400 F. They have also been used for intermittent exposure as high as 1700 F. Typical high-temperature uses are chemical and metal processing pots, bolts, air heater tubes, and parts for furnaces, steam superheaters and oil and gas polymerizers.

Carburized

Carburizing allows steels to retain high internal strength and toughness and at the same time have high surface hardness. The hardened surface is produced by introducing carbon into a steel surface by heating the metal above the transformation temperature while it is in contact with a carbonaceous material which may be a solid, gas or liquid.

In general, carburizing is limited to steels low enough in carbon (below about 0.45%) to take up that element readily. Plain carbon steels are generally used if surface hardness is the principal requirement and core properties are not too critical. Alloy steels must usually be used if high strength and toughness are needed in the core. Typical applications are gears, cams, pawls, racks, and shafts.

Chromized

Chromizing is the process of diffusing chromium into ferrous metals to improve their resistance to corrosion, heat, and wear. Typical of the chromizing methods that have been developed is one in which the parts to be treated are packed in a proprietary powdered chromium compound and heated to 1500 to 1900 F. This method produces a high chromium-iron alloy on ferrous metals with a low carbon content. The case (3 mils thick) exhibits good resistance to scaling and corrosion at high temperatures. A chromium carbide case is produced on high carbon materials such as cast iron, iron powder, tool steel, and plain carbon steels containing over 0.40% carbon. This case (½ to 2 mils thick) has a hardness of 1600 to 1800 vpn.

Cyanided, carbonitrided

Both cyaniding and carbonitriding produce a hard and wear-resistant surface on low carbon steels. Both methods cause carbon and nitrogen to diffuse into the surface of the base metal. The case developed has high hardness after quenching. The methods differ in that a liquid bath is used in cyaniding, whereas a gas atmosphere is used in carbonitriding.

In general, cyaniding and carbonitriding are used with the same base metals and for the same applications as carburizing. Warpage is usually less serious than in carburizing. Quenching is usually required for full hardness but file hardness can be obtained without quenching.

Nickel-phosphorus coated

With some exceptions, nickel-phosphorus coatings can be roughly classified as diffusion coatings. The coatings are prepared from nickel oxide, dibasic ammonium phosphate and water, and are applied to ferrous surfaces just like a paint. Subsequent heat treatment in a controlled atmosphere produces coatings with a degree of corrosion resistance approaching that of stainless steel and the high nickel alloys. The coatings have little porosity and high resistance to heat and abrasion.

Nitrided

Nitriding is a means of improving wear resistance. In the most widely used process, steel is exposed to gaseous ammonia at a temperature (about 1000 F) suitable for the formation of metallic nitrides. The hardest cases are obtained with aluminum-bearing steels such as the Nitralloys. Where lower hardness is acceptable, steels containing no aluminum, such as medium carbon steels containing chromium and molybdenum, can be used.

Stainless steels can also be casehardened by nitriding (e.g., Malcolmizing). Straight chromium steels are more readily nitrided than nickel-chromium steels, although both are used. Tool steels can also be given a thin hard case.

Nitriding produces minimum distortion. Some growth occurs, but this can be allowed for. In general, nitriding is used for the same applications as carburizing.

Sherardized

Sherardizing is the process of applying zinc coatings to ferrous and nonferrous metals to improve their corrosion resistance. The coatings are applied by heating parts in a zinc powder at 650 to 700 F for 3 to 12 hr.

Sherardized zinc coatings are not as protective as plated or hot dip zinc coatings; however, they can be applied in more uniform thicknesses. Ordinarily, they are quite satisfactory for mild atmospheres. Principal applications are small steel parts such as nuts, bolts and washers, or castings that must resist atmospheric corrosion.

Siliconized

Substantial improvements in the wear resistance and hardness of steel and iron parts can be obtained by impregnating with silicon (about 14%). The most wear-resistant cases are formed on low carbon, low sulfur steels. High carbon, low sulfur steels can also be impregnated, although treatment time is longer. White and malleable iron can also be siliconized; siliconizing of gray irons is not recommended.

The case of a siliconized surface (about 5 to 10 mils) is rather brittle, hardness varying from Rockwell B80 to B85. Siliconized surfaces are virtually nongalling and are especially effective in resisting combined wear and corrosion.

DUCTILE IRONS

Ductile iron, also variously called spheroidal graphite iron, S. G. iron or nodular iron, is a graphite-containing ferrous metal in which the graphite appears as rounded particles or spheroids rather than the usual plates, flakes or clumps typical of other graphitic ferrous metals. In spheroidal form, the graphite exerts a minor influence on the properties of the steel-like matrix since in that form the graphite to metal contact area is at a minimum. It is relatively economical to produce and finds application in many industries.

The occurrence of graphite in spherulitic form rather than plates or flakes is thought to be due to the nearly complete elimination of oxygen and sulfur by the spheroidizing element, which is usually magnesium. The absence of oxygen and sulfur alters the surface tension of the liquid metal and changes other solidification characteristics which possibly accounts for the occurrence of spherulitic graphite in the frozen metal. The properties of ferrous metals are determined largely by the character of their matrices, and, if graphite is present, by the amount, form and distribution of the graphite. Graphite contrib-

utes certain desirable properties such as easy machining, increased damping capacity and resistance to galling in metal-to-metal contact.

Ductile iron is similar to other high carbon-high silicon alloys with respect to corrosion resistance, sliding wear and machinability except insofar as the shape of the graphite alters these characteristics. The spheroidal form of the graphite, having lower surface area, alters corrosion resistance where the graphite is cathodic to the surrounding material. In comparison with flake graphite iron, greater power will be required to remove metal in machining, while in metal-to-metal contact the greater toughness resists removal of metal particles and therefore increases resistance to certain types of wear.

Principal types are listed in Table 1.

TABLE 1—PRINCIPAL TYPES OF DUCTILE IRON

Type [a]	Bhn	Characteristics
80–60–03	200–270	Essentially pearlitic matrix, high strength as-cast. Responds readily to flame or induction hardening.
60–45–10	140–200	Essentially ferritic matrix, excellent machinability and good ductility.
60–40–15	140–190	Fully ferritic matrix, maximum ductility and low transition temperature.
100–70–03	240–300	Uniformly fine pearlitic matrix, normalized and tempered or alloyed. Good combination of strength, wear resistance, and ductility.
120–90–02	270–350	Matrix of tempered martensite. May be alloyed to provide hardenability. Maximum strength and wear resistance.

[a] The type numbers indicate the minimum tensile strength, yield strength, and percent of elongation. The 80–60–03 type has a minimum of 80,000 psi tensile, 60,000 psi yield, and 3% elongation in two in.

Composition

The carbon and silicon contents of ductile iron are usually maintained in a range to provide essentially eutectic compositions which furnish the process advantages of low melting point and low liquid shrinkage. The greatest tonnage of ductile iron is melted in the cupola furnace which is the most economical means of melting the required high-carbon base composition. Carbon content usually ranges from about 3.50% to 4.00% and silicon from 1.00% to 2.50% but both elements have wider ranges in special purpose compositions. These are discussed later under properties. Manganese is usually in the range of 0.10% to about 0.60% and phosphorus 0.020 to 0.050%. A sulfur content of below 0.020% is normal. Alloys intensify the effect of carbon and their effect on hardenability of ductile iron is great since it is a high-carbon alloy. Nickel and molybdenum are used principally to increase strength and hardness in heavy sections. Because of its tendency to form stable carbides, chromium is little used except as a hardener for extremely heavy sections such as are found in metalworking rolls and dies. Magnesium content varies from about 0.020% to about 0.10%, the normal range being 0.045% to 0.060%. A few elements are deleterious to the formation of good graphite: arsenic, antimony, lead and titanium.

High-alloy types. A discussion of the physical and mechanical characteristics of the spheroidal graphite-containing irons should certainly also consider the high alloy or austenitic type of materials. These are the 18 to 36% nickel alloys which are manufactured for their corrosion- and heat-resisting characteristics and other special properties such as controlled expansion. Their relationship to the low-alloy variety of graphite-containing materials can be likened to the relationship of the austenitic steels to low-alloy steels. For this discussion, it is sufficient to point out that the benefits received in changing the graphite from flake to spherulitic is as great for the austenitic types of irons as for the unalloyed or low alloy variety which have been discussed.

Heat treatment

The microstructure of ferrous materials determines, to a large extent, their physical properties. In the spheroidal graphite irons, the matrix structure is determined by the cooling rate of the castings, the precasting treatment (inoculation), the type and amount of alloy present and the heat treatment. By suitable process control and alloy content, structures ranging from substantially ferritic to pearlitic and acicular may be obtained in the as-cast condition in medium section sizes. Structure control may also be accomplished by thermal treatment. Some prefer this method to produce the required structures since thermal treatments have the advantage of producing more uniform structures, especially in components having wide variations in section thickness. Broadly, these treatments are used to accomplish: (1) softening for better machining, greater ductility or greater resistance to impact, (2) hardening for increased mechanical properties or wear resistance and (3) increasing combined carbon for better response to surface hardening.

The change of structure by heat treatment is accomplished in three ranges of temperature. At the upper range of temperature, 1350 to 1700 F, the crystallographic form of iron existing at that temperature, called austenite, contains carbon in solid solution. The hard, brittle constituent, iron carbide, if present, tends to dissociate at this temperature into iron and carbon. The latter deposits on the spheroids as graphite. Conversely, if the austenite is not saturated with carbon at that temperature, it tends to dissolve carbon from spheroids. The rate of cooling from the austenitic range determines the microstructure of the metal as it reaches normal temperatures.

The second important range of heat-treating temperatures is about 1250 to 1325 F. On cooling to this temperature from the upper temperature range, austenite has transformed to another crystallographic form which can no longer hold the carbon in solid solution. On holding in this rather narrow temperature range the carbon formerly in solution migrates to the graphite spheroids and is deposited thereon as additional graphite. At this temperature the carbide which had existed as a constituent of pearlite also dissociates and that carbon also migrates to the existing graphite. By these mechanisms, it is then possible to produce a fully ferritic structure free of carbide and pearlite. Upon cooling from the austenitic temperature range at a rate too fast to decompose the pearlite, such as cooling in air, pearlite will be the microconstituent that exists at room temperature. Upon very fast cooling from the austenitic temperature range, such as oil or water quenching, acicular structures are formed which are the hardest and strongest possible to produce in the material.

The third important temperature range in the heat treatment of these alloys is from about 800 to 1200 F. Stress relieving is holding in this range, usually at about 1100 to 1150 F. Stresses sometimes accumulate in complicated castings when cooling from the casting temperature or in fast cooling from high heat-treating temperatures. Drawing or tempering is holding in this range to modify the as-quenched hardness. Oil quenching, for example, will produce hardnesses on the order of 600 Bhn and tempering or drawing is the means used to reduce this hardness to one desired, with corresponding reduction in tensile strength and increase in ductility. The higher the tempering temperature, the greater the reduction in as-quenched hardness achieved. Fully quenched structures containing martensite may be tempered at 400 to 500 F to relieve quenching stresses without appreciably reducing the hardness.

In summary, the heat treatments common to the ductile irons are *annealing:* heating to 1650 F, holding for one hr per in. of section thickness, cooling to 1275 F and holding for 4 hr plus one hr per in. of section thickness followed by any rate of cooling to room temperature. An alternate method is to modify the second step of the cycle: heat to 1650 F and hold one hr per in. of section thickness, cool to 1350 F at any rate, then cool from 1350 F to 1200 F at a rate not exceeding 35 F per hr. *Normalizing:* heating to 1650 F, holding one hr per in. of section, re-moving from furnace and cooling in air. Some components, because of their complicated nature, are cooled in the furnace to 1550 F and cooled in air from that temperature in order to reduce the tendency for producing high residual stresses in the component. *Oil quenching:* heating to 1650 F holding one hr per in. of section thickness, cooling in oil. As in normalizing, it may also be desirable to oil-quench from 1550 F because of the complicated nature of the castings. Following the normalizing or oil-quenching operations, it is usual to draw or temper the castings in order to remove stresses imposed by the quick cooling or to reduce the quenched hardness. The usual temperature range for tempering is 800 to 1200 F and the higher the range chosen, the lower will be the final hardness of the casting. Simple stress relieving is also done in this temperature range but the stresses are relieved most readily in the 1100 to 1150 F temperature range. The order of properties of heat-treated ductile iron is shown in Table 2. The localized hardening treatments, flame and induction hardening, are used especially in such applications as crankshaft bearing surfaces and gear teeth, giving hardnesses in the neighborhood of 60 R_c. The material also responds well to carbonitriding when carried out in the upper temperature ranges although case depths after carbonitriding are somewhat shallower than that of steel for equal periods of holding at temperature.

TABLE 2—TYPICAL MECHANICAL PROPERTIES: HEAT-TREATED DUCTILE IRON

Heat Treatment	Ten Str 1000 psi	Yld Str 1000 psi	Elong, %	Hard.	Charpy Impact [e] ft-lb
As-cast, moderate [a] to heavy section	80–100	60–75	3–10	200–270 Bhn	15–65
Anneal [b]	60–80	45–60	10–25	140–200 Bhn	60–115
Normalize and temper [c]	100–120	70–90	3–10	240–300 Bhn	35–50
Quench and temper [d]	120–150	90–125	2–7	270–350 Bhn	25–40
Martemper and temper	—	—	—	40–50 R_c	—
Flame hardening	—	—	—	Up to 60 R_c	—
Induction hardening	—	—	—	Up to 60 R_c	—

[a] ASTM 80–60–03
[b] ASTM 60–45–10
[c] ASTM 100–70–03
[d] ASTM 120–90–02
[e] Unnotched

Critical temperature quenching refines the ferrite grains and produces higher mechanical properties for a given hardness. Fully annealed material, having all the carbon present as graphite spheroids, is reheated to just above the critical temperature, then given a rapid quench such as in brine at 100 F. This causes a recrystallization of the ferrite grains but with little absorption of carbon from the spheroids. Tempering at 1200 F will reduce the as-quenched hardness from around 500 Bhn to 155 to 190 Bhn. As a result of the grain refinement, the mechanical properties produced are over 90,000 psi tensile strength, with over 10% elongation.

Machinability

The spheroidal graphite irons have very good machining characteristics by virtue of their contained graphite and compare favorably with other materials at equivalent hardnesses and strengths. The soft annealed grade provides a strong, tough material that can be turned at very high speeds and feeds. Maximum turning speeds of well over 400 sfpm are attainable in this grade. The cutting speed is, however, determined by the microstructure with maximum turning speed of the pearlitic grade reduced to about 150 sfpm and of the hardest oil-quenched grade to 100 sfpm. The chip produced in machining operations tends to be long and continuous which complicates such operations as deephole drilling and internal broaching and reaming. For fast metal removal in these operations, special tool configurations are required to provide access for chip removal. Cutting fluids are recommended for all machining operations and, of course, production machining requires the use of carbide-tipped tools.

Welding

It is possible and practical to weld the spheroidal graphite irons by the several welding processes. Most commonly used are the metallic arc and oxyacetylene processes. Due to the high carbon content and consequent high hardenability, the material must be handled very carefully in order to prevent excessive iron carbide from forming at the welding temperature. If maximum ductility and machinability are desired in the heat-affected zone, the welding operation should be followed by annealing. Filler rods which undergo no phase changes, such as the

60% nickel—40% iron rods, are the easiest and most practical to use. The operation requires the lowest practical input of amperage and the use of welding techniques which prevent overheating of the base metal in any location. The annealed grade, containing little or no combined carbon (to harden on cooling) is the grade on which welding is most easily performed.

Brazing

The joining of ductile iron castings to similar or dissimilar metals can be accomplished with silver or copper brazing alloys, usually without resorting to special techniques in preparing the surfaces to be brazed. Overlaying with hard surfacing alloys also presents no special problems.

Available forms

Because its normal composition makes it eminently suited, ductile iron is essentially a casting alloy. In this condition a wide variety of control is available on its properties through composition and heat treatment. Ferritic castings may be processed further by hot bending to shapes too difficult to cast conveniently. Tubular shapes may be formed by centrifugal casting and by the hot extruding of billets. The forging, rolling, extruding and stamping processes may be used to produce bars, shafts, pipe, plate, angles, beams, gear blanks, etc., in order to utilize the good machinability and corrosion resistance of this high-silicon, high-carbon material. Certain shapes are being produced by explosive forming.

Mechanical properties

Tensile. The various grades or types of ductile iron are usually designated by their tensile properties. For example, the 60-45-10 grade has a minimum tensile strength of 60,000 psi, 45,000 psi yield strength, and 10% elongation. The percent reduction of area is usually not specified but has a numerical value approximately the same as that of the elongation.

The properties of ductile iron castings are primarily dependent upon their microstructure which depends upon the cooling rate and composition of the metal. These factors affect the amount and composition of ferrite and the amount, fineness and hardness of the pearlite in the microstructure. Acicular structures are developed by faster cooling such as in oil quenching or by higher alloy contents. The presence of more and finer pearlite produces stronger and harder matrices at the expense of ductility while the hardest and strongest structures are the acicular structures which are characterized by the least ductility.

A definite relationship exists between the hardness, elongation, tensile strength and yield strengths for the various grades of the metal. In the softer as-cast and annealed grades, the tensile strength in psi is about 420 times the Brinell hardness number with the yield strength about 75% of the tensile strength. In the normalized and oil-quenched grades the tensile strength in psi is about 470 times the Brinell hardness number. Because of the generally good founding characteristics, castings have quite uniform properties in varying section thicknesses. The effect of alloys upon the tensile properties is similar to their effects upon other ferrous metals. Silicon has a softening effect upon as-cast material but raises tensile and yield strength in annealed material. Manganese is a pearlite stabilizer and, therefore, hardens as-cast castings but has little effect upon annealed properties. The alloys nickel and molybdenum raise tensile and yield strength, hardness, hardenability and are most important for increasing strength in heavy sections. Their effect is much greater upon heat-treated properties. Nickel is often used as a ferrite strengthener and as such, will increase the yield strength without great sacrifice of ductility. Molybdenum, in addition to strengthening and increasing hardenability, promotes strength at elevated temperature. The carbide-forming elements, chromium and vanadium, are seldom used except to promote hardness for wear resistance in very heavy sections.

The spheroidal graphite irons have a definite modulus of elasticity which is dependent upon their graphite contents. The matrix, being steel-like, has a modulus of elasticity of 29×10^6 psi which value is reduced by the area of the cross section occupied by the graphite. The modulus of the irons in the ordinary carbon ranges is 23 to 25×10^6 psi but this value can be increased or decreased by varying the amount of graphite in the microstructure.

Impact. Impact properties of the spheroidal graphite irons are affected by the microstructure and composition. As in steels, impact tests are conducted over a range of temperature and vary from ductile fractures with high energy absorption characterized by a dark, fibrous appearance, to brittle fractures having light crystalline appearance and low energy absorption. At lower temperatures, the fracture changes from ductile to brittle and the range of temperature where this occurs is called the transition temperature. The softer, alloy-free grades absorb more energy in fracturing and remain ductile to lower temperatures. Nickel has slight effect on energy absorbed while silicon and most other alloys adversely affect this property. The preferred microstructure for high resistance to impact is unalloyed ferrite. In the iron-carbon alloys, pearlite has a transition temperature above room temperature, and for maximum impact resistance this microconstituent should not be present in ductile iron. Increasing silicon raises the transition temperature and lowers impact values for any given temperature. For ductile fractures and high energy absorption at freezing temperatures, a maximum silicon content of 2.50% is usually specified. Below 2% silicon, the transition temperature range is lowered to as much as − 75 F. The effect of phosphorus is even more detrimental to this property than that of silicon and for very low temperatures the phosphorus content should be held to a maximum of 0.05%. Values of energy absorbed in impact tests range from about 30 ft-lb in a Charpy test at 32 F on a specimen 0.50 in. × 0.75 in., 3 in. span, and V-notched, to 90 ft-lb at − 90 F in an unnotched bar 0.375 in. square, 2.25 in. span.

Elevated temperature. Ductile irons exhibit useful elevated temperature properties up to about 1200 F and like most of the other properties, are dependent upon microstructure. The pearlitic irons will have good strength and little growth up to about 800 F. Above this temperature, the pearlite has a tendency to decompose into ferrite and graphite with a change in volume, or growth. Above 1200 F pearlite decomposes very rapidly. The ferritic irons have little tendency for growth up to about 1500 F because of the absence of pearlite. Resistance to scaling is very good within the useful load-carrying temperatures because the shape of the graphite does not offer paths for penetration of the heat effects and the natural resistance of silicon-alloyed ferrite to such phenomena. High silicon compositions (up to 6%) are as scale resistant as the stainless steels up to 1800 F but these compositions are brittle at room temperature. They are also not as resistant to thermal shock as lower silicon compositions. The hot tensile properties of the ductile irons are summarized in Fig 1. However, the elevated temperature strengths of these materials are greatly increased by the addition of molybdenum and copper. For instance, at 1100 F the 1000-hr rupture stress of an iron of conventional silicon content is increased from about 4200 psi to over 7000 psi by the addition of about 0.75% molybdenum.

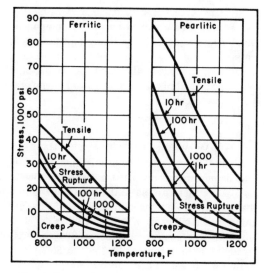

FIG. 1. Comparison of tensile, stress rupture and creep strength (0.0001% per hr) of ferritic and pearlitic ductile iron.

Endurance. Endurance properties of this class of irons vary with the hardness and microstructure, the harder and stronger types having a somewhat lower ratio of endurance limit to tensile strength. Alloys seem to have little effect on this property although shot peening or rolling of the surface increases the endurance limit of a part. The softer grades are improved to a greater extent than the harder grades by putting their surfaces under compression by rolling treatment.

The ratio of tensile strength to endurance limit (endurance ratio) is about 0.45 for the as-cast or annealed grades and about 0.35 for the normalized and oil quenched grades. The notch sensitivity factor (ratio of unnotched to 45 deg V-notched endurance limits) is about 1.45 for as-cast and annealed material, 1.60 in the normalized condition and 1.65 in the quenched and tempered grade.

Applications

The incorporation into one material of high strength in all section sizes, excellent castability, wear resistance, machinability and relatively simple control of properties suggests that such material would find application in many different industries and applications. It is lighter than nongraphitic ferrous metals and, with its high strength and good castability, permits the design of thin-section, lighter-weight machine components. The largest tonnages are used in automobile and diesel engines for crankshafts, rocker arms, and other engine and clutch components; in farm implements and tractors for gears, sprockets, brackets, transmission casings, housings and many structural components; in earth-moving machinery for wheels, gears, rope drums and in heavy machinery for load-carrying components requiring good wear, easy machining, high strength and antideflection properties. Other applications include: fluid-handling devices such as valves and fittings; paper-mill machinery—rolls, roll heads, gears; steel mill rolls and mill equipment, as well as many others in a variety of industries.

E

ELASTOMERIC LININGS

Elastomeric lining materials are available in natural rubber, GR-S, butyl, neoprene, nitrile and "Hypalon" synthetic rubbers as well as in polyvinyl chloride and polyvinylidene chloride (saran) plastics and the new fluorocarbon rubbers. These basic materials, when formulated and processed into linings, are normally sold and applied as heavy sheets at 3/16 to ¼ in. thick.

Elastomeric linings are manufactured by plying up thin calendered films of 0.015 to 0.030 in. thickness to the full thickness of the lining. Linings are handled in uncured rolls about 42 to 48 in. wide and 15 to 25 yds long, depending upon thickness. They are applied in the soft and plastic uncured state to the base metal and are cured in place.

Heavy linings differ from organic coatings and other protective films which are much thinner and are customarily applied by spraying or dipping procedures. Because of their plied construction, it is almost impossible for any pinholes or abnormalities to be present in the same spot in the six or eight calendered plies. Therefore it is almost impossible to have lining failures due to pinholes or other abnormalities. Such causes for failure are a serious problem in other types of protective films.

When conflicting properties are needed, a suitable lining with the best balance of properties can normally be obtained by proper compounding. The basic elastomeric lining materials as a group are noted for their chemical resistance, toughness, flexibility, dielectric properties and resistance to abrasion and reasonable temperatures (see Table 1).

Natural rubber

About 70% of the applied linings are estimated to be based upon natural rubber. This is due to the fact that natural rubber can be formulated in an extremely wide range of physical properties. Linings from natural rubber are classified as soft (sulfur content at 2% to 3%), semihard (sulfur content at 15% to 30%) and hard (sulfur content at 30% to 45%). As the sulfur content is increased, general chemical resistance increases from good to excellent; but at

TABLE 1—PROPERTIES OF ELASTOMERIC LINING MATERIALS

Type of Lining	Hardness (Shore A)	Tensile Str, psi	Elongation, %	Water Pickup oz/sq ft of face	Volt Breakdown, v/mil	Temp Range, F
Natural Rubber						
Soft	35–70	2000–3500	400–800	0.3–0.4	to 600	−65–160
Semihard	90–95	1200–2000	100–200	0.5–0.6	to 600	30–210
Hard	60–80 [a]	1800–6000	0.2		to 600	35–220
Neoprene						
Zinc Cure	50–70	1800–3000	300–500	0.9–1	to 600	−40–250
Lead Cure	50–70	1800–3000	300–500	0.6–0.7	to 600	−40–250
Butyl	50–70	1200–2200	300–500	0.5	to 600	−100–300
Nitrile	50–70	1500–2500	300–500	0.4	to 300	−60–250
Hypalon	60–80	—	250–400	0.3	to 700	−30–300
PVC	80–90 [b]	—	—	0.04	750–900	−40–230 [c]

[a] Shore D.
[b] Plasticized.
[c] Special adhesives used.

132

the same time the lining also becomes harder and less flexible.

Soft natural rubber linings are extremely flexible and tough above −65 F, but are limited to a maximum operating temperature of 160 F. Hard rubber linings give satisfactory service to approximately 210 F but are not suitable for use at low temperatures. Linings may be protected from high temperatures and mechanical damage by facing the inside of vessels with an acid-resistant insulating brick.

Because of brittleness at low temperatures and the difference in coefficient of expansion between steel and hard rubber, large tanks lined with hard rubber are not suitable for use outdoors in northern climates. Furthermore, hard rubber linings are not suitable for use with abrasive slurries where specifically compounded soft rubbers give excellent service. Because of these opposites in the mechanical properties of soft and hard rubbers, the semihard rubbers often present the best compromise of properties, and for this reason are extensively used.

Lining properties can be improved by sandwiching a layer of hard or semihard rubber between layers of soft rubber. In this way, the hard rubber provides chemical resistance, and the soft rubber provides the resistance to mechanical shock and/or abrasion necessary to make the lining serviceable at low temperatures. The soft layer on the back of the lining also improves adhesion to the base structure where temperature variations could result in thermal expansion which would rupture the bond of a semihard or hard rubber if it were laid directly to the metal. The thermal coefficient of expansion for steel is about 12×10^{-6} as compared to 84×10^{-6} for hard rubber. The electrical resistance of both soft and hard rubbers is excellent.

Linings from natural rubber are among the lowest in cost, and all natural rubber linings can be applied without difficulty. Even hard rubber can be used satisfactorily on large vessels if the vessel is designed in sections which are then bolted together and the joints covered with lap straps. With proper usage and suitable repair when damaged, they can be expected to give service for twelve to fifteen years or more, which makes for a very low over-all cost.

Natural-rubber linings give excellent service for many water solutions of chemicals and can be used for practically all plating solutions (excepting chromium). Hard rubber linings will also withstand paraffenic oils, greases and fats at moderate temperatures. The linings will not withstand the solvating action of aniline, benzene, carbon disulfide, carbon tetrachloride and other chlorinated or halogenated hydrocarbons. They are not resistant to the chemical attack of the oxidizing acids, nitric acid above 10% or sulfuric acid above 50% at 160 F. However, lower operating temperatures will increase the upper concentration of chemicals which can be safely used with a lining.

GR-S

With the exception of gum (nonfilled) rubber linings, GR-S can be used to duplicate practically all of the linings which are now made from natural rubber and for the same type of service. However, there is a slight quality advantage in linings made from natural rubber and accordingly, linings are not made from GR-S as long as natural rubber is available at prices reasonably close to GR-S.

Butyl rubber

Butyl rubber linings are available in physical characteristics approaching those of the soft natural rubbers. Although the linings are not used extensively now, they could well develop into large volume applications in the future.

Butyl linings have a higher degree of general chemical resistance than natural rubber, and are suitable for weak oxidizing acids and for organic chemicals which deteriorate soft rubber linings over a long period of time (e.g., fats, greases and soaps).

Butyl linings are suitable for use above 160 F and it has been reported that butyl rubber may be compounded for continued use at 300 F as a lining material. These characteristics indicate that butyl will find extensive use in tank trucks and railroad tank cars where hot solutions are to be transported, and where the empty car can be returned without fear of rupture of the lining by shock in cold climates. Also, butyl is used where the continual vibration and shock encountered in railroad hauling eliminates the semihard and hard rubbers which would normally be used for such solutions. Hydrofluoric acid and ethylene chlorohydrin can be handled with butyl linings.

Neoprene

Neoprene linings are suitable for use with a very broad range of chemical solutions and are satisfactory for most of the same applications where natural rubbers are used. There are two basically different formulations for neoprene linings, one of which (litharge cure) is suitable for use with low-pressure steam and will also withstand boiling solutions of many acids, such as sulfuric acid at 230 F for a period of approximately two years.

Neoprene linings have gained acclaim for their resistance to caustic solutions and are recommended for use with fluorides and phosphates. Flexible neoprene linings also have good resistance to a large group of chemicals such as all fatty acids, oils, greases and alphatic hydrocarbons; they are superior to butyl rubber in this respect. Such resistance enables neoprene linings to give good service in chemical processes where mixtures of acids with kerosene, oils or other organic materials are involved. There is a question about the suitability of neoprene linings for use with hydrochloric acid because this acid is reported to permeate the lining down to the metal and cause blistering and loss of adhesion.

Because neoprene is more expensive than natural rubber it is normally used only where it gives service superior to natural rubber. Neoprene linings are specifically used where flexibility is needed above 160 F.

Nitrile rubber

Nitrile rubber linings, because of their relatively high cost, are used only with specific organic solvents, where organic solvents are used in conjunction with acids and other water-base corrosive systems and where superior oil resistance is needed. The general chemical resistance of nitrile rubbers is only fair, but their resistance to oils is good. They can be compounded with sulfur to a hard stage and make very good linings for processes involving wet chlorine.

Hypalon

Hypalon linings are flexible and are useful because of their resistance to heat as well as to oxidizing chemicals. Hypalon linings are espe-cially suitable for use with hypochlorites and sulfuric acid; cold solutions of sulfuric acid up to 90% exhibit very little attack upon hypalon. It probably is the best lining available for hot solutions of sulfuric acid above 50% concentration.

Polyvinyl chloride

PVC linings are available both in the flexible and rigid state. Flexible PVC is used almost exclusively where chromic acid solutions are involved in chromium plating, and where nitric acid or mixtures of nitric and hydrofluoric acids are involved in the electropickling of stainless steels. No other linings are suitable for these operations. PVC linings are also suitable where mixtures of water solutions of corrosive chemicals with oils or specific organic solvents are involved. The high cost of these linings practically limits their use to the kinds of processes and applications mentioned.

A notable advance in the performance of PVC linings has been the development of high-strength adhesives which bond at temperatures up to at least 230 F, thus eliminating the need for insulating brick to protect the lining. At this temperature strong acids will permeate the lining if it is not thick enough and destroy it by attacking the base metal.

Similar to PVC are the saran (polyvinylidene chloride) linings which look like and are applied in much the same manner as PVC. They are good for use with hydrochloric acid.

Application and cost

A lining system consists of the following:

1. A primer which provides specific adhesion to the surface to be lined.

2. A cover coat which provides adhesion both to the primer and to the base linings.

3. The lining itself which has been selected to give the best balance of characteristics for services required.

The adhesives used to adhere the lining to the tank or in forming joints must be so chosen as to meet the service demands on the lining itself. Adhesion to the substrate must be maintained at temperatures at which the lining is to be used, and seaming cements must also be so chosen as to provide chemical resistance equivalent to the lining itself.

Elastomeric linings are applied in much the same manner as wallpaper. In the application of rubber linings, it is important that the lining be applied when the rubber is uncured and is plastic and workable. After cure, a complete envelope or solid sheet covers the area completely. Although it is not advisable, the heavy linings can be applied over irregular surfaces, rough edges and sharp ridges. Compound curves are easily covered with large uncut sheets. When the rubber is lapped back upon itself to provide a 1½ to 2 in. joint, its cohesion is tenacious and will give leakproof joints of the same strength as the lining itself.

Peel strengths of 60 to 100 lb per lb per inch of width are readily obtained; straight pull (tensile) adhesive values of 1200 to 1500 psi are also common. Both of these values are far in excess of what is needed.

Polyvinyl chloride and polyvinylidene chloride linings are applied similarly to rubber linings but they are not cured. They are applied with butt joints covered with a narrow sealing strip. The strip is welded (melted) to the lining with a hot-air gun or electric knife to give the joint. The bond is as strong as the lining and chemical resistance is not changed. Adhesion values at room temperature for peel strength normally are above 20 lb per in. but strength declines rapidly to give negligible adhesion at 150 F.

Where warranted, PVC linings can be applied for use to at least 230 F with adhesives that cure and give good adhesion at this temperature. Polyvinylidene linings are applied in much the same manner.

Uses

Elastomeric linings can be compounded to provide specific corrosion resistance for many applications and can be applied to almost any shapes of equipment or surface under almost any set of conditions and at any location.

Those industries using the highest volume of linings include:

1. Chemical process industry, where practically every type of lining is utilized for a wide variety of chemicals.

2. The steel and aluminum industries, where the linings are used in pickling and anodizing operations.

3. In the automotive and appliance industries, for plating.

4. In ordnance, where missile fuels are involved.

5. In the food processing industry, where specifically compounded linings are used in smaller volume.

ELECTRICAL CERAMICS

Electrical ceramics, like all other ceramics, consist of randomly oriented small crystallites, bonded together by either a glassy matrix or by close interlocking of one or several crystalline phases. The atomic structures of the crystalline and glassy phases determine the physical properties of the finished ceramic. A detailed knowledge of crystal chemistry is essential for the compounding and processing of electrical ceramics. The transition of ceramics from an art to a science came about through the need of developing specific properties for very exacting electrical requirements. With the aid of crystal chemistry, physics of the solid state, and knowledge of atomic structures, it is possible to anticipate or explain essential electrical properties.

Raw materials of electrical ceramics. Conventional raw materials for ceramic production are earthy, natural minerals, used in finely ground form, such as clays, talc, feldspar and flint. These are used for the production of electrical insulators, often carefully selected and refined from gross impurities by flotation or filtering processes. In many instances, these naturally occurring minerals are replaced by synthesized inorganic compounds which are prepared from various oxides. This may be done by melting, recrystallizing and grinding, or more frequently by high-temperature solid state reactions.

These compounds may be composed of oxides of practically any metal, single or in combination, or they may be refractory compounds such as carbides and silicides. The use of these inorganic synthetic raw materials has opened the field of likely ceramic compositions to practically infinite numbers, and the possibility of variation in physical and electrical properties is therefore equally staggering.

Principal fabricating techniques

The compositions, which may consist either of finely ground minerals or synthetic compounds or combinations of both, are blended into so-called "bodies." These are then shaped into desired forms and fired. The resultant products are hard, dense and brittle. Further shaping is possible only by grinding. Other compositions can be formed by glass-shaping techniques, such as casting in the molten state or hot pressing.

The demand for complex shapes, close dimensional tolerances, and controlled physical and chemical properties has resulted in new developments in forming techniques. The older methods of slip casting and jiggering are still used for forming electrical porcelain which contains a considerable amount of plastic clays. High quality of product is assured by evacuation of the "slip" to remove entrapped air; casting under pressure is frequently done to arrive at a dense, uniform structure of high dielectric strength.

Other methods of forming, particularly adaptable to nonplastic compositions and to line production techniques, are automatic dry pressing, extruding, and injection or compression molding. Film casting and stamping of ceramic sheets is another method used for the production of accurately formed thin ceramic shapes, such as capacitor dielectrics or vacuum tube spacers. For these forming processes it is necessary to plasticize the nonplastic ceramic composition by addition of organic binders or plasticizers. These are burned out during the firing process and the resultant product is refractory and completely inorganic.

During the high-temperature sintering process, ceramic materials undergo a considerable amount of dimensional shrinkage. This makes the control of dimensions rather difficult. Electronic ceramics are usually supplied to dimensional tolerances of $\pm 1\%$; if closer tolerances are required, grinding or lapping is necessary. It has to be borne in mind that ordinary machining techniques are not possible for ceramics, because of their brittle and hard nature. Grinding of flat surfaces or outside diameters is, however, quite economical and can be done with the same precision as the grinding of metals.

Electrical porcelain

Porcelain is the outstanding ceramic insulator, because it combines mechanical stability and strength with heat and arc resistance and the ability to resist the passage of an electric current.

Conventional electrical porcelain is made of the natural minerals clay, flint, and feldspar and is very similar in composition to porcelain used for high-quality vitrified dinnerware. After forming, the ceramic is fired to complete vitrification in order to develop maximum mechanical and dielectric strength. For outdoor use and at locations exposed to humidity, it is desirable to use insulators with glazed surface. The glaze consists essentially of the same components as the porcelain itself, but in different proportions. Sometimes additional fluxes such as calcium carbonate or boric acid are added. A glaze not only serves as protection against dust and moisture, but it also improves mechanical strength if it is so designed that it exerts a compressive action on the insulator body. This can be achieved by selecting a glaze composition which has a slightly higher thermal expansion than the underlying ceramic body. Glazes can be colored by the addition of certain oxides to the glaze formula. The well-known brown color of high voltage suspension insulators is obtained by using iron and chromium oxides.

Electrical porcelain is a satisfactory insulator for low tension electrical wiring systems, for outlets and switches, lamp sockets and electrical appliances. It is unsurpassed for insulation of outdoor high-voltage power transmission and distribution systems, and is used extensively for suspension and pin type insulators. Porcelain high-tension insulators also find numerous applications in transformers and as lead-in insulators.

High frequency insulation

A dielectric material for use at high frequencies must have the additional characteristic of low dielectric loss and its properties must not be affected by changes over the required temperature range. Porcelain has a rather high dielectric loss under high frequency conditions, but a number of low loss ceramic materials have been developed especially for high frequency insulation. It is generally agreed that a low loss ceramic body should consist chiefly of uniformly small crystallites bonded with only a very small amount of glassy matrix. Typical special ceramics are frequently named according to the predominant crystalline phase in the ceramic structure, which is the chief contributor to the specific properties of the ceramic product.

Among the best-known high-frequency ceramics are *Steatite* ceramics ($MgO \cdot SiO_2$). They are based on the mineral talc or steatite, a hydrous magnesium silicate. The finely ground talc powder is fluxed with small amounts of feldspar or alkaline earth oxides. The choice of flux has a deciding effect on the dielectric loss of the fired steatite ceramic product. Other low loss ceramic compositions are *Forsterite* ($2MgO \cdot SiO_2$), *Wollastonite* ($CaO \cdot SiO_2$), *Zircon* ($ZrO_2 \cdot SiO_2$), *Mullite* ($3Al_2O_3 \cdot 2SiO_2$), and *Spinel* ($MgO \cdot Al_2O_3$).

Sintered alumina

Highest physical and dielectric strength and low dielectric loss over a wide temperature range is found in sintered aluminum oxide or alumina ceramics. The main crystalline phase is corundum (Al_2O_3). Finely ground particles of alumina bonded with a very small amount of flux are sintered at the very high temperature of 3200 F, or even higher and a network of interlocking corundum (Al_2O_3) crystals is developed. Sintered alumina ceramics are impervious to gases and meet the very exacting requirements of spark plug insulation, of envelopes for ultrahigh frequency receiving and power vacuum tubes. Paper-thin wafers of sintered alumina are used as wire supports in vacuum tubes, as a substitute for stamped natural mica spacers.

Thermal shock-resistant electrical ceramics

Numerous applications in the low and high frequency field demand ceramics which will withstand sudden heat shock. For instance, insulation for electrical appliances such as toasters, ovens, also electric arc chambers and switch gears, thermocouple insulation, and supports for electrical resistors, are subjected to sudden thermal changes. Low thermal expansion, combined with high heat conductivity is desirable in ceramics for such applications. The outstanding material in this group is *Cordierite* ($2MgO \cdot 2Al_2O_3 \cdot 5SiO_2$). Other ceramics of low thermal expansion are based on the mineral β-Spodume,

a complex lithium alumino silicate. Both groups of ceramics have lower mechanical strength than steatite and sintered alumina, but their excellent thermal shock resistance makes these materials very useful for the applications just mentioned.

Glass

A unique combination of properties has made glass an indispensable material of ocnstruction in the electrical industry. Among its most important properties are transmission of light, imperviousness to gases, ability to seal readily to metals, good dielectric properties, and ease of fabrication into many shapes. The chemical compositions of glasses vary widely, but in general it can be stated that silica (SiO_2) is the foundation of most commercial glasses, fused with such metallic oxides as soda, potash, calcium, lead, barium, magnesium and boron.

The properties of glasses are primarily determined by their composition. The most widely used glasses for electrical insulation are of the soda-lime-silicate type, the lead glasses, and the boron alumino silicate (Pyrex) glasses. The latter type's major characteristics are good mechanical strength, low thermal expansion, good weathering stability and good dielectric properties.

Glass insulators are used extensively as line insulators for open wire lines, radio antennas, and for envelopes of incandescent lamps, vacuum tubes, mercury switches, and other devices. In the electronic field, glass serves as an insulation basis for electronic components, such as resistors and inductors and as a dielectric for capacitors.

A primary benefit of glass is the versatility of glass-forming techniques available. Glass can be drawn into sheets or ribbons in thicknesses down to 0.001 in. and widths up to 2 in. Glass ribbon of this type is quite flexible and adaptable to the manufacture of fixed capacitors, by alternately stacking metal and glass foils and molding into a single body under heat and pressure. Powdered glass is formed into vitreous articles by slip-casting or dry pressing and firing, methods which are useful for forming small insulators, such as spacers for coaxial cables, bushings and insulation beads.

Certain glasses can be converted to finely crystalline bodies bonded together with a vitreous matrix or by fusion of the crystallites at their grain boundaries (Pyroceram). Articles to be converted from a glass to a crystalline ceramic are first fabricated in the glassy state with special nucleating agents added to the glass composition to form crystallites in the body. After forming, crystallization is effected by either exposure to ultraviolet radiation and temperature treatment, or by heat treatment alone. The outstanding advantage of this new group of materials is that they can be fabricated into a variety of shapes by conventional glass-forming methods and have the higher strength and stability of crystalline ceramics.

Fibrous glass is an important insulating material for electrical equipment. It may be used in the form of yarns, tapes, sleeving, or for reinforcement of organic plastics. On the other hand, plastics improve abrasion resistance and dielectric strength of fibrous glass constructions.

Ceramic papers

There are a number of ceramic papers on the market which serve as high-temperature electrical insulation in a similar way as fibrous glass. These papers are made from reconstituted mica flakes (Samica, Mica Mat), clays, or glass and mineral (mullite) fibers.

Glass-bonded mica

As the name suggests, this group of materials consists of either natural or synthetic mica flakes which are bonded under heat and pressure by a glassy matrix. Vitrified ceramics can only be lapped or ground with the hardest abrasives, such as silicon carbide, corundum, or diamond powder. In contrast, glass-bonded mica can be subjected to all normal machining operations. The material is impervious to moisture, has high dielectric strength, and can be molded to accurate dimensions. Metal inserts can be molded directly into mica bonded articles. This is a distinct advantage over other ceramics, which require assembly or attachment of metal parts after processing.

Electronic ceramics

Great progress has been made during the past decades in the area of solid state research. Solid state components are finding many applications in electronics. Some of these, such as diodes, rectifiers, or transistors, are based on single crystals, germanium and silicon, grown from a melt and cut into required shape. These are not considered ceramics, but others based on polycrystalline systems and fabricated by ceramic processes, are considered among electric ceramics.

Ferroelectric ceramics. Ferroelectricity, the spontaneous alignment of electric dipoles under the influence of an electric field, is an outsanding property of certain ceramic materials, especially those based on barium titanate. This unusual property was discovered in ceramics about 20 years ago. Besides barium titanate, additional ferroelectric materials were discovered in the field of niobates, tantalates and zirconates.

Ferroelectric ceramics exhibit a high dielectric constant which makes them attractive for dielectrics in capacitors. The dielectric constant of these ceramics can be varied through compositional changes; values as high as 10,000 at room temperature can be obtained. Very high dielectric constant ceramics have rather steep negative temperature coefficients of capacity.

The capacitance of special barium titanate capacitors is sensitive to applied voltage which suggests their use in dielectric amplifiers. Nonlinear capacitors of this type have also been employed for frequency modulation and remote tuning devices.

Ferroelectric ceramics can be made to show piezoelectric effects, i.e., the ability to convert mechanical strain into electric charges and, conversely the ability to transform a voltage into mechanical force. This is produced by exposing the material to an orienting electric field during the cooling period after firing. Piezoelectric ceramics of the barium titanate type have found applications for phonograph pickups, ultrasonic thickness gauges, accelerometers, ultrasonic cutting tools and even in electromedical instruments for measuring heart conditions.

Ferromagnetic ceramics. Ferromagnetic ceramics, also known as ferrites, are compounds of various metal oxides and have the general formula $MO \cdot Fe_2O_3$, where M stands for a bivalent metal ion, such as Zn, Ni, Mg, and others. They are ceramic materials with a crystalline structure of the spinel $(MgO \cdot Al_2O_3)$ type. The mineral magnetite $(FeO \cdot Fe_2O_3)$ is the only naturally occurring mineral of this type and has been well-known and used for ages as lodestone for its magnetic properties.

Ceramic ferrites were originally proposed in 1909, but their development and industrial use did not take place until about 1942. In contrast to metallic magnetic materials, ceramic ferrites have high volume resistivity and high permeability. Their specific gravity is between 4 and 5, considerably less than iron (8). They can be made both into "soft" and "hard" permanent magnetic materials.

The properties of soft ferrites can be varied over a wide range to meet specific application requirements. For cores of radio and television loop antennas they are made with emphasis on a high quality (Q) factor, to attain optimum in reception quality and selectivity. For memory cores in electron computers, they are so compounded that they exhibit a square hysteresis loop of magnetization and have an extremely low switching time between magnetic saturation and demagnetization.

Permanent ceramic magnets of barium ferrite ($BaFe_{12}O_{19}$) composition are considered as the most important advance in permanent magnets since the development of "Alnico" materials. Some of these ceramic magnets have the highest coercive force of any commercial magnetic material. They have very high resistance to demagnetization from vibration and shock. Since they are electrical nonconductors, they can be used in applications where metallic magnets would cause short circuits or eddy current losses in high frequency alternating fields. Ceramic magnets are about 35% lower in density than metallic magnets, an important factor for military and airborne applications.

Resistor ceramics

Certain nonmetallics pass limited amounts of electricity and therefore are useful as electrical resistors. Some of these have advantages over metals in that they are more resistant to oxidation and are therefore useful at higher temperatures.

Carbon and graphite can be formed according to ceramic methods and are used for electronic and electrothermic applications. Pyrolytically deposited carbon on porcelain rods, so-called carbon-deposited resistors, can be made to very precise resistor tolerances and in many cases have replaced more expensive wire-wound resistors.

Silicon carbide, either self-bonded or bonded with silicon in form of rods or tubes, is used for resistor heating elements in electric furnaces. These can be heated to 1400 C in air for indefinite periods of time and may be heated as high as 1600 C for shorter periods.

Silicon carbide as a resistor element shows nonlinear characteristics and is used where a nonohmic variation of current is desired (Thyrite resistors). Nonlinear resistors of this type are chiefly used for voltage regulation and current suppression, such as in lightning arrestors.

Many oxides become electrically conductive at elevated temperatures. The electrical conductivity of oxides is governed to a large degree by the amount of impurities present. For instance, the electrical resistance of thoria is greatly reduced by the presence of such oxides as ceria, yttria, erbia, etc. (e.g., Nernst glower). Thoria and zirconia resistors have been used as electric furnace heating elements up to 2000 C. These oxides are nonlinear with temperature change and therefore have found applications as thermistors. These components are used as temperature sensors in pyrometers, temperature bridges, and microwave power meters.

Other semiconductor ceramics

Semiconductive ceramics of silicide, telluride and oxide compositions are being considered for thermoelectric devices, to convert heat directly into electrical energy (Seebeck effect) and for refrigeration (Peltier effect).

ELECTROFORMED PARTS

Electroforming is essentially an electrolytic plating process for manufacturing metal parts. In general, it is best to consider electroforming for applications where a part is impossible or difficult to make by any other standard method or if the tooling required by another method such as forging or die casting is extremely expensive. Electroforming is not generally used for large-quantity production because it is a batch operation. It is, however, valuable for short-run, simple parts or long-run, complicated parts because of its low tooling costs.

Briefly, the procedure for making parts by electroforming is as follows:

A male mandrel is made which creates the inside dimensions of the part to be made. The mandrel or core is placed in a plating bath and metal is deposited on its surface. The metal is allowed to continue depositing as long as necessary; the longer it is allowed to deposit the thicker it will be. After the required amount of deposition is reached, the assembly is removed from the plating bath and the core and metal deposit are separated. This provides a self-supporting metal structure with inside dimensions matching those of the mandrel or shape upon which the deposition was made, and with a wall thickness corresponding to the buildup of plating metal.

The forms upon which the deposition may be made may be metal such as stainless steel, aluminum, zinc and nickel, or nonmetallic materials such as plaster, wax, plastic materials of all types and even the leaves of a tree. The process differs essentially from that of electroplating in that the base or core upon which the metal is deposited is generally not treated in order to get adhesion of the deposition to the base material. Because electroforming is used to produce metal parts, it is not necessary that adhesion be obtained to the base material.

Self-supporting parts can be made from such metals as nickel, copper, iron, silver, chromium and gold. In all instances the pure metal has been found to be the only practical deposit obtainable. There are, however, a number of institutions at present working on bimetallic deposits such as cobalt and/or nickel and tungsten.

The electroforming baths used are generally similar to the standard types of electroplating solutions. For instance, a copper plating solution is made up of copper sulfate and sulfuric acid. A nickel bath may be of the nickel sulfamate or watts nickel type.

Advantages

Following are the chief advantages of electroforming:

1. Initial tooling costs are extremely low. This has several advantages. Short-run parts can be produced to determine market acceptability and, as is the case in missile and aircraft applications, small numbers of parts may be produced economically. The low initial tooling cost makes it economically possible to try various shapes and configurations without undue tooling charges.

2. Since this is an electrochemical process the

surface finish of the mandrel is duplicated exactly, thereby providing a means of producing parts to any surface finish—from a high polish to a surface resembling gravel. Many examples of these are found in industry. Among them are molds to make records, plastic tile resembling different fabric textures, and surface finish standards.

3. Complex shapes such as ducting and tubing may be made in one piece, thereby eliminating welding and soldering.

4. Parts requiring extremely close tolerances are easily made by the electroforming process, inasmuch as the mandrel or shape upon which the metal is deposited may be machined or produced to extremely close tolerances, inspected easily and thereby duplicated exactly. Examples of extremely close tolerance parts that can be made are radar plumbing, hot-air ducting and tubing, wind-tunnel test nozzles and liners, reflectors, collectors, mirrors, nose cones, etc.

Sizes

Parts produced by electroforming range in size from miniature to extremely large. Typical miniature parts such as small electronic devices are 0.020 to 0.030 in. in dia and $\frac{1}{8}$ in. long with a wall thickness of 0.001 ± 0.0001 in. Large sizes may be as big as 16 ft in dia. Thicknesses also may vary from 0.001 to 1 in. or even thicker.

Parts have been made in all sizes and shapes, an example being wind-tunnel nozzles 16 ft long and 30 in. in dia and varying in thickness from 0.06 to 1 in. Tolerances on a part like this are extremely close and are usually ± 0.001 in. in the throat area and ± 0.002 in. in the downstream sections. Tolerances on waveguide plumbing may be as close as ± 0.0005 in.

In many cases where an attachment is required to the duct, a machined flange or boss may be made integral with the mandrel and electroformed to the part. Examples of this are waveguide plumbing with irises and flanges and aircraft ducting with attaching flanges on each opening.

The range of mechanical properties of the

materials which may be deposited by electroforming are shown in the following table:

Material	Brinell Hardness (Bhn)	Ult Tens Str, psi $\times 10^3$	Yield Str, psi $\times 10^3$	Elong, % in 2 in.
Nickel	140–500	55–225	40–125	2–20
Copper	51–170	36–80	12–40	21–39

ELECTROPLATED COATINGS

Electroplating may be defined as the electrodeposition of an adherent metallic coating upon an electrode for the purpose of securing a surface with properties or dimensions different from those of the basis metal. It must not be confused with electroforming, which is the production or reproduction of massive objects by electrodeposition. Nor should it be confused with "electroless" plating, a type of immersion plating in which no electric current is required, and in which the metal is deposited by chemical reduction of the metal ions in the electrolyte, together with some catalytic reduction. "Brush" plating, on the other hand, is a special method of electroplating which will be discussed briefly later.

The plating process

The basis metal (out of which the object to be plated is made) may be any of the common metals or their alloys. As our definition implies, the basis metal, in copper plating, for example, is made an electrode by immersing in a copper plating bath as a cathode, i.e., by connecting it to the negative terminal of a low-voltage source of direct current. One or more copper electrodes are then immersed in the plating bath as anodes, i.e., by connecting them to the positive terminal of the direct-current source. A controlled quantity of current is then allowed to flow through the plating bath and, in accordance with Faraday's Laws, copper goes into solution from the anodes and is electrodeposited from the bath upon the basis metal object serving as the cathode.

In "brush" plating the anode may be soluble or insoluble. It is covered with a cloth or similar spongelike material and moistened with the plating solution while it is gently moved back and forth across the surface to be plated. It is usually used for specialized applications.

Electroplating is an electrochemical process (not a "dunking" operation as many believe) in which the quantity of current flowing as ampere hours per square foot of cathode surface determines the weight or thickness of metal deposited. Table 1, giving electrochemical equivalents and related data, presents the necessary information to determine the ounces per square foot or the ampere hours per square foot to deposit 0.001 in. of an element, assuming 100% current efficiency. From these data the time of plating at a given current density, i.e., amperes per square foot, to obtain a given thickness of metal can be determined.

In applying the above data, allowance must be made for any variation in current efficiency in a given plating process, because the efficiency will vary with the bath composition as well as with the operating conditions of temperature, current density and agitation. Also, the distribution of current over a large irregularly shaped object, or a plating rack load of a number of pieces, will cause further variation in metal distribution from point to point. Because most plating specifications define the minimum metal thickness required at any point, it is most important that the plating installation and racking of parts be properly engineered to give the optimum current and metal distribution. By failing to do so, quality specifications may be difficult if not impossible to meet and the cost will be increased.

Limitations and scope

Electroplating is applied to objects varying greatly in size. Small screws, costume jewelry, the inside surfaces of large (7000 gallon) tanks and the continuous plating of strip steel up to 42 in. in width while traveling at speeds up to 2000 fpm are but a few examples.

The thickness of plated coatings may also vary widely. Gold or chromium coatings may be no more than 10 μin. in thickness, while in other cases nickel and chromium coatings may exceed 0.01 in. Electrodeposited thicknesses of copper or nickel in electroforming may be as great as 0.1 to 1.0 in. Thickness in general is limited by such factors as the specified metal to be electrodeposited, the plating bath employed, and the intended use of the plated object. Specifications for electrodeposited coatings of the common

TABLE 1—ELECTROCHEMICAL EQUIVALENTS AND RELATED DATA [a]

Element	Atomic Weight	Valence	oz/sq ft for 0.001 in.	Amp hr to deposit 0.001 in/sq ft
Aluminum	26.97	3	0.225	19.05
Antimony	121.76	5	0.557	17.4
		3	0.557	10.4
Arsenic	74.91	5	0.475	24.1
		3	0.475	14.5
Bismuth	209	5	0.816	14.8
		3	0.816	8.93
Cadmium	112.4	2	0.72	9.73
Chromium	52.01	6	0.591	51.8
		3	0.591	25.9
Cobalt	58.94	2	0.74	19
Copper	63.57	2	0.74	17.7
		1	0.74	8.84
Gallium	69.72	3	0.491	16
Germanium	72.6	4	0.445	18.6
		2	0.445	9.31
Gold	197.2	3	1.61	18.6
		2	1.61	12.4
		1	1.61	6.2
Indium	114.76	3	0.608	12.1
Iridium	193.1	4	1.869	29.4
		3	1.869	22.1
Iron	55.84	2	0.66	17.9
Lead	207.2	2	0.94	6.91
Manganese	54.93	2	0.598	16.5
Mercury	200.61	2	1.129	8.55
		1	1.129	4.27
Nickel	58.69	2	0.742	19.2
Palladium	106.7	4	0.998	28.6
		3	0.998	21.4
		2	0.998	14.2
Platinum	195.23	4	1.78	27.6
		2	1.78	13.85
Rhodium	102.9	4	1.04	30.8
		3	1.04	23.1
		2	1.04	15.37
Rhenium	186.31	7	1.71	48.8
Selenium	78.9	4	0.400	15.4
Silver	107.88	1	0.875	6.16
Tellurium	127.61	4	0.52	12.4
		2	0.52	6.19
Thallium	204.39	1	0.986	3.82
Tin	118.7	4	0.61	15.63
		2	0.61	7.82
Zinc	65.38	2	0.59	13.7

[a] All figures based on 100% current efficiency. From Graham, K., "Electroplating Engineering Handbook," Second Edition, Reinhold Publishing Corp., New York, 1962.

metals are available for varying degrees of service severity.

Most any of the common metals and a number of their alloys can be electrodeposited from aqueous plating baths (other than aluminum and magnesium). Of the more uncommon metals, plated coatings of gold, silver, platinum, palladium, rhodium and chromium are extensively used (see separate articles on these metals).

Types

Nickel. One of the most widely used electroplates, nickel plates have excellent corrosion resistance. Proprietary plating baths are available for depositing nickel with a fully bright, semi-bright or satin finished surface. Thicknesses up to 60 mils or more are used on equipment subject to severe corrosive environments, whereas lower thicknesses—0.2 to 3 mils—are used for protection and appearance on steel, copper, brass, zinc, aluminum and magnesium. Nickel deposits up to ¼ in. thick are used in building up worn parts and for producing parts by the electroforming process.

Chromium. Decorative chromium plating systems, consisting of a top layer of chromium applied over layers of copper, nickel or copper plus nickel, have an attractive blue-white appearance. In addition, chromium plates can be produced in blue, black and gray. The underlayers of copper and nickel provide a nonporous undercoat for the relatively brittle and porous chromium layer. Besides colored and bright chromium plates, no gloss finishes can be obtained by using satin nickel undercoats.

Hard chromium, also known as industrial chromium, consists of thick, hard layers adding up to coating thicknesses ranging up to 20 mils, and in some cases up to 100 mils or more. Applied usually to steel, zinc and aluminum hard chromium plates provide a combination of hardness, corrosion resistance and low coefficient of friction. Porous chromium plates, produced by special techniques, are used on piston rings where oil retaining surfaces are desired.

Zinc and cadmium. These plates, with good resistance to many atmospheres, are often used as anodic coatings on steel and as a corrosion resistant paint base. Zinc is lower in cost than cadmium. Lacquered or chromate-treated zinc plates are sometimes used to simulate chromium plate. Cadmium is sometimes used as a base for zinc plating. It is seldom used for decorative purposes.

Tin. Tin is a relatively low cost plate with good resistance to corrosion and tarnish. Because its corrosion products are not toxic or objectionable to the taste, it is used on "tin cans" and copper kitchen ware. Thick tin plates are used to resist special chemical environments.

Tin-copper plates, containing about 45% tin and known as speculum, resemble polished silver when buffed to a high luster. Tin-zinc (about 80% tin) under certain conditions offer better outdoor protection to steel than zinc or cadmium plates. Tin-nickel plates are bright, tarnish resistant coatings that are sometimes competitive with chromium plate.

Copper and brass. Copper is used primarily as an undercoating for deposits of other plating materials, such as nickel and chromium. Lacquered, bright copper plates sometimes serve as an inexpensive decorative finish for steel. Because brass plates have low corrosion resistance they are used chiefly for decorative purposes. Applied as thin coatings (from 0.07 to 0.3 mil) they are usually protected with a clear organic coating.

Lead. Lead plates provide good protection for steels exposed to industrial atmospheres. They also provide a good paint base and can be severely deformed without stripping off the base metal. Lead-tin plates are used on bearings and as a base for soldering. Plates with about 5 to 6% tin have excellent corrosion resistance and are competitive with terne plate (hot dipped).

Precious metals. Most precious metals can be plated. However, being expensive, they are used only where their high cost can be justified and then in thin layers only. Silver plate is most common. It has a pleasing appearance, high chemical resistance, good resistance to high temperature oxidation, high electrical conductivity and good bearing qualities. Because gold plates are fine grained and dense in structure they can be used in extremely thin layers—for example, as thin as 0.0001 in. on brass. Hardness of gold plates is considerably increased by alloying them with cobalt and other metals. Such plates have hardnesses of over 300 Vickers.

Brass (copper-zinc). Brass is plated for decorative applications and for the bonding of

rubber to metal. High-zinc "white brass" is used for finishing some tubular furniture and for miscellaneous decorative uses. Solutions are based on zinc and copper cyanides; free cyanide, carbonates and alkali are also present, and proprietary brighteners are available.

Copper tin. Red bronze (10 to 15% tin) is superior to copper as an undercoat for nickel-chromium systems; alone it is used as a golden finish. It is also used as a stop-off in selective nitriding of steel. Stannate-cyanide baths are used (with some proprietary addition agents) and proprietary processes are also available. A 45% tin alloy (speculum) alloy has a silverlike appearance.

Tin-lead consisting of 7 to 10% tin is used in the graphic arts and a 50 to 60% tin alloy is used for solderability, particularly in printed circuits. A fluoborate solution is used.

Tin-zinc (20% Zn) is plated from a stannate-cyanide solution; it has good corrosion resistance and solderability.

Tin-nickel (35% Ni) is a decorative finish somewhat resembling chromium. It differs from other finishes of its type in being solderable, and the solution is characterized by unusually good throwing power.

Cobalt-nickel (1–5% Co) was one of the first "bright nickel" processes; its use has lessened somewhat in recent years owing to the increasing popularity of other types of nickel-plating solutions, but it still has many proponents.

Other alloy systems which have been mentioned include gold-copper, gold-silver, silver-cadmium, tin-cadmium, silver-lead, zinc-cadmium, tin-antimony, zinc-iron, tin-indium, nickel-iron and tin-cobalt.

Tungsten and molybdenum cannot be deposited from aqueous solution, but their alloys with cobalt, nickel, iron and some other metals are plateable, and the tungsten-cobalt system in particular seems to have interesting possibilities.

EPOXY PLASTICS

Epoxy resins (developed in the 1940's) comprise an extremely broad and diverse family of materials. They are used in the form of protective coatings, adhesives, reinforced plastics, molding compounds, casting and potting compounds, and foams.

Epoxies, perhaps best known as adhesives, are premium thermosetting plastics, and are generally employed in high-performance uses where their high cost is justified. They are available in a wide variety of forms, both liquid and solid, and are cured into the finished plastic by a catalyst or with hardeners containing active hydrogen. Depending on the type, they are cured at either room temperature or at elevated temperatures.

Liquid epoxies are used for casting, for potting or encapsulation, and for laminating. They are used unfilled or with any of a number of different mineral or metallic powders. Molding compounds are available as liquids, and also as powders with various types of fillers and reinforcements.

General properties

As a class, epoxies provide: (1) outstanding adhesion to both metallic and nonmetallic surfaces, (2) excellent chemical resistance, (3) low moisture absorption (0.01 to 0.2% in 24 hr), (4) excellent dielectric properties, (5) high strength in reinforced laminate form (in filament wound structures, they provide strength-weight ratios unexcelled by any engineering material), and (6) relatively good heat resistance.

Epoxies are thermosetting resins, usually cured to form rigid materials. They can also be produced in flexible, resilient form. Maximum recommended service temperatures generally range about 300 to 350 F. Epoxies can be cured at either room or elevated temperatures, though heat cure provides maximum properties. Several basically different curing systems can be used to provide different characteristics in the cured material.

In comparison with other thermosets, unmodified epoxies are relatively expensive.

Casting and foaming resins. Casting resins are primarily used for potting or encapsulating electrical or electronic equipment. Epoxies' excellent adhesion and extremely low shrinkage, coupled with high dielectric properties, provide a well-sealed, voidless, well-insulated component.

Foaming resins have also been used for electronic potting. More recently developed low density foams (2 lb per cu ft) are finding application as thermal insulation for refrigeration units.

Liquid resins are also formulated with a variety of fillers, such as metal powder, to provide effective patching and repair putties or pastes. Such compounds can be used to patch both metal and plastic surfaces.

Laminating resins. Liquid resin systems are used to produce low-pressure reinforced laminates and moldings, high-pressure industrial thermosetting laminates (NEMA Grade G-10 glass cloth-base, as well as paper-base laminates for electrical uses), and filament wound shapes.

Properly prepared glass-reinforced epoxy laminates offer the highest mechanical strength of any reinforced plastics material, as well as outstanding chemical resistance. Price has generally limited use of such laminates to critical parts, such as high strength-weight aircraft parts, chemical tanks, pipe and bases for printed circuits.

Epoxy laminates are also widely used in plastics tools. They are used for drilling, checking and locating fixtures, where dimensional accuracy is critical. They are also used to provide durable surfaces for metal-forming tools, such as draw dies for short-run production.

Filament wound structures are used for such applications as rocket-motor booster cases, pressure vessels, and chemical tanks and pipe.

Molding compounds. Molding compounds provide the performance characteristics of epoxy resins with the automated speed and economy of compression and transfer molding. They are being used primarily for electrical components.

Composition

Most epoxy resins are reaction products of epichlorohydrin and polyhydric phenols, commonly, "bisphenol A," that when mixed with a crosslinking agent (hardener) polymerize to a thermosetting solid. They are produced as amber-colored, solvent-free, liquids and solids, of which the lower molecular weight liquids are better known because they offer a lower viscosity beneficial to mixing, pouring and de-airation. As they come to the eventual user, however, epoxies are supplied in system form, i.e., a base with a companion hardener, and these are most often modified to meet specific application requirements.

These modifications normally take two directions, the first being chemical where the actual molecular structure is changed, the second being mechanical where fillers are added that make for changes primarily in performance.

Hardeners. The hardeners for epoxies have the most important role in creating the versatility found in the compounds and, unlike a catalyst, actually take part in the reaction and then become a part of the cured piece. By changing the hardener, but retaining the same base resin, such important characteristics as pot life, viscosity, flexibility, and resistance to heat can be modified. Commonly used hardeners include aliphatic amines (e.g., diethylenetriamine), aromatic amines (e.g., metaphenylenediamine), acid types (e.g., phthalic anhydride and methyl nadic anhydride), and borontrifloride complexes. All of these are liquids with the exception of certain anhydrides, such as phthalic, which are supplied in solid form and must be melted before use.

While the class of hardeners that promotes cure at low temperatures (room or slightly higher) are used more widely, in general those hardeners that require heat for cure produce a tighter chemical linkage, and offer superior performance. One fact of importance here, particularly to new users, is that random substitution of hardeners should not be made unless a chemical compatibility with the base resin is first established, and mixing ratios properly determined. Another often overlooked cause of poor cure is that weighing and mixing procedures are not given enough attention. For good results proportioning must be precise and mixing thorough. Table 1 illustrates the differences in both handling and performance characteristics of three separate hardeners when used with the same unmodified epoxy base resin.

Fillers. Mechanical modification is usually provided by adding fillers to the base resins, though in some cases, to adjust mixing ratios or flow characteristics fillers are included in the hardener.

Most fillers are of an inorganic nature and range all the way from mica, silica, zirconium silicate, hydrated alumina, and iron powder to cork and hollow glass beads. They may serve to decrease costs, improve heat transfer, impart thixotropism, and induce electrical conductivity in an otherwise electrical insulator. In addition, the specific filler and percentage used has a determinable effect on such factors in the compound as the coefficient of thermal expansion, specific gravity, electricals, heat transfer, and resistance to impact.

Certain fillers also increase resistance to burning, although within the last few years this feature has been improved considerably by the development of the chlorendic anhydride hardeners, and the chlorinated and brominated base resins.

TABLE 1—EFFECT OF HARDENER ON EPOXY SYSTEM [a]

Hardener	Aliphatic Amine	Poly-amide	Anhy-dride [c]
Mix Ratio, PBw.	100:11	100:75	100:170
Pot Life (100 gm), min	30	105	60 [b]
Viscosity (@ 70 F), cps	3000	20,000	3000 [b]
Cure, hr; °F	24; 70 or 2: 140	2; 140	3; 250
Linear Shrinkage, %	1.0	1.2	0.9
Max Operating Temp, F	240	220	270
Hardness (Shore)	D89	D74	D60
Tensile Strength, psi	10,000	7,000	4,000
Dielec Const (@ 105 C, 1000 cps)	3.7	3.7	4.3
Dissipation Factor (@ 105 C, 1000 cps)	0.08	0.04	0.10
Volume Resist (@105 C) ohm-cm	3×10^{11}	6×10^{11}	1×10^{10}
Type	Rigid	Flexible	Flexible

[a] Same base resin used in each formulation (Conap #1200).
[b] At 160 °F.   [c] Modified.

TABLE 2—TYPICAL PROPERTIES OF EPOXIES

Type	Bisphenol A, Cast, Rigid	Bisphenol A, Molded, Mineral	Novalac, Cast, Rigid	Novolac, Molded, Mineral	Cyclo-aliphatic, Cast, Rigid
Specific Gravity	1.15	1.6–2.1	1.24	1.7	1.22
Mod. Elast. in Tens., 10^5 psi	4.5	–	4–5	–	5
Elong.,%	4.4	–	2–5	–	2.2–4.8
Hardness, Rockwell M	106	101	107–112	–	–
Tens. Str., 1000 psi	9.5–11.5	5–7	8–12	5.3	9.5–12
Imp. Str., ft-lb/ in. notch	0.2–0.5	0.25–0.45	0.5	0.3–0.5	–
Max. Svc. Temp., F	175–190	300–500	450	450–500	450–500
Vol. Res., ohm-cm	6×10^{15}	9×10^{15}	2×10^{14}	3×10^{14}	$> 10^{16}$
Dielect. Str., v/mil	> 400	350–400	–	280–400	444

Handling

Since the cure of an epoxy is brought about by a chemical reaction, and not by a simple process of solvent evaporation, it is quite important that users recognize at the outset that handling (particularly for the two component, low-temperature curing compounds) requires more care than do older materials such as asphalt. However, because this is, and probably will be, a problem for some time to come, procedures have come about that keep difficulties in handling to a minimum. One simple and inexpensive way of eliminating frequent weighing is to use calibrated mixing containers where it is only necessary to fill to the first calibration with the base, to the second with hardener, mix and apply. Disposable unwaxed Dixie cups are acceptable for this purpose although polyethylene jars are superior.

Formulators meet this problem by supplying the compounds packaged in preweighed containers, while equipment manufacturers recognizing the need introduced automatic mixers and dispensers about five years ago. It is possible through the use of these machines, of which there is a broad selection, to eject preweighed and premixed shots of compound as the operator requires. There is little danger of the material setting up in lines and mix chambers as ways of cleaning, through the automatic interjection of solvents or by the purging with single shots of either the base or hardener, can be utilized. Certain of these machines are operated manually and are capable of weighing only, but others are power-driven and incorporate such desirable features as vacuum de-airation.

Guns also are marketed that operate on the principle of the hand-powered caulking tools. The cartridges for these are made from polyethylene which must be filled with the mixed epoxy system immediately prior to inserting into the gun particularly if the pot life is thirty minutes or less. Refrigerating multiple cartridges, and keeping them under refrigeration until use, minimizes the problem of frequent handling. Many types of nozzles are available for these guns, some for narrow aperture potting, and others for sealing such structures as decorative building facings, decks and cement runways. Of course, there is a broad selection of two component systems available that offers longer pot lives. With these, materials handling time varies from two hours to three months, or even longer, but such compounds normally require cure schedules of from one hour at 140 F to eight or more hours at 350 F.

Certain of the two-component, heat-cured systems can also be cast in large masses without the destructive exotherm found in the low-temperature cure group. To illustrate, a 200-gm cube of an unmodified epoxy/aliphatic amine system, providing it is cast without the use of a heavy metal mold or metal insert to draw off exothermic heat, will gel within minutes and result in an unsatisfactory casting. On the other hand, anhydride systems have been cast without trouble in blocks weighing fifty pounds. The only problem is the danger of strains, which can be relieved either by a careful annealing or through a chemical flexibilization process.

One-component systems. The one-component systems, as the name indicates, eliminate problems of weighing, mixing and pot life, and are subdivided into solids, most often "B" staged epoxy powders, and liquids containing latent hardeners such as borontrifloride, or anhydride/solvent solutions. In each case the hardener is included by a different method but heat, serving as a catalyst, is usually necessary for complete cure.

Powders have been used to some extent as adhesives and in potting, but are employed most widely for the fluidized bed coating method or in compression and transfer molding. The epoxy fluidized bed approach provides a relatively easy way of applying encapsulant coatings in uniform thicknesses over contours but is not suitable for impregnation, or for coating units which cannot be heated.

This process is essentially one, where the dry powders are kept in a state of suspension within an open-top chamber by air pressure and exhibit flow characteristics much like a liquid. The units to be coated are preheated and dipped into this chamber and the particles then touching the surface cling to it and fuse. Following withdrawal and further oven heating the epoxy cures to form a tough, conforming coat. Fluidizing powders eliminate the need for molds but impose certain application problems because of the difficulty in masking leads on small components.

The molding powders, also supplied as preforms, are very similar in make-up to the fluidized bed materials but are adapted to use in compression and transfer presses. The initial investment in equipment here is considerably higher than for the other materials, but the press cycle can be as short as a minute and a half at 375 F, coating tolerances can be maintained precisely, and operating qualities are exceptional.

Resin selection

Epoxy compounds then can be categorized (although there are exceptions) as (1) two-component liquids that will cure at temperatures from 70 F to 140 F, but which have short pot lives; (2) as two-component liquid systems requiring cure at temperatures up to 350 F, but which offer longer pot lives and improved operational characteristics; and (3) as one-component liquids and powders that reduce handling problems, offer optimum operating characteristics, but that require slightly higher curing temperatures (in the range of 350 to 400 F).

With respect to possible methods of application, with the exception of the powders already explained, epoxies are supplied for use by spray, brush, spatula, roller coat, knife coat, dipping, filament winding, laminating, and casting. All of these are self-descriptive but casting and this is simply the method of producing a defined shape by pouring a compound into a mold where it cures and from whence it can later be removed. On the other hand, if an object has been placed in the mold and the epoxy cast around it, the process is called potting.

In either case, insurance against entrapped air bubbles can be obtained by vacuum de-airation at approximately 29 in. of mercury, both after mixing and after casting. Molds can be made from metals (generally brass, steel, or aluminum) and these are recommended whenever extended production is called for, or when cure temperatures are to be above 140 F. Other lower cost materials such as plaster, plastic and wood, if the occasion demands it, can also be used but these are normally considered semipermanent. Plaster should always be oven-dried and sealed with a parting agent such as a vinyl paint, whereas wood should be sealed with two or more coats of a high-grade patternmakers' lacquer. All molds must be treated with a release agent (e.g., silicone grease) before casting takes place.

ETHYLENE-PROPYLENE ELASTOMER

Ethylene-propylene elastomer is a completely saturated copolymer made by solution polymerization. The material's remarkable properties include exceptional ozone resistance, excellent electrical properties, good high (300 to 325 F) and low temperature properties, good stress-strain characteristics and resistance to chemicals, light and other types of aging.

The raw polymer has an amber appearance and a Mooney viscosity of 35 to 45 at 212 F after 8 min. Typical physical properties for a 60 phr HAF carbon-black vulcanizate are 2700 psi tensile strength, 500% ultimate elongation, 1600 psi modulus at 300% extension, and a Shore A hardness of 65. The first commercially available copolymer contains 43% by weight ethylene. It is stabilized with a nonstaining type of antioxidant.

Ethylene-propylene rubber, at present, requires a peroxide or peroxide-sulfur modified curing system. Sulfur improves the peroxide curing efficiency and assists in chemical cross-linking of the polymer chains, thereby imparting better physical properties to the vulcanizate.

Some plasticizers used in other rubbers are not suitable for ethylene-propylene rubber. Most acceptable are saturated materials of relatively low polarity such as parafinic hydrocarbon oils and waxes.

For most applications, antioxidants are not needed although they may be used where extreme heat resistance is required. Calcium stearate is recommended as a replacement for stearic acid to improve release from processing equipment and maintain an alkaline medium desired for peroxide cure.

The polymer mixes well in a Banbury mixer and the batch dumps easily without sticking. It does not break down, and the mill cycle is short.

On the open mill, ethylene-propylene elastomer tends to split unless rolls are heated to 100 to 200 F in the early stages. Preheating the polymer is helpful. Once the stock becomes warm, compounds with black-to-oil ratios equal to or greater than 1.5 are processible. With the proper amount of carbon black, extrudability is excellent, with low die swell and excellent definition. Very little processing oil is needed. A recommended die temperature is 220 F; barrel temperature, 160 F; screw temperature, 120 F.

In calendering operations, there is little sticking at roll temperatures between 180 and 210 F. The polymer tends to be dry and somewhat lacking in building tack. Tackifier resins may be used to improve raw compound tack.

EUROPIUM. See Rare Earth Metals.

EXTRUDED METALS

Extrusion, in its most common form, may be defined as the conversion of a cylindrical billet of metal into a continuous length of uniform reduced cross section by forcing the metal under compression to flow through a die of a desired shape. In practice, a preheated billet placed in a rigid, thick-walled, cylindrical container closed at one end by a die is forced to flow through the die by the action of a ram moving through the other end. These three major components of an extrusion press—the container, die, and ram—are shown in Fig 1, and illustrate the principles of the basic process known as *direct extrusion*. Other less common forms of the process differ in the relative movement between the billet and tools during deformation, but operate on essentially the same basic principles.

For direct extrusion of hollow shapes, a mandrel positioned in the die through a hollow ram is generally used with a cored or pierced billet, as shown in Fig 2. In large presses (over 2500-ton capacity), the mandrel may have sufficient force for piercing and solid billets can be used, but this method is applicable only to relatively large-diameter, thick-walled tubular shapes.

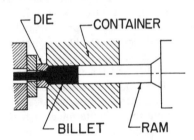

FIG. 1. Direct extrusion process.

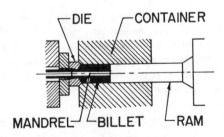

FIG. 2. Extrusion of hollow shapes.

Process characteristics

Because the metal is deformed completely by the action of compressive stresses, greater amounts of deformation in a single operation can be achieved by extrusion than by any other metalworking process. This stress condition also makes it possible to work materials that exhibit brittle behavior and tend to crack when deformed by other methods that impose tensile stresses on the workpiece. The amount of deformation in extrusion is usually expressed as the *extrusion ratio,* which is the ratio between the cross-sectional areas of the billet and the ex-

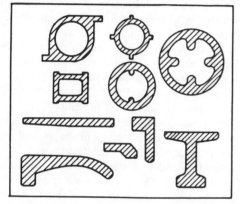

Fig. 3. Typical extruded steel sections.

truded section. Generally, extrusion ratios in practice will range between about 10:1 to 60:1,

but certain intricate and thin sections of some materials have been extruded at ratios greater than 100:1, or more than 99% reduction in area. Extrusion ratios below 10:1 are seldom used other than for preliminary working of ingots to break down brittle cast structures for subsequent fabrication of finished products by extrusion or other methods.

The most outstanding feature of the extrusion process is its ability to produce a wide variety of section configurations. Structural shapes can be extruded that have complex nonuniform and nonsymmetrical sections that would be difficult or impossible to roll, such as the steel shapes shown in Fig 3. In many instances, extrusions can replace bulky assemblies made up by joining, welding, or riveting rolled structural shapes, or sections previously machined from bar, plate or pipe. Some examples of structural designs—improved in both strength and cost—that can be realized by the use of aluminum and magnesium extrusions are depicted in Fig 4.

An extrusion die is relatively simple to make and inexpensive when compared to a pair of rolls or a set of forging dies. The low cost of dies and the short lead time for die changes make it possible to extrude small quantities more economically than by most other methods.

Metals available as extrusions

Extrusion can be used to fabricate practically

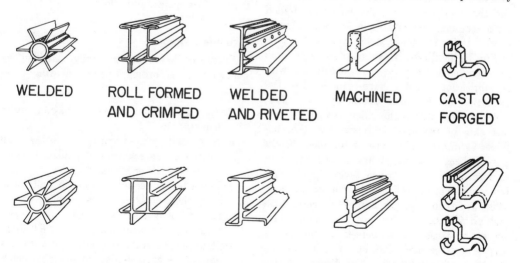

SECTIONS PRODUCED DIRECTLY BY EXTRUSION

Fig. 4. Examples of improved structural designs possible with light alloy extrusions.

all of the structural metals and alloys. Among the more common materials extruded on a commercial or semicommercial basis are alloys in the following metal systems:

Magnesium	Carbon and alloy steels
Aluminum	Stainless steel
Brass	Iron superalloys
Copper	Nickel superalloys
Titanium	Columbium
Zirconium	Molybdenum
Beryllium	Tantalum
Nickel	Tungsten

These materials span a range of working temperatures from about 600 to 4000 F, in approximately the order shown above.

Production limitations

The wide range of extrusion temperatures gives rise to the major differences in processing which center around such variables as extrusion lubricants, die materials, die design, billet preparation, and extrusion speed. Present tool materials are capable of maintaining adequate strength and wear resistance for extrusion at temperatures only slightly higher than 1000 F. At higher temperatures, lubricants are necessary not only to reduce friction but to insulate and protect the tooling surface from overheating. Also, the speed of extrusion must be more rapid to avoid prolonged contact between the tools and the hot billet. Thus, the extrusion method for magnesium and aluminum is quite different from that for the other metal systems.

The light alloys. Magnesium and aluminum are extruded at temperatures below 1000 F with no lubrication and flat sharp-cornered dies. Deformation of the billet occurs by *shear flow*, which is from within the billet so that the surface skin of the billet is retained in the container as discard. This type of turbulent flow is possible because of the ability of these materials to form sound welds when severely deformed, but requires comparatively slow pressing speeds, often less than 5 ft per min. With clean tools and no lubricants there are no contaminants present to cause internal defects or laminations, and several sections can be extruded at one time by using multihole dies. Precise dimensional control is attained with ordinary hot-work tool-steel dies which last for hundreds of extrusions.

Other metals. With higher temperature materials, e.g., titanium, steels, refractory metals, it is necessary to use lubricants and die designs so that deformation occurs by *uniform flow*. In this case, the surface of the billet becomes the surface of the extrusion; otherwise, laminations and inclusions could occur. Graphitic lubricants are suitable for producing relatively short lengths at temperatures up to about 2000 F if the operation is performed at high speeds. The Ugine-Sejournet process in which molten glass serves as a lubricant is most widely used for high-temperature extrusion. Because of the insulating as well as lubricating properties of glass, overheating of tools does not occur and die life is increased. For titanium and steels, dies are usually made of tungsten hot-work tool steels. Ceramic coatings (alumina or zirconia) on the dies are necessary at the temperatures required for refractory metals. Pressing speeds are usually in the range of 25 to 100 ft per min.

As extrusion temperatures increase, processing costs increase and the range of shapes and section sizes available becomes narrower. Tolerances are broader on both dimensions and surface finish, unless a subsequent sizing and finishing operation by drawing is performed. As improvements are made in lubricants and tool materials for high temperatures, however, more complex configurations with thinner sections and closer tolerances will be available in the high-temperature metals and alloys.

Extruded shapes

Extruded shapes are generally classified by configuration according to the following categories:

Rod: A round-solid section ⅜ in. or more in diameter.

Bar: A solid symmetrical section other than round such as square, rectangular, hexagonal, octagonal, or elliptical.

Tube: A hollow section of uniform wall thickness, whose cross section is completely symmetrical and may be round, square, rectangular, hexagonal, octagonal, or elliptical.

Hollow Shape: An extruded shape with any part of its cross section completely enclosing a void.

Semihollow Shape: An extruded shape with any part of its cross section partially enclosing a void, in which the area of the void is greater than the square of the width of the gap.

Solid Shape: Any extruded shape other than a hollow or a semihollow.

Shape limitations. While many asymmetrical shapes can be produced, probably the most important factor in the extrudability of a shape is symmetry. Hollow and semihollow shapes cost more than solid shapes and usually cannot be extruded with as thin sections. Semihollow shapes with long thin voids should be avoided. For best extrudability the length-to-width ratio of partially enclosed voids, channels, or grooves should not exceed 3:1 for aluminum and magnesium, 2:1 for brass, or 1:1 for copper, titanium, and steels. Wall thickness surrounding the voids should be as uniform as possible.

Adjacent section thicknesses should be as nearly equal as possible to permit uniform metal flow through the die. In shapes with varying section thicknesses, the length of thin protruding legs should not exceed ten times their thickness. Very thin sections in shapes with large circumscribing circles are undesirable because of straightening difficulties.

Size limitations. The size and weight of extruded shapes are limited both by the section configuration and by the material properties. The maximum size which can be extruded on a press of given capacity is determined by the *circumscribing circle,* which is the smallest circle that will enclose the shape. The circumscribing circle size controls the die size, which in turn is limited by the press size. Press capacities range from several hundred tons to 14,000 tons. The larger presses are capable of extruding aluminum shapes with 25-in.-dia circumscribing circles, and steel and titanium shapes with 22-in. circles.

The minimum cross-sectional area which can be extruded on a given size press is a function of the pressure required to overcome the resistance to flow. The pressure will depend on the properties of the material, the extrusion ratio, and the complexity of the shape. Minimum cross-

section sizes for steels and titanium alloys are approximately as follows:

Material	Minimum Cross Section, sq in.
Carbon and alloy steels	0.40
Stainless Steels	
Type 410	0.45
Types 304, 321, 347	0.50
Type 316	0.60
Type 310	0.70
19-9 DL	0.90
Hot-work tool steel	0.90
High-speed tool steel	1.25
A-286	1.25
Titanium alloys	0.50

Aluminum and magnesium alloys can be extruded in sizes with much smaller cross sections to fit circumscribing circles as small as ½ in. in diameter.

Thickness limitations. These are related to the size of the cross section as well as the type of material. As a rule, thicker sections are required with increased section size. The table below shows thickness limitations for several alloy groups:

Material	Minimum Thickness, in.
Carbon and alloy steels	0.120
Type 410 stainless	0.120
Type 304 stainless	0.187
A-286	0.250
Titanium alloys	0.150
Aluminum	0.040
Magnesium	0.040

Sharp corners and edges are usually possible with aluminum and magnesium alloys, but 1/64-in. corner and fillet radii are preferred. Minimum fillet radii of 0.125 in. for steel and 0.188 in. for titanium are suggested by most extruders. Typical minimum corner radii are 0.031 in. for steel and 0.055 in. for titanium.

Surface and tolerances

Smooth surfaces with finishes better than 30 μ in. rms are readily attainable in magnesium and aluminum alloys. High-temperature alloys are characteristically rougher; an extruded finish of 125 μ in. rms is generally considered acceptable for most steels and titanium alloys. Improved

TABLE 1—TYPICAL SECTION THICKNESS TOLERANCES FOR SOLID EXTRUDED SHAPES

Specified Dimension, in.	Tolerance, ± in.					
	Aluminum Alloys	Magnesium Alloys	Extruded Brass	Drawn Brass & Copper Alloys	Titanium Alloys	Steels
Under 0.125	0.006	0.006	0.005–0.010	0.003–0.005	—	—
0.125 to 0.250	0.007	0.007	0.005–0.010	0.003–0.005	0.020	0.020
0.250 to 0.500	0.008	0.008	0.005–0.015	0.003–0.005	0.020	0.020
0.500 to 0.750	0.009	0.009	0.0075–0.020	0.005–0.0075	0.020	0.020
0.750 to 1.000	0.010	0.010	0.0075–0.020	0.005–0.0075	0.030	0.020
1.000 to 1.500	0.012	0.012	0.010–0.025	0.0075–0.010	0.030	0.030
1.500 to 2.000	0.016	0.016	0.010–0.025	0.0075–0.010	0.030	0.030
2.000 to 3.000	0.024	0.024	0.015–0.030	0.010–0.015	0.040	0.030
3.000 to 4.000	0.024	0.024	0.020–0.035	0.015–0.020	0.050	0.045
4.000 to 5.000	0.034	0.034	0.030–0.045	0.020–0.030	0.060	0.060
5.000 to 6.000	0.034	0.034	0.035–0.060	0.030–0.035	—	—
6.000 to 8.000	0.044	0.044	—	—	—	—
8.000 to 10.000	0.054	0.054	—	—	—	—
10.000 to 12.000	0.064	0.064	—	—	—	—
12.000 to 14.000	0.074	0.074	—	—	—	—
14.000 to 15.000	0.080	0.080	—	—	—	—

surface finishes can be produced by a cold-draw finishing operation.

Although extruded shapes minimize and often eliminate the need for machining, they do not possess the dimensional accuracy of machined parts. The tolerances of any given dimension vary somewhat depending on the size and type of shape, and the relative location of the dimension. Detailed standard tolerances covering straightness, flatness, twist, and cross-sectional dimensions such as section thickness, angles, contours, and corner and fillet radii have been established for magnesium, aluminum, copper, and brass by most extruders and are published in handbooks. Standard tolerances also have been established for steels and titanium alloys in simple sections, but in many instances these are subject to mill inquiry. The typical tolerances on section thickness for solid shapes of several materials are summarized in Table 1.

Special shapes. Recent developments in extrusion of light metals have led to a variety of specialty shapes in aluminum and magnesium alloys. Stepped extrusions that have one or more changes in cross-sectional area at intervals along the length can be produced in a variety of shapes including tubing. Extrusion of longitudinally tapered structural sections and tubing is possible by special split die and moving mandrel tech-

niques. Wider, thinner, and flatter shapes can be produced by extrusion of a slab-shaped ingot in presses with rectangular containers. Flat aluminum-alloy panels with integral stiffeners have been extruded in widths of 18 in. with thicknesses as low as 0.035 in. Application of this method to larger presses would extend the maximum widths to about 42 in.

EXTRUDED PLASTICS

Extrusion is a process for making articles of constant cross section, called "continuous shapes," by forcing softened material through a hole approximating the desired shape. With plastics the process is carried out by one of two methods: *ram extrusion* or *screw extrusion*.

In ram extrusion, the softened mass fills a cylinder to which the die—the shaped hole—is attached at one end. A closely fitting piston, the ram, enters the cylinder and pushes the mass through the die at pressures ranging up to 10,000 psi. The product, or extrudate, is cooled or otherwise hardened shortly after leaving the die. Subsequent handling depends on the material and the shape. Ram extrusion is used chiefly for extruding TFE fluorocarbon (tetrafluoroethylene) resin (Teflon), which is damaged by the shearing action of screw extrusion, and for cellulose nitrate, whose extreme heat sensitivity and inflammability make screw extrusion dangerous.

Screw extrusion, by far the more economical and commercially important process, centers around the screw extruder. This consists of a heavy cylindrical barrel inside which turns a motor-driven screw, or worm. The screw is essentially a thick shaft with a helical blade, or flight, wrapped around it. At the rear end of the barrel, a feed hopper admits cold plastic particles that normally fall into the screw channel by gravity. As the screw rotates, the particles are dragged forward by frictional action between screw, plastic and barrel. Electric band heaters on the outside of the barrel heat the plastic which is further heated by the frictional action of the screw. Soon the particles coalesce into a voidless mass which softens further to become a melt. This plastic melt is very viscous (a million times as viscous as water), so considerable pressure, on the order of 500 to 10,000 psi, must be developed to force it through the die at the front end of the extruder at economical rates.

Screw extruders are specified by the inside diameter of the barrel, by the L/D ratio (the length of the barrel measured in diameters), and by the drive horsepower. Most commercial extruders range in diameter from 1.5 to 8 inches, though smaller and larger ones are in use, while L/D ratios range from 8 to 15 in older machines, 15 to 30 in recent models. A 2-inch extruder delivers about 50 lb/hr of polyethylene sheet, requires a 7.5-hp drive. Corresponding figures for a 6-incher are 500 lb/hr and 75 hp. Variations on the basic machine include: multiple screws; two-stage screws with an opening in the barrel between stages for extracting volatile matter from the melt; screws with special mixing sections; two or more extruders feeding into a single die so as to make extra-large or multicolor products; intermittent extruders; multiflighted screws.

Capabilities and limitations of process

In most screw extruders a large fraction of the heat required to melt the plastic comes from the frictional action of the screw. Since plastics conduct heat very slowly, and are highly viscous, it is difficult to heat them simply by putting them in contact with a hot surface. The shearing action of the screw thus presents an economical

and compact method for melting these materials, making the extruder's output per investment and labor dollar very high in comparison with those of other plastics processing machines. It is possible to extrude a variety of materials and shapes from a single machine with a very low changeover cost. On the other hand, where production volume warrants, it is possible to arrive at an optimum design for a given material and shape.

There is a proven body of theory that is a valuable aid to the extrusion engineer who studies and understands it. At present extrusion is limited mostly to making continuous shapes, though some noncontinuous shapes are possible with special die designs and post-forming techniques. Typical plastics extrusions are: films and sheets ranging in thickness from 0.25 mil to 0.5 inch and in widths up to 20 ft (in polyethylene film); filaments and rods from a few mils to 6 in. in diameter; structural shapes—channels, L's, H's, etc.—and gaskets of many shapes and sizes; pipe and tubing from $\frac{1}{8}$ to 8 in. o.d. in a variety of wall thicknesses; wire coverings on single and multiple strands ranging in diameter from a few mils to 8 in.; coatings on papers, metal foils and plastic films. Any plastic object that is technically feasible to make in quantity by extrusion is most economically made by that process.

A problem in the extrusion of complex shapes is the estimation of the die dimensions required to achieve the desired extrudate dimensions. Plastic melts tend to swell as they emerge from the die, and this viscoelastic recovery must be compensated for, either by drawing down the emerging melt (i.e., taking it away at a higher lineal speed than it is emerging) or by making the die aperture smaller than the desired extrudate dimension. Since the melt will also flow faster through a thick opening than a thin one, careful attention must be given to land length (the thickness of the die metal in the direction of flow) in designing the die. The normal procedure is one of trial and error: the die is constructed according to the best estimates from previous experience with similar shapes extruded from the given material. A test run is made, the extrudate is examined as it emerges, and the die aperture is adjusted by filing and/or peening the metal until the desired extrudate dimensions are obtained. With simpler shapes, e.g., sheeting, wire covering, and pipe, die design is more systematic and confident.

Highly accurate extrusions are possible under best conditions, and variations in diameter and wall thickness of extruded plastic pipe are well within those found in commercial steel pipe, while surface quality is considerably better. Control of film thickness crosswise to the direction of flow is probably the most difficult dimensional control problem commonly encountered in plastics extrusion. Slight variations from point to point in die aperture, melt pressure and temperature, drawdown, and cooling conditions can strongly influence the local thickness of thin films. In tubular polyethylene film nominally 1 mil thick, for example, crosswise caliper varies from 0.7 to 1.3 mils in typical commercial products.

Extrudable materials

All thermoplastics can be extruded by either ram or screw extrusion. For many years it was difficult to extrude any but the simplest shapes from type 66 nylon because of its low melt viscosity. Today, however, high-viscosity grades of this and other nylons are available and their extrusion presents no special difficulties. In extruding rigid polyvinyl chloride, extreme care must be taken not to overheat the resin since thermal decomposition, once started, snowballs. This simply means being careful to avoid extreme temperatures everywhere and to meticulously streamline all passages through which the melt must pass. To a lesser degree, CFE fluorocarbon (trifluorochloroethylene) resin (Kel-F, et al.), cellulosics, nylons, and acetal (polyoxymethylene) are similarly heat-sensitive.

Some thermosets can be extruded provided they are formulated to flow at temperatures safely below the curing temperatures. The process has been used to make pipe and structural shapes, to coat wire with thermosetting compositions and to prepare "rope" and pellets for compression molding. The extrusion of rubbers closely resembles plastics extrusion.

F

FABRICS, NONWOVEN BONDED

While there are several types of fabrics which are not woven, the term "Nonwoven Fabrics" is recognized in the textile trade as applying to those materials composed of a fibrous web held together with a bonding agent to obtain fabric-like qualities. These fabrics may be of a uniform, close-bonded fibrous structure or of a foraminous unitary construction.

They may be formed by processing on modifications of either textile type machines or of papermaking equipment. In either case the fibers as laid up in the basic web prior to bonding or to postforming may be oriented in one or more prescribed directions, or distributed in a completely random fashion. They are secured in place by suitable adhesives incorporated in the web. The application of these adhesives may be controlled to completely coat and bond the fibers, or to bond them only in selected areas, or at points of individual fiber contact.

Nonwovens may be thick or thin and of either low or high density. The conditions under which nonwoven fabrics are manufactured and the possible combinations of fibers and adhesives permit the production of structures offering a wide range of physical and chemical properties.

Production methods

Production of a nonwoven fabric may be divided into two basic steps: (1) formation of the web and (2) bonding the web. The most widely used means for forming the web is a series of cotton cards feeding to a common conveyor belt to build up a unidirectional composite web of the desired weight. The number of cards per line will depend upon the maximum weight product to be produced. Each card in the line may be geared to produce webs ranging from some 35 grains up to 100 grains per yard at speeds of 200 fpm down to 45 fpm for the heavier material. Material from these lines is usually limited to 40 in. widths, with strength favoring the machine direction over the cross machine direction in ratios from 3 to 1 to as

much as 20 to 1. Where wider material of heavier weight, or with the strength balanced in the machine and cross machine directions is required a web production line consisting of a single breaker and a finisher garnett equipped with a cross lapper may be used to advantage. Production rates with garnetts run considerably higher than with cotton cards though the quality of the web frequently suffers.

The most versatile type of web for nonwoven fabrics is that produced by the air disposition of precarded fibers which are collected with a minimum of orientation as a uniform mat. Several machines have been specially developed for this purpose which will handle a wide range of fiber descriptions, producing webs up to 84 in. in width of some $\frac{1}{2}$ oz to 10 oz/sq yd in weight. Production rates with synthetic fibers of 1.5 lb per hr per in. of width have been quoted.

In contrast to nonwoven fabrics made from webs which have been dry processed on modified textile equipment, are those produced from wet-laid webs using papermaking machines. The chief departure from papermaking technique is in the use of longer unbeaten nonhydrated fibers, generally ranging from $\frac{1}{4}$ to $\frac{1}{2}$ in. in length. Such webs usually depend upon adhesive additives or post bonding to impart the necessary physical properties.

Once the web has been formed, by either the dry or wet laid process, it may be further modified by techniques such as needle punching, aeration or impingement with gaseous, liquid or other means, to produce a patterned configuration of desired characteristics. Nonwovens of such postformed webs may be characterized by added resistance to delamination, superior drape, flexibility, porosity, abrasion and flame resistance or other desirable properties.

Types of fibers

In the production of nonwoven fabrics most every type of natural and man-made fiber can be used. Price, equipment and quality, as well as chemical and physical requirements of the product will govern the particular fiber used as well as the bonding agent. While reworked and off-quality fiber is quite satisfactory for some of the less critical industrial uses, the trend is toward the use of more virgin first quality fiber, especially the man-made cellulosics, in everything from diapers to casket liners.

Rayon is the predominant fiber used for both utility as well as aesthetic appeal. It is made in a wide variety of descriptions to fit the different manufacturing methods and end use requirements. The finer 1 to 3 denier fibers are generally of advantage in the lighter weight, denser materials where softness, flexibility and draping properties are important. The coarser deniers ranging from 5.5 through 50 are generally more desirable in the stiffer, thicker, lower density and more open structure materials. With other factors remaining constant, the finer deniers give the best tensile, tear and bursting strength values. Lengths within the range of 1 to 2 in. are also generally best for ease of processing as well as physical properties in the product. Little seems to be gained by using lengths over 2 in. Blends of several different fibers frequently may be used to advantage, such as a blend of rayon and nylon for added wet strength, abrasion resistance or flex resistance. For special applications calling for particular chemical or electrical resistance, more use of the expensive synthetic fibers such as Acrilan, nylon or Dacron may be warranted.

Bonding agents

Properties of nonwoven fabrics are as dependent upon the bonding agent as they are upon the fiber which forms the foundation of the material. Both are selected with the end use in mind and each must be compatible with the other.

Bonding agents may be grouped into three broad classifications: (1) liquid dispersions, (2) powdered adhesives, and (3) thermoplastic fibers.

Liquid dispersions. Liquid dispersions are the most extensive type used. Among these are polyvinyl alcohol, generally used as a preliminary binder or where high strength and permanence are not essential; polyvinyl acetate for good strength and flexibility where freedom from odor and taste are important; polyvinyl chloride for good wet and dry strength, and toughness; synthetic latices of butadiene-acrylonitrile or butadiene-styrene for good adhesive and elastic properties where strength and a high degree of permanence are more important than color stability and odor; the acrylics for good strength, soft hand, color stability and permanence. These dispersions are applied by (1) spraying, generally used for low density materials, (2) saturator, for denser more durable material, and by (3) printing, usually for selective bonding of localized areas in soft absorbent products.

Powdered adhesives. These are usually of thermosetting or thermoplastic resin types and are sifted into the fiber web as formed. They are used especially in the low density, high bulk nonwovens where wetting by the binder or the application of pressure might cause excessive matting and compression of the material. Bonding is effected by heating either with or without the use of pressure.

Thermoplastic fibers. The thermoplastic fiber binders have the advantage of constituting an integrated structural part of the fiber web which forms the fabric. To bond the web they may be activated by solvents or by heat and pressure. By regulation of the amount of heat and pressure as well as the amount of thermoplastic fiber present, a wide variety of characteristics may be built into these nonwoven fabrics.

Applications

Construction and performance of nonwoven fabrics have not been standardized. They are usually constructed to fit a particular end-use requirement or are built around particular specifications. The various manufacturers may use entirely different equipment or composition of material for the same general class of application unless the user is specific as to these. This could limit the source of supply to perhaps one or two manufacturers as no two have identical production methods. A greater responsibility is therefore placed upon the consumer in selecting the proper material or supplier. Once he has established these to his satisfaction, he should expect to be supplied with material of consistent quality.

Industrial products for which nonwoven fabrics have been used include: acoustical curtains; artificial leather and chamois; automotive plumpers; backing for adhesive tapes; base for vinyl and rubber coatings; bagging; buffing wheels; cable and wire wrappings; electrical tapes; filters for air, gases, and liquids; insulation; laminate reinforcements; polishing and wiping cloths and wall coverings.

FABRICS, WOVEN

By far the greatest volume of textile materials is used in consumer textiles, such as apparel. But textiles are extremely versatile materials which have been applied to a large number of engineered uses, e.g., thermal, acoustical and electrical insulation; padding and packaging; barrier applications; filtration, both dry and wet; upholstery and seating; reinforcing for plastics or rubber; and various mechanical uses such as fire hose jackets, tenting, tarpaulins, parachutes and marine lines.

Textiles are highly complex materials. Their properties depend not only on the fiber but on the form in which it is used—whether the form be a felt, a bonded fabric, a woven or knit fabric or cordage. Properties such as heat, chemical and weather resistance depend primarily on the type of fiber used; properties such as mechanical strength, thermal transmission, and air or liquid permeability depend both on the fiber and the textile form.

The versatility of textiles stems from (1) the wide range of fibers which can be used, and (2) the range of complicated textile structures which can be formed from the fibers.

Two important factors which should be considered in discussing textile needs with textile suppliers, but which are not discussed here in detail are as follows:

1. The types of finishes that can be applied to the finished textile product can substantially alter or modify the stability, "hand," and/or durability of the textile.

2. Combining of textiles with other materials, such as resins or rubber, either by impregnation or by coating, will substantially alter performance characteristics of the final composite.

Textile constructions

Textile engineering materials can be classified generally as (1) nonwoven fabrics, including both felts and bonded fabrics; (2) woven or knit fabrics; and (3) cordage. Nonwoven fabrics are discussed elsewhere.

Woven and knit fabrics. The basic building blocks of woven or knit fabrics are fibers. Fibers may be either filament (essentially continuous) or staple (relatively short). From these fibers either filament or staple yarns are produced.

Filament yarns are primarily used for maximum strength. Staple yarns are produced by spinning staple fibers into long continuous strands or yarns. Friction caused by the twist holds the staple fibers together in the yarn. Two or more of these so-called "single yarns" may then be twisted together to form "ply yarns." The latter, in turn, can be twisted together to form "cabled yarns" or "cord."

Yarns are either knit or woven to form fabrics. Knitting is a process of interlocking one or more yarns, and except for some knit textiles used for coated fabrics, is little used for industrial fabrics. Knitting generally provides a fabric of greater bulk and extensibility than does the weaving process.

Weaving is the process of interlacing lengthwise "warp" yarns with crosswise "filling" yarns in any of a variety of constructions. All woven fabrics are variations of three basic weaves:

1. Plain weave is simply "one yarn up and one down." Tightly woven, a plain weave can provide the most yarn interlacings per square inch, and thus, maximum "cover" and impermeability. The weave can also be "opened up" to practically any desired degree.

2. Twill weave is characterized by a sharp diagonal "twill line" produced by the warp yarn crossing over two or more filling yarns, with interlacing advancing one filling yarn with each warp. More porous than an equivalent plain weave, twills are used in drill fabrics and jeans.

3. Satin weave is characterized by regularly spaced interlacings at wide intervals. Satins provide porosity, good cover, and a smooth surface. Originally woven from silk yarn, satin weave is called "sateen" when woven from cotton.

Cordage. The term cordage includes all types of threads, twine, rope and hawser. Essentially all cordage consists of fibers twisted together, plied, and in many cases cabled to produce essentially continuous strands of desired cross section and strength.

In addition to the type of fiber used, the most important determinants of the end properties of cordage are the type and degree of twist employed. The two major types of twist are: (1) cable twist, in which the direction of twisting is alternated in each successive operation, i.e., singles may be "S" twisted, plies "Z" twisted and cables "S" twisted (a yarn or cord has "S" twist if, when held in a vertical position, the spirals conform in direction of slope to the central portion of the letter "S," and "Z" twist if the spirals conform in direction of slope to the central portion of the letter "Z"), and (2) hawser twist, in which the singles, plies, and cables are twisted "SSZ" or "ZZS." Hawser twist generally provides higher strength and resilience.

Specifications

Textile specifications contain two important types of information: (1) descriptive information and (2) service property requirements.

Specifications which physically describe the textile fabric usually include (1) width, in inches, (2) weight, usually in ounces per square yard, (3) type of weave, such as twill, broken twill, leno or satin, (4) thread count, both in warp and filling (e.g., 68 × 44 denotes 68 warp yarns per in. and 44 filling yarns per in.), (5) type of fiber and whether the yarn is to be filament or staple, (6) crimp, in percent, (7) twist per inch, and (8) yarn number both for warp and fill.

Yarn number designations are somewhat complex, as they have been developed in a relatively unorganized fashion over the years, and different systems are used in different types of fibers. (Filament yarns are usually stated simply in denier, which is the weight in grams of 9000 meters of yarn.)

Essentially, yarn numbers provide a measure of weight per unit length, or length per unit weight. A typical yarn designation on a specification may appear as "210 (denier)/1 × 20/2 (cotton system)." This means that (1) the warp yarn is a 210-denier single yarn, and (2) the filling yarn contains 2 plies, each of which is a 20 singles yarn (determined by the cotton numbering system).

A number of fabric-designation systems have been formalized by tradition. For example, sheetings, drills, twills, jeans, broken twills and sateens are designated only by width in inches, number of linear yards per pound, and number of warp and filling threads per inch (e.g., 59 in. 1.85 68 × 40 drill). "Specs" for equivalent synthetic fabrics also include fiber type, whether staple or filament.

FELTS

There are two classes of felts in use today as engineering materials: wool felts and synthetic fiber felts. Both of these are obtained without use of spinning, weaving, or knitting, and in a sense are nonwoven fabrics.

Wool felt is a fabric obtained as a result of the interlocking of wool fibers under suitable combinations of mechanical work, chemical action, moisture and heat, alone or in combination with other fibers.

Synthetic fiber felt is a fabric obtained as a result of interlocking of synthetic fibers by mechanical action. In some cases chemical action or heat is utilized in addition to further enhance or stabilize the mechanical interlocking or to achieve greater strength or density.

General properties

Neither wool nor synthetic fiber felts require binders and exist as 100% fibrous materials. Felts generally exhibit the same chemical properties as do the fibers of which the felt is composed. Wool felts are characterized by excellent resistance to acids but are damaged by exposure to strong alkalies. They exhibit remarkable resistance to atmospheric aging and as a class are probably the most inert of all nonmetallic engineering materials with respect to nonaqueous liquids, oils, or solvents. Wool felts are not generally recommended for stressed dry uses at temperatures in excess of 180 F, due to changes in physical properties, but are used as dry spacers and gaskets in many applications up to 300 F, and the use of the materials as oil wicks and lubricating system components at ambient temperatures up to 300 F is common. A natural resistance to burning is common to wool felt but where additional resistance is required this can be achieved by chemical treatment. The latter approach is used also to obtain resistance to water absorption and biological and insect pest attack when required.

Synthetic fiber felts are available in virtually all classes of fiber composition including regenerated cellulose, cellulose acetate, cellulose triacetate, polyamide, polyester, acrylic, modacrylic, olefin, and TFE fluorocarbon (tetrafluorethylene). The variety of types available provides a virtually infinite range of physical and chemical properties for application beyond the natural versatility of wool fiber felts. Outstanding among the properties of this class of engineering material are chemical, solvent, thermal, and biological stability as well as low moisture absorption, quick drying after aqueous wetting, abrasion resistance, and frictional and dielectrical features.

Forms available

Wool felts are produced as "sheet" stock in a standard 36 by 36-in. size in thickness ranging from 1/16 to 3 in. There are a number of density classifications based upon the weight for a 36 by 36-in. sheet in 1 in. thickness from 12 to 32 lb with the weight for any given thickness in the density class proportioned to the weight per square yard at the 1-in. thickness. "Roll" felts are produced in either 60 or 72-in. widths in lengths up to 60 yards. Standard thicknesses range from ½₂ to 1 in. and densities from 8 lb per square yard for 1-in. thickness to 18 lb per sq yd for 1 in. thickness.

In either sheet or roll form normally available thickness increments are based upon 1/32 of an inch but virtually any thickness can be produced on order within the broad thickness ranges given above.

Specifications and general properties of wool felts are detailed in U.S. Department of Commerce, Commercial Standard 185-52 "Felt, Wool;" Federal Specification C-F-206A "Felt, Sheet, Wool;" ASTM Standard D 1114 Specification for Sheet Wool Felt; and in the Society of Automotive Engineers Handbook under "Felt." In addition, complete details for testing wool felt are provided in ASTM Standard D-461, "Standard Methods of Test for Wool Felt."

Synthetic fiber felts have only been commercially available as such since 1955 and no detailed standards for weight, thickness and density classes have been developed as in the case of wool felt. However, widths are available from 54 to 72 in. in thicknesses from 1/32 to ¾ in. 60 yards is the generally accepted piece length although special size sheets are available on order. At the present time, 8 pounds per square yard is the maximum weight produced but this provides a broad range of felted densities controlled by the thickness range mentioned before.

Both wool and synthetic fiber felts are nonfraying and nonravelling and thus provide considerable ease of cutting and fabricating. Most of the major producers maintain up-to-date cutting and fabricating divisions for the production of cut and fabricated parts to blueprint specification. In addition, wool felt lends itself to most grinding, cutting, shaping, extruding and other machining operations so that special shaped parts such as polishing laps, round wicking, ink rollers and others are produced.

Applications

The structural elasticity of felts as a class make these materials suitable for molding and forming and parts of these types are produced to provide special shaped gaskets, seals, fillers, instrument covers, and the like. In many cases these shaped parts are stabilized through the use of resinous or rubber impregnants and this modified class of felt is finding ever-increasing use as flat stock for special gasketing, sealing, and other applications. Combinations of felt and plastic and elastomer sheet materials are also produced for application where resilience plus nonpermeability is desired as in sealing and frictional uses.

Use of felt as an engineering material covers a broad spectrum of application. Major contributing properties include resilience, mechanical, thermal and acoustic energy absorption; high porosity-to-weight ratio; resistance to aging; thermal and chemical stability; high effective surface area per unit volume; and solvent resistance.

Applied uses include wet and dry filtration, thermal and acoustical insulation, vibration isolation, impact absorption, cushioning and packaging, polishing, frictional surfacing, liquid absorption and reservoirs, wicking, gasketing, sealing, and percussion mechanical dampening.

FERROUS METAL POWDERS

Iron-carbon powders contain up to 1% graphite. When pressed and sintered internal carburization results and produces a carbon steel structure, although some free carbon remains. In general, iron-carbon steel P/M parts have densities of around 6.5 g per cc. However, densities of over 7.0 are used to produce higher mechanical properties. These carbon steel P/M parts have higher strength and hardness than those of iron, but they are usually more brittle. As-sintered strengths range from about 35,000 psi to 70,000 psi depending on density. By heat treatment, strengths up to 125,000 psi are achieved.

The mechanical properties of ferrous powder

parts can be considerably improved by impregnating or infiltrating them with any one of a number of different materials, both metallic and nonmetallic, such as oil, wax, resins, copper, lead and babbitt.

In addition to the above powders, which are used for the bulk of P/M parts applications, a number of specialty ferrous alloys are available. These include 3 to 9% silicon irons, iron-nickel alloys, alloy steels (2, 4 and 7% nickel steels and 4600 series steels). Also a range of stainless steel powders are finding increasing use. These include types 302, 303, 304, 316, 330, 410, 430 and 71-4 PH.

Ferrous base P/M parts can range in size from about 0.10 in. thick and 1/8 in. diameter to 2 in. thick and over 2 ft. in diameter. Because they can be mass produced at relatively low cost, iron base P/M parts find a wide variety of uses in such high volume products as appliances, business machines, power tools and automobiles. Typical parts are gears, bearings, rotors, valves, valve plates, cams, levers, ratchets and sprockets. Also recent developments have made possible the production of forging preforms or blanks out of low alloy steel powders and some of the superalloys.

TABLE 1–MECHANICAL PROPERTIES (AS-SINTERED) OF TYPICAL FERROUS P/M PARTS

Type	Density, g/cc	Tens. Str., 1000 psi	Elong., %	Hardness, Rockwell B	Imp. Str., ft-lb
99 Iron (Sponge)	5.7–6.1	19	5	20 (H)	4
	7.3	40	12	20	
99 Iron, 1 Carbon	6.1–6.5	35	1.0	50	1
	7.0	60	3.0	–	2
90 Iron, 10 Copper	5.8–6.2	30	0.5	–	–
9 Iron, 7 Copper,	5.8–6.2	50	0.5	70	3
1 Carbon	6.8	83	1.0	73	4

FIBER REINFORCED PLASTICS

In recent years the fastest growing engineering material has been fiber reinforced plastics (FRP). These materials are extremely versatile composites with relatively high strength-to-weight ratios and excellent corrosion resistance. In both these respects they out-perform most metals. In addition, they can be formed economically into virtually any shape and size part. In size, FRP products range from tiny electronic components up to boat hulls of 70 ft. and longer. In between these extremes, there are a wide variety of FRP gears, bearings, housings and other parts used in all the product manufacturing industries.

Fiber reinforced plastics are composed of three major components—matrix, fiber and bonding agent. The plastic resin serves as the matrix in which are embedded the fibers. Adherence between matrix and fibers is achieved by a bonding agent or binder, sometimes called coupling agent. Most plastics, both thermosets and thermoplastics, can be the matrix material. In addition to these three major components, a wide variety of additives—fillers, catalysts, inhibitors, stabilizers, pgiments and fire retardants—can be used to fit specific application needs.

Fibers

Glass is by far the most used fiber in FRP. Plastics composites reinforced with it are referred to as GFRP or GRP. Asbestos has some use, but is largely limited to applications where maximum thermal insulation or fire resistance is required. Other limited use fibrous materials are paper, sisal, cotton and nylon. Metals can be used as wire mesh or cloth for special applications. High performance and costly fibers, such as boron and graphite, and metal fibers are also used in advanced technology products. In recent years other fibers have been developed that will find increasing use in the years ahead, in commodity FRP plastics. One of these is polyvinyl alcohol fiber (PVA) developed in Japan.

The standard glass fiber used in GRP, is a borosilicate type, known as E-glass. The fibers are spun as single glass filaments with diameters ranging from 0.0002 to 0.001 in. These filaments collected into strands, usually around 200 per strand, are manufactured into many forms of reinforcement. The E-glass fibers have a tensile strength of 500,000 psi. A relatively new glass fiber, known as S-glass, is higher in strength, but because of its higher cost its use is limited to advanced, high performance applications. In general, in reinforced thermoplastics, glass content runs between 20 and 40%; with thermosets it runs as high as 80% in the case of filament wound structures.

There are a number of standard forms in which glass fiber is produced and applied in GRP.

1. Continuous strands of glass supplied either as twisted, single-end strands (yarn) or as

untwisted multistrands (continuous) roving.

2. Fabrics woven from yarns in a variety of types, weights and widths.

3. Woven rovings—continuous rovings woven into a coarse, heavy, drapable fabric.

4. Chopped strands made from either continuous or spun roving cut into 1/8 to 1/2 in. lengths.

5. Reinforcing mats made of either chopped strands or continuous strand laid down in a random pattern.

6. Surfacing mats composed of continuous glass filaments in random patterns.

FRP resins

Although a number of different plastic resins are used for reinforced plastics, thermosetting polyester resins are the most common. The combination of polyester and glass provide a good balance of mechanical properties as well as corrosion resistance along with low cost and good dimensional stability. In addition, curing can be done at room temperature without pressure, thus making for low processing equipment costs. For high volume production, special sheet molding compounds are available in continuous sheet form for use in the matched die process. In recent years resin mixtures of thermoplastics with polyesters, have been developed to produce high quality surfaces in the finished molding. The common thermoplastics used are acrylic, polyethylene and styrene.

Other glass reinforced thermosets include phenolics and epoxies. GRP phenolics are noted for their low cost and good overall performance in low strength applications. Because of their good electrical resistivity and low water absorption they are widely used for electrical housings, circuit boards and gears. Epoxies, being more expensive than polyesters and phenolics, are limited to high performance parts where their excellent strength, thermal stability, chemical resistance and dielectric strength are required.

Up until about a decade ago GRP were largely limited to the thermosets. Today, however, more than 1000 different types and grades of reinforced thermoplastics are commercially available. Leaders in volume use are nylon and the styrenes. Others include sulfones and ABS. Unlike thermosetting resins, GRP thermoplastic parts can be made in standard injection molding machines. The resin can be supplied as pellets

containing chopped glass fibers 1/8 to 1/2 in. long. As a general rule, a GRP thermoplastic with chopped fibers at least doubles the plastics tensile strength and stiffness.

Glass reinforced thermoplastics are also produced as sheet materials for forming on metal stamping equipment. Produced in sheet and coil form, the materials are preheated at about 400 F and then formed and trimmed. The mechanical properties are comparable to GRP parts made by other methods.

FRP processing methods

Matched Metal Die Molding—This is the most efficient and economical method for mass producing high-strength parts. Parts are press-molded in matched male and female molds at pressures of 200 to 300 psi and at heats of 235 to 260 F.

Four main forms of thermosetting resin reinforcement are used: 1. Chopped fiber preforms, shaped like the part, are saturated with resin at the mold. They are best for deep-draw, compound curvature parts. 2. Flat mat, saturated with resin at the mold, is used for shallow parts with simple curvature. 3. Sheet molding compound, a pre-impregnated material, has advantages for parts with varying thickness. 4. Bulk molding compound, a premix made up of short fibers pre-impregnated with resin. It's used for parts similar to castings.

Injection Molding—In this high-volume process, a mix of short fibers and resin is forced by a screw or plunger through an orifice into the heated cavity of a closed matched metal mold. It is the major method for forming reinforced thermoplastics and is beginning to be used for thermoplastic-modified, thermosetting bulk molding compounds.

Hand Lay-Up—This is the simplest of all methods of forming thermosetting composites. It is best employed for quantities under 1000, for prototypes and sample runs, for extremely large parts and for larger volume where model changes are frequent, as in boats. In hand lay-up, only one mold is used, usually female, which can be made of low-cost wood or plaster. Duplicate molds are inexpensive. The reinforcing mat or fabric is cut to fit, laid in the mold and saturated with resin by hand, using a brush, roller or spray gun. Layers are built up to the required thickness, then the laminate is cured to permanent hardness, generally at room temperature.

Spray-Up—Like hand lay-up, the spray-up method uses a single mold, but it can introduce a degree of automation. This method is good for complex thermoset moldings, and its portable equipment eases on-site fabrication and repair. Short lengths of reinforcement and resin are projected by a specially designed spray gun so they are deposited simultaneously on the surface of the mold. Cure is usually accomplished by a catalyst in the resin at room temperature.

Filament Winding—This method produces moldings with the highest strength-to-weight ratio of any reinforced thermoset because of its high glass-to-resin ratio. It is generally limited to surfaces of revolution—round, oval, tapered or rectangular—but it can achieve a high degree of automation. Continuous fiber strands are wound on a suitably-shaped mandrel or core and precisely positioned in predetermined patterns. The mandrel may be left in place permanently or removed after cure. The strands may be preimpregnated or the resin may be applied during or after winding. Heat is used to effect final cure.

Centrifugal Casting—This is another method of producing round, oval, tapered or rectangular parts. It offers low labor and tooling costs, uniform wall thicknesses, and good inner and outer surfaces. Chopped fibers and resin are placed inside a mandrel and uniformly distributed as the mandrel is rotated inside an oven.

Continuous Laminating—Continuous laminating is the most economical method of producing flat and corrugated panels, glazing and similar products in large volume. Reinforcing mat or fabric is impregnated with resin, run through laminating rolls between cellophane sheets to control thickness and resin content, then cured in a heating zone.

Pultrusion—Pultrusion produces shapes with high unidirectional strength such as "I" beams, flat stock for building siding, fishing rods and shafts for golf clubs. Continuous fiber strands, combined with mat or woven fibers for cross strength, are impregnated with resin and pulled through a long heated steel die. The die shapes the product and controls resin content.

Properties

The mechanical properties of fiber composites are dependent upon a number of complex factors. Two of the dominating ones in glass reinforced plastics are the length of fibers and the glass content by weight. In general, strength increases with fiber length. For example, reinforcing a thermoplastic with chopped glass fibers at least doubles the plastic's strength, whereas long-glass fiber reinforced thermoplastics exhibit increases of 300 and 400%. Also heat distortion temperatures usually increase by about 100 F, and impact strengths are raised appreciably. Similarly, as a general rule, an increase in glass content results in the following property changes:

Tensile and impact strength increase.
Modulus of elasticity increases.

TABLE 1—TYPICAL PROPERTIES OF GLASS-REINFORCED THERMOPLASTICS

Resin	Specific Gravity	Tens. Str., 1000 psi	Elong., %	Tens. Mod., 10^5 psi	Imp. Str., (ft-lb/in.)	Deflection Temp. at 264 psi, F
ABS	1.19–1.36	14.5–18	1.5–2.0	8.0–10.0	0.4–3.0	210–230
Acetal	1.54–1.69	9–18	1.0–3.0	8.0–14.5	0.8–3.0	310–335
Nylon	1.32–1.52	15–30	1.8–3.0	8.0–20.0	1.0–4.5	390–500
Polycarbonate	1.34–1.58	13–21	1.2–4.0	8.1–17.0	1.5–4.0	285–295
Polyethylene (linear)	1.09–1.28	7–11	1.5–3.0	7.0–10.5	1.2–3.5	240–260
Polypropylene	1.04–1.22	6–9	2.0–3.0	4.5–9.0	1.2–3.0	270–300
Polystyrene	1.20–1.34	10–15	1.0–1.6	8.0–19.0	0.4–2.2	210–220
Polysulfone	1.31–1.47	11–17	1.5–4.0	6.5–10.0	0.9–2.5	340–350
Styrene-Acrylonitrile	1.22–1.35	13.5–18	1.1–2.0	9.5–18.0	0.6–3.0	210–220
Vinyl-Chloride	1.49	12	1.5	12.0	1.0	170

Ranges shown represent various amounts of glass content, most being glass content of 20 to 40% (Plastics/Elastomers Issue, *Machine Design*, Feb. 15, 1973).

TABLE 2—TYPICAL PROPERTIES OF GLASS FIBER REINFORCED THERMOSETS

Type	Phenolics, Fibers	Melamines, Fibers	Alkyds, Fibers	Epoxy, Laminate	Epoxy, Fil. Wound	Silicone, Woven Fabric
Specific Gravity	1.75–1.90	1.9–2.0	2.0–2.1	1.8	2.17	1.75–1.8
Mod. Elast. in Tens., 10^5 psi	30–33	24	20–25	33–36	64–72	28
Tens. Str., 1000 psi	5–10	5–10	5–9	50–60	230–240	30–35
Elong., %	0.2	–	–	–	–	–
Hardness, Rockwell	E50–70	–	–	M115–117	M98–120	–
Imp. Str., ft-lb/ in. notch	10–33	5–10	8–12	12–15	–	10–25
Max. Svc. Temp., F	350–450	300–400	300	250–350	250–350	450–500
Vol. Res., ohm-cm	7–10×10^{12}	1–7×10^{11}	10^{14}	–	–	2–5×10^{14}
Dielect. Str., v/mil	200–370	250–300	300–350	450–550	–	725

Heat deflection temperature increases, sometimes as much as 300 F.
Creep decreases and dimensional stability increases.
Thermal expansion decreases.

More specifically, moldings of glass reinforced polyesters range in strength from around 15,000 to 55,000 psi, compared to strengths of 8,000 to 12,000 psi for non-reinforced polyesters (see "Polyester Plastics"). GFR thermoplastic moldings, such as nylon, have strengths up to about 30,000 psi compared to about 10,000 psi for unreinforced parts (Table 1). In the high performance area, GFR epoxies using glass cloth laminates, go as high as 60,000 psi. Filament wound structures can exceed 240,000 psi (Table 2). Typical GRP epoxy applications of this kind are rocket motor cases, chemical tanks and pressure bottles.

There are similar improvements in impact strength and stiffness. For example, impact strength of GRP polyesters run from 10 to 30 ft lb compared to less than 1 ft lb for unreinforced moldings; and modulus of elasticity ranges up to 2.5 and 3.0 million psi in contrast to ½ million psi. And the modulus elasticity of most thermoplastics is usually doubled or tripled by the addition of glass fibers.

Another outstanding characteristic of GRP materials is their chemical stability, a property which has been used to advantage in the construction of a large variety of tanks, containers and piping for use in corrosive environments.

Improvements in properties of GRP are usually accompanied by an increase in material cost. The added cost varies widely depending on the plastic, glass content, form of glass and other factors. The increased material cost is generally at least 15%, but it can be several times this figure. However, as we have seen, material cost is only one consideration. Even though a reinforced plastic may have a higher raw material cost, processing costs and other considerations may make it more economical than unreinforced plastic.

FIBROUS GLASS

. The primary engineering benefits of glass fibers are their (1) inorganic nature which makes them highly inert, (2) high strength-to-weight ratio, (3) nonflammability, and (4) resistance to heat, fungi and rotting.

Glass fibers are produced in both filament and staple form. Their major engineering uses are (1) thermal and/or acoustical insulation and (2) as reinforcements, primarily for plastics.

Types

The largest volume of glass fibers used for engineering applications are so-called "E" type, made from a lime-alumina borosilicate glass that is relatively soda-free. Although its initial strength at the bushing may be about 400,000 to 500,000 psi, surface damage to fibers (both mechanical damage in handling and effects of moisture) reduces usable strength to 150,000 to 200,000 psi. But at 200,000 psi tensile strength, the relatively low density of glass (0.092 lb per cu ft) produces a strength-to-weight ratio of about 2,170,000 in., superior to that of a 450,000 psi tensile strength steel. Modulus of E glass fibers is about 10 million psi. Although essentially unaffected by low temperatures, E glass is limited to a maximum continuous operating temperature of about 600 F.

Other specialized types of glass (primarily used in specialty reinforced plastics applications) include:

1. High silica, leached glass fiber—Fibers with silica content of 96 to 99% are produced by leaching glass fibers. Such fibers provide excellent heat resistance, but relatively low strengths. They are usually used in short-fiber form for molding compounds.

2. Silica or quartz fibers—Fibers of pure silica provide optimum heat resistance (to about 200 F), though strength is somewhat lower than that of conventional E glass.

3. High modulus fibers—Fibers of a beryllia-containing glass have been developed (primarily for filament winding use) with modulus of about 16 to 18 million psi.

Production methods

Most fibrous glass is produced either by air, steam or flame blowing, or by mechanical pulling or drawing. Blowing produces relatively short staple fibers; mechanical drawing produces continuous monofilaments.

In blowing, steam or air jets impinge upon, and break up molten streams of glass, forming fibers. The type of fiber produced depends on the pressure of the steam or air, and the temperature and viscosity of the molten glass. As pressure and temperature are increased, the action changes from the smooth attenuation of relatively long, uniform, textile grade staple fibers to a violent action which draws the streams out turbulently in many directions and produces shorter, finer fibers ("wool").

After the fibers are blown, they form a tangled mass whose properties are determined by such factors as size and length of the fibers, density of packing and the type of lubricant or binder used.

In the mechanical drawing process, the molten glass is fed into a "bushing" which contains a number of orifices through which the glass flows. Continuous filaments are then drawn from the molten glass stream. During the early stages of cooling, the stream is attenuated into filaments by being pulled at very high speeds—usually ranging from 5,000 to 10,000 fpm—from the bushing. Resulting filaments have diameters of 0.00020 to 0.00075 in., depending on drawing speed, orifice size, molten glass temperature and other variables.

For efficient production, a number of filaments are pulled simultaneously from several orifices in the bushing. These filaments (usually numbering about 204) are collected into a bundle, called a "strand," at a gathering device where a "size" is applied to the filament surfaces. The strand is then wound into a forming package called a "cake." From this cake, shippable forms of fibrous glass are produced.

Thermal insulation

In general, fibrous glass insulation is available in densities ranging from 0.5 to 12 lb per cu ft. Maximum operating temperature is about 600 to 2000 F, depending on type of glass. Thermal conductivity values range from about 0.23 to 0.30 Btu/hr/sq ft/°F/in. It provides high sound absorption, relatively high tensile strength, and resistance to moisture, fire, rotting, fungi and bacteria growth. It is available in either flexible or rigid form.

The excellent insulating properties of fibrous glass are due to the large pockets of air between the fibers. These air pockets take up considerable volume. At a given temperature, thermal conductivity is a function of fiber diameter and orientation, density and the type and amount of bonding agent, if used. Bonding agents consisting of phenolic or high-temperature resins are usually used with the semirigid or rigid types of insulation.

Fibrous glass is not affected by low temperatures and has been used satisfactorily at temperatures as low as −350 F. Heat resistance depends on type of glass: borosilicate glass is generally limited to operating temperatures of 600 to 1000 F; high silica glasses are capable of operating at 1830 F; silica (quartz) fibers are usable up to 2000 F. The heat resistance of bonded insulations is normally limited by the heat resistance of the binder (maximum of about 450-600 F).

Forms available. Types of glass insulations

used include: unbonded, bonded, blanket, block and preformed insulations.

Unbonded insulations can be cut and shaped to insulate curved or irregularly shaped surfaces. They are frequently used in heating appliances because of their higher heat resistance than bonded insulations.

Bonded glass wool insulations provide greater rigidity; temperature of use is generally limited to 450 to 600 F maximum.

Blanket insulations are frequently faced on one or both sides with a metal mesh, for use on equipment such as boilers, cylinders, piping and ovens. They may consist of either bonded or unbonded insulations.

Block insulations are fabricated from high-density bonded wools, and are sometimes used instead of blankets for applications below about 450 F.

Preformed insulations consist primarily of bonded glass fibers, preformed to the shape of piping.

Glass-fiber reinforcements

Fibrous-glass reinforcements for plastics are usually of E glass type, although special types are used for special applications. Some of the special uses include: (1) heat resistance where the high silica leached glass provides superior heat resistance, but somewhat lower strength, and the silica or quartz fibers provide the optimum in heat resistance (among glasses); (2) high strength or modulus where special high modulus glass fibers and high tensile-strength glass fibers have been developed (primarily for filament winding).

Forms available. Following are the various forms in which fibrous glass is used in reinforced plastics:

1. Rovings consist of a number of strands (usually 60) gathered together from cake packages and wound on a tube to form a cylindrical package. Rovings have very little or no twist. They are used either to provide completely unidirectional strength characteristics, such as in filament winding, or are chopped into predetermined lengths for preform matched metal or spray molding.

2. Chopped strand consists of strands which have been cut into short lengths (usually ½ to 2 in.) in a manner similar to chopped roving, for use in preform matched metal or spray

molding, or to make molding compounds. It is the least expensive form of fibrous-glass reinforcement.

3. Milled fibers are produced from continuous strands which are hammer-milled into small modules of filamented glass (nominal lengths of $\frac{1}{32}$ to $\frac{1}{4}$ in.). Largely used for filler reinforcement in casting resins and in resin adhesives, they provide greater body and dimensional stability.

4. Yarns are twisted from either filaments or staple fibers on standard textile equipment. Although primarily an intermediate form from which woven fabrics are made, yarns are used for making rod stock, and for some very high-strength, unidirectionally reinforced shapes. A common form in which yarn is available is the "warp beam" where many parallel yarns are wrapped on a mandrel.

5. Nonwoven mats are available both as reinforcing mats and as surfacing or overlay mats. Reinforcing mats are made of either chopped strands or swirled continuous strands laid down in a random pattern. Strands are held together by resinous binders. In laminates, mats provide relatively low strength levels, but strengths are isotropic.

Surfacing or overlay mats are both thin mats of staple monofilaments. They provide practically no reinforcing, but serve to stabilize the surface resin coat, providing better appearance.

6. Woven fabrics and rovings provide the highest strength characteristics to reinforced plastics laminates (except for filament wound structures), although strengths are orthotropic. A wide variety of fabrics and weaves are available both in woven yarns and woven rovings. Probably the most common types used are plain, basket, crowfoot satin, long shaft satin, unidirectional, and leno weaves.

Fabric finishes. When glass fibers are first produced, a "size" is applied to the fiber to protect it from abrasion and moisture, and to serve as a lubricant. After the fibers are woven into a fabric, the size is removed because it prevents good coupling action with most resins. When used with silicone resins, glass fabrics need only size removal; in laminating with other resins, a finish or coupling agent is applied to improve the bond between resin and glass.

Probably the most common finish for glass

cloth is "Volan A," a chrome finish which provides good bonds with either polyester or epoxy resins, and reasonably good bonds with phenolics. Another commonly used family of finishes is the silanes, which are generally used to improve wet strength of the laminate. Among commonly used silanes are "Garan," "NOL 24," and "A-1100."

In most cases the end user need not concern himself with finish, as it is ordinarily applied by the glass textile supplier.

FILAMENT WOUND REINFORCED PLASTICS

The true fiber-glass filament-wound structure may be more appropriately termed a resin-bonded filament-wound structure as it comprises approximately 80% glass fibers by weight and 20% bonding resin. Fibers are generally oriented to resist the principal stresses and the resin protects while secondarily supporting the fiber system. Filament winding is well adapted to the fabrication of internal pressure vessels; it has also performed well under external pressure and can be designed to function efficiently as a column or beam. This structure has the tensile strength of moderately heat-treated alloy steel and one-quarter of the weight.

The winding process

Bands of parallel glass filaments (usually in the form of roving) are wound over a mandrel following a precise pattern in such a way that subsequent bands lie adjacent, progressively covering the mandrel in successive layers, thus generating a shell structure. Liquid resin is simultaneously applied, generally by passing the filament band through a bath of catalyzed resin.

Tension generates a running load between the curved work surface and filament band which forces out air and excess resin and allows each successive layer to ultimately rest on solid material while the remaining interstices are filled with resin. Precision of filament placement plus tension and viscosity control are primary controlling factors in the attainment of high fiber content which is generally desired for high strength.

Preimpregnation of the fiber strands is occasionally used as a means of applying the resin binder. Such prepregs must fuse on contact with the work in order to accomplish a bond so the fundamental relationships remain the same. The fiber bands are parallel strands only, since a cross weave or other structural filler would not bear primary loads and would preclude the true maintenance of equal tension on fibers in the band or roving when winding over crowned surfaces.

Tension serves only the purpose of accomplishing high fiber content and cannot be considered as accomplishing any prestressing. This is primarily true as the dry strength of glass fibers, as wound, is only about one-quarter of the ultimate resin consolidated strength. Also unless some structural component is to remain within the wound shell (rather than a removable mandrel), there is no member against which a prestress can be maintained.

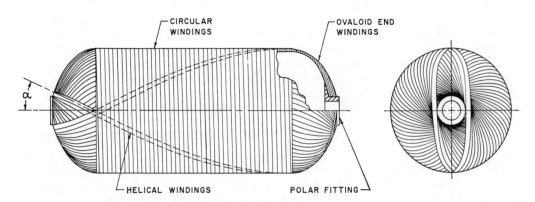

FIG. 1. Filament-wound pressure cylinder.

Structural behavior

These structures have highly differentiated physical properties as the glass fiber is roughly thirty times stronger and fifteen times stiffer than the consolidating resin. A filament-wound ring designed for girth strength only, may have an ultimate tensile strength in this direction of 250,000 psi and a dependable interlaminar shear strength of only 1000 psi. In spite of such considerable differences, the well-informed designer will be able to create well-proportioned high performance structures.

Products fall into three general categories: (1) balanced pressure vessels, (2) tubular structural members in which the stresses are not in balance, and (3) postmolded shell structures which may be formed into a variety of shapes following combination of the fiber and resin by winding. In the last case, the uncured sheet is removed from a cylindrical mandrel, draped over a form in full thickness to the desired contour and cured, generally in open heat.

Balanced structures. The balanced structure is so called because the netting system of its fibers are so arranged that all forces due to internal pressure are balanced by uniform tension in all fibers, and there are no other stresses. This statement can only be strictly true if certain constraints are assumed, such as zero stiffness of the resin and a very large cylinder radius-to-wall-thickness ratio. The filament winding process also requires that all filament paths will closely approximate geodesic paths to avoid lateral slippage in the liquid resin. The geodesic line on any surface is the shortest distance between two points. A straight line on a plane and a great circle on a sphere are examples.

The closed-end cylinder is one clear example of a balanced netting structure. While the cylinder and the end are two distinct netting systems, they may be integrally fabricated and can be examined together.

Fig 1 illustrates a typical filament-wound pressure cylinder. It consists of a system of low helix-angle windings carrying the longitudinal forces in the cylinder shell and forming integral end closures that retain their own polar fittings. Circular or circumferential windings are also applied to the cylindrical portion of this vessel, yielding a balanced netting system. Such a netting arrangement is said to be balanced when the membrane generated contains the appropriate combination of filament orientations to balance exactly the combination of loadings imposed. In the case of the cylindrical shell, the girth load is two times the axial load, and the helical system is so designed that its longitudinal strength is exactly equal to the task. Such a low-angle helical system possesses, in combination with high longitudinal strength, a limited girth strength. Therefore, the circular windings are designed to carry the balance of the girth load.

It has been found in practice that the helical windings, because of shorter radius of curvature and imperfect packing over the end shapes, must operate at somewhat lower fiber stress than the circular windings. In this respect, a practical balanced structure differs somewhat from a theoretical one in which all fibers would operate at identical stresses.

The end dome contains no circular windings, as the profile is designated to accommodate the netting system generated by the terminal windings of the helical pattern. The end dome shape is termed an ovaloid; specifically, it is that surface of revolution whose geometry is such that it is amenable to winding along geodesic paths and simultaneously fiber stress is uniform throughout. A hemisphere satisfies the first requirement but not the second.

The ovaloid netting system is the natural result of the reversal of helical windings over the end of the vessel. The windings become thicker as they converge near the polar fittings, and it therefore follows that when resisting internal pressure by constant filament tension only, the radius of curvature must increase in this region. It can also be reasoned that the radius of curvature at the cylinder junction will be equal to one-half the cylinder radius when the helix angle $\alpha = 0$, and equal to the cylinder radius when $\alpha = 45°$. The profile will also be affected by the presence of an external axial force. Derivation of this profile was initially accomplished by an analog computing device and has now been done analytically.

The netting analysis is not strictly precise for real articles but the errors are generally decidedly secondary and the method is superior to that in which the material is regarded as homogeneous and properties are derived by test.

A brief description of the method is given for clarification.

Considering the helical winding system in Fig 1, it may be seen that the continuous overlay of right and left helices generates a multiple diamond pattern. Now, if the end load due to internal pressure is to be resolved into the direction of the helical filaments, it is necessary to divide the running load (in lb per in. across an imaginary girth cut) by the cosine of α.

It is also true that the number of helical fibers which can cross a girth cut (thickness remaining constant) varies directly as cosine α. Diminishing the number of fibers again increases stress, and it follows that the longitudinal strength of a helically wound cylinder is proporticnal to the cosine squared of α. The same relation can be shown to be true for girth strength and the sine squared of α. It is to be remembered that the loading of the helical system in such a design as illustrated in Fig 1 all derives from end load and the girth load carried by the helical system is independent of the stress in that system. This is because girth deflection is restrained by circular windings.

In the following example, let

the cylinder radius, $R = 3$ in.
the helix angle, $\alpha = 20°$
the internal pressure, $P = 2,000$ psi
the unidirectional or band
 strength of structure
 (glass and resin) in the
 helical system $= 150,000$ psi
and in the circular system $= 180,000$ psi

Now end load $= \dfrac{PR}{2} = \dfrac{2,000 \times 3}{2} = 3,000\,\text{lb/in.}$

The longitudinal strength of the helical system $= 150,000 \times \text{COS}^2\alpha = 150,000 \times 0.94^2 = 132,500$ psi and the thickness of helicals $=$

$$\frac{3,000}{132,500} = .0226 \text{ in.}$$

The girth load carried by the helical system $=$

$3,000 \times \dfrac{\text{SIN}^2\alpha}{\text{COS}^2\alpha} = 3,000 \times \dfrac{0.342^2}{0.94^2} = 398$ lb/in.

Now the total girth load $= PR = 3 \times 2,000 = 6,000$ lb/in. and the net girth load to be carried by circular windings $= 6,000 - 398 = 5,602$ lb/in.

and the thickness of circular windings $=$

$$\frac{5602}{180,000} = 0.0311 \text{ in.}$$

Now the total cylinder wall $= 0.0311 + 0.0226 = 0.0537$ in. and the composite wall stress at

$$2000 \text{ psi} = \frac{2000 \times 3}{0.0537} = 113,000 \text{ psi.}$$

The ovaloid end closure is designed when the profile is selected from tabulated data now available but not reproduced here. Helix angle is the only required parameter. The thickness of helical windings established in the example is correct for the ovaloid.

Based on currently available materials and methods, the cylinder in the example would sustain twenty to fifty cycles at 2000 psi before failure and has a strength-to-density quotient of

$$\frac{113,000}{0.072} \text{ or } 1,550,000 \text{ in.}$$

Unbalanced structures. Following the principles outlined above, a netting analysis can be made of a cylindrical tube in bending, compression or tension. The primary difference is that, although axial load generates a resulting girth load as in the pressure vessel, the circular windings are designed to balance this load only. The lower the helix angle, the fewer girth windings required. A zero helix angle is not the ultimate solution, however, as a system of parallel fibers alone has low shear strength and tends to split.

Size range

Large or small structures are easily fabricated by adhering to the basic principles. Winding precision is important as is the relation of filament tension, resin viscosity and radius of curvature of the filament path on the work surface. Tanks twelve feet in diameter are presently in production and structural tubes thirty-three feet long have been produced. Small tubes have been made down to one-quarter inch in diameter, and there appear to be no fixed size limitations in either direction. Wall thicknesses may be several inches or more as the normal bonding resins contain no volatile components and the glass content is so high that there is little danger of the exothermic heat becoming excessive.

Dimensional accuracy

Dimensional control in winding depends upon

mandrel accuracy as well as both material and process control. The bands of filaments are generally 0.005 in. thick while a full layer requires coverage by both right and left hand helices making the layer thickness 0.010 in. The fiber deposition is recorded by a counter, and the band width and thickness are established in the winding machine setup. Glass-fiber thickness is subject to some variation, and resin content variation will also affect thickness. Wall thickness can generally be held to plus or minus 5%. Length and diameter are easily held to one-tenth of 1% as there is little resin shrinkage or wound-in strain due to winding tension.

Machining may be accomplished by carbide tools or grinding techniques. Tolerances can be held as closely as in metals. The inner surface as-wound against a good steel mandrel can have a finish of approximately thirty microinches, and normal machining or grinding will produce a forty to sixty microinch finish. Cutting of surface fibers in machining does not weaken the structure.

Component materials

There are three primary materials in this composite: the glass fiber, fiber finish and the bonding resin. Glass fibers are continuous and each "end" contains 204 monofilaments approximately 0.0003 in. in diameter. These ends are plied together without twist to form a strand of "roving." Twelve-end, equally tensioned, roving is generally used for the winding of high performance structures.

Metallic and organic fibers have also been used in filament winding but the lower strength-to-weight ratio of both and the low elastic modulus of organic fibers relegate their use to special applications.

The fiber finish generally includes compounds having a chemical affinity for the glass surface and the bonding resin. These are called coupling agents. Other functional components are lubricants and "film formers" for the generation of strand integrity. Both improve handling properties but do not contribute to the performance of filament structures.

The resin binder materials are generally liquid at room temperatures or are wound hot for liquid integration of the system. Best results are obtained with strong tough resins such as the epoxies. Polyesters have also been extensively

used (primarily for radomes where electrical properties are critical) and any bonding resin should be free of polymerization products of a volatile nature which would have to escape through the cured structure.

Chemical resistance and electrical properties of the constituents are usually critical in selecting the proper resin, as applications of the filament-wound structure are found in both the electrical and chemical industries.

Production considerations

Winding is an economical and efficient process by which structural fibers and bonding resins may be combined. The machinery is automatic in that it places the fibers precisely, and resin content is easily controlled. Production rate depends upon product size and may vary from one lb to one thousand lb per hr.

Production winding machines are generally designed for a limited range of products and, if quantities are large, small articles can be produced economically.

FLAME-SPRAYED COATINGS
Process and equipment

Flame-spraying methods are used to produce coatings of a wide variety of heat-fusible materials, including metals, metallic compounds, and ceramics (see Tables 1 and 2). There are three basic flame-spray processes. Following is a brief description of the equipment used in carrying out these processes:

Wire type guns are used to produce coatings of metals, alloys, and in some cases, ceramics. With this type of equipment wire or rod from $\frac{1}{32}$ to $\frac{3}{16}$ in. in dia is fed axially through the center of a fuel gas/oxygen flame at a controlled rate. The flame is surrounded by an annular blast of air, which imparts high velocity to the burning gases and provides the kinetic energy needed to atomize the metal as it melts. Acetylene is the most commonly used fuel gas, although propane, natural gas, and hydrogen are also used.

Power for feeding the wire through the gun may be supplied by an electric motor, usually electronically controlled, or by an air motor provided with a speed control. Wire feed rate varies from a few inches per minute to over 20 fpm, depending on the melting point of the metal

TABLE 1—PROPERTIES AND USES OF FLAME-SPRAYED METALS

Metal	Density	Rockwell Hardness	Tensile Str, psi	Uses
Aluminum	2.41	H 72	19,500	Corrosion resistance (salt, atmospheric, and high temp)
Babbitt A	6.67	H 58	—	Bearing linings
M Bronze	7.47	B 50	26,500	Bearing surfaces, bushings
A Bronze	7.06	B 78	29,000	Bearing surfaces, bushings
Copper	7.54	B 32	—	Electrical and heat conductivity
Molybdenum	8.86	C 38	—	Bond, wear resistance
Monel	7.67	B 39	—	Wear and corrosion resistance
Nickel	7.55	B 49	—	Corrosion resistance
Stainless (300 series)	6.93	B 78	30,000	Wear and corrosion resistance
Stainless (400 series)	6.74	C 29	40,000	Wear and abrasion resistance
Steel, 1010	6.67	B 89	30,000	Wear resistance
Steel, 1080	6.36	C 36	27,500	Wear and abrasion resistance
Tin	6.43	H 10	—	Corrosion resistance
Zinc	6.36	H 46	—	Corrosion resistance

and the diameter of the wire or rod. Precise control of the wire speed is essential to proper atomization.

Guns of this type vary from two or three pounds in weight to over twenty pounds. The smaller types are hand held, and the heavy duty types are usually machine mounted. Capacity of wire guns ranges from one or two pounds per hour for fine wires of high-melting metals to over one hundred pounds per hour for low-melting alloys such as lead, solder or babbitt. Steels can be sprayed at speeds of twenty or more pounds per hour.

Wire type guns are the most widely used, because of their versatility and ease of operation. Hand-held guns are used to coat large areas such as bridges, ship hulls and tanks. Machine-mounted guns are used principally for machine element applications, as in surfacing rolls or salvaging journals by building up worn areas. In many production applications one or more electronically controlled guns are operated and cycled automatically by a central console.

Powder type guns are used with a variety of materials which cannot be readily produced in the form of wire or rod. They are also used for spraying low-melting metals which are readily available in powder form.

With the powder gun, metal or ceramic powders are fed axially through a fuel gas/oxygen flame at a controlled rate. The powders are entrained in a carrier gas which may be air, acetylene or oxygen. Since the powders are finely divided and dispersed as they enter the flame, further atomization is not required and annular air blast is not needed. However, for some purposes an air blast is used to increase particle velocity and to avoid excessive temperature at the work surface.

The powder supply for guns of this type may be a reservoir connected to the gun by a powder tube or hose, or it may be a cannister attached directly to the gun. In either case, the powder is fed into the carrier gas stream at a controlled rate, and is kept in suspension until it enters the flame. As the carrier gas diffuses in the flame zone, the particles are melted or heat-softened and projected against the work surface.

Powder guns are used for flame spraying a wide variety of ceramics, and for coating with "self-fluxing" alloys which are subsequently fused to the base material. They have also been used for many years to apply zinc and aluminum coatings for corrosion prevention.

Plasma guns are the most recent development in the flame-spraying field. The plasma flame is produced by passing suitable gases through a confined arc, where dissociation and ionization occur. The ionized gases form a conductive path within a water-cooled nozzle, so

TABLE 2—FLAME-SPRAYED INORGANIC COATING MATERIALS

Material	Method of Application [a]	Melting or Softening Temp, F	Uses
Alumina	P, O	3700	Heat and abrasion resistance, flame-sprayed shapes
Barium Titanate	P, O	3010	Dielectric properties Superior to Al_2O_3
Calcium Zirconate	P, O	4250	Thermal barriers, abrasion resistance, molten metal resistance
Ceric Oxide	P, O	4712	Thermal barriers, combustion catalyst
Chromium Carbide	P, O	3435	Wear resistance (unfused) or as ingredient in cermets
Columbium (Niobium)	P	4379	Hard (over Rc 60) coatings Self-bonding
Columbium (Niobium) Carbide	P	6330	Hard, wear-resistant coatings
Hafnium Oxide	P, O	5030	Hard, dense, high barns number—thermal barrier
Magnesia Alumina Spinel	P, O	3875	Hard, dense abrasion-resistant coatings. Resists attack by molten glass
Magnesium Zirconate	P, O	3830	Thermal barriers, molten metal resistance, coating graphite
Misc. Glasses	P, O	800–2000	Dielectrics, seals
Molybdenum	P	4760	Bonding, wear resistance
Nickel	P, O	2650	Corrosion resistance (unfused, sealed) oxygen barrier, cermet mixtures
Rare-Earth Oxide	P, O	4000	Thermal barriers, flame catalyst
Tantalum	P	5425	Catalytic uses Self-bonding
Tantalum Carbide	P	7020–8730	Hard, wear-resistant coatings
Titanium Oxide	P, O	3490	Abrasion resistance where low porosity is needed, and for blends with other refractories
Tungsten	P	6170	High-temp abrasion resistance
Tungsten-Carbide Crystalline	P	5035	High-temp abrasion resistance
Tungsten Carbide + Cobalt	P, O	2715	Unfused-wear resistance. Fused-blended with Ni and Co Alloys for abrasion resistance
Zirconium Oxide	P, O	4700	Thermal barriers
Zirconium Oxide (Hafnium-free)	P, O	4700	Hard, low barns number—thermal barrier
Zirconium Silicate	P, O	4390	Thermal barriers

[a] P—Plasma flame equipment, O—Oxyacetylene equipment.

that an arc of considerable length is maintained. The gases most commonly used are nitrogen, hydogen and argon.

Temperature of the plasma flame depends on the type and volume of gas used, the size of the nozzle, and the amount of current used. For flame-spraying purposes temperature ranges of 10,000 to 15,000 F are generally employed, though much higher temperatures may be attained if desired. Plasma flame-spray guns usu-ally operate at 20 to 40 kw, using 100 to 300 cfh of gas.

In addition to their high-temperature capabilities, plasma guns have other advantages. Extremely high velocities are possible, and favorable environmental conditions can be obtained by proper selection of gases. Spraying within a controlled atmosphere chamber permits the production of oxide-free coatings.

With the plasma gun, metal or ceramic powders are fed into the flame at a point downstream from the actual arc path. Current, gas flow, and powder flow must be adjusted for different coating materials, many of which would otherwise be completely vaporized.

Surface preparation

Regardless of the flame-spray method used or the type of coating material being applied, some sort of surface preparation is usually required. Bond to the base is often largely mechanical, and in general the greater the degree of surface roughness the better the bond. Thin coatings require less elaborate preparation than thick coatings, and some coating materials require much more thorough preparation than other materials.

Bonding methods used include abrasive blasting, rough threading, molybdenum bonding coats, heating, or combinations of these steps. Abrasive blasting is probably the most widely used, and the abrasive may run from G-16 angular steel grit for maximum bond, to 100 mesh (or finer) alumina for bonding very thin coating.

Coating characteristics

Studies of flame-sprayed deposits show that several mechanisms are responsible for the cohesion of particles and adhesion to the base. Mechanical interlock, oxide cementation, molecular attraction, and microscopically small areas of fusion are all involved, and the significance of each of these factors varies with the type of coating material. With metallic coatings produced with wire or powder type guns, it has been observed that metals and alloys (e.g., stainless steels or aluminum bronzes) which develop dense, tenacious oxide films provide sound coatings having relatively good physical properties. On the other hand, metals (e.g., coppers and brasses) that develop soft, weakly adherent oxides produce coatings which show rather poor physical properties.

Flame-sprayed coatings have certain characteristics in common, regardless of coating material or type of equipment used. Careful consideration must be given to these characteristics if these coatings are to be used advantageously:

Porosity. All flame-sprayed deposits are porous to a greater or lesser degree. Although porosity can be controlled to some extent by variations in spray technique, it cannot be entirely eliminated except by subsequent treatment of the coating. Porosity is an advantage in some applications, and a disadvantage in others.

Sprayed metal wearing surfaces can have outstandingly good wearing properties, particularly when lubricant is retained in the pores. Ceramics such as alumina and zirconia are more effective heat barriers and have better thermal shock resistance if applied in a manner designed to provide high porosity. Where ceramic coatings are used for abrasion resistance the reverse is true, and application technique is adjusted to produce hard, dense deposits. Sprayed zinc and aluminum coatings used for corrosion prevention protect steel perfectly even though the pores are not sealed, since these metals are anodic to steel and provide electrolytic protection. On the other hand, coatings of monel, nickel, stainless steel, bronze or other metals which are cathodic to steel must be thoroughly sealed if used over steel in a corrosive environment. Otherwise traces of moisture penetrating through to the interface will initiate rapid electrolytic attack on the steel substrate, with complete loss of adhesion.

Tensile stress. Nearly all sprayed metals and ceramics tend to shrink as they are applied. The amount of shrink or contraction varies greatly with different coating materials. With any given material, tensile stress in the coating increases with increased thickness. If the geometry of the work to be coated is such that the stress in the coating will be imposed as a shear or tensile stress on the bond, then steps must be taken to prevent loss of adhesion. Such steps may include preheating, additional roughening or keying, the use of special bonding coats, or various combinations of these steps.

Tensile strength. The tensile strength of flame-sprayed deposits varies with the material and the mode of application, and is further influenced by the magnitude of the locked-in stresses which may exist. These stresses, in turn, depend on the geometry and thickness of the deposit. Since tensile strength is a function of the cohesion between particles, and since erosion resistance requires that surface particles be firmly adherent, it is obvious that some degree of tensile strength is important. However, flame-sprayed coatings are not ordinarily used for purposes where tensile strength *per se* is critical.

Ductility. Nearly all flame-sprayed deposits

lack ductility. In most cases ultimate elongation is less than 1%. For this reason, heavy coatings should not be used on parts that are flexed or distorted in service or on raw material which will be bent or deformed in subsequent processing steps. Although flame-sprayed coatings may contribute to the stiffness or rigidity of thin sheet or strips, they should not be depended upon to add strength to machine parts.

Hardness. Hardness is frequently of primary interest in dealing with materials intended to resist erosion and abrasion. In the case of flame-sprayed deposits, both the over-all or macrohardness and the particle hardness or microhardness must be considered. Conventional hardness tests such as the Rockwell or Brinell tests are based on a relatively large-scale penetration of the material under test. With porous materials, the collapse of pores under and adjacent to the penetrator results in lower readings than with dense materials. Microhardness tests, on the other hand, measure the actual hardness of the particles which make up the structure. Thus, for example, flame-sprayed 1080 steel applied with a wire type gun will show a macrohardness of Rockwell C40, but a microhardness equivalent to Rockwell C 65-67. This latter figure is the true particle hardness, and 1080 steel quenched rapidly from above the critical temperature will show this hardness. The type of erosion or abrasion to which a flame-sprayed coating will be subjected will determine whether macrohardness or microhardness is the more important.

Finishing methods

Flame-sprayed deposits may be finished by machining or by grinding, depending on the hardness of the particular coating material. Sintered carbide tools are generally used for machining since even the softer materials contain some amount of abrasive oxides which may cause rapid wear of tool steel. For those materials which must be finished by grinding, specific wheel recommendations are available from flame-spray equipment manufacturers.

Flame-sprayed coatings may require sealing, depending on the type of coating and service requirements. A wide variety of impregnants are used to reduce porosity, enhance physical properties, or to improve friction characteristics. Equipment manufacturers should be consulted for specific sealing recommendations.

FLUIDIZED BED COATINGS

The fluidized bed process is used to apply organic coatings to parts by first preheating the parts and then immersing them in a tank of finely divided plastic powders which are held in a suspended state by a rising current of air. In this suspended state the powders behave and feel like a fluid. The method produces an extremely uniform, high quality, fusion bond coating that offers many technical and economic advantages.

Fusion bond (the basic process is covered in U.S. patent 2,844,489) coatings are generally applied to metal parts, although other substrates have been successfully coated. The major fields of application are electrical, chemical, and mechanical equipment as well as household appliances. The process is now being used in every major industrial country in the world for applying many different plastics to a variety of parts.

How coatings are applied

Objects to be coated are preheated in an oven to a temperature above the melting point of the plastic coating material. The preheat temperature depends upon the type of plastic used, the thickness of the coating to be applied, and the mass of the article to be coated.

After preheating, the parts are immersed with suitable motion in the fluidized bed. Air used to fluidize the plastic powders enters the tank through a specially designed porous plate located at the bottom of the unit.

When the powder particles contact the heated part, they fuse and adhere to the surface, forming a continuous and extremely uniform coating. In many cases, the part is postheated to coalesce the coating completely and improve its appearance. Thickness of the coating depends on the temperature of the part surface and how long it is immersed in the fluidized bed.

Cellulosic. Cellulosic fluidized bed coatings are noted for their all-around combination of properties and are especially popular for decorative-protective applications. They combine good impact and abrasion resistance with outstanding electrical insulation. The coatings have excellent weathering properties, salt spray resistance, high gloss, and can be made in an almost unlimited range of colors. They can be solvent-etched to

TABLE 1—PROPERTIES OF FLUIDIZED BED COATINGS [a]

Type	Cellulosic	Vinyl	Epoxy	Nylon	Poly-ethylene	Chlorinated Polyether
CHEMICAL RESISTANCE						
Exterior Durability	E	E	F	F	F	F
Salt Spray	E	E	VG	G	E	E
Water (salt, fresh)	VG	E	G	F	VG	E
Solvents						
Alcohols	F	E	E	G	E	E
Gasoline	G	E	E	E	VG	VG
Hydrocarbons	G	G	E	E	VG	E
Esters, Ketones	P	P	F	G	G	VG
Chlorinated	P	P	E	E	F	P
Salts	VG	E	E	VG	E	E
Ammonia	P	E	P	G	E	E
Alkalies	F	E	VG	G	VG	E
Mineral Acids						
Dilute [b]	G	E	E	F	E	E
Conc [c]	P	G	G	P	VG	E
Oxidizing Acids						
Dilute [b]	P	E	G	P	VG	E
Conc [c]	P	G	P	P	P	G
Organic Acids						
Acetic, Formic, etc.	P	F	F	P	VG	E
Oleic, Stearic, etc.	F	E	E	VG	VG	E
MECHANICAL AND PHYSICAL PROPERTIES						
Abrasion Resis	VG	G	VG	E	F	VG
Flex	G	E	F [e]	G	E	F
Impact Resis	E	E	G	VG	F	G
Max Svc Temp, F	180	200+	350+	180 [d]	160	250
Dielectric Str	VG	VG	E	G	E	VG
DECORATIVE PROPERTIES						
Color Range	E	E	P	F	G	P
Color Retention	E	VG	F	VG	VG	G
Initial Gloss	E	VG	G	G	VG	G
Gloss Retention	E	G	P	—	—	—

[a] These data are intended only as a preliminary selection guide. Final selection should be made after consulting with coating formulator and after suitable testing. Data are based on Corvel® fusion bond coating materials as supplied by Polymer Corp. Key: E = excellent, VG = very good, G = good, F = fair, P = poor.

[b] Dilute = 10%.   [c] Con = over 30%.

[d] Up to 300 F in nonoxidizing environment.

[e] Ranges from good to poor depending upon composition.

provide a satin finish and heat-embossed for additional decorative effects.

The economy of cellulosic coatings combined with their excellent appearance and durability are particularly useful for such applications as indoor and outdoor furniture, kitchen fixtures, home and marine hardware, metal stampings, fan guards and sporting goods. Major uses of cellulosic fusion bond finishes are coated transformer tanks and covers, reclosure tanks and covers, outdoor electrical equipment housings and many pole line hardware parts.

Vinyl. Vinyl fluidized bed coatings have a good combination of chemical resistance, decorative appeal, flexibility and toughness, and low-frequency insulating properties. They have excellent salt-spray resistance, outstanding outdoor weathering characteristics and can be used for general-purpose electrical insulation. The vinyl fusion bond coatings are claimed to have better uniformity and edge coverage than plasticol coatings.

Fusion bond vinyl coatings have been especially successful on wire goods for applications such as dishwasher racks, washing-machine parts and refrigerator shelves. They are also being used on wire furniture and hardware, and in industrial applications such as bus bars, pump impellers, transformer tanks and covers, auto battery brackets, conveyor rollers and other material-handling equipment. Cast iron, die castings and expanded metal parts can be readily coated.

Epoxy. Both thermoplastic and thermosetting materials can be applied. Epoxy coatings have a smooth, hard surface and exceptionally good electrical-insulation properties over a wide temperature range. They are available in rigid and semiflexible variations with different combinations of electrical, physical and chemical properties. Epoxy coatings on electrical motor laminations provide good dielectric strength and uniform coverage over sharp edges. When properly applied, epoxy coatings have good impact resistance and do not sacrifice toughness for surface hardness. Other electrical applications include torroidal cores, wound coils, encapsulated printed circuit boards, bus bars, watt hour meter coils, resistors, and capacitors.

Nylon. The combination of properties which have made the polyamide plastics unique for molded and extruded parts are obtainable in coatings. Until a few years ago, nylon could not be applied by any practical process, including solution coating techniques, but now it can be applied by the fluidized bed process and offers many advantages to the design engineer.

Nylon fusion bond finishes combine a decorative, smooth and glossy appearance with low surface friction and excellent bearing and wear properties. They minimize scratching and cut down undesirable noise. The frictional heat developed on nylon is dissipated more rapidly when used as a coating over a conductive metal surface than when used as a solid plastics member. Coated metal parts offer increased dimensional stability. Due to the unique properties of nylon coated metal parts, many users of the process are able to reduce the number of metal component parts at substantial savings.

By an additional immersion of the heated and coated part into a fluid bed of whirling molybdenum sulfide or graphite, an impregnated nylon surface can be produced with unusual bearing, frictional and wear characteristics.

Nylon fusion bond finishes are effectively used in machine shop fixtures, modern indoor furniture to simulate a wrought iron finish, aircraft instrument panels, ball-joint suspensions, collars, guards, and slide valves used in textile and farm equipment, knitting machine parts, switch box cover panels, tractor control handles, and radar and calculator component parts.

Polyethylene. Polyethylene coatings combine low water absorption and excellent chemical resistance with good electrical insulation properties. They can be aplied successfully in thicknesses of 10 to 60 mils by the fluidized bed process. The primary uses for polyethylene coatings are for protecting chemical processing equipment and on food-handling equipment. Typical chemical applications include pipe and fittings, pump and motor housings, valves, battery holddowns, fans and electroplating jigs.

Chlorinated polyether. This relatively new fluidized bed coating is attaining rapid recognition due to an excellent combination of mechanical, chemical, thermal and electrical properties. It has good resistance to wear and abrasion. It provides good electrical insulation even under high humidity and high temperature conditions, and has very low moisture absorption. It can be used continuously at 250 F and even up to 300 F.

Chlorinated polyether coatings have excellent

chemical resistance and are widely used to coat equipment for the chemical industry such as valves, pipe and pipe fittings, pump housings and impellers, electroplating jigs and fixtures, cams and bushings. However, chlorinated poly-ether coatings should not be used in contact with some chlorinated organic solvents, or with fuming sulfuric and nitric acids.

FLUOROCARBON PLASTICS, CFE

Chlorotrifluoroethylene (CFE) plastic, as the chemical name implies, is made up of chlorine and fluorine atoms attached to a carbon-carbon skeleton to form a true thermoplastic resin. The properties of CFE plastic are, in effect, determined by the combined characteristics of these elements.

General nature

Chlorine contributes a powerful cohesive force between the polymer chains which results in low cold flow and resistance to cut-through not present in completely fluorinated resins. The presence of chlorine also results in a very dense plastic which is more impermeable to gaseous matter. Typical of all fluorocarbon resins, CFE has good electrical properties with high dielectric strengths, low dielectric constants and high volume and surface resistivities.

The presence of fluorine in the polymer contributes excellent chemical resistance, high temperature stability, and zero moisture absorption.

CFE fluorocarbon plastics are available in basic forms generally adapted to thermoplastic processing methods. These forms include pelletized and unpelletized resins, dispersions, copolymerized resins and some plasticized resins for special applications. The basic materials are then processed by extrusion, injection molding, transfer molding, compression molding and spray or dip coating into a variety of useful products. Standard sheet, rod, tubing and film can be formed or machined into any desired shape. Coil forms and other electronic components made of CFE plastic have achieved wide acceptance due to zero moisture absorption and excellent electrical properties.

Cryogenic gasketing and valve seals rely heavily on the low temperature flexibility and compressive strength properties of the material. CFE film and thin sheets work well as infrared and visible light-transmission media while low permeability to most "hard-to-contain" chemicals is responsible for many other uses.

Although the glass transition temperature of CFE plastic is approximately 137 F (which indicates a gradual softening of the plastic from 137 F up to the 414 F melt temperature), use temperatures of this material vary from 400 F to −400 F. Liquid hydrogen seals fabricated from CFE, for example, fill the requirement of flexibility at −420 F while electrical insulating components have proven successful at 400 F.

Some typical properties of CFE plastic are shown in Table 1 and Table 2.

Crystallinity

CFE fluorocarbon plastic may well be described as crystallizable but never completely crystalline. The degree and kind of crystallinity in a given sample is a function of its thermal history. It is this variation in crystallinity which gives rise to so many of the interesting properties of the plastic. It is common to speak of the "quick-quenched" resin as "amorphous" and of the "slow-cooled" resin as "crystalline." At best, these terms are relative since only fairly thin samples can be quenched rapidly and completely enough to inhibit crystal growth. In all cases some crystal nuclei develop even when molded or extruded parts are cooled rapidly. These nuclei may later grow into large, spherulitic domains if the part is subjected to prolonged aging at 300 to 390 F.

At higher degrees of crystallinity, CFE is a denser, less transparent material with higher tensile modulus, lower elongation, and greater resistance to the penetration of liquids and vapors. The "amorphous" plastic, on the other hand, is less dense, more elastic, optically clear and tough.

Molecular weight plays a vital role in the crystallization process. Long chain molecules are slow to form crystal nuclei and reluctant to rearrange themselves into large spherulites. Consequently, components made from high molecular weight CFE are strong and tough and therefore retain amorphic properties even after extended aging at high temperatures.

In contrast, parts produced from resin degraded to low molecular weight during processing are susceptible to excess crystallization and embrittlement. Molding conditions must be carefully controlled, therefore, to make sure that

TABLE 1—TYPICAL PROPERTIES OF CFE
FLUOROCARBON PLASTIC [a]

Properties (@77 F)	ASTM Test	Physical State of Plastic	
		Crystalline [b]	Amorphous
Sp Gr	D-792-50	2.1312	2.1047
Ten Str, psi	D-638-56T	4900	5070
Yld Str, psi	D-638-56T	5710	4650
Elongation, %	D-638-56T	105	175
Impact Str, ft-lb/in. notch	D-256-56	3.1	7.3
Compr Yld Str (0.2% offset), psi	D695-54	5440	—
Hardness (Shore D)	D-1706-59T	78	74
Shear Str, psi	D-732-46	5440	6010
Water Abs (21 days)	D-570-59T	0.00	0.00
Dielec Const			
10^3 cps	D150-54T	2.45	2.50
10^5 cps		2.40	2.34
Dissip Factor			
10^3 cps	D150-54T	0.0247	0.0271
10^5 cps		0.0143	0.0144
Arc Resis, sec	D459-58T	360	>360
Volume Resis, ohm-cm	D257-58	2.5×10^{16}	4×10^{16}
Surface Resis, ohm	D257-58	5×10^{15}	$>10^{18}$
Infared Transmission (6 mil film)	1 to 4 microns wavelength		90%
	4 to 7 microns wavelength		70%
Radiation Resis (Gamma)			
Ten Str Retained (16 Megarads), %		75	70%
Yld Str Retained, %		80	70%
Elong Retained, %		20	80%
Flame Resistance		Nonflammable	

[a] Minnesota Mining and Mfg. Co.'s Kel-F.
[b] Semi-Crystalline.

TABLE 2—CHEMICAL RESISTANCE OF CFE
FLUOROCARBON PLASTIC [a] (ASTM D543;
7 DAYS AT 77 F)

	Crystalline		Amorphous	
	% Weight Change	% Volume Change	% Weight Change	% Volume Change
Acetone	0.1	−1.1	0.3	0.2
Benzene	0.1	−1.2	0.3	−0.3
Ethanol, 95%	0.0	−0.8	0.0	−0.7
Methyl Ethyl Ketone	0.2	−0.5	0.4	0.1
Trichloroethylene	1.8	1.5	3.0	1.9
Ammonium Hydroxide, 10%	0.0	−0.7	0.0	−0.3
Ethyl Ether	4.4	7.6 [b]	5.8	11.2 [b]
Freon 113	0.9	0.6	0.9	0.4
Hydrochloric Acid, 10%	0.0	−0.5	0.0	−0.6
Sodium Hydroxide, 10%	0.0	−0.9	0.0	−0.6
Nitric Acid, 90% Fuming	0.0	−0.2	0.0	−0.4
Hydrofluoric Acid, Anhyd.	0.0	−0.1	0.0	−0.8
Nitrogen Tetroxide (5 C)	8.2	9.4 [b]	9.9	10.5 [b]
Unsym. Dimethyl Hydrazine	0.0	0.8	0.1	0.1

[a] Minnesota Mining and Mfg. Co.'s Kel-F.
[b] Becomes rubberlike.

finished parts retain to the fullest extent possible all the properties inherent in the resin. A simple test (ZST, zero strength time) has been developed to enable the molder to check molecular weight of finished parts during production (see ASTM D1430-58T).

Processing

As is true with many vinyl polymers, CFE plastics are generally processed under conditions approaching thermal degradation. Because of this, and because contaminants can so grossly affect end-product performance, it is mandatory that part design and processing be accomplished by those having a thorough understanding of the material and its processing techniques.

CFE is thermally stable well above its melting point of 414 F. Compression molding at 500 F causes little or no degradation, whereas injection molding and extrusion at much higher temperatures (up to 650 F) degrade the resin to some extent. In this connection, it is important to distinguish between *reversible* changes which occur in the resin through heat treatment, i.e., through "quick quenching" or "slow cooling," just below its melting point, and *permanent* changes which may occur because of overheating above the melting point (chain scission or degradation to a lower molecular weight).

Improperly processed, contaminated, or thermally degraded CFE plastic cannot and will not yield fabricated items with the ultimate in expected properties. Properly processed, however, CFE will perform where the utmost in corrosion resistance, electrical insulation reliability, high-low temperature stability, radiation resistance, and excellent physical properties are all or singly required, coupled with relative processing ease.

CFE plastic was originally developed for use in atomic-energy installations as gasketing and containers. The discovery of other uses in the fields of cryogenics, electronics, space exploration, medicine and chemical processing quickly established this material as an important aid in man's conquest of his environment.

FLUOROCARBON PLASTICS, PVF

Polyvinylidene fluoride is a fluorine-containing thermoplastic resin. It is unlike other resins currently available, being a crystalline, high molecular weight polymer of vinylidene fluoride ($CH_2 = CF_2$) containing 59% fluorine.

Resin forms

The material has the trade name Kynar, and is available as a base resin in the following forms:

1. *Powder:* White, free flowing, nonhygroscopic; bulk density, 30 lb per cu ft.

2. *Pellets:* Cylinders, approximately $\frac{1}{8} \times \frac{1}{8}$ in.; translucent, off-white; bulk density 60 lb per cu ft.

3. *Dispersions:* Stable suspensions of polymer (1-10 microns) in high boiling organic liquids; solids (by weight) 45%; stable in storage from −20 to 120 F.

4. *Solution:* Colloidal solution of polymer in strongly polar organic liquid; solids (by weight) 20%; stable in storage from −20 to 120 F.

5. *Fabricated forms* such as sheet, slab rod, tubing, pipe, valves and custom-molded shapes are available from a number of fabricators.

Further, PVF fluorocarbon resin is made in three grades. One is a general purpose, high molecular weight resin. Another is designed for specific resistance to liquid propellants for use in the aerospace industry. The third is specifically designed for electrical applications requiring high insulating properties.

Physical properties

Properties of the material are summarized in Table 1. Important mechanical properties of PVF include high tensile and compressive strength, high impact strength, and low level of creep. It is flexible in thin sections for film and tubing aplications; it is rigid in heavy sections where load bearing is important.

The material has a broad useful temperature range, from below −80 to above 300 F. Its crystalline melting point of 340 F permits fabrication at reasonably low temperatures.

PVF is self-extinguishing. When exposed to a flame, it degrades essentially to hydrogen fluoride gas and a carbon residue, without dripping. It is classified, "Self-Extinguishing, Group I," by Underwriters Laboratories, Inc.

Weatherability. One of the outstanding properties of PVF is its stability to weather. Its combination of low moisture absorption, high chemical resistance and stability to ultraviolet radiation makes it attractive for applications requiring long service life under severe atmospheric conditions. The tensile properties of thin film before and after one year exposure to an industrial atmosphere are shown below:

	Before Exposure	After Exposure
Tensile Strength, psi	6500	6800
Elongation, %	220	190
Stress at Yield, psi	6000	6500

Uv and gamma radiation. Stability to ultraviolet radiation (2000–4000Å) is demonstrated by the fold endurance of thin film after accelerated aging:

	Before Exposure	After One Year Exposure (GE S-1 lamp ASTM D-795)
MIT Fold Endurance Test, cycles to failure for 0.002 in. film, 0.5 kg load	500,000	450,000

Comparative endurance tests with other plastic films of the same thickness show that only the highly fluorinated polymers possess this degree of stability to ultraviolet radiation.

Resistance to gamma radiation is exceptional. At a dosage level exceeding 300 million roentgens (Co^{60} source), the material shows no change in tensile strength or elongation, though a darkened color is observed.

TABLE 1—PHYSICAL PROPERTIES OF PVF [a]

Property	Measurement
Clarity	Transparent to translucent
Melt Pt (crystalline), °F	340
Spec Gr	1.76
Spec Vol, cu in./lb	15.7
Refractive Index, n_D^{25}	1.42
Molding Temp, °F	400–550
Mold Shrinkage (average), in./in.	0.020
Color Possibilities	Unlimited
Machining Qualities	Excellent
Flammability	Self-extinguishing, nondripping
Ten Str (77 F), psi	7000
Ten Str (212 F), psi	5000
Elong (77°F), %	300
Elong (212°F), %	400
Yld Pt (77°F), psi	5500
Yld Pt (212°F), psi	2500
Compr Str (77 F), psi	10,000
Mod of Elast (77 F), psi	
Tension	1.2×10^5
Flexure	2.0×10^5
Compr	1.2×10^5
Izod Impact (notched; 77 F), ft-lb/in. notch	3.8
Izod Impact (unnotched; 77 F), ft-lb	30
Durometer Hardness	D80
Heat Dist Temp (66 psi), F	300
Heat Dist Temp (264 psi), F	195
Abrasion Res (Tabor CS-17, ½ kg load), mg/1000 cycles	17.6
Coef of Sliding Friction on Steel	0.14–0.17
Ther Coef of Linear Exp, /°F	8.5×10^{-5}
Ther Cond (RT-325 F), Btu/hr/sq ft/°F/ft	0.14–0.11
Specific Heat, Btu/lb/°F	0.33
Ther Degradation Temp, °F	>600
Low Temp Embrittlement, °F	<−80
Water Abs (24 hr), %	0.04

[a] Kynar grade 18.

Electrical properties. PVF has high dielectric strength in thin films, an unusually high dielectric constant, and a high dissipation factor. Its dielectric strength in combination with mechanical strength, flexibility, thermal stability, and inertness to chemicals and weather suggest such applications as thin-wall primary insulation and jacketing for specialty hook-up and control wire and motor windings. Its high dissipation factor and the attendant power losses may limit its application as an insulating material for high-frequency circuits.

DIELECTRIC PROPERTIES OF PVF

	60 cps	10^3 cps	10^6 cps	10^9 cps
Dielectric Constant	8.40	7.72	6.43	2.98
Dissipation Factor	0.0497	0.0191	0.159	0.11
Volume Resistivity, ohm-cm				2×10^{14}
Dielectric Strength, volts per mil				
0.125 in. thickness, short time				260
0.008 in. thickness, short time				1280

Low-temperature properties. No evidence of crazing or rupture is observed when 1/16-in. thick sheets of the material are cooled to −80 F and bent around a ⅛-in. radius. Although its stiffness increases sharply at extremely low temperatures, it retains a moderate degree of impact strength at −300 F. (Izod unnotched value of 3.2 ft-lb per in.), and should be serviceable in applications at cryogenic temperatures.

High-temperature stability. PVF is thermally stable (no significant dehydrofluorination or chain scission) under the following conditions:

Short Periods (½ hour), F	500
Intermediate Periods (16 hours), F	400
Long Periods (2 years), F	300

The material can be used continuously at temperatures up to 300 F. Use temperatures are generally limited by changes in mechanical strength and not by chemical decomposition of the polymer.

Chemical resistance. PVF is resistant to attack or penetration by most corrosive chemicals and organic compounds, including acids, alkalies, strong oxidizers and halogens.

Resistance to penetration has been determined by immersing PVF-covered metal panels in corrosive media. For example, a steel test panel encased in 0.040-in. sheet shows no metal attack after exposure (vapor and liquid) to boiling 15% nitric acid for three months.

PVF is, however, attacked by fuming sulfuric acid at room temperature and by strong sulfuric acid and other sulfonating agents at high temperatures. It can be swelled and partially dispersed by acetone and will form colloidal solutions with such strongly polar solvents as dimethylacetamide. Certain strongly basic primary amines such as n-butylamine, darken and embrittle the resin.

PVF film

Film has been produced from dispersion on commercial scale equipment, in thicknesses from 0.75 to 2.0 mil, and by laboratory procedures to 10.0 mil.

The properties shown below relate to quickly quenched, unoriented film. Biaxially oriented film is under study.

PROPERTIES OF PVF FILM

Transparency	
1 to 3 mil thickness	Clear
3 to 10 mil thickness	Slightly hazy
Gloss	High
Tensile Strength, psi	6000–6500
Elongation, %	150–500
Burst Strength (Mullen), psi/mil	15–20
Tear Strength (Elmendorf), gm/mil	40–60
Flex Life (MIT Endurance, 3-mil film) cycles	75,000
Flammability	Self-ext
Thermal Stability (1 year, 300 F)	
Weight loss	None
Change in color	None
Stability in Ultraviolet	Excellent
Water Vapor Permeability, gm/mil-24 hr-1000 sq in.	0.6
Gas Permeability, cc/mm-sec-cm^2 cm Hg	
Oxygen	24×10^{-12}
Nitrogen	5.5×10^{-12}
Carbon Dioxide	9×10^{-12}

Fabrication

PVF can be fabricated by most conventional plastics molding and forming techniques.

Compression molding. PVF is readily compression-molded into sheets, gaskets, rods and other shapes. Its low thermal conductivity and relatively high melt viscosity are factors to be considered when choosing molding conditions. The resin's high stability at molding temperatures allows complete thermal conversion without degradation.

Transfer molding. Transfer molding may show advantages in economy over compression or injection molding dependent on size and shape of parts, production rate, and other considerations.

Conventional molds and equipment suitable for other thermoplastics (polyvinyl chloride or modified styrenes, for example) may be used. Because PVF sets at a relatively high temperature, transfer molding cycles are much shorter than those for most other thermoplastics.

Injection molding. PVF can be injection-molded without making major modifications to standard machines. Materials of construction in contact with molten polymer (cylinder, nozzles and mold cavity) should be chrome-plated hardened steel or a high nickel stainless type (316). It is not necessary to dry the resin. Mold design will vary with the size and complexity of the cavity. In general, gate lands should be short, 0.015 to 0.020 in., and the gate openings should be flared. Runner diameters may vary from ⅛ to 5/16 in.

To minimize shrinkage of molded parts, it is advisable to keep temperatures low and use reasonably high pressures. Molding cycles will be longer than those experienced with low melt viscosity plastics, and may vary from 20 to 80 sec depending upon the size and shape of the molded part.

Extrusion. PVF can be extruded into shapes, rod, tubing, sheet, wire coating and cable jackets. Long barrel (L/D ratios of 20/1) extruders, using nickel alloy liners ("X alloy"), and stainless or chrome plated steel dies and screws are recommended. The conventional polyethylene screw design (constant pitch, varying root diameters leading to a shallow metering section) has given satisfactory results. Elongated die assemblies, employing die lands of ½ to 1 in., and long core tubes are recommended. Barrel temperature should be held between 400 and 500 F, and the temperature in the extrusion head and die-holder zone should not exceed 500 F. At the tip or die outlet a separate controlled zone should be maintained at 25 to 30 F above the head temperature.

The extrusion of PVF sheet (0.020 to 0.125 in.) is most readily carried out on a flat die assembly integral with a three roll sheet mill. Jaw openings should be set at the same dimension as the desired caliper. The extruded sheet should pass directly into the pressure nip of the sheet mill, with the roll temperatures maintained at 220 F and the pressure rolls set at 5% less than the finished caliper. The cooling roll should be held at 175 to 190 F.

Extrusion of PVF pipe can be carried out using an internally cooled sizing mandrel with a taper of 0.007 in./in.

Machining. PVF sheet and rod can be machined to fine tolerances. High-speed polishing or buffing should be accompanied by adequate cooling.

Hot welding and heat sealing. PVF sheet and rod can be hot air-welded to form intricate shapes and strong seals for pipe, tubing and sheet.

Film is readily heat-sealed by conventional hot impulse sealers, dielectric sealers and by sealers using ultrasonic impulses to generate heat in the polymer.

Applications

PVF finds its most important applications where long service life and high performance are required, and where unfavorable chemical and physical environments may preclude the use of most other materials. Following are typical potential uses.

Equipment for chemical processing. Seals and gaskets; valve and pump parts, diaphragms, impellers, seats and plugs; translucent tubing, sight glasses and flowmeter tubes; heavy-wall unsupported pipe and fittings; lining for pipe, fittings, valves and pumps; tank and autoclave linings or coatings; gears, cams and bearings; laboratory ware.

Packaging. Blown containers for corrosive chemicals; drum and container coatings and liners; films for packaging and hermetically sealing chemicals, lubricants and propellants; sterilizable packaging.

Electric insulation and protective coatings. Molded components, terminal boards, coil forms, printed circuit boards; film laminated to metals, to wood and to other plastics for protection from weather and chemical attack; extruded jackets and prime insulation for high temperature wire insulation; coatings and saturants for fabrics and braids for tarpaulins, hose and wire insulations.

FLUOROCARBON PLASTICS, TFE, FEP

Due to their excellent electrical properties, high-temperature capabilities, chemical resistance and useful mechanical properties, TFE and FEP fluorocarbon resins (e.g., "Teflon") are used in a wide variety of applications. These include gaskets and packings, linings for tanks, valves, fittings and pipe, electrical insulation for wire, motors, etc., and for mechanical applications, such as bearings, seals and piston rings, and for antistick surfaces.

Types and composition

There are now available two basic types of completely fluorinated resins:

1. TFE fluorocarbon resin is the fully saturated fluorinated polymer polymerized from tetrafluoroethylene. This material must be fabricated using techniques similar to those used in powdered metallurgy (i.e., compacting and sintering). Although originally this process was expensive and difficult, improvements in the past few years have reduced costs and permitted a wide range of products to be made that heretofore were not considered possible.

2. FEP fluorocarbon resin is a copolymer (fluorinated ethylene propylene) made from tetrafluoroethylene and hexafluoropropylene monomers. This resin can be fabricated in conventional molding and extruding equipment used for thermoplastics resins. Its melting point is about 540 F, compared with the melting or gel point of 621 F for the TFE resin. Except for the difference in upper service temperature the properties of the FEP resin and TFE resin are essentially identical for most practical purposes.

Reinforcing agents to enhance physical properties are used in most seals, bearings, and piston rings where resistance to wear and deformation is important. Typical additives include glass fiber, graphite, MoS_2, asbestos, bronze powder, etc. In general, the use of reinforcing materials increases stiffness, hardness, compressive strengths, and resistance to deformation under load. Sheets, rods, and tubes, as well as molded parts, can be supplied with a variety of reinforcing agents.

Aqueous dispersions of both TFE and FEP resins are available for coating and impregnating items of varying shape and size.

Properties

Following are the outstanding properties of these fluorocarbons (they are summarized quantitatively in Table 1).

Chemical inertness. They are inert to practically all chemicals and solvents. In fact, the only commercial chemicals and solvents that affect TFE fluorocarbon resins are molten alkali metals and fluorine at elevated temperatures. TFE and FEP resins have been found entirely satisfactory in missile fuels and oxidizers (such as nitrogen tetraoxide, hydrogen peroxide, unsymmetrical diemethyl hydrazene) for seals, gaskets, bladders, hose, etc. TFE resins are safe for use with liquid oxygen.

Heat resistance. They are capable of continuous service (10 to 20 years) at 500 F for TFE, and 400 F for FEP.

Excellent dielectric properties. These properties are essentially constant at all frequencies, and over the wide service temperature range of the materials.

Toughness and flexibility. These qualities are retained even at very low temperatures. They are more flexible than most materials at −110 F, and have been used successfully at −450 F (liquid helium service).

Low coefficient of friction. Coefficients are lowest of any solid, comparable to ice rubbing against ice. They do not show "slip-stick" characteristics—static coefficient is less than dynamic coefficients of friction.

Nonadhesive characteristics. Very few materials tend to stick to surfaces.

Moisture resistance. They do not absorb moisture.

Weather resistance. They are completely unaffected by outdoor weathering. Samples exposed in Florida for over twelve years are completely unchanged.

Permeability to gases and liquids. Permeability is generally very low with the exception of the halogenated solvents.

Radiation resistance. Radiation resistance in air of the TFE resins indicates an exposure limit of approximately 5×10^6 rads before serious impairment of physical properties. The FEP resins are slightly superior to the TFE resins in this respect. However, under vacuum conditions such as those found in outer space the radiation resistance rises to a value of 5×10^7 rads which places TFE and FEP fluorocarbon resins above the service levels for common electronic hardware, such as transistors and capacitors.

Ablative properties. The high temperature resistance, low thermal conductivity, and high energy absorption on decomposition of TFE resins provide superior coatings for missiles and satellites. As TFE decomposes cleanly and evenly to monomer, accurate calculations from thermal and thermodynamic data available on the TFE resins, can be made of the thickness required to provide thermal protection. Thus, minimum quantities of resin can be used permitting savings in weight.

Service at elevated temperatures. Although TFE resins are rated for use at 500 F for continuous service, they can withstand much higher temperatures for limited periods and exhibit fair mechanical strength at temperatures above the so-called melting point of 621 F. Electrical properties remain excellent at temperatures up to at least 600 F. The TFE resins in combination with ceramic or mineral fibers, such as asbestos, are used in electrical insulation at temperatures up to 1200 F for short periods of time. At 750 F, elongation is over 100%, tensile strength over 200 psi. These properties are maintained for at least six hours at this temperature.

Fabricating techniques

Following are fabricating techniques used with TFE and FEP film, sheet, rod, tubes and tapes.

Machining. Stock shapes of TFE fluorocarbon resins can be readily machined using standard equipment and techniques. When working to extremely close tolerances stress relieving may be necessary. This is done by heating a piece prior to machining for one hour per inch of thickness at a temperature below the melt point but above the intended use temperature.

For best results cutting tools should be kept sharp and vibration of machinery should be minimized. The use of coolants such as water or oils is recommended to permit higher cutting speeds.

Table 1—Properties of TFE and FEP Fluorocarbon Resins [a]

Properties	ASTM Test Method	TFE Fluorocarbon	FEP Fluorocarbon
Injection-Molding Qualities	—	—	Excellent
Injection-Molding Temp, F	—	—	625–760
Injection-Molding Pressure, 1000 psi	—	—	5–20
Mold Shrinkage, in. per in.	—	—	0.03–0.06
Specific Gravity	D 792	2.13–2.2	2.14–2.17
Specific Volume, cu in./lb	D 792	13.2–12.6	12.8–13.0
Refractive Index	D 542	1.35	1.338
Ten Str, psi [b]	D 638, D 651	2500–4500	2700–3100
Elongation, %	D 638	250–450	250–330
Mod of Elast in Tension, 10^5 psi	D 747	0.58	0.5
Impact Str, ft-lb per in. of notch ($\frac{1}{2} \times \frac{1}{2}$ in. notched bar, Izod test)	D 256	3.0	No break
Hardness, Rockwell	D 785	D 50–D 65 [c]	R 25
Ther Conductivity Btu/sq ft/hr/°F/in.	C 177	1.7	1.4
Specific Heat, Btu/lb/°F	—	0.25	0.28
Ther Expansion, 10^{-5} per °C	D 696	10	8.3–10.5
Max Resis to Heat (continuous), F	—	550	400
Volume Resis (50% RH, 23 C), ohm-cm	D 257	$>10^{18}$	$>2 \times 10^{18}$
Dielec Str, v/mil [d]	D 149	480	500–600
Dielec Str, v/mil [e]	D 149	430	—
Dielec Constant, 60 cycles	D 150	2.2	2.2
Dielec Constant, 3×10^9 cycles	D 150	2.2	2.2
Dissipation (power) Factor, 60 cycles	D 150	<0.0002	<0.0003
Dissipation (power) Factor, 3×10^9 cycles	D 150	<0.0002	0.0003
Arc Resis, sec.	D 495	>300	>300
Water Absorption, 24 hr $\frac{1}{8}$ in. thick., %	D 570	<0.01	<0.01
Burning Rate	D 635	None	None
Effect of Sunlight	—	None	None
Effect of Weak Acids	D 543	None	None
Effect of Strong Acids	D 543	None	None
Effect of Weak Alkalies	D 543	None	None
Effect of Strong Alkalies	D 543	None	None
Effect of Organic Solvents	D 543	None	None
Machining Qualities	—	Excellent	Excellent
Clarity	—	Opaque	Transparent to translucent

[a] Teflon TFE and Teflon FEP respectively, produced by E. I. du Pont de Nemours & Co., Inc.
[b] Tensile strength of oriented fibres may be as high as 47,000 psi.
[c] Shore D scale.
[d] Short time, $\frac{1}{8}''$ thickness.
[e] Step by step, $\frac{1}{8}''$ thickness.

Small flat parts, such as gaskets, can be produced by punching or stamping from sheet stock or cutting from rod.

Like other plastics, fluorocarbon resins have a coefficient of linear thermal expansion about ten times greater than that of metals. In addition the TFE resins (not the FEP resins) undergo a transition between 64 and 74 F, which results in an abnormally high thermal expansion (1%) within this range. This means that parts machined above 74 F, and then measured at a temperature below 64 F, will have a marked change in dimensions. Because of these two factors, where tolerances are critical, measurements should be made at the temperature at which the part will be used, and preferably above or below the 64 to 74 F transition range.

Forming. The high temperature flow properties of FEP fluorocarbon film permits thermoforming into more complex shapes and contours than is possible with film of TFE resins. Film of FEP resins can be distended and formed at 480 to 545 F, using matched metal molds or by standard vacuum-forming techniques. Deeper draws can be vacuum-formed using a heated metal plug assist.

Tape and sheet stock of TFE resin can be formed by drawing between male and female dies. Cold forming is adequate in many cases. However, preheating to temperatures of about 450 F will help where extensive drawing is necessary. A depth of draw equivalent to one-half the thickness of the original sheet can be obtained quite readily. Thin wall tubing can be flared with standard flaring tools, such as are used with copper tubing.

Heat sealing. Due to the melt properties of FEP resins, thin films (less than 10 mils) can be readily heat-sealed in commercial impulse sealers. Heavier films in the range of 10 to 20 mils can be heat-sealed to form strong bonds using slight pressure (1 to 2 psi maximum). Heavy FEP film (20 to 60 mils) can best be heat-sealed by use of a hot air gun, with a technique similar to that used on PVC sheeting.

Sheeting of TFE resin can be heat-sealed or bonded, utilizing temperatures of about 700 F, with the aid of a fluxlike material, under contact pressure of about 35 psi. This technique is most suitable for thicknesses above ⅟₁₆ in., and has been used to make large gasket rings with bonds that are as strong and as chemically resistant as the original sheet.

Cementing. Due to the antistick properties of TFE and FEP fluorocarbon resins, special treatment is required to enable them to be used with conventional cements and adhesives. The FEP fluorocarbon film is available with such a treatment on one side so that it can be used with common adhesives, such as the epoxies, polyesters, and rubber-based cements. The ease of fabrication of FEP film by thermoforming techniques is unchanged by the cementability process.

Sheets, rods, tubes, molded parts, etc., of TFE resins can be bought with a prepared surface for cementing to various substrates using conventional adhesives. It is now possible to buy the "resin-etching" solutions for treating surfaces.

These etching solutions presumably extract fluorine atoms, leaving a carbonaceous film on the resin which changes only the surface. This treatment will slightly lower the surface resistivity but has no effect on other properties. The surface treatment will deteriorate when exposed to direct sunlight, thus reducing the bond strength.

Other less frequently used methods for preparing the surface are graft polymerization onto the surface and the bonding of fine silica particles onto the surface.

Yarns of the TFE fibers can be interwoven with cotton yarns so that the back surface is cotton, which can be cemented, and the front surface is TFE fiber. Coated glass fabric is available with a coating only on one side permitting cementing of the glass fabric to the substrate. Reinforced sheeting using asbestos or ceramic fibers can be obtained with the fibers exposed on one side which permits cementing by conventional means.

Bonding. Due to the thermoplastic nature of the FEP film, bonding to metallic substrates can be accomplished via heat and pressure without the use of an adhesive. This is of particular interest for printed circuits where the presence of an adhesive may detract from the electrical characteristics at ultrahigh frequencies. The FEP film can be used to cement TFE film to metallic substrates or to cement together films of TFE resins.

Fluidized bed coating. Materials to be coated are heated to about 650 to 900 F, and are then dipped into a fluidized bed of suitably powdered FEP resin. After a 1 to 2 sec dip, followed by low-pressure air blasting to remove excess powder, the component is reheated. This fuses adhering particles of FEP into a continuous 1 to 10 mil film. Additional heat-dip-heat cycles can effectively increase covering laydown to 20 to 40 mils.

Printing and striping. The nonreactive surface of fluorocarbon resins presents a problem when marking or identification is required, for instance, in wire insulation. These problems have been overcome by the use of striping inks based on fluorocarbon resins. The stripes for wire insulation are put on by conventional equipment and then are "cured" or "sintered" by heating to 700 to 800 F. Printing can be accomplished by using a hot stamp with pressure on a fluorocarbon-resin inked tape. These stripes and prints have the properties of the fluorocarbon resins, have excellent bonds, and reasonably good abrasion resistance.

Applications

Following are some of the major uses for TFE and FEP resins.

Gaskets. All types of gaskets are made from TFE resins: Solid, envelope, spiral wound, and reinforced. If bolt loadings must be low, an envelope gasket with a rubber inner core or an impregnated felt of TFE fibers is used. Where pressures are high, and there is danger of extrusion or cold flow, spiral wound, coated glass fabric or reinforced gaskets are used. This reinforcement may be in the form of asbestos or glass fibers.

Unreinforced gaskets of TFE resins will exhibit cold flow, thus it is desirable to take up on the bolts 24 to 48 hr after initial tightening. After this time most of the cold flow has occurred and retightening will not be required. Gaskets of fluorocarbon resins can be easily removed and used over again.

Packings. Braided packings of asbestos or TFE fluorocarbon fibers impregnated with TFE fluorocarbon dispersions are available in many sizes and constructions. Unlike the oil, grease and graphite lubricants that ooze from conventional packings under operating temperatures and pressures, the lubricant in asbestos packings of TFE will not evaporate or extrude from the packings. Therefore, after the initial break-in period, which is very important, the stuffing box or gland need not be retightened, and the packings will maintain their size and lubricity indefinitely.

Valve packings should be tightened initially just enough to provide a seal. After heat and pressure have been applied to the line for a day or two, the glands or expansion joints should be retightened to complete the seal. Overtightening pump packings can be disastrous. Equally important are a smooth shaft finish and small shaft runout. Packings for high-speed pumps should be externally lubricated. Many of the failures of pump packings of fluorocarbon resins are due to improper installation and lack of lubrication. Experience has shown that speeds up to 2000 ft/min on pump shafts can be tolerated by correctly installed braided packings impregnated with TFE fluorocarbon resins.

Hose. Hose lined with TFE is completely inert to virtually all chemicals and is rated for continuous service up to 450 F. Liners of TFE are easy to clean and do not contaminate the fluids handled by the hose. Such hoses are tough and flexible; they can be flexed at liquid oxygen temperatures. Hose outer constructions include wire and fiber braiding, elastomeric coverings and convoluted metal tubing.

Lining of chemical process equipment. Metal pipe lined with TFE fluorocarbon resins has been a commercial product for several years. Piping systems including fittings have successfully handled such difficult fluids as hydrochloric acid and organic liquids in combination at 275 F. Sizes available range from 1 to 10 in. In lined pipe a gasket of fluorocarbon resin is an integral part of the lining.

Wire and cable. Wires and cables insulated with TFE fluorocarbon resins are used extensively in electronic, aircraft, and missile wiring where the high temperature performance, and excellent dielectric characteristics offer the ultimate in performance and reliability as well as weight savings. In addition, the nonflammability and chemical resistance are important attributes.

Insulations of the fluorocarbon resins can be directly extruded onto wire or as jackets on coax and large cables. Various types of tapes are also available for wrapping wire. Tapes of FEP come as thin as one-half mil and can be readily fused to form a smooth insulation. Unsintered TFE tape is also available for wrapping with a minimum thickness of approximately 2 mils. This tape can also be fused onto a smooth jacket although not as readily as the FEP film. Coated glass tape and skived tape (sintered tape) of TFE is also available for wrapping.

Electronic components. Laminates containing asbestos or glass fiber or impregnated glass fabrics are available as rigid insulating boards in thicknesses from $\frac{1}{16}$ up to 1 in. Thin flexible laminates are also available as a base for printed circuits. Laminates utilizing the FEP resins can be adhered to the conducting surface without any adhesive, insuring the optimum in dielectric properties for critical circuits.

Other components include insulators, tube sockets, coil forms, electrical contacts, cams, brush holders, radomes and radar components. In most of these applications the low dielectric loss and the high temperature resistance are generally the most important properties for design considerations. The ability to injection-mold the FEP resins makes many intricate electromechanical parts now feasible. High performance, high temperature capacitors utilizing metallized films of FEP and TFE resins are also available.

Electrical machinery. Coated fabrics, laminates, and other reinforced forms provide electrical insulation in excess of Class H requirements (> 180 C). These forms are used for armature and field coil wrappings, support pad insulation, armature spiders, phase separators, slot liners, slot stick and commutator insulation.

Seals. Seals of TFE fluorocarbon resins are desirable due to their low coefficient of friction, good mechanical strength over wide range of temperatures, and unparalleled resistance to chemicals and solvents. As these resins are not elastomeric, designs must overcome the lack of resiliency. Such designs are now available in any type of seal including backup rings, cup seals, rod seals, "U" and "V" seals, mechanical seals and floating rings. Many mechanical seals now utilize TFE on the sealing surfaces. As with packing materials, more and more original equipment manufacturers are turning to seals of the fluorocarbon resins due to their universal application.

Bearings. Due to the exceptional frictional properties of TFE, it is used extensively for nonlubricated bearings. Frictional properties of TFE are superior to those of FEP resins. Forms used include woven fabric of TFE fibers, impregnated porous bronze, reinforced molded sleeves, reinforced resin molded on or around metallic screens, and reinforced tapes. Reinforcement is used to improve wear and load-bearing capabilities.

Woven cloth bearings are used at loads up to 60,000 psi at low speeds. PV values of the various forms range from 20,000 to 50,000, and many of the compositions will operate above 2000 ft/min. The operable PV value is determined by the wear and load-carrying capacity.

Piston rings. Piston rings of TFE fluorocarbon resins up to 15 in. in diameter are being used in place of carbon-graphite rings due to the superior strength and longer life of the TFE rings. Nonlubricated rings of TFE resins are used in compressors processing gases, such as chlorine, oxygen, air, hydrogen, and liquids which must not be contaminated. Rings of TFE resins have been used in lubricated compressors to reduce the amount of lubrication needed, reduce cylinder wear and cylinder damage due to ring breakage.

Nonadhesive uses. One of the earliest applications of TFE resins was as a lining in bread-baking pans to permit easy removal of the baked loaf. Other nonadhesive uses include bread-sheeting rolls, heat-sealer plates, guides and plates in packaging machinery, troughs and hoppers in fertilizer and feed plants, candy-making equipment, etc. For antistick purposes thin coatings can be applied by spraying and baking. Heavy coatings can be applied by fluidized bed techniques provided the part can be removed for processing. For plant installations and where the thickest coating is required, sheeting or tape of the fluorocarbon resins can be cemented on to the surfaces. Materials will stick to fluorocarbon surfaces but almost without exception the adherence is much weaker than the adherence of the sticky substance to other surfaces. For instance, paints will stick to the surface of a sheet of TFE resin, but the adhered paint can be readily removed by a mild scraping.

Filtering. TFE in the form of woven fabric is used as filtering media to separate solids from gases and liquids, and liquids from gases. Due to the hydrophobic nature of the fully saturated fluorocarbon resins, felts made of TFE fibers have been used to separate water from hydrocarbons. The water is retained upon the felt of the fluorocarbon resin while the hydrocarbon which wets the surface passes readily through. In absorption towers, separators utilizing filaments of the fluorocarbon resins have been used to separate entrained liquids from the gas stream. Due to the chemical inertness and heat resistance these filaments are replacing a similar construction made of stainless steel and more exotic alloys. Filters with very small pore size made from molded sheets of the TFE fluorocarbon resins are available.

Miscellaneous applications. Coatings and linings for food processing equipment; flexible bellows for expansion joints; thread sealant for pipe; laboratory ware: beakers, stopcocks, stirring rods, etc.; "puffer" nozzles, bearings, and seals for high-voltage switchgear; antistick coatings on frying pans and other cooking utensils;

raschig rings; dry film lubricants for submarine periscopes, rifles, 20 mm ammunition shells; spaghetti tubing for "slip-on" wire insulation; lacing from TFE yarn for wire cabling; brush holders for DC motors and generators; conveyor belting for sticky materials; radar "windows" for missiles.

FLUOROELASTOMERS

Fluoroelastomers are rubberlike polymers which, when compounded with other materials and vulcanized, form useful engineering materials. At modest temperatures, vulcanizates of fluoroelastomers exhibit mechanical properties which are similar to those of the more familiar hydrocarbon elastomers. However, in comparison, vulcanizates of fluoroelastomers are much superior in resistance to: (1) attack by active chemicals, (2) swelling by a variety of fluids, and (3) deterioration on exposure to elevated temperatures.

General characteristics

Typical fluoroelastomer vulcanizates exhibit high densities; specific gravity is usually between 1.85 and 2.00. The materials are self-extinguishing when removed from a flame. Thermal degradation is accompanied by the evolution of hydrogen fluoride. Hydrocarbon solvents, fuels, and lubricants, as well as halogenated solvents, do not swell fluoroelastomer vulcanizates excessively; whereas some esters, ketones, and ethers cause a high degree of swelling which is accompanied by a drastic loss of strength.

Hence, a small specimen of fluoroelastomer vulcanizate can be identified, or distinguished from other types of vulcanizates by the following simple tests: (1) it will sink but not swell in trichloroethylene; (2) it is immediately and excessively swollen in acetone, and (3) it will etch glass on a brief exposure at temperatures above 700 F.

While the properties of any elastomer end item can be widely varied by the use of differing amounts and types of compounding ingredients, the basic chemical, solvent, and heat-resistance properties are characteristic of the type of polymer. A knowledge of these basic properties of each fluoroelastomer is essential, because they impose the limits of service for fabricated end items.

Kel-F Elastomers

The Kel-F Elastomers 3700 and 5500 are copolymers of trifluorochloroethylene and vinylidene fluoride ($CF_2 = CFCl/CF_2 = CH_2$) containing approximately 50% fluorine. The Kel-F Elastomers were among the earlier fluoroelastomers and helped to solve many critical problems.

Items fabricated from these polymers possessed outstanding resistance to high-temperature jet fuels, hydraulic fluids and lubricating oils, in addition to an excellent retention of properties in red fuming nitric acid. Difficulties of flow and knitting during the molding of precision parts and poor compression set resistance precluded the use of these materials as seals, gaskets, and other parts where these properties are very important.

Viton A

Viton A was the trade name assigned by du Pont to a copolymer of hexafluoropropylene and vinylidene fluoride ($CF_3 - CF = CF_2/CF_2 = CH_2$). At a 65% fluorine content, there are about nine vinylidene fluoride units to every two of the hexafluoropropylene units in the polymer chain. The copolymer is a water-white elastomeric gum which processes and molds without difficulty. The gum is usually compounded with a large particle-size furnace carbon black, magnesium oxide and a blocked diamine. Viton A compounded with fine silica, zinc oxide and benzoyl peroxide also produces a useful vulcanizate, but lacks the excellent compression set properties of the black-diamine vulcanized compound.

Chemical resistance. Viton A is chemically resistant to a wide variety of concentrated acids and salt solutions. Strong oxidants such as N_2O_4 and permanganates chemically attack the hydrocarbon amine crosslinks of the more common Viton A vulcanizates. The nonblack, peroxide-vulcanized Viton A materials are much more resistant to oxidizing chemicals. The most severe chemical attack is by hydrocarbon amines and high pH solutions, which cause a removal of hydrogen fluoride from the polymer chain, leaving a site for further chemical attack and ultimately the loss of all useful properties.

Solvent resistance. Viton A resists the swelling action of a wide variety of aliphatic, aromatic and halogenated solvents. Swelling of less than 10% is common even at elevated temperatures. A selected group of solvents (esters, ketones, and ethers) can cause swelling as high as 300%. These fluids are often used to make solvent cements of compounded, but not yet vulcanized, stocks. Other fluids which usually cause high swelling are phosphate esters, glacial acetic acid, fuming nitric acids, aldehydes, and many nitrogen-containing solvents and chemicals.

TYPICAL VITON A FORMULATION

	Parts
Viton A	100
Magnesium Oxide	20
Medium Thermal Black	25
Hexamethylenediamine Carbamate	1.3

Press Mold: 30 min at 320 F
Oven Postcure: 24 hr at 400 F

TYPICAL PHYSICAL PROPERTIES OF VITON A VULCANIZATES

Tensile Strength, psi	2000–2400
Elongation, %	200–220
Hardness, Shore A	70–75
Low-Temperature Brittle Point, F	−30
Lower Limit of Rubberlike Properties (TR-10 Value), F	0 to +4
Compression Set (25% Compression), %	
70 hours at 250 F	20–25
24 hours at 400 F	45–50
Air Aged 16 hr at 600 F	
Tensile Strength, psi	1400–1600
Elongation, %	100–130
Hardness, Shore A	75–80
Weight Loss, %	>10

Heat resistance. Viton A is stable to many fluid environments such as jet fuels, lubricating oils, and hydraulic fluids for 1000 hours or more at 400 F. Short-term exposures at higher temperatures can be tolerated; 600 F is considered to be the maximum short-term operating temperature for such parts as seals, hose, and gaskets. The features which limit service at high temperatures are rapidly increasing compression set above 400 F and a loss of tensile (or tear) strength and ultimate elongation. The loss of tensile strength with increasing temperature is common with all elastomers to some extent, but becomes more critical for the fluoroelastomers because of their use at higher operating temper-

atures. The extent of this loss is illustrated below:

Temperature, F	Tensile Strength, psi	Elongation, %
75	2200	220
400	300	70
500	150	60

The careful design of elastomer parts can do much to overcome the deficiencies of poor resistance to compression set in the 500 to 600 F range and the low tensile strength at these elevated temperatures.

A heat-related environment is high-energy nuclear radiation such as the gamma rays from Cobalt 60 and mixed gamma rays and high-speed neutrons from nuclear reactors. Elastomers in general are very sensitive to this harsh environment. In addition to radiation resistance, heat, solvent and chemical resistance are often required. While the radiation resistance of all of the present commercial fluoroelastomers is considered poor, the other requirements of nuclear service may dictate the use of these materials for short-time exposure to this severe environment.

FLUOROELASTOMER PROPERTIES AFTER 1 ×10¹⁰ ERGS/GRAM (GRAPHITE) WHILE IMMERSED IN 400 F POLYPHENYLETHER FLUID

	Tensile Strength, psi	Elongation, %	Volume Increase, %	Shore A Hardness
Viton A	1210	30	4.0	90
Viton A-HV	1210	35	4.5	90
Viton B	2140	50	3.8	88

The property often limiting the service life of seals and similar parts which are under compression is radiation-induced compression set. A 50% set is taken by Viton A at room temperature after an exposure to 5 x 10^8 ergs/gm (graphite) and almost 100% set after an exposure of 5 x 10^9 ergs/gm (graphite). Again good design can do much to overcome such deficiencies, while the search is on for newer, more stable materials.

Fluorel

Fluorel, a product of the Minnesota Mining and Manufacturing Co., is also a copolymer of hexafluoropropylene and vinylidene fluoride ($CF_3 - CF = CF_2/CF_2 = CH_2$) and approximately the same monomer ratios, and hence fluorine content, as Viton A. The polymer appears to be less branched and of higher viscosity, i.e., intermediate between Viton A and Viton A-HV. The chemical, solvent and heat resistance as well as mechanical properties are very similar to those of Viton A.

Viton A-HV

The second fluoroelastomer in the Viton series is a high viscosity version of Viton A designated as Viton A-HV by du Pont. The copolymer is tougher than Fluorel and considerably more so than Viton A. This toughness can cause difficulty in processing; however, with care, moldings and extrusions of high quality can be made.

While slight improvements in heat and fluid resistance are obtained with vulcanizates of Viton A-HV, its primary advantage over the lower viscosity polymer is an increased tensile strength (2800 vs 2200 psi). The high viscosity polymer is often blended with Viton A to obtain a tensile strength at some desired level intermediate to either elastomer alone.

Viton B

The trade name for this fluoroelastomer implies that it is significantly different from Viton A. The chemical composition has not yet been released; however, the physical and chemical properties indicate a slightly higher fluorine content and fewer sites for chemical attack. The vulcanizates of this fluoroelastomer have improved chemical, solvent, and heat resistance over the prior Vitons and Fluorel. The clear white polymer processes easily and has excellent mold flow characteristics. Viton B requires a higher level of vulcanizing agent, otherwise the typical formulation is the same as the other Vitons and Fluorel.

The low temperature properties of Viton B are significantly different from Viton A and Fluorel. The low-temperature brittle point is about −50 F, 20 F lower than Viton A; however, the lower limit of rubberlike properties (as measured by a retraction of 10% from the frozen stretched state) is +15 F, about 15 F higher than Viton A. This rubberlike property is important for low-temperature seal performance. If it were not for this feature, Viton B could be expected to replace Viton A and Viton A-HV as du Pont's commercial fluoroelastomer.

TYPICAL VITON B FORMULATION

	Parts
Viton B	100
Magnesium Oxide	20
Medium Thermal Black	25
Ethylenediamine Carbamate	1.0

Press Cure: 30 min at 320 F
Oven Postcure: 24 hr at 400 F

TYPICAL PROPERTIES OF VITON B FLUOROELASTOMER VULCANIZATES

Tensile Strength, psi	2400–2800
Elongation, %	230–240
Hardness, Shore A	70–75
Low-Temperature Brittle Point, F	−50
Lower Limit of Rubberlike Properties (TR-10 Value), F	+15
Compression Set (25% Compression), %	
70 hr at 250 F	20–26
70 hr at 400 F	60–65
Air Aged 24 hr at 600 F	
Tensile Strength, psi	1000–1200
Elongation, %	200–250
Hardness, Shore A	80–85
Weight Loss, %	>5

Chemical resistance. Viton B is slightly more resistant to harsh chemicals than Viton A and Fluorel; however, the organic amines and other high pH chemicals attack the polymer in a similar manner.

Solvent resistance. Viton B vulcanizates are more resistant to the swelling action of solvents than those of Viton A, Viton A-HV, or Fluorel. However, the same general classes of low swelling and high swelling fluids apply to Viton B. In some cases this increased solvent resistance is sufficient for Viton B to be used where the other Vitons and Fluorel would be unsuitable. For example, Viton A vulcanizates swell over 30% after 72 hours at 400 F in trimethylolpropane esters (−65 to 420 F gas turbine lubricants) whereas those of Viton B swell under 20%.

Heat resistance. Viton B vulcanizates are more stable than those of Viton A or Fluorel in high temperature environments. After twenty-four hours at 650 F Viton B vulcanizates suffer only slightly greater loss of mechanical properties than do those of Viton A after 16 hr at 600 F. Thus, Viton B represents a substantial improvement in heat resistance over Viton A. The change in elongation on high-temperature aging can be regulated by the choice of the amount and type of vulcanizing agent. High levels of vulcanizing agent result in a decrease from the original elongation on aging, while lower levels result in an increased elongation on aging. The tendency to evolve hydrogen fluoride at elevated temperatures is also reduced, and Viton B vulcanizates can be used as coatings for glass fabrics for high-temperature service.

The high-temperature tensile strength of Viton B is somewhat improved over the other fluoroelastomers with 440 psi tensile strength and 130% elongation at 400 F. The compression set properties are similar to Viton A, hence, the long-term operating temperature of such parts as seals and gaskets is limited to about 400 F. Short-term use up to 650 F can be obtained in some applications.

The radiation resistance of Viton B is significantly better than Viton A (as shown before). Viton B is the best choice of all commercial elastomers for service in a nuclear environment where fluid resistance above 400 F is involved. This level of resistance is still far below that desired of elastomers for service in nuclear environments.

Other types

Development of other fluorelastomers continues. One of the newer types is trifluoronitrosomethane/tetrafluoroethylene. The outstanding feature of this type copolymer is a phenomenal chemical and solvent resistance. Selected perfluorocarbon fluids are among the very few fluids that can be used as a solvent for the base polymer. The pressed raw gum is reported to be stable to chlorine trifluoride, possibly fluorine gas, nitrogen tetroxide, and similar highly reactive chemicals. The gum polymer is also very flame resistant.

Perfluoroalkyl triazine is another relatively new fluoroelastomer. It has excellent resistance to chemicals, fluids, heat, and high energy radiations. Lower limit of rubberlike properties is around room temperature, with a glossy brittle point near 10 F. High temperature limit has been reported to be 800 F where a gum vulcanizate loses about 20% of its weight in five hours. Properties such as tensile strength, elongation, modulus and color appear unchanged after such rigorous exposure.

FOAMED GLASS

Foamed or cellular glass is produced from borosilicate-type glass. The resulting material is completely inorganic, containing millions of tiny closed cells.

Fabrication and properties

Foamed glass is easily fabricated by cutting with a hand or band saw. Sections to be used for pipe insulation can be formed by working a rectangular block of the material over a rotating cylinder.

In general, physical properties range as follows:

TABLE 1

Composition	Borosilicate glass
Density, lb/cu ft	9
Permeability	0
Coefficient of Ther Expans, per °F	0.0000046
Therm Conductivity, Btu/hr/ sq ft/°F/in.	
At 50 F	0.38
At 75 F	0.40
Thermal Diffusivity, sq ft/day	0.42
Specific Heat, Btu/lb	0.20
Sound Insulation Values:	
Sound Absorption	No value
Sound Reduction Factor, Decibels	28.3
Dielectric Constant	
At 1000 cps	83
At 10⁴ cps	68
Capillarity	0
Moisture Absorption (ASTM C240), %	0.2
Hygroscopicity (246 days at 90% RH)	No increase in weight
Acid Resistance	Impervious to common acids and acid fumes
Combustibility	Will not burn
Ultimate Compressive Str (avg), psi	100
Flexural Strength, psi	75
Modulus of Elasticity, psi	180,000
Shear Strength, psi	40
Tensile Strength, psi	50

Available shapes and sizes

The material is available for roof insulation in sheets of 1¾ × 12 × 18 in. Other thicknesses available are 2, 2½, 3, and 4 in., in sizes of 12 × 18 or 18 × 24 in. For wall use, sheets 1½ in. thick are available in 12 × 18 in. dimensions. Special thicknesses ranging from 1 to 3 in. in increments of ⅟₁₆ in. are produced for curtain wall applications.

Pipe insulation is available in standard thicknesses from 1 to 4 in. nominal, to fit ¼- to 24-in. pipe sizes. The 1-in. thickness is not made in pipe sizes larger than 6 in.

Pipe covering is also available in factory-applied jackets of glass fabric, aluminum, or white kraft paper-aluminum foil.

In standard curved segments, the material is available in 12 × 18 in. sections, in beveled lags of 6 × 18 in., flat lags of 6 × 18 in., and special head segments.

Applications

Foamed glass is used as insulating material at temperatures ranging from −450 to 500 F. For cooler and freezer rooms the material is used for lining floors, walls and ceilings. Its rigid nature permits it to be used as free-standing walls to provide partitions between freezer and cooler spaces in the same building.

For normal temperature insulation, foamed glass is used as wall-lining insulation applied directly to the masonry. For curtain wall construction it is used as an insulating core between two sheets of metal; the outer skin is usually a colored porcelain enamel. The material is also used as: (1) a backup spandrel insulation for glass-clad buildings, (2) roof insulation over most any type deck, (3) core walls between outer brick face and the backup masonry, and (4) perimeter insulation for slab-on-ground construction.

Industrial uses include insulation for pipe and equipment, large tanks and underground pipe. The only protection foamed glass usually requires is glass cloth coated with tar enamel.

FOAM PLASTICS

A broad range of polymeric materials are available in foam or cellular form. In general, the cellular structure provides one or a combination of the following characteristics: light weight, thermal and/or electrical insulation, energy absorption, and/or flexibility.

Foam plastics, which can be classified as rigid and flexible, are produced by a variety of methods.

1. Gas expansion is the most widely used method. Gas can be used by (a) introducing gas under pressure directly into a soft plastic mass, (b) incorporating under pressure a low boiling point solvent, which volatilizes when the pressure is released at a temperature above the solvent's boiling point, (c) aeration or frothing of the soft plastic mass, (d) incorporating a chemical or group of chemicals which when heated decompose, releasing gas; and (e) reacting chemicals in the formulation which during polymerization release a gas.

2. Soluble solids can be incorporated in the soft plastic mass. After the mass has been formed the material is leached out by solvents.

3. Bulky fillers can be incorporated in a binding resin. Partial fusion of the mass produces a porous material.

4. Bonding of loose fibrous structures with resinous materials forms a lightweight porous mass.

Rigid foam structures have been produced of urethanes, polystyrenes, epoxies, phenolics, silicones and cellulose acetate. So-called "syntactic" rigid foams have also been produced by incorporating such cellular materials as microballoons and glass microspheres in a liquid thermosetting resin and allowing the resin to cure.

Of the rigid foams, urethanes, polystyrenes and epoxies are currently of most interest. Primary uses are thermal insulation and energy absorption. Other uses include structural cores for sandwich panels and filtration for certain types of urethanes.

Thermal insulation. The biggest potential application for rigid foams is in thermal insulation, both in building panels, and in refrigerated storage, transportation and home units. Here the major interest is in the rigid fluorocarbon-blown polyether type urethane foams, and the newer low-density, fluorocarbon-blown epoxy foams. Both materials are foamed as liquid systems and provide essentially the same performance characteristics, e.g., densities of about 1.5 to 2 lb per cu ft, k factors of 0.1-0.15 Btu/hr/sq ft/°F/in., good structural strength and good adhesion to surfaces against which the materials are foamed, and the ability to be foamed in place.

Both materials also appear promising for sprayed-on foam insulation, used for insulating storage tanks and large surfaces.

Packaging. Foamed polystyrene has found its largest volume use in packaging (as well as building insulation). It is available as a pre-foamed board or sheet, or as beads which expand on heating. Densities range from 1 to almost 5 lb per cu ft.

In foam sheet or film, the material can be embossed or thermoformed to provide light-weight decorative and protective packaging. As expandable beads the material can be molded by a variety of techniques to form protective packaging. As a rigid crushable material, it can offer nearly ideal one-time protection to delicate equipment.

Rubber, urethanes, vinyls and polyethylenes provide flexible, resilient foams. Of these, urethanes comprise the largest volume use in auto and home seating and bedding. Vinyls have been primarily used for institutional seating applications, and polyethylenes for wire insulation and flotation uses.

Insulated clothing is a large potential volume for flexible foams. At present, both thin-sheet, flexible urethane foams, either flame or adhesive bonded to fabrics, and thin-sheet flexible foam vinyl are contenders.

Silicone foams, both rigid and flexible, are specialty high-cost materials, particularly useful for critical electrical applications where exposure to heat and cold are problems. Higher-density epoxy foams are being used for some critical electrical potting applications.

Phenolics, cellulose acetates and ureas are in only limited use. Phenolics are friable and are relatively low in strength, but are finding limited use as cores for plastics tools; cellulose acetate, a relatively high-strength material, is limited by relatively high cost; urea formaldehyde is extremely low in density, but also extremely low in strength.

FOAM RUBBER

Also called latex foam or foamed latex, foam rubber is a cellular cushioning material made from latex. It differs from most other cellular products in that its cells are almost completely interconnecting and permit ready passage of air throughout the structure. This feature eliminates the "hot feel" of most rubber products. Sponge rubber and the flexible plastic foams differ considerably from foam rubber not only in the raw materials used in their manufacture but also in properties.

Production method and materials

Latex, a colloidal dispersion of elastomeric particles in water, is compounded with antioxidant, vulcanizing ingredients and soap, and then frothed to a creamy, low density foam. The liquid foamed latex is cast or shaped prior to gelation, which is carried out through a pH-lowering mechanism such as the addition of a slowly soluble acid salt. Sodium silicofluoride is most widely used as a gelling agent.

The gelled froth or foam is then exposed to elevated temperatures using live steam, hot air or the like to effect the sulfur cross-linking reaction known as vulcanization. The product can then be stripped from molds or casting belt and is usually washed to remove most of the water solubles and reduce odor. It is dried and is then ready for use.

Although SBR or natural rubber latex is most generally used separately or in combinations, neoprene or acrylonitrile types are sometimes used to make foams which have special properties typical of these elastomers. Neoprene can be used to make foams with outstanding flame resistance and also resistance to oils which swell and deteriorate natural rubber and SBR. Acrylonitrile latex is used to make foam with outstanding solvent resistance.

Properties

Foam rubber is typified by its interconnecting cells which permit passage of air or water easily. It exhibits the good elastic memory of vulcanized elastomers and retains its original shape and cushioning comfort even after severe use over long periods of time. These properties are demonstrated by compression set, flexing, oven aging and other tests described in ASTM D-1055.

The degree of softness (or firmness) of foam is controlled during its manufacture by adjusting the density of the froth so that a wide range of cushioning properties can be obtained. Density of the final product is ordinarily in the range of 4 to 10 lb per cu ft, but foam rubber is purchased on the basis of a compression reading. The standard machine for making this measurement deflects the foam with a circular, 50 sq in. indentor foot to 75% of its original height. The result in pounds is listed as the RMA (Rubber Manufacturers Assn.) compression. Standard compression ranges are as follows:

A. Molded goods

RMA Grade	RMA Compression, lb per 50 sq in.
RC- 5	5 ± 3
RC- 10	10 ± 3
RC- 15	15 ± 4
RC- 20	20 ± 4
RC- 25	25 ± 5
RC- 30	30 ± 6
RC- 40	40 ± 7
RC- 50	50 ± 8
RC- 60	60 ± 9
RC- 70	70 ± 12
RC- 90	90 ± 14

B. Slab or uncored stock

RMA Grade	RMA Compression, lb per 50 sq in.
RU- 11	11 ± 4
RU- 20	20 ± 5
RU- 35	35 ± 10
RU- 55	55 ± 10
RU- 80	80 ± 15
RU-150	150 ± 55

Cushioning properties are retained over a range of temperatures from −40 to +250 F, although service life is impaired upon continuous exposure at high temperatures. At low temperatures the stock gets progressively firmer, but does not show embrittlement.

Fabrication and forming

Foam rubber can be cut with scissors, band saw or clicker die and can be cemented (usually with a neoprene solvent or latex adhesive) and joined to create special shapes.

The product is available as a "slab" in sheets or rolls in thickness ranges of from one-quarter inch to two inches and in widths of five to six feet. Slab stock is generally made in a continuous ribbon which is cut according to needs. It may be cast directly on fabric or paper and in such cases a tight bond is achieved.

The material is perhaps in widest demand be-cause it can be molded into a great variety of shapes and forms. These range from parts as large as mattresses to articles weighing only a few ounces. Through coring, special cushioning effects are achieved such as providing firm, solid edges and soft, cored-out seating areas where desired.

Although most foam rubber is off-white or natural colored, substantial amounts are made pure white and in colors.

The fields of application are varied and a partial list follows: mattresses, pillows, and furniture cushions; aircraft, train, bus, truck, and automobile seats and backs; journal-box lubricators for railroad freight cars (the oil-resistant foam acts as an oil retainer and also serves to maintain a light continuous contact against the journal); mops; ink stamping pads; pressing pads for dry cleaning industry (open structure permits passage of steam); powder puffs; brassiere pads; kneeling pads; shoe insoles; bunion pads; packaging; arm and head rests; and carpet underlay.

FORGINGS

Forging is a process of plastic deformation (under pressure or a blow) of a cast or sintered ingot, bar or billet, mill product or metal powder to produce a desired shape and mechanical properties. Forging develops a metallurgically sound, uniform and stable material which will have optimum properties in the operating component after being completely processed and assembled.

During the forging process, metal is distributed within the required shape as it is needed according to function and stress requirements for purposes of improved design, material utilization, producibility and cost. Forged integral components permit reduction in the number of mechanical and welded joints and elimination of welds in critical areas, with resultant reduction in stress concentrations, increased reliability and weight savings.

Forging is a fine blend of art and science. Furthermore, the dynamic nature of the industry illustrated by the rapid and constant development of new forging alloys with their attendant problems, the need to develop new die materials and methods of die sinking, the research in die lubricants and the construction of equipment allowing the forger to make larger and more complex parts, makes it almost mandatory for the component designer to consult with the forging designer and metallurgist. It is difficult, even for the forging expert, to predict what the forging industry may be offering in the near future and, therefore, to set limits on design geometry, weight, tolerances, alloy and strength is to be short-sighted. The forging industry has developed with the demand, and alloys unheard of five years ago, unforgeable three years ago, are in production today.

At the time, presses having 50,000-ton capacity, hammers over 50,000-lb rated capacity and counterblow hammers to 125,000 mkg (meter-kilogram) capacity are available. Closed-die forgings weighing at least 15,000 lb are being made with every probability of larger sizes in the future. Of course, open die forgings much heavier than this are currently in production.

Two types

All methods of forging are basically related to hammering or pressing, the main difference between the two being the speed of pressure application. At times, the two methods may be interchangeable, depending on the availability of equipment and forging characteristics of the alloy. Certain metals and alloys resist rapid deformation and require the normally slower pressing operation.

Hammers are energized by gravity, air or steam, and repeated blows of a vertically guided ram on metal (usually heated) resting on the anvil, cause the metal to change shape. The steam hammer is generally used in the modern forge shop because of its flexibility of operation. Modern modifications of the conventional hammer are currently being operated. The "counterblow" hammer is one example: this has an anvil rising counter to the direction of the moving ram. Another related type of equipment—Dynapak—operates by suddenly released energy of highly compressed gas.

Pressing causes metal to move as the result of a slowly applied force. Presses operate by hydraulic, air or steam action or by mechanical means such as crank or screw. Hydraulic presses have slower action than hammers, and their pressure application may be more closely controlled, allowing the maintenance of sustained pressures. Mechanical action presses, particu-

larly of the crank type, may closely approach the speed of hammer blows and cannot maintain sustained pressures.

While presses are usually assumed to operate in a vertical plane, *upsetters,* which fall into the category of mechanical presses, operate in a horizontal plane with side or gripper dies moving in coordination and at 90 deg to the basic press movement. Ring rolling machines come in various designs, but fundamentally operate as presses, with moving rolls forcing the ring to the desired shape and size. Extrusion presses may operate either vertically or horizontally.

Die types

In forging, force is transmitted to the work piece through dies generally made of chromium-molybdenum-vanadium steels, sometimes modified by addition of nickel or tungsten. In open dies, the primary force (compression) is applied locally, and different parts of the forging are progressively worked. In closed dies, the primary force is applied on the entire surface and the metal is forced into a cavity for forging to shape.

Open die (or hand) forgings are produced either in a hammer or a press, using a minimum of tooling. This method offers low tool cost and fast initial delivery, but relatively poor utilization of material and slow production rates. Closed die forgings involve higher tool cost for small quantities of finished parts, but offer relatively good utilization of material, generally better properties, close tolerances, good production rates and good reproducibility.

Flat die forgings are produced on the anvil and offer simple configurations. *Blocker* type forgings have fairly generous design tolerances and are usually made in one set of closed dies. *Normal (or commercial) dies* impart a more exact configuration to the work piece which may be machined to required tolerances. *No draft-close tolerance dies* impart a finished shape and size to the forging which requires little or no machining. Generally, precision is inversely proportional to size, melting point of the metal, forging temperature and the reactive tendencies of the metal surface with the atmosphere and lubricants at the forging temperature. Precision is directly proportional to the number of dies used.

With most configurations, therefore, economics

dictate that it is cheaper to machine a commercial forging than to continue to approach the finished tolerances with die forging alone.

Forging specifications

Specifications for forging can be grouped into two classifications, material and dimensional.

Material specifications deal with chemistry, strength, ductility, impact resistance, conductivity, soundness, etc. These specifications can apply to the "as-forged" condition, but generally apply to condition after heat treatment. There are many standard specifications applied to the physical qualities of a forging, such as the *ASTM Standards* and the *Aircraft Material Specifications* (*AMS*) of the SAE.

Often the customer writes his own specifications based upon his particular need. The forging vendor reviews these specifications, and, based upon his experience or the experience of the mill from whom he will purchase material, either accepts or asks for modification of the specification.

Property guarantees. Physical property guarantees are based on experience with an alloy and the property level the alloy can be expected to exceed when properly melted and processed at the producing mill as well as forged and heat-treated. The guaranteed mechanical properties are based upon the alloy specified, and influenced somewhat by forging size, shape and test location. When no alloy is specified, and only minimum physical properties are known, then the forging vendor will determine the proper alloy based upon his experience. Mechanical properties are generally certified by the forging vendor through the destructive evaluation of a forging after heat treatment and prior to shipment. Testing frequency and method is generally agreed upon at the time of quotation.

Dimensions. Dimensional specifications refer to the size of the forging. Certain standards exist in the forging industry for such characteristics as mismatch (shifting of one die with respect to the other), oversize, warpage, etc. These tolerances are generally applied unless a customer requests more stringent dimensional requirements. Ordinarily a ⅛-in. envelope is placed around a finish machined part plus draft when designing a forging. However, many forgings today are designed with unmachined surfaces, especially in nonferrous alloys. Ferrous

alloys, because of their tendency to decarburize at ordinary forging and heat-treating temperatures without atmosphere protection, may require a machine envelope. Many forgings are made ± 0.010-in. tolerances through coining operations or special metal-removal techniques such as electro-chemical milling. Blade forgings of titanium, aluminum and austenitic stainless steels are produced regularly by these methods which require little or no machining on complex blade surfaces. These auxiliary techniques also make possible much thinner sections than ordinary forging procedures will permit, since in smaller forgings metal chilling against die surfaces results in loss of plasticity while in larger forgings, the plane area-thickness ratio is the limiting factor.

Forging alloys and applications. While all ductile metals can be forged, forgeability varies widely. In general, forgeability of a metal at the forging temperature depends on the crystallographic structure of the material, melting point, yield strength and ductility, recovery from forging stresses, surface reactivity and on strain rate during forging and on die friction.

In practice, relative forgeability might be interpreted in terms of: (1) resistance to flow into a standard die cavity; (2) force required for forging; (3) the number of forging cycles and reheats necessary; (4) amount of die wear; (5) crack susceptibility; and (6) cost of producing a forging of the metal in question. Materials most frequently forged, in the approximate order of forgeability based primarily on flow into a standard die cavity at proper forging temperature, are copper, aluminum, magnesium, carbon steels, low-alloy steels, stainless steels, tantalum, iron-base superalloys, titanium, columbium, molybdenum, tungsten, nickel-base alloys and refractory metal alloys. Beryllium, chromium and vanadium have limited forgeability because of inherent low ductility in presently available material. Depleted uranium, zirconium and other reactor metals are as easily forged as stainless steels, but offer special problems, particularly due to their reactivity.

Various alloys of these metals may be less (or more) difficult to forge than this list would indicate. Forgeability improves with quality of material. Quality is determined by chemical purity and composition-structure control and uniformity which is a result of the raw material and the melting-solidification or other consolidation processes.

Copper and its alloys (including brass and bronze) are eminently suited to forging. Copper can be forged cold, but usually is forged at 1400 to 1750 F. It is important for its corrosion resistance and for its electrical and thermal conductivity.

Aluminum, forged at 700 to 850 F, is employed for temperatures ranging up to 500 F, primarily for structural applications in the aircraft and transportation industries. With low density and good strength-to-weight ratio, aluminum forgings are used for airframe structural components, aircraft propellers, supercharger impellers, radial aircraft engine crankcases, internal combustion engine pistons and cylinder heads.

Magnesium, forged at 600 F, is employed within the same temperature range as aluminum, although certain alloys (magnesium-thorium, for example) are employed for short-time service up to 700 F. Magnesium forgings are used efficiently in lightweight structures and have the lowest density of any commercial metal.

Low-carbon and low-alloy steels compose the largest volume of forgings produced. These steels, forged at 2000 to 2300 F, are employed within the temperature range to 750 F because of low material cost, ease of fabrication and good mechanical properties. The design flexibility of steel is due in great part to its varied response to heat treatment, giving the designer a wide choice of properties in the finished forging.

Special alloys offer up to 240,000 psi yield strength and higher at room temperature. Steel forgings are indispensable in the transportation, power, petroleum and mining fields, and for industrial tools and equipment, farm implements and aircraft and missiles in highly stressed applications.

Stainless steel forgings are important for use at moderate temperatures and/or wherever corrosion is a problem. Forging temperature is usually 2000 to 2250 F. Pressure vessels and steam turbines and other applications in the chemical, food processing, petroleum and hospital service industries employ stainless steel forgings for stressed parts used up to 1250 F and unstressed to 1800 F and higher.

Titanium, within its primary temperature use range (up to 1000 F), is important for high strength, low density and excellent corrosion resistance. Forged at 1350 to 1950 F, depending on the alloy, alloys of titanium offer yield strengths in the 200,000 psi range with elongation up to 5% at room temperatures. Configurations nearly identical to steel parts are forgeable, and are 40% lighter than steel in weight. Titanium forgings are useful not only in compressor discs and blades and airframe construction, but also in submarine structures, ship propeller blades, valves and pipes and transportation and chemical applications.

Nickel-base superalloys are the ranking metals in terms of use for the temperature range between 1200 and 1800 F. Forging temperature for most alloys is 1900 to 2100 F. They are noted for long time creep-rupture strength and oxidation resistance. While present capability in forging these alloys is limited to 50- to 60-in. diameter because of size and ingots available, pressing capability is available for forgings up to 100-in dia. Steam and gas turbines gave impetus to forgings in both nickel and cobalt base alloys. Structural shapes, turbine wheels and buckets, pipe fittings and valves are commonly forged. Turbine wheels, for example, may be forged in these materials offering up to 150,000 psi yield strength at room temperature and 130,000 psi yield strength at 1400 F with 10% elongation at room temperature and above, and stress rupture strength up to 20,000 psi for 35 hours at 1800 F.

For cryogenic applications, certain aluminum, titanium, austenitic stainless steel and nickel-base alloys have good toughness and high strength-weight ratios, and do not exhibit ductile-brittle transition behaviour.

Advanced chemical, electrical and nuclear propulsion systems, and advanced flight vehicles, with their attendant high temperatures, have made *refractory metal* forgings important. Columbium, molybdenum, tantalum and tungsten, in order of melting point, are now forged for applications requiring enhanced resistance to creep in high thermal environments. With metallurgical stability during rapid temperature rise from room to operating temperatures, various refractory metals are usable for rocket nozzles, liners and throats, domes, leading edges, nose caps and hot structures. Parts forged of these metals are of medium to high density, hard, somewhat brittle and difficult to machine. Pure tantalum and columbium can be forged cold;

but, like molybdenum, are usually forged at 2100 to 2500 F. Tungsten, with the highest melting point (6170 F) of any metal, is forged at temperatures above 2500 F. These metals and their alloys are prepared for forging by either a melting or a powder metallurgical process. Size of forgings is currently restricted by the maximum volume of ingot material available.

Columbium, tantalum and other reactive metals are susceptible to contamination from the atmosphere during heating which causes cracks in forged surfaces. Proper coatings can minimize this problem.

Beryllium is among the newest metals in the forging field. Sintered ingots are forgeable to a limited extent; methods have been developed to forge such ingots at a temperature below 1500 F where rapid oxidation occurs. This method is currently restricted to cases where forging design or special economic advantages exist. For flexibility and economy, however, beryllium is being press-forged directly from powder contained in a steel can. Press-forging utilizes pressures of up to 75,000 psi within the range of sintering temperatures, and parts are produced in short cycles (minutes). Plastic deformation takes place during the latter part of the cycle. Capability for beryllium forgings up to 6 to 7 ft in diameter exists. Facilities and techniques in this area will grow in scope as designs dictate.

Zirconium and *hafnium*, used almost exclusively in nuclear applications, are readily forged.

Advantages of forgings

Orientation of crystal structure of the base metal and the flow pattern distribution of secondary phases (nonmetallics and alloy segregation) aligned in the direction of working is called grain flow. Metals in the solid state are a crystalline aggregate and, therefore, anisotropic to a degree in properties such as strength, ductility, impact and fatigue resistance. This anisotropy or directionality can be employed to the desired extent by orienting the metal during forging (through die design) in the direction requiring maximum strength.

It is possible to develop the maximum strength potential of a particular alloy in the forging process. Only quality-controlled rolled bar or cast ingot stock is used. Quality is determined by chemical analysis, macrostructure, microstructure, ultrasonics and mechanical testing. Further improvement comes during the forging process, since work on the material itself achieves recrystallization and grain refinement to produce material for optimum heat-treatment response.

Forgings are better than cast material for many applications because of their greater strength and ductility in a particular alloy as well as greater soundness, uniformity in chemistry, and finer grain size. There is no drastic change of state or volume in forging as there is in castings during solidification.

Forgings are highly reproducible because of this, and because of the use of carefully controlled material, controlled working temperatures and controlled metal flow in specially designed permanent die cavities. Forged components are also stronger than welded fabrications because weld efficiencies rarely equal 100%. Ordinary sintered products do not develop the full strength potential of an alloy because of porosity in the sintered part. As a result, a larger section is required for equivalent strength, but even here, ductility will be appreciably lower than in a part forged from material of the same chemistry.

Forgings, designed to approach the desired finished part configuration, make better utilization of material than parts machined from plate or bar stock. The closeness of this approach will be determined by the amount of money it is desired to spend on forge tools rather than machining capacity. Also, bar or plate stock has only one direction of grain flow. As a result, changes in section size cut across flow lines and render the material more notch, fatigue and stress-corrosion sensitive.

G

GALLIUM

Pure gallium is a silvery-white metal having a slight blue luster in incandescent light. Because it lies between aluminum and indium in Group IIIb of the periodic system, its chemical properties are close to those of aluminum and indium. Gallium shows a principal valence of three but compounds with lower valence can be formed. As with its sister metals, the nature of gallium compounds is a function of valence, size of ions, and type of bonding. Its crystal structure is orthorhombic.

Polycrystalline gallium shatters readily. However, gallium shows good malleability in single-crystal form. Gallium alloys readily with most metals at elevated temperatures. It alloys with tin, zinc, cadmium, aluminum, silver, magnesium, copper and others. Tantalum resists attack up to 450 C, and tungsten to 800 C. Gallium does not attack graphite at any temperature and silica-base refractories are satisfactory up to about 1000 C.

Gallium has a melting point of 29.78 C. Only two metals, mercury (−38.9 C) and cesium (28.5 C), melt at lower temperatures. Its boiling point at 2403 C gives it an extremely high liquid range. Gallium, like aluminum, forms an oxide film upon exposure to air.

Gallium reacts with sulfuric, nitric, perchloric, and hydrochloric acids, and sodium hydroxide. However, as the purity is increased, attack by acids and bases is reduced markedly. Aqua regia and concentrated caustic attack pure gallium most rapidly. Gallium is nonreactive with boiling water. The most common gallium compound is the trichloride ($GaCl_3$) which may be used in place of aluminum chloride in Friedel-Craft type syntheses.

Gallium, like water, bismuth, and germanium, expands on freezing. High-purity gallium super-cools very greatly and can remain liquid below the freezing point indefinitely.

Gallium is extremely anisotropic. Electrical resistance varies as 1:3.2:7 and the coefficient of thermal expansion varies as 31:16:11. The variation in electrical resistivity is thought to be greater than for any other known metal.

In general, gallium and gallium salts are considered to be nontoxic.

PHYSICAL PROPERTIES OF GALLIUM

Atomic Number	31
Atomic Weight	69.72
Atomic Structure	K shell—2
	L shell—2, 6
	M shell—2, 6, 10
	N shell—2, 1
Isotopes	
Ga-69	60.2%
Ga-71	39.8%
Crystal Structure	Orthorhombic
	a = 4.5258A
	b = 4.5198A
	c = 7.6602A
Melt Pt, C	29.78
Boil Pt, C	2403
Sp Gr, gm/ml	
Solid (29.6 C)	5.904
Liquid (29.8 C)	6.095
(32.38 C)	6.093
(301 C)	5.095
(600 C)	5.720
(806 C)	5.604
(1100 C)	5.445
Solidification Exp, %	3.2
Lin Coef of Ther Exp of Solid, $10^{-5}/°C$	
Crystallographic Axis	
c	1.6
a	1.1
b	3.1
Vol Coef of Ther Exp, $\times 10^{-5}$	
0 to 30 C (solid)	5.8
100 C (liquid)	12.0
900 C (liquid)	9.7
Vapor Pressure, mm Hg	
600 C	0.0000000044
800 C	0.0000059
1000 C	0.00082
1200 C	0.030

PHYSICAL PROPERTIES OF GALLIUM

1400 C	0.45
1600 C	3.8
1800 C	21
2000 C	86
2200 C	280
2403 C	760

Sp Ht, cal/gm/°C

−268.9 C (solid)	0.0000291
−257.1 C (solid)	0.0046
−213.1 C (solid)	0.042
0 to 24 C (solid)	0.089
12.5 C to 200 C (liquid)	0.095

Latent Ht of Fusion, cal/gm	19.16

Latent Ht of Vaporization at
Boil Pt, cal/gm   930.4

Viscosity, poises (cgs units)

97.7 C	0.01612
1100 C	0.00578

Vol Resis of Liquid, μ ohm-cm

0 C	25.2
20 C	25.6
40 C	26.0

Vol Resis of Solid, μ ohm-cm

	c Axis	a Axis	b Axis
29.7 C	54.3	17.4	8.1
0 C	48.0	15.4	7.16
−195.6 C	10.1	3.08	1.43
−268.9 C	0.00138	0.00068	0.00016
−272.06 C	Superconducting Transition Temp.		

Magnetic Suscept, cgs units,
$\times 10^{-6}$

Solid (18 C)	−0.24
Liquid (100 C)	−0.04

Standard Electrode Potential, V
$Ga \leftrightarrows Ga^{+3}$ −0.52

Hydrogen Overvoltage, V

Solid	0.31
Liquid	0.44

Ionic Rad, A	0.62
Covalent Rad, A	1.22
Hardness, Mohs scale	1.5–2.5

Applications

Semiconductors. In the initial development of semiconductors, germanium was the first material used. This was supplanted in part by silicon principally in military applications because germanium cannot be used in ambient temperatures much above 100 C. Silicon, however, has certain severe upper frequency limitations.

The temperature limitations on germanium and frequency limits on silicon led to a search for materials which might be used at both high temperatures and high frequencies. Certain combinations of elements of Group III and Group V of the periodic system looked promising.

Gallium arsenide is an interesting new material because both gallium and arsenic are available in the state of extreme purity required for semiconductor applications and because the finished gallium arsenide in the proper state of purity can be used in transistors at high frequencies and high temperatures.

Another device using gallium arsenide is the tunnel diode. Basically a tunnel diode is a heavily doped junction diode that displays a quantum-mechanical tunneling effect under forward bias. This effect leads to an interesting negative resistance effect.

Several applications for tunnel diodes are: replacement for phase-locked oscillators; switching circuits; FM transmitter circuits; and amplifiers.

Gallium arsenide is also used in increasing quantities for solar cells. A "paddle wheel" satellite in orbit demonstrates a dramatic commercial application of solar batteries.

In addition to gallium arsenide, several gallium compounds have found application in the semiconductor field. Gallium oxide has been used for vapor-phase doping of other semiconductor materials, and the oxide and halides have application in epitaxial growth of GaAs and GaP. Gallium itself has been used a a dopant for semiconductors. Gallium ammonium chloride has been used in plating baths for the electro-deposition of gallium onto whisker wires used as leads for transistors.

Other gallium alloys are suitable as dental alloys, and gallium is used in gold-platinum-indium alloys for dental restoration. Because of its low vapor pressure, gallium is being used as a sealant for glass joints in laboratory equipment, particularly mass spectrometers. Certain alloys (principally with cadmium and zinc) are used as cathodes in specialized vapor-arc lamps. Hard gallium alloys are used as low-resistance contact electrodes for bonding thermocouples and other wires to ferrites and semi-conductors.

GERMANIUM

Germanium is a hard, silver-white metallic element which at present finds predominant use as a semiconductor in transistors, diodes, rectifiers and infrared optics. Minor quantities are consumed in precious metal brazing alloys, color correcting fluorogermanates in mercury arc lamps and in intermetallic telluride compounds for thermoelectric devices.

It does not occur in nature in the elemental state. Although its abundance in the earth crust is in the same range as zinc and lead, economic concentrations occur only in complex copper ores at the Leopold Mine in Katanga Province and at Tsumeb, Southwest Africa. Other sources are zinc ores mined in the Tri-State area in the United States and some zinc ores mined in Italy. Germanium also occurs in certain coals or near-coals and some recovery is made from fly ashes from burning of these coals in England.

Production

Germanium recovery from ores involves considerable upgrading by hydro and pyrometallurgical methods to produce a germanium concentrate. Chlorination is generally used to extract the germanium followed by extensive processing to obtain a pure germanium tetrachloride. Hydrolysis with ultrapure water yields germanium dioxide. Reduction of the dioxide to powder and subsequent melting yields metallic germanium, which still contains considerable impurities and is called "first reduction metal" by the industry. Zone refining of this material produces polycrystalline semiconductor grade germanium metal. Due to the extremely high purity demanded of semiconductor materials, chemical analytical methods known at the present time are inadequate for detection of impurities. Several electrical methods have, however, been devised and one basic indication of the degree of purity is the resistivity of the metal measured under controlled conditions. First reduction metal is generally available with a minimum resistivity of 5 ohm-cm. Zone refined germanium has a minimum resistivity of 40 ohm-cm when measured at 25 C.

Properties

The important position germanium has achieved in the semiconductor industry is based on several factors. It has excellent semiconducting properties, it is easier to purify, and has a lower melting point than other semiconductors, specifically silicon, which results in greater ease in fabrication of finished devices. Metallic scrap produced during fabrication can also be recovered for re-treatment resulting in lower over-all material cost. Since germanium has a lower energy gap than silicon, it cannot operate in as high a temperature range as silicon. Its use in rectifiers is also restricted to low voltage, high amperage applications. Pertinent properties of metallic germanium are shown below.

PHYSICAL CONSTANTS

Atomic Weight	72.60
Density, 25 C, gm/cu cm	5.35
Melting Point, C	937 ± 1
Boiling Point, C	~2700
Hardness—Mohs scale (s)	6.25
Thermal Energy Gap (0 K), ev	0.75
Mobility Electrons 20 C	3950
Holes 20 C	1930
Vol Resis, 25 C, microhm-cm	60×10^6
Index of Refraction	4.068–4.143
Specific Heat, 25 C, cal/gm	0.086

Germanium is not attacked by oxygen to any appreciable extent below 600 C. Above that temperature oxidation proceeds rapidly. The halogens combine readily with the metal, with chlorine the most active and iodine the least reactive. Hydrochloric and sulphuric acid do not react with metallic germanium at room temperature. Potassium and sodium hydroxide solutions have little or no effect on the metal; however, molten alkali salts readily dissolve germanium. In addition, both nitric acid and aqua regia will attack the metal.

Germanium dioxide is reduced to the metal by hydrogen at 650 C. Several allotropic forms of the dioxide exist, depending on the method of formation; however, all forms melt at about 1100 C.

Germanium dioxide and metal show no evidence of toxicity.

GLASS

Glass, one of our oldest and most extensively used materials, is made from the most abundant of the earth's natural resources—silica sand. For centuries considered as a decorative, fragile material suitable for only glazing and art objects, today glass is produced in thousands of compositions and grades for a wide range of consumer and industrial applications.

Composition and structure

As just stated, the basic ingredient of glasses is silica (silicon dioxide), which is present in various amounts, ranging from about 50 to almost 100%. Other common ingredients are oxides of metals, such as lead, boron, aluminum, sodium and potassium.

Unlike most other ceramic materials, glass is non-crystalline. In its manufacture, a mixture of silica and other oxides is melted and then cooled to a rigid condition. The glass does not change from a liquid to a solid at a fixed temperature, but remains in a vitreous, noncrystalline state, and is considered as a supercooled liquid. Thus, with the relative positions of the atoms being similar to those in liquids, the structure of glass has short-range order. However, glass has some distinct differences compared to a supercooled liquid. Glass has a three-dimensional framework and the atoms occupy definite positions. There are covalent bonds present the same as those found in many solids. Therefore, there is a tendency towards an ordered structure in that there is present in glass a continuous network of strongly bonded atoms.

Glass production and processing

The major steps in producing glass products are 1. melting and refining 2. forming and shaping 3. heat treating and 4. finishing. The mixed batch of raw materials along with broken or reclaimed glass are fed into one end of a continuous type furnace where it melts and remains molten at around 2730 F. Molten glass is drawn continuously from the melting furnace and runs in suitable troughs to the working area where it is drawn off for fabrication at a temperature of about 1830 F. Where small amounts are involved, glass is melted in pots.

Forming Methods—Most glass products are manufactured on automatic high speed equipment by either blowing, pressing, rolling, drawing, or casting. Pressing, usually the lowest cost fabrication method, is used in manufacture of table and ovenware, insulators, lenses and reflectors. In this process, gobs of glass are fed into molds on a rotating press. The molds are moved beneath a plunger which forces the glass into final shape.

Glass blowing is used to produce hollow products such as bottles, jars and light bulbs. In glass blowing machines, molten glass in ribbon form sags through holes as air is blown in from above. At the same time, molds move up from below and clamp around the molten glass. More puffs of air force the glass into the mold to form the final shape.

The drawing process, which forms glass into tubing and rod, is performed at up to 40 miles per hour. For tubing, the molten glass passes around a ceramic or metal cone-shaped mandrel as air is blown through the center of the mandrel. A special drawing process is used to produce continuous glass fibers, or filaments. Molten glass is fed into a bushing that has a number of orifices through which the glass flows. The drawn filaments are collected into bundles called strands.

Sheet glass can be manufactured by either drawing, rolling or floating methods. In drawing, glass is drawn from molten pool through or over rollers. In rolling, molten glass passes between cooled rolls. And in floating, the molten glass is formed into sheet on the surface of a pool of molten tin in a controlled atmosphere.

Stationary casting is perhaps the most difficult method of forming glass, and is usually restricted to large, simple shapes, such as astronomical telescope lenses. Centrifugal casting in which a gob of glass is spread over the inside of a rapidly spinning mold by centrifugal force is used to form products such as the funnel portion of TV tubes and missile radomes.

Other fabricating methods include pressing and fusing powdered glass into shapes, and forming molten glass to produce cellular materials.

Heat Treating—There are two glass heat treating processes—annealing and tempering. When glass cools from the forming range to room temperature, thermal stresses develop that adversely affect strength properties. Therefore, almost all glass products are annealed to eliminate the stresses. The treatment involves heating the glass to its annealing temperature range, holding it there for a period of time and then cooling slowly to room temperature.

In contrast, tempering of glass involves heating it to around the softening point and then cooling it rapidly with blasts of air or by quenching in oil. Tempering produces glass with a rigid surface layer that is in compression and an interior that is in tension. Therefore, in service, compression stresses in the outer skin of the glass resist imposed tensile stresses and thereby greatly enhance over-all strength. Heat tempered glass is three to five times stronger than annealed glass while still retaining its initial clarity, hardness and expansion coefficient.

Special glass compositions can also be tempered by chemical treatments. The strength achieved is dependent on the specific treatment applied and the glass composition used. Sheets of chemically tempered glass can undergo repeated flexing without failure.

Finishing—Glass surfaces can be treated or finished in a number of ways. Chemical methods using hydrofluoric acid are used for polishing and etching. Glass can also be stained by copper or silver compounds. Metallizing is often used to form a base for sealing, for decoration, or to provide electrical conductivity. Fired-on films are used on such products as instrument windows and electronic components. Mechanical finishing of glass includes grinding to square and smooth edges. Polishing with materials such as ferric oxide or cerium oxide produces smooth, accurately-finished surfaces.

Glass can be sealed to most metallic materials. The principal problem is matching the coefficients of expansion of the two dissimilar materials. The wide range of expansion rates in glass help solve this problem, as do a number of special metal alloys developed principally for sealing glass.

Types of glass

There are a number of general families of glasses, some of which have many of hundreds of variations in composition (Table 1).

Soda-Lime Glasses—The soda-lime family is the oldest, lowest in cost, easiest to work and most widely used. It accounts for about 90% of the glass used in this country. Soda lime glasses have only fair to moderate corrosion resistance and are useful up to about 860 F, annealed, and up to 480 F tempered. Thermal expansion is high and thermal shock resistance is low compared to other glasses. They are the glass of ordinary windows, bottles and tumblers.

Lead Glasses—Lead or lead alkali glasses are produced with lead-contents ranging from low to high. They are relatively inexpensive and are noted for high electrical resistivity and high refractory index. Corrosion resistance varies

with lead content, but they are all poor in acid resistance compared to other glasses. Thermal properties also vary with lead content. Coefficient of expansion, for example, increases with lead content. High lead grades are heaviest of the commercial glasses. As a group, lead glasses are the lowest in rigidity. They are used in many optical components, for neon sign tubing and for electric light bulb stems.

Borosilicate Glasses—Borosilicate glasses are most versatile and are noted for their excellent chemical durability, for resistance to heat and thermal shock, and for low coefficients of thermal expansion. There are six basic kinds. The low expansion type is best known as the Pyrex brand ovenware. The low electrical loss types have a dielectric loss factor only second to fused silica and some grades of 96% silica glass. Sealing types, including the well-known Kovar, are used in glass-to-metal sealing applications. Optical grades, which are referred to as crowns, are characterized by high light transmission and good corrosion resistance. Ultraviolet transmitting and laboratory apparatus grades are two other borosilicate type glasses. Because of this wide range of types and compositions, borosilicate glasses find use in such products as sights and gages, piping, seals to low expansion metals, telescope mirrors, electronic tubes, laboratory glassware, ovenware, and pump impellers.

Aluminosilicate Glasses—These glasses are roughly three times more costly than borosilicate types, but are useful at higher temperatures and have greater thermal shock resistance. Maximum service temperature for annealed condition is about 1200 F. Corrosion resistance to weathering, water and chemicals is excellent, although acid resistance is only fair compared to other glasses. Compared to 96% silica glass, which it resembles, in some respects, it is more easily worked and is lower in cost. It is used for high performance power tubes, traveling wave tubes, high temperature thermometers, combustion tubes and stove-top cooking ware.

Fused Silica—Fused silica is 100% silicon dioxide. If natural occurring, the glass is known as fused quartz. There are many types and grades of both glasses, depending on impurities present and manufacturing method. Because of its high purity level, fused silica is one of the most transparent glasses. Also the most heat resistant of all glasses, it can be used up to 1650 F in continuous service and to 2300 F for

TABLE 1—COMPOSITION OF TYPICAL GLASSES

Designation	Composition, %								
	SiO_2	Na_2O	K_2O	CaO	MgO	BaO	PbO	B_2O_3	Al_2O_3
Silica Glass (fused silica)	99.5+								
96% Silica Glass	96.3	<0.2	<0.2					2.9	0.4
Soda-Lime (window sheet)	71–73	12–15		8–10	1.5–3.5				0.5–1.5
Soda-Lime (plate glass)	71–73	12–14		10–12	1–4				0.5–1.5
Soda-Lime (containers)	70–74	13–16		10–13		0–0.5			1.5–2.5
Soda-Lime (electric lamp bulbs)	73.6	16	0.6	5.2	3.6				1
Lead-Alkali Silicate (electrical)	63	7.6	6	0.3	0.2		21	0.2	0.6
Lead-Alkali Silicate (high-lead)	35		7.2				58		
Borosilicate (low-expansion)	80.5	3.8	0.4					12.9	2.2
Borosilicate (low-electrical loss)	70.0		0.5				1.2	28.0	1.1
Aluminosilicate	57	1.0		5.5	12			4	20.5

short term exposure. In addition, it has outstanding resistance to thermal shock, maximum transmittance to ultraviolet and excellent resistance to chemicals. Unlike most glasses, its modulus of elasticity increases with temperature. Because fused silica is high in cost and difficult to shape, its applications are restricted to such specialty applications as laboratory optical systems and instruments and crucibles for crystal growing. Because of a unique ability to transmit ultrasonic elastic waves with little distortion or absorption, fused silica is used in delay lines in radar installations.

96% Silica Glass—These glasses are similar in many ways to fused silica. Though less expensive than fused silica, they are still more costly

TABLE 2—TYPICAL PROPERTIES OF GLASSES

Designation	Viscosity Data			Coeff of Expan per °C (0 to 300 C)	Specific Gravity	Refract Index (sod. D)	Young's Modulus, psi
	Strain Pt., C	Anneal Pt., C	Soft Pt., C				
Silica Glass (fused silica)	1070	1140	1667	5.5×10^{-7}	2.20	1.458	10×10^6
96% Silica Glass (7900)	820	910	1500	8×10^{-7}	2.18	1.458	9.7×10^6
Soda-Lime (window sheet)	505	548	730	85×10^{-7}	2.46	1.510	
Soda-Lime (plate glass)	510	553	735	87×10^{-7}	↓ to ↓	↓ to ↓	10×10^6
Soda-Lime (containers)	505	548	730	85×10^{-7}	2.49	1.520	
Soda-Lime (elect. lamp bulbs)	470	510	696	92×10^{-7}	2.47	1.512	9.8×10^6
Lead-Alkali Silicate (electrical)	395	435	626	91×10^{-7}	2.85	1.539	9.0×10^6
Lead-Alkali Silicate (high-lead)	395	430	580	91×10^{-7}	4.28	1.639	7.6×10^6
Borosilicate (low expansion)	520	565	830	32×10^{-7}	2.23	1.474	9.8×10^6
Borosilicate (low electrical loss)	455	495		32×10^{-7}	2.13	1.469	6.8×10^6
Aluminosilicate	670	715	915	42×10^{-7}	2.53	1.534	12.7×10^6

than most other glasses. Compared to fused silica they are easier to fabricate, have a slightly higher coefficient of expansion, about 30% lower thermal stress resistance, and a lower softening point. They can be used continuously up to 1470 F. Uses include chemical glassware and windows and heat shields for space vehicles.

Other Glasses—Nonsilicate glasses include borate glasses, which have very low light dispersion and high refractive index; phosphate glasses that are resistant to hydrofluoric acids, but low resistance to water attack; calcium aluminate germanate, and arsenic trisulfide glasses that are useful in infrared transmission systems.

Colored glasses, made by adding small amounts of colorants to glass batches, are used in lamp bulbs, sunglasses, light filters and signalware.

Opal glasses, contain small particles dispersed in transparent glass. The particles disperse the light passing through the glass, producing an opalescent appearance.

Laminated or safety glass is composed of two or more layers of glass with a layer or layers of transparent plastic, usually vinyl, sandwiched between the glass. Bullet resisting plate glass is similarly made up of multiple layers of glass which are bonded together with plastic film.

Photosensitive glass is sensitive to ultraviolet light and heat. This property permits reproduction of images on it from a photographic negative. Upon immersion in an acid bath, the image is etched on the surface. The glass can be intricately shaped and patterned without use of mechanical tools.

Cellular or foam glass, made by heating a mixture of pulverized glass and a foaming agent, is almost as light as cork and is used as an insulating material.

Coated glass has a thin, metallic oxide surface coating which can conduct electricity. In sheet or panel form it is used in lighting applications. In rod and tube form it is used for resistors in electronic devices.

Physical properties

Glasses and ceramics have similar, but not identical, mechanical properties. The same broad generalizations, with some important differences, can be made of the behavior of the vitreous and crystalline states. They are perfectly elastic, and return to their original shape (at ordinary temperatures) after the release of any applied forces. They do not exhibit the phenomenon of plastic flow, and consequently have no yield point. Their behavior differs so much under applied loads from that of metal that a few generalizations need to be made.

1. Glasses and ceramics are brittle materials. They are elastic to fracture; they fail without previous yield or plastic deformation.

2. Failure always results from a tensile component of stress, even when the load is applied in compression. The strength of these materials, then, can be described by an appropriate value of tensile strength.

3. The effect of static fatigue is relatively large. The extrapolated infinite time modulus of rupture for glasses is usually ½ to ⅓ that for short times. For ceramics this reduction in breaking stress with time is not so great; it may be nearer 70 to 80%. The cumulative effect, integrating the load over the total time, depends little for all ceramics (including glass) on whether the load is static or cyclic.

Optical properties

Most ceramics, because of their randomly oriented crystalline character, are opaque; their colors are those of reflection. The seemingly limitless range and variety of colors of glasses, from the deepest red to the most intense blue, are the result of selective absorption and transmission resulting from the various constituents of the glasses. This occurs not only in the visible region of the spectrum, but at longer and shorter wave lengths also. The low transmittance for many glasses in the infrared at about 2.7 microns is the result of absorption by hydroxyl ions found in the structure as the result of water present during melting of the glass.

Glasses may be designed to interact with radiation impinging on them. One special glass composition contains phosphates which are sensitized by gamma-ray irradiation. If the glass is then bombarded with X-rays, some of the electrons raised to higher energy states by interaction with the gamma rays will fall back to lower energy states, with emission of fluorescence proportional to the total amount of radiation. Such a glass can therefore be used as a dosimeter;

the sensitivity can be adjusted over wide limits by deliberate changes in composition. The lovely soft purple color of glass in old windows or bottles after long exposure to sunlight is the result of the interaction of the ultraviolet light with one of the minor constituents, manganese dioxide, of the glass. Proper compositional changes can greatly increase the sensitivity of glasses to UV. This is the basis of the photosensitive family of glasses.

GLASS-CERAMICS

In May 1957, Corning Glass Works announced the discovery of a new family of fine-grained crystalline materials made from glass. These materials are called glass-ceramics, and products made from them are designated by the trademark "Pyroceram." They are made by a process of controlled crystallization, from special glass compositions containing nucleating agents.

Properties

Glass-ceramics are nonporous and generally have an opaque white appearance in the finished state, though some compositions may also be transparent like glass. They are resistant to high temperatures in terms of both strength and oxidation. They are like glasses in their ability to resist chemical attack with durabilities measured in weight loss, comparable to those of durable chemical glasses. They are lighter in weight, yet harder than most metals. In thermal conductivity they are classed as heat insulators. The thermal expansion coefficients range from negative values through zero to positive values of 110×10^{-7} per degree F. Hence, extreme heat shock resistance and dimensional stability with temperature may be had by choice of composition to provide coefficients of expansion near zero.

In electrical properties glass-ceramics are classed as electrical insulators having dielectric constants from 5 to 10 and power factors at high frequencies ranging down to extremely low values. In mechanical properties these materials are generally higher in Young's modulus than glasses, ranging from 12.5 to some 20×10^6 psi with strengths higher than glasses and most ceramics. Glass-ceramics are not ductile or malleable as are most metals, but are classed as brittle materials.

Various compositions can be tailor-made to provide desired properties, with the final control depending on special heat treatment.

Properties of Pyroceram Code 9606 indicate that it is suitable for high-temperature, high-frequency applications in the electronics field. The maximum usable temperature under load is 1560 F for 1000 hours. This glass-ceramic has been successfully used as a missile radome. Its hardness and ability to withstand high temperatures made it suitable for consideration as an abrasive binder and for bearing parts.

Special glass-melting techniques are used to assure uniform composition, constant density, freedom from bubbles and striations, and uniform electrical properties. Pyroceram Code 9606 is most easily formed by gravity and centrifugal casting. With certain limitations, it can be pressed and rolled.

Applications

The first use of glass-ceramics has been in radomes for supersonic missiles, where a radar-transmitting material having a combination of strength, hardness, temperature and thermal shock resistance, uniform quality, and precision finishing is required.

The second large-scale application is in "heatproof" skillets and saucepans, again taking advantage of the thermal shock resistance.

One of the products of potential interest to mechanical engineers is bearings for operation at high temperatures without lubrication, or in corrosive liquids. Another is a lightweight, dimensionally stable honeycomb structure which has promise for use in heat regenerators for gas turbines operating at high temperatures. Still a third possibility is precision gauges, and machine-tool parts whose dimensions do not change with temperature.

Bearings. While metal bearings are perfectly satisfactory for most applications, there are conditions such as high temperature, dry (unlubricated) operation, and presence of corrosive media in which even the best metal bearings may perform poorly. Also, the metals which do perform best are expensive, difficult to fabricate, and heavy.

Since glass-ceramic bearings might be expected to be stable at high temperatures, resistant to oxidation and to corrosive conditions as well as light in weight, a thorough evaluation of their characteristics is being made by a number of laboratories and bearing manufacturers. One of the surprising findings has been that glass-ceramic bearings can be finished almost as easily

TABLE 1—COMPARATIVE PROPERTIES OF GLASS-CERAMICS

	Glass-Ceramic (Pyroceram)		96% Silica Glass (Code 7900)	Borosilicate Glass (Code 7740)	Ceramic (High Purity Aluminas)
	9606	9608			
GENERAL					
Specific Gravity (77 F)	2.60	2.50	2.18	2.23	3.6
Water Absorption, %	0.00	0.00	0.00	0.00	0.00
Porosity (gas permeability)	gas-tight	gas-tight	gas-tight	gas-tight	gas-tight
THERMAL					
Softening Temp, F	2282	2282	2732	1508	3092
Specific Heat (77 F)	0.185	0.190	0.178	0.186	0.181
Mean Specific Heat (77 to 572 F)	0.230	0.235	0.224	0.233	0.241
Thermal Conductivity (77 F mean temp), Btu/hr/sq ft/ °F/ft	2.10	1.13	—	0.626	1.25–1.40
Linear Coeff. of thermal expansion (77 to 572 F) $\times 10^7$ per °F	31.6	3.9–11.1 [a]	4.45	17.8	40.5 [b]
MECHANICAL					
Modulus of Elasticity, 10^6 psi	17.3	12.5	9.6	9.5	40
Poisson's Ratio	0.245	0.25	0.17	0.20	0.32
Modulus of Rupture (abraded), 10^6 psi	20	16–23	5–9	6–10	40–50 [c]
Hardness					
Knoop 100 gm	698	703	532	481	1880
500 gm	619	588	477	442	1530

[a] Depending on heat treatment.

[b] 68–932 F.

[c] Unabraded values.

as steel bearings and in the same general type of equipment.

High-temperature heat exchangers. The thermal shock resistance and dimensional stability of the low expansion glass-ceramics make them useful in various kinds of heat exchangers.

One interesting type has been developed for use in high-temperature turbine engines.

Precision uses. As requirements for precision instruments and machines become more stringent, the gauges and machine tools required to make them must become still more precise in dimensions. If these are made of relatively high expansion metals such as steel, the dimensions vary with temperature.

The expansion coefficient of Pyroceram Code 9608 between 0 C and 100 C is so nearly zero that no measurable change in length occurs in this temperature range. For this reason, it should be a useful material for precision gauges, for the beds of special machine tools, and other parts requiring constant dimensions.

GOLD AND ALLOYS

Gold is a soft, very ductile, yellow metal. Commercially pure gold is 99.97% pure, and higher purity material is available.

The outstandingly useful property of gold is its oxidation resistance. Gold does not oxidize in air, even on heating. It is not attacked by the common acids when used singly. It does, however, dissolve in aqua regia (nitric acid plus hydrochloric acid) and cyanide solutions, and is attacked by chlorine above 80 C. It is resistant to dry fluorine up to about 300 C, hydrogen fluoride, dry hydrogen chloride and dry iodine. It is also resistant to sulfuric acid, sulfur and sulfur dioxide.

Table 1 gives some of the properties of gold.

TABLE 1—PROPERTIES OF GOLD

Atomic Number	79
Atomic Weight	197.0
Crystal Structure	face-centered cubic a = 4.08A at 25 C
Density, 25 C, gm/cu cm	19.3
Melting Pt, C	1063
Lin Ther Exp Coef, 0–100 C, per °C	14.2×10^{-6}
Sp Ht, 20 C, cal/gm/°C	0.031
Therm Cond, 20 C, cal/cm/ sec/sq cm/°C	0.71
Electrical Resis, 20 C, μ ohm-cm	2.44
Temp Coef of Elec Resis, 0–100 C, per °C	0.0039
Mod of Elast, 10^6 psi	11.4
Mod of Torsion, 10^6 psi	4.0
Poisson's Ratio	0.42
Ten Str (hard, 60% red), 1000 psi	30
Elong, in 2 in., %	0.5
Ten Str (ann), 1000 psi	17
Elong, in 2 in., %	40
Bhn (60% red)	60
Bhn (ann)	25

Alloys and applications

Gold is a very soft, easily deformed metal, and in most uses it is alloyed to increase its hardness without appreciable loss of oxidation resistance. Copper is a common alloying element along with silver and small amounts of the platinum metals. Some of these alloys can be heat-treated to relatively high strengths. Gold and its alloys are worked into all the usual forms of sheet, wire, ribbon and tubing. Precision casting is also used to form gold alloys, particularly for jewelry. The expense of gold and its low hardness are often offset by using it as a laminate or plating on base metals. It can also be applied to metals, ceramics and some plastics by the thermal decomposition of certain gold compounds.

Gold alloys (Au-Ag-Cu and Au-Ni-Cu-Zn) can be made in a range of colors from white to many shades of yellow. For this reason, gold is widely used for jewelry and other decorative applications. Similar alloys (Au-Ag-Cu-Pt-Pd) are used in dentistry, the nobility of the alloys and their response to heat-treatment hardening being of concern here. Gold-silver alloys have been used as low current electrical contacts (under 0.5 amp). Gold is often used in electrical and other equipment which is used for standards, and where stability is of prime concern.

The corrosion resistance and melting point of gold also make it useful as a brazing material. It is used as well in chemical equipment, where its susceptibility to chlorine attack is not a problem. In particular, it is used to line reaction vessels, and as a gasketing material.

Gold may be readily applied by electroplating from cyanide and other solutions. It may also be applied to some metals by simple immersion in special plating solutions. Plating has, of course, many decorative applications.

Gold plating is also used to make reflectors, particularly for the infrared wavelengths. Electrical components are often gold plated, especially for high frequencies, because of the low electrical resistance of gold. Vacuum tube grids may be gold plated to reduce electron emission and some electrical contacts are gold plated.

Gold may also be applied by using "liquid bright golds." These are varnishlike solutions of gold compounds which may be applied in any suitable manner—brushing, spraying, printing, etc.—to metals, ceramics, and some plastics. After being applied, the material is heated to decompose the compound, depositing a tightly adhering gold layer. Some gold alloys may also be applied in this manner. This method is used for the decoration of china and glassware, as well as for printed circuits and electrical resistance elements.

GRAY IRONS

Gray iron is characterized by the presence of flakes of graphite supported in a matrix of ferrite, pearlite, austenite, or any other matrix attainable in steel. The major dimension of the flakes may vary from about 0.002 to 0.04 in. Because of their low density, the graphite flakes occupy about 10% of the metal volume. The flakes interrupt the continuity of the matrix, and have a large effect on the properties of gray iron. In addition, the flakes give a fractured surface that is gray. This is responsible for the name, "gray iron."

High carbon content and the flakes of graphite give gray iron unique properties as follows:

1. Lowest melting point of the ferrous alloys,

so that low-cost refractories can be used for molds.

2. High fluidity in the molten state so that complex and thin designs can be cast.
3. Excellent machinability, better than steel.
4. High damping capacity and ability to absorb vibrations.
5. High resistance to wear involving sliding.
6. Low ductility and low impact strength when compared with steel.

Gray iron is by far the most common and widely used cast iron. Average production in the United States is about 12,000,000 tons per year. Gray iron is encountered almost exclusively as shaped castings used either with or without machining. Typical common applications include:

1. Pipe for underground service for water or gas.
2. Ingot molds into which steel and other metals are cast.
3. Cylinder blocks and heads for internal combustion engines.
4. Frames and end bells for electric motors.
5. Bases, frames and supports for machine tools.
6. Sanitary ware such as sinks and bathtubs (usually coated with porcelain enamel).
7. Pumps, car wheels and transmission cases.

The major industries that consume gray-iron castings are as follows: automotive, building and construction, utilities, machine tools, architectural, rolling mills (steel plants), general machinery, household appliances and heating equipment.

Gray-iron castings commonly are ordered to meet ASTM specifications. Some of the most common are:

ASTM Designation	Title
A 48	Gray Iron Castings
A 159	Automotive Gray Iron Castings
A 278	Gray Iron Castings for Pressure-Containing Parts for Temperatures up to 650 F
A 377	Cast Iron Pressure Pipe
A 436	Austenitic Gray-Iron Castings

For engineering applications where tensile strength is important, gray iron usually is classified on the basis of minimum tensile strength in a specimen machined from a separately cast test bar. These are the three most common sizes of test bars:

Test Bar Designation	Nominal As-Cast Dia, in.	Minimum Length, in.
A	0.88	5.0
B	1.20	6.0
C	2.00	7.0

Each class of gray iron is designated by a number followed by a letter. The number indicates the minimum tensile strength of the iron in thousands of psi in a separately cast test bar, and the letter indicates the size of the test bar. Examples of proper designations are as follows:

1. Gray Iron Castings, ASTM Designation A 48, Class 30B

   (Minimum tensile strength of 30,000 psi in a test bar cast with a diameter of 1.20 in.)

2. Gray Iron Castings, ASTM Designation A 48, Class 40C

   (Minimum tensile strength of 40,000 psi in a test bar cast with a diameter of 2.0 in.)

ASTM Designation A 48 provides for irons with minimum tensile strengths of 20,000, 25,000, 30,000, 35,000, 40,000, 45,000, 50,000 and 60,000 psi.

Gray irons in Classes 20A, 20B, 20C, 25A, 25B, 25C, 30A, 30B, 30C, 35A, 35B and 35C are characterized by excellent machinability, high damping capacity, low modulus of elasticity, and comparative ease of manufacture. Castings in Classes 40B, 40C, 45B, 45C, 50B, 50C, 60B and 60C are usually more difficult to machine, have lower damping capacity, a higher modulus of elasticity, and are more difficult to manufacture than the lower strength alloys.

The relationship between tensile strength and Brinell hardness of gray iron is illustrated by the following series of irons in the *Standard for Automotive Gray Iron Castings* of the Society of Automotive Engineers:

SAE No.	Ten Str, 1000 psi, min.	Bhn
110	20	187 max.
111	30	170 to 223
120	35	187 to 241
121	40	202 to 255
122	45	217 to 269

Other mechanical and physical properties of gray iron are summarized below:

Compressive strength. Unusually high; at least three times the tensile strength.

Modulus of elasticity. Increases with tensile strength. About 12×10^6 psi for a tensile strength of 20,000 psi, and up to about 20×10^6 psi for a tensile strength of 60,000 psi.

Endurance limit. About 35 to 50% of tensile strength. Gray iron is relatively insensitive to the effect of notches.

Damping capacity. Very high, especially in irons of high carbon content. Specific damping capacity is about 10 times that of steel.

Specific gravity. Varies from about 6.8 for high-carbon, low-strength irons to about 7.6 for low-carbon, high-strength irons.

Coefficient of thermal expansion. Slightly lower than that of steel; averages about 6×10^{-6} per °F.

Coefficient of thermal conductivity. About the same as many other ferrous alloys; about 0.11 to 0.14 in CGS units. Can be lowered appreciably by adding alloying elements.

Gray iron heat treatment

Gray cast iron is amenable to many of the types of heat treatment applied to steel. Because of its high content of carbon and other elements, gray iron has high hardenability.

Residual stresses in the casting can be removed and dimensional stability improved by heating the casting to about 1000 F, holding it at least 1 hr, and cooling it to 400 F at a rate not higher than 100 F per hr. This treatment has little effect on the strength or hardness of the casting.

Machinability can be improved by a subcritical spheroidizing anneal at about 1200 F, followed by slow cooling.

If a gray-iron casting contains hard edges or spots of chilled white iron, complete softening can be obtained by a suitable anneal at about 1650 F.

Wear resistance and strength can be improved by austenitizing the casting at about 1650 F, quenching to martensite in oil or water, and tempering to the desired hardness. Some designs or types of iron castings may crack when heat-treated in this way. One common solution to this problem is to flame-harden or induction-harden only those surfaces or sections where improved properties are needed.

H

HAFNIUM AND ALLOYS

Hafnium, the heaviest of the three metals comprising the Group IV transition metals, has developed from the stage of a laboratory curiosity to the production stage within the past decade. Because of the startling similarity in their chemical properties, zirconium and hafnium always occur together in nature. In their respective ability to absorb neutrons however, they differ greatly, and this difference has led to their use in surprisingly different ways in nuclear reactors. Zirconium, with a low neutron-absorption cross section (0.18 barn), is highly desirable as a structural material in water-cooled nuclear reactor cores. Hafnium, on the other hand, because of its high neutron-absorption cross section (105 barns) can be used as a neutron-absorbing control material in the same nuclear reactor cores. Thus, the two elements, which occur together so intimately in nature that they are very difficult to separate, are used as individual and important but contrasting components in the cores of nuclear reactors. In view of the chemical similarity of zirconium and hafnium and the lack of analytical X-ray techniques, it is not surprising that hafnium went undetected until 1922.

Properties

Pure hafnium is not so ductile nor so easily worked as zirconium; nevertheless, hafnium can be hot- and cold-rolled on the same equipment and with similar techniques as those used for zirconium. Hafnium is characterized by a high thermal-neutron-capture cross section, excellent strength up to 540 C (which makes it useful in the unalloyed form in nuclear reactors), and excellent resistance in a wide range of corrosive environments.

TYPICAL PROPERTIES OF ANNEALED IODIDE
PROCESS HAFNIUM

Density, gm/cu cm	13.09 ± 0.01
Melting Point, °C	2222 ± 30
Crystal Structure	Below 1760 C, close-packed
	hexagonal Above 1760 C, body-centered cubic
Coef of Ther Exp, 25 C, in./in./°C	5.9×10^{-6}
Ther Cond, watts/cm/°C	0.223
Sp Ht, 25 C, cal/mol/°C	6.25
Ther Neutron Cross Section, barns	105 ± 5
Hardness, Bhn	150–180
Ten Str, 1000 psi	66.5
Yld Str, 1000 psi	35.3
Elong, in 2 in., %	35
Red in Area, %	38

Hafnium alloys and compounds

Hafnium forms refractory compounds with carbon, nitrogen, boron and oxygen. Hafnium carbide has a density of 12.20 gm per cu cm and a melting point of 3890 C, the highest melting point of any simple carbide. Hafnium nitride, with a melting point of 3300 C, has the highest melting point of any nitride and hafnium boride, with a melting point of 3260 C, has a melting point higher than any other boride. The refractory nature of these hafnium compounds is opening up new areas of application in missiles and high-temperature nuclear reactors.

Fabrication

Hafnium is not as ductile as its sister element zirconium, but it is fabricated in much the same way.

Formability is proportional to the purity of the metal (the iodide-process or equivalent material is quite easy to work). Hafnium that contains over 1500 ppm oxygen is nearly impossible to fabricate. The metal may be joined to itself or to zirconium by welding in an inert atmosphere in which tungsten, zirconium or hafnium electrodes are used.

The recommendations for fabrication with zirconium also apply to hafnium.

Corrosion properties

Very little reliable information has been accumulated on the corrosion-resistant properties of hafnium in various chemical environments. The information that is available indicates that hafnium is equivalent to or superior to zirconium. For example, in pressurized water reactors, unalloyed hafnium exhibits a corrosion rate approximately one-third that of the most satisfactory zirconium alloys developed to date. As hafnium supplies become more generally available, corrosion data for many severe conditions will become available.

The rapid development of hafnium from a laboratory curiosity to a commercially available metal of importance in modern technology has made the most up-to-date literature on the subject obsolete within a relatively short period of time. For a more detailed presentation on hafnium, reference should be made to *The Metallurgy of Hafnium*, edited by Thomas and Hayes and available from the Superintendent of Documents, Washington 25, D. C.

HARD FACINGS

Hard facing is a technique by which a wear-resistant overlay is welded on a softer and usually tougher base metal. The method is versatile and has a number of advantages: (1) wear resistance can be added exactly where it is needed on the surface; (2) hard compounds and special alloys are easy to apply; (3) hard facings can be applied in the field as well as in the plant; (4) expensive alloying elements can be economically used; (5) protection can be provided in depth; and (6) a unique and useful structure is provided by the hard-surfaced, tough-core composite.

Many of the merits of hard facing stem from the hardness of the special materials used. For example, ordinary weld deposits range in hardness up to about 200 Brinell, hardened steels have a hardness up to 700 to 800 Vickers, and special carbides have hardness up to about 3000 Vickers. However, it is important to note that the hardness of the materials does not always correlate with wear resistance. Thus, special tests should be performed to determine the material's resistance to impact, gouging abrasion, grinding (high-stress abrasion), erosion (low-stress scratching abrasion), seizing or galling, and hot wear.

Another important point is that durable overlays are not necessarily hard. Most surfacing

is used to protect base metals against abrasion, friction and impact. However, many "hard facings" such as the stainless steels, related nickel-base alloys and copper alloys are used for corrosion-resistant applications where hardness may not be a factor. Also, the relatively soft leaded bronzes may be used for bearing surfaces. Other facings are also used for heat- and oxidation-resistant applications.

Methods of application

Hard facings can be applied by: (1) manual, semiautomatic and automatic methods using bare or flux coated electrodes; (2) submerged-arc welding; (3) inert-gas shielded arc welding (both consumable and tungsten electrode types); (4) oxyacetylene and oxyhydrogen gas welding; (5) metal spraying; and (6) welded or brazed on inserts. Gas welding and spraying usually provides higher quality and precise placement of surfaces; arc welding is less expensive. Automatic or semiautomatic methods are preferred where large areas are to be covered, or where repetitive operations favor automation.

Surfacing filler metals are available in the form of drawn wire, cast rods, powders, and steel tubes filled with ferroalloys or hard compounds (e.g., tungsten carbide). The electrodes may take the form of filled tubes or alloyed wires, stick types, or coils specially designed for automated operations. The stick type electrodes may have a simple steel core and a thick coating containing the special alloys. In submerged-arc welding the alloys may be introduced through a special flux blanket. In spray coating, the materials are used in the form of powders or bonded wire.

Sprayed facings are advantageous in producing thin layers and in following surface contours. With this method it is usually necessary to fuse the sprayed layer in place after deposition in order to obtain good abrasion resistance. However, under boundary lubrication conditions the as-sprayed porosity of the facings may aid against frictional wear.

Hard facings are used in thicknesses from $\frac{1}{32}$ to 1 in. or more. The thinnest layers are usually deposited by gas welding, usually with low melting alloys that solidify with many free carbide or other hard compound crystals. The thick deposits are usually made from air-hardening or austenitic steels.

Hard overlays are usually strong in compression but weak in tension. Thus, they perform better in pockets, grooves or low ridges. Edges and corners must be treated cautiously unless the deposit is tough. Brittle overlays should be deposited over a base of sufficient strength to prevent subsurface flow under excessive compression.

Gas welding is a useful method for depositing small, precisely located surfacings in applications where the base metal can withstand the welding temperatures (e.g., steam valve trim and exhaust valve facings). On the other hand, heavy layers and large areas may be impossible to surface without cracks with the harder, more wear-

resistant alloys because of the severe thermal stresses that are encountered (e.g., usually in arc welding). Thus, the opposing factors of wear resistance and freedom from cracking frequently require a compromise in process and material selection.

Materials selection

Table 1 shows the principal hard-surfacing materials and their important properties and applications. These materials are roughly listed in order of decreasing abrasion resistance and increasing toughness. Although many transition and proprietary materials are available, the

TABLE 1—PROPERTIES AND APPLICATIONS OF HARD-FACING MATERIALS

Types of Hard Facing	Important Features	Applications
Tungsten Carbide Deposits Granules or Inserts Coarse Granule Tube Rods	Maximum abrasion resistance. —— Worn surfaces become rough.	Wide range of severe abrasive applications. Oil well rock drill bits and tool joints.
Fine Granule Tube Rods	Best performance when gas-welded.	Plows.
High Chromium Irons Multiple Alloy Martensitic Austenitic	Excellent erosion resistance. Oxidation resistant. Hot hardness from 800–1200 F with W and Mo. Can be annealed and rehardened. ——	Erosion by catalysts (1000 F) in refineries. Abrasion by hot coke (coke pusher shoes). —— Agricultural equipment in sandy soil such as plowshares and cultivator sweeps.
Martensitic Alloy Irons Chromium-Tungsten Chromium-Molybdenum Nickel-Chromium	Excellent abrasion resistance. High compressive strength. Good for light impact.	General abrasive conditions with light impact. Machine parts subject to repetitive wear and impact such as engine rocker arm faces. Metal to metal wear. Same.
Austenitic Alloy Irons Chromium-Molybdenum Nickel-Chromium	More crack resistant than martensitic irons.	General erosion conditions with light impact such as crusher rolls, dipper teeth and lips dredge shells and cutter heads.
Chromium-Cobalt-Tungsten Alloys High Carbon (2.5%) Medium Carbon (1.4%) Low Carbon (1%)	Hot strength and creep resistance. Brittle; abrasion resistant. Tough and oxidation resistant.	— Hot wear and abrasion above 1200 F, as in blanking and forming dies. Exhaust valves of gasoline engines; valve trim of steam turbines.

TABLE 1—PROPERTIES AND APPLICATIONS OF HARD-FACING MATERIALS (*Continued*)

Types of Hard Facing	Important Features	Applications
Nickel Base Alloys	Good hot hardness and erosion resistance.	—
Nickel-Chromium-Boron	Good hot hardness and erosion resistance.	Oil well slush pumps; coal conveyor screws.
Nickel-Chromium-Moly-Tungsten	Corrosion resistant.	—
Nickel-Chromium-Molybdenum	Resistant to exhaust gas erosion.	Engine exhaust valves.
Nickel-Chromium	Oxidation resistant.	—
Copper Base Alloys	Antiseizing; resistant to frictional wear.	Pumps, shafts and bearing surfaces.
Martensitic Steels		
High Carbon (0.65–1.7%)	Fair abrasion resistance.	General abrasive conditions with medium impact. Crushing rolls; dredge cutter heads.
Medium Carbon (0.30–0.65%)	Good resistance to medium impact.	
Low Carbon (below 0.3%)	Tough, economical.	Hot working dies.
Semiaustenitic Steels	Tough, crack resistant.	General low cost hard facing; crusher jaws; sand pump impellers.
Pearlitic Steels	Crack resistant and low in cost.	Base for surfacing or buildup to restore dimensions.
Low Alloy Steels	Suitable for buildup of worn areas.	Tractor rails, rollers and idlers; hammer mill hammers.
Simple Carbon Steels	Good base for hard facing.	—
Austenitic Steels	Tough; excellent for heavy impact.	General metal-to-metal wear under heavy impact, railway trackwork, dipper teeth and bucket fronts.
13 Manganese-1% Moly	Fair abrasion and erosion resistance.	
13 Manganese-3% Nickel	Lower yield strength.	Same.
13% Manganese-Nickel-Chromium	High yield strength for aust. types.	Same.
High Carbon Nickel-Chromium Stainless	Oxidation and hot wear resistant.	Frictional wear at red heat; furnace parts.
Low Carbon Nickel-Chromium Stainless	Oxidation and corrosion resistant.	Corrosion-resistant surfacing of large tanks.

materials listed will satisfy most industrial applications (see also AWS-ASTM Specification A5.13-56T).

Basically, hard-facing materials are alloys that lend themselves to weld fusion and which provide hardness or other properties without special heat treatment. Thus, for hard surfacing, the steels and the matrices of high carbon irons may contain enough alloys to cause the hardening transformation during weld cooling, rather than after a quenching treatment.

The properties of the iron, nickel and cobalt base alloys are strongly affected by carbon content and somewhat by the welding technique

used. For example, gas welding usually provides superior abrasion resistance, although carbon pickup may lower corrosion resistance. Arc welding tends to burn out carbon and alloys, thereby lowering abrasion resistance but increasing toughness; high thermal stresses from arc welding may also accentuate cracking tendencies.

The martensitic irons, martensitic steels and austenitic manganese steels are suited for light, medium and heavy impact applications, respectively. Gouging abrasion applications usually require an austenitic manganese steel because of the associated heavy impact. Grinding abrasion is well resisted by the martensitic irons and steels. Erosion is most effectively resisted by a good volume of the very hard compounds (e.g., high-chromium irons). Tungsten carbide composites have outstanding resistance to abrasion where heavy impact is not present, but deposits may develop a rough surface.

Selection of materials for hot wear applications is complicated by oxidation, tempering, softening and creep factors. Oxidation resistance is provided by using a minimum of 25% chromium. Tempering resistance (up to 1100 F) is provided by chromium, molybdenum, tungsten, etc. Creep resistance is provided by the austenitic structure in nickel or cobalt-bearing alloys. The chromium-cobalt-tungsten grade of materials usually provides a good combination of properties above 1200 F.

HARD RUBBER

Hard rubber is a plastic. It is a resinous material mixed with a polymerizing or curing agent and fillers, and can be formed under heat and pressure to practically any desired shape. The original hard rubber, "Ebonite," was made from natural rubber. It was hard, horny, and jet black in appearance. The bulk of today's hard rubber is made with SBR synthetic rubber. Other types of synthetic rubbers, such as butyl or nitrile or, in rare cases, silicone or polyacrylic, can also be used.

Once it has gone through the process of heat and pressure, hard rubber cannot be returned to its original state and therefore falls into the class of thermosetting plastics, i.e., those which undergo chemical change under heat and pressure. It differs, however, from other commercial thermosetting plastics such as the phenolics and the ureas in that after it has gone through the thermosetting process it will still soften somewhat under heat. In this characteristic it most resembles the thermoplastic acetates, polystyrenes, and vinyls. It differs from all others in that it is available in pliable sheet form before vulcanization and is therefore adaptable to many shapes for which molds and presses are not necessary.

Because of this feature and because it can be softened again after vulcanization, it falls into a class by itself in the field of plastics.

The term "hard rubber" is self-descriptive. The hardness is measured on the Shore D scale, which is several orders of magnitude higher than the Shore A scale used for conventional rubbers and elastomers. Similar in composition to soft rubber, it contains a much higher percentage of sulfur, up to a saturation point of 47% of the weight of the rubber in the compound. If sulfur is present in rubber compounds in amounts over 18% of the weight of rubber in the compound when the material is completely vulcanized, the product will be generally known as hard rubber.

Besides rubber and sulfur, a rubber compound may consist of any or all of the following materials:

Lime	Mineral fillers
Clay	Metallic fillers
Carbon black	Synthetic resin
Mineral carbon	Natural resin
Mineral oil	Cotton and other fibers
Mineral rubber	Pulps
Softening agents	Anti-oxidants
Waxes	Other less commonly used
Wood flour	ingredients
Organic accelerators	

Hard rubber compounds are not standardized and will differ to some extent with each producer. Thus the hard rubber producer should be consulted in the initial design stages to advise on the design requirements.

Properties

The most important properties of hard rubber are the combination of relatively high tensile strength, low elongation and extremely low water absorption. Property ranges shown in Table 1 are typical for low, medium and high grades.

Values for these properties can be altered by specialized compounding or by changing the processing conditions. In general, hard rubbers have good thermal and electrical insulating characteristics. Since cold flow is negligible, they usually have excellent dimensional stability. Resistance to aging, acids, alkalies and other chemicals is good, and special compounds can be made where good oil resistance is required.

TABLE 1—TYPICAL PROPERTIES OF HARD RUBBER

	Type	High Grade	Medium Grade	Low Grade
	ASTM			
Ten Str, 1000 psi	D530	5–10	2.5–5	0.9–2.5
Elong, %	D530	4–50	4–50	2–20
Durometer Hardness (Shore D)	D676	55–95	50–85	50–85
Max Fiber Stress, 1000 psi	D530	44	15.2	1.75
Impact Str (Ball Drop), in.–lb	D639	10–100	10–100	10–100
Sp Gr	D297	1.13	1.20	1.34
Heat Distortion Temp (264 psi), F	D648	155	110	115
Max Continuous Svc Temp, F [a]	—	190–300	135	140
Water Absorption, %	D570	0.010	0.020	0.030
Acid Absorption, %	D639	0.050	0.100	0.20
Dielec Str, v/mil	—	420–450	360–380	320
Cost, $/lb		0.32	0.16	0.10 or less

[a] This is the temperature at which the material can usually be used continuously in an unstressed condition.

Fabrication

Hard rubber may be compression-, transfer- or injection-molded. In sheet form it can be hand-fabricated into many shapes. Its machining qualities are comparable to brass, and it may be drilled and tapped. The material lends itself readily to permanent or temporary sealing with hot or cold cements and sealing compounds.

The size and shape of a hard rubber part is dependent only upon the size of press equipment and vulcanizers available. Parts are presently made from as little as 1 oz to 40 or 50 lb.

Uses

Perhaps the largest application for hard rubber is in the manufacture of battery boxes. The water-meter industry is also a large user. Hard rubber linings and coatings either molded or hand laid-up account for large amounts of material. In the electrical industry, hard rubber is used for terminal blocks, insulating materials and connector protectors. The chemical, electroplating and photographic industries use large quantities of hard rubber for acid-handling devices.

HEAT-RESISTANT ALLOYS (CAST)

Cast alloys intended for service in which the metal temperature will exceed 1200 F are classed as "heat-resistant" or "high-temperature" alloys. They have the characteristic of corroding at very slow rates compared with unalloyed, or low-alloy cast iron or steel in the atmospheres to which they are exposed, and they offer sufficient strength at operating temperature to be useful as load-carrying engineering structures. Iron-base and nickel-base alloys comprise the bulk of production, but cobalt-base, chromium-base, molybdenum-base and columbium-base alloys are also made.

Although some cast heat-resistant alloys are available in compositions similar to wrought alloys, it is necessary to differentiate between them. Cast alloys are made to somewhat different chemical specifications than wrought alloys; physical and mechanical properties for each group are also somewhat different. For these reasons it is advisable to follow the standard Alloy Casting Institute (ACI) designations for cast high alloys, just as AISI designations should be used for wrought materials. There are, moreover, a number of heat-resistant cast alloys that are not available in wrought form; this is frequently of advantage in meeting special conditions of high-temperature service. In addition to the grades listed in Table 1, the industry produces special heat-resistant compositions. Many of these are modifications of the standard types, but some are wholly different and are designed to meet unique service conditions.

Selection of a particular alloy, of course, is dependent upon the application, and in this article composition, structure and properties of the various cast heat-resistant alloys are discussed from this point of view.

Proper selection of an alloy for a specific high-temperature service involves consideration of some or all of the following factors: (1) required life of the part, (2) range and speed of temperature cycling, (3) the atmosphere and its contaminants, (4) complexity of casting design and (5) further fabrication of the casting. The criteria that should be used to compare alloys depend on the factors enumerated, and the designer will be aided in his choice by providing the foundry with as much pertinent information as possible on intended operating conditions before reaching a definite decision to use a particular alloy type.

TABLE 1—STANDARD DESIGNATIONS FOR
HEAT-RESISTANT CASTINGS

Cast Alloy Designation	Wrought Alloy Type [a]
HA	—
HC	446
HD	327
HE	—
HF	302B
HH	309
HI	—
HK	310
HL	—
HN	—
HT	330
HU	—
HW	—
HX	—

[a] Wrought-alloy type numbers are listed only for the convenience of those who want to determine corresponding wrought and cast grades. Because the cast-alloy chemical composition ranges *are not*

the same as the wrought composition ranges, buyers should use cast-alloy designations for proper identification of castings.

Physical and mechanical properties of the various grades are listed in Tables 2 and 3. For high-temperature design purposes a frequently used design stress is 50% of the stress that will produce a creep rate of 0.0001% per hr at the maximum operating temperature. Such a value should be applied only under conditions of direct axial static loading and essentially uniform temperature or slow temperature variation. Where impact loading or rapid temperature cycles are involved, a considerably lower percentage of the limiting creep stress should be used. In the selection of design stresses, safety factors should be higher if the parts are inaccessible, nonuniformly loaded, or of complex design; they may be lower if the parts are accessible for replacement, fully supported or rotating, and of simple design with little or no thermal gradient.

TABLE 2—SELECTED PHYSICAL PROPERTIES OF HEAT-RESISTANT GRADES

Alloy Type	Density, lb/cu in.	Sp Ht, at 70 F, Btu/lb/°F	Ther Cond at 212 F, Btu/hr/sq ft/ ft/°F	Ther Exp, 70–1800 F, in/in/°F $\times 10^6$	Mag Perm	Elec Res, microhm-cm, at 70 F
HA	0.279	0.11	15.2	7.5 [b]	Ferromagnetic	70.0
HC	0.272	0.12	12.6	7.4	Ferromagnetic	77.0
HD	0.274	0.12	12.6 [a]	8.9	Ferromagnetic	81.0
HE	0.277	0.14	10.0	10.8	1.3 to 2.5	85.0
HF	0.280	0.12	9.0	10.5	1.00	80.0
HH	0.279	0.12	8.2	10.5	1.0 to 1.9	75 to 85
HI	0.279	0.12	8.2	10.5	1.0 to 1.7	—
HK	0.280	0.12	8.2	10.0	1.02	90.0
HL	0.279	0.12	8.2	9.9	1.01	94.0
HN	0.283	0.11	—	—	1.10	—
HT	0.286	0.11	7.7	9.8	1.10 to 2.00	100.0
HU	0.290	0.11	—	9.6	1.10 to 2.00	105.0
HW	0.294	0.11	7.7	8.8	16.0	112.0
HX	0.294	0.11	—	9.2	2.0	—

[a] At 1500 F.
[b] 70–1200 F.
Source: ACI Data Sheets.

TABLE 3—REPRESENTATIVE MECHANICAL
PROPERTIES, HEAT-RESISTANT GRADES [a]

Alloy	Room Temp [a]			1800 F	
	Ten Str, 1000 psi	Yld Str, 0.2% offset, 1000 psi	Elong, %	Creep Stress, 0.0001% 1000 psi	100 hr Rupture Stress, 1000 psi
HA [b]	95	65	23	16.0 [c]	37 [c]
HC	70	65	2	—	—
HD	85	48	16	0.9	2.5
HE	95	45	20	1.4	2.5
HF	85	45	35	3.2 [d]	6.0 [d]
HH₁	80	50	25	1.1	3.1
HH₂	85	40	15	2.1	4.0
HI	80	45	12	1.9	4.1
HK	75	50	17	2.7	4.5
HL	82	52	19	2.2	5.2
HN	68	38	17	3.1	4.9
HT	70	40	10	2.0	4.5
HU	70	40	9	2.2	4.5
HW	68	36	4	1.4	3.6
HX	65	36	9	1.6	3.5

[a] As-cast.

[b] Annealed.

[c] 1000 F. This grade not recommended for service above 1200 F.

[d] 1600 F. This grade not recommended for service above 1600 F.

Source: *ACI Data Sheets,* Alloy Casting Institute, Garden City, N.Y. These sheets contain additional data for temperatures in range 1000 to 2150 F.

Individual alloy characteristics

Group I—Iron Chromium Alloys. These alloys contain 8 to 30% chromium, and under 7% nickel (types HA, HC and HD) and are ferritic in structure. They have excellent resistance to oxidation and sulfur-containing atmospheres.

Type HA (8 to 10 Cr) is recommended for use only up to 1200 F. Types HC (26 to 30 Cr, 4 max. Ni) and HD (26 to 30 Cr, 4 to 7 Ni) can be used for light load-bearing applications up to 1200 F and where only light loads are involved, up to 1900 F. Both HC and HD are especally useful in high sulfur-containing atmospheres, and in applications where high nickel content cannot be used. The HC alloy is classified in two carbon ranges: up to 0.50% for normal high-temperature service, and up to 3.0% where an abrasion-resistant material is required.

Group II—Iron-Chromium-Nickel Alloys. These alloys contain 19 to 32% chromium and 8 to 22% nickel, with higher chromium than nickel content (types HE, HF, HH, HI, HK, and HL). Since they are partially or completely austenitic, they have greater high temperature strength and ductility than iron-chromium alloys. They can be used in either oxidizing or reducing atmospheres.

Type HF (19 to 22 Cr, 9 to 12 Ni) is similar in composition to the 18–8 stainless steels used for their corrosion resistance, except that carbon content is higher. Castings of this alloy operate in the 1200 to 1600 F temperature range.

Type HE (26 to 30 Cr, 8 to 11 Ni) is suitable for service up to 2000 F. It has excellent corrosion resistance at high temperatures. Its relatively low nickel content makes it readily applicable for use in high sulfur-bearing atmospheres, where moderate strength is required.

Type HH (24 to 28 Cr, 11 to 14 Ni) exhibits high strength and resistance to oxidation at temperatures up to 2000 F. These properties make it an extremely useful alloy, and it accounts for about one-third of the production of all heat-resistant castings. Depending upon composition balance, the alloy can be partially ferritic or wholly austenitic. The austenitic type should be used if operation will be in the range 1200-1600 F. For service above 1600 F, either of the compositions will serve—the ferritic for highest hot ductility, and the austenitic for highest hot strength. They are not generally recommended for service where severe temperature cycles are encountered (such as quenching fixtures).

Type HI (26 to 30 Cr, 14 to 18 Ni) is more resistant to oxidation than type HH and can be used up to 2150 F. Similar to the HH alloys in mechanical properties, the HI grade has been used mainly for cast retorts operating above 2100 F in magnesium production.

Type HK (24 to 28 Cr, 18 to 22 Ni) is also similar to a fully austenitic HH alloy; it has high resistance to oxidation and is one of the strongest heat-resistant cast alloys at temperatures above 1900 F. It can be used in structural applications up to 2100 F. It is not recommended for high sulfur-bearing atmospheres, or where severe thermal shock is a factor, but is widely used for parts where high creep strength is needed.

Type HL (28 to 32 Cr, 18 to 22 Ni) is similar to Type HK but has higher chromium content. The composition of this alloy is one of the most resistant to corrosion in high sulfur-containing atmospheres up to 1800 F. It is used where higher strength is required than is obtainable with straight chromium alloys.

Group III—Iron-Nickel-Chromium Alloys.

These alloys contain 23 to 68% nickel and 10 to 23% chromium, and employ nickel either as the predominant alloying element or as the base metal (Types HN, HT, HU, HW, and HX). As a result, they have a stable austenitic structure and are not as sensitive to variations in composition as the chromium predominant grades. They can be used for almost all applications up to 2100 F, have excellent hot strength, do not carburize readily, and give excellent service life where subject to rapid heating and cooling. The formation of an adherent scale makes them practical for use where a loose, flaking scale would be detrimental, such as in the enameling industry. Because of their high nickel content, however, they are not recommended for use in high sulfur-bearing atmospheres.

In selecting alloys from this group, consider the following factors:

1. Increasing nickel content increases resistance to carburization, decreases hot strength somewhat, and increases resistance to thermal shock.

2. Increasing chromium content increases resistance to corrosion and oxidation.

3. Increasing carbon content increases hot strength.

4. Increasing silicon content increases resistance to carburization, but decreases hot strength.

Type HN (23 to 27 Ni, 19 to 23 Cr) has been used successfully at temperatures up to 2100 F. Although adequate field data are not yet available for this alloy, it appears to have very high strength at high temperatures.

Type HT (33 to 37 Ni, 13 to 17 Cr) can be used satisfactorily at temperatures up to 2100 F in oxidizing atmospheres, and 2000 F in reducing atmospheres.

Type HU (37 to 41 Ni, 17 to 21 Cr) is substantially the same as HT. However, it is often recommended for severe service conditions because the increased chromium content increases hot strength. The higher alloy content also increases resistance to corrosive attack.

Type HW (58 to 62 Ni, 10 to 14 Cr) is not as strong at elevated temperatures as the HT and HU alloys, but is used in applications in oxidizing atmospheres up to 2050 F, and in reducing atmospheres up to 1900 F. It is extensively used for parts that require resistance to oxidation, carburization, nitriding, and thermal shock. Its high electrical resistivity makes it suitable for use as cast electrical heating elements. It is not recommended for use in high sulfur-containing hot gases. Typical uses for Type HW involve service at high temperatures coupled with either cyclic heating or thermal shock.

Type HX (64 to 68 Ni, 15 to 19 Cr) is more resistant to corrosion at elevated temperatures than Type HW because of its higher nickel and chromium content; it is also considerably more resistant to reducing gases containing up to 100 grains of sulfur per 100 cu ft at 1800 F. Both HW and HX alloys are highly resistant to carburization when in contact with tempering and cyaniding salts; they are not recommended for use with neutral salts, however, or salts used *in* hardening high-speed steels. Type HX is used for the same applications as HW when an improvement in resistance to hot gas corrosion is required.

Applications

Iron-chromium alloys containing from 8 to 32% chromium and iron-chromium-nickel alloys containing up to 32% chromium and up to 80% nickel are widely used in the cast form for industrial processing equipment at temperatures up to 2200 F. Parts for heat-treating furnaces such as trays, fixtures, rails, rollers and radiant tubes are a major use. Furnaces for oil refining, glass making, and cement kilns also employ many heat-resistant castings for supports, conveyor mechanisms, and other structural parts. In addition to furnace and process equipment applications, these alloys are used in gas turbines, jet engines and missiles, as well as power-plant superheaters, steel mill economizers, and municipal incinerators.

Other alloys containing large amounts of molybdenum or cobalt, with additions of columbium, tungsten, aluminum, and titanium, are used for cast blades in jet engines and other applications in the 1200 to 1600 F range where very high creep strength is essential.

HEAT RESISTANT PLASTICS (SUPERPOLYMERS)

Several different plastics developed in recent years that maintain mechanical and chemical integrity above 400 F for extended periods, are frequently referred to as superpolymers. They are polyimide, polysulfone, polyphenylene sulfide, polyarylsulfone and aromatic polyester.

In addition to their high temperature resistance, all of these materials have in common high strength and modulus, and excellent resistance to solvents, oils, and corrosive environments. They are also among the highest priced plastics, and a major disadvantage is processing difficulty. Molding temperatures and pressures are extremely high compared to conventional plastics. Some of them, including polyimides and arooatic polyester, are not molded conventionally. Because they do not melt, the molding process is more of a sintering operation. Because of their high price, superpolymers are largely used in specialized applications in the aerospace and nuclear energy field.

Indicative of their high temperature resistance, the superpolymers have a glass transition temperature well over 500 F as compared to less than 350 F for most conventional plastics. In the case of polyimides, the glass temperature is greater than 800 F, and the material decomposes rather than softens when heated excessively.

Polysulfone has the highest service temperature of any melt-processable thermoplastic. Its flexural modulus stays above 300,000 psi at up to 320 F. At such temperatures it does not discolor or degrade:

Aromatic polyester does not melt, but at 800 F can be made to flow in a nonviscous manner similar to metals. Thus, filled and unfilled forms and parts can be made by hot sintering, high-velocity forging and plasma spraying. Notable properties are high thermal stability, good strength at 600 F, high thermal conductivity, good wear resistance and extra-high compressive strength.

HIGH ENERGY RATE FORMED PARTS

In high energy rate forming, parts are shaped by the extremely rapid application of high pressures. Pressures as high as 2,000,000 psi and speeds as high as 3000 fps may be used.

The principal advantages of high energy rate forming are:

1. Parts can be formed which cannot be formed by conventional methods.
2. Exotic metals, which do not readily lend themselves to conventional forming processes, may be formed over a wide range of sizes and configurations.
3. The method is excellent for restrike operations.
4. Springback after forming is reduced to a minimum.
5. Dimensional tolerances are generally excellent.
6. Variations from part to part are held to a minimum.
7. Scrap rate is low.
8. Less equipment and fewer dies cut down on production lead time.

Explosive forming

There are three different explosive forming techniques now being used: free forming, bulkhead forming and cylinder forming. All are shown schematically in Fig 1. Both free forming and bulkhead forming allow the workpiece to be heated before forming. Although air can be used as a coupling medium between the explosive and the workpiece, in most cases water is used. Efficiency in air is approximately 4%; in water, 33%.

TYPICAL PROPERTIES OF HEAT RESISTANT PLASTICS (SUPERPOLYMERS)

Type	Polyimide	Polysulfone	Polyphenylene Sulfide	Polyarylsulfone
Specific Gravity	1.43–1.47	1.24	1.34	1.36
Mod. Elast. in Tens., 10^5	5.4–7.5	3.6	4.8	3.7
Tens. Str., 1000 psi	10.5		–	13
Elong., %	1.2	50–100	3	15–20
Hardness, Rockwell	H85	R120	R124	M110
Imp. Str., ft-lb/ in. notch	0.5	1.3	–	5.0
Max. Svc. Temp., F	500	340	500	500
Vol. Res., Ohm-cm	4×10^{15}	5×10^{16}	–	3.2×10^{16}
Dielect. Str., v/mil	310	425	595	350

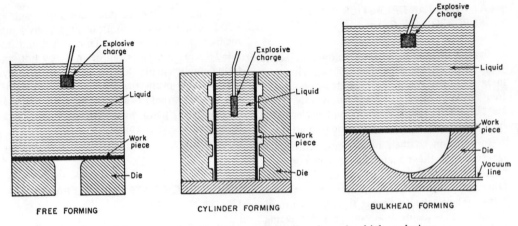

FIG. 1. Three general methods for explosive forming using high explosives.

TABLE 1—ROCKWELL HARDNESS OF EXPLOSIVELY
FORMED PARTS STRETCHED 30–35%

Condition → Metal ↓	Before Forming	After Forming	After 35% Cold Work [a]
"A" Nickel	62B	99B	98B
Inconel	80B	102B	95B
Inconel "X"	18C	30C	—
Monel	68B	95B	90B
AISI 1020 Steel ...	72B	100B	85B

[a] Data added by author.
Source: National Northern Div., American Potash & Chemical Corp.

Changes in mechanical properties caused by explosive forming correlate closely with those obtained in material cold-worked to the same degree. Table 1 shows how hardness of several nickel alloys is affected by explosive forming and Table 2 shows the effect of explosive forming on the mechanical properties of three stainless steels. In both cases, strength and hardness are in-

creased and, as expected, ductility is decreased.

Studies indicate that explosive impact hardening is useful with materials hardened by cold work, e.g., austenitic stainless steel, Hadfield steel, nickel, molybdenum, etc. An interesting application is the possibility of restoring mechanical properties of parts that have been welded or heat-treated, and thus softened.

Simple forgings can be made by explosive forming techniques. One study showed that aluminum alloys could be explosive-forged if the design had no extreme contours.

Copper has been welded to copper by the application of explosive force. The joint was a metallurgical, not a mechanical one.

Expanding gas

Gases generated by the burning of propellant powders in a closed container produce the pressures required to form metals. Most of the work that uses propellant powder gases as the energy source is classified as bulge-type forming. Sev-

TABLE 2—EXPLOSIVE IMPACT TREATMENT AFFECTS MECHANICAL PROPERTIES OF STAINLESS STEEL

	Before Explosive Treatment			After Explosive Treatment		
Stainless Type →	302	304	316	302	304	316
Yld Str, 1000 psi	27.5	30.5	26.7	142	154.0	127.0
Ten Str, 1000 psi	88.0	95.0	77.7	154	180.0	147.0
Elong, %	84	81	72	19	25	17
Red. in Area, %	80	78	79	62	46	60

Source: E. I. du Pont de Nemours & Co., Inc.

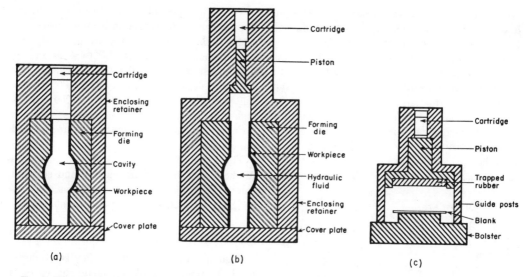

FIG. 2. Three general types of cartridge actuated devices: (a) gases act directly against the workpiece; (b) pressure acts through a hydraulic system; (c) gases act on piston.

eral types of cartridge-actuated dies are shown in Fig 2.

In each die, which is completely closed, the gases expand directly against the workpiece, forcing it to accept the contour of the female (or be bent over a male die). There are two variations: the expanding gases may press against some intermediate medium which, in turn, presses against the workpiece; the expanding gases actuate a piston that acts on the intermediate medium which ultimately forms the workpiece to the proper configuration.

Dynapak

The Dynapak machine uses a gas-powered ram to form a part and the operation is under close control. The Dynapak system has the following advantages:

1. Low cost. A $3 cylinder of nitrogen will operate the machine for one day.

2. Rapid production. Cycle time ranges from 30 to 60 sec.

3. The machine can generate high energy levels and velocities above 200 fps. Velocity can be controlled within 2 fps.

4. Dies can be made of ordinary steel.

5. Parts can be extruded, forged, formed and compacted.

6. Parts can have zero draft angles with minimum curve and fillet radii.

Dynapak can form low-alloy steels such as AISI 4340, austenitic steels of the 200 and 300 series, titanium, and the refractory metals, to name a few examples. It can also be used to compact powders to a density higher than normally obtained with conventional powder-metallurgy processes.

Parts have been extruded with excellent surface finish and close dimensional tolerances. Web thicknesses of 0.01 in. can be obtained and wire 0.020 in. in dia has been extruded directly from a 1 in. billet.

Hot and cold forgings of various materials can be produced with zero draft angles and minimum radii. The smooth, close tolerance surfaces that are produced often minimize finish machining requirements.

Capacitor discharge techniques

Explosives and compressed gases are not the only means of achieving high deformation rates. In one type of device, a spark is discharged in a nonconducting liquid medium and generates a shock wave that travels at the speed of sound from the spark source to the workpiece. A schematic diagram of this system is shown in Fig 3. This forming technique has several advantages:

1. Explosives, with their potential safety hazard, are eliminated.

2. Parts can be sized into a die by several

applications of energy impulses. Since the device is electrical, components of the system do not have to be repositioned after each shot.

3. A standard machine tool, based on this principle, can be constructed for about one-tenth the cost of a conventional hydraulic press and occupy only a fraction of the floor space.

One problem encountered with capacitor discharge techniques is the containment of high voltages, since stored electrical energy increases with the square of voltage ($E = CV^2$, where $C =$ capacitance and $V =$ voltage). Two other prob-

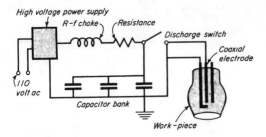

FIG. 3. Schematic diagram of hydrospark forming system. (*Republic Aviation Corp.*)

lems: corona and arcing. Normal safety procedures for handling high voltage must be followed.

Magnetic forming

Use of the pressure generated by a magnetic field permits parts to be formed in 6 sec. Of this time, only 10 to 20 μ sec may be needed for the forming operation; the balance is taken up by setup and removal of the part from the apparatus.

The many possible magnetic coil configurations permit wide variety in forming operations, as Fig 4 shows. Coils usually are massively supported since they must be able to withstand the high pressures generated by the magnetic field. For example, at a flux density of 300,000 gausses, the pressure is approximately 50,000 psi. At higher fields (up to a million gausses) magnetic forming devices may generate pressures exceeding 500,000 psi.

The value of magnetic forming methods lies in the ability to perform quickly and economically many conventional operations such as swaging, bulging, expanding and assembly.

HIGH-SPEED STEELS

High-speed steels are those alloy steels developed and used primarily for metal-cutting tools. They are characterized by being heat-treatable to very high hardness (usually Rockwell C 64 or over) and of retaining their hardness and cutting ability at temperatures as high as 1000 F, thus permitting truly high-speed machining. Above 1000 F they rapidly soften and lose their cutting ability.

All high-speed steels are based on either tungsten or molybdenum (or both) as the primary heat-resisting additive, with carbon for high hardness, chromium for ease of heat treating, vanadium for grain refining and, in amounts over 1% for abrasion resistance, and sometimes cobalt for additional hardness and resistance to heat softening.

Popular compositions of high-speed steels read-

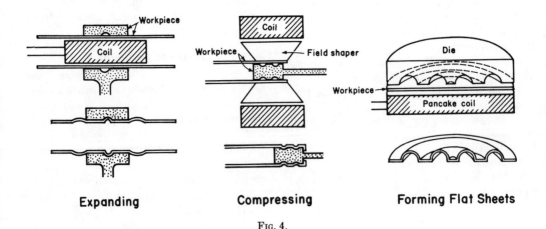

Expanding   **Compressing**   **Forming Flat Sheets**

FIG. 4.

ily available in the United States and the most common characteristics are shown in Table 1.

By far the most important of these are M1, M2 and M10. Nearly all twist drills, taps, chasers, reamers, saw blades, and high-speed steel hand tools are made from M1 or M10, while the more complex tools such as milling cutters, gear hobs, broaches, and form tools utilize the M2 type. Very little tungsten-base high-speed steel is used in the United States today, although it is popular abroad. Wartime restrictions on tungsten and its high price started the trend. The ready availability of molybdenum in America and the generally better properties developed by the molybdenum steels are other factors working against the use of high-tungsten types.

Special applications, such as machining of hard heat-treated materials, call for a more abrasion-resistant type (such as M3 or T15) while extra heavy-duty cutting involving max-

arate types) but they are usually designed for a specific purpose, and have limited use.

Some of the common grades in Table 1 are also available with sulfide additives to provide improved machinability.

Properties

High-speed steel possesses the highest hardness after heat treating of any well-known ferrous alloy. The value of high-speed steel lies in its ability to retain this hardness under considerable exposure to heat, and to retain a sharp cutting edge when exposed to abrasive wear.

The hardness of high-speed steel when heat-treated is usually Rockwell C 64 to 66, equivalent to Brinell 725 to 760. It is brittle at this hardness, particularly in the cobalt bearing grades, and must be sharpened and handled carefully. This high hardness is obtained by somewhat special heat-treating techniques as compared with lower alloyed steels. Temperatures much in excess of normal steel heat-treating temperatures are employed.

TABLE 1

AISI Type	Composition, %					Characteristics	Relative Cost
	C	Cr	W	Mo	V		
T1	0.75	4	18	—	1	General purpose	1.5
T5 a	0.85	4	18	—	2	Extra heat resistance	2.4
T15 b	1.50	4	14	—	5	Heaviest duty	2.3
M1	0.80	4	1.5	8	1	General purpose	1.0
M2	0.85	4	6	5	2	General purpose	1.1
M3	1.25	4	6	5	3	Heavy-duty, abrasive materials	1.3
M7	1.00	4	1.75	8.75	2	Special applications	1.0
M10	0.88	4	—	8	2	General purpose	1.0
M36 a	0.90	4	6	5	2	Extra heat resistance	1.9

a Contains 9% Co.
b Contains 5% Co.

Heat treatment

Table 2 lists recommended temperature ranges for the various grades for austenitizing (hardening) and tempering. In general, for maximum hardness and heat resistance it is necessary to heat-treat at as high a temperature as possible, short of the point of initial fusion or grain growth. This would normally result in severe surface damage when done in conventional furname atmospheres, so surface protection by special gaseous atmospheres or by molten salt (usually $BaCl_2$) is required for production of quality tools.

After quenching in oil, or a salt eutectic (KCl-NaCl-BaCl$_2$) held at about 1050-1100 F, and air cooling to 125 F, high-speed steel is tempered in the "secondary" hardening range (975-1050 F) to develop maximum hardness and cutting life. Such tempering is usually done two or three times successively for best results.

Occasionally a shallow surface treatment (under 0.001 in.) is imparted by nitriding in salt, or by one of several proprietary methods, to elevate surface hardness and reduce friction. These treatments are often very useful in improving cutting life.

imum heat generation leads to the use of the high cobalt types M36 or T5.

Many other compositions exist in the United States (the AISI Tool Steel Manual lists 24 sep-

TABLE 2

AISI Type	Austenitizing Temp, °F	Tempering Temp, °F	Resis to Decarb [a]
T1	2300–2350	1025–1075	Exc
T5	2325–2375	1025–1075	Fair
T15	2225–2275	1000–1050	Good
M1	2150–2225	1025–1050	Poor
M2	2175–2250	1025–1050	Fair
M3	2200–2250	1000–1025	Fair
M7	2175–2240	1000–1050	Poor
M10	2150–2225	1000–1050	Poor
M36	2225–2275	1025–1075	Fair

[a] During heat treatment.

Use and selection

High-speed steels are used for all types of cutting tools, particularly those powered by machines, such as drills, taps, reamers, milling cutters of all types, form cutting tools, shavers, broaches, and lathe, planer and shaper bits. They have preference over cemented carbide when the tool is difficult to form (high-speed steel is machinable before heat treating), when subject to shock loading or vibration (high-speed steel is tough and resistant to fatigue, considering its hardness level), or when the machining problem is not particularly difficult. High-speed steel is considerably less expensive than carbides and much simpler to form into complex tools but it does not have the high hardness, abrasion resistance, or tool life in severe high-speed cutting applications associated with cemented carbide. On the other hand, high-speed steel, having good heat resistance, consistently cuts far better than carbon steel, or one of the "fast-finishing" types.

Other uses for high-speed steel are in forming dies, drawing dies, inserted heading dies, knives, chisels, high-temperature bearings and pump parts. In these applications use is made of the combination of high hardness, heat resistance and abrasion resistance rather than cutting ability.

Among the types of high-speed steel listed in Table 1, common mass-produced tools are made from M1 or M10. These grades have the lowest cost, and are easiest to machine, heat-treat and sharpen. They also are the toughest when hard and thus withstand the abuse often given common tooling—drills, taps, threading dies, etc.

More complex, expensive tools are usually made from M2 high-speed steel. It has better abrasion resistance and is easier to heat-treat in complex shapes. Most milling cutters, gear hobs, broaches, and similar multiple-point tools are in this category. M7 is also becoming popular for specialized applications.

Occasionally, extremely difficult machining operations are encountered, such as cutting plastic, synthetic wood and paperboard products, or hardened alloy steels. Better tool life can then be obtained by use of M3 or T15, the high-vanadium high-speed steels. They are more expensive and considerably more difficult to sharpen and maintain because of resistance to grinding, but these factors are often outweighed by the superior tool life developed.

The high-cobalt high-speed steels, T5 and M36, have the best heat resistance, and therefore are particularly suited for tools cutting heavy castings or forgings, where cutting speeds are relatively slow but cuts are deep and the cutting edge gets very hot. T5 and M36 are more expensive than other grades and thus have limited use, but are the most economical for some operations.

Fabrication

High-speed steels in the annealed condition are machinable by all common techniques. Their machinability rating is about 30% of Bessemer screw stock, and they must be cut slowly and carefully. The recent development of free-machining high-speed steels has eased this situation, but considerable care is still required.

Ordinarily tools are machined from bar stock or forgings, either singly for complex tools or in automatic screw machines for mass production items (taps, twist drills, etc.). After finishing almost to final size, the tools are heat-treated to final hardness, then finish-ground. The manufacture of unground tools machined to final size before heat treating is growing because of improvements in heat-treating facilities and better cutting ability of a properly hardened unground surface.

After high-speed steel tools become dull they

can easily be resharpened, with some care given to selection of the proper grinding wheel and technique. They are rarely softened by annealing and re-heat-treated, since this may produce a brittle grain structure in the steel unless great care is employed. High-speed steels are never welded after hardening, and tools are seldom repaired by welding because of extreme brittleness in the weld. Often high-speed steel inserts are brazed to alloy-steel bodies, or flash-welded to alloy-steel shanks (heavy drills, taps and reamers).

HOT-DIP COATINGS

A hot-dip coating is produced by immersing a base metal in a bath of the molten coating metal. Adhesion results from the tendency of the coating metal to diffuse into the base metal and form an alloy layer. Most hot-dip coatings consist of at least two distinct layers: an alloy layer and a layer of relatively pure coating metal. The alloy layer is usually a brittle intermetallic compound. Hot-dip coatings in which the alloy layer is relatively thick are not readily deformable, but modern techniques make it possible to keep the alloy layer quite thin.

Fairly thick coatings of inexpensive metals can be obtained more cheaply by hot dipping than by electroplating. Except on simple shapes, however, hot-dip coatings are nonuniform and wasteful of material. The nature of the process is such that coating metals are restricted to relatively low-melting metals, and base metals are limited to high-melting metals such as cast iron, steel, and copper.

Zinc

Properties and uses. Hot-dipped or galvanized zinc coatings (see also Zinc Coatings) have been popular for many years for protecting ferrous products because of their ideal combination of high corrosion protection and low cost. Their corrosion protection stems from three important factors: (1) zinc has a slower rate of corrosion than iron, (2) zinc corrosion products are white and nonstaining, and (3) zinc affords electrolytic protection to iron.

The amount of protection against corrosion depends largely upon coating weight—the heavier the coating the longer the life of the base metal. For example, a coating 1.8 mils thick (1 oz per sq ft) is estimated to have a life of 25 years in rural atmospheres, whereas, a 3.6 mil coating (2 oz per sq ft) will last 50 years. The life of zinc coatings may be five to ten times greater in rural atmospheres than in industrial atmospheres containing sulfur and acid gases. Nevertheless, the coatings are still popular for industrial use because of their low cost.

Hot dipping is particularly valuable for zinc coating parts that cannot conveniently be made of galvanized sheet. Thus it is quite popular for structural parts, castings, bolts, nuts, nails, pole-line hardware, heater and condenser coils, windlasses, and many other products.

How they are applied. Hot-dip zinc coatings must be applied to absolutely clean metal. Consequently, surfaces are usually cleaned in a caustic or degreasing medium and then pickled. After rinsing, parts must be fluxed to promote bonding of the zinc coating. The coating itself is applied at 840 to 860 F. A zinc-iron alloy layer is formed between the base metal and coating. Immersion time depends on the thickness of coating desired, most coatings being applied in less than 1 min.

Lead

Properties and uses. Hot-dip lead coatings (see also Lead Coatings) provide many important advantages on ferrous metals. They are relatively inexpensive, provide very good protection against indoor and outdoor atmospheric corrosion, and can be used in contact with many chemicals such as sulfuric, hydrochloric, hydrofluoric, phosphoric, and chromic acids. Their atmospheric corrosion resistance stems from the formation of a superficial oxide film which is relatively impervious to corrosion.

Because of their softness lead coatings can withstand severe deformation. Poor adhesion may be a problem since the bond is mechanical, but this problem can be minimized by adding alloying elements. Pinholes formed during application may be potential sources of corrosion but they can be eliminated by slight working or burnishing. Typical successful applications for the coatings are wire, pole-line hardware, nuts, bolts, washers, tanks, barrels, cans and miscellaneous air ducts.

How they are applied. Since pure lead will not alloy with ferrous metals, it is usually combined with an alloying agent such as tin which alloys with iron, forming an interface between the base metal and the coating. A typical sequence of operations for coating small parts involves: solvent cleaning, electrocleaning, pickling, predipping, fluxing, coating, and quenching.

Tin

Properties and uses. Hot-dip coatings (see also Tin Coatings) can be applied to fabricated parts made of mild and alloy steels, cast iron, and copper and copper alloys to improve appearance and corrosion resistance. Like zinc, the coatings consist of two layers—a relatively pure outer layer and an intermediate alloy layer.

An invisible surface film of stannic oxide is formed during exposure which helps to retard, but does not completely prevent, corrosion. The coatings have good resistance to tarnishing and staining indoors, and in most rural, marine, and industrial atmospheres. They also resist foods. Corrosion resistance in all cases can be markedly improved by increasing thickness and controlling porosity. Typical applications where they can be used are: milk cans, condenser and transformer cans, food and beverage containers, and various items of sanitary equipment such as cast iron mincing machines and grinders.

How they are applied. Steel products first have to be thoroughly cleaned and fluxed. They are then immersed in a preliminary tinning pot, followed by immersion in a second pot at lower temperature. Finally, the parts are withdrawn through palm oil or are dipped in a separate oil pot. Small parts are handled in one pot, centrifuged to remove excess tin, and then quenched or air-cooled. Thickness of the coatings varies from 0.3 to 0.5 mil. The above treatments are typical and variations are used for cast iron and copper and copper alloy products.

Aluminum

Aluminum hot-dip coatings (usually about 1 mil) are more expensive but much more atmospheric-resistant than zinc. The coatings (see also Aluminum Coatings) are also highly heat-reflective and the aluminum-iron alloy layer is highly refractory. Although these coatings seem promising for more general use in outdoor (especially industrial) atmospheres, they are currently used primarily to protect steel from high-temperature oxidation; typical applications include aircraft fire walls, toasters, auto mufflers, and water-heater casings. Hot-dip aluminum-coated (aluminized) steel sheet, strip and wire are commercially available.

I

IMMERSION AND CHEMICAL COATINGS

Immersion coatings are applied without electricity by immersing parts in a chemical solution or bath containing the metal to be deposited. Deposition can take place either by displacement, where the metal in solution displaces the base metal, or by reduction, where the base metal does not enter into the reaction.

Although many metals can be deposited in immersion baths, comparatively few have proved acceptable for decorative or functional applications. These are: nickel, tin, copper, gold and silver.

Electroless nickel

Properties. Electroless nickel is generally more expensive than electroplated nickel. For this reason it is used primarily for its functional properties, although a very smooth, bright deposit can be obtained on buffed ferrous and nonferrous metals.

Because of its amorphous structure and phosphorous content (8 to 10%) the coatings are said to have better corrosion resistance than electrolytic or wrought nickel. Hardness of the coatings is relatively high—about 50 R_c—and can be raised to 64 R_c by heat treatment. Thickness of the coatings ranges from 1 to 5 mils, depending on end use.

Applications. The most important uses for electroless nickel are to protect parts from corrosion and to prevent product contamination. The coatings are widely used on tank-car interiors to protect caustic soda, ethylene oxide, tetraethyl lead, tall oil and many other liquids from contamination. Other similar applications include: oil refinery air compressors, missile fuel injector plates, gas storage bottles for liquid rockets, and pumps for petroleum and related products.

The hardness of the coatings is particularly valuable in increasing the life of rotating and reciprocating surfaces in gas compressors, pumps, hydraulic cylinders, sheaves and armatures. The coatings are also used on aluminum electronic devices to facilitate soldering, on stainless steel to facilitate brazing, on moving metal parts to prevent galling, and on stainless steel equipment to prevent stress corrosion cracking.

Tin

Properties. Tin immersion coatings are especially noted for their low cost, bright appearance, good frictional properties, and ease of application to many common metals such as copper, brass, bronze, aluminum and steel. However, their corrosion resistance is only fair.

As with some other immersion coatings, plating usually stops when the base metal is completely covered. Thus, thickness for common decorative uses is limited to about 0.015 mil. However, thicknesses up to 2 mils for heavy-duty applications have been produced by placing the base metal in contact with a dissimilar metal, thereby generating current and promoting additional plating.

Applications. Tin immersion coatings are popular for decorative finishing of small parts such as safety pins, thimbles and buckles. They are also applied to copper tubing to prevent discoloration from water, and to aluminum engine pistons to provide lubrication during break-in periods.

Copper

Properties. The most important characteristics of copper immersion coatings are their high electrical conductivity, good lubrication properties, and unique appearance. In addition to steel, they can be applied to brass and aluminum and to printed circuit boards. Usual thickness range is 0.1 to 1 mil.

Applications. Because of their conductivity, copper immersion coatings have proved particularly useful for printed circuits. They are not especially noted for their decorative appeal, but can be used in applications where a particular appearance is required; e.g., inexpensive, decorative hardware such as casket parts. Because of their good lubrication properties they can also be used on steel wire in die-forming operations.

Gold

Properties. Gold immersion coatings are relatively inexpensive because of their extreme thinness—about 0.001 mil. The coatings have good electrical conductivity and emissivity characteristics, and a bright, attractive appearance. As deposited they are not especially resistant to discoloration and abrasion; however, they can be protected with a clear lacquer finish. They are used on a wide variety of ferrous and nonferrous metals, and on copper printed circuit boards.

Applications. Because of their good appearance gold immersion coatings are principally used on costume jewelry, trophies, auto trim and inexpensive novelties. Their conductivity and solderability are used to advantage in electrical applications such as printed circuits, transistors and connectors. Also, the unique emissivity properties of the coatings have proved useful in missile applications.

Silver

Properties. Like gold, silver immersion coatings are relatively inexpensive because of their extreme thinness. The coatings have a bright, attractive appearance when first deposited. Their resistance to tarnishing and abuse is poor; however, they can be protected somewhat with a clear lacquer coating.

Silver immersion coatings can be applied to most base metals except lead, zinc, aluminum and very active metals. They perform best on copper, nickel and steel. Usual thickness is about 0.001 mil, but 0.03 mil can be deposited in some cases.

Applications. Because of their poor durability silver immersion coatings are not too popular, the only applications being cheap decorative products, minor electronic parts and maintenance plating.

IMPACT EXTRUSIONS

Impact extrusion consists of subjecting metallic materials to very high pressures at room temperature. Under these pressures the metals become "plastic" and assume predetermined shapes. Whereas in coining this process takes place within a closed die, typical impact extrusions allow a portion of the metal to be "squirted" or "squeezed" out of the die cavity, thereby forming an integral part of the desired shape.

From a practical point of view impact extrusion of suitable parts may result in substantially lower unit manufacturing costs because it permits:
1. High production rates.
2. Substantial material savings (up to 75%).
3. Little or no machining.
4. Low initial tool costs and long tool life.
5. Bright and smooth surface finish ready for decorating.

Two types. Impact extrusion of the slug takes place between punch and die. Under pressure the metal may be forced to flow counter to the direction of punch travel (backward extrusion) or in the direction of punch travel (forward extrusion). Frequently, parts require a combination of both.

Starting material

The raw material for the process is usually referred to as a *slug*. Slugs may be blanked from sheet or plate, sawed from bar stock, or cast. The cross section of a slug—round, oval, square or rectangular—fits into the die bottom. Its height is determined by the volume of metal required to produce the part.

Equipment selection

The pressures necessary for impact extrusion are available on mechanical or hydraulic presses. Mechanical presses are usually preferred when higher production rates are required. Both toggle and crank presses are used—depending on load and performance characteristics required. Hydraulic presses have primarily been used for forward extrusions requiring particularly high pressures, and heavy cross sections. Factors bearing on the selection of equipment for impact extrusion are: dimensional characteristics of the part to be produced, tonnage and speed.

Dimensional characteristics of the part (diameter, height, etc.) determine die size of the press and length of stroke required. Most commercially extruded parts range in diameter from ½ to 4 in. and in length from 2 to 12 in.

Tonnage required for metal flow must be carefully determined in advance. It is predicated on the relative plastic deformation required for the part, i.e., the ratio of the cross-sectional area of the extruded part to the area of the slug. Maximum limits of plastic deformation vary from

90 to 95% for 99.5% aluminum, lead and tin, to 70 to 80% for mild steel and brasses

Design criteria

Impact extrusion should be specified by the designer primarily for:

1. Parts that are essentially hollow shells consisting of a wall and a bottom or partial bottom section. While in drawing the ratio between height and bottom diameter is limited, to approximately 2 to 3 to 1, it is possible to impact-extrude (backwards) up to 8 to 1. Forward extrusions many times as long as the diameter have also been made.
2. Parts with straight, no-draft walls.
3. Parts requiring high-strength characteristics.
4. Parts with longitudinal ribs, flutes, splines, etc. or with bosses, cavities, etc., in the bottom section.
5. Parts made in large quantities calling for low unit cost.

Applications

It is significant for this process that the flow of metal takes place primarily in a direction parallel to punch travel. Parts produced by impact extrusion are essentially longitudinally oriented, e.g., collapsible tubes, cans, etc.

Originally used only on soft materials (lead, tin, etc.) to make collapsible tubes, the process has, in recent years, found rapidly increasing applications in the field of metallic containers, as well as in the production of a wide range of automotive, electrical and hardware components. Aluminum and its alloys, copper, high brasses and mild steel are impact-extruded commercially in large quantities today.

Since it is a "chipless" metalworking process, impact extrusion competes in many applications with the automatic screw machine, deep drawing, die casting, hot forging, cold upsetting and other operations. To achieve optimum results, it is important that likely parts be designed with an understanding of characteristics of metal flow, die design and proper distribution of pressures in impact extrusion.

IMPREGNATING MATERIALS FOR CASTINGS

Castings are impregnated for several reasons, the most obvious of which is to prevent leakage. Other important reasons for impregnating are to prevent corrosion, improve surface finish, remove sites that may lodge food particles and cause bacterial growth, and to prevent back seepage of occluded fluids. Leakage may be avoided by blocking the pores at any point along their length, whereas all of the other aims can be met only when the pores are blocked at the surface as well as in depth.

Impregnating processes

Shown in the sketch are the stages in the impregnation process. "A" shows a section of a porous casting with different types of porosity (continuous; noncontinuous or blind; and isolated). After evacuation and impregnation under pressure, the condition shown in "B" is produced. If neither bleeding nor contraction occur before or during the curing operation, then "B" will also represent the final condition. However, this ideal condition is rarely attained, and some condition between "B" and the extreme represented by "C" is more likely to occur.

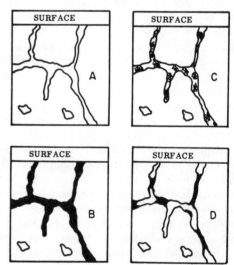

Stages in casting impregnation process.

Bleeding will lead to incompletely filled pores near the surface, and contraction in curing will lead to incomplete filling along the length of the pore (condition "C"). The latter condition may arise from: loss of solvent from impregnants introduced as a solution; loss of water from vehicles such as sodium silicate; and decomposition of organic impregnants by exposure to excessive temperatures (carbonization). Some compensation for the loss of solvent or water can be obtained if the residue can be made to expand on curing, e.g., by oxidation of metallic particles carried in sodium silicate.

It has been demonstrated that the sealing action of an impregnant generally decreases as the amount of volatile in the formulation increases. One exception to this rule is demonstrated by condition "D," in which a self-fluxing brazing alloy powder is fused in the pores after evaporation of the vehicle. The molten braze alloy runs into the narrowest part of the pore under capillary action and gives optimum sealing in spite of incomplete filling of the pores.

Impregnating materials

The oldest method of reducing leakage, particularly in cast iron, is to rust or oxidize the pores, sometimes after "impregnation" with mud. Natural drying oils such as tung and linseed were also among the first impregnant materials to be used. Another early type of impregnant was based on the readily available water glass (sodium silicate).

The contraction of sodium silicate impregnants during drying was reduced by the addition of inert particles such as asbestos, chalk, or oxides. Further development has led to the so-called "metallic" impregnants in which a large proportion of the solids are metal powders (e.g., copper or iron). Special agents are added to the sodium silicate to reduce sedimentation. During the curing of these "metallic" impregnants, the metal particles oxidize so that a certain amount of expansion occurs to offset the considerable contraction. However, these impregnants have several disadvantages: their resistance to high temperatures is poor (e.g., 300 F max); the impregnants are attacked by steam; and the abrasive oxides make them unsuitable for bearings or machining.

There are several differences between the sodium silicate and plastics impregnants. In general, because plastics or oils do not wet castings readily, a vapor degreasing is usually necessary, followed by pumping under a vacuum of at least 28 in. of Hg for periods of 30 min (one impregnator has achieved success in difficult cases by holding under a vacuum of 10 microns for two hr). A second difference results from the higher viscosity of plastics. During the pressure

stage of impregnation, a minimum of 30 min under pressure may be necessary to force the plastics to flow deeply into the pores. One solution to the problem of high viscosity is to thin the plastic by solvents, but this leads to "C."

The cost of impregnating with sodium silicate is usually less (about one-quarter less) than with plastics because prior cleaning of castings need to be less thorough and excess impregnant is washed away with water.

The most important types of plastics impregnants are based on the styrene monomer. These were developed to replace tung oil when the latter was in short supply during the Sino-Japanese War. A typical composition of this substitute is 80% styrene monomer with 20% linseed oil and a small quantity of organic oxidizer as a catalyst. The instability of the catalyst in the presence of lead, zinc and copper restricts the use of this type of impregnant primarily to the light metals. The castings are heated for two hr at 275 F to polymerize the impregnant.

The thermosetting plastics formed by copolymerization of the styrene monomer with polyesters represents an advance on the styrene-linseed oil polymers because it is possible to introduce liquid bath curing. The surfaces are cured rapidly when they make contact with the liquid curing agent, so that subsequent bleeding (which occurs on curing in ovens) is prevented. Other improvements that have been introduced with the styrene-polyester plastics include detergent washing (does not wash the impregnant from pores), freedom from inhibition by copper, and low contraction on curing (e.g., 6 to 7%). A typical curing cycle is 1 hr at 275 F. These impregnants will withstand 500 F for short periods and 400 F in continuous service.

Phenolics, epoxy and furfural plastics can also be used for impregnation. The phenolics are dissolved in a solvent such as alcohol, and require curing in two stages; the first to remove the alcohol (e.g., 1 hr at 175 F) and the second to cure the plastic (e.g., 3 hr at 375 F). Phenolic resins do not have good adhesion to metal and they seal according to condition "C." Therefore, they do not provide the highest quality seals, although they may be suitable for many types of work. Epoxy resins have good adhesion to metals and resistance to chemicals, but many have to be thinned with solvents to achieve low viscosity so that the impregnation occurs readily. Furfural type plastics have been investigated for resistance to alkalies.

Polyester-styrene and sodium silicate are the most widely used impregnants. The plastics give the higher quality seal, but sodium silicate continues to be used because of its low cost.

INDIUM

Indium is a silver-white metal with a color similar to that of platinum. It is one of the softest of metals and can be scratched easily. Highly plastic, the metal can be deformed almost indefinitely in compression.

Physical properties

The following table presents the physical properties of indium:

Atomic No.	49
Atomic Wt	114.76
Absorption Coef, barns	190 ± 10
Crystal Structure	Face centered tetragonal, a = 4.583, c = 4.936
Density (20 C), gm/cu cm	7.31
Sp Ht (20 C), cal/gm/°C	0.057
Electrochem Equiv, mg/coulomb	0.39641
Melt Pt, C	156.17
Boil Pt, C	2000
Coef Lin Exp (20 C), 10^{-6} in./in./°C	33
Therm Cond, cal/sq cm/cm/°C/sec	0.057
Elec Resis, μ ohm-cm	
20 C	9
156 C	29
Latent Ht of Fusion, cal/gm	6.8
Ht of Vaporization, cal/gm	468
Vapor Press, mm Hg	
1249 C	1
1466 C	10
1756 C	100
1863 C	200
Mod of Elast, 10^6 psi	1.57

Mechanical properties

Typical properties of annealed indium are as follows: tensile strength, 380 psi; hardness, 0.9 Bhn; elongation in one in., 22%; compressive strength, 310 psi.

Because indium does not work-harden and almost all of the deformation occurring in the tensile test is localized, deformation is very low for such a low-strength material.

Reactivity

Indium is stable in dry air at room temperature. The metal boils at 2000 C but sublimes when heated in hydrogen or in vacuum. Because a thin, tenacious oxide film forms on its surface, indium resists oxidation up to, and a little beyond, its melting point. The film, however, dissolves in dilute hydrochloric acid.

Indium can be slowly dissolved in dilute mineral acids and more readily in hot dilute acids. It unites with the halogens directly when warm. Concentrated mineral acids react vigorously with indium but there is no attack by solutions of strong alkalies. The metal dissolves in oxalic acid, but not in acetic acid.

There is no evidence that the metal is toxic and it has no action as a skin irritant.

Applications

The three largest uses of indium are in semiconductor devices, bearings and low melting point alloys.

Semiconductors. Indium is used to form p–n junctions in germanium. Two characteristics—the fact that indium readily wets the germanium and dissolves it at 500 to 550 C, as well as the fact that after alloying, the indium does not set up contraction stresses in the germanium—make it suitable for this application.

If a piece of n-type germanium is dissolved in indium, germanium containing excess indium recrystallizes after subsequent cooling. The excess indium changes the germanium from n-type to p-type.

Low melting point alloys. Addition of indium to Wood's metal alloys lowers their melting points by 1.45 C for each 1% of added indium. The lowest melting point—47 C—is found at 19.1% indium. The eutectic alloy 76% gallium–24% indium melts at 16 C and is liquid at room temperature. Low melting point alloys are used for fusible safety plugs, foundry patterns, etc.

Bearings. Impregnating the surface of steel-backed lead-silver bearings increases the strength and hardness, improves resistance to corrosion by acids in the lubricants and permits better retention of the bearing oil film. Indium-coated bearings can be used for high-duty service such as found in aircraft engines and diesel engines.

Alloying. A common glass sealing alloy contains approximately 50% tin–50% indium. A solder alloy containing 37.5% lead, 37.5% tin and 25% indium has greater resistance to alkalies than the 50% lead–50% tin solder.

Adding indium to gold for use in dental alloys increases the tensile strength and ductility of gold, improves resistance to discoloration and improves bonding characteristics.

INDUSTRIAL THERMOSETTING LAMINATES

So-called "industrial thermosetting laminates" are those in which a reinforcing material has been impregnated with thermosetting resins and the laminates have usually been formed at relatively high pressures.

Resins most commonly used are phenolic, polyester, melamine, epoxy and silicone; reinforcements are usually paper, woven cotton, asbestos, glass cloth or glass mat.

NEMA (National Electrical Manufacturers' Association) has published standards covering over 25 standard grades of laminates. Each manufacturer, in addition to these, normally provides a range of special grades having altered or modified properties or fabricating characteristics. Emphasis here is on the standard grades available.

Laminates are available in the form of sheet, rod and rolled or molded tubing. Laminated shapes can also be specified; these are generally custom-molded by the laminate producer.

General properties

Resin binders generally provide the following characteristics: (1) phenolics are low in cost, have good mechanical and electrical properties, and are somewhat resistant to flame; (2) polyesters used with glass mat provide a low-cost laminate for general-purpose uses; (3) epoxy resins are more expensive, but provide a high degree of resistance to acids, alkalies, and solvents, as well as extremely low moisture absorption, resulting in excellent mechanical and electrical property retention under humid conditions; (4) silicones, high in cost, provide optimum heat resistance; and (5) melamines provide resistance to flame, alkalies, arcing and tracking, as well as good colorability.

Mechanical properties. Tensile strengths of paper, asbestos and cotton fabric-base laminates vary from about 8000 to over 20,000 psi; flexural strengths are somewhat higher. Moduli of elasticity are about 1 to 2 million psi.

Glass cloth reinforcement provides tensile strength values in the 13,000 to 37,000 psi range; modulus of elasticity can approach 3 million psi. Typical flatwise compressive strength values fall in the 20,000 to 50,000 psi range.

Impact strength values range from about 1.0 to 4.5 ft-lb per in. notch for cotton and asbestos-base laminates, to 4.5 to 15 ft-lb per in. notch for glass-base grades.

Electrical properties. Dielectric strengths (perpendicular to laminates) generally range between 400 and 700 v per mil, and up to 1000 v per mil for paper-base phenolics. Dissipation factor, as-received and at 10^6 cps, ranges from a high of about 0.055 for cotton-base phenolics to a low of about 0.0015 for a glass-reinforced silicone laminate.

Other properties. In general, moisture absorption (ASTM D 570) is about 0.3 to 1.3% (24 hr, $\frac{1}{16}$ in. thickness), depending on resin and reinforcement.

Heat resistance depends on both resin and reinforcements. Most standard laminates are designed for AIEE Class A or B insulation requirements, though glass silicones are suitable for Class H insulation. General maximum continuous service temperatures are about 250 F for phenolics reinforced with organic reinforcements, 300 F for phenolics, melamines and epoxies reinforced with inorganic materials, and 500 F for silicones reinforced with inorganic materials.

Flame retardance of most standard grades, other than melamines is relatively poor, but special flame-resistant or self-extinguishing grades are available.

Mechanical grades

Following are brief characterizations of NEMA grades intended primarily for mechanical use. Remember that laminate manufacturers offer a variety of specialized grades with substantially altered properties.

Grade X. High-strength kraft paper with a low phenolic resin content. It has the best impact strength of the paper-base laminates, though values are not as high as those of fabric-base grades.

Grade P. A paper-base plasticized phenolic suitable for hot punching (at 200 to 250 F) in thicknesses up to ⅛ in., and cold punching in thicknesses up to ¹⁄₁₆ in. It is more flexible but not as strong as Grade X; it has relatively good dielectric properties under dry conditions, but does not have the dimensional stability of Grades XXP or XXXP.

Grade PC. A paper-base phenolic laminate with properties similar to those of Grade P, but intended for cold-punching applications.

Grades ES-1, 2, 3. Paper-base laminates used as engraving stock. ES-1 is usually a melamine laminate with a black or gray surface and a white core. ES-2 is usually a phenolic and has a black or gray surface, a white melamine subcore and a black phenolic core. ES-3 usually consists of a black phenolic core with a white or gray melamine surface.

Grade C. A cotton canvas-base phenolic laminate, intended for uses requiring toughness and high impact strength. Fabric weighs over 4 oz per sq yd.

Grade L. Fine weave cotton linen-base phenolic laminate for use where a fine machined finish is desirable. Fabric weighs not over 4 oz per sq yd. It has higher tensile but lower flexural and impact strengths than Grade C.

Grade MC. A specialty grade made from purified cotton fabric with a melamine resin. Used for plating barrels.

Electrical grades

Following are brief descriptions of NEMA grades intended primarily for electrical uses.

Grade XX. A paper-base phenolic laminate intended for general-purpose electrical use. It has good dimensional stability and moderate mechanical properties.

Grade XXP. A paper-base phenolic laminate suitable for hot punching (at 200 to 250 F), has better moisture resistance and electrical properties than Grade XX. It is intermediate between Grades P and XX in resistance to cold flow and in punching characteristics.

Grade XXX. A paper-base laminate with a high phenolic resin content. It has good electrical properties, low miosture absorption and low cold flow under high humidity conditions.

Grade XXXP. A paper-base laminate with a plasticized phenolic resin to permit hot punching. It has better dielectric properties than Grade XXX and is intermediate in punching characteristics between Grades XXP and XX.

Grade LE. A cotton linen-base phenolic laminate recommended for electrical applications requiring greater toughness than Grade XX

provides, and better machining properties and finer appearance in thinner sections than Grade CE.

Grade G-2. A staple-fiber glass cloth-base phenolic laminate with good electrical properties under high humidity conditions. Although good in dimensional stability, it is the weakest of the glass-base grades.

Grade G-6. A continuous-filament glass cloth-base silicone laminate. It is recommended for Class H insulation applications. Dimensional stability and impact strength are high, although other mechanical properties are only fair.

Grade G-7. A continuous-filament glass cloth-base silicone laminate combining the thermal and electrical properties of Grade G-6 with higher tensile and flexural strengths.

Grade N-1. A nylon-fabric base phenolic laminate with exceptionally high insulation resistance and excellent electrical properties under high humidity conditions. Good impact strength and easy to machine and punch.

General-purpose grades. Following are grades which cannot easily be classified either as primarily electrical or primarily mechanical. This group also includes some specialty grades which have been developed for specific qualities other than mechanical or electrical properties.

Grade CE. A medium weave, canvas-base phenolic laminate, used for electrical applications where greater toughness than Grade XX is required, or for mechanical applications requiring greater moisture resistance than Grade C provides.

Grade A. An asbestos paper-base phenolic laminate, it is more flame resistant and somewhat more resistant to heat than cellulose-reinforced grades. It retains strength relatively well after exposure at fairly high temperatures.

Grade AA. An asbestos cloth-base phenolic laminate, it is stronger, tougher and more heat resistant than Grade A. It is made with AA grade asbestos fabric, containing a minimum of 90% asbestos by weight.

Grade G-3. A continuous-filament glass cloth-base phenolic laminate, combines good electrical and mechanical properties with excellent heat resistance.

Grade G-5. A continuous-filament glass cloth-base melamine laminate that has the highest mechanical strength of all NEMA grades; it also has excellent electrical properties under dry conditions.

Grade G-10. Continous-filament glass-base epoxy laminate has the highest insulation resistance and dielectric strength of the glass-base grades, an extremely low dissipation factor and the lowest water absorption of all NEMA grades.

Grade GPO-1. A random fiber glass mat-base polyester laminate designed for use as a moderate cost, general-purpose material. It has good mechanical and electrical properties and good cold-punching characteristics.

INJECTION MOLDINGS

Injection molding of plastics is analogous to die casting of metals. The plastic is heated to a fluid state in one chamber then forced at high speed into a relatively cold, closed mold where it cools and solidifies to the desired shape. This method of processing plastics became a commercial reality in the early 1930's.

It is the fastest and most economical of all commercial processes for the molding of thermoplastic materials, and is applicable to the production of articles of intricate as well as simple design.

A slightly modified version of injection molding known as "jet molding" is applicable to the molding of thermosetting materials. The principal difference between the two processes is the function of the temperature of the nozzle and mold on the material. Injection molding employs a relatively cold mold to solidify thermoplastics by chilling the mass below the melting point. Jet molding employs a relatively hot nozzle and mold to harden the thermosetting material by completing the cure.

The process

The sequence of operation known as the molding cycle is as follows:

1. Two mold halves, which, when closed together, combine to form one or more negative forms of the article to be molded, are tightly clamped between the platens of an injection-molding machine.
2. The closed mold is brought into contact with the nozzle orifice of a heating chamber. The heating chamber, known as the plastifying cylinder, is of sufficient size to carry an inventory of material equal to

several volumes required to fill the mold. This permits gradual heating of the plastic to fluidity.

3. An automatically weighed or measured quantity of granulated thermoplastic, sufficent to fill the mold cavity, is fed into the rear of the plastifying cylinder.

4. A reciprocating plunger actuated by a hydraulically operated piston forces the material into the plastifying cylinder. An equal quantity of fluid plastic is thus forced out of the front of the cylinder through the nozzle orifice and into the mold.

5. A pressure of several thousand pounds per square inch is maintained on the material within the mold until the plastic cools and solidifies.

6. After the molded item has hardened sufficiently to permit removal from the mold without distortion, the mold is opened and the part ejected.

Injection molding offers several advantages over other methods of molding. Some of the more important of these are: (1) The process lends itself to complete automation for the molding of a great number of parts; (2) molded parts require little if any post-molding operations; (3) high rates of production are made possible by the high thermal efficiency of the operation and by the short molding cycles possible; (4) low ratio of mold-to-part cost in large-volume production; (5) long tool and machine life requiring a minimum of maintenance and relatively low amortization costs, and (6) possible reuse of material in most applications.

Molding machines vary considerably in design as well as capacity. For example, the clamp mechanism may be hydraulically operated, hydraulic-mechanically operated or entirely mechanical. The injection of the material into the mold may be accomplished by a rotating screw as well as a plunger.

Machines are rated according to the weight of plastic which can be injected into the mold (the shot) with one stroke of the injection plunger. The capacity of commercially available machines covers a range from a fraction of an ounce to 400 oz.

Limitations

The injection-molding process is subject to the following limitations:

Material. Any thermoplastic material may be a candidate for injection molding, if upon heating it can be rendered sufficiently fluid to permit injection into a mold and the resulting molded article retains all the desired properties.

Geometry of part. Any part, regardless of geometry, which can be removed from a mold without damage to the mold or part is moldable.

Weight of part. The weight of the part must be within the "shot" rating of the particular machine being used.

Projected area and wall thickness of part. The limitation on these dimensions will be governed by several factors including the relative fluidity of the plastic being molded, the pressure necessary to fill the mold, the rigidity of the part to permit ejection from the mold without deformation, and sufficient clamp force available to hold the mold in a closed position when the necessary injection pressure is developed within the cavity.

Production economy. Economy can be realized only when relatively large quantities are produced. In addition to the material consumed and the cost of the molding operation, the number of pieces produced must also bear the cost of the mold in which they are cast. Depending on the size and complexity of the geometry of the part to be molded, mold costs will vary from a few hundred to many thousands of dollars. Again depending on size and complexity of the part, the rate of production will vary from a few minutes for a single part to several hundred parts per minute.

In addition to the high production rates made possible by injection molding, articles of high quality and relatively precise dimensions can be produced. The production of industrial parts held to dimensional tolerances of $\pm$ 0.002 in. per in. is quite common.

The injection-molding process has made possible the development of an extremely large family of plastics. The basic types of thermoplastics and the many modified compositions available are analogous to metals and alloys. New com-

positions are constantly being developed to meet the demands of particular situations.

INVESTMENT CASTINGS

The investment-casting process derives its name from one operation of the over-all process: that of investing, enclosing or encasing a disposable pattern or pattern cluster within a refractory slurry which subsequently hardens to form a mold. This method of mold-making and casting in a hot mold (1600 to 1900 F) distinguishes investment casting from other casting procedures. Other casting processes such as sand casting, die casting, and shell mold casting, employ a split mold-making technique.

The investment-casting process is also known as the lost wax, or precision casting process. (Another technique akin to it is the frozen mercury process). Ornamental jewelry was produced in ancient times by investment casting. And in the early 1900's, the process gained prominence in the production of dental inlays and dentures. Investment castings are now used in many industries.

General description

Investment casting consists of three major operations—pattern production, mold production and casting.

1. Castings are produced by using disposable patterns. Patterns are formed of wax or plastic by injecting these materials, which are fluid, into a die cavity or cavities. The shape and detail of the part intended for reproduction is cast or machined into the die.

A pattern die is constructed in two parts so that the pattern, or, in the case of a multiple-cavity die, patterns can be removed. Machined-pattern dies may be cut from a solid piece of steel or may be built in components for assembly. Cast-pattern dies require that a metal master pattern first be produced, which is in turn used to cast a soft metal die cavity or cavities. These dies are usually cast from a low melting alloy such as tin-bismuth. Steel dies are normally used for injection of plastics. In many instances, a part of the foundry gating system, (e.g., feeders and runners) is built into the die cavity to eliminate subsequent assembly work in making up a pattern cluster. A pattern die can be made to produce a single pattern, multiple patterns, or only a portion of a pattern of a given part. In the latter case, more than one die would be used and these components would be put together to make a complete pattern.

After the patterns are produced, the next step is making the pattern assembly or pattern cluster which subsequently forms the mold cavity. The pattern assembly consists of the part pattern or patterns and the gating system which includes feeders, runners and metal reservoir.

2. The first operation in mold making is to dip the pattern cluster into a liquid mix of silica flour to wet the patterns and then cover them with a dry mix of flour and sand. This forms the dip coat of the patterns and is the start of the mold buildup. There are now two major methods of completing the molds.

The first method is to seal the pattern cluster to a mold board and place a container on the board which encloses the pattern cluster. A refractory slurry is poured into the container which invests (encloses) the pattern cluster except at that area at the bottom of the mold board where the pattern cluster is stuck to the board. The fluids in the slurry harden chemically and the result is a very dense, hard, heat-resistant mold with the pattern cluster embedded. The mold board is then removed, exposing the end of wax pattern clusters.

An alternate method of completing the mold after the initial dip coating is to continue with the procedure of alternately dipping in a heavier slurry and coating with dry refractory particles until a ceramic shell is formed around the pattern cluster. Drying is necessary between cycles and the number of cycles depends upon the strength requirements to contain the molten metal when poured into the mold.

3. The casting operation includes three steps: removing the pattern cluster from the mold, heating the metal to a molten state and pouring it into the mold, and cooling the mold and removing the cast parts from it. The pattern material (wax or plastic) is removed from the mold by a combination of melting and burning. Approximately an 8-hr heating cycle at temperatures of about 1600 to 1900 F is used to fire the ceramic mold and remove the pattern material.

This, then, leaves a cavity in the mold. The molds are removed from the furnace shortly before casting so they are near furnace temperature when poured.

Depending upon the alloy from which the castings are to be made, various melting-casting techniques are used. Alloys are melted while exposed to normal atmosphere, while covered protectively by an inert gas, or while the melting crucible and mold are in a vacuum chamber. Melting is usually accomplished by indirect arc or induction heating. Molds may be cast by "floor pouring" to fill the mold directly from a furnace or from furnace to ladle to mold. The floor-pouring method is used primarily for large molds. The more prevalent way is to secure the mold on top of the melting furnace and invert both furnace and mold so that the alloy enters directly into the mold from the crucible, normally under slight pressure.

After solidification and cooling have taken place, the mold material is mechanically or hydraulically broken away from the castings. The gates are cut from the casting and the casting processed through other operations of finishing and inspection as required.

Materials

The majority of the alloys cast by this process are iron-, nickel-, or cobalt-base alloys. The iron-base alloys include the complete range from plain carbon steels to the stainless and high-alloy steels. Many nonferrous alloys can also be used, such as gold, brasses and bronzes, as well as gold, aluminum and copper alloys.

The alloy to be cast is generally not considered as a restriction, as the process is sufficiently versatile to accommodate almost all types. However, not all manufacturers cast all alloys. The use of inert gas and vacuum protection for melting and pouring has enabled the industry to cast the ultra-high-temperature iron- and nickel-base precipitation-hardening alloys.

Sizes and configurations

The size of investment castings is often thought of in terms of castings weighing less than 1 oz to castings weighing about 3 lb; this size range includes the majority of the castings poured. However, the capability of the process permits the successful casting of parts over 100 lb and over 20 in. in diameter with considerable accuracy and detail. Small castings have been produced with edges down to 0.015 in., holes ⅛ in. in diameter by 6 to 8 in. long, shorter holes to 0.03 in. The type of alloy selected and the corresponding pouring practice may modify the maximum or minimum sizes and tolerances.

It is difficult to place a restriction on the configurations that can be produced. The use of coring techniques, pattern assembly, and other processing details make for great latitude in design. As a general statement, it can be said that as a part becomes more difficult or costly to finish by machine operations or fabrication, the more feasible it will be to produce the part as an investment casting.

Quality

The investment-casting process is noted for its inherent quality and reliability.

Dimensional quality. The accompanying figure shows a tolerance prediction curve based on the statistical analysis of the dimensional variance incurred in many different types and sizes of parts. The mean tolerance is an arithmetical average. The limits represent statistically three standard deviations from that mean; i.e., 99.6% of all of the parts having a dimension on it of a given size would be expected to fall within the limits shown. This chart then represents a grand average.

It is definitely possible to work within closer tolerances on a specific dimension or a few critical dimensions by closer quality-control procedures, finishing and inspection. On small dimensions up to about an inch and a half, it is sometimes possible to work as closely as ± 0.003 in. but general tolerance capabilities of ± 0.003 in. or ± 0.005 in. per in. of length for various in-

crements of length over an entire part which may contain from 10 to 75 dimensions is definitely misleading if not somewhat incorrect.

Casting and metallurgical quality. The melting and casting techniques employed in the process make possible accurate, reproducible metallurgical qualities. Alloys can be cast to very close chemical analysis which, in turn, insures consistent physical and mechanical properties. Proper application of the techniques of investment-casting practice produces parts with minimum amounts of dirt, gas, shrinkage and other imperfections which adversely affect properties and performance.

Surface finish and appearance. Investment castings normally have a smooth surface finish as cast which ranges about 80 to 100 rms. One of the features of the process is that this good quality finish and appearance enable the user to eliminate or minimize finishing.

Applications

One of the major uses of investment-cast parts is in gas turbines. Investment-cast parts have proven very successful here because the shapes required (airfoils) are particularly difficult to machine and the high-strength alloys being used are difficult to finish, even for fairly simple operations. Many of the alloy compositions develop their outstanding properties in the cast form and may not be amenable to working in the wrought form.

No single industry, other than the gas-turbine

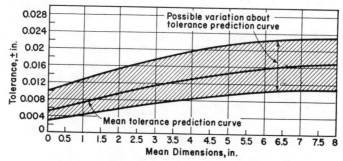

Example: on a ½ in. dimension the "mean" line falls on approx. 0.006 in. indicating a regular tolerance of plus or minus 0.006 in. for the represented ½ in. dimension, or a total spread of 0.012 in.

Tolerance prediction curve for investment castings.

industry, uses a large volume of investment castings. However, practically all industries are using some investment castings in their products or processes.

Advantages of the process

There are several important advantages in the mass production of metal parts by the investment-casting process, not only by comparison with other casting processes, but also with other production methods such as machining or forging.

Compared with other casting processes (such as sand or shell molding) investment casting provides:

1. An improved casting from the standpoint of cleanliness of the alloy and metallurgical soundness. This is especially true when an inert gas or vacuum is used to protect the alloy.

2. Considerable more versatility in producing intricate shapes and cast detail. This feature opens the possibility of combining two or more components of an assembly into a single casting.

3. Greater dimensional accuracy which in many cases eliminates or minimizes the need for finishing. This certainly reduces the necessary stock allowance.

4. Better cast surfaces which are beneficial from the standpoint of quality and appearance, and may further eliminate or minimize finishing operations.

The advantages of investment casting compared with other manufacturing processes are:

1. Many alloy compositions develop their outstanding properties in the cast form only. Also, many of the compositions cannot be produced other than in cast form.

2. There has always been controversy as to whether cast parts could compete with the consistent high quality and uniformity of forged parts. The investment casting industry has proven this can be done, primarily, by consistently producing to the stringent quality requirements of the aircraft industry.

3. There have been many instances on both small and large parts where castings have been very beneficial in eliminating great amounts of machining, either starting with a wrought mill stock or from forged parts. Castings also have

greater versatility in accommodating changes in section and shape.

IRIDIUM. See Platinum Metals.

IRON (PURE)

"Armco Ingot Iron" is a trade name for commercially pure iron produced by the basic open-hearth process. The refining operations are carried farther than for making mild steel. Metalloids are reduced to low values by oxidizing the molten steel with iron ore. This oxidizes the manganese, phosphorus and carbon and the highly basic slag reacts with the manganese, phosphorus and sulfur present to effect their removal from the metal. The carbon is eliminated as a gas.

A typical chemical analysis of this material is: carbon, 0.015%; manganese, 0.025%; phosphorous, 0.005%; sulfur, 0.025%; silicon-trace. The microstructure of Armco Ingot Iron consists of a simple network of grain boundaries characteristic of all pure metals since no phases other than ferrite are present. The size and shape of the grains depends on the mechanical and thermal treatments.

Physical properties

Physical properties vary according to form and treatment but typical values are: specific gravity, 7.866; specific heat at 25 C, 0.108; melting point, 1539 C; heat of fusion, 65 cal per gm; thermal conductivity, 0.16 cal per sec per sq cm per °C per cm at 100 C; thermal coefficient of expansion, 12.6×10^{-6} per °C (20 to 300 C); electrical resistivity, 10.7 microhm-cm. at 20 C; temperature coefficient of electrical resistivity, 0.0056 per °C (0 to 100 C).

Mechanical properties

Typical mechanical properties after various treatments are shown in Table 1.

Young's modulus is 29.3×10^6 psi in both tension and compression and shear modulus is 11.8×10^6 psi. Poisson's ratio is generally considered to be 0.28.

TABLE 1—MECHANICAL PROPERTIES OF INGOT IRON

Property	Hot-Rolled Rods or Plates	Ann. Rods or Plates	Dead Soft	Cold-[a] Worked	Finished Cold	Finished Hot	Finished Hot [b]
Compression							
Elastic Limit, 1000 psi		19.4					
Prop Limit, 1000 psi		19.2					
Yield Str, 1000 psi		20.6					
Tensile							
Yield Str, 1000 psi	26–32		19		26.9	19.1	30.3
Tensile Str, 1000 psi	42–48		38.5	100	43.8	42.2	47
Elongation, %	22–28		43–48		40	44	36
Gauge Length, in.	8		2		8	2	2
Reduction of Area, %	65–78		70–77	65–70	75.6	77.3	70.0
Torsion							
Prop Limit, 1000 psi	12–15						
Yield Str, 1000 psi	14–23						
Hardness							
Bhn	82–100		67	220	101	90	110
Rockwell B	39–55						
Fatigue							
Rev Bend, 1000 psi		26					
Rev Torsion, 1000 psi		12.8					
Axial Stress, 1000 psi		26.7					

[a] Approximate max.

[b] Quenched from 1725 F to water.

Fabrication

Cold working. Because of its high ductility, Armco Ingot Iron can be bent and formed more easily than steel thus making it possible in some cases to substitute cold-forming operations for hot forming with steel. Deep-drawing operations of a wide variety of forms and degrees of intricacy are carried out in cold stamping presses. It is very suitable for cold-spinning operations.

Hot working. Armco Ingot Iron is workable above 1920 F and below 1560 F but between these two temperatures it will check or break due to hot shortness resulting from the very low manganese content. For forging or hammer welding, it should be heated to a white heat from which it can be worked until temperature drops to about 1920 F or a bright orange color. If more working must be done, reheating is necessary. The material can be worked below the lower limit of the range if hammer welding is not involved. This is generally better than working above the critical range because it avoids the difficulty of some portions of the piece cooling into the critical range. When worked at temperatures either above or below this range, satisfactory results will be obtained.

Machining

The high purity of Armco Ingot Iron makes it very tough and ductile and this requires slight changes in the cutting angles of machine tools in order to get best results. Otherwise it machines much the same as dead soft steel. More metal can be removed with less power and a better finish can be obtained by taking fine feeds and deeper cuts. All cutting tools should have a side slope angle of about 28 to 32 deg and a back slope of about 15 deg and 8 to 10 deg clearance. A sulfur base mineral oil of about 70 to 80 viscosity is recommended for cutting operations where machine speeds are high. For drilling the clearance angle of the drill should be increased slightly and the drill ground to cut continuous spirals rather than chips. A light feed of only 0.002 to 0.004″ per revolution should be used.

For tapping use a tap drill a size or two larger than standard.

Joining

Brazing and soldering. The usual procedures may be used without difficulty on Armco Ingot Iron.

Riveting. The high ductility of Armco Ingot Iron makes it possible to cold-drive rivets up to about 3/8 in. Larger sizes that are driven hot should never be heated above the lower range of critical working temperature in order to avoid the breakage described under hot working.

Annealing

Annealing results in a change in the grain structure and physical properties of the material. At temperatures below 1650 F the recrystallization of the grains and change in properties that occur depend upon previous cold work. Certain combinations of annealing temperature and degree of cold working may cause excessive grain growth and inferior physical properties. This is characteristic of low-carbon steels as well as "Armco Ingot Iron" and should be considered when parts must be shop annealed during the manufacturing process. Optimum physical properties and microstructure in the original material can be entirely destroyed by ill-advised annealing which may cause excessive grain growth after cold working. These difficulties can be avoided by normalizing or annealing above 1650 F. One advantage of Armco Ingot Iron is that its high ductility frequently makes it possible to complete the fabrication in one operation so that no intermediate annealing is required.

Forms

Armco Ingot Iron is available in the form of hot and cold rolled sheet and strip in all variations of surface and temper, in plate, wire rods, bars and spiral welded pipe.

Applications

Armco Ingot Iron was originally developed to resist corrosion and in sheet form it has been galvanized for roofing, siding, tanks and signs. As corrugated pipe it is used for culverts. Its high purity makes it suitable for oxyacetylene welding both as material to be welded and as welding rods. It is widely used in the form of sheets for vitreous enameling because of its freedom from enamel defects such as bubbles and blisters.

It is is very ductile and is used for severe deep-drawing operations such as refrigerator parts, washing machine tubs, etc. Superior drawing properties and enameling quality are thus combined in the same material.

Relatively low electrical resistance compared to steel has led to its use as telegraph wire and bond wire for rail signal circuits for railroads. High magnetic permeability and low retentivity has stimulated its use for generator fields, frames and poles, solenoid plungers, magnetic parts for relays, magnetic brakes and clutches and for other parts of electrical equipment.

K

KNITTED AND WOVEN METALS

Although they are commonly referred to in the same context and are used for some of the same applications, there is a considerable difference between knitted and woven metals. As their name implies, knitted metals are knitted into a mesh structure in much the same way as stockings or sweaters. The structure of woven wire, on the other hand, is usually simpler, consisting of interwoven strands of wire. Whereas weaving usually produces a symmetrical mesh, usually with square openings and parallel wires, knitting produces an asymmetrical mesh of interlocking loops.

Knitted metals

An important advantage of knitting is that it produces a mesh of interlocking loops each of which acts as a small spring and provides resiliency. Because of this resiliency knitted metals are usually able to withstand greater loads and deflections without being permanently deformed.

Knitted wire also has both a large surface area and a high percentage of free space. Thus, knitting permits construction of a mesh using wire with a maximum surface area and with interstices of almost any desirable size, regardless of wire diameter. Fine wire, for example, has been knitted into a mesh with as few as three to five openings to the inch.

The free volume of knitted wire can be controlled between 50 and 98%, depending on the interstice size, and regardless of the wire size used. Even when the wires are widely spaced to produce a free volume of 98% the structure retains its shape; a similar spacing with woven wire would result in a shape that would be almost impossible to handle.

Another important property of knitted metals is their ability to maintain their dimensions during temperature cycling. When the meshes are slightly stretched in every direction and expansion occurs, the sides of the loops are merely forced closer together without changing overall dimensions or the plane of the surface.

Knitted metals can be produced in wire diameters of 0.0005 to 0.025 in. in such materials as steel, copper, brass, aluminum, stainless steel, and various nickel alloys including monel. In fact, almost any metal that can be drawn into wire can be knitted. Thus, knitted parts can be made in a wide range of strength, corrosion resistance, wear resistance, electrical shielding and heat-resistance properties.

The asymmetrical mesh of knitted parts is advantageous for electronic shielding because the continuous loop structure apparently causes induced currents to cancel themselves. Apart from this application, however, the biggest use for knitted wire is to remove one material from another. Thus, it can be used to separate two phases of the same material, to separate two immiscible liquids, or to separate a solid from a liquid or a gas. The degree of separation can be controlled by varying the compression used to form or shape the knitted part, and by controlling the size and shape of the wire and the size of the loops.

A good example of this kind of application is a mist eliminator in which the knitted mesh separates the liquid phase from the gas phase. Gases bearing droplets from such processes as distillation, evaporation, scrubbing, cleaning or absorption are passed through built-up layers of knitted fabrics up to 6 in. thick. The droplets collect on the loops and slowly run to the bottom where they accumulate. Liquid particles as small as 5 to 20 microns can be separated with efficiencies as high as 98 to 100%. Although small eliminators are usually made in one piece, eliminators can be made up to 28 ft in dia by building up layers of crimped mesh.

Knitted metals are also used in many other applications where their special structure and properties are useful. Fuel line filters, for example, of knitted wire are resilient and do not require precise machining for sidewall fit. Gaskets of knitted wire have excellent conductivity and sufficient resiliency to produce tight joints on uneven surfaces, thereby preventing RF leakage. Heat dissipation sleeves of knitted metal for subminiature glass tube envelopes provide high cooling efficiency. Knitted metals also provide good shock and vibration control when used as mountings for airborne and industrial equipment.

Woven metals

Woven wire cloth is used in a wide range of applications for grading materials, filtering, straining, washing, guarding, reinforcing and decoration. A variety of meshes in different materials and sizes is available to meet these applications.

Like knitted wire, woven wire cloth can be produced in almost any metal that can be drawn into wire, including carbon and stainless steel, copper, brass, bronze, nickel, monel, Inconel and aluminum. Plain steel wire cloth is one of the most economical and generally used materials. However, it may require a protective coating to prevent rusting.

Corrosion can be prevented by using tinned or galvanized wire; tinned wire is preferred for handling food products, galvanized for all other applications. Where required, the cloth can also be provided with a protective electroplate of cadmium, chromium or tin. Phosphate coatings and paints can also be used.

Where severe corrosive conditions are encountered, the weave can be made from such materials as stainless steel, monel, phosphor bronze and silicon bronze; some of these materials are available only in limited sizes. Abrasive resistant steels are available for applications where abrasive materials have to be handled. Many grades of stainless can also be used to withstand high temperatures, e.g., 347 stainless (up to 1400 F), 309 and 310 (up to 1600 F), and 310 stainless as well as Inconel and Nichrome (up to 2000 F).

Woven wire cloth is widely used for filtering and straining, particularly in automotive and aviation applications for carburetors, air screens and oil and fuel strainers. They are also used to grade materials, and for wire baskets, insect screening and safety guards.

L

LEAD AND ALLOYS

Lead is a heavy, bluish-gray metal soft enough to be easily cut with a knife. The exposed lustrous face acquires a silvery gray patina in air. It has a face-centered cubic crystal structure and, compared with most structural metals, has low tensile strength and a high rate of creep. Lead's melting point is 621 F.

Design properties

Lead is the most *dense* of the common, inexpensive metals. It has a specific gravity of about 11.3 to 11.4 depending on its form—equivalent to 707 lb per cu ft.

A member of the fourth group of the chemical periodic table, lead has a chemistry which makes it inherently *corrosion resistant* to many problem chemicals for other structural materials. Its resistance to most acids is excellent and it is among the most permanent metals in terms of weathering resistance or exposure to most natural environments such as burial, exposure to sea water, etc.

The softness and low tensile strength of lead are as often an asset as a liability. Many of lead's important uses hinge on its *ability to flow* or "give." Lead and most of its alloys present a soft "greasy" surface which is used in bearings and often contributes useful lubricating power even in applications where *lubricity* is not the key requirement—in deep drawing of lead-coated stock or in the machining of leaded ferrous and nonferrous alloys, for example. Also, because it is a soft material, lead is quite easily formed by almost any technique.

Where *radiation shielding* is a problem, lead is probably the first material to be considered by the designer. Design manuals dealing with X-ray or gamma radiation shielding often make the point that comparisons of shielding efficiency are usually expressed in terms of lead because it represents the best practical gamma shield material for most applications. Despite its high density, lead almost invariably saves weight and space when used for shielding. The

AEC *Reactor Shielding Design Manual* points out that "where large attenuations of gamma-radiation intensities are desired, the use of lead instead of iron results in a weight saving of at least 30%." This corresponds to shield thickness reduction of more than 50%. Where water is the shielding material, the savings in weight, size—and cost—are even more dramatic: a reactor employing water shielding (without lead) having a designed weight of 683,000 lb can be built with the same shielding effectiveness to a weight of 137,000 lb by using lead shielding.

The role of lead in lead-acid storage batteries is well-known, but there are newer electrical uses. Lead and some of its alloys, for example, are *superconductive* at low temperatures and have been employed in computers, gyrocompasses and other devices making use of superconductivity. Lead's semimetals such as lead telluride are being used commercially as *thermoelectric* generators. They are employed in SNAP III nuclear power package and have been used in domestic and industrial temperature control instrumentation where no secondary power is needed to augment the output of the temperature-sensing element. Still another growing use of lead is in cathodic protection systems using impressed current. *Lead anodes* develop a protective lead peroxide coating in such service and show excellent life and current-handling performance.

Available forms

Sheet and foil. Because of its malleability, lead and its alloys are readily rolled to any desired thickness down to 0.0005 in. Sheets are easily fabricated by burning or soldering. Standard widths run up to 8 ft or more for sheet and sheets may be cut to any desired size. Blanks for impact extrusion, gaskets, washers or other purposes may be stamped out. Tin-coated lead can be produced by rolling lead and tin together.

Extrusions. Lead is easily extruded in the form of pipe, rod, wire or any desired cross section like window cames (H-shaped), rounds, hollow stars, rectangular duct. Commercially available extrusions range in size from 24 in. pipe down to solder wire 0.010 in. in diameter. Lead is extruded over paper, rubber or plastic in making electrical cable and around steel bars. Common flux-cored solder is a lead extrusion; toothpaste tubes are impact extrusions.

Castings. One of the simplest metals to cast, lead is used in tiny die castings and massive cast counterweights. Type metal, renowned for its ability to reproduce minute detail, is a lead alloy. Lead grids for most batteries are die-cast. Casting temperature (usually about 600 F) is moderate. Arsenic, antimony, or tin are frequently alloyed to impart strength or special properties. Small die castings can have wall thicknesses as low as 0.05 in. and "as-cast" dimensions are reproducible to 0.001 in.

Coatings. Protection of underlying iron and steel is the main objective in most lead coating. In the purely protective class one finds terneplate for roofing, fireproof frames and doors, automotive parts, and containers for paint and oil. Lubricity imparted by the coating eases drawing and stamping operations and produces an excellent surface for soldering—hence, television chassis and automotive gas tanks are made of terne. Other hot-dip processes as well as electroplating and flame spraying are also used for outdoor hardware, automotive mufflers, bearings, bushings, nuts, bolts, and for maintenance as well.

Laminations. Developed originally for X-ray protection, a large family of laminated lead materials now exists. In addition to their original niche these are finding increasing use in sound isolation and noise control. Typical examples include lead-plywood, lead-gypsum board, lead-cinder block, leaded plastic-fabric laminates, leaded plastic and glass-fiber combinations.

Cladding. Metallic lead in thicknesses from ⅛ in. to 12 in. or more may be bonded to other metals. Thus, for example, lead and steel may be combined for corrosion resistance and strength or lead and copper for gamma shielding and heat transfer. In many instances a product such as a tank or chemical reactor is completely or partially fabricated in steel and then clad with bonded lead—as a unit.

Powder. Spheres, irregular grains, and flakes of lead from 4 μ dia up find use in special greases, as a constituent of bearings, brake, and clutch facings, in filling plastics and rubber and

in paints and pipe-joint compounds. Wire rope is usually treated with such a powder to lubricate it and to fill any nicks in the fiber, thus renewing it with self-lubricating surfaces.

Shot. This form of lead is produced in abundance—about 33,000 tons go into shotgun ammunition each year. Ammunition sizes range from 0.04 to 0.44 in.; small shot is made for other uses. Easily handled, it is a preferred form when mass or shielding is required inside an irregular enclosure. It is also used in making free-machining steels.

Wool. By passing molten lead through a fine sieve and allowing it to solidify in the air, a loose rope of fibers is produced. This weighs about 0.4 lb per ft. Under pressure, usually by being driven into a crevice with a calking iron and hammer, the fibers weld into a homogeneous mass. This permits the forming of a solid metal seal where temperature or explosion hazards prohibit jointing procedures requiring heat. Continuous lead fiber is also produced by being spun on textile machines.

Alloys. The extremely wide range of lead alloys makes it impossible to treat them in detail here. There are scores of fusible alloys and solders with melting temperatures from 117 to 588 F. A number of bearing alloys, babbitt being the most familiar of them, make use of lead. Antimony and arsenic are frequent additions to lead since small quantities of them harden and strengthen it. Type metal is a lead-tin-antimony alloy sometimes carrying a little copper. Newer developments include alloys made by powder metallurgy. Some of these show tensile and creep strengths several times as high as those normally associated with lead.

Lead is frequently added to alloys such as steel, brass, and bronze to promote machinability, forgeability, corrosion resistance, or other special qualities.

Compounds. The highest tonnage complex organic compound, tetraethyl lead, is produced at a rate of over a quarter of a million tons per year. By controlling the flame propagation rate it promotes even combustion, hence better efficiency in internal-combustion engines. Another compound for the same purpose is tetramethyl lead. Litharge (PbO) is used in storage batteries, crystal glass (and leaded-glass radiation shielding windows), in ceramics, and to vulcanize rubber. While litharge is used to cure many synthetic rubbers, today newer chemical derivatives of lead with the same action—tribasic lead malate, for example—are substituted for it. Many modern plastics are stabilized with other lead compounds. Another lead oxide—Pb_3O_4—is the ubiquitous red lead, valued for its ability to prevent corrosion. Lead chromates, yellow, orange, and red, are similar pigments. Basic lead carbonate is the familiar white lead for house paint. Lead azide is a standard explosive detonator; lead arsenate an insecticide and herbicide; lead silicates find use as pigment and enameling frits.

Mixtures. Quite a variety of useful materials made with lead can only be described as "mixtures." Lead powder in paraffin, for example, combines lead's gamma shielding ability with the neutron capturing ability of an organic. And further, it can be melted for casting or removal from a shield structure at a temperature of about 200 F. Many lead-organic mixtures are used similarly—some with exotic chemicals, some with many of the common plastics. Lead-plastic mixtures show unusual sound attenuation properties. A mixture of lead and polyethylene has been used to mold precision small parts with substantial mass—small flywheels, for example. A combination acoustical, antifouling, and "porpoise skin" (antiturbulence) coating for submarines employs lead shot in neoprene.

Fabricating methods

Because the designer has such a wide selection of forms of metallic lead and its alloys, he has, in effect, a partial prefabrication of any product or component he wishes to make of lead.

Where lead must be joined to lead, soldering and welding are the two most common techniques. A properly made joint by either technique will be as strong as or stronger than the metal itself. Soldering is exceptionally easy with lead—mild noncorrosive fluxes are used and the metal is cleaned at the joint mechanically by scraping or wire brushing. Tallow or rosin can be used as flux. Rosin dissolved in alcohol is often employed. No pretinning is necessary though the cleaning should be thorough and a good fit is essential to producing a solid joint. Where a closely fitted joint—lap or lock seam, for example—is used, 50-50 lead-tin solder is the most popular.

Many solder joints in lead and especially in lead pipe and cable sheathing are "wiped." This involves using a solder with a wider plastic range, usually 37 to 40% tin, the balance, lead. Lead-silver solders cannot be used successfully with lead because their melting point is too close to that of the metal itself.

Lead welding, commonly called lead burning, produces a true weld by fusing the parts together without the addition of any different metal. A torch is used to fuse the metals. Since little heat is required to fuse lead, only the smallest tip should be used on the torch. Two gases are normally employed—oxygen and hydrogen are the most popular, but oxyacetylene or oxypropane and even city gas with oxygen are sometimes employed. The flame should be neutral. An oxidizing flame will produce a noticeable formation of brown oxide on the surface of the molten metal. Joint preparation is much the same as in soldering except that no flux is employed. Lap joints are preferred except in very heavy lead and the lead welding rod used as a filler in the joint should have the same composition as the lead being joined.

Lead liners. Often lead lining is used for corrosion protection of tanks or to block X-ray or gamma radiation. Because of its low strength, lead is usually supported by some other stronger material such as steel, wood or concrete. The thickness of these linings depends largely on the degree of corrosive attack or intensity of radiation to be blocked. Quite a common lead thickness for these applications is 8-lb sheet lead (⅛ in. thick).

In tanks, conical linings are often held by 2-in. half-oval steel bars fastened at frequent intervals inside the lead and through it to the outer steel shell. These straps are covered with strips of lead welded to the lead lining on either side. These vertical straps are usually about 18 in. apart and never more than 24 in. Horizontal strapping in lieu of vertical strapping is seldom used, particularly when under-hand welding would be involved.

If the lining is to be bolted in place it is advantageous to use a large steel cut washer and a round head bolt to fasten the sheet lead to the steel or wooden shell. The bolt and washer are then covered with a lead cap which is burned to the lining. This tends to distribute the load over a large area and aids considerably in minimizing fatigue. The bolts used in this manner are evenly spaced over the entire lining on about 18-in. centers.

Where radiation shielding is the object, there is no need to preserve watertight integrity but it is important to leave no straight-line path for the radiation to pass through the shielding. Smaller pieces of lead sheet are thus often lapped, like shingles, so that the tail of each sheet covers the nail holes in the next. Normally these joints have 2-in. overlaps.

The availability of many forms of lead such as lead laminated to sheet materials like gypsum board or plywood, and in the form of lead tile "sandwiches," offers another route to designing radiationproof structures or linings.

Typical uses of lead

	Uses
Sheet and Foil	Tank linings, packaging, blanks for impact extrusions, gaskets, washers.
Lead Extrusions	Pipe, rod, wire, cames (H-shaped extrusion for leaded glass windows), rounds, rectangular duct, toothpaste tubes (by impact extrusion).
Castings	Die castings, bearings, metal type, lead grids for lead-acid storage batteries, counterweights, flywheels.
Coatings	Terneplate for roofing, automotive parts, cans for paint and oil, hardware, bearings, bushings.
Bonded Sheath	Nuclear shielding, chemical processing tanks, piping, equipment.
Laminations	Lead-plywood, lead-gypsum board, lead cinder block, leaded plastic fabric laminates, leaded plastic in glass-fiber combinations for X-ray protection and noise and sound isolation and control.
Powder	A constituent of greases, bearing faces, brake linings, clutch facings; also used in filling plastics and rubber, in paints and pipe joints compounds.
Shot	Major uses in ammunition; also used in shielding or for adding mass in irregularly shaped voids.
Lead Wool	Calking, "cold welding."
Alloys	Solder, fusible alloys, bearing materials, type metal.

	Uses
Mixtures	Lead powder in paraffin for combination gamma and neutron shielding, lead plastic mixtures for sound attenuation or gamma and X-ray shielding, lead and polyethylene mixtures for precision moldings of flywheels, for example, lead in neoprene as a special-purpose, antifouling, porpoise skin, and sound-deadening sheathing for submarines.

LEATHER

Leather is made from the hides or skins of animals, birds, reptiles and fish. Two main steps are involved in this process. First the hides or skins are cured or dressed to prepare them for tanning. This curing process removes all the flesh, hair and foreign matter. The hides are then tanned to produce durable, useful leather.

There are two general tanning methods—bark or vegetable, and chrome tanning. In chrome tanning, salts of chromium are used to process the hides. Chrome-tanned leather cannot be tooled, but it can be dyed and embossed.

Vegetable tanning makes use of the tannic acid which is found in bark, leaves and other vegetable products. Leather made by vegetable tanning is toolable and is, in general, the most suitable for leathercraft.

Tanned leather which is about the color of human skin is smoother grained, but can be further finished by dyeing, glazing, buffing, graining or embossing.

While the better leathers are usually left smooth, those having scratches, flaws or other surface defects after tanning are generally embossed with a grain.

Leather can be purchased in many sizes and thicknesses. The thickness of leather is expressed either in ounces or fractions of an inch. When ounce is used it is really a measure of thickness, and not of weight. Leathers range from about one to eight ounces. The weight to use depends on the end service requirements.

Quality

Leather varies widely in quality depending on what part and what layer of the hide it comes from. The term "genuine leather" gives no measure of quality, but merely refers to a skin or hide that has been dressed and tanned for use.

The highest or prime quality leather comes from the center back portion of the hide which is commonly called the back or bend. The shoulder portion also is of relatively good quality. Belly leather with its loose grain is soft and flabby and is usually considered lowest in quality. However, the belly section leather from alligator and lizard is fine quality.

Thin split-off parts of thick hides are called skivers. They are usually available in either smooth or rough finish and in various colors. Skivers are not toolable and are widely used for linings.

Leather from larger animals such as cowhide and steerhide is split into layers and divided into three grades accordingly.

Top Grain is the top or outside layer of the hide and of the best quality. It has close fibers and is more flexible, durable and attractive than the other grades. Top grain is most often used "natural" and is seldom embossed.

First-split leather comes from the layer next to the top layer and is second to top grain in quality. It is not as flexible or durable as top grain and is sometimes coated and embossed to resemble the better-grade leather.

Second split, the inner layer of the hide, is lowest in quality, and generally considered waste.

Leather can be readily tested for quality by making a small cut into it and then tearing it. Split-grade leather will tear easily and show long, loose fibers. Another common test is to fold and crease the leather and then crease it again at right angles to the first fold. This will usually separate the finish from the leather surface and give the appearance of cracking.

Kinds of leather

Cowhide. This is a strong, tough, durable leather available in a wide variety of grades, types, weights, finishes and colors. Heavy vegetable-tanned cowhide back leather, ranging from 6 to 8 oz, is widely used for tooling, carving and stamping. Heavy cowhide sides, sometimes called strap leather, are also suitable for tooling deep designs. Shoulder pieces are widely used for belts. Thinner cowhide, both vegetable- and chrome-tanned, is used in leathercraft kits. Cowhide belly, lowest in cost and quality, can be used where appearance and durability are not important.

Calfskin. Calf has the finest grain of all tooling leathers and is therefore best for tooling. It ranges in weight from 1½ to 3½ oz; it is more expensive than cowhide. When moistened and tooled on the grain side, the designs will remain permanently. Calfskin is available in natural finish or in a variety of colors. Chrome-tanned grades are less expensive than the tooling grades and are excellent for lining. (Vegetable-tanned calfskin is known as saddle leather.)

Steerhide. Steerhide has a crinkly surface, ranges in weight from about 2½ to 4½ oz, tools easily, and is not readily marred. It is available in many colors and in three-toned combination finish.

Sheepskin. Sheepskin is available in both tooling and nontooling grades. Largest nontooling use is for linings. Several types are available: suede, the best of lining leathers; skivers, thin and lightweight; glazed, with a fine, smooth glossy finish and firm texture. All these are available in many different colors, and are generally inexpensive.

The tooling grade is economical and tools without difficulty. Relatively deep tooling is also possible, but the leather should be dampened only very lightly. Sheepskin is often embossed or dyed to resemble calfskin.

Goatskin or "Morocco." Although goat is naturally a fine-grained, smooth leather, it is most commonly seen with the pebbly or crinkly "morocco" grain produced by boarding. It is stronger and more attractive than sheepskin of the same finish. In the heavier weights it is suitable for tooling line design and initials. Morocco leather is used extensively for book covers. Glazed goatskin has a high gloss, firm finish and is excellent for lacing things.

Kidskin. Exceptionally thin, smooth and strong, kidskin is characterized by tiny holes on the grain side. Excellent for lining, and available in various colors, it is used for linings, handbags, book covers and wallets.

Pigskin. This leather has a fine grain and smooth surface with visible holes in groups of three. The finest grades are usually imported. Tooling of pigskin is very difficult and not recommended. It is an expensive leather and therefore its use is confined largely to small parts such as billfolds, card holders, and pocket secretaries.

Suede. Suede can be made from many kinds

of leather, but most commonly from sheepskin. It has a soft velvet finish, is nontooling, and is suitable as a lining as well as for garments. Velvet Persian is a high-quality suede from skins of Persian sheep.

Alligator. Alligator is very expensive and is therefore restricted to small products. Besides genuine alligator, simulated alligator-grain leather is also available. It is usually made from calfskin or cowhide. It is attractive, strong, nontooling, available in many colors, and relatively inexpensive.

Snake and lizard. Reptile skins are specialty leathers noted for the beauty of their markings. The skins are durable as well as beautiful. Since they are quite expensive they are used only for very small items.

LITHIUM

Lithium is a soft, silver-white metal. Fresh surfaces tarnish quickly at room temperature through the formation of nitrides, oxides and other compounds. The metal reacts with water to form hydrogen and lithium hydroxide. The heat evolved is not sufficient to ignite the hydrogen and the lithium therefore does not present the explosion hazard common to other alkali metals, such as sodium.

Physical properties

Melt Pt, C	186
Boil Pt, C	1336–1372
Density (20 C), gm/cu cm	0.534
Sp Ht, cal/gm/°C	
0 C	0.784
50 C	0.884
100 C	0.905
186 C	1.010
Lin Coef Therm Exp (20 C), 10^{-6} in./in./°C	56
Therm Cond, cal/sec/cm/°C	0.17
Ht of Fusion, cal/mole	1100
Ht of Vaporization, cal/mole	36,100

Applications

Lithium has several uses in organic reactions and as an alloying element in aluminum and magnesium. In addition, silver-lithium alloys can be brazed without the use of a flux in inert atmospheres.

Organic catalyst. Lithium functions as an organic catalytic control in synthesizing organic preparations.

Alloying. The aluminum alloy 2020, containing 1% lithium, is an important high-strength aluminum alloy which is being applied in the aerospace industry. In the −T6 condition, 2020 exceeds the strength of 7075–T6 above 200 F.

Alloys containing up to 10 to 15% lithium are being investigated for possible use in aircraft and missiles.

LOW ALLOY CARBON STEELS

Low alloy steels are roughly defined as those steels that do not have more than 5% total combined alloying elements. They are designated by the same AISI system used for plain carbon steels. The last two digits show the nominal carbon content. The first two digits identify the major alloy element(s) or group. For example 2317 is a nickel alloy steel with a nominal carbon content of 0.17%.

Alloy effects

Table 1 lists the standard AISI low alloy steels along with the principal alloying elements. As the table shows one or more of the following elements are present: manganese, nickel, chromium, molybdenum, vanadium and silicon. Of these nickel, chromium and molybdenum are most frequently present.

Whereas surface hardness attainable by quenching is largely a function of carbon content, the depth of hardness depends in addition on alloy content. Therefore, a principal feature of low alloy steels is their enhanced hardenability compared to plain carbon steels. As is the case with plain carbon steels, low alloy steels' mechanical properties are closely related to carbon content. In heat treated, low alloy steels there is the added factor of alloying elements that contribute to the mechanical properties through a secondary hardening process that involves the formation of finely divided alloy carbides. Therefore, for a given carbon content, tensile strengths of low alloy steels can often be double those of comparable plain carbon steels.

TABLE 1 – LOW ALLOY
STEEL GRADES

AISI No.	Alloying Elements, %
13XX	Manganese, 1.75
23XX	Nickel 3.5
25XX	Nickel 5
31XX	Nickel 1.25; chromium 0.65
33XX	Nickel 3.50; chromium 1.55
40XX	Molybdenum 0.25
41XX	Chromium 0.95; Molybdenum 0.20
43XX	Nickel 1.80; chromium 0.80; molybdenum 0.25
46XX	Nickel 1.80; molybdenum 0.26
47XX	Nickel 1.05; chromium 0.45; molybdenum 0.20
48XX	Nickel 3.50; molybdenum 0.25
50XX	Chromium 0.30 or 0.60
51XX	Chromium 0.70–1.05
52XX	Chromium 0.40–1.60; carbon 0.95–1.10
61XX	Chromium 0.70–1.10; vanadium 0.10–0.15
81XX	Chromium 0.40; nickel 0.30; molybdenum 0.12
86XX	Nickel 0.55; chromium 0.50; molybdenum 0.20
87XX	Nickel 0.55; chromium 0.50; molybdenum 0.25
92XX	Silicon 2.00; chromium 0.10–0.40
93XX	Nickel 3.25; chromium 1.20; molybdenum 0.12
98XX	Nickel 1.00; chromium 0.80; molybdenum 0.25

Standard grades

A majority of the low alloy steel types are produced in both surface hardening (carburizing) and through-hardening grades. The former are comparable in carbon content to low carbon steels. Grades such as 4023, 4118, and 5015 are used for parts requiring better core properties than obtainable with the surface hardening grades of plain carbon steel. The higher alloy grades, such as 3120, 4320, 4620, 5120 and 8620 are used for still better strength and toughness in the core.

Most through, or direct, hardening grades are of medium carbon content and are quenched and tempered to specific strength and hardness levels. They can be divided into three classes:

Tensile Strength, psi:	Hardness, Bhn:
275,000–300,000	550–600
175,000–225,000	350–450
125,000–170,000	260–350

Alloy steels also can be produced to meet specific hardenability limits as determined by end quench tests. Identified by the suffix letter *H*, they afford the steel producers more latitude in chemical composition limits. The boron steels, a group containing very small amounts of boron, are also H steels. They carry the letter *B* after the first two digits of their designation.

TABLE 2 – MECHANICAL PROPERTIES OF TYPICAL
LOW ALLOY CARBON STEELS

Type & Condition	Tens. Str., 1000 psi	Elong., % in 2 in.	Hardness	Imp. Str., Izod, ft-lb
Carburizing Grades:				
Carb., Quench, Temp.				
4320	215	12	60 (Case) R (C)	28
4620	118	20	60 (Case) R (C)	56
8620	175	13	62 (Case) R (C)	28
Through Hardening Grades:				
Quench & Temp.				
1340	100–140	20–26	212–285 Bhn	50–90
4130	100–165	16–25	195–330 Bhn	40–90
8740	115–180	13–23	230–250 Bhn	35–90

Note: Range of values represents differences in heat treatment

A few low alloy steels are available with high carbon content. These are principally spring steel grades (such as 9260, 6150, 5160, 4160 and 8655), and bearing steels (52100 and 51100). The principal advantages of alloy spring steels are their high degree of hardenability and toughness. The bearing steels, because of their combination of high hardness, wear resistance and strength are used for a number of other parts, in addition to bearings.

Special grades

Special low temperature service alloys steels are also available. The three most common grades have a carbon content of 0.12 to 0.20% and nominal nickel contents of 2.25, 3.50 and 9%. They have relatively high strength and very good toughness at temperatures from −75 to as low as −320 F, and therefore find wide use for pressure vessels and gas storage tanks.

Steels with exceptionally good magnetic properties can be classified as low alloy steels. Known as electrical steels, they contain from 0.5 to 4.5% silicon, they have high magnetic permeability, high electrical resistance and low hysteresis loss. Grain-oriented and non-oriented grades are available, with the latter grades subdivided into low, medium and high silicon grades. The standard AISI designation is the letter *M* followed by a number that originally stood for the specified core loss for the grade.

Nitriding steels are low alloy steels developed specifically for developing optimum properties by nitriding. They are low and medium carbon with combinations of chromium and aluminum or nickel, chromium and aluminum. These steels after nitriding have extremely high surface hardnesses of about 92 to 95 Rockwell N. The nitride layer also has considerable resistance to certain types of corrosion. It is extremely resistant to alkali, atmosphere, crude oil, natural gas combustion products, tap water and still salt water. Nitrided parts usually grow about 0.001 to 0.002 in. during nitriding. The growth can be removed by grinding or lapping, which also removes the brittle surface layer. Most uses of nitrided steels are based on resistance to wear. The steels can also be used up to as high as 1000 F for long periods without softening. The slick, hard and tough nitrided surface also resists seizing, galling and spalling. Typical applications are cylinder liners for aircraft engines, bushings, shafts, spindles and thread guides, cams, and rolls.

Quenched and Tempered Low Alloy Steels— As contrasted to the HSLA steels, quenched-and-tempered steels are usually treated at the steel mill to develop optimum properties. Generally low-carbon, with an upper limit of 0.2%, they have minimum yield strengths from 80,000 to 125,000 psi. Some two dozen types of proprietary steels of this type are produced. Many of these are available in three or four strength or hardness levels. In addition, there are several special abrasion resistant grades.

The mechanical properties of these steels are significantly influenced by section size. Hardenability is chiefly controlled by the alloying elements. Roughly, an increase in alloy content counteracts the decline of strength and toughness as section size increases. Thus, specifications for these steels take section size into account.

In general, the higher strength grades have endurance limits of about 60% of their tensile strength. Although their toughness is acceptable, they do not have the ductility of HSLA steels. Their atmospheric corrosion resistance in general is comparable, and in some grades it is better. Most of them are readily welded by conventional methods.

LOW-ALLOY, HIGH-STRENGTH STEELS

The steels that are the subject of this discussion have been termed "High-Tensile Steels" and "Low-Alloy Steels," but the name "High-Strength Low-Alloy Steels" is gradually being adopted as the more acceptable designation. For the sake of brevity, they are referred to in this text as high-strength steels.

TABLE 1—IMPORTANT CHARACTERISTICS OF ALL-PURPOSE HIGH-STRENGTH STEEL

Property	Method of Determination
High Strength	Yield point in tension test.
Good Corrosion Resistance	Atmospheric corrosion resistance judged from weight loss in exposure rack test, and useful life judged from service performance.
Good Formability	Bend test, tensile elongation, and fabrication performance.
Good Weldability	Weldment performance judged by various weld-bending tests.
Good Toughness under Adverse Conditions	Temperature of transition from tough to brittle behavior in notched-bar impact test.
Good Resistance to Repeated Loading	Endurance tests on simple specimens or built-up structures.
Good Abrasion Resistance	Service performance.

High-strength steels are categorized on the basis of their mechanical properties; for example, within certain thickness limitations, they have a minimum yield point of 50,000 psi as compared with about 33,000 psi for structural carbon steel. This method of classifying is in contrast to the usual method, in which steels are designated as "plain-carbon steels," "alloy steels," and "stainless steels" on the basis of the presence (or absence) of added alloying elements in the particular steel. Thus, steels of a particular range of composition may be specified so that the purchaser will be able to obtain certain desired properties, including mechanical properties, in a part to be heat-treated. When chemical composition is the primary controlling factor in obtaining these required properties, such steels automatically fall into one of the above compositional classifications.

Because enhanced mechanical properties can be obtained with several different alloying elements and combinations of these elements (other than carbon), high-strength steels may have different chemical compositions but each steel in the group must meet essentially the same minimum mechanical-property requirements. The high-strength steels, which are available as sheet, strip, bar, plate, and shapes, are intended for general structural applications and are not, therefore, to be considered as special-purpose steels or as steels requiring, or adapted for, heat treatment.

Fundamental characteristics

To be of interest as construction materials, high-strength steels must have characteristics and properties that result in economies to the user when the steels are properly applied. They should be considerably stronger, and in many instances tougher, than structural carbon steel. Also, they must have sufficient ductility, formability, and weldability to be successfully fabricated by customary shop methods. In addition, sufficient resistance to corrosion is often required so that equal service life in a thinner section or longer life in the same section is obtained when compared to that of a structural carbon steel member.

Strength. The yield point of a constructional member determines the stress to which a structure may be subjected without permanent deformation. Therefore, the unit working stresses of each structure are based upon this important property. The minimum yield point of structural carbon steel (ASTM A 7 steel) is 33,000. psi; that of high-strength steel is generally 50,000 psi. Hence, on the basis of the proportionality of their yield points, the unit working stress employed with most high-strength steels may be increased to 1½ times that used with structural carbon steel. The use of higher unit working stresses generally permits reduction in the thickness of section in the structure and this results in a decrease in weight. Frequently, high-strength steels are substituted for structural carbon steel without change in section, the sole purpose being to produce a stronger and more durable structure with no increase in weight. Savings in weight are of utmost importance in mobile structures when these structures are enabled to carry greater payload.

Corrosion resistance. When high-strength steels are employed to save weight, it is desirable to make the maximum reduction in section. However, the structure must be considered not only as it is at the time it is built, but also as it will exist perhaps 20 or 30 years later when the members may be thinned by corrosion. Thus, for members initially relatively thin by virtue of desired weight saving it is necessary to prevent thinning in service insofar as possible, or the structure will be too weak to serve its intended purpose. For applications in which corrosion resistance is as important as increased strength, high-strength steels that exhibit 4 to 6 times the atmospheric corrosion resistance of structural carbon steel are available. The superior atmospheric corrosion resistance of these high-strength steels has been shown by exposure tests in different atmospheres and has been confirmed by the performance of these steels in a variety of applications. In other exposure tests, it has been shown that the paint coatings applied to corrosion-resistant high-strength steels exhibit an appreciably longer service life than when applied to structural carbon steels. Fig 1 shows the relative resistance to corrosion by an industrial atmosphere of structural carbon steel, structural copper steel, and a "corrosion-resistant" high-strength steel.

Formability. High-strength steels must have suitable properties so that they may be hot- or cold-worked readily and economically into various commodities for engineering structures. These operations and others such as shearing, punching, and machining, can generally be performed on high-strength steels with almost as much ease as on structural carbon steels. Despite their high yield points, high-strength steels can be satisfactorily formed in the same press brakes, draw benches, presses, and other equipment used for cold forming structural carbon steels.

There are some inherent differences between the cold-forming characteristics of high-strength steels and those of structural carbon steels. First, more force is required to produce a given amount of permanent set in a high-strength steel section than in a structural carbon-steel section of *the*

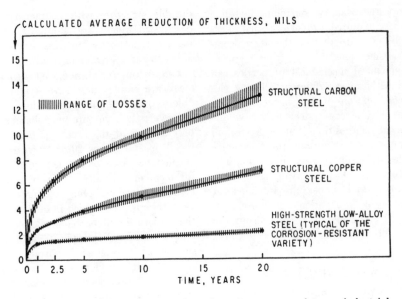

FIG. 1. Comparative corrosion of steel specimens exposed to an industrial atmosphere.

same thickness. Second, a somewhat greater allowance for springback should be provided when forming the high-strength steels. When lighter sections of high-strength steel are used, more springback is obtained even though the force to produce the part is little or no greater than that required for the original part of structural carbon steel.

Experience has shown that a more liberal bend radius should be used with a high-strength steel than with structural carbon steel for successful cold forming. This increased bend radius is ascribed to the greater strength of the high-strength steel.

Weldability. Since welding is often employed in fabricating structural steel, it is important that high-strength steels for these applications be readily weldable by metal-arc welding and by gas welding in plate thicknesses and by all the resistance-welding processes in sheet and strip thicknesses. It is equally important that the welds in fabricated structures have the required strength and ductility to withstand the most adverse conditions anticipated in the contemplated service. The development of the present-day high-strength steels has paralleled the growth of the various welding processes and particular care has been exercised to make sure that these steels possess suitable welding characteristics. Most high-strength low-alloy steels are considered to be readily weldable by conventional processes.

Notch toughness. As measured in a notched-bar impact test, high-strength steels exhibit toughness superior to that of structural carbon steels. This superiority is shown both when the notch toughness is expressed in terms of the amount of energy absorbed in breaking a specimen at room temperature and in terms of the low temperatures to which they preserve their toughness.

Fatigue resistance. Laboratory tests on polished specimens indicate that the resistance to repeated loading, or the fatigue resistance, of high-strength steels is superior to that of structural carbon steels. It should be noted, however, that the results of tests on polished specimens do not necessarily provide a measure of the fatigue resistance of full-size structures, because fatigue failures of structural members generally initiate at a notch or other surface discontinuity that acts as a stress raiser.

Abrasion resistance. There is general agreement that the resistance of various steels to abrasive action increases with strength or hardness, and, to some extent, with carbon content.

Service tests have demonstrated that the abrasion resistance of the high-strength steels with their inherently greater strength is somewhat higher than that of structural carbon steels containing 0.15 to 0.20% carbon.

Applications

High-strength steels can be used advantageously in any structural application where their greater strength can be utilized either to decrease the weight or increase the durability of the structure.

Although high-strength steels find application in all recognized market classifications, the largest single field of aplication has been in the manufacture of construction machinery and transportation equipment. From 1934 to 1959, one of the leading grades of high-strength steel has been used in the construction of 340,000 railroad freight cars and 9000 railroad passenger cars. If all the prominent grades were considered, the number of freight cars in service constructed partly of high-strength steels would doubtless exceed 400,000 or 25% of the current car ownership by American Class I railroads.

Previously, the main emphasis in railroad freight car use was on the savings in operating costs obtained by using high-strength steels in somewhat reduced thicknesses to decrease the dead weight of the cars, or to increase the capacity of the car without increasing the dead weight. In recent years, however, the emphasis has been shifted in many applications from weight reduction to that of obtaining stronger, more durable equipment, with little or no increase in weight. Similar benefits have also been obtained with other mobile equipment, such as various types of trucks, trailers and buses.

In bridges, designers are giving increased recognition to the importance of reducing dead weight by using high-strength steels, particularly for bridges involving long spans in which a reduction of weight at the center permits additional savings in the weight of supporting members. High-strength steels also lend themselves to economical tower construction where the properties permit the use of sections smaller than would be required in structural carbon steel. This advantage is important in tall television towers where dynamic loading due to wind resistance is lessened by use of smaller sections, and in transmission towers where lighter weight is a substantial advantage in reducing freight and handling costs.

A recent new use of high-strength steels has been for columns in high-rise buildings. Judicious use of high-strength steels in place of, and in combination with, structural carbon steel can result in substantial cost savings and an increase in usable floor area. High-strength steels are also being used to advantage in framing members of industrial and farm buildings.

The weight of containers for liquefied petroleum gas has been reduced appreciably by the use of high-strength steel, making them easier and less costly to handle and ship. Almost all such containers are now made of high-strength steel.

Other applications of high-strength steels include: the inner bottoms, floors, tanks, and hatch covers of ore boats; hulls and other structural members of small tankers, barges, tugs, launches, and river boats; coal bunkers; street-lighting poles; portable oil-drilling rigs; jet-blast fences; cable reels; automobile bumpers; pole-line hardware; air-conditioning equipment; stokers; agricultural-machinery parts; earth-moving equipment; military and domestic shipping containers; and air-preheater units.

M

MAGNESIUM AND ALLOYS

General characteristics

Magnesium is probably best known for its light weight. This silvery-white metal has a specific gravity of only 1.74 and is the lightest structural metal. Aluminum weighs 1½ times more, iron, 4 times more, and nickel alloys, 5 times more. Magnesium also has other desirable properties which give it a place among our common metals.

The excellent machinability of magnesium makes its use economical in parts where weight saving may not be of primary importance, but where much costly machining is required. Such parts, when made of magnesium, can be machined at higher speeds and with greater economy than would be possible with most other commonly used metals. Chemical milling can be used on magnesium.

Magnesium can be cast and fabricated by practically every method known to the metal worker. The metal is cast in sand or permanent molds to obtain lightweight castings with good strength, stiffness and resistance to impact or shock loading. Magnesium sand and permanent-mold castings are heat-treatable to further improve mechanical properties.

The die-casting process is likewise applicable to magnesium and this method of casting should always be considered when the quantities desired are in the range that indicates its use. Both hot and cold chamber processes are usable. An automatic metering device has been developed for cold chamber machines which permits high production rates equivalent to hot chamber.

The metal can also be cast by some of the less common methods, including plaster mold, centrifugal, shell molding, and investment processes.

Magnesium is rolled into sheet and plate and can be extruded into rods, bars, tubing, and an almost endless variety of structural and special shapes.

Sheet and extrusions are very easily formed using techniques that have been developed especially for magnesium. Stamping, deep and shallow drawing, blanking, coining, spinning, and impact extrusion are just a few of the production forming operations regularly used on magnesium and which indicate the metal's adaptability to a large variety of metalworking procedures.

The forging of magnesium is accomplished by methods much the same as those used for forging other metals. Both press and hammer equipment are used, but the former is most commonly employed because the physical structure of magnesium makes the metal better adapted to the squeezing action of the forging press. Magnesium forgings are chosen when a high strength-to-weight ratio, rigidity, or pressure tightness is required. The selection of forging, however, is governed by the fact that, like permanent mold and die castings, a sufficient number of parts must be needed to justify the cost of die equipment.

Magnesium parts can be joined by any of the common methods. Arc and electric resistance welding, adhesive bonding, and mechanical fastening are in daily production use. Brazing and gas welding, although not as frequently used as the other methods, are also suitable ways of joining magnesium.

Magnesium possesses relatively high thermal and electrical conductivities, very high damping capacities, and is nonferromagnetic. It has good stability to atmospheric exposure and good resistance to attack by alkalies, chromic and hydrofluoric acids, and many organic chemicals, including hydrocarbons, aldehydes, alcohols (except methyl), phenols, amines, esters, and most oils. Bare magnesium surfaces are nonsticking (snow, ice, sand, etc.) and nonmarking. Magnesium also has a low sparking tendency.

Magnesium has a close-packed hexagonal crystal structure as do other metals such as zinc, cadmium, titanium, zirconium, etc. The fact that magnesium has this structure allows a mode of plastic deformation called "twinning" to take place under compressive loading at room temperature and slightly above. This can result in wrought magnesium products (which have preferred orientation of their grains) having a compressive yield strength at these temperatures which is below the tensile yield strength. At elevated temperatures twinning becomes less influential and compressive and tensile yield strengths are equal, as they are for metals having cubic crystal structures. Twinning does not result in a lower compressive yield strength at room temperature in cast magnesium because of the random orientation of the grains in castings.

Another result of the hexagonal structure of magnesium is its limited formability at room temperature. This is because of the limited number of "slip planes" available for plastic deformation at room temperature in close-packed hexagonal crystals in comparison to the relatively large number available in cubic crystals.

Above 400 F, however, more slip planes become available and magnesium becomes very workable. In fact, its formability increases so much that greater deformations are possible in magnesium at elevated temperatures in some forming operations than in the cubic metals at room temperature. Also significant is the fact that even when cubic metals are heated, they still do not attain the same excellent workability possessed by magnesium at elevated temperatures. One advantage of hot working magnesium over cold working other metals is that parts can be deep-drawn in one operation without the necessity for repeated annealing and redrawing and for multiple sets of dies. Likewise, there is very little or no springback.

Alloys

The American Society for Testing Materials' alloy designation system is the only one in current common use and is used in this article. The ASTM designations consist of two letters representing the two major alloying elements followed by numbers representing the weight percentage of the two elements present with a serial letter at the end. The serial letter indicates some variation in composition.

In common with other metals, magnesium is not used in its pure state for stressed applications, but is alloyed with certain other metals such as aluminum, zinc, rare earths, thorium, zirconium, manganese and silver in order to obtain the strong, lightweight alloys needed for structural uses. These alloys, with their nominal chemical compositions, are listed in Table 1 and are discussed below.

TABLE 1—NOMINAL CHEMICAL COMPOSITION AND DENSITY OF MAGNESIUM ALLOYS

Cast Alloys [a]	Nominal Composition, %						
	Ag	Al	Mn	RE [b]	Th	Zn	Zr
AZ63A	—	6.0	—	—	—	3.0	—
AZ81A	—	7.6	0.2	—	—	0.7	—
AZ91A & B	—	9.0	0.2	—	—	0.6	—
AZ91C	—	8.7	0.2	—	—	0.7	—
AZ92A	—	9.0	0.2	—	—	2.0	—
AM100A	—	10.0	0.2	—	—	—	—
ZE41A	—	—	—	1.2	—	4.2	0.7
ZK51A	—	—	—	—	—	4.6	0.7
ZK61A	—	—	—	—	—	6.0	0.7
ZH62A	—	—	—	—	1.8	5.7	0.7
EK30A	—	—	—	3.0	—	—	0.3
EK41A	—	—	—	4.0	—	—	0.7
EZ33A	—	—	—	3.0	—	2.7	0.7
QE22A	2.5	—	—	2.0	—	—	0.6
HM11XA	—	—	1.2	—	1.2	—	—
HK31A	—	—	—	—	3.0	—	0.7
HZ32A	—	—	—	—	3.0	2.1	0.7
K1A	—	—	—	—	—	—	0.7

Wrought Alloys [c]	Nominal Composition, %					
	Al	Mn	RE [b]	Th	Zn	Zr
M1A	—	1.5	—	—	—	—
AZ10	1.0	0.2	—	—	0.4	—
AZ31B & C	3.0	—	—	—	1.0	—
AZ61A	6.5	—	—	—	1.0	—
AZ80A	8.5	—	—	—	0.5	—
ZE10A	—	—	0.2	—	1.2	—
ZK21A	—	—	—	—	2.3	0.6
ZK60A	—	—	—	—	5.7	0.7
ZK60B	—	—	—	—	6.2	0.6
HK31A	—	—	—	3.0	—	0.7
HM21A	—	0.5	—	2.0	—	—
HM31A	—	1.5	—	3.0	—	—

[a] Density varies between 0.063–0.067 lb/cu in.
[b] Rare-earth metals.
[c] Density varies between 0.063–0.066 lb/cu in.

Sand and permanent-mold casting alloys

The magnesium alloy systems of commercial importance are magnesium-aluminum-zinc, magnesium-zinc-zirconium (with and without thorium), magnesium-rare earth metal-zirconium (with and without zinc or silver), and magnesium-thorium-zirconium alloys (with and without zinc).

Magnesium-aluminum-zinc system. AZ92A, AZ63A, AZ91C, and AM100A are the principal Mg-Al-Zn sand and permanent mold casting alloys. AZ92A provides the optimum combination of high yield strength and moderate elongation with good pressure tightness in this group. AZ63A, AZ91C, and AZ81A are used where greater elongation and toughness are required. AZ91C and AZ81A also provide good pressure tightness. AM100A is used principally for permanent mold casting.

The properties of alloys in the Mg-Al-Zn system are stable up to about 200 F and often give satisfactory service at temperatures as high as 350 F. The particular alloy and heat treatment desired depends on the application.

Magnesium-zinc-zirconium system. In the Mg-Zn-Zr system, ZK51A (equivalent to the British Z5Z) and ZK61A have interesting mechanical properties. ZK51A develops high yield strength and maintains high ductility after an artificial aging treatment (–T5). ZK61A is usually given the –T6 solution heat treatment plus artificial aging to develop its properties fully. Tests show that these alloys have a fatigue strength at least equal to the Mg-Al-Zn alloys. ZK51A and ZK61A are suggested for simple, highly stressed parts of uniform cross section. These alloys are somewhat higher in cost than the Mg-Al-Zn alloys and have foundry castability problems associated with shrinkage, incidence of microporosity, and cracking. ZK51A and ZK61A alloys are not readily weldable. The addition of thorium or rare-earth metals to this alloy system results in decreased porosity problems and improved weldability. The properties of ZH62A (British TZ6) are equivalent to ZK51A and ZK61A, while those of ZE41A (British RZ5) are somewhat lower.

Magnesium–rare-earth metal–zirconium system. The addition to magnesium of rare-earth metals and zirconium, with or without zinc, results in casting alloys for use at temperatures falling roughly between 350 to 500 F. Use of these alloys for this temperature range usually allows a design lighter than is possible with the Mg-Al-Zn alloys. This is true particularly where high operating stresses require wall thicknesses greater than the minimum that can be cast.

Relatively small differences in properties exist among EK30A, EK41A, and EZ33A. EZ33A has more strength stability when exposed to elevated temperatures than do EK30A or EK41A alloys. Strength stability is defined as the ability to resist deterioration of strength due to extended exposure to elevated temperature.

The magnesium-rare earth metal-silver-zirconium alloy, QE22A (British MSR), has a higher yield strength at temperatures up to 600 F than any other magnesium sand casting alloy. In contrast to other magnesium alloys, it is quite sensitive to quenching rate subsequent to solution heat treatment and requires a relatively rapid quench in order to develop optimum properties. Mg-RE-Zr castings usually are quite free of porosity, but are more prone to superficial shrinkage defects and to dross inclusions than are the Mg-Al-Zn alloys. For these reasons they are more difficult to cast in some designs than the Mg-Al-Zn alloys. Mg-RE-Zr castings have excellent pressure tightness.

Magnesium - thorium - zirconium system. Mg-Th-Zr alloys (with and without zinc) are intended primarily for elevated temperature applications at 400 F and above, where properties superior to those of the Mg-RE-Zr system are required. Castings of this alloy group have interesting properties up to 650-700 F.

Two compositions are available: HZ32A, a Mg-Th-Zn-Zr alloy and HK31A, a Mg-Th-Zr composition. HZ32A (British ZT1) requires only the –T5 artificial aging treatment, while HK31A is given the –T6 solution heat treatment plus artificial aging to develop its properties fully.

Mg-Th-Zr alloys are less castable than Mg-RE-Zr alloys in that oxide inclusion and defects attributable to gating turbulence are harder to control. The tendency toward inclusions observed in the Mg-Th-Zr alloys is particularly marked in thin walled parts that require a rapid pouring rate. These alloys are found to have quite adequate castability for the production of complex parts having moderate to heavy wall thickness.

In general, at temperatures of 500 F and higher, HZ32A is equal to or better than HK31A in long time creep at all total extensions, and is equal to or better than HK31A in very small total extensions of creep for most times and temperatures. HK31A is superior to HZ32A in tensile and yield strengths up to 700 F and in short time higher total extension values of creep. The strength stability of HZ32A is less affected by exposure to elevated temperature than is HK31A alloy.

General considerations. Except in elevated

temperature applications, the very few reported service failures in magnesium alloy castings are of fatigue origin and result from inadequate design allowance for stress concentrations. These failures take place most frequently around fillets, flanges, ribs, bolt holes, etc. Failures may occur at these points even in sound metal when proper design features are not used.

Creep is an important consideration in designs for elevated temperature operation. For example, loads occurring in jet engines persist over several hundred hours. These stresses arise from engine weight and plane maneuvers as well as from differential thermal expansion. Over a long period the material deforms gradually until satisfactory functioning is lost. At 400 F a typical creep value at 0.2% total extension for 100 hr would be 1500 psi for the Mg-Al-Zn alloys, compared to about 7000 psi for Mg-Rare Earth-Zr alloys and about 11,000 psi for Mg-Th-Zr alloys.

A choice of an elevated temperature alloy cannot be made on the basis of operating temperature alone. Stresses and times of operations should also be known for both peak and normal conditions. Thus, the limits of stress, time and allowable deformation during operation at given temperatures determine the specific alloy composition for each application.

Die-casting alloys

The most commonly used magnesium die-casting alloy is AZ91B. In certain military applications AZ91A, a premium-purity version of this same basic alloy, is used to assure maximum corrosion resistance. Both of these alloys possess a desirable combination of good casting characteristics and good mechanical properties. These alloys are not heat treated, but used in the as-cast temper.

An experimental Mg-Th-Mn alloy for elevated temperature service, HM11XA, is also now available. It is the only die-casting alloy in any metal that is especially designed for good elevated temperature properties. It is stronger than AZ91B and AZ91A above about 300 F and can be used as high as 700 F or above. In addition, it has good castability and good room temperature strength and ductility. Other die-casting alloys with special properties are also available.

High-damping casting alloy

The damping capacity of pure magnesium is outstandingly high, but castings of pure magnesium have little practical application due to their low strengths and the poor castability of pure metals. Therefore, K1A, a Mg-Zr alloy, was developed to give the greatest damping capacity in an alloy which has improved castability and better strength. Both sand and die castings have been produced.

Sheet and plate alloys

The most commonly used magnesium sheet and plate alloy is AZ31B, a Mg-Al-Zn-Mn alloy. It has good strength, toughness, formability, and weldability. In addition, a Mg-Zn-Rare Earth alloy, ZE10A, is now available. It is generally comparable to AZ31B, but in addition it is free from stress-relief requirements after welding in contrast to AZ31B. Both of these alloys have excellent dent resistance.

Two magnesium-thorium alloys are also available for elevated temperature service. HK31A is used up to about 500 F, while HM21A is normally used from about 500 to 800 F. They are useful because of their good short-time strength, creep, strength, and modulus of elasticity at these temperatures. The alloys also have good formability and weldability.

AZ31B, ZE10A, and HK31A are available in both strain-hardened tempers and the annealed temper. HM21A is available only in the solution-heat-treated, strain-hardened, and aged temper (–T8), as this temper develops the optimum combination of properties for the service temperatures in which it is used.

Extrusion alloys

The most commonly used magnesium extrusion alloy is AZ31B. It has the same good strength, toughness, formability, and weldability in the extruded form as it has in the form of sheet and plate. It is available in all the extruded forms: bars, rods, shapes, and tubes. This same basic alloy is also available in a commercial grade, AZ31C. It has the same properties as AZ31B but has higher impurity limits. Where higher strength than that offered by AZ31B is required, AZ61A can be used. For still higher strength AZ80A extrusions are available. AZ80A can also be artificially aged for further improvement of its properties. AZ80A is not available as hollow shapes or tube.

AZ31B, AZ31C, AZ61A, and AZ80A alloys all require stress relief after welding to prevent stress corrosion cracking. In contrast, M1A, AZ10, and ZK21A alloys are available as extrusions and these have no stress relief requirements. M1A and AZ10 have strengths lower than AZ31B, while the strength of ZK21A is more like AZ80A-F extrusions.

ZK60A extrusions are used where high strengths over a broad range of sizes combined with good toughness is required. It is available in the as-extruded temper, but it is normally used in the artificially aged condition. A variation of this same basic alloy is used in extrusions that are extruded from pellets, rather than from a solid billet. These extrusions are designated (P)ZK60A-T5 and have the highest combination of properties now available in magnesium alloy extrusions, especially in the larger sizes. Both versions of ZK60 alloy have very limited weldability.

A magnesium-thorium alloy, HM31A, is available in extrusions to be used at elevated temperatures. It is normally used in the artificially aged condition. It has good short-time strength, creep strength, and modulus of elasticity at temperatures up to about 800 F.

Forging alloys

M1A, AZ31B, AZ61A, AZ80A, and ZK60A alloys are all used for forgings. AZ31B is suitable for hammer forging, while M1A is both hammer and press forged. AZ61A, AZ80A, and ZK60A are press forged only. The properties and characteristics of these alloys in forgings is about the same as in extrusions.

HM21A alloy is available in forgings for elevated temperature service and has properties and characteristics similar to HM21A plate and HM31A extrusions. Other magnesium-thorium and magnesium-rare-earth metal alloys are being developed for use in forgings for various elevated temperature requirements.

Properties

The values for various physical properties of magnesium and magnesium alloys are listed in Table 2. Note that some apply only to pure magnesium and others apply to all forms and tempers of the commercial magnesium alloys as well as to pure magnesium. (The chemical composition and densities of the commercial magnesium alloys are listed in Table 1.) Other physical properties as well as the mechanical properties of the alloys vary with product form and

TABLE 2—SOME PHYSICAL PROPERTIES OF
MAGNESIUM AND MAGNESIUM ALLOYS

Pure Magnesium	
Atomic Number	12
Atomic Weight	24.32
Sp Gr, 68 F	1.738
Melting Pt, F	1202
Latent Ht of Fusion, cal/gram	88
Elec Resis, 68 F, microhm-cm	4.45
Elec Cond, 68 F, Volume % IACS ..	38.6
Therm Cond, 68 F, Btu/sq ft/ hr/°F/in.	1070
Pure Magnesium and Magnesium Alloys	
Sp Ht, 68 F	0.245
Coef of Therm Exp, 68–212 F, in./in./°F	0.0000145
Mod of Elas, 68 F, 10^6 psi	6.5
Mod of Rigidity, 68 F, 10^6 psi	2.4
Poisson's Ratio	0.35

temper. The typical values for these properties are listed in Table 3. In addition to the properties listed, other properties of the alloys are important in the design of structural parts. Some of the most important are discussed below.

Elevated temperature properties. As stated in the discussion of the various magnesium alloys, Mg-Al-Zn alloys often have sufficient strength to give satisfactory service at temperatures as high as 350 F. For still higher temperature services (and sometimes lower temperature service) the alloys especially developed for such service are usually required. These are the magnesium-rare earth metal and the magnesium-thorium alloys. They have good short-time mechanical strength, good strength under long-time loading (creep strength) and good modulus of elasticity, at these higher temperatures. In addition, they retain these good properties throughout long exposures to elevated temperatures.

Fatigue. Magnesium alloys are similar to other non-ferrous alloys in that they do not exhibit a definite endurance limit when subjected to fatigue loading. Instead, their fatigue strength continues to drop as the required life of the part

TABLE 3—TYPICAL PROPERTIES OF MAGNESIUM ALLOYS AT ROOM TEMPERATURE

Alloy Form	Alloy	Temper	Elec Resis, microhm-cm	Therm Cond, Btu/sq ft/ hr/°F/in.	Ten Str, 1000 psi	Yld Str, 1000 psi [a]	Elong, in 2 in., %
Sand and Perm. Mold Casting	AZ63A	—T4	14.0	363	40	13	12
	AZ81A	—T4	15.0	340	40	12	15
	AZ91C	—T6	12.9	389	40	19	5
	AZ92A	—T6	12.4	404	40	21	2
	AM100A	—T6	12.4	404	40	22	1
	ZE41A	—T5	5.6	856	30	20	3.5
	ZK51A	—T5	6.4	758	40	24	8
	ZK61A	—T6	—	—	45	28	10
	ZH62A	—T5	6.4	749	40	25	6
	EK30A	—T6	6.3	766	23	16	3
	EK41A	—T6	7.3	665	25	18	3
	EZ33A	—T5	7.0	691	23	15	3
	QE22A	—T6	6.8	711	40	30	4
	HK31A	—T6	7.7	630	32	15	8
	HZ32A	—T5	6.5	740	30	14	7
	K1A	—F	5.6	865	25	7	19
Die Casting	AZ91A & B	—F	12.9	389	34	23	3
	HM11XA	—F	6.6	763	30	21	4
	K1A	—F	5.3	912	24	12	8
Sheet and Plate	AZ31B	—H24	9.2	534	42	32, 26	15
	AZ31B	—O	9.2	534	37	22, 16	21
	ZE10A	—H24	5.2	906	38	29, 26	12
	ZE10A	—O	5.0	961	33	20, 16	23
	HK31A	—H24	6.1	787	38	30, 23	9
	HK31A	—O	6.6	732	33	20, 14	23
	HM21A	—T8	5.0	949	35	23, 19	11
Extrusion	M1A	—F	5.4	897	37	26, 12	12
	AZ10	—F	6.4	751	35	22, 11	10
	AZ31B & C	—F	9.2	534	33	28, 15	14
	AZ61A	—F	12.5	406	46	33, 18	17
	AZ80A	—T5	12.2	412	55	38, 34	8
	ZK21A	—F	5.4	883	42	33, 25	10
	ZK60A	—T5	5.7	840	53	44, 36	11
	(P)ZK60B	—T5	5.5	870	49	38, 40	17
	HM31A	—T5	6.6	732	44	38, 27	8
Forging	M1A	—F	5.4	897	36	23, 16	7
	AZ31B	—F	9.2	534	38	28, 16	9
	AZ61A	—F	12.5	406	43	26, 17	12
	AZ80A	—T5	12.2	412	50	34, 28	6
	ZK60A	—T5	5.5	840	47	33, 28	13
	HM21A	—T5	5.0	949	38	23, 17	10

[a] First value is tensile yield strength, second is compressive yield strength. When both are equal, only single value is given.

increases. But the "S-N curves" for magnesium are much more horizontal at long lives than the curves of aluminum alloys for example. Therefore, magnesium alloys are especially suited to applications requiring a high number of cycles of fatigue loading.

Damping. All magnesium alloys have excellent damping capacity compared to the same form in other metals. As with other metals, sand castings have the highest, die castings have lower, and the wrought forms have the lowest damping capacity. As discussed in the section on alloys, K1A magnesium casting alloy gives the greatest damping capacity in both sand and die castings. Where greater strength is required, especially creep strength, the elevated-temperature casting alloys can be used as they have about the next highest damping capacity. EZ33A-T5 sand castings are often used for their high damping capacity combined with good strength. They have damping capacity comparable to that of gray cast iron. The magnesium-zinc-zirconium and the magnesium-aluminum-zinc casting alloys have lower, but still good, damping capacity. No special high-damping wrought alloys have been developed as the common alloys, AZ31B and ZE10A sheet and AZ31B extrusions, already have as high damping capacity as appears possible for wrought magnesium alloys.

Dimensional stability. An important advantage in using magnesium alloy parts and assemblies is the ability to obtain good dimensional stability. The correct use of the common procedures used in their production leave the parts and assemblies with such low internal stresses that warpage is no problem, even in relatively thin unsupported sections. Another aspect of dimensional stability is growth or shrinkage of alloys due to metallurgical changes, such as compound precipitation or solution. Dimensional changes due to metallurgical changes, even after years of service, has been found to be either nil or negligible in all magnesium alloys, forms and tempers, except the magnesium-aluminum-zinc alloys of high aluminum content in the as-fabricated or solution-heat-treated temper.

Some small growth has been measured on these latter alloys with these tempers after long exposures to elevated temperature so they are normally used in the artificially aged or solution-heat-treated and artificially aged temper whenever dimensional stability at elevated tempera-

ture is required with these particular alloys. The magnesium alloys which contain rare earth metals or thorium contract rather than grow after exposure to elevated temperature. Like growth, however, the degree is so small as to be negligible for most purposes.

Applications

Because of its many attractive properties and characteristics, magnesium has been successfully used in a great variety of applications. Its extreme lightness alone makes it attractive in all parts that are moved or lifted during their manufacture or use. In addition, the low density of magnesium allows thicker sections to be used in parts, eliminating the need for a large amount of stiffening, and thus simplifying the part and its manufacture. While light weight is probably the most important factor in the choice of magnesium in most structural applications, in many applications other properties are also significant. Examples of these are elevated temperature properties in missiles and aircraft, fatigue in wheels, damping in electronic housings for aircraft and missiles, dimensional stability in electronic housings and jigs and fixtures, dent resistance in luggage, non-marking in textile machinery, etc.

In the field of *missiles, satellites,* and both military and commercial airplanes and helicopters, magnesium is used extensively in airframes, engines, and gear boxes, wheels, auxiliary equipment, flooring, seating, electronic and instrument cases, etc. Other military uses are in ground handling equipment for aircraft and missiles, ordnance equipment such as vehicles and weapons, and many uses in the Quartermaster and Signal Corps such as portable shelters and hand carried communications equipment.

In *ground transportation* vehicles, both military and commercial, magnesium is used in the engines, transmissions, differentials, pumps, and other parts of the power plant. It has also been used in the floors and body panels of trucks and trailers, while for many years magnesium wheels have been used on all the Indianapolis "500" racing cars. In addition, magnesium automotive interior body parts are also being used.

Magnesium is widely used in the *materials handling* field in such items as handtrucks, dockboards, barrel skids, grain shovels, gravity conveyors, platform trucks, foundry equipment, brick tongs and many others. Other industrial uses are storage tanks and hoppers, plasterer's tools, concrete tools and forms, ladders and scaffolds, and portable tools such as electric drills, chain saws, power hammers, sanders, etc. It is also used in the moving parts of manufacturing machinery such as textile machines and printing equipment and is also used for magnetic-tape reels.

In *consumer goods* magnesium is used in a wide variety of equipment. Among the household goods in which magnesium is used are portable appliances, furniture, luggage, griddles, ladders, and lawn mowers. Among types of office equipment which use magnesium are typewriters, dictating machines, adding machines, calculators, and furniture. In sporting goods magnesium is used in such items as sleds, high jump and pole vault cross bars, and baseball masks. Magnesium is also used in instruments and for such items as binoculars and camera bodies.

Because of its rapid, yet controlled, etching characteristics as well as its lightness, strength and wear characteristics, magnesium finds usage in the photoengraved printing plates.

In addition to the many structural uses, magnesium has several *non-structural uses*. A large non-structural use is in the cathodic protection of other metals from corrosion. It also functions well in dry-cell battery construction. Magnesium also has chemical and metallurgical uses. Among these are the Grignard reaction, pyrotechnics, high energy fuels, alloying in aluminum, zinc, and lead alloys, and as an additive in the manufacture of nodular cast iron, lead, nickel alloys, and copper alloys. It is also used in the production of titanium, zirconium, beryllium, uranium, and hafnium.

MAGNETIC IRONS AND STEELS

All magnetic materials fall into two general classes: they are either permanently or nonpermanently magnetized. The permanently magnetized materials retain their magnetization after being placed in a magnetic field and can be used as a constant source of magnetic field. On the other hand, the nonpermanently magnetized, or soft magnetic materials, retain their magnetization only while a magnetic field is applied to the material. It is this latter class of materials, the soft magnetics, which will be discussed here. The

discussion will be further limited to those magnetic materials which are pure metals or alloys. However, for the sake of completeness there are many nonmetallic permanent and soft magnetic materials. These materials are usually oxides or intermetallic compounds.

Basic concepts

Before a magnetic material can be considered for any application the designer must refer to two basic curves, the magnetization curve, shown in Fig 1, and the hysteresis loop, shown in Fig 2.

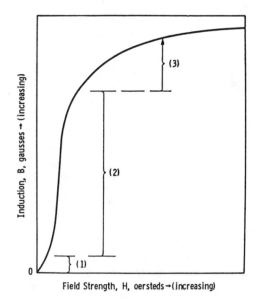

FIG. 1. Typical magnetization curve.

Fig 1 is a schematic representation of a magnetization process in a d.c. magnetic field. For a soft magnetic material, the intrinsic magnetic induction, B_i, in gauss, is plotted versus the field H, in oersteds. In some of the literature on magnetic materials, B is plotted against H. B is related to B_i by Eq. (1).

$$B = H + B_i. \tag{1}$$

For most soft magnetic materials H is much smaller than B_i so that B is approximately equal to B_i. The magnetization curves can be usually divided into three segments as shown on Fig 1. The first segment gives nonlinear reversible magnetization, the second segment a linear nonreversible segment, and the third segment above the knee is again reversible. In order to qualitatively understand the shape of the magnetization curve it is of interest to study the magnetization process.

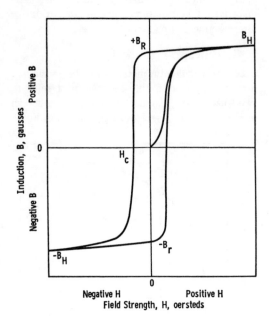

Fig. 2. Typical hysteresis loop.

When a magnetic material is cooled from above the Curie temperature to a lower temperature it becomes spontaneously magnetized even in the absence of an applied external magnetic field. Yet material cooled in this way has no external moment. The reason for this lies in the fact that the material forms into small regions called magnetic domains (illustrated in Fig 3). Each domain is magnetized to saturation, but since magnetization is a vector quantity, the sum of the magnetic vectors is zero. If a magnetic field is applied at some angle θ, to the specimen shown in Fig 3, the domains which have a component of magnetization parallel to the applied field will enlarge at the expense of the oppositely oriented domains. The distance that the domain walls move determines whether the magnetization is reversible or irreversible. The process of domain wall movement accounts for region 1 and 2 of the curve shown in Fig 1. When all domain wall motion has ceased, so that the material is essentially a single domain, further magnetization must occur by rotation of the magnetic vector into the direction of the field.

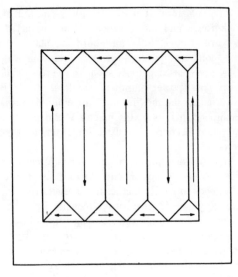

Fig. 3. Domain structure of magnetic material.

The difficulty of this rotation is determined by the composition and crystal orientation. For most magnetic materials, the magnetic domains will lie along a preferred crystal direction. In iron for example, the preferred axis is parallel to the <100> direction, or cube edge. Therefore, magnetic materials are generally anisotropic and to minimize the energy consumed in the rotation process, they are often oriented. The difference in the energy required to magnetize material along various crystal direction is called the magnetocrystalline anisotropy.

In the application of soft magnetic materials, it is often desirable to optimize the permeability, μ, which is defined in Eq. (2).

$$\mu = \frac{B}{H}. \tag{2}$$

To do this it is necessary to reduce the obstructions to domain wall motion and to eliminate the rotational process as much as possible. The highest quality soft magnetic alloys are usually crystallographically oriented, of very high purity and stress-free. The effect of stress on the magnetization curve is schematically illustrated in Fig 4 and 5. In Fig 4, curve a is the magnetization curve under zero tension and curve b is for the same material when a small tensile stress is applied. It is observed that the magnetization

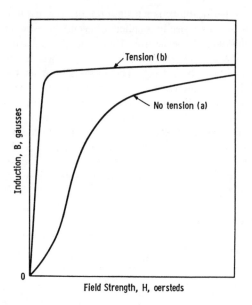

FIG. 4. Magnetization curve of a material with positive magnetostriction.

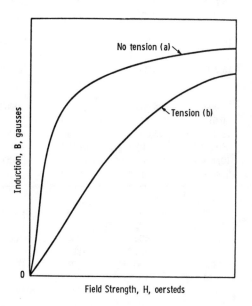

FIG. 5. Magnetization curve of a material with negative magnetostriction.

curve shows, for the tensile case, a higher permeability. The reason for this behavior is the fact that most magnetic materials change their lengths when they are magnetized. For the case illustrated in Fig 4, the material elongates during magnetization. In Fig 5 tensile stress has been applied to a magnetic material which normally contracts during magnetization. For this case, the effect of the tensile stress is to severely reduce the permeability. The dimensional change occurring during magnetization is called magnetostriction. Since stress has been shown to be so effective in changing the magnetic properties, care is usually taken to avoid the presence of internal stress, and correct practice requires that magnetic materials be stress-relief-annealed after punching and other mechanical operations which may introduce strain.

The nonreversibility of the magnetization process has been referred to in the discussion of Fig 1. The effect is shown in the hysteresis loop schematically shown in Fig 2. When a material has been magnetized from O to B_H and the field removed, the material does not demagnetize along the original curve, OB_H, but instead traces the curve $B_H B_r H_c$. If a still more negative field is applied, the magnetization follows from H_c to $-B_H$. If the direction of the field is changed again, the loop is formed as shown in the figure. The area included by the curve is the energy loss during one cycle of magnetization. This energy is usually expressed as watts per lb/cycle and is commonly referred to as the hysteresis loss. The area of this loop is a function of the magnitude of H_c, the coercive force, B_r, the residual induction or remanence and, of course, the value B_H. Generally, in order to estimate the magnetic quality, it is satisfactory to report the peak magnetization, coercive force and remanence.

The basic magnetization process and hysteresis cycle is best understood in terms of the static domain theory which is applicable under d.c. excitation. However, most applications of magnetic materials are for a.c. use. The predominant frequency commercially used in the United States is 60 cycles. For many special applications particularly in the aircraft and military field, 400 cycles is a common frequency. The a.c. losses are not simple multiples of the d.c. hysteresis loss, for in addition to the basic magnetic losses, eddy currents are generated in the material increasing the loss. The eddy current loss in a perfectly homogeneous material of uniform permeability is given by Eq. (3).

$$W_e = \frac{\pi^2 \delta^2 B^2 f^2}{6 \rho \times 10^{16}} \qquad (3)$$

where W_e is in watts per cu cm, δ is in cm, f is cycles per sec, B is in gausses and ρ is in ohm-cm. Unfortunately this equation cannot be applied directly to calculate the a.c. losses because

the permeability is not uniform and the material is microscopically nonhomogeneous because of the domain structure. However, the equation indicates several important factors which are qualitatively correct. The losses go up as the thickness increases and down as the resistivity increases. For these reasons, magnetic alloys used for a.c. applications are used in sheet form and are most often alloyed so that maximum resistivity is obtained commensurate with the induction required, ability to hot- and cold-work the material, minimum hysteresis loss and cost.

TABLE 1—D.C. MAGNETIC PROPERTIES

Alloy	Sheet Thick, 0.001 in.	H_c	μ_1	μ_{10}	μ_{100}	B_r
Iron and Iron Silicon						
Comm Pure Fe	62.5	0.83 [a]	3600	1530	180	8362
Fe–0.35% Si	25.0	2.0 [a]	1000	1340	176	8500
Fe–0.70% Si	25.0	0.92 [a]	4300 [c]	1400	175	8250
Fe–1.60% Si	25.0	1.00 [a]	3900 [c]	1380	175	7800
Fe–2.80% Si	18.5	0.75 [a]	5000	1380	171	7500
Fe–3.25% Si	18.5	0.50 [a]	6400	1380	171	6600
Fe–3.25% Si (oriented)	12.0	0.095 [b]	16,000	1800	198	12,700
Fe–3.25% Si (oriented)	14.0	0.095 [b]	15,700	1760	196	12,200
Fe–3.25% Si	5.0	0.70 [a]	5200	1400	—	7600
Fe–3.25% Si (oriented)	4.0	0.28 [b]	16,000	1800	—	14,200
Fe–3.25% Si (oriented)	2.0	0.40 [b]	13,200	1630	—	13,800
Nickel-Iron						
50% Ni–50% Fe	14	0.04	13,000 (0.01) [d]	85,000 (0.1) [d]	12,000 (1.0) [d]	9000
50% Ni–50% Fe (oriented)	2 to 6	0.07	13,000 (0.1) [d]	15,000 (1.0) [d]	1500 (10.0) [d]	14,000
4% Mo–79% Ni, Bal. Fe	20	0.015	40,000 (0.001) [d]	200,000 (0.01) [d]	60,000 (0.1) [d]	5000
Cobalt-Iron						
27% Co–Fe	14	2.0	2000 (3.0) [d]	1500 (10) [d]	208 (100) [d]	10,500
35% Co–Fe	17	1.8	2000 (3.0) [d]	1500 (10) [d]	230 (100) [d]	9000
50% Co–Fe	4	1.0	4400 (1.0) [d]	2000 (10) [d]	226 (100) [d]	10,000
50% Co–2% V	4	0.3	47,000 (0.17) [d]	19,500 (1.0) [d]	216 (10) [d]	19,000

[a] Hysteresis loop measured from $B_{max.} = 10,000$ gausses.

[b] Hysteresis loop measured from $B_{max.} = 15,000$ gausses.

[c] Permeability at $H = 2$ oersteds.

[d] Field at which permeability is measured.

TABLE 2—A.C. MAGNETIC PROPERTIES

Alloy	Sheet Thick, 0.001 in.	60 Cycles			400 Cycles			Applications
		W_a	W_b	W_c	W_a	W_b	W_c	
Comm Pure Fe	62.5	—— [a]			——			D.c. devices, i.e., electromagnets, relays, pole pieces.
Fe–0.35% Si	25.0	0.80 (5) [b]	2.8 (10) [b]	7.0 (15) [b]	——			Small motors, intermittently used electrical apparatus.
Fe–0.70% Si	25.0	0.54 (5) [b]	1.9 (10) [b]	4.3 (15) [b]	——			Small motors, fractional horsepower motors.
Fe–1.6% Si	25.0	0.49 (5) [b]	1.6 (10) [b]	3.5 (15) [b]	——			High-quality motors, medium efficiency motors and generators.
Fe–2.8% Si	18.5	0.33 (5) [b]	1.1 (10) [b]	2.4 (15) [b]	——			Motors and generators, reactors, small transformers.
Fe–3.25% Si	18.5	0.26 (5) [b]	0.90 (10) [b]	1.95 (15) [b]	——			High-efficiency motors and generators, reactors, motors.
Fe–3.25% Si (oriented)	12.0	0.15 (7) [b]	0.25 (10) [b]	0.57 (15) [b]	1.0 (4) [b]	5.0 (10) [b]	12.5 (15) [b]	High-quality, high-power, continuous duty transformers.
Fe–3.25% Si	5.0	——			0.11 (1) [b]	1.8 (5) [b]	6.0 (10) [b]	Aircraft motors and transformers, television transformers.
Fe–3.25% Si (oriented)	4.0	——			0.035 (1) [b]	0.70 (5) [b]	2.5 (10) [b]	Magnetic amplifiers, television transformers, power aircraft transformers.
Fe–3.25% Si (oriented)	2.0	——			0.035 (1) [b]	0.84 (5) [b]	3.0 (10) [b]	Magnetic amplifiers, pulse transformers.
50% Ni–50% Fe	14.0	0.0035 (1) [b]	0.054 (5) [b]	0.21 (10) [b]	0.64 (1) [b]	1.1 (5) [b]	5.1 (10) [b]	Instrument transformers, magnetic shields, sensitive low-current relays.
50% Ni–50% Fe	2–6	Specialized use at 60 and 400 cycles requires *Constant Current Flux Reset Test*						Magnetic amplifiers, current transformers, pulse transformers.
27% Co–Fe	0.014	3.8 (15) [b]	5.1 (18) [b]	6.6 (21) [b]	50 (15) [b]	62 (17) [b]	76 (19) [b]	High-temperature generators, high-temperature transformer transducers, aircraft equipment, pole pieces.
27% Co–Fe	0.004	2.9 (15) [b]	3.9 (18) [b]	5.0 (21) [b]	24 (15) [b]	29 (17) [b]	34 (19) [b]	
35% Co–Fe	0.017	3.0 (15) [b]	3.9 (18) [b]	5.4 (21) [b]	52 (15) [b]	69 (18) [b]	110 (21) [b]	
50% Co–2% V, Bal Fe	0.004 [c]	0.67 (18) [b]	0.80 (20) [b]	1.00 (22) [b]	8.5 (18) [b]	12.0 (20) [b]	17.0 (22) [b]	

[a] Blank space indicates alloy is not usually used at this frequency.
[b] Induction in kilogauss at which loss is measured.
[c] Field annealed.

Applications

The number of applications of magnetic materials is so large that a complete listing is not feasible. Furthermore, even for the same general application, such as a motor, many acceptable designs could be made, each requiring different-quality levels of magnetic material. Therefore, no attempt will be made to evaluate all the uses for a given alloy. However, several typical uses will be given. In some instances very special properties of a magnetic material may be required which cannot be covered in detail here.

Tables 1 and 2 list the most common magnetic materials with their characteristic a.c. and d.c. properties. The reader is referred to Fig 1 and 2 for the relation of the d.c. properties to the characteristic magnetization curve and hysteresis loop for each material. The table is designed to give the reader a ready reference source to identify the kinds of materials available and a brief resumé of their most important properties. After preliminary selection, reference should be made to the detailed curves available from reputable suppliers before any design is anticipated. Many of the materials listed require special handling and annealing techniques in order to achieve optimum properties.

MAGNETIC MATERIALS (HARD)

These are materials capable by themselves of producing and maintaining relatively high magnetic fields without the aid of external sources of energy. The magnetic parameters most useful in the selection of permanent magnet materials include: the *coercive force*, which should be high if short magnets are required; the *residual induction*, which needs to be high when magnets of small section are to be specified; and the *maximum energy product*, which must be high if one wishes magnets having the smallest volume for a given output of magnetic flux.

As compared with high permeability materials, many of these materials are hard physically and all are hard to magnetize and demagnetize. Hard magnetic materials may be classified as hardened steels, carbon-free nonmachinable alloys, carbon-free machinable alloys, pressed- or sintered-metal powders, and pressed-metal oxides. Data for some of the more widely used compositions of commercially available materials in these groups are given in Tables 1 to 4.

Hardened steels

Hardened steels used for making permanent magnets include carbon steel, chrome and tungsten steels and several cobalt steels. They all have relatively low energy products, but they are attractive for certain uses because of their high physical strength, workability, and in some cases, low cost. These steels are used mainly in the form of hot-rolled rods, bars, or strips. Typical sizes are, ¼ in. to 1 in. dia rods, bars ⅛ in. to 1 in. thick in widths up to about 2 in., and strips 0.010 in. to ⅛ in. thick in widths up to 2 in. or more. Tolerances for thickness and width are the same as for any hot-rolled steel.

In the as-rolled state, the Rockwell hardness may vary from approximately C23 to C40, depending upon size and previous processing history. All of these materials can be hot-formed, but for cold forming and certain types of machining, it may be desirable to anneal the steel. Hardness values as low as C15 for chromium and tungsten steels and C25 for cobalt steels may be obtained by annealing for a short time in the range 1500 to 1725 F. If possible, however, annealing should be avoided since it tends to reduce magnetic properties.

Carbon steel, similar to that used in permanent magnets before the year 1900, is still made and sold in appreciable quantities despite the later development of many materials with greatly improved, permanent magnet properties. Although the coercive force of carbon steel is one of the lowest for permanent magnet materials, the combination of low raw-material cost and relatively high residual induction makes carbon steel suitable for magnets for toys, compass needles, latching relays, certain types of meters, and other applications where high magnetic flux output is not required.

Tungsten steel was developed as an improvement over carbon steel. The most popular tungsten steel contains about 5% tungsten and 0.7% carbon. It has found use in magneto magnets, d.c. meter magnets, and in other devices where comparatively large size is permissible. A unique aplication was in the flywheel of outboard motor engines where it supplied both the weight for flywheel action and the magnetic field for the magneto. Tungsten steel is seldom used now because of its relatively higher cost per unit of magnetic flux output.

TABLE 1—MAGNET STEELS

Subject	Carbon Steel	3.5% Chrome Steel	17% Cobalt Steel	36% Cobalt Steel
Approximate Composition, % (balance Fe)	1 Mn 0.9 C	3.5 Cr 0.9 C 0.5 Mn	17 Co 0.7 C 2.5 Cr 8 W	36 Co 0.7 C 4 Cr 5 W
Heat Treatment	1470 F W.Q.	1520 F O.Q.	1720 F O.Q.	1720 F O.Q.
Magnetizing Force Used, oersteds	300	300	1000	1500
Coercive Force, H_c, oersteds	50	60	160	240
Residual Induction, B_r, gauss	10000	10000	9500	10100
Maximum Energy Product, gauss-oersteds, $\times 10^6$	0.2	0.3	0.65	0.97
Flux Density for Maximum Energy, gauss	6500	6500	6000	6500
Density, lb/cu in.	0.28	0.28	0.30	.030
Ten Str, 1000 psi	>200	>200	>200	>200
Rockwell C Hardness (as used)	63	63	63	63
Coef of Ther Exp, 10^{-6} in./in./°C	12.0	13.5	16.0	17.0
Resistivity, microhm-cm, 80 F	20	30	29	28

Chromium steels containing 1 to 6% chromium and 0.6 to 0.9% carbon were developed during World War I when tungsten was difficult to obtain. These steels are quenched in oil to produce the desired permanent magnet properties, and thus they are much less susceptible to cracking than the carbon and tungsten steels which are quenched in water. The 1% chromium steel has been found to be a good substitute for carbon steel and the 3.5% chromium steel can be used in place of 5% tungsten steel. The 3.5% chromium steel is still one of the best permanent magnet materials from the standpoint of low cost per unit of magnetic flux output. Its relatively low coercive force and low energy product, however, require that magnets made of it be long and rather large.

Cobalt steels were introduced by the Japanese shortly after World War I. These steels were the first of the high coercive force materials.

Alloys containing 3 to 38% cobalt combined with 2 to 8% chromium and tungsten are included in this group. For many years 36% cobalt steel with an energy product of about 1.0×10^6 gauss-oersteds was the best permanent magnet alloy obtainable. Although "Alnico" alloys have largely replaced 36% cobalt steel, the high tensile strength and wide range of magnetic properties obtainable with cobalt steels makes them desirable for certain applications.

The 17% and 38% cobalt steels have been found particularly useful for making magnets for hysteresis motors. In comparison with chromium steel, the cobalt steels are much more difficult to form and machine.

Magnets made from hardened steels are very likely to lose magnetic flux output with time. This effect may be reduced to a negligible amount by heating at 212 F for about 24 hr before final magnetization. Typical magnetic data for hardened steels are given in Table 1 for samples magnetized approximately to saturation.

Carbon-free nonmachinable alloys

This group is comprised of materials in the "Alnico" class which are produced in magnet form by casting and grinding processes. As the name implies, Alnicos contain aluminum, nickel, cobalt, and iron with the exception of "Alnico 3," which has no cobalt. Other elements such as copper and titanium are also found in some types. Although hot and cold forming or machining are not feasible, the high energy products of these alloys are so attractive that their use has dominated the permanent magnet market for some time.

Alnico type alloys have been produced under many different names and compositions. The data given below will be confined, however, to the general types most used in the United States. These include Alnicos 2, 3, 4, 5, 6 and 7 and some of their variations. Some of these alloys are heat-treated in a magnetic field during the course of their manufacture; a few are cast in such a way as to produce a grain-oriented structure. These latter materials are anisotropic and should be used accordingly.

The recommended tolerances for Alnico sand castings are ± 1/64 in. up to 2 in. and ± 1/64 in. for each additional 2 in. Approximately 0.025 in. should be allowed for grinding when such finishing is required. Holes or sections less than ⅛ in. should be avoided as should bars whose length to equivalent diameter ratio is greater than about 16 to 1. For all normal uses in the range − 40 to + 140 F, the variations due to temperature, shock, and aging range from about ½ to 2%. The greatest variation is caused by temperature and this effect is practically reversible in all well-designed magnets.

Cost is the factor which most frequently determines the choice of the Alnico used. Meaningful cost comparisons are difficult to make, however, since in addition to magnetic requirements, the required size and shape of the magnet, raw material cost and processing also are factors. In many cases, the very high energy product of Alnico 5 makes possible magnets having the highest flux output per dollar although the raw-material cost is higher than that of many other alloys.

Alnicos 2, 3, and 4 have lower raw-material costs than other Alnicos. They may be magnetized in any direction since they are for all practical purposes isotropic.

Alnico 2 is a general-purpose cast "Alnico" combining relatively low cost and a good energy product. Its comparatively low tensile strength must be considered, however, in those applications where physical strength is important.

Alnico 3 contains no cobalt and has the lowest raw material cost of all the Alnicos. It is used where its comparatively low energy product is adequate for the application.

Alnico 4 also has a relatively low cost per pound. Its high coercive force makes it especially desirable for designs requiring a short magnet.

Alnico 5 has one of the highest energy products of any of the available materials. In many cases this makes it the most suitable material both from a performance and a cost standpoint. Energy products ranging from an average 5×10^6 gauss-oersteds for the original alloy to as high as 7.5×10^6 gauss-oersteds for a directional grain type are obtainable. Trade names for this type include "Alnico VB-DG," "Alnico V-7," "Alnicus," "Alnico V-DG," etc.

Alnicos 6 and 7 are high-energy product materials having higher coercive forces than the average for Alnico 5. They are thus useful in cases where demagnetizing factors are high and where the length of the magnet must be kept minimum.

Alnicos 5, 6 and 7 are heat-treated in a magnetic field during processing and are thus made anisotropic. The desired direction of magnetization should therefore be specified for magnets made of these alloys. In addition, the directional grain type Alnico 5 is more difficult to produce in irregular shapes and, for this reason, some manufacturers offer it in straight bars or cylinders only.

Alnico alloys find application in all cases where a cast magnet can be used. Their magnetic properties are very stable with time, temperature and shock. Representative data for the Alnico alloys are given in Table 2.

Carbon-free machinable magnet alloys

In addition to being machinable, some of these alloys have good cold-punching and forming characteristics. The permanent magnet properties (see Table 3) are produced by precipitation hardening and by severe cold-working ef-

TABLE 2—CARBON-FREE NONMACHINABLE MAGNET ALLOYS

Subject	Alnico 2	Alnico 5 (Normal)	Alnico 6	Alnico 7
Approximate Composition, % (balance Fe)	10 Al 17 Ni 12.5 Co 6 Cu	8 Al 14 Ni 24 Co 3 Cu	8 Al 15 Ni 24 Co 3 Cu 1.25 Ti	8.5 Al 18 Ni 24 Co 3.25 Cu 5 Ti
Magnetizing Force Used, oersteds	2000	3000	3500	5000
B_r, Residual Induction, gauss	7500	12500	10000	7200
H_c, Coercive Force, oersteds	550	650	750	1050
Maximum Energy Product, gauss-oersteds, $\times 10^6$	1.6	5.5	3.5	2.9
Flux Density for Maximum Energy, gauss	4700	10400	7300	4000
Density, lb/cu in.	0.256	0.265	0.268	0.259
Ten Str, 1000 psi	3	5.5	—	—
Rockwell C Hardness (as used)	45	50	56	60
Resistivity, microhm-cm, 80 F	65	47	50	58

fects. Their maximum energy products are in most cases better than those of the steels and they are relatively more stable with respect to time, temperature and mechanical shock. They are not generally as available as the steels and their costs per pound are in some cases much higher. The comparatively high raw material cost sometimes is more than offset by the low cost of fabrication.

The 68% iron, 12% cobalt, 20% molybdenum alloy, "Remalloy," produced by relatively low cost hot-rolling or hot-forming operations, has been used extensively in telephone receiver magnets and other applications. Another type, 17, Remalloy, having 71% iron, 12% cobalt, 17% molybdenum has a relatively high residual induction which makes possible magnets of small section. "Comol" and "Indalloy" are other trade names by which 17 Remalloy is known.

A very ductile permanent magnet material, called "Cunife" is made from copper, nickel, and iron. Cunife, because of its exceptionally good cold-punching qualities, has been used for speedometer magnets and in many other devices where magnets punched from thin strip are desired.

Since the properties of this material are produced in part by severe cold reduction, the magnets made of it are anisotropic and should be magnetized in the direction of the cold reduction. Although it is obtainable in wire sizes of the order of ¼ in. dia and strips up to about ⅛ in. thick, its best properties are obtained in smaller sizes. Even after heat treatment to produce the best permanent magnet properties it can be machined and worked like mild steel.

"Cunico" is a copper-nickel-cobalt alloy somewhat similar to Cunife. Its magnetic properties are not as dependent upon the degree of previous cold working, however, and it can be obtained in a wider variety of sizes. The low values of residual induction for both Cunife and Cunico are limiting factors in their use.

"Vicalloy I" is one of a number of iron-cobalt-vanadium alloys having good ductility and good permanent magnet properties. Although an energy product of only 1×10^6 gauss-oersteds is usually obtained, much higher energy products are possible after heavy cold working. Although "Vicalloy" has good ductility when properly processed, it is not as readily worked as Cunife,

TABLE 3—CARBON-FREE MACHINABLE MAGNET ALLOYS

Subject	17 Remalloy Indalloy Comol	20 Remalloy	Cunife I	"P-6" Alloy
Approximate Composition, % (balance Fe)	12 Co 17 Mo	12 Co 20 Mo	60 Cu 20 Ni	45 Co 6 Ni 4 V
Magnetizing Force Used, oersteds	1500	1500	2500	1000
H_c, Coercive Force, oersteds	250	320	500	60
B_r, Residual Induction, gauss	10500	9000	5400	14200
Maximum Energy Product, gauss-oersteds, $\times 10^6$	1.1	1.4	1.5	.41
Flux Density for Maximum Energy, gauss	7000	6000	4000	11500
Density, lb/cu in.	0.295	0.297	0.310	0.285
Ten Str, 1000 psi	125	125	100	—
Hardness (as used)	60 R_c	60 R_c	200 Bhn	—
Resistivity, microhm-cm, 80 F	45	—	18	30

and it tends to become brittle if heat-treated to produce the highest coercive force. The high cobalt content of Vicalloy makes raw material costs high, but the desirable mechanical properties of Vicalloy I are used to advantage in such applications as long-wearing recorder tape.

"P-6" alloy, another alloy containing cobalt and iron plus nickel and vanadium, was developed primarily for hysteresis motor use. A combination of high hysteresis loss and relatively high permeability is desirable for this type use. To achieve these characteristics, "P-6" alloy was developed to have a very high residual induction and a comparatively low coercive force. Its unusually high residual induction makes it worthy of consideration also for uses where small sectional area is important.

"Silmanal," made of silver-manganese-aluminum, has the highest intrinsic coercive force of all available permanent magnet materials. Very low residual induction has limited its use, however. It is most useful in applications where very high demagnetizing forces are present.

In spite of very high raw material costs, platinum-cobalt containing aproximately 77% platinum and 23% cobalt has been found very desirable for certain specialized applications. The unusually high coercive force and very high energy product obtainable in this alloy has made it suitable for very small but powerful magnets for watches, meters, and special instruments, where compactness is essential.

Pressed or sintered metal-powder magnets

Pressed-powder magnets made of finely divided iron and iron-cobalt powders have inherent advantages over other types of permanent magnet materials. They require no heat treatment to produce their permanent magnet qualities; they have the desirable close tolerances typical of die-pressed powders; they are capable of relatively high-energy products, and yet they are machinable without requiring the use of special tools. High manufacturing costs have been a problem in making them competitive with other permanent magnet materials.

Tolerances on magnets made of pressed fine powders without heat treatment are said to be of the order of ± 0.001 in. except parallel to the pressing direction where the tolerance is given as ± 0.003 in. The powders used for making such magnets are highly pyrophoric and adequate protection from oxidation is therefore a must.

TABLE 4—PRESSED POWDER MAGNET MATERIALS

Subject	Gecalloy	Sintered "Alnico" 2	Sintered "Alnico" 5	"Indalloy"
Approximate Composition, % (balance Fe)	Fine Fe Particles	10 Al 17 Ni 12.5 Co 6 Cu	8 Al 14 Ni 24 Co 3 Cu, 1 Ti	12 Co 17 Mo
Magnetizing Force Used, oersteds	2000	2000	3000	1500
Coercive Force, H_c, oersteds	330	550	575	240
Residual Induction, B_r, gauss	7000	7000	10000	9000
Maximum Energy Product, gauss-oersteds, $\times 10^6$	1.0	1.4	3.5	0.9
Density, lb/cu in.	0.145	0.243	0.241	0.288
Ten Str, 1000 psi	—	65	—	125
Rockwell C Hardness (final)	—	40	45	60
Resistivity, microhm-cm, 80 F	—	68	—	48

The change in magnetic flux output with temperature for relatively long magnets of some of these materials is given as being about 0.017%/°C over the range −48 C to 150 C. This compares with 0.034%/°C for cast Alnico 5.

"Gecalloy" is typical of the powdered iron and iron-cobalt materials made in Europe; ESD (elongated single domain particle) material called "Lodex" is in production in the United States.

Sintered powder magnets also are made by sintering pressed-metal powders at high temperatures using conventional powder-metallurgy techniques. Very small magnets can be made in this manner more readily than by the use of casting methods. In general, the sintered magnets do not have as high energy products as their cast counterparts, but they are more uniform and much stronger physically. Sintered Alnico 2, Alnico 5, and "Indalloy" ("Remalloy") magnets find many uses in instruments and other devices where very small magnets are desirable. The relatively high tensile strength of these materials sometimes makes them the best choice for such things as high-speed rotors for motors and generators. Magnets weighing less than about 20 gm are usually made more economically by the sintering process rather than by casting.

Accepted tolerances for sintered magnets are: ± .005 in. up to 1/8 in., ± .010 in. for 1/8 in. to 5/8 in., and ± .015 in. from 5/8 in. to 1¼ in. Table 4 gives typical properties for some of the pressed powder and sintered powder magnet materials.

Oxide magnets

Magnets made from iron oxides or compounds of iron oxides and the oxides of other metals can be made to combine good permanent magnet properties with high chemical stability, lower than average density, and very high electrical resistivity.

Iron oxides in the form of Fe_2O_3 or Fe_3O_4 powder having certain sizes and shapes of particles are produced primarily for use in making magnetic recording media. Coercive forces in the range of 150 to 300 oersteds are characteristic. Bars or other forms of magnets are not usually made from these oxide powders as such.

However, bars, disks, and various other shapes are produced from "Vectolite," a compound of iron and cobalt oxides and also of "Ferroxdure," a material made from barium and iron oxides.

Vectolite is composed of 30% Fe_2O_3, 44% Fe_3O_4 and 26% Co_2O_3 powders pressed into the desired shapes and sintered at a high temperature. The sintered product is heat-treated finally in a magnetic field and consequently must be magnetized along the same axis for best results. Typical properties are: coercive force, 900 oersteds; residual induction, 1,600 gauss; maximum energy product, 0.6×10^6 gauss-oersteds; density, 0.113 lbs/cu in., and a resistivity of 225 ohm-cm per sq cm at 77 F. Its high resistivity permits it to be used at high frequencies without incurring high eddy current loss penalties. Its light weight and high coercive force are used to advantage in small motor rotors and other similar applications.

Ferroxdure, sometimes called "Barium Ferrite," is a ferritelike compound of barium and iron oxides ($BaFe_{12}O_{19}$). It is marketed under many other trade names such as: "Arnox," "Index," "Ferroba," "Ferrimag," "Ceramag," etc. It is made from a pressed powder mixture fired at a high temperature to produce a hard, brittle, ceramiclike substance having good permanent magnet properties and an electrical resistivity of 1×10^{12} ohm-cm at 80 F. Two types are available. One has an oriented structure produced by processing in the presence of a strong magnetic field; the other is a nonoriented type and thus, more nearly isotropic. Typical properties are given as follows:

FERROXDURE PROPERTIES

Type	H_c, Coercive Force, oersteds	B_r, Residual Induction, gauss	Max. Energy Product, gauss-oersteds, $\times 10^6$	Density, lb/cu in.
Non-oriented	1700	2200	0.9	0.17
Oriented, Normal	2000	3700	3.2	0.18
Oriented, High Coercive Force	2500	3200	2.4	0.16

The raw material costs for Ferroxdure are low. Processing costs also are low for the nonoriented variety. Dimensional tolerances for unground magnets are as follows: In direction of pressing, $\pm 1/64$ in. up to 5/8 in., $\pm 1/32$ from 5/8 in. to 1 1/2 in.; at right angles to pressing $\pm 1 1/2 \%$ or ± 0.010 in. whichever is greater; warp tolerance

is 0.011 in. per in. of length.

Ferroxdure magnets are more subject to change in magnetic properties with temperature than Alnico magnets. From -40 F to $+140$ F, the output flux of oriented Ferroxdure magnets is likely to decrease at the rate of about $0.11\%/°F$. This effect is almost completely reversible for magnets designed to operate at a flux density above the "knee" of the demagnetization curve.

Because of their high coercive force and relatively low cost, nonoriented Ferroxdure type magnets are being used for holding magnets, magnetic coupling devices, small loudspeakers, and in many other applications where a very high-energy product is not essential. In one unique application the Ferroxdure in powder form is mixed with rubber or plastic to form flexible permanent magnets suitable for door closures and other holding devices. The higher energy product of the oriented type permits it to compete with the Alnicos for loudspeaker magnets, traveling wave tube magnets, generator magnets, etc. Both types of Ferroxdure have very high resistivity which makes them suitable for high-frequency applications. Because of their relatively low residual inductions, sectional areas for Ferroxdure magnets are necessarily large.

MALLEABLE IRONS

The malleable irons are a family of cast alloys —consisting primarily of iron, carbon, and silicon—which are cast as hard, brittle white iron and then rendered tough and ductile through a controlled heat conversion process. Because of their unique metallurgical structure, they possess a wide range of desirable engineering properties including: strength, toughness, ductility, resistance to corrosion, machinability and castability. The malleable irons have been established as an excellent material for a wide range of applications and have been employed in virtually every branch of American industry since 1826 because of the advantages inherent in the material and the valuable properties it consequently offers the designer.

Three principal types of malleable iron are in wide use in this country: ferritic, pearlitic, and alloy malleable iron. A fourth type, called cupola malleable because of the method of manufacture, is also produced but only in small tonnages. Most important from the standpoint of production volume and use is standard malleable iron, which has a ferritic matrix. Pearlitic malleable, which, as the name implies, has a pearlitic matrix, is being produced in ever-increasing quantities. Alloy malleable iron is basically a specialty type iron, with higher strength and corrosion resistance, finding primary use in railroad parts.

By far the largest tonnage of malleable castings normally is consumed by the automotive industry. The railroad, agricultural implement, electrical line hardware, pipe fittings, and detachable chain industries, and most other basic industries, use standard and pearlitic malleable castings.

All malleable iron produced in this country is of the blackheart type, i.e., the normal fracture shows a velvety black appearance. The whiteheart type, produced in Europe, also derives its name from the appearance of the fracture surface. This type of material depends, to a large extent, on decarburization for its mechanical properties.

Advantages

Important attributes of malleable iron can be summarized as follows:

1. Malleable iron can be produced with a high yield strength, which is the static mechanical property upon which most mechanical design is based.

2. Ferritic and pearlitic malleable irons have a high ratio of yield strength to tensile strength. This means that the engineer can design to high applied strength values in service for materials of construction, concomitant with good machinability and low production cost for the final part.

3. Pearlitic malleable irons can be produced to a wide range of mechanical properties through carefully controlled heat treatments.

4. Malleable irons have a high modulus of elasticity and a low coefficient of thermal expansion, compared with the nonferrous metals.

5. Malleable and pearlitic malleable irons exhibit a low nil ductility transition temperature for brittle fracture.

6. Compared with steel, malleable and pearlitic malleable irons have considerably better damping capacity, which makes operation of moving components less prone to noise because of resonant vibration.

7. Pearlitic malleable irons have good wear resistance and can be selectively hardened by flame, induction, or the carbonitriding process.

8. Pearlitic malleable irons will take a high-quality finish. Honed surfaces with two to three μin. finish at hardness values of 197 to 207 Bhn have been reported.

9. The uniformity of properties from surface to center is excellent, particularly in oil-quenched and tempered pearlitic malleable iron.

10. All malleable irons are substantially free from residual stresses as a result of long heat treatments at high temperatures.

11. Pearlitic malleable iron provides the properties of medium to high carbon steel coupled with a machinability rating unequalled for a material of similar hardness.

With respect to mechanical properties, minimum specification values are generally exceeded by a comfortable margin in the better-controlled malleable foundries.

Brinell hardness of ferritic malleable irons varies from about 110 to 145. Pearlitic malleable and alloyed malleable iron grades have higher values, ranging usually between 160 and 280 Bhn. Both hardness and tensile strength increase with combined carbon content.

Since the final properties of malleable iron castings are the result of thermal treatments, section thickness has no appreciable effect on strength. Therefore, mechanical properties will be essentially the same throughout the entire cross section.

Manufacture

The manufacture of malleable iron castings is fundamentally a two-phase operation. Phase one consists of producing the white iron castings, and the second phase involves the controlled heat treatment of these castings to obtain the desired finished product.

The first step in phase one is the selection of varying amounts of sprue, malleable iron and steel scrap, and pig iron, which make up the charge. The charge can be placed in an air furnace, cupola, or electric furnace for melting. When the melting is carried out in a combination of cupola and air or electric furnace, the melting operation is called duplexing. In the duplex operation, both coke and limestone are added to the cupola charge. When only an air (open hearth) furnace or an electric furnace is used, the operation is termed batch melting.

After the metal has reached the molten state in the melting unit, impurities which are found in the charge, ash from the fuel, and any refractory which has reacted with the charge, separate and form a slag on top of the liquid iron. Since the slag is comprised of impurities, it is skimmed off with the molten metal bath and discarded.

Batch melting. Most of the batch, or cold, melting in this country is performed in acid-hearth reverberatory type furnaces having capacities of 20 to 50 tons. The fuels for air furnace melting are either pulverized coal or oil with quantities of air so that the melting flame can be adjusted to be oxidizing or reducing.

Electric furnaces of 2 to 10 ton capacities are also employed for malleable iron melting. These furnaces are generally of the direct-arc type, and melting is accomplished by the radiation of heat from the arcs, from metal contact with the arcs and to some extent from the passage of current through the charge.

Duplex melting. The first step of duplex melting is carried out in a cupola with a rating of from 6 to 45 tons of molten metal per hr. Coke is used as the fuel and limestone or some other flux is added to insure a slag of low viscosity. The molten metal flows directly from the cupola to the second unit of the duplexing operation, either an air or electric furnace. This second unit is designed to refine and superheat the liquid iron so that the proper composition and temperature are attained to insure uniformity of the white iron castings.

Regardless of the method of melting, the chemical composition of the white iron plays a very important role in the manufacture of malleable iron. Typical composition ranges for the major constituents are: carbon 2.00 to 2.65%, silicon 1.40 to 0.90%, manganese 0.25 to 0.55%, sulfur 0.05 to 0.18%, and phosphorus less than 0.18%, and it is essential for these values to be maintained if castings of good quality are to be produced.

After the malleable castings have solidified and cooled sufficiently in the mold they are shaken out of the mold for spruing and trimming. Because the parts are very brittle as-cast, gates and risers may be removed by impact. Some foundries clean the hard iron castings by tumbling or abrasive blasting, while others proceed to subsequent operations without the use of a cleaning step.

Hard iron castings are annealed in the second phase of malleable iron production in either continuous or batch furnaces. Generally speaking, the continuous furnaces are constructed so that atmosphere control is possible. In some batch furnaces controlled atmospheres permit loose packing of castings, while others use slag, sand or gravel to support the castings and preserve their surface condition during annealing.

Subsequent to annealing, castings are cleaned and finished. Malleable iron's ductility permits straightening or coining to close dimensional tolerances, thus reducing the necessary machining allowances substantially.

Structurally, malleable iron castings consist essentially of carbon-free iron (ferrite) and uniformly dispersed nodules of temper carbon. This combination of soft, ductile ferrite and nodular temper carbon accounts for the desirable mechanical properties of malleable iron. In pearlitic malleable the matrix is essentially pearlitic, resembling that of a medium-carbon steel. Minimum ASTM specifications for the two grades of ferritic and seven grades of pearlitic malleable iron are given in Table 1.

TABLE 1—TENSILE PROPERTIES—ASTM MINIMUM
SPECIFICATIONS [a]

Designation (Grade)	Ten Str, 1000 psi	Yld Str, 1000 psi	Elong (in 2 in.), %
Ferritic			
35018	53	35	18
32510	50	32.5	10
Pearlitic			
45010	65	45	10
45007	68	45	7
48004	70	48	4
50007	75	50	7
53004	80	53	4
60003	80	60	3
80002	100	80	2

[a] Strength up to 135,000 psi tensile and 110,000 psi yield are produced commercially under individual producers' specifications.

Properties

Effect of temperature. Studies of the behavior of ferritic malleable iron at both high and lower temperatures demonstrate, in general, that this material is well suited to applications in a temperature range from −60 to 1200 F.

Low-temperature investigations have been concerned primarily with impact resistance and notch sensitivity; high temperature studies have focused principally on tensile strength, yield point, elongation, stress rupture and creep behavior.

Results of research sponsored by the Malleable Founders Society have indicated a high level of performance at elevated temperatures, equal or superior to other ferritic materials for which data are available, particularly at 800 F. Strength at 1000 F is adequate for many applications and strength is retained even at 1200 F. No evidence was found in any of the investigations of changes in structure or performance during the test periods which extended from 1 to over 2,000 hr.

Surface hardness. Many structural parts require high surface hardness backed up by a strong, tough core. In steel components this can be accomplished by carburizing or nitriding after machining, followed by a suitable heat treatment. In pearlitic malleable iron the combined carbon content is adequate for production of high surface hardnesses through quenching after either induction or flame heating. Many parts are preferentially hardened on wearing surfaces.

MANGANESE

Manganese, as one of the transition elements of the first long period, belongs to Group VII of the periodic system. It has an atomic number of 25, an atomic weight of 54.93 and one stable and four unstable isotopes, respectively 55 and 51, 52, 53 and 56. The metal has four allotropic modifications. Alpha, stable at ordinary temperature, and beta are hard and brittle, body-centered cubes, while gamma is a soft and flexible face-centered tetragonal crystal. Chemically, the element resembles iron, but it has all valences up to and including 7, the most stable being the divalent salt and the tetravalent oxide.

Composition and general chemical nature

The most common method for producing commercial high-purity manganese is the electrolysis of sulfate or chloride solutions. Other methods, particularly the distillation of manganese in vacuum, have been used but so far have not led to a commercial product.

A typical analysis of electrolytic metal, in per cent, is as follows: Fe, 0.0015; Cu, 0.001; As, 0.0005; Co, 0.0025; Ni, 0.0025; Pb, 0.0025; Mo, 0.001; sulfide sulfur, 0.017; sulfate sulfur, 0.014; C, 0.002 and H, 0.015.

Physical properties

The pure metal cannot be used for constructional purposes, since it is too brittle. The principal physical properties of manganese and its modifications are given below. The transformation temperatures shown below are now generally accepted, but may be again modified in the future. (See Table 1.)

TABLE 1

	Alpha	727 C $\rightleftarrows$ 692-622	Beta	1100 C $\longleftrightarrow$	Gamma	1138 C $\longleftrightarrow$ Delta	1245 C $\longleftrightarrow$ Liquid
Crystal Lattice	Body Centered Cubic		Body Centered Cubic		Face-Centered Tetragonal	?	—
Density: gm/cu cm at 20 C	7.44		7.29		7.21	?	6.54
Specific Heat, cal/gm/°C at 25 C	0.114		0.154		0.148	0.191	—
Lin Coef of Ther Exp $\times 10^{-6}$ per °C (8-100°)	22		14		—	—	—
Elec Res, 20 C, microhm-cm	185		91		45.1	—	—
Heat of Transformation, cal/gm atom		535		545		430	—
Heat Capacity Equations cal/gm atom	$Cp = 5.70 + 3.38 \times 10^{-3} T - 0.375 \times 10^{5} T^{-2}$		$Cp = 8.33 + 0.66 \times 10^{-3} T$		$Cp = 10.7$	—	$Cp = 11.0$

The brittleness of pure manganese prevents the metal from being used unalloyed. Also, it oxidizes easily and rusts rapidly in moist air. In the liquid phase manganese dissolves carbon, and absorbs substantial amounts of hydrogen and numerous metals to form a wide range of alloys.

Alloys

A ductile manganese alloy results when 2% Cu and 1% Ni are added. Manganese-copper alloys, e.g. 80/20, have excellent vibration damping properties. A 72% Mn, 18% Cu and 10% Ni composition has a high thermal expansion rate, while a low expansion alloy contains 50% Mn, 54% Fe and 5% Al. 60% Cu, 20% Mn and 20% Ni is an age hardening composition.

Applications

Electrolytic manganese is used extensively in the production of high-quality alloys in which special advantage is derived from the purity of this melting stock. Among these are the austenitic iron-chromium-manganese stainless steels, which may or may not additionally contain nickel or nitrogen. This form of manganese is also frequently used in the production of manganese-bearing aluminum alloys. Specialty nonferrous alloys, such as copper-manganese, special manganese-containing brasses, bronzes, and some nickel silvers also employ the electrolytic grade.

Virtually all steels contain at least a small amount of manganese, while in some steels massive amounts are added to achieve specific effects. The principal functions of manganese in steel are those of controlling and counteracting the deleterious effects of sulfur and of mildly increasing hardenability. Since its carbide-forming tendency is somewhat greater than that of iron, manganese is sometimes added to a greater extent (pearlitic manganese steels contain over 0.80% Mn). Apart from the manganese-bearing stainless steels mentioned above, the element has been used for many years at levels above 12% to produce austenitic wear-resistant steel. In the making of ordinary steels the use of electrolytic manganese is of course precluded on the basis of price, and one of the many grades of ferromanganese is usually employed.

Also, most commercial aluminum alloys contain manganese, usually in amounts of less than 1%, to raise the recrystallization temperature, slightly improve mechanical properties and, sometimes, the corrosion resistance. Many producers use a special, low-iron, high-carbon ferromanganese rather than electrolytic metal in the making of manganese-containing aluminum alloys.

MARAGING STEELS

The maraging steels develop unique combinations of properties which have not been obtained in conventional low-alloy steels. Some of these properties are: (a) useful yield strengths to and above 300,000 psi; (b) high toughness and impact energy even at the 300,000 psi yield strength level; (c) low nil ductility temperature (NDT); (d) exceptional stress-corrosion resistance; (e) through hardening without quenching; (f) simple heat treatment; (g) good formability without prolonged softening treatments; (h) good machinability; (i) low distortion during maraging after forming or machining; (j) good weldability; and (k) freedom from decarburization problems.

The key to this combination of high strength and toughness and the revolutionary characteristics of heat-treatment response is the martensitic structure having a low carbon content and high nickel content. This martensite is tough and has a low work-hardening rate, which allows a high degree of cold working. This, of course, is in contrast to the higher-carbon/lower-nickel types of low-alloy steels, which have a hard brittle martensite. Furthermore, the maraging treatment in this system tends to harden the material, rather than soften it, as tempering does in the conventional low-alloy steel martensite.

18% and 20% Ni maraging steels

Heat treatment. The 18% Ni and 20% Ni maraging steels are annealed at 1500 F to dissolve the hardening elements. During subsequent cooling to room temperature the alloys transform to martensite. Because ferrite, pearlite and bainite do not form in these alloys during cooling at slow rates, quenching is unnecessary to ensure the complete transformation to martensite. Therefore, hardenability (section size) limitations are absent and air cooling may be used, even for heavy sections.

Although the martensite obtained after cooling from the annealing temperature has a hardness of 28 to 35 Rc, its work-hardening rate is low. The alloys may be cold-reduced by 90% with little increase in hardness. As shown in Table 2, cold reduction can be used between annealing and maraging to increase maraged strength.

Maraging is accomplished at 850 to 900 F, followed by air cooling. Although the sequence of austenitizing, cooling and reheating is the same as for carbon steels, their effects are different. In the maraging steels, the untempered martensite is most suitable for forming or machining operations. The steels can be machined easily at 28 to 35 Rc and even as maraged, the steels still can be machined more easily than a carbon steel of the same hardness, probably because carbides are absent. At no stage of the heat treatment is quenching required and distortion during hardening is reduced to a minimum. In some forms, 18% Ni maraging steel can be treated by direct maraging of hot-rolled stock.

18% Ni maraging steel for 200,000 psi yield strength. This is an air-melting steel which can be melted in the basic electric-arc furnace and processed in the mill in the same manner as low-alloy hardenable steels. This steel has high toughness for its 200,000 psi yield strength level and fits into critical applications in plate, billet and other heavy sections. Reduction of area in the conventional tensile test is in excess of 65% and the ratio of notch tensile strength to tensile strength (with K_t of 12) is 1.6. The Charpy V-notch impact energy on bar stock is greater than 50 ft-lb at room temperature, 30 ft-lb at −320 F and 20 ft-lb at −423 F. In addition, this alloy passes the drop-weight test at −320 F and is very resistant to sea-water stress corrosion cracking. High notched toughness has been found in sheet testing (with machined edge notch specimen with K_t of 18) at −320 F.

TABLE 1—CHEMICAL COMPOSITION OF MARAGING STEELS

Type	Composition, %						
	Ni	Co	Mo	Ti	Al	Cb	Fe
18% Ni (200,000 psi)	17/19	8/9	3.0/3.5	0.15/0.25	0.1	—	Bal
18% Ni (250,000 psi)	17/19	7/8.5	4.6/5.1	0.3/0.5	0.1	—	Bal
18% Ni (300,000 psi)	18/19	8.5/9.5	4.7/5.2	0.5/0.7	0.1	—	Bal
20% Ni (250,000 psi)	18/20	—	—	1.3/1.6	.15/.35	.3/.5	Bal
25% Ni (250,000 psi)	25/26	—	—	1.3/1.6	.15/.35	.3/.5	Bal

TABLE 2—HEAT TREATMENT CYCLES

Type	Sol Ann 1 hr/in. of section at 1500 F, A.C.	Ausage 4 hr at 1300 F, A.C.	Cold Work	Refrigerate 8 hr at −100 F	Marage
18% Ni	X (28-32 Rc)	—	—	—	3 hr at 900 F
	—	—	—	—	3 hr at 900 F
	X	—	X	—	3 hr at 900 F
20% Ni	X (30-35 Rc)	—	—	—	4 hr at 850 F or 1 hr at 900 F
	X	—	X	—	
25% Ni	X (10-15 Rc)	—	X	X	1 hr at 850 F
	X	X	—	X	1 hr at 850 F

TABLE 3—MECHANICAL PROPERTIES OF MARAGING STEELS

Type	Melting	Heat Treat	Yld Str, 0.2% offset, 1000 psi	G_c[a] in.-lb/$\sqrt{in}$		NTS/TS[b] Sheet		Bar[c]	Impact Str. Charpy-V, ft-lb	
				R.T.	—320 F	R.T.	—320 F	R.T.	R.T.	—320 F
18% Ni (200,000 psi)	Air	1500 + 3/900 ..	200	>1800	>2700	1.1	1.0	1.6	50	30
	CVM	1500 + 50% CW + 3/900	235	>1680	>1800	1.1	0.95	—	—	—
18% Ni (250,000 psi)	Air	1500 + 3/900 ..	250, 209 d	>2000	—	0.93	—	1.4	18/26	12/20
	CVM	1500 + 3/900 ..	250	>2240	920	1.0	.74	1.5	25/30	—
	CVM	1500 + 50% CW + 3/900	280	1600	1000	0.90	0.75	—	—	—
18% Ni (300,000 psi)	CVM	1500 + 3/900 ..	280, 240 d	1610	—	.77	—	1.4	—	—
		3/900	300	—	—	—	—	1.5	22	14
		1500 + 50% CW + 3/900	310	1100/>1800	—	.85	—	—	—	—
20% Ni (250,000 psi)	CVM	1500 + 1/900 ..	250, 180 d	1900	—	.82	—	1.3	15/20	10
		1500 + 50% CW + 1/900	270	1000	—	.68	—	—	—	—
25% Ni (250,000 psi)	CVM	1500 + 4/1300 + Ref. + 1/850	250	500	—	.48	—	1.1	—	—
		1500 + 50% CW + Ref. + 1/850	250, 187 d	1700	—	.72	—	—	—	—

a Longitudinal, 0.06-0.08 in. sheet, ASTM-NASA 1-in edge notch specimen, $K_t = 18$.
b NTS = notched tensile strength based on machined notch area.
c With $K_t = 12$ and root dia. of 0.212 in.
d Last value determined at 800 F.

TABLE 4—OTHER PROPERTIES OF MARAGING STEELS

	18% Ni			20% Ni	25% Ni
	(200,000 psi)	(250,000 psi)	(300,000 psi)		
Mod of Elast, 10^6 psi	27.5	26.5	27.5	25.5	24.5
Density, gm/cu cm	—	8	—	7.9	7.9
Av. Lin Coef of Ther Exp, 70–900 F, per °F	—	5.6×10^{-6}	—	6.2×10^{-6}	6.2×10^{-6}
Length Change After Maraging, %	—	−0.04	—	−0.12	−0.10
Stress Corrosion at Yld Str, Min Life, days					
U-Bends, Sea Water	No failure, 12 mo	35	2 mo a	1	<1
Bayonne Atm, 3 or 2 Pt. Ld.	3 mo a	24 mo a	3 mo a	7	—
Magnetic Properties					
Intrinsic Induction, k gauss					
H-250 oersted	—	16.55	—	17.1	13.5
H-500 oersted	—	18.5	—	18.375	14.750
Remanence, k gauss	—	5.5	—	5.0	3.7
Coercive Force, oersteds	—	28.1	—	15.6	25.0

a No failure.

18% Ni maraging steel for 250,000 psi yield strength. Outstanding for this steel as air-melted at the yield strength level of 250,000 psi are:

1. Passing of the drop weight at −80 F. For equivalent NDT temperature, this represents a jump in yield strength of about 80,000 psi over low-alloy steels.

2. High notch tensile strengths of 372,000 to 386,000 psi. This is 80,000 psi higher than low-alloy steels at the same yield strength level.

3. Excellent weldability. No heat-affected zone cracking has been found even in welding fully hardened restrained plate of up to 4-in. sections with no preheat. The heat-affected zone can be restored to full strength and deposited weld metal brought to a high strength level by a simple post-weld marage.

4. Relatively low coefficient of expansion.

Additional data are provided in Table 5.

TABLE 5—ADDITIONAL PROPERTIES OF 18% NI
MARAGING STEEL (250,000 PSI)

Elec Resis, μ-ohm cm	
Ann 1500 F	60–61
Ann plus maraged 3 hr at 900 F	38–39
Mod of Rigidity, 10^6 psi [a]	10.2
Poisson's Ratio	0.30
True Stress-Strain	
Strength Coef (true stress at true strain of 1.0), 1000 psi [a]	272
Strain Hardening Exponent,[a] n	0.039
Shear Str, 1000 psi [a]	143
End Lim, 10^8 cycles, 1000 psi	95
Comp Yld Str, 0.2% offset, 1000 psi [a]	247

[a] Determined on specimens from 1-in. plate having a 0.2% offset yld str = 230,000 psi and tensile strength = 240,000 psi.

This steel has been air and vacuum melted and can be used in plate, bar, sheet, rod, wire and extruded forms. An advantage of vacuum melting is noted at this strength level in higher notched tensile strength and impact energy.

18% Ni maraging steel for 300,000 psi yield strength. This steel, which is capable of developing a 300,000 psi yield strength, uses higher cobalt and titanium levels. At this strength level, the benefits of vacuum melting are pronounced. Therefore, this melting procedure is used to ensure adequate toughness. Yield strengths of 300,000 psi with over 440,000 psi notched tensile strength, have been found on Research Laboratory and commercial heats. G_c values in excess of 1000 in lb per sq in. are available in sheet form at this yield-strength level. This grade retains the best hot strength of those listed in Table 3 and has scaling resistance superior to H-11 steel at 1000 F. It can be considered for applications in billet, plate, sheet, bar, rod, wire and extruded forms.

20% Ni maraging steel. Consumable-arc vacuum melting has been employed for this steel. Compared to 18% Ni maraging steel, it has the advantage of lower alloy content and freedom from cobalt and molybdenum, which may be desirable for some applications and environments. At the present stage of commercial development, the cost falls between that of air-melted and consumable-arc vacuum melted 18% Ni maraging steel. The 20% Ni maraging steel can be used in sheet, bar, wire, and extruded tubing. Compared to 18% Ni maraging steel, the 20% Ni maraging steel has lower toughness, hot strength, modulus of elasticity, stress corrosion cracking resistance, weld toughness, and less dimensional stability during heat treatment.

25% Ni maraging steel

This alloy is intended for use when a material is required which must be softened below 28 Rc for severe fabrication processes such as deep drawing. The 25% Ni maraging steel is largely austenitic after annealing at 1500 F. The softened material has a yield strength of 40,000 psi, an ultimate strength of 132,000 psi, with 30% elongation, and 72% reduction of area.

The steel must be transformed virtually completely to martensite to reach high-strength levels subsequently. This transformation may be achieved by two methods.

1. The M_s temperature (temperature at which martensite starts forming during cooling) of the austenite may be raised by precipitating some of the hardeners in combination with nickel. This is achieved conveniently by aging the austenite at 1300 F (ausaging). In addition to raising the M_s temperature of the austenite, the ausaging treatment also causes some hardening

which is inherited by the martensite. In alloys containing about 1.5% titanium the martensite transformation is nearly completed by cooling and holding at room temperature. Alloys containing less than about 1.4% titanium require refrigeration (at −100 F) to complete the transformation after ausaging because less nickel is removed from solution during treatment at 1300 F.

2. The austenite may be transformed directly by a combination of cold working and refrigeration to about −100 F.

Once the structure has been transformed to martensite it is hardened by a simple "maraging" treatment as described previously. From the mechanical properties standpoint, the second treatment of cold working, refrigeration and maraging is preferable.

This alloy is considered for sheet and other applications where severe cold forming is necessary in manufacturing the part. Consumable-arc vacuum melting has been used on this version.

The unusual properties give maraging steels a wide range of applications where high-strength density ratio with high toughness is required. Typical products and parts for which they are suitable include cryogenic structural parts, landing gear for aircraft, hot extrusion dies, rotors, cold-headed bolts, mortar and rifle tubing, shafting, tubing, and various tools and fixtures.

MECHANICAL FINISHES

Finishing with coated abrasives

Coated abrasives or "sandpaper" can be used in many ways on many materials to produce a wide variety of finished surfaces. The method of application as well as the type of abrasive used greatly influences the results obtained. Following is a description of the three principal methods of using endless abrasive belts and the applications for each method:

Contact wheel application. The contact wheel method (Fig 1) produces the best abrasive efficiency, including the fastest rate of cut and best abrasive life. The finish produced by this method is characterized by short individual scratches which are uniform across the width of contact and in line with the running direction of the belt. This method can be used on all types of parts having flat or lightly contoured

surfaces such as hand tools, small household appliances, cutlery, and stainless-steel sheet for decorative trim in kitchens, hospitals, etc.

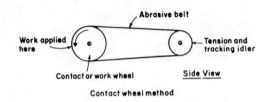

Contact wheel method

Fig. 1. Contact wheel method produces fast rate of cut and provides long abrasive life.

Platen method. This method (Fig 2) is used for finishing work with a flat surface that is either continuous or interrupted. Because individual abrasive grain remains in contact with the work from top to bottom the method produces a continuous scratch running the length of the part in contact with the belt. The advantage of the platen method is that the whole surface is contacted simultaneously and it easily produces flatness with a high degree of accuracy.

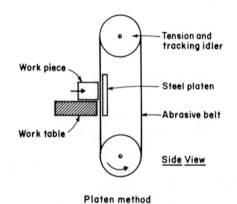

Platen method

Fig. 2. Platen method produces continuous scratches running length of part.

Slack belt. This method (Fig 3) is used on work that is highly contoured such as metal furniture, light fixtures and plumbing fixtures. The flexibility of the belt is useful in polishing hard-to-reach surfaces where contours and shapes are involved. The length of scratch pattern produced is between the short scratch pattern of the contact wheel method and the long scratch pattern of the platen.

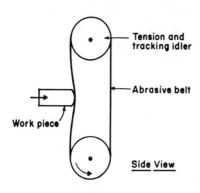

Slack belt method

Fig. 3. Slack belt method works well on parts with highly contoured surfaces.

Types of abrasives and finishes. Two principal types of abrasives are used for coated-abrasive types of finishing operations—aluminum oxide and silicon carbide. The former has a blocky shape and is very tough. It is used for grinding and finishing materials having a high tensile strength. Silicon carbide is harder and sharper but is more friable or brittle; consequently it is used for grinding and finishing materials having low tensile strength.

Both grain types, when used without a lubricant, customarily produce a dull finish with little or no reflectivity. However, when a suitable lubricant is used, the finish produced is approximately one grit size finer than the finish from the same grit size when used dry. In addition to being finer, the lubricated finish is much brighter and, when fine grit sizes are used, a high degree of reflectivity is obtained without buffing.

Silicon carbide abrasive belts are often used for the final polishing operation when a sequence of grit sizes is used to polish the surface of high tensile strength materials. Because silicon carbide grain is more friable, it breaks down quickly when used on ferrous materials and produces a burnishing effect on the work with a very bright and highly light reflective finish. When used on stainless steel, it also imparts a faint bluish color or cast to the finish that has considerable eye appeal.

For a given grit size, the more firmly the abrasive is supported or backed up the coarser the finish will be. Hard contact wheels and platens will produce substantially coarser finishes than will soft contact wheels and platens. With

a coarse grit size, such as 36, used with a hard contact wheel and heavy grinding pressure, it is possible to produce finishes as coarse as 250 to 400 μin. rms (root mean square) on mild steel. Using a soft contact wheel under identical conditions can improve the finish to the point of reducing it to 100 to 125 μin. rms.

The use of fine grit sizes in the range of 240 mesh and finer, makes it relatively easy to produce finishes under 20 μin. and finer. On ferrous materials it is usually necessary to start with a finish of 20 rms or finer in order to subsequently buff satisfactorily to a scratch-free finish. On most nonferrous materials buffing can be accomplished from a starting finish of approximately 30 rms.

The surface speed of the abrasive materially affects the fineness of the finish produced. For belts the recommended speed range is 3000 to 7000 sfpm and for discs, 5000 to 11,000 sfpm. The lower end of the range will provide a faster cut and coarser finish; the upper end of the range produces some sacrifice in the rate of cut, but provides a substantially finer finish, especially when light grinding pressure is used.

Coated abrasives are used extensively for a vast multitude of grinding and finishing operations because of their inherent versatility and economy. The finishes produced with coated abrasives are primarily decorative but can also be used for functional purposes in cases where geometry of the finished parts is important.

Finishing with loose abrasives

Finishes produced by loose abrasives, such as aluminum oxide and silicon carbide, are available in many variations of surface roughness and appearance, depending on the methods used. The methods used to finish or polish a surface may include polishing wheels (referred to as "set-up" wheels), lapping, pressure blasting, and barrel finishing (often referred to as tumbling).

"Set-up" wheels can be likened to bonded grinding wheels. The basic difference is that set-up wheels are resilient because they are made of cloth laminations which are stitched so as to give a certain rigidity or resiliency. Abrasive grain is cemented on the periphery of set-up wheels with cold cement or hot glue. Because of its "cushioning" effect, the resiliency of set-up wheels simplifies the blending of contoured surfaces.

As with grinding wheels, the coarser the grit the deeper the furrow plowed; but grit size for grit size, the set-up wheel produces shallower furrows than bonded grinding wheels because of its cushioning effect. Fine finishes can be generated with set-up wheels, and for this reason the surfaces of many metal parts are polished by this method preparatory to plating. Set-up wheels are not intended for so-called precision, surface or cylindrical grinders but are intended to eliminate surface imperfections in metal without regard to establishing or retaining dimensional accuracy.

Buffing with loose abrasives usually follows the polishing operation, the latter being done with either set-up wheels or coated belts. Buffing generates smooth surfaces which are essentially free of scratches and have high reflectivity. It is accomplished by bringing the surface into contact with the periphery of a buffing wheel (usually cloth laminations) to which abrasive materials have been applied in composition form. Buffing generates smooth and reflective surfaces by removing or displacing relatively small amounts of metal. It is not designed to establish or retain dimensional accuracy.

Pressure blasting is the process in which solid abrasive particles are propelled against a surface by means of expanding compressed air. The method always produces a matte finish. The finer the abrasive the finer the finish; the coarser the abrasive the faster the rate of stock removal and the rougher the finish.

Pressure blasting can be used for many finishing purposes. These include: removal of heat-treat scale and oxides from metallic surfaces; preparation of surfaces for protective coatings; removal of contamination from nuclear devices; generation of decorative finishes; preparation of surfaces for visual inspection for surface defects; and removal of small burrs.

Lapping can be used to produce very smooth finishes and geometrically true surfaces. Like grinding, the degree of finish obtained by lapping can be controlled by the particle size of the abrasive. Coarse grits plow deep furrows and therefore remove stock faster than fine particles. Fine particles, conversely, remove stock at a lower rate but produce finishes with a lower rms. To reduce the surface roughness from say, 30 to 5 rms, usually requires using two to three progressively finer abrasives.

Factors other than abrasive particle size can also influence the finish and appearance. Such factors include speed of lap; type of vehicle, such as water, kerosene, grease, lard, etc.; and the type of material from which the lap is made, such as cast iron, bronze or plastic. A bronze lap will produce a finer finsh than one made of cast iron, all other factors remaining constant.

The finishes generated by lapping are functional rather than decorative. The surface smoothness dictates a low rms where wearing parts are involved. Flatness and scratch-free parts are essential for components such as high-pressure valves, and germanium and silicon transistor wafers.

Barrel finishing (also covered in a separate article in this Encyclopedia) is a surface-conditioning operation in which a mixture of metallic or nonmetallic parts, abrasive media and various compounds are placed in a receptacle and rotated for the purpose of rounding corners, deburring, improving surface finish, cleaning, derusting, burnishing for high luster and producing low μin. finishes. Many of these finishing operations are accomplished in simultaneous operations.

Unlike grinding wheel, set-up wheel or abrasive belt operations, tumbling produces a random scratch pattern. This is because the media and parts have a random motion while the barrel is rotating. But like grinding wheels and other types of abrasive finishing methods, the types and size of abrasive media (sometimes called nuggets, pellets, chips, etc.) influences the roughness and rate of stock removal.

The appearance and finish of barrel-finished parts are also influenced by the ratio of media to parts; compounds or additives used; water level in barrel; speed of barrel; height of load, and length of cycle.

Rough finishes and rapid stock removal are produced with a large medium and low water level. Finer finishes are produced with a small-size medium and usually higher water level. A burnished finish (one with high luster) is accomplished by using well-rounded media and sufficient detergent and water to provide the necessary cushioning effect to produce a scratch-free reflective surface.

Finishing with bonded abrasives

Finishes produced by grinding wheels vary over a wide range both in surface roughness and in surface appearance. These finishes can be divided into two major classifications, those produced by rough grinding and those produced by finish grinding.

Rough-grinding operation is primarily intended for stock removal without any particular regard for appearance. The finish produced can be described as having the appearance of a plowed field with the direction of the furrows in the same direction as the path of the grinding wheel. The depth of the furrows is in direct proportion to the roughness of the grind. The rougher the grind, the deeper the furrow; the less severe the grind, the shallower the furrow. The surface roughness produced by rough grinding varies over a wide range, beginning in the 30 μin. area and becoming progressively rougher.

Finish grinding, as its name implies, us used for the production of finish and, in addition, the generation of geometry. This finish may be one that has a low surface roughness or a high luster and, in some cases, both a low roughness and a high luster.

The appearance of a finely ground surface can either be a dull matte finish or it can be a highly reflective surface. A distinction is made between the roughness of the dull matte ground finish and that of a highly reflective ground surface. A dull matte finish is generally as smooth and, many times, smoother than a highly reflective surface. The matte finish is a truly ground surface and can be described as a plowed field where the furrows are very shallow in depth and all the ridges have been removed.

A highly reflective surface on the other hand can be described as a plowed field where some of the ridges have fallen into the adjacent furrow and are pushed into that furrow by the grinding wheel as it passes over it. This produces a luster but not necessarily a low surface roughness.

Grinding wheels can produce a surface roughness as fine as 1 μin. and a surface so reflective that it can be used as a mirror. These surfaces can be produced on both cylindrical and flat surfaces and are produced by using: (a) fine grit size grinding wheels; (b) light infeeds; (c) fast work speeds; (d) slow table speeds; (e) a copious supply of grinding fluid.

In a finish-grinding operation, any combination of grinding wheel, speeds, feeds and grinding fluid can produce a variety of finishes. It is common practice to produce finishes of about 15 μin. with a 46 grit size grinding wheel, and of about 10 μin. with a 60 grit size grinding wheel. Progressively finer finishes are produced by using finer grit size grinding wheels and taking lighter feeds and a longer time to produce the finish.

MELAMINE FORMALDEHYDE PLASTICS

The outstanding features of melamines, one of the amino resins, include excellent colorability, high hardness, and good electrical characteristics, including exceptional arc resistance.

Primary uses for melamine plastics as engineering materials are in decorative molded dinnerware, decorative thermosetting laminates, molded housings and wiring devices, closures and buttons.

The resins are produced by reaction of melamine and formaldehyde. Crystals of melamine, an amino chemical, are reacted with formaldehyde and mixed with highly purified filler, dried and ground. Color pigments, plasticizers, lubricant and accelerators are added to produce the molding compound. Melamine is similar in many respects to urea-formaldehyde. Melamine resists water absorption and chemical attack to a greater extent than urea and its cost is higher.

Molding compounds

For molding applications, melamine's hardness and colorability are outstanding advantages. Melamine is the hardest of plastics (Rockwell M118–124) and has unlimited colorability. Parts can be produced in any color in the spectrum (both translucent and opaque). Color quality ranges from pastels to bright, jewel-like tones, as well as two tones in the same object. Products such as dishes and tableware can be decorated by foil inlays.

Among other useful characteristics of melamine are its nonelectrostatic nature (does not attract dust), heat resistance (210 F), dimensional stability, low moisture absorption (0.4 to 0.6% in 24 hr), flame resistance, excellent dielectric properties (arc resistance, 120 to 140 sec.; dielectric strength, 300 to 400 v per mil.), resistance to organic solvents, oils and greases.

Melamine products can be molded by most of the common processes, i.e., compression, transfer and plunger. In compression molding, the molding compound is usually preformed into a "pill," preheated and molded. Mold temperatures range from 280 to 350 F at pressures of 2000 to 8000 psi. Because finish is usually important in melamine products, dies are highly polished.

Melamine compounds can be molded at higher temperatures than urea and are not as sensitive to overcure.

Dinnerware (dishes, plates, cups and saucers) make up 90% of the uses for melamine molding compounds. Hardness and colorability as well as resistance to attack by foods, detergents and hot water account for its popularity in the home. In addition to an unlimited range of solid colors, melamine tableware is now available in two colors. The top of a plate or cup may be blue while the bottom may be white with full fusion of the two. By the use of melamine-impregnated foil, decorative designs are made an integral part of the plate.

To combat the tendency of melamine to stain, attention is being given to a stain-resistant glaze.

Melamine molding compounds are widely used as handles for flatware and kitchen utensils because of hardness, color and resistance to heat and water. Handles can now be decorated with patterns applied integrally by foil.

Agitators for washing machines are being molded of melamines because of their resistance to hot water and detergents and excellent appearance. A colored agitator is more attractive than a black one.

Nondecorative applications for molding compounds include wiring fixtures and other electrical uses, where the arc resistance and nontracking characteristics of melamine, coupled with its nonflammability, make it useful. For such uses, strength and impact resistance is improved by incorporating fabrics and glass-fiber reinforcements. Although dimensional stability has been lacking in such compounds, more recent grades have greatly improved dimensional stability.

High pressure laminates

The majority of top-quality Decorative Thermosetting Laminates, or high pressure laminates (e.g., "Formica," "Micarta" and "Textolite") use melamine resins impregnated in alpha-cellulose paper or other clear reinforcements to provide their decorative surfaces. With the proper resin selection, such laminates can be burn and stain resistant. Further the high hardness of melamines makes them scuff and mar resistant.

Melamines cure by a condensation type polymerization, thus are not practical for wet lay-up fabrication of reinforced plastics. But they are suitable for the "dry lay-up" techniques commonly used in producing high-pressure industrial thermosetting laminates.

Several standard grades of NEMA Industrial Thermosetting Laminates consist of melamine resin binders. There are two standard common grades: (1) Grade MC, a specialty grade consisting of purified cotton fabric with a melamine resin binder. It is intended primarily for use in plating barrels and has high alkali and arc resistance. (2) Grade G-5 is a glass-reinforced melamine laminate which provides high strength and hardness. It has good flame resistance, and is second only to silicone laminates in heat and arc resistance. It has excellent electrical properties under dry conditions, and low insulation resistance under high humidities. Grade G-5 is used for switchboard panels, arc barriers and circuit-breaker parts, armature and slot wedges.

METAL POWDERS

By definition a metal powder is an aggregate of discrete metal particles that are usually in the size range of 1 to 1000 μ.

The metal powders in common use are lead-tin

TABLE 1–TYPICAL PROPERTIES OF MELAMINE

Type	Melamine, Cellulose, Elect.	Melamine, Glass Fiber
Specific Gravity	1.43–1.50	2.0
Mod. Elast. in Tens., 10^5 psi	10.5×10	24
Tens. Str., 1000 psi	5–9	5–10
Elong., %	0.6	–
Hardness, Rockwell M	110	120
Imp. Str., ft-lb/ in notch	0.25–0.35	0.5–10
Max. Svc. Temp., F	280	300–350
Vol. Res., ohm-cm	$10^{12} \times 10^{13}$	2–7×10^{11}
Dielect. Str., v/mil	350–400	250–300

alloys for solders; iron, nickel and cobalt base alloys for hard facing; copper, silver and nickel base alloys for brazing; aluminum, bronze and stainless steel for paint pigments; magnesium for pyrotechnics; iron for welding rods, torch cutting and scarfing and metal powder parts; copper for metal-powder parts and carbide for tools. Used in smaller quantities are iron powders for radio and television tuning cores; nickel-iron and silicon-iron alloy powders for other soft magnetic parts; aluminum, iron, nickel and cobalt powders for small permanent magnets in the "Alnico" series; nickel and cobalt powders as binders in the production of carbide tools and alloy and stainless-steel powders for high strength and special property parts.

Of these uses, the fabrication of structural parts or machinery components accounts for the largest single use of metal powders and competes with such other metal-forming methods as machining, casting, stamping and forging. The powder-metallurgy technique may be selected by the designer as the best way to make a particular part for one of several reasons:

1. The process is ideal for mass-producing machine components at low unit cost.
2. Residual porosity can be controlled to provide long-wearing qualities through the self-lubricating feature of the "oil-less" sleeve bearing.
3. Metal combinations are possible through powder metallurgy that cannot be melted.
4. Powder-metallurgy techniques provide the only practical method of forming high melting point metals.

The good surface finish and close dimensional tolerances possible are additional reasons why the designer may specify metal-powder parts.

Although most metal powder parts are made from iron, copper or mixtures of these primary powders with or without graphite additions, many other powders are used to develop special properties like high strength, magnetic or electrical properties, corrosion resistance or oxidation resistance. These special powders include brass, bronze, alloy steels, stainless steel and various nickel base alloys. Many of these special powders are also employed in the production of sintered metal filters.

Production methods

The methods used to produce these powders are varied. Most iron powder is the product of *gaseous reduction* of either selected iron ore or mill scale. The iron oxide is ground and mixed with coke dust and ground limestone. When this mixture is heated either in a tunnel kiln or in saggers in a car furnace, the carbon monoxide formed reduces the iron oxide to a spongy mass of relatively pure iron. This sponge iron is then ground to the desired size, magnetically separated, annealed and screened to yield sponge iron powder noted for its soft and spongy characteristics.

Electrolytic methods are used to make most of the copper powder used for metal-powder parts and a special, high purity iron powder often used for making high density iron powder parts. The electrolytic process begins with the production of an anode—by casting in the case of copper or rolling in the case of iron. The current density of the cell is arranged so that the metal deposited on the cathode is both porous and brittle. When the porous cathode deposit has built up to the desired thickness, the deposit is stripped from the cathode and forms a powder slurry. This powder slurry is washed free of electrolyte, dried, milled to size and given a reduction-annealing treatment. Screening and blending complete the process. The resulting electrolytic powder is very pure and soft.

Atomization of molten metal is a production method applied to a wide range of powders including aluminum, magnesium, brass, bronze, lead, tin, nickel-silver and stainless and low alloy steels. Other more complex iron, nickel and cobalt base alloys are also made by atomization. In the atomization process a stream of molten metal of the desired chemical composition is dispersed by the impact of a high velocity stream of air, steam, inert gas or a suitable liquid. The resulting tiny metal droplets are collected, annealed and screened to size.

Although metal powders made by machining, milling and crushing operations are not widely used for structural parts, such powders make important contributions to the economy. Among the powders made by *pulverization* are metal carbides, tungsten, molybdenum and manganese. All are brittle enough to mill readily. Soft and ductile powders, such as bronze and aluminum, tend to become quite flaky when milled and so they find ready application as metallic paint pigments.

Each production method imparts special physical characteristics to the powder. The method used will be decided on the basis of economics and the powder properties desired.

Because the metal powder fabricator has at his disposal primary metal powders, mixtures of these primary powders and prealloyed powders, a wide range of strength levels can be provided by selecting the desired combination and varying density and heat treatment.

Through control of the many variables possible, metal-powder parts can have a wide range of unique and interesting properties. Although mechanical property levels for metal-powder parts are generally lower than for wrought parts of equivalent composition, the strength of metal powder parts is seldom a problem. Close cooperation between the design engineer and the metal-powder fabricator will insure maximum realization of the desired properties.

MICA

Mica has been known as an electrical material of excellent insulating and fabricating properties for many years. Almost everyone has seen or used mica in one form or another, yet it is surprising how little even the average technically trained person knows about its composition, origin and properties.

It is, of course, a naturally occurring mineral having a wide range of possible chemical compositions and properties. All true mica, however, belongs to one mineral class of silicates having a sheet type of structure, and found in certain areas of the world growing from pegmatite deposits in "book" form. In recent years the increasing demands for high-quality sheet mica have led to the development and commercial production of synthetic mica as an electric furnace product which will be discussed later.

Nature and composition

Only two types of natural mica are considered to be of commercial importance for most engineering applications. These are muscovite, known generically as "ruby mica," and phlogopite, or "amber mica."

Almost 70% of the high-quality natural muscovite mica used in this country comes from overseas sources. Muscovite, which has the composition $KAl_3Si_3O_{10}(OH)_2$, generally comes from India, Brazil and Argentina, although some lesser deposits are found in the United States, Russia, Rhodesia and Tanganyika.

Ruby mica is generally colorless or has tinges of gray, brown, green or red. It is the most desirable mica for electrical applications and is of strategic importance for military requirements.

Phlogopite mica is a higher-temperature-resistant natural mica having the formula $KMg_3,AlSi_3O_{10}(OH)_2$. It is usually darker in color than muscovite, ranging from brown to greenish-yellow. In general, natural phlogopite is not as desirable as natural muscovite for electrical purposes but finds use in many subcritical applications.

The synthetic mica which is being produced commercially today is a fluor-phlogopite $KMg_3AlSi_3O_{10}F_2$ made by the internal-resistance melting process. This method utilizes graphite electrodes, and melts the mica at 1365 C in its own raw batch so that no crucible is required. Standard production melts made by some producers are on the order of 12 to 15 tons in weight. While the mica crystallizes in book fashion, the intergrowth of the resultant crystals makes economic recovery of single sheets impractical. Major research efforts are in progress to effect growth of larger-area synthetic mica sheets.

Properties

The most singular outstanding property of mica is its physical structure. As a sheet mineral it can be split into strong, flexible films having good high-temperature resistance and electrical insulating properties.

The natural muscovite and phlogopite micas all contain chemically combined water in the form of hydroxyl $(OH)^-$ ions. Muscovite contains about 4.5% water, which at temperatures of 600 C and above, volatilizes, causing breakdown of the mica with resultant blistering delamination. Phlogopite contains about 3% water and is stable at temperatures up to 800 C. Synthetic fluor-phlogopite contains no water. During its formation from the inital oxide and fluoride raw materials the $(OH)^-$ ion positions are filled by fluorine $(F)^-$ as an ionic substitute. The result is that fluor-phlogopite is temperature stable to 1000 C. Above this temperature evolution of fluorine begins and continues until the melting point is reached.

Table 1 summarizes the properties of the three principal types of micas.

Forms available

Natural mica is sold in many forms and is found in a widely varying range of quality. The American Society for Testing and Materials has published a specification (D351) which establishes the quality of muscovite mica on the basis of visual evaluation. Mica is classified into four forms: (1) Full-trimmed natural block mica of 0.007 in. minimum thickness; (2) partially trimmed natural block mica of 0.007 in. minimum thickness; (3) films (cut or uncut) split from natural block mica to any range of specified thickness, and (4) scrap.

Block mica is graded according to size from "No. 6" which is 1 to 2¼ sq in. in area up to "No. A-1" special which is 36 to 48 sq in. Visual quality is classified into 12 groups ranging from "V-1" clear mica down to "V-10a" which is densely black and red-stained. Among the qualities considered in addition to discolorations are inclusions, stains, waviness, hardness, cracks, holes, etc.

The cost of sheet mica also varies widely from around $1.00 per lb for 1-sq-in. sheets to $100.00 per lb for 100-sq-in. sheets. The price, of course, is strongly dependent on the quality of the mica. A general rule for economy is to select the smallest area sheet of lowest quality mica which will perform in the application.

Mica film split from block mica is generally grouped into three categories of quality ranging from "V-3" fair-stained and above, to "V-5" stained-A quality. In general, the mica used in electronic tube and capacitor applications is first quality mica. The lower qualities of mica find their place in commercial appliance applications.

Scrap and flake mica are used widely in the manufacture of roofing, wallpaper, paints, plastics, etc.

Products and applications

Sheet mica is used primarily in electrical equipment and appliances in the form of washers, spacers, sleeves, tubes, etc., where high voltages must be withstood. It is used widely as a dielectric in stable high-reliability capacitors and in the majority of types of receiving tubes. It is also used in microwave windows and X-ray tube applications.

In addition to use in the form of flexible single sheets, both natural and synthetic mica are available in many other forms, primarily for electrical applications. Some of these include the following:

Pasted mica. This product is prepared by bonding loose mica splittings with various resins and glues into the form of hard plates or flexible sheets. The sheets are in many cases reinforced with glass cloth, paper or plastic film to increase tensile strength and improve tear resistance.

Pasted mica products are used in many applications where sheet mica would not be economical but where many of the desirable properties of mica are required. It is used as flat segment

TABLE 1

Physical Properties	Natural Muscovite	Natural Phlogopite	Synthetic Fluor-Phlogopite
Formula	$KAl_3Si_3O_{10}(OH)_2$	$KMg_3AlSi_3O_{10}(OH)_2$	$KMg_3AlSi_3O_{10}(F)_2$
Sp Gr	2.6–3.2	2.6–3.2	2.8–3.0
Sp Heat	0.207	0.207	0.20
Ther Conductivity, cal/cm²/sec °C/cm	0.0008–0.0014	0.0008–0.0014	0.0012
Ther Expansion, 10^{-6} in./in./°C	9–12	12–15	8–13
Elast Mod, 10^6 psi	25	25	26
Ten Str, 1000 psi	40–50	40–50	40–50
Dielec Str, v/mil	3000–6000	3000–4200	2500–3500
Dielec Constant, (10^6 cps)	6.5–8.7	5–6	5–6
Dissip Factor (10^6 cps)	0.0001–0.0004	0.004–0.07	0.0004–0.0007
Vol Resis, ohm-cm	2×10^{13}–1×10^{17}	—	10^{16}–10^{17}

plate in rotary machinery and as molding plate where more complex shapes are required such as in commutators, V-rings and channels. These materials are generally punched, sawed, cut or pressed to the required shapes.

Mica tape and wrappers. These materials are similar to pasted mica but are generally much thinner. Mica tape and wrapper are used for insulating high-voltage coils, motor armatures and other areas of rotating machinery.

Heater plate. This material is made by using small-size loose splittings with a minimum amount of bond. The bond is burned out leaving a heat-resistant plate which finds use in percolators and similar commercial appliances.

Reconstituted mica paper. Natural mica paper is made by laying down fine scrap mica flake by processes utilizing heat, pressure and chemical reaction. It can be made without a bond to form a continuous flexible sheet. By the introduction of various binders such as silicones and epoxy resins it can be given increased handling strength.

Recent government-sponsored studies have led to the successful development of reconstituted synthetic mica paper which has extreme heat resistance and high tensile strength. The manufacturing process is similar in nature to that used in paper making and results in a material suitable for use in high-temperature (500 C) transformers, capacitors and motors. By a subsequent hot pressing at 1300 C, this material can be converted to a recrystallized sheet having many of the properties of single-crystal synthetic mica.

Glass-bonded mica and ceramoplastics. For many years, glass-bonded mica has been used in every type of electrical and electronic system where the insulation requirements are preferably low-dissipation factor at high frequencies, a high-insulation resistance and dielectric-breakdown strength along with extreme dimensional stability. Glass-bonded micas are made in both machinable grades and precision-moldable grades. Basically, the material consists of natural mica flake bonded with a low-loss electrical glass. This material is heated to the plastic state at temperatures up to 600 C with subsequent compression or injection molding, depending on the specific part requirement.

The availability of synthetic mica resulted in the development of so-called ceramoplastics,

consisting of high-temperature electrical glass filled with synthetic mica. Ceramoplastics provide an increase in the electrical characteristics over those of natural mica, and, in addition, are more easily molded and have greater thermal stability.

Glass-bonded mica and ceramoplastics have found use in many advanced components such as telemetering commutation plates, molded printed circuitry, high-reliability relay spacers and bobbins, coil forms, transducer housings, miniature-switch cases, and innumerable other component applications.

MOLYBDENUM AND ALLOYS

Molybdenum metal

Molybdenum metal is produced in a high state of purity with only very small amounts of trace elements and gases. The arc-cast product is deoxidized with carbon and therefore may contain up to 0.040% carbon residual.

Physical properties. Outstanding among the physical properties of molybdenum metal (Table 1) are the high melting point, high thermal conductivity, high modulus values and low expansivity.

TABLE 1—PHYSICAL PROPERTIES OF
MOLYBDENUM METAL

Melt Pt, F	4730
Density, 68 F, lb/cu in.	0.369
Elec Cond, 32 F, % IACS	34
Sp Ht, Btu/lb/°F	
70 F	0.06
2000 F	0.07
Mod of Elast, (static), 10^6 psi	
70 F	46
2000 F	29
Mod of Rigidity, 10^6 psi	
80 F	17
1600 F	15
Poisson's Ratio	
80 F	0.324
1600 F	0.321
Therm Cond, Btu/ft/hr/°F	
70 F	81
2000 F	58
Lin Therm Exp, %	
70 to 1000 F	0.29
70 to 2000 F	0.63
Thermal Neutron Absorption Cross Section, (2200 m/sec), barns	2.5

Mechanical properties. The mechanical properties of molybdenum depend to a large degree on the amount of working done below its recrystallization temperature. Room-temperature ductility and toughness are greatly affected by the temperature of transition from ductile to brittle behavior, since for many types of stressing this occurs around room temperature.

The mechanical properties of most interest today are those at high temperatures:

STRESS-RELIEVED BAR STOCK

| | Short-Time Tensile Properties | | | |
| | | | | 100-hr Stress-Rupture Str, 1000 psi |
	Ten Str, 1000 psi	Elong, %	Red of Area, %	
1200 F	65.2	22	86	56
1600	52.4	24	89	32
2000	44.1	19	88	13
2400	11.6	71	96	—

The elevated-temperature properties are strongly influenced by the recrystallization temperature, which sets a limit to the operating temperatures where strain-hardened material offers a strength advantage. The minimum recrystallization temperature reported for pure molybdenum is 1650 F; commercial products normally require higher temperatures. Fully recrystallized molybdenum has lower strength than stress-relieved material; at 1600 F, the 100-hr rupture strength is about 40 to 50% lower. This difference lessens as the testing temperature approaches the recrystallization temperature. With unalloyed arc-cast molybdenum, recrystallization may take place in stress-relieved bars tested at 2000 F, so stress-relieved and recrystallized bars may have essentially the same strength and structure at the conclusion of testing.

Corrosion resistance. Molybdenum has particularly good resistance to corrosion by mineral acids provided oxidizing agents are not present. For instance, molybdenum components have been giving good service in contact with 50% sulfuric acid at 400 F where nickel-molybdenum alloys, tantalum and glass-lined equipment were not satisfactory. In gases, molybdenum is relatively inert to hydrogen, ammonia and nitrogen atmospheres up to about 2000 F; a superficial nitride case may be formed at higher temperatures in the last two gases. Molybdenum has ex-

cellent resistance to liquid metals such as molten bismuth, sodium, potassium and lithium. It is used for electrodes in mercury switches because it does not react in any manner with mercury even at elevated temperatures. Also of commercial significance is its resistance to many types of molten glass and nonferrous slags.

Protection from oxidation. At temperatures over about 1000 F, unprotected molybdenum oxidizes so rapidly in air or oxidizing atmospheres that its continued use under these conditions is impractical. Uncoated molybdenum is, however, used satisfactorily where very short lives are involved (as in some missile parts) or where the surrounding atmosphere is nonoxidizing (as in hydrogen and vacuum furnaces). Protective coatings seem to be the answer where oxidation is a problem. Various coatings differing in maximum time-temperature capabilities and in physical and mechanical characteristics are available. Selection of the proper coating for a specific application involves consideration of a number of factors, foremost among which is the service temperature. For temperatures up to 2200 F, nickel-base alloys applied as cladding or sprayed coatings, and chromium-nickel electroplates appear most generally suitable. For temperatures up to 2800 F, or short periods at higher temperatures, modified chromized coatings and sprayed aluminum-chromium-silicon are predominant. For longer periods at higher temperatures, the choice would probably rest between siliconizing and ceramic coatings. Component tests have been found most reliable for final selection of suitable coatings.

Fabricating characteristics. Molybdenum cannot be hardened by heat treatment but only by working below the recrystallization temperature. For optimum ductility, it should be given at least a 50% reduction in area by warm working. Relative to its high melting and recrystallization temperatures, molybdenum is much "colder" at room temperature than steel. Consequently, the toughness and ductility are considerably higher at temperatures somewhat above room temperature (for example, 400 F) than at room temperature. Therefore, except for fine wire and sheet, at least a moderate amount of heating is recommended for all fabricating operations. Under these conditions, molybdenum is being fabricated satisfactorily by all standard methods, including the following: forming, bending, punching, stamping, deep drawing, spinning and power roll forming.

No particular problems are posed by mechanical joining or brazing. Molybdenum can also be welded by a number of different processes, such as arc, electrical resistance, percussion, flash and electron beam. Arc-cast molybdenum can be arc-welded without porosity or cracks, but this is generally not true of commercially available powder-metallurgy molybdenum. The ductility of molybdenum is adversely affected by even minute amounts of oxygen or nitrogen, not only in the parent metal but also on the surface and in the welding atmosphere. Therefore, for optimum ductility, fusion welding must be carried out on carefully cleaned material in a closely controlled atmosphere. With optimum arc-welding conditions, welds that will take a longitudinal free bend of 180 deg at 80 F have been made in arc-cast molybdenum sheet. These welds have a strength equal to the recrystallized strength of the base metal.

Molybdenum can be machined and ground on standard equipment. The choice between high-speed and sintered-carbide tools depends largely on individual preference. In either case, tool life is shorter than would be expected with steel at the same hardness. The preferred carbide grades are the straight tungsten-carbide types recommended for cast iron. Tool angles and rakes are usually similar to those used for cast iron. Many machining operations are done without lubrication. When a lubricant is used during machining, various highly chlorinated oils have proved satisfactory.

Availability. Molybdenum metal is available in many forms: forging billets, sheet bar, bars, wire, plate, sheet, strip, foil and seamless tubing. The maximum size available depends on the method of production as the arc-cast process produces heavier ingots than are feasible with powder metallurgy.

Applications. The applications of molybdenum are generally those based on its high melting point; high modulus of elasticity; high strength at elevated temperatures; high thermal conductivity; high resistance to corrosion; low specific heat and low coefficient of expansion. Long-established uses are found in the electric and electronic industries for applications such as mandrels and supports in lamp fabrication; anodes and grids in electronic tubes; resistance

elements and radiation shields in high-temperature furnaces; electrical contacts and electrodes. In recent years molybdenum has become increasingly important in the missile field for nozzles, nozzle inserts, leading edges of control surfaces, support vanes and jetavators. Uses in the metalworking industry include boring bars, grinding quills, resistance-welding electrodes and thermocouples. The nuclear-energy industry has been active in developing molybdenum heat exchangers, piping, heat shields and structural parts. The glass industry is a major user of large molybdenum parts, especially for melting electrodes and stirrers. The chemical and petrochemical industry is beginning to use molybdenum for corrosion resistance. Molybdenum's usefulness as a metallized coating, either to improve bonding between the base metal and a sprayed top coating, or alone for improved wear resistance, is no longer confined to maintenance but is extending to original equipment.

Molybdenum-base alloys

The three major molybdenum-base alloys today are: molybdenum-0.5% titanium; molybdenum-0.5% titanium-0.08% zirconium (TZM); and molybdenum-30% tungsten. A fourth alloy, molybdenum-1.25% titanium-0.3% zirconium-0.3% carbon (TZC) is still in the experimental stage. All appear to be produced more satisfactorily by the arc-cast process than by powder metallurgy.

Molybdenum-titanium and molybdenum-titanium-zirconium alloys. The main advantage of these alloys, as compared with unalloyed molybdenum, is their higher recrystallization temperature and better hot strength as shown in these data for stress-relieved bar stock:

	100-hr Str-Rupture Str, 1000 psi		Approx Temp for Complete Recryst of Bars (1 hr at Temp), F
	1800 F	2000 F	
Unalloyed Molybdenum	22	13	2100
Mo—0.5% Ti	53	34	2450
Mo—0.5% Ti—0.08% Zr	70	52	2600
Mo—1.25% Ti—0.3% Zr—0.3% C	67.5 (est)	—	3200

Physical properties, resistance to corrosion and oxidation and fabrication characteristics are not significantly different from those of unalloyed molybdenum.

Applications are generally those that require higher hot strength than molybdenum metal and include extrusion dies, die-casting cores, piercer points, and structural parts of experimental spaceships, rockets and planes.

Molybdenum-30% tungsten alloy. While the molybdenum-30% tungsten material has higher hot strength than unalloyed molybdenum (47,000 psi vs. 22,000 psi for 100-hr stress-rupture strength at 1800 F), its main applications depend either on its melting point or on its resistance to molten zinc. This alloy has a solidus temperature of about 5115 F and a liquidus of 5145 F. Therefore, it is being used in rocket nozzles and similar parts where melting point is the critical factor. The other field of application comprises pumps and other equipment for handling molten zinc.

The density and modulus of the molybdenum-30% tungsten alloy are higher than those of molybdenum metal (0.43 lb per cu in. and about 50×10^6 psi). Its fabrication into finished parts requires more care than is needed with molybdenum metal or the other molybdenum-base alloys.

N

NATURAL RUBBER

Rubber is characterized as being a highly elastic or resilient material, and the natural product is obtained mainly as a latex from cuts in the trunks of the *Hevea brasiliensis* tree. The latex consists of small particles (averaging about 2500 Å units in diameter) of rubber suspended in an aqueous medium (at about 35% solids content). The system also contains about 6 to 8% nonrubber constituents of which some are emulsifiers, naturally occurring antioxidants and proteins.

The rubber part of the composition, which is obtained from the latex by coagulation, washing, and drying, consists of long linear polymeric molecules of high average molecular weight. One report states that natural rubber contains molecules of molecular weights from 50,000 to 3,000,-000 with 60% of it over 1,300,000. The rubber is built up of C_5H_8 units, each containing a double bond; and the over-all structure is called cis-polyisoprene. Although a product almost identical to natural rubber has been made by polymerizing isoprene, there is no indication that a tree makes rubber from isoprene.

On long standing at low temperature, the solid rubber will tend to crystallize, and also, uncrystallized raw rubber will show some crystallization on stretching. Frozen or crystallized bales of rubber must be heated so that they can be processed in the standard procedures.

Vulcanization

The raw rubber as obtained from the latex is not useful directly in many commercial applications because of stiffening at low temperatures and softening at high temperatures. It must be vulcanized in order to give a wider temperature range for application and good physical properties. The most usual vulcanization is by the use of sulfur and organic accelerators of vulcanization in the presence of zinc oxide at temperatures in the neighborhood of 212 to 400 F depending upon the curing system.

Also, for many applications the high rubber vulcanizates are not tough enough for good service. In order to improve this situation, reinforcing fillers are used for compounding the rubber, such as finely divided carbon black, silica, and silicates including clays. In addition, many other nonreinforcing fillers, such as whiting (calcium carbonate) and barytes, are used for cheapness or to lend special properties to finished articles.

Properties

Some of the important properties of soft vulcanized natural rubber are as follows:

Tensile strength. Tensile strengths vary from about 2000 pounds to 6000 psi depending upon composition, with elongations-at-break varying from around 200 to around 1000%.

Modulus. Strength at various percent elongations can be varied over a wide range up to the tensile at break by controlling the degree of vulcanization and additives such as fillers.

Hardness. This property, for soft rubber, can be varied over a wide range by the amount of softener or by the amount of fillers or reinforcing agents used in the compounding. Also by using large quantities of sulfur for vulcanization, a very tough material known as hard rubber can be produced.

Resilience. One of the important features of natural rubber is good resilience and especially rapid recovery from a deformation. A ball from a good rubber compound, when dropped from a given height, can rebound in the range of 70% of the height from which it dropped.

Hysteresis. Low hysteresis or heat buildup in a rubber compound (i.e., a tire compound) under dynamic conditions such as repeated flexing is another of the good features of natural rubber. Most of the synthetic rubbers with the exception of synthetic cis-polyisoprene and cis-polybutadiene are appreciably inferior to natural rubber in this respect.

Abrasion resistance. Rubber compounds reinforced with carbon black give quite good abrason resistance, and this property has led to their use in many applications where this property is important. For many years it was the only material for making pneumatic tires, although it has been displaced in the United States to a large extent, particularly in passenger tires, by a number of synthetic rubbers.

Tear resistance. Natural rubber, and particularly the reinforced product, gives vulcanizates with good resistance to tear, and this property is better than that obtained with many of its synthetic competitors.

Other properties. Natural rubber compounds have also been made with good electrical resistance, particularly those of purified rubber, which are resistant to water absorption. Also, rubber compounds can be made resistant to dilute acids and dilute alkalies. Natural rubber, due to its hydrocarbon nature, is not resistant to swelling by many organic chemicals.

Flame resistance of natural rubber is not particularly good due to its hydrocarbon nature. However, it may be compounded with flame retardants to greatly improve its resistance to burning.

Environmental effects. Under service conditions natural rubber articles are subject to some deteriorating effects, and the rubber must be compounded so as to minimize these effects. For instance it is subject to deterioration by the action of oxygen, ozone, heat, and light and to fatigue cracking under dynamic conditions. Antioxidants are used in compounds to retard air and heat deterioration, and also special chemicals are added in addition to protect against light deterioration.

Certain antioxidants, called antiflex-cracking agents, have great protective value against deterioration caused by repeated flexing of rubber in the presence of air, heat and light.

Articles from unsaturated rubbers such as natural rubber react very rapidly with ozone to form cracks when in the stretched condition. In fact, stretched rubber is one of the best tests for the presence of small amounts of ozone in the atmosphere. However, rubber can be compounded to eliminate or greatly minimize this reaction. For many years waxes have been incorporated in rubber to protect it against weather or ozone cracking when under tension under static conditions. Recently, chemicals called antiozonants have been developed to retard both static and dynamic cracking of rubber influenced by the ozone in the air.

Processing

Natural rubber is, for most applications, masticated and compounded on mills or in Banbury mixers; the masticated material is then passed through a tubing machine to form various articles or parts of articles which are then vulcanized at an elevated temperature in a mold or under other suitable conditions.

Compounded raw rubber may also be calendered into sheets which are built into various types of articles and then vulcanized. In addition to this, a number of articles are made directly from aqueous dispersions of the rubber such as from the latex. In this case the latex is compounded as such. The rubber article may be made by a dipping procedure or by some coagulating procedure such as is used in making latex thread or by foaming the latex and gelling it as is done in a foam sponge process. The articles are subsequently vulcanized with rapid accelerators at relatively low temperatures (around 212 F).

Natural rubber is furnished to fabricators in the form of large bales or as a latex. The bale rubber comes in a number of different grades which are priced according to quality. The latex is sold at different solids contents, the higher solids contents being preferred partly because of shipping costs from the plantations.

Applications

Natural rubber is used for making many types of articles. Because of its abrasion-resistant quality and low hysteresis in reinforced compounds, it is used in truck-tire tread stocks and in conveyor belts which are employed in conveying abrasive material such as coal, crushed rock, ores and cinders. In large-size tires, it has found application in carcass compounds because of the tack and building qualities of the raw polymer. It has also been used in carcass compounds because of the low heat buildup (low hysteresis) of the carcass compound vulcanizate during severe service conditions in tire usage.

Some of the many other applications for natural rubber include waterproof clothing and footwear, and wire insulation; as well as mechanical goods and sundries, products such as hot water bottles, surgical goods, rubber bands and thread, engine mounts, tank tread blocks, gaskets, sound and vibration damping equipment, rug underlay, etc. In fact, it can be used almost any place where a highly resilient, elastic, durable material is required. A few applications in the latex field are (1) in making latex thread by coagulating a fine stream of latex compound coming from a nozzle and then vulcanizing it; (2) in making foam sponge such as mattresses by gelling and vulcanizing a foamed latex; (3) in tire cord dipping; (4) in making such articles as surgeon's gloves by the latex dipping method and (5) in carpets where it is used as a backing material to provide anchorage of the yarns and to give dimensional stability.

NEOPRENE RUBBER

Essentially, the treatment of monovinylacetylene with hydrochloric acid results in chloroprene, which can be polymerized to produce a rubber-like solid, known generically as neoprene.

General characteristics

The name neoprene describes a family of chemical rubbers consisting of polymerized chloroprene. Neoprene and neoprene latex both possess outstanding resistance to deterioration by oils, solvents and chemicals, sunlight and weathering, oxidation, heat, flexing, abrasion and flame. Each type of neoprene has its own special engineering characteristics which contribute to its usefulness in specific applications.

The family of elastomers is highly versatile, in that there are a number of neoprene polymers available, and each can be further modified by choice of fillers, softeners and vulcanizing agents. Optimum values for all properties, of course, cannot be obtained in one compound; compromise is essential.

Mechanical properties

Neoprene vulcanizates can be produced with hardness ranging from less than 30 to more than 90 durometer A. Typical neoprene compositions have tensile strengths in the 2000 to 3500 psi range. Tear resistance is of the same order of magnitude as that of similar compounds of other elastomers.

Neoprene is generally characterized by excellent resistance to compression set. The following data compare compression set values of a general-purpose neoprene with those of a natural rubber compound designed for minimum set. Note that the neoprene compound is superior at all conditioning temperatures:

COMPRESSION SET: NEOPRENE VS. RUBBER
(Expressed in % set)

Conditioned: 30% deflection for 70 hr at:	Neoprene	Natural Rubber
70 C	13	20
100 C	18	28
121 C	20	38

In dynamic characteristics, neoprene is generally acknowledged to approach natural rubber more closely than does any other synthetic rubber within practical hardness ranges. Resilience of natural rubber is superior to that of neoprene in the lower hardness range (40 to 50 durometer A); as hardness increases, neoprene becomes superior to natural rubber, the point at which the curves cross depending largely on the type of filler used to adjust hardness.

Energy loss in a cycle of deformation (i.e., hysteresis) is in the form of heat and since elastomers are not sufficiently good thermal conductors to dissipate heat as rapidly as it is generated, rapid vibration may result in a substantial temperature rise within a compound. At constant amplitude vibration, natural rubber has a lower degree of heat rise than neoprene, up to a hardness of about 50 A. However, in hardnesses above 50 A, neoprene is superior to natural rubber in this respect.

Initial creep of neoprene compounds is substantially higher than that of natural rubber. The slope of the creep curve, however, is considerably higher for natural rubber, so that creep values approach those of neoprene after prolonged conditioning.

Effect of temperature

Neoprene's resistance to low temperatures is not as good as that of natural rubber (minimum recommended service temperature for neoprene is about −40 C, compared with −50 C for natural rubber). On cooling from room temperature to −34 C, neoprene increases in hardness by about 30 durometer points.

Special compounding techniques can be used to improve the low temperature flexibility of neoprene. Low-temperature plasticizers can be added to reduce the brittle point from −40 to −54 C.

Neoprene is substantially superior to natural

rubber in heat resistance. Maximum recommended service temperature of neoprene is about 116 C, compared with about 80 C for natural rubber.

Physical properties

Neoprene is inherently flame resistant. Although it can be made to burn, it will not propagate flame provided it is not compounded with excessive amounts of flammable materials (such as hydrocarbon softeners).

Neoprene has excellent resistance to permeation by gases. Permeability is one-fourth to one-tenth that of natural rubber, depending on the gas.

Dielectric properties of neoprene are somewhat inferior to those of natural rubber and SBR. Neoprene compounds are adequate, however, for insulation at low voltages. Volume resistivity of about 10^{12} ohm-cm can be obtained by compounding. Conductive compounds can also be made with resistivities as low as 100 ohm-cm. Dielectric strength of typical commercial neoprene compounds ranges from 150 to 550 volts per mil.

Resistance to deteriorating environments

In general, neoprene is noted for its outstanding resistance to chemicals, sunlight and weathering, aging and ozone. This combination of properties has probably been responsible for the greatest volume of use of the material.

Resistance to specific reagents is difficult to state meaningfully in general terms. Effects will depend on the type and concentration of the reagent, service temperature and ratio of exposed area to total volume of part. Further, the degree of importance of such effects depends on the demands on the part in service.

Very small differences in chemical composition of the reagent can have important effects on resistance of the elastomer. For example, the following data show the equilibrium volume increase of a neoprene composition after immersion at 100 C in SAE 20 lubricating oils from different crude stocks. The data show that substantial differences are obtained with oils of different chemical type and even with different oils from crudes of the same general class.

SAE 20 Oil, Chemical Type:	Volume Increase, %
Paraffinic A	13
Paraffinic B	29
Naphthenic	62

Whether or not a neoprene part is in continuous or intermittent service is particularly important in applications involving contact with the many highly volatile materials, such as benzene and trichloroethylene, which are widely used as chemical process intermediates, solvents and degreasing agents. Even though continuous contact with trichloroethylene, for example, may result in substantial swelling of the neoprene, the same compound may be satisfactory for only intermittent contact with this solvent. Intermittent contact allows time for evaporation of highly volatile hydrocarbons so that the compound rapidly regains its original properties. For example, tensile strength of a neoprene part measured immediately after removal from the solvent was only 50% of its original value; after 2 hr of drying, tensile strength returned to 90% of its original.

In general, neoprene compounds have excellent resistance to (1) all straight-chain hydrocarbons even when substantial amounts of aromatic constituents are added, as in some types of aviation gasoline, (2) all aliphatic hydroxy compounds such as methyl and ethyl alcohols and ethylene glycol, (3) animal and vegetable fats and oils, and (4) fluorinated hydrocarbon compounds, such as Freon refrigerants.

Neoprene compounds are either unsuitable or have limited serviceability in contact with (1) chlorinated hydrocarbons such as chloroform and benzylchloride, (2) organic esters such as butyl acetate and methyl salicylate, (3) aromatic hydrocarbons such as toluene and benzene, (4) aromatic hydroxy compounds such as phenol and cresol, and (5) some ketones, such as methyl ethyl ketone.

Generally, neoprene renders excellent service even at elevated temperatures in contact with (1) dilute mineral acids, except those of a strong oxidizing nature, (2) all alkalies including concentrated sodium and potassium hydroxides, and (3) all solutions of inorganic salts except those of a strong oxidizing nature.

Neoprene is unsuitable or has limited serviceability in contact with strong oxidizing agents such as nitric acid, potassium dichromate, concentrated sulfuric acid and hydrogen peroxide. In cases where the suitability of neoprene is doubtful, use of simulated or actual service tests is recommended.

NICKEL AND ALLOYS

The element nickel, along with the elements iron and cobalt, constitute the transition group in the fourth series of the periodic table. It has an atomic number of 28 and an atomic weight of 58.71. This atomic weight represents a composite of 5 stable isotopes that are found in the natural state. The abundance of the natural isotopes is 67.8% for Ni^{58}, 26.23% for Ni^{60}, 1.25% for Ni^{61}, 3.66% for Ni^{62}, and 1.16% for Ni^{64}. Seven radioactive isotopes are also known and have been produced.

The normal crystallographic system of nickel is face-centered cubic at all temperatures. At 25 C, its lattice constant is 3.5238 Å and the distance of closest approach of the atoms is 2.491 Å.

The usual commercial grade of wrought nickel ("A" nickel) nominally contains 99.0% nickel plus up to 0.4% cobalt, which traditionally are combined and reported as 99.4% nickel. A purer grade, obtained by electrolytic refining, contains 99.55% nickel plus up to 0.4% cobalt and is usually reported as 99.95% nickel. Nickel made by a third process, the decomposition of nickel carbonyl vapor, contains negligible cobalt.

Wrought nickel is available as the standard "A," a lower carbon variety for improved workability, and an electronic grade with closely controlled residuals. It can be obtained in all of the usual forms such as plate, bar, sheet, strip, rod, wire, and tubing. Electrolytic nickel is deposited as 145 lb cathodes with the dimensions $38 \times 28 \times \frac{3}{8}$ in. These cathodes are sheared to approximately $1 \times 1 \times \frac{3}{8}$ in. pieces for alloy manufacture and are melted and cast into anodes for electroplating and into shot for alloy additions. Electrolytic nickel powder, principally for powder-metallurgical applications, is made by a suitable adjustment in the conditions of electrodeposition. Carbonyl nickel is available in powder form with three different purities: HPM, Grade A, and Grade B, containing 99.9+, 99.8, and 99.7% nickel, respectively. These powders are hot- and cold-worked to form high purity plate, bar, sheet, strip, rod and wire.

Fabrication

With suitable modifications in temperatures, tools, pressures, rates, etc., wrought nickel is amenable to most of the fabrication processes used for mild steel. It can be hot-formed (forged, rolled, bent, extruded); cold-formed (sheared, punched, spun, deep drawn); machined with high-speed steel and carbide tools; and ground, polished and buffed. It can be annealed to remove the effects of cold work by heating above the recrystallization temperature, which is somewhat in excess of 600 C for commercial nickel and at lower temperatures for purer nickel. Generally, temperatures of 820 to 930 C are specified for full annealing and 540 to 600 C for stress-relief annealing. A general precautionary measure for all nickel and many high nickel alloys is to avoid heating them in the presence of sulfur. Sulfur contaminates nickel by forming nickel sulfides at grain boundaries and in the body of grains, which adversely affects mechanical properties. Sulfur can be picked up from fuels used for heating, some salt baths, and even from cutting fluids used in machining and lubricants used in rolling if they are not completely removed prior to heating the nickel.

Physical properties

Several of the physical properties of nickel are shown in Table 1. The values are for nickel of higher purity than the best commercial grade.

Nickel is ferromagnetic at ordinary and low temperatures but becomes paramagnetic at elevated temperatures. The exact temperature of the magnetic transformation (Curie point) varies with the purity and prior treatment of the metal and occurs within the range 350 to 360 C. Specific heat and the coefficient of thermal expansion increase with increasing temperatures up to the Curie point and then decrease sharply between the Curie point and 500 C; thereafter they increase again. Electrical resistivity is also temperature dependent and shows a sharp increase, but no maximum, in the region of the Curie point. Thermal conductivity becomes irregular in the region of the Curie point.

TABLE 1—PHYSICAL PROPERTIES OF NICKEL

Density, gm/cu cm, 20 C	8.908
Melting Pt, C	1453
Latent Heat of Fusion, cal/g	73.8
Boiling Pt, C	2730 (est.)
Sp Ht, 0 C, cal/g/°C	0.1025
Coef of Ther Exp, 25 to 100 C, $\times 10^{-6}$/°C	13.3
Ther Cond, 100 C, cal/cm²/cm/°C/sec	0.198
Elec Resis, 20 C, microhm-cm	9.5
Magnetic Saturation, gauss	6500
Magnetic Perm (mag. field strength = 1 oersted)	2,000 to 3,000 max
Thermal-Neutron-Absorption Cross Section, barns	4.5 ± 0.2

Mechanical properties

The mechanical properties of nickel are influenced by purity, temperature of test, and prior thermal and mechanical history. The metal can only be hardened by cold work. Mechanical properties at room temperature are given in Table 2.

TABLE 2

Grade → Mechanical Property ↓	99.4%			99.95% Ann.
	Ann.	Hot Rolled	Cold Drawn	
Ten Str, 1000 psi	75	55–80	65–115	46
Yld Str, 0.2% offset, 1000 psi	25	15–45	40–90	8.5
Prop. Limit, 1000 psi	—	18	—	4.5
Elong, in 2 in., %	50	65–30	35–15	30
Red. of Area, %	70	75–60	70–50	88
Mod of Elast, 10⁶ psi	30	30	31	30
Hard., Rockwell B	60	40–65	70–100	30
Impact Str, ft-lb [a]	204	—˙	195	—
Endurance Limit, 10⁸ cycles, 1000 psi	24	30	42.5	—

a Charpy V notch.

Corrosion resistance

As a commercially pure metal, one of its principal uses is for corrosion resistance, and for this application nickel is very frequently used as an electroplated coating or cladding on some other material. It visibly tarnishes in the atmosphere, but its rate of corrosion is low. It is rapidly attacked by most acids, although it resists sulfuric and phosphoric acids under favorable conditions. Nickel is highly resistant to solutions of sodium hydroxide and many other alkaline salt solutions, subject to limitations on composition and temperatures of the solutions. It finds important use as a catalyst, where it is used in about 40 different fields including the hydrogenation of oils, the cracking of ammonia, and the artificial aging of liquors. Other uses include thin films for resistance thermometers and filters for photoelectric cells, wire and strip for cathode bases in electronic tubes, and anodes for electroforming and building up of worn parts by electrodeposition.

Ferrous alloys

The largest commercial application for nickel is as an alloying element in both ferrous and nonferrous alloys. In unhardened nonaustenitic steels, nickel has the effect of slightly strengthening ferrite and mildly improving corrosion resistance. In balanced combinations with small amounts of Cr and Mn; or Cr, Mo, and Mn; or Mo and Mn (AISI 3100, 3300, 4300, 4600, 4800, 8600, 8700, 9300, 9800 series) hardenabilities of steels are greatly increased. In carburizing steels (AISI 2300 and 2500 series) nickel is used to strengthen and toughen the core.

Nickel is a strong austenite stabilizer and with chromium is used to form the important AISI 300 series of nonmagnetic austenitic stainless steels. In these steels, chromium is the principal alloying element and is the element responsible for their excellent corrosion resistance. Nickel contributes to this corrosion resistance, but its primary function is to promote and retain an austenitic structure at all recommended temperatures of use and under most conditions of plastic deformation. The mechanical properties of an austenitic steel are generally much less adversely affected by moderately elevated temperatures and very low temperatures than are ferritic or martensitic steels. Wrought stainless steels (AISI listing) with chromium and nickel contents up to 22 and 26%, respectively, and in some cases with the addition of molybdenum, have been designed for specific heat and corrosion resisting service over and above the standard 18-8 type. Some cast stainless steels with extremely high chromium and nickel contents (up to 28% Cr, 10% Ni and 17% Cr, 66% Ni) are used for severe corrosion or heat-resisting applications. Nickel is also widely used as a property improving addition to cast iron and as one of the many alloying elements in some ferrous high-temperature, high-strength superalloys.

Magnetic alloys

Nickel and iron form a series of alloys with thermal-expansion and magnetic characteristics of commerical importance. The alloy of 36% nickel ("Invar") has a thermal coefficient of expansion of almost zero over a small temperature range. The addition of 12% chromium, in lieu of some of the iron, produces an alloy ("Elinvar") with an invariable modulus of elasticity over a considerable temperature range as well as a fairly low coefficient of expansion. Variations in content of nickel and other metals have produced a large number of alloys with particular coefficients for specific applications such as geodetic tapes, balance wheels in watches, tuning forks, and glass-to-metal seals. The alloy with 78% nickel (permalloy) and many modifications of this composition have very high magnetic permeabilities at low field strengths and find application in the electronics field. Nickel also plays an important role in permanent magnets of the Alnico type which contain 14-18% nickel plus aluminum, cobalt, and sometimes copper.

Nonferrous alloys

The principal groups of nonferrous alloys are those of slightly alloyed nickel, nickel and copper, nickel and chromium, and the superalloys. The slightly alloyed nickels ("Duranickel," "Permanickel," E nickel, electronic nickel, etc.) in most cases contain more than 94% nickel. These alloys retain much of the corrosion resistance and physical properties of nickel, but they are alloyed to make them heat treatable by age hardening, more resistant to a specific form of corrosion (Mn to reduce susceptibility to sulfur embrittlement) or to produce desirable electronic characteristics.

With copper. The major nickel-based alloy with copper is "Monel," which nominally contains 66% Ni, 31.5% Cu, 1.35% Fe, 0.90% Mn, plus residuals. It has a brighter appearance than nickel, is stronger and tougher than mild steel, has excellent resistance to atmospheric and seawater corrosion, and generally is more resistant than nickel to acid, less resistant to alkalies, and equally resistant to salts. It is used in architectural and marine applications where appearance plus corrosion resistance is important, and in specialized equipment used by the food, pharmaceutical, paper, oil and chemical industries. Several variations have been produced in which compositions have been adjusted to obtain an additional characteristic. These include an age-hardenable grade (K "Monel"), a free-machining grade (R "Monel"), a hard-casting grade (H "Monel"), an age-hardenable casting grade (S "Monel") and several others.

Most copper-based alloys with nickel, the cupronickels, can be classified into three groups according to their nickel content: 30%, 20% and 10%. They are valued for their resistance to atmospheric and water corrosion (sea and fresh) and for their high thermal conductivity. This combination of properties has led to their accepted usage in marine equipment and particularly for heat exchanger tubes and condensers. The alloy used for the United States five-cent coin at the present time is a cupronickel with 75% copper and 25% nickel. The alloy with 45% nickel, constantan, is widely used as a thermocouple element and to some extent as an electrical-resistance material.

Nickel silver. The copper-nickel-zinc alloys are known as nickel silver, although in actuality they are brasses with sufficient nickel added to give a silvery white color, improved corrosion resistance, and high strength. These alloys are used as low-cost substitutes for silver in tableware and jewelry, usually with a silver or gold electroplate on the surface. Nickel silvers are also construction materials for many musical, drafting, and scientific instruments, and are used for marine and architectural applications.

With chromium. The alloys of nickel with chromium, with and without iron, form a series of corrosion- and heat-resistant materials. In this group the 80% Ni, 14% Cr, 6% Fe alloy ("Inconel") with many modifications resists progressive oxidation below 2000 F, and is used in applications such as furnace chambers, salt pots, aircraft exhaust manifolds, and high-temperature springs. It was originally developed for use in the milk industry as a corrosion-resistant alloy and now is much used in many chemical industries because of its excellent corrosion resistance. The 80% Ni, 20% Cr alloy ("Chromel A," "Nichrome V," "Tophet A") and the 60% Ni, 16% Cr, 24% Fe alloy ("Nichrome," "Chromel C," "Tophet C") form the bulk of materials used for heater elements. The 90% Ni, 10% Cr

("Chromel P") alloy in combination with alumel is much used as a dependable base-metal thermocouple.

Superalloys. The nonferrous superalloys are complex materials in that their compositions usually contain a minimum of four principal alloying elements. They were developed to produce greater strengths and corrosion resistance at higher temperatures than were obtainable with existing materials, including the ferrous superalloys. Many are modifications of the nickel-chromium alloys previously mentioned ("Inconel X").

There are three major groups of nonferrous (less than 25% iron) superalloys: cobalt base, nickel base, and nickel-chromium base. The actual number of these alloys reaches the hundreds.

About one percent of the total production of nickel is used in the formation of nickel-based chemicals. Of these, nickel sulfate and nickel-ammonium sulfate, nickel chloride, and nickel carbonate are used in electroplating baths; nickelous hydrate for electrical storage batteries, and green and black nickel oxide for coloring glazes on pottery.

NITRIDES

Only in recent years have nitrides been of more than academic interest. In general, they are less stable than the oxides, carbides and sulfides, and their use in air at elevated temperature is limited because of their tendency to oxidize. However, in several instances, the oxide film is protective and deterioration is slow. Despite their limitations, nitrides have interesting properties and are sure to find many specialized uses as technology becomes more complex.

Table 1 gives the melting points and volatilities of the more stable nitrides. To be useful for extended periods in refractory applications, a material should have a vapor pressure below 10^{-6} atm at the operating temperature. The use of a static atmosphere of nitrogen to prevent dissociation of the nitride can extend its usefulness.

Aluminum nitride

Aluminum nitride is conveniently prepared by an electric arc between aluminum electrodes in a nitrogen atmosphere. Crucibles of the pressed powder, sintered at 3600 F, are resistant to liquid aluminum at 3600 F, to liquid gallium at 2400 F and to liquid boron oxide at 2000 F. Aluminum nitride has good thermal shock resistance and is only slowly oxidized in air (1.3 percent converted to Al_2O_3 in 30 hr at 2600 F). It is inert to hydrogen at 3100 F but is attacked by chlorine at 1100 F.

Boron nitride

The crystal structure of boron nitride is similar to that of graphite, giving the powder the same greasy feel. The platy habit of the particles and the fact that boron nitride is not wet by glass favors use of the powder as a mold wash, e.g., in the fabrication of high-tension insulators. It is also useful as thermal insulation in induction heating.

Boron nitride can be hot-pressed to strong ivory-white bodies which are easily machined. The apparent density is about 2.1 g/cc or 92 percent of theoretical, and the purity is 97 percent. The 2.5 percent boron oxide in the finished product is termed an "inadvertent impurity." Unfortunately, its presence leads to weakening of bodies exposed to hot water. However, this small amount of oxide is probably necessary to bond the nitride.

Hot-pressed boron nitride is stable in air to about 1300 F. From 1300 to 1800 F the rate of oxidation increases moderately. It is also stable in chlorine up to 1300 F. However, at 1800 F it is attacked rapidly.

Like commercial graphite, hot-pressed boron nitride is anisotropic. Thermal expansion parallel to the direction of pressing is ten times that in the perpendicular direction. The ratio

TABLE 1—MELTING POINTS AND VAPOR PRESSURES OF REFRACTORY NITRIDES

Nitride	Melting Point, F	Vapor Pressure (at indicated temperatures), atmospheres		
		2250 F	3150 F	4050 F
AlN	>4400 (sublimes)	—	10^{-6}	—
BN	>4400 (sublimes)	$<10^{-7}$	10^{-4}	($<10^{-6}$)
CbN	3600	—	—	—
HfN	6000	$<10^{-7}$	$<10^{-7}$	10^{-5}
ScN	4800	$<10^{-7}$	—	—
Si_3N_4	>3450 (sublimes)	10^{-4}	—	—
TaN	5350	$<10^{-7}$	—	—
TiN	5350	$<10^{-7}$	$>10^{-6}$	10^{-3}
ThN	—	$<10^{-7}$	$<10^{-7}$	10^{-5}
UN	4770	$<10^{-7}$	$<10^{-7}$	10^{-5}
YN	>4800	$<10^{-7}$	$>10^{-6}$	—
ZrN	5400	$<10^{-7}$	$<10^{-7}$	10^{-5}

TABLE 2—PROPERTIES OF BORON AND SILICON NITRIDES COMPARED WITH GRAPHITE AND ALUMINA

Property	Boron Nitride	Graphite	Silicon Nitride	Alumina
Melt Pt, F	>4400 (subl.)	>6500 (subl.)	>3450 (subl.)	3722
Sp Gr				
Crystal	2.25	2.25	3.18	3.96
Body	2.10	1.6–2.0	1.5–2.7	2.6–3.9
Hardness, DPH [a]	30	20	1100	2800
Mod of Rupture (room	16 $\parallel$	2.2 $\parallel$ [c]		
temp), 1000 psi [b]	7.3 $\perp$	1.8 $\perp$	16–20	38
Coef of Ther Exp (Avg at	4.17 $\parallel$	4.5–12 $\parallel$ [c]		
70–1800 F),/°F $\times$ 10^6 [b]	0.43 $\perp$	1.5–4 $\perp$	1.37	4.3
Ther Cond (room temp),	105 $\parallel$ [d]			
Btu-in./ft^2 hr/°F [b]	199 $\perp$ [d]	120–360	130	20–30
Elec Res, ohm-cm				
At Room Temp	1.7×10^{13} $\parallel$ [b]	10^{-3} [c]	10^{13}	10^{16}
At 900 F	2.3×10^{10} $\parallel$		10^{13}	10^{12}
Dielec Const	4.1–4.8	—	9.4	12.3

[a] Diamond pyramid hardness.
[b] $\parallel$ = Parallel to molding pressure; $\perp$ = perpendicular.
[c] Varies widely with type of graphite.
[d] At 570 F.

for modulus of rupture is two to one.

Although boron nitride resembles graphite in many respects, it differs uniquely in electrical characteristics, having high resistivity and high dielectric strength even at elevated temperatures. This feature, combined with easy machinability, should lead to extensive use in high-temperature electronics.

Table 2 compares the properties of boron nitride and other materials.

Boron nitride is available as minus 325-mesh powder and as hot-pressed bars up to 9 in. dia by 8 in. long.

A cubic form of boron nitride (Borazon) similar to diamond in hardness and structure has been synthesized by the high-temperature–high-pressure process for making synthetic diamonds. Any uses it may find as a substitute for diamonds will depend on its greatly superior oxidation resistance.

Considerable interest is developing in boron nitride formed in place by chemical vapor-deposition reactions. The highly anisotropic structure has many of the properties of pyrolytic graphite with the notable exceptions of lower electrical conductivity and greater oxidation resistance.

Silicon nitride

Silicon nitride is most easily prepared by direct reaction of nitrogen at about 2400 F with finely divided elemental silicon ($\leq$ 150 mesh), either as loose powder or as a slip-cast or otherwise preformed part. Conversion of silicon particles to the nitride Si_3N_4 is accompanied by the growth of a felt of interlocking needles in the void space between particles. Despite an over-all porosity of 15 to 25%, silicon nitride bodies are effectively impervious in many applications because of the microscopic size of the pores.

Although silicon nitride is not machinable in its final form except by grinding, the partially converted body can be machined by conventional methods after which conversion can be completed without dimensional change.

Silicon nitride is indefinitely resistant to air oxidation up to 3000 F, but begins to sublime at about 3500 F. It is not attacked by chlorine at 1650 F or hydrogen sulfide at 1800 F nor by the common acids. However, it is slowly attacked by boiling 50 percent sodium hydroxide. Due to a low coefficient of thermal expansion, resistance to thermal shock is relatively good. Table 2 includes properties of silicon nitride for comparison.

An outstanding characteristic of silicon nitride is its resistance to molten nonferrous metals as indicated in Table 3. Slip-cast thermocouple wells and other parts are finding increasing use in the aluminum industry.

TABLE 3—RESISTANCE OF SILICON NITRIDE TO ATTACK
BY MOLTEN METALS

Metal	Temperature, F	Time, hrs	Remarks
Aluminum	1800	3000	No attack
Lead	752	144	No attack
Tin	572	144	No attack
Zinc	1022	500	No attack
Magnesium	1382	20	Slightly attacked
Copper	2120	7	Badly attacked

Silicon nitride is available as a powder and in the form of small crucibles and plates for experimental use. Parts are custom-made by slip casting. Tubes $3\frac{1}{2}$ in. in diameter by 25 in. can be obtained. Tiles 6 in. $\times$ 12 in. $\times$ $\frac{3}{4}$ in. have been produced.

Titanium and zirconium nitrides

Titanium and zirconium nitrides for use in refractory bodies are most conveniently prepared by treating the corresponding metal hydrides with ammonia at 1000 C. The ground powder ($< 4 \mu$), after pressing and final firing in NH_3 (2900 F for TiN; 3700 F for Zr), yields bodies which are resistant to attack by molten iron and molten cerium, but are attacked rapidly by molten beryllium. Sintered TiN can be heated to a bright red heat with only superficial oxidation, and then plunged into water without cracking. ZrN is less resistant to oxidation.

NITRIDING STEELS

Nitriding steels are alloy steels designed particularly for optimum results when they are subjected to the nitriding operation. The composition is such that the required microstructure for optimum nitriding is produced after heat treatment. Nitrided parts made from nitriding steels have extremely high surface hardnesses of about 92 to 95 Rockwell N scale, wear resistance, and resistance to certain types of corrosion.

The nitriding process is one in which steel is heated in a nitrogen atmosphere at temperatures usually between 925 F and 1050 F for periods up to 100 hr, but more usually for periods of 24 hr to 48 hr, depending on the depth of nitrided case desired. The nitrogen is generally provided by dissociated ammonia or sodium and/or potassium cyanide salt. The process forms a very

hard layer of nitrides on the surface of the steel, which is responsible for the unique properties produced by this treatment.

Composition

There are several basic compositions of nitriding steel available, all of which will produce a definite set of core properties, and the same basic case properties.

	\	%					
	C	Mn	Si	Cr	Ni	Mo	Al
135 Type G	0.35	0.55	0.30	1.15	—	0.20	1.00
Modified 135 [a]	0.40	0.60	0.30	1.60	—	0.35	1.15
Type N	0.24	0.60	0.30	1.15	3.50	0.25	1.25

[a] Standard AISI composition.

Modifications are generally offered to affect machining characteristics (such as sulfur or selenium) or hardening and physical properties (such as obtained with aluminum-nickel combinations).

The nitriding operation is done below the critical temperature of the steel and offers advantages when critical machining, tolerances, or distortion problems are involved.

Treatment and properties

Generally, nitriding steels are heat-treated prior to machining. This also produces a sorbitic structure which is ideal for the nitriding operation. A tempering temperature of 1100 to 1300 F (about 200 F above the nitriding temperature) is used following an oil quench from 1700 to 1750 F. The following properties of a 1 in. rod are typical following an 1100 F temper:

	Ten Str, 1000 psi	Yld Str, 1000 psi	Elong, in 2 in., %	Red of Area, %	BhN
135 Type G	155	137	15	52	310
Modified 135	181	165	15.5	54.3	368
Type N	145	130	21	58	302

The following case depths are typical when nitriding in a normal atmosphere for the time indicated:

10 hr	0.006 to 0.008 in.
30 hr	0.014 to 0.016 in.
60 hr	0.025 to 0.026 in.

A typical cycle in using nitriding steel might be as follows: rough machining; heat-treat to desired core properties; finish-machine; stress-relieve at 1100 F; nitride at 950 F for 10 hr; grind and/or lap the surface.

While the nitride layer on nitriding steel cannot be classed as stainless steel, it does exhibit remarkable resistance to certain types of corrosion. It is extremely resistant to alkali, atmosphere, crude oil, natural gas combustion products, tap water and still salt water. It is slightly attacked by aerated salt water or under conditions of alternate wetting and drying. It is not suited to mineral acid media such as hydrochloric or sulfuric acid.

The fabrication characteristics of nitriding steels are basically the same as those of other steels of similar alloy content. They can be drilled, broached, tapped, milled, sawed, or ground. Light feeds and depth of cuts are recommended. Welding is done with rod or wire of similar composition. Flash welding is permissible.

If very heavy cuts are involved, they usually are made prior to heat treatment. Normal machining is done on heat-treated material, and is followed by a stress-relieving treatment of not less than 100 F above the nitriding temperature, before finish machining or grinding. It is essential that in machining, sufficient removal be allowed to remove all decarburization from the surface prior to nitriding. The surface also must be clean and free of any surface contamination.

Since nitriding is a low-temperature treatment, little or no warpage is encountered. If it is necessary to straighten because of residual stress, the part should be heated to 1000 to 1100 F to prevent surface cracking.

Nitrided parts normally grow about 0.001 to 0.002 in. during nitriding. This may be removed by grinding or lapping. This also has the advantage of removing a brittle layer on the surface and exposes a slightly harder layer immediately beneath it. This operation, however, will reduce corrosion resistance to a large degree.

Nitriding steels are available in all standard steel forms. They can be purchased heat-treated or annealed to desired physical properties.

Most uses of nitriding steels are based on resistance to wear. An outstanding property is that these steels can be heated to as high as 1000 F for long periods without softening.

The slick, hard surface produced also makes it ideal to prevent seizing, galling, and spalling, and it is not readily attacked by combustion products.

Typical applications include: cylinder liners and barrels of aircraft engines, bushings, shafts, piston pins, spindles and thread guides, cams, rubber and paper-mill product rolls, special oil tool equipment, bearings, rollers, etc.

NITRILE RUBBER

The discovery and early investigation of nitrile rubber (ASTM designation, NBR), a copolymer of butadiene and acrylonitrile, occurred in the Central Research Laboratories of I. G. Farbenindustrie. The general term "Buna N" was the early German designation for such copolymers, which were sold under the trade name of "Perbunan."

First commercial production of nitrile rubber in the United States was undertaken in 1939. The outstanding oil and solvent resistance of nitrile rubber was of immediate interest in the expanding national defense program and was used in the manufacture of bullet-sealing tanks, fuel hose, seals and many other military applications. When the United States entered World War II, all nitrile rubber was placed on strict allocation for military uses only. During the war nitrile rubbers were among the most critical of all elastomeric materials. Major commercial development has occurred only since late 1945.

General characteristics

The most important property of nitrile rubber is its high degree of resistance to oils and fuels, both at normal and elevated temperatures. Oil and fuel resistance refers to the ability of a vulcanized nitrile rubber part to retain its original physical properties such as modulus, tensile strength, abrasion resistance and dimensions while in contact with these fluids. Other properties such as heat resistance, permeability to gases, compatibility with resins and resistance to a wide range of solvents and chemicals are important in certain applications.

Essentially the same techniques of emulsion polymerization employed in manufacture of general-purpose synthetic rubber may be used for production of nitrile polymers. Nitrile rubbers are supplied in various physical forms including sheet, crumb, powder and liquid. The sheet is the most widely used type, with the other varieties being offered for specialty applications.

Blends. An outstanding feature of nitrile rubber is its compatibility with many different types of resins permitting it to be easily blended with them. In combination with phenolic resins it provides adhesives with especially high strengths. Other resins used include resorcinol formaldehyde, urea formaldehyde, alkyd, epoxy and polyvinyl chloride (to produce Type 2 rigid PVC). Both slab and crumb type nitrile rubber are used in this type of application, the crumb type being directly soluble and of special interest to adhesive manufacturers who do not have rubber-mixing equipment. Nitrile rubber-phenolic resin solvent solutions are used in shoe sole-attaching adhesives, for structural bonding in aircraft, adhering automotive brake lining to brake shoes and many other industrial applications.

The powder type rubbers were also developed for blending with phenolic resins, primarily for the manufacture of improved impact phenolic molding powders.

The liquid nitrile polymer finds use as a tackifier and nonextractable plasticizer in molded rubber parts, cements, friction and calendered stocks.

Both the liquid and powder are of interest as curing type plasticizers in vinyl plastisols. Nitrile rubber-polyvinyl chloride blends of various types are used in many other fields including cable jacket, retractable cord, abrasion-resistant shoe soles, industrial face masks, boat bumpers and fuel lines.

Composition

Ratio of butadiene to acrylonitrile in the commercially available rubbers ranges from a low of about 20% to as high as 50% acrylonitrile. The various grades are usually referred to as high, medium-high, medium-low and low acrylonitrile content.

The high acrylonitrile polymers are used in applications requiring maximum resistance to aromatic fuels, oils and solvents. This would include oil-well parts, fuel-cell liners, fuel hose and other similar applications. The low acrylonitrile grade finds use in those areas requiring good flexibility at very low temperatures where oil resistance is of secondary importance. The medium types are most widely used and are satisfactory for all oil-resistant applications between these two extremes. Typical applications include conveyor belts, flexible couplings, soles, heels, floor mats, printing blankets, rubber rollers, sealing strips, aerosol bomb gaskets, milking inflations, seals, diaphragms, O-rings, packings,

hose, washing machine parts, valves and grinding wheels. These established uses give only a slight idea of products which are made of nitrile rubbers.

Physical properties of cured nitrile rubber parts are directly related to the ratio of butadiene and acrylonitrile in the polymer, as indicated below:

As acrylonitrile content increases:
1. Oil and solvent resistance improve.
2. Tensile strength increases.
3. Hardness increases.
4. Abrasion resistance improves.
5. Gas impermeability improves.
6. Heat resistance improves.

As acrylonitrile content decreases:
1. Low-temperature resistance improves.
2. Resilience increases.
3. Plasticizer compatibility increases.

Properties

Table 1 presents in more detail the properties of four nitrile polymers, covering the range of high to low acrylonitrile content, compounded and cured in a typical general-purpose molding compound formulation.

Analysis of the above data indicates the general trend in physical properties based on acrylonitrile content. The polymer having the highest acrylonitrile content produces the highest tensile strength and hardness; it also exhibits the best resistance to fuels and oils. As the percentage of acrylonitrile decreases, there is a corresponding decrease in resistance to fuels and oils; at the same time low-temperature flexibility characteristics are improved. Resiliency also increases. The lowest acrylonitrile polymer exhibits only moderate resistance to swelling in aromatic fluids but remains flexible at very low temperatures in the range of -70 to -80 F.

Thus, properly compounded nitrile polymers will provide high tensile strength, excellent resistance to abrasion, low compression set, very good aging under severe operating conditions and excellent resistance to a wide range of fuels, oils and solvents. They are practically unaffected by alkaline solutions, saturated salt solutions and aliphatic hydrocarbons, both saturated and unsaturated. They are affected little by fatty acids found in vegetable fats and oils or by aliphatic alcohols, glycols or glycerols.

Nitrile rubber is not recommended, generally,

for use in the presence of strong oxidizing agents, ketones, acetates and a few other chemicals.

TABLE 1

Nitrile Type	High	Med-High	Med-Low	Low
Approximate Acrylonitrile Content, %	40	32	27	20
Tensile Strength, psi	2800	2600	2250	2100
Elongation, %	500	500	425	450
Hardness, Durometer A	68	65	62	60
Compression Set (70 hr @ 250 F), %	41	40	45	49
Rebound, %	30	50	51	59
Low-Temperature Brittleness, F				
Pass	−15	−35	−55	−75
Fail	−20	−40	−60	−80
% Change in volume after 70 hr immersion:				
At 212 F				
ASTM Oil No. 1	−9	−8	−6	−4
ASTM Oil No. 3	−1	+6	+19	+45
At Room Temperature				
Diisobutylene	+1	+6	+13	+23
40% aromatic fuel	+26	+38	+63	+103
Air Aged 70 hr @ 250 F				
% Change in tensile strength	−1	−3	−2	−30
% Change in elongation	−40	−45	−41	−61
% Change in hardness	−1	+3	+3	+8

NYLON PLASTICS

Although nylon polymers are most familiar in fiber form, their combination of excellent chemical and mechanical properties, plus their ability to be molded and extruded into precise forms, have permitted their use in a wide variety of nontextile applications. These range from hammerheads, gears, and rifle stocks to miniature coil forms and delicately colored personal products.

Characteristics

Basically, nylons are linear polyamides, with recurring amide linkages an integral part of the molecular chain. They may be made by the amidation of diamines with dibacic acids, for example, hexamethylenediamine plus adipic acid (Type 66 nylon), or hexamethylenediamine and sebacic acid (Type 610 nylon). They can also be made by the polymerization of amino acids or their derivatives, for example, polycaprolactam (Type 6 nylon) and polymerized 11-aminoun-decanoic acid (Type 11 nylon).

All these nylons have many similar properties. All are strong, tough, and resistant to abrasion, fatigue, and to the effect of most chemicals.

Nylons rank among the strongest of plastics although their toughness is more difficult to measure. Izod impact values range from 0.9 to 5.0 ft lb/inch notch, but nylons rank much better in actual service than do many other plastics. Moreover, they are particularly resistant to repeated impact.

Abrasion resistance is excellent but depends on the type of wear involved. In Taber or ball mill tests, nylons surpass other plastics and even most metals. When a cutting action is involved, nylons will not equal the stronger metals but, in sliding wear, they normally outwear unlubricated steel.

Nylons also have excellent fatigue properties. When constant strains are repeatedly applied, as in the case of tubing connected to fabricating machinery, they will often outperform metals. Repeated stress levels of 2,500 to 3,000 psi can be withstood almost indefinitely.

Nylons resist electrolytic corrosion, hydrolysis, fungi, bacteria and most chemicals. Of the common organic solvents, only phenols and formic acid dissolve nylons. However, while strong mineral acids, oxidizing agents, and concentrated solutions of certain salts, such as zinc chloride and potassium thiocyanate will attack them, nylons are resistant to lubricants, fuels, phosphate esters, fluorocarbons and most other compounds of practical concern.

Nylons are considered to be biologically inactive and neither support nor inhibit the growth of bacteria or fungi. Several grades meet FDA requirements for applications where contact with food is involved.

The higher melting nylons have excellent properties at high temperatures. Unlike many polymers, they remain rigid until the melting point is reached (but with decreasing stiffness and increasing tendency to creep). On a short-term basis, and where loads are not excessive, practical service temperatures for 66 nylon approach 400 F. Thermally stabilized compositions, capable of resisting oxidative embrittlement, are available and are recommended for use where long-term exposure over 200 F is involved.

The weathering of unstabilized nylon must be rated as only "fair." While it performs satisfac-

TABLE 1—TYPICAL PROPERTIES OF REPRESENTATIVE NYLONS

Property	ASTM Method	Units	66 Nylon		610 Nylon		6 Nylon		11 Nylon
			Dry-As-Molded	Equilibrium with 50% RH	Dry-As-Molded	Equilibrium with 50% RH	Dry-As-Molded	Equilibrium with 50% RH	Dry-As-Molded
Ten Str	D-638	psi	11,800	11,200	8,500	7,100	11,800	10,000	8,500
Yld Stress	D-638	psi	11,800	8,500	8,500	7,100	11,800	6,400	—
Elongation	D-638	%	60	300	85	220	200	300	100
Flex Mod	D-790	psi	410,000	175,000	280,000	160,000	395,000	140,000	185,000
Impact Str, Izod	D-256	ft-lb/in.	0.9	2.0	0.6	1.6	1.2	4.0	3.5
Shear Str	D-732	psi	9,600	—	8,400	—	7,600	6,200	—
Def Under Load (2000 psi)	D-621	%	1.4	—	4.2	—	1.8	—	—
Hardness, Rockwell	D-785		R118 M79	R108 M59	R111	—	R119 M75	R97 M50	A50
Melting Point	D-789	F	482-500	—	405-430	—	420-435	—	367
Water Absorption	D-570	%	1.5	—	0.4	—	1.6	—	0.4
Sp Gr	D-792		1.14	—	1.09	—	1.13	—	1.04

torily in many outdoor applications, it will not last indefinitely. When long-term retention of properties out-of-doors is needed, weather-stabilized grades should be used, particularly those filled with carbon black.

Nylons differ among themselves primarily in melting point, stiffness, and in their tendency to absorb moisture. They are usually processed in an essentially dry state but subsequently absorb moisture from the atmosphere. This, in effect, plasticizes the resin. Important properties, both dry and as they exist at equilibrium in an atmosphere of 50% RH, are listed in Table 1.

Generally, 66 nylons are selected where stiffness, low creep, high yield strength, and marginally better chemical resistance are needed. Extracted polycaprolactam offers greater flexibility and, in some cases, higher impact strength. Its lower melting point gives it a wider range of processing temperatures.

The 610 and 11 nylons are higher cost resins of substantially reduced moisture absorption, and hence, better dimensional stability.

In addition to these homopolymers, an increasing variety of copolymers are being offered to achieve special effects, particularly flexibility and transparency. Also, many nylons can be plasticized to obtain the flexibility needed for certain tubing and wire-jacketing constructions. One effective plasticizer for polycaprolactam is the equilibrium monomer left after the polymerization. Unextracted polycaprolactam possesses the right combination of toughness and flexibility for spin fishing lines, bowling-pin bases, etc.

Processing

Nylons are generally fabricated by injection molding or by extrusion. Precise, intricate shapes of a variety of colors can be molded with little or no finishing required. These can often replace an assembly of several metal parts. Thus, even where nylons cost more in a per-volume basis than the common die-casting metals, economies in finishing and assembly often result in lower ultimate costs.

Tubing and rod stock manufacture, plus the coating of wire and cable, are the major forms of nylon extrusion. Film and relatively complex cross sections are also made, but in less volume. In general, tubing, film, and other unsupported shapes require higher melt viscosity than is desirable for injection molding. Most manufacturers of nylon supply these high viscosity grades.

Applications

Nylons are usually specified because of their combination of properties. Gears, bearings, cams, clutch facings, and similar mechanical parts require their strength, stiffness, low coefficient of friction, and resistance to fatigue and abrasion. In cases where oiling or greasing is apt to be neglected, as in home appliances, or is undesirable from a contamination standpoint, as in textile and food handling machinery, nylon parts usually perform satisfactorily without any lubrication whatsoever.

Designers often utilize the mechanical properties of nylons, plus one or more characteristics of particular value. For example, the nylon housing for an electric drill must be tough, stiff, dimensionally stable, and resistant to commonly encountered lubricants and solvents. However, its electrical nonconductivity and safety are the critical advantages.

Similarly, one manufacturer makes an entire rifle stock, plus many of the rifle's moving parts, out of nylon. It is far lighter and tougher than wood and provides moving surfaces which do not need lubrication. Nylon's ability to be molded into precise sections thus permits custom-quality guns to be mass-produced.

Utilizing different properties, marine electrical stuffing tubes of nylon capitalize on the resin's durability, lightness, resistance to corrosion, and cost advantage over machined brass. Washing machine mixing valves and valves for the dispensing of hot beverages require nylon's mechanical properties and excellent resistance to the effects of hot water.

The coating of wire and cable construction is nylon's most important extrusion application. While it is most commonly used as a jacket over a primary insulator such as polyvinyl chloride or polyethylene to impart resistance to lubricants, abrasion, and to the effects of high temperature, its electrical properties are adequate for low voltage uses.

O

ORGANIC COATINGS

Organic coatings are chiefly additive type finishes that find use on almost all types of materials. They can be monolithic consisting simply of one layer, or coat, or they can be composed of two or more layers. Total thickness of coating systems varies widely. Some run less than 1 mil thick. Others go as high as 10 and 15 mils thick. Generally, by definition, coatings that are more than 10 mils thick are referred to as linings, films or mastics.

To function as a protective barrier against corrosion and oxidation, organic coatings depend principally on their chemical inertness and impermeability. In addition, however, some coatings provide protection with the use of inhibiting pigments that have a passivating action, particularly on metal surfaces. Also, some coatings contain metallic pigments that give electrochemical protection to metals.

Coating application and drying

Organic coatings are commonly applied by the following methods: brush, spray, dip, roller, flow coat, knife, tumbling, silk screen and electrostatic means. Of all these, application by brushing is the slowest. All the others are production methods.

Organic coatings dry, or cure, by one or more of the following mechanisms: 1) evaporation or loss of solvent, 2) oxidation, and 3) polymerization. After application of the coating the volatile ingredients, which are almost always present in at least a small amount, evaporate. Some finishes, such as lacquers, dry completely by evaporation of solvents. After evaporation, coatings that do dry solely by evaporation are still in a semi-fluid state, and depend upon oxidation or polymerization, or a combination of both to convert to their final form.

Drying by oxidation, which is usually done at room temperature, is the slowest of the three methods. Polymerization, which involves polymer chain forming mechanism, can be done at normal or at elevated temperatures. Polymerization is speeded up by use of heat. In recent years radiation curing involving the use of an electron beam to polymerize the coating in a few seconds has found increasing use.

Coating types and systems

Coating composition. An organic coating is made up of two principal components: a vehicle and a pigment. The vehicle is always there. It contains the film-forming ingredients that enable the coating to convert from a mobile liquid to a solid film. It also acts as a carrier and suspending agent for the pigment. Pigments, which may or may not be present, are the coloring agents, and, in addition, contribute a number of other important properties.

Organic coatings are commonly divided into about a half dozen broad categories based on the types and combinations of vehicle and pigment used in their formulation. They are paints, enamels, varnishes, lacquers, dispersion coatings, emulsion coatings and latex coatings. However, with the complexity in modern formulations, distinctions between these various types are often difficult to make.

As mentioned earlier, organic finish systems are frequently composed of more than one layer or coat. These various layers are commonly classified as primers, intermediate coats, and finish coats.

Primers. They are the first coatings placed on the surface, (except for fillers, in some cases). Where chemical pretreatments are used, primer coats may often be unnecessary. Primers for industrial or production finishing are of two types: air-dry types and baking types. The air-dry types have drying oil vehicle bases and are usually referred to as paints. They may or may not be modified with resins. They are not used as extensively as the baking type primers, which have resin or varnish vehicle bases and dry chiefly by polymerization. Some primers, known as flash primers, are applied by spraying, and dry by solvent evaporation within 10 minutes. In practically all primers, the pigments impart most of the anti-corrosion properties to the primer and, along with the vehicle, determine its compatibility and adherence with the base metal.

Intermediate coats. These are fillers, surfacers, and sealers. They can be applied either before or after the primer, but more often after the primer and sometimes after the surfacer coat. Their function is to fill in large irregularities in the surface or local imperfections. They are usually putty-like substances, and a variety of materials is used. Their chief characteristics are: (1) must harden with a minimum of shrinkage; (2) must have good adhesion; (3) must have good sanding properties; and (4) must work smoothly and easily.

Surfacers are often similar to primers; they usually have the same composition as the priming coat, except that more pigment is present. Surfacers are applied over the priming coat to cover all minor irregularities in the surface. Sealers as a rule are used either over the fillers or surfacers. The chief function of sealers is to fill up the pores of the undercoat to avoid "striking in" of the finish coats. This filling-in of the porous surfacer or sealer also tends to strengthen the entire coating system. The sealers when used over surfacers are usually formulated with the same type pigment and vehicle as used in the final coat.

Finish coats. Finish or top coats are usually the decorative and/or functional part of a paint system. However, they often also have a protective function. The primer coats may require protection against the service conditions, because although the pigments used in primers are satisfactory for corrosion protection of the metal, they are frequently not satisfactory as top coats. Their color retention upon weathering, or their physical durability may be poor. There are also one-coat applications where the finish coats are applied directly to the base material surface, and, therefore, provide the sole protective medium.

Vehicles

Vehicles are composed of film-forming materials and various other ingredients, including thinners (volatile solvents) which control viscosity, flow and film thickness, and driers which facilitate application and improve drying qualities. We will be concerned here chiefly with the film-forming part of the vehicle, because it is that part of the vehicle that to a large extent determines the quality and character of an organic finish. It determines the possible ways in which the finish can be applied and how the

"wet" finish will dry to a hard film; it provides for adhesion to the metal surface; and it usually influences the finish's durability.

Vehicles can be divided into three main types: (1) oil, (2) resin, and (3) varnish. The simplest and among the oldest vehicles are the straight drying oil types. Resins, as a class, can serve as vehicles in their own right, or can be used with drying oils to make varnish type vehicles. Varnish vehicles are composed of resins and either drying or non-drying oils, together with required amounts of thinners and driers. They are often used alone as a full-fledged organic finish.

Drying oils. Vehicles consisting of oil only are used to a limited extent in industrial finishes. Linseed oil is probably the most widely used oil. There are a number of different kinds that differ in rate of drying, and in such properties as water resistance, color and hardness.

Tung oil or China wood oil, when properly treated, excels all other drying oils in speed of drying, hardening, and water resistance. Oiticica oil is similar to tung oil in many of its properties. Dehydrated castor oil dries better than linseed oil, but slower than tung oil. Some of its advantages are good color and color retention, and flexibility. The oils from some fish are also used as drying oils. If processed properly they dry reasonably well and have little odor. They are often used in combination with other oils. Perilla oil is quite similar in properties to fast drying linseed oil. Its use is largely dependent upon its price and availability. Soybean oil is the slowest drying in the drying oil classes, and is usually used in combination with some faster drying oil such as linseed oil.

Resins. Although, both natural and synthetic resins can serve as organic coating vehicles, today the natural types, such as rosin, have been largely replaced by plastic resins. Nearly all the plastic resins—both thermosets and thermoplastics as well as many elastomers can be used as film formers, and frequently two or more kinds are combined to give the set of properties desired. Typical thermoplastics used in vehicles are acrylics, acetates, butyrates and vinyls. Commonly used thermosets for vehicles include phenolics, alkyds, melamines, ureas and expoxies. The properties of these and other plastics in the form of coatings are similar to those of the bulk form as covered in other articles in this book.

Pigments

Pigments are the second of the two principal components that make up most organic finishes. They contribute a number of important characteristics to a coating. They, first of all, serve a decorative function. The choice of color and shades of color by use of one or combinations of pigments is practically unlimited. Closely associated with color is the hiding power function or their ability to obscure the surface of the material being finished. In many primers the principal function of pigments is to prevent corrosion of the base metal. In other cases they may be added to counteract the destructive action of ultra-violet light rays. Pigments also help give body and good flow characteristics to the finish. And, finally, some pigments may give to organic coatings what is termed package stability—that is, they keep the coating material in usable condition in the container until ready for use.

Pigments can be conveniently divided into three classes as follows: (1) white hiding pigments, (2) colored pigments, and (3) extender or inert pigments. White pigments are used not only in white paints and enamels, but also in making white bases for the tinted and light shades. Colored pigments furnish the finish with both opacity and color. They may be used by themselves to form solid colors, or in combination with whites to produce tints, and often provide rust inhibitive properties. For example, red lead, certain lead chromates, zinc chromates, and blue lead are used in iron and steel primers as rust inhibiters. There are two general classes of colored pigments—earth colors, which are very stable and are not readily affected by acids and alkalies, heat, light, and moisture, and chemical colors, which are produced under controlled conditions by chemical reaction. Under this class the metallic pigments can also be included. Aluminum powder is perhaps the best known.

The chief functions of extender pigments are to help control consistency, gloss, smoothness and filling qualities, and leveling and check resistance. Thus, particle size and shape, oil absorption and flatting power are important selection considerations. Extender pigments are for the most part chemically inactive. They usually have little or no hiding power.

Enamels

By definition, enamels are an intimate dispersion of pigments in a varnish or a resin vehicle, or in a combination of both. Enamels may dry by oxidation at room temperatures and/or by polymerization at room or elevated temperatures. They vary widely in composition, in color and appearance, and in properties, and are available in all colors and shades. Although they generally give a high-gloss finish, there are some that give a semi-gloss or eggshell finish and still others that give a flat finish. Enamels as a class are hard and tough and offer good mar and abrasion resistance. They can be formulated to resist attack of most commonly encountered chemical agents and corrosive atmospheres.

Because of their wide range of useful properties, enamels are probably the most widely used organic coating in industry. One of their largest fields of use is for coating household appliances—washing machines, stoves, kitchen cabinets and the like. A large portion of refrigerators, for example, are finished with synthetic baking enamels. These appliance enamels are usually white, and therefore must have a high degree of color and gloss retention when subjected to light and heat. Other products finished with enamels include automotive products, railway equipment, office equipment, toys and sport supplies, industrial equipment, and novelties.

Lacquers

The word lacquer comes from lac resin, which is the base of common shellac. Lac resin dissolved in alcohol was one of the first lacquers and has been in use for many centuries. Nowadays, shellac is called spirit lacquer. It is only one of several different kinds of lacquers; these, except for spirit lacquer, are named after the chief film forming ingredient. The most common ones are cellulose acetate, cellulose acetate butyrate, ethyl cellulose, vinyl, and nitrocellulose.

A distinguishing characteristic of lacquers is that they dry by evaporation of the solvents or thinners in which the vehicle is dissolved. This is in contrast to oils, varnishes, or resin base finishes, which are converted to a hard film chiefly through oxidation and/or polymerization.

Because many modern lacquers have high resin content, the gap between lacquer and synthetic type varnishes diminishes until finally you have what might be called modified synthetic air drying varnishes. They may dry chiefly by oxidation and/or polymerization.

Lacquers normally dry hard and dust-free in a very few minutes at room temperature. In production line work, forced drying is often used. It is possible, therefore, to do a multi-coat job without having to lose time between coats. Because of the speed of drying and the fact that they are permanently soluble in the solvents used for application, lacquers usually are not applied by brush. Spray application or dipping are the usual procedures.

Lacquers can be either clear and transparent or pigmented, and their color range is practically unlimited. Lacquers in themselves have good color retention, but sometimes the added pigments, modifying resins and plasticizers may adversely affect this property. They are hard and mar-resistant. Inherently, they lack good adhesion to metal, but modern lacquer formulations have greatly improved their adhesion properties. Lacquers can be made to be resistant to a large variety of chemicals, including water and moisture, alcohol, gasoline, vegetable, animal and mineral oils, mild acids and alkalies. Because of the volatile solvents, lacquers are inflammable in storage and during application, and this sometimes limits their application.

Because of their fast drying speeds, lacquers find wide application in the protection and decoration of products which can be dipped, sprayed, roller coated, or flow coated. They are especially advantageous for coating metal hardware and fixtures, toys and other articles which, because of volume production, must dry hard enough to handle and pack in a short period of time. Lacquers are widely used in automobile finishing and especially for refinishing autos and commercial vehicles where fast drying without baking equipment is a requirement. Lacquers also compete with enamels for coating metal stampings and castings, including die castings.

Varnishes

Varnishes consist of thermosetting resins and either drying or non-drying oils. They are clear and unpigmented and can be used alone as a coating. However, their major use in industrial finishing is as a vehicle to which pigments are added, thus forming other types of organic coatings.

The drying mechanism of varnishes all follow the same general pattern. First, any volatile solvents that are present evaporate; then, drying by oxidation and/or polymerization takes place, depending on the nature of the resin and oil. At high temperatures, of course, there is more tendency to polymerize. So varnishes can be formulated for either air or bake drying. Varnishes may be applied by brushing or by any of the production methods.

It is evident that with the large variety of raw materials to choose from and the unlimited number of combinations possible that varnishes have an extensive range of properties and characteristics. They range from almost clear white to a deep gold; they are transparent, lacking any appreciable amount of opacity. Japan, a hard baked black-looking varnish, is an exception. It is opaque, due to carbon and carbonaceous material being present.

There are some distinctions in properties between oil-modified alkyd varnishes and the other types. In general, oil-modified alkyds have better gloss and color retention and better resistance to weathering. They form a harder, tougher, more durable film and dry faster. On the other hand, they have less alkali resistance than the other varnishes. In such things as adhesion and rust inhibitiveness there is no distinctive difference.

The major use of varnishes, as coatings in their own right, is for food containers, closures such as bottle caps, and bandings of various kinds. Another large application is as a clear finish coat over lithographic coatings.

Paints

The word paints is sometimes used broadly to refer to all types of organic coatings. However, by definition a paint is a dispersion of a pigment or pigments in drying oil vehicle. They find little use these days as industrial finishes. Their principal use is for primers. Paints dry by oxidation at room temperature. Compared to enamels and lacquers their drying rate is slow; they are relatively soft and tend to chalk with age.

Other organic coatings

Dispersion and emulsion coatings. In recent years these coatings have become known as water-base paints or coatings because many of them consist essentially of finely divided ingredients, including plastic resins, fillers and pigments, suspended in water. An organic media may also be involved. There are three types of water base coatings. Emulsions, or latexes, are aqueous dispersions of high-molecular weight resins. Strictly speaking, latex coatings are dispersions of resins in water, whereas emulsion coatings are suspensions of an oil phase in water.

Emulsion and latex coatings are clear to milky in appearance, have low gloss, excellent resistance to weathering and good impact resistance. Chemical and stain resistance varies with composition. Dispersion coatings consist of ultrafine fine insoluble resin particles present as a colloidal dispersion in an aqueous medium. They are clear or nearly clear. Weathering properties, toughness and gloss are roughly equal to those of conventional solvent paints.

Water soluble types, which contain low-molecular-weight resins are clear finishes and they can be formulated to have high gloss, fair to good chemical and weathering resistance and high toughness. Of the three types, they handle and flow most like conventional solvent coatings.

Plastic powder coatings. Several different methods have been developed to apply plastic powder coatings. In the most popular process—fluidized bed—parts are preheated and then immersed in a tank of finely divided plastic powders, which are held in a suspended state by a rising current of air. When the powder particles contact the heated part, they fuse and adhere to the surface, forming a continuous, uniform coating.

Another process, electrostatic spraying, works on the principle that oppositely charged materials attract each other. Powder is fed through a gun, where an electrostatic charge is applied opposite that applied to the part to be coated. When the charged particles leave the gun, they are attracted to the part where they cling until fused together as a plastic coating. Other powder application methods include flock coating, flow coating, flame and plasma spraying, and a cloud chamber technique.

Although many different plastic powders can be applied by the above techniques, vinyl, epoxy and nylon are most often used. Vinyl and epoxy provide good corrosion and weather resistance as well as good electrical insulation. Nylon is used chiefly for its outstanding wear and abrasion resistance. Other plastics fre-

quently used in powder coating include chlorinated polyether, polycarbonate, acetal, cellulosics, acrylic and fluorocarbons.

Hot melt coatings. These consist of thermoplastic materials that solidify on the metal surface from the molten state. The plastic is applied either in solid form and then melted and flowed over the surface, or is applied molten by spraying or flow coating. Since no solvent is involved, thick single coats are possible. Bituminous coatings are also commonly applied by the hot melt process.

Lining and sheeting. Sheet, film and tapes of various plastics and elastomers cemented to material shapes and parts are used to provide corrosion and abrasion resistance. Thicknesses usually range from 1/8 to 1/2 in. Most widely used materials are polyvinyl chloride, polyethylene butadiene-styrene rubber and neoprene.

Specialty finishes. An almost infinite number of specialty or novelty finishes are available. Most of them are really lacquers or enamels to which special ingredients have been added or which are processed in some unique way to give the effects desired.

One of the most common types are those giving a roughened or wrinkle appearance, which is obtained by use of high percentages of driers causing wrinkling when the finish is baked. Another group of specialty finishes give a crystalline effect. They are enamels in which impurities are purposely introduced during the baking process by retaining the products of combustion in the oven while the coating dries. The wrinkle and crystalline finishes are widely used on instrument panels, office equipment and a variety of other industrial and consumer products.

Other unusual finishes are obtained by adding special ingredients to lacquers to give them a stringy or "veiled" appearance when applied by spraying. The application of the silk-screen process to organic finishing of metals has also resulted in unique finishes with multi-colored effects.

OLEFIN COPOLYMERS

The principal olefin copolymers are the polyallomers, ionomers and ethylene copolymers. The polyallomers, which are highly crystalline, can be formulated to provide high stiffness and medium impact strength; moderately high stiffness and high impact strength; or extra-high impact strength. Polyallomers, with their unusually high resistance to flexing fatigue, have "hinge" properties better than those of polypropylenes. They have the characteristic milky color of polyolefins, are softer than polypropylene, but have greater abrasion resistance. Commonly injection-molded, extruded and thermoformed, polyallomers are used for such items as typewriter cases, snap clasps, threaded container closures, embossed luggage shells and food containers.

Ionomers are nonrigid plastics characterized by low density, transparency and toughness. Unlike polyethylenes, density and properties are not crystalline-dependent. Their flexibility, resilience and high molecular weight combine to provide high abrasion resistance. They have outstanding low temperature flexural properties but upper temperature use is limited to 160 F. Resistance to attack from organic solvents and stress cracking chemicals is high. Ionomers have high melt strength for thermoforming and extrusion-coating, and a broad temperature range for blow-molding and injection molding. Representative ionomer parts include injection molded containers, housewares, tool handles and closures; extruded film, sheet, electrical insulation, and tubing; blow molded containers and packaging.

There are four commercial ethylene copolymers, of which ethylene vinyl acetate (EVA) and ethylene ethyl acrylate (EEA) are the most common.

Ethylene-vinyl acetate (EVA) copolymers approach elastomers in flexibility and softness, although they are processed like other thermoplastics. Many of their properties are density-dependent, but in a different way from that of polyethylenes. Softening temperature and modulus of elasticity decrease as density increases, which is contrary to the behavior of polyethylene. Likewise, the transparency of EVA increases with density to a maximum that is higher than that of polyethylenes, which become opaque when density increases above around 0.935 gm/cc. Although EVA's electrical properties are not as good as those of low-density polyethylene, they are competitive with vinyl and elastomers normally used for electrical products. The major limitation of EVA plastics is their relatively low resistance to heat and solvents, the Vicat softening point being 147 F. EVA copolymers can be injection, blow,

compression, transfer and rotationally molded; they can also be extruded. Molded parts include appliance bumpers, and a variety of seals, gaskets and bushings. Extruded tubing is used in beverage vending machines and for hoses for air-operated tools and paint spray equipment.

Ethylene-ethyl acrylate (EEA) is similar to EVA in its density-property relationships. It is also generally similar to EVA in high temperature resistance, and like EVA it is not resistant to aliphatic and aromatic hydrocarbons as well as chlorinated versions thereof. However, EEA is superior to EVA in environmental stress cracking and resistance to ultraviolet radiation. Similar to EVA, most of EEA'S applications are related to the plastic's outstanding flexibility and toughness. Typical uses are household products such as trash cans, dish washer trays, flexible hose and water pipe, and film packaging.

Two other ethylene copolymers are ethylene hexene (EH) and ethylene butene (EB). Compared to the other two, these copolymers have greater high temperature resistance, their useful service range being between 150 and 190 F. They are also stronger and stiffer, and therefore less flexible than EVA and EEA. In general, EH and EB are more resistant to chemicals and solvents than the other two, but their resistance to environmental stress cracking is not as good.

TYPICAL PROPERTIES OF SOME
OLEFIN COPOLYMERS

	Polyallomer	Ionomer	EVA
Specific Gravity	0.898–0.905	0.94	0.94
Tens. Str., 1000 psi	3–4.5	3–5	0.5–1.0
Elong., %	350	450	650
Hardness, Phase D	–	60	35
Imp. Str., ft-lb/ in. notch	1.5	9–14	–
Softening Pt., Vicat, F.	250–275	162	147
Dielect. Str., V/mil	500–650	1000	525
Vol. Res., ohm-cm	10^{15}	10^{15}	–

ORGANOSOL COATINGS

Organosol coatings are coatings in which the resin (usually polyvinyl chloride) is suspended rather than dissolved in an organic fluid. The dispersion technique permits the use of high molecular weight, relatively insoluble resins without the use of expensive solvents. In organosols, the fluid, or dispersant, consists of plasticizers together with a blend of inexpensive volatile diluents selected to give the desired fluidity, speed of fusion, and physical properties. The dispersant provides little or no solvating action on the resin particles until a critical temperature is reached at which point the resin is dissolved in the dispersant to form a single-phase solid solution. Since a portion of the liquid is made up of volatile diluents, the fusion process results in a proportional shrinkage. (Nonvolatile dispersions using only plasticizer dispersants are termed plastisols and are covered under Plastisol Coatings.)

Organosols are available with a wide range of flow characteristics and consequently may be formulated for application by any of the conventional techniques. Because they contain substantial quantities of volatile diluents their thickness is limited to 10 to 12 mils per coat. Single-coat applications of greater thickness blister during bake due to trapped solvents.

Baking is generally accomplished in two stages. The volatiles are removed in the first stage at temperatures of from 200 to 250 F, but fusion does not occur until a temperature of 300 to 375 F is reached. The fusion stage accomplishes the union of the discrete vinyl particles into a single-phase solid. In addition to polyvinyl chloride, the organosol technique can be used with acrylonitrile-vinyl and polychlorotrifluoroethylene resins. A balance must be achieved in the baking operation between the removal of the volatiles and the solvation of the resins. Rapid heating results in solvent blistering, whereas the reverse causes a mud-cracking effect.

Application methods

Organosol coatings can be applied by several methods including:

Spread coating. The bulk of the organosols are applied by spread-coating methods to fabrics and paper. Several plants are using spread coaters for the application of organosols and plastisols to strip steel to provide materials competitive with the light metals and plastics. Two basic processes are available: the knife coater and the roll coater. Knife coating is simple and fast but the product lacks the uniformity afforded by roller coating. Fusion is accomplished in a tunnel oven and embossing rolls may be employed at the oven exit to impart a texture or pattern to the hot gelled coating.

Strand coating. Wires or filaments may be coated with organosols by first passing the strand through a dip tank, then through a wiping die to set the thickness. Fusion is accomplished in a drying tower. As many as 9 to 10 passes may be required to build a thickness of 20 mils.

Dip coating. One of the major problems encountered in dip coating with organosols is the tendency of dried organosol to fall back into the dip tank and cause coating rejects. For this reason much of the dip coating is done with modified plastisols rather than organosols. A dipping formulation must have low viscosity with high yield value to prevent sags and drains. The rate at which the article is withdrawn from the dip tank is a determining factor in the thickness and quality of deposit. Withdrawal rates generally range between 4 to 18 in. per min. Special techniques such as inversion of the dipped article just prior to fusion may be employed to alleviate the drip problem when it is particularly troublesome.

Spray coating. Organosols are readily handled in either suction or pressure spray equipment. For production-line spraying, pressure systems afford rapid delivery and are generally preferred. There is an increasing tendency to use electronic spray processes for handling organosols. A number of electrostatic processes are available including several "hand guns."

Properties

Organosols have the characteristic vinyl properties of toughness and moisture resistance. However, although the coatings possess good electrical resistance and are frequently used as secondary insulation to reduce shock hazards, they seldom meet the needs of primary wire insulation. Adhesive primers are required to bond the materials to metals and other dense substrates. Prolonged exposure to temperatures greater than 200 F causes thermal degradation which is generally evidenced by a gradual darkening. Typical properties are shown in Table 1.

TABLE 1—PROPERTIES OF ORGANOSOL COATINGS	
Physical Properties	
Density of Liquid, lb/gal	8–10
Therm Cond, Btu/hr/ sq ft/°F/ft	0.07–0.1
Max Rec Svc Temp, F	150–200
Mois Vapor Trans g/100 sq in./mil (24 hr, 95 F, 100% RH)	10
Water Abs, % in 24 hr	0.1–0.15
Odor	None
Weatherability	Good
Flammability	Nonflammable
Colors	Complete range
Dried Film Density, lb/sq ft/mil	0.006–0.008
Mechanical Properties	
Tensile Strength, psi	200–4000
% Elongation	100–400
Shore A Hardness	50–100
Cold Temp Flex, F	−65
Electrical Properties	
Spec Resis, ohm-cm	10^{14}–10^{15}
Dielec Str, v/mil	500
Chemical Properties [a]	
Dilute Min Acids	Resistant
Lower Fatty Acids	Nonresis
Higher Fatty Acids	Lim Resis
Alkalies	Resistant
Aliphatic Hydrocarbons	Resistant
Aromatic Hydrocarbons	Nonresis
Alcohol	Gen Resis
Esters, Ethers, Ketones, Chlorinated Hydrocarbon	Nonresis
Application Properties	
Surface Prep	Special primer
Diluent	Hydrocarbon type
Baking Req	200–250 F prebake 300–375 F fusion-bake
Max Thick per coat, mils	10–15
Coverage, sq ft/gal/mil	100–250
Approximate Cost (incl mat and labor), ¢/sq ft/mil	5 to 7
Recoatability	Fused film may be recoated or touched up
Stripping	Fused film removed by burning or with chemical strippers

[a] Other chemical properties are given under Plastisol Coatings.

TABLE 2—APPLICATIONS OF ORGANOSOL COATINGS	
Application	Related Properties
Automotive Interiors: station-wagon flooring, roof liners, dashboards, cowling, kick plates	Ease of application; uniformity of color and texture; resistance to gasoline, grease and polishes; resistance to impact damage.
Commercial Vehicles: seat backs, trim, interior paneling, luggage, racks, sill plates	Durability: resistance to abrasion and scuffing; cleanability: resistance to moisture and detergents.
Appliance Finishes: television and radio cabinets, slide projectors, refrigerator panels	Esthetic qualities: novelty of appearance; durability: resistance to abrasion and scratching; resistance to moisture and staining.
Business Machines: typewriters, calculators, electronic computers, laboratory instruments	Durability: resistance to abrasion, scuffing and chipping; resistance to chemicals and perspiration; sound-deadening qualities.
Architectural Applications: paneling, partitions, shower stalls, elevator doors, bathroom wall sections	Cleanability: resistance to moisture and detergents; esthetic qualities; durability: resistance to abrasion and scuffing; sound-deadening qualities.
Office Furniture: desks, file cabinets, showcases, counters, waste baskets, chair finishes	Durability: resistance to abrasion, impact and scuffing; resistance to moisture and staining.
Luggage	Durability: toughness, resistance to abrasion and scuffing; esthetic qualities.
Paper and Fabrics: floor and wall covering, place mats, bottle-cap liners, containers for food packaging, bandage dressings, upholstery fabrics, safety clothing, glove coatings	Ease of application: economy and simplicity of application equipment; resistance to abrasion and tearing; resistance to moisture and staining.
Glass Coatings: perfume bottles, bleach and chemical reagent bottles, photo flash bulbs	Resiliency, feel, cohesion strength; resistance to alcohol, moisture and chemicals; ease of application.

Uses

Vinyl organosols have found their widest usage in the coating of paper and fabric stock where their ease of handling has permitted the use of simplified low-cost application equipment. Where fabric coating strike-through is to be avoided, as in open weave, thixotropic plastisols are employed rather than organosols. Closer weaves require some penetration.

Textured finishes for metals are a growing use and offer competition to the vinyl laminates and other decorative finishes.

Typical applications and related properties are shown in Table 2.

OXIDE CERAMICS

Oxide ceramics can be divided into two groups—single oxides that contain one metallic element, and mixed or complex oxides that contain two or more metallic elements. From Tables 1 and 2, it is evident that they differ widely among themselves. As a class they are low in cost compared to other technical ceramics, except for thoria and beryllia. Each of them can be produced in a variety of compositions, porosity, and microstructure, to meet specific property requirements. Thus, the property data given in the tables are only typical values.

Oxide ceramic parts are produced by slip casting or pressing or extrusion and then fired at about 3270 F. They are more difficult to fabricate than other types of ceramics, because of the usual requirement to obtain a high-density body with minimum distortion and dimensional error, except in the case of porous bodies for use as thermal insulation. Powder pressing produces bodies with the lowest porosity and highest strength, because of the high pressures and the small amount of binder required.

Single oxides

Aluminum Oxide (Alumina)—Alumina is the most widely used oxide, chiefly because it is plentiful, relatively low in cost and equal to or better than most oxides in mechanical properties. Density can be varied over a wide range, as can purity—down to about 90% alumina—to meet specific application requirements. Alumina ceramics are the hardest, strongest, and stiffest of the oxides. They are also outstanding in electrical resistivity, dielectric strength, are resistant to a wide variety of chemicals, and are unaffected by air, water vapor and sulfurous atmospheres. However, with a melting point of only 3700 F, they are relatively low in refractoriness, and at 2500 F retain only about 10% of room temperature strength. Besides wide use as electrical insulators, and chemical and aerospace applications, alumina's high hardness and close dimensional tolerance capability make this ceramic suitable for such abrasion resistant parts as textile guides, pump plungers, chute linings, discharge orifices, dies and bearings.

Beryllium Oxide (Beryllia)—Beryllia is noted for its high thermal conductivity, which is about ten times that of a dense alumina (at 930 F), three times that of steel, and second only to that of the high conductivity metals (silver, gold and copper). It also has high strength and good dielectric properties. However, beryllia is costly and is difficult to work with. Above 3000 F it reacts with water to form a volatile hydroxide. Also, because beryllia dust and particles are toxic, special handling precautions are required. The combination of strength, rigidity, dimensional stability make beryllia suitable for use in gyroscopes; and because of high thermal conductivity, it is widely used for transistors, resistors and substrate cooling in electronic equipment.

Magnesium Oxide (Magnesia)—Magnesia is not as widely useful as alumina and beryllia. It is not as strong, and because of high thermal expansion, it is susceptible to thermal shock. Although it has better high temperature oxidation resistance than alumina, it is less stable in contact with most metals at temperatures above 3100 F in reducing atmospheres or in a vacuum.

Zirconium Oxide (Zirconia)—There are several types of zirconia: a pure (monoclinic) oxide and a stabilized (cubic) form, and a number of variations such as yttria and magnesia stabi-

TABLE 1—PROPERTIES OF SOME SINGLE OXIDE CERAMICS

	Alumina	Beryllia	Magnesia	Zirconia	Thoria
Melting Point, F	3700	4620	5070	4710	6000
Mod. Elast. in Tens., 10^6 psi	65	35	40	30	20
Tens. Str., 1000 psi	38	14	20	21	7.5
Compr. Str., 1000 psi	320	300	120	300	200
Hard., Micro (Knoop)	3000	1300	700	1100	700
Max. Svc. Temp. (Oxid. Atm.), F	3540	4350	4350	4530	4890

lized zirconia and nuclear grades. Stabilized zirconia has a high melting point, about 5000 F, low thermal conductivity, and is generally unaffected by oxidizing and reducing atmospheres and most chemicals. Yttria and magnesia stabilized zirconias are widely used for equipment and vessels in contact with liquid metals. Monoclinic nuclear zirconia is used for nuclear fuel elements, reactor hardware and related applications where high purity (99.7%) is needed. Zirconia has the distinction of being an electrical insulator at low temperatures, gradually becoming a conductor as temperatures increase.

Thorium Oxide (Thoria)—Thoria, the most chemically stable oxide ceramic, is only attacked by some earth alkali metals under some conditions. It has the highest melting point (6000 F) of the oxide ceramics. Like beryllia, it is costly. Also, it has high thermal expansion and poor thermal shock resistance.

Mixed oxides

Except for zircon the principal mixed oxides are composed of various combinations of magnesia, alumina and silica.

TABLE 2—PROPERTIES OF SOME MIXED
OXIDE CERAMICS

	Cordierite	Forsterite	Steatite	Zircon
Melting Point, F	2680	3470	2820	2820
Mod. Elast. in Tens., 10^6 psi	7	–	13–16	21
Tens. Str., 1000 psi	4–8	9	5–10	5–11
Compr. Str., 1000 psi	50–95	80–85	65–90	60–100
Hard., Mohs	7	7.5	7.5	8
Vol. Res., ohm-cm	$> 10^{14}$	$> 10^{14}$	$> 10^{14}$	$> 10^{14}$
Dielect. Str., v/mil	140–230	250	150–280	60–300
Max. Svc. Temp., F	1830	1830	1830	2000

Cordierite $(2MgO \cdot 2al_2O_3 \cdot 55iO_2)$—Cordierite, is most widely used in extruded form for insulators in such parts as heating elements and thermocouples. It has low thermal expansion and excellent resistance to thermal shock, and good dielectric strength. There are three traditional groups of cordierite ceramics: 1. Porous bodies that have relatively little mechanical strength due to limited crystalline intergrowth and absence of ceramic bond. With long thermal endurance and low thermal expansion, they are used for radiant elements in furnaces, resistor tubes and rheostat parts. 2. Low porosity bodies, developed principally for use as furnace refractory brick. 3. Vitrified bodies used for exposed electrical devices that are subjected to thermal variations.

Forsterite $(2Mg O \cdot SiO_2)$—This mixed oxide has high thermal shock resistance, but good electrical properties and good mechanical strength. It is somewhat difficult to form and requires grinding to meet close tolerances.

Steatite—Steatites are noted for their excellent electrical properties and low cost. They are easily formed and fired at relatively low temperatures. However, compositions containing little or no clay or plastic material present fabricating problems because of a narrow firing range. Steatite parts are vacuum tight, can be readily bonded to other materials, and can be glazed or ground to high quality surfaces.

Zircon $(ZrO_2 \cdot SiO_2)$—This mixed oxide provides ceramics with strength, low thermal expansion, and relatively high thermal conductivity and thermal endurance. Its high thermal endurance is used to advantage in various porous type ceramics.

OXIDE COATINGS

The black oxide finish on steel is one of the most widely used black or blue-black finishes. Some of the advantages of this type of finish are: (1) attractive black color; (2) no dimensional changes; (3) corrosion resistance, depending upon the final finish dip used; (4) nongalling surface; (5) no flaking, chipping or peeling because the finish becomes an integral part of the metal surface; (6) lubricating qualities due to its ability to absorb and adsorb the final oil or wax dips; (7) ease and economy of application; and (8) nonelectrolytic solutions and a minimum of plain steel tank equipment required.

The black oxide finish produced on steel is composed essentially of the black oxide of iron Fe_3O_4, and is considered by many to be a combination of FeO and Fe_2O_3. It can be produced by several methods: the browning process, carbonia process, heat treatment and the aqueous alkali-nitrate process. Each of these processes produces a black oxide of iron finish, although the finish produced by each particular process differs in some characteristics. Inasmuch as the chemical dip aqueous alkali-nitrate process is the most widely used to apply a black oxide finish on steel, this process will be discussed in detail and only a brief description of the others will be given.

Aqueous alkali-nitrate process

In this process a blackening solution is used which is highly alkaline and which also contains strong oxidizing chemicals. Refinements such as penetrants and rectifiers are also used to promote ease of operation, faster blackening and trouble-free processing. At specific concentrations and boiling temperatures these solutions will react with the iron in the steel to form the black oxide of iron (Fe_3O_4).

Because the reaction is directly with the iron in the steel, the finish becomes an integral part of the metal itself and, therefore, cannot flake, chip or peel. For all practical purposes there are also no dimensional changes. In transforming iron to black iron oxide, it is rightful to assume that there is a change in volume. However, because of the highly alkaline nature of the blackening solution and because of the operating temperature, the blackening solution will dissolve a small amount of iron. Therefore, the amount of iron lost in this manner is compensated for by the buildup in volume from the change of iron to iron-oxide—resulting in, for all intents and purposes, no dimensional changes. Extremely close measurements have shown that the actual change amounts to a buildup of only about five millionths of an inch.

In order to apply a black oxide finish on steel by the alkali-nitrate method, a minimum of five steps is usually required, as follows:

1. Hot alkali clean at 180 F to boiling for about 5 min.
2. Water rinse.
3. Immersion in blackening solution (boiling at 290 F) until a good black color is obtained.
4. Rinse in clean water.
5. Final oil or wax dip.

In some cases the first two steps can be eliminated and an organic solvent degreaser used. The black oxide solution operates at a boiling temperature at 285 to 300 F and it must be maintained at the recommended operating boiling temperature. This is accomplished by adding fresh water to compensate for that which is lost through evaporation.

There are types and conditions of steel and many different products which may require special procedures or additional steps other than those listed above. There are occasions when, in order to insure proper rinsing, a warm or hot water rinse is used as a second rinse after the blackening solution. In some cases, if the steel or steel parts have rust or oxidation on them prior to blackening, it is necessary to remove this rust or oxidation by pickling, followed by one or more water rinses. However, in most cases the steps above are usually adequate.

Because a blackening solution operates only as a boiling solution, electrical current is not required. Most work can be processed in baskets or barrels and only in a few instances does the work have to be racked. If racking is required, parts can be racked much closer together than during electroplating. A large volume of work can be processed in a relatively small volume of solution and automatic equipment can be used very satisfactorily.

If the black oxide surface is to be used as a bond for paint, lacquer or enamels, after step 4 the pieces must be immersed for approximately 5 min in a solution of $\frac{1}{2}$ to 1 oz of chromic acid per gal of water at around 150 F, after which the pieces are removed, rinsed and dried. The pieces can then be painted, enameled or lacquered.

The blackening solution is normally maintained at operating strength by replenishing the solution which has been dragged out during processing and is determined by the loss of depth of solution below the original 6-in. level. Assuming that each inch of solution depth represents 3 gal of solution and a loss of 2 in. is noted after a day's processing, it would be necessary to replenish 6 gal of solution. Since it would require 6 lb of black oxide salts to make up each gallon of blackening solution, 36 lb of black oxide salts should be added to the blackening solution and then sufficient water to obtain the 6-in. level from the top of the tank.

It is important that only steel or steel alloys be immersed in the solution because other metals such as copper, zinc, cadmium and aluminum will contaminate it. Many improvements have been incorporated into some proprietary black oxide salts and the latest one will rectify approximately fifty times more contaminants than heretofore. This particular product causes the contaminants to boil to the top, from which they can be skimmed, or dragged out and rinsed away during processing.

After a black oxide blackening solution has been mixed, there is a "breaking-in" period which can run from 24 to 48 hr, depending upon the volume of blackening solution and the amount of work being processed. During this period, if it occurs, erratic blackening may be encountered, resulting in some work being blackened and some remaining partially or totally unblackened.

The usual reason for this condition is that when the new blackening solution is made up, all of the water in the solution is new, fresh water and contains carbon dioxide. This carbon dioxide has to react with the alkali in the blackening solution to form sodium carbonate, until all of the carbon dioxide is used up. During this period not only is the reaction trying to produce the black finish but also the carbon dioxide is reacting to form sodium carbonate. Because two chemical reactions are trying to take place at the same time each will hinder the other and cause erratic blackening.

In a newly mixed blackening solution there is approximately 2% sodium carbonate. During the "breaking-in" period the percentage of sodium carbonate will increase from 5 to 7%. At this point, the reaction is usually satisfied and continuous, uninterrupted blackening will be obtained.

Chemical black oxide finishes today are being used on a wide variety of consumer and military parts. Some of the most important applications are: guns, firearms and components, metal stampings, toys, screws, spark plugs, machine parts, screw machine products, typewriter and calculating machine parts, auto accessories and parts, tools, gauges and textile machinery parts. A black oxide finish can normally be used for indoor or semioutdoor applications on metal parts or fabrications that require an economical attractive finish, nominal corrosion resistance and, in many cases, where "dimensional changes" cannot be tolerated. For military use, the chemical black oxide finish meets the following government specifications: MIL-C-13924A, Class I; 57-0-2C Type 3; 3-51-70-1A; MIL-P-12011 (ORD) FINISH 22.04; MIL-STD-171 (ORD) FINISH 3.3.1: Aeronautical Specification AMS 2485C.

Heat-treatment methods

These methods can be divided into three classes: (1) oven or furnace heating; (2) molten

salt bath immersion; and (3) steam heat process.

In the oven or furnace method the parts are heated to a temperature of 600 to 700 F at which temperature the metal surface is oxidized to a bluish-black color. The shade of color depends on the temperature and the analyses of the steel.

In the molten salt bath method the oxide finish can be obtained in several ways, depending upon the manufacturing or processing requirements:

1. In a molten salt bath composed essentially of nitrate salts maintained at 600 to 700 F the pieces are first cleaned of any oils, greases or objectionable oxides and then immersed in the molten bath. They will take on a blue-black finish, after which they are quenched in clean water and given a final oil dip.

2. In a molten nitrate bath in an austempering operation, the pieces are heated to the hardening temperature in a neutral salt hardening bath, after which they are quenched in the molten nitrate bath at 600 to 700 F. They are then removed, cleaned and given a final oil dip.

Steam process. In this method the steel is placed in a retort and heated to a minimum temperature of 600 F. The retort is then purged with steam. Under these conditions a black oxide is formed on the metal surface.

Browning process

The browning process is commonly known as a rusting process. The pieces are first thoroughly cleaned and then swabbed with an acidic solution. After drying in a dry atmosphere at approximately 170 F, the pieces are then placed in an oven with an atmosphere of 100% humidity and temperature at around 170 F for about 1½ hr. They then become quite rusty and the surface is rubbed down to remove the loose rust. This procedure is carried out three or four times, after which the surface will have taken on a bluish-black finish. The pieces are then given a final oil dip.

Carbonia process

In order to apply a black oxide finish by this method the pieces are placed in a rotary furnace heated to around 600 to 700 F. Charred bone or other carbonaceous material is placed in the furnace along with a thick oil known as carbonia oil. The door or cover of the furnace is occasionally opened and closed to allow circulation of air. After the parts have been treated for approximately 4 hr, they are removed and immersed in oil. This finish is essentially a black oxide of iron, but because the metal surface is in contact with carbonaceous material and oil, some black carbon penetrates into the metal surface.

Processes for nonferrous metals

Following are typical methods for applying black oxide coatings to nonferrous metals:

1. *Stainless steel:* aqueous alkali-nitrate and molten dichromate methods.

In the aqueous alkali-nitrate method a solution is made up by using approximately 4½ to 5 lb of blackening salt mixture to make up a gallon of blackening solution which is operated at a boiling point of 255 to 260 F. The parts, after cleaning and acid pickling, are immersed in the boiling solution and will take on a black color. They are then given a rust-preventive oil dip.

In the molten dichromate method a molten bath of sodium dichromate or a mixture of sodium and potassium dichromates are used at a molten temperature of 600 to 750 F. The parts are immersed in the molten bath until they take on a blue or blue-black color, after which they are removed and cooled in oil or water. They are then cleaned to remove the salt and oil (if cooled in oil), after which they are given a dip in a clean, rust-preventive oil.

2. *Zinc, zinc plate, zinc-base die castings and cadmium plate:* hot molybdate blackening method, chromate and black dye method, and the black nickel plate method.

3. *Copper and copper alloys (brasses and bronzes):* alkali-chlorite aqueous solution method, cuprammonium carbonate method, anodic oxidation method and aryl-sulphone mono-chloramide-sodium hydroxide method.

4. *Aluminum and aluminum alloys:* anodize and dye method.

P

PAPER

Paper can provide a variety of combinations of engineering properties, in some cases unique, and in some cases equivalent to those provided by more expensive materials.

According to the *Dictionary of Paper,* paper is a term used to describe "all kinds of matted or felted sheets of fiber (usually vegetable, but sometimes mineral, animal or synthetic) formed on a fine wire screen from a water suspension." In addition to such materials, paperlike sheets can be produced from platelets or flakes of mica or glass. These are not covered here.

Although there is no distinct line to be drawn between papers and paperboard, paper is usually considered to be less than 0.006 in. thick. Most all fibrous sheets over 0.012 in. thick are considered to be board. In the borderline range of 0.006 to 0.012 in., most are considered to be papers, though some are classified as board.

General features and uses

The major attribute of paper is its extreme versatility. A wide range of end properties can be obtained by control of the variables in (1) original selection of the type and size of fiber, (2) the various pulp processing methods, (3) the actual web-forming operation, and (4) the treatments which can be applied after the paper has been produced.

Papers have been specifically developed for a number of engineered applications. These include gasketing; electrical, thermal, acoustical and vibration insulation; liquid and air filtration; composite structural assemblies; simulated leathers and backing materials; cord or twine; and as yarns for paper textiles.

Paper production

Most papers are made from crude fibrous wood pulps.

Types of pulp. The type of pulp to a large extent determines the type of paper produced. Pulps are generally classified as mechanical or chemical wood pulps.

Mechanical wood pulps produced by mechanical processes include:

1. Ground wood, which is used in a number of papers where absorbency, bulk, opacity and compressibility are primary requirements, and permanence and strength are secondary.

2. Defibrated pulps, which are used for insulating board, hardboard or roofing felts where good felting properties are required.

3. Exploded pulps, used for building and insulation hardboards, or so-called "wood composition materials."

Chemical wood pulps are produced by "cooking" the fibrous material in various chemicals to provide certain characteristics. They include:

1. Sulfite pulps, used in the bleached or unbleached state for papers ranging from very soft or weak to strong grades. There are about 12 grades of sulfite pulps.

2. Neutral sulfite or monosulfite pulps, used for strong papers for bags, wrappings and envelopes.

3. Sulfate or kraft pulps, providing high strength, fair cleanliness and, in some instances, high absorbency. Such pulps are used for strong grades of unbleached, semibleached, or bleached paper (called kraft paper) and board.

4. Soda pulps, used principally in combination with bleached sulfite or bleached sulfate pulp for book-printing papers.

5. Semichemical pulps, used for specialty boards, corrugating papers, glassine and greaseproof papers, test liners and insulating boards and wallboards.

6. Screenings, used principally for coarse grades of paper and board such as millwrapper, and as a substitute for chipboard, corrugating papers and insulation board.

Stock preparation. The various operations performed on the pulp prior to the actual making of the paper can have a critical effect on properties of the paper produced. Beating, refining and curlation are three techniques commonly used to modify the fibers of the pulp.

Beating fluffs out the fibers and transfers hydrogen bonds from within fibers to between fibers, permitting them to interlock to a greater degree within the paper, thus improving such properties as tensile strength, burst strength, density, folding endurance, stiffness and transparency.

Refining cuts the fibers into shorter lengths providing better distribution of fibers throughout the paper web.

Curlation actually curls and crinkles the fibers and breaks up any fiber bundles. Curlation is used to improve such properties as tear strength, bulk, and brightness as well as to provide an even-textured matte finish.

So-called "beater additions" can be made at the beater stage in pulp treatment. A wide variety of nonfibrous materials can be added at this stage to obtain special properties in the final paper. Commonly added materials are sizes, mineral fillers, starch, silicate of soda, wet-strength resins, and coloring dyes and pigments.

Papermaking. The two basic types of papermaking machines are the Fourdrinier and the cylinder machines, the Fourdrinier machine being the most commonly used.

In Fourdrinier papermaking the pulp, mixed to a consistency of 97.5 to 99.5% water, is fed continuously to Fourdrinier, which consists of an endless belt of fine mesh screen called the "wire." This wire is usually agitated to help the fibers criscross, felt and mat together to provide a degree of isotropicity in properties.

As the pulp web travels along the wire, water drains from it into suction boxes. As it leaves the wire (at about 83% water) it passes through presses, and usually a variety of other types of equipment, such as dryers and calender rolls, depending on the type of paper produced.

Cylinder machine papermaking differs from Fourdrinier in that the web of pulp is formed on a cylindrical mold surface instead of a continuous wire covered with fine wire cloth, which revolves in a vat of paper stock or pulp. The felting occurs on the face of the cylinder and water drains through the cylinder. A felt conveyor carries the resulting web to the press and dryers.

The cylinder machine is used to produce a greater variety of paper thicknesses, ranging from the thinnest tissue to the thickest building board.

Types of papers

In the broadest classification, there are three basic types of papers: cellulose fiber, inorganic fiber and synthetic organic fiber papers.

Cellulose fiber papers. These papers, made from wood pulp, constitute by far the largest number of papers produced. A great many of the engineering papers are produced from kraft or sulfate pulps. The term "kraft" is used broadly today for all types of sulfate papers, although it is primarily descriptive of the basic grades of unbleached sulfate papers, where strength is the chief factor, and cleanliness and color are secondary. By various treatments, kraft can be altered to produce various grades of condenser, insulating and sheathing papers.

Other types of vegetable fibers used to produce papers include:

1. Rope, used for strong, pliable papers, such as those required in cable insulation, gasketing, bags, abrasive papers and pattern papers.

2. Jute, used for papers possessing excellent strength and durability.

3. Bagasse, used for paper for wallboard and insulation, usually where strength is not a primary requirement.

4. Esparto, used for high grade book or printing papers. A number of other types of pulps are also used for these papers, not discussed here.

Inorganic fiber papers. There are three major types of papers made from inorganic fibers:

1. Asbestos is the most widely used inorganic fiber for papers. Asbestos papers are nonflammable, resistant to elevated temperatures, and have good thermal insulating characteristics. They are available with or without binders and can be used for electrical insulation or for high-temperature reinforced plastics.

2. Fibrous glass can be used to produce porous and nonhydrating papers. Such papers are used for filtration and thermal and electrical insulation, and are available with or without binders. High purity silica glass papers are also available for high-temperature applications.

3. Ceramic fiber (aluminum silicate) papers provide good resistance to high temperatures, low thermal conductivity, good dielectric properties and can be produced with good filtering characteristics.

Synthetic organic fiber papers. A great deal of research has been carried out on the use of such synthetic textile fibers as nylon, polyester, and acrylic fibers in papers. Some of the earliest appear highly promising for electrical insulating uses. Others appear promising for chemical or mechanical applications. They are most commonly combined with other fibers in a paper, primarily to add strength.

Paper treatments

Papers can be impregnated or saturated, coated, laminated, or mechanically treated. The discussion here covers the major treatments used, to indicate the extent of treatments available. It is not intended to be comprehensive, as it would be impossible to cover all the various specialty treatments.

Impregnation or saturation. Impregnation or saturation can be carried out either at the beater stage in the processing of the pulp, or after the paper web has been formed. Beater saturation permits saturation of nonporous or *ad*sorbent papers, whereas papers saturated after manufacture must be of the *ab*sorbent type to permit complete impregnation by the saturant.

Papers can be saturated or impregnated with almost any known resin or binder. Probably the most commonly used are asphalt for moisture resistance; waxes for moisture vapor and water resistance; phenolic resins for strength and rigidity; melamine and certain ureas for wet strength (not to be confused with moisture resistance); rubber latexes, both natural and synthetic, for resilience, flexibility, strength and moisture resistance; epoxy or silicone resins for dielectric characteristics or dielectric characteristics at elevated temperatures; and ammonium salts, or other materials for flameproofing.

A number of proprietary beater saturated papers are currently available. They are used primarily for gasketing, filtration, simulated leathers and backing materials. Most of these consist of cellulose or asbestos fibers blended with natural or synthetic rubbers. In some cases cork is added to the blend for increased compressibility. Another type of proprietary beater saturated paper consists of leather fibers blended with rubber latexes.

Coatings. Papers may be coated either by the paper manufacturer or by converters. Coating materials, which also impregnate the paper to a greater or lesser degree, include practically every known resin or binder and pigment used in the paint industry. Coatings can be applied in solvent or water solutions, water emulsions, hot melts, and extrusion coatings, or in the form of plastisols or organisols.

The most important properties provided by coatings are (1) gas and water vapor resistance, (2) water, liquid and grease resistance, (3) flexibility, (4) heat sealability, (5) chemical resistance, (6) scuff resistance, (7) dielectric properties, (8) structural strength, (9) mold resistance, (10) avoidance of fiber contamination, and (11) protection of printing.

Coating materials range from the older asphalts, waxes, starches, casein, shellac and natural gums, to the newer polyethylenes, vinyl copolymers, acrylics, polystyrenes, alkyds, polyamides, cellulosics and natural or synthetic rubbers.

Flock coatings can be applied to papers for decoration, sound absorption, vibration cushioning and surface protection. They consist of extremely short (usually 0.015 to 0.060 in.) fibers, usually of rayon, cotton, hair or wool, and are applied to adhesive coated paper by spraying, vibrating or electrostatic techniques.

Laminations. Paper can be laminated to other papers or to other films to provide a variety of composite structures. (Paper-based industrial thermosetting laminates are covered elsewhere.)

Probably the most common types of paper laminates are those composed of layers of paper laminated with asphalt to provide moisture resistance and strength. Simple laminations of paper can be so oriented that over-all characteristics of the composite are isotropic.

Laminating paper with plastics or other types of films or with metal foils will, in many cases, combine the desirable properties of the film or foil with those of the paper.

Scrim is a mat of fibers, usually laminated as a "core" material between two faces of paper. It is usually used to provide strength but can also provide bulk for cushioning, or a degree of "hand" to the composite material.

Mechanical treatments. Several mechanical treatments can be applied to papers to provide particular special properties.

Crimping, which can be done either on the paper web or on the individual fibers of the paper, essentially adds stretch or extensibility. Crimping the paper web results in crepe paper with improved strength, stretch, bulk and conformability and texture similar to that of cloth. The creping process usually consists of "crowding" the paper into small pleats or folds with a "doctor." Typical range of elongation or stretch obtainable is 20 to 300%. Cross-creping can provide controllable stretch in directions perpendicular to each other, further improving drapability.

A high degree of stretch, conformability and flexibility is produced by a patented process which differs from creping in that the individual fibers in the paper web are crimped, rather than the web itself. Amount of stretch is variable, but about 10% stretch in the machine direction seems to be optimum for most industrial applications. The major advantages of this type of paper (trade name "Clupak") are reported to be a high degree of toughness, combined with a smooth surface, and a high resistance to tearing or punching.

Twisting. Twisting is used to convert paper to twine or yarns. Such yarns have substantially higher strength than the paper from which they are made.

Twisting papers are usually sulfate papers, either bleached or unbleached, with basis weight usually varying from 12 to 60 lb. High tensile strength is required in the machine direction, and the heavier weight papers should usually be soft and pliable. Treatments to impart such characteristics as wear and moisture resistance can be applied during or after the spinning operation.

Embossing and other techniques. Decorative papers can be produced by embossing in a variety of patterns. Embossing does not usually improve strength significantly. Embossing or "dimpling" in certain patterns, followed by lamination, can produce composites with added strength as well as bulk and thermal insulation.

Other mechanical methods include: (1) shredding for bulk or padding, (2) pleating, used as a forming aid and for strength in paper cups and plates, (3) die cutting and punching, and (4) molding, which consists of compressing the wet pulp web in a mold to form a finished shape, such as an egg crate.

Applications

As an engineering material, paper has several important applications.

As filtration material, paper can be used either as a labyrinth barrier material to guide the fluid or gas to be filtered or, more commonly, as the filtering medium itself. Paper is used for filtering automotive air and oil, and air in room air conditioners, as well as for industrial plant filtration, filtering liquids in tea bags, in addition to machine-cooling oil and domestic hot water filters.

Papers are used for both light- and heavy-duty gaskets, for such applications as high and low pressure steam and water, high and low temperature oil, aromatic and nonaromatic fuel systems and for sealing both rough and machined surfaces.

Electrical insulation represents one of the largest engineering uses of papers. For electrical uses, special types include coil papers or layer insulation, cable paper or turn insulation, capacitor papers, condenser papers, and high-temperature, inorganic insulating papers.

A large and growing use for paper is in structural sandwich materials, where papers are impregnated with a resin such as phenolic and formed in the shape of a honeycomb. The honeycomb is used as a core between facing sheets of a variety of materials including paperboard, reinforced plastics and aluminum.

Another large-volume application of papers is as backing material for decorative films or other surfacing materials. These provide bulk and depth to the product in simulating leather or fabrics.

PERMANENT MOLD CASTINGS

Permanent mold casting is performed in a mold, generally made of metal, that is not destroyed by removing the casting. Several types of casting can be included in the description: pressure die casting, centrifugal casting and gravity die casting. This article is concerned with the latter process.

Advantages and disadvantages

Gravity die castings are dense and fine grained and can be made with better surfaces and to closer tolerances than sand castings. Tolerances are wider than for pressure die castings and plaster mold castings, but narrower than for sand castings. Production rates are lower than those obtainable by die casting.

The process stands somewhere between sand

and die casting with respect to possible complexity, dimensional accuracy, mold or die casts, etc. For many parts it provides an attractive compromise when the ultimate—whether in complexity of one-piece construction, narrow tolerances or ultra high production rates—need not be met. Table 1 analyzes cost and design factors for permanent mold casting.

TABLE 1—PERMANENT MOLD CASTINGS: FACTORS TO CONSIDER

Cost Factors

Raw-Material Costs	Medium cost. Aluminum, copper and magnesium alloys most used; iron and steel sometimes used.
Tool and Die Costs	Moderate. Less than in die casting, but more than in other methods.
Optimum Lot Size	Large quantities—in the thousands.
Direct Labor Costs	Medium, compared with sand, plaster mold or precision casting.
Finishing Costs	Low to moderate, as little machining allowance is necessary.
Scrap Loss	Low. Most scrap reusable in other castings.

Production Design Factors

Choice of Materials	Wide (see above).
Complexity of Part	Limited. Mold and lack of pressure feeding are controlling factors.
Maximum Size	Practical limit is under 50 lb (for aluminum).
Minimum Size	About 1 oz. Sections as thin as 0.1 in. often cast; 0.060 in., rarely.
Mechanical Properties	Fair.
Precision, Tolerances	High; ±0.015 in. −0.010 in.
Surface Smoothness	Good. Often obviates finishing.
Surface Detail	Fair.

Getting into Production	Several days to a few weeks.
Rate of Output	Moderate. Up to 100/hr customary. Sometimes higher.
General	Gravity die castings fall somewhere between sand castings and die castings as far as complexity, tolerances, cost and production rates are concerned.

Design considerations

The general design rules are:

1. Maintain section thickness at or above ⅛ in.
2. Use generous fillets to aid metal flow.
3. Avoid undercuts to eliminate expensive coring procedures.
4. Minimum draft angle on inside surfaces is 2 deg and on outside surfaces, 3 deg.
5. Avoid sharp section changes.
6. Know the position of locating points and chucking procedure for gate and riser design.
7. Allow ¹⁄₁₆ to ¹⁄₃₂ in. for machining stock.

Minimum diameter of cored holes is ⅜ in.

Mold design. The tolerances in Table 2 should be followed when producing the mold cavity.

TABLE 2—PERMANENT MOLD CASTING TOLERANCE ALLOWANCES.

Dimension	Casting Tolerance
Across Parting Line	±0.015 in. for 1 in. or less; over 1 in. add 0.002 in./in.
Between Points Produced by One Part of the Mold	±0.015 in. for 1 in. or less; over 1 in. add 0.001 in./in.
Between Points Produced by the Core and the Mold	±0.015 in. for 1 in. or less; over 1 in. add 0.002 in./in.
Maximum Length of Core Supported to One End	Core diameter × 10
Outside Draft	1° min, 3° desirable
Draft in Recesses	2° min, 5° desirable
Draft on Cores	½° limited, 2° desirable

Casting alloys

Lead, zinc, aluminum, magnesium and copper base alloys, as well as gray cast iron, can be cast by permanent molding. Less than 1% of total gray iron production is permanent mold cast, 5 to 7% of copper and magnesium alloy, and up to 40% of total aluminum casting production. Here is a breakdown of problems to expect:

Gray iron. Mechanical properties depend on thickness.

Aluminum. Gates must be made larger to take the low specific gravity of aluminum into account.

Magnesium. Extreme caution is needed when removing metal from the furnace. Ladles must be kept at red heat to exclude moisture and prevent an explosion.

Copper. Aluminum bronze is the most popular permanent-mold-cast copper base alloy. Turbulence must be prevented in the die cavity to reduce oxidation of the molten metal during solidification.

PHENOLIC PLASTICS

Phenolic resins are used most extensively as thermosetting plastic materials, there being only a few uses as thermoplastics. The polymer is composed of carbon, hydrogen, oxygen and sometimes nitrogen. Its molecular weight varies from a very low value during its early state of formation to almost infinity in its final state of cure. The chemical configuration, in the thermoset state, is usually represented by a three-dimensional network in which the phenolic nuclei are linked by methylene groups. The completely crosslinked network requires three methylene groups to two phenolic groups. A lesser degree of crosslinking is attainable either by varying the proportions of the ingredients or by blocking some of the reactive positions of the phenolic nucleus by other groups, such as methyl, butyl, etc. Reactivity can be enhanced by increasing the hydroxyl groups on the phenolic nuclei, for example, by the use of resorcinol.

General characteristics

Phenolic resins are manufactured commercially by reaction of phenols with aldehydes, of which there are many varieties; consequently, a great many variations are possible. The basic chemistry can best be illustrated by the products formed from phenol and formaldehyde. Under alkaline conditions, one mole of phenol will react with more than one mole of formaldehyde to form products known as resoles. These will cure upon heating to form infusible, insoluble polymers, commonly identified as single-stage, or one-step resins.

When, under acid conditions, less than one mole of formaldehyde is reacted with one mole of phenol, low molecular weight, linear, thermoplastic polymers are formed. These are called novolaks and require a crosslinking agent, such as hexamethylenetetramine, to form the three-dimensional network polymer, the reaction usually being carried out at elevated temperatures. The mixture of novolak resin and hexamethylenetetramine is called a two-step or two-stage resin. In contrast to one-step resins, a phenolic two-step resin is stable indefinitely when stored at room temperature in its normal powdered state.

The outstanding characteristics of phenolics are: good electrical properties, very rigid set, good tensile strength, excellent heat resistance, good rigidity at elevated temperature, good aging properties; also, good resistance to water, organic solvents, weak bases, and weak acids. All of these characteristics are coupled with relatively low cost.

Phenolics are used in applications that differ widely in nature. For example, wood is impregnated to make "impreg" and "compreg"; paper is treated to make battery separators and oil and air filters; specific chemical radicals can be added to the molecule to make an ion-exchange material. Phenolics are also widely used in protective coatings. These are in addition to the more common applications described below.

Molding compounds

The largest single use for phenolic resins is in molding compounds. To make these products, either one- or two-stage resins are compounded with fillers, lubricants, dyes, plasticizers, etc. Wood flour is used as an inexpensive reinforcing agent in the general-purpose type of compounds. Cotton flock, chopped fabric, and sisal and glass fibers are used to improve strength characteristics; mineral fillers such as asbestos and mica are used where improvements in dimensional stability, heat resistance, or electrical properties are desired. The compounds are usually produced in granular, macerated, or nodular forms, depending on type of filler used. Since the color of the base resin is not stable to light, molding compounds are commonly produced only in dark colors such as black and brown.

Molding compounds are usually processed in hardened steel molds at a temperature between 270 F and 400 F, and at a pressure of from 500 to 10,000 psi. Molds can be designed to operate using the compression, transfer, or plunger-molding techniques, depending on the design of the article to be fabricated. Molded parts can be drilled, tapped, or machined. Properties of molded materials are summarized in Table 1.

The hundreds of different phenolic molding compounds can be divided into the following groups on the basis of major performance characteristics.

General purpose phenolics are low cost compounds with fillers, such as wood flour and flock, and are formulated for non-critical functional requirements. They provide a balance of moderately good mechanical and electrical properties and are generally suitable in temperatures up to 300 F.

Impact resistant grades, higher in cost, are designed for use in electrical and structural components subject to impact loads. The fillers are usually either paper, chopped fabric, or glass fibers.

Electrical grades, with mineral fillers, have high electrical resistivity plus good arc resistance, and they retain their resistivity under high temperature and high humidity conditions.

Heat resistant grades are usually mineral or glass filled compounds that retain their mechanical properties in the 375–500 F range. Special grades, such as phenyl silanes, provide long term stability at up to 550 F.

Special purpose grades are formulated for service applications requiring exceptional resistance to chemicals or water, or combinations of conditions such as impact loading and a chemical environment. The chemical resistant grades are inert to most common solvents and weak acids and alkali resistance is good.

Nonbleeding grades compounded specially for use in container closures and for cosmetic cases.

About one-third of all phenolic resins produced is processed into parts by molding. Compression and transfer molding are the principal processes used but they can also be extruded and injection molded.

Molded phenolic parts are used in bottle caps, automotive ignition and engine components, electrical wiring devices, washing machine agitators, pump impellers, electronic tubes and components, utensil handles and a multitude of other products.

Adhesives

The thermosetting nature and good water- and fungus-resistant qualities of phenolic resins make them ideal for adhesive applications. Almost all exterior grade plywood is phenolic resin-bonded. This constitutes the second largest market for phenolics. The essential ingredient in many metal-to-metal, and metal-to-plastic adhesives is a phenolic resin. One-step phenol-formaldehyde resins are used predominantly for hot-pressed plywood. Special resorcin-formaldehyde resins curing at room temperature are employed for fabricating laminated timber.

Laminates

The third largest use for phenolic resins is in the manufacture of laminated materials. Many variations of paper, from cheap kraft to high quality alpha cellulose, besides asbestos, cotton, linen, nylon and glass fabrics are the most commonly used reinforcing filler sheets. The laminate is formed by combining under heat (350 F) and pressure (500-2000 psi) multiple layers of the various reinforcing sheets after saturation with phenolic resin, generally of the one-step type dissolved in alcohol.

Paper laminates are used most extensively in the electrical and decorative fields. A large number of the laminates for the electrical industry are of the punching grade, making it possible to fabricate all kinds of small parts in a punch press. The laminate used for decorative purposes usually contains a surface sheet of melamine resin-treated paper for providing unlimited color or design configurations. Other fillers are used for special applications where superior dimensional stability, or water, fire, or chemical resistance, or extra strength is required. Properties are published in NEMA (National Electrical Mfrs' Assn.) Standards for Industrial Thermosetting Laminates.

TABLE 1—TYPICAL PROPERTIES OF PHENOLICS

Type	No Filler	General, Wood Flour and Flock	Shock, Fabric or Cord	Heat, Res., Glass Fiber	Electrical, Mineral
Specific Gravity	1.28	1.34–1.46	1.36–1.43	1.75–1.90	1.6–3.0
Mod Elast. in Tens., 10^5 psi	7.5–10	8–13	9–14	30–33	10–30
Tens. Str., 1000 psi	7.5	5.0–8.5	5–9	5–10	6
Elong., %	–	–	0.40–0.55	0.2	–
Hardness	M126	E85–100	E80–90	E50–70	E80–90
Imp. Str., ft-lb/ in. notch	Nil	0.24–0.50	0.6–8.0	10–33	0.32
Max. Svc. Temp., F	–	300–350	250–300	350–450	400
Vol. Res., ohm-cm	10^{11}	10^9–10^{12}	$> 10^{10}$	7–10×10^{12}	6×10^{12}
Dielect Str., v/mil	–	200–425	200–350	200–370	380

Casting resins

Plastic parts can also be manufactured by pouring resin into molds and heat-curing without pressure. Two basic grades of casting resins are manufactured commercially. Both are of the one-stage type and are cured under neutral to strongly acidic conditions depending on the application. The first grade is manufactured for its variegated color and artistic possibilities and is used primarily in the cutlery and decorative field. Since this type of cast material is noted for ease of machining, it is well adapted for small production runs or where machined prototypes are desired.

The second grade of casting resins includes all those modified by fillers and reinforcing agents. Designed primarily to have low-shrinkage characteristics during cure, they are usually set with a strong acid catalyst to obtain a low temperature set. Uses include containers, jigs, fixtures and metal-forming dies.

Bonding agents

Phenolic resins are noted for their excellent bond strength characteristics under elevated temperature conditions, and thus are used in such applications as thermal and acoustical insulation, grinding wheels, coated abrasives, brake linings and clutch facings. Glass wool insulation is manufactured by spraying water-soluble one-stage resins on the glass fibers as they are formed. Heat given off by the fibers as they cool is suffi-

cient to set the resin. Where organic fibers are used, finely pulverized, single-stage resins are distributed between the fibers, either by a mixing or a dusting operation, followed by an oven treatment.

Grinding wheels bonded with phenolic resin are commonly known as resinoid-bonded wheels. By combining a liquid single-stage resin and a powdered two-stage resin, the material can be evenly distributed with abrasive grit and fillers so that the mixture can be pressed into wheels and baked in ovens. Some wheels are also hot-pressed and cured directly in a press. Resinoid-bonded wheels are used primarily where the application requires a bond of exceptional strength such as in cut-off and snagging wheels.

Coated abrasives are manufactured by bonding abrasive grit to paper, fabric, or fiberboard by means of phenolic resins of the liquid one-stage type. Sander discs and belts are common applications.

Brake linings and clutch facings are made by bonding asbestos, fillers, metal shavings, and friction modifiers with phenolic resin, usually performed by a mixing, forming and baking operation. Most resins for these applications are specially formulated, but simple two-stage resins are sometimes used. Formulation for each friction lining or clutch face must be determined carefully. Wood composition products of all descriptions are manufactured by hot-pressing sawdust, wood chips, or wood flour containing 8 to

20% resin. Two-step resins are most frequently used in composition boards. One-step resins are employed for special applications where the yellow color or the slight ammonia odor of two-step resins is undesirable.

Foundry use

A relatively new application holding great promise for phenolic resins is in the shell mold process for the foundry industry. The basic principle involves binding sand grains with resin. Molds made in this manner are permeable shells with an over-all thickness of $\frac{1}{8}$ to $\frac{5}{16}$ in. The process is equally well adapted to making cores and is satisfactory for practically all metals, including magnesium and high chrome alloys. Invented by Johannes Croning of Germany, variations have since been developed. All of them basically use a two-stage resin, although in many different forms and modifications. The following illustrates the basic principles of the process.

A mixture of powdered resin and dry sand is made in a suitable mixer, then dumped onto a hot metal pattern or blown into a hot core box, and allowed to remain there a few seconds. The excess is dumped off or dropped out. The pattern or core box, with shell mold or core, is then immediately placed in a heated oven until the resin binder is thoroughly hardened. The shells are then removed from the pattern or core box and are ready for use.

The process has many intriguing possibilities. Molds thus made can be reproduced in exact composition, detail and size; they are rigid and, having no affinity for water, can be racked and stored indefinitely, with or without cores. Castings from these molds have excellent surface finish and detail and can be held to close dimensional tolerances. The process can be completely mechanized, yields more castings per ton of melt than sand casting, simplifies cleaning of castings, minimizes problems with sand control and handling, and is a relatively clean operation. It is revolutionizing the foundry industry.

Consumption of resin in the foundry industry, while still relatively small, is growing rapidly and conceivably this could become the largest single outlet for phenolics. However, markets for phenolic resins in plywood adhesive, insulation, and wood composition board applications can also be expected to expand.

PHOSPHATE CONVERSION COATINGS

Phosphate coatings have been used commercially for approximately fifty years. They are used on iron, steel, zinc and aluminum surfaces to increase corrosion protection, provide a base for paint, reduce wear on bearing parts, and aid in the cold forming and extrusion of metals.

Phosphate coatings are formed by chemically reacting a clean metal surface with an aqueous solution of a soluble metal phosphate of zinc, iron or manganese, accelerating agents and free phosphoric acid. For example, when steel is treated, the surface is converted into a crystalline coating consisting of secondary and tertiary phosphates, adherent to and integral with the base metal.

Properties of the coating such as size, weight and uniformity of the crystals, are influenced by many factors such as composition and concentration of the phosphating solution, temperature and processing time, type of metal and the condition of its surface due to previous treatment.

The performance of a phosphate coating depends largely on the unique properties of the coating which is integrally bound to the base metal and acts as a nonmetallic, adsorptive layer to hold a subsequent finish of oil or wax, paint, or lubricant. Heavy phosphate coatings are normally used in conjunction with an oil or wax for corrosion resistance. The combination of the coating with the oil film gives a synergistic effect, which affords much greater protection than that obtained by the sum of the two taken separately. The stable, nonmetallic, nonreactive phosphate coating provides an excellent base for paint. It is chemically combined with the metal surface which results in increased adsorption of paint and materially reduces electrochemical corrosion normally occurring between the paint film and the metal.

The oil absorptive phosphate coatings are useful in holding and maintaining a continuous oil film between metal-to-metal moving parts. They also permit rapid break-in of new bearing surfaces. Even after the coatings have been worn away, the controlled etched condition of the metal surface continues to hold the oil film between the moving parts. The ability of a well-anchored coating to hold a soap or oil type lubricant is used in the cold forming and drawing of metals.

Processes

Most phosphate coatings are formed from heated solutions following a hot cleaning cycle. Effective coatings, both the zinc phosphate and the iron phosphate types, are now produced by the relatively cold system with up to 70% savings in heating costs.

This new system, which is designed for cleaning and coating steel by spray application, has the following sequence of operations:

1. Cleaning, 60 sec, 80-100 F.
2. Water rinsing, 60 sec, 80-90 F.
3. Phosphating, 60 sec, 105-120 F.
4. Water rinsing, 20-60 sec, unheated.
5. Acidified rinsing, 15-45 sec, unheated-160 F.
6. Drying, to remove surface moisture.

All steel must be thoroughly cleaned prior to phosphate coating to remove grease, oil, rust and undesirable soils on the steel surface which prevent or alter the formation of a satisfactory phosphate coating and interfere with paint adhesion. The cleaner used to remove oily soil prior to the formation of the zinc phosphate coating is a light-duty, mildly alkaline material especially formulated to function effectively at low temperature. The cleaner which is best suited for use in connection with the low temperature iron phosphate coating process may be either a light duty, low foaming, mildly acidic mixture, or the above mild alkaline cleaner, depending upon the conditions of operation.

Proper formulation of the phosphating solutions allows the coatings to form rapidly at low temperature on the cleaned steel surface. Simple, on-the-job, chemical controls enable the operator to adjust the addition of coating chemicals to the requirements of the steel being treated.

Following the coating operation, a cold water rinse is used to remove excess coating chemicals. The flow of water through the rinse is regulated with the rate of production so that contamination of the main body of the rinse is minimized. An acidified rinse containing hexavalent chromium compounds follows the water rinse. This rinse has the specific effect of enhancing the corrosion resistance of the coating. An oven dry-off to remove surface moisture completes the process.

Typical products being treated by the cold phosphate system are automotive body and sheet metal parts, refrigerator cabinets, office furniture, lighting fixtures, commercial air conditioners, home heating equipment, home laundries, kitchen cabinets, desk and filing cabinets, steel drums and window sash.

The advantages of the cold phosphate system over conventional methods which require higher temperature operation are:

1. Direct heat savings—up to 70%.
2. Less heat-up time.
3. Less maintenance due to decreased load on heating coils, steam traps, etc.
4. Reduced downtime. Maintenance personnel can enter the units immediately after shutdown to make adjustments or repairs.
5. Increased worker comfort near the installation.
6. Reduced use of water through decreased evaporation.
7. Elimination of exhaust fans.

Applications

Phosphate conversion coatings have a wide range of application. For example:

Base for plastisol coatings. Recently, the domestic appliance industry has pointed towards the use of plastisol (polyvinyl chloride) coatings as a replacement for more costly porcelain enamel. During this period several major manufacturers of domestic dish washers have standardized on plastisol films for coating tubs, lids, dish racks, etc. In the preliminary development work it was established that the steel base metal would have to be cleaned and treated in a manner that would give maximum adhesion of the plastisol coating to the metal and at the same time provide maximum moisture or blister resistance in the exacting and severe vapor life tests.

The metal preparation method meeting this requirement was a zinc-phosphate coating system. The fabricated parts are cleaned in a conditioned cleaner, phosphated and rinsed thoroughly. The rinse procedure includes an acidified chromic-phosphoric acid rinse followed by a closely controlled deionized water rinse and thorough drying to remove surface moisture. This treatment produces a continuous, uniform, fine-grained coating of approximately 150 to 250 mg of zinc-phosphate coating per square foot of surface area treated. Some manufacturers have incorporated additional stages to provide a phosphoric acid pickle which is controlled to a constant composition by use of an ion-exchange unit. The phosphoric pickle removes steel mill contaminants from the surface of steel and, at the same time, levels off the "pickle-lag variable" between different heats of steel or steels from different suppliers.

After the steel has been zinc phosphate coated, an especially formulated primer is applied to a very thin and closely controlled film build and followed by the subsequent application of 12 to 15 mils of plastisol with an intermediate and final baking operation.

Bond for vinyl coatings. Vinyl films of 5 to 15 mils thickness are applied to both steel and aluminum sheets by lamination of calendered and decorated films or by spray or roller coating. For the vinyl laminate as with the plastisol application, the base metal must be chemically treated to provide the necessary bond for laminate to metal surface. A controlled, accelerated iron phosphate treatment followed by an acidified chromate rinse has proved to be best for the preparation of steel for vinyl lamination to sheet or coil on a continuous line. This method is now being used by several fabricators and rollers of steel.

Aluminum can also be finished with vinyl laminates and plastisols. Chromate conversion coatings of the "gold" oxide type provide excellent adhesion of the vinyl film as well as excellent corrosion resistance on the unprotected surface.

Metal preparation for both steel and aluminum is usually done by spray application in 5-stage equipment in the following sequence:

1. Alkali clean—30 sec (smutty steel may require brush scrubbing).
2. Water rinse—10 sec.
3. Phosphate coating—10 sec.
4. Water rinse—5 sec.
5. Acidified chromate rinse—5 sec.

The same sequence and time cycles are used for aluminum except that an appropriate chromate conversion coating solution is in stage 3. Consequently, installations have been engineered to prepare both steel and aluminum for subsequent application of vinyl laminates by providing interchangeable coating solution tanks at that stage, thereby providing efficient and versatile "in-line" operation for this new decorative treatment.

Bolt making. Phosphate coatings are just now finding wide acceptance in the fastener industry as applied to rod to facilitate the forming of bolts. The coating is produced from a dilute, especially accelerated, zinc acid phosphate solution which reacts chemically with the surface of the rod to form an insoluble nonmetallic phosphate coating integral with the surface of the rod. The coating forms a porous bond to carry the extruding and heading lubricant and prevent metal-to-metal contact in subsequent heading operations. The result is longer tool life, increased percentage of reduction, and improved surface appearance of the finished product.

The methods of application vary from the conventional pickle-house immersion method to the newly developed in-line strand processes. In the immersion method, the coils of scaled rods are dipped in the treating solutions in the following sequence of operations:

1. Acid pickle—10 min to 2½ hr.
2. Water rinse—cold, overflowing or spray.
3. Hot water rinse.
4. Phosphate coating—5 to 15 min, 160 to 200 F.
5. Cold water rinse.
6. Neutralizing rinse—lime, etc.
7. Bake dry.

It has been found that a dip in a high-strength soap solution improves die life and reduces knockout pressures. The soap solution follows the neutralizing rinse, or in some instances, it is combined with the neutralizing rinse.

The in-line strand process eliminates intermediate handling from scaled rod to drawn wire and the necessity of acid disposal. The coils are handled continuously. Line speeds up to 600 fpm are required with this process so that the phosphate solution must deposit the desired heavy coating in about 15 sec.

Even at this extremely short processing time, a line speed up to 600 fpm requires a special coating tank. Compact, prerinse and neutralizing rinse stages are incorporated in this unit from which it is possible to obtain better than 750 mg per sq ft of coating in 15 sec by use of specially formulated phosphate solutions adapted to strand processing.

A recent refinement is the adoption of the strand principle "in-line" with the boltmaker. The combination produces finished bolt from coiled, scaled rod in one operation with an absolute minimum of floor space. The material is delivered to the unit as hot rolled rod and leaves as a finished bolt without intermediate handling.

All operations are handled by the boltmaker operator including the phosphating operation which includes cleaning, rinsing, coating, rinsing and lubricating.

Wire produced by the strand method of processing produces wire superior in quality to that produced by conventional cleaning and drawing methods and does so at lower cost. As a consequence of the better quality, phosphate base lubricated material will produce close tolerance bolts with full heads and sharp shoulders.

Wire drawing. Phosphate coatings are also gaining acceptance as a lubricant carrier in the wire industry to permit increased drawing speeds and prolonged die life. This is particularly true in connection with both the dry and wet drawing of high carbon wire. Other advantages which the zinc phosphate coatings afford to the wire industry are increased corrosion resistance after drawing and closer dimensional tolerances. Improved dimensional tolerance is a major advantage in forming springs from spring wire.

The zinc phosphate coating may be applied by immersion in the conventional pickle house installation, or by the newly developed fast coating, continuous strand method. The continuous strand method fits in well with fast, in-line travel of the wire. This adaptability and the lower labor costs involved warrant the recommendation of strand phosphate lines in wire processing.

Extrusion. Another important development is the use of zinc phosphate coatings as an aid in the rapidly developing field of cold extrusion of steel and aluminum. In this application the phosphate coating solution is formulated to deposit a considerably heavier zinc phosphate coating than heretofore mentioned. The coating is then chemically reacted with a soap base lubricant to form a water-insoluble lubricating film.

The use of the zinc phosphate coating is considered a basic requirement in the cold extrusion field where the pressure exerted by the tools on the steel being formed may be in excess of 300,000 psi. The phosphate and lubricant coating withstands the high unit pressure and temperatures developed in this type of cold forming. At the same time, it maintains the required separating film between the tools and workpiece being extruded to prevent scoring, galling and tool breakage. A specially formulated phosphate

coating solution produces a zinc phosphate coating on aluminum and is gaining wide acceptance as an aid in the cold extrusion of the heat-treatable alloys of this metal.

Facilitate processing

Attempts to box anneal cold-reduced steel sheets in the mill at higher temperatures, and greater stack height than conventional can produce a welding or sticking of sheets. A new development to prevent "stickers" in annealing is the application of a specifically adjusted phosphate coating to the steel prior to the cold-reduction process. The coating must be uniformly distributed over the steel surface and controlled within a narrow coating weight range without being adversely affected by line slow-down or stoppage. The sequence of operations for the application of this process is as follows:

1. Pickle—6 to 25% H_2SO_4, 200 F, 1 to 2 min.
2. Water rinse—3 sec.
3. Phosphate coat, spray or immersion—10 sec, 175 F.
4. Cold water rinse—3 sec.
5. Neutralizing rinse—175 F, 3 sec.

This phosphate coating aids in the cold reduction of steel strip by: (1) acting as a lubricant carrier; (2) reducing friction; (3) increasing mill roll life; (4) reducing power requirements and heat generation in the cold-rolling equipment, and thereby requiring fewer passes to final gauge by permitting greater reductions.

The phosphate coating is retained by the steel surface even after cold reduction and acts as a parting medium to prevent sticking of sheets during annealing. Additional advantages afforded by the phosphated steel are the increased ease of cleaning the sheet, increase in corrosion protection obtained by oiling the treated sheet, and resistance to abrasion during recoiling operation.

PLASMA-ARC COATINGS

In this process, a flow of gas, such as argon, is directed through the nozzle of a device called the plasma arc torch. When a high-current electric arc is struck within the torch between a negative tungsten electrode and the positive water-cooled copper nozzle, electrical and aerodynamical effects force the arc through the nozzle which concentrates and stabilizes it. A substantial portion of the gas flows through the arc and is heated to temperatures as high as 30,000 F and accelerated to supersonic speeds to form an ionized gas jet called plasma. A cool layer of gas next to the nozzle wall effectively insulates the torch from the tremendous heating effect of the arc column.

Particles of refractory coating material, introduced into the plasma in either powder or wire form, are melted and accelerated to high velocity. When these molten particles strike the workpiece, they impact to form a dense, high-purity coating. Sprays of cold carbon dioxide gas, played on the workpiece, keep it from overheating during the process and protect the purity of the coating from air oxidation.

Characteristics

The primary advantage of the process is its ability to combine the bulk properties of a base material with the surface properties of a refractory material. Furthermore, the application of the thin, tenacious coatings can be limited to the specific areas of the base material where a coating is needed, and warpage or distortion of precision parts is eliminated because of the low base material temperature maintained during coating.

Whether as-coated or finished, the refractory coatings have extremely good resistance to wear, abrasion, and corrosion and erosion, even under the adverse conditions of high temperature, high load and lack of lubrication and cooling. When ground and lapped, the coatings give superior performance under conditions of fretting corrosion. Finished coatings, when mated with proper materials, have generally lower coefficients of friction than most metal-to-metal combinations. This ratio is also true at elevated temperatures. The coatings have a porosity of less than 1% and an as-coated surface finish of approximately 150 μ-in. rms which can be finished down to better than 1 μ-in. rms.

Fabrication

Coatings can be applied in practically any desired thickness. But only areas which allow the particles free access will be coated evenly. This limitation excludes narrow holes, blind cavities and deep V-shaped grooves. All corners and edges should be rounded by a minimum 0.015-in. rad or have a minimum chamfer of 0.015 in. by 45° to prevent weak spots. The sizes of the parts which can be coated are:

1. Long external cylindrical parts, 0.2 to 28 in. in dia by 76 in. long.
2. Short external cylindrical parts, 0.2 to 60 in. in dia by 6 in. long.
3. Internal diameters, greater than 0.38 in., open at both ends, coating depth limited to 1½ times the diameter.
4. Rectangular flat surfaces, 24 in. wide by 76 in. long.
5. Circular flat surfaces, 0.1 to 60 in. in dia.

Since plasma-arc coatings can be deposited in practically any desired thickness, it is also possible to fabricate parts by this method. The required thickness is built up on a mandrel formed to the desired internal shape of the finished part, and the mandrel is then removed chemically from the part with acid or caustic.

This method allows intricate shapes to be made of materials which are normally difficult to fabricate. But, as with flame spraying, only areas which allow the particles of coating material sufficient access will be plated evenly. Although the shape of the part or mandrel to be plated is limited to a surface of revolution or a contour that deviates no more than 45 deg from the axis of revolution if uniform thickness is desired, more extreme asymmetrical shapes can be coated with hand-operated equipment which allows somewhat less control over thickness.

Materials

Almost any base material can be coated, even certain reinforced plastics, and any known inorganic solid which will melt without decomposition can be used as a coating material. Many basic coatings have already been established including tantalum, palladium, platinum, molybdenum, tungsten, alumina, zirconium diboride and oxide, and three combinations of tungsten with additives to improve its properties. These additives are zirconia, chromium and alumina.

Development work is continuing on coatings of the refractory metals such as columbium; some of the refractory metal compounds such as the borides of tungsten, columbium, tantalum, titanium and chromium; the refractory carbides of columbium, hafnium, tantalum, zirconium, titanium, tungsten and vanadium; the refractory oxides of thorium, hafnium, magnesium, cerium and aluminum; and other pure metals such as aluminum, copper, nickel, chromium and boron.

Properties

The properties of the coatings or parts made from all of these materials are equivalent to those of the pure materials themselves. For example, in the as-coated state, tungsten parts have a density of 86 to 90% theoretical, a modulus of rupture of 49,000 psi, and a Young's modulus of 24×10^6 psi. Furnace treatment at 2550 F in an argon-hydrogen atmosphere increases the coating density to 91 to 95%, the modulus of rupture to 57,000 psi, and Young's modulus to 40×10^6 psi. The volumetric and linear shrinkage upon firing are 2.2% and 0.7% respectively.

Cost

The cost of depositing the high-melting-point coatings with the plasma-arc process is comparable with that of the flame-sprayed coatings. Power costs are much higher, but the process is much faster and there are no carbonization products to affect purity. In fact, since the material cost is lower, the process is inherently the cheaper of the two. There is no comparison possible for the fast and accurate mass production of refractory parts by this process since these materials have been virtually unworkable by conventional means up to now.

PLASTER MOLD CASTINGS

Plaster mold casting is primarily used for producing parts in quantities that are too small to justify the use of permanent molds, yet large enough to outweigh the machining costs of sand castings. The process is noted for its ability to produce parts with high dimensional accuracy, smooth and intricate surfaces and low porosity. On the other hand, it is limited to nonferrous metals (aluminum and copper alloys) and relatively small parts. Also, production times are relatively high because the molds take relatively long to make and are not reusable.

The process

The plaster used for molding generally consists of water mixtures of gypsum or plaster of Paris (calcium sulfate) and strengthening binders such as asbestos, magnesium silicate, silicate flour and others. Impurities such as salts and hydrochloric acid are also added to accelerate setting.

After mixing, the plaster is poured over the pattern (usually brass) which is fixed in a suitable flask. Groupings of more than one part can be made on the same pattern plate. After the plaster has air-hardened for about 20 to 30 min the pattern is removed and the mold is then baked at at least 400 F for several hours (10 to 20 hr at 400 F in the case of aluminum castings).

After baking, the mold is removed from the flask and suitable inserts, cores and guide pins are installed along with the cope and drag. The mold is then ready for casting, after which the metal is allowed to cool and the mold is destroyed.

A variation of plaster mold casting known as the Antioch process can also be used to produce highly accurate parts. The materials used in this process combine the advantages of sand and gypsum plaster and consist of a mixture of bulk silica sand, a gypsum binder and control ingredients such as talc, sodium silicate and magnesium oxide. This mixture sets around the pattern in 5 to 7 min and is air-dried for 5 to 6 hr before being placed in an autoclave where it is subjected to steam at about 2 atm. The mold is then air-dried again for 12 hr and rebaked for 12 to 20 hr at 450 F. This procedure is designed to make the mold permeable and improve the quality of the castings during pouring.

Materials

Plaster-mold casting is limited by the melting temperature of the mold material—about 2200 F. Thus, it is principally used to cast metals such as yellow brass (Muntz metal or naval brass), manganese bronze, aluminum bronze, silicon-aluminum bronze, nickel brass and aluminum. It has been reported that red brass and naval bronze do not have a good surface finish when cast in plaster molds.

It is possible that the development of more heat-resistant molds will permit the casting of higher melting materials. Variations of the process are suitable for casting metals such as the stainless steels with melting points up to 3000 F.

Sizes and tolerances. Although plaster-mold casting can be used for making parts weighing up to 100 lb it is usually limited to parts of less than 15 lb, and even down to 1 oz. Minimum section thickness is about 40 to 60 mils and bosses and undercuts can be incorporated into the design.

Dimensional tolerances are about 5 mils per in. for the first inch and 1 mil is added for each additional inch. A common total of dimensional tolerance on parts is 5 to 10 mils; draft allowance is ½ to 1 deg. The surface finish of as-cast surfaces is 30 to 50 μin. (rms) and about $\frac{1}{32}$ in. should be allowed for finish machining.

Applications

Plaster-mold castings are usually used for medium-production applications and their cost falls between sand castings and permanent-mold castings. Typical parts where the process has been used include: gears, ratchet teeth, cams, handles, small housings, pistons, wing nuts, locks, valves, hand tools, and radar parts for aircraft, railroad, household and electrical uses.

PLASTICS ALLOYS

Plastics, like metals, can be alloyed. And like metal alloys, the resulting materials have different, and often better, properties than those of the base materials making up the alloys.

There are about a half-dozen plastic alloys commercially available. However, a number of others are in use that have not been announced for various reasons. Some have been developed by end-users who do not want to reveal any information about them. And others have been "tailor-made" by resin suppliers for large-volume special applications.

The plastics most widely used in alloys today are polyvinyl chloride (PVC), ABS and polycarbonate. These three plastics can be combined with each other or with other types of polymers.

ABS, besides being used with polycarbonate, can also be alloyed with polyurethane. Commercially available in two grades, these alloys combine the excellent toughness and abrasion resistance of the urethanes with the lower cost and rigidity of ABS. The materials can be injection molded into large parts but cannot be extruded. Typical applications for which they are suitable include such parts as wheel treads, pulleys, low load gears, gaskets, automotive grilles and bumper assemblies.

ABS is also being successfully combined with polyvinyl chloride and is available commercially in several grades. One of the established grades provides self-extinguishing properties, thus eliminating the need for intumescent (non-burning) coatings in present ABS applications, such as power tool housings, where self-extinguishing materials are required. A second grade possesses an impact strength about 30% higher than general-purpose ABS. This improvement, plus its ability to be readily molded, has resulted in its use for automobile grilles.

ABS-PVC alloys also can be produced in sheet form. The sheet materials have improved hot strength which allows deeper draws than with standard rubber-modified PVC base sheet. They also are non-fogging when exposed to the heat from sunlight. This is an important advantage where transparent materials are needed. Some properties of ABS-PVC alloys are lower than the base resins. Rigidity, in general, is somewhat lower and tensile strength is more or less dependent on the type and amount of ABS in the alloy.

Another sheet material, which is an alloy of about 80% PVC and the rest acrylic plastic, combines the nonburning properties, chemical resistance and toughness of vinyl plastics with the rigidity and deep drawing merits of the acrylics. The PVC-acrylic alloy approaches some metals in its ability to withstand repeated blows. Because of its unusually high rigidity, sheets ranging in thickness from 0.60 to 0.187 in. thick can be formed into thin-walled, deeply drawn parts. Typical thermoformed products include luggage, truck cargo liners and a wide variety of machine and equipment housings.

PVC is also alloyed with chlorinated-polyethylene (CPE) by end-users to gain materials with improved outdoor weathering or to obtain better low-temperature flexibility. Applications include wire and cable jacketing, extruded and molded shapes, and film sheeting.

Acrylic-base alloys with an additive of polybutadiene have also been developed, chiefly for blow-molded products. The acrylic content can range from 50 to 95 pct depending on the application. Besides blow-molded bottles, the alloys are suitable for thermoformed products such as tubs, trays and blister pods. The material is rigid and tough and has good heat distortion resistance up to 180 F.

Another group of plastics, polyphenylene-oxide (PPO) can be blended with polystyrene to produce an alloy with improved processing traits and lower cost than non-alloyed PPO. The addition of polystyrene reduces tensile strength and heat deflection temperature somewhat and increases thermal expansion.

PROPERTIES OF TYPICAL PLASTICS ALLOYS

	ABS-Polycarbonate	PVC-Acrylic
Specific Gravity	1.04	1.35
Tens. Str., psi	6200	6500
Tens. Mod., 10^5 psi	3.7	3.4
Imp. Str., ft lb/in. notch	10	15
Heat Deflection Temp., F @ 255 psi	250	160

PLASTISOL COATINGS

Vinyl plastisols, or pastes, as they are described in Europe, are suspensions of vinyl resin in nonvolatile oily liquids known as plasticizers. They vary in viscosity from a motor oil consistency to a puttylike dough. In the more viscous state the plastisol is termed plastigel, while the more fluid materials, to which volatile diluents have been added, are known as modified plastisols. Modified plastisols differ from organosols in the function of the volatile components. In organosols, the volatiles are used as resin dispersants, whereas in modified plastisols they serve as diluents to adjust fluidity and are generally present in small quantities.

The polyvinyl chloride resin, resembling confectioner's sugar, is blended into a mixture of one or more plasticizers to form a suspension. This fluid remains essentially unchanged until heat is applied. During the heating process, the dispersion first sets or gels; this is followed by solution or fusion of the resin in the hot plasticizer to form a single phase solid solution. Upon cooling, the coating assumes the properties of a tough, rubbery plastic.

These plastisols have no adhesion to metals or dense, nonporous substrates and consequently require the use of adhesive primers for bonding. To a large extent, the nature of these primers determines the suitability of plastisol coatings for specific applications.

Application methods

The fluidity or absence of fluidity in the liquid plastisol is sometimes deceiving. These materials are supplied at very high solids content and consequently exhibit non-Newtonian flow (viscosity varies with applied shear). While the viscosity cup is satisfactory for many paints and lacquers, and may even suffice for organosols, it can only serve to mislead the plastisol user. Viscosity of plastisols should be specified and measured using a viscosimeter capable of operating over a range of shear rates preferably within the area of use.

Spread coating. Roller and knife coaters are the two major types of spread-coating equipment used for handling plastisols. Fabric, paper and even strip steel are all being coated with this type of process. Compound viscosity characteristics, speed of coating, clearance between the web and the knife or roll, type and angle of the knife are all factors in determining the quality of the coating. Heavy paper and fabric coatings, which will withstand folding and forming, may be applied to porous stock without danger of penetration or strike through. The momentary application of heat to the coated side of the stock will fuse the plastisol with a minimum of thermal action on the paper.

Plastisol-coated strip steel is currently being produced by roller-coating processes for use by appliance and other manufacturers.

Dip coating. Two dip-coating processes are available. In hot dipping, the object for coating is prebaked, prior to immersion in the plastisol. The heat content in the article serves to gel a deposit on the surface of the object. This gelled coating must then be fused by baking. Plastisol formulation and temperature, dipping rate, mass, shape and heat content of the article to be coated all serve to determine the thickness and nature of deposit.

Cold-dip processes permit the application of from 1 to 60 mils per coat without the necessity of a prebake operation. Cold dips lend themselves to conveyor line coating of products. These coatings have a high yield value and permit controlled film thicknesses without the presence of sags or drips to mar the appearance.

Spray coating. Plastisols may be spray applied through either pressure or suction type guns, but generally the pressure equipment is preferred since it permits faster delivery with a minimal use of volatile diluents. Airless spray equipment is currently available which operates at fluid pressures in excess of 2000 psi and requires no atomizing air. This airless-spray process is reported to give extreme smoothness to highly thixotropic formulations.

Molding coating. There are a number of applications in which equipment is lined using a molding technique. In this process, a polished steel core is employed and the void between the casting and core filled with plastisol. After baking, the steel core is removed leaving behind a molded lining. This process is particularly adaptable to regular-shaped articles such as piping.

Properties

Although plastisols may be modified with slight additions of volatile diluents, their fluidity is mainly due to the presence of large quantities of plasticizer. Unlike the fluid phase of the organosol, which is largely volatile, the plasticizers remain behind after baking, as a portion of the fused film. Thus, the plastisol tends to be softer, and more resilient than the organosol. Its low volatile content (or its absence altogether) permits wide baking latitude by eliminating the problem of mud cracking and reducing the solvent entrapment tendency found in organosols. Film thicknesses may range from 2 to 250 mils per coat.

Prolonged exposure to elevated temperatures causes thermal degradation of the polyvinyl chloride by splitting off HCl. Maximum recommended service temperature ranges between 150 and 200 F. The following tabulation outlines the

PHYSICAL PROPERTIES

Density of Dried Film, lb/sq ft	0.185 to 2.25
Tensile Strength, psi	200 to 4000
% Elongation	100–600
Hardness (Shore A)	20–100
Cold Temperature Flexibility, °F	−65°
Specific Resistivity (ohm-cm)	10^{14} to 10^{15}
Dielectric Strength, v/mil	500

properties of plastisols for coating application. It should be noted that verification by test is suggested for marginal applications since performance in these areas is a function of the particular formulation examined. The indiscriminate use of plastisols for chemical application is never advisable and the recommendations of the lining supplier should be followed.

A more detailed tabulation of physical, me-

chanical and electrical properties may be found under Organosol Coatings.

CHEMICAL PROPERTIES

Chemical	Conc, %	Max Temp, F	Limitations
Acids			
Acetic	10	70	Verify for conc >2%
Chromic	40	140	
Hydrochloric	25	70	Verify for conc >10%
Lactic	Conc	90	
Nitric	35	70	
Oleic	Conc	90	
Phosphoric	85	150	Verify for conc >75%
Sulfuric	50	150	
Alkalies			
Ammonium Hydroxide	Conc	90	
Calcium Hydroxide	Conc	150	
Sodium Hydroxide	35	90	
Salt Solutions	Conc	150	
Solvents			
Aliphatic Hydrocarbon		150	
Alcohol		90	Verify
Formaldehyde	Conc	70	Verify
Fats and Oils		90	Verify
Moist Gases		90	Verify
Water		150	Verify for contamination if water is distilled

Uses

Plastisols, because of their ability to be applied readily in heavy thicknesses, found early success in the electroplating field as rack coatings. The plastisol serves as an insulation, confining current to the work being plated, and is resistant to chemical attack by plating solutions. The use of plastisols as linings for tanks, chemical equipment and steel drums followed.

In fabric coating, they replaced solution coatings by eliminating the need for expensive solvents lost in the baking operation. Rubber has been replaced by vinyl plastisols and organosols as coatings for wire baskets because of their superior resistance to moisture and detergents.

One of the most dramatic applications of plastisols is as a lining for kitchen dishwashers. The use of a plastisol lining permitted a lightweight tub design not possible with porcelain enameling in which firing resulted in buckling and warping of light-gauge steel. Plastisols also served to reduce scrap units since defects may be readily patched and repaired. The resistance to impact damage, etching and enamel erosion are other factors which prompted manufacturers to select plastisols for this application.

A list of applications and the properties related to the specific application follow:

Application	Related Property
Industrial: Tool handles, stair treads, conveyor hooks, conveyor rollers, railings.	Resiliency, thermal and electrical insulating qualities, resistance to abrasion.
Electrical: Bus bars, conduit boxes, battery clamps and cases, toggle switches, electroplating racks and plating barrels.	Dielectric strength, electrical resistivity, resistance to moisture and chemicals.
Linings: Tanks, ductwork, pumps, filter presses, centrifugal cleaners, dishwasher tubs, piping, drums and shipping containers.	Resistance to abrasion and impact, resistance to moisture and chemicals.
Wire Goods: Dish-drain baskets, egg baskets, deep-freeze baskets, refrigerator shelves, record racks, clothes hangers.	Resistance to moisture, detergents and staining, resiliency.
Miscellaneous: Bottles and glassware, glove coating, bobby-pin coatings.	Resiliency and esthetic qualities, abrasion resistance, softness.

PLASTISOLS

Plastisols are dispersions of high molecular weight vinyl chloride polymer or copolymer resins in nonaqueous liquid plasticizers which do not dissolve the resin at room temperature. Plastisols are converted from liquids to solids by fusing under heat, which causes the resin to dissolve in the plasticizers.

There are many advantages in molding with plastisols each varying in importance according to the particular type of molding application. Vinyl plastisol is supplied as a liquid and consequently is easy to handle. The material requires no catalysts or curing agents to convert it to a solid, only moderate heat in the range of 300-400 F. Vinyl plastisol does not require a long baking cycle nor high pressures to fuse and shape it. Consequently, lightweight inexpensive molds are suitable for molding. It can be formulated to have a virtually indefinite shelf life. Plastisols are usually 100% solids and shrinkage from the mold is at an absolute minimum, thus assuring that the molded object is exact and consistent.

Properties

Chemical and physical properties of plastisols can be varied throughout a wide range. This versatility makes plastisols adaptable to a multitude of end uses.

The following general ranges indicate the properties which may be compounded into plastisols:

Specific Gravity: 1.05—1.35
Tensile Strength: As required to 4,000 psi
Elongation: As required to 600%
Flexibility: Good to a temperature as low as −65 F
Hardness: From 10 to 100 on the Shore A Durometer Scale; up to 80 on the Shore D Durometer Scale.
Chemical Resistance: Outstanding to most acids, alkalies, detergents, oils and solvents.
Heat Resistance: Can resist 225 F for as long as 2,000 hours and 450 F for over 2 hours.
Electrical Properties: Dielectric strength at a minimum of 400 v/mil in thicknesses of 3 mils and over.
Flammability: Slow burning to self-extinguishing.
Colors: All colors available including phosphorescent and fluorescent shades.

Molding methods

There are several different methods by which plastisols may be molded.

Pour and injection. Two of the simplest are pour molding and low-pressure injection molding. The first method entails merely pouring plastisol into a cavity until it is filled and subsequently fusing the compound. This system is used in manufacturing products such as plastic doilies, sink stoppers and display plaques.

If the mold is closed, a low-pressure injection system such as a grease gun can be used to inject the liquid plastisol into the cavity. A low-pressure injection mold should be designed with bleeders at the extremities of the cavity to insure complete filling of the mold, as well as to relieve the minor pressure on the mold surface caused by expansion during the heating. Laboratory models, novelties and electrical harnesses are products which are commonly low pressure injection molded with vinyl plastisol.

The proper fusing time for a quantity of plastisol varies with the formulation and the thickness of the material to be fused. It has been determined that the fusing time for a ⅛-in. thickness should be at least seven minutes at the fusing temperature of the compound, which normally is in the range of 300 to 400 F. In the case of many sections being fused at once, undoubtedly a longer heating period would be necessary in order to bring the entire mass of material up to the proper temperature.

Heating sources used for molding plastisols vary according to the particular product under consideration. For shallow, open molds, such as those for plastic doilies, radiant heat would be satisfactory. When this type of heat is used, the material thickness should not be so great that the open surface exposed to the heat overfuses in the time taken for the temperature to reach the mold surface of the part. Conductive heat is another source for fusing plastisol, in particular when closed molds are employed. Immersing the mold in a hot bath or using cartridge heaters are two conductive heating methods. A more commonly used method of fusion is convection heat. The advantage of a convection oven, in particular a forced air type, is that the entire inner area of the oven, and consequently the entire surface of the mold, is maintained at a constant temperature, ensuring a more even heat transfer through the mold.

Pour and low-pressure injection molds usually are made of aluminum, electroformed copper, brass or steel. The thickness of the metal should be kept to a minimum for good heat transfer yet should be thick enough to withstand expansion pressure during fusing. Two-piece molds should be machined to provide a tight fit at the parting line in order to minimize flashing, and the finish on the interior of the mold should be as smooth as the texture desired on the surface of the molded piece.

In-place molding. Another molding process, which is somewhat similar to the foregoing methods, is in-place molding. This method permanently attaches plastisol to another component during the fusing process. The combination serves an important functional purpose and usually eliminates the need for several assembly steps. In-place molding is most commonly used in forming seals and gaskets of various sorts. For example, the paper and metal screen elements of an automotive air cleaner are placed in a shallow, ringlike mold in which has been poured a measured quantity of plastisol. Upon fusing, the plastisol seals the filter elements together into a permanent shape with good structural strength and no additional gaskets are needed for use with the filter. Gaskets are applied to vitrified clay pipe by pouring plastisol into special molds on the bell and spigot ends. Such gaskets compensate for inherent out-of-roundness of the pipe. Flowed in gaskets also are applied to bottle caps and jar lids.

Dip molding. When a hollow object is to be molded and the internal dimensions are of importance, many times a dip molding process is employed. The metal molds are shaped according to the interior design of the molded object. They are usually solid and are made of cast or machined aluminum, machined brass, steel or ceramic. These molds are preheated to a temperature in the range of 300 to 400 F, dipped into the plastisol and allowed to dwell until the proper thickness has gelled on the mold. In order to eliminate drips or sags, the mold is withdrawn at a rate that does not exceed the rate at which the liquid residue drains from the gelled coating. The thickness of this coating can be varied by altering the preheating time and temperature, as well as the dwell time. The mold sometimes is inverted after withdrawal in order to allow any excess to flow back evenly onto the coating. It is then placed in the oven in order to fuse the gelled coating, cooled upon removal and the molded piece stripped from the mold. In some cases, the stripped piece is turned inside out so that the designs of the mold can be displayed on the exterior of the molded piece.

A dip molding system can easily be conveyorized. In such a system mandrels holding the molds would be conveyed through a preheat oven, dipping station, fusing oven, cooling station and stripping station.

Slush molding. Another method for molding pieces which are hollow is slush molding. In this process an open-end metal mold is heated to a temperature in the range of 300-400 F and then filled with plastisol. The plastisol is allowed to dwell until the desired thickness has gelled on the inner surface of the mold and then the remaining liquid in the mold is poured back into a reservoir for use again. The mold with the gelled inner coating is placed in an oven where the plastisol is fused. Upon cooling, the plastisol part is stripped from the mold, retaining the design of its inner surface on the exterior of the piece. Molds of electroformed copper or fine sandcast aluminum are usually used for slush molding.

The above process is generally known as the single-pour system of slush molding. For a mold of intricate detail, a two-pour method often is used. In this process, the mold is filled when it is cold, vibrated to remove bubbles, and then emptied, leaving a thin film of plastisol on the inner surface. In this way, the plastisol does not have a chance to gel before flowing into the mold extremities. The first skin is gelled when the mold is heated. It is refilled again and another thicker film of plastisol gels over the first skin. The mold is then returned to the oven for the final fusing. In both of the foregoing cases, the thickness of the molded piece is determined by the time and temperature used in preheating the mold and the dwell time of the plastisol in the mold.

Rotational molding. Completely enclosed hollow parts can be produced by rotational molding. A measured amount of plastisol is poured into one-half of a two-piece mold. The mold is closed and rotated in two or more planes while being heated. During this rotation, the plastisol flows, gels and fuses evenly over the interior walls of the mold. Molds for this operation are either electroformed copper or cast or machined aluminum, and the molds are arranged in clusters or "gangs" so that the maximum number of molds can be operated per spindle.

In rotational molding it is possible to vary the thickness of the walls of the molded piece. One way this can be accomplished is by rotating the mold more in one plane than in another. An alternate way is to construct the mold so that the metal is thicker in some sections than in others. When such a mold is rotated the heat transfer will be slower through the thicker sections and, consequently, the wall of the molded piece will be thinner in this area.

A few familiar products manufactured by this process are toys and novelties such as dolls and beach balls, swimming pool floats, and artificial fruit.

Combinations. In many cases several of the above molding methods are combined in order to produce a product made from plastisol. For example, vinyl foam products such as armrests, toys or electrical harnesses are manufactured by first forming a tough vinyl skin by spraying, slush molding or rotational molding. The interior then is formed by casting, low-pressure injection, or rotational molding a vinyl plastisol foam within the pregelled skin.

PLATINUM METALS

The platinum group metals—ruthenium, rhodium, palladium, osmium, iridium, and platinum are found in the second and third long periods in Group VIII of the periodic table. Platinum and palladium are the most abundant of the group although all are generally found together.

The outstanding characteristics of platinum, the most important member of the group, are its remarkable resistance to corrosion and chemical attack, high melting point, retention of mechanical strength and resistance to oxidation in air, even at very high temperatures. These qualities, together with the ability of the metal to greatly influence the rates of reaction in a large number of chemical processes, are the basis of nearly all of its technical applications. The other five metals of the platinum group are also characterized by high melting points, good stability and resistance to corrosion. Addition of these metals to platinum forms a series of alloys that provide a wide range of useful physical properties combined with the high resistance to corrosion that is characteristic of the parent metals.

Platinum group metals

Table 1 shows some physical and mechanical properties of the platinum group metals. Each possesses specific characteristics which distinguish it from the others.

Platinum (Pt) is a white, malleable and ductile metal that takes a very high permanent polish. When heated to redness it softens and is easily worked. It is virtually nonoxidizable and is soluble only in liquids generating free chlorine, such as aqua regia. At red heat platinum is attacked by cyanides, hydroxides, sulfides and phosphides. When heated in an atmosphere of chlorine, platinum volatizes and condenses as the crystalline chloride. Reduction of platinum chloride with zinc gives platinum black which has a high adsorptive capacity for hydrogen. Platinum sponge is finely divided platinum.

Palladium (Pd) is silvery white, very ductile and slightly harder than platinum. It is readily soluble in aqua regia and is attacked by boiling

TABLE 1—PHYSICAL AND MECHANICAL PROPERTIES OF PLATINUM GROUP METALS

	Platinum (thermopure)	Palladium	Iridium	Rhodium	Ruthenium	Osmium
Density (68 F), lb/cu in.	0.775	0.434	0.813	0.449	0.441	0.813
Melting Point, F	3216	2826	4449	3571	4530±180	4900±360
Ther Cond (32–212 F), Btu/ft/in./°F/hr	480	493	406	618	—	—
Ther Coef of Exp (32–212 F), 10^{-6} per °F	4.94	6.50	3.61	4.6	5.33	3.33
Spec Ht, Btu/lb/°F						
32 F	0.0315	0.0584	0.0307 (68 F)	0.058	0.057	0.0309
212 F	0.0325	—	—	—	—	0.0314
Elec Res, ohms/cir mil-ft						
32 F	58.86	60.0	—	30.1	47	54
68 F	63.60	64.8	—	—	84	57.1
212 F	81.90	—	—	—	—	—
Mean Temp Coef of Res (32–212 F), per °F	0.00392	0.0037	0.00392	0.00457	—	0.0042
Tensile Strength, 1000 psi						
Annealed	17–19 [a]	28	21	80	—	—
50% Reduction	34 [b]	47	—	360	—	—
Vickers Hardness						
Cast	45	49	183	147	231	362
Annealed	42	46	189	144	310	381
50% Reduction	106	118	351	401	—	—

[a] Comparable value for mechanical grade: 20–23,000 psi.   [b] Comparable value for mechanical grade: 35,000 psi.

nitric and sulfuric acids. Palladium has the remarkable ability to occlude large quantities of hydrogen. When properly alloyed it can be used for the commercial separation of and purification of hydrogen. Palladium and platinum can both be worked by normal metalworking processes.

Iridium (Ir) is the most corrosion-resistant element known. It is a very hard, brittle, tin-colored metal with a melting point higher than that of platinum. It is soluble in aqua regia only when alloyed with sufficient platinum. Iridium has its greatest value in platinum alloys where it acts as a hardening agent. By itself it can be worked only with difficulty.

Rhodium (Rh) serves an important role in high temperature applications up to 3000 F. Platinum-rhodium thermocouple wire makes possible high temperature measurement with great accuracy. Rhodium and rhodium alloys are used in furnace windings and in crucibles at temperatures too high for platinum. It is a very hard, white metal and is workable only under certain conditions, and then with difficulty. Applied to a base metal by electroplating, it forms a hard, wear-resistant, permanently brilliant surface. Solubility is slight even in aqua regia.

Ruthenium (Ru) is hard and brittle with a silver-gray luster. Its tetraoxide is very volatile and poisonous. When alloyed with platinum its

effect on hardness and resistivity is the greatest of all the metals in the group. It is unworkable in the pure state.

Osmium (Os) has the highest specific gravity and melting point of the platinum metals. It oxidizes readily when heated in air to form a very volatile and poisonous tetraoxide. Application has been predominantly in the field of catalysis. As a metal it is also practically unworkable.

Platinum alloys

Platinum is alloyed to obtain greater hardness, strength and electrical resistivity. Because most applications require freedom from corrosion, the other platinum metals are usually employed as alloying agents.

Platinum-iridium. Iridium is the addition to platinum most often used to provide improved mechanical properties. It increases resistance to corrosion while the alloy retains its workability. Up to 20% iridium the alloys are quite ductile. With higher iridium content fabrication becomes difficult.

Platinum-rhodium. The addition of rhodium to platinum also provides improved mechanical properties to platinum and increases its resistance to corrosion. For applications at high temperatures the platinum-rhodium alloys are preferred because of retention of good mechanical properties including good hot strength and very little tendency toward volatilization or oxidation.

Platinum-ruthenium. Alloying platinum with ruthenium has the most marked effect upon both hardness and resistivity. However, the limit of workability is reached at 15% ruthenium. The lower cost and the lower specific gravity of ruthenium offer an appreciable economic benefit as an alternate to other platinum alloys.

Platinum-gold. Platinum-gold alloys cover a wide range of compositions and provide distinct chemical and physical characteristics.

Palladium alloys

Electrical resistivity of palladium is the highest of the platinum group metals and its temperature coefficient of resistance is the lowest of the group with figures closely approximating those of platinum. Palladium is sometimes used as a substitute for platinum but begins to tarnish at above 750 F. It will maintain a low and steady contact resistance but is inferior to platinum in this respect.

Palladium is the principal constituent in several contact alloys, the main consideration being an improvement in hardness and strength without affecting the corrosion-resistant properties of the parent metal.

More recently alloys of palladium with silver have been developed as membranes for the selective separation of ultrapure hydrogen from hydrogen containing gas mixtures. Unlike palladium, these alloys are not destroyed by alternating phase changes that occur when the pure metal is heated and cooled a number of times in hydrogen. In addition to this remarkable stabilizing effect, the permeability of an alloy membrane to the passage of hydrogen is over twice that of palladium at 600 F and nearly 25% greater at 1000 F.

Applications

In an age when metals must perform under greater stresses and disintegrating exposure, platinum and its alloys fill the requirements where no other metal would endure. Frequently this is accompanied by a substantial cost saving since the noble metals are completely recoverable and can be used again and again. In some instances, because of price increases, a platinum user has received more for his worn equipment than its original cost.

Due to its long life, high purity, outstanding corrosion resistance, ability to withstand high temperatures and selective catalytic properties, industrial uses of platinum and its alloys are expanding. Engineers, designers and researchers have learned that when conditions are beyond the capabilities of other materials the platinum metals may offer the least expensive solution. Chemical processing equipment, corrosion-resistant electrodes, catalysis, temperature measurement, electrical contacts, high temperature furnaces, glass manufacture, synthetic fiber, dentistry, medicine, rocketry and missiles are fields in which the platinum group metals are becoming increasingly useful.

PLYWOOD

Plywood is a term generally used to designate glued wood panels made up of layers, or plies, with the grain of one or more layers at an angle, usually 90°, with the grain of the others. The outside plies are called faces or face and back, the center plies are called the core, and the plies immediately below the face and back, laid at right angles to them, are called the crossbands.

The core may be veneer, lumber, or various combinations of veneer and lumber; the total thickness may be less than 1/16 in. or more than 3 in.; the different plies may vary as to number, thickness, wood species. Also, the shape of the members may vary. The crossbands and their arrangement generally govern both the properties (particularly warping characteristics) and uses of all such constructions.

Composition

The composition of a plywood panel is generally dependent upon the end use for which it is intended. The number of plywood constructions is almost endless when one considers the number of wood species available, the many thicknesses of wood veneers used in the outer plies or cores, the placement of the adjacent plies, the types of adhesives and their qualities, various manufacturing processes and more technical variations.

Conventional plywood generally consists of an odd number of plies with the grains of the alternate layers perpendicular to each other. The use of an odd number permits an arrangement that gives a substantially balanced effect; that is, when three plies are glued together with the grain of the outer two plies at right angles to that of the center ply, the stresses are balanced and the panel tends to remain flat with changes in moisture content. These forces may be similarly balanced with five, seven or some other uneven number of plies. If only two plies are glued together with the grain of one ply at right angles to the other, each ply tends to distort the other when changes in moisture content occur; cupping will result.

Grades and types

Broadly speaking, two classes of plywood are available—hardwood and softwood. Most softwood plywood is composed of Douglas fir, but western hemlock, white fir, ponderosa pine, redwood and other wood species are also used. Hardwood plywood is made of many wood species.

Various grades and types of plywood are manufactured. "Grade" is determined by the quality of the veneer and "Type" by the moisture resistance of the glue line. For example, there are two types of Douglas fir plywood—interior and exterior. The interior type is expected to retain its form and strength properties when occasionally subjected to wetting and drying periods. It is commonly bonded with urea formaldehyde resin type adhesives. On the other hand, the exterior type is expected to retain its form and strength properties when subjected to cyclic wetting and drying and to be suitable for permanent exterior use. It is commonly bonded with hot-pressed phenolic-resin glues.

Several grades of plywood are established within each type depending on the quality of the veneer on the two faces of the panel. The veneer is designated "A," "B," "C," or "D" in descending order of quality. For example, "Grade A-D" plywood has Grade A veneer on one face and Grade D on the other. The types and grades in general use are listed in the various commercial standards established by the plywood industry in cooperation with the Department of Commerce. These are commercial standards applicable to the various grades and types of hardwood plywoods also.

Engineering properties

The mechanical and physical properties of plywood are dependent upon the particular construction employed. Plywood may be designed for beauty, durability, rigidity, strength, cost or many other properties. With practically an unlimited variety of constructions to choose from, there is a wide range of differing characteristics in any given plywood panel. Most important among properties are the following.

High strength-weight ratio. Perhaps the most notable feature of plywood is its high strength-weight ratio. Plywood is given special consideration whenever lightness and strength are desired. Plywood is widely used for concrete form work, floor underlayment, roof decks, siding and many other applications because of its high strength-weight ratio. A comparison between birch plywood and other structural materials shows that its strength-weight ratio is 1.52 times that of 100,000-lb test heat-treated steel and 1.36 times that of 10,000-lb test aluminum.

TABLE 1—STRENGTH PROPERTIES OF SOME PLYWOODS

Species	Density, lb/sq in.	[a]Ult. Tens. Str., 1000 psi	[a]Mod. Elast. in Bend., 10^6 psi	[b]Max. Crush. Str., 1000 psi
Douglas Fir				
Three-ply	0.31	6.5	1.39	3.51
Five-ply	0.77	5.7	1.14	3.13
Birch				
Three-ply	0.39	9.06	1.74	4.02
Five-ply	0.98	7.97	1.42	3.57
Mahogany				
Three-ply	0.30	7.02	1.29	3.93
Five-ply	0.75	6.17	1.06	3.52
Yellow Poplar				
Three-ply	0.28	4.82	1.15	2.38
Five-ply	0.66	4.65	.94	2.30
Sweet Gum				
Three-ply	0.31	6.72	1.26	2.98
Five-ply	0.77	5.91	1.03	2.65

Three-Ply: Face and back, 0.030 in.; core, 0.040 in.
Five-Ply: Face and back, 0.047 in.; core, 0.040 in.; cross bands, 0.060 in.
[a]Parallel to surface.
[b]Parallel to face.

Bending properties. A most desirable characteristic of plywood is its flatness but it can and will support substantial curvatures without appreciable loss of strength. Standard construction plywood can be bent or shaped to nominal radii and held in place with adhesives, nails, screws or other fixing methods.

The radius of curvature to which a panel can be formed varies with panel thickness and the specie or species of wood employed in the panel construction. The arc of curvature is limited by the tension force in the outer plies of the convex perimeter and by the compression forces in the outer plies of the concave perimeter. A ¼-in. dry Douglas fir panel can be bent to a radius of 24 in. lengthwise and 15 in. crosswise.

Waterproofed plywood, soaked or steamed before bending, exhibits approximately 50% greater flexibility than panels bent when dry.

Resistance to splitting. Because plywood has no line of cleavage, it cannot split. This is an exceptional property when one considers its effect on fastening. The crisscross arrangement of wood plies in plywood construction develops extraordinary resistance to pull-through of nail or screw heads.

Resistance to impact. The absence of a cleavage line has a pronounced effect on the impact resistance of plywood. Plywood will fracture only when the impact force is greater than the tensile strength of the wood fibers in the panel composition. Under an impact force, the side of the panel opposite the impact point will rupture along the long grain fiber followed by successive shattering of the various plies. Splintering usually does not take place because pressure is dispersed throughout the panel at the point of impact. Under similar conditions, solid lumber will show complete rupture.

Beauty. Plywood has certain intrinsic qualities that add much to any structure or construction in which it is used. Because of improved modern methods of manufacture, there is practically no limit to the decorative potential of plywood.

The entire range of fine woods is at the designer's disposal; they vary in shade from golden yellow to ebony, from pastels to reds and browns. The use of bleaches, toners and stains in manufacturing procedures gives the designer even greater latitude with respect to design freedom.

The fine woods employed in the manufacture of plywood offer warmth and charm to any decorative scheme because the surface of the wood variously absorbs, reflects and refracts light rays, giving the wood pattern depth and making it restful to the eye. This phenomenon accounts for the play of color and pattern when a plywood panel is viewed from different angles.

Dimensional stability. The absorption of water causes wood to swell and this movement is much greater across the grain than along the grain. The alternating layers of veneers in standard plywood construction inhibits this cross-grain movement because the cross-grain weakness is reinforced by the long-grain stability. Therefore, the dimensional stability of a plywood panel can be controlled by controlling its moisture content. In the field, this control can be achieved by applying coatings such as paints, lacquers and sealers of various types.

Thermal insulating qualities. The thermal insulating qualities of plywood are the same as those of the wood of which it is composed. For example, a Douglas fir panel has a coefficient of heat transmission of 0.78 Btu per in., the same as Douglas fir lumber.

The use of plywood as an insulating material can be attributed to two factors: (1) The use of large sheets reduces the numbers of cracks and joints and thereby inhibits wind leakage; and (2) the resistance of plywood to moisture vapor transmission stabilizes the moisture content of the trapped air and maintains its insulating qualities.

Fire resistance. Fire-resistant plywood is manufactured by impregnating the core stock with a salt solution which upon evaporation leaves a salt deposit in the wood. Plywood or wood treated in this manner will not support combustion but will char when heated beyond the normal charring point of the wood.

Nonimpregnated plywood can be made fire resistant by applying surface coatings such as intumescent paint and chemicals such as borax. An intumescent paint has a silicate of soda base which bubbles or intumesces in the presence of heat, thus forming a protective coating. At high temperatures, borax releases a gas such as carbon dioxide, which blankets the fire. It must be noted that fire-resistant coatings are effective only in direct ratio to their thickness. Highly resistant plywood can be manufactured by using an incombustible core such as asbestos.

Resistance to borers. Plywood panels are subject to attack by borers to the same extent as the wood species of which they are composed but phenol-formaldehyde-resin glue lines are fairly effective barriers against further penetration. Panels may also be treated with pentachlorophenol for increased resistance to these pests.

Fatigue resistance. Plywood has the same resistance to fatigue as the wood of which it is composed.

Fabrication

The fact that almost anyone can use plywood has contributed greatly to its wide acceptance in many varied applications. The utilization of plywood does not require special tools, special skills or safeguards and practically anyone capable of handling a saw and hammer can make use of its inherent engineering properties.

Since plywood does not exhibit the typical cross-grain weakness of lumber, it can very often be used in place of lumber for various applications. For example, $5/16$-in. thick plywood replaces conventional $3/4$-in. sheathing; $1/4$-in. plywood can be used for interior wall paneling without sheathing and $3/8$-in. plywood can be used for shipping containers, furniture and case goods instead of $3/4$-in. lumber. The use of the thinner plywood reduces weight, bulkiness and is less fatiguing for tradesmen to handle. The use of large plywood sheets instead of narrow boards also reduces the amount of cutting, fitting and fastening involved in a particular job.

Plywood is especially adaptable to the portable power-driven saws, drills, and automatic hammers normally used on production or construction jobs. Multiple cutting with band saws may be done with assurance because even the thinnest plywood has strength in all directions and the danger of splitting or chipping is reduced to a minimum. This quality is particularly important when fine fitting is required.

Whenever plywood is employed in a structure, fewer and smaller fastenings can be specified because consideration need be given only to the holding power of the fastener and the tensile strength of the fastening itself.

Available forms, sizes, shapes

Plywood is available in practically any size, but the 4 by 8 ft panel has become the standard production unit of the industry. Larger panels usually demand a price premium and are available on special order. Panels with continuous cores and faces can be produced in one piece up to 12 ft long.

Oversize panels up to 8 ft in width and of unlimited length can be manufactured by scarf jointing. (A scarf is an angling joint, made either in veneers or plywood, where pieces are spliced or lapped together. The length of the scarf is usually 12 to 20 times the thickness. When properly made, scarf joints are as strong as the adjacent unspliced material.)

A considerable quantity of plywood is produced with overlays of paper or fabric impregnated with synthetic resins, plastic film or plastic sheet. An overlay of this type is sometimes applied to one face to provide a desired decorative effect combined with good wearing characteristics. A product of this type is widely used for counters and table tops. If a decorative overlay is applied on one face, a plain overlay is usually applied on the back to reduce the tendency to warp. In many cases, an overlay is applied primarily to improve serviceability and painting characteristics, or to reduce face roughness. As a general rule, decorative overlays are applied in a separate operation after the plywood panel has been fabricated; plain overlays are laid at the time of gluing the veneer into plywood.

Decorative plywood is now commercially available with a plant-applied finish. Prefinished wall paneling is supplied with the finish varying from offset printing to polyester film.

General fields of application

There are many uses and applications for plywood in industry today. Owing to the wide diversity of plywood applications, only the more prominent ones are mentioned here:

Architectural	Marine construction
Aviation	Mock-ups, models
Boatbuilding	Paddles
Building construction	Panel boards
Cabinet work	Patterns
Concrete forms	Prefabrication
Containers, cases	Remodeling
Die boards	Sheathing
Display	Signs
Fixtures	Sporting goods
Floor underlayment	Table tops
Furniture	Toys
Hampers	Trays
Luggage	Truck floors, bodies
Machine bases	Wall paneling

POLYACRYLIC RUBBER

The development of polyacrylic rubbers was pioneered by the Department of Agriculture's Eastern Regional Research Laboratory, the B. F. Goodrich Company and the University of Akron Government Laboratories. The first types were identified as Lactoprene EV and Lactoprene BN, and were proposed as oxidation-resistant elastomeric materials. The first commercial products were made available in 1948. Chemically, they were polyethyl acrylate, and a copolymer of ethyl acrylate and 2-chloro ethyl vinyl ether.

The development of polyacrylic rubbers was accelerated by the interest expressed throughout the automotive industry in the potential applications of this type of polymer in special types of seals. An effective seal for today's modern lubricants must be resistant not only to the action of the lubricant but to increasingly severe temperature conditions. It must also resist attack of highly active chemical additives which are incorporated in the lubricant to protect it from deterioration at extreme temperature.

Polyacrylic rubber compounds were developed to provide a rubber part which would function in applications where oils and/or temperatures as high as 400 F. would be encountered. These were also very resistant to attack by sulfur-bearing chemical additives in the oil. These properties have resulted in general use of polyacrylic rubber compounds for automotive rubber parts as seals for automatic transmission fluids and extreme pressure lubricants.

Cured polyacrylic rubber will exhibit the following properties:

1. Temperature resistance to + 400 F.
2. Resistance to oxidation at normal and elevated temperatures.
3. Good flex life.
4. Resistance to sunlight fading.
5. Excellent ozone resistance.
6. Good resistance to swelling and deterioration in oils, particularly sulfur-bearing oils at high temperatures.
7. Resistance to permeability by many gases.
8. Permanence of color in white or pastel shades.
9. Physical properties in the following range:

Tensile strength, psi	500–2400
Elongation, %	100–400
Hardness, Durometer A	40–90

Polyacrylic rubber will prove most useful in fields where these special properties are used to the maximum. It is recommended for products such as automatic transmission seals, extreme pressure lubricant seals, searchlight gaskets, belting, rolls, tank linings, hose, o-rings and seals, white or pastel colored rubber parts, solution coatings, and pigment binders on paper, textiles and fibrous glass.

Curing

A typical polyacrylic rubber, such as the copolymer of ethyl acrylate and chloroethyl vinyl ether, is supplied as a crude rubber in the form of white sheets having a specific gravity of approximately 1.1. It may be mixed and processed according to conventional rubber practice.

However, polyacrylic rubber is chemically saturated and cannot be cured in the same manner as conventional rubbers. Sulfur and sulfur-bearing materials act as retarders of cure and function as a form of age resistor in most formulations. Polyacrylic rubber is cured with amines; "Trimene Base" and triethylene tetramine are most widely used. Aging properties may be altered by balancing the effect of the amine and the sulfur. Widely used curing systems are: (a) 3.0 parts of "Trimene Base" with 0.5 parts of sulfur or (b) 2 parts benzothiazyl disulfide and 1.5 parts triethylene tetramine.

Like other rubber polymers, reinforcing agents such as carbon black or certain white pigments are necessary to develop optimum physical properties in a polyacrylic rubber vulcanizate. Selection of pigments is more critical in that acidic materials, which would react with the basic amine curing systems, must be avoided. The SAF or FEF carbon blacks are most widely used, while hydrated silica or precipitated calcium silicate are recommended for light-colored stocks. It is possible to make white compounds which really stay white in service.

Typical curing temperatures are from 290 to 330 F at cure times of 10 to 45 min. depending on the thickness of the part. Polished, chromium-plated molds are recommended. For maximum over-all physical properties, the cured parts should be tempered in an air oven for 24 hours at 300 F.

Forming

In order to obtain smooth extrusions, more loading and lubrication are necessary than for molded goods, because of the inherent nerve of the polymer. Temperatures of 110 F in the barrel and 170 F on the die are recommended.

Generally, those compounds which extrude well, are also good calendering stocks. Suggested temperatures for calendering are in the range of

100 to 130 F. Higher temperatures will result in sticking of the stock to the rolls. Under optimum conditions, 15 mil films may be obtained.

Polyacrylic rubber may be coated on nylon either by calendering or from solvent solution. It also has excellent adhesion to cotton and is often used as a solvent solution applied to cotton duck to be used as belting. Solvents generally used include methylethyl ketone, toluene, xylene or benzene.

Polyacrylic rubber is most widely used in many types of seals because of its excellent resistance to sulfur-bearing oils and lubricants. This is illustrated in the accompanying table which compares resistance to Hypoid Oil of polyacrylic rubber and an excellent oil-resistant nitrile rubber. The nitrile rubber exhibits almost complete loss of tensile strength and elongation; it becomes hard and brittle. The polyacrylic rubber, under the same conditions, remains flexible and would continue to function satisfactorily as a seal.

RESISTANCE OF POLYACRYLIC RUBBER TO SULFUR-STABILIZED OILS

Recipe No.	1	2*
Hycar 1041	100	—
Hycar 4021	—	100
Zinc oxide	5	—
FEF black [1]	40	40
Tetramethylthiuram disulfide [2]	3.5	—
Stearic acid	1	1
Sulfur	—	0.5
Trimene Base	—	3
Totals	149.5	144.5

[1] Philblack A   [2] Methyl Tuads

ORIGINAL PROPERTIES—All Cures at 310°F.

Modulus at 100% elongation (psi)		
Minutes cured—30	380	870
Ultimate tensile strength (psi)		
Minutes cured—30	2880	1670
Ultimate elongation (%)		
Minutes cured—30	570	170
Hardness (Durometer A)		
Minutes cured—30	70	66
Block cured 45 minutes at 310°F.		
Compression set—ASTM Method B		
70 hours at 212°F. (%)	29	22
70 hours at 250°F. (%)	45	26
70 hours at 300°F. (%)	—	54

* Recipe No. 2 tempered 24 hours at 300°F.

IMMERSIONS—Samples cured 30 minutes at 310°F.

Hypoid Oil—70 hours at 300°F.		
Ultimate tensile strength (psi)	0	1080
Tensile change (%)	—100	—35
Ultimate elongation (%)	10	120
Elongation change (%)	—98	—29
Hardness (Durometer A)	90	73
Hardness change (points)	+20	+7
Volume change (%)	+1.6	+9.4
180° bend	Cracked	Pass

In general, polyacrylic rubber vulcanizates are resistant to petroleum products and animal and vegetable fats and oils. They will swell in aromatic hydrocarbons, alcohols and ketones. Polyacrylic rubber is not recommended for use in water, steam, ethylene glycol, or in alkaline media.

Laboratory tests indicate that polyacrylic vulcanizates become stiff and brittle at a temperature of —10 F. But in actual service, these same polyacrylic rubbers have been found to provide satisfactory performance at engine start-up and operation in oil at temperatures as low as —40 F.

For those applications requiring improvement in low temperature brittleness by as much as 25 F and which can tolerate considerable sacrifice in over-all chemical oil and heat resistance, a copolymer of butyl acrylate and acrylonitrile may be used. Compounding and processing techniques for these polymers are, in general, the same as described above.

POLYBUTADIENE RUBBER

Polybutadiene may be prepared in several ways to yield different products. The method of polymerization can have a marked effect on polymer structure which in turn controls the properties of the polymer and thus its ultimate end use.

Composition

When polybutadiene is made, the butadiene molecule may enter the polymer chain by either 1,4-addition or 1,2-addition. In 1,4-addition, the unsaturated bonds may be either of *cis* or of *trans* configuration. Polybutadienes containing a more or less random mixture of these polymer units can be prepared with alkali metal catalysts or with emulsion polymerization systems but these have not achieved commercial significance as general-purpose rubbers in the United States.

In recent years, new catalysts have been developed which allow the structure of the polymer chain to be controlled. Polybutadienes containing in excess of 85% of either *cis, trans* or vinyl unsaturation have been studied and, in the case of *cis* and *trans* polymers, produced commercially. It is also possible to prepare a number of other polymers with various combinations of these three component structures. Of current importance are those types commercially available which are: (1) polymers of high *cis* 1,4-content (more than 85%) with low *trans* and vinyl content, (2) polymers of more than 80% *trans*, low *cis* and low vinyl content and (3) polymers of intermediate *cis* content (approximately 40% *cis,* 50% *trans*) with low vinyl content.

High-*cis* polybutadiene

Outstanding properties of high-*cis* polybutadiene are high resilience and high resistance to abrasion. The high resilience is indicative of low hysteresis loss under dynamic conditions, i.e., low heat rise in the polymer under rapid, repeated deformations. In this respect *cis*-polybutadiene is similar to natural rubber. The combination of good hysteresis properties and resistance to abrasion makes this polymer attractive for use in tires, especially in heavy-duty tires where heat generation is a problem.

Another favorable factor, particularly for use in tire bodies, is that high-*cis* polybutadiene imparts good resistance to heat degradation under heavy loads in dynamic applications. Tensile strength (in reinforced stocks) is lower than for natural rubber or SBR but is adequate for many uses. Modulus depends on the degree of cross-linking but *cis*-polybutadiene generally requires relatively low stress to reach a given elongation (at slow deformation). Hardness may be varied depending on the compound formulation and is similar to that of SBR and natural rubbers. Ozone resistance is typical of unsaturated polymers and is inferior to that of saturated rubbers. Oil resistance is comparable to that of SBR and natural rubber. Permeability to gases is higher than that of most other rubbers which may be either an advantage or a disadvantage, depending on the application. Freeze point is quite low; *cis*-polybutadiene tread compounds do not become brittle until very low temperatures are reached, lower than −150 F.

Typical properties of one of the high-*cis* polybutadienes ("Cis-4" polymer with a *cis* content

of about 95%) are shown in Table 1 in com-

TABLE 1—PROPERTIES OF HIGH-*cis* POLYBUTADIENE

	cis-Poly-butadiene	Natural Rubber	SBR 1500
300 % Modulus, psi	1100	1900	1500
Tensile Strength, psi	2650	3900	3600
Elongation, %	550	500	550
Tensile Strength at 200 F, psi	1500	3000	1700
Heat Rise (T-100 F), F	44	40	63
Resilience, %	75	71	60
Effective Dynamic Modulus, psi	1220	940	1170
Hardness, Shore A	61	64	61
Brittle Point, F	<−160 [a]	−72 [a]	−75

[a] Crystallizes at higher temperature.

parison with SBR 1500 and natural rubber. Each compound is reinforced with 50 phr (parts per 100 parts rubber) HAF carbon black in a tire-tread formulation (also contains other common ingredients such as plasticizing oil, sulfur, accelerator, zinc oxide and organic acid).

Abrasion resistance of the high-*cis* polybutadienes varies with the type of service and in nearly all cases is superior to that of natural rubber or SBR. The advantage for these polybutadienes in relation to natural rubber and SBR increases as the severity of the service increases. From 30% to as high as 100% improvement in wear resistance has been reported through the substitution of these polybutadienes for natural rubber in tire treads.

Processing. Polybutadiene rubbers in general are more difficult to process in conventional equipment than natural rubber, particularly with regard to milling and extrusion operations. The processing problems have been overcome in many cases by treatment of the polymer, changes in compounding formulations or by blending with other rubbers. Blends of natural rubber and high-*cis* polybutadiene are particularly attractive. The presence of natural rubber alleviates the processing difficulties and improves tensile and tear properties of the polybutadiene while the latter improves abrasion resistance and complements the already good hysteresis properties of natural rubber.

Processability can also be improved by the use of higher levels of reinforcing fillers and oils than normally employed. Consideration must also be given to changes in quality, but extension with 35 to 70 phr oil is possible with retention of properties suitable for tires and many rubber goods. Even in blends with natural rubber, where processing is not a problem, increases in carbon black and oil content have proved practical and often desirable.

As a general rule, internal mixers are the most suitable for mixing polybutadiene rubbers with the necessary compounding ingredients such as reinforcing fillers and plasticizers. High-*cis* polybutadienes are more resistant to breakdown during mixing than many rubbers and may require high temperatures and selected peptizers to achieve a reduction in viscosity. To obtain satisfactory processability during mixing, milling or extrusion operations it may be desirable to use blends or extension with oil and fillers as discussed above or to seek specific recommendations of the manufacturer.

Applications. For most uses, vulcanization is necessary to develop the desired strength and elastic qualities. The polymer chain is unsaturated and can be readily vulcanized with sulfur (in conjunction with the usual activators) or with other crosslinking agents such as peroxides. With some exceptions, admixture of the rubber with a reinforcing pigment is required to obtain high strength. Antioxidants and/or antiozidants should be used for most applications. Lower than normal sulfur levels provide a better balance of properties in many instances, particularly with blends of high-*cis* polybutadiene and natural rubber.

In large tires, the use of blends of high-*cis* polybutadiene and natural rubber in treads improves both abrasion resistance and resistance to tread groove cracking compared to natural rubber alone. Substitution of high-*cis* polybutadiene for a portion of the natural rubber in the tire body has improved resistance to blowout or other heat failures (in some cases to a remarkable degree).

Examples of tire tread stocks containing high-*cis* polybutadiene with large amounts of black and oil or blends of this polymer and natural rubber are shown in Table 2.

Processing	Fair	Good	Good	Good
Tensile Strength, psi	2400	3000	3600	4100
Elongation, %	500	550	600	570
Heat Rise (T-100 F), F	66	56	46	41
Resilience, %	60	63	71	67
Hardness, Shore A	61	61	62	64
Typical Abrasion Index (Road Tests)	160	145	115	100

[a] Phillips Cis-4 polybutadiene (*cis* content about 95%).

The high-*cis* polybutadienes are also used in increasing quantity in blends with styrene-butadiene rubber (SBR) for the production of tires for passenger cars and small trucks. In this use, advantages include improved resistance to abrasion and cracking as well as adaptability to extension with large amounts of oil and carbon black.

As indicated, properties of high-*cis* polybutadiene are well-suited for tire use and this is expected to be the major application. Its use should be considered in other areas where high resilience, resistance to abrasion or low-temperature resistance is required. Sponge stocks, footwear, gaskets and seals and conveyor belts are possible applications. In shoe heels high-*cis* polybutadiene may be used to provide good resilience and better abrasion resistance than realized with other commonly used rubbers. In shoe soles designed to be soft and resilient, high-*cis* polybutadiene has been used as a partial or total replacement for natural rubber to maintain high resilience, improve abrasion resistance and to provide better resistance to crack growth.

An important use for high-*cis* polybutadiene is in blends with other polymers to improve low-temperature properties. Replacement of 35 percent of an acrylonitrile (nitrile) rubber in a compound with *cis*-polybutadiene can reduce the brittleness failure temperature from −40 F to −65 F. Similarly, it is possible to reduce the brittle point of neoprene compounds by some 20 F with the substitution of *cis*-polybutadiene for one-fourth of the neoprene rubber. Such substitutions usually reduce resistance to swelling in hydrocarbons but this effect can be lessened by compounding with oil-resistant resins or by using *cis*-polybutadiene as a replacement for plasticizer in the compound rather than as a replacement for the polymer. In many cases, of course, oil resistance is not required and *cis*-polybutadiene may be blended with various polymers to give low-temperature properties approaching those of special arctic rubbers such as

TABLE 2—HIGH-*cis* POLYBUTADIENE TIRE STOCKS

cis-Polybutadiene, parts [a]	100	60	50	—
Natural Rubber, parts	—	40	50	100
Carbon Black (ISAF), parts	70 (HAF)	60	45	45
Petroleum Oil, parts	30	18	5	5

butadiene-styrene copolymers with low styrene content.

Partial substitution of *cis*-polybutadiene for other rubbers used in light-colored or black-reinforced mechanical goods can provide a product that displays better snap. Also, resilience, abrasion resistance and low-temperature properties are usually improved; tensile strength and tear strength may be reduced but show less decline after aging.

High-*cis* polybutadienes have been used successfully as base components of caulking compounds and sealants. Another use is as a base polymer for graft polymerization of styrene to produce high-impact polystyrene.

Commercially available *cis*-polybutadiene rubbers are supplied in bale form. The rubber contains a small amount of antioxidant to provide stability during storage.

High-*trans* polybutadiene

In contrast to the soft, rubbery nature of *cis*-polybutadiene, polybutadiene of high *trans* content (available from Phillips Chemical Co. under trade name of "Trans-4" polybutadiene) is a hard, horny material at room temperature. It is thermoplastic and thus can be molded without addition of vulcanization agents. The softening point, hardness and tensile strength increase with increasing *trans* content. A polymer of approximately 90% *trans* content, for example, displays a softening point near 200 F, Shore A hardness of about 98 and tensile strength in excess of 1000 psi without addition of other ingredients.

High-*trans* polybutadiene can be readily vulcanized with sulfur or peroxides. It remains a hard, partially thermoplastic material when lightly vulcanized but can be made rubbery by increasing the curative level. Tensile strength is increased by the addition of reinforcing pigments (carbon black, silica or clay). Compounded in such a manner, high-*trans* polybutadienes are characterized by high modulus, tensile strength, elongation and hardness, by moderate resilience and by excellent abrasion resistance. Properties are shown in Table 3 for a 93% *trans*-polybutadiene reinforced with carbon black (50 phr black, 5 phr softener, 1.75 phr sulfur plus small amounts of zinc oxide, stearic acid and accelerator).

In vulcanized stocks properties such as hardness, resilience and heat buildup can be modi-fied considerably by variation in curative level. A high-*trans* polybutadiene containing carbon black and 2.0 phr sulfur may display hardness of 96, resilience of 65 percent and heat buildup of 90 F. The same compound with 6.0 phr sulfur rather than 2.0 phr will have a hardness of 76, resilience of 74% and heat buildup of 38 F. Very low compression set values can be realized with a peroxide cure.

Uses. Applications include those where balata has been used such as golf ball covers and wire and cable coverings. Properties have also been demonstrated to be suitable for shoe soles (high hardness, good abrasion resistance), floor tile (high hardness, low compression set), gasket stocks, blown sponge compounds and other molded or extruded items. High-*trans* polybutadiene requires processing temperatures above its softening point but is relatively easy to mix, mill or extrude at these temperatures. Care should be taken to avoid scorch or precure when handling stocks containing curatives at elevated temperatures.

TABLE 3—PROPERTIES OF HIGH-*trans* POLYBUTADIENE

	High-*trans* Polybutadiene	SBR 1500	Natural Rubber
300% Modulus, psi	1850	1500	2000
Tensile Strength, psi	3200	3500	4200
Elongation, %	690	530	500
Resilience, %	61	61	71
NBS Abrasion, rev/mil	197	11	12
Shore A Hardness, 80 F	97	59	64
212 F	58	56	59
300 F	59	56	59

Polybutadiene of intermediate *cis* content

This type of product has a structure of approximately 40% *cis*, 50% *trans* and 10% (or less) vinyl content (available from Firestone Tire and Rubber Co. under trade name of "Diene"). The raw polymer is soft and somewhat waxlike in character. This polybutadiene displays high resilience, high dynamic modulus and a low brittle point. Properties of a polybutadiene of intermediate *cis* content ("Diene"), published by the manufacturer, are listed in Table 4 (50 phr HAF carbon black).

TABLE 4—POLYBUTADIENE OF INTERMEDIATE *cis* CONTENT

	40% *cis*-Polybutadiene	Natural Rubber
300% Modulus, psi	1525	1925
Tensile Strength, psi	2550	3650
Elongation, %	440	480
Rebound (Steel Ball) (73 F), %	69	54
Rebound (Steel Ball) (212 F), %	71	72
Dynamic Modulus (212 F), psi	312	202
Hardness, Shore A	65	62
Bell Brittle Point, F	−141	−72

Processing difficulties in milling and extrusion operations may be encountered and it is usually recommended that polybutadienes of intermediate *cis* content be used in blends with other rubbers. In tire treads substitution of 40% *cis*-polybutadiene for a portion of the natural rubber or SBR 1500 improves resilience and abrasion resistance and gives some reduction in operating temperature. Other applications include products where similar changes in properties to those above are desired or in products where improvements in low temperature properties are desired.

Liquid polymers

Liquid polybutadienes can be prepared in most of the systems used to make solid rubbers. Such polymers may be crosslinked with chemicals or solidified with heat. One type of liquid polybutadiene (available from Phillips Chemcal Co. under the trade name Butarez) has been manufactured on a small scale utilizing sodium catalyst. Uses for this type polymer include coatings, binders, adhesives, potting agents, casting and laminating resin or vulcanizable plasticizer for rubber.

POLYCARBONATE PLASTICS

Polycarbonate resins offer a combination of properties that extends the usefulness and fields of application for thermoplastic materials. This relatively new plastic material is characterized by very high impact strength, superior heat resistance and good electrical properties.

In addition, the material's low water absorption, high heat distortion point, low and uniform mold shrinkage, and excellent creep resistance result in especially good dimensional stability.

Of value for many applications is the material's transparency, shear strength, stain resistance, colorability and gloss, oil resistance, machinability, and maintenance of good properties

over a broad temperature range from less than −100 F to 270 to 280 F. The fact that polycarbonate resin is self-extinguishing is important in many applications.

Composition

The polycarbonate name is taken from the carbonate linkage which joins the organic units in the polymer. This is the first commercially useful thermoplastic material which incorporates the carbonate radical as an integral part of the main polymer chain.

There are several methods by which polycarbonates can be made. One method involves a bifunctional phenol, bisphenol A, which combines with carbonyl chloride by splitting out hydrochloric acid to give a linear polymer consisting of bisphenol groups joined together by carbonate linkages. Bisphenol A, which is the condensation product of phenol and acetone, is the basic building block used also in the preparation of epoxy resins.

Other members of the polycarbonate family may be made by using other phenols and other ketones to modify the isopropylidene group, or to replace this bridge entirely by other radicals. Substitutions on the benzene ring offer further possibilities for variations.

Properties

Parts molded of polycarbonate resin have very high impact strength. Notched Izod impacts, using the ⅛ in. test specimen, average 12 to 16 ft-lb/in. of notch. Unnotched Izod value is more than 60 ft-lb/in. In the newer tensile impact test, polycarbonate resin molded parts give values from 600 to 900 ft-lb/cu in. These high impact values are one of the significant properties of this new resin, being exceptionally high for a rigid thermoplastic.

High heat resistance is another characteristic of the resin. Under standard ASTM conditions, with the test specimen loaded to a fiber stress of 264 psi, polycarbonate thermoplastic resin has a heat distortion point of 270 to 280 F. The material exhibits little load sensitivity, and essentially the same value is obtained with a lower stress of 66 psi. Because of its essentially all-aromatic constitution, the resin also shows excellent resistance to thermal-oxidative degradation at temperatures up to 300 F.

Low water absorption of molded parts is important because moisture often affects dimensional and chemical stability, as well as mechanical and electrical properties. Exposed to 50% relative humidity at room temperature, polycarbonate parts (1/8-in. thickness) reach equilibrium in about ten days' time at a low 0.15% water absorption. When immersed in water at room temperature, a maximum of 0.35% water is absorbed. In boiling water, the maximum water absorption of 0.59% is rapidly attained and no further absorption occurs.

Molded polycarbonate parts show excellent dimensional stability under a variety of conditions. Four of the aging conditions in Table 1 produced no dimensional change. After aging 90 days at 125 C, a molded part showed a change in dimension of 1 mil/in.

Under the conditions of the heat deformation test, polycarbonate parts show little deformation, indicative of the material's excellent resistance to creep. Standard 1/2-in. test pieces under 1000 lb load for 24 hr at 75 F and at 158 F, per ASTM D621-51, show dimensional changes no greater than 0.3%. These measurements characterize the resin as a rigid plastic approaching the thermosets in this property at these temperatures.

TABLE 1—DIMENSIONAL STABILITY [a]
(Avg dimensional change in mils per in.)

30 days at 70 C	0
90 days at 125 C	−1
24 hours in water at RT	0
4 days at 100% RH at 100 F	0
7 days in oil at 160 F	0

[a] G.E.'s Lexan polycarbonate.

Significant among the thermal properties of polycarbonate resin is its high heat distortion temperature of 270 to 280 F. Also of importance in many applications is the fact that the material is self-extinguishing, as determined by ASTM method D-635.

Tables 2 and 3 summarize the properties of molded polycarbonate resin.

In chemical resistance, polycarbonates are stable to water, dilute mineral and organic acids, aliphatic hydrocarbons and some alcohols. They are attacked and decomposed by alkaline solutions, ammonia and amines. They are easily soluble in chlorinated hydrocarbons and par-

tially soluble in aromatic hydrocarbons and ketones. They are resistant to coffee, grape juice, orange juice, tomato juice, catsup, tincture of iodine, merthiolate, milk, wine, beer and whiskey.

In outdoor weathering tests at Pittsfield, Mass., polycarbonate specimens have shown no change in appearance after one year. Prolonged outdoor exposure produces surface dulling and gradual degradation of exposed surfaces.

Fabrication

Polycarbonate resin has been molded in standard injection equipment using existing molds designed for nylon, polystyrene, acrylic, or other thermoplastic materials. Differences in mold shrinkage must be considered. And, in fabrication, the polycarbonate does have its own unique processing characteristics. Most important among these are the resin's broad plastic range and high melt viscosity. Production runs in molds designed for nylon or acetal resin are not recommended.

Like other amorphous polymers, polycarbonate resin has no precise melting point. It softens and begins to melt over a range from 420 F to 440 F. Optimum molding temperatures lie above 520 F. The most desirable range of cylinder temperatures for molding the resin is in the area of 525 to 600 F.

Mold design must take into consideration the material's high melt viscosity. Large sprues, large full-round runners, and generous gates with short lands usually give best results. Tab gating is good for filling large thin sections. The gate to the tab should be large.

In injection molding polycarbonate, the following conditions are desirable:

1. Heated molds. Generally, hot water heat is adequate for heating molds, with typical mold temperatures ranging from 170 to 200 F. Molds for large areas, thin sections, or complex shapes or multiple cavity molds having long runners may require higher temperatures. Some molds have run best at 230 to 250 F.

2. Cylinder temperatures. The most usual cylinder temperatures for molding polycarbonates are in the range of 525 to 600 F. Few parts require cylinder temperatures above 600 F. A heated nozzle and adequate mold temperature are helpful in keeping cylinder temperatures below 600 F. In most cases rear cylinder tem-

TABLE 2—PROPERTIES OF MOLDED POLYCARBONATE RESIN [a]

Physical Properties	Units	Value	ASTM Test
Color	—	Light Amber, transparent	—
Spec Grav	—	1.20	D 792
Odor	—	None	—
Taste	—	None	—
Refractive Index at 77 F	—	1.586	—
Rockwell Hardness	—	M70, R118	D 785
Abrasion Res, Taber Abraser with CS-17 Wheel	mg/1000 cycles	7–11	D 1044
Impact Strength, Notched Izod, ⅛ in. Specimen	ft-lb/in. of notch	12–16	D 256
Impact Strength, Unnotched Izod, ⅛ in. Specimen	ft-lb/in.	>60	D 256
Tensile-Impact	ft-lb/cu in.	600–900	—
Tensile Yield Strength	psi	8,000–9,000	D 638
Tensile Ultimate Strength	psi	9,000–10,500	D 638
Tensile Mod	psi	340,000	D 638
Elongation	%	60–100	D 638
Compressive Strength	psi	11,000	D 695
Compressive Mod	psi	240,000	D 695
Flex Strength	psi	11,000–13,000	D 790
Flex Modulus	psi	340,000	D 790
Shear Yield Strength	psi	5,400	D 732
Shear Ultimate Strength	psi	9,200	D 732
Light Transmission (⅛ in. thick disc)	%	84–86	—
Water Vapor Permeability	g. cm/hr cm^2 mm Hg	3–4×10^{-8}	—
Nitrogen Permeability	cc (STP) mm/sec cm^2 cm Hg	0.012×10^{-8}	—
Carbon Dioxide Permeability	cc (STP) mm/sec cm^2 cm Hg	0.32×10^{-8}	—
Bulk Factor of Pellets	—	1.74	—
Thermal Properties			
Heat Distortion Temp			
264 psi	°F	270–280	D 648
66 psi	°F	283–293	D 648
Mold Shrinkage	in./in.	0.005–0.007	D 955
Thermal Conductivity, °F	Btu/hr/sq. ft/°F/ft.	0.1112	—
Coef of Linear Ther Expansion			
−103 to 32 F	in./in./°F	3.6×10^{-5}	D 696
32 to 136 F	in./in./°F	3.8×10^{-5}	D 696
140 to 284 F	in./in./°F	4.1×10^{-5}	D 696
Crystalline Melt Point	°F	514	—
Flammability	—	Self-extinguishing	D 635
Brittle Temperature	°F	<-211	D 746
Electrical Properties			
Arc Resis			D 495
Stainless-Steel Strip Electrodes	sec	10–11	
Tungsten Electrodes	sec	120	
Dielec Strength, S/T, 77 F			D 149
3.0 mils	v/mil	3,080	
23.0 mils		1,130	
125.0 mils		400	
Dielec Strength, S/T, 212 F			D 149
3.0 mils	v/mil	3,380	
23.0 mils		1,250	
125.0 mils		600	
Res to Electron Beam Radiation		No significant change up to 5×10^7 r.	—

[a] G.E.'s Lexan polycarbonate.

TABLE 3—ELECTRICAL PROPERTIES AT VARIOUS TEMPERATURES [a]

	−22 F	9 F	73 F	212 F	257 F	ASTM Test
Dielec Constant						
60 cycles	3.12	3.14	3.17	3.15	3.13	D 150
10^6 cycles			2.96			D 150
Power Factor						
60 cycles	0.005	0.004	0.0009	0.0009	0.0011	D 150
10^6 cycles			0.0100			D 150
10^8 cycles			0.0100			
Volume Resistivity, ohm-cm	$>10^{17}$	$>10^{17}$	2.1×10^{16}	2.1×10^{15}	2.7×10^{14}	—

[a] G.E.'s Lexan polycarbonate.

peratures higher than front cylinder temperatures give best results.

3. Heated nozzle. In general, nozzle temperature equal to front cylinder temperature gives good results.

4. Adequate injection pressure. Injection pressures used in molding polycarbonate resin range from 10,000 to 30,000 psi. Most usual range is in the 15,000 to 20,000 psi range. Typical pressure setting is ¾ to full pressure capacity of the press.

5. Fast fill time. For most parts molded, a fast ram travel time has been found desirable for thick as well as thin sections. For very thick sections, it is better to utilize somewhat slower ram speed.

Polycarbonate must be well dried to get optimum properties in the molded part. For this reason the resin is packaged in sealed containers. Recommended drying conditions for regrind material and material left exposed to air after the sealed container has been opened are 250 F for at least 12 hours, and preferably overnight, with the resin placed in trays to a depth of 1 in. or less. Preheating pellets in the can to 250 F for 4 to 8 hours and use of hopper heaters at 250 F are recommended for production operations to prevent moisture pickup.

While polycarbonates are fabricated primarily by injection molding, other fabricating techniques may be used. Rod, tubing, shapes, film and sheet may be extruded by conventional techniques. Films and coatings may be cast from solution. Parts can readily be machined from rod or standard shapes. Cementing, painting, metalizing, heat sealing, welding, machining operations and other standard finishing operations may be employed. Film and sheet can be vacuum-formed or cold-formed.

Applications

The properties of polycarbonate resin make this new plastic suitable for a wide variety of applications. It is now being used in business machine parts, electrical and electronic parts, military components, and aircraft parts; and is finding increasing use in automotive, instrument, pump, appliance, communication equipment, and many other varied industrial and consumer applications.

One of the applications is in molded coil forms which take advantage of the resin's electrical properties, heat and oxidation resistance, dimensional stability, and resistance to deformation under stress.

A transparent plastic having the heat resistance, the dimensional stability and the impact resistance of polycarbonate resin has created considerable interest for optical parts, such as outdoor lenses, instrument covers and lenses and lighting devices.

Housings make use of the material's impact resistance and its attractive appearance and colorability. In many cases, also, heat resistance and dimensional stability are important.

An interesting application area for plastic materials is the use of polycarbonate resin for fabricating fasteners of various types. Such uses as grommets, rivets, nails, staples, and nuts are in production or under evaluation. The ability of polycarbonate parts to be cold-headed has developed considerable interest in rivet applications.

Terminal blocks, connectors, switch housings,

and other electrical parts may advantageously be molded of polycarbonate resin to take advantage both of the electrical properties of the material and the unusual physical properties, which give strength and toughness to the parts over a range of temperatures. Because of polycarbonate's heat resistance, and the fact that it is self-extinguishing, molded parts can be used in current-carrying support applications.

Another application is in bushings, cams and gears. Here dimensional stability is important as is the high impact strength of the resin. Good physical properties over a broad temperature range, low water absorption, resistance to deformation under load and resistance to creep suggest its use for many applications of this type. However, the resin has a higher coefficient of friction and a lower fatigue endurance limit than do some other plastics used in these types of applications. For this reason, it should not be considered a general purpose gear and bearing material, but might be considered for applications subject to light loading, or to heavier but intermittent loading.

POLYESTER FILM

Polyester film is a transparent, flexible film, ranging from 0.15 to 14 mils in thickness, used as a product component, in industrial processes, and for packaging.

Several types have been announced by producers. The most widely used is produced from polyethylene terephthalate (i.e., Mylar) and this article is based on characteristics obtained with this polymer. Polyester films based on other polymers or copolymers or manufactured by other methods are not identical, though they are similar in nature.

The polyethylene terephthalate polymer (a saturated polyester) is a product of the condensation reaction between ethylene glycol and terephthalic acid. The film is oriented in such a way that properties are uniform in all directions of the sheet, and many of the properties are dependent upon this orientation. The film contains no plasticizers and consequently does not tend to lose desirable properties as a result of age or environmental conditions.

Engineering properties

It is the strongest of all plastic films and strength is probably the outstanding property.

However, it is useful as an engineering material because of its combination of desirable physical, chemical, electrical and thermal properties. For example, strength combined with heat resistance and electrical properties makes it a good material for motor slot liners; strength plus dimensional stability under most environmental conditions makes it an excellent base for reliable magnetic recording tapes.

Tensile strength is 23,000 psi; tensile modulus, 550,000 psi; break elongation, 100 percent. Impact strength and bursting strength are high. Tear propagation strength is moderate, but this is offset by the fact that tears are difficult to start. The flex life of one-mil film is over 100,000 cycles. Density is 1.38 and the area factor is 20,000 sq in./lb per mil.

Typical electrical property values are shown below:

Dielectric strength (25 C, 60 cps), v/mil	7000
Dielectric constant (25 C, 60 cps)	3.25
Power factor (25 C. 60 cps)	0.002
Insulation resistance (100 C), megohm-mfds	5000
Volume resistivity (25 C), ohm-cm	1×10^{19}
Surface resistivity (25 C, 0% RH), ohm	$>10^{12}$

Polyester film has excellent resistance to attack and penetration by solvents, greases, oils and many of the commonly used electrical varnishes. At room temperature, permeability to such solvents as ethanol, ethyl acetate, carbon tetrachloride, hexane, benzene, acetone and acetic acid is very low. It is degraded by some strong alkali compounds and embrittles under severe hydrolysis conditions.

Moisture absorption is less than 0.8% after a week's immersion at 25 C. The hygroscopic coefficient of expansion is 11×10^{-6} in./in./%RH. Water-vapor permeability is similar to that of polyethylene film and permeability to gases is very low. The film is not subject to fungus attack and copper corrosion is negligible.

An outstanding feature of the film is the fact that good physical and mechanical properties are retained over a wide temperature range. Service temperature range is − 60 to 150 C. The effect of temperature is relatively small between − 20 and 80 C. No embrittlement occurs at as low as − 60 C, and useful properties are retained up to 150 to 175 C. Tensile modulus drops off sharply at 80 to 90 C.

Melting point is 250 to 255 C; thermal coefficient of expansion is 15×10^{-6} in./in./°F; and shrinkage at 150 C is 2 to 3%.

Fabrication and forms

Polyester film can be printed, laminated, metalized, coated, embossed and dyed. It can be slit into extremely narrow tapes ($\frac{1}{64}$ in. and narrower) and light gauges can be wound into spiral tubing. Heavy gauges can be formed by stamping or vacuum (thermo-) forming. Matte finishes can be applied, and adhesives for bonding the film to itself and practically any other material are available.

Because of its desirable thermal characteristics, polyester film is not inherently heat sealable. However, some coated forms of the film can be heat-sealed and satisfactory seals can be obtained on the standard film by the use of benzyl alcohol, heat and pressure.

Polyethylene terephthalate film is available in several different types:

A. General-purpose and electrical film for wide variety of uses.

C. Special electrical applications requiring high insulation resistance.

D. Highly transparent film, minimum surface defects.

K. Coated with a polymer for heat sealability and outstanding gas and moisture impermeability.

HS. Shrinks uniformly about 30% when heated to approximately 100 C. After shrinking, it has substantially the same characteristics as the standard film.

T. A film with high tensile strength (available in some thin gauges) with superior strength characteristics in the machine direction; designed for use in tapes requiring high-strength properties.

W. For outdoor applications; resistant to degradation by ultraviolet light.

Eleven gauges are available (not in all types) ranging from 0.00015 to 0.014 in. Standard types are available in both rolls and sheets. Maximum width is approximately 72 in.

Applications

The balance of properties outlined above has made polyester film functional in many totally different industrial applications and suggests its use in numerous other ways.

Largest current user of the film is the electrical and electronic market, which uses it as slot liners in motors and as the dielectric for capacitors, replacing other materials which are less effective, bulkier and more expensive. It is found in hundreds of wire and cable types, sometimes used primarily as an insulating material, sometimes for its mechanical and physical contributions to wire and cable construction. Reduced cost of materials and processing and improved cable performance result.

Magnetic recording tapes for both audio and instrumentation uses are based on polyester film. In audio applications they contribute toughness, durability and long play; for instrumentation tapes, the film insures maximum reliability. The film has proved to be a highly successful new material for the textile industry since it can be used to produce metallic yarns that are nontarnishing and unusually strong. They can be run unsupported, knit, dyed at the boil, and either laundered or dry-cleaned. Yarns are made by laminating the film to both sides of aluminum foil or by laminating metalized and transparent film. The structure is then slit into the required yarn widths.

As a surfacing material, polyester film is used on both flexible and rigid substrates for both protective and decorative purposes. Metalized, laminated to vinyl, and embossed, the film becomes interior trim for automobiles, for example.

With a special coating, the film becomes a drafting material that is tougher and longer lasting than drafting cloth. It is used for map making, templates, and other applications in which its dimensional stability becomes a significant factor.

Strength in thin sections makes the film advantageous for sheet protectors, card-holders, sheet reinforcers, and similar stationery products. Pressure-sensitive adhesives make the film's properties available for uses ranging from decorative trim to movie splicing.

The weatherable form (Type W) has a life of four to seven years. It is principally used in greenhouses, where it cuts construction costs by as much as two-thirds since a simple, inexpensive structure suffices and maintenance costs are at a minimum.

In the packaging field, polyester film serves in areas where other materials fail or have functional disadvantages. In window cartons it lasts longer and does not break as other materials do; its toughness permits transparent packaging of heavy items; and, coated with polyethylene, it has made possible the "heat-in-the-bag" method of frozen-food preparation.

An optically clear form of polyester film in thicknesses of four to seven mils is used as a base for the coating of light-sensitive emulsions in the manufacture of photographic film. The outstanding qualities of toughness and dimensional stability make this film especially well suited as a base for graphic arts films, motion picture film, engineering reproduction films, and microfilm. Other advantages include excellent storage and aging characteristics.

POLYESTER PLASTICS

The materials to be discussed here are commonly called polyester resins but this simple name does not distinguish between at least two major classes of commercial materials. Also, the same name is used within the unsaturated class to designate both the cured and uncured state. The subject covered here may be defined by identifying the materials as "unsaturated polyester resins which, when cured, yield thermoset products as opposed to thermoplastic products." The latter, as exemplified by Dacron and Videne, are saturated polyesters.

Composition

Unsaturated polyester resins of commerce are composed of two major components, a linear, unsaturated polyester and a polymerizable monomer. The former is a condensation type of polymer prepared by esterification of an unsaturated dibasic acid with a glycol. Actually most polyesters are made from two or more dibasic acids.

The most commonly used unsaturated acids are maleic or fumaric with limited quantities of others being used to provide special properties. The second acid does not contain reactive unsaturation. Phthalic anhydride is most commonly used but adipic acid and, more recently, isophthalic acid, are employed. The properties of the final product can be varied widely, from flexible to rigid, by changing the ratios and components in the polyester portion of the resin.

The polyesters vary from viscous liquids to

hard, brittle solids but, with a few exceptions, are never sold in this form. Instead, the polyester is dissolved in the other major component, a polymerizable monomer. This is usually styrene; diallyl phthalate and vinyl toluene are used to a lesser extent. Other monomers are used for special applications. These solutions of unsaturated polyesters in the polymerizable monomers are sold by resin manufacturers as "polyester resins." They are liquids which vary in viscosity from thin syrups to very thick honey depending partly on the amount of monomer present. Conventional resins contain from 20 to 40% monomer, and have specific gravities ranging from 1.10–1.20.

Polyester resins may contain numerous minor components such as light stabilizers and accelerators, but must contain inhibitory components so that storage stability is achieved. Otherwise polymerization will take place at room temperature. Most resin manufacturers guarantee six months or longer storage stability.

The most widely used polyester resins are supplied with viscosities in the range of 300 to 5000 centipoises. They are clear liquids varying in color from nearly water-white to amber. They can be colored with certain common pigments which are available ground in a vehicle for ease of dispersion. Inert, inorganic fillers such as clays, talcs, calcium carbonates, etc., are often added, usually by the fabricator, to reduce shrinkage and lower costs. Resins with thixotropic properties are also available.

Curing

The liquid resins are cured by the use of peroxides with or without heat to form solid materials. During the cure the monomer copolymerizes with the double bonds of the unsaturated polyester. The resulting copolymer is thermoset and does not flow easily again under heat and pressure. Heat is evolved during the cure. This must be considered when thick sections are made. Also, a volume shrinkage occurs and the density increases to 1.20–1.30. The amount of shrinkage varies between 5–10% depending on monomer content and degree of unsaturation in the polyester. The most popular catalyst is benzoyl peroxide but methylethyl ketone peroxide, cumene hydroperoxide and others find application.

Reinforcement

The development of this field of commercial resins owes a great deal to the commercial production of two other products: The first was the production of low-cost styrene for the synthetic rubber program. Other cheap monomers will work but not nearly as well. Styrene and its homologues have relatively high boiling points and fast copolymerization rates. Both properties are important in this field.

The second development was the commercial production of fibrous glass. Polyester resins are not widely used as cast materials nor are the physical properties of such castings particularly outstanding.

The unique property of polyesters is their ability to change from a liquid to a hard solid in a very short time under the influence of a catalyst and heat. This property was not available in any of the earlier plastic materials. Polyesters flow easily in a mold with little or no pressure so that expensive, high-pressure molds are not required. Alternatively, very large parts can be made because the total pressure required to form the material is low.

Glass fibers of fine diameter have high tensile strengths, good electrical resistance properties, and low specific gravity when compared to metals. When such fibers are used as reinforcement for polyester resins (like steel in concrete) the resulting product possesses greatly enhanced properties. Specific physical properties of the polyester resins may be increased by a factor varying from 2 to 10. Naturally the increase in physical strengths obtained will depend upon both the amount of glass fibers used and their form. Strengths approaching those of metals on an equal weight basis are obtained with some constructions.

Fabrication techniques

Hand lay-up. This method involves the use of either male or female molds. Products requiring highest physical properties are made with glass cloth. The cloth may be precoated with resin but usually one ply is laid on or in the mold, coated with resin by brushing or spraying, and then a second ply of cloth laid on top of the first. The process is continued until the desired thickness is built up. Aircraft parts, such as radomes, usually require close tolerances on dimensions and resin-to-glass weight ratio. After the lay-up has been completed, pressure is applied either by covering the assembly with a flexible, extensible blanket and drawing it down by vacuum or the mold is so made that a rubber bag can be contained above the part. This is then blown up to apply pressure on the laminate. Thereafter the cure is accomplished by heating in an oven, by infrared lamps or heating means built into the mold.

Corrugated sheet molding. Glass-reinforced polyester sheets are sold in large volume and are made by a relatively simple intermittent or continuous method. The process consists of placing a resin-impregnated mat between cellophane sheets, rolling or squeegeeing out the air, placing the assembly between steel or aluminum molds of the desired corrugation and curing the assembly in an oven.

Matched die molding. This method produces parts rapidly and generally of uniform quality. The molds used are somewhat similar to those employed in compression molding, usually of two-piece, mating construction. The process consists of two steps. A "preform" of glass fibers is prepared by collecting fibers on a screen which has the shape of the finished article. Suction is used to hold the fibers on the screen; fibers are either blown at it or fall on it from a cutter. Commercial equipment is available for this operation. When the desired weight of glass fibers has been collected, a resinous binder is applied. The preform and screen are then baked to cure the binder, after which the preform is ready for use. The second step is the actual molding. The preform is placed in the mold, the catalyzed resin poured on the preform in correct amount, the mold closed and the cure effected by heat. Design of the mold is very important and some features are different from those in other molding fields.

The economic features previously mentioned are important in this process. The molds are commonly required to withstand only 50 to 200 psi pressure so that they are less expensive than those used for compression molds for comparably sized parts. The liquid resin flows easily with heat and pressure before it cures. This makes it possible to mold parts as large as the complete Corvette underbody in one piece without the use of tremendous molds and pressures.

Molding cycles vary from 2 to 5 min at temperatures from 230 to 300 F. Trimming, sanding and buffing are usually required at the flash line. Costs of the molded part fall roughly into $0.60–1.00/lb range.

Premix molding. This branch of the field provides parts for automotive and similar end uses. The parts are strictly functional and are usually pigmented black. The strength properties required are relatively low except that impact strength must be good. As the name infers, the unsaturated resin is first mixed with fillers, fibers, and catalyst to provide a nontacky compound. The mixers used are of the heavy duty Day or Baker-Perkins type. The "premix" is usually extruded to provide a "rope" or strip of material easily handled at the press. The fibers used most extensively are cut sisal, but glass and asbestos are also used, and frequently all three are present in a compound. The fillers are clays, carbonates and similar cheap inorganic materials. A typical premix will contain 38% catalyzed unsaturated resin, 12% total fiber and 50% filler. However, rather wide variations in composition are practiced to obtain specific end use properties.

The premix is molded at pressures of 150 to 500 psi and at temperatures ranging from 250–310 F. Cure cycles are short, usually from 30 to 90 seconds. Again the fact that the resin starts as a liquid makes possible the molding of intricate parts because of ease of flow in the mold. Heater housings for autos are the largest use, but housings of many types and electrical parts are produced in volume. This process provides a cheap but very serviceable molded material.

Properties

The strengths obtainable in the finished product are of prime interest to the engineer. However, the numerous forms of reinforcement materials and the variations possible in the polyester constituent present a whole spectrum of obtainable properties. As an example, products are available that will resist heat for long periods of time at temperatures varying from 150 to 350 F; but the higher the temperature, the more expensive the resin.

In general, the commercial resins have good electrical properties and are resistant to dilute chemicals. Alkali resistance is poor, as is resistance to strong acids. The strength-to-weight ratio of polyester parts and their impact resistance are outstanding physical properties. The data in Table 1 are intended to be illustrative of the properties obtainable by different fabrication techniques.

Available forms

Probably over 95% of the unsaturated polyester resins sold are liquids in the uncured state. However, certain types are available in solid

TABLE 1—GENERAL PROPERTIES OF POLYESTERS

Type of Reinforcement	None	Glass Cloth	Mat or Preform	Mat [a]	Parallel Yarn or Roving	Premix
Glass Content, % by Wt	—	60–70	35–45	20–30	60–80	10–40 [b]
Sp Gr Range	1.20–1.30	1.7–1.9	1.5–1.6	1.4–1.5	1.7–1.95	1.6–1.9
Rockwell Hardness	M100–110	M100–110	M90–100	M85–95	M90–110	M55–75
Flex Str, 1000 psi	13–17	40–85	25–35	15–25	80–115	5–20
Ten Str, 1000 psi	8–12	30–55	15–25	10–15	70–100	3–6
Compr Str, 1000 psi	18–23	20–45	17–28	20–25	50–75	10–16
Flex Mod, 10^6 psi	0.50–0.60	2.0–3.0	1.0–1.8	0.8–1.5	3.0–6.0	0.8–1.2
Ten Mod. 10^6 psi	0.45–0.55	1.8–3.0	0.8–1.6	0.7–1.4	3.0–6.0	0.8–1.2
Impact, Notched, ft-lb/in. notch	0.17–0.25	15–30	10–20	6–10	—	1.5–3.0
Shear Str, 1000 psi	—	15–25	12–18	8–12	—	—
Water Absorption (24 hr), %	0.15–0.25	0.10–0.20	0.2–0.5	0.2–0.4	0.15–0.30	0.3–1.0

[a] Corrugated sheet-type laminate.
[b] Total fiber content.

or paste form for special uses. The cured resins are available in laminate form as corrugated sheeting which is sold widely for partitions, windows, patio roofs, etc. Rod stock may be purchased for fishing rods and electrical applications. Paper and glass cloth laminates are sold to fabricators. The boat end use is making the material more familiar to the general public. However, a large part of the industry's production is concerned with custom molded parts. These are perhaps best classified by industries rather than specific products. The aircraft industry uses substantial quantities but automotive end uses are larger. The chemical industry is using increasing amounts in fume ducts and corrosion-resistant containers. The electrical industry uses the material in laminate and molded forms and as an encapsulation medium. Furniture applications are growing. The machinery industry uses moldings as housings and guards; latest volume use is motor boat shrouds. There are few fields which have not found the material useful in some application.

POLYETHYLENE PLASTICS

Polyethylene is the generic term for a variety of thermoplastic resins made by polymerization of essentially pure ethylene. The small molecules of ethylene containing two carbon atoms and four hydrogen atoms ($CH_2 = CH_2$) are polymerized or joined end-to-end, forming very long molecules which are called polyethylene. It is these giant molecules which give polyethylene its unusual and useful properties. The final product is not a single chemical compound made up of molecules identical in size and shape, but rather a mixture of molecules which are similar in structure but vary considerably in length, with the variations in length following a probability curve. In addition to length variation, the molecules generally have branches and the number and length of branches vary with different compositions.

Composition and structure

By controlling the synthesis conditions a wide variety of materials are made which differ in three important ways, namely: (1) average size of molecules or molecular weight; (2) distribution of size of molecules or molecular weight distribution; and (3) branching or nonlinearity of the molecules.

In the molten state, polyethylene resins are masses of entangled long chain molecules. When they solidify, the molecules tend to align themselves into three-dimensional ordered structures or crystals. However, the entanglements of the molecules and the presence of branches on the sides of the chains prevent perfect alignment over large distances, and forestall the growth of large crystals. As a result, the structure of solid polyethylene is composed of a large number of crystals, commonly called crystallites, which are connected by material which is not crystalline and is called amorphous.

The length of the crystallites is small compared with the length of the individual molecules, so one polymer molecule often passes through several crystallites. By controlling the molecular weight and structure during manufacture, it is possible to produce a series of polyethylene resins, each designed to meet specific end use requirements in film, pipe, bottles, coated paper, and wire and cable applications.

Polyethylene resins are best described in terms of three parameters, namely melt index, density and molecular weight distribution. The melt index is an arbitrary measure of the fluidity of the molten polymer under certain specified conditions (ASTM D1238–57T). Density is a measure of the crystallinity of polyethylene. Molecular weight distribution (MWD) gives an indication of the range of sizes and relative number of each size of molecule in the resin. One way MWD can be expressed is as the ratio of the weight average molecular weight to the number average molecular weight, or M_W/M_N.

The three parameters just mentioned have certain general effects on the properties of polyethylene. For example, if melt index and molecular weight are held constant, and density or crystallinity is increased, an increase in stiffness, yield strength, Graves tear strength, and resistance to creep will be observed; whereas there will be a decrease in Elmendorf tear strength and permeability to liquids. If density and molecular weight distribution are held constant an increase in melt index causes an increase in melt drawability, often an increase in film coefficient of friction, decrease in tensile strength, Elmendorf tear strength, and resistance to environmental stress cracking. If density and melt index are held constant, a decrease in the breadth of molecular weight distribution increases tensile strength, resistance to environmental stress cracking, and film brittleness.

Specifications

ASTM specification D1248–58T covers the requirements for polyethylene molding and extruding compounds. This specification provides for the identification of three types of polyethylene molding and extrusion materials furnished in powder, granule or pellet form.

Type 1 is polyethylene in a density range of 0.910 to 0.925 gms/cc (often called "low density" or "branched").

Type 2 is polyethylene in a density range of 0.926 to 0.940 gms/cc (intermediate density).

Type 3 is polyethylene in a density range of 0.941 to 0.965 gms/cc (high density, or "linear").

Polyethylene resins should be annealed by conditioning near their crystalline melting points prior to density determination. Each of these three types is subdivided into three classes according to composition and use as follows: Class A, natural color only (general purpose and dielectric); Class B, colors and white and black (this is also general purpose and dielectric); and Class C, black (weather resistant) containing not less than 2% carbon black.

Mechanical properties

Measured tensile properties of polyethylene are dependent on the rate of strain and temperature. Many polyethylene compositions have an ultimate tensile strength higher than the yield strength. However, some of the compositions with very high density have an ultimate tensile strength lower than the yield strength at customary loading rates. The effect of temperature on the yield strength of polyethylene is shown in Fig 1.

Stiffness of polyethylene varies with density from flexible (14,000 psi flexural modulus at 0.914 density) to semirigid (about 130,000 psi flexural modulus at 0.960 density). The effect of temperature on the stiffness of polyethylene is shown in Fig 2.

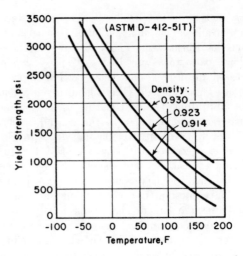

FIG. 1. Temperature *vs.* yield strength of polyethylene.

When polyethylene is subjected to a constant stress below the yield point, the strain in the material increases with time. With a long continued stress there is a very slow continuous yielding of the material. This slow yielding of the material under steady load is called creep.

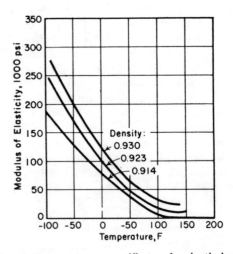

FIG. 2. Temperature *vs.* stiffness of polyethylene.

Data on deformation vs. time for various stresses are given in Fig 3.

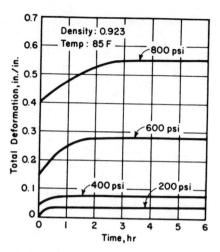

FIG. 3. Total deformation of polyethylene (0.923 density) @ 85 F.

Polyethylene has excellent impact resistance. Even at temperatures in the range of − 40 F, the Izod impact test results range from 0.8 ft-lb/in. for the more dense materials to 2.2 ft-lb/in. for less dense materials.

Electrical properties

Because polyethylene consists of long hydrocarbon chains and exhibits a high degree of electrical symmetry, it is almost completely nonpolar. In addition, because of its moisture resistance, the electrical properties of polyethylene are not significantly affected by wet conditions. Polyethylene containing no modifiers other than an antioxidant has a power factor of 0.0002 and

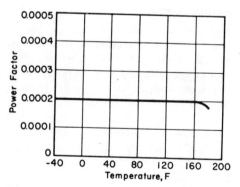

FIG. 4. Temperature *vs.* power factor of polyethylene.

a dielectric constant of 2.27 over the complete range of frequencies measured to date.

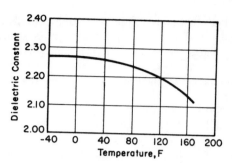

FIG. 5. Effect of temperature on dielectric constant of polyethylene.

The loss characteristics of polyethylene do not vary appreciably over a wide temperature range. The effects of temperature on the power factor and dielectric constant of polyethylene are shown in Figs 4 and 5 respectively. The volume resistivity of polyethylene is extremely high, greater than 10^{15} ohms-cm, even at temperatures as high as about 185 F. The dielectric strength at various thicknesses is given in Fig 6.

Thermal properties

Like most polymeric materials, polyethylene has a relatively high coefficient of linear thermal expansion, particularly as compared with metals. This coefficient for polyethylene is 9×10^{-5} inches/inch/°F by ASTM D696–44. However, in most uses of polyethylene, thermal expansion or contraction does not present a problem because of the material's resilience.

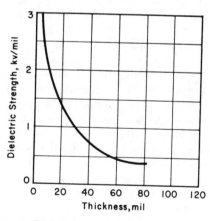

FIG. 6. Dielectric strength of polyethylene at various thicknesses.

Polyethylene is an excellent thermal insulator and as such has many applications where a thermal barrier is needed. Thermal conductivity of polyethylene is about 2 Btu/hr/sq ft/°F/inch as determined by the Cenco-Fitch apparatus.

Chemical resistance

All polyethylene resins have excellent resistance to chemicals and solvents. Polyethylene is essentially inert to the action of food chemicals and has good resistance to alkalies and mineral acids, including hydrofluoric acid. Sea water and salt spray, which are so highly corrosive to many other materials, have no effect whatsoever on polyethylene. Only the chlorinated hydrocarbons, aliphatic and aromatic hydrocarbons, a few esters and certain oils have any appreciable effect on polyethylene at room temperatures. These cause a slight swelling on immersion but do not dissolve it. At temperatures above 160 F polyethylene is dissolved by a number of solvents, including xylene, toluene, amyl acetate, trichloroethylene, petroleum ether, paraffin, turpentine and lubricating oils.

Environmental stress cracking

Polyethylene will crack in certain environments at points where stresses are present. Although the conditions necessary to produce stress cracking are rather critical and probably will not be encountered frequently, knowledge of this phenomenon is important in the use of polyethylene.

This phenomenon is not directly related to chemical or solvent attack, since unstressed polyethylene is highly resistant to many of the materials that are the most active agents in causing stress cracking. Environmental stress cracking does not occur immediately in most cases and frequently does not develop for many months. The rate depends upon the amount of stress and the temperature. In general, surface active agents such as metallic soaps and sulphated and sulphonated alcohols produce the most active stress-cracking environments. Less active environments are aliphatic and aromatic liquid hydrocarbons, alcohols and some organic acids. Fortunately, environmental stress cracking can be overcome to a large degree by the use of the correct composition of polyethylene. Compositions are available which have excellent resistance to this effect.

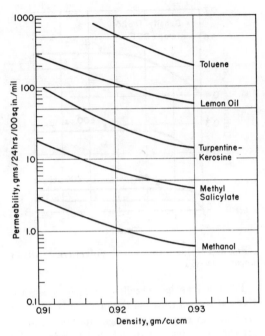

FIG. 7. Permeability of polyethylene to liquids at 73°F.

Permeability

Polyethylene is permeable to some materials that show little sign of chemical attack on it. This property varies with the density of the polyethylene composition. The more dense materials are more resistant to permeation. The permeability of films of polyethylene to a number of liquids is shown in Fig 7.

Moisture resistance

Polyethylene has excellent resistance to moisture and to the transmission of moisture vapor. The water absorption of polyethylene as determined by standard ASTM methods, is less than 0.01 percent. Articles made of polyethylene are unaffected by changes in atmospheric humidity. The moisture vapor transmission rates of different thicknesses of polyethylene are shown in Fig 8. These values are not appreciably altered by sealing, cracking, crumpling, or direct contact with water.

Weathering properties

Black pigmented polyethylene supplied for outdoor use has excellent resistance to the effects of outdoor weathering. Specimens exposed in

Florida show no change in properties after three years. These compositions contain finely divided carbon black intimately dispersed. Unpigmented polyethylene is not generally recommended for prolonged outdoor use. The effects of outdoor weathering on unmodified polyethylene vary considerably with exposure conditions. The general effects are increase in color, loss of tensile strength and elongation, and impairment of low temperature properties. In thin sections, such as film, this may occur in a few weeks.

Abrasion resistance

The abrasion resistance of polyethylene varies with the type of abrasive action. Because of its resiliency it tends to deform under load, then spring back to its original form when the load is removed. When the load is of a type that can be distributed over the surface, there is little tendency to wear. A sharp edge will, however, cut into articles made of polyethylene.

Flammability

Polyethylene has a relatively low flammability rate, particularly for a hydrocarbon. In thin sections it can be made to burn, but usually at a slow rate. In thick sections polyethylene is

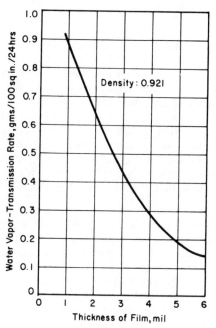

FIG. 8. Moisture vapor-transmission of film made from polyethylene (0.921 density).

slow to ignite and usually extinguishes itself as soon as the source of heat is removed. For applications in which nonflammability is imperative, certain types of modified polyethylenes which do not support combustion and do not drip when heated above the melting point have been developed. However, certain physical and processing properties are impaired.

Methods of processing

Polyethylene is readily injection-molded in any commercial machine. It has a wide molding range and excellent flow characteristics and is not sensitive to moisture. Polyethylene is readily extruded with standard equipment into film, heavy sheets, thin-wall tubing, pipe, and other shapes. It can be extruded as a coating over wire or as a coating for paper and other base stocks. Other processing techniques in which polyethylene performs well include blow molding, compression molding and thermoforming.

Applications

The general field of application of polyethylene resins includes packaging, wire and cable, housewares, chemical equipment, and pipe. Polyethylene is used as a packaging material in the form of film, coatings, bottles and other containers, closures, liners for metal and fiber drums, and multiwall bags. Film extruded from polyethylene is tough and strong, transparent, moistureproof, resistant to greases, flexible at low temperatures and heat sealable. It is used for the packaging of foods, chemicals, textile items, metal parts and many other products. The squeeze bottle is one of the best known uses of polyethylene. Bottles made of polyethylene are lightweight and unbreakable. They can be made in an endless variety of distinctive shapes and colors. Closures of screw-on and snap-on types are molded of polyethylene for both functional and decorative reasons.

Polyethylene resins are outstanding in the field of electrical insulating and jacketing materials. For these uses polyethylene offers low power loss and dielectric constant, toughness and flexibility even at extremely low temperature, light weight and resistance to moisture. Its excellent electrical properties are utilized in such high frequency applications as coaxial cable, microphone cable, spiral four cable, appliance wire, hook-up wire, and signal and control cable. In addition to its uses in wire and cable, polyethylene is also used in molded electrical components. One such use is the molded cone-shaped shield for color television picture tubes. The storage battery makes use of slotted tubes and tube sealers molded of polyethylene.

Polyethylene compositions containing relatively high carbon content for superior resistance to outdoor weathering are used for weatherproof line wire and service drop wire. Polyethylene compositions for jacketing offer excellent resistance to outdoor weathering and to environmental stress cracking.

Polyethylene, because of its durability, flexibility, lightness of weight, and low cost, enjoys many uses in the kitchen and household. Many large items such as garbage pails, wash basins, and clothes baskets have been introduced in recent months. These items are colorful, light in weight, long lasting, and cannot corrode. Such items as refrigerator bowls manufactured from low density resins can be squeezed to form pouring spouts. Ice cube trays of polyethylene remain tough and flexible at subzero temperatures and permit easy removal of ice. Bowls, cups, pitchers, juice containers and other similar articles are colorful, light in weight and practically unbreakable. The flexibility of low density polyethylene also makes possible tight sealing covers.

Many applications of polyethylene utilize its chemical inertness, usually in combination with other properties. Beakers, funnels, graduates and other chemical equipment made of polyethylene are extremely durable and light in weight. A variety of pouring spouts takes advantage of the chemical resistance of polyethylene. Spouts also resist adhesion to sticky materials. Some of these spouts molded of polyethylene also double as closures.

Pipe and tubing of polyethylene are used in the handling of chemicals because of the material's toughness, chemical resistance, plus flexibility and light weight. Pipe made of low density polyethylene is flexible, lightweight and corrosion resistant. It can be extruded in any length and diameters up to three inches may be coiled. It is outstanding in the flexible pipe field for long-term burst strength. It is exceptionally tough, even at low temperatures. It has excellent resistance to corrosive soils and atmospheres and to outdoor weathering and is approved by sanitation authorities for drinking water. A major

use of pipe made of polyethylene is in jet well installations. In this use the durability and ease of installation are particularly important. One man can install the full length of pipe in a single operation. Other major uses include the golf course and lawn sprinkler systems, irrigation, sewage and waste installations, industrial installations, land drainage and ice skating rinks.

Polyethylene is an ideal material for use in toys. It can be economically molded into intricate shapes and is strong even in thin sections. It can be made in bright colors and is tasteless and odorless. Baby bottles of high density polyethylene will not break if dropped and can be sterilized in boiling water.

Other important uses of polyethylene include sporting goods, shoe components, protective coverings, interlayer for cured rubber, medical and surgical appliances, flashlight cases, transparent flange shields and frames for face shields.

POLYPHENYLENE OXIDE PLASTICS

This plastic is notable for its high strength and broad temperature resistance. There are two major types: phenylene oxide (PPO) and modified phenylene oxide (Noryl). These materials have a deflection temperature, which ranges from 212 to 345 F at 264 psi. Their coefficient of linear thermal expansion is among the lowest for engineering thermoplastics. Room temperature strength and modulus is high and creep is low. In addition they have good electrical resistivity. Their ability to withstand steam sterilization and their hydrolytic stability makes them suitable for medical instruments, electric dishwashers and food dispensers. They are also used in the electrical and electronic field and for business machine housings.

Tensile strength and modulus of phenylene oxides ranks high among engineering thermoplastics. They are processed by injection-molding, extrusion and thermoforming techniques. The foam grades, with their high rigidity, are suitable for large structural parts. Because of good dimensional stability at high temperatures and under moisture conditions, these plastics are readily plated without blistering.

TYPICAL PROPERTIES OF PHENYLENE OXIDE PLASTICS

	PPO	Noryl
Specific Gravity	1.06	1.06
Mod. Elast. in Tens., 10^5	3.8	3.6
Tens. Str., 1000 psi	11.0	9.6
Elong., %	60	60
Hardness, Rockwell	120	120
Imp. Str., ft-lb in. notch	1.5–1.9	1.3
Max. Svc. Temp., F	212	212
Vol. Res., ohm-cm	10^{17}	10^{17}

POLYPROPYLENE PLASTICS

Polypropylene is made by polymerization of propylene using catalysts similar to those developed by Ziegler for the low-pressure polymerization of ethylene. It is a linear polymer, more than 95% of which has a spatially ordered structure. The commercial polymer contains over 99% C_3H_6, the remainder being stabilizing additives. Molded pieces or molding pellets of unpigmented material may be identified by density (0.905) and by maximum crystalline melting point (168 to 170 C) when observed with a hot-stage microscope using crossed Nicol prisms. The unpigmented material burns with a colorless, bluish-tinged flame.

Several grades of polypropylene are sold with stabilizers especially selected to offer maximum utility under high service temperatures, outdoor exposure, or general purpose uses. A broad range of molecular weight resins are supplied to provide optimum performance during fabrication and use. A special grade, designed for use in contact with foods, is available and uses a stabilizing system acceptable to the U. S. Food and Drug Administration. Modified formulations are also available to provide high impact properties at low temperatures.

Physical properties

Physical properties shown in Table 1, with the exception of electrical properties, apply generally to all grades of polypropylene. Incorporation of the stabilizers used in the different grades alters the electrical properties shown; pigmentation can also make a difference. The low moisture-absorption values indicate why polypropylene retains its good electrical properties over a wide range of temperature and humidity conditions. The material is nonmagnetic.

The physical properties of polypropylene, like those of many other thermoplastics, will vary with the rate at which it is stressed. At high loading rates, values for tensile modulus and ultimate tensile strength tend to be higher, and ultimate elongations lower, than under slowly applied loads. Behavior in flexure is similar; however, a semirigid material such as polypropylene does not fracture or show a definite yield point in standard flexural tests, but simply continues to deform. The maximum fiber stress at which such continuous deformation occurs ranges from 7000 to 9000 psi at 73 F, depending on the nature of the test piece and the rate of loading. Under compression, the material deforms continuously as load is increased, hence results of such tests can be interpreted to give values of elastic modulus only.

The values of tensile modulus at 212 F, and the low values of deformation under load (2000 psi at 122 F) further indicate its load-bearing properties at elevated temperatures.

In addition to the purely physical factors involved in heat resistance, the chemical behavior of polypropylene in hot air must also be taken into account; hence, stabilizers or antioxidants are used in the plastic, just as they are used with rubber. Natural colored, general purpose material is estimated to have a continuous service life of 15 to 18 years in air at 180 F, six to seven years in air at 194 F, and two to three years in air at 212 F. Age-life at these elevated temperatures can be expected to be extended as new stabilization systems now in development are established. Special stabilizers to resist wet-heat environments now assure satisfactory performance for several years in such uses.

Where intermittent exposure to high temperature air is encountered, and the physical properties are the limiting factor in design, non-load-bearing parts are form stable to 270 to 300 F, with load-bearing pieces being designed as indicated by creep data.

In general, polypropylene at subzero temperatures has good shock resistance to loads applied at moderate rates. Cracks that may appear under rapid loading require special consideration. Proper attention to mold design and technique of molding, moldings from standard polypropylene will minimize impact problems, however low-temperature impact-resistant grades are also available. Electrical connectors and steering

TABLE 1—PROPERTIES OF GENERAL-PURPOSE POLYPROPYLENE [a]

Test	Method	Range of Properties
Melt Index	ASTM D1238–57T, Procedure E	4 g/10 min
Density, 73 F	ASTM D792–50	0.905 g/cc
Refr Index	Oil immersion, white light, 25 C	1.53
Crystalline Melt Pt	Disappearance of birefringence—hot-stage microscope	334 F
Burning Rate	ASTM D635–56	1 in./min
Mechanical Properties at 73 F [b]		
Rockwell Hardness, R	ASTM D735–51	98
Izod Impact, notched, at 73 F	ASTM D256–56	0.9 ft-lb/in. notch
Ultimate Ten Str, 20 in./min	ASTM D638–58T	5500 psi
Ultimate Elong, 0.2 in./min	ASTM D638–58T	220%
Ten Yld Stress, 0.2 in./min, "drop beam"	ASTM D638–58T	4500 psi
Ten Yld Stress, 0.2 in./min, 1% offset	ASTM D638–58T	3300 psi
Elastic Mod in tension, 0.2 in./min, tangent	ASTM D638–58T	203,000 psi
Ten Elong at yield, 0.2 in./min	ASTM D638–58T	11%
Elastic Mod in flex, 0.2 in./min	ASTM D790–59T Procedure A	140,000–200,000 psi
Compr Stress at 1% offset	ASTM D695–54	6000 psi
Compr Strain at 1% offset	ASTM D695–54	3.2%
Elastic Mod in compression, 0.05 in./min	ASTM D695–54	270,000 psi
Permanence Properties		
Water Abs	ASTM D570–54T	0.03%
Environ. Stress Cracking—Igepal C0630	Bell Labs. Test	None observable
Abr Res, mg loss/1000 cycles, Taber CS-17 wheel	ASTM D1044–54T	18–28
Properties at Other Temperatures		
Deflection Temp	ASTM D648–56	221 F (66 psi)
Deflection Temp	ASTM D648–56	130–150 F (264 psi)
Ten Yld Stress, 0.2 in./min, "drop beam," 212 F	ASTM D638–58T	1720 psi
Ten Yld Stress, 0.2 in./min, 1% offset, 212 F	ASTM D638–58T	1000 psi
Elastic Mod in tension, 0.2 in./min, 212 F	ASTM D638–58T	37,000 psi
Ten Elong at yld, 0.2 in./min, 212 F	ASTM D638–58T	17%
Def under load of 2000 psi, 6 hrs, 122 F	ASTM D621–59	4.5%
Ten Yld Stress, 2.0 in./min, 25 F	ASTM D638–58T	7000 psi
Thermal Properties		
Coef of Lin Exp	ASTM D864–52	5.8×10^{-5} in./in./°F −22 to 86 F
Density at 140 F, at 212 F, at 284 F	From dilatometry	0.89, 0.872, 0.853, resp.
Ther Cond, 80 to 250 F	—	1.3 Btu/hr/sq ft/°F/in.
Sp Heat, 32–192 F; over 374 F	From calorimetry	0.476; 0.639 Btu/lb/°F
Electrical Properties at 73 F (and 50% RH)		
Dielec Const	ASTM D150–54T	2.25 at 50 megacycles
D.c. Vol Res	ASTM D257–58	2×10^{17} ohm-cm
Dielec Str, step-by-step	ASTM D149–55T	480 v/mil, ⅛ in. thick
Arc Res [c]	ASTM D495–48T	185 sec, does not track
Dissip Factor	ASTM D150–54T	0.0005 at 60 cycles
" "	"	0.0008 at 10^4 cycles
" "	"	0.0013 at 5×10^4 cycles
" "	"	0.0015 at 10^6 cycles
" "	"	0.0005 at 5×10^7 cycles

[a] Hercules Powder Co.'s Pro-Fax 6523. Data on modified and special types available from suppliers.

[b] Injection-molded test specimens.   [c] At 90 F and 90% RH.

wheels molded with metal inserts have withstood manufacturers' service tests at −30 F without failure. Wire coatings, monofilament and film from polypropylene have given satisfactory performance at −40 F; in these forms, the material is somewhat oriented, and in this condition its shock resistance is good at subzero temperatures.

If mold design and technique do not provide the subzero properties sought, recourse may be made to blends of polypropylyene and other polymers. These provide adequate resistance at low temperatures to high-speed shock at some small sacrifice of thermal resistance, stiffness and surface hardness. In the present state of the art of polypropylene plastics, design for low-temperature performance is most realistically done on the basis of simulated service testing of prototype models or sections.

The proper design of load-bearing plastic parts for resistance to creep in the range 70 to 212 F is just as important for their proper functioning as it is in the design of steel members for use above 1000 F.

Compressive creep tests made at 73 F on general purpose polypropylene, using bars 0.5 in. × 0.5 in. × 1.75 in. loaded to apply 2000 psi to the square section, showed a deformation in one minute of 0.89%, increasing to 1.44% in one day, and to 1.89% in 21 days; on unloading, the irreversible deformation was 0.52%.

In the molding and cooling of molded parts made of polypropylene, the fact that it goes through a melting range between 280 and 330 F is significant; a heat of fusion must be imparted to melt it, and a heat of crystallization must be removed in cooling.

The heat of fusion of polypropylene polymer, obtained from enthalpy measurements over the range 248 to 392 F, is 42.1 Btu per pound of polymer.

For optimum toughness, polypropylene should be molded under conditions leading to the formation of a large number of small crystals. As the temperature is lowered below the melting point, rate of nucleation and crystallization increases rapidly. By cooling quickly to at least 240 F, rapid crystallization and the toughness associated with small crystals can be obtained. By contrast, slow cooling below the melting point leads to the formation of large spherulitic aggregates which impart brittle characteristics to

the plastic. It is for this reason that methods of molding test specimens are specified in Table 1; the varied rates of cooling, and the state of internal stress imparted by compression or injection molding, or extrusion, to the same batch of polymer formed in different ways can make it show test values in mechanical tests differing as much as 40% from each other.

Methods of fabrication

Compression molding. This is seldom used with polypropylene except for making heavy slab in multiple daylight presses. For this, positive molds, arranged so heat can be applied from all sides, are needed. The stock should be heated to 380 F and cooled from top and bottom only under pressure, with at least 2500 psi on the plastic, to a core temperature of about 212 F, to get sound pieces free of porosity.

Injection molding. Standard techniques for molding apply to polypropylene. Cylinder temperatures of 550 F or less, and fast rams operating in the neighborhood of half the available machine pressure, generally give good moldings at fast cycles. No special metals are needed for cylinders or molds. Since polypropylene shows a definite change in melt viscosity between 450 and 525 F, the indicated cylinder temperature should be balanced with the machine heater capacity, the size of each shot, and the cycle time to maintain such melt temperatures.

Control of mold temperature is necessary for best molding. On sections below 0.125 in. thickness, good surface may be attained at 100 F mold temperature if the melt is hot enough. Heavy sections may require mold temperatures above 212 F so that ram pressure can be kept effectively on the piece during cooling to minimize shrinkage and sinks. Heavy sections are prone to bubble during cooling unless the contraction that occurs is controlled by control of molding conditions and proper application of pressure during cooling. Thinner sections which can be more nearly shock-quenched minimize this problem.

Molds for polypropylene should embody the best techniques used with thermoplastics: uniform wall thicknesses; avoidance of heavy ribs, bosses and fillets; use of channels or curved walls instead of ribs to increase rigidity. Balance of cavity layout to provide equal runner length and equal volume discharge per runner is essential. Runners are preferably unpolished, and making them ¼ to ⅜ in. diameter minimizes problems in filling. Gates should be of minimum size, with about 0.040 in. land length, and should direct the material to a mold surface to enhance the smooth flow of plastic in a cavity. This is especially helpful in molding heavy sections; jump or tab gates will also prevent "worming" of the melt during fills.

The best flow patterns for a mold may require a choice between center gating to a heavy central section and multiple gating. Details of mold design may often be clarified by consultation with technical service departments of manufacturers of the plastic.

Mold shrinkage of injection-molded polypropylene varies with section thickness. Commercial experience with general-purpose material indicates the following maximum shrinkage allowances, in inches/inch, for mold design, in the flow direction: For parts 0.0625 in. thick, 0.014; 0.125 in. thick, 0.018; 0.250 in. thick, 0.025. In the direction perpendicular to flow, these values will be 5 to 10% smaller.

Proper gating avoids air traps, sinks, and weld lines, with due regard for venting the mold.

In long runs of heavy section, differences in plastic hardness may be noted near the gate and at points furthest away from it; multiple gating minimizes this.

Objects ranging in size from small valve and pen parts to bookshelves and trash can covers are being molded today from polypropylene.

Extrusion. Heavy sheet, shapes, thin film, and monofilament are all being produced commercially by extrusion. Equipment required can be made of the usual steel alloys with no danger of corrosive degradation products. For even melting, screws of length:diameter ratio of 20:1 are best. The screws should have a compression ratio of 3.1-3.5:1, the compression preferably being applied in the length of one or two flights at the end of the melting zone, followed by a few metering flights. Screen packs ranging from 60 to 200 mesh are common. Good streamlining of all flow channels past the screw is essential. Dies are made preferably with lands of length 10 to 50 times the thickness of the shape being made to increase back pressure on the stock and to avoid turbulent flow. Approach channels should be streamlined with tapers of 5-30°.

In extruding heavy sheet sections, best melt temperatures are 450 to 490 F, with temperature control provided all the way across the slot die. Three-roll sheet take-off devices used for other plastics handle propylene polymer if each roll is individually temperature controlled near 180 F. Best practice calls for no more than 10% drawdown of the sheet, with the top roll adjusted to press down on it uniformly without building up a nip of material behind the rolls. Uniform cooling, protecting the extrusion from stray air currents, is essential.

Shape extrusion of tubing, pipe, or profiles is done using melt temperatures similar to those for heavy sheet, with temperatures rising from the hopper to the end of the extruder. Dies are set at melt temperature and are provided with lands 30 to 50 times as long as the wall thickness being made. Standard methods, such as outside sizing rings, sleeves, or internal mandrels are used to provide size control on closed sections. Open profiles may be formed and supported by cooled tools or air streams located to minimize warpage before water cooling is applied. Here also drawdown should not exceed 10%. Cooling arrangements in existing setups may be adapted for use with polypropylene.

Sheets of 0.005 to 0.250 in. thickness may be vacuum-formed. Equipment with sandwich heaters is best for driving the heat into both sides. After forming, mist cooling is recommended to speed up setting. Exact control of heating, drawing, and cooling cycles is necessary to make perfect draws. Chemical-resistant tanks, hoods, and tank linings are fabricated from polypropylene in thicknesses of 0.125 to 0.500 in. Joints are made by butt welding or applying a gas torch (with a welding rod made of the plastic) to chamfered sheet.

Propylene polymers are well adapted to chemical-resistant work, being insoluble in all solvents at room temperatures and being softened only by aromatic and chlorinated solvents. They are insensitive to nonoxidizing aqueous acids and alkalies up to 180 F. Where oxidizing solutions are encountered, careful pretests of the materials and the joints proposed, under simulated-use conditions, are necessary to arrive at sound bases for equipment design.

Film

Thin films of polypropylene are best made by

extruding melts at 550 to 575 F from a flat die at 10 to 15-mils die opening directly onto quench rolls. Films 0.5 to 5.0 mils thick may be made in this manner; best transparency and toughness are obtained if the die lips are within 0.5 in. of the chill roll, which should operate at 55 to 65 F. Direct extrusion into cold water located the same distance below the die lips will also give similar results. Production rates over 400 fpm have been attained. The extreme quench of the film so made results in a product of density about 0.885 when first produced; after aging a few days in storage, this will rise and stabilize at about 0.895, due to partial recrystallization.

Film has gloss and transparency equivalent to that of cellophane; its permeability is as follows:

H_2O permeability (73° F, 1 mil):
  .48-.72 g/100 sq in/day
Gas permeability (73° F, 1 mil):
  Oxygen—100-300 ml/100 sq in/24 hrs.
  Nitrogen—15-20 ml/100 sq in/24 hrs.

Films have elastic moduli ranging between 100,000 and 120,000 psi, tensile strengths of 3000-4000 psi, and breaking extensibility over 400% at 1 mil thickness. They may be surface treated for printing by oxidative processes and may be used for overwrap and heat-sealed on commercial machines.

Tubular blowing processes do not give film of as good clarity or toughness, since they do not provide sufficient quench. Polypropylene may be uniaxially or biaxially oriented to provide tensile strengths as high as 25,000 psi. Such films may be heat set or heat shrinkable. High speed coatings of polypropylene are now being applied to paper and similar substrates using conventional extrusion-lamination techniques and relatively low molecular weight polymer.

Filaments

Monofilament of 4 to 20 mils diameter is spun from polypropylene in equipment suitable for polyethylene; an extruder conforming to the general requirements listed, quench tank, and an orientation section are required. Melt temperatures of 480 to 570 F are used, depending on the type of material being processed, and dies have tapered approach sections to lands of length five to 15 times that of the opening in the spinning capillary. Cold-water quench of the

strands puts them in form for orientation, which is preferably carried out at 220 to 250 F at draw ratios up to 8:1. Such oriented yarn may shrink 5 to 7% on immersion in boiling water; this can be minimized by an after-treatment. A suitable treatment is provided by winding the yarn on a crushproof bobbin and heating for 10 to 20 min at 250 F.

Typical properties of commercially available yarns are:

Ten Strength, 75 F	35,000–70,000 psi
Tenacity, 220 F	2.3 g/denier
75 F	4.8–6.1 g/denier
−4 F	5.8 g/denier
Breaking elongation, 75 F	25–15%
Knot strength, 75 F	3.6 g/denier
−4 F	4.7 g/denier
Loop strength, 75 F	3.0 g/denier
Resilience, at 75 F:	nearly 100% recovery of elongations of 5 to 15%; 90% recovery of 20% elongation.
Light res, light stabilized grade:	colors retain 47 to 70% of original strength after six months of direct sun exposure in Florida, and 50% of original strength after 1500 Fade-O-Meter hours.

Uses of monofilament include rope and cordage, woven tape for furniture, woven fabric for auto seat covers, windlaces, etc. In rope, it is particularly useful in mooring lines for ocean-going ships.

Through the use of long land dies, wires and cables can be coated with polypropylene. Because of the stiffness of the coating, this is done commercially only on the smaller gauges of wire, where its advantages of heat resistance, hardness, and low creep are predominant. Such coatings have replaced ethyl cellulose on detonator wire, tests showing them to be more flexible at subzero temperatures. By mixing small quantities of chemical foaming agents with molding pellets of polypropylene, and extruding these over wire, foamed coatings of good dielectric properties can be made. Resistant to crushing over a wide temperature range, these coated wires are in commercial use in communications work.

Bottles that withstand the highest heats of sterilization can readily be blown from plastic grade polypropylene on equipment in use with polyethylenes. As yet, blowing technique has not overcome the shock sensitivity of these items at low temperatures, but use of special blends of polypropylene and other polymers overcomes this defect. Several automotive and industrial applications for polypropylene use the blow molding technique.

POLYSTYRENE PLASTICS

Polystyrenes comprise one of the largest and most widely used families of plastics. Often called the "workhorse" of the thermoplastics, polystyrenes consist of the basic general-purpose materials, plus a wide variety of modified grades of polymers, copolymers and blends.

All polystyrenes generally have in common: (1) low cost, present price ranging around 16 to 25¢ per lb, (2) unexcelled electrical insulating properties, (3) virtually unlimited colorability, (4) ability to be made crystal clear (in general-purpose grades), and (5) high hardness and gloss. Table 1 lists typical property ranges of several types.

Polystyrenes have excellent molding and extrusion characteristics, and are formed readily and inexpensively by any of the thermoplastic forming methods. Types available include molding and extrusion grades, foams, sheet and film, and fiber.

General-purpose types

So-called GP polystyrenes are characterized by clarity, luster, colorability, rigidity, unexcelled dielectric properties and moldability. They are used where rigidity and appearance are important, but where toughness is not required. Typical uses include wall tile, container lids, and brush backs.

Grades are available with a wide range of processing characteristics. Higher heat-resistance types, which also have improved toughness, are available in crystal and a full range of transparent, translucent and opaque colors.

TABLE 1—TYPICAL PROPERTIES OF SEVERAL GRADES OF POLYSTYRENES

	General Purpose	Medium-High Impact	Extra-High Impact	Chemical and Heat Resistant
MECHANICAL PROPERTIES				
Mod of Elast in Ten, 10^5 psi	4–5	3.0–4.5	2.0–2.5	4–6
Ten Str, 1000 psi	5–8	3.5–6.8	3.0–5.5	10–11
Impact Str, ft-lb/in. notch	0.2–0.4	0.6–3.0	6–11	0.2–0.5
Hardness, Rockwell	M68–80	M15–80	M15–60	M78–88
Flex, Str, 1000 psi	8–15	No failure	No failure	11–17
PHYSICAL PROPERTIES				
Specific Gravity	1.04–1.07	1.04–1.08	1.0–1.1	1.05–1.11
Heat Dist. Temp (264 psi), F	165–190 [a]	155–180	185–190	200–220
Max Rec Svc Temp, °F	140–160 [b]	125–165	120–160	175–190
Water Absorp (24 hr), %	0.03–0.05	0.03–0.08	0.05–0.20	0.1–0.3
Flammability, ipm	1.0–1.5	0.5–2.0	0.5–2.0	0.4–1.0
ELECTRICAL PROPERTIES				
Volume Resistivity, ohm-cm	10^{18}–10^{19}	10^{14}–10^{16}	10^{12}–10^{17}	10^{13}–10^{17}
Dielec Str (short time), v/mil	>500	>450	300–650	400–600
Dielec Const				
60 cps	2.5–2.6	2.45–2.70	2.5–4.0	2.5–3.4
10^6 cps	2.45–2.65	—	2.5–4.0	2.5–3.31
Dissip Factor (60–10^6 cps)	0.0001–0.0005	0.0004–0.01	0.003–0.0095	0.001–0.008

[a] 180–205 F for heat-resistant grades.
[b] 160–180 F for heat-resistant grades.

Impact grades

To overcome the relatively low impact strength of general-purpose polystyrene, combinations of polystyrene and rubber provide grades whose impact strength depends on the proportion of rubber added. Grades are generally characterized as medium, high and extra-high impact types (i.e., 0.6–1, 1–3, and 4–15 ft-lb per in. notch Izod impact strengths, respectively). As impact strength increases, rigidity or modulus decreases. Such materials are available in virtually unlimited colors, but cannot be produced crystal clear.

Medium-impact polystyrenes are used where moderate toughness plus good translucency is required. They are used for such products as containers, closures and table-model radio cabinets. Types of medium impact polystyrenes are available with improved heat resistance, surface gloss and moldability.

High-impact polystyrenes include special grades with improved heat resistance, moldability and surface gloss. They are used for refrigerator inner-door liners, crisper trays, containers, appliance housings and toys. Higher heat-resistant high-impact grades (generally suitable for sections greater than 0.080 in.) are used for television masks and housings, portable radio cabinets, auto heater ducts and automatic washer soap dispensers.

Extra-high impact strength grades have relatively low moduli (about 200,000 to 250,000 psi) and are used primarily where resistance to high-speed loading is required.

Chemical-resistant grades

Copolymers of styrene and acrylonitrile provide resistance to chemicals such as carbon tetrachloride, aliphatic hydrocarbons, and food stains, and provide much better stress-crack resistance than polystyrene. The copolymers are transparent and haze-free, but are slightly yellow; they are available also in a wide variety of colors.

Primary uses for the copolymers are in drinking tumblers and cups which have high resistance to crazing by butter fat and staining by coffee oils. Industrial uses include water filter parts, oil filter bowls, storage battery containers and washing machine parts.

Other special grades

Light-stabilized types styrene-methyl methacrylate copolymers, and glass-reinforced molding materials are other special grades of polystyrene.

Light stabilized grades were developed specifically for applications involving exposure to intense fluorescent-light radiation, e.g., "egg crate" light diffusers. Such formulations prevent the yellowing which occurs in unstabilized polystyrene when it is exposed to fluorescent light.

Styrene-methyl methacrylate copolymers were developed to provide a material with weatherability approaching that of acrylics, but at a lower cost. They have been used primarily for escutcheons, instrument panels, decorative medallions, brush blocks, auto tail light lenses and advertising signs.

Glass-reinforced polystyrenes (incorporating chopped glass fibers) are available for injection molding or extrusion, and provide higher strength, greater durability and higher strength at elevated temperatures than do unreinforced polystyrenes.

Foams

Foamed or cellular polystyrenes have been produced both from GP polystyrene and styrene-acrylonitrile copolymer. They are available both in the form of prefoamed logs or planks, and as expandable beads. Resulting foams are all of the closed-cell, relatively rigid type.

Boards or slabs are available in densities ranging from 1 to 5 lb per cu ft. They are low in moisture absorption and permeability and not subject to decay. They are used for buoyancy, low-temperature insulations and roof insulation. One type is self-extinguishing.

Expandable beads can be placed in a mold and by heating expanded to form a foamed structure of the desired final shape. Densities ranging from 1 to 10 lb per cu ft are readily obtainable. Extruded and molded expandable polystyrene beads are produced in the form of film, sheet and rigid moldings.

POLYSULFIDE RUBBER

Basically, polysulfide rubbers are chemically saturated polymers with reactive terminals through which conversion to a thermoset, elastomeric state can be effected by means of suitable catalysts or curing agents. More specifically, these rubbers are the products of a condensation polymerization in which one or more organic dihalides are reacted with an aqueous solution of sodium polysulfide giving high molecular weight polymers with the general structure:

$$(RS_x)_n \text{ or } (R_1S_xR_2S_x)_n.$$

Such polymers may be di-, tri-, or tetrasulfides; however, with one exception, present commercial polymers are all essentially disulfides. The exception is polysulfide Type A, a tetrasulfide, which has only limited industrial use.

The reactive terminals are developed as a result of cleavage of disulfide linkages by various reduction techniques, and while polymers may be prepared with various functional end groups, those now available are either thiol ($-SH$) or hydroxyl ($-OH$) terminated. The organic dihalides which are employed to provide the polymer backbone structure are either dichlorodiethyl formal or ethylene dichloride, or a combination of both. In order to impart some branch chain or cross-linking effect, up to 2 mole % of an aliphatic trihalide (trichloropropane) is introduced into the polymer reaction along with the regular dihalide.

Properties

As with a number of other synthetic elastomers, polysulfide rubbers require reinforcing fillers to achieve optimum physical properties. However, the high tensile values possible with unsaturated rubbers are not obtainable with polysulfides. For practical purposes, values of 1500 to 1800 psi are the upper maximum limit.

The primary assets of these rubbers are outstanding oil, gasoline, and solvent resistance, as well as very low permeability to gases and solvent vapors. Also because of their saturated structure, polysulfides possess excellent resistance to oxidation, ozone and weathering. Performance as regards temperature is somewhat dependent on polymer structure, but with few exceptions the serviceability range is from -60 to 250 F and intermittently up to 300 F. The principal limitation of polysulfide rubber compounds is relatively low resistance to compression set.

Available forms

Polysulfide rubbers are novel in that they are available not only as solids but also as liquids. The liquids are 100% polymer, unadulterated with solvents or diluents, which also can be converted to highly elastic rubbers with properties closely approaching those of the cured, solid polysulfides. Viscosities range from a very fluid 5 poises to a heavy molasses-like 700 poises.

Fabrication and applications

Design with both polysulfide crudes and liquids is quite similar, but because of the difference in physical form, the products made from these materials are processed and fabricated somewhat differently.

Crudes or solid polymers. Polysulfide crude rubbers are processed in the same way as other synthetic rubbers, on conventional mixing and fabricating equipment. The incorporation of reinforcing fillers, curing agents, and other additives is accomplished on 2-roll mills or Banbury type internal mixers. Subsequent operations such as extruding, calendering, molding, or steam vulcanization can be carried out in the normal manner, except that somewhat closer factory control must be exercised than might be necessary with the larger volume, general-purpose rubbers.

Since polysulfide crudes may be classified as specialty rubbers their use is usually limited to those applications which demand exceptional solvent resistance. Such products include gaskets, washers, diaphragms, various types of oil and gasoline hose, and other mechanical rubber goods items. However, the solvent combinations encountered in various paints, coatings, and inks are responsible for the major consumption of these rubbers. A great number of the rollers employed for can lacquering, wood and metal coating, and the application of quick drying inks are fabricated with a polysulfide rubber covering. Much of the hose used with hot lacquer and paint spraying equipment is made with a polysulfide tube or inner liner. One other unique commercial use of crudes is in the form of nonhardening putties, which make effective solvent-resistant seals for static type joints.

Liquid polymers. Liquid polymer compounds are mixed by a three roll paint mill, colloid mill, ball mill, or internal mixer. The resulting products may be applied by brushing, spraying, casting, or caulking gun depending on the characteristics of the specific compound involved and the type of application for which it was designed. Curing or conversion of these materials to highly elastic rubbers can be accomplished over a relatively wide temperature range, but their ability to cure at room temperature has been a primary factor in the employment of these polymers for a diversity of industrial applications. Despite their similarity in performance properties there has been little overlapping use of the polysulfide crudes and liquid polymers in the same application areas; however, because of versatility in end-use product fabrication, liquids are threatening to intrude into solid-rubber fields of application.

One of the major uses of liquid polysulfide polymers is in the manufacture of sealants for the aircraft, building and marine industries. Such products may be compounded to bond to most building materials, and provide flexible, elastic seals for joints which are subject to a high degree of movement and vibration. Curing or conversion to the rubbery state can be regulated to occur in minutes or hours at normal atmospheric temperatures depending on the demands of the application techniques involved.

Other applications include cold setting casting, potting, and molding compounds which exhibit flexibility, very low shrinkage, and excellent dimensional stability. The impregnation of leather with polysulfide liquid polymers imparts water and solvent resistance without loss of pliability. The liquid polysulfides may also be employed as coatings and adhesives, but more commonly they are used as modifiers of epoxy resins for these and many other industrial applications. Modification is chemical rather than physical and results from an addition reaction between the polysulfide thiol terminals and epoxide groups. Greatly improved flexibility and impact resistance, lower shrinkage, less internal strain, better wetting properties and lower moisture vapor transmission are the advantages gained.

POLYVINYL FLUORIDE FILM

Polyvinyl fluoride film is a flexible, transparent film, which, unlike most films, is inherently weatherable as well as tough, inert and easy to fabricate. All of its properties derive from its chemical structure and are not dependent on additives or plasticizers.

Engineering properties

Two types of the film are available which vary slightly in properties. The principal difference lies in the % elongation at break and the dimensional stability at higher fabricating temperatures. The manufacturer is able to recommend specific types for specific uses. Properties listed below are for one particular type, but those of the other type are of the same order.

Polyvinyl fluoride film is strong, flexible and fatigue-resistant. Tensile strength is 13,700 psi; tensile modulus, 280,000 psi; break elongation, 185% at 68 F. Flex life is 200,000 cycles (one to two mils, at 68 F). Toughness and flexibility are retained over a wide range of temperatures. The surface is stain-resistant, easily cleaned, and highly resistant to abrasion.

The film is impermeable to greases and oils, and retains its film form and strength even when boiled in strong acids and bases. At ordinary temperatures it is not affected by many classes of common solvents including hydrocarbons and chlorinated solvents. It is partially soluble in a few highly polar solvents at temperatures above 300 F. Its resistance to hydrolysis is excellent. Strength, yield stress and elongation are not measurably affected after 60 hr exposure in 85 psi steam (325 F).

Impermeability to gases, water vapor, and organic vapors (with the exception of ketones and esters) is good. Resistance to thermal embrittlement is excellent. The film remains flexible at 400 F and resists flexural fatigue at zero F. Service temperature range is −100 to 225 F.

Its outstanding electrical characteristic is a high dielectric constant—6.8 at 72 F, one kilocycle; for the pigmented form, 8.5 under the same conditions.

Resistance to degradation by sunlight is an outstanding characteristic—the result of chemical inertness and the fact that the material is essentially transparent to and unaffected by the near ultraviolet, visible, and near infrared regions of the spectrum. Unsupported film is not discolored and is still flexible and strong with 50% residual tensile strength after ten years of Florida exposure. Outdoor life of two or three times that period is predicted under less severe exposure conditions and for film used as the surface in laminates.

The film can be colored as desired by the addition of pigments and will be supplied in standard colors.

The film can be sealed by electronic or impulse methods and some types by the hot-bar process as well. The film can be metallized, embossed, vacuum-formed, delustered, printed, postformed as a laminate, and fabricated in simple processes using existing equipment.

Applications

The film is used for metal prefinishing, building board prefinishing, and for roofing—applications in which its weatherability permits it to make a unique contribution as an exterior finish. It can be applied to galvanized steel and to aluminum as well as to plywood, hardboard, cement-asbestos, fibrous glass and similar building materials to provide both a decorative and long-lived protective surface. Laminated to flexible substrates the material offers promise as an easily installed, highly reflective, maintenance-free roofing material.

Chemical, physical and thermal properties combine to make the film a unique material for bag molding and as a parting sheet. Shapes can be fabricated by pressure or vacuum-forming.

Polyvinyl fluoride film in either transparent or pigmented form provides a weatherable, soil-resistant surface layer for flexible and rigid plastics. It can be combined with glass reinforced polyester panels during manufacture and remains as a permanent skin on the panel. A special polymer containing permanent UV absorber is used to produce this film.

The material appears suitable for some difficult packaging applications and offers useful properties in tape form. Used as a jacketing material over insulation on tanks and pipelines, it gives superior performance at lower cost than other materials in use.

Outdoor glazing is another application for the film because of its strength, transparency and weatherability, combined with ease of fabrication through heat sealing. Solar energy structures, transparent tarpaulins, air-supported structures, storm windows, greenhouses, poultry sheds and crop covers can be made from it.

In the electrical industry, capacitors, transformers, motors, wire and cable are all uses in which the film's high dielectric constant, high dielectric strength, outstanding resistance to thermal degradation and the effects of hydrolysis, and low moisture absorption offer substantial benefits.

The film can be furnished in thicknesses ranging from one-half to four mils and in widths up to about twelve feet. Area factor is 20,000 sq in. per lb per mil for clear film and approx. 18,200 for pigmented film.

PORCELAIN ENAMELS AND CERAMIC COATINGS

Earlier definitions of ceramic materials usually stressed their mineral origin and the need for heat to convert them into useful form. As a consequence, only porcelain enamels and glazes were recognized as ceramic coatings until recently, when the principles of phase relations, bonding mechanisms, and crystal structure were applied to ceramic materials and to coatings made from them. In consequence, ceramic materials can now be most safely defined as solid substances which are neither metallic nor organic in nature, a definition which is somewhat more inclusive than older ones, but more accurately reflects modern scientific usage.

Most ceramics are metal oxides, or mixtures and solutions of such oxides. Certain ceramic materials, however, contain little or no oxygen. As a whole, ceramic materials are harder, more inert, and more brittle than organic or metallic substances. Most ceramic coatings are employed to exploit the first two properties while minimizing the third.

Low-temperature coatings

The outstanding resistance to corrosion of certain metals, notably aluminum and chromium, is attributable to the remarkable adherence of their oxide films. Aluminum does not corrode because its oxidation product, unlike that of iron, is a highly protective coating. It was once believed that some mysterious kinship between a metal and its own oxide was needed for this protection, but recent knowledge relating to the structure of metals and metal oxides has enabled metallurgists to develop alloys that form even more stable and adherent films. Methods for thickening or stabilizing these oxide coatings by heat treatment, electrolysis, or chemical reaction are widely accepted.

The presence of such coatings on metals and alloys strongly influences such properties as their emission and absorption of radiation, frictional and wetting characteristics, and electrical and electrochemical properties.

Usually, ceramic coatings contain compounds of metals other than the substrate. Most of these are oxides which are amorphous or crypto-crystalline in nature. They can be divided into two groups: chemical reactants (in which a new compound or complex is formed), and inorganic colloids.

Chemical reactants. These coatings usually involve the chemical modification of the natural metal oxide into a coating which is more stable or more dense. This classification includes the treatment of aluminum, tin, and zinc, with chromates, electrolytic "anodizing," the treatment of iron with soluble phosphates, etc. The films that are formed are more inert to most chemicals than normal oxide films, and the process is termed "passivation."

Inorganic colloids. Colloidal particles are sufficiently small so that their surface energy is sufficient for bonding. These particles can be made into colloidal suspensions called "slurries" or "slips." They contain natural colloids such as hydrophilic clays, or finely ground materials of a fibrous or platelike nature such as asbestos, mica, graphite, potassium titanate, alumina monohydrate, zirconia hydrate or molybdenum sulfide.

Some inorganic colloids are so finely dispersed as to be true sols. "Water glass" is a familiar example of the colloidal sol; depending on its sodium content, alkalinity, and dilution it may be a thin fluid, a sticky, viscous liquid, or a translucent semisolid. Newer aqueous sols include other alkali silicates, aluminum acid hydrates and phosphates, alkyl titanates and silicates, and lime hydrate.

These coatings are usually processed by drying or flocculation. The dried film and substrate are usually heated sufficiently to drive off all traces of moisture and irreversibly "set" the coating for its final use, as for lubricants, thermionic emitters and fluorescent lamp coatings. The relatively weak bonding power of these dried films can be used to hold them in place for further heating; this is the basis of "wet process" porcelain enameling.

Moderate temperature processes

In the temperature range between 1000 and 2000 F, most silicate glasses melt. In this range the fusion of a vitreous powder (usually called a frit) or the formation of a glass from its component materials is then used as the basis for porcelain enameling and glazing processes.

Glazes differ from enamels principally in the substrates on which they are applied. When the substrate is metallic, the vitreous coating is called an enamel; when applied to ceramic bodies such as porcelain, china, terra cotta, pottery, and electrical ceramics, the coating is termed a glaze. Since these coatings are vitreous they may be transparent, but most enamels and many glazes contain finely divided crystalline materials which color them and make them more or less opaque.

Enamels are used primarily to provide resistance to corrosion, heat, and/or abrasion and wear; they are frequently employed for their attractive appearance as well.

Vitreous enamels cannot easily be classified; a typical enameling slip may contain ten or more ingredients, including those that are glassy or form a vitreous network (feldspars, frits, borax, and other mineral sources of silicates, phosphates, and borates). The modifiers usually consist of alkali and alkaline earth compounds or lead. They also contain fluxing agents, opacifiers, suspending agents and clays, as well as refractory oxides intended to dissolve in the melting glass and increase its viscosity. The oxides or other compounds of cobalt, nickel, manganese, arsenic, or antimony may be included to promote adhesion between glass and metal.

Most porcelain enamels consist of two or more layers of glass, separately applied and fused. The first layer is called the ground coat or base coat; its purpose is to attach firmly to the metal substrate and prevent undesirable interactions between substrate and enamel or the evolution of gas from the metal. It is in this ground coat that the adherence-promoting additives are used, and it is these additives which produce the blue, brown, or black color of most ground coats. Where light colors are not required (range parts, heat-resisting coatings, and certain abrasion-resistant applications) a ground coat may suffice over the base metal.

Formerly, ground coats were fired to about 1600 to 1700 F but newer formulations require firing to 1500 to 1550 F and reduce distortion or sagging of steel substrates. Such "soft" ground coats can be applied to relatively thin steel.

Although "cover coat" may be required to resist chemical attack, abrasion, impact, heat, or weathering, enamels are most commonly used to provide a durable and attractive finish for steel and ferrous alloys. Most of these are white, hence the name "porcelain" enamels. Earlier enamels contained antimony or zircon to provide opacity and whiteness, but zirconium oxide and titanium oxide are now most frequently used. The latter not only provides superior opacity but usually improves acid resistance as well. As a consequence, titania-bearing enamels may be applied in a single coating, not only over a suitable base coat but directly upon special steels. Such enamels can now be fused at less than 1400 F to titanium-bearing steels, to steels precoated with a thin nickel film, or to steels pretreated with iron phosphate.

On cast iron, enamels are used for range parts and high-grade sanitary ware. Certain chemical ware (tanks, pumps, etc.) are also made with cast iron. The rigidity and good acoustical damping of cast iron, together with its resistance to distortion by heat, permits heavier layers of the protective glass to be used than are possible on steel or enameling iron.

Special enamels have been developed for the chemical industry and hot water tanks. Such equipment is frequently called "glass lined," and can contain all acids except hydrofluoric and hot phosphoric, moderately alkaline solutions up to their boiling point, and water under pressures up to 500 psi.

Enamels are not limited to use on ferrous substrates; the earliest decorative enamels were used on precious metals and copper artware and jewelry. Enamels have recently been developed suitable for aluminum. Such enamels can be fired at temperatures well below 1000 F. While not so hard or corrosion resistant as sheet-steel enamels, these aluminum enamels provide attractive and durable finishes for sheet, extrusions and castings.

While most enamels are applied to the substrate by a wet-process technique, some coatings may be applied by dusting or sieving the powdered composition directly upon the heated surface. This "dry process" is principally used with chemical ware and cast iron sanitary ware.

The mechanical properties of enamels are strongly influenced by the composition, thickness, and geometry of the substrate as well as by the kind, thickness, method of application and firing conditions of the enamel layer or layers. In general, thin enamel coatings are best able to resist thermal and mechanical shock and stress. Very thin (1 to 4 mils thick) coatings on

steel or aluminum may even be bent, punched, sheared, or drilled without damage. Thicker coatings are usually required for ordinary applications (appliances, curtain walls or structural panels, and signs); the thickness of the glassy layers being usually between 3 and 20 mils. For cast iron sanitary ware the vitreous coating may be 40 mils thick or more, and some chemical tanks employ $1/4$ in. of protective glass.

For maximum resistance to chipping, the enamel must be supported by a hard, relatively thick substrate. Where the enameled article must resist bending or twisting, thinner and more ductile substrates are indicated.

The selection of metal for porcelain enamel and its preparation strongly influence the properties of the composite. "Enameling irons" are essentially very low carbon, basic open-hearth, rimmed steels. Regular SAE and AISI low-carbon steels can seldom be perfectly enameled especially when hot-rolled. Premium enameling stock may contain titanium sufficient to further lower the available carbon content and improve resistance to warping or sagging.

For thicker substrates a basic open-hearth plate steel of flange quality may be used. Higher impurity levels, however, may require that the enamel be fired in an inert atmosphere to eliminate "boiling" defects. In some cases aluminized steel can be used in moderately oxidizing furnace atmospheres.

Cast steel and cast iron can be enameled acceptably; carbon is oxidized from the surface during the relatively long firing period needed to fuse and consolidate the ground coat. Purity of the metal is not so stringent a requirement for cast iron as for steel.

Surface treatment of the metal usually requires the removal of all scale and dirt (this may require sand blasting for heavy gauge stock) followed by a light pickling for ultimate control of the oxide layer thickness. For some metals and cast irons, sand or grit blasting alone may be sufficient pretreatment. Special enamels may require phosphate bath treatment or the deposition of a nickel, copper, or aluminum base coating over the metal.

Metals to be enameled should be reasonably strain-free. Burrs, sharp edges or small external radii and large variations in substrate thickness should be avoided. Welds must be sound and metallurgically similar to the parent area.

Refractory enamels and glazes

Since the fused porcelain enamel cools with its substrate, it is important that the total contraction which occurs during cooling be approximately equal in both metal and glass. If the metal contracts more than the enamel, the latter will be forced into compression and may shatter or chip easily; if the enamel contracts more than the metal it will crack or craze.

Among the silicate glasses, the most refractory generally have a low thermal expansion coefficient, and they must therefore be used on metals which have low expansion. Most of the refractory enamels contain finely powdered silica and chromium oxide which dissolve in the glass on fusion, further lowering expansion. In consequence such enamels are largely restricted to refractory substrate metals such as certain stainless steels and nickel and chromium alloys. The service temperature of enamel-coated stainless steel and nickel alloys is about 1750 F in aircraft engine exhaust manifolds, turbosupercharger linings, jet engine combustors and commercial burners.

Thin, electrically conductive ceramic coatings can be used as resistance heaters on aircraft windshields and the like. Most of these consist of a mixture of the oxides and suboxides of tin with traces of bismuth, antimony, cadmium, or arsenic. They are applied by heating the glass to a dull red heat and spraying it with a fusible tin halide under slightly reducing conditions. As it cools in air, the tin halide decomposes into the conductive complex. When sufficiently thin, these coatings are quite transparent.

High-temperature coatings

To attach a ceramic coating to any substrate by enameling, both substrate and glass must be heated to the fusion temperature of the glass. However, fusion methods have not been successful for the more refractory materials. Because most refractory ceramic coatings are amorphous or crystalline in nature, they have to be applied by relatively novel techniques.

While most ceramic materials are refractory, some of them can be vaporized in an electric arc or hot vacuum. Thin coatings of amorphous silica can be applied readily to relatively cool substrates by vaporizing metallic silicon or silicon halides in the presence of small quantities of oxygen. Apparently the transfer is accomplished largely as silicon monoxide, which recombines with oxygen on cooling. The process is used to obtain thin, protective, optically transparent films on lenses, certain electrical components and metal reflectors.

Other ceramic coatings may be produced by vaporizing one or more components of the coating. In this way coatings of the respective carbides of silicon, boron, aluminum, and chromium can be deposited on graphite, silicon nitride can be formed on metallic silicon, and silicide coatings can be deposited on metals such as tungsten and molybdenum. These processes are necessarily expensive and are poorly adaptable to large specimens or complex shapes.

Flame- or arc-spraying. Many metallic oxides and interstitial compounds can be heated to or above their melting temperatures with a chemical flame or electric arc. The coatings obtained by directing a spray of nearly molten ceramic particles toward an otherwise unheated substrate are interesting and useful for a variety of purposes. Most of the processes are proprietary and differ chiefly in the form in which material is fed into the heat source. (See also Flame Sprayed Coatings.)

The coatings obtained in this way may be quite porous or they may approach the theoretical density of the material being sprayed. Since the substrate need not be heated, the need for a close match of thermal expansion is less important than with vitreous coatings. The porous coatings are surprisingly immune to thermal and mechanical shock, but confer little chemical protection.

Substrates must usually be roughened before application of these coatings, but heat-resistant glasses, glazes, and some procelain enamels make excellent substrates. Adherence seems to be largely mechanical, and the adherence tests used for porcelain enamels are not applicable to these coatings. No standards for testing or performance have yet been established.

Coatings obtained by flame spraying may consist of pure ceramic materials, metals, and some organic polymers; a modification of flame spraying can be used to produce pyrolytic graphitic coatings of high density and resistance to hot gas erosion. Mixtures of ceramic and metal powders may be used to produce cermet coatings (see also Cermet Coatings) or even graded coatings, and unusual electrical, magnetic, and dielectric properties can be obtained with such mixtures or with multilayer application.

Because each particle is suddenly and individually chilled, the structures of flame-sprayed ceramics may be unlike those of the bulk materials. Nonstoichiometry is common, and the stresses in the coating and between the coating and substrate are complex. Nevertheless, certain flame- and arc plasma-sprayed coatings have already found acceptance in missile technology, metal working and foundry applications, and for heat- or wear-resistant coatings on metallic, ceramic and polymer substrates.

POWDER METALLURGY (P/M) PARTS

Powder metallurgy parts, commonly referred to as P/M parts, are produced by the powder metallurgy process which involves blending of powders, pressing the mixture in a die, and then sintering or heating the compact in a controlled atmosphere to bond the contacting surfaces of the particles. Where desirable, parts can be sized, coined or repressed to closer tolerances; they can be impregnated with oil or plastic or infiltrated with a lower melting metal; and they can be heat treated, plated and machined. Production rates range from several hundred to several thousand per hour.

Shapes that can be fabricated in conventional P/M equipment range up to about 35 lbs. Parts of over 1000 lb. can be produced with special techniques such as isostatic compacting and extrusion. However, most P/M parts weigh less than 5 lb. While most of the early P/M parts were simple shapes, such as bearings and washers, developments over the years in equipment and materials now make economical the production of more intricate and stronger parts. And shapes with flanges, hubs, cores, counterbores, and combinations of these are fairly commonplace.

P/M parts are made from a wide range of materials, including combinations not available in wrought or cast form. And these materials can be processed by P/M techniques to provide tailored densities in parts ranging from porous components to high density structural and mechanical parts. In addition, almost any conceivable alloy system under equilibrium or non-equilibrium conditions can be achieved, and segregation effects (non-homogeneities) are avoided or minimized.

Most metal powders are produced by atomization, reduction of oxides, electrolysis, or chemical reduction. Metals available include iron, nickel, copper and aluminum, as well as refractory and reactive metals. These metals can be blended together to form different alloy compositions during sintering. Also, prealloyed such as low alloy steels, bronze, brass, nickel-silver and stainless steel are produced in which each particle is itself an alloy, thus ensuring a homogeneous metallurgical structure in the part. And it is possible to combine metal and nonmetal powders to provide composite materials with the desirable properties of both in the finished part.

The P/M process is being used to produce many thousands of different parts in most product and equipment manufacturing industries, including automotive, business machines, aircraft, consumer products, electrical and electronic, agricultural equipment, machinery, ordnance and atomic energy.

The applications of P/M parts in these industries fall into two main groups. The first are those applications in which the part is impossible to make by any other method. For example, parts made of refractory metals like tungsten and molybdenum or of materials such as tungsten carbide cannot be made efficiently by any other means. Porous bearings and many types of magnetic cores are exclusively products of the powder metallurgy process. The second group of uses consists of mechanical and structural parts that compete with other types of metal forms such as machined parts, castings and forgings.

Part Size—Although there is no known theoretical limit to the size to which P/M parts can be pressed, maximum practical size is governed chiefly by available press equipment and powder characteristics. The majority of conventionally pressed P/M parts range in projected area from about 1/8 to 25 sq. in. and run between 1/32 and 6 in. in length. The maximum projected surface area possible depends on the material, part density and press capacity. Also, it is evident that for a given metal powder, the lower the density required, the larger the size of the part that can be produced on a given press.

A concern when designing parts that are relatively long in the direction of pressing is to obtain adequate density through the total length of the part, and particularly in the center sections. Another factor limiting part length is the apparent density of metal powders. In general, compression ratio is about 2 to 1. This means that the depth of mold must be more than two times the length of the pressed compact. Sometimes the thickness limitation is not a function of the press but of the part design, such as where difficulty in filling thin wall sections (see below) may occur or core rod length may get beyond a practical limit.

Shapes—While a variety of shapes, sections and profiles can be produced by the P/M process, the most suitable ones are those that have uniform dimensions in the direction of pressing. These include simple cylindrical, square and rectangular shapes as well as odd shapes in which the contour is in a plane at right angles to the direction of pressing. For example, with radial projections and contours, parts like cams and gear with fixed thickness are relatively simple to press.

Since the tooling is subject to enormous pressures, shapes that require fragile tools should be avoided. Also, the contour of parts must allow ejection of the part from the die by means of an upward motion of the bottom punch.

Because powders must be compressed and do not flow hydraulically, perfect spheres cannot be made by the P/M process. Therefore, spherical P/M parts are designed with straight or flat areas around the equator. Parts that must fit into ball sockets are repressed to produce a more spherical shape. Hemispheres, such as those used in automotive ball joints can be readily compacted. Spherical depressions up to a hemisphere are also possible.

Multi-Level Shapes—As previously mentioned, P/M parts are pressed from the top and bottom and not from the sides. This, in addition to the fact that metal powders have almost no lateral flow, imposes certain conditions on a part having variations in thickness, such as steps, flanges, slots or grooves. Difference in thickness of up to 10 or 15% are feasible, if variations in density are acceptable. Variations of this magnitude can be formed in the punch faces. When only limited nonuniformity in density can be tolerated or when the overall thickness variation is too great, multiple punch tooling or steps in the die or both, must be used to provide the needed variations in fill and compression. In many such cases, the part is held captive in the tools and special motions are required for proper ejection. Of course, multiple punches increase tool and equipment complexity and thus increase both original and maintenance costs.

POWDER PLASTICS MOLDINGS

The term *powder molding* broadly describes a technique of sintering or fusing finely divided thermoplastic materials so as to conform to the surface of a mold.

Processing

There are a number of processes and techniques loosely termed powder molding. The Engel or Thermofusion process was developed in Europe and sublicensed in this country in 1960. This patented technique employs inexpensive sheet-metal molds and a hot air oven heated by either gas or electricity. In essence, the process consists of filling the mold with powdered thermoplastic material, fusing the layer of material next to the mold walls, removing the excess powder, and then smoothing the inside surface with heat.

Another older technique, known as the Heisler process, involves rotating a heated mold. Both of these patented techniques employ an *excess* of powdered material over that required for the object being fabricated, the extra material being removed as one of the processing steps.

More recently, it was discovered that powdered polyethylene behaves enough like a liquid to permit use of rotational casting equipment designed for use with vinyl plastisols. The technique involves multiaxial rotation of a closed mold filled with a *measured* charge of material, all of which is fused during the heating step. This permits fabrication of totally enclosed articles. A hobby horse for children is the most striking example of the versatility of the rotational casting process. There is presently a variety of rotational casting apparatus available. As soon as it is engineered specifically for use with powdered thermoplastic materials, it is anticipated that this technique will find widespread acceptance.

The selection of a particular powder-molding technique depends on the application under consideration. The selection depends on factors such as size, wall thickness, geometric configuration and quantity of parts desired. Rotational casting with a measured charge is obviously indicated where automation for a large volume of production is desired, due to the reduction of material-handling requirements.

Advantages

Because of low tooling costs, all of these various powder-molding processes offer significant economies over conventional fabricating techniques such as injection or blow molding in small or moderately large quantities (up to 250,000 pieces, depending on the application concerned).

Typically, inexpensive sheet metal and/or cast aluminum molds may be used, as compared with expensive matched metal molds made of tool steels. The heat source is usually an electric or gas-fired air oven, a small capital cost in comparison to an injection-molding machine or extruder. Moreover, the large size of some objects being fabricated by powder molding defy production by most other techniques. The size of an object to be fabricated by this particular process is limited only by the size of the oven. Large tanks, refuse containers, and even boats are included in the wide variety of products being fabricated by this method. These parts range in wall thickness from 0.060 to 0.250 in. with a tolerance of ± 10% considered reasonable.

Materials used

While powder-molding processes are theoretically applicable to any thermoplastic material, current practice has been restricted primarily to polyethylenes of low and intermediate density. As soon as effective stabilizing systems are found for other materials, particularly the broader range of polyolefin materials, it is anticipated that such materials will also find use where their particular properties offer advantages in specific applications. The particles of powder range in size from 30 to 100 mesh (U. S. Standard), with the peak of the bell-shaped distribution at approximately 60 mesh. The powder is made by grinding the material in pellet form as an additional manufacturing step.

Production cycles can be cited only generally, as they depend on the application concerned. Cycles range from roughly three minutes for small, thin-walled parts made by rotational casting, to as much as 30 minutes for very large, heavy-walled parts made by the Engel process.

PRASEODYMIUM. See Rare Earth Metals.

PREFINISHED METALS

Prefinished metals are sheet metals that are precoated or treated at the mill so as to eliminate or minimize final finishing by the user. The metals are made in a ready-to-use form with a decorative and/or functional finish already applied. Prefinished metals provide many advantages including: (1) better product appearance; (2) lower product cost; (3) greater product uniformity; and (4) improved product function.

A very large number and variety of prefinished metals have been developed in recent years. (See Table 1.) It is possible to obtain base metals preplated with almost any decorative or functional metal, from bright, shiny chromium to dull, rich-looking brass. Similarly, it is possible to obtain sheet with prepainted surfaces in almost every color and in a wide variety of special-purpose plastics resins. Also, where extra durability or a special decorative effect is needed some of these resins, notably polyvinyl chloride, are available in sheet or film, laminated to several different metals.

Furthermore, almost every metal is available in a limitless number of textured, patterned and embossed finishes right from the mill. These textured metals can be used as is, but even they can be supplied preplated, prepainted, or even with a colored preanodized finish, as in the case of aluminum.

Many sheet metals are also available with galvanized, aluminized, tin and terne coatings. These materials, especially galvanized, were among the first prefinished metals to be developed and remain today as basic prefinished metals.

Appearance

In many cases appearance alone is not the dominant reason for selecting a prefinished metal. In the case of preplated and prepainted metals the user is usually seeking good appearance in addition to something else such as lower cost, easier fabrication, or greater product uniformity. If appearance alone were the dominant selection factor, it might be just as advantageous to plate or paint the part after fabrication. Except for uniformity, a plain plated or painted surface looks the same whether it is made of prefinished metal or finished after fabrication.

This does not apply to brushed or textured effects. Many preplated metals can be obtained with a brushed mechanical finish over the entire surface, or in the form of stripes. Such effects are costly and in many cases impossible to apply after fabrication. Similarly, heavy prepainted coatings can be mechanically embossed with a variety of patterns.

Vinyl (PVC)-metal laminates can be produced in just about every pattern, texture and color. Textile effects, wood grains and leather surfaces have all been closely duplicated. Outside of using the real material, many of these effects cannot be duplicated on metal by any other method except in plastics laminates. Designers are devising new patterns and textures every day and many unusual patterns can be made on special order.

Textured, patterned and embossed metals also have a unique appearance. In almost all cases it is impossible to obtain these effects after fabrication. Here again, appearance possibilities are unlimited and custom patterns can be obtained by having special embossing dies made.

Except for galvanized sheet, the conventional hot-dipped and electroplated sheets such as aluminized, tin and terneplate are not ordinarily used for decorative reasons. However, although the appearance of galvanized metal is not as pleasing as chromium and nickel electroplates, in some applications it is quite satisfactory for many functional products.

Cost

Prefinished sheet is not the ultimate answer for lowering production costs, as borne out by the fact that most electroplated and painted parts made today are still finished after fabrication. Prefinished metals—notably preplated and prepainted—are most effective in lowering the cost of mass-produced small parts. Small sheet metal parts can be relatively expensive to electroplate or paint in small quantities.

Comparative costs can be determined by totaling up conventional finishing costs and metal costs and comparing this figure with the cost of an equivalent amount of prefinished metal. Because the scrap loss of prefinished metal is higher than that of unfinished metal this factor should also be accounted for. Fabrication costs can usually be ignored because in most cases it

does not cost any more to fabricate prefinished metal than unfinished metal.

An important benefit of using preplated and prepainted metal is that smaller shops are able to eliminate completely the need for maintaining a plating or painting department. This applies only when the user is dealing exclusively with prefinished materials for his products.

Greater uniformity

An important advantage of prefinished metals is their high uniformity. This uniformity is obtainable because the coatings are machine-applied under constantly controlled conditions. This does not mean too much from an appearance standpoint as most consumers would be hard-pressed to tell the difference between a prefinished surface and a surface coated by ordinary methods. The only difference that might be apparent is that the coating on prefinished metal products does not have any tendency to build up at bends and corners.

The greater uniformity of prefinished metals can be beneficial in functional applications, especially where an electroplate of constant thickness is needed. As is well-known, electroplates tend to build up on high spots and thin out in low areas; this can cause problems where good corrosion and wear resistance is needed. Preplated metals are plated flat and uniformly.

Functional advantages

Many prefinished metals are used for functional applications. Practically all zinc-plated sheet, for example, is used in functional applications where good corrosion resistance, rather than a bright, decorative finish is wanted (e.g., condenser cans and hidden parts in door locks). The zinc coating also provides a good paint base, provided the surface is first given a chemical conversion treatment.

Copper-plated steel is another good example of a functional finish. It provides a good lubricating surface for deep-drawing operations and also makes a good base for further electroplating. The material is also used for its electrical conductivity, its usefulness in low-temperature tinning and high temperature brazing operations, and as a stop-off coating in carburizing operations.

Prepainted metals are also popular for functional applications; this is borne out by the wide

TABLE 1—AVAILABLE PREFINISHED METALS

MATERIAL ↓	Surface Composition, Appearance	Base Metals
PREPLATED METALS	**NICKEL, CHROMIUM** Excellent appearance; available in dull, satin and bright (sometimes provided with a clear lacquer coating for added protection) finishes, many of which can be embossed with wide range of patterns.	Steel, zinc, brass, copper, aluminum
	BRASS, COPPPER Excellent appearance; available in dull, satin and highly polished finishes, many of which can be embossed with wide range of patterns.	Steel, zinc
	ZINC Natural grayish finish which can be used to improve product appearance by providing with semilustrous finish and coating with clear lacquer.	Steel
PREPAINTED METALS	Almost every organic coating is available or can be ordered in prepainted form. Selection of coating resin depends on end use requirements. Five most popular prepainted metals now in use are alkyds, acrylics, vinyls, epoxies and epoxyphenolics. Many coatings can be pigmented to provide metal-like appearance.	Most common ferrous and nonferrous metals. For added protection reverse side of ferrous metals is usually given a rust-preventive treatment or provided with an organic or metallic coating. Selection of base metal depends on cost, appearance and product life requirements. Most popular base metals are: cold-rolled steel, tinplate, tin mill blackplate, hot-dipped and electrogalvanized steel, and standard aluminum alloys.
PLASTIC-METAL LAMINATES	**VINYL (PVC)-METAL** Polyvinyl chloride sheet laminated to base metal with thermosetting adhesives under heat and pressure (25 to 60 psi)	Can be applied to most metals; popular are: Steel—Provides strength at low cost. Reverse side can be painted or provided with corrosion resistant coating. Aluminum—Light weight and/or corrosion resistance. Magnesium—Light weight.
	VINYL (PVF)-METAL Polyvinyl fluoride film laminated to base metal with thermosetting adhesive under heat and pressure.	Cold-rolled steel—Provides strength at low cost. Galvanized, aluminized and tinplated steel—Corrosion resistance. Aluminum—Light weight and corrosion resistance.
	POLYESTER-METAL Polyester film laminated to base metal.	Usually steel.

TABLE 1—AVAILABLE PREFINISHED METALS (*Continued*)

MATERIAL ↓	Surface Composition, Appearance	Base Metal
TEXTURED AND EMBOSSED METALS	Surfaces available in hundreds of different textures and patterns. Available with texture on one surface only or with pattern that extends completely through cross section of metal. Also available in perforated form. Surfaces can be provided with dull satin or highly polished finish or combinations thereof (e.g., dull background with polished highlights). Can also be painted, porcelain enameled, or oxidized; these finishes can be buffed off high spots to provide two-tone effect.	All common sheet metals, including: Carbon steel—Strength at low cost. Stainless steel—Corrosion resistance plus strength. Aluminum—Light weight plus corrosion resistance. Copper—Pleasing appearance plus strength and corrosion resistance.
HOT-DIPPED OTHER PLATED METALS	GALVANIZED (ZINC-COATED) Zinc surface with intermediate zinc-iron alloy layer.	Steel, ingot iron.
	ALUMINIZED Aluminum surface. Intermediate aluminum-iron alloy layer forms above 900 F.	Steel.
	TIN-COATED Hot-dipped or electroplated tin.	Mild carbon steel.
	TERNE OR LEAD-COATED Lead-tin alloys, pure lead.	Steel.
SPECIALLY FINISHED ALUMINUM	COLOR ANODIZED Anodized aluminum available in clear, yellow gold (70:30 brass) red gold, rich low brass (85:15 brass), copper, blue, green, red and black. Colors are obtainable over standard mill, satin and bright finishes, as well as over embossed and perforated textures.	All commercial aluminum alloys and tempers.
	SPANGLED Uncoated surface containing large grains which stand out in relief and facets which break up and reflect light. Available in wide variety of colors and mill finishes.	Wrought aluminum alloys.

number of functional resins that are now available. In addition to providing decorative appearance, these coatings prevent the base metal from corroding and can provide good resistance to chemicals and foods (epoxies), toughness and resistance to forming damage (vinyls), and good resistance to outdoor exposure (acrylics).

PREIMPREGNATED DECORATIVE FOIL

Special papers preimpregnated with melamine resin make possible a broad range of molded-in decoration for thermoset plastic products. By far the leading application to date is the use of melamine-impregnated rayon paper. It is used to decorate about 70% of all melamine dinnerware. Development of new materials and techniques to shorten the production cycle is extending use of prepreg paper in closures, cutlery handles and other markets.

The first successful prepreg overlay was achieved experimentally in 1945 with decoration of a plate and an ashtray.

Development work and continuing quality control at every level—paper production, pre-impregnation, printing, and molding—have been essential to the growing use of prepreg foils. The paper found most successful for dinnerware foils is a rayon paper based on pure bright white ¾-in. rayon staple with a blend of wet-strength additives. No fillers, pigments or additives are used that would affect clarity of the paper, since the foil must disappear in molding.

How they are made

The rayon paper is impregnated with resin under closely controlled conditions at the impregnator's plant to provide a foil with precise resin content, even weight and resin distribution, specified penetration and absorption. The melamine resin is partially cured, then held at a stable, flexible B-stage, cut into sheets convenient for printing ($23\frac{1}{2} \times 30\frac{1}{2}$ in. is a common size), and shipped to printers for decoration.

Designs are printed either by lithography or silk screen with special inks and techniques developed by specialty printers. Virtually any design can be reproduced, including photographs in black and white or full color. Even moldable silver and gold inks have been perfected.

The printed foil is cut to final size and shape and delivered to the molder ready for the mold.

Normally the product to be decorated is formed and partially cured, then the mold is opened, the foil inserted and the mold closed to fuse the foil to the product and complete the cure. Except when metallic inks are used, the paper is inserted so the designs face the product. The foil becomes transparent but forms a protective melamine surface.

Stretch of the foil is carefully limited to avoid distortion of the design, but the effect of "deep draw" can be achieved by shaping the printed foil. Thus, "doughnut" shapes are used effectively to decorate the inside of plates and bowls. Wrap-around shapes put molded designs on the outside of cups and pitchers. Mold design must allow for the decoration of side surfaces.

Decorating closures

A relatively new development is the use of prepreg foils to decorate closures such as cosmetic jar covers. Previously, decoration was limited to the one or two colors possible with hot stamping and direct screening or to lithography on metal, which is subject to rust. Molding the foil into a thermosetting plastic closure gives the package designer full color range and combination, and later permits change of design without altering the molds.

The demand of the closure producer for high-volume, high-speed molding encouraged a research program which has yielded a new one-shot decorative foil. Instead of preforming the product and opening the mold for insertion of the foil, the new one-shot foil is put into the mold first, the molding powder poured on top of it, and the mold closed for a complete, uninterrupted cycle. As yet, the new technique has been used only for closures, but the same principles are believed to be technically feasible for larger products such as dinnerware. More immediate markets for the one-shot foil are control knobs and drawer pulls.

Materials

Prepreg foils can be used to apply decoration to a wide variety of thermosetting resins, including all melamines, all ureas, and most phenolics.

Though the self-effacing rayon foil is most popular for dinnerware and most melamine and urea products, opaque prepreg paper is favored for such products as cutlery handles, clock faces, switch plates, and trays. The opaque paper shields the darker resins and even permits use of vari-colored odds and ends in a molder's inventory for such applications as a clock face. Often, of course, the opaque paper is used to provide a decorative color contrast.

The opaque print-base paper is made with a high purity alpha pulp, fillers, pigments, and other additives. Like the rayon foils, this paper is preimpregnated with resin and advanced to a stable B-stage for shipment and printing.

Among other products for which the opaque prepreg papers are suitable are handles for pots and pans, organ keys and wall tile.

PREIMPREGNATED MATERIALS FOR REINFORCED PLASTICS

So-called *prepregs* are ready-to-mold reinforced plastics with the resin and reinforcement pre-combined into one easy-to-handle material. They are fabricated by what is called dry lay-up techniques; the alternative is the wet lay-up method of combining a liquid resin and reinforcement at the mold.

How they are made

Prepregs are produced by impregnating continuous webs of fabric or fiber with synthetic resins under close control. The resins are then partially cured to the B-stage, or partly polymerized. At this stage the preimpregnated material remains stable, and may be shipped and/or stored, ready for forming into the final shape by heat and pressure.

Most of the controllable variables in a reinforced plastics structure are taken into account in the production of a prepreg, e.g., resin type and content, reinforcement and finish, which are largely determined by the requirements of the final laminate. Prepregs are pre-engineered to meet performance and processing requirements.

Major advantages

The major advantages of using preimpregnated materials for molding reinforced plastics products are: (1) high and more uniform strength, (2) uniform quality, (3) simplified production, and (4) design freedom.

Key to the majority of prepreg's advantages is the close control maintained through all stages of the preimpregnation process. For example, the amount of resin solution picked up by the reinforcing web as it passes through one or more resin baths is influenced not alone by resin viscosity and resin solids, but by web tension and speed (which regulate the time the reinforcement is in the bath) and by the temperature of the resin. Because of the high degree of control afforded by preimpregnating equipment over tension, speed and temperature, resin content can be controlled ±2%.

Such controls also permit the production of prepregs with a reinforcement content from as high as 85 to as low as 20%. It is the high reinforcement content possible with prepregs that accounts for the high strength of these materials. Further, since the reinforcement is thoroughly saturated during preimpregnation, little air is entrapped in the reinforced plastic material, assuring better quality in the molded part.

Handling characteristics of prepregs can also be closely controlled by controlling temperature and speed of the material as it passes through the second stage of the process where the resin is partially cured. Temperature of the drying oven, speed of travel, and resin type and content determine the degree of polymerization or B-stage, which, in turn, determines such handling and molding characteristics as resin flow, gel time, tack and drape. In prepregs these factors can be tailored to suit a variety of production molding requirements.

Prepregs also facilitate use of such resins as phenolics, melamines, and silicones, which when available in liquid form, are usually solvent solutions (except for newer solventless silicones). Wet lay-up of solvent solutions of resins is usually unsatisfactory, at best, because solvents are released during cure. As B-staged prepregs such resin systems are virtually solvent-free.

Prepregs can also be made using resins whose viscosities are so low that they would be difficult, or impossible to handle by wet lay-up techniques.

Prepregs can be particularly advantageous to the smaller reinforced plastics fabricator, eliminating the need for resin formulation. Scrap loss can be minimized by chopping up or macerating leftover prepreg to form molding compound.

Because the resin is already properly distributed throughout the reinforcement, prepregs offer distinct advantages in molding odd-shaped parts, with varying thicknesses, undercuts, flanges, etc. They preclude the problem of excessive flow of resin, eliminating resin-rich or resin-starved areas of such parts. Cut to the proper pattern, the prepreg can be accurately pre-positioned in the mold to provide optimum finished parts.

Possibly the greatest advantage of prepregs is that they lend themselves to automated, high production molding. The continuous prepreg web can be prepared for molding by cutting, slitting or blanking operations. Often parts can be molded directly from such blanks. Die-cut parts can also be used to pre-assemble a complete lay-up in advance of molding, thus minimizing press time.

Materials used

A wide range of reinforcing materials and resins are available in prepreg form. Principal resins are polyesters, epoxies, phenolics, melamines, silicones and several elastomers. The most commonly used reinforcements are glass cloth, asbestos, paper and cotton. Specialty fibrous reinforcements include high-silica glass, nylon, rayon and graphite.

Reinforcing materials. Reinforcing materials available in prepregs today include the following:

1. Glass fiber is by far the largest volume reinforcing material in use today. Prepregs incorporating glass fiber are available in the form of preimpregnated roving (for filament winding), cloth and mat. It provides a good balance of properties: outstanding strength-to-weight characteristics, high tensile strength, high modulus of elasticity compared to other fibers, resilience and excellent dimensional stability. It is used for such products as airplane and missile parts, ducts, trays, electrical components, truck body panels and construction materials.

2. Asbestos is used in prepregs in the form of paper, felt and fabrics. All provide good thermal insulation. Papers provide cost advantage where structural characteristics are not critical. Felts provide maximum tensile and flexural strength, and good ablation resistance. Asbestos fabrics are used principally in flat laminates. Prepregs employing asbestos webs have won an important place in rocket and missile parts in good part because of their remarkable short-term resistance to extremes of temperature and flame exposure.

3. Paper, a leading reinforcement for high-

pressure prepreg laminates, is inexpensive, prints well and has adequate strength for its principal uses as counter surfacing, furniture and desk tops, and wall boards.

4. The specialty reinforcements, like high-silica glass, quartz, graphite, nylon, have their main use in missile and rocket parts. Some exhibit unusual high-temperature resistance, some are good ablative materials.

Resins

The major plastic resins used in prepregs are the polyesters, epoxies, phenolics, melamines and silicones, although any thermosetting resin or elastomer may be used.

1. Polyester resin prepregs are comparatively low in cost and easy to mold at low temperatures. They have good mechanical, chemical and electrical properties, and some types are flame-resistant.

2. Epoxy prepregs provide high mechanical strength, excellent dimensional stability, corrosion resistance and interlaminar bond strength, good electrical properties, and very low water absorption.

3. Phenolic prepregs have high mechanical strength, excellent resistance to high temperature, good thermal insulation and electrical properties, and high chemical resistance.

4. Melamine prepregs have excellent color range and color retention, high abrasion resistance, good electrical properties, and are resistant to alkalies, and flame.

5. Silicone prepregs provide the highest electrical properties available in reinforced plastic and are the most heat stable. They retain strength and electrical characteristics under long-term exposure to 500 to 600 F.

Actually any thermosetting resin or elastomer may be used and any continuous-length reinforcement. Even sheet material not capable of supporting its own impregnated weight in a drying oven can be used. The web may be a woven fabric, nonwoven sheet or continuous strand or roving. After impregnation the prepreg may be slit into tape, chopped or macerated, or cut to specified pattern, depending on the requirements of the part and the desire of the molders.

Molding

The one requirement for using prepregs is that

some heat and pressure must be used to mold them to final shape. They can not be room-temperature cured. Prepregs can be formulated to fit all other methods of processing, the four major methods being vacuum-bag molding, pressure-bag molding, matched metal-die molding, and filament and tape winding.

The choice of method of forming prepregs is most often governed by production volume. The exception is the critical part, like a missile component, where the higher performance qualities achieved by matched metal-die molding justifies use of this process even for a few hundred parts.

The vacuum-bag molding method uses the simplest and most economical molds and equipment and is used where the number of parts needed is low or where frequent design changes occur.

The pressure-bag method of molding uses pressures up to 100 psi to achieve greater strength than is possible with the vacuum-bag method. Mold and equipment costs remain comparatively low. In both vacuum-bag and pressure-bag molding, additional pressure with resulting higher strengths can be achieved through use of an autoclave.

Matched metal-die molding is used for long runs where it gives the lowest cost and the high-est production rates. This method permits accurate control of dimensions, density, rate of cure and surface smoothness.

Tape and filament winding are used for cylindrical, spherical and conical shapes and can produce high physical properties and high strength-to-weight characteristics. Techniques range from simple winding to highly automated operations. In fact, it comes closest to automation of current fabrication methods and achieves homogeneous, void-free products. It does not involve the expense of matched metal dies.

PREMIX MOLDINGS

Premix molding materials are physical mixtures of a reactive thermosetting resin (usually polyester), chopped fibrous reinforcement (usually fibrous glass, asbestos or sisal) and powdered fillers (usually carbonates or clays). Such mixtures, when properly formulated, can exhibit a wide range of performance properties at variable costs. In general, the resin type will determine the corrosion-resistance properties of the premix; increasing amounts of glass reinforcement will increase strength and increasing amounts of filler will reduce cost and corrosion resistance. Properties of typical premix compounds are shown in Table 1.

INDICATIVE PROPERTIES OF PREMIX COMPOUNDS

Property	ASTM Method	Low Cost [a]	High Performance	
			Intermediate Strength [b]	High Strength [c]
Izod Impact Str, ft-lb/in. notch....	D256	1–5	4–7	12–24
Flex Str, 1000 psi	D790	4–9	6–17	11–25
Flex Mod of Elas, 10^5 psi	D790	6–9	7–11	16–20
Rockwell Hardness	D785	M35–60	—	M98–100
Heat Dist Temp (264 psi), °F	D648	200–350	200–350	300–550
Water Abs (24 hr), %	D570	0.7–2.3	0.1–0.4	0.1–0.4
Spec Gr	D792	1.5–1.8	1.8–2.0	1.8–2.0
Mold Shrinkage in./in.		0.004	0.001–0.004	0.001–0.004
Cost, ¢/cu in.[d]		0.7–1.0	1.6–2.9	2.6–4.3

[a] Range covered by sisal, sisal-glass, and low glass content.
[b] Range generally provided by glass contents of 15–20%, and various filler loadings.
[c] Range generally indicated for high glass (25–30%) by compound suppliers.
[d] Cost figures based on compound cost estimate of 12–15¢ per lb for low cost grades; compounders' large volume price range of 40–60¢ per lb for high-glass grades; and intermediate estimate of 25–40¢ per lb for intermediate strength types. Actual costs depend on specific compound and whether purchased or premixed.

Comparative benefits

In comparison with such corrosion-resistant metals as brass, bronze and stainless steels, premixes offer, primarily, cost reductions due to ease of manufacture of complex shapes, and reductions in actual material costs. In some cases, design latitude, colorability, resistance to abrasion and reduction in weight are also important factors in favor of the premix parts.

In comparison with aluminum, premixes can offer superior corrosion resistance, especially to alkaline detergents, in addition to the advantages noted above.

In comparison with thermoplastic resins, the premix materials offer considerably greater hardness, rigidity and heat resistance. The premix materials are also stronger. The thermoplastics usually have superior colorability, surface smoothness and gloss, and a broader range of corrosion resistance. Thermoplastics are also somewhat lower in cost for the molded item when the number of parts is large. In general, tooling costs for injection-molded thermoplastics are higher, but molding, material-handling and finishing costs are lower for the thermoplastics.

Phenolics, ureas and melamines are considerably less strong than the glass-fiber reinforced premixes and in general, they do not have as good a range of resistances to aqueous solutions as do properly formulated premixes. The phenolics, of course, also have limited color possibilities.

In comparison with most other plastics, premixes are usually more difficult to handle because they are more difficult to meter or preweigh automatically. Generally, they must be preweighed by hand to an exact mold charge. Recently, however, there has been a trend toward extrusion and chopping of premix to logs of predetermined charge weight.

Molding

Because premixes are essentially heterogeneous materials, certain well-known molding phenomena such as weld lines and orientation during flow take on added significance. Fiber orientation tends to give somewhat greater effects of anisotropy and more care must be taken with premix molded parts to prevent excessive orientation. Similarly, weld lines in premix parts tend to be proportionately weaker than weld lines in thermoplastics. Much of the "art" of premix molding is concerned with reducing or eliminating orientation and weld-line effects by proper mold design, molding conditions (closing speed, pressure, temperature) and by selection of charge shape and location in the mold. Again, recent advances by glass-fiber manufacturers has resulted in superior glass fibers which reduce orientation and weld-line effects.

Premixes are generally molded in conventional compression molds and presses using pressures in the order of 500 to 2000 psi and temperatures in the range of 250 to 350 F.

Premixes are also occasionally molded in transfer presses, usually at somewhat higher pressures. Transfer molding tends to reduce strength properties by degrading the glass fibers in the compounds.

Premixes are suitable for molding very complex shapes over a wide range of sizes. Filter plates and frames weighing over 100 lb have been molded successfully.

A large volume of premix is consumed in automotive air-conditioner and heater housings and ducts. These parts are low-cost, low-strength, items requiring high rigidity and good heat resistance. Other large-volume applications include electrical insulators and housings of various types. Most premixes are manufactured captively, i.e., the molder prepares his own premix material. The advantages of captive premix are lower initial raw-material costs, reduced packaging costs, ability to formulate faster curing compounds and ability to tailor-make special compounds for each part. The major disadvantage of captive premix very often is lack of good quality control. Another disadvantage of the captive operation is that not enough compound development effort is spent in working out the "bugs" in a new formulation before such a formulation becomes commercial.

Compounding

Premixes are prepared by mixing glass, filler and resin mix in a 180° spiral double-arm dough mixer. Blade clearance should be about $\frac{1}{4}$ to $\frac{1}{2}$ in. The resin mix is usually prepared separately and includes resin, lubricant, catalyst, inhibitor and other minor additives. The resin mix is charged to the mixer, followed by the filler. After the filler is thoroughly blended (usually 10 minutes are required) the fibrous material can be added. If the fibrous material is glass, care must be taken to disperse the fiber to prevent clumping; close control of mixing time must be maintained to prevent fiber breakdown. Recently, a high strand integrity fiber has been introduced which reduces fiber-breakdown tendency.

After mixing the premix must be stored in airtight containers to prevent styrene monomer loss. Cellophane bags may be used to store small units of premix (15 to 20 lb) while larger bins or hoppers may be used for larger quantities.

Premix of high filler loadings may be extruded through screw type or ram type extruders and automatically chopped to logs of predetermined weight.

The storage life of premix which is sealed against styrene loss may vary from one or two days to one year depending upon formulation. The short storage life premix can be molded at lower temperatures and shorter cycles than the more stable premix.

PYROLYTIC MATERIALS

Essentially, pyrolytic deposition (literally, deposition by thermal decomposition) is a form of so-called gas or vapor plating.

Gas or vapor plating can be accomplished by (1) hydrogen reduction, (2) displacement, or (3) thermal decomposition. Pyrolytic deposition is accomplished by the last mechanism.

The process involves passing the vapors of a compound over a surface maintained at a temperature above the decomposition temperature of the compound, in a vacuum furnace. The surface provides a source for nucleation of the desired material which is built up to the desired section thickness.

Elemental materials are deposited from single compound vapors, e.g., carbon from a hydrocarbon, metals from their halides; compound materials are deposited from mixtures of compounds, e.g., boron nitride from boron halide and ammonia.

In producing coatings, the substrate serves as the surface on which the coating is deposited. In producing self-supporting parts, the substrate serves as a mandrel or mold from which the pyrolytic material is removed after deposition.

The pyrolytic deposition process is used to produce pyrolytic graphite, as well as coatings or self-supporting structures of an extremely broad range of materials. Theoretically, the only limitation on the type of material that can be produced is that (1) it must be available in the form of a compound whose vaporization temperature is below its decomposition temperature, and (2) the desired material must separate cleanly from the compound's vapor.

General properties

The major benefits of pyrolytic materials are:

1. Highly directional properties are obtained in some materials by the substantial degree of orientation of crystals or grains. (Note: Highly directional properties are only obtained in materials such as pyrolytic graphite and boron nitride which possess the unique and anisotropic graphite crystal structure.)

2. High densities, equivalent to theoretical densities, are obtainable.

3. High purity of material and close control of ingredients in "alloys" are obtainable by control of the reactant gases.

All initial work has been aimed at producing high-temperature materials, primarily for aerospace use. The unique properties of these materials make them attractive for a number of commercial applications.

Materials and forms

To date, pyrolytic graphite has been the largest volume material produced. But newer materials include the following:

1. Graphite-boron compounds: pyrolytic graphite to which less than 2% boron has been added for increased strength and oxidation resistance as well as lower electrical resistivity.

2. Other graphite compounds: pyrolytic graphite to which varying percentages of columbium, molybdenum or tungsten have been added.

3. Boron nitride: pyrolytic BN containing 50 atm percent of boron and nitrogen.

4. Carbides: pyrolytic carbides of tantalum, columbium, hafnium and zirconium.

5. Tungsten: pyrolytic tungsten has been produced in the form of coatings and parts such as crucibles and tubes.

Coatings can be deposited on complex surfaces, providing gas impermeability in extremely thin sections. On the other hand, only those surfaces of the shape which can be exposed to the flow of gases will be coated. Also, differences in coefficients of thermal expansion between substrate and coating materials must be carefully considered.

Self-supporting shapes are limited by the fact that the material must be produced by deposition on a mandrel that must be removed after the part is formed.

Directionality depends on crystal structure. The directional properties of the produced part or coating depend on the inherent crystal structure of the material. In general, materials are either highly anisotropic or nearly isotropic. Properties of each are shown in the accompanying table.

Pyrolytic deposition of material with hexagonal graphite type crystal structures (i.e., pyrolytic graphite and boron nitride) results in preferred crystal orientation producing a high degree of directionality of properties. The hexagonal or layer plane alignment of the grains is essentially parallel to the substrate's surface. Directionality results from strong atomic bonds within layer planes and weak bonds between layer planes, and also from the mode of heat transfer through the material which is predominantly by lattice vibration.

Pyrolytic deposition of materials with face-centered or body-centered cubic structures (i.e., carbides or tungsten) results in relatively isotropic properties. Although such materials can have a high degree of crystal orientation, orientation does not necessarily produce directional properties.

Properties of anisotropic materials

Pyrolytic graphite and its compounds and boron nitride all have substantial directionality of properties. The ratio of the number of crystallites having layer planes parallel to the deposition surface (i.e., a axis) to the number normal to the surface (i.e., c axis) can be varied by process control. For example, in graphite, ratios may range from 100 to 1000 to 1, compared with ratios of 3 or 4 for some commercial graphites. Orientation obtained in boron nitride has been as high as 1900 to 1.

Conductivity. One of the most useful properties of PG (pyrolytic graphite) is its insulating ability. In the direction normal to the deposition surface, PG is a better insulator than the most refractory ceramic materials. In addition, the thermal conductivity parallel to the deposition surface is comparable to the more *conductive* metals, tungsten and copper. This high conductivity evens out hot spots over the total surface.

The high destruction temperature of PG combined with low conductivity normal to the surface allows the surface temperature to become very high. This cuts down heat absorbed by the component by reradiating heat back to the at-

ANISOTROPIC MATERIALS

Material	Composition	Structure	Lattice Constant C/2	M.P. Subl.	Density	Young's Modulus	Resistivity ohm-cm "a"	"c"	Bend Strength "c"
Pyrolytic Graphite	C	Hex.	3.35 3.42	6600 F	2.20 2.23	$4.26(10)^6$	.0005	0.55	15,000
Boron Pyralloy	C+<2%B	Hex.	3.35 3.42	6600 F	2.20 2.23	$4.6(10)^6$	.00025	0.02	25,000
Boron Nitride	BN	Hex.	3.36	5430 F	2.20 2.2	$4.4(10)^6$	$>10^6$	$>10^6$	12,000

ISOTROPIC MATERIALS

Material	Composition	Structure	Lattice Constant C/2	M.P. Subl.	Density	Young's Modulus	Resistivity ohm-cm "a"	"c"	Bend Strength "c"
HfC	—	FCC	4.63	7030 F	12.2	—	$109(10)^{-6}$		—
TaC	—	FCC	4.43	7020 F	14.5	41.5	$30(10)^{-6}$		10,000–40,000
NbC	—	FCC	4.46	3500 F	7.8	49.4	$74(10)^{-6}$		—
ZrC	—	FCC	4.68	3200 F	6.4	45	$63(10)^{-6}$		—
W	—	BCC	3.16	3380 F	19.3	60	$5.48(10)^{-6}$		95,000

mosphere, thus acting as a "hyperinsulator."

Electrical resistivity is highly directional in pyrolytic graphite and graphite-boron compounds, but not in boron nitride. PG has relatively high electrical resistivity in the c direction, but low resistivities on the order of 0.0005 ohm-cm in the a axis.

Tensile strength improved. Tensile strengths in the a axis are orders of magnitude higher than in the c axis. For example, at room temperature, pyrolytic graphite has an a axis average tensile strength of about 14,000 psi, compared with about 500 psi in the c direction. The graphite-boron compound has room temperature tensile strengths of 16,500 and 700 psi in the a and c axes; boron nitride has a and c tensile strengths of 12,000 and 650 psi respectively.

Pyrolytic graphite (like conventional graphite) offers the singular advantage of increasing in strength with increasing temperature. Preliminary data indicate that the c axis strength decreases with temperature.

Addition of alloying elements has been found to improve c-axis strength. Additions of low concentration of less than 1% of tungsten and molybdenum have increased c-axis tensile strength by 50 to 90%.

Oxidation resistance improved. PG has somewhat greater oxidation resistance than normal graphite due largely to its imperviousness. Addition of boron improves oxidation resistance of pyrolytic graphite by a factor of ⅓. Oxidation resistance of boron nitride is superior to other pyrolytic materials produced to date specifically at temperatures below 2000 C.

Other properties. Owing to the atom-by-atom deposition process, pyrolytic materials are all near theoretical density, are impervious, and have extremely high purity levels. The materials do exhibit substantial directionality of thermal expansion which must be carefully considered in designing components.

Specific heat increases with temperature from about 0.23 Btu per lb per °F at room temperature to a maximum of about 0.50 at about 2000 F. The emissivity of pyrolytic graphite is relatively high. More important for uses such as radiating heat shields is the fact that as a shell material pyrolytic graphite will radiate a greater percent of heat input than other materials because of its high thermal conductivity parallel to the surface, which effectively increases the radiating surface area.

Properties of isotropic materials

Performance data on the isotropic pyrolytic materials, tungsten, and the carbides of tantalum, hafnium and columbium are much more limited than those for the anisotropic pyrolytic materials.

As mentioned before, although such materials are considered to be isotropic (in comparison with the anisotropic materials) crystal growth does tend to provide a preferred orientation in a plane normal to the deposition surface. Thus, generally speaking, the strength of such materials is greater in the plane normal to the surface than in the plane parallel to the surface.

Carbides. The pyrolytic deposition process can be controlled to produce carbides of varying metal-to-carbon ratio, resulting in carbides of differing microhardness. Following are Knoop hardness number (K_{100}) for three carbides, hardness increasing with increasing carbon-to-metal ratio:

TaC	1400–3500
HfC	2000–3600
NbC	1700–4000

The very hard grades of carbide are extremely brittle and difficult to handle. Their strength is low but they promise to be useful as thin, well-bonded coatings.

As hardness decreases, the carbides' ductility and strength increase. Bend strengths, except for the high hardness grades, were found to fall in the range of 10,000 to 40,000 psi; a value as high as 150,000 psi was observed for tantalum carbide low in carbon.

Tungsten. Pyrolytic tungsten which can now be produced in thicknesses over ¼ in. as coatings on components to 1½ ft in dia, is being evaluated for missile application. The increased strength and lower impurity level of pyrolytic tungsten is believed to be a major factor permitting production of sound coatings on large rocket nozzles.

R

RARE EARTH METALS

The rare earths are a closely related group of highly reactive metals comprising about one-sixth of the known elements. They form a transition series including the elements of atomic number 57 through 71, all having three outer electrons and differing only in the inner electronic structure. Since chemical properties are determined by the outer electronic structure, it is evident why these metals are chemically alike. Although not truly members of this series, scandium and yttrium (atomic numbers 21 and 39, respectively) are frequently included with this grouping shown in Table 1. They occur together in nature with the rare earths and are similar in properties.

TABLE 1—THE RARE-EARTH METALS

Element	Atomic No.	Density, lb/cu in.	Melt Pt, F
Scandium	21	0.108	2860 [a]
Yttrium	39	0.161	2820 [a]
Lanthanum	57	0.221	1690
Cerium	58	0.244	1480
Praseodymium	59	0.233	2820
Neodymium	60	0.253	1870
Promethium	61	0.263	1895 [a]
Samarium	62	0.272	1920
Europium	63	0.189	1650
Gadolinium	64	0.284	2460 [a]
Terbium	65	0.296	2490
Dysprosium	66	0.308	2550 [a]
Holmium	67	0.316	2730 [a]
Erbium	68	0.326	2770 [a]
Thulium	69	0.336	2910 [a]
Ytterbium	70	0.250	1515
Lutetium	71	0.355	3090 [a]

[a] Best estimated values.

The rare earths are neither rare (even the scarcest are more abundant than cadmium or silver) nor earths. The term *earth* stems from the oxide mineral in which these elements were first discovered. More suitable names such as lanthanons (after lanthanum, the first member of the group) have been proposed, but the term rare earths persists today.

Two groups. This series of metals is frequently broken into two groups based upon atomic weight and chemical properties. The "light" rare earths consist of elements with atomic number 57 to 63 and may be called the cerium group. The "heavy" rare earths consist of elements 64 through 71 as well as scandium and yttrium because of similar chemical behavior. The metals of the heavy or yttrium group are harder to separate from each other and, as a result, have found commercial interest only recently. Other historical terms will be found in commercial usage and can lead to confusion. For example, *didymium* is not actually another metal in this series, but refers to the neodymium-rich rare-earth mixture left after lanthanum and cerium are removed.

Separation now simpler. Separation of the individual elements, particularly of the heavy or yttrium group, was extremely difficult in the past and resulted in high costs and incompletely separated metals. With the development and application of ion-exchange techniques, a great change occurred. During the period 1949 to 1957 the prices of many of the separated rare-earth oxides decreased by a hundredfold or more.

Misch metal is an alloy of the cerium-group rare earths which are roughly in the same proportion as found in the ore. Costing about $3 per lb, it is the cheapest rare-earth material in alloy form. Cerium, being relatively abundant and easy to separate, is available in an impure (95%) grade for about $20 per lb. The price of separated material is dependent upon purity, and caution must be exercised in considering purity. Some rare earths on the market have a purity designation of 99.9% but this refers to the content of other rare-earth elements only and does not include impurities such as oxygen or tantalum which might be present in amounts up to 1% or even more. At a nominal purity of 99%, the price of the majority of the rare earths falls in the range of 100 to 300 dollars per lb. Several of the materials such as europium, terbium, thulium, and lutetium are much more costly, however. The metals are generally available in lump or ingot from a number of suppliers.

Properties

Recent work with separated rare-earth elements of good purity has shown that the metals are not so much alike in the metallic state as earlier data had indicated. Consistent property trends are readily noted; with the exception of europium and ytterbium, density and melting point increase with higher atomic numbers as shown in the table. The melting point of yttrium, as well as its chemical behavior, places it with the highest atomic number metals.

Mechanical. For the various metals of 99.5% purity, the room-temperature ultimate tensile strength ranges from 15,000 to 40,000 psi with elongations of 5 to 25%. Strength at 800 F is about one-half that at room temperature. Therefore, the rare earths do not appear to offer any outstanding mechanical properties that would indicate their use as a base for structural alloys. Scandium has a density similar to aluminum, but the potential for high strength-to-density materials is unknown. Yttrium provides an interesting combination of properties: density similar to titanium, melting point of 2820 F, transparency to neutrons, and formation of one of the most stable hydrides.

Fabricability. The rare earths may be hot-worked, and some of them can be fabricated cold. Small arc-cast ingots of yttrium have been reduced 95% at room temperature. The metals are poor conductors of electricity; specific resistances are in the range of 70×10^{-6} ohm-cm. All the metals are paramagnetic, and some are strongly ferromagnetic below room temperature.

Oxidation resistance. Yttrium and the higher atomic number rare earths maintain a typical metallic appearance at room temperature. Lanthanum, cerium, and europium oxidize rapidly under ordinary atmospheric conditions; the other metals of the light group form a thin oxide film. Some of the rare earths, notably samarium, form a stable protective oxide in air at temperatures up to at least 1100 F.

Applications

Applications of the rare earths may be divided into two general categories: the long-established uses and the newer developments that frequently require the higher purity separated elements. Three of the older applications that still account for three-fourths of total output are: rare-earth-cored carbons for arc lighting; lighter flints which are misch metal-iron alloys; and cerium oxide for polishing of glass and also salts for coloring or decolorizing of glass.

The use of rare earths to improve the properties of other metals is not new, but the data are still only qualitative. Rare earths have a strong tendency to combine with oxygen and other impurities so the addition may serve either as a scavenger or a true alloy constituent. Many beneficial effects have been noted in cast iron and steels with the addition of misch metal and various rare-earth compounds. Their function is generally to serve as a very potent deoxidizer or desulfurizer, depending upon alloy content. Conflicting data, probably caused by unreproducible retention of rare earths, have limited their application to a very small percentage of the ferrous alloy production.

Improved magnesium alloys containing rare earths have been developed, but currently this is not an important usage. Small additions to a variety of alloys improves the protective oxide film. A minor cerium content in nickel-chromium heating elements will greatly lengthen service life at high temperatures. The addition of 1% yttrium to 25% chromium steel increases the oxidation-resistant service temperature from 2000 to 2500 F. Included in chromium, yttrium greatly improves the barrier properties of chromium oxide to oxygen and nitrogen.

Yttrium oxide and cerium sulfide are high-melting refractories. Yttria-stabilized zirconia ceramic materials with excellent properties are commercially available. Good catalytic properties have been reported; assessment of their ability to compete with current materials is difficult.

A number of new industrial applications, still in the developmental stage, appear to offer considerable promise. Rare-earth compound semiconductors exist, and gadolinium selenide may become a valuable thermoelectric material. Oxides have been useful in assisting emission from cathodes in electron tubes. Rare earth-iron garnets (commonly called ferrites) may be particularly valuable for microwave magnetic cores because of special magnetic properties.

Applications as nuclear-reactor control materials are growing in importance. For this application a high neutron absorption capability (thermal neutron-capture cross section) is desired. Gadolinium, samarium, and europium are extremely efficient in this respect. Whereas considerable past effort has been devoted to europium (as an oxide dispersant), the importance of samarium may increase because it has a higher neutron absorption and is much cheaper.

REFRACTORIES, SPECIALTY

Specialty or superrefractories cover a group of materials that, because of their properties, are quite different from common furnace refractories made from natural minerals such as fire clay, diaspore, kyanite, silica or chrome. Primary usage of the latter is to resist or contain moderate to high heats.

Superrefractories are made from electric furnace crystals that include silicon carbide, fused aluminum oxide, electric furnace mullite, stabilized zirconia or fused magnesium oxide. These crystals are crushed, graded, bonded, formed and fired at high temperature. Usually a ceramic bond is employed although some silicon carbide compositions use a chemical bond (silicon nitride).

Also included here are fusion-cast refractories that are produced by pouring electric-furnace melts directly into either bonded alumina or graphite molds and subsequently annealed.

Most significantly, these materials retain to

TABLE 1—TYPICAL PROPERTIES OF BONDED AND FUSED-CAST REFRACTORIES

Chemical Analyses	Silicon Carbide [a]	Nitride-Bonded Silicon Carbide [a]	Fused Alumina Low Silica	Fused Alumina Regular	Lightweight Fused Alumina	Electric Furnace Mullite	Beta Alumina Fused Cast	Alpha-Beta Alumina Fused Cast	Zirconia Alumina Silica Fused Cast
Ignition	1.37	—	Nil	0.15	Nil	Nil	0.18	0.07	0.10
SiC	89.16	—	—	—	—	—	—	—	—
Al_2O_3	0.35	—	98.99	92.03	81.02	75.57	93.80	94.56	47.56
SiO_2	7.72	—	0.50	6.75	16.80	23.47	0.04	0.80	14.65
Fe_2O_3	0.87	—	0.13	0.27	0.09	0.27	0.11	0.12	0.10
CaO	<0.10	—	Nil	0.10	Tr	<0.10	<0.10	<0.10	Nil
MgO	—	—	Tr	0.10	0.10	<0.10	<0.10	<0.10	Nil
Na_2O	—	—	0.37	0.53	1.59	0.40	5.62	4.07	1.19
TiO_2	—	—	0.01	0.07	.40	0.09	<0.05	<0.05	0.05
ZrO_2	—	—	—	—	—	—	—	—	36.35
Physical Properties									
Pyrometric Cone Equivalent	38	—	39–40	38–39	38–39	38–39	39–40	39–40	38–40
Weight of 9" × 4½" × 2½" Straight in lbs	9.2	10.3	10.4 .	10.3	4.8	9.0	10.2	11.6	12.6
Average Bulk Density C.G.S.	2.55	2.87	2.90	2.87	1.30	2.50	2.80	3.20	3.46
Porosity—%	13.2	7.9	21.9	22.6	66.6	22.7	7.5	3.5	2.1
Mean Specific Heat 0° to 1400°C	.285	—	—	.330	.320	.230	—	—	—
Mean Coefficient of Expansion per °C, 25°–1400°C	4.4×10^{-6}	4.4×10^{-6}	7.4×10^{-6}	7.4×10^{-6}	8.6×10^{-6}	5.9×10^{-6}	6.6×10^{-6}	8.2×10^{-6}	4.7–8.4 [c] $\times 10^{-6}$
Thermal Conductivity at 2200°F in British Units	109	113.5	24	24	7	15	24	31	20
Modulus of Rupture @ 1350°C	2000	5640	200	658	63	433	—	—	—
1500°C Load Test @ 25 lb/sq in.—Contraction Cold	Nil	0.0%	0.50%	0.35%	0.65% [b]	0.15%	Nil	Nil	Nil
Abrasion Resistance	Excellent	Excellent	Fair	Fair	Poor	Fair	Poor	Excellent	Excellent
Relative Resistance to Spalling	High	Very High	Good	Good	Good	Very Good	Good	Low	Fair

[a] This covers a variety of silicon carbide refractory commonly used.. Where required, higher SiC content varieties can be supplied.
[b] @ 12.5 lb/sq in.
[c] Zirconia crystals change at 1200°C.

a maximum degree the properties of their respective electric-furnace crystals. They offer excellent thermal conductivity, abrasion resistance, hot strength, thermal-shock resistance, impact resistance and/or chemical inertness. Types and general properties are shown in Table 1.

A new development in fused cast refractories is a chrome alumina spinel fusion containing 27% Cr_2O_3. Its density and physical properties are similar to zirconia alumina silica fusions, but its corrosion resistance against slags and glasses is two to three times greater. Chrome-magnesia and chrome-magnesia-alumina fusions hold promise in metallurgical applications.

Properties

Following are the primary engineering properties of specialty refractories.

Thermal conductivity. Fig 1 shows that specialty refractories have a wide range of thermal conductivity, ranging from insulation to conduction. While cast iron and alloy steels possess exceptional ability to transmit heat, the ability decreases at higher temperatures. Silicon carbide, usable to about 3000 F, transmits 109 Btu/hr/sq ft/°F/in. at 2200 F or, roughly, 11 times that of fireclay and 70% that of chrome-nickel steels.

High-thermal-conductivity refractories, used in indirect or muffled furnaces to insure complete separation of heated materials and heating media, serve to lower combustion-chamber temperatures, improve fuel efficiency and combustion and minimize refractory problems. This property allows occasional substitution for metals at high temperatures.

Fused alumina and silicon carbide bodies also dissipate heat. Alumina panels, for example, are used in cooling sections of tunnel kilns to remove heat rapidly. Silicon-nitride-bonded silicon carbide rocket-nozzle throats withstand destructive heats developed in uncooled motors because of fast heat dissipation.

Where high-thermal conductivity is a disadvantage, refractories are available with low conductivity.

Fig 1 shows that some of these superrefractories make good insulators. One such material is made of hollow spheres of fused-aluminum oxide. Another is lightweight stabilized zirconia.

Ceramic fibers, blown from electric furnace melts, are made into a variety of forms with

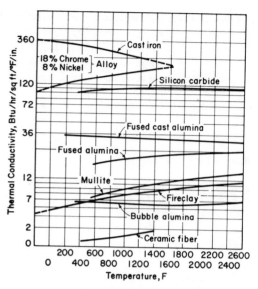

FIG. 1. Comparative thermal conductivity curves for heat-resistant metals, silicon carbide and various refractories. Silicon carbide conducts heat rapidly in a range far above temperatures that metals withstand.

extraordinarily low K factors. They are available in bulk, blanket, rope, tape, roving, yarn, wicking, cloth and wire inserted cloth.

Load-bearing strength. This is a far more important property in high-temperature work than refractoriness, or ability to withstand temperature. Hot load-bearing strength indicates degree of volume stability at temperature; in other words, retention of volume under hot load. Load may come from external weight, such as a furnace charge; superimposed, such as a furnace sidewall; or even result from thermal-expansion stresses within the structure itself. Many ceramic and metallic materials give satisfactory service in definite ranges of temperature and load. Above these points they become increasingly susceptible to deformation and ultimate failure.

Table 2 provides comparative data on sustained and short-duration, high-temperature load tests. Both are informative in themselves. In addition, they indicate relative thermal-shock resistance of the body. Note that silica refractories show excellent load-bearing characteristics in the 100-hour test but a tendency to spall during the rapid cycle test.

A 5% contraction is considered failure since

TABLE 2—COMPARATIVE DATA ON SUSTAINED AND SHORT-DURATION, HIGH-TEMPERATURE LOAD TESTS

Refractory Compositions	100 hour 25 psi Load Test		1½ hour 25 psi Load Test	
	Hold Temperature °F	% Linear Cold Contraction	Hold Temperature °F	% Linear Cold Contraction
Silicon Nitride-Bonded Silicon Carbide	2732	0.0	2732	0.0
Ceramic-Bonded Silicon Carbide	2732	0.58	2732	Nil
			3146	0.78
Electric Furnace Mullite	2732	1.23	2732	0.11
			3128	1.31
Bonded Fused Alumina	2462	0.25		
	2732	8.68	3130	8.86
Fused-Cast Beta Alumina	2732	0.0	2732	Nil
Converted Kyanite	—	—	3137	14.3
Chrome Magnesia	—	—	Crushed at 3002	Sheared
Magnesia	—	—	Crushed at 2552	Sheared
Silica	2732	0.13	Crushed sharply at 2948	Sheared
High alumina	2462	7.7	Squashed at 3011	25.46
Fireclay	2462	11.3	Squashed at 3002	25.0
Fireclay	2462	15.1	Squashed at 2993	26.6

this represents a 7/16-in. slump in a 9-in.-long brick; this is sufficient to lose the keying, wedging or arching action on a modular brick unit.

Heat-shock resistance. Rapid and repeated temperature changes prevalent in many furnacing and processing operations cause temperature gradients that stress refractory parts sometimes so severely that they crack and spall. Economics of a particular operation may preclude steps to lessen thermal shock in which case a material hàving good thermal-shock resistance should be employed.

Thermal conductivity, tensile strength, modulus of elasticity and thermal expansion have major influence on thermal-shock resistance. Low-thermal expansion obviously helps prevent a differential expansion across the cooled and uncooled portions of the structure. High-heat conductivity minimizes thermal gradients. Hot strength is essential to meet the inevitable strains that develop. Possession of these unique qualities enables silicon carbide refractories—both ceramic and nitride bonded—to offer high resistance to thermal shock with electric-furnace mullite materials a close second.

On the other hand their high coefficient of expansion eliminates fused-cast refractories from consideration for applications where severe shock will be encountered. Ordinarily, they find use in furnaces such as glass-melting tanks that do not go through rapid temperature fluctuations.

Chemical resistance. Chemical action often destroys refractories. Attack may be by materials charged into a furnace, by slags or fluxing agents, by fuel ash, by evolved furnace gases, by dust or even by other construction materials.

Refractories made of electric-furnace crystals are generally far more resistant to corrosion than those made of natural raw materials—largely because of the extraordinary temperatures at which the crystals are formed. This lessens tendency to change shape or to go through conversion phases at lower temperatures. These classes of material are governed by the general conditions that acids react with bases and vice versa and that reaction rates usually increase with temperature. Silicon carbide materials are acid whereas other bonded and fused-cast refractory types may be acid, neutral or basic.

Permeability. As a group, bonded refractories are porous, containing interstices between crystals or grog and bond.

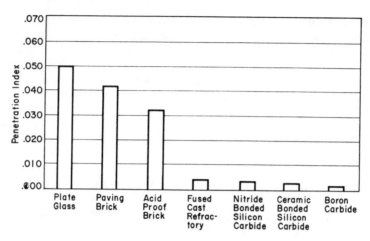

FIG. 2. Abrasion resistance of several nonmetallics after being blasted for the same period in a Zeiss impingement test. Aluminum oxide grain was the abrading material.

Fused-cast refractories, however, may be extremely dense as they are solidified from complete fusion. Dense silicon carbide is capable of limiting the passage of cold air to one or two cubic feet per hour. This is important when the material is used as externally heated tubes carrying gases to be cracked or reformed; leakage would minimize effectiveness and perhaps set up explosive conditions.

Abrasion resistance. Abrasion resistance of these materials is important at all temperatures from subzero to over 3000 F.

There are two basic types of abrasion. The first is impingement by fine solids that are suspended in air, gas or fluids. Fig 2 gives results of Zeiss impingement tests on superrefractories using fused-aluminum-oxide grain (Mohs scale hardness: 9) as the impinging medium. Impingement of suspended fine solids tends to erode containing material; here, bonded silicon carbide and some varieties of bonded alumina refractories are the most resistant.

Direct abrasion by a mass of wet or dry solids flowing across or at a surface is the second basic type. Where there is little or no impact, silicon carbide materials are recommended. Severe continuing impact by heavy solids may cause progressive fracture of the ceramic bond without which refractory grains are removed mechanically. Fused-cast refractory materials are recommended here because they depend on interlocking crystalline structures rather than bonded crystals.

Many times refractories are chosen on a basis of temperature resistance, without enough consideration to these other key characteristics. Only in cases of stabilized zirconia (melting point 4710 F) and fused magnesia (melting point 5072 F) is refractoriness significant; nevertheless even these materials must be applied with careful attention to their other properties.

Design considerations

The properties discussed indicate where care must be taken in design and application. Although high in modulus of rupture, as well as hot strength, superrefractories do possess relatively poor tensile strengths. Nitride-bonded silicon carbide has a compressive strength of approximately 20,000 psi at 2400 F, but a tensile strength of only 3500 psi. Being essentially ceramics, these materials do not have the same stress-strain ratio or yield point that is expected of less refractory metals. Although strong, they are brittle and break before bending.

Overly large pieces or ones with thin sections reduce heat-shock resistance. When there is impact or thermal shock, a small compact shape is less susceptible to breakage or spalling than a large piece. Similarly, irregular cross sections should be avoided.

Ceramic-bonded refractories should not be made under ¼ in. or over 3 in. thick. With fusion-cast materials, preferred and most economical shapes are of comparatively heavy cross sections and roughly rectangular in design.

Available forms, shapes and sizes

Ceramic-bonded superrefractory materials are available in all brick sizes as well as a great variety of shapes far more complex than normally furnished in common refractories. Special cylinders, tubes, checkers, hearths and other furnace parts are produced regularly.

These materials fail to match the intricate designs possible with metals, although nitride-bonded silicon carbide materials come close. The latter have low volume changes between drying and firing. Tolerances of plus or minus 0.003 to plus or minus 0.01 in. per in. are possible up to a foot, with comparable tolerances above that. When cast, this material has a surface of approximately 250 microinches; compacted and pressed parts attain a finish of 500 microinches.

It is possible to make silicon-nitride-bonded refractories with standard threads of good accuracy. Other techniques permit making separate pieces that, on second firing, provide an assembly hitherto impossible to make by ceramic-forming methods.

RHENIUM

Rhenium is a rare element which probably does not occur in high concentrations anywhere. It has been estimated to constitute about 4×10^{-9} parts of the earth's crust, making it comparable to platinum in abundance.

The primary source of rhenium is the by-product molybdenite obtained from porphyry copper deposits.

Virtually all of the wrought products of pure rhenium are produced from bars made by powder metallurgy. Although rhenium can be arc-melted in an inert atmosphere, the resultant metal is not well-suited for fabrication due to the coarse as-melted grain size and the possible segregation of small amounts of rhenium oxide at the grain boundaries. Even with vacuum melting, the workability has not equalled that of bars produced by powder metallurgy. The latter bars have a small grain size and are much more amenable to fabrication.

Rhenium powder produced by hydrogen reduction of ammonium perrhenate is consolidated by pressing in split rectangular dies at pressures of 25 to 30 tsi using nonlubricated powder. These pressed bars are vacuum presintered at 1200 C and a pressure of 0.5 to 1μ. They are subse-

quently resistance sintered in dry hydrogen at a maximum temperature of approximately 90% of the melting point in a manner almost identical to that used for tungsten. The sintered bars are normally 90 to 96% of the theoretical density, having undergone 15 to 20% shrinkage in each dimension. These bars are then ready for fabrication.

Pure rhenium is fabricated by cold working with frequent intermediate recrystallizing anneals. Hot working of rhenium results in "hot short" failures, probably due to the presence of rhenium heptoxide at the grain boundaries. This volatile oxide has a melting point of 297 C and a boiling point of 363 C.

The fabrication of rhenium wire requires processing by swaging and drawing. Work hardening is so great in cold swaging that reductions are limited to one 10% reduction between anneals to avoid serious damage to the swaging dies. The reductions are somewhat greater in cold drawing beginning with 10% but increasing to about 40% between anneals. Diamond drawing dies are used for all pure rhenium wire drawing.

The production of rhenium strip is likewise accomplished by cold rolling. After several light initial reductions on the sintered bars, reductions up to 40% can be affected between anneals. Strip thinner than 0.005 in. is rolled using small diameter tungsten carbide work rolls in a "4 high" mill. Strip can be processed in this manner to thicknesses of 0.001 in.

Due to the rapid rate of work hardening which accompanies the cold working of rhenium, annealing constitutes a very important part of the fabrication processes. Frequent anneals are necessary at temperatures of 1550 to 1700 C for times varying from 10 to 30 min in an atmosphere of dry hydrogen or a hydrogen-nitrogen mixture. Furnaces utilizing molybdenum heating elements and aluminum oxide (Al_2O_3) refractories are used most extensively. Anneals performed as indicated result in complete recrystallization with the grain size usually between 0.010 and 0.040 mm and a hardness of about 250 to 275 VHN.

Rhenium-molybdenum alloys. The benefit of rhenium additions to both molybdenum and tungsten was first reported by Geach and Hughes of the Associated Electrical Industries Laboratory, Aldermaston, England and confirmed by groups at Battelle Memorial Institute and Chase Brass & Copper Co. Its effect on fabricability will be considered later.

The greatest interest in rhenium-molybdenum alloys lies in those containing approximately 30 to 50 wt % rhenium. Rod, wire, and sheet of these alloys can be fabricated from either melted ingots or pressed and sintered bars made from premixed powders. Bars or ingots can be warm-forged or warm-swaged at approximately 800 to 1200 C. Heating must be done in hydrogen or a hydrogren-nitrogen mixture, although the working can be done in air. The rhenium-molybdenum alloys are also amenable to cold working so that wire can be drawn either warm or cold.

A similar procedure can be used for the fabrication of Re-Mo alloy strip. Vacuum-melted ingots can be forged flat at approximately 1000 C to permit warm rolling or pressed and sintered bars can be directly warm-rolled at 800 to 1200 C. Sizes thinner than about 0.050 in. can be cold-rolled with good results.

Rhenium-tungsten alloys. As with the Re-Mo alloys, the most promising range of Re-W alloys extends from a few percent to 30% of rhenium which takes in the solid solution range of rhenium in tungsten. The greatest improvement in ductility seems to occur with additions of 20 to 26% rhenium.

The fabrication of wrought Re-W alloys is best accomplished using pressed and sintered bars. Sintered bars in excess of 90% of the theoretical density can be obtained with or without the use of binders. These bars are subsequently warm-swaged to approximately 0.100 in. dia and warm-drawn to the desired sizes. In like manner sintered bars can be warm-rolled to produce strip.

Properties

Rhenium is a refractory metal with properties similar in many ways to tungsten and molybdenum. Pure rhenium has significantly greater room-temperature ductility combined with good high-temperature strength. When used as an alloying addition with tungsten or molybdenum forming body-centered cubic solid solutions, rhenium produces significant improvements in the ductility of those metals. Some of the properties of rhenium for comparison with tungsten and molybdenum are given in Table 1. Mechanical properties of rhenium and rhenium alloys

are given in Table 2.

It is significant to note that the melting point of rhenium is second highest of the metals, exceeded only by tungsten. Its density is surpassed only by osmium, iridium and platinum and its modulus of elasticity only by osmium and iridium. Moreover, the hardness and strength of rhenium are very high and its rate of work hardening is greater than that for any other metal.

The resistance of rhenium to the water-cycle

TABLE 1—TYPICAL PHYSICAL PROPERTIES OF RHENIUM

Atomic No.	75
Atomic Weight	186.31
Crystal Structure	HCP
Density, 20 C, gm/cu cm	21.0
Melt Pt, °C	3180 C
Lin Coef of Exp, 20 C, $10^{-6}/°C$	6.7
Vapor Pressure, mm	
2000 C	3×10^{-8}
2200 C	7.5×10^{-7}
2500 C	4.5×10^{-5}
3000 C	8×10^{-3}
Elec Resis, 20 C, μ-ohm-cm	19.0
Elec Cond, % IACS	9.1
Temp Coef of Elec Resis, 20–100 C, per °C	0.00395
Electronic Properties	
Work Function, ev	4.8
Mod of Elas, 10^6 psi	67

effect as encountered in a tube or lamp is far superior to that of tungsten. However, the oxidation resistance of rhenium is very similar to that of molybdenum. It is good to approximately 500 C but at higher temperatures oxidation becomes catastrophic due to the formation of the volatile oxide, Re_2O_7.

The combination of high strength and ductility exhibited by rhenium and rhenium alloys is of great interest. Furthermore, these materials can be inert-arc-welded with a resultant weld which is ductile and possesses corrosion resistance as good as the parent metal. No ductile to brittle transition is observed in pure rhenium or the rhenium-molybdenum alloys containing 30 to 50 wt % Re as occurs in molybdenum at about room temperature and in tungsten at approximately 300 C. Although rhenium additions lower the transition temperature of tungsten it

TABLE 2—MECHANICAL PROPERTIES OF RHENIUM
AND RHENIUM ALLOYS

	Yld Str, 0.2% offset, 1000 psi	Ten Str, 1000 psi	Elong in 2 in., %
Rhenium			
0.005 in. Strip			
Ann.	39	150	19
Cold Rolled 20%	274	287	2
0.020 in. dia Wire			
Ann.	45	160	15
Cold Drawn 20%	280	290	1
50Re–50Mo (wt %)			
0.005 in. Strip			
Ann.	123	150	22
Cold Rolled 50%	210	240	1
0.005 in. dia Wire			
Ann.	—	175	22
Cold Drawn ...	—	250	2
75W–25Re (wt %)			
0.010 in. dia Wire			
Ann. 1600 C	—	225	10
Ann. 2000 C	—	200	7
As Drawn	—	475	2

still occurs above room temperature. However, even below the transition temperature for the recrystallized material, the ductility of the Re-W alloys is higher than that of pure tungsten.

Many of the properties of the rhenium-bearing alloys are still unknown while others (Table 2) are just coming into focus. Some of the rhenium-molybdenum alloys have been observed to possess excellent superconductivity. A great deal of interest has developed in this characteristic of these alloys and extensive work is under way to advance the technology in this important and promising field. Likewise, the surface has just been scratched in the W-Re alloy field. The thermoelectric properties of rhenium and especially W-Re alloys when coupled with tungsten are outstanding. These same W-Re alloys (3 to 26% Re) have been observed to possess excellent strength at both room and elevated temperatures. All of these alloys seem to have appreciable room-temperature ductility when not exposed to temperatures above 1800 C. Even after heating to 2000 C in a protective atmosphere, alloys containing 20 to 26% rhenium have ap-

proximately 10% elongation and alloys containing as little as 10% rhenium appear to possess at least 4% elongation.

Applications

The oustanding properties of rhenium and rhenium alloys suggest their use in many specialized applications.

Thermocouples. Re *vs.* W thermocouples can be used for temperature measurement and control to approximately 2200 C whereas previous thermocouple use was limited to temperatures below 1750 C. Indeed, 74 W-26 Re *vs.* W thermocouples can be used to temperatures of at least 2750 C with a high emf output insuring accurate precise temperature measurements with excellent reproducibility and reliability.

Electronic. Rhenium is now widely used for filaments for mass spectrographs and for ion gauges for measuring high vacuum. Its ductility, chemical properties, including the fact that it does not react with carbon to form a carbide, and its emission characteristics make it superior to tungsten for these applications.

The elevated temperature properties of rhenium and rhenium alloys, their weldability, and electrical resistivity suggest their use for various components in electronic tubes.

Electrical. The superconducting properties of Re-Mo alloys has prompted their use for coils in compact-size electromagnets with high field strengths. While this application is still in the development stages it looks most promising.

Rhenium has received considerable acclaim as an electrical contact material. It possesses excellent resistance to wear as well as arc erosion. Furthermore, the contact resistance of rhenium is extremely stable due to its good corrosion resistance in addition to the fact that possible formation of an oxide film on the contacts would not cause any appreciable change in the contact resistance since the resistivity of the oxide is almost the same as that of the metal. Extensive tests for some types of make and break switching contacts has shown rhenium to have 20 times the life of platinum-palladium contacts currently in use.

Heating elements for resistance heated vacuum or inert atmosphere furnaces seems to be a potential application for rhenium and especially rhenium-tungsten alloys. These materials would not undergo the embrittlement which occurs in tungsten upon heating, and they would allow for usage in vacuum, hydrogen, or inert atmospheres.

Welding filler rod. Rhenium and Re-Mo alloys can be readily welded, permitting fabrication of the welds. In fact Re-Mo alloy can be used as a filler material for obtaining ductile welds in molybdenum. Although not all of the problems associated with the heat-affected zone are readily overcome, the properties of welds made with Re-Mo alloy filler wire are very much better.

Nuclear. Investigations are currently under way to evaluate rhenium for various uses in the nuclear field. The elevated temperature properties of these alloys as well as their high density and moderately high nuclear cross section suggest consideration for radiation shields as well as other reactor components.

RUST PREVENTIVES

Petroleum rust preventives have been used for the temporary preservation of corrodible metal surfaces ever since Colonel Drake discovered the first oil well over one hundred years ago. From this first oil production waxes were separated, refined and made into "petroleum jelly," bringing into being the first petroleum rust preventives. The use of petroleum products as protective coatings for oil-well equipment spread into all areas requiring low-cost, temporary rust protection.

The raw materials used today are produced in modern petroleum refineries in large volume at low cost. Petroleum has provided the wide range of physical properties required to meet the exacting and varied engineering specifications of our complex industrial and diversified military rust-preventive requirements. These physical properties are viscosity, melting point, pour point, consistency, adhesion, volatility and density. Chemical additives are usually incorporated with petroleum rust preventives to improve lubricity, to modify viscosity-temperature relationships, to reduce pour point, to displace water from metal surfaces, to dissolve salts, to prevent foaming, to prevent oxidation of the compound, to impart antiwear properties, to improve detergency and dispersancy, and to prevent corrosion and rusting.

The importance of rust prevention can be judged by considering one authoritative estimate which reported that rust and corrosion of metals in the United States result directly in an annual loss of six billion dollars. Atmospheric corrosion of ferrous metals accounts for a substantial part of this figure. The rust prevention of machined parts presents an over-all cost saving because the cost of protection is much less than the cost of cleaning to remove rust, of reprocessing, or of scrapping parts which cannot be salvaged.

Petroleum rust preventives provide temporary but complete protection against rusting, are easy to apply, easy to remove and are low in cost. The results of rusting may result in dimensional change, structural weakening and loss of decorative finish.

Rust-preventive types

Petroleum rust preventives may be conveniently separated into four major types: (1) solid or semisolid waxes; (2) oils; (3) solvent cutbacks; and (4) emulsions.

Solid or semisolid waxes. These compounds are generally prepared from combinations of petrolatums and microwaxes with heavy lube oils added to obtain the desired consistency. Three consistencies are specified for military use: hard, medium and soft.

1. The hard-film grade is applied in the molten state and is intended for long-term, temporary outdoor storage of heavy arms, jigs and forms, for protecting small metal parts, either packaged or unpackaged, and for long-term indoor storage protection of brightly finished surfaces.

2. The medium-film grade is applied in the molten state or at about ambient temperature by dipping or brushing. It is intended for unshielded outdoor storage in relatively moderate climates at temperatures not exceeding the flow point temperature of the compound.

3. The soft-film grade is applied either by brushing or dipping at room temperature or from the molten state. It is intended primarily for the preservation of friction bearings and for use on machined surfaces indoors or for packaged parts.

Additives may or may not be present, depending on the type of service for which the compound is intended. The advantages claimed for the solid type compounds are:

1. They form coatings of controllable thick-

ness, they do not drain off during storage, and they provide a substantial physical barrier to both moisture and corrosive gases.

2. They form coatings resistant to mild abrasion, air-borne contaminants and ultraviolet light (sunlight) for extended periods.

3. The fast-setting wax coatings eliminate prolonged draining or drying periods during processing.

Some of the disadvantages found for these compounds are:

1. Special thermostatically controlled dip tanks must be used to control thickness.

2. The coating, once removed accidentally during handling, is not self-healing and must be reapplied.

3. The compounds are difficult to remove by wiping or rinsing with solvents. Special facilities must be available for the efficient removal of these hard grades. They are not to be applied to intricate parts or inaccessible surfaces.

Preservative oils. These are available in as many types as there are special uses for oil-type products. For example:

1. Preservative lubricants may contain detergents, pour depressants, bearing corrosion inhibitors and antioxidants, as well as antirust additives for gasoline or diesel engine use.

2. Special preservative oils for both reciprocating and turbojet aviation engines are available containing nonmetallic (nonashing), antirust additives.

3. Light lubricants for automatic rifles, machine guns and aircraft, suitable for operation at temperatures as low as −65 F, are available in several viscosity grades.

4. General-purpose preservatives are produced specifically for marine use where salt spray as well as high humidity may rust machinery and shipboard cargo.

5. Preservative hydraulic fluids are compounded to preserve hydraulic systems during storage, shipment and manufacture. These oils contain additives that will permit temporary operational use of the hydraulic systems. In addition to antirust additives, oxidation inhibitors, viscosity index improvers, antiwear agents and pour depressants are present in hydraulic fluids.

6. Slushing oils are used extensively for the preservation of rolled steel and aluminum sheet rods and bar stock. These oils require careful selection of additives to prevent "black stain-

ing" of stacked or rolled sheet. This form of corrosion is a common problem in steel mills.

Metal-conditioning oils are used to loosen and soften scale and rust from heavily corroded surfaces such as marine bulkheads, decks, and structural supports of ballast tanks. These oils appreciably reduce the pitting and general corrosion resulting from the combination of humid atmospheres and frequent immersion in salt water.

Solvent cutbacks. These are generally of three types and can be applied by spraying, dipping or slushing. The volatile solvent used in these compounds normally has a flash point above 100 F and will evaporate completely in several hours leaving the residual protective coating. They include:

1. Asphalt cutbacks formulated to dry "to handle" in several hours leaving a smooth black coating approximately two mils thick. These coatings are recommended for severe outdoor exposure. They are resistant for several years to salt spray, sunlight, rain and high humidity.

2. Petrolatum and waxlike coatings which can be either hard and "dry to touch," or soft and easily removed. The "dry to touch" or tack-free coating is usually transparent and provides an excellent preservative coating for spare parts which are stored for long periods indoors and are handled frequently. The soft coatings provide excellent protection against corrosive atmospheres, high humidity and airborne contamination. These coatings are easily removed by wiping or solvent degreasing but must be protected by protective packaging or storage arrangements. These coatings are not fully self-healing if damaged.

3. Oil coatings are used largely for short-term, in-shop use where parts are still being processed. Humidity protection is provided, with additional emphasis being placed on suppression of fingerprint rusting and on displacement of the water remaining from processing.

Emulsifiable rust preventives. The emulsifiable rust preventive is an oil or wax concentrate which on proper dilution and on mixing with water will provide a ready-to-use product in the form of an oil-in-water emulsion system. These emulsion systems are fire resistant, provide excellent protection and are economical. They are intended for use whenever a low-cost, fire resistant, soft-film, corrosion preventive can be utilized. The compound is formulated to be emulsified in distilled or deionized water, normally at a one to four ratio, reducing the volatile content of the corrosion preventive to a minimum. Its fire resistant properties make it especially adaptable as a substitute for volatile solvent cutbacks where fire hazards increase the expense of application. It has proven efficient for use on small hardware or on parts of uniform contour. Caution should be exercised in applying it to parts of radical design having cracks, crevices or depressions, which would prevent adequate drainage and cause subsequent corrosion by trapping excess amounts of water.

Selection of rust preventives

The selection of petroleum rust preventives is based upon their temporary nature, ease of application and removal, the economy of providing protection against corrosion, and the type of protection desired (e.g., indoor protection or outdoor exposure).

Indoor protection. Preservatives of this group must provide a high degree of protection against high humidity, corrosive atmospheres and moderate amounts of airborne contaminants. Resistance to abrasion is occasionally required where spare parts are stored in bins or on uncovered racks. Normally, however, the more easily removed products are used, such as oils,

soft petrolatums or soft-film solvent cutbacks.

Outdoor exposure. Preservatives of this group must withstand the effects of rain, snow, and sunlight as well as provide good resistance to abrasion. The products used for this service are the hard-film, asphalt solvent cutbacks and the hard grades of hot-dip waxes.

Short-term storage. These preservatives are commonly used to protect parts between machining operations or prior to final preservation with a heavy-duty product. These light oils or solvent cutbacks usually have fingerprint-suppressing properties to protect the parts during production handling.

Special-purpose lubricants. Lubricants for many purposes are formulated to provide primary or secondary rust-prevention properties. These special lubricants include nonashing lubricants for aircraft engine use, high-detergency oils for motor vehicles, low-temperature machine gun or rifle oils, and special salt-spray resistant lubricating oils for shipboard use.

Functional uses. Hydraulic oils possessing rust-preventive properties are used during the shipment or storage of equipment containing hydraulic power-transmission systems. This equipment is found in aerospace systems, submarines, aircraft carriers and industrial machine tools. Various viscosity grades are available.

RUTHENIUM. See Platinum Metals.

S

SAND CASTINGS

Sand casting can be broken down into three general processes: green-sand molding, dry-sand molding and pit molding. Green-sand molding, in all probability, produces the largest tonnage of castings but is limited in that long, thin projections are very difficult to cast. Dry-sand molds, on the other hand, can produce castings of any desired intricacy (within the normal dimensional limits of the process) because the sand is baked with a binder, increasing the strength of the thin mold sections. Pit molding is the sand process used to make very large, heavy castings.

The processes

Green-sand molding. The essential parts of a green-sand mold are shown in Fig 1. The pattern is placed in the flask (consisting of the cope and drag) and molding sand is tightly rammed around it. Usually, the mold is made in two halves, with the pattern lying at the parting line. Removing the pattern from the rammed sand leaves the desired cavity from which the casting is produced. The molder then produces a sprue for pouring metal into the cavity and an opening for a riser to permit air in the cavity to be expelled. Both the sprue and riser also act as a source of hot metal during solidification and help eliminate shrinkage cavities in the casting.

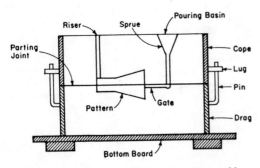

FIG. 1. Cross section of a sand casting mold.

Cores can be added to the mold cavity to shape internal casting surfaces.

Among the advantages of green-sand molding are the following: one pattern can be used to produce any number of castings; most nonferrous alloys, gray irons, ductile irons, malleable irons and steel can be cast; and finally, the fragility of the sand cores permits them to collapse after metal is cast around them, thereby eliminating or reducing casting stresses and tendency for hot tearing.

The limitation of the process is its inability to support long, thin projections in the sand.

Dry-sand molding. Core boxes, not patterns, are used to make the various parts of the mold. A mixture of sand and a binder is formed in a core box and baked at 400 to 500 F in order to harden the sand. The various pieces are then assembled into a mold.

Dry-sand molds, because of the way in which they are produced, can be used to make intricate castings. The baking operation strengthens the sand and permits thin projections to be cast without danger of collapsing the mold walls. Cores used in dry-sand molding are collapsible and help reduce hot tearing tendencies.

Pit molding. If large, intricate parts must be produced, pit molding is considered the most economical production method. (Castings weighing up to 500,000 lb have been produced in pit molds.) Pit molding is a highly specialized operation and the equipment used—sand slingers, molding machines, etc.—precludes the use of much hand labor. The mold usually is dried, increasing sand strength and, consequently, the ability to resist mold erosion during pouring as well as the weight of the casting being poured.

Complexity and dimensions

The maximum size casting that can be produced in green-sand molds is 20 to 30 tons; limitation for dry-sand molds is 5000 to 6000 lb. Both processes can produce castings as small as 1 oz. There is no weight limitation for pit molding. Minimum section thickness for both green- and dry-sand castings, in inches, are as follows:

aluminum, $\frac{3}{16}$; copper, $\frac{3}{32}$; gray and malleable iron, $\frac{1}{8}$; steel, $\frac{1}{4}$ to $\frac{1}{2}$.

Dimensional tolerances in inches per foot are as follows: gray iron, $\frac{3}{64}$; malleable iron, $\frac{1}{32}$; steel, $\frac{1}{16}$; aluminum and magnesium, $\frac{1}{32}$; copper, $\frac{3}{32}$. The draft allowance for green- and dry-sand castings is 1 to 2 deg.

All processes produce a surface finish range of 100 to 1000 μ in. rms.

SANDWICH MATERIALS

Sandwich construction refers generally to those types of composite constructions consisting of two high-strength, thin facings, rigidly attached to either side of a low-density, shear-resistant and relatively thick core material. Typical structures are made of aluminum, steel, or plywood skins adhesive bonded or brazed to cores made of various honeycombs, foam, wood, plastic, or other low-density material.

Theory

The theory of sandwich materials and functions of the individual components may best be described by making an analogy to an I-beam. The high density facings of a sandwich correspond to the flanges of the I-beam, the object being to place a high-density, high-strength material as far from the neutral axis as possible to increase the section modulus without adding much weight. Honeycomb in a sandwich is comparable to the I-beam web which supports the flanges and allows them to act as a unit. The web of the I-beam and honeycomb of the sandwich carry the beam shear stress. Honeycomb in a sandwich differs from the web of an I-beam in that it maintains a continuous area support for the facings, allowing them to carry stresses up to or above the yield strength without crippling or buckling. The adhesive which bonds honeycomb to its facings must be capable of transmitting shear loads between these two components, thus making the entire structure an integral unit.

When a sandwich panel is loaded as a beam, the honeycomb and the bond resist the shear loads while the facings resist the moments due to bending forces, and hence carry the beam bending as tensile and compressive loads. When loaded as a column, the facings alone resist the column forces while the core stabilizes the thin facings to prevent buckling, wrinkling, or crippling.

Materials

Most of the materials common to conventional methods of construction have found their way into sandwich construction. Facing materials commonly used, in the order of their volume of usage, are:

1. Aluminum alloys (widely used in aircraft and missile structures).

2. Mild steel facings (commonly used in toilet partitions, commercial desks, and furniture).

3. Porcelain enameled steel (the most common of the spandrel panel facing materials used in building construction).

4. Gypsum board (widely used for interior partition panels in buildings).

5. Plywoods (a common flush door facing material).

6. Reinforced plastic laminates (standard material in aircraft radomes and antenna panels).

7. Stainless steel (used as facing in all-stainless sandwich in high performance aircraft).

8. Various other specialized stainless steels, plastics, superalloys, titanium, magnesium, reinforced plastic, particle boards, asbestos fiberboards, and other less common materials.

Core materials most commonly occurring in sandwich construction are the honeycomb types of cores used in most aircraft construction and building construction. The most frequently used single honeycomb material is a family of paper honeycombs used in building construction, furniture construction, etc. Closely following these are the aluminum honeycombs used in high-performance aircraft and automotive tooling. Third most commonly used honeycomb is the family of reinforced plastic honeycomb materials used in radome construction for both aircraft and ground-based radar antenna systems and in many aircraft primary structures. The least commonly used is the stainless-steel honeycomb used in supersonic aircraft having skin temperatures too high to permit use of standard aluminum alloys or high-temperature reinforced plastic materials. Although relatively few aircraft have been built using stainless-steel honeycomb, the high cost of the material and the extreme complexity of fabricating stainless-steel honeycomb into brazed sandwich panels has resulted in a nationwide technological effort of large magnitude over the past few years.

Following the honeycomb core materials, the most commonly used cores are the urethane foams, CCA foams, balsa and plywood cores. In nearly all cases, the selection of both the facing and core materials is dictated principally by cost, ease of fabrication and suitability for end use.

Advantages of sandwich construction

The largest single reason for the use of sandwich construction and its rapid growth to one of the standard structural approaches during the past 10 years is its high strength or stiffness-to-weight ratio. As an example consider a 2-ft span beam having a width of 1 ft and supporting a load of 3600 lb at the mid span. This beam, if constructed of solid steel, would have a deflection of 0.058 in. and weigh 68.6 lb. A honeycomb-sandwich beam using aluminum skins and aluminum cores and carrying the same total load at the same total deflection would weigh less than 8 lb. As an interesting further comparison, a magnesium plate to the same specifications would weigh 26 lb and an aluminum plate 34.2 lb. Although such clear and simple cases of comparative strength, weight and stiffness are not normally found in actual designs, it has generally been found that equivalent structures of sandwich construction will weigh from 5% to 80% less than other minimum weight structures and frequently possess other significant advantages.

Other advantages of sandwich construction include: extremely high resistance to vibration and sonic fatigue, relatively low noise transmission, either high or low heat transmission depending upon the selection of core materials, electrical transparency (varying from almost completely transparent in the case of radome structures to completely opaque in the case of metal sandwich structures), relatively low-cost tooling when producing complex aircraft parts, ability to mass-produce complicated shapes, ability to absorb damage and absorb energy while retaining significant structural strength and flexibility of design available.

SAPPHIRE

Sapphire has a combination of properties that makes it unique among ceramic materials. It was first successfully synthesized by A. V. L. Vernueil in France, in 1904, by a flame-fusion process. Since 1940, the original process has been significantly developed to provide various large synthetic sapphire shapes that have proved to be economical in a variety of technical applications.

General nature

Synthetic sapphire, chemically and physically, is the same as the natural mineral, which is now uneconomical to mine because of the low cost of the synthetic ($.05 per gram). It is a single crystal of very high-purity aluminum oxide and crystallizes in the hexagonal-rhombohedral crystal system. It has a hardness which is second only to diamond among naturally occurring minerals. It has unusual chemical stability, high thermal conductivity, excellent high-temperature strength, broad band optical transmission and low electrical loss. Table 1 lists typical properties.

Chemical properties

Sapphire is inert to most chemical agents. For example, it is not attacked by any acids, including hydrofluoric acid. It is resistant to alkalies and is not attacked by uranium hexafluoride up to 600 C. Sapphire will dissolve in water at the rate of 1 mg per day at 30,000 psi and 700 C. It is also attacked by anhydrous sodium borate-sodium carbonate at 1000 C, by silicon at 1500 C, by sodium hydroxide at 700 C and by boiling phosphoric acid.

Radiation effects

When irradiated by X-rays, it develops two absorption bands at 2300 Å and 4000 Å, which reach maximum optical density with 3×10^3 roentgens exposure. At 2300 Å, the transmission is decreased by 14% and a slight amber color is produced. There is no significant effect on infrared transmission. Irradiation by fast neutrons in a reactor produces two absorption bands at 2040 Å and 2600 Å. The transmission is decreased by 60 and 20%, respectively, at the two wave lengths. No coloring is visible due to the irradiation. A decrease in physical density of about 0.13% occurs during irradiation. No significant effect is produced on transmission in the infrared.

Forms available and uses

Synthetic sapphire is available in the form of rods, e.g., 0.010 in. dia × 6 in. long, 0.125 in. dia

TABLE 1—PROPERTIES OF SAPPHIRE

PHYSICAL PROPERTIES

Melting Point, C	2040
Boiling Point, C	3500
Density, gm/cu cm	3.98
Hardness	
Vickers	2000
Mohs Scale	9
Heat of Fusion, Kcal/mol	26
Thermal Conductivity, cal/sec/sq cm/ °C/cm	
At 2.5 K	0.08
At 50 K	15.0
At 100 K	1.0
At 200 C	0.024
At 600 C	0.016
At 1200 C	0.021
Thermal Expansion, per °C	
90° to optic axis:	
20–500 C	8.34
20–1000 C	9.08
20–1700 C	9.95
Parallel to optic axis:	
20–500 C	7.10
20–1000 C	7.35
20–1700 C	8.45 [a]
56° to optic axis:	
20–1700 C	9.35

MECHANICAL PROPERTIES

Young's Modulus, 10^6 psi	52.6
Modulus of Rupture, 1000 psi	
At 25 C	70
At 400 C	50
At 600 C	45
At 900 C	70
At 1000 C	90
Creep Yield Stress, psi [b]	
At 900 C	11,300
At 1000 C	5670
At 1100 C	3540
At 1400 C	2120
Tensile Stress to Initiate Creep (in rods), psi [c]	
For 45 deg angle	
At 900 C	22,100
At 1200 C	5250
At 1400 C	3540
For 30 deg angle	
At 900 C	25,500
At 1200 C	6100
At 1400 C	4100
Compressive Strength, 1000 psi	300

OPTICAL PROPERTIES

(1 mm thickness, transmission uncorrected for Fresnel losses.)

Wavelength:	Transmission, %:	Refractive Index:
1450 Å	15	6.2 [d]
1500 Å	22	5.5 [d]
1800 Å	53	3.0 [d]
2500 Å	78	1.84
1 micron	86	1.76
2.8 microns	92	1.71
4 microns	90	1.67
5 microns	86	1.63
6 microns	50	—
7 microns	15	—

ELECTRICAL PROPERTIES

Volume Resistivity, ohm-cm

At 900 C	10^8–10^{11}
At 1000 C	10^6–10^9
At 1150 C	10^6
At 1300 C	10^5

Dielectric Constant and Loss Tangent:

Temperature, C	Frequency, cps	Loss Tangent	Dielectric Constant
24	10^2 to 8.5×10^9	—	9.3 [e]
600	10^2	0.021	10.6 [e]
600	10^4	0.006	10.1 [e]
600	8.5×10^9	0.00013	10.1 [e]
24	10^2 to 8.5×10^9	—	10.1 [f]
600	10^3	0.02	12.8 [f]
600	10^5	0.0018	12.8 [f]
600	8.5×10^9	0.00018	12.8 [f]

[a] Extrapolated.

[b] Resolved shear stress measured on rods with optic axis 45 deg to rod axis.

[c] For 90° and 0° no detectable plastic deformation occurs up to 1400 C; slip plane is .0001 and direction is 1120 at 1300 to 2000 C.

[d] Calculated from absorption data.

[e] Field perpendicular to optic axis.

[f] Field parallel to optic axis.

$\times$ 20 in., and 0.250 in. dia $\times$ 12 inch. Boule is available $\frac{3}{8}$ in. dia $\times$ 8 in., $\frac{5}{8}$ in. dia $\times$ 4 in., 1 in. dia $\times$ 3 in. Discs are available 2 in. dia $\times$ 1 in. thick, 6 in. dia $\times$ $\frac{1}{2}$ in. Optical lens shapes are available to 5 in. in diameter. Sapphire is fabricated with diamond grinding techniques into a variety of surface roughnesses, can be optically polished, and can be chemically or flame polished.

Over 60 tons of sapphire are utilized internationally in a year. Most applications are for watch and instrument bearings, gem stones, phonograph needles, ball point pen balls, infrared and ultraviolet optics, as well as electronic insulators and masers. Others include: orifices, arc lamp tubes, radiation pyrometry, crucibles, film and textile guides, solar cell covers, optical flats, liquid nitrogen heat sinks, klystron windows and optical filter substrates.

By deliberately adding impurities to the starting material, other modifications of sapphire crystals can be obtained. The most important today is ruby, which is alumina with 0.05% chromium oxide added and is the basic solid state crystal in microwave and optical maser systems.

The submicron size high-purity alumina powder, which is used as a starting material for producing crystals, is also used as a polishing agent in dentifrice formulation, metallurgical polishing, as supports for high-purity chemical catalyst carriers, and as the raw material in specialty ceramic manufacture.

SBR RUBBER

SBR synthetic rubbers are versatile, general-purpose elastomers, developed as a substitute for natural rubber during World War II. Known as Buna-S (BU-tadiene, NA-trium or sodium catalyzed, S-tyrene) in Germany, they were designated GR-S (G-overnment R-ubber S-tyrene) by the United States. The materials have been designated SBR (S-tyrene B-utadiene R-ubber) by the ASTM (American Society for Testing and Materials). They constitute about 80% of domestic synthetic-rubber consumption.

SBR types

SBR polymers are generally available as dry rubbers (bale or crumb form) and as latices. So-called "hot" polymers (polymerized at 122 F) have been steadily replaced in recent years by the 20 to 30% higher tensile, longer-wearing "cold" rubbers (polymerized at 41 F). Cold rubbers generally accept higher pigment loadings than the hot types. The remaining 12% usage of hot polymers is maintained by certain demands for the inherently easier processing characteristics of these materials.

Styrene content varies from as low as 9% in so-called "Arctic" or low temperature-resistant rubbers up to 44% in a rubber designed for its excellent flow characteristics. Materials above an arbitrary 50% level are termed plastics, butadiene-styrene, or high-styrene resins. These resins are used as stiffening agents for SBR rubbers in applications such as shoe soles, and latices in this composition range are used in the paint industry and paper coating.

Latices vary in styrene content from 0 to 46%, the higher levels being used where greater strength is needed. Solids vary from 28 to 60%. The higher solids content, large particle size, cold types provide the largest single outlet for SBR latices as they are rapidly taking over the major share of the natural-rubber foam market. Other applications for SBR latices include adhesives, coatings, impregnating or spreading compounds for rug backing, textiles and paper saturation.

In addition to modifications previously noted as possible through variation of polymerization temperature and styrene content, specific desired characteristics may be built into the dry SBR's through the polymerization formula. For example, rosin soaps are used for added tack and fatty acid soap types are somewhat faster curing. Cross-linking the polymer chain with 1% of divinyl benzene yields a rubber which promotes ease of processing, smooth extrusions and absence of shrinkage, i.e., maximum dimensional stability when it is used in the range of only 10 to 20% with other synthetics or natural rubber.

Alum or glue-acid coagulation and special finishing techniques in contrast to the usual salt-acid coagulated types, yield low moisture absorption polymers with good electrical properties.

Staining type antioxidants are added to polymers to be blended with natural rubber and where color is not important, e.g., dark compounds or those incorporating carbon black. Nonstaining, nondiscoloring antioxidants are employed for light-colored SBR's used: (1) in light or brightly colored stocks, (2) in contact with white stocks where "bleeding" would be detrimental, (3) where the rubber is used as the elastomer component of high-impact polystyrene or (4) where the rubber is used in contact with foods.

Although regular SBR is basically a low-cost rubber, further economies were realized with the introduction of oil-extended SBR polymers in

1950. Very high Mooney viscosity polymers (e.g., 120-150) incapable of being processed on standard rubber equipment, are blended with 25 to 50 or more parts of nonstaining naphthenic or staining aromatic oils in an emulsified state at the polymer plant. The coagulated dried material (Mooney viscosity now about 50) is used widely in tires, camelback, mechanical goods, etc., not only due to its cost advantage but because of actually superior wearing qualities.

The oil-extended types now constitute 50% of all SBR produced in the United States.

The introduction of new mechanical mixing techniques in 1958 to supplant the latex blending methods for incorporating carbon black in SBR brought about renewed interest in SBR-carbon black masterbatches. The favor of black masterbatches for tires, retread stock, mechanical goods, etc., is expected to continue to increase as smaller particle-size blacks are developed and employed. These finer blacks will in turn permit the use of higher oil extension, resulting in high abrasion-resistant stocks, possibly softer stocks and further economies.

Physical properties

With natural and SBR compounds of the same hardness, it is found that the synthetic formulations not only extrude faster, smoother and with a greater degree of dimensional stability, but also with less danger of sticking if the article is to be coiled during curing.

In molded items SBR compounds are more nearly free of taste and odor than those made with natural rubber.

The inherently superior physical properties of SBR over natural rubber (or those superior characteristics obtainable with special polymers or through compounding of SBR) include the following:

1. Enhanced resistance to attack by organisms (e.g., soil bacteria)
2. Lower compression set
3. Less subject to scorch in curing operations
4. Superior flex resistance
5. Better resistance to animal and vegetable oils
6. Higher abrasion resistance
7. Superior water resistance
8. Lower freeze temperatures

9. Better dampening effect
10. Increased filler tolerance
11. Less high-temperature discoloration
12. Superior aging, including:
    a) Heat resistance (steam or dry heat)
    b) Sunlight checking
    c) No reversion or tackiness
    d) Resistance to oxidation
    e) Ozone cracking
13. Enhanced electrical insulation properties

Natural rubber excels SBR in building tack, tensile strength, tear resistance, elongation, hysteresis (heating on flexing) and resilience properties.

Although SBR has a higher thermal conductivity than natural rubber it is not sufficient to overcome its higher hysteresis value. While natural rubber's low conductivity is an engineering advantage in gaskets or seals used in low-temperature equipment, it becomes a distinct disadvantage in an application such as electrical insulation where overheating at the surface of the conductor is experienced during an overload condition.

Another drawback of natural rubber's low thermal conductivity is encountered when thick slabs or sections are molded, as the time required to bring the stock up to temperature and reduce it again is excessive. This is particularly significant when the compound used is a hard rubber, containing large amounts of sulfur which liberate internal heat as well. Although the internal heating factor is not important with soft compounds containing small amounts of sulfur, the polymer content of hard rubber may actually be destroyed due to excessive internal temperatures.

Compounding

After selection of the proper polymer from the wide diversity of types available, the compounder is afforded a further latitude in building in desired properties through selection of vulcanizing agents, fillers, deodorants, age resistors, blends of other rubbers and resins, etc.

One of the outstanding differences between natural rubber (*Hevea*) and SBR is the need for reinforcement in the latter. Whereas *Hevea's* tensile strength might be increased 25% with the addition of carbon black from its uncompounded value of about 3500 to 4000 psi, SBR's tensile is increased to above 3000 psi from an initial value of less than 1000 psi.

Although not as effective as the carbon blacks, other inorganic fillers including clays, whiting, coated calcium carbonate and precipitated silicas, barytes, zinc oxide, magnesia, lithopone, and mica are used. Titanium dioxide pigment is used where covering power is desired in white or very brightly colored stocks.

Various degrees of oil resistance may be imparted to SBR compounds (or put conversely, lower-cost, oil-resistant rubbers can be produced) through blending SBR with the higher-costing neoprene or butadiene-acrylonitrile copolymer rubbers.

Organic fillers and extenders such as cumarone-indene resins as well as hydrocarbon plasticizers are often used by a compounder to soften or further extend SBR elastomers. Azelate and sebacate esters are utilized as plasticizers for stocks subjected to very low temperatures, e.g., -10 to -35 F.

The durometer hardness of SBR stocks can be varied from very soft, solid materials or foams or sponges, to very hard or ebonite type materials. The rubber sponge stocks are made from low Mooney rubbers by chemically "blowing" them. Very soft solid materials may be compounded by using a high degree of plasticization or through the use of low Mooney rubbers. SBR hard rubber is made by utilizing enough sulfur in the compound to saturate all the double bonds and through the incorporation of high carbon-black loadings.

SCANDIUM

Scandium, element number 21, was discovered by L. F. Nilson in 1879, while he was making separations of the rare-earth erbium group. It was recognized as the Ekaboron predicted by Mendeleff. Although as plentiful in the earth's crust as tungsten and more than cadmium, it is considered a rare element because it is so widely distributed that commercial quantities are difficult to obtain.

Chemically, scandium is somewhat similar to aluminum but is more nearly like yttrium and the rare-earth metals. It is quite reactive, forming very stable compounds. The halides, however, can be reduced by calcium, sodium, and potassium, and it is in this manner that scandium metal is produced. For example, scandium fluoride is generally reduced with calcium metal

in a tantalum or tungsten crucible in an inert atmosphere. The scandium is purified by distillation at 1650 to 1700 C in a vacuum of 10^{-5} mm Hg. This produces a material with a metallic luster that has a slight yellow tinge but the general appearance of aluminum.

Physical properties

Atomic Vol, cu cm/gm-atom 15.03
Atomic Weight 44.96
Density, gm/cu cm 3.01
Crystal structure,
    below 1335 C—hexagonal close packed
    above 1335 C—body-centered cubic
Lattice constants, Å ....   a = 3.308,
                                        c = 5.267,
                                     c/a = 1.592
The lattice constants are not known for
    b.c.c. structure.
Melting Pt, C 1540
Boiling Pt, C 2725

Vapor pressure, $\log P_{mm} = \dfrac{1.718 \times 10^4}{T^\circ K} + 8.298$

Heat of Sublimation 298 K, kcal per mole   80.79
Compressive strength is 76,700 psi.

Electrical and magnetic properties

Elec Resis, ohm-cm
  -194 C .... 21.1×10^{-6}
  -76 C .... 39.3×10^{-6}
    0 C .... 50.5×10^{-6}
   22 C .... 54.3×10^{-6}
Therm Coef of Elec Résis,
  per °C .... 2.97×10^{-3}
Magnetic Susceptibility, 25 C, emu 315×10^{-6}

Chemical properties

Scandium reacts very rapidly with dilute acids but reacts slowly in a mixture of concentrated nitric and hydrofluoric acids owing to the formation of an insoluble layer of scandium fluoride on the surface of the metal.

Scandium metal is quite stable in air at room temperature but oxidizes fairly rapidly at higher temperatures. Scandium oxidizes at a rate of 0.0187, 0.304, and 0.421 mg per sq cm per hr at 400, 600 and 800 C, respectively. The oxides at 400 and 600 C are very dense, adherent films, while the coatings at 800 C or above are loose and nonprotective.

Mechanical properties

Hardness, Brinell (500-kg load)—50
Sonic Mod of Elast, 10^6 psi—11.5

Because of the scarcity and difficulty in purification, the tensile properties have been determined for only a small amount of material of at least 99.0% purity:

Temp, F	Yld Str, 1000 psi	Ten Str, 1000 psi	Elong, %	Red in Area, %
70	25.2	37	5.0	8.0
800	25.1	32.2	4.0	7.6
1600	—	1.78	0.5	1.7

Therm Coef of Expansion, per °C

25 to 300 C	12.80×10^{-6}
25 to 950 C	16.04×10^{-6}
300 to 600 C	15.27×10^{-6}
600 to 950 C	17.86×10^{-6}

The only mechanical properties determined on alloys have been on scandium-titanium alloys:

Titanium, %	Temp, F	Yld Str, 1000 psi	Ten Str, 1000 psi	Elong, %	Red in Area, %
3.94	70	22.3	37.1	9.7	15.4
3.94	800	21.3	37.8	8.2	11.2
11.35	70	39.7	46.2	1.0	12.8
11.35	800	26.5	47.95	8.5	11.2

Fabrication

Scandium metal can be arc-melted in an inert atmosphere such as argon at pressures of 50 mm of Hg without loss by evaporation. It is extremely active, however, and will "getter" the furnace atmosphere, producing a dark film on the surface of the metal.

Pure scandium can be cold-rolled or swaged into thin pieces without annealing if the oxygen content is kept below 200 ppm. Excessive oxygen occurs as dispersed oxides in the grain boundaries and makes fabrication difficult by causing intergranular cracking.

Uses

Owing to its scarcity and high cost, no commercial uses have been found for scandium.

SCREW-MACHINE PARTS

Among the many screw-machine parts are such things as bushings, bearings, shafts, instrument parts, aircraft fittings, watch and clock parts, pins, bolts, studs and nuts. Screw-machine parts can be produced at rates up to 4000 parts per hour, and tolerances of ±0.001 in. are common. In addition, these parts can be made of practically any material that is machinable and

that can be obtained in rod form. In comparison, die casting is restricted to a limited number of alloys of relatively low melting point, and cold heading requires materials that are ductile at room temperature.

However, screw-machine parts do have certain limitations:

1. Since the material must be in the form of bar or rod, the cross section of the part is generally limited to a circle, hexagon, square or other readily available extruded cross section, although a special extruded cross section can sometimes be ordered. The stock used must be as large as the greatest cross section of the part; depending on the shape, this may or may not result in large scrap losses. In general, irregular and nonsymmetrical parts are not good screw-machine parts, although some parts of this type are produced.

2. Parts can become quite expensive unless certain restrictions are observed in selection of material and in design of parts. These restrictions and other design suggestions are given in this article.

Materials

Selection of the material is very important. Cost of producing a part is influenced greatly by ease of machining; poor machinability reduces tool life and causes frequent shutdown to replace tools. If service requirements permit, readily machinable materials, such as resulfurized steels or leaded brasses, should be given preference.

Some of the many metals used in screw-machine parts are: carbon and low alloy steels; stainless and heat-resisting steels; cast and malleable irons; copper alloys, particularly brass, bronze and nickel silver; aluminum and magnesium alloys; nickel and cobalt superalloys; and gold and silver. Among the nonmetallic materials used are vulcanized fiber, hard rubber, nylon, "Teflon," methyl methacrylate, polyethylene, polystyrene and phenolics.

Lot size should be considered in selecting the material, as it influences not only the cost of the part but the material that can be used. Generally, for a lot size of less than 500 lb, materials are limited to those stocked in warehouses. For a lot size of a ton or more, orders can be placed with a mill and a wider range of materials is available. However, delivery may be considerably slower than from warehouse stocks, and delays of three or more months are not uncommon.

Price per pound is a function of lot size. The per-pound price of a material ordered in 75-lb lots can be twice that of the same material in 500-lb lots.

Design

Screw-machine parts should be designed to be made from standard compositions, standard stock sizes and standard shapes. Specifying nonstandard bar not only increases costs but may result in delays in obtaining the parts.

Tolerances. Commercial tolerances are given below:

Diameters
    Fractional ± 0.005 in.
    Decimal ± 0.003 in.
Lengths
    Fractional ± ⅟₆₄ in.
    Decimal ± 0.010 in.
Angles and holes
    Angles 2½ deg
    Drilled holes −0.001 to +0.010 in.
Fillets
    0.020-in. max radius
Corners
    Dia <½ in.: 0.020-in. radius or 0.005 in. x 45-deg chamfer
    Dia ½ in. or > : ⅟₆₄ in. x 45-deg chamfer

Closer dimensions can be held but should be specified only when necessary, since they increase the cost of producing and inspecting the part.

Holes. Standard drill sizes should be specified if possible. When holes require reaming, sizes that can be finished with standard reamers should be specified.

Threads. A major item in the design of a screw-machine product is the selection of a proper standard thread, the aim being to provide interchangeability. In new designs, use the Unified Screw Thread System, adopted by the United States, Canada and Great Britain. For existing designs, however, continue to use the American National System. Indicate threads by size, pitch, series and class, only; using a special pitch and specifying major and minor diameters is often unnecessary and always expensive.

On screw-machine products having tapped holes, adopt a thread depth that is no greater than the application demands. Tests have shown that accurately made threads that are 65% of full thread depth, and equal or greater than the screw diameter, have ample strength to meet most commercial requirements. Excessive percentage of thread in tapped holes can cause increased tap breakage, machine downtime and inspection cost.

Concentricity. When concentricity is required, specify it in terms of total indicator reading (TIR) rather than as a dimension. It is good engineering practice to indicate the diameters that must run true by an arrow leading to a note reading "diameters marked A must be concentric to within —— TIR measured at points ——." Specifying concentricity when it is not required increases cost of the part unnecessarily.

Burrs. Do not specify removal of burrs unless this operation is essential. Since there is some doubt about the definition of a burr, the screw-machine industry has established certain criteria:

1. Sharp corners are not considered burrs unless they have ragged edges and interfere with operation of the part.

2. Slight tears or roughness on the first two threads of a tapped hole or the male thread are not burrs unless they interfere with assembly.

3. A projection is not a burr if it must be found with a magnifying glass.

Burrs can be eliminated in most cases without additional cost by specifying chamfered or rounded corners on the parts.

Finish. In general, cost of finishing will depend on the degree of finish required; the finer the finish, the higher the cost. A finish for appearance only need not be as fine as a finish required on mating or bearing surfaces. In any case, finish should be specified in terms of microinch rms units. Microinch finish depends on material and operation and without secondary operations can range from the usual 125 rms to as low as 16 rms. The latter finish requires special care and is a very high cost operation on a production basis.

Dimensions. If a part requires heat treatment or plating, dimensions supplied to the part producer will depend on where these operations are to be done. If the supplier is to finish the part, give dimensions to apply after heat treatment or plating; if the buyer is to finish the part, give dimensions to apply before heat treating or plating.

SELENIUM

Selenium, discovered in 1817 by Berzelius, is the third member of Group VIB of the periodic arrangement of the elements, being more metallic than sulfur but less metallic than tellurium, the two adjacent elements of this group.

It is rarely found in its native state but is usually associated with lead, copper, and nickel from which it is recovered as a by-product. There are at present seven producers of selenium in the United States and Canada with an output of between 1 and 1.4 million lb annually. At the present time and for the foreseeable future the supply is adequate to take care of any reasonable demand based on stocks in hands of producers and production potential.

Physical properties

Selenium exists in several allotropic forms, and Table 1 gives some of the more common physical constants of the various allotropes where known.

Chemical properties

The chemical properties of selenium are very similar to other Group VIB elements such as sulfur and tellurium. It reacts readily with oxygen and halogens to form their respective compounds, i.e., oxides and halides. Selenous and selenic acids are easily prepared and in many respects resemble sulfurous and sulfuric acids. Selenium is not very soluble in aqueous alkali solutions, but selenites and selenates are readily prepared by fusion with oxidizing salts of the alkalis.

Forms available

Commercial grades (99.5%) of selenium are usually sold as a powder of various mesh sizes packed in steel drums. High-purity selenium (99.99%), used primarily in the electronics industry, is available as small pellets (approximately $\frac{1}{8}$ in. dia), packed in polyethylene bags. Ferroselenium containing 50 to 58% selenium and nickel-selenium containing 50 to 60% selenium are available in the form of metallic chunks packed in wooden barrels or boxes.

Applications

The most important uses for selenium are in the electronics industry, in such components as current rectifiers, photoelectric cells, xerography plates, and as a component in intermetallic compounds for thermoelectric applications.

A selenium dry-plate rectifier consists of a base plate of aluminum or iron, either nickel plated or coated with a thin layer of bismuth, a layer of 0.002 to 0.003 in. of halogenated selenium, an artificial barrier layer, and a counter-electrode of cadmium or a low-melting alloy. The purpose of the controlled amount of halogen in the selenium is to accelerate the transformation from amorphous to the hexagonal form which is done at temperatures slightly above 210 C during the manufacturing process.

Xerography is a dry printing method for the production of images by light or other rays. A thin layer of selenium is given a strong positive electrical charge and the design projected on it which discharges the selenium in proportion to the light intensity, producing a latent image. This image is developed by dusting with negatively charged particles which adhere to the plate where it was not struck by the projected light. Some further simple steps allow the image to be used as a master for many additional prints.

One of the oldest uses of selenium is in the glass and ceramic industries. As a decolorizer selenium is added to glass to counteract the greenish tint caused by the presence of iron. The familiar ruby red which is commonly seen in glass is produced by cadmium-selenide. These same cadmium-selenide reds find broad application in the field of ceramics and enamels.

Selenium (usually in the range of 0.15 to 0.25%) imparts excellent machinability to some types of austenitic stainless steels and at the same time acts as a degasifier. In the nonferrous industry selenium is used to improve the machinability of copper and copper alloys.

A minor use of selenium is as a vulcanizing agent in the rubber industry to increase resistance to heat, as well as abrasion resistance and resilience.

TABLE 1—PROPERTIES OF SELENIUM

Form	Amorphous		Crystalline	
	Powder	Vitreous	Monoclinic	Hexagonal
Color	Red	Black to red	Deep red	Gray to dark gray
Sp Gr	4.25	4.28	4.46	4.79
Melt Pt, C	Softens at 40–50, changes state	Softens at 40–50	170–180	217
Boiling Point, C	684.9 ± 1.0	684.9 ± 1.0	684.9 ± 1.0	684.9 ± 1.0
Latent Ht of Fusion, cal/gm	—	—	—	16.5
Latent Ht of Vaporization, cal/gm ...	79.6	79.6	79.6	79.6
Therm Cond	Depends on crystal structure and impurities.			
Elec Cond	Depends on crystal structure and impurities.			
Lin Coef of Therm Exp, μ-in./°C	—	—	—	37

SEMICONDUCTING MATERIALS

Semiconductors may be defined as materials which conduct electricity better than insulators, but not as well as metals. An enormous range of conductivities can meet this requirement. At room temperature, the conductivities characteristic of metals are on the order of 10^4 to 10^6 ohm^{-1} cm^{-1}, while those of insulators range from 10^{-25} to 10^{-9} ohm^{-1} cm^{-1}. The materials classed as semiconductors have conductivities which range from 10^{-9} to 10^4 ohm^{-1} cm^{-1}. The conductivity of metals normally decreases with an increase of temperature, but semiconductors have the distinctive feature that in some range of temperature their conductivity increases rather than decreases with an increase of temperature.

Another criterion for a semiconductor is that the conduction process be primarily by electrons and not ionic, since the latter involves the transfer of appreciable mass as well as charge.

Performance parameters

The unique properties of semiconductors depend upon the number of carriers of electrical current. These carriers can be of two kinds: electrons and holes. The number of carriers can be changed by (1) temperature, (2) injection by photons, nuclear particles, or from electric fields, or (3) crystal imperfections in the form of foreign atoms, deviations from stoichiometry, or lattice defects. Semiconductors containing large numbers of foreign atoms or impurities correspondingly have large numbers of free carriers which cannot be increased appreciably by temperature or injection. Such materials are sometimes called "semimetals" because their electrical conductivity is similar to, but less than, that of metals.

In addition to the number of carriers that are free to conduct, two other parameters in a semiconductor are important. The first is the velocity per unit electric field or mobility of the carriers in a semiconductor. Electrons (or holes) of different but unique effective masses are sometimes present in semiconductors. Light mass carriers have high mobilities and heavy mass carriers have low mobilities. Carrier mobility values ranging from 1 to 10^6 cm^2/volt sec are commonly observed.

The mobility of the free carriers is reduced by impurities or other imperfections. Carrier mobilities usually increase with decreasing temperature. However, at low temperatures this trend is reversed and the carrier mobilities decrease with decreasing temperature. The temperature of reversal point or the maximum mobility increases with increasing impurity content.

The operation of many semiconductor devices depends upon the influence of injected carriers on the electrical conductivity. Therefore, the time these carriers remain free, i.e., the mean carrier lifetime, is another important parameter in a semiconductor.

Energy bands

The quantitative features of semiconduction can be explained in terms of "energy bands," which constitute allowed electronic energy levels in crystalline solids. The atoms forming each crystal are held together in a rigid lattice by "chemical bonds." These bonds consist of electron pairs which are "shared" by adjacent atoms. Electrons in bonds are in a lower energy state than would be the case if the bond were broken, i.e., it requires work to pull apart the electrons in a bond. Actually the electrons cannot be individually identified with a particular atom, but the interaction of the electrons from an assembly of a large number of atoms can be described by quantum mechanics. Such a treatment shows that there are allowed electronic energies which fall into "energy bands" of closely spaced levels with "forbidden gaps" between these bands.

When the allowed bands are partially filled or overlap, as is the case in most metals, a large number of electrons are free to conduct. This is shown diagrammatically in Fig 1. In a semiconductor the bands are separated, but conduction is still possible by either of two mechanisms. The first is to supply sufficient energy to electrons in a lower or valence band to raise them to an energy corresponding to a higher or conduction band. This energy corresponds to that required to break one of the covalent bonds between two of the atoms. This energy can be supplied by

phonons, photons, or nuclear particles.

The second way in which conduction is possible in a semiconductor is by the introduction of impurities which form steps or levels in the forbidden region. A relatively small amount of energy is required to raise an electron from a level just below the conduction band to an energy level within the conduction band where it is free to conduct. Fig 1 shows the positions of a number of the impurity levels in germanium and the energy required to free an electron from the impurity to the conduction band. The distinction between a semiconductor and many insulators is merely that the forbidden region or band gap in the insulator is larger than that in the semiconductor, and that the impurity levels are correspondingly farther from the allowed energy bands. Thus the distinction is one of quantity and not mechanism of conduction.

Types

While some of the chemical elements are semiconductors, most semiconductors are chemical compounds. Table 1 lists the elemental semiconductors and some of their characteristic properties. Compound semiconductors can be simple compounds, such as compounds formed from elements of the III-V, V-VI, II-IV, II-VI, I-VI, II-V, II-VI groups of the periodic table, as well as alloys of these compounds, ternary compounds, complicated oxides, and even organic complexes.

Characteristics of III-V compounds are shown in Table 2 and examples of other compound semiconductors are given in Table 3.

Organic semiconductors and some of their properties are shown in Table 4. Most of the materials that have been investigated in detail are monoclinic and belong to the C^5_{2h} space group (twofold screw axis normal to a glide plane).

For organic semiconductors it is generally true that the larger the molecule, the higher the conductivity. Inclusion of atoms other than carbon, hydrogen, and oxygen in the molecule often appears to increase the conductivity. The phthalocyanine molecule, which is much like the active center of several biologically active compounds, e.g., chlorophyll, can accommodate a metallic atom in its center.

An organic semiconductor may be either

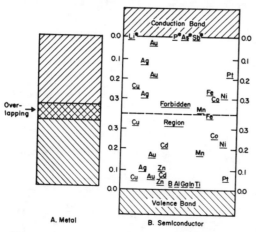

FIG. 1. Energy-level diagrams. Germanium is used as the example for a semiconductor and the positions of a number of impurity levels indicated. For levels above the dashed line (gap center), energy is measured from the conduction band; for levels below, from the valence band. Asterisks indicate donor levels; all others are acceptors.

TABLE 1—PROPERTIES OF ELEMENTAL SEMICONDUCTORS

Element	Crystal Structure	Density, gms/cm³	Melting Point, C	Linear Coeffi-cient of Expansion, 10^{-6}/°C	Energy Band Gap at 300 K, ev	Electron Mobility, cm²/volt sec	Hole Mobility Light Mass cm²/volt sec	Heavy Mass cm²/volt sec
B	—	2.34	2075	—	1.4	1	—	2
C (dia-mond)	Cub. (f.c.), O^7_h	3.51	3800	1.18	5.3	1800	—	1600
Si	Cub. (f.c.), O^7_h	2.33	1417	4.2	1.09	1500	1500	480
Ge	Cub. (f.c.), O^7_h	5.32	937	6.1	0.66	3900	14,000	1860
α–Sn	Cub. (f.c.), O^7_h	5.75	231.9	—	0.08	144,000 [a]	—	1600 [a]
As	Hex. (rhomb.), D^5_{3d}	5.73	814	3.86	1.2	—	—	—
Sb	Hex. (rhomb.), D^5_{3d}	6.68	630.5	10.88	0.11	—	—	—
α–S	Rhomb. (f.c.), V^{24}_h	2.07	112.8	64.1	2.6	—	—	—
Se	Hex., D^4_3	4.79	217	36.8	1.8	—	—	1
Se	Amorphous	4.82	—	—	2.3	0.005	—	0.15
Te	Hex., D^4_3	6.25	452	16.8	0.38	1100	10,000	700

[a] Values for carrier mobilities are experimental values obtained at 300 K, except in the case of gray tin where the 77 K values are given.

TABLE 2—III-V COMPOUNDS AND THEIR PROPERTIES

Com-pound	Crystal Structure	Density, gms/cm³	Melting Point, C	Linear Coeffi-cient of Expan-sion, 10^{-6}/°C	Energy Band Gap at 300 K, ev	Electron Mobility [a] Light Mass cm²/volt sec	Heavy Mass cm²/volt sec	Hole Mobility [a] Light Mass cm²/volt sec	Heavy Mass cm²/volt sec
BN	Hex. (graphite), D^4_{6h}	2.2	3000	—	—	—	—	—	—
BN	Cub. (ZnS), T^2_d	—	—	—	4.6	—	—	—	—
BP	Cub. (ZnS), T^2_d	—	>3000	—	—	—	—	—	500–1000
BAs	Cub. (ZnS), T^2_d	—	—	—	—	—	—	—	—
AlN	Hex., C^4_{6v}	3.26	>2700	—	—	—	—	—	—
AlP	Cub. (ZnS), T^2_d	—	>2100	—	2.42	—	—	—	—
AlAs	Cub. (ZnS), T^2_d	—	1600	—	2.16	—	—	—	—
AlSb	Cub. (ZnS), T^2_d	4.28	1065	—	1.6	—	180–230	—	420–500
GaN	Hex., C^4_{6v}	—	1500	—	3.25	—	—	—	150–250
GaP	Cub. (ZnS), T^2_d	4.13	1450	5.3	2.25	—	120–300	—	70–150
GaAs	Cub. (ZnS), T^2_d	5.32	1238	5.7	1.43	8600–11000	1000	3000	426–500
GaSb	Cub. (ZnS), T^2_d	5.62	706	6.9	0.70	5000–40000	1000	7000	700–1200
InN	Hex., C^4_{6v}	—	1200	—	—	—	—	—	—
InP	Cub. (ZnS), T^2_d	4.79	1062	4.5	1.27	4800–6800	—	—	150–200
InAs	Cub. (ZnS), T^2_d	5.67	942	5.3	0.33	33000–40000	—	8000	450–500
InSb	Cub. (ZnS), T^2_d	5.78	530	5.5	0.17	78000	—	12000	750

[a] Since InSb is the only III-V compound which has been prepared with an impurity concentration low enough that the characteristic lattice carrier mobility can be determined, the best experimental value is given followed by the theoretical estimate for higher-purity material. All mobility values are for 300 K.

TABLE 3—COMPOUND SEMICONDUCTORS

Formula	Typical Compounds
I-V	KSb, K_3Sb, $CsSb$, Cs_3Sb, Cs_3Bi
I-VI	Ag_2S, Ag_2Se, Cu_2S, Cu_2Te
II-IV	Mg_2Si, Mg_2Ge, Mg_2Sn, Ca_2Si, Ca_2Sn, Ca_2Pb, $MnSi_2$, $CrSi_2$
II-V	$ZnSb$, $CdSb$, Mg_3Sb_2, Zn_3As_2, Cd_3P_2, Cd_3As_2
II-VI	CdS, $CdSe$, $CdTe$, ZnS, $ZnSe$, $ZnTe$, HgS, $HgSe$, $HgTe$, $MoTe_2$, $RuTe_2$, $MnTe_2$, BeS, MgS, CaS
III-VI	Al_2S_3, Ga_2S_3, Ga_2Se_3, Ga_2Te_3, In_2S_3, In_2Se_3, In_2Te_3, GaS, $GaSe$, $GaTe$, InS, $InSe$, $InTe$
IV-IV	SiC
IV-VI	PbS, $PbSe$, $PbTe$, TiS_2, $GeTe$
V-VI	Sb_2S_3, Sb_2Se_3, Sb_2Te_3, As_2Se_3, As_2Te_3, Bi_2S_3, Bi_2Se_3, Bi_2Te_3, Ce_2S_3, Gd_2Se_3
$A^IB^{II}C_2^{VI}$	$CuFeS_2$
$A^IB^{III}C_2^{VI}$	C_4AlS_2, $CuInS_2$, $CuInSe_2$, $CuInTe_2$, $AgInSe_2$, $AgInTe_2$, $CuGaTe_2$
$A^IB^VC_2^{VI}$	$AgSbSe_2$, $AgSbTe_2$, $AgBiS_2$, $AgBiSe_2$, $AgBiTe_2$, $AuSbTe_2$, $Au(Sb,Bi)Te_2$
$A_3^IB^VC_3^{VI}$	Cu_3SbS_3, Cu_3AsS_3
$A_3^IB^VC_4^{VI}$	Cu_3AsSe_4
$A^{II}B^IC_2^V$	$ZnSnAs_2$
Oxides	SrO, BaO, MnO, NiO, Fe_2O_3, $BaFe_{12}O_{19}$, Al_2O_3, In_2O_3, TiO_2, BeO, MgO, CaO, CdO, ZnO, SiO_2, GeO_2, $ZnSiO_3$, $MgWO_3$, CuO

n-type or p-type. For example, among the organic dyes, the cationic dyes appear to be n-type semiconductors and the anionic dyes p-type. The quinolinium salt of tetracyanoquinodimethane has a room-temperature resistivity of 0.01 ohm cm, a remarkably low value for a pure organic compound.

Magnetic ferrites are usually polycrystalline ceramics, composed of mixed semiconducting oxides, and possess useful magnetic properties combined with high electrical resistivities. The magnetically soft ferrites have the general form, MFe_2O_4, where M represents one or more of the following divalent metals, Cu, Mg, Mn, Ni, Fe and Zn. The crystallites have the cubic structure of the spinel system.

Fabrication

Although some electronic devices require extremely small semiconductor elements, others require large masses of material. For example, some diodes and thermistors have volumes, exclusive of the lead wires, of less than one ten-millionth of a cubic inch, while some silicon domes require as much as a thousand cubic inches of material, and xerographic and luminescent panels cover many square feet. Semiconductors are prepared in the form of single crystals (as large as several inches in diameter and several feet in length), polycrystalline ingots, thin films, or sintered ceramic shapes. Semiconductors are generally brittle materials and some make excellent abrasives. However, at higher temperatures (near their melting points) they can be deformed plastically in a manner analogous to engineering metals at room temperature.

Single crystals and polycrystalline ingots can be cut into the desired shapes by using diamond or other appropriate abrasive wheels or by using ultrasonic cutting tools and a boron carbide abrasive. A continuous wire and an abrasive slurry are sometimes used for cutting semiconductors. Lapping and etching are commonly used for preparing small semiconductor elements of the desired shape and surface condition. Large-area selenium xerographic plates are prepared by vacuum evaporation, while ceramic molding and sintering techniques are used in the preparation of ferrites and some luminescent materials.

Single crystals are grown by the horizontal Bridgman, the vertical Bridgman, Czochralski, floating-zone, and vapor-phase deposition techniques. The impurities may be added to the starting materials or to the melt during growth. Controlled regions of impurities are created by diffusion, alloying, or vapor-phase decomposition to form epitaxial layers on a single-crystal substrate.

Some semiconductor elements are formed by use of a semiconductor powder and an appropriate binder. In the manufacture of magnetic ferrites the oxides are mixed in required proportions and milled to obtain a fine particle size. The powder is then partially sintered in a pre-firing kiln, and subsequently crushed, milled, and granulated before being pressed or extruded into the required shape. The main firing cycle takes place at temperatures around 1250 C in a controlled atmosphere. As a result the parts are sintered into a dense, homogeneous ceramic. During sintering the parts shrink some 20% in linear dimensions.

TABLE 4—ORGANIC SEMICONDUCTORS

Substance	Formula	Resistivity, ohm-cm	Band Gap	
			Conductivity, ev	Photo Conduct, ev
POLYACENES Athracene		300	0.83	—
Tetracene		10	0.85	3.6
Pyrene		300	1.01	3.2
Perylene		10	0.98	—
Chrysene		100	1.10	3.2
Coronene		0.2	1.15	—
Pyranthrene		10^7	0.54	0.85
POLYACENES WITH QUINONOID ATTACHMENTS Violanthrone		1000	0.39	0.84
Pyranthrone		10^6	0.54	1.14
AZO-AROMATIC COMPOUNDS Indanthrone black		3000	0.28	—
1,9,4,10-Anthradipyrimidine		1000	1.61	—

TABLE 4—ORGANIC SEMICONDUCTORS (*Continued*)

Substance	Formula	Resistivity, ohm-cm	Band Gap	
			Conductivity, ev	Photo Conduct, ev
PHTHALOCYANINES		10^4	1.2	1.56
FREE RADICALS α,α-Diphenyl β-picryl hydrazyl		10^6	0.74	—

Applications

The technology and practical importance of semiconductor devices have been growing steadily in the past decade. The major applications of semiconductors can be divided into ten categories: diodes and transistors, luminescent devices, ferrites, special resistors, photovoltaic cells, infrared lenses and domes, thermoelectric and piezoelectric devices, xerography, and electron emission. The materials most commonly used for these applications are listed in Table 5.

TABLE 5—SEMICONDUCTOR APPLICATIONS AND
MATERIALS USED
(including some insulators)

1. TRANSISTORS, DIODES, RECTIFIERS, and RELATED DEVICES

Transistors	Ge, Si, GaAs
Diodes	
Switching Diodes	Ge, Si, GaAs
Varactor Diodes	Ge, Si, GaAs
Tunnel Diodes	Ge, Si, GaAs, GaSb
Photodiodes	Ge, Si, GaAs
Zenner Diodes	Ge, Si, GaAs
Microwave Diodes	Ge, Si, GaAs
Magnetodiodes	InSb
Power Rectifiers	Cu_2O, Se, Si, Ge
Varistors	SiC, Cu_2O

2. LUMINESCENT DEVICES

Electroluminescence	ZnS, ZnO
Phosphors	ZnS, CdS, ZnO, ZnSiO_3, MgWO_3, SrWO_4
Lasers	Al_2O_3, CaF_2, $CaWO_4$, BaF_2, $SrMoO_4$, SrF_2, LaF_2, LaF_3, As_2S_3, $CaMoO_4$, $KMgF_3$, $(Ba,Mg)_2P_2O_7$

3. FERRITES

Soft	ZnO, MnO, NiO, Fe_2O_3
Permanent	$BaFe_{12}O_{19}$

4. SPECIAL RESISTORS

Thermistors	B, U_3O_8, Si, $(Ni-Mn)O_2$
Photoconductors	Ge, Se, CdS, CdSe, GaAs, InSb, PbSe, PbTe
Particle Detectors	CdS, Diamond, Si, GaAs
Magnetoresistors	InSb, InAs
Piezoresistors	Si, PbTe, GaSb
Cryosars	Ge
Bokotrons	Ge
Helicons	Fe_2S
Oscillistors	Ge, Si, InSb
Chargistors	Ge, Si

5. PHOTOVOLTAIC AND HALL EFFECT	
Photovoltaic Cells	Se, Si, GaAs
Photoelectro-magnetic Cells	InSb, InAs
Hall Effect Devices	InSb, InAs, GaAs
6. OPTICAL MATERIALS	
Infrared Lenses and Domes	Ge, Si, Se, Al_2O_3, As_2S_3, MgO, TiO_2, $SrTiO_3$
7. THERMOELEC-TRIC DEVICES	
Generators	PbTe, Bi_2Te_3, ZnSb, GeTe, MnTe, CeS
Refrigerators	Bi_2Te_3
8. PIEZOELECTRIC DEVICES	$BaTiO_3$, $PbTiO_3$, $(PbZr)TiO_3$, Pb-NbO_2, CdS, GaAs
9. XEROGRAPHY	Se, ZnO
10. ELECTRON EMISSION	BaO, SrO

SHEET METAL PARTS

Stamping and pressing make up a large family of metal forming processes. Included in this group are blanking, pressing, stamping and drawing, all of which are used to cut and/or form metal plate, sheet and strip. The steps common to all stamping and pressing operations are the preparation of a flat blank and shearing or stretching the metal into a die to attain the desired shape.

In drawing, the flat stock is either formed in a single operation or progressive drawing steps may be needed to reach the final form. In spinning, flat disks are dished by a tool as they revolve on a lathe.

Stamping involves placing the flat stock in a die and then striking it with a movable die or punch. Besides shaping the part, the dies can perform perforating, blanking, bending and shearing operations. Almost all metals can be stamped. In general, stampings are limited to metal thicknesses of 3/8 in. or less. Pressing and drawing operations can be performed on cold metals up to 3/4 inch thick and up to about 3½ inches on hot metals.

In recent years many new press forming and drawing techniques have been developed. A number of them make use of rubber pads, bags and diaphragms as part of the die or forming elements. Some involve stretch forming over dies. Others combine forming and heat treating operations. And still other methods, known as high energy rate forming, employ explosives, electrical or magnetic energy to produce shock waves that form the material into the desired shape.

Sheet-metal shops

Some industries, such as the building industry, use sheet metal for corrugated roofing, heating and ventilating duct, spouts, hoods, gutters, etc. Such parts are often made in sheet-metal shops, many of which are quite small.

The operations done in such shops would include shearing and notching; forming in bending rolls, or in roll formers and lock seamers (the long way); corrugating and flattening, piercing, riveting, flanging or putting collars on duct, and often, circle shearing, welding and brake forming. A good deal of art is involved in getting precise fits as the tooling is rather simple, or applicable to a range of parts. As a result, many folded-up boxes, cabinets, and some chassis can be economically made in somewhat limited quantities.

Of course, such equipment may be combined with additional equipment to accomplish more precise ends. Sheet-metal operations are also involved in much larger components such as aircraft frames, steel lockers, railroad cars, trailers, large office machines, machine tools, etc.

A number of nonintegrated makers of welded tubing, boiler tanks, signs, conduit moldings, etc., are examples of specialized sheet-metal forming shops.

Spinning

Spinning is a smaller branch of metal forming. With the application of power to rolls, it has grown beyond the production of limited quantities of round shapes like lamp bases, tank ends, or diver's helmets, which call for low tool costs and perhaps a good finish, free of any draw marks. With power rolls, large items such as missile cones of refractory metals, and items where material can be saved by varying metal thicknesses in walls have increased the scope of spinning. As an example of the latter, television picture tubes are spun and then "bopped" into shape on a large press. Where several beads, or rolled threads, are called for, spinning lathes are often more practical than stamping, using an expanding, rubbery insert and split dies.

Stamping

The independent stamping industry is a two-billion dollar industry in the United States, and, at that, many components made on presses are not classified as stampings. Frequently, secondary operations, such as annealing in furnaces, trimming in lathes or rolls, brake bends, tapping, etc., make it difficult to define a part as a stamping. Such operations, as well as finishing operations, may cost more than comparatively fast and economical press operations.

Sizes and materials. Presses, and presslike machines, are not necessarily limited to sheet-metal forming. They punch paper doilies and cut uppers for shoes. Multislide machines form round or flat wire; some presses impact-extrude aluminum, zinc and steel into deep shells for toothpaste tubes or shell bodies, using slugs cut from bars, or cast for the purpose, or sometimes punched from sheets, and plates, in bellmouth dies to burnish the edges. Presses forge from billets and they compress powdered metal and carbon into compacts. In size, they run from small kick presses, for staking on a bench, to giants with a squeeze of 10,000 tons (the largest forging presses can exert 50,000 tons pressure to form the wing spar of a plane). Although most perform one operation per stroke, there are transfer presses doing as many as 24 operations to complete a door knob each stroke.

Stampings are available from bassinet brackets to coffin hinges, and in size from shoe lace eyelets to freight car ends. The first presses were drop presses built in Germany in the late 18th century. Presses blank at up to 800 strokes per min, but even large deep-draw operations at 90 per hr are much faster than other ways of forming. All types of materials are stamped from 0.002 in. thick to 5/8 in., or more. Rolled raw materials are strong, dense, uniform, and non-porous, as well as inexpensive and available.

Types of work. Presses perform such operations as blanking (some blanks are made on shears) and cutting off; piercing (punching) holes, cutouts, or extruding holes; bending to almost any angle (this would include lance forming of tabs); embossing, or forming strengthen-ing ribs or shallow pockets and hemming (bending edges up and then back flat on themselves). The edges of shells may be curled in or out. Coining changes thicknesses, for the raw material of stamping is almost always uniform.

Related to blanking are trimming operations to remove excess stock, and shaving, a slight removal of stock. Examples are the first step of broaching in order to get close tolerances, small teeth, and straight edges where the breakout on stampings is objectionable, or to improve edge appearance to about 125 μ in. on thinner materials, and about 150 μ in. on stock 3/32 in. and over. Sometimes, two or more shaves are called for. An example of stamping accuracy is a die plate for a key punch machine with rectangular holes held accurate within ±0.0001 in. These are flattened, punched and shaved twice most carefully.

The operation of stamping itself usually refers to placing identifying marks on a surface for up to 20% of stock thickness. Some of these marks are embossed, as are locating extrusions and stops—little buttons forced out of the stock thickness with a depression, or pit, on the back side of the material. Press flattening is not uncommon, although its success is dependent on the material's not having too much springback. Of course, two or more operations can be combined in one die.

Drawing

The operations related to drawing call for softer material than in stamping. There must be an adequate spread between yield and tensile strength of the material. The material may work-harden, but annealing restores ductility.

First operations are often called cupping, and use either a double-acting press, which has a hold-down, or a press with a cushion below, supporting a pressure pad, so that the metal can be held back from rushing in and forming wrinkles. There are straight draw-throughs, and also necking operations when a shell with a flange returns up through the die. Necking is also used to gather metal for a small diameter boss in a larger part, and the smaller its diameter in relation to the whole part, the more operations are needed. Necking can also be defined as drawing small diameters below larger ones, the type of operation which precedes a slant walled striking operation. Striking is simply drawing in a die with a bottom; the purpose may be to flatten slanted sides or bottoms, to sharpen radii or to emboss a shape in the bottom, or to reduce radii under a flange and flatten the flange.

Walls of drawn parts tend to thicken at the top. To develop straighter walls and to hold tighter tolerances, sizing operations are used. If the wall is to be reduced below the thickness of the bottom of the cup, ironing operations are used. To make the top thicker, to a limited degree, it is upset.

Swaging reduces the diameter of an open-ended part or tube about 15% per stroke. Stretching around a radius or around an outside corner is called flanging. Shaped beads, or double outward flanges, can be placed in a part like a bellows by bulging and then striking flat.

Design characteristics

It may be well to review a few characteristics of stampings. The rolled metal surfaces, unless there are many operations, are bright, but edges tend to be sheared (frequently with burrs which can be expensive to remove if the part cannot be tumbled). Pinch trim edges are most economical on stock 0.025 to 0.078 in. thick, but they are sharp. Multiple thickness variations and undercuts most often must be machined.

Flatness can be a problem—about 0.005 in. per linear in. is the commercial tolerance. Hole diameters are held on the punch side only; below that is breakout. Holding specific angles after stamping is difficult because of springback, particularly with harder materials such as stainless steel. If practicable, specify hole locations on legs from the ends of parts, or from the material on the hole side (so that material tolerances do not affect locations). Do not put holes too close to edges or long slots next to bends.

The bottoms of deep-drawn parts are frequently a little smaller than the open ends to help strip the part off the punch. Cups are measured twice at right angles, to arrive at the same dimension, since they tend to be oval. If one set of tools is to be used for various materials, decide early so that overlapping radius ranges are used on the draw dies. Consider tolerances carefully because too-tight tolerances are expensive. Substantial savings are being made by specifying prefinished stock (galvanized, for instance).

Quantities

Metals are formed in presses using punches and dies, and these are costly. In the last few years, savings have been developed in blanking and piercing operations by the use of steel rule dies and thin short-run dies. Striking punches and dies made on profilers, elaborate cam piercing tools for sides of cups, and close tolerance shaving tools, progressive and eyelet machine tools, are particularly expensive. Using special dies on runs of a few thousand pieces is on the borderline of being economical. It may pay to form boxes and weld corners, rather than draw and trim them. Round tools made in lathes cost much less than other shapes. Hydroform tooling, explosive tooling, and rubber dies for shallow parts all promise to reduce the cost of draw tooling of the more expensive types. Brake bending, and frequently piercing, is often done with standard tools and simple inserts at little or no extra charge.

Many blanks have been sheared, sawed, routed, or flame-cut, although cheaper blank tools are changing economies. Quantities play an important part in selecting a particular form of sheet-metal fabrication. To deep-draw a few hundred rowboats would involve astronomical tool charges. Therefore, while plastic boats may be vacuum-formed with no mating punch needed, a metal boat would be made in several pieces and riveted together.

SHELL MOLD CASTINGS

Shell mold casting is a process that uses a relatively thin-wall mold made by bonding silica or zircon sand with a thermosetting phenolic or urea resin. It has gained widespread use because it offers many advantages over conventional sand castings. Shell mold casting is a practical and economical way to meet the demand for weight reduction, thinner sections, and closer tolerances. Table 1 lists the general limitations of the shell mold process and compares it with green sand casting.

Advantages

There are five basic advantages:

1. Lower costs. High production rates and fewer finishing operations result in a lower unit cost for applicable parts.

2. Closer tolerances. Shell castings have closer tolerances than sand castings (as Table 1 shows). Draft allowances (see Table 2) are also reduced.

3. Smoother surface finish. Shell molding pro-

TABLE 1—COMPARISON OF SHELL MOLDING AND
GREEN SAND CASTING

Process	Green Sand	Shell
Casting Weight	1 oz to several tons.	1 oz to several hundred pounds; usually under 25 lb.
Tolerances (avg, ±)	In in./ft: aluminum, ⅓₂; copper, ³⁄₃₂; gray iron, ³⁄₆₄; malleable iron, ⅓₂; steel, ¹⁄₁₆; magnesium, ⅓₂.	0.003 to 0.005 in./in.
Across Parting Line	Included above.	Add 0.005 to 0.015 in./in. to above.
Surface Finish μin. rms	250–1000.	50–250.
Min Section Thickness, in.	Aluminum, ³⁄₁₆; copper, ³⁄₃₂; gray iron, ⅛; malleable iron, ⅛; steel ¼–½; magnesium, ⁵⁄₃₂.	Same as green sand.
Max Section Thickness, in.	No limit in floor molds.	—

vides an improved surface compared with sand casting (250 to 1000 μ-in. rms for sand, 125 to 250 for shell).

4. Less machining. The precision of shell molding reduces, and in many cases eliminates, machining or grinding operations.

5. Uniformity. Insulating properties of the molds produce casting surfaces free of chill and with a more uniform grain structure.

Although a shell mold is more expensive than a green sand mold, the possibility of reducing weight, minimizing machining and eliminating cores often results in savings sufficient to offset the mold cost.

TABLE 2—DRAFT ALLOWANCES (Deg)

Process	Green Sand	Shell
Normal	2	1
Minimum	1	¼
In Pockets	3–10	1–2

The process

The shell molding process can be broken down into five operations:

1. A match plate pattern is made from tool steel with dimensions calculated to allow for subsequent metal shrinkage.

2. The resin-sand mixture is applied to the metal pattern which is then heated to 425 to 450 F. The hot pattern melts the resin which flows between the grains of sand and binds them together. Thickness of the mold increases with time. After the desired thickness is reached, excess unbonded sand is poured from the pattern.

3. The pattern, with the soft shell adhering, is placed in an oven and heated to 1050 to 1200 F for 30 sec to 1 min. This cures the shell and produces a hard, smooth mold that reproduces the pattern surface exactly. The shell is stripped from the pattern by ejector pins. The other half of the mold is produced in the same way.

4. Sprues and risers are opened and cores inserted to complete the cope and drag halves of the mold.

5. The two shell halves are glued together under pressure to form a tightly sealed mold which can be stacked and stored indefinitely.

Shell molded cores

Shell molded cores are an offshoot of the shell molding process. These cores have several advantages over the sand cores which they replace: the shell core usually costs less than the sand core; strength and rigidity permit handling without damage or distortion; sharp details, including threads, are accurately reproduced; core weight is reduced; cured cores are unaffected by moisture and can be stored for long periods of time; most shell cores are hollow and can function as vents during casting.

One manufacturer of pipe fittings found that use of a shell core reduced the weight of one core from 350 lb in green sand to 100 lb. In addition, the greater accuracy of the shell cores permitted a 5 to 10% reduction in casting weight. Tool life is usually increased because of the reduction in the amount of burned-in sand embedded in the casting.

SILICA

Silica, as the dioxide of silicon is commonly termed, is the principal constituent of the solid crust of the earth. Consequently, it is a major ingredient of most of the nonmetallic, inorganic materials used in industry. Consideration of this vast field of silicate materials would require more space than is here available. This article will be limited to the relatively pure forms of silica which are applicable to design engineering. The use of silica as a raw material for glass and other ceramics will also be omitted, although quartz sand employed in this way is one of the major industrial commodities and is unique in its high natural purity, abundant supply and low cost.

Silica occurs in a number of allotropic forms which have different properties and different uses. Omitting two high density forms, coesite and keatite, which were only recently discovered and for which industrial uses do not yet appear to have been found, the principal modifications of silica are: quartz, silica glass, cristobalite and tridymite. The last two are sometimes combined under the term of inverted silica. The valuable physical properties of these four principal modifications lead to a wide variety of applications for silica in industry and technology.

Quartz

The common low-temperature form of silica, quartz, is strong and insoluble in water. Consequently, sandstone, which is composed of grains of quartz held together by a siliceous cement, is an excellent building stone. Huge quantities of silica stone are used as aggregate for concrete. One curious variety has a structure which renders the rock flexible and is known as itacolumite or flexible sandstone. Quartzite, in which the quartz grains of an original sandstone have recrystallized and grown to a compact mass, and vein quartz are too hard and difficult to shape for use as building stones.

Microfibrous varieties of silica are frequently colored and are worked into beautiful decorative articles often being considered as semiprecious stones. Chalcedony; green crysoprase, prase, plasma and heliotrope; red carnelian and sard; banded agate, onyx and sardonyx and opaque jasper are examples of these gem stone varieties of silica. Hydrated silica containing up to 12% of water occurs as hyalite. Sometimes this material shows iridescent colors and is known as gem opal or fire opal.

Rock crystal, as the euhedral quartz found in nature is termed, is often carved into ornaments of great beauty because of its perfect transparency and because of the high luster given it by polishing. The "crystal ball" of story and romance is a polished quartz sphere. Colored varieties of quartz are much valued as semiprecious stones. The purple amethyst, blue sapphire quartz, yellow citrine or false topaz, red or pink rose quartz, smoky quartz and the dark brown morion are examples.

Sandstone is also useful as an abrasive. It is used for grindstones and pulpstones. "Berea" grit from northern Ohio and novaculite from Arkansas are examples of coarse and fine-grained natural stones which are still preferred for some purposes to the artificial abrasives made from silicon carbide or aluminum oxide.

The strength and hardness of various natural forms of silica leads to use for crushing and for grinding. Flint pebbles are used in ball mills for grinding all sorts of materials. Agate mortars are invaluable in the chemical laboratory for pulverizing minerals before analysis. Sandstone and quartzite are used for millstones and buhrstones to crush and grind grain, paint pigments, fertilizers and many other products. Quartz sand, driven by compressed air, constitutes the useful sand blast for cleaning metal, decorating glass and refacing stone buildings. Coated on paper, quartz grains provide the carpenter and cabinet maker with the familiar "sandpaper"; although much modern sandpaper is really coated with crushed glass containing only about 70% silica.

Fine-grained quartz sand is used in scouring soaps and compositions for polishing metal. Diatomite, which consists of the silica skeletons or shells of prehistoric water organisms with diameters of about ten microns, has been applied to scouring and polishing since early times. It is often called "tripoli" since beds near Oran in Tripoli, North Africa, were an early source of the material.

The abrasive properties of quartz grains are also employed by locomotives to gain traction on steel rails, in billiard cue chalk and in the chicken gizzard, the latter being a kind of natural ball mill.

The thermal properties of the various forms of silica also lead to important uses. Its high melting point (3110 F) makes it a good refractory while its low cost and low thermal expansion have brought about a wide use in industrial furnaces, particularly for melting steel and glass.

Properties

In general, the thermal expansion of all of the common forms of silica is low at high temperatures. This makes silica refractories capable of withstanding sudden temperature changes and large thermal gradients very well in these high-temperature ranges. At lower temperatures, large volume changes, due to rapid inversions from the high-temperature crystalline forms of quartz, tridymite and cristobalite to corresponding low-temperature forms, make these materials rather sensitive to sudden temperature changes. The low-temperature forms of quartz, tridymite and cristobalite also have higher thermal expansions than the high-temperature forms and the amorphous form of silica. This form of silica, variously known as silica glass, vitrosil and fused quartz is thus the only form of silica which may be heated rapidly from room temperature without fear of breakage. A further limitation on the thermal behavior of silica refractories is occasioned by the large volume increase which occurs as quartz changes slowly to tridymite or cristobalite at temperatures above 870 C. This may lead to swelling and warping of silica bricks if

they are not first converted to the forms stable at high temperatures by prolonged firing. Since the volume changes which occur when tridymite changes from its high temperature form to the forms stable at lower temperatures are less than the corresponding change for cristobalite an effort is made to convert the silica refractory as completely to tridymite as is feasible before using it.

Table 1, which shows the densities and volumetric coefficients of thermal dilation of the common forms of silica at various temperatures, throws light on this behavior of silica refractories. The data for the table, as well as much of the other information in this article, are taken from Sosman's authoritative work, "The Properties of Silica," a volume of the American Chemical Society Monograph Series published by the Chemical Catalogue Company.

The inversion temperature of cristobalite, at which a volume change of about 2.8% occurs, depends upon the previous history of the material, being higher the higher the temperature at which the cristobalite was formed from quartz. This inversion temperature ranges from about

TABLE 1—DENSITIES AND VOLUMETRIC THERMAL DILATION

Temperature, C	Quartz		Tridymite		Cristobalite		Vitreous	
	d, g/ml	$10^6 a/°C$	d, g/ml	$10^6 a/°C$	d, g/ml	$10^6 a/°C$	d, g/ml	$10^6 a/°C$
−200	2.664	12					2.203	−2.6
−100	2.659	25.2					2.203	−0.9
0	2.651	33.6	2.262	73	2.320	17	2.203	1.1
100	2.641	40.0	2.249	49	2.313	56	2.203	1.7
117 [a]			2.247					
			2.242					
150	2.636	43.3	2.235	89	2.305	81		
163 [a]			2.233					1.7
			2.228					
200	2.630	46.6	2.221	72	2.292	180	2.202	1.7
300	2.616	54.9	2.209	45	2.222	58	2.202	1.8
400	2.601	67.4	2.201	31	2.212	35	2.202	1.8
500	2.581	100	2.195	21	2.208	22	2.201	1.7
573 [b]	2.554							
	2.533							
600	2.533	0	2.192	12	2.203	13	2.201	1.6
800	2.535	−4	2.189	0			2.200	1.5
1000	2.541	−12	2.190	−7			2.200	1.9

[a] Transition temperatures for tridymite.
[b] Transition temperature for low-high inversion of quartz.

200 to 280 C. Recent evidence indicates that tridymite is not a pure form of silica but depends for its existence on the incorporation of traces of alkali in the crystals.

Uses

Silica bricks are made from crushed quartzite rock, known as ganister, which is bonded with 1.5 to 3.0% of lime. They are molded smaller than the dimensions desired in the finished brick to allow for an expansion of about $3/8$ in. per ft as the quartz inverts to tridymite during firing. A firing schedule of about 20 days at 1450 C (about cone 16) is required. Study of the phase diagram of the silica-lime system shows why considerable quantities of lime may be used to bond the quartzite in silica bricks without loss of refractoriness. The lime is taken up in an immiscible lime-rich glass phase of which only a small amount is formed because of its high lime content.

In use, silica bricks are characterized by retention of rigidity and load-bearing capacity to temperatures above 1600 C, without the slow yield characteristic of fireclay brick. If a cold silica brick is heated suddenly it spalls and disintegrates owing to the sudden volume changes taking place at the high-low inversions of tridymite and cristobalite. If heated cautiously through this sensitive temperature region silica brick is very resistant to temperature shock.

The refractory properties of quartz sand are also employed in molds for cast iron. Huge quantities of sand, with varying amounts of clay impurity to bond the grains, are used for this purpose in the foundries of the world. A recent innovation is the use of 96% silica glass for precision-casting molds. These "Glascast" molds are a development of the Corning Research and Development Laboratories. Tubes of "Vycor Brand" 96% silica glass are used as mold inserts to produce castings with fine holes.

The fine-grained diatomite silica is used in large quantities as a heat insulator in ovens and refrigerators. Southern California, as well as Tripoli, is a source of this material, which is also bonded into lightweight silica insulating brick.

The optical and electrical properties of crystal quartz lead to important uses which employ grams rather than tons of silica. Formerly, industry depended upon natural rock crystal from Brazil or Madagascar for crystals of "optical"

grade. The principal source is near Diamantina in the State of Minas Geras. In recent years, however, a method of growing large transparent crystals from aqueous solutions of silica under high pressure in steel bombs has been developed in the laboratories of the Bell Telephone Company. This technological triumph comes just in time to avert a serious shortage of satisfactory crystal quartz for piezoelectric oscillators which have become indispensable for the control of radio transmitters, radar equipment and other electronic "timing" gear.

This use and other electrical and optical uses of crystal quartz depend upon the symmetry of the material. Sosman states that "low quartz is the most important and most familiar example of the trigonal enantiomorphous hemihedral or trigonal trapezohedral class of symmetry. This class is characterized by one axis of threefold symmetry with 3 axes of twofold symmetry perpendicular thereto and separated by angles of 120°. This class has no plane of symmetry and no center of symmetry.

"The trigonal trapezohedral symmetry class to which low-quartz belongs has the quality of enantiomorphism. This is the name given to the relation between two objects which are similar but not superposable—the relation between our right and left hands or between an object and its image in a mirror." (Sosman, pp. 183–186.)

Whether a given quartz crystal is "right-handed low-quartz" or "left-handed low-quartz" is revealed by small faces which bevel some of the corners where the six prism faces intersect the terminal pyramid faces.

Insulating crystals with the enantiomorphous, right- and left-hand symmetry characteristic of low quartz, generally display the phenomenon of piezoelectricity. If a plate cut from such a crystal is compressed, electrical charges appear upon its opposite faces and conversely, if electrical potentials are applied to opposite faces of such a plate, the plate responds by changing its thickness.

A quartz plate, furnished with metallic electrodes on its surfaces, may be used to measure pressure changes. Such a pressure gauge or piezometer has the advantage of an extremely rapid response and can be used to measure the pressures developed in a cannon barrel when it is fired or in a blast wave from a nuclear explosion.

In 1922 Cady invented a device for stabilizing

the frequency of electrical oscillator tube circuits (*Proc. Inst. Radio Eng.*, **10**, 83–114, 1922) which has become of enormous importance in modern electronic gear. A quartz plate furnished with a pair of metallic electrodes may be set into mechanical vibration by application of an alternating voltage of suitable frequency and in turn feeds back an alternating voltage due to its piezoelectrical property which exerts control on the electronic oscillator. By cutting the quartz plate at a suitable angle from the original crystal and by controlling its temperature in a carefully thermostated oven, the frequency of the resulting electromechanical oscillator may be held constant to about 1 in 100,000,000. Such extreme precision is possible only because of the excellent electrical and mechanical properties of the quartz. Its elasticity is so perfect that the mechanical vibrations take place with almost zero loss of energy and it is such a perfect insulator that there is equally small electrical loss in the vibrating crystal. By dividing frequencies, it is possible to construct electrical clocks controlled by oscillating quartz crystals which keep time with an error of less than one second per year.

Optical properties

The transparency of crystal quartz, particularly to the short waves of the ultraviolet, makes it very useful in optical instruments. Quartz prisms are employed in spectographs for analysis of light waves varying in length from almost 5 microns in the infrared to almost 0.2 microns in the ultraviolet. Lenses made of crystal quartz are also useful over this range for photography and microscopy. Crystal quartz is more transparent than fused quartz for these purposes but its birefringence introduces complications in design.

Quartz is a positive, uniaxial, birefringent crystal. That is, polarized light vibrating in the direction of the trigonal axis (prism axis) of quartz is retarded more than light vibrating at right angles to this axis. The refractive indices for the D line of sodium are $N\varepsilon = 1.553$ for light vibrating parallel to the axis and $N\omega = 1.544$ for light vibrating perpendicular to it. A thin, wedge-shaped piece of quartz cut parallel to the axis is useful as a standard for measuring the birefringence (difference in refractive indices) of other birefringent substances. Quartz is also used for constructing polarizers for ultraviolet light.

Quartz also has the property of rotating the plane of vibration of polarized light traveling along its axis. This property is associated with its left- or right-handed character. Sugar solutions have the same optical rotating power and quartz plates or wedges are employed as standards in optical devices used for analyzing sugar solutions by means of this rotary power. Many other organic chemicals which have asymmetric right- or left-handed molecules can be studied and assayed by this optical means.

The variation of this rotary power of quartz with wave length or color of light makes it possible to construct monochromators capable of selecting light of a desired color from a white source with very large optical apertures.

Amorphous silica

The amorphous form of silica also has mechanical, optical, thermal and electrical properties which make it very useful in hundreds of technical applications. Vitreous silica is made by three different processes. The earliest process, which is still used, consists simply of melting quartz by application of high temperature. The oxyhydrogen flame was the first source of heat for this process and is still used to melt fragments of rock crystal which are combined to form rods and other shapes of silica glass. Electrical heat from graphite resistors is more commonly employed at present, however, since there is less trouble from volatilization of the silica. Large masses can be made by this technique and shaped into various useful forms although the very high temperature required for working silica glass makes this method of manufacture difficult and expensive.

In order to obtain a clear transparent product, crystal must be melted. If sand is used the product is white and opaque because of the numerous air bubbles trapped in the very viscous glass. It is impossible to heat silica glass hot enough to drive out the bubbles or, in glass parlance, to "fine" it. This opaque "vitrosil" made by fusing sand, however, is very useful for chemical apparatus, when the low thermal expansion and insolubility of the silica glass play a role.

Recently, the Corning Glass Works developed a new process for making silica glass by hydrolysis of silicon tetrachloride in a flame. The resulting silica is deposited directly on a support in the form of transparent silica glass which is more homogeneous and purer than the glass made by fusing quartz.

Large pieces can be made by this process and the unusual homogeneity of the product makes it especially useful for optical parts and for sonic delay lines where striations in the older fused quartz are detrimental. The sonic delay lines are polygons of the silica glass in which acoustic waves travel on a long path, being reflected at the numerous polygonal faces.

Finally silica glass of 96% purity or better is made by an ingenious process invented by Mr. H. P. Hood and Dr. M. E. Nordberg of the Corning Glass Works. In this third process for making vitreous silica, an object is shaped first from a soft glass containing about 30% of borax and boric acid as fluxes. A suitable heat treatment causes a submicroscopic separation of two glass phases. One phase, which is composed chiefly of the fluxes, is then leached from the glass in a hot, dilute nitric acid bath. If the composition and heat treatment are exactly right, the high silica phase is left as a porous "sponge" having the shape and size of the original soft glass article. This is carefully washed and dried and fired by heating to about 1200 C.

In the firing step, the porous silica sponge shrinks to a dense transparent object of glass containing 96% SiO_2 or more which has the desirable properties of silica glass made by either of the other processes. The 4% of impurity in the glass made by this process consists chiefly of boric oxide. Because of its presence this glass is appreciably softer than pure silica glass. For corresponding viscosities the "Vycor" 96% silica glass requires about 100 C lower temperature.

Silica glass made by any of these processes has many uses because of its unique combination of good properties.

Fibers of silica glass have very high tensile strength and almost perfect elasticity. This makes them useful in constructing microbalances, electrometers and similar instruments. Fibers estimated to be as fine as 0.025 microns in diameter may be made by a bow and arrow method devised by Boy's *Phil. Mag.*, **23**, 489–499, 1887. Bordon gauges which may be constructed of thin-walled, flattened, silica glass tubing, are very sensitive and accurate for measurement of gas pressures.

Because of its small thermal expansion, fused silica is very resistant to sudden temperature changes. It is also a very hard glass so that it may be used in the laboratory for crucibles and combustion tubes to much better advantage than ordinary glass. Vitreous silica makes possible the construction of thermometers operating up to 1000 C. The small thermal expansion and durability of fused silica have led to its use in fabrication of standards of length. Fused silica plates, ground to optical flatness, are used in interferometers for measurement of thermal expansion.

The fact that fused silica is an excellent insulator with little or no tendency to condense surface films of moisture makes it valuable in the construction of electrical apparatus. It has a very high dielectric strength and low dielectric loss.

Silica glass also has very good transmission for visible and ultraviolet light. This makes it useful for the construction of mercury lamps and other optical equipment. In the mercury lamps the strength and heat resistance of silica glass make it possible to operate with high internal pressures, producing very high light intensities and great efficiency. The 96% silica glass made by the Vycor process may be fired in a reducing atmosphere or in vacuum to modify and improve its light-transmitting properties.

Silica glass is among the most chemically resistant of all glasses. This makes it particularly useful in the analytical laboratory where there is the added advantage that any contamination of contained solutions can only be by the one oxide, silica. Crucibles of fused silica glass may be used for pyrosulfate fusions. Condensers of fused silica are extremely useful for distilling acids (except hydrofluoric) and for preparation of extremely pure water.

Another useful form of silica is obtained by dehydrating silicic acid. In this way a porous "gel" is obtained which has an enormous surface area and is capable of adsorbing various gases and vapors, particularly water vapor. Silica gel is used in dehumidifiers to remove water vapor from the air. It is also used as a catalyst and as a support for other catalysts.

The porous 96% silica skeleton obtained in the Vycor process also has a very large surface and is useful as an adsorbent and drying agent. It has less capacity than silica gel but better mechanical strength.

A very finely divided form of silica known as silica soot is obtained by hydrolizing silicon tetrachloride in a flame without heating the product hot enough to consolidate it as a glass. This material is valuable as a thermal insulator and as a white filler for rubber and plastics.

SILICIDES

Silicides are a group of substances, usually compounds, comprising silicon in combination with one or more metallic elements. These hard, crystalline materials are closely related to inter-metallic compounds and have, therefore, many of the physical and chemical characteristics and some of the mechanical properties of metals.

Silicides are not natural products. They received but little attention prior to the development of the electrical furnace which provided the first practical means of attaining and controlling the high temperatures generally required in their preparation. In the twenty-year period immediately preceding World War I, Moissan, Honig-schmid and their contemporaries prepared a large number of silicides and studied their properties. The practical utilization of silicides has attracted increasing interest since World War II. A small but slowly expanding market has developed for silicide products.

Composition

Although a majority of the metals react with silicon, many of the resulting silicides do not have the properties usually required in engineering materials. The silicides which appear most promising for practical utilization in engineering and structural applications are, with few exceptions, limited to those of the refractory, or high melting, metals of Groups IV, V and VI of the periodic table. Included in this category are the silicides of titanium, zirconium, hafnium, vanadium, columbium, tantalum, chromium, molybdenum and tungsten.

Silicides can be prepared by direct synthesis from the elements, by reduction of silica or silicon halogenides and the appropriate metal oxide or halogenide with silicon, carbon, aluminum, magnesium, hydrogen, etc., and by electrolysis of molten compounds. They are also obtained as by-products in many metallurgical processes. High purity silicides of stoichiometric composition are difficult to prepare.

The chemical composition of silicides cannot, in general, be predicted from a consideration of the customary valences of the elements. The zirconium-silicon system, for example, is reported to include the compounds Zr_4Si, Zr_2Si, Zr_3Si_2, Zr_4Si_3, Zr_6Si_5, $ZrSi$ and $ZrSi_2$. The disilicide composition (MSi_2) occurs in all of the refractory metal-silicon systems and probably will prove the most important, particularly in those applications requiring high temperature stability.

General properties

Silicides resemble silicon in their chemical properties with the degree of similarity being roughly proportional to the silicon content. At normal temperatures the refractory metal-disilicides are inert to most ordinary chemical reagents. The compounds are not thermodynamically stable in the presence of oxygen and in the finely pulverized state they oxidize readily. However, in massive form they are oxidation resistant due to the formation of a protective surface layer of silica. Bodies of $MoSi_2$ and WSi_2 are highly resistant to oxidation even at temperatures approaching their melting points.

The physical and mechanical characteristics of silicides are, to a large extent, determined by the properties of the component metal. The refractory metal silicides have highly crystalline structures and moderate densities. Their melting points are intermediate to relatively high. They have low electrical resistance, high thermal conductivity and fair thermal shock resistance. They have high hardness, high compressive strength and moderate tensile strength at both room and elevated temperatures. Their elevated temperature stress-rupture and creep properties are good. Brittleness and low impact resistance are the most serious disadvantages of these materials.

Excellent oxidation resistance has prompted detailed studies of molybdenum disilicide. Quantitative information on the properties of most silicides is scarce and incomplete. The data for $MoSi_2$ (see Table 1) can, however, be used in estimating the properties of the other refractory metal disilicides provided adjustments are made on the basis of the characteristics of the metal components.

Fabrication

The general methods used for consolidating

powders can be applied to the silicides. High density parts are obtained by cold pressing and sintering and also by hot pressing. Slip casting and extrusion are convenient methods for preparing certain shapes and sizes. Casting in the molten state is difficult due to the partial decomposition of silicides at their melting points. Silicide coatings can be prepared by vapor deposition techniques. Dense, fully sintered silicide parts are extremely difficult to work. They can be cut using silicon carbide or diamond wheels. Grinding has shown some promise but it is a slow process. "Green" or presintered compacts can, however, be shaped by conventional methods.

Welding of silicides has not proved successful because of the decomposition which accompanies melting. Brazing techniques which provide strong, ductile bonds have not yet been developed.

Availability

Although the availability of silicides is gradually increasing the varieties, quantities and shapes are limited. Molybdenum disilicide is commercially available in powder form and as furnace heating elements, and certain shapes and sizes have been produced on a custom basis. Some of the other compositions can be obtained on special order. The small market which has developed for $MoSi_2$ can be expected to stimulate general interest in the silicides and to foster their commercial development and production.

The practical utilization of silicides, with few exceptions, notably $MoSi_2$ furnace heating elements, has not progressed beyond the exploratory stage. Molybdenum disilicide is a promising structural material for gas turbine and missile components which do not require high impact and thermal shock resistance. Igniter elements, thermocouple shields, gas probes and nozzles are other potential applications. The high hardness of these materials suggests their use in metal working dies and tooling.

Silicide coatings, prepared by vapor phase deposition, afford excellent oxidation protection to molybdenum and tungsten. This method of providing oxidation resistance is versatile because fabrication can be completed before the coating is applied. Similar coatings can be produced on other materials such as graphite by using silicide powders. Silicide coatings are commercially available and are a major item in the present market for silicide products.

TABLE 1—PROPERTIES OF MOLYBDENUM DISILICIDE

Chemical Composition	$MoSi_2$
Crystal Structure	Tetragonal
	$a = 3.200$ Å,
	$c = 7.861$ Å
Theoretical Density, g/cm³	6.27 ± 0.06
Melting Point, C	2030 ± 50
Electrical Resistivity, microhm-cm	
-80 C	18.9
Room temperature	21.5–28.3
1100 C	193
Temperature Coefficient, %/°C	
0–300 C	0.0281
600–1100 C	0.0411
Electrical Conductivity, % IACS	5.9–6.1
Thermal Conductivity, cal/cm²/cm/	
°C/sec	
Room temperature	0.14
150 C	0.13
540 C	0.09
Linear Thermal Expansion, per °C	
27–1500 C	9.2×10^{-6}
Heat Capacity (absolute),	
joules/g/°C	
0 C	0.419
900 C	0.535
Thermoelectric Emission	Very small
Chemical Stability	Very good
Oxidation Resistance (1700 C and	
lower)	Excellent
Modulus of Elasticity, 10⁶ psi	59
Hardness (Rockwell)	
Room temperature	A87
650 C	A80
Modulus of Rupture, 1000 psi	50–65
Compressive Strength, 1000 psi	300–350
Tensile Strength (1315 C and lower),	
psi	40,000
Stress-Rupture and Creep	

Temperature (C)	Stress (psi)	Time to Rupture (Hours)	Creep Rate (in./in./hr)
870	35,000	107	0.000024
980	20,000	224	0.000028
1040	12,000	110	0.00073
1095	10,000	85	0.0018

Impact Resistance	Poor
Thermal Shock Resistance	Fair

SILICON

Silicon, the most abundant solid element in the earth's crust (28%), was first produced in the amorphous form by Gay-Lussac and Thenard in 1811. Crystalline silicon was first prepared by Deville in 1854. The metal has been prepared by reducing the tetrachloride by hydrogen using a hot filament, or by aluminum, magnesium or zinc. The fluoride or alkali fluosilicates have been reduced with alkali metals or aluminum. Silica can be converted to metal by reduction in the electric furnace with carbon, silicon carbide, aluminum, or magnesium. The element has also been produced by the fusion electrolysis of silica in molten alkali oxide-sodium chloride-aluminum chloride baths. Extremely pure metal has been made by the treatment of silanes with hydrogen.

Silicon has an atomic number of 14, and its electron configuration is $(1s)^2$, $(2s)^2$, $(2p)^6$, $(3s)^2$, $(3p)^2$. It is located in Group IV of the periodic table, and exhibits a similarity to carbon in many of its compounds. Its atomic radius is 1.17 Å.

The atomic weight of silicon is 28.08, and it is composed of the isotopes Si^{27}, Si^{28} (92.28%), Si^{29} (4.67%), Si^{30} (3.05%), and Si^{31}.

It has the diamond structure and forms octahedral crystals with eight atoms to the unit cube, with a side of 5.42 to 5.43 Å. It is also found in the amorphous form.

The density of 99.95% silicon is 2.32 gm/cu cm.

Thermal properties

Melt Pt, C—1410
Ht of Fusion, kcal/mol—12.1
Ht of Vaporization, kcal/mol—71
Vapor Pressure, mm Hg
  1920 K—1
  2120 K—10
  2360 K—100
Boil Pt, C—2480
Entropy cal/mol/°K
  Solid
    298 K—4.5
    1500 K—15.48
  Liquid
    2000 K—23.75
  Gaseous
    298 K—40.13
    2000 K—49.66
Heat Content, cal/mol/°K
  Solid $(H_t - H_{298})$
    500 K—1.06
    1500 K—7.37
  Liquid
    2000 K—22.92
  Gaseous
    500 K—1.065
    1500 K—8.700

Free Energy, cal/mol/°K $\left[-\dfrac{(F-H_{298})}{T} \right]$
  Solid
    298 K—4.50
    1500 K—10.56
  Liquid
    2000 K—12.29
  Gaseous
    298 K—40.13
    1500 K—44.12
Ht of Combustion, Si to SiO_2, kcal—209,750
Sp Ht, °C, cal/gm—0.162
  $C_P = 5.74 + 0.617 \times 10^{-3}\ T - 1.01 \times 10^5\ T^{-2}$
  (0–900 C)

Mechanical properties

Coef of Compressibility, 20 C, 100 to 500 megabars, 10^{-6} kg/sq cm—0.32
Hardness
  Moh's Scale—7
  Bhn—240
Mod of Elast, 10^6 psi—15.49
Mod of Rup, psi—9,046
Plasticity No., $Pt = 116$—89.3
Trans Strength, lb—35
Deflection, 3-in. span, 12-lb load, in.—0.0018

Electrical and magnetic properties

Dielectric Constant—12
Energy Gap, ev—1.12
Elec Cond, ohm-cm
  −190 C—2×10^3
  350 C—15
  500 C—0.4
Magnetic Suscept, cgs— -13×10^{-6}

Optical properties

Reflectance, %—26 to 35 (highest in the blue)
Refractive Index
  μ—3.87
  μK (absorp. coef.)—0.47
Infrared
  1.2 μ—3.5
  2.6 μ—3.4

Miscellaneous properties

Cathodoluminescence (minimum voltage for excitation), V
  Amorphous Silicon—1600
  Crystalline Silicon—2400
Coef of Ther Exp, 10^{-6} in./in./°C
  Mean Value, 15 to 1000 C—4.68
  −157 C—0
  −190 C—Negative
Ther Cond, cal/cu cm/sec/°C—0.20
Vibration Frequency

Red Ray—13.57 × 10^{12}
Violet Ray—3.96 × 10^{15}
Gamma Ray—9.60 × 10^{12}

Chemical properties

The resistance of silicon to various reagents is listed below:

Agent	Resistance	Remarks
HCl	Resistant	Dilute or conc., cold or boiling
HF	Resistant	Same as above
HNO_3	Resistant	Same as above
H_2SO_4	Resistant	Same as above
Air	Resistant	—
NH_3	Resistant	Reacts at bright red heat
Br	Resistant	Burns at 500 C
CO_2	Resistant	—
Cl_2	Resistant	Burns at 340 C
$CuSO_4$	Resistant	10% solution
$FeCl_3$	Resistant	10% solution
H_2S	Resistant	—
I	Resistant	—
O_2	Resistant	Reacts at red heat
KOH	Attacked	—
NaOH	Attacked	—
S	Resistant	Reacts at elevated temperature
SO_2	Resistant	—
Water	Resistant	—

Silicon combines with many elements including boron, carbon, titanium, and zirconium in the electric furnace. It readily dissolves in molten magnesium, copper, iron and nickel to form silicides. Most oxides are reduced by silicon at high temperatures.

Fabrication

Silicon can be cast by melting in a vacuum furnace and cooling in vacuum by withdrawing the element from the heated zone. Single crystal ingots have been prepared by drawing from the melt and by the "floating zone" technique. It is claimed that silicon exhibits some workability above 1000 C but at room temperature it is very brittle. Metals can be coated with a silicon-rich layer by reducing the tetrachloride with hydrogen on the hot metal surface.

Uses

Due to its inherent brittleness, there are no engineering applications of silicon. The chief use of highly purified metal is as a semiconductor in transistors, rectifiers, and solar cells. It may be fired with ceramic materials to form heat-resisting articles. Silicon can serve as an autoxidation catalyst and as an element in photocells. Mirrors for dental use are formed with a reflecting surface of silicon. The metal is also employed to prepare silicides and alloys, and to coat various materials. In commercial quantities it is also used as a starting material for the synthesis of silicones.

SILICONE RESINS

Silicone resins are synthetic materials capable of crosslinking or polymerizing to form films, coatings, or molded shapes with outstanding resistance to high temperatures.

Composition

Silicones are made by first reducing quartz rock (SiO_2) to elemental silicon in an electric furnace, then preparing organochlorosilane monomers ($RSiCl_3$, R_2SiCl_2, R_3SiCl) from the silicon by one of several different methods. The monomers are then hydrolyzed into crosslinked polymers (resins) whose thermal stability is based on the same silicon-oxygen-silicon bonds found in quartz and glass. The properties of these resins will depend on the amount of crosslinking, and on the type of organic groups (R) included in the original monomers. Methyl, vinyl and phenyl groups are among those used in making silicone resins.

Properties

Common characteristics shared by most silicone resins are outstanding thermal stability, water repellency, general inertness and electrical insulating properties. These properties, among others, have resulted in the use of silicone resins in the following fields:

1. Laminating (reinforced plastics), molding, foaming, and potting resins.
2. Impregnating, cloth coating, and wire varnishes for Class H (high performance) electric motors and generators.
3. Protective coating resins.
4. Water repellents for textiles, leather and masonry.
5. Release agents for baking pans.

Laminating resins

Laminates made from silicone resin and glass cloth are lightweight, strong, heat-resistant materials used for both mechanical and dielectric applications. Silicone-glass laminates have low moisture absorption and low dielectric losses, and retain most of their physical and electrical properties for long periods at 500 F.

Laminates may be separated into three groups according to the method of manufacture: high pressure, low pressure, and wet lay-up. Typical properties of the three types are shown in Table 1.

TABLE 1—TYPICAL PROPERTIES OF SILICONE-GLASS LAMINATES

Type	High Pressure	Low Pressure	Wet Lay-up
Applications	Dielectric	Mechanical	Prototypes
Resin Content, %	40	30	40
Spec Grav	1.7	1.8	—
Flex Str, 1000 psi			
At 77 F	27	40	35
At 77 F after 200 hr at 500 F	24	38	—
At 500 F	6	14	5
At 500 F after 200 hr at 500 F	11	16	—
Tensile Strength (77 F), 1000 psi	30	40	25
Compr Str (77 F), 1000 psi	13	20	20
Water Absorption, %	0.10	0.05	0.2
Arc Resistance, sec	240	240	40
Electric Str, v/mil	320	110	160 [a]
Dielec Constant (10^6 cps)			
As supplied	3.6	4.0	3.6
After 24-hr immersion	3.7	4.1	3.7
Dissipation Factor (10^6 cps)			
As supplied	0.002	0.002	0.002
After 24-hour immersion	0.007	0.009	0.010

[a] After 370 hours at 480 F.

High pressure. Silicone-glass laminates (industrial thermosetting laminates) have excellent electric strength and arc resistance, and are normally used as dielectric materials.

To prepare high pressure laminates, glass cloth is first impregnated by passing it through a solvent solution of silicone resin. The resin is dried of solvent and precured by passing the fabric through a curing tower. Laminates are prepared by laying up the proper number of plies of preimpregnated glass cloth and pressing them together at about 1000 psi and 350 F for about one hour. They are then oven-cured at increasing temperatures, with the final cure at about 480 F. The resulting laminates can be drilled, sawed, punched, or ground into insulating components of almost any desired shape. Typical applications include transformer spacer bars and barrier sheets, slot sticks, panel boards and coil bobbins. Electrical grade silicone-glass laminates can be specified by MIL-P-997B, or by NEMA codes G-6 and G-7.

Low pressure. Silicone-glass laminates made by low-pressure reinforced plastics molding methods usually provide optimum flexural strength, e.g., about 40,000 psi even after heat aging. They are used for mechanical applications such as radomes, aircraft ductwork, thermal barriers, covers for high-frequency equipment, and high-temperature missile parts.

In making low-pressure laminates, glass cloth is first impregnated and laid up as described above. Since the required laminating pressure can be as low as 10 psi, matched-metal-molding and bag-molding techniques can be used in laminating, making possible greater variety in laminated shapes. Lamination should be after-cured as already described.

Wet lay-up. Silicone-glass laminates can now be produced by wet lay-up techniques because of the solventless silicone resins recently developed. Such laminates can be cured without any pressure except that needed to hold the laminate together. This technique should prove especially useful in making prototype laminates, and in short production runs where expensive dies are not justified.

Laminates are prepared by wrapping glass cloth around a form and spreading on catalyzed resin, repeating this process until the desired thickness is obtained. The laminate surface is then wrapped with a transparent film, and air bubbles are worked out. Laminates are cured at 300 F and, after the transparent film is removed, postcured at 400 F.

Molding compounds

Silicone molding compounds consist of silicone resin, inorganic filler, and catalyst which, when molded under heat and pressure, form thermosetting plastic parts. Molded parts retain exceptional physical and electrical properties at high temperatures, resist water and chemicals, and do not support combustion. Specification MIL-M-14E recognizes two distinctly different types of silicone molding compounds: type MSI-30 (glass-fiber filled), and type MSG (mineral filled). Requirements for both are as follows:

Property (min. values unless noted)	Type MSI-30 (fiber filled)	Type MSG (mineral filled)
Arc Resist, sec	175	210
Dielec Const at 1 kc (max)	5.0	5.0
Dissip Factor at 1 kc (max)	0.015	0.015
Elec Str (short time), v/mil	160	325
Compr Str, 1000 psi	10	15
Flex Str, 1000 psi	7	6
Impact Str, ft-lb/in. notch	3.2	0.25
Tensile Str, 1000 psi	2	2.5
Water Absorp, % (max)	0.50	0.50

Type MSI-30. Glass-filled molded parts have high strengths which become greater the longer the fiber length of the glass filler. Parts compression-molded from ¾ in. fiber length material, for example, may have an Izod impact strength of 15 ft-lb/in. notch (11 at 400 F) and a flexural strength of 12,000 psi (over 6,000 at 800 F). Where simple parts can be compression-molded from a continuous fiber length compound, strengths will be approximately twice as great.

Properly cured glass-fiber-filled molded parts can be exposed continuously to temperatures as high as 700 F, and intermittently as high as 1000 F. The heat-distortion temperature after postcure is 900 F. Because of the flow characteristics of these fiber-filled compounds, their use is generally limited to compression molding.

Type MSG. Mineral-filled compounds are free-flowing granular materials. They are suitable for transfer molding, and can be used in automatic preforming and molding machines. They are used to make complex parts which retain their physical and electrical properties at temperatures above 500 F, but which do not require impact strength.

Silicone molding compounds are excellent materials for making Class H electrical insulating components such as coil forms, slot wedges and connector plugs. They have many potential applications in the aircraft, missile and electronic industries.

Foaming powders

Silicone foaming powders are completely formulated, ready-to-use materials that produce heat-stable, nonflammable, low density silicone foam structures when heated. Densities vary from 10 to 18 lb per cu ft, compressive strengths from 100 to 325 psi. Electrical properties are excellent (dissipation factor at 10^5 cps is only 0.001), and water absorption after 24-hr immersion is only 2½%. The maximum continuous operating temperature of these foams is about 650 F.

Foams are prepared by heating the powders to between 300 and 350 F for about 2 hours. The powder can be foamed in place, or foamed into blocks and shaped with woodworking tools. Foams are normally after-cured to develop strength, but can often be cured in service.

Silicone foams are being used in the aircraft and missile industries to provide lightweight thermal insulation and to protect delicate electronic equipment from thermal shock. They can also be bonded to silicone-glass laminates or metals to form heat and moisture resistant sandwich structures.

Potting resins

Solventless silicone resins can be used for impregnating, encapsulating and potting of electrical and electronic units. Properly catalyzed, filled, and cured, they form tough materials with good physical and electrical properties, and will withstand continuous temperatures of 400 F and intermittent temperatures above 500 F.

Typical physical properties of cured resins include flexural strength of 7,000 psi, compressive strength of 17,000 psi, and water absorption of 0.04 per cent. Electrical properties include electric strength of about 350 volts per mil, dissipation factor of 0.003, dielectric constant of 2.9, and volume resistivities on the order of 10^{15} ohm-cm.

Before use, resins are catalyzed with dicumyl peroxide or ditertiary butyl peroxide. Resins can be simply poured in place, although vacuum impregnation is suggested where fine voids must be filled. Fillers such as glass beads or silica flour are added to extend the resin; their use increases physical strength and thermal conductivity, but decreases electrical properties. The resin is polymerized by heating it to about 300 F, and postcured, first at 400 F, then at the intended operating temperature if higher.

Electrical varnishes

Silicone varnishes (solvent solutions of silicone resins) have made possible the new high temperature classes of insulation for electrical motors and generators. The hottest spot in conventional Class B insulation should not be above 275 F; Class H insulation systems based on silicone resins operate at hottest spot temperatures of 428 F (and systems based on silicone rubber at 356 F). Electrical equipment that operates at higher temperatures makes possible motors, generators, and transformers that are much smaller and lighter, or equipment that delivers 25 to 50 percent more power from the same size and still has a much longer service life.

The resinous silicone materials used in Class H electrical insulating systems include the following:

1. Silicone bonding varnish for glass-fiber-covered magnet wire.

2. Silicone varnishes for impregnating and bonding glass cloth, mica and/or asbestos paper. Sheet insulations made of these heat-resistant materials are used as slot liners for electric motors and as phase insulation.

3. Silicone dipping varnish which impregnates, bonds, and seals all insulating components into an integrated system.

Other silicone materials used in electrical equipment include silicone rubber lead wire, silicone-adhesive-backed glass tape, and temperature-resistant silicone bearing greases. Silicone insulated motors, generators, and transformers are now being produced by the leading electrical equipment manufacturers.

SILICONE RUBBER

Silicone rubbers are a group of synthetic elastomers noted for their (1) resilience over a very wide temperature range, (2) outstanding resistance to ozone and weathering, and (3) excellent electrical properties.

Composition

The basic silicone elastomer is a dimethyl polysiloxane. It consists of long chains of alternating silicon and oxygen atoms, with two methyl ($-CH_3$) side chains attached to each silicon atom. By replacing a part of these methyl groups with other side chains, polymers with various desirable properties can be obtained. For example, where flexibility at temperatures lower than -70 F is desired, a polymer with about 10% of the methyl side chains replaced by phenyl groups ($-C_6H_5$) will provide compounds having brittle points below -150 F. Side-chain modification

can also be used to produce elastomers with lower compression set, increased resistance to fuels, oils or solvents, or to permit vulcanization at room temperature.

Curing, or vulcanization, is the process of introducing crosslinks at intervals between the long chains of the polymer. Silicone rubbers are usually crosslinked by free radical-generating curing agents, such as benzoylperoxide, which are activated by heat, or the crosslinking can be accomplished by high energy radiation beams. Room temperature vulcanized compounds are crosslinked by the condensation reaction resulting from the action of metal-organic salts, such as zinc or tin octoates. Pure polymers upon crosslinking change from viscous liquids into elastic gels with very low tensile strength. In order to attain satisfactory tensile strength, reinforcing agents are necessary. Synthetic and natural silicas and metallic oxides are commonly used for this purpose. In addition to the vulcanizing agents and reinforcing fillers described, other additives may be incorporated into silicone compounds to pigment the stock, to improve processing, or to reduce the compression set of certain types of silicone gum.

Types

Silicone rubber compounds can be conveniently grouped into several major types according to characteristic properties. Typical properties of several types are shown in Table 1. According to one such classification system, types are: (1) general purpose, (2) extremely low temperature, (3) extreme high temperature, (4) low compression set, (5) high strength, (6) fluid resistant, (7) electrical, and (8) RTV rubbers.

(1) General-purpose compounds are available in Shore A hardnesses from 30 to 90, tensile strengths of 600 to 1200 psi, and ultimate elongations of 100 to 500%. Their service temperature range extends from -65 to 500 F, and they have good resistance to heat and oils, along with good electrical properties. Many of these compounds contain semireinforcing or extending fillers to lower their cost.

(2) Extremely low temperature compounds have brittle points near -180 F and are quite flexible at -120 to -130 F. Their physical properties are usually about the same as those of the general purpose stocks, with some reduction in

TABLE 1—PROPERTIES OF SOME TYPICAL SILICONE RUBBERS

	General Purpose	Extreme Low Temp	Extreme High Temp	Low Compression Set	High Strength	Fluoro-silicone (General Purpose)	RTV [a]
Hardness (Shore A)	50 ± 5	25 ± 5	50 ± 5	60 ± 5	50 ± 5	60 ± 5	65 ± 5
Tens Str, psi	1000	1000	1200	900	2000	950	750
Elongation, %	400	600	300	130	600	225	110
Tear Str, lbs/in.	80	120	150	60	300	75	40
Compression Set (22 hr at 300 F), %	20	20	15	10	30	15	13 [b]
Max Service Temp, F							
Continuous	500	500	550	500	500	500	500
Intermittent	600	600	700	600	600	550	600
Low Temp Flex, F	−65	−130	−130	−65	−130	−65	−65
Volume Swell (ASTM No. 1 Oil; 70 hr at 300 F), %	7	10	+9	+5	+10	+1	+3

[a] Room temperature vulcanizing silicone rubber.
[b] After additional postcure 24 hours at 480 F.

oil resistance.

(3) Extreme high temperature compounds are considered serviceable for over 70 hr at 650 F and will withstand brief exposures at higher temperatures; for example, 4 to 5 hr at 700 F and 10 to 15 min at 750 F. In comparison, general-purpose compounds are limited to about 500 F for continuous service and 600 F for intermittent service.

(4) Low compression set compounds provide typical values of 10 to 20% compression set after 22 hr at 300 F. They have improved resistance to petroleum oils and various hydraulic fluids, and are particularly suitable for use in O-rings and gaskets.

(5) High-strength compounds, in Shore A hardness of 25 to 70, provide tensile strengths from 1200 to over 2000 psi and elongations from 400 to 700%, with tear strengths from 150 to 325 lb/in. Compounds of this class may operate over a service temperature range from −130 to 600 F.

(6) Excellent resistance to a wide range of fuels, lubricants and hydraulic fluids is offered by compounds based on a silicone polymer with 50% of its side methyl groups substituted by trifluoropropyl groups. Physical properties are similar to properties of other types of silicone compounds. However, service temperature range is somewhat limited. Its low-temperature properties are about the same as those of the dimethyl polymer, with a brittle point around −90 F, while the upper service temperature is around 500 F.

(7) In general, silicone-rubber compounds have excellent electrical properties which, along with their resistance to high temperatures, make them suitable for many electrical applications. With proper compounding, dielectric constant can be easily varied from about 2.7 to 5.0 or higher, while the power factor can be varied from 0.0005 up to 0.1 or higher. The volume resistivity of a typical silicone compound will be in the range 10^{14} to 10^{16} ohm-cm and its dielectric strength will be about 450-550 volts per mil thickness (measured on a slab 0.075 in. thick). Resistance to corona is excellent and water absorption is low. In most cases, excellent electrical properties are retained over a wide temperature and frequency range. Compounds can also be prepared with very low resistivity, as low as about 10 ohm-cm, for special applications.

Insulated tapes for cable-wrapping applications can be prepared from electrical grade compounds with a partial cure or from a completely cured self-adhering silicone compound.

(8) RTV (room temperature vulcanizing) silicone rubbers are available to provide most of the performance characteristics of silicone rubbers in compounds that cure at room temperature.

In addition to having excellent heat resistance, the silicone rubbers retain their properties to a much greater extent at high temperatures than do most organic rubbers. For example, a silicone compound having a tensile strength at room temperature of 2000 psi will have a tensile strength at 600 F of 300 psi or over. Most organic rubbers, although their initial properties are much higher, will be virtually useless at 500 F (except for the fluoroelastomers). In applications where a silicone rubber part operates in a low oxygen atmosphere, such as sealing on high altitude aircraft, its heat resistance will be still further improved.

In using silicone rubber at high temperatures, care must be taken to prevent reversion or depolymerization, which may occur where a part is required to operate in an enclosed environment. Here again, where this problem cannot be eliminated by the design engineer, the silicone-rubber fabricator can produce compounds with relatively high resistance to reversion.

Fabrication and uses

In general, silicone rubbers may be handled on standard rubber-processing equipment. Their fabrication differs from that of organic rubbers chiefly in that uncured silicone compounds are softer and more tacky and have much lower green strength. Also, an oven postcure in a circulating air oven is often required after vulcanization in order to obtain optimum properties. Silicone rubbers can be extruded, molded, calendered, sponged and foamed. Since compounding of the rubber stock determines to a great extent the processing characteristics of the material, the fabricator should be consulted before the material is specified, to determine whether compromises are necessary to obtain the best combination of physical properties and the most desirable shape. Silicone-rubber compounds can also be applied to fabrics by calender coating, knife spreading or solvent dispersion techniques.

For certain applications, very soft or low durometer materials are required. Suitable for such applications are low durometer solid silicone rubbers, closed or open cell expanded silicone rubbers (designated here as sponge and foam, respectively) and fibrous silicone rubber. Their properties are compared in Table 2.

Silicone-rubber sponge is available in molded sheets, extrusions and simple molded shapes. As in the case of solid silicone rubber, improved resistance to fluids or abrasion can be obtained by bonding molded or extruded sponge to a fabric or plastic cover. Silicone foam rubber can be fabricated in heavy cross sections and complex shapes, and can be foamed and vulcanized either at ambient or elevated temperature. It is suitable for use where an extremely soft, low density silicone material is required. Like sponge, it can be bonded to fabrics and plastics.

For some applications a solid, low durometer material has advantages over sponge or foam. For example, it should probably be specified for gaskets or seals where the low compression set of foam, the compression-deflection characteristics of sponge and the higher tensile and tear strength of solid silicone rubbers must be combined.

The last highly compressible material, fibrous silicone rubber, consists of hollow rubber fibers sprayed in a random manner and bonded into a low-density porous mat. Its properties include excellent compression set combined with good tear and tensile strength, and very high porosity. As manufactured at present, it is serviceable from −65 F to over 500 F, and is available in mats ¼ in. thick and 9 in. wide.

TABLE 2—COMPARISON OF SOFT SILICONE RUBBERS

	Slab, 1/16 in.	Sponge Slab, 1/2 in.	Extrusion	Foam (1 in.)	Fibrous Silicone Rubber, 1/4 in.	Low Durometer Solid Stack, 1/16 in.
Density, lb/cu in.	0.057	0.025	0.035	0.014	0.015	0.064
Compr-Defl (stress at 25% compr), psi	12	7	6	1	1	20
Compr Set, % orig. deflection						
22 hr at 70 F	25	25	25	—	—	—
22 hr at 350 F	80	80	80	20	15	30

Silicone rubbers are most widely used in the aircraft, electrical and automotive industries, although their unique properties have created many other applications. Specific examples would include seals for aircraft canopies or access doors, insulation for wire and cable, dielectric encapsulation of electronic equipment and gaskets or O-rings for use in aircraft or automobile engines. An example of another field in which they are useful is the manufacture of stoppers for pharmaceutical vials, since silicone rubbers are tasteless, odorless and nontoxic. In the future, an even greater variety of uses will certainly be found for these versatile elastomers.

SILVER AND ALLOYS

Silver is a white metallic element, having the symbol Ag. Its atomic number is 47 and its atomic weight, 107.88.

Properties

The properties of silver are:

Melting Pt, F	1761
Boiling Pt, F	3551+
Density, lb/cu in.	0.379
Elec Cond, % IACS [a]	105.2
Elec Resis, μ ohm-cm	1.59
Sp Ht, 20 C, cal/gm/°C	0.0562
Therm Cond, cal/sec/sq cm/cm/°C	1.0
Therm Expan, 10^{-5} in./in./°F	
(32–212 F)	1.97
Mod of Elast, 10^6 psi	10.3
Crystal Structure	face-centered cubic
Lattice Parameter	4.0778 A
Ten Str, ann., 1000 psi	25
Ten Str, hard (50% red), 1000 psi .	45

[a] Depending on purity and temper, the conductivity will vary from 100% to 108% (copper = 100%). Cold working lowers the conductivity of the metal.

Pure silver has the highest thermal and electrical conductivity of any metal, as well as the highest optical reflectivity. Next to gold it is the most ductile and most malleable of any metal. Silver can be hammered into sheet 0.00001 in. thick or drawn out in wire so fine 400 ft would weigh only one gram. Classified as one of the most corrosion-resistant metals, silver, under ordinary conditions, will not be affected by caustics or corrosive elements, unless hydrogen sulfide is present, causing silver sulfide to form. Silver will dissolve rapidly in nitric acid and more slowly in hot concentrated sulfuric acid. Unless oxidizing agents are present, the action of diluted or cold solutions of sulfuric acid is negligible. Organic acids generally do not attack the metal and caustic alkalies have but a slight effect on pure silver.

Although silver tarnishes quickly in the presence of sulfur and sulfur-bearing compounds, it oxidizes slowly in air and the oxide decomposes at a relatively low temperature.

Classification

Silver is classified by grades in parts per thousand based on the silver content (impurities are reported in parts per hundred). Commercial grades are Fine Silver and High Fine Silver. As ordinarily supplied, fine silver contains at least 999.0 parts silver per thousand, and may go as high as 999.3 parts per thousand. Any of the common base metals may be present, although copper is usually the major impurity. Any silver of higher purity than commercial fine contains its purity in its description, i.e., 999.7 High Fine Silver. The purest silver obtainable in quantity is 999.9 plus; the impurities are less than 0.01 part per thousand. Fine silver may also contain small percentages of oxygen or hydrogen; deoxidized silver is available for applications where these elements may be a detriment.

Fabricability

Silver can be cold-worked, extruded, rolled, swaged and drawn. It can be cold-rolled or cold-drawn drastically between anneals, and can be annealed at relatively low temperatures. To prevent oxidation when casting by conventional methods, silver should be protected by a layer of charcoal or by melting under neutral or reducing gas. Deoxidation by adding lithium or phosphorus can be obtained leaving a residual content of 0.01% max. Silver's excellent ductility makes it readily workable hot or cold.

Molten silver will absorb approximately 20 times its own volume of oxygen. Most of this oxygen is given up when the silver solidifies in cooling, but care should be taken in melting and casting because any oxygen left in the cast bars will cause cracking when they are fabricated and the castings may have blow holes.

Galling, seizing of the tool, and surface tearing are problems encountered when machining fine silver. This can be somewhat alleviated by using material cold-worked as much as possible.

Joining. Fine silver can be soldered without difficulty using tin-lead solders. Boron-silver filler metal can be used in brazing, and welding can be done by resistance methods and by atomic hydrogen or inert-gas shielded arc processes. A range of 400 to 800 F is recommended for annealing, with best strength and ductility being achieved between 700 and 800 F. Little additional softening occurs at higher temperatures, which may induce welding of adjacent surfaces. The lighter the gauge, the lower should be the annealing temperature.

Applications and alloys

Pure silver has many uses in electronic and electrical equipment where maximum conductivity is of prime importance: in fine silver contacts which are suited for light pressure, light-duty devices or for intermittent service; fluorescent lamp controls; electromagnetic counters; protective devices for motors; refrigerator thermostats, and telephone relays. Because of silver's resistance to many corrosive elements, it has applications in tubing, piping and linings for equipment used in the chemical industry.

Usually alloyed. Because pure silver is so soft, it is usually alloyed with other metals for strength and durability. The most common alloying metal is copper, which imparts hardness and strength without appreciably changing silver's desirable characteristics. Sterling silver, which is 92.5% silver and 7.5% copper, is perhaps the best known silver-copper alloy, having popular use in tableware, jewelry, etc. However, sterling silver also has applications in manufacturing processes. For example, sterling silver plus lithium, a recently developed alloy, is currently being used in the aircraft industry for brazing honeycomb sections. Other silver-copper alloys are coin silver, 90.0% silver, 10.0% copper, and the silver-copper eutectic, 72% silver, 28% copper. This latter alloy has the highest combination of strength, hardness and electrical properties of any of the silver alloys.

Silver filler metals, commonly called *silver brazing alloys*, are widely used for joining virtually all ferrous and nonferrous metals, with the exception of aluminum, magnesium and some other lower melting point metals. While

pure silver melts at 1761 F, silver alloys, having compositions of 10% to 85% silver—the alloying metals being copper, zinc, cadmium and/or other base metals—have melting points of 1145 F to 1760 F. These alloys have ductility and malleability and can be rolled into sheet or drawn into wire of very small diameter. They may be employed in all brazing processes and are generally free flowing when molten. Recommended joint clearances are 0.002 in. to 0.005 in. when used with flux. Whereas fluxes are usually required, zinc and cadmium free alloys can be brazed in a vacuum or in reducing or inert atmospheres without flux. Joints made with silver brazing alloys are strong, ductile and highly resistant to shock and vibration. With proper design there is no difficulty in obtaining joint strength equal to or greater than that of the metals joined. The strongest joints have but a few thousandths of an inch of the alloy as bonding material. Typical joints made with silver brazing alloys, giving the greatest degree of safety, are scarf, lap and butt joints.

For *electrical contacts*, silver is combined with a number of other metals, which increase hardness and reduce the tendency to sulfide tarnishing. Silver-cadmium, for instance, is extensively used for contacts, with the cadmium ranging from 10% to 15%. The advantages of these alloys are resistance to sticking or welding, more uniform wear and a decreased tendency for metal transfer.

Alloys recently developed for contact and spring purposes are the silver-magnesium-nickel series (99.5% silver), which are used where electrical contacts are to be joined by brazing without loss of hardness, in miniature electron tubes for spring clips where high thermal conductivity is essential and for instruments and relay springs requiring good electrical conductivity at high temperatures. These are unique, oxidation hardening alloys. Before being hardened, the silver-magnesium-nickel alloy can be worked by standard procedures. After hardening in an oxidizing atmosphere, the room temperature tensile properties are similar to those of hard rolled sterling silver or coin silver.

Gold and palladium are also combined with silver for contact use because they reduce welding and tarnishing and, to some extent, increase hardness.

When certain base metals do not combine with

silver by conventional methods, powder metal processes are employed. This is particularly true of silver-iron, silver-nickel, silver-graphite, silver-tungsten, etc. These alloys are used in electrical contacts because of silver's desirable conductivity and the base metals' mechanical properties. They can be pressed, sintered and rolled into sheet and wire which is ductile and suitable for forming into contacts by heading or stamping operations.

Other silver products are those produced chemically—powder, flake, oxide, nitrate and paint. Silver powder and flake are composed of large amounts of silver with 0.03% or 0.04% copper and traces of lead, iron and other volatiles.

Silver paints, which are used as conductive coatings are pigmented with metallic silver flake or powder and bonding agents that are specially selected for the type of base material to which they are applied. These coatings are used to make conductive surfaces on such materials as ceramics, glass, quartz, mica, plastics and paper, as well as on some metals. They are used for making printed circuits, resistor and capacitor terminals and in miniature electrical instruments and equipment. Silver paints fall into two classifications: (a) fire-on types for base materials which can withstand temperatures in the 750 to 1700 F range, and (b) air-dry or bake-on types for organic base materials which are dried at temperatures ranging from 70 F to 800 F. The bonding agent in the fire-on type of coating is a powdered glass frit, while in the air-dry or bake-on type of coating, organic resins are used. The viscosity and drying rate of each type varies, depending on the method of application, such as spraying, dipping, brushing, roller coating or screen stenciling.

Because they are six times lighter and five times smaller than other batteries of similar capacity, *silver-zinc batteries* have found wide use in guided missiles, telemetering equipment, guidance control-circuits and mechanisms. Where longer life and ruggedness are more important than the weight, silver-cadmium rechargeable batteries are specified. Where sea-water activation is required, silver chloride-magnesium couples are used. Another silver type of battery is the solid electrolyte type made with silver, silver iodide and vanadium pentoxide. This battery, designed for low current applications, weighs less than one ounce and has almost unlimited shelf ilfe.

Decoratively, silver powder is used for coating watch and clock dials and for the ornamentation of glass. In the first instance, silver powder is made into a paste with salt and tartatic acid, which is then brushed on the dial. The beautiful white surface obtained cannot be duplicated by any other metal. For the ornamentation of glass, silver powder is mixed with borate of lead and lavender oil, brushed on the surface of the glass to cover the design, and when the article is heated, the mixture fuses on the glass.

Silver is also used to enhance cloth and leather, and it is employed as the backing for mirrors and reflectors, automobile lamps, etc., because it will reflect up to 95%.

The *photosensitive halides* used in photography, the cyanides used in electroplating and most of the minor silver salts are prepared from silver nitrate which is a salt made by treating silver with nitric acid.

In *silver plating,* the plating bath contains double salt of cyanide with either sodium or potassium cyanide. The silver anodes are supplied in rolled plates of proper size. The silver used for plating must be free from impurities if the formation of a black film on the anode is to be prevented. Anodes and articles acting as cathodes to be plated are suspended in the bath and a direct current is passed through the bath, causing the silver to adhere to the base-metal articles.

SOLDER ALLOYS

Solder alloys are available in a wide range of sizes and shapes, enabling the user to select that one which best suits his application. Among these shapes are pig, slab, cake or ingot, bar, paste, ribbon or tape, segment or drop, powder, foil, sheet, solid wire, flux cored wire and preforms. There are eleven major groups of solder alloys:

Tin-antimony. Useful at moderately elevated operating temperatures, around 300 F, these solders have higher electrical conductivity than the tin-lead solders. They are recommended for use where lead contamination must be avoided. A 95Sn-5Sb alloy has a solidus of 452 F, a liquidus of 464 F, and a resulting pasty range of 12 F.

Tin-lead. Constituting the largest group of all solders in use today, the tin-lead solders are used for joining a large variety of metals. Most are not satisfactory for use above 300 F under sustained load. Some standard tin-lead solders are given in the table following:

ASTM Solder Class.	Composition, %		Temperature, F		Pasty Range
	Sn	Pb	Solidus	Liquidus	
5A	5	95	572	596	24
10A	10	90	514	573	59
15A	15	85	437	553	116
20A	20	80	361	535	174
25A	25	75	361	511	150
30A	30	70	361	491	130
35A	35	65	361	477	116
40A	40	60	361	455	94
45A	45	55	361	441	80
50A	50	50	361	421	60
60A	60	40	361	374	13
70A	70	30	361	378	17

Tin-antimony-lead. They may normally be used for the same applications as tin-lead alloys with the following exceptions: aluminum, zinc, or galvanized iron. In the presence of zinc, these solders form a brittle intermetallic compound of zinc and antimony. Some characteristics of tin-lead-antimony alloys are as follows:

ASTM Solder Class.	Composition, %			Temperature, F		Pasty Range
	Sn	Sb	Pb	Solidus	Liquidus	
20 C	20	1.0	79	363	517	154
25 C	25	1.3	73.7	364	504	140
30 C	30	1.6	68.4	364	482	118
35 C	35	1.8	63.2	365	470	105
40 C	40	2.0	58.0	365	448	83

Tin-silver. They have advantages and limitations similar to those of tin-antimony solders. The tin silvers, however, are easier to apply with a rosin flux. Relatively high cost confines these solders to fine instrument work. Two standard compositions: 96.5Sn-3.5Ag, the eutectic; 95Sn-5Ag, with a solidus of 430 F and liquidus of 473 F.

Tin-zinc. These are principally for soldering aluminum since they tend to minimize galvanic corrosion. The table lists five tin-zinc solders:

Composition, %		Temperature, F		Pasty Range
Sn	Zn	Solidus	Liquidus	
91	9	390	390	0
80	20	390	518	128
70	30	390	592	202
60	40	390	645	255
30	70	390	708	318

Lead-silver. Tensile, creep and shear strengths of these solders are usually satisfactory up to 350 F. Flow characteristics are rather poor and these solders are susceptible to humid atmospheric corrosion in storage. The use of a zinc chloride base flux is recommended to produce a good joint on metals uncoated with solder.

ASTM Solder Class.	Composition, %			Temperature, F		Pasty Range
	Pb	Ag	Sn	Solidus	Liquidus	
2.5S	97.5	2.5	—	579	579	0
5.5S	94.5	5.5	—	579	689	110
1.5S	97.5	1.5	1.0	588	588	0

Cadmium-silver. The primary use of cadmium-silver solder is in applications where service temperature will be higher than permissible with lower melting solder. Improper use may lead to health hazards. The solder has a composition of 95Cd-5Ag. Solidus is 640 F and liquidus is 740 F.

Cadmium-zinc. These solders are useful for soldering aluminum.

Composition, %		Temperature, F		Pasty Range
Cd	Zn	Solidus	Liquidus	
82.5	17.5	509	509	0
40	60	509	635	126
10	90	509	750	241

Zinc-aluminum. The solder 95Zn-5Al is specifically for use on aluminum and develops joints with the highest strength as well as satisfactory corrosion resistance. This eutectic alloy melts at 720 F.

Low-temperature solders. Solders containing bismuth are useful when a soldering temperature below 361 F is required. Some particular examples of the need for low-temperature soldering are: soldering heat-treated surfaces where higher soldering temperature would cause softening of the part; soldering joints where the adjacent material is very temperature sensitive. The table presents some low-temperature solders:

Alloy	Composition, %				Temperature, F	
	Pb	Bi	Sn	Other	Solidus	Liquidus
Lipowitz	26.7	50	13.3	10Cd	158	158
Wood's Metal	25	50	12.5	12.5Cd	158	165
Eutectic	40	52	—	8Cd	197	197
Eutectic	32	52.5	15.5	—	203	203
Rose's	28	50	22	—	204	229
Matrix	28.5	48	14.5	9Sb	217	440
Mold	44.5	55.5	—	—	255	255

Indium. These solders possess certain properties, such as improved wetting characteristics, which make them valuable for some special applications. Their usefulness for any particular application should be checked with the supplier. Composition and melting characteristics of a representative group of indium solders are given below:

Composition, %					Temperature, F		Pasty
Sn	In	Bi	Pb	Cd	Solidus	Liquidus	Range
8.3	19.1	44.7	22.6	5.3	117	117	0
12	21	49	18	—	136	136	0
12.8	4	48	25.6	9.6	142	149	7
50	50	—	—	—	243	260	17
48	52	—	—	—	243	243	0

SPUN PARTS

Metal spinning, essentially, involves forming flat sheet metal disks into seamless circular or cylindrical shapes. It is a useful processing technique when quantity does not warrant investment needed for draw dies.

The process

The first step in the spinning process is to produce a form to the exact shape of the inside contours of the part to be made. The form can be of wood or metal. This form is secured to the headstock of a lathe and the metal blank is, in turn, secured to the form. In manual spinning, an operator forces the blank against the form by means of a spinning tool pressed against the spinning blank. Mechanical spinning lathes usually can be set up to mechanically force the blank against the form.

In addition to manual or power spinning, hot spinning is sometimes used to either anneal a spun part, eliminating the need to remove a partially formed blank from the lathe, or else to increase the plasticity of the metal being formed. In the latter category, some metals such as titanium or magnesium must be spun hot because their normal room-temperature crystal structure lacks ductility. Heavy parts (up to 5 in. in some cases) can also be spun with increased facility at elevated temperature.

Shapes and tolerances

Basically, a component must be symmetrical about its axis in order to be adaptable to spinning. The three basic spinning shapes are: the cone, hemisphere, and straight-sided cylinder. The shapes are listed in order of increasing difficulty to be formed by spinning.

Available spinning equipment is the limiting

factor in determining the size of parts. Parts can be made ranging in diameter from 1 in. to almost 12 ft. Thickness ranges from 0.004 in. to 5 in. Most commonly, spun parts range in thickness from 0.024 to 0.187 in.

Tolerances depend strongly on the metal being spun, thickness of the part, spinning method, etc. Generally stated, for parts up to 24 in. in dia, tolerances range from $\pm\frac{1}{64}$ to $\frac{1}{32}$ in.; between 24 to 48 in. dia, $\pm\frac{1}{32}$ to $\frac{1}{16}$ in.; between 48 and 96 in. dia, $\pm\frac{1}{16}$ to $\frac{1}{8}$ in.; and over 96 in. in dia, $\pm\frac{1}{8}$ to $\frac{1}{4}$ in.

STAINLESS STEEL (CAST)

Cast alloys intended for continuous or intermittent service in corrosive environments at temperatures less than 1200 F are classed as "corrosion-resistant" alloys. These iron-chromium and iron-nickel-chromium alloys have a minimum of 8% alloy content, and are commonly referred to as "cast stainless steels."

There are corresponding grades of cast and wrought corrosion-resistant (stainless) alloys, but chemical compositions are not identical. Although superficially minor, differences in chemistry are metallurgically important in proper balancing of the compositions to provide workability on the one hand and castability on the other. In general, corrosion resistance of corresponding cast and wrought alloys is equivalent. Designations for the cast corrosion-resistant alloys are shown in Table 1 together with the related wrought type numbers. These designations—not those for wrought forms—should be used for specifying castings.

TABLE 1—STANDARD DESIGNATIONS FOR CORROSION-RESISTANT CASTINGS

Cast Alloy Designation	Wrought Alloy Type [a]
CA–15	410
CA–40	420
CB–30	431
CB–7Cu	—
CC–50	446
CD–4MCu	
CE–30	—
CF–3	304L
CF–8	304
CF–20	302
CF–3M	316L

CF–8M	316
CF–12M	316
CF–8C	347
CF–16F	303
CG–8M	317
CH–20	309
CK–20	310
CN–7M	—

^a Wrought alloy type numbers are listed only for the convenience of those who want to determine corresponding wrought and cast grades. Because the cast alloy chemical composition ranges *are not the same* as the wrought composition ranges, buyers should use cast alloy designations for proper identification of castings.

Individual alloy characteristics

Group I—Iron-chromium alloys. The iron-chromium alloys are generally highly resistant to oxidizing solutions and are applied in chemical plants processing nitric acid and nitrates. De-aerated or reducing conditions are unfavorable for these alloys. The degree to which they become passive in oxidizing media in general increases with increasing chromium content.

The alloys are divided into two subgroups: those hardenable by heat treatment and those virtually nonhardenable. All of the alloys exhibit

good resistance to erosion and are utilized for valve trim.

Type CA-15 (11.5 to 14 Cr) is hardenable by heat treatment and mildly corrosion resistant. CA-15 has many uses in the chemical and petroleum industries and in power plants. A wide range of mechanical properties can be obtained by heat treatment.

Type CA-40 (11.5 to 14 Cr) has the same general corrosion properties as CA-15. The major difference is a higher carbon content which produces greater hardness after heat treatment. Uses are similar to the applications mentioned for CA-15, except that greater hardness is required.

Type CB-30 (18 to 22 Cr, 2 max. Ni) has greater resistance to most corrosive environments than the CA types, but the material is only slightly hardenable by heat treatment. The alloy is used in nitric acid service, but in recent years, the CF type (19 Cr, 9 Ni) has replaced the CB grade in many applications.

Type CC-50 (26 to 30 Cr, 4 max. Ni) is resistant to oxidizing corrodents, and is used extensively in contact with acid mine waters, alkaline liquors and in nitrocellulose production. The alloy has a ferritic structure at all temperatures and cannot be hardened by heat treatment.

TABLE 2—PHYSICAL PROPERTIES OF CORROSION-RESISTANT GRADES

Alloy Type	Density, lb/cu in.	Sp Ht, at 70 F, Btu/lb/°F	Ther Cond at 212 F, Btu/hr/sq ft/ ft/°F	Ther Exp, 70–1000 F, in/in/°F $\times 10^6$	Mag Perm	Elec Res, microhm-cm, at 70 F
CA–15	0.275	0.11	14.5	6.4	Ferromagnetic	78
CA–40	0.275	0.11	14.5	6.4	Ferromagnetic	76
CB–30	0.272	0.11	12.8	6.5	Ferromagnetic	76
CC–50	0.272	0.12	12.6	6.4	Ferromagnetic	77
CD–4MCu	0.277	0.12	8.8	6.5	Ferromagnetic	75
CE–30	0.277	0.14	—	9.6	1.5	85
CF–3 ⎱ CF–8 ⎰	0.280	0.12	9.2	10.0	1.0 to 2.0	76
CF–20	0.280	0.12	9.2	10.4	1.01	78
CF–3M ⎫ CF–8M ⎬ CF–12M ⎭	0.280	0.12	9.4	9.7	1.5 to 2.5	82
CF–8C	0.280	0.12	9.3	10.3	1.2 to 1.8	71
CF–16F	0.280	0.12	9.4	9.9	1.0 to 2.0	72
CG–8M	0.281	0.12	9.4	9.7	1.5 to 2.5	82
CH–20	0.279	0.12	8.2	9.6	1.71	84
CK–20	0.280	0.12	8.2	9.2	1.02	90
CN–7M	0.289	0.11	12.1	9.7	1.01 to 1.10	90

Source: ACI Data Sheets.

In the CC-50 type containing more than 2% Ni, strength and ductility are improved by increasing the nitrogen content to 0.15% or more.

Group II—Iron-chromium-nickel alloys:

Type CB-7Cu (Cr 15-17, Ni 3-5, Cu 2-3, C max. 0.07)—a proprietary alloy commonly known as 17-4 PH—is a precipitation-hardening grade having good resistance to mild corrodents combined with high strength properties. It is readily machinable in the solution annealed state, and, when hardened, has an appreciable amount of ductility and toughness.

Type CD-4MCu (25 to 27 Cr, 4.75 to 6.00 Ni, 1.75 to 2.25 Mo, 2.75-3.25 Cu) is a precipitation-hardening grade with twice the strength of the CF-type alloys with equivalent corrosion resistance. It is especially resistant to stress corrosion cracking in chloride-containing solutions, and is highly resistant to sulfuric acid as well as to nitric acid and other strongly oxidizing media. CD-4M Cu is readily machined in either the water-quenched or aged conditions.

Type CE-30 (26-30 Cr, 8-11 Ni) is used for digester fittings, pumps, and valves in sulfite service in the paper industry, and other applications involving mixed sulfuric and nitric acids. Because of its high chromium content it is not susceptible to intergranular attack and it can be used, therefore, where heat treatment after welding is not feasible. Although the CE-30 alloy is often used in the as-cast condition, the ductility and resistance to corrosion may be improved somewhat by quenching the material from about 2000 F.

CF alloys (19 Cr, 9 Ni) constitute the most widely used group of corrosion-resistant alloys. When properly heat-treated, the alloys are resistant to a great variety of corrodents, and are usually considered the best "general-purpose" types.

Type CF-3 (17 to 21 Cr, 8 to 12 Ni, 0.03 max. C) is an extra-low-carbon variety used where castings must be welded without subsequent heat treatment.

Type CF-8 (18 to 21 Cr, 8-11 Ni, 0.08 max. C) resists attack by strongly oxidizing media such as boiling nitric acid. Type CF-8 castings have excellent subzero properties, retaining high impact strength at temperatures below −400 F.

Type CF-20 (18 to 21 Cr, 8-11 Ni, 0.20 max. C) is similar to CF-8, except that it is used for less severe service.

Type CF-8M (18 to 21 Cr, 9 to 12 Ni, 2 to 3 Mo, 0.08 max. C) is a modification of CF-8 in which molybdenum is added to enhance resistance to reducing corrosive media. The alloy has improved resistance to pitting in chloride-containing solutions. An extra-low-carbon type, CF-3M, is used where castings cannot be heat-treated after welding.

Type CF-12M (18 to 21 Cr, 9 to 12 Ni, 2 to 3 Mo, 0.12 max. C) is similar to CF-8M, but because of its higher carbon content, is used for less severe service.

Type CF-8C (18 to 21 Cr, 9-12 Ni, 1.0 max. Cb, 0.08 max. C) is stabilized with columbium (8 × carbon content) or with columbium-tantalum (10 x C) to avoid intergranular corrosion. Applications are similar to CF-8, but the alloy is used where parts cannot be heat-treated after welding, and for parts subjected to temperatures in the carbide precipitation range (800 to 1600 F). The aircraft industry uses castings of this composition to handle ethyl gasoline exhaust gases and other products of combustion.

Type CF-16F (18 to 21 Cr, 9-12 Ni, 0.16 max. C) alloys contain additions of selenium or sulfur and molybdenum to improve machinability. They are used in services similar to CF-8 and CF-20.

Type CG-8M (18 to 21 Cr, 9-13 Ni, 3-4 Mo, 0.08 max. C) is a modification of the CF-8M alloy, containing more molybdenum. It is especially useful in resisting liquors encountered in pulp processing.

Type CH-20 (22 to 26 Cr, 12 to 15 Ni, 0.20 max. C) has considerably better resistance to many corrosive media than Type CF-8 because of higher nickel and chromium contents. It is used for specialized applications in the chemical processing industries. CH alloys are made in two carbon grades: 0.10 and 0.20 C max with the lower carbon type used for more severe corrodents. Molybdenum is added in the low-carbon type for resistance to reducing chemicals.

Type CK-20 (23 to 27 Cr, 19-22 Ni, 0.20 max. C) is used for agitators and fittings to handle sulfite, liquor and cold dilute sulfuric acid. It is similar in application to the CE and CH compositions.

Type CN-7M (27 to 30 Ni, 19-22 Cr, 1.75 to 2.50 Mo, 3 min. Cu, 0.07 max. C) is used for resistance to various concentrations of hot sulfuric acid, dilute hydrochloric acid, and many reducing chemicals, as well as nitric acid and nitric-hydrofluoric pickling solutions.

Applications

Iron-chromium alloys containing from 11.5 to 30% chromium and iron-chromium-nickel alloys containing up to 30% chromium and 31% nickel are widely used in the cast form for industrial process equipment at temperatures from −430 F to 1200 F. The largest area of use is in the temperature range from room temperature to the boiling points of the materials handled.

Typical stainless castings are pumps, valves, fittings, mixers and similar equipment. Chemical industries employ them to resist nitric, sulfuric, phosphoric and most organic acids, as well as many neutral and alkaline salt solutions. The pulp and paper industry is a large user of high alloy castings in digesters, filters, pumps and other equipment for the manufacture of pulp. Fatty acids and other chemicals involved in soap-making processes are often handled by high alloy castings. Bleaching and dyeing operations in the textile industry require parts made from high alloys. These corrosion-resistant alloys are also widely used in making synthetic textile fibers. Pumps and valves cast of various high alloy compositions find wide application in petroleum refining. Other fields of application are food and beverage processing and handling, plastics manufacture, preparation of pharmaceuticals, atomic-energy processes, and explosives manufacture. Increasing use is being made of cast stainless alloys for handling liquid gases at cryogenic temperatures.

STAINLESS STEELS, WROUGHT

The stainless steels are a large group of iron-base alloys characterized by their corrosion resistance. This property of resisting chemical attack is attained through surface passivity which is commonly attributed to the presence of a thin oxide film on the surface of the metal. This film is considered insoluble, self-healing and nonporous. A chromium content of at least 12% is essential to produce passivity. In conventional classification, therefore, the stainless steels are defined as iron-base alloys with at least 12% chromium. Alloys with more than 30% chromium content are generally used for high-temperature application and are classified as heat-resisting alloys.

While the corrosion resistance is a common characteristic of all the stainless steels, the outstanding mechanical properties of the various classes of stainless steels have further contributed to their usefulness in engineering applications throughout industry. The hardenable stainless steels can develop high strength in thick sections. The austenitic steels have forming and welding qualities unequaled by other steels. They can also work-harden to extremely high strength. The austenitic steels retain their strength at high temperature and have excellent resistance to oxidation.

Stainless steels were discovered at the beginning of this century in France, Germany and England. Leon Guillet and Albert Portevin of France, W. Giessen of England and P. Monnartz of Germany are credited with the original discoveries. Industrial usefulness was soon found and was followed by commercial development of these steels both in Europe and the United States. Intensive research in this field resulted in development of a considerable variety of new steels for diversified purposes. The stainless steels are available commercially today as sheet, plate, strip, billet, bar, tubing, wire and castings. With the exception of castings all other products are in wrought form.

Classification

The stainless steels are classified in four groups according to composition and structure: martensitic, ferritic, austenitic and semi-austenitic or precipitation hardening. The standard American Iron and Steel Institute (AISI) specifications are more extensively used and have been adopted here when referring to specific grades. The AISI divisions for stainless steels are the 200, 300 and 400 series. Typical physical and mechanical properties are shown in Table 1.

Martensitic. The martensitic stainless steels are hardenable by heat treatment. This treatment consists in heating above 1650 F for a sufficient time to become fully austenitic. Upon cooling, the austenite transforms to martensite. A tempering treatment between 400 and 1300 F brings the mechanical properties to the desired level. The martensitic stainless steels are straight chromium steels with the chromium content between 11.5 and 18%. Certain grades have small additions of sulfur or selenium to improve machinability or vanadium, molybdenum and tung-

TABLE 1—SOME TYPICAL PHYSICAL AND MECHANICAL PROPERTIES OF WROUGHT STAINLESS STEELS

Properties	Type						
	410 [a]	420 [a]	430 [b]	304 [b]	316 [b]	AM–355 [a]	15–7 Mo [a]
Physical							
Melt Range, °F	2700/2790	2650/2750	2600/2750	2550/2650	2500/2550	2500/2550	—
Density, lb/cu in.	0.28	0.28	0.28	0.29	0.29	0.282	0.277
Ther Cond, 200 F, Btu/sq ft/hr/°F/ft	14.4	14.4	—	9.4	9.4	9.18	—
Coef of Ther Exp, 68–212 F, in./in./°F × 10^{-6}	5.5	5.5	5.6	9.2	9.2	6.4	5.0
Mechanical							
Yld Str, 0.2% offset, 1000 psi	35–180	50–220	35	30	30	181	220
Ten Str, 1000 psi	60–200	90–270	60	80	75	219	238
Elong, in 2 in., %	25–2	15–2	20	40	40	12.5	5
Hardness, Rockwell	B70–C45	B90–C55	B95 max	B90 max	B95 max	C45	C46

[a] Data for heat-treated (hardened) condition.
[b] Data for annealed condition.

sten added to enhance the high-temperature properties or the resistance to tempering. Up to 2.5% nickel is used in certain grades in order to improve the mechanical properties. Compositions of the martensitic grades are shown in Table 2. The martensitic steels can be hot-worked or forged easily and can also be cold-worked with ease since they have a low rate of work hardening. The corrosion is adequate for atmospheric exposure, water or some chemicals. The martensitic steels can be welded by all common processes, but because of their thermal

TABLE 2—MARTENSITIC STEELS

Type	Composition, %		
	Cr	C	Other
403	11.5/13	0.15 max	—
410	11.5/13.5	0.15 max	—
414	11.5/13.5	0.15 max	Ni, 1.25/2.5
416	12/14	0.15 max	S, 0.18/0.35 Mo, 0.60 max
418	12/14	0.20 max	Ni, 1.8/2.2 W, 2.5/3.5
420	12/14	0.25/0.35	—
422	11.5/13	0.20/0.25	Mo, 0.75/1.25 V, 0.17/0.27 W, 0.75/1.25
431	15/17	0.20 max	Ni, 1.25/2.5
440A	16/18	0.60/0.75	Mo, 0.75 max

hardenability require special preheating and postheating in order to prevent cracking. A final heat treatment after welding is even more desirable as it can bring the properties in the weld and heat-affected zone to the desired level.

Type 410 has just enough chromium to attain stainlessness and is for this reason a low-priced, general-purpose steel hardenable to a wide range of properties. Type 403 is used for turbine blades. Type 416 is a general-purpose, free-machining steel. Type 414 is used primarily for knife blades while 422 and 618 are used for high-temperature applications, principally in jet engines.

Ferritic. Ferritic stainless steels are chromium steels with 11.5 to 30% chromium content. They cannot be hardened by heat treatment. Like the martensitic grades, they are ferromagnetic. They can be hot- or cold-worked with ease and are readily welded. Because of the high chromium content the ferritic steels have good corrosion resistance, which improves with increasing chromium content, and excellent resistance to oxidation at high temperature, although their high-temperature strength is not as good as that of the austenitic stainless steels. The so-called sigma phase, a brittle compound, can form, especially in the higher chromium steels. A second embrittling phenomenon, "885 brittleness," occurs because of prolonged heating in the

TABLE 3—FERRITIC STEELS

Type	Composition, %		
	Cr	C	Other
405	11.5/14.5	0.08 max	Al, 0.10/0.30
430	14/18	0.12 max	—
442	18/23	0.25 max	—
446	23/27	0.20 max	N, 0.25 max

TABLE 4—AUSTENITIC STEELS

Type	Composition, %			
	Cr	Ni	C [b]	Other
301	16/18	6/8	0.15	—
302	17/19	8/10	0.15	—
303	17/19	8/10	0.15	S, 0.18/0.35
304 [a]	18/20	8/12	0.08	—
308	19/21	10/12	0.08	—
309	22/24	12/15	0.20	—
310	24/26	19/22	0.25	—
316 [a]	16/18	10/14	0.08	Mo, 2/3
321	17/19	9/12	0.08	Ti, 5 × C min
347	17/19	9/13	0.08	Cb, 10 × C min
201	16/18	3.5/5.5	0.15	Mn, 5.5/7.5 N, 0.25 max
202	17/19	4/6	0.15	Mn, 7.5/10 N, 0.25 max

[a] In both Types 304 L and 316 L, C = 0.03% max.
[b] Maximum.

750 to 1000 F range. The composition of the ferritic steels is shown in Table 3. Typical properties are shown in Table 1. Of these grades, Type 430 is primarily used for automotive trim. Type 446 is used in application requiring high resistance to corrosion or oxidation.

Austenitic. The austenitic stainless steels are chromium-nickel or chromium-nickel-manganese steels with sufficient nickel or manganese to suppress the transformation of austenite to martensite or ferrite. For this reason these steels are nonmagnetic and cannot be hardened by heat treatment. They can be hot-worked or cold-worked with ease. They work-harden considerably during cold rolling, especially the lower grades, such as Type 301, which transforms partially to martensite during cold working. Standard cold-rolled tempers of Type 301 have tensile strength up to 200,000 psi and nonstandard tempers up to 300,000 psi. The austenitic chromium-nickel steels have excellent toughness at all temperatures including subzero temperatures. This makes them useful in cryogenic application. These steels can be welded without difficulty and have high ductility in the welds. With proper control the corrosion properties of the weld are not adversely affected. Most of the austenitic grades can be "sensitized" when heated between approximately 1000 and 1500 F. Sensitizing is caused by precipitation of chromium carbides at the grain boundaries and can be prevented by using "stabilized" grades containing columbium or titanium. The high-temperature properties of the austenitic steels are superior to the martensitic or ferritic grades and the nickel improves the oxidation resistance. Table 1 shows typical physical and mechanical properties of the austenitic steels. Sulfur or selenium is added in some grades in order to improve machinability; titanium or columbium for stabilizing; molybdenum in order to improve the corrosion properties. Types 201 and 202 have been developed

in recent years primarily to save nickel. Type 301 is a general-purpose steel used for trim, household utensils and structural applications. Type 302 is used in decorative and corrosion applications. Type 303 is a free-machining grade. Types 321 and 347 are the "stabilized grades," 309 and 310 are used at high temperatures and 316 is for severe corrosion conditions.

Semiaustenitic or precipitation hardening. The semiaustenitic or precipitation-hardening stainless steels are the most recent additions and were developed primarily for aircraft and missile applications. They have an intermediate position between the austenitic and martensitic grades. In the annealed condition they are austenitic and display the good forming and welding properties of this group. The stability of the austenite can be upset by a heat treatment in the carbide precipitation temperature range (either longer time at 1400 F or short time at 1750 F) and the steel transforms to martensite while cooling to room temperature or by a subzero treatment at −100 F. A tempering or aging treatment in the 800 to 1100 F range brings the properties to the desired level. In the hardened condition these steels are, therefore, martensitic and compare in strength with the martensitic steels. Titanium or aluminum have been added to some of these grades to promote precipitation hardening. Table 1 shows typical properties of these steels. AM-350, 17-7 PH and

15-7 Mo are used almost exclusively in sheet form mostly in aircraft and missiles. AM-355, 17-4 PH and Stainless W are used primarily in bar form in aircraft and jet engines. (Stainless W and 17-4 PH are, in effect, precipitation hardening martensitic rather than semi-austenitic stainless steels.) Table 5 gives the composition of the semiaustenitic grades.

Table 5—Semiaustenitic Steels

| Type | Composition, % | | | | |
	C	Cr	Ni	Mo	Other
AM-350	0.10	16.5	4.3	2.75	N, 0.10
AM-355	0.13	15.5	4.3	2.75	N, 0.10
17-7 PH	0.09	17	7	—	Al, 1.3
15-7 Mo	0.09	15	7	2.5	Al, 1.2
17-4 PH	0.07	16.5	4	—	Cu, 4
Stainless W	0.07	16.5	7	—	Al, 0.30 Ti, 0.75

Corrosion resistance

As mentioned in the introduction, the stainless steels owe their resistance to corrosive attack to a phenomenon called "passivity" produced by the presence of an oxide film at the surface of the metal. There are other theories on the nature of passivity, but the protective film theory is more popular, possibly because of its simplicity.

The state of passivity is not a general condition. It exists only under certain well-defined conditions. Outside the range of these favorable conditions the stainless steels can lose their passivity and react like ordinary iron. It is, for this reason, very important to know well the factors which favor passivity in order to take advantage of this remarkable property.

The corrosion resistance of stainless steel is attained by chromium additions. A minimum chromium content of approximately 12% is, however, required in order to attain passivity. With further increases in chromium the corrosion resistance continues to improve. In parallel with this the resistance to oxidation at high temperature also benefits from chromium addition. Aluminum and silicon contribute strongly in improving the oxidation resistance. Nickel added to chromium steels broadens the passivity range, especially in nonoxidizing environments. Manganese, on the other hand, is rather ineffec-

tive in this respect though it can be helpful by producing a fully austenitic structure with lower nickel content such as in the AISI 200 series. Molybdenum is beneficial when added to straight chromium steels but particularly to chromium-nickel austenitic steels since it improves the resistance to pitting and sea-water corrosion. Molybdenum bearing steels also have better resistance in sulfurous and sulphuric acids. Copper is believed to help make molybdenum more effective.

Because of their corrosion resistance the stainless steels have great industrial importance in oxidizing environments which tend to promote passivity but cannot be used in severe reducing media which tend to destroy passivity. Nitric acid and chromic acids are an example of oxidizing agents; hydrofluoric and hydrochloric acids are reducing; sulfuric acid is about borderline.

One characteristic of stainless steels is that they are attacked by chloride ion in aqueous solutions. The attack is localized in nature and is referred to as pitting. Special alloy additions of molybdenum, copper or silicon to the austenitic stainless steels (300 series) considerably reduce susceptibility to pitting.

In order to attain optimum conditions for resisting corrosive attack the stainless steels are heat-treated to produce the microstructure which has the most suitable properties. The correlation between microstructure and corrosion has been extensively investigated and has contributed to solving many corrosion problems. To attain maximum corrosion resistance it is desirable to have all of the carbon in solution rather than precipitated as chromium carbides. Martensitic stainless steels exhibit the best corrosion-resisting properties in the as-hardened state or with a low stress-relieving treatment up to approximately 650 F. Tempering treatments above this temperature promote precipitation of carbides and a gradual reduction in resistance to corrosion. The heat treatment of ferritic steels is less critical than that of the hardenable martensitic types and they do not show carbide precipitation like the austenitic grades; however, annealing is beneficial after welding. The austenitic stainless steels have optimum corrosion resistance when annealed from 1850 to 2050 F and cool rapidly to prevent any carbide precipitation. These steels may be intergranularly attacked in a chemical environment when sensitized in the

900 to 1500 F temperature range in which carbide precipitation occurs. Sensitizing can occur in welding, stress relieving or slow cooling from the annealing temperature. It can be prevented by suitable fabrication methods or by using the columbium and titanium stabilized grades (Type 347 and 321). The extra-low carbon (0.03% max) Type 304L and Type 316L also have found increased use because with this low carbon content no carbide precipitation can take place.

Stress corrosion is a brittle type of failure which can affect the austenitic stainless steels exposed under stress in solutions containing chlorides.

Because the desirable passive condition of stainless steels must exist for their good corrosion resistance, any condition which disturbs or prevents the passive film from forming in service should be avoided. Contamination of the surface by deposits of foreign substances, dirt, grease, oxide scale, and iron particles is undesirable since these may lead to localized pitting attack and rusting at the areas where they are present, even under very mild ordinary atmospheric conditions. Care in fabrication and handling to avoid contamination is advisable. Oxide scales created by heat treatment or welding should be removed either by mechanical or chemical means (pickling). Passivation with warm dilute nitric to remove iron particles from machining or forming operations is frequently recommended before stainless steels are placed in service. A periodic cleaning or maintenance program for stainless steels in actual service is usually very beneficial in preventing heavy accumulated deposits from remaining on surfaces for long times. With good clean surfaces which permit the passive condition to be attained and maintained continuously in service, the stainless steels are able to offer maximum corrosion resistance.

STRIPPABLE COATINGS

The need for permanent protection of building equipment, tools and other equipment during storage for long periods of time has led to the development of strippable coatings which can be removed quickly when required. A wide series of strippable coatings are available for commercial and industrial products.

Vinyl types. Vinyl coatings like those described in MIL-C-3254 specification were first developed for ships. These are called cocooning systems and are applied over chicken wire or a similar frame over the object to be protected. The interstices of the chicken wire are coated by a process known as webbing. This consists of spraying a specially designed vinyl coating in a web fashion so that it coats the interstices with a very thin, spider-web-like, fragile covering. This in turn is coated with a material similar to an ordinary strip coat vinyl by spraying. A more rigid protective coat of asphalt is then applied, following which coats of vinyl or aluminum enamel are applied. The advantages of this coating system are long life, ability to cover irregular surfaces, and easy removal. Disadvantages include a cumbersome structure which is expensive.

Strippable, sprayable, vinyl coatings. These materials are generally applied by spraying in thicknesses of 30 to 40 mils. Their tensile strength runs 500 psi minimum with an elongation of 200%, minimum. Adhesion on the Scott Tester 1-in. cut is ¾ to 2¼ lb. These materials are designed to be strippable after years of protective service. They are suitable for bright steel, aluminum, painted surfaces, wood, etc. They can also be used to protect spray booths, for aircraft protection, and on tanks, trucks, ships and similar equipment.

Special colors in vinyls are available and from 1900 to 2000 pounds psi tensile strength can be obtained with 100% elongation. Coatings can also be produced in a translucent, colored effect or in a clear form so as to show any defects in the substrate. These coatings are generally sprayed from 1 to 2 mils thick and are designed for protection in covered storage or in the transportation or fabrication of tools. They can be readily removed even though film thickness is low.

Ethyl Cellulose, Type I and Cellulose Acetobutyrate, Type II. These 100% coatings are designed for dip application in a hot melt bath of 350 F. Thickness ranges from 100 to 200 mils, depending upon the protection desired. The mineral oil generally present in these coatings exudes and coats the metal surface to keep it from corroding and in a strippable condition for long periods. Tensile strength of the coatings is about 300 psi minimum, with an elongation of about 90% when originally made, and 70% when aged.

The strippable material hardens by cooling when it is removed from the bath and is immediately ready for packaging. Generally, coated parts are packed with a protective wrapping of some sort, although parts can be shipped directly without any packaging because of the thickness and toughness of the film.

Coated parts will withstand a cycling test of 16 hr at 100 F and 100% RH; 3 hr at minus 40 F; 2 hr at 160 F and 3 hr immersion in 5% sodium chloride solution at room temperature. They will also withstand 720 hr of 100% RH at 100 F without any attack of the bright metal substrate.

The coatings are generally used on tools, steel and aluminum parts, and many other parts which can withstand the temperature of dipping. Variations of specification types can be formulated that do not exude oil, which can be objectionable in handling, particularly with electrical equipment. Pourable variations are also commercially available.

Degradation of the material on long use at 350 is a problem and exterior durability of the cellulosic or cellulose acetobutyrate types is not as good as that of the vinyl types.

Some special types of ethyl cellulose strippable materials can be used on painted surfaces and are formulated so that these surfaces are not affected. They can be applied by spraying. Other specially designed strippable materials are also practical; some of them can be used for packaging.

SUPERALLOYS

The term "superalloy" is broadly applied to iron-base, nickel-base and cobalt-base alloys, often quite complex, which combine high-temperature mechanical properties and oxidation resistance to an unusual degree. Alloy requirements for turbo-superchargers and, later, the jet engine, largely provided the incentive for superalloy development.

This article deals with superalloys but principal emphasis is on those containing more than 40% nickel.

Strengthening mechanisms

Superalloys of the nickel-chromium and iron-nickel-chromium types usually contain sufficient chromium to provide the needed oxidation resistance and are further strengthened by the addition of other elements. The strengthening mechanisms include solid solution, precipitation and carbide hardening. Most of the superalloys combine at least two and frequently all three of the above mechanisms.

Solid solution. This is accomplished by introducing elements having different atomic sizes than those of the matrix elements, to increase the lattice strain. In addition to chromium, needed also for oxidation resistance, the elements molybdenum, columbium, vanadium, cobalt and tungsten are effective in varying degree as solid-solution strengtheners when added in proper balance.

Superalloys of this type, such as 16-25-6, were used in gas turbines of older design for discs, employing "hot-cold work" to obtain the required yield strength in the hub area. However, hot-cold work is not effective as a means of getting high strength at temperatures much above about 1000 F. Therefore, as gas turbine operating temperatures increased and turbine disc rim temperatures appreciably exceeded 1000 F, other approaches were needed.

Cobalt-base alloys are strengthened principally by solid solution hardening, usually combined with a dispersion of stable carbides.

Precipitation hardening. Precipitation hardening is the method now employed to impart high strength at high temperatures to most of the superalloys used for critical components of aircraft gas turbines. This includes alloys ranging from the 25% nickel A-286 wheel alloy to such recent and complex high-temperature nickel-base wrought and cast turbine blade alloys as "Udimet" 700, "Nimonic" alloy 115 and IN-100.

The presence of aluminum and titanium, usually jointly, in a nickel-chromium base, with or without iron, imparts unique age-hardening characteristics through the precipitation of gamma prime phase ($Ni_3(Al,Ti)$).

The outstanding difference between the previously used age-hardening systems, typified by duralumin or beryllium-copper, and those utilizing gamma prime hardening lies in the fact that an alloy of the latter type can be heated in service appreciably above the optimum aging temperature without permanent loss of strength. In the conventional critical dispersion type age-hardening alloys, strength can only be restored by a complete cycle of heat treatment involving a high-temperature solution treatment and reaging.

Fig 1 shows the way in which the two types of age-hardening systems differ, with the aluminum-titanium hardened alloy (curve B) regaining most of its initial hardness (and high-temperature properties) when the overheat is removed. However, an alloy typical of the other type system (such as beryllium-copper or beryllium-nickel) overages and does not regain its hardness when the overheat is removed. It is this "reversible" aging behavior, along with

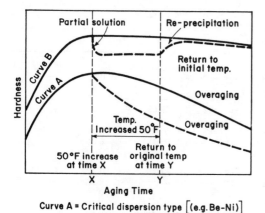

Curve A = Critical dispersion type [(e.g. Be-Ni)]
Curve B = Reversible hardening [(e.g. Ni₃(Al Ti)]

FIG. 1. Effect of aging time on hardness varies with type of age-hardening system.

high resistance to overaging (agglomeration of gamma-prime hardening phase), that has led to the widespread use of aluminum-titanium age-hardened nickel-base alloys for first stage turbine blades in commercial and advance design jet engines. The somewhat less complex iron-nickel-chromium alloys, such as A-286 and "Incoloy" alloy 901, used for turbine discs, are similarly age-hardened, with Ni_3Ti comprising most of the age-hardening component in these two alloys. Sufficient aluminum is also present in the gamma-prime precipitate to improve structural stability.

Carbide hardening. Carbides provide a major source of dispersed phase strengthening in cobalt-base alloys. These alloys do not respond to age hardening with aluminum and titanium since the gamma-prime phase does not form unless substantial amounts of nickel are present. While gamma prime ($Ni_3(Al,Ti)$) is the principal strengthener in the age-hardenable

nickel-base alloys, important auxiliary strengthening can be obtained by precipitation of quite complex carbides. The nature of the carbides formed and their mode of distribution can usually be controlled by alloy formulation and heat treatment.

Deoxidizers and malleabilizers. In addition to the major alloying elements, small but effective amounts of malleabilizers such as boron and zirconium must be present to neutralize the effects of impurities that adversely affect hot ductility. As little as 0.005% boron is highly effective in nickel-base alloys, and zirconium, in a concentration about ten times that of boron, is also useful. In air melting of nickel-base alloys, magnesium is also added to "fix" any sulfur picked up during melting.

The high-temperature properties of the age-hardenable alloys are governed to a large degree by the hardener (aluminum plus titanium) content. However, as the hardener increases, the temperature of incipient fusion decreases and the lower temperature limit of forgeability rises, because of increased high-temperature strength, until forging is no longer practical by conventional procedures (Fig 2). The use of cast turbine blade alloys to meet very high temperature requirements is a natural consequence.

While other co-present elements and the amount of hot and cold workability required are also factors, it may be considered that alloys

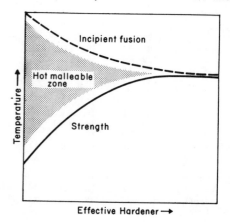

FIG. 2. Effect of hardness content on lower temperature limit of forgeability.

with up to 4% total hardener (Al + Ti) are generally available in all common mill forms and are fabricable by conventional methods. As the hardener increases to about 8 or 9%, wrought alloys are still available, but in progressively fewer forms, ultimately being limited to small forgings such as turbine blades, at increasingly greater cost. The commercial cast turbine blade alloys contain from about 7 to 11% total hardener and are often chosen over forgings for reasons of cost or necessity or both.

Compositions and properties

While factors such as tensile properties, corrosion resistance, fatigue strength, expansion characteristics, etc., are important, the creep-rupture characteristics are usually the prime requisite in the selection of a superalloy.

Experimental work on nickel-chromium base age-hardening alloys of the Al-Ti type led to the development of the "Inconel" and Nimonic series of alloys. This early work led the way to the development of a large number of age-hardenable alloys of proprietary nature, bearing such designations as Inconel alloy X-750, Nimonic

TABLE 1—NOMINAL COMPOSITION OF SOME SUPERALLOYS, %

No.	Alloy	Ni	Cr	Fe	Co	Mo	Cb	W	Al	Ti	B	Zr	C	Other
Iron-Base														
1.	16–25–6	25	16	Bal	—	6.0	—	—	—	—	—	—	0.06	—
2.	19–9 DL	9	19	Bal	—	1.3	0.4	1.2	—	0.3	—	—	0.30	—
3.	15–15 N	15	16	Bal	—	1.6	1.1	1.4	—	—	—	—	0.10	—
4.	A-286	26	15	Bal	—	1.3	—	—	0.2	2.0	0.003	—	0.06	—
5.	Discaloy 24	26	13	Bal	—	2.8	—	—	0.15	1.8	—	—	0.03	—
6.	Incoloy alloy 901	42	13	35	—	6.0	—	—	0.3	2.8	0.02	—	0.05	—
7.	D 979	45	15	Bal	—	4.0	—	4.0	1.0	3.0	0.01	—	0.05	—
High Cobalt														
8.	N 155	20	21	—	20	3.0	1.0	2.5	—	—	—	—	0.10	0.15 N
9.	S 816	20	20	3.5	Bal	4.0	4.0	4.0	—	—	—	—	0.38	—
10.	X 40, HS 31	11	26	1.5	Bal	—	—	7.5	—	—	—	—	0.50	—
11.	L 605, HS 25	10	20	3 max	Bal	—	—	15.0	—	—	—	—	0.10	—
12.	HS 151	—	20	—	Bal	—	—	12.7	—	—	0.05	—	0.50	—
13.	SM 302	—	21	1.5 max	Bal	—	—	10.0	—	—	—	0.2	0.85	9 Ta
14.	Nivco 10	22.5	—	—	Bal	—	—	—	—	1.8	—	1.1	0.05	—
15.	WI-52	3 max	21	—	Bal	—	—	11.0	—	—	0.09	—	0.4	Cb & Ta-1.5
16.	J-1570	28	20	—	Bal	—	—	6.0	—	4.0	—	—	0.2	—
17.	J-1650	26	19	—	Bal	—	—	12.0	—	4.0	0.02	—	0.2	Ta-2
Nickel-Base														
18.	Inconel alloy 600	76	16	7.0	—	—	—	—	—	—	—	—	0.04	—
19.	Nimonic alloy 75	78	20	Low	—	—	—	—	—	0.4	—	—	0.12	—
20.	Inconel alloy 721	71	16	7.2	—	—	—	—	—	3.0	—	—	0.04	2.25 Mn
21.	" " 722	74	15	7.0	—	—	—	—	0.6	2.4	—	—	0.04	—
22.	" " X-750	73	15	7.0	—	—	0.85	—	0.8	2.5	—	—	0.04	—
23.	" " 751	73	15	7.0	—	—	1.00	—	1.2	2.5	—	—	0.04	—
24.	" " 700	46	15	1.0	29	3.7	—	—	3.0	2.2	0.005	0.05	0.14	—
25.	" " 702	79	16	Low	—	—	—	—	3.4	0.7	—	—	0.03	—
26.	" " 718	53	19	18.0	—	3.0	5.2	—	0.6	0.8	—	—	0.04	—
27.	" " 713	72	13	1.0	—	4.5	2.0	—	6.0	0.6	0.01	0.1	0.12	—
28.	M-252	56	19	2.0	10	9.7	—	—	1.0	2.5	0.005	0.06	0.15	—
29.	René 41	55	19	1.0	11	9.8	—	—	1.7	3.2	0.003	—	0.09	—
30.	Astroloy	56	15	—	15	5.3	—	—	4.4	3.5	0.03	—	0.06	—
31.	GMR 235	65	15	9.0	—	5.5	—	—	3.0	2.0	0.05	—	0.15	—
32.	GMR 235 D	68	16	4.5	—	5.0	—	—	3.5	2.5	0.08	—	0.15	—
33.	Waspaloy	56	20	1.0	14	4.5	—	—	1.3	3.0	0.005	0.06	0.07	—
34.	Waspaloy-mod.	57	19	1.0	11.5	7.0	—	—	1.2	2.5	—	—	0.05	—
35.	Udimet 500	50	19	4 max	19.5	4.0	—	—	2.9	2.9	0.008 max	0.01	0.08	—
36.	Udimet 700	53	15	1 max	19	5.2	—	—	4.3	3.5	0.1 max	—	0.12	—
37.	IN-100	60	10	1 max	15	3.0	—	—	5.5	5.0	0.015	0.05	0.18	1.0 V
38.	Nimonic alloy 80	75	20	5 max	—	—	—	—	1.0	2.3	—	—	0.08	—
39.	" " 90	57	20	5 max	18	—	—	—	1.4	2.4	—	—	0.08	—
40.	" " 105	53	15	1 max	20	5.2	—	—	4.6	1.5	0.002	0.05	0.18	—
41.	" " 115	57	15	—	15	3.5	—	—	4.0	5.0	—	—	0.18	—
42.	SM 200	Bal	9	Low	10	—	—	12.5	5.0	2.0	0.015	0.05	0.15	—

alloy 90, M-252, Nimonic alloy 115, "Waspaloy," Udimet 500, Udimet 700, Inconel alloy 700, alloy 713C, "René" 41, "Astroloy," IN-100, SM-200, and others. While the alloys vary considerably in the composition of the matrix with respect to solid solution strengtheners such as molybdenum, tungsten, cobalt, etc., all depend predominantly on aluminum plus titanium to attain their unique high temperature properties.

Table 1 shows the composition of a number of typical superalloys. The 100- and 1000-hour rupture properties for a few temperatures are shown in Table 2. Most of the rupture data in the tables are taken from an article by W. F. Simmons.

Cast alloys such as IN-100 are generally stronger at high temperatures than the best wrought alloys. This is partially due to the fact

TABLE 2—RUPTURE PROPERTIES OF SOME SUPERALLOYS, 1000 PSI

No.	Alloy	1200 F		1350 F		1500 F	
		100 hr	1000 hr	100 hr	1000 hr	100 hr	1000 hr
Iron-Base							
1.	16–25–6	45	34	25	17	13.5	9
2.	19–9 DL	44	37	22	17	13	8.6
3.	15–15 N	—	33	—	18	—	10
4.	A-286	63	46	35	21	13	8
5.	Discaloy 24	52	41	30	20	15	—
6.	Incoloy alloy 901	94	78	55	40	24	15
7.	D 979	94	76	63	44	35	22
High Cobalt							
8.	N 155	52	43	30	23	20	16
9.	S 816	60	46	38	29	25	18
				1500 F		1800 F	
10.	X 40, HS 31	56	51	28	22	11.3	9.8
11.	L 605, HS 25	70	54	24	17	7	3.8
12.	HS 151	73	68	37	33	14	11.5
13.	SM 302	—	—	—	—	—	—
14.	NIVCO 10	54	43	—	—	—	—
15.	WI-52	—	—	—	—	11.5	7.8
16.	J-1570	95	78	33	24	—	—
17.	J-1650	—	82	46	33	13	—
Nickel-Base							
18.	Inconel alloy 600	23	14.5	8	5.6	2.8	1.8
19.	Nimonic alloy 75	25	—	12	—	1.9	—
20.	Inconel alloy 721	55	35	17	10	—	—
21.	" " 722	74	54	19	11.5	3.2	1.4
22.	" " X-750	80	68	28	16	3.3	2.3
23.	" " 751	79	60	34	21	3.3	—
24.	" " 700	100	87	43	30	6	3.4
25.	" " 702	54	41	15	9	3.1	2.5
26.	" " 718	100	85	—	—	—	—
27.	" " 713	101	92	68	47	20	15
28.	M-252	102	88	37	23	—	—
29.	René 41	110	102	45	29	11	—
30.	Astroloy	—	—	50	39	14	—
31.	GMR 235	86	65	38	29	11	4.5
32.	GMR 235 D	110	100	56	39	17.5	14
33.	Waspaloy	110	86	40	25	6.5	—
34.	Waspaloy (mod.)	—	—	—	—	—	—
35.	Udimet 500	120	94	46	33	12	8
36.	Udimet 700	—	100	58	43	16	6
37.	IN-100	—	—	75	59	25	18
38.	Nimonic alloy 80	66	46	25	16	4	—
39.	" " 90	79	66	28	17	3.1	—
40.	" " 105	112	89	44	31	9.6	4.5
41.	" " 115	—	—	53	37	16.4	9.4
42.	SM 200	—	—	73	—	26	—

Source: Principally W. F. Simmons, "What Alloy Shall I Use For High Temperature Applications Above 1200 F," *Metal Progress*, Oct. '61, p. 84.

that higher hardener contents can be employed in the cast alloys than are feasible in wrought products, and partially due to the fact that the cast structure has an inherently better high-temperature rupture strength for a given composition than the wrought structure. The data in Table 2 are included to permit a broad comparison of a number of superalloys, and should not be used for design purposes.

Vacuum melting. Vacuum induction-melted superalloys are generally more ductile than air-melted material, permitting the use of higher aluminum plus titanium levels while still retaining adequate ductility. Vacuum induction melting accomplishes refining and also permits closer control of composition than does air melting. The vacuum arc (consumable electrode) method is widely used for superalloys, especially those employed for turbine discs. Often the electrode for the vacuum arc-melting charge is obtained from a vacuum induction melt, the end product thus combining the benefits of refining occurring in vacuum induction melting with the controlled solidification and sound ingot structure associated with the vacuum arc process. Some refining is also accomplished in the vacuum arc process, but to a lesser degree than in vacuum induction melting.

Fabrication

Forging and hot working. Many of the commercial nickel-base age hardenable superalloys can be forged or hot-worked with varying degrees of ease. As has been indicated, the top side of the forging temperature range is limited by such considerations as incipient fusion temperature, grain size requirements, tendency for "bursts," etc., and the lower side by the stiffness and ductility of the alloy. The recommendations of the metal producer should be sought for optimum forging practice for a given alloy.

The hot extrusion process increases the gamut of superalloy compositions that can be hot-worked. By the use of a suitable sheath on the extrusion billet, otherwise unworkable alloys can be reduced to bar by hot extrusion. In some instances, mild steel has been used for a sheath while in others nickel-chromium alloys such as Nimonic alloy 75 and Inconel alloy 600 have been employed. Some of the very high-speed hot extrusion processes may be of value in hot-working the more refractory superalloys.

Heat treatment. The conventional heat-treating equipment and fixtures generally suitable for nickel alloys and austenitic stainless steels are also applicable to the nickel-base high-temperature alloys. Nickel-base alloys are more susceptible to sulfur and lead embrittlement than iron-base alloys. It is therefore essential that all foreign material, such as grease, oil, cutting lubricants, marking paints, etc., be removed by suitable solvents, vapor degreasing, or other methods, before heat treatment.

When fabricated parts made from thin sheet or strip of age-hardening alloys such as Inconel alloy X-750 must be annealed during and after fabrication, it is desirable, especially in light gauges, to provide a protective atmosphere such as argon or dry hydrogen to lessen the possibility of surface depletion of the age-hardening elements. This precaution may not be as necessary in heavier sections, since the surface oxidation involves a much smaller proportion of the effective cross section.

It is usually necessary after severe forming, or after welding, to apply a stress relief anneal (above 1650 F) to assemblies fabricated from Al-Ti age-hardenable nickel-base alloys prior to aging. It is vitally important to heat the structure *rapidly* through the age-hardening temperature range of 1200 to 1400 F (which is also the low ductility range) so that stress relief can be achieved before any appreciable aging takes place. This is conveniently done by charging into a furnace at or above the desired annealing temperature. It has been found at times that the efficacy of this procedure has been vitiated in large welded structures by charging on to a cold car, resulting in a slower and nonuniform heating of the fabricated part when run into the hot furnace. Contrary to expectations, little difficulty has been encountered with distortion under the above rapid heating conditions. In fact, distortion of weldments of substantial size has been reported to be less than by conventional slow-heating methods.

Forming. All of the wrought nickel-base alloys available as sheet can be formed successfully into quite complex shapes involving much plastic flow. The lower strength Inconel alloy 600 and Nimonic alloy 75 offer few problems. The high-strength age-hardening varieties, processed in the annealed condition, can be subjected to a surprising amount of cold work and deformation, provided sufficient power is available. Explosive forming has also been successfully employed on a number of nickel-base alloys.

Machining. All of the alloys discussed can be machined, the strongest and highest hardener content materials causing the most difficulty. The recommendations of the metal producer should be followed with respect to optimum condition of heat treatment, type of tool, speed and feed, cutting lubricant, etc. Wrought alloys of quite high hardener content, such as Inconel alloy 700 and Udimet 500, though difficult to handle, can be machined with reasonable facility using high-speed-steel tools of the tungsten-cobalt type, and cemented carbide tools of the tungsten-cobalt and tungsten-tantalum-cobalt type.

Various electroerosion processes have been successfully used on a number of the age-hardened superalloys and, at high hardener levels, may be necessary for some operations such as drilling.

Welding. Inconel alloy 600, Nimonic alloy 75 and other nickel-base alloys of the predominantly solid solution strengthened type offer no serious problems in welding. All of the common resistance and fusion welding processes (except submerged arc) are regularly and successfully employed.

In handling the wrought superalloys age-hardened with gamma prime (Ni_3 (Al,Ti)), it is necessary to observe certain precautions. Material should be welded in the annealed condition to minimize the hazard of cracking in weld or parent metal. If the components to be joined have been severely worked or deformed they should be stress-relief annealed before welding by charging into a hot furnace to insure rapid heating to the stress-relieving temperature. Similarly, weldments should be stress-relieved before attempting to apply the 1300 F age-hardening treatment.

Where subassemblies must be joined in the age-hardened condition, the practice of "safe ending" with a compatible nonaging material prior to age hardening can be usefully employed. The final weldment joining the fully age-hardened components is then made on the "safe ends."

Such new welding processes as "short arc" and "electron beam" are coming into increasing use and may be helpful in joining some of the very high hardener content alloys.

Brazing. The solid solution type chromium-containing alloys, such as Inconel alloy 600, are quite readily brazed, using techniques and brazing alloys applicable to the austenitic stainless steels. Generally speaking, it is desirable to braze annealed (stress-free) material to avoid embrittlement by the molten braze metal. Where brazing alloys are employed that melt above the stress-relieving temperature, a prior anneal is usually not needed. As with the stainless steels, dry hydrogen, argon, and helium atmospheres are used successfully, and vacuum brazing is also quite widely employed.

The age-hardened nickel-base alloys containing titanium and aluminum are rather difficult to braze, unless some method of fluxing, solid or gaseous, is used. Alternatively, the common practice is to preplate the areas to be furnace brazed with 0.0005 to 0.001 in. of nickel, which prevents the formation of aluminum or titanium oxide films and permits ready wetting by the brazing alloy.

Silver brazing alloys can be used for lower temperature applications. However, since the nickel-base superalloys are usually employed for high-temperature applications, the higher melting point and stronger and more oxidation-resistant brazing alloys of the Ni-Cr-Si-B type are generally used. The silver-palladium-manganese, palladium-nickel-manganese, and palladium-nickel alloys also provide useful brazing materials for intermediate service temperatures.

SYNTHETIC NATURAL RUBBER (ISOPRENE)

Stereoregular polyisoprene and polybutadiene elastomers, high in *cis*-1,4 content, are of growing interest to the engineer, both because of engineering performance, and their competitive price. IR (*cis*-polyisoprene) has been called "synthetic natural rubber" because chemically and physically it is similar to *Hevea*.

General properties and examples of end-use performance show it to be a satisfactory supplement to natural rubber in a wide variety of products. Molecular weight can be controlled within quite wide limits and linearity can be maintained even with the longest chains. Higher molecular weight materials have been satisfactorily extended with oil to give compositions having a desirable combination of low cost and attractive properties.

The development of IR latex is a noteworthy advance in latex technology. The low emulsifier level, stereoregularity of the polymer, large particle size, and low viscosity have not hitherto been available in a general-purpose synthetic latex. These properties, combined with high gum strength and elongation, offer advantages for many latex applications.

Vulcanizate properties

IR can be processed in a manner similar to that used for natural rubber. Vulcanization can be carried out by means of curatives commonly used with natural or SBR. Properties of gum vulcanizates are quite similar to those of natural rubber, although IR has somewhat lower modulus and higher extensibility.

Table 1 compares properties of IR with those of other elastomers reinforced with 50 phr HAF

TABLE 1—TREAD VULCANIZATE PROPERTIES [a]
(AT 77 F)

	IR	NR	SBR	BR
Ten Str, psi	3600	4000	3800	2900
300% Modulus, psi	1650	2000	1700	1300
Elongation at Break, %	520	550	550	500
Shore Hardness	56	63	58	56
Angle Tear, lb/in.	380	600	320	300
Yerzley Resilience, %	71	72	60	74
Heat Buildup (Goodrich Flexometer), ΔT, °F	34.2	33.3	43.2	32.4

[a] 50 phr HAF; IR and NR 5 phr oil; SBR and BR 8 phr oil. Conventional recipes, optimum cures.

(each compound mixed with conventional recipes and given optimum cures). The data indicate the excellent hysteresis properties, low heat buildup and high resilience of both the IR and the polybutadiene tread vulcanizates. Note that IR is second only to natural rubber in tear strength.

Processing and compounding procedures for oil-extended IR are similar to those for unextended polymer, except that lower curative levels are recommended for maximum tensile strength and flat curing characteristics. Properties of extended vulcanizates approach those of the non-extended materials.

Applications

Similarity of performance between IR and natural rubber has permitted use of IR as a supplement for natural rubber in uses such as tire treads, carcasses, and white sidewalls.

In nontire uses IR's low ash content, light color and good mold flow characteristics are of particular advantage. Good electrical properties and low moisture absorption make it suitable for a number of electrical insulating uses. Parts molded in IR exhibit sharp definition and excellent color stability to light.

The low cost and good performance of oil-extended IR are promising for tire carcass compounds, molded mechanical goods, and footwear.

In latex form, IR is the first synthetic which possesses an average particle size as large and a particle size distribution as broad as that of natural rubber latex. It is highly promising for a number of coating and dipping applications, as well as foaming.

T

TANTALUM AND ALLOYS

Tantalum is a high-density, ductile, refractory metal which exhibits exceptional corrosion resistance and good high-temperature strength over 3000 F. The annealed wrought metal in its pure form is easily worked and can be cold-worked in much the same manner as fully annealed mild steel.

Tantalum belongs to Group V of the periodic table and has an atomic weight of 180.95. Density is 0.6 lb/cu in. Tantalum possesses negligible corrosion rates in most environments with the exception of hydrofluoric acid, alkalies, and hot concentrated sulfuric or phosphoric acid. But tantalum is a reactive metal and must be protected from oxidation when heated in air above 500 F. Tantalum will become embrittled when heated in hydrogen below 2200 F and will form nitrides when heated in the presence of nitrogen. Heated in the presence of hydrocarbons, tantalum will react to form carbides.

Fabricability

Hot working. High-purity tantalum sintered bar, cast ingot, and annealed wrought forms can be worked at room temperature, although the working of large ingots and billets is sometimes performed at elevated temperature to permit working within equipment strength capacity. Cast ingots, protected by canning or coating materials, have been forged and extruded at temperatures up to 2400 F. Cold-worked tantalum can be stress-relieved or annealed at a variety of time-temperature schedules depending upon stress level and chemical purity of the material. A temperature of at least 2200 F is generally used for full annealing while temperatures between 1500 and 1700 F can be used for stress relieving. Annealing atmosphere must be either high purity argon, helium, or preferably a vacuum of one-tenth of a micron or less.

Cold working. The excellent room-temperature ductility of stress-relieved and fully annealed tantalum makes the forming of tantalum comparatively simple. But the grain size of the material must be carefully considered for requirements where surface finish is of importance. Fully annealed tantalum has a tendency to "neck" and fold rapidly during some forming operations and care must be exercised to prevent breakage. The combination of the higher tensile strength and fair uniform elongation of stress-relieved tantalum sometimes makes it more satisfactory for drawing and forming than fully annealed material. It is very important to consider the temper properties of the material in the design of forming tools.

Joining. Tantalum can be joined by electron beam, TIG, and spot and resistance welding, but must be carefully protected from the effects of oxidation during welding. Uncontaminated welds are ductile and usually can be worked at room temperature.

Electron beam melted 90% tantalum-10% tungsten alloy can be formed and joined in a similar manner as pure tantalum provided that the higher strength and more rapid work-hardening characteristics of the alloy are considered.

Properties

Typical properties of high-purity tantalum and electron beam melted (EBM) 90% Ta-10% W alloy sheet are shown in Table 1. Properties of several developmental alloys are given in Table 2.

Applications

The corrosion resistance, electrical properties, gettering ability, and high melting point of tantalum have been the principal properties responsible for its past and present use. Tantalum is used extensively in the chemical industry where its excellent fabrication and joining properties permit the application to acid-resistant heat exchangers, condensers, duct-work, chemical lines, and other chemical process equipment. Tantalum also finds use in the medical profession. Because of its nontoxic properties and immunity to body chemicals, tantalum is used for sutures, gauze, pins and plates.

The electrical characteristics of anodic films

TABLE 1—PROPERTIES OF TA AND 90 TA-10 W.

MECHANICAL	R.T.	
	Ten Str, 1000 psi	Elong, %
Tantalum		
Annealed	35–50	25–40
As Rolled, 90% C.W.	90–110	1–3
90 Ta-10 W (EBM)		
Annealed	70–80	20–25
Str Rel	110–130	7–15
As Rolled, 90% C.W.	160–180	0–3

	2200 F	
Tantalum		
Annealed	12–15	40–50
90 Ta-10 W (EBM)		
Annealed	35–40	—

	4000 F	
Tantalum		
Annealed	2–3	30–40
90 Ta-10 W (EBM)		
Annealed	6–10	—

THERMAL (OF TANTALUM)

Melt Pt, F 5393
Vapor Pressure, atm
  4264 F .... 6.216×10^{-9}
  4847 F .... 3.655×10^{-7}
Lin Exp, %
  2912 F 1.04
  4352 F 2.04
  5072 F 2.99
Ther Cond cal/sq cm/°C/cm
  68 F 0.130
  2606 F 0.174
  3326 F 0.198
Sp Ht, cal/gm
  212 F 0.03364
  2012 F 0.03823
  2912 F 0.04078

ELECTRICAL (OF TANTALUM)

Elec Cond, % IACS, 64.4 F 13.9
Elec Res, μohm-cm
  77 F 12.4
  1832 F 54
  2732 F 71
  3632 F 87

on tantalum permit its use in rectifiers and capacitors. The anodic film acts as a dielectric. The combination of high melting point, low vapor pressure, and gettering properties are useful for applications in the electronic vacuum tube industry.

TABLE 2—PROPERTIES OF TANTALUM ALLOYS

Alloy and Condn	Temp, F	Yld Str, 1000 psi	Ten Str, 1000 psi	Elong	Density, lb/cu in.
Ta-30Cb-10V	−320	189	217	21	0.413
Recryst	75	127	148	21	
strip	2190	35	64	18	
	2400	30	42	76	
	2600	19	27	104	
	2730	12	20	114	
Ta-30Cb-7.5V	75	125	139	27	0.426
Recryst	2200	48	60	35	
strip	2600	22	36	76	
	3000	6.2	10	99	
Ta-10W-5V	75	137	153	19	0.56
Recryst	2200	59	75.5	10	
Ta-4Mo-4Hf	−320	197	199	16	0.58
Strip, 0.050	−100	151	155	16	
in. thk,	75	139	144	16	
95% red,	2200	78	85.0	26	
stress rel	2500	35	39.6	59	
2000 F,	2700	28	28.7	79	
1 hr					
Ta-8W-4Hf	−320	204	205	11	0.60
Strip, 0.050	75	140	147	15	
in. thk,	2200	80	91.0	23	
95% red,	2500	37.7	43.0	50	
stress rel	2700	30.3	32.2	67	
2000 F,					
1 hr					
Ta-10HF-5Mo	75	104	106	3	0.567
Recryst	2200	58.6	80.8	7	
Ta-10HF-5W	75	105	115	23	0.589
Sheet, 0.040	2400	31	41	10	
in. thk,	2600	25	37	17	
Recryst	3000	12	18	29	

Availability

Pure tantalum is commercially available as billet, bar, plate, sheet, strip, foil, tubing, rod, and wire processed from both cast and sintered material. Sheet is available up to 36 in. wide and foil is rolled down to 0.0002 in. thick. Fine wire is drawn down to 0.003 in. thick.

TELLURIUM

Tellurium, discovered by Muller von Richenstein in 1782, is the fourth member of Group VIB of the periodic arrangement of the elements, being more metallic than selenium, the next adjacent element. It is rarely found in its elemental state but is usually associated with gold, silver, lead and copper.

Physical properties

Atomic No.	52
Atomic Weight	127.61
Crystal Structure	Hexagonal
Density, 20 C, gm/cu cm	6.25
Elec Resis, 19.6 C, μohm-cm	436,000
Latent Ht of Fusion, cal/gm	32
Latent Ht of Vaporization, cal/gm	107
Lin Coef of Therm Exp, μ in./°C	16.75
Melt Pt, C	449 ± 0.3
Sp Ht, 20 C, cal/gm	0.047
Therm Cond, 20 C, cal/sq cm/cm/ °C/sec	0.014

Chemical properties

The chemical properties of tellurium are more basic than either sulfur or selenium. The oxidation states of tellerium are:

-2, in the tellurides
+2, TeO is known to exist
+4, in TeO_2 and $TeCl_4$
+6, in TeO_3, H_6TeO_6, and TeF_6.

Tellurium dioxide can readily be made and is a white odorless crystal which is only slightly soluble in water and dilute acids.

Tellurous acid, H_2TeO_3, is a white crystalline powder, slightly soluble in water, more soluble in dilute acids, and very soluble in alkalies. Telluric acid, H_6TeO_6, is an orange-colored solid that is insoluble in water, dilute acids and alkalies.

Tellurium combines readily with the halogens to form numerous compounds, and over thirty halides and oxyhalides are mentioned in the literature. In general, the halogen compounds of tellurium hydrolyze with water to form hydrogen halide plus an acid of tellurium.

Tellurium and its compounds are considered to be less toxic than selenium compounds. The accepted limit concentration for eight-hour daily exposure is 0.1 mg per cu in. of air. While this concentration is a safe one, it will probably produce garlic breath which will disappear in a few days, if no further exposure is experienced. Large doses of Vitamin C, of the order of 10 mg per kg of body weight, have been found to alleviate the garlic odor.

Forms available

Commercial Grade, 99.5% Te:
  Powder, −200 mesh.
  Sticks, approximately 1 in. dia and 6 in. long weighing approximately 1 lb.
  Slabs, 7 in. × 2 in. × 1½ in., 5 lb.
  Tablets, in various sizes which are a combination of tellurium powder (95%) and carnauba wax (5%)
High Purity Grades, 99.99% and 99.999% Te:
  Random size fragments and 5 lb cast cakes.

Applications

Like selenium, tellurium improves the machinability of stainless steels and acts as a degasifier but, because of the fume problem, it has usually been avoided. Small amounts of tellurium are used in the mold dressing of cores or as ladle additions to control the depth of chill in the manufacture of hard-chilled iron castings.

The addition of tellurium to copper, while not materially reducing the electrical conductivity or hot- and cold-working properties, greatly increases machinability. Small amounts of tellurium, added to lead in the range of 0.05 to 0.1%, restrict grain growth, raise the annealing temperature, and markedly improve fatigue resistance.

Tellurium is an important component of many thermoelectric devices, and such devices can be used for both power generation and cooling. The requirements of a good thermoelectric element are high thermoelectric power, low thermal conductivity, and low electrical resistivity. Lead telluride (PbTe), bismuth telluride (Bi_2Te_3) and silver antimony telluride meet these requirements better than any other presently known materials. By the addition of various other elements these compounds can be made either p-type or n-type semiconductors.

TEXTILE FIBERS

Natural fibers constitute one of man's oldest sources of building materials. There is evidence to indicate that weaving and probably spinning were not unknown to our Stone Age ancestors. It is important to realize that there is no such thing as a natural textile fiber, although today there are man-made textile fibers. There are only natural fibers which have been diverted from their original function by mankind for use in textiles.

In man's search for fibers which can be used

to further his own ends, literally dozens of naturally occurring fibers have been investigated. In the 1960 issue of the ASTM Committee D-13 Manual on Textile Materials, a total of 161 vegetable fibers, 29 animal fibers, and one mineral fiber are listed. Yet of this vast number, only 23 are readily recognized by most textile authorities as being of commercial importance, and one fiber alone, cotton, accounts for approximately 70% of the total fibers consumed by the world's population for textile purposes. If, however, this listing of natural fibers is carefully reviewed, it will be found that all the fibers therein can be grouped into six different types of spinnable fibers, each differing fundamentally with respect to molecular and morphological structure. The distinctive characteristics possessed by the fibers in these six groups are such that the groups may be subjectively described as cottonlike, linenlike, sisal-like, wool-like, silklike, and asbestoslike.

This same ASTM Manual lists some 33 different man-made fibers, ranging from a regenerated fiber such as rayon to a true synthetic fiber such as nylon. Yet of this number, only ten can be classed as being of some commercial importance. Here again, the traditions of centuries are revealed in the name of "artificial silk" given to the first fibers produced by man. Only in the past score of years or so have traditions started to crumble with the introduction of true synthetic fibers designed and engineered for specific functions.

Form characteristics of fibers

Fibers possess a number of form characteristics, the important ones being length and cross-section shape or diameter. Other factors, such as crimp and surface character may be of considerable importance in certain applications.

Length. ASTM Standard D123 contains a specific definition for a fiber, wherein the length is stated as being at least 100 times its diameter or width, with this length being at least 5 mm. From a length viewpoint, fibers are classed as either staple or filament, wherein filament implies continuous length. The length-diameter ratios for common staple fibers range from about 500 to 15,000, with the various grades of cotton covering the low range (500 to 1300). Silk is the only natural fiber existing in filament form, while man-made fibers may be obtained in both staple and

TABLE 1—RANGE OF LENGTHS FOR SELECTED STAPLE FIBERS
(Length range in inches)

Cotton, Indian	½–1¼₁₆
Cotton, Am. Upland	¾–1⁵⁄₁₆
Cotton, Sea Island	1½–2
Camel Hair	2–2½
Wool, Apparel	2–15
Ramie	3–10
Mohair	6–12
Flax	12–36
Jute	50–80

Source: Harris' *Handbook of Textile Fibers*, Harris Research Laboratories, Incorporated, 1954.

filament forms. With respect to staple fibers, most processing is done with fibers ranging from just under 1 to 6 in. long; however, lengths of several feet in the rope industry are not uncommon.

Cross section. The direct expression of the cross section of a fiber is not common to the textile industry, as only a few fibers ("Dacron," glass, nylon, "Dynel") have essentially circular cross sections. Some few others (hair fibers) have nearly circular or oval cross sections, but most fibers have irregular cross sections. It is accordingly a common practice to express this characteristic of a fiber as fineness, which is defined in terms of a simple weight-length or linear density relationship. An exception to this is wool, which is graded or classed for commercial purposes in terms of its diameter expressed in microns.

Linear density is a convenient basis for comparing fibers in this respect because it is not influenced by variations in fiber density and cross-sectional shape. A common measure of linear density is the "denier," which is the weight in grams of a fiber length equal to 9000 m (meters). Another measure, sponsored by the International Organization for Standardization and receiving considerable acceptance, is the "tex," defined as the grams per 1000 m or 1 km (kilometer). In the tex system, it is recommended that fiber fineness be reported in millitex units, which is the number of grams per 1000 km. Actual fiber diameters range from about 7 to 40 μ.

The related diameter and cross-sectional area of a fiber for a given denier vary considerably because of differences in fiber density. Also these characteristics have a major affect upon yarn packing, weave tightness, yarn twist acceptance,

TABLE 2—MEAN DIAMETERS FOR SELECTED
STAPLE FIBERS

Fiber	Width, microns Range	Mean	Denier	Millitex
Silk	3–23	12	1.1	122.
Cotton, Am. Upland	8–30	18	2.4	266.
Camel Hair	9–40	18	2.4	266.
Cotton, Indian	10–33	21	3.3	366.
Mohair	14–75	33	10.9	1210.
Wool, Apparel	17–33	25	4.7	522.

Source (except millitex): Harris' *Handbook of Textile Fibers*.

yarn and fabric stiffness, fabric thickness and weight, and cost relationships. Comparative engineering data for various fibers involving fineness, length, and density can be easily calculated if it is assumed that such fibers are essentially circular in cross section, an assumption which is reasonable for many purposes. On the basis of this assumption, the following relationships hold:

$$\text{Diameter, in.} = 1.13 \sqrt{\text{Area}} \tag{1a}$$

$$= 0.469 \times 10^{-3} \sqrt{\frac{\text{Denier}}{\text{Density}}} \tag{1b}$$

$$\text{Diameter, microns} = 11.91 \sqrt{\frac{\text{Denier}}{\text{Density}}} \tag{2}$$

$$\text{Denier} = \frac{(\text{Density})(\text{Diameter, microns})^2}{142} \tag{3}$$

$$\text{Cross-Section Area, in.}^2 = \frac{1.722 \times 10^{-7}\,\text{Denier}}{\text{Density}}. \tag{4}$$

The denier/density value is conveniently referred to as "denier per unit density," and as such, it represents the equivalent denier for a fiber with the same cross-sectional area and a density of 1.0.

In blending operations, it may be desired to blend two fibers, each of essentially the same diameter. Diameters are inverse functions of the square roots of the densities involved, and this relationship is expressed as follows:

$$\text{Diameter A} = \frac{\text{Diameter B} \sqrt{\text{Density 2}}}{\sqrt{\text{Density A}}} \tag{5}$$

The shape of the cross section of a fiber has a bearing on its ability to pack in a yarn structure, on the fiber-to-fiber functional characteristics of the yarn structure, on its soiling properties, and on its stiffness or ease of bending and twisting characteristics.

The cross-sectional shape also relates to the surface area, which in turn has an important bearing on fiber-to-fiber contact area, air permeability, and other properties. If the assumption of a circular shape is maintained, then the surface area may be calculated from the following relationships.

$$\text{Surface Area, in.}^2/\text{oz} = \frac{31558\,\text{Mils}}{\text{Denier}} \tag{6a}$$

$$= \frac{1212\,\text{Microns}}{\text{Denier}} \tag{6b}$$

$$= \frac{6936}{(\text{Mils})(\text{Density})}. \tag{6c}$$

While not strictly an area relationship, the number of fibers per unit weight of fibers can provide useful information.

$$\text{Fibers/oz} = \frac{10045188}{(\text{Denier})(\text{Fiber Length, in.})}. \tag{7}$$

Obviously, the number of fiber ends per ounce is simply twice the number of fibers per ounce.

TABLE 3—GENERALIZED CROSS SECTIONS OF
SELECTED FIBERS

Round, Round to Oval	Irregular and Serrated-Round	Flat Tube	Dog Bone	Tri-angular
"Acrilan"	Acetate	Cotton [a]	"Orlon"	Silk
Rayon, cupra	"Arnel"	Ramie	"Verel"	
"Dacron"	"Fortisan"	Flax		
"Dynel"	"Velon"			
Glass	Rayon,			
Nylon	viscose			
Polyethylene				
Saran				
"Teflon"				
"Terylene"				

[a] Also convoluted.
Source: ASTM Committee D-13 *Manual on Textile Materials*, 1960.

Physical characteristics of fibers

Density. The relationship of fiber density to such characteristics as denier, diameter, and cross-sectional and surface areas has been expressed previously. Fibers in common use range from a density of 0.93 g/cm³ for polyethylene to 2.80 g/cm³ for asbestos, as illustrated in Table 4.

The denier/density value is of use in blending operations. It is generally desirable to blend fibers of essentially the same diameter, all other factors being equal. For example, the finest wools are relatively coarse compared to fine man-made fibers, and blending such fine man-made fibers with wool will usually result in softer, fuzzier fabrics. A fiber of 1 denier/density has a cross-sectional area of 1.722×10^{-7} in.^{2} and a diameter of 11.91 μ.

Moisture absorption. Changes in the relative humidity of the surrounding air result in significant changes in the hygroscopic moisture content

TABLE 4—DENSITY VALUES FOR SELECTED FIBERS

By Density		Alphabetically	
Density, g/cc	Fiber	Fiber	Density, g/cc
0.93	Polyethylene	Acetate	1.31–1.33
1.12–1.15	Nylon	"Acrilan"	1.14
1.12–1.19	"Orlon"	"Arnel"	1.30
1.14	"Acrilan"	Asbestos	2.10–2.80
1.28–1.31	"Dynel"	Cotton	1.50–1.55
1.28–1.33	Wool	"Dacron"	1.38–1.40
1.30	"Arnel"	"Dynel"	1.28–1.31
1.31–1.33	Acetate	"Fortisan"	1.52
1.34–1.37	Silk	Glass	2.47–2.57
1.37	"Verel"	Nylon	1.12–1.15
1.38–1.40	"Dacron"	"Orlon"	1.12–1.19
1.38–1.40	"Terylene"	Polyethylene	0.93
1.48–1.53	Rayon, viscose	Rayon, cupra	1.54
1.50–1.55	Cotton	Rayon, viscose	1.48–1.53
1.52	"Fortisan"	Saran	1.62–1.75
1.54	Rayon, cupra	Silk	1.34–1.37
1.62–1.75	Saran	"Terylene"	1.38–1.40
2.10–2.80	Asbestos	"Verel"	1.37
2.47–2.57	Glass	Wool	1.28–1.33

Source: ASTM Committee D-13 *Manual on Textile Materials*, 1960; Harris' *Handbook of Textile Fibers*.

of many of the textile fibers, which may be broadly classed as either hydrophilic or hydrophobic in nature. For the hydrophilic fibers, this moisture content can affect not only the weight, but also the dimensions, most of the mechanical properties, and the ease with which the fiber may be manipulated in processing. The extent to which a fiber absorbs moisture depends upon the presence of polar groups and the availability of these groups within the amorphous regions of the fiber. The principal water binding groups in fibers include the hydroxyl, carboxyl, carbonyl, and amino groups.

The moisture component in textile materials is customarily expressed as a percentage of the dry weight of the fiber, and this relationship is defined as moisture regain. For the hydrophilic fibers, the relationship between moisture regain and relative humidity may be expressed by an S-shaped curve, with moisture regain increasing rapidly for the first 10 to 20% relative humidity, rising slowly and in an essentially linear fashion up to 80 to 90% relative humidity, and finally again rising rapidly from this point to 100% relative humidity.

The initial portion of the curve represents the absorption of moisture by freely available un-

coordinated polar groups. The middle portion results from a slow spreading apart of the remaining polar groups within the amorphous regions of the fiber and the saturation of the binding potential of all polar groups. The final portion of the curve probably represents multilayer moisture condensation, with the moisture being held rather loosely by the fiber. The moisture held by the fiber as represented by the middle portion of this S-curve exerts its greatest effect in reducing fiber rigidity.

The hydrophilic fibers are represented by the natural fibers as a group and by the regenerated man-made fibers. The true synthetic fibers are, in general, hydrophobic.

TABLE 5—MOISTURE REGAIN AND WATER
SWELLING OF FIBERS

Fiber	Moisture Regain, % [a]	Water Swelling, % [b]
Acetate	6.3–6.5	6–30
Cotton	8.5	44–49
"Dacron"	0.4	Nil
"Fortisan"	10.0–11.5	18–26
Glass	0	0
Nylon	4.0–4.5	About 2
"Orlon"	Nil	Nil
Rayon, cupra	11	99–134
Rayon, viscose	11.5–16.6	45–82
Saran	0	0
Silk	11.0	30–41
Wool	17.0	32–38

[a] Regain at 21 C and 65% RH.
[b] Swelling in water based on cross-sectional area increase.

Source: Harris' *Handbook of Textile Fibers*.

Stress-strain behavior

Textile fibers, with but a few exceptions, exhibit plastic and elastic deformation under a stress, and are commonly referred to as "viscoelastic" materials. The plastic or viscous component in a textile fiber stress-strain diagram is much greater in magnitude than in other engineering materials.

The ultimate or breaking strength of fibers is customarily expressed in terms of breaking load per unit fineness, such as grams per tex or grams per denier, and measure of strength is referred to as "tenacity." Tensile strength in psi units is a calculated quantity based on the assumption of a circular cross section. When wet, cotton, as representative of the natural vegetable fibers, shows an increase in breaking strength, while the regenerated cellulosic fibers show an appreciable loss. The synthetic fibers, as a group, show no significant loss when wet.

Because of the viscoelastic nature of essentially all textile fibers under stress, when reference is made to the modulus value for a fiber, this term is not interchangeable with the classic Young's modulus of elasticity. Although the modulus of elasticity may be determined by static and velocity-of-sound or sonic modulus techniques, the modulus values commonly reported in the literature are simply the ratio of the tenacity, usually at the breaking point, to the corresponding strain. Actually, this is the slope of a straight line, usually from the origin of the tenacity-strain curve to the terminal point. De-Witt Smith defined this property as "stiffness" so as not to create confusion with the conventional engineering concept of modulus. ASTM Committee D-13 has defined this ratio as "secant modulus," and this expression, or more simply "modulus," seems to have general acceptance in textile work. All other things being equal, fibers

having low secant modulus values exhibit a high degree of compliance.

A useful measure for ranking the relative performance abilities of the various fibers with respect to their ability to absorb energy is their "toughness index," which DeWitt Smith defined as the area under the secant modulus line. Because of the simplifying assumption with respect to the secant modulus, the toughness index may be easily calculated from tenacity-strain data. Due to the actual shape of the tenacity-strain curve for nearly all fibers, this toughness index is a conservative measure of the ability of a fiber to do work.

The foregoing relates to the one-time loading of a textile fiber. In actual use, a textile fiber is subjected to numerous loading and unloading cycles, usually to limits appreciably below the maximum limits of breaking strength and elongation. The majority of fibers exhibit a pronounced strength-elongation hysteresis effect, and the behavior after a few loading-unloading cycles is usually significantly different from the behavior for a one-time loading-unloading cycle. Generally, from five to ten cycles will mechanically condition a fiber so that the hysteresis loop

TABLE 6—BREAKING STRENGTH AND ELONGATION FOR SELECTED FIBERS [a]

Fiber	Break Tenacity,[b] g/denier	Break Strength, 1000 psi	Wet Strength, % of conditioned	Break Elongation, %
Acetate	1.1–1.4	18–23	60–65	23–30
"Acrilan"	3.5	50	90	17.5
Cotton	2.1–6.3	42–125	110–130	3–10
"Dacron"	4.2–5.0	74–89	100	22–30
"Dynel"	3.0	50	100	31
"Fortisan"	7.0	138	85	6
Glass	7.7	· 250–315	99	2.5–3.2
Nylon, reg	4.5–6.0	65–88	85–90	25–31
Nylon, hi ten	6.0–7.8	88–114	85–90	15–23
"Orlon"	4.8–5.8	73–88	95	15–18
Polyethylene	1.0–2.5	11–30	100	20–60
Rayon, cupra	1.7–2.3	33–42	59	10–17
Rayon, viscose, reg	1.5–2.4	29–47	44–54	15–30
Rayon, viscose, hi ten	3.0–4.0	58–78	55–65	9–20
Saran	1.1–2.9	25–60	100	20–35
Silk	2.8–5.2	45–83	75–95	13–31
Wool	1.0–1.7	17–28	76–97	20–50

[a] Man-made fiber values are for filament form except for "Acrilan" and "Dynel." The staple form yields significantly different values. Values reported are for results obtained at 21 C and 65% RH.
[b] To convert g/denier to g/tex, multiply by 9.

Source: Harris' *Textile Fibers Handbook*.

becomes closed at zero load, thus indicating the absence of secondary creep, which has been removed during the mechanical conditioning cycles.

DeWitt Smith applied the term "degree of resilience" to the ratio of the energy recovered to the energy put in for a one time loading-unloading cycle. Hamburger extended this concept in his "elastic performance coefficient" so as to include the ratio of the energy recovered to input for the mechanically conditioned state of the fiber. In many industrial and consumer applications, the relative amounts of immediate elastic deformation, creep deformation and the primary and secondary components of creep have an important bearing, along with the total elongation of the fiber. As an illustration, both nylon and rayon possess high amounts of elongation, but rayon is noted for its large amount of secondary creep or nonrecoverable deformation when mechanically conditioned, while nylon exhibits but a small amount of secondary creep, and if the strain is small enough, essentially none. Accordingly, cold-worked or mechanically conditioned nylon yarn will still be essentially the same length as it was before conditioning, but rayon will be appreciably longer in the mechanically conditioned state. It follows that nylon is an ideal fiber for women's hose, while rayon is not when it is considered that the wearer of such hose is mechanically conditioning the yarns during wearing.

As the relative amounts of elastic and creep deformation and recovery are time dependent, it follows that any quoted values are dependent upon the experimental conditions employed. Results reported in the literature are voluminous, and for this reason, reference should be made to appropriate sources, such as Harris' *Textile Fibers Handbook*, for more detailed information.

Thermal behavior. Fibers used for textile purposes may be grouped into three categories in terms of their behavior to heat. In the first category are those few fibers which do not burn in the presence of flame, and which are noted for their resistance to high temperature. The inorganic fibers, represented by asbestos, glass, some experimental quartz and ceramic fibers, and metal threads, fall in this grouping.

In the second category may be grouped most of the true synthetic fibers, for most such fibers are thermoplastic in nature. Fibers in this group, such as "Acrilan," "Dacron," "Dynel," nylon, "Orlon," polyethylene, and saran, are difficult to classify precisely, for they tend to soften gradu-

TABLE 7—SELECTED STRESS-STRAIN PROPERTIES FOR FIBERS [a]

Fiber	Modulus of Elasticity (Sonic method), g/denier	Secant Modulus, g/denier	Toughness Index, g/denier	Recovery from Strain, %
Acetate	52	4.7	0.17	94
"Acrilan"	30	20	0.31	80
Cotton	—	65	0.14	74
"Dacron"	120 app	18	0.60	97
"Dynel"	7.9	9.7	0.47	97
"Fortisan"	—	117	0.21	82
Glass	331	270	0.11	100
Nylon, reg	60 app	19	0.74	100
Nylon, hi ten	—	36	0.66	100
"Orlon"	26–39	32	0.44	97
Polyethylene	—	4.4	0.35	—
Rayon, cupra	—	15	0.14	—
Rayon, viscose, reg	86–96	9	0.22	82
Rayon, viscose, hi ten	150–252	24	0.25	82
Saran	—	7	0.28	—
Silk	130	18	0.44	92
Wool	44	3.9	0.24	99

[a] Recovery from strain at 21 C, 65% RH; rate of loading 10 g/denier/minute; loading 30 sec, recovery 60 sec, strain 2%. These values and modulus of elasticity values from Harris' *Textile Fibers Handbook*; other values calculated from data in Table 6.

ally on heating. "Teflon," du Pont's TFE fluoro-carbon fiber, exhibits maximum resistance to heat for organic base fibers, retaining some degree of serviceability to temperatures up to 500 F, but being badly degraded between 500 and 700 F. The melting point of nylon, which is of the order of 480 F, is used in many specifications as an identification test. (However, a new heat-resistant nylon fiber is reported to retain about 60% of its tenacity at 480 F.)

Some of these synthetic fibers, such as "Acrilan," "Dynel," polyethylene and saran, shrink significantly at temperatures from 160 to 250 F; advantage is taken of such heat shrinkage in the production of roller applicators for paints and in artificial furs. While the fibers in this second group will burn in the presence of an open flame, the burning action is characterized by melting and fusing, the release of considerable gaseous by-products, and a general tendency to be self-extinguishing when the flame is removed.

The natural fibers and regenerated man-made fibers make up the third category. There is some overlap between this group and the thermoplastic group with respect to useful temperature ranges. The cellulose fibers burn readily with little or no ash, while the protein-base fibers, silk and wool, leave a black crumbly ball-like ash and produce a strong odor of burning hair or feathers. The effect of heat on such fibers varies widely with temperature, duration of heating, test atmosphere, and previous history. Various finishes may be employed to greatly alter the burning characteristics of these fibers, as well as fibers in the second group.

The effect of low temperatures on the physical properties of most fibers has not been extensively investigated. The mechanical properties of fibers in the first category are essentially unaffected by low temperatures. The thermoplastic fibers tend to become stiff and brittle, with resulting losses in elongation and gains in strength. As the actual shape of the tenacity-strain curve may be drastically altered by low temperatures, it follows that the area involved may be of greater interest than merely the change in breaking point.

Thermal-transmission measurements exist primarily in the form of measurements for fabrics. Herein, the relative thermal transmission of a fabric is but only slightly affected by the particular fiber employed in the fabric, but is in-

TABLE 8—EFFECT OF HEAT AND FLAME ON SOME SELECTED FIBERS

Fiber	Effect of Heat, F	Rate of Burning
Acetate	Delustered 170, softens 400, melts 500	Burns moderately fast
"Acrilan"	Softens 464, 5% shrinkage at 490	Burns slowly
Asbestos	Useful to 1490, wt loss above 700	Nonflammable
Cotton	Useful to 275	Burns rapidly
"Dacron"	Sticks 455, melts about 480	Burns slowly
"Dynel"	Shrinks above 240	Self-extinguishing
"Fortisan"	Similar to cotton	Burns moderately fast
Glass	Softens 1380–1550, 50% str loss at 685	Nonflammable
Nylon	Sticks 455, melts about 480	Self-extinguishing
"Orlon"	Sticks 455, no str loss after 32 days at 250	Self-extinguishing
Poly-ethylene	5% shrinkage at 165, melts 225	Burns slowly
Rayon	Str loss above 300, decomposes 350–400	Burns rapidly
Saran	Shrinks above 240, softens 240–280	Self-extinguishing
Silk	Disintegrates at 340	Self-extinguishing to slow
"Teflon"	No effect at 400, sublimation above 550	Self-extinguishing
Wool	Good to 212	Self-extinguishing to slow

Source: Harris' *Textile Fibers Handbook* and others.

stead dependent upon the geometry of the fabric, that is the amount and configuration of air spaces within the fabric. Thus, it is primarily a function of fabric thickness for a given air permeability value. This relationship is essentially linear, work by Schiefer, Stevens, Mack, and Boyland (*J. Research*, Nat. Bur. Standards **32**.261, 1944) showing thermal transmission values ranging from 1.52 to 0.41 Btu/hr/F/sq ft/in. for thicknesses from 0.010 to 0.600 in. when measured at a load of 0.10 psi. Thermal-transmission values for a few fibers are available: loose asbestos 1.07, packed asbestos 1.62, cotton 0.29, glass wool 0.27, wool felt 0.27, and silk 0.31. The specific heats for a number of natural fibers range from 0.317 to 0.331, with asbestos at 0.251 and glass wool at 0.157.

Effect of chemicals and environments on fibers

In many industrial applications, the response of a fiber to various chemicals and dyestuffs or to environmental factors such as sunlight and weathering, to wear, or to microorganisms, becomes a deciding factor in the selection of one fiber over another. While the resistance of some fibers to degradation by a particular condition, such as cotton to attack by microorganisms, can be altered by suitable finishing treatments, inherent resistance is always to be preferred. Because actual use conditions vary so greatly, tabulations of chemical resistances and related properties should only be used for screening purposes, with the final selection of a particular fiber being dependent upon actual field trials.

Chemical resistance. In resistance to the action of chemicals, fibers range all the way from essentially inert materials, such as "Teflon" and glass, down to fibers such as cotton, which has poor resistance to most acids and good resistance to alkalis, and the protein fibers, which are easily damaged by alkalis. The mercerization process for cotton is based upon its reaction to a sodium hydroxide solution under controlled tension conditions. Cotton can be hydrolized by strong alkalis, and "Fortisan" is basically a saponified cellulose ester. Wool is so sensitive to alkali damage that precise control of both pH and ion concentration in the raw-wool scouring bath must be maintained. Glass fibers, as a group, resist attack by all acids except hot phosphoric and hydrofluoric; however, they have poor resistance to strong alkalis, especially at elevated temperatures. The amount of silica in the glass influences the chemical resistance of common glass fibers. Thus silica or quartz fibers are the most resistant, 96% silica is next, followed by borosilicate, with soda lime lowest in chemical resistance.

Dyeability. The ability of a fiber to accept a given dyestuff in a manner so as to exhibit satisfactory resistance to color-degrading conditions such as laundering, dry cleaning, light, weathering, and other end-use conditions is af-fected by the chemical structure of the fiber. Many methods are available for dyeing a textile material, some methods being peculiar to a given fiber. A fiber may be dyed in the fiber state (stock dyeing) or in the solution or melt state for some man-made fibers, or the fibers may be dyed after they have been processed into a yarn (yarn dyeing), or into a fabric (piece dyeing), or the fabric may be colored by various printing techniques.

Chemicals used as dyestuffs are classed in terms of their chemical composition or the nature of the dyeing process employed. In Table 10, the major classes of dyestuffs commercially employed for dyeing some common fibers are tabulated.

Resistance to environments. Here, environmental factors have been separated from chemical factors, although in many applications the chemical resistance of a fiber may dictate its use in a chemical environment. Generally, environment is interpreted as consisting of light, weathering, microorganisms, and wear conditions. In industrial and consumer uses, one or more of these conditions may be of primary importance, and thus determine the selection of a particular fiber. In some instances, selection may be dictated by economic factors. In this latter sense, cotton is widely used under conditions which make it susceptible to attack by microorganism, attack that it is poorly equipped to resist unless specially treated, yet it is used in place of fibers better equipped because of the cost factor.

The inorganic fibers are generally unaffected by weathering. The poor weatherability of some of the man-made fibers can be overcome by the use of special finishing and/or coating treatments when other considerations require the use of such a fiber. The natural bast and leaf fibers, such as flax, hemp, hennequin, jute, manila, ramie, and sisal, are essentially unaffected by weathering. Nylon and Dacron lose strength on exposure to sunlight, being only slightly better than silk in this respect. Cotton is better than silk, and wool is better than cotton. Teflon is regarded as being completely inert to weathering, and Orlon is also noted for its resistance to weathering. While polyethylene loses appreciable strength on exposure to weathering, the incorporation of colored pigments improves the weathering ability of this fiber, with black providing maximum improvement.

TABLE 9—CHEMICAL RESISTANCE OF SELECTED FIBERS

Fiber	Effect of Chemicals	Fiber	Effect of Chemicals
Acetate	Deteriorated by cold conc and hot dilute acids.		Excellent resistance to other chemicals.
	Swollen by strong alkalis, strength reduced; good resistance to weak alkalis.	Nylon	Dissolves in cold conc mineral acids. generally good resistance to weak acids.
	Dissolves in acetone. not affected by dry cleaning and related solvents, attacked by strong oxidizing agents. not affected by hypochlorite or peroxide bleaches.		Excellent resistance to strong and weak alkalis.
			Soluble in some phenolic compounds and conc formic acid, good resistance to most other chemicals.
"Acrilan"	Good to excellent resistance to strong acids, excellent to weak acids.	"Orlon"	Very resistant to most mineral acids, disintegrated by 96% sulfuric acid. excellent resistance to weak acids.
	Poor resistance to strong alkalis, fair to good to weak alkalis.		Moderate resistance to cold strong alkalis, disintegrates at boil, good resistance to weak alkalis.
	Not affected by solvents, oils, greases, and some acid salts.		Excellent resistance to common solvents.
Asbestos	Good resistance to cold acids, poor to hot, all conc.	Polyethylene	Excellent resistance to all but strong oxidizing acids.
	Good resistance to strong and weak alkalis.		Excellent resistance to strong and weak alkalis.
	Excellent resistance to other chemicals.		Some swelling and weakening in benzene, toluene, soluble above 160 F in some solvents.
Cotton	Poor resistance to strong and to hot weak acids.	Rayon	Acids, similar to cotton.
	Excellent resistance to strong and weak alkalis, can be hydrolized.		Swollen and weakened by strong alkalis, good resistance to weak alkalis.
	Generally good resistance to organic solvents, bleached by hypochlorites and peroxides, swells and disintegrates in cuprammonium hydroxide.		Other chemicals, similar to cotton.
"Dacron"	Dissolves in sulfuric acid, good resistance to weak acids.	Saran	Excellent resistance to strong acids except only fair to sulfuric, excellent resistance to weak acids.
	Disintegrates when boiled in strong alkalis, good resistance to weak alkalis.		Excellent resistance to strong and weak alkalis except ammonia.
	Good resistance to most other chemicals.		Swollen or softened by oxygen-bearing solvents at elevated temperatures.
"Dynel"	Resistance to acids and alkalis, similar to "Acrilan."	Silk	Dissolved by strong acids, fairly resistant to weak acids.
	Good resistance to other chemicals, softened by acetone and some ketones.		Dissolved by strong alkalis, attacked by hot weak alkalis.
"Fortisan"	Acids, alkalis, other chemicals, similar to cotton.		Good to excellent resistance to other chemicals.
E-Glass	Attacked only by hydrofluoric and hot phosphoric acids.	"Teflon"	Inert to all acids and alkalis.
	Poor resistance to hot strong alkalis, good resistance to weak alkalis.		Affected by alkali metals and some hot halogenated hydrocarbons.
		Wool	Acids, alkalis, and other chemicals, similar to silk.

Sources: Harris' *Textile Fibers Handbook; Man-Made Textile Encyclopedia,* Textile Book Publishers, Incorporated. 1959; and other selected sources.

TABLE 10—DYE CLASSES ACCEPTED BY SELECTED
FIBERS

Fiber	Dye Classes Accepted
Acetate	Special acetate classes, selected vats and azo, some acid and basic, pigment, and solvent dyes.
"Acrilan"	Acetate classes, acid dyes.
Cotton	Basic, direct, mordant, naphthol, sulfur and vat classes.
"Dacron"	Acetate classes, azo, and vat with a carrier or at high temperatures.
"Dynel"	Selected acetate classes, acid, direct, and some vat types.
"Fortisan"	Generally same classes as used for cotton.
Glass	Resin-bonded pigments, also other dyes used on coated fibers.
Nylon	Acetate classes and acid types usually preferred, but most other classes are used also.
"Orlon"	Acetate classes, acid by copper technique, basic, naphthol and vat.
Polyethylene	Colored before extrusion.
Rayon	Generally same classes as used for cotton.
Saran	Colored before extrusion.
Silk	Acid, basic, direct, mordant, naphthol, and vat classes.
Wool	Acid, basic, direct, mordant, and vat types.

Source: Harris' *Textile Fibers Handbook; Man-Made Textile Encyclopedia;* and other selected sources.

Wear is a complex phenomenon, and most workers speak of only one aspect of wear with respect to textile fibers, that aspect being abrasion resistance. Abrasion is the wearing away of the textile material by rubbing or attrition. The abrasion resistance of a fabric is greatly influenced by the geometry of the yarns and of the fabric itself, and can be significantly altered by various finishing treatments. In connection with fabric geometry, the incorporation of a wear-resistant fiber in the structure so that it lies on the face to take the abrasive action will improve the abrasion resistance of the fabric, as will the use of smooth (filament) yarns and fabrics as opposed to spun (staple) fiber yarns and fabrics.

While the literature is replete with papers pertaining to the abrasion resistance of fabrics of different constructions, much yet remains to be done in terms of a mathematical study of abrasion. Consequently, few objective data are available as to the absolute abrasion-resistance properties of the common fibers. Hard, less resilient fibers such as glass have very poor abrasion resistance. Nylon is generally recognized as having the best abrasion resistance of any of the textile fibers, with other fibers ranking in the following descending order: cotton, wool, medium high-tenacity viscose rayon, cupra rayon, regular viscose rayon and acetate.

Fiber classification

The accompanying fiber classification chart is

TABLE 11—ENVIRONMENTAL RESISTANCE OF SELECTED FIBERS

Fiber	Sunlight	Effect of Exposure To: Mildew [a]	Moths [a]
Acetate	Strength loss	Resistant	Not attacked
"Acrilan"	Slight loss	Not attacked	Not attacked
Cotton	Strength loss, yellow	Attacked	Not attacked
"Dacron"	Strength loss	Not attacked	Not attacked
"Dynel"	Strength loss	Not attacked	Not attacked
"Fortisan"	Slight loss	Resistant	Not attacked
Glass	None	Not attacked	Not attacked
Nylon	Strength loss	Not attacked	Not attacked
"Orlon"	Very slight loss	Not attacked	Not attacked
Polyethylene	Depends on pigment	Not attacked	Not attacked
Rayon	Strength loss	Attacked	Not attacked
Saran	Darkens slightly	Not attacked	Not attacked
Silk	Strength loss, great	Not usually attacked	Better than wool
"Teflon"	None	Not attacked	Not attacked
Wool	Better than cotton	Better than cotton	Attacked

[a] Depends upon presence of finish, some finishes will support attack while fiber itself is not attacked.
Sources: Harris' *Textile Fibers Handbook;* and other selected sources.

NATURAL FIBERS

Cellulose Base

- Bast
  - Flax (Linen)
  - Hemp
  - Jute
  - Ramie
- Fruit
  - Coir
- Leaf
  - Abacá (Manila)
  - Cantala
  - Henequen
  - Istle
  - Maguey
  - Sisal
- Seed
  - Cotton
  - Kapok

Protein Base

- Staple
  - Hair
    - Alpaca
    - Camel
    - Cashmere
    - Llama
    - Mohair
    - Rabbit
    - Vicuna
  - Wool
    - Sheep
- Filament
  - Silk

Mineral Base

- Asbestos

MAN-MADE FIBERS

Organic Base

Natural Polymer Base

- Acetate
  - ACELE
  - ESTRON
- Azlon
  - CASLAN
  - MERINOVA
- Rayon
  - Cupra
  - Saponified cellulose acetate FORTISAN
  - Viscose
- Rubber
- Triacetate
  - ARNEL

Synthetic Polymer Base

- Acrylic
  - ACRILAN
  - CRESLAN
  - ORLON
  - ZEFRAN
- Monacrylic
  - DYNEL
  - VEREL
- Nylons
- Nitrile
  - DARVAN
- Olefin
  - Polyethylene
  - Polypropylene
- Polyester
  - DACRON
  - FORTREL
  - KODEL
  - TERYLENE
  - VYCRON
- Saran
  - VELON
- Spandex
  - LYCRA
  - VYRENE
- Vinal
  - VINLON
- Vinyon
  - RHOVYL

Inorganic Base

- Glass
  - FIBERGLAS
- Metallic

FIG. 1. Fiber Classification Chart.

Textile Fiber Products Identification Act of 1958 generic names given for man-made fibers group only. Trademark names given in upper case letters only. The trademark names listed are not offered as the only fibers within a group, but are offered as representative names.

given to provide a general guide to the generic names of fiber groups as defined by the Federal Textile Fiber Products Identification Act of 1958 (effective March 3, 1960) and of the trademarks or commercial names for some fibers within each generic class. In this chart, the Textile Fiber Product Identification Act generic names apply only to the man-made fibers. The natural fibers are grouped along traditional lines. In considering the generic names given by the Act, it will be noted that no provision has been made for listing "Teflon," a polytetrafluoroethylene fiber, or ceramic (aluminum silicate) fibers such as "Fiberfrax." An excellent chart wherein the fibers are classed in terms of their basic chemical composition is given in the *ASTM Standards on Textile Materials* published by Committee D-13 of ASTM.

Some new fibers

Two new generic fiber families have been introduced within the past few years—the aramids and novoloids. Aramids are produced from long chain polyamides in which 85% of the amide linkages are attached directly to two aromatic rings. The fibers are exceptionally stable and have good strength, toughness and stiffness, which is retained well above 300 F. The three initial aramids were Nomex, Kevlar 29 and Kevlar 49. Kevlar 29, with high strength intermediate stiffness, is suitable for cables, ropes, webbings and tapes. Kevlar 49, with high strength and stiffness, is used for reinforcing plastics. Nomex, best known for its excellent flame and abrasion resistance, is used for protective clothing, air filtration bags and electrical insulation.

The novoloids are fibers containing at least 85%, by weight, cross-linked novalacs (epoxies). One novoloid, trade named Kynol, is noted for its exceptionally high temperature resistance. At 1920 F the fiber is virtually unaffected. The fiber also has high dielectric strength and excellent resistance to all organic solvents and nonoxidizing acids.

THALLIUM

Thallium is a dull gray metal resembling lead in its softness and high malleability. It is commercially available in the metallic form in purities of 99.95% (ingots, 34 lb, and sticks, 7 oz) and 99.999+% (sticks, 1½ oz). During the last

decade the price per lb for the 99.95% grade has ranged between $7.50 and $15.

Physical properties

Some of its physical properties are as follows:

Atomic No.	81
Atomic Weight	204.39
Melting Pt, C	303
Boiling Pt, C	1457
Density, 20 C, gm/cu cm	11.85
Latent Heat of Fusion, cal/gm	5.04
Latent Heat of Vaporization, 20 C, cal/gm	189.9
Sp Ht, 20 C, cal/gm	0.031
Therm Cond, 20 C, cal/sq cm/cm/°C/sec	0.093
Lin Coef of Therm Exp, μ in./°C	28
Electrical Resis, 0 C, μ ohm-cm	18

The commercial uses of metallic thallium are very limited. Although a number of thallium alloys have been investigated for high corrosion resistance, bearings and electrical contacts, they have not found commercial acceptance. Some use is made of the thallium-mercury amalgam (melting point, −60 C) when alloy that is liquid at very low temperatures is required.

Some low-melting alloys formed by the addition of thallium to other elements are as follows:

	Melt Pt, C
39% Tl, 19% Cd, 42% Sn	129.5
11.5% Tl, 46.5% Bi, 28% Pb, 14% Sn	93
11.5% Tl, 33.3% Pb, 55.2% Bi	90.8
8.9% Tl, 44.3% Bi, 11% Cd, 35.8% Pb	81

No extensive use has been found for these alloys.

Chemically, thallium resembles the elements of Group III. Both thallous (Valence 1) and thallic (Valence 3) compounds are stable.

The metal oxidizes rapidly in air forming thallous oxide, which has a very high solubility in water.

The best solvent for thallium metal is nitric acid.

Applications

The major use for thallium is as a rodenticide and insecticide. The sulfate compound is most commonly employed for this application; therefore, the largest commercial sale of the element is in the form of the sulfate. Thallium sulfate is a heavy white crystalline powder, odorless, tasteless and soluble in water. The advantage of this compound over many other rodenticides is that it is not detected by the rodent. Killing time with a lethal dose is about 72 hr and small doses are said to halt the reproduction processes, thus restricting the rat population. An example of such a poison for extermination of rodents and insects would be a mixture of starch, sugar, glycerine and water with the thallium sulfate added.

Other commercially available thallium chemicals are thallous nitrate and thallic oxide. Further uses of thallium compounds are as follows: (a) thallium oxisulfide, employed in a photosensitive cell which has high sensitivity to wavelengths in the infrared range; (b) thallium bromide-iodide crystals, which have a good range of infrared transmission and are used in infrared optical instruments; and (c) alkaline earth phosphors, which are activated by the addition of thallium.

Other minor uses for thallium are in glasses with high indices of refraction, in the production of tungsten lamps as an oxygen getter, in high-density liquids used for separating precious stones from ores by flotation, and in mercury thermometers for lowering the solidification point.

Toxicity

Thallium and thallium compounds are toxic to humans as well as other forms of animal life. Therefore, special care must be taken that thallium is not touched by persons handling it. Rubber gloves should be used in handling both the metal and its compounds. Proper precautions should be taken for adequate ventilation of all working areas.

THERMOFORMED PLASTIC SHEET

Thermoplastic sheet-forming consists of the following three steps: (1) a thermoplastic sheet or film is heated above its softening point; (2) the hot and pliable sheet is shaped along the contours of a mold, the necessary pressure being supplied by mechanical, hydraulic or pneumatic force or by vacuum; and (3) the formed sheet is removed from the mold after being cooled below its softening point.

Sheet materials

The following five groups of thermoplastic materials account for the major share of the thermoforming business:

1. *Polystyrenes:* High impact polystyrene sheet, ABS (acrylonitrile-butadiene-styrene) sheet, biaxially oriented polystyrene film and polystyrene foam.

2. *Acrylics:* Cast and extruded acrylic sheet, and oriented acrylic film.

3. *Vinyls:* Unplasticized rigid PVC, vinyl copolymers and plasticized PVC sheeting.

4. *Polyolefins:* Polyethylene, polypropylene and their copolymer films.

5. *Cellulosics:* Cellulose acetate, cellulose acetate butyrate and ethyl cellulose sheet.

A sixth group of increasing importance is the linear polycondensation products, such as polycarbonates, polycaprolactam (type 6 nylon), polyhexamethylene adipamide, oriented polyethylene terephthalate (polyester) and polyoxymethylene (acetal) films.

General process considerations

Thermoplastic sheets soften between 250 and 450 F. It is important that the sheets be heated rapidly and uniformly to the optimum forming temperature. The fastest heating is brought about with infrared radiant heaters. Some thermoplastics cannot tolerate such intense heat and require convection heating in air circulating ovens or conduction heating between platens. In a few instances, the sheets are formed "in line," making use of the heat of extrusion.

Four basic forming methods and more than twenty modifications are known:

1. *Matched mold forming* (Fig 1): The process in which the hot sheet is formed between a registering male and female mold section, employs mechanical or hydraulic pressure.

It is used for corrugating flat rigid sheeting either "in line" or in a separate operation. For continuous longitudinal corrugation, the hot sheet is pulled through a matched mold with registering top and bottom teeth. Transverse corrugation is accomplished with matched top and bottom rolls or molds mounted on an endless conveyor belt or chains.

For stationary molding operations, a rubber blanket, backed with a liquid or inflated by air, frequently replaces the male section. Another modification is the *plug and ring* technique in which the ring acts as a stationary clamping device and the moving plug resembles the top portion of the male mold.

Elastic and oriented thermoplastic sheets possess a plastic memory, i.e., they tend to draw tight against the force which stretches them. This property occasionally permits forming against a single mold only.

2. *Slip forming* (Fig 2): A loosely clamped sheet is allowed to slip between the clamps and is "wiped" around a male mold. This process has been in use to avoid excessive thinning when forming articles with deep draws.

3. *Air blowing* (Fig 3): A hot sheet is blown with preheated compressed air into a female mold.

Variations of this process are: *free blowing* without a mold into a bubble, *plug-assist blowing* in which a cored plug pushes the hot sheet ahead before blowing, and *trapped sheet forming* in which a clamping ring slides over the mold before applying the compressed air. The last process is employed in automatic roll-fed packaging machines with biaxially oriented films.

4. *Vacuum forming* (Fig 4): This is the most common sheet forming process with many modifications.

Modifications of vacuum forming. In *straight vacuum forming*, the hot thermoplastic sheet is

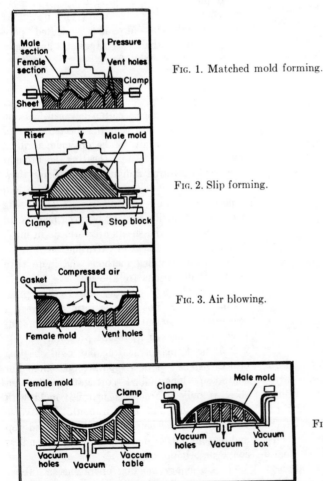

Fig. 1. Matched mold forming.

Fig. 2. Slip forming.

Fig. 3. Air blowing.

Fig. 4. Vacuum forming.

The four basic methods of thermoplastic sheet forming.

clamped tight to the top of a female mold or of a vacuum box which contains a male mold. Drawing sheets into a female mold results in excellent replica of fine details on the outer surface of the drawn articles.

Straight vacuum forming has proved excellent for shallow draws; however, in articles having small radii or deeper dimensions, the corners and bottom are excessively thinned out. Employing a female mold with multiple cavities is more economical than forming over a number of male molds, since it permits smaller spacing between cavities (without bridging) which in turn allows more pieces per sheet.

Drawing sheets over a male mold produces articles with the thickest section on top. It is used for the production of three-dimensional geographical maps, since it gives a greater accuracy of registration due to restricted shrinkage.

Free vacuum forming into a hemisphere, without a mold, is similar to free air blowing and is employed with acrylic sheets where perfect optical clarity has to be maintained.

Vacuum snap-back forming makes use of the plastic memory of the sheet. The hot sheet is drawn by vacuum into an empty vacuum box, while a male mold on a plug is moved from the top into the box. The vacuum is released and the sheet, still hot and elastic, snaps back against the male mold and cools along the contours of the mold. This method is employed with ABS and plasticized vinyl sheets for the production of cases and luggage shells.

Drape forming is a technique which allows deeper drawings. After clamping and heating, the sheet is mechanically stretched over a male mold, then formed by vacuum, which picks up the detailed contours of the mold. Acrylic and polystyrene sheets slide easily over the mold, whereas polyethylene sheets tend to freeze on contact with the mold, causing differences in thickness. To overcome the problem of thinning, *vacuum plug-assist forming* into a female mold has been developed. Its principle is to force a heated and clamped sheet into a female mold using a plug-assist before applying the vacuum. This technique may be considered a reverse of the snap-back method.

Vacuum air-slip forming represents a modification of drape forming and is designed to reduce thinning on deep-drawn articles. It consists of prestretching the sheet pneumatically prior to vacuum forming over a male mold. Prestretching is accomplished either with entrapped air by moving the male mold like a piston in the vacuum chamber or by compressed air.

There are at least three variations of the *reverse-draw technique.* All employ the principle of blowing a bubble of the hot plastic sheet and pushing a plug in reverse direction into the outside of the hot bubble. This accomplishes a folding operation permitting deeper draws than any other common practice.

The technique of *reverse-draw with plug-assist* consists of heating the clamped sheet, raising a female mold so that a sealed cavity is formed while a bubble is blown upward. The preheated plug-assist is lowered and pushes the sheet into the cavity. The final shaping is accomplished with vacuum.

A variation of this method is *reverse-draw with air-cushion.* The plug-assist is furnished with holes through which hot air is blown downwards and pushes the hot sheet ahead of the plug-assist, minimizing mechanical contact. This technique is used in forming materials with sharp softening ranges and limited hot strength, such as polyethylene and polypropylene. *Reverse-draw on a plug* uses a male mold on the plug to preserve the finish of the sheet.

Design and construction of vacuum molds

Depth of draw is a prime factor controlling the wall thickness of the formed article. During straight vacuum forming into a female mold, the depth of draw should not exceed one-half of the cavity width. For drape forming over a male mold, the height to width ratio should be 1:1 or less. With plug-assist, air slip or one of the reverse-draw techniques, the ratio may exceed the 1:1 ratio.

Proper air evacuation assists material flow in the desired direction and in uniform wall thickness. In general, deep corners require intensified evacuation. The diameter of vacuum holes should be 0.010-0.025 inch for polyethylene sheets, 0.025 to 0.040 in. for other thin-gauge materials and may increase to 0.060 in. for heavier rigid materials. Sharp bends and corners should be avoided, because they result in excessive stress concentration and in reduction of strength. The forming cycle can be accelerated and maintained by the use of mold temperature controls. Molds for permanent use are cast from aluminum or magnesium alloy.

Finishing

After the article has been formed, it must be cooled, removed from the forming machine and separated from the remainder of the sheet. Trimming of thin-walled articles can be carried out hot or cold, in the forming machine or after removal. Heavy-gauge articles should be trimmed only after cooling. Clicker dies, high dies and Walker dies are frequently used. Decorating formed articles is generally accomplished by printing the flat sheet before the forming operation. Formed articles may be spray-coated with

TIN AND ALLOYS

Tin, Sn, atomic number 50, atomic weight 118.70, is in Group IV of the periodic table. It forms stannous (Sn^{2+}) and stannic (Sn^{4+}) compounds as well as complex salts of the stannite (M_2SnX_4) and stannate (M_2SnX_6) types. It alloys readily with nearly all metals.

Tin is a nontoxic, soft and pliable metal adaptable to cold working such as rolling, extrusion and spinning. It melts at a low temperature, is highly fluid when molten and has a high boiling point, which facilitates its use as a coating for other metals. It can be electrodeposited readily on all common metals.

Properties

The most important physical, mechanical and electrical properties are shown in Table 1. Tin reacts with strong acids and strong bases but is relatively inert to nearly neutral solutions. In indoor and outdoor exposure it retains its white silvery color because of its resistance to corrosion. A thin film of stannic oxide is formed in air which provides surface protection.

Two allotropic forms exist: white tin (β) and gray tin (α). Although the transformation temperature is 13.2 C, the change does not take place unless the metal is of high purity, and only when the exposure temperature is well below 0 C. Commercial grades of tin (99.8%) resist transformation because of the inhibiting effect of the small amounts of bismuth, antimony, lead and silver present as impurities.

Alloying elements such as copper, antimony, bismuth, cadmium and silver increase its hardness. Tin tends rather easily to form hard, brittle

intermetallic phases, which are often undesirable. It does not form wide solid solution ranges in other metals in general, and there are few elements which have appreciable solid solubility in tin. Simple eutectic systems, however, occur with bismuth, gallium, lead, thallium and zinc.

TABLE 1—PROPERTIES OF TIN

Melt Pt, °C	231.9
Boil Pt, °C	2270
Sp Gr	
α-Form (gray tin)	5.77
β-Form (white tin)	7.29
Liquid at Melt Pt	6.97
Transformation Temp, °C	13.2
Sp Ht, cal/gm	
White Tin, 25 C	0.053
Gray Tin, 10 C	0.049
Latent Ht of Fusion, cal/gm	14.2
Latent Ht of Vaporization, cal/gm	520 ± 20
Ht of Transformation, cal/gm	4.2
Therm Cond, white tin, 0 C,	0.150
cal/sec/sq cm/°C/cm	0.150
Coef of Lin Exp, 0 C, 10^{-6} in./in.	19.9
Solidification Shrinkage, %	2.8
Vol Resis, white tin, μ-ohm/cu cm	
0 C	11.0
100 C	15.5
Bhn, 10 kg/5 mm, 180 sec	
20 C	3.9
220 C	0.7
Ten Str, as-cast, 1000 psi	
200 C	0.65
15 C	2.1
−40 C	2.9
−120 C	12.7

Tin is rarely used alone, generally as a coating for a baser metal or as a constituent of an alloy. The range of useful alloys is extensive and extremely important.

Fabrication characteristics

Small amounts of mechanical working result in an increase in hardness but because of their low recrystallization temperature, most tin-base alloys soften spontaneously at room temperature. Reduction of more than about 20% by rolling or drawing induce softening rather than hardening. It is difficult to obtain a permanent degree of hardening of tin-base alloys by heat treatment. Diffusion occurs at normal temperatures so that quenched and tempered alloys tend to overage and soften.

Available forms

Tin is usually sold in ingots of 28, 56 and 100 lb and bars of 1 lb and up. The bulk of the tin sold on the common market is Grade A which contains a minimum of 99.8% tin.

Special electrolytic grades of tin (99.99+%) and zone refined tin, containing a few ppm of impurities, are available from at least two sources.

Tin can be obtained in a number of forms: granulated, mossy, fine powder, sheet, foil and wire. Tin-base alloys are available in many forms: solder can be obtained in 1-lb bars, solid and cored wire, powder, sheet and foil; babbitt type metal and casting alloys in bars and ingots; bronze in ingot, continuously cast bars and shapes, sheet and foil.

Applications

Full use is made of the ductility, surface smoothness, corrosion resistance and hygienic qualities of tin in the form of foil, pipe, wire and collapsible tubes. Tin foil is devoid of springiness and is ideal for wrapping food products, as liners for bottle caps and for electrical condensers. Heavy walled tin pipe and tin-lined copper pipe is used by the food and beverage industries for conveying distilled water, beer and soft drink syrups. Tin wire is used for electrical fuses and for packing glands in pumps of food machinery. Collapsible tubes, made by impact extrusion from discs of pure tin, are used for pharmaceutical and food products.

More than 40% of the world's tin is used as a coating for steel and copper. Tin-coated steel strip (tinplate) is a major outlet for tin. Present production of tinplate for the container industries is about 7 million tons. The United States produces over 46 billion tinned steel containers annually of which 60% are food cans and 40% non-food cans. Tinplate manufacture is now largely a continuous electrolytic process with only a small percentage of production in hot-tinning machines. The coating thickness may be less than 0.0001 in. Heavily coated tinned steel sheet is used in making gas meters and automotive parts such as filters and air coolers.

The electrical industry is a large user of tin coated steel and copper in the form of connectors, capacitor and condenser cans and tinned copper wire. Many kinds of food-handling machinery, including holding tanks, mixers, separators, milk cans, pipes and valves are made of tinned steel, cast iron, copper or brass. The tin coating may be applied by dipping in molten tin or by electrodeposition.

Electrodeposited tin-alloy coatings (tin-copper, tin-cadmium, tin-lead, tin-zinc and tin-nickel) have advantages over single metal plates. They are denser and harder, more corrosion resistant, brighter or more easily buffed, and more protective for the base metal. Tin-copper coatings serve as an attractive finish for jewelry, handbag frames, wire goods and hardware. Tin-lead coatings and tin-zinc coatings have excellent corrosion resistance and solderability and find use as a plating for printed circuits and electronic parts, radio and television chassis. Commercial applications for the tin-nickel (66% tin) coating include watch parts, surgical and scientific instruments, household appliances, and electrical fittings and components.

Alloys

Tin alloys account for about one-half of the world's consumption of tin. These cover a wide composition range and many applications because tin alloys readily with nearly all metals.

Soft solders, used for the sealing and joining of metals, are one of the most widely used series of tin-containing alloys. Common solder is an alloy of tin and lead usually containing 20-70% tin. The eutectic alloy, with 63% tin, melts sharply at 361 F and is used extensively in the electrical industry. General-purpose solder, containing equal parts of tin and lead, has a melting range of 56 F. Plumber's solder, containing 30% tin, has a long pasty range which is useful in making wiped joints. Lower tin solders (15-20% tin) are used as dipping solders for sealing automotive radiator cores. Solders for special uses are alloys of tin, antimony, silver, indium and zinc. A 95 tin-5 antimony solder and a 97 tin-3 silver solder are used extensively for higher temperature service than is possible with tin-lead solders. Tin-zinc solders are used in soldering aluminum.

A combination of bismuth and cadmium with tin and lead gives a range of alloys with low melting point. The *fusible alloys* are used as fuses for fire-extinguishing equipment, boiler plugs, solders and seals, for tube bending and for tearable strip on can lids.

Copper-tin alloys, with or without modifying

elements such as zinc, lead or manganese, are classed under the general name of bronzes. The so-called phosphor bronzes, containing from 8-12% tin with small additions of phosphorus, added chiefly for deoxidation purposes, are used for springs, condenser tubes, wire screen, bearings and bushings and diaphragms. The gun metals, tin-bronze casting alloys modified with 1-6% zinc, are used for valves and fittings for water and steam lines, and for bearing backs. The leaded bronzes are excellent for heavy-duty bearings, steam valves and fittings where gas-free, pressuretight, corrosion-resistant castings are required. Many brasses contain 0.75-1.0% tin for additional corrosion resistance.

Babbitt bearing metals are essentially tin-base alloys containing 4-8% each of copper and antimony and the lead-base alloys, containing 12-18% antimony and 10-12% tin. These alloys can be cast easily to steel, cast iron and bronze backings for all types of bearing applications.

Aluminum-tin alloys have been developed with mechanical and physical characteristics which meet the requirements of bearing applications. The 6% *tin-aluminum* alloys can be cast or bonded to steel by rolling. The chief merit of the low-tin aluminum alloys is their high fatigue strength which enables them to carry higher fluctuating loads than either the tin-base or copper-lead alloys. Seizure resistance of aluminum-tin bearing alloys is reduced if the tin content is increased to 20%. A continuous network of tin is achieved by recrystallizing the aluminum matrix. The high tin-aluminum bearing alloys backed to steel are standard in some small foreign cars.

Modern *pewter* is a tin-base alloy with antimony and copper added to harden it. Ancient pewter contained lead and darkened readily. Modern pewter is tarnish resistant, contains about 7% antimony and 2% copper and is lead-free. Pewter metal is easy to cast and to work into complicated shapes by spinning and hammering. Pewter is readily available in the form of coffee and tea services, plates, trays, bowls, tankards, candelabra, pitchers and vases.

Type metals are lead-base alloys containing 3-15% tin and a somewhat larger proportion of antimony. Tin adds fluidity and gives a structure that reproduces fine detail.

Alloys of titanium containing tin and aluminum are being used industrially in the aircraft industry. They contain 2.5% tin with 5% aluminum or 13% tin with 2.75% aluminum. Their use depends on the fact that they are stable α-alloys at temperatures below the α-β transition temperature, have high strength and good creep resistance. A series of corrosion-resisting alloys developed commercially for use in water-cooled nuclear reactors are: "Zircaloy" 1, containing 2.5% tin; Zircaloy 2, containing 1.5% tin; and Zircaloy 3, containing 0.25% tin. Zircaloy 2 has the best strength and corrosion resistance.

Tin is a strong pearlitic stabilizer in both gray iron and nodular iron and a residual tin content of 0.05% is usually sufficient to render the matrix of these irons completely pearlitic. Additions of tin to the ladle assure an iron with uniform hardness values, improved wear resistance and resistance to heating at elevated temperatures. Among the applications that are suited to this development are pistons and piston rings, clutch plates, flywheels, brake drums, machine tool beds, cylinder liners, compressor bodies and cylinders. One large automobile manufacturer is using tin-treated cast iron for engine blocks.

TITANATES

The high dielectric constants, high refractive indices, and ferroelectric properties of titanates contribute primarily to their commercial importance.

Ferroelectricity may be described as the electric analog of ferromagnetism. As a field is applied to a ferroelectric material, a nonlinear relationship between polarization and field (similar to the magnetization curve for iron) is observed. This increase in polarization is a function of the orientation of ferroelectric domains within the crystal.

As these domains become aligned, a saturation point is reached. If the field is now removed, the domains tend to remain aligned and a finite value of polarization (called remanent polarization) can be measured. Extrapolation of the polarization at high field strength to zero field gives a somewhat higher value (spontaneous polarization).

In order to eliminate the remanent polarization, the field must be applied in the opposite direction, and the field required to return to the original state is called the coercive field. On further increase in the electric field, polarization in the opposite direction is achieved. This behavior leads to a characteristic ferroelectric hysteresis loop as the field is alternated.

Preparation: titanium dioxide

Titanates, for the most part, are prepared by heating a mixture of the specific oxide or carbonate with titanium dioxide. Titanium dioxide has an exceptionally high refractive index for a white oxide (2.6 to 2.9 for the rutile form, 2.5 for anatase) and due to its high refractive index finds wide application as a white pigment of high reflectance for the opacification of paint, plastics, rubber, paper and porcelain enamels.

For electronics, polycrystalline (ceramic) titanium dioxide with its moderately high dielectric constant (ca. 95) has been used as a capacitor; it does not show ferroelectric behavior. Titanium dioxide is normally an insulator, but in the oxygen deficient state where some of the Ti^{4+} sites are occupied by Ti^{3+} ions, it becomes an "n-type" semiconductor with conductivities in the range 1 to 10 (ohm-cm)$^{-1}$. A similar semiconducting behavior is found when ions with a valence greater than four (e.g., Nb^{5+}, Ta^{5+}, W^{6+}) are substituted for Ti^{4+}; to preserve electroneutrality, some of the titanium must exist as Ti^{3+} ions.

Titanium dioxide is commonly prepared by solution of ilmenite ($FeTiO_3$) ore in sulfuric acid, crystallization of the ferrous sulfate and hydrolysis of the resulting titanyl sulfate solution to hydrated titanium oxide. Calcination of this hydrate at temperatures in the neighborhood of 900 C yields titanium dioxide. Ceramic components are readily fabricated by dry pressing or slip casting powdered titanium dioxide in a suitable mold, followed by firing at elevated temperatures to yield a dense sintered compact. As a ceramic, titanium dioxide normally exists as rutile, the modification stable at high temperature.

Rutile single crystals have been grown by the flame fusion technique. The high refractive index and high dispersion have led to the use of rutile as a gem stone of high brilliancy and fire. Rutile transmits from a wavelength of about 0.42 to about 5.8μ and has been used in conjunction with a photoconductor such as lead sulfide as a lens for the concentration and detection of infrared radiation.

Small amounts of impurities may be readily incorporated in the crystal during growth, and the use of rutile modified in this way as a maser crystal has been suggested.

As a dielectric, rutile crystal shows a dielectric constant parallel to the c-axis of about 190 with a loss tangent of about 0.017.

Titanates are usually prepared by heating a mixture of the specific oxide (or carbonate) and titanium dioxide according to the scheme:

$$\left.\begin{array}{r} MO \\ or\ MCO_3 \end{array}\right\} + TiO_2 \longrightarrow \left\{\begin{array}{l} MTiO_3 \\ or\ MTiO_3 + CO_2 \end{array}\right.$$

The specific temperature required will depend on the particular titanate being prepared, but as a general rule, temperatures of 1000 C or higher are required. As with titanium dioxide, the titanates prepared in this way may be dry-pressed or slip-cast to a desired shape and fired at elevated temperatures to yield a dense sintered compact.

Barium titanate

Reportedly, five different compounds exist between BaO and TiO_2, including Ba_2TiO_4, $BaTiO_3$, $BaTi_2O_5$, $BaTi_3O_7$ and $BaTi_4O_9$. Of these, the metatitanate, $BaTiO_3$, is of primary interest and the term "barium titanate" in the following discussion will refer to the metatitanate.

Above 120 C barium titanate exists as a cubic crystal (perovskite structure), but on cooling below this temperature the structure changes to tetragonal. On further cooling the structure shifts again, at about 0 C, to orthorhombic and, at −90 C, to monoclinic. A fifth (hexagonal) modification is stable at high temperatures, between about 1460 C and the melting point (1612 C).

At room temperature ceramic barium titanate (tetragonal) has a very high dielectric constant (in the order of 1700). The dielectric constant shows an extremely high sharp maximum (in the order of 8,000) at 120 C (tetragonal to cubic transition) with smaller, less well-defined maxima at the two lower temperature transitions. The power factor of barium titanate at room temperature is of the order of one percent and its resistivity is about 10^{12} ohm-cm.

The exact value of the dielectric constant of barium titanate is strongly influenced by the method of fabrication, impurities, stoichiometry of the compound, firing temperature and time, and oxidation state.

The variation in dielectric constant with temperature is a disadvantage where a minimum temperature dependence is required. Modification of barium titanate with various additions results in capacitors in which the peak in the dielectric constant vs. temperature curve has been flattened so that the temperature dependence is considerably less pronounced.

Another modification is obtained by the addition of strontium titanate to form a barium-strontium titanate solid solution with a Curie temperature below room temperature so that a relatively low loss capacitor with a negative temperature coefficient is obtained. The decrease in the Curie temperature is about 4°C for each mole percent of $SrTiO_3$ substituted for $BaTiO_3$.

The nonlinearity of the dielectric constant of barium titanate with applied field suggested its use in dielectric amplifiers. Because of the hysteresis loss, barium titanate is usually modified with strontium titanate to yield a material with a Curie temperature slightly below room temperature but which still behaves like a nonlinear dielectric.

Barium titanate is also piezoelectric, i.e., electric polarization results from the application of a mechanical stress and vice versa, and as a true piezoelectric compound there is a one-to-one correspondence between the direct and inverse effect. Ceramic barium titanate can be made piezoelectric by applying a polarizing field of about 30 kv per cm at room temperatures. The remanent polarization after removal of the polarizing voltage is permanent unless the material is overheated or subjected to high reverse voltages.

The advantages of ceramic barium titanate as a transducer lie in its mechanical strength, chemical durability and ease of fabrication into virtually any shape desired. Barium titanate transducers, as ultrasonic generators, are used in various applications (emulsification, mixing, cleaning, drilling); other applications include such things as phonograph pickups and accelerometers.

In order to raise the operating temperature and to avoid heating of the transducer (either from external or internal sources of heat) to a temperature above its Curie point, the addition of lead titanate (normally about five percent which raises the Curie temperature about 15°C)

as a solid solution may be used. While the addition of lead titanate lowers the piezo response slightly, its presence increases the ability of the material to resist depolarization, i.e., it stabilizes the remanent polarization.

By substitution of a small amount of lanthanum for barium in barium titanate a thermistor with a large positive temperature coefficient of resistance (12 to 20%/°C) in the neighborhood of the Curie point is obtained. The range of temperatures over which this rapid increase in resistance takes place may be shifted toward lower temperatures by the substitution of strontium for part of the barium in order to lower the Curie temperature.

Barium titanate has been grown as a single crystal by the Stockbarger technique, by flame fusion (with the addition of a small amount of $SrTiO_3$ to prevent the formation of a hexagonal crystal) and by the slow cooling of a potassium fluoride-barium titanate melt. The last technique has apparently been the most successful to date and crystals of this type show a reasonably rectangular hysteresis loop which suggests their application in memory circuits. Drawbacks to their use in such an application appear to be the fields required for switching, switching times and aging effects which tend to reduce the reliability of the system.

Calcium titanate

Calcium titanate ($CaTiO_3$) is orthorhombic (although nearly cubic) and occurs in nature as the mineral perovskite. As a ceramic it has a room temperature dielectric constant of about 160. It is frequently used as an addition to $BaTiO_3$ or by itself as a temperature compensating capacitor.

Single crystals have been grown by the flame fusion technique; calcium titanate crystals show a strong tendency toward twinning and, although the material is not ferroelectric, the twinning in the crystal shows a marked resemblance to the domain structure observed in barium titanate crystals. By rather careful heat treatment during and after growth, crystals free from twinning may be obtained. The crystals show a room temperature dielectric constant of about 200 with a loss tangent of about 0.001. The crystals transmit over a wavelength region of 0.365 to 5.5μ.

Strontium titanate

Strontium titanate ($SrTiO_3$) has a cubic perovskite structure at room temperature. At about −90 C, on cooling, it changes from cubic to tetragonal but the latter form does not seem to be ferroelectric. It has a dielectric constant of about 230 as a ceramic, and it is commonly used, as indicated earlier, as an additive to barium titanate to decrease the Curie temperature. By itself it is used as a temperature compensating material with even greater negative temperature characteristics than calcium titanate.

Single crystals of strontium titanate have been grown by the flame fusion process. Strontium titanate is essentially colorless and because of its high index of refraction ($n_D = 2.41$) and high dispersion ($n_F - n_C = 0.108$) has been utilized as a gem stone ("Fabulite") with considerable brilliance and fire. Strontium titanate transmits from about 0.395 to 6.2μ; this transmission together with its high index makes it usable as a lens for the concentration of and, in conjunction with a photoconductive lead sulfide film, detection of infrared radiation. Because of its cubic structure, it is somewhat more satisfactory as an optical material than rutile (tetragonal). Strontium titanate crystals show a room temperature dielectric constant of about 300 with a loss tangent of 0.0003.

Magnesium titanate

As a result of the relatively small size of the magnesium ion, magnesium titanate ($MgTiO_3$) crystallizes as an ilmenite rather than a perovskite structure. Ilmenite itself is ferrous titanate ($FeTiO_3$).

Magnesium titanate is not ferroelectric and is used with titanium dioxide to form temperature compensating capacitors. It has also been used as an addition agent to barium titanate.

Magnesium metatitanate does not appear to have been grown as a single crystal although the orthotitanate (Mg_2TiO_4, spinel structure) apparently has.

Lead titanate

Lead titanate ($PbTiO_3$) crystallizes in the perovskite structure and exists as a tetragonal crystal at room temperature. The structure shifts to cubic at 490 C (Curie temperature). Good ceramic specimens of lead titanate are somewhat difficult to prepare owing to the volatility of lead oxide at the firing temperature. A dielectric constant of the order of 100 with a sharp peak of about 1000 at the Curie temperature has been measured. As indicated earlier, lead titanate is commonly used as a solid solution additive to increase the Curie temperature of barium titanate.

By substitution of zirconium for titanium in lead titanate a solid solution, lead zirconium titanate, may be produced. Volatility of the lead oxide is minimized by firing in a lead oxide atmosphere. Compositions of the ratio of about 55 $PbZrO_3$:45 $PbTiO_3$ have room temperature dielectric constants as high as 1300, are piezoelectric up to about 350 C and have a high coupling coefficient (measure of the conversion of mechanical to electrical energy).

Miscellaneous titanates

The metatitanates of cadmium, manganese, iron, nickel and cobalt all have an ilmenite rather than perovskite structure. None of these, as far as is known, is ferroelectric and they are not particularly important electrically other than perhaps as addition agents to barium titanate. Nickel titanate has been grown as a single crystal, and its use as a rectifier has been suggested after the addition of suitable impurities during growth.

TITANIUM AND TITANIUM ALLOYS

Although the basic metallic element, titanium, was discovered over 150 years ago, and is present in abundance in the earth's crust (of the structural metallic elements, only iron, aluminum, and magnesium are more abundant), it has not been until the last decade that significant amounts of this metal and its alloys have been produced commercially. The initial slow development of titanium as a production material was due to the inability of separating it from its ores in a relatively pure and ductile state.

However, in the 1930's, W. J. Kroll developed a process which successfully produced titanium of a high degree of purity and ductility. The Kroll process involves the reduction of titanium tetrachloride (produced by reaction of TiO_2 with chlorine at high temperatures) with a highly reactive metal such as magnesium or sodium in a protective atmosphere of argon or helium at relatively low pressures. The U. S. Bureau of Mines further expanded the Kroll process to a pilot plant scale and demonstrated its feasibility as a commercial process for producing titanium sponge in the 1940's. The resultant product is relatively porous and spongy, from which is derived the term "titanium sponge."

From the late 1940's until the present, great strides have been made in the production of high-purity sponge on a production basis and from it, the melting of titanium ingots. The initial laboratory melts were on the order of a few grams, whereas today 9000-lb ingots are being produced, with larger ingots a future prospect.

Melting. Due to the fact that titanium has a great affinity for oxygen, nitrogen, and hydrogen at elevated temperatures, particularly when molten, all melting operations must be conducted in a vacuum and/or an inert-gas atmosphere. To avoid contamination resulting from the use of a nonconsumable electrode material, such as tungsten or carbon, and to improve chemical homogeneity, practically all of the melting today is done by the "consumable-electrode double melting process." Briefly, this process involves the following steps:

1. Titanium sponge of a uniform particle size and alloying elements (when required) are compacted or briquetted.
2. The compacts or briquettes are joined together by welding to form the desired size initial electrode.
3. The initial electrode is then melted to form the first melt of a predetermined size.
4. Generally, two first melt ingots are joined together and then remelted in a mold of larger diameter to form the final ingot.

Because all known refractory materials will react with titanium, virtually all commercial melting is done in a water-cooled copper receptacle which acts as both the melting chamber and mold. No pouring of hot metal takes place in the production of titanium ingots.

Structure and properties

Titanium is allotropic in that it possesses two distinct crystallographic structures, alpha and beta. Alpha (close-packed hexagonal structure), which is the low-temperature stable constituent, transforms to beta (body-centered cubic structure) at 1625 F in the absence of metallic alloying elements. The fundamental concept in alloying

titanium is that the transformation temperature, % beta present at lower temperatures, and concomitant changes in properties can be controlled by the addition of alloying elements such as Al, Mn, V, Mo, Cr, Fe, Sn and Zr. Although seldom purposely added, O_2, N_2, H_2, and C are generally present as residuals and also have a significant effect on the mechanical properties of titanium, particularly at room temperature.

Of these elements, Al, O_2, N_2, and C are alpha stabilizers; Sn and Zr are more or less neutral in the range of percentages normally added to titanium, and the remainder are beta stabilizers.

Titanium-base alloys are classified into three main categories: alpha, beta and alpha + beta. The main characteristics of each of these alloy groups are as follows:

Alpha alloys. Weldable, good stability up to 1000 F, strong and relatively tough at low temperatures down to −423 F, non-heat-treatable and somewhat more difficult to form at room temperature than the other two alloy groups.

Beta alloys. Weldable, good stability up to 600 F, strong at higher temperatures for short times, brittle at temperatures below −100 F, heat-treatable and very formable at room temperatures.

Alpha + beta alloys. Depending on specific alloy, weldable or nonweldable, good stability up to 800 F, strong and relatively tough at low temperatures down to −320 F, heat-treatable and generally more formable than alpha alloys.

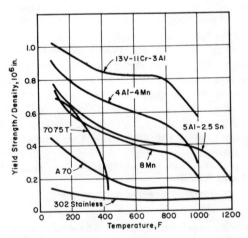

Fig. 1. Effect of temperature on strength-to-weight ratio of several titanium alloys as compared with aluminum and stainless steel.

TABLE 1—TITANIUM ALLOYS, FORMS AND CHARACTERISTICS

Composition—Type	Yld Str, 0.2% offset, 1000 psi, min.	Forms Available						General Characteristics
		Billet	Bar	Sheet	Strip	Plate	Wire	
Commercially Pure (alpha)	40	√	√	√	√	√	√	For applications requiring maximum ductility and formability and general corrosion resistance.
Commercially Pure (alpha)	55	√	√	√	√	√	√	General-purpose grade for moderate to severe forming and general corrosion resistance.
Commercially Pure (alpha)	70	√	√	√	√	√	√	For higher strength applications such as moderately stressed aircraft parts and general corrosion resistance.
5% Al, 2.5% Tin (alpha)	110	√	√	√	√	√	√	Combines high strength at elevated temperatures with excellent welding characteristics.
8% Mn (alpha-beta)	110	—	—	√	√	√	—	Good elevated-temperature strength and good bend ductility.
4% Al, 4% Mn (alpha-beta)	130	√	√	—	—	√	√	Excellent high strength/weight ratio, creep resistance, stability and good forgeability.
6% Al, 4% V (alpha-beta)	120	√	√	√	√	√	√	Excellent high-strength/weight ratio, creep resistance as well as good low-temperature impact.
7% Al, 4% Mo (alpha-beta)	135	√	√	—	—	√	√	Heat-treatable type providing higher room and elevated temperature strength than other alpha-beta types.
13% V, 11% Cr, 3% Al (beta)	120	√	√	√	√	√	√	Heat-treatable type providing highest strength combined with optimum ductility and formability.
4% Al, 3% Mo, 1% V (alpha-beta)	115	—	—	√	√	√	—	Heat-treatable providing high strength with good ductility and formability.

Table 1 lists most of the titanium and titanium alloys currently available, together with the general characteristics of each grade.

Titanium has a number of characteristics which, individually or in combination, make it useful for certain industrial applications. To date, the physical characteristic of titanium which has led to its most extensive use has been its relatively low density in conjunction with high strength in the intermediate temperature range of 200 to 1000 F, as illustrated in Fig 1.

This unique combination of properties has made titanium and its alloys a major structural material in the aircraft industry and has evolved considerable interest in the missile industry.

Other properties of particular interest are the high melting point, low thermal conductivity, low coefficient of expansion, and high electrical resistivity. Table 2 presents the physical properties of commercially pure titanium.

Corrosion resistance and commercial applications

By far, the greater majority of titanium production during the past ten years has gone into air frames and jet engines. Recently, it has been used more and more extensively in nonmilitary applications particularly because of its excellent

TABLE 2—PHYSICAL PROPERTIES OF COMMERCIALLY
PURE TITANIUM

Melting Pt, F	3135
Density, lb/cu in.	0.1628
Atomic Number	22
Atomic Weight	47.90
Crystal Structure Type	H.C.P.
	below 1625 F
	B.C.C.
	above 1625 F
Lattice Constant	2.9505
	4.6833
Ther Cond, Btu/ft^2/hr/°F/in.	105
Ther Exp, 10^{-6} in./in./°F,	
32–212 F	5
Sp Ht (68 F), Btu/lb/°F	0.130
Elec Cond, % IACS	3.1
Elec Res (68 F), ohms/cir mil ft	370
Tensile Mod of Elast, 10^6 psi	15.5
Torsional Mod of	
Elast, 10^6 psi	6.5
Poisson's Ratio	0.34

corrosion resistance. The inertness of titanium can be attributed to a layer of oxygen which is absorbed upon exposure of air. Although the film can be mechanically damaged, it is readily repaired by absorption in natural media, in oxidizing acids and in most organic acids. Hence, titanium exhibits outstanding resistance to nitric acid, chloride compounds, urea and sea water in addition to many other media.

This protective oxide film generally is not stable in strong reducing acids such as hydrochloric, hydrofluoric, sulfuric, oxalic, trichloroacetic, and in acids containing fluoride ions. However, corrosion of titanium in reducing atmospheres is markedly inhibited when oxidizing agents or metal ions are present or when an *anodic* current is applied to the titanium. Alloying titanium with as little as 0.1% palladium or platinum also inhibits corrosion in reducing media.

Examples of these inhibiting techniques are listed below:

Environment	Temp, F	Remarks	Corrosion Rate
20% HCl	70		50.0 mils/yr
20% HCl	70	+0.2 gm/1 potassium dichromatic	0.1 mils/yr
15% H$_2$SO$_4$	95		50.0 mils/yr
15% H$_2$SO$_4$	95	Impressed current on Ti	Nil
5% HCl	212		1120.0 mils/yr
5% HCl	212	Ti + 0.2% Pd	7.0 mils/yr

The major portion of titanium's commercial applications have been in the chemical processing industries in the form of reactors, vessels and heat exchangers. The pulp and paper industry has used it in bleaching equipment primarily as chlorine dioxide mixers, while the electrochemical industry has utilized heating and cooling (tubing) coils and anodizing and plating racks. Pumps, valves, thermowells and other miscellaneous items are additional examples of commercial applications of titanium.

TOOL AND DIE STEELS

Tool and die steels might best be defined as special steels which have been developed to form, cut or otherwise change the shape of a material into a finished or semifinished product. Because of the performance demanded of tool steels, only steel which has been melted and processed to rigorous procedures has found acceptance in industry. Quality tool steels of the past and present have been melted in electric-arc furnaces. The latest improvement in the quality of tool steels has come through vacuum melting. Although presently employed only to a limited degree, this method of melting offers exceptionally clean, gas-free steels.

Tool-steel classification

A great variety of tool steels are needed to meet the numerous conditions and special requirements in present day manufacturing. Certain operations require wear resistance, others require toughness, and still others require hot hardness or combinations of these three properties to varying degrees. While these three fundamental properties are common to all tool steels, they are not complementary. The degree to which they are present depends upon the chemical composition and heat treatment applied. Consequently, numerous compositions have been developed so that optimum properties can be supplied for specific applications.

Brand names have been assigned to each steel by the manufacturer. With each producer manufacturing as many as a hundred different tool steels, the brand names, although associated with steel quality, have become numerous and confusing. Recognizing this situation, the American Iron and Steel Institute and the Society of Automotive Engineers jointly devised a classification chart which has received wide acceptance.

An abbreviated form of this chart is given in Table 1.

Basically, excluding high-speed steels, there

TABLE 1—CHEMICAL COMPOSITION OF THE MOST POPULAR TOOL STEELS [a]

Steel	Nominal Alloying Elements, %							
	C	Mn	Si	Cr	W	Mo	V	Ni
W1	[b]	—	—	—	—	—	—	—
W2	[b]	—	—	—	—	—	0.25	—
P6	0.10 max.	—	—	1.50	—	—	—	3.50
P20	0.30	—	—	0.75	—	0.25 (pre-treated)	—	—
L3	1.00	—	—	1.50	—	—	0.20	—
L6	0.70	—	—	0.75	—	0.25 (opt)	—	1.50
S4	0.55	0.80	2.00	—	—	—	—	—
O1	0.90	1.00		0.50	0.50	—	—	—
A2	1.00	—		5.00	—	1.00	—	—
D2	1.50	—		12.00	—	1.00	—	—
H11	0.35	—		5.00	—	1.50	—	—
H21	0.35	—		3.50	9.00	—	—	—

[a] Excluding high-speed steels.
[b] Varying carbon content, usually offered in 0.90 to 1.10% carbon range.

are six general classes of tool steels listed on the chart. These are: water hardening, class "W"; special purpose, class "L"; mold steels, class "P"; shock resisting, class "S"; cold-work die, class "O," "A," and "D", and hot-work die, class "H."

Principal characteristics and uses of popular tool steels

Water hardening (W-1, W-2). Water hardening or carbon tool steels were the forerunner of all tool steels and are still widely used in industry today. They have good edge strength and wear resistance if not heated above 350 F in service; consequently, inexpensive taps, drills, reamers, chasers, etc., are made from water-hardening tool steels. In section sizes over about ½ in., these steels develop a tough core and hard case. This combination is particularly desirable for many cold-working applications.

Inherent disadvantages are: susceptibility to cracking during hardening, particularly when a complicated shape is encountered; low heat resistance; poor predictability of dimensional change in hardening and poor hardenability for applications requiring through hardening.

Special purpose (L-3, L-6). Special-purpose tool steels have low alloy content and are used in a wide variety of applications that do not necessarily fall within the other classifications. Type L-3 is universally used as a bearing steel. It is further used in cold-work applications requiring a tough core and a deeper case than can be obtained with W-1.

Type L-6 is a deeper hardening steel than L-3. This grade is widely used for pins, jigs, shims, hand stamps and in any application requiring a machinery steel capable of heat treatment to high hardness.

Mold steels (P-6, P-20). Mold steels were developed and are used almost exclusively for molding plastics and zinc. Two prerequisites for a mold steel are: (1) machinability and (2) the ability to be polished to a high luster.

Types P-6 and P-20 are the most popular in the manufacture of machined cavities. Type P-6, which is easily carburized and has high hardenability, is used generally for large, long-run molds. P-20 finds greater application for short runs.

Shock resisting (S-4). The shock-resisting tool steels were developed for applications in which toughness and strength are of prime importance, while hardness and wear resistance are of lesser importance. Although these steels vary considerably in composition, the S-4, silicon-manganese type, is the most widely used. These steels are particularly suitable for pneumatic tools, hand chisels, track chisels, cold cutters, punches and heavy-duty shear blades.

Cold work (O-1, A-2, D-2). Cold-work tool steels are used in manufacturing operations in which wear resistance and nondeformation in heat treatment are of greatest importance. Steels in this classification are among the most important and widely used of all tool steels. Of the cold-work steels, types O-1, A-2 and D-2 are most frequently employed.

Type "O" (oil hardening) is the least expensive of the three, and is widely used. However, intricate shapes and dies with close tolerances are generally made from Type "A" steels which are air-hardening and less apt to distort in heat treatment. When greater wear resistance is required, Type "D" steels, which are also air-hardening, are employed.

The cold-work tool steels are adapted to a wide variety of tools and dies. Examples are blanking, forming, coining and trimming dies, thread rolling dies, punching, cold-forming rolls, gauges, etc.

Hot work (H-11, H-21). Hot-work die steels, as the name implies, are used in manufacturing operations involving the forming, shearing or punching of metals at comparatively high temperatures. Whereas low-alloy tool steels tend to soften and wash out when subjected to combined heat, pressure and abrasion, hot-work steels give much longer life. The most widely used are H-11 and H-21. Typical applications for these steels are forging dies, hot heading punches, extrusion dies, hot-drawing dies, sheer blades, hot cutoff tools, gripper dies and die casting dies.

Properties

Toughness. Toughness in *tool steels* is best defined as the ability of a material to absorb energy without fracturing rather than the ability to deform plastically without breaking. Thus, a high elastic limit is required for best performance since large degrees of flow or deformation are rarely permissible in fine tools or dies. Hardness of a tool has considerable bearing on the toughness since the elastic limit increases with an increase in hardness. However, at very high hardness levels, increased notch sensitivity and brittleness are limiting factors.

In general, lower carbon tool steels are tougher than higher carbon tool steels. However, shallow hardening carbon (W-1) or carbon-vanadium (W-2) tool steels with a hard case and soft core will have good toughness regardless of carbon content. The higher alloy steels will range between good and poor toughness depending upon hardness and alloy content.

Abrasion resistance. Some tool steels exhibit better resistance to abrasion than others. Attempts to measure absolute abrasion resistance are not always consistent but, in general, abrasion resistance increases as the carbon and alloy contents increase. Carbon is an influential factor. Additions of certain alloying elements (chromium, tungsten, molybdenum and vanadium) balanced with carbon have a marked effect on increasing the abrasion resistance by forming extremely hard carbides.

Hardness. Maximum attainable hardness is primarily dependent upon the carbon content except possibly in the more highly alloyed tool steels. Tool steels are generally used somewhat below maximum hardness except for deep-drawing dies, forming dies, cutting tools, etc. Battering or impact tools are put in service at moderate hardness levels for improved toughness.

Hot hardness. The ability to retain hardness with increasing temperature is defined as hot hardness or red hardness. This characteristic is important in steels used for hot-working dies. Generally, as the alloy content of the steel is increased (particularly in chromium, tungsten, cobalt, molybdenum and vanadium which form stable carbides), the resistance to softening at elevated temperatures is improved. High-alloy tool steels with a properly balanced composition will retain high hardness up to 1100 F. In the absence of other data, hardness after high-temperature tempering will indicate the hot hardness of a particular alloy.

Heat treatment

Hardenability. Carbon tool steels are classified as shallow hardening, i.e., when quenched in water from the hardening (austenitizing) temperature, they form a hardened case and a soft core. Increasing the alloy content increases the hardenability or depth of hardening of the case. A small increase in alloy content will result in a steel which will harden through the cross sections when quenched in oil. If the increase in alloy content is great enough, the steels will harden throughout when quenched in still air. For large tool or die sections, a high-alloy tool steel should be selected if strength is to be developed throughout the section in the finished part.

For carbon tool steels which are very shallow in hardening characteristics, the P/F test, Disc test, and PV test are methods for rating this characteristic. Oil-hardening tool steels of medium-alloy content are generally rated for hardenability by the Jominy End Quench test.

Dimensional changes during heat treatment. Carbon tool steels are apt to distort because of the severity of the water quench required. In general, water-hardening steels distort more than oil hardening, and oil hardening more than air hardening steels. Thus, if a tool or die is to be machined very close to final size before heat treatment and little or no grinding is to be performed after it, an air-hardening tool steel would be the proper selection.

Resistance to decarburization. During heat treatment, steels containing large amounts of silicon, molybdenum and cobalt tend to lose carbon from the surface more rapidly than steels containing other alloying elements. Steels with extremely high carbon content are also susceptible to rapid decarburization. Extra precaution should be employed to provide a neutral atmosphere when heat-treating these steels. Otherwise, danger of cracking during hardening will be present. Also, it would be necessary to allow a liberal grinding allowance for clean-up after heat treatment.

Machinability

Since most tool steels, even in the annealed state, contain wear-resistant carbides, they are generally more difficult to machine than the open-hearth grades or low-alloy steels. In general, the machinability tends to decrease with increasing alloying content. Microstructure also has a marked effect on machinability. For best machinability, a spheroidal microstructure is preferred over pearlitic.

The addition of small amounts of lead or sulfur to the steels to improve machinability has gained considerable acceptance in the tool-steel industry. These free machining steels not only machine easier but give a better surface finish than the regular grades. However, some caution is advised in applications involving transverse loading since lead or sulfur additions actually add longitudinal inclusions in the steel.

Available forms

Tool steels are available in billets, bars, rods, sheets and coils. Special shapes can be furnished upon request. Generally the material is furnished in the soft (or annealed) condition to facilitate machining. However, certain applications require that the steel be cold drawn or pre-hardened to a specified hardness.

A word of caution: Mill decarburization is generally present on all steel except that guaranteed by the producer to be decarburization-free. It is important that all decarburized areas be removed prior to heat treating or the tool or

die may crack during hardening.

TUNGSTEN AND ALLOYS

Tungsten is best known for its high melting point and excellent high-temperature strength. Until recently, the pure metal in the form of wire and strip has been used primarily in lamp, electron tube, and electrical contact applications. But, the severe high-temperature requirements for present and future space-vehicle components has generated much interest in tungsten sheet and forgings for structural application. The density of the metal has led to development and application as the base metal in high-density components for aircraft control systems and high-temperature nuclear shielding. Tungsten powder is also used as the base for the production of tungsten-carbide tools.

The conventional production of tungsten wrought metal is by the working of sintered-powder metallurgy billets, but the development of vacuum arc-cast and electron-beam cast ingots indicates that future commercial production of wrought tungsten and tungsten alloys will utilize the cast ingot as well as the sintered-powder metallurgy billet.

The oxidation resistance of tungsten, although better than molybdenum, is not good enough to permit elevated temperature use without protection from the atmosphere with the exception of short-time exposure of massive sections. The oxide produced by exposure to the air will not effectively volatilize until a temperature of approximately 900 C.

Tungsten exhibits the highest ductile to brittle transition temperature of the body-centered cubic system. The high transition temperature makes it necessary to conduct most forming operations at 700 C or over.

Fabrication

Full-density wrought tungsten can be hot-forged, swaged, extruded, rolled and drawn. Working temperature is usually 1100 C or above depending upon the grain size and type of deformation.

Sintered billets are forged, swaged or rolled initially at temperatures in excess of 1400 C. Working temperature can be progressively lowered as the amount of work increases, but consideration must be given to equipment capacity because of the high strength of tungsten.

Heat treatment

Heat treatment of wrought forms of tungsten to provide increased ductility just above and below the transition temperature is still an art and recommendations for stress-relieving treatments will vary from temperatures of 1000 to 1350 C and at even higher temperatures for very short exposure. Any heat treatment should be preferably conducted in pure dry hydrogen, argon, helium, or a vacuum of less than 1 μ.

Properties

Typical properties of tungsten are shown in Table 1.

TABLE 1—PROPERTIES OF TUNGSTEN

Form	Test Temp, C	Ten Str, 1000 psi	Elong, %
MECHANICAL			
Rod, Swaged			
and Recryst	350	60	48
	450	53	44
	900	43	40
	1370	32	—
	1650	19	—
	1980	9.8	—
Sheet, Cold-Worked			
0.010 in.	25	300	—
0.020 in.	25	200	—

PHYSICAL

Mod of Elast, 10^6 psi		
25 C	....	49–53
1000 C	....	35
Mod of Rigidity, 10^6 psi		
25 C	....	~ 20
727 C	....	10

Melt Pt, C	....	3410 ± 20
Boil Pt, C	....	5500

THERMAL

Vapor Press, mm Hg		
1227 C	....	2.1×10^{-18}
2227 C	....	1.3×10^{-7}
2727 C	....	6.8×10^{-5}
4727 C	....	9
Sp Ht, cl/gm/°C		
900 C	....	0.04
1800 C	....	0.045
Ther Cond, cal/sq cm/sec/°C/cm		
20 C	....	0.31
1027 C	....	0.271
1527 C	....	0.253
Coef of Ther Exp, 10^{-6}/°C		
20 C	....	4.4
1000 C	....	5.2

ELECTRICAL

Elec Res, μ-ohm-cm		
20 C	....	5.5
1227 C	....	39.5
1727 C	....	55.7
2727 C	....	90.4

Available forms

Tungsten metal is commercially supplied in the form of sintered billet, forged bar, swaged rod, rolled rod, drawn rod and wire, plate, sheet, strip and foil. Development work indicates the possibility of producing extruded, drawn, and welded tubular products. Sheet is commercially produced up to 24 in. wide and has been experimentally produced over 24 in. wide. Foil has been rolled down to 0.001 in. thick. Fine wires are drawn to 0.001 in. thick.

U

ULTRAHIGH STRENGTH STEELS

Arbitrarily, steels with tensile strengths of around 200,000 psi or higher are included in this category, and surprisingly, more than one hundred alloy steels now can be thus classified. They differ widely among themselves in composition and/or the way in which the ultrahigh strengths are achieved.

Medium carbon low alloy steels were the initial ultrahigh strength steels, and within this group a chromium-molybdenum (4130) grade and a chromium-nickel-molybdenum (4340) grade were the first. Later others of this type were developed. They have yield strengths as high as 240,000 psi and tensile strengths approaching 300,000 psi. They are particularly useful for thick sections because they are moderately priced and have high hardenability.

Several types of stainless steels are capable of strengths above 200,000 psi, including a number of martensitic, cold-rolled austenitic, and semiaustenitic grades. The typical martensitic grades are types 410, 420, and 431 as well as certain age-hardenable alloys. The cold rolled austenitic stainless steels work harden rapidly and can achieve 180,000 psi yield and 200,000 ultimate strength. The strength can also be increased by cold working at cryogenic temperatures. Semi-austenitic stainless steels can be heat treated for use at yield strengths as high as 220,000 psi and ultimate strengths of 235,000 psi.

Maraging steels, a relatively new family, contain 18 to 25% nickel plus and substantial amounts of cobalt and molybdenum. Some newer grades contain somewhat less than 10% nickel and between 10 and 14% chromium. Because of the low carbon (0.03% max.) and nickel content, maraging steels are martensitic in the annealed condition, but still readily formed, machined and welded. By a simple aging treatment at about 900 F, yield strengths of as high as 300,000 and 350,000 psi are attainable, depending on specific composition. In this condition, although ductility is fairly low, the material is still far from being brittle.

Two types of ultrahigh-strength, low carbon, hardenable steels have been developed in recent years. One, a chromium-nickel-molybdenum steel, named Astralloy, with 0.24% carbon is air hardened to a yield strength of 180,000 psi in heavy sections when it is normalized and tempered at 500 F. The other type is based on the iron-chromium-molybdenum-cobalt system and is strengthened by a precipitation hardening and aging to levels of up to 245,000 psi in yield strength.

Finally, high-alloy, quenched-and-tempered steels are another group that have extra high strengths. They contain 9% nickel, 4% cobalt and carbon ranging from 0.20 to 0.30% carbon and develop yield strengths close to 300,000 psi and ultimate strengths of 350,000 psi. Another group in this high alloy category resemble high speed tool steels, but are modified to eliminate excess carbide, thus considerably improving ductility. These so-called matrix steels contain tungsten, molybdenum, chromium, vanadium, cobalt and about 0.5% carbon. They can be heat treated to ultimate strengths of over 400,000 psi—the highest strength presently attainable in steels, except for heavily cold worked plain high carbon steel strip used for razor blades and drawn wire for musical instruments, both of which have tensile strengths as high as 600,000 psi.

UREA FORMALDEHYDE PLASTICS

The chemical reaction of urea and formaldehyde under certain carefully controlled conditions results in the formation of a clear, water white, thermosetting resin. Resins of this type are used in a wide variety of commercial applications where they impart strength, hardness, water resistance and durability within a broad color range. Applications include foundry resins, adhesives, textile resins, paper treating resins, coating resins, laminating resins and plastic molding compounds. This article is primarily concerned with urea formaldehyde molding compounds as engineering materials.

Chemistry

In the initial formation of the resins the mol

ratio of the reactants may be varied to produce certain variations in resin properties. Usually two or three mols of formaldehyde are reacted with one mol of urea first producing methylol ureas as indicated by the following equations:

$$NH_2—CO—NH_2 + HCHO \longrightarrow$$
$$NH_2—CO—NH—CH_2OH$$
$$NH_2—CO—NH_2 + 2HCHO \longrightarrow$$
$$CO(NH—CH_2OH)_2$$

These intermediates will condense under mildly acidic conditions to produce sirups which may be used directly in compounding or which may be dried to form a water-soluble white powder for shipment or storage.

Conversion to the final insoluble, infusible form occurs under conditions of acidity. The rate at which this change occurs is dependent upon temperature. Materials which will react with formaldehyde to produce acid under certain temperature conditions may be utilized to accomplish the final reaction rapidly at the desired time.

Molding compounds

In the preparation of molding compounds, alpha cellulose pulp or wood flour is first impregnated with the resinous sirup then carefully dried under controlled conditions and ground into a fine powder. This powder is compounded with pigments, lubricants, accelerators and other additives and densified into a solid mass to reduce bulk. The densified product is reground into the form of coarse granules and packaged in drums or bags for use by the molders who convert it into an endless variety of useful products. Typical properties are shown in Table 1.

Alpha cellulose-filled urea molding compound is supplied in an almost limitless range of light-fast colors and a wide range of translucency. These properties, combined with good mechanical strength, electrical characteristics, chemical resistance, hard, non-static surface, and relatively low cost, have made the material popular in the following typical applications: electrical wiring devices; closures; buttons; stove and cabinet hardware; bathroom accessories; lighting fixtures; radio, appliance and other housings; cosmetic and jewelry containers. A recently developed, highly fire-resistant type shows great promise in completely illuminated ceilings.

Wood flour-filled urea molding compounds offer the same mechanical, electrical and chemical properties in a more limited range of colors and translucency at lower cost. The colors are also lightfast. Typical applications include closures, wiring devices and electrical switchgear.

Processing

Both the alpha cellulose-filled and wood flour-filled materials are fabricated by compression molding in hot matched steel molds. Mold temperatures in the range 275 to 325 F and pressures of 2000 to 8000 psi on the projected area of the piece are commonly employed. In general, compression rather than transfer or plunger molding is recommended.

During the molding process, the material softens into a plastic mass which flows into every unrestricted area of the mold. This flow period is followed by the final chemical reaction which converts the material into a hard infusible form having a specific gravity of about 1.5. This complete molding process is accomplished in a matter of a few seconds to a few minutes depending on the thickness of the molded part.

Besides determining the final configuration and surface luster of the piece, the mold must serve to contain the material under the required pressure during the plastic phase in order to insure good density. Molds for urea are often constructed from oil hardening tool steel which are hardened to 54 to 56 Rockwell (C Scale). After polishing to the required finish they are sometimes chrome plated to increase mold life and insure the best possible flow characteristics.

Multicavity molds are customarily used on higher volume parts. For example, it is quite common to see 50 to 100 cavities in a 150-ton concerning these details. In building molds for the majority of applications, total shrinkage of 0.008 to 0.010 in. from cold-mold dimensions should be allowed.

Urea poses no unusual requirements for plastic molding presses. Most modern presses have closing speeds of 250 in. per min or more and closure press. The type of mold, whether positive, semipositive, flash or flash with loading well, will be determined by the shape of the part to be molded, the number of cavities to be constructed and the number of pieces to be run. Molds are heated by steam or electricity. A competent plastic toolmaker should be consulted

TABLE 1—PROPERTIES OF UREA MOLDING COMPOUNDS

	ASTM Test Method	Alpha Cellulose-Filled	Wood Flour-Filled
MOLDING PROPERTIES			
Bulk Factor	D392–38	2.2–3.0	2.2–2.5
Preformability		Excellent	Excellent
Compr Molding			
Temp, F		275–325	275–325
Pressure, psi		2,000–8,000	2,000–8,000
Mold Shrinkage, in./in.		0.006–0.014	0.006–0.014
PROPERTIES MOLDED			
Electrical			
Arc Res, Tungsten Electrodes, sec	D495–58T	110–130	110–130
Dielec Str, V/mil			
Short Time, $\frac{1}{8}$ in.	D149–55T	300–400	300–400
Step-by-Step, $\frac{1}{8}$ in.		250–300	250–300
Dielec Cons			
60 Cycles	D150–54T	7.0–9.5	7.0–9.5
10^6 Cycles	D150–54T	6.4–6.9	6.4–6.9
Dissip Factor			
60 Cycles	D150–54T	0.035–0.040	0.035–0.040
10^6 Cycles	D150–54T	0.028–0.032	0.028–0.032
Loss Factor			
60 Cycles	D150–54T	0.24–0.38	0.24–0.38
10^6 Cycles	D150–54T	0.18–0.22	0.18–0.22
Physical			
Sp Gr	D792–50	1.47–1.52	1.47–1.52
Ht Res, F Max		170	170
Defl Temp Under Load (Heat Distortion)			
264 psi, F	D648–56	270–280	270–280
Flamm	D635–56T	Nonburning	Nonburning
Water Asborption, %			
24 hr @ 25 C	D570–57T	0.5–0.7	0.8–1.1
Mechanical			
Impact Str, Izod ft-lb/in. of Notch	D256–56	0.25–0.35	0.25–0.35
Compr Str, psi	D695–54	25,000–35,000	25,000–35,000
Flex Str, psi	D790–58T	10,000–16,000	10,000–16,000
Mod of Elast, psi	D790–58T	$1.3–1.6 \times 10^6$	$1.3–1.6 \times 10^6$
Tens Str, psi	D638–58T	5,000–10,000	5,000–10,000
Rockwell Hardness	D785–51	M116–M120	M116–120
Chemical			
Effect of Light		Negligible	Negligible
Res to Organic Solvents, Oils and Greases		Excellent	Excellent
Res to Weak Acids and Alkalies		None to marked	
Res to Strong Acids and Alkalies		Poor	Poor

pressing speeds of 10 in./min or more. High volume parts are often molded on automatic presses of either rotary or flat bed design. In these the material is loaded into hoppers from which it is dispensed automatically into the mold cavities each cycle.

Urea preforms with ease on standard preform machines. This often represents the most economical way to measure the individual mold charge. Round ball preforms are frequently used with automatic molding equipment. A preform density of about 18 gm per cu in. is considered average.

Preheating the material before charging it into

the mold is not always necessary but is frequently done to eliminate entrapment of gas on heavy sections. The proper preheat temperature range is 145 to 165 F. This is accomplished most efficiently by high frequency dielectric preheating equipment. The material may be preheated in either preformed or granular form. When preheating granular compound on dielectric preheating equipment, it is customary to use linear polyethylene containers.

Urea molding compounds are furnished in a variety of plasticities and cure speeds to meet individual requirements. A small closure or button may be molded from a fast-curing, relatively stiff type while a large radio cabinet will require a soft, relatively slow-curing formulation. The former case may require a cure cycle of only 45 sec while the latter may require several minutes. It may be necessary to incorporate a "breathe" or degassing cycle a few seconds after the mold is closed, especially on heavy sections.

Certain tests have been developed to determine whether or not urea moldings have a commercially acceptable degree of cure. These include 15 min immersion in boiling water without chalking or gain in weight of more than 0.030 gm per sq in. of surface. In the case of heavy sections, baking in a circulating air oven at 140 F for 16 to 24 hr will detect internal porosity.

Very small parts such as buttons and closures are often deflashed and polished in tumbling barrels. Larger parts are either individually deflashed or placed in commercial machines where a blast of ground nutshell or plastic grit quickly removes all flash. Heavy grinding operations on flash lines may ultimately result in surface checking or crazing and are to be avoided.

It is generally recommended that unmolded urea formaldehyde molding compound be stored at temperatures below 70 F. At such temperatures plasticity changes are very slow and most types may be stored safely for many months. Low temperatures, even below freezing, have no effect on the material. However, prior to opening the container, the material should be allowed to reach approximately room temperature.

URETHANE MATERIALS

The term urethane at present refers to several distinct types of engineering materials, produced by essentially a common type of chemical reaction. Types available include urethane rigid and flexible foams, elastomers, coatings and adhesives.

Composition

In brief, the chemistry involved is the reaction of a diisocyanate with a hydroxyl-terminated polyester or polyether to form a higher molecular weight prepolymer which in turn is chain-extended by adding difunctional compounds containing active hydrogens, such as water, glycols, diamines or amino alcohols, etc. The use of water leads to the evolution of carbon dioxide and under proper conditions (catalyst, emulsifier, temperature, time, etc.) a stable flexible or rigid foam. The use of each of the above-mentioned components produces various types of linkages such as urethane, substituted ureas, etc. The systems using water can be degassed or water can be left out, in which case a solid elastomer having rather unique properties is formed.

In both cases described above, the systems can also be handled as a one-shot process, wherein all components are reacted simultaneously to produce the final product, either a foam or elastomer. The final form and the final properties desired dictate the initial components of a formulation, that is, the type of isocyanate and polyol used, as well as extender or blowing agent and catalyst to produce the various chemical linkages which in turn affect the physical properties.

The urethanes are block-polymers and one can think of them as being capable of being formed by literally an indeterminate number of combinations of the components listed above.

Urethane foams

The principal isocyanate used for both flexible and rigid foam at this time, is a mixture of the isomers of tolylene diisocyanate (80% 2,4-isomer, 20% 2,6-isomer). Various linear and branched polyols can be used such as polyesters of the diethylene glycol-adipic acid type as well as polyethers, notably the adducts of propylene oxide and glycerin, poly(oxypropylene)triols, having molecular weights of 400 to 5000.

Depending on the application and dictated by cost as well as properties, either of the two basic systems above is used. The cheaper polyether system predominates the flexible foam field today especially because of the cost, but also because of its improved hydrolytic stability and superior "comfort" characteristics as delineated in its compression-deflection curve.

In addition various catalysts such as tertiary amines ("Dabco," T. M. Houdry) and tin compounds (i.e., stannous octoate, Metal and Thermit Corp.) and stabilizers such as silicones (Dow-Corning, Union Carbide Chemical, General Electric) are used. Fluorinated hydrocarbons (e.g., trichlorofluoromethane) are also used to produce foams of low thermal conductivity or special properties.

Foams can be classified somewhat according to modulus, being flexible, semiflexible or semirigid as well as rigid. No sharp lines of demarcation have been set on these different classes as the gradation is continuous from the flexibles to the rigids.

Flexible foams. The techniques of manufacture of flexible urethane foam vary widely, from intermittent hand mixing to continuous machine operation, from prepolymer to one-shot techniques, from slab-forming to molding, from stuffing to foamed-in-place. Densities of foams range from about 1.0 lb per cu ft at the lightest to 4-5 lb/cu ft depending on the end use.

Applications of flexible foams are many, ranging from comfort cushioning of all types, e.g., mattresses, pillows, sofa seats, backs and arms, automobile topper pads, rug underlay and so on, to clothing interliners for warmth at light weight. Future applications envision the flexible foam not as a substitute for latex rubber foam or cotton, but as a new material of construction allowing for design of furniture, for instance, that is essentially all foam with a simple cloth cover and a very simple metal supporting framework (a Karpen Furniture-General Tire development).

Rigid foams. Densities from about 1.5 to 50 lb per cu ft on the semirigid side have been produced with corresponding compression strengths again for particular end uses ranging from insulation to fully supporting structural members. The usefulness of the urethane system has been in the foam-in-place principle using a host of containing wall materials.

Applications in the more rigid foam field have been thermal insulation of all types (low-temperature refrigeration ranging from liquid nitrogen temperatures up to the freezing point of water and high temperature insulation of steam pipes, oil lines, etc.); shock absorption such as packaging, crash pads, etc. where the higher hysteresis values produce either a better one-time high impact "crash" use or, more often, lower amplitude but higher frequency container end use; filtration (air, oil, etc. where a large surface to volume ratio is needed with a simple technique to produce a reusable filter to allow for its initially higher cost factor); structural (building applications of all kinds combining a good thermal as well as structural behavior, filling of building voids, and curtain walls are some basic applications); flotation (boats, buoys and every other imaginable object afloat represents some possible application using urethane foams) and finally, general-purpose applications which include all other uses such as decorative applications.

Rigid foams can be produced using a simple spray technique and a number of machines are sold on the market for this technique. Time-consuming lay-up of foam is eliminated using this method. Insulation of walls, tanks, etc., are applications in use today. With the use of low vapor pressure isocyanates such as MDI (4,4'-diphenylmethane diisocyanate), the potential irritant hazard during spraying is greatly lowered. Self-adhesion of the sprayed foam is a valuable asset of this type of system.

Urethane foams offer advantages over many of the better-known foams such as latex foam rubber, polystyrene, polyethylene with the combination of excellent properties and lower installed costs. Depending on the application, a lower foam density can be used with similar load-bearing properties, also one having an extremely low thermal conductivity can be fabricated. The oil resistance, high temperature resistance, good high tensile properties, good permanence properties, resistance to mildew, resistance to flammability, and so on are in general the types of properties which, combined with foamed-in-place technology, put urethane foam far ahead in use over competitive materials.

Urethane elastomers

Urethane solid elastomers are made with various isocyanates, the principal ones being TDI (tolylene diisocyanate) and MDI (4,4'-diphenylmethane diisocyanate) reacting with linear polyols of the polyester and polyether families. Various chain extenders such as glycols, water, di-amines or aminoalcohols, are used in either a prepolymer or one-shot type of system to form the long chain polymer.

Fabrication. Urethane elastomers can be further characterized by the method of fabrication of the final article. Three principal types of fabrication are possible: (1) casting technique wherein a liquid prepolymer or a liquid mixture of all initial components (one-shot) is cast into the final mold, allowed to "set" and harden and is then removed for final cure; (2) millable gum technique wherein conventional rubber methods and equipment are used to mill the gum, add fillers, color, etc., and/or banbury, extrude, calender and compression mold the final shaped item; (3) thermoplastic processing techniques wherein the resin can be calendered, extruded,

TABLE 1—TYPICAL PROPERTIES OF FLEXIBLE URETHANE FOAMS

	Polyester-Base	Polyether-Base
Density, lb/cu ft	1.5–6.0	1.0–3.0
Ten Str, psi	20–40	15–25
Elong, %	250–500	200–300
Tear Str, lb/in.	3–7	1.5–3
Compr Set (158 F), % of orig. thick.	2–15	2–10
Compr Defl, psi		
at 25%	0.55–0.85	0.15–0.50
at 50%	0.60–1.3	0.20–0.75
Resilience (falling ball), %	20–35	35–50
Hydrolysis Res	Fair to Good	Excellent
Oxidation Res	Excellent	Fair to Good
Ozone Res	Excellent	Fair to Good
Solvent and Chemical Res	Excellent	Good
Heat Res (284 F)	Excellent	Excellent
Flex Fatigue Res	Good	Good
Ther Conductivity, Btu/sq ft/hr/°F/in.	0.2–0.25	0.2–0.25
Flammability (with additive)	Self-extinguishing	Self-extinguishing
Dielec Const, 10^2–10^6 cps	1.0–1.3	1.0–1.3
Dissip Factor	0.001–0.01	0.001–0.01

TABLE 2—TYPICAL PROPERTIES OF RIGID URETHANE FOAMS

Density, pcf	1.0–5.0	5.0–25	25–50
Ten Str, psi	15–80	80–600	600–1000
Shear Str, psi	15–50	50–600	600–1500
Flex Str, psi	25–125	125–1000	1000 and up
Compr Str, psi at yield	15–80	80–1000	1000–3000
Impact Str, in.-lb/ sq in.	0.5–1.5	1.5–8	8 and up
Mod of Elast, psi	500–2000	2000–8000	8000 and up
Heat Res. F	190–225	225–350	350 and up
Ther Cond, Btu/sq ft/hr/°F/in.	0.11–0.25	0.15–0.30	0.30
Water Abs (in volume), %	2.5–1.7	1.7–0.5	0.5
Hydrolysis Res	Excellent	————————————→	
Oxidation Res	Excellent	————————————→	
Ozone Res	Excellent	————————————→	
Solvent Res	Excellent	————————————→	
Dimens Stability	Good	Excellent	Excellent
Flammability (with additive)	Self-extinguishing ————————————→		
Dielec Const, 10^2–10^6 cps	1.05	1.20	1.40
Dissip Factor	0.001	0.002	0.003

and injection- or blow-molded on conventional plastic machinery in final form (an important benefit here is that scrap can be reground and reused in fabricating other parts).

The choice of the proper method of fabrication largely depends upon the economics of the process, since the properties of the final product may be about the same regardless of the method of fabrication. If a few large-volume items are needed, casting these into a single mold is usually more economical. However if many thousands of small, intricate pieces are needed, usually injection molding is the preferred, more economical method of fabrication.

Uses. Applications of urethane elastomers have been developed wherein high abrasion resistance, good oil resistance and good load-bearing capacity are of value, as in solid tires and wheels, especially of industrial trucks, the shoe industry, drive and belting applications, printing rolls, gasketing in oil, etc. Other applications include vibration dampening; for example, in hammer heads, air hammer handles, shock absorption underlays for heavy machinery, etc.; low coefficient of friction with the addition of molybdenum disulfide for self-lubricating uses as ball and socket joints, thrust bearings, leaf spring slide blocks, etc. In the electrical industry, cable jacketing and potting compounds are developing as important uses. Various systems of urethane

TABLE 3—TYPICAL PROPERTIES OF URETHANE ELASTOMERS

	Elastomers	Printing Rolls and Potting Compounds Caulks and Sealants
Trade Names ⟶	"Multrathanes," "Texin," "Neothane," "Adiprene," "Genthane," "Vibrathanes," "Estane"	"Multrathanes," "Solithanes"
Durometer Range (Shore)	A50–D70	A30–80
Sp Gr	1.1–1.3	1.0–1.3
Ten Str, psi	4000–8000	250–5000
Ult Elong, %	400–700	150–500
Resilience	Good	Fair–Excellent
Compr Set (158 F), %	20–50	5–20
Tear Res (split method), pli	50–400	25–400
Abrasion Res	Excellent	Excellent
Flex Mod, psi	8000–90,000	—
Compr-Defl (at 5%), psi	400–6000	—
Low Temp Brittle Point, F	−60 to −90	−65 to −80
Clash Berg Glass Transition, F	−20 to −50	
Flame Res	Fair	Fair
Heat Res, F	250	200–350
Heat Aging (212 F)	Good	Fair–Good
Oxidation Res	Excellent	Excellent
Ozone Res	Excellent	Excellent
Radiation Res	Good	Fair–Good
Acid Res	Poor–Fair	Fair
Alkali Res	Poor–Fair	Fair
Resis to Aliphatics Solvent	Good	Good–Excellent
Resis to Aromatic Solvent	Good	Good
Resis to Chlorinated Solvents	Fair	Fair
Resis to Water (below 160 F)	Good	Fair–Excellent
Insulation Res, ohm	—	10^9–10^{13}
Volume Res, ohm-cm	2–3×10^{11}	10^9–10^{13}
Surface Resis, ohm	0.4–1×10^{11}	10^9–10^{13}
Dielec Const (10^2–10^6 cps)	6.5–7.1	3–8
Dielec Str, v/mil	450–500	250–600
Power Factor (60 cps)	0.015–0.017	0.01–0.5

elastomers with specific fillers have been developed into an important class of caulks and sealants which are just beginning to take hold in applications such as concrete road-expansion joints, building caulking and so on, in direct competition with such older materials as the polysulfides but at a much lower price and superior properties.

A host of other applications vary from adhesive bonding of fibers of all kinds to rocket fuel binders of the more exotic variety which are becoming so important in our national defense picture. Application development of urethane elastomers has just begun. Therefore, it is imperative that the design engineer understand fully the material he is using and how he intends to utilize it in the final piece of equipment. For instance, one recommendation is to limit the use of urethanes to below 180 F in water for continuous exposures. Dry uses can go somewhat higher, e.g., to 225 F for certain systems. In oil, exposures can be up to 250 F. Disregard of such limitations can result in failures, but the design engineer can eliminate these by the proper choice of material. On the other hand, he should choose the urethanes for their virtues, such as hardness and elasticity, where other materials such as natural and other synthetic rubbers may fail.

Properties. The urethanes have excellent tensile strengths and elongation, good ozone resistance, and good abrasion resistance. Knowledge of these properties is mandatory for good engineering design. Table 3 shows typical properties of urethane elastomers. Though particular formulations vary in specific properties, these typical values are indicative of the ranges obtainable. Combinations of hardness and elasticity unobtainable with other systems are possible in urethanes, ranging from Shore hardnesses of 15 to 30 on the "A" scale (printing rolls, potting compounds) through the 60 to 90 "A" scale for most industrial or mechanical goods application, to the 70 to 85 Shore "D" scale, comparable to polystyrene hardnesses, which begin to fall right into the plastic hardness range of applications.

The greater load-bearing capacity of urethanes as compared to other elastomers is noteworthy, for it leads to smaller, less costly, lower weight parts in equivalent applications. Tear strength is extremely high which may be of importance in particular applications along with the very high tensile strengths. The high abrasion resistance has made possible driving parts for which no other materials could compete. However in every such dynamic application, the engineer must design the part to allow for the higher hysteresis losses in the urethanes. Whereas in some applications such as dampening, the higher hysteresis works to advantage, in others, hysteresis will lead to part failure if the upper temperature limit is thereby exceeded. Redesign of the part (thinner walls, etc.) to allow for greater dissipation of the heat generated will permit the part to operate successfully. This has been proven to be the case many times.

Urethane elastomers generally have good low-temperature properties. The same hysteresis effect works in reverse here so that a part in dynamic use at temperatures as low as −60 F, while stiff in static exposure, immediately generates enough heat in dynamic use to pass through its second order transition and does not show any brittleness but becomes elastic and usable. By proper choice of the polyester or polyether molecular backbone, lower use temperatures (as far as −80 F) have been formulated in urethane elastomers.

In addition to good mechanical properties, urethanes have good electrical properties which suggest a number of applications. Oxygen, ozone and corona resistance of this system are generally excellent.

Urethanes are fairly resistant to many chemicals such as aliphatic solvents, alcohols, ether, certain fuels and oils, etc. They are attacked by hot water, polar solvents, and concentrated acids and bases. The best way to make certain of a new application with an unknown environment, is to actually run a swelling test under end-use conditions, which would prevent misapplication of this kind.

V

VACUUM COATINGS

The process of vacuum coating is used to modify a surface by evaporating a coating material under vacuum and condensing it on the surface. It is normally carried out under high vacuum conditions (at approximately one millionth of an atmosphere pressure). The material to be evaporated is heated until its vapor pressure appreciably exceeds the residual pressure within the vacuum system.

Vacuum coating can be used for many applications. For example, optical lenses are coated with magnesium fluoride to a fraction of a wavelength to prevent glare and provide much better transmission of light and a more reliable optical system. The deposited film is extremely adherent and will withstand normal cleaning.

Silicon monoxide is frequently used as an abrasion-resistant coating material. As deposited, it is soft and requires postheat treatment in air to convert it to silicon dioxide which is transparent and extremely hard. It is frequently used to protect front surface mirrors and increases abrasion resistance by a factor of 5000 to 10,000, while maintaining equal or higher reflectivity. Similarly, titanium is sometimes used for coating and is subsequently oxidized to give a titanium dioxide abrasion-resistant surface.

By far the most common type of vacuum coating is the process of vacuum metallizing. In this process metal is evaporated and used as deposited without further treatment as opposed to the evaporation of compounds or materials which require posttreatment.

Vacuum metallizing has generally been used as a decorative process whereby costume jewelry, toys, etc., are given a metallic sheen and are made highly reflective. The base material may be either plastic or metal. In either case the part is frequently lacquered before metallizing to prevent the evolution of gas from the base and to provide a smooth surface without mechanical buffing. Because the metal deposit is only about two or three millionths of an inch thick, the smooth surface is necessary to give a specular reflection.

When the metal is on the outside of the coated part, it is referred to as front surface. However, in applications where it is used on the back of a transparent plastic (e.g., dashboards and taillight assemblies on automobiles) it is referred to as a second surface coating. The advantage of second surface coating is provided by using the plastic as the exposed surface. Front surface coatings must generally be protected with a transparent lacquer overcoat (applied after metallizing) because the thin decorative coatings are not wear resistant in themselves.

Aluminum is the most popular vacuum metallizing coating material for most applications. However, other metals may be used, such as zinc, cadmium, copper, silver, gold or chromium. Of all these metals, aluminum has the best general combination of reflectivity, conductivity and stability in air. By adding color to the top-coat lacquer, the aluminum deposit may be made to appear like copper or gold as well as metallic sheens of blues, reds, yellow, etc. By using separate sources for each constituent, it is possible to deposit alloys as well as pure metals.

The above applications are for parts produced by batch metallizing, i.e., the individual parts are mounted on racks inserted in the vacuum system and after the necessary vacuum and evaporation temperatures are obtained, the parts are rotated so as to be uniformly coated by the evaporating metal. In batch metallizing the aluminum is evaporated from tungsten filaments which are heated by direct resistance. Because of this, the amount of aluminum which can be charged is limited and only thin coatings can be produced. Similarly, only small surfaces (a few square feet), such as can be exposed within a matter of seconds, can be metallized.

When it is desirable to coat larger surfaces, e.g., rolls of flexible material, a semicontinuous metallizing process must be employed. For semicontinuous metallizing a roll of material is mounted in the vacuum chamber and unrolled under vacuum so as to coat either or both sides of the web which is subsequently rewound in vacuum. This process is currently in use for coating rolls of plastic sheeting and paper. In order to coat continuously over a period of hours, it is necessary to have larger volumes of aluminum available for evaporation than can be held on resistance-heated tungsten filaments. Therefore, the aluminum is generally heated by induction in crucibles.

Coating of rolls of materials provided one of the first functional applications for coatings that used the electrical conductivity of the metal deposited. This conductive layer was deposited on thin insulating layers of either paper or plastic and could be used for winding miniature condensers. The electrical conductivity is also used in the metallizing process itself as a means of measuring the amount of metal deposited. Since the conductivity is a function of the thickness of the metal, continuously measuring conductivity gives a control for the amount of metal deposited. Other functional uses of the coating are based on its reflectivity (e.g., reflective insulation).

Vacuum metallizing has recently been extended to include thick films, i.e., in the range of 1 to 3 mils. Such coatings serve as corrosion-resistant barriers, particularly on high tensile strength steel exposed to marine atmospheres. Where the temperature requirements of steel are less than 500 F, cadmium deposits can be used. For temperatures in excess of this, aluminum shows much better protection and does not react with the base steel as cadmium does.

Truly continuous operation is necessary for coating rolled steel. Here the rolls are unwound and rewound in air with the strip passing through seals into the vacuum chamber where it is coated. The process has been satisfactorily performed on a pilot plant scale, but to date (1962) is not in commercial production. The metallizing of rolled stock allows separate control on each side of the web and the composition of the coating, as well as thickness, may be changed from one side to the other.

Typical advantages of vacuum metallizing include:

1. Close control of coating thickness and composition.

2. Uniform deposits without buildup at sharp discontinuities.

3. High coating rate.

4. Low coating costs in volume production.

5. Long life of equipment since few moving parts.

Disadvantages of the process are as follows:

1. Part must be extremely clean.

2. Surfaces to be metallized must not evolve gas under vacuum.

3. Parts must not be temperature sensitive, i.e., must be stable to about 255 F.

4. Deposits form well only on a surface exposed to hot metal; re-entrant angles are not well coated.

The cost for metallizing in production lots for corrosion-resistant coatings is comparable to electroplating. Decorative metallizing is generally much less expensive than electroplating.

VANADIUM

Vanadium, an element belonging to Group V B in the periodic system, has an atomic number of 23, an atomic weight of 50.95 and six isotopes from 47 to 52. It is a gray-colored metal, crystallizing as a body-centered cube. Chemically very active, the element is close to its daughter elements in Group V B, columbium and tantalum, and to its sisters, titanium and chromium.

Composition and general chemical nature

The principal production method for commercial high-purity vanadium consists in the reduction of vanadium pentoxide by means of calcium, using a small amount of iodine as a booster.

The composition, in percent, of calcium-reduced metal typically is as follows: 0.08 O; 0.03 N; 0.005 H; 0.04 C; 0.03 Fe and 99.75 V. This material is soft and ductile, and has a tensile strength of about 80,000 psi. As purity increases, tensile strength decreases, since the impurities are interstitial and offer resistance to deformation. The principal physical properties are set forth below:

Density, gm/cu cm	6.1
Melting Pt, C	1000 ± 25
Elec Resis, 20 C, μohm-cm	24.8–26
Ther Cond, 100 C, cal/sq cm/ °C/cm/sec	0.074
Ther Neutron Capture Cross Section, barns/atom	4.5 ± 0.9
Sp Ht, 0–100 C, cal/gm/°C	0.119

Ther Exp, 10^{-6} in./in./°C

23–100 C	8.3
23–500 C	9.6
23–900 C	10.4
23–1100 C	10.9
Mod of Elast, 10^6 psi	18–20
Shear Mod, 10^6 psi	6.73

Mechanical properties

Vanadium metal shapes have relatively low strength, combined with high ductility, while their tensile-yield ratio is consistently advantageous. The following table indicates their most important mechanical properties.

Material	Size, in.	Ten Str, 1000 psi	Yld Str, 1000 psi	Elong, in 2 in., %	Hard, Rockwell
Bar [a, b]	1	68.4	63.7	27	85B
Wire [c, d]	0.154	78	67.1	25	48A
Wire [d, e]	0.154	132.1	111	6.8	54A
Sheet [c, f]	0.075	77.8	65.9	20	83B
Sheet [f, g]	0.075	120.1	112.6	2	100B

[a] Hot rolled.
[b] Test specimen, 0.505 in. dia x 2 in.
[c] Vacuum annealed.
[d] Test specimen, 0.154 in. dia x 2 in.
[e] Cold drawn 80%.
[f] Test specimen, 0.075 x 0.5 x 2 in.
[g] Cold rolled 84%.

At 1025 C, vanadium metal has a tensile strength of 7700 psi and an elongation in 1 in. of 37%. At 1000 C, 672 psi will produce 1% deformation in 24 hr.

Vanadium of good purity (Rockwell 85B) has a transition temperature of approximately −70 F.

Principal fabricating characteristics

Hot working. Since vanadium oxidizes rapidly at hot-working temperatures, forming a molten oxide, it must be protected during heating. This is most easily accomplished by heating in an inert-gas atmosphere. Other common practices have been found less suitable.

Vanadium ingots up to 6 in. in size have been successfully hot-worked, but the degree of contamination is a modifying factor. Generally, the procedures used in working alloy steels apply.

In view of the difficulties involved in heating the metal, reheating is generally avoided and the starting temperature is a function of the amount of hot work to be accomplished and of the desired finishing temperature. Starting can range as high as 1260 C and finishing limited by the beginning of recrystallization. Straightening is performed between 371 to 427 C but not at room temperatures.

Cold working. Vanadium has excellent cold-working properties, provided its surfaces are uncontaminated. They are therefore machined clean by removing between 0.02 to 0.04 in.

Strip can be readily made from hot-rolled sections ¾ × 6 in. in cross section, and 0.010 in. material has been produced without, and 0.001 in. with intermediate annealing. Where incipient cracking is observed, vacuum annealing at 899 C becomes necessary.

Extrusion is one of the most suitable fabricating methods for vanadium, since warm extrusion followed by cold rolling or drawing avoids hot working with the troublesome heating step. At temperatures below 538 C, tube blanks 2 in. o.d. × ¼-in. wall thickness have been produced from hot-rolled and turned bars as well as from ingots.

Wires can be drawn from ⅜ in. dia stock down to 0.010 in., especially after copper-plating. Reductions are usually 10% per pass.

In machining, vanadium resembles the more difficult stainless steels. Low speeds with light to moderate feed are used and very light finishing cuts at higher speeds are possible.

Welding is not difficult but contamination of the metal must be avoided by shielding from air by means of an inert gas, i.e., argon.

Available forms, sizes, shapes

Vanadium is available in a wide range of forms. But the present high cost of the metal limits its use to special applications.

Applications

Vanadium is used in the cladding of fuel elements in nuclear reactors because it does not alloy with uranium and has good thermal conductivity as well as satisfactory thermal neutron cross section.

As the metal alloys with both titanium and steel, it has found application in providing a bond in the titanium-cladding of steel. Also, the good corrosion resistance of vanadium offers interesting possibilities for the future; it has excellent resistance to hydrochloric and sulfuric acids and resists aerated salt water very well. But its stability in caustic solutions is only fair and in nitric acid, inadequate.

VAPOR DEPOSITION COATINGS

The terms "vapor plating," or "vapor deposition" are occasionally applied to the coating process involving the evaporation of metals in high vacuum. This process is better known as "evaporative coating," or "vacuum metallizing." As used here, "vapor plating," "gas plating," or "vapor deposition" refer to those processes in which solid or liquid coatings are produced on a substrate by chemical reactions induced in a gas or vapor in contact with the substrate. The inducing agent is generally heat, the substrate being hotter (in a few instances, colder) than the gas, thus disturbing the chemical equilibrium in the gas to the extent that a reaction occurs, depositing a solid or liquid reaction product upon the substrate.

In some processes, the material of the substrate itself enters into the reaction by displacing certain components from the vapor in contact with the substrate. Materials can be applied by vapor plating at much higher rates than they can be applied by evaporative coating, and at temperatures far below those at which they develop an appreciable vapor pressure.

Vapor plates can be applied to virtually any material which is not melted or decomposed at the processing temperature. Materials which have been successfully coated include iron, steel, alloy steel, copper, molybdenum, tungsten, chromium, titanium, zirconium, uranium, aluminum, Pyrex glass, Vycor, quartz, porcelain, graphite, alumina and glass fibers. Also, easily decomposable metal carbonyls, hydrides, and organometallic compounds can be used to plate heat-sensitive materials such as paper, plastics and rubber.

Three vapor plating processes

Most vapor-plating processes are comprised of one or more of three basic types of chemical reactions: (1) hydrogen reduction of metal halide vapors; (2) thermal decomposition of halides, hydrides, carbonyls, organometallics, or other volatile compounds; and (3) displacement reactions between the material of the base and a metal halide vapor.

Any material which meets the following requirements can be applied as a coating by vapor deposition: (a) the material, or its components, must form a compound which can be vaporized at a relatively low temperature without appreciable decomposition; (b) the volatile compound must be sufficiently unstable to be capable of decomposition or chemical reduction at a temperature somewhat higher than its vaporization temperature; and (c) the material to be deposited must not have an appreciable vapor pressure at the deposition temperature.

Hydrogen reduction. Materials which have been vapor-deposited by hydrogen reduction of their halide (fluoride, chloride, bromide or iodide) vapors include boron, silicon, titanium, zirconium, vanadium, columbium, tantalum, chromium, molybdenum, tungsten, iron, cobalt, nickel, germanium, tin, arsenic, antimony and bismuth.

Thermal decomposition. Most of the above metals, in addition to hafnium, uranium and rhenium, can also be deposited by thermal decomposition of their halide vapors. Thermal decomposition of the metal hydrides has been used in depositing boron, silicon, germanium, arsenic, antimony and bismuth. Iron, cobalt, nickel, chromium, molybdenum and tungsten have also been deposited from their respective carbonyls, and copper and the platinum group metals from their carbonyl chloride compounds.

Thermal decomposition of organometallic compounds, such as the acetylacetonates and the biscyclopentadienyl derivatives, has been used in depositing copper, iron, nickel, cobalt and some of the platinum group metals. The metal alkyls, such as tetraethyl lead, diethyl zinc and triethyl aluminum, have been used in depositing these metals.

More than one plating compound can be used in the plating atmosphere. This permits the deposition of alloys and of compounds such as refractory carbides, silicides, borides, and some oxides, as well as sulfides, selenides, tellurides and arsenides. Vapor plating is thus seen to be a process of potentially great versatility.

Displacement plating. In the displacement type of vapor plating, the material of the base displaces a constituent of the vapor which then alloys or otherwise combines with the remaining base material. For example, iron will displace chromium from chromous chloride vapor; the formation of ferrous chloride vapor and chromium metal converts the surface of the iron object to a high-chromium stainless alloy. A sufficiently high temperature is maintained to permit fairly rapid interdiffusion of the depositing metal and the base metal, thus maintaining enough of the base metal at the surface to sustain the plating reaction. This type of process has been used in applying chromium, aluminum, silicon, boron, titanium, vanadium, niobium, antimony, cerium, manganese, zinc, tin, molybdenum and tungsten alloy coatings on iron and steel; in applying chromium, silicon, and boron coatings on molybdenum; and in applying titanium, zirconium, niobium, and tantalum carbide coatings on graphite.

Processing variables

Vapor plating processes of the reductive type are carried out by saturating hydrogen with 1 to 10% (volume) of the plating vapor and passing the mixture over the heated object to be coated. In a few instances, the plating vapor may be reducible with hydrogen at the vaporization temperature; thus an inert carrier gas such as argon is used to transport the vapor to the plating chamber where it is mixed with hydrogen.

The thermal decomposition plating process and the displacement plating process are carried out either in an atmosphere of plating vapor and inert carrier gas, or in the plating vapor alone.

Vapor plating is carried out at atmospheric pressure where possible. However, in some systems, such as those involving hydrogen reduction of molybdenum or tungsten halides, or thermal decomposition of the metal hydrides and carbonyls, the plating compounds are sufficiently unstable so that plating must be carried out at reduced pressures (of the order of 0.001 to 0.02 atmosphere) to prevent premature reaction of the plating mixture and resultant formation of powdery, nonadherent deposits. Low pressures are also used in plating by thermal decomposition of the metal iodides to increase the extent of iodide decomposition and thus decrease the required deposition temperature. Low pressures are sometimes used in the displacement type of coating process to increase the rate of diffusion of the gaseous reactant to the surface of the object and to remove the plating by-products (e.g., lower halides of the base metal) as fast as they are formed.

Heating of the object can be done by induction, by electric resistance, by internal radiant heaters (i.e., "hot fingers"), or by placing the object and coating chamber in an electric or gas furnace. The heating method used is determined largely by the nature or shape of the object to be plated. Heating in an external furnace is generally the least desirable, except in the case of displacement coatings, since the bulk of the plating action occurs on the hot walls of the plating chamber instead of on the desired object. This results in very low plating efficiency and gross nonuniformity in the plate due to depletion of the coating atmosphere at the walls of the plating chamber before it can contact all parts of the object.

Deposition temperature used in the various vapor plating processes are as follows: (a) hydrogen reduction of metal halides, 300 to 1400 C (570 to 2550 F); (b) thermal decomposition of hydrides, carbonyls, and organometallic compounds, 150 to 600 C (300 to 1110 F); (c) thermal decomposition of metal iodides, 900 to 1400 C (1650 to 2550 F); of metal bromides and chlorides, 1200 to 2000 C (2190 to 3630 F); (d) displacement coatings on iron and steel alloys, 700 to 1200 C (1290 to 2190 F); on refractory metals such as molybdenum, 1000 to 1800 C (1830 to 3270 F); displacement carbide coatings on graphite, 1400 to 2500 C (2550 to 4530 F).

Deposition rates of most vapor plating processes range from 0.1 to 20 mils per hr, the average rate usually being 1 to 2 mils per hr. The efficiency of utilization of the plating compound generally ranges from 5 to 80%. Efficiency can vary widely with the method of heating the object, with the type of material used as plating compound, and with the rate of gas flow through the plating chamber. It can usually be improved by making provision for recovering and recycling unused plating compound.

The rate of production of coated pieces in vapor plating varies chiefly with the thickness of coating desired. Small objects plated with only a few tenths of a mil of metal by the carbonyl decomposition process can be coated in continuous fashion with a treatment period of only a few seconds or minutes per piece. Wire has been plated with 0.0005 to 0.001 in. coatings at speeds of about 120 feet per hr, and glass fibers have been plated, with much thinner coatings, at considerably higher speeds. Molybdenum articles can be siliconized with treatment periods of one hour or less. Other displacement-diffusion coating processes, such as pack-chromizing, generally use a three- to six-hour treatment period, with the remainder of an eight-hour day being devoted to loading, unloading and cleaning of articles plated. An additional eight-hour period is required for heating and cooling the treatment container or box. This long, over-all coating cycle does not require continuous supervision or attention, however, and since hundreds or thousands of pieces can be treated in one batch, the over-all cost per piece is low.

Coating structure

Vapor plates are more or less crystalline. In low temperature deposits the size of the crystallites is so small that the plate is virtually amorphous. The size of the crystals in high temperature deposits may range up to ¼ in. or more in diameter, depending upon the coating thickness. Crystallinity also tends to increase with plate thickness. Deposits under one mil thickness usually approximate the finish of the underlying metal in smoothness. The density of the coating is usually close to theoretical.

Plate thicknesses may range from a few millionths of an inch to a quarter of an inch or more, and are determined primarily by the intended usage. Coatings intended to provide protection from corrosion or oxidation generally range from 5 to 10 mils in thickness. Techniques are being developed to produce pore-free vapor plates in thicknesses of 1 to 2 mils. Coatings intended to provide only hardness and resistance to scoring or abrasion usually are not over 1 mil thick, in order to obtain some flexibility in the coating.

The bond and adhesion of the plate depends upon the particular combination of plate and base material used, and upon the temperature of application. Low temperature deposits resemble electroplates and are mechanically bonded. Careful cleaning and degassing of the surface of the substrate before plating are necessary to obtain strongly adherent coatings in low-temperature plating. Roughening of the surface by etching or grit-blasting before plating likewise improves adhesion.

High temperature deposits may develop a diffusion layer at the bond if the coating material and substrate are capable of alloying or mutual reaction. Since such a layer usually improves bonding, initial surface preparation is not too important. In some cases, however, the coating and base material may interact to form brittle intermetallic, or other type compounds which make the bonding layer susceptible to easy fracture by mechanical or thermal stresses. This may occur, for example, with aluminum coatings on iron, and with tantalum and columbium coatings on iron or carbon steel (in the latter instance, a brittle carbide or high carbon interlayer is formed). When this is likely to occur, a thin intermediate coating of another material, such as nickel, is applied, either by vapor plating or by other techniques, before applying the main vapor plate. Displacement type coatings are always bonded by a diffusion layer, and, because of the gradation of properties within this layer, are generally the best bonded and most nonporous of vapor deposited coatings.

Applications

Vapor plating is not considered to be competitive with electroplating. Its chief use (present and future) is to apply coating materials that cannot be electroplated, or cannot be applied in a nonporous condition by other techniques. Such materials include titanium, zirconium, columbium, tantalum, molybdenum and tungsten, and refractory compounds such as the transition metal carbides, nitrides, borides and silicides. Vapor plating will also continue to be useful in the preparation of ultrahigh-purity metals and compounds for use in electronics applications and in alloy development.

A few of the main commercial uses of vapor plating are:

1. The application of high-chromium alloy coatings to iron and steel articles by the displacement-diffusion coating process (known as pack chromizing), for abrasion resistance and for protection from corrosion by food products, strong oxidizing acids, alkalies, salt solutions, and gaseous combustion products at temperatures up to about 800 C (1470 F).
2. The application of molybdenum disilicide coatings to molybdenum by gas-phase siliconizing, for protection against air oxidation at temperatures between 800 and 1700 C (1470 to 3100 F).

3. The preparation of ultrahigh-purity titanium, zirconium, chromium, thorium, and silicon by iodide vapor decomposition processes.
4. The preparation of junction transistors by the controlled diffusion of boron from boron halide into the surface of silicon or germanium wafers.
5. The preparation of oriented graphite plates and shapes (pyrolytic graphite), by the high-temperature pyrolysis of hydrocarbon gases, for use in rocket and missile applications.

In addition, the following coatings have been prepared in development work:

1. Tantalum coatings on iron and steel for corrosion resistance.
2. Vanadized, tungstenized, and molybdenized iron for wear resistance.
3. Tungsten coatings on copper X-ray and cyclotron targets.
4. Conductive metallic coatings on glass, porcelain, alundum, porous bodies, rubber, and plastics.
5. Refractory metal coatings on copper wires.
6. Oxidation-resistant carbide coatings on graphite tubes, nozzles and vanes.
7. Metallic coatings of all types on metallic and nonmetallic powders.
8. Decorative, colored coatings on glass.
9. High purity boron, rhenium, vanadium, germanium and aluminum.

At the present time, vapor plating is generally applicable to approximately isodiametric objects having a maximum dimension of 8 to 10 in. with no deep undercuts or blind recesses. Objects such as nozzles, thermometer wells, rings, bushings, crucibles, cups and dies have been successfully plated. Wire in diameters up to at least 0.1 in. can be plated in a continuous process. Rods and tubing ranging in length up to several feet and in diameter from 0.15 in. to 2 to 3 in. can be plated externally, internally (in the case of tubing) or both, by the expedient of heating only a small section of the rod or tubing to plating temperature and moving this plating zone slowly over the entire length of the rod or tubing.

The displacement-diffusion plating processes, such as pack-chromizing, can plate uniformly somewhat larger pieces and more complex shapes

having inaccessible areas. Sheets and rod up to 2 to 3 ft in dimension have been coated, and no technical obstacles are seen to scaling the processes up to coat even larger pieces. The pack coating processes have the advantage of minimizing the problems of specimen support and warpage during plating.

Plating uniformity varies somewhat with the particular process used, with the shape of the object being plated, and with the attention given to providing proper gas flow around the object. A variation in thickness of 10 to 25% is usually obtained. However, some coating processes can be made self-limiting so that the variation in coating thickness is much less than this range.

Disadvantages

Vapor plating has the following disadvantages:

1. Relative instability and air and moisture sensitivity of most of the compounds used as plating agents.
2. A tendency to produce nonuniform deposits due to unfavorable gas flow patterns around the work, or to uneven specimen temperature.
3. Alteration of physical properties of the substrate due to the elevated processing temperatures.
4. The possibility of poor coating quality arising from undesired side reactions in the plating process.

In general, the materials used as plating compounds in vapor plating are relatively unstable and easily decomposed by air and moisture, thus rendering them more expensive and more difficult to store and handle than the compounds used in other plating techniques. Also, some of the metal carbonyls, hydrides, and organometallic compounds are highly toxic; and some of the hydrides and metal alkyls inflame spontaneously upon contact with air.

To develop optimum properties in deposits of many materials, the plating compounds, particularly the moisture-sensitive metal halides, must be purified and used without contamination from the atmosphere. This apparent disadvantage is sometimes put to good use, however, when intentional contamination of the coating atmosphere with nitrogen or moisture is used to produce harder deposits (e.g., of titanium or tantalum), or to reduce the codeposition of carbon (e.g., with molybdenum from the carbonyl).

Nonuniform plating may result in all vapor-plating processes, except the displacement-diffusion process, if consideration is not given to the gas-flow pattern around, or through, the article being coated. The shape factor may also have to be taken into account in selecting the method of heating the article, to avoid nonuniform deposition due to nonuniform heating of the part. These difficulties can be overcome in extreme cases by applying more than one coating and using a different direction of gas flow over the specimen for each application.

If necessary, the displacement-diffusion type of coating process can be carried out at very low gas flow rates (since solid-state diffusion is the rate-controlling factor), and still produce uniform coatings. For this reason, this type of coating process is ideally suited for coating large, or highly irregular objects, or large numbers of small objects.

The elevated processing temperatures required in vapor plating may produce undesired physical changes in the article, such as loss of temper, grain growth, warping, dimensional change or precipitation or solution of alloying constituents. However, in many instances a vapor plating procedure can be selected which will avoid marked undesirable change.

Undesirable side reactions in vapor plating processes must be watched for and avoided. A particularly troublesome one in plating from metal halide vapors is the interaction of the base material and the halide vapor to form lower-valent halides, either of the base material or of the coating vapor. If the substrate temperature is too low, or the plating atmosphere too rich in plating vapor, these lower-valent halides will condense at the surface of the substrate, producing a plate underlaid or contaminated with halide salts. Such deposits are always poorly adherent, porous and sensitive to moisture. Incomplete reduction or decomposition of the plating vapor alone can produce the same result. Contamination of this type is less likely to occur when plating inert base materials such as graphite, glass and some ceramics.

After having been plated in a hydrogen atmosphere, some metals such as tantalum, columbium, and titanium, having a strong affinity for hydrogen, must be vacuum annealed, or at least cooled in an inert-gas atmosphere to avoid excessive hydrogen absorption and embrittlement.

Also, carburization of the substrate may occur in processes employing the metal carbonyls to coat metals having a strong affinity for carbon, if the substrate temperature is too high. The metal of the deposit itself may be partially carburized in some cases, as when depositing molybdenum, chromium and tungsten from their carbonyl vapors.

VINYL PLASTICS

Polyvinyl chloride (PVC) is a thermoplastic polymer formed by the polymerization of vinyl chloride. A variety of copolymers of vinyl chloride and other vinyl monomers, notably vinyl acetate and vinylidene chloride, are of commercial importance. Resins of different properties can be made by variations in polymerization techniques. These resins can be compounded with plasticizers, colors, mineral fillers, etc., and processed into usable forms, varying widely in physical and electrical properties, and chemical resistance, and possessing versatility in coloring and design. Compared with other themoplastics of comparable cost, articles produced from the vinyl chloride plastics have outstanding chemical, flame and abrasion resistance, tensile properties, and resistance to heat distortion.

PVC homopolymer resins, the largest single type of vinyl chloride-containing plastics, are produced by several methods of polymerization:

1. Suspension: The largest volume method which produces resins for general-purpose use, processed by calendering, injection molding, extrusion, etc.

2. Mass or solution: Produces fine particle size resins used principally for calendering and solution coating.

3. Emulsion: Produces extremely fine particle size resin used for the preparation of liquid plastisols or organosols for use in slush molding, coatings and foam.

The two largest volume members of the family of vinyl-chloride polymers are the pure polyvinyl chloride or homopolymer resins and the vinyl chloride-vinyl acetate copolymers containing approximately 5 to 15% vinyl acetate.

Table 1 lists some of the pertinent properties of several types of plastics made from vinyl-chloride polymers. The following sections describe specific types of vinyl-chloride plastics and some of the characteristic properties and applications in the engineering field.

Rigid vinyls

Products made from rigid vinyls are perhaps of most interest in the engineering field. Rigid materials can be prepared by calendering, extrusion, injection molding, transfer molding, and solution casting processes. Rigid polyvinyl-chloride products are available in sheets, films, rods, pipes, profiles, valves, nuts and bolts, etc. The products can be machined easily with wood and metalworking tools.

Sheets or other forms can be conventionally welded by hot-air guns, using extruded welding rods of essentially the same composition as that of the sheet. The welded joints have strength equal to that of the base material. Rigid sheets can be thermoformed into many intricate shapes by several different thermoforming techniques such as vacuum forming, ring and plug forming, etc. Rigid pipe can be threaded and joined like steel pipe or sealed with adhesives in a manner similar to the sweating of copper pipe. Vinyl pipe is being used increasingly in water works,

TABLE 1—PROPERTIES OF POLYVINYL CHLORIDE AND COPOLYMERS

Property	ASTM Method	Rigid Polyvinyl Chloride		Rigid Vinyl Chloride-Vinyl Acetate Copolymer	Vinyl Chloride-Vinylidene Chlor. Copolymer	Flexible Polyvinyl Chloride	
		Type I Normal Impact	Type II High Impact			Filled	Unfilled
Physical Properties							
Sp Gr	D-792	1.38	1.35	1.37–1.45	1.65–1.72	1.3–1.7	1.2–1.35
Ther Cond, Btu/hr/sq ft/°F	D-325	0.07–0.10	0.09–0.13	0.08–0.11	0.053	0.07–0.10	—
Coef of Ther Expan, 10^{-5} in./in./°F	D-696	2.8–3.3	5	3.5	8.8	—	—
Mechanical Properties							
Ten Str, psi	D-638	8500	6000	7500	3000–5000	1000–3500	1500–3500
Mod of Elast in Tension, 10^5 psi	D-638	3.5–4.0	2.5	3.0	0.5–0.8	—	—
Elongation, %	D-638	5–25	100	30	50–240	200–400	200–450
Flex Str	D-790	13,500	12,000	12,000	4300–6100	—	—
Mod of Elast in Flex, 10^5 psi	D-790	4.5–5.4	3.5	4.5–5.0	0.55–0.95	—	—
Impact Str, ft-lb/in. of notch	D-256	0.8	15	0.5	0.3–1.0	—	—
Hardness, Rockwell	D-785	R110–120	R100–105	R65	M50–60	—	—
Heat Distortion Temp (264 psi), °F	D-648	165	155	145	130–150	—	—
Electrical Properties							
Vol. Resistivity (23 C), ohm-cm	D-257	10^{14}–10^{16}	—	10^{13}–10^{15}	10^{14}–10^{16}	10^{11}–10^{14}	10^{11}–10^{13}
Dielec Str (short-time), v/mil	D-149	700–1400	1100	425	350–450	250–280	300–1000
Dielec Constant at 1000 cps	D-150	3.5	3.3	3.2	3.5–5.0	4.0–7.0	4.0–8.0
Power Factor at 1000 cps	D-150	0.009–0.017	—	0.011–0.013	—	0.112	—
Miscellaneous Properties							
Resis to Acids	D-543	Exc	Good	Good	Exc	Good	Good
Resis to Alkalies	D-543	Exc	Good	Good	Exc	Good	Good
Machining Qualities	—	—	Exc	Exc	Exc	Good	—	—

the petroleum industry, in natural gas distribution, irrigation, hazardous chemical applications and food processing.

Rigid vinyl made from vinyl acetate-vinyl chloride copolymers is prominent in sheeting used for thermoforming for such items as maps, packaging, advertising displays, toys, etc. Rigids made from homopolymer vinyl chloride resins are used in heavier structural designs, for example: pipe, pipe valves, heavy panels, electrical ducting, window and door framing parts, architectural moldings, gutters, down-spouts, automotive trim, etc. In these fields, rigid vinyls compete with aluminum and other metals. Rigid vinyl products made from homopolymer resins are available in two types: Type I, or unmodified PVC, is approximately 95% PVC and has outstanding chemical resistance but low impact strength; Type II PVC, containing 10 to 20% of a resinous or rubbery polymeric modifier, has improved impact strength but reduced chemical resistance.

Flexible vinyls

Flexible vinyl products are produced by the same general methods used for rigid vinyl products. Flexibility is achieved by the incorporation of plasticizers (mainly high boiling organic esters) with the vinyl polymer. By proper choice of plasticizer type, flexible products can be obtained which excel in certain specific properties such as gasoline and oil resistance, low temperature flexibility, flame resistance, etc. The flexible sheeting and film can be fabricated by heat sealing to itself or other substrates by induction or high-frequency methods, solvent sealing, sewing, etc. Flexible vinyl film and sheeting find application in upholstery, packaging, agriculture, etc. The corrosion resistance of vinyl sheeting makes it ideal for a pipe wrap to prevent corrosion of underground installations. Flexible extrusions in many different shapes and forms have application as insulating and jacketing on electrical wire and cable, refrigerator gaskets, weatherstripping, upholstery and shoe welting. Injection-molded flexible vinyl products are used as shoes, electrical plugs, and insulation of various sorts.

Abrasion and stain resistance, coupled with unlimited coloring and design possibilities, have made flexible vinyl flooring one of the largest items in the floor covering field. The major revolution in vinyl flooring is the greater emphasis on the use of relatively low to very low molecular weight homopolymer resins in place of the more expensive vinyl chloride-vinyl acetate copolymers.

Coatings

Coatings, based on polyvinyl chloride polymers and copolymers, can be applied from solutions, latex, plastisols or organosols. Plastisols are liquid dispersions of fine particle size emulsion PVC in plasticizers. Organosols are essentially the same as plastisols but contain a volatile liquid organic diluent to reduce viscosity and facilitate processing. In many coating operations, conventional paint-spraying equipment is used. In other techniques, articles can be dip, knife or roller coated. Plastisols and organosols are used extensively for dip coating of wire products such as household utensils and knife coating of fabrics and paper. Plastisol and organosol products can be varied from hard (rigidsols) to very soft (vinyl foam). Vinyl coatings are used for a variety of applications requiring corrosion and/or abrasion resistance.

A recent development in coating with vinyl plastics involves a fluidized bed technique. A metallic object, heated to 400 to 500 F, is immersed in a bed of finely ground plastic which is "fluidized" by air entering the bottom of the container. Articles can be coated by this method to a thickness of 7 to 60 mils. Both rigid and plasticized vinyls can be applied by this method.

Vinyl-metal laminates

These products are made by direct lamination of preprocessed, embossed, and designed vinyl sheet to metal, or continuous plastisol coating of metal sheet followed by fusing or curing of the plastisol and subsequent embossing. In the former process, the vinyl can be laminated to both sides of the metal sheet. Steel, aluminum, magnesium, brass and copper have been used. The hardness, elongation, general properties and thickness of the vinyl can be modified within wide limits to meet particular needs.

These laminates are dimensionally stable below 212 F and combine the chemical and flame resistance, decorative and design possibilities of vinyl with the rigidity, strength and fabricating attributes of metals. The vinyl-metal laminates can be worked without rupture by many of the metalworking techniques, such as deep-drawing, crimping, stamping, punching, shearing and reverse-bending. Disadvantages are inability to spot-weld, and lack of covering of metal edges, which is necessary where severe exposure conditions are encountered. Vinyl-metal laminates find use in appliance cabinets, machine housings, lawn and office furniture, automotive parts, luggage, chemical tanks, etc. The cost of these laminates is comparable to some lacquered metal surfaces.

Vinylidene chloride copolymers

The vinylidene chloride-vinyl chloride copolymers are more resistant chemically and have better vapor barrier characteristics than the vinyl-chloride homopolymers or vinyl chloride-vinyl acetate copolymers. Vinylidene chloride-vinyl chloride copolymers, containing 50% or more vinylidene chloride, are known generically as "saran." The two largest volume applications are in upholstery made from monofilaments and film for food packaging. Other uses are in window screening (monofilaments), paper and other coatings, pipe and pipe linings, and staple fiber.

Saran pipe and saran-lined metal pipe are of interest to the engineering field. Saran-lined pipe is prepared by swaging an oversize metal pipe on an extruded saran tube. These products can be installed with ordinary piping tools. Fittings and valves lined with saran and flange joints with saran gaskets are available. Vinylidene chloride-acrylonitrile copolymers have applications as coatings for tank car and ship-hold linings. Lacquers of these polymers are used in cellophane coatings giving a product having the low moisture vapor permeability of vinylidene chloride polymers plus the handling ease of cellophane. The lacquers are also used for paper coatings, dip coatings and sprayed packaging. Vinylidene chloride copolymers in latex form are used in paper coatings and specialty paints.

VULCANIZED FIBRE

Vulcanized fibre is a pure, dense, cellulosic material with good electrical insulating properties and high mechanical strength. It is one-half the weight of aluminum, easily machined and formed, and is used for parts such as arc barriers, abrasive-disc backing, high-strength bobbin heads, materials-handling equipment, railroad-track insulation and athletic guards.

Forms

Most manufacturers provide vulcanized fibre in the form of sheets, coils, tubes and rods. Sheets are made in a thickness of 0.0025 to 2 in., approximately 48×80 in. in size, or in rolls and coils from 0.0025 to $\frac{3}{32}$ in. thick. Tubes are made in the outside diameter range of $\frac{3}{16}$ to $4\frac{3}{8}$ in., and rods are produced $\frac{3}{32}$ to 2 in. in diameter.

Sheets can be machined and formed to produce a variety of useful shapes for insulating or shielding purposes. Sheets, tubes and rods can be machined using standard practices for cutting, punching, tapping, milling, shaping, sanding, etc. NEMA (National Electrical Manufacturers' Association) has a publication on recommended practices for fabricating vulcanized fibre.

Properties

Vulcanized fibre possesses a versatile combination of properties, making it a useful material for practically all fields. It has outstanding arc resistance, high structural strength per unit area and can be formed and machined. In thin sections it possesses high tear strength, smoothness and flexibility. In heavier thicknesses it resists repeated impact and has high tensile, flexural and compressive strength.

The material is unaffected by normal solvents, gasoline and oils, and therefore, is recommended for applications where a structural support is required in the presence of these materials.

Moisture absorption is high and dimensional stability is affected by conditions of humidity when not protected by moisture-resistant coatings.

Vulcanized fibre is produced in 13 basic grades, and numerous special grades to meet specific application requirements. Typical values for four grades, electrical insulation, commercial grade, bone grade and special white fibre are shown in Table 1.

Applications

Vulcanized fibre serves as the insulating material in a signal block for railroad track insulations. At the end of a signal station, the rails are completely insulated from the next adjoining section to form what is termed a "block." The two meeting rails and the coupling fixtures are insulated with formed parts made from vulcanized fibre. This junction is effective while absorbing the repeated impact from trains under all weather conditions.

Vulcanized fibre offers durability, ease of fabricating, excellent wear characteristics and lightness of weight for materials-handling and luggage applications. The materials-handling equipment resists scuffing, battering, denting, rusting and other general wearing conditions and provides protection by its hardness and resilience.

Formed pieces of vulcanized fibre offer outstanding service as arc barriers in circuit breakers. The arc-resistant properties of vulcanized fibre prevent a breakdown when the circuit breaker is subjected to an overload. The formed barrier is tested to take higher electrical loads than the maximum which can be produced by the circuit and, since repeated circuit breaks will not affect the performance of vulcanized fibre, the need for replacement is negligible.

Peerless control tape is used for programming data, processing equipment or automatic machining equipment. The special properties which give this material outstanding service life are high tensile strength, high tear strength, low stretch and good abrasive resistance.

Flame-resistant vulcanized fibre gives designers a structural material which can be used in those applications requiring a nonburning material and reduces fire hazards by containing a fire at its source. Flame-retardant parts serve as barriers in electrical equipment, materials-handling equipment and waste baskets.

Chemistry

Vulcanized fibre is produced by the chemical action of zinc chloride solution on a saturating grade of absorbent paper when processed under heat and pressure. The action of the zinc chloride converts the cellulosic fibers to a dense, homogeneous structure producing a laminated material which is refined to a chemically pure form. Final processing consists of drying to the proper moisture content and applying the proper calender to give smoothness and uniformity of thickness.

Grades

There are many grades of vulcanized fibre. The 13 basic grades are defined by NEMA as follows:

Electrical insulation grade. Primarily intended for electrical applications and others involving difficult bending or forming operations. It is sometimes referred to as "fishpaper."

Commercial grade. Considered to be the general-purpose grade, sometimes referred to as "mechanical and electrical grade." It possesses good physical and electrical properties and fabricates well.

Bone grade. Characterized by greater hardness and stiffness associated with higher specific gravity. It machines smoother with less tendency to separate the plies in the machining operations.

Trunk and case grade. Conforms to the mechanical requirements of "commercial grade," but has better bending qualities and smoother surface.

Flexible grade. Made sufficiently soft by incorporating a plasticizer, it is suitable for gaskets, packings, and similar applications. It is not recommended for electrical use.

Abrasive grade. Designed as the supporting base for abrasive grit for both disk and drum sanders. It has exceptional tear resistance, ply adhesion, resilience, and toughness.

White tag grade. It has smooth clean surfaces and can be printed or written on without danger of ink feathering.

Bobbin grade. Used for the manufacture of textile bobbin heads. It punches well under proper conditions, but is firm enough to resist denting in use. It machines to a very smooth surface.

Railroad grade. Used as railroad track joint, switch rods, and other insulating applications for track circuits. It conforms to A.A.R. Signal Section Specification for Hard Fibre.

Hermetic grade. Used as electric-motor insulation in hermetically sealed refrigeration units. High purity and low methanol extractables are essential because it is immersed in the refrigerant.

White grade. Recommended for applications where whiteness and cleanliness are essential requirements.

Shuttle grade. Designed for gluing to wood shuttles to withstand the repeated pounding received in textile power looms.

Pattern grade. Made to provide maximum dimensional stability and minimum warpage for use as patterns in cutting cloth, leather, and similar materials.

TABLE 1—TYPICAL PROPERTIES OF VULCANIZED FIBRE

		Thick-ness	Units	"Peer-less" Insula-tion	Bone	Commercial	Special White
MECHANICAL							
Ten Str	MD	$\frac{1}{16}$	1000 psi	21	17	16	16
	CD	$\frac{1}{16}$	1000 psi	10	10	9	8
Mod of Elast in Tension	MD	$\frac{1}{16}$	10^5 psi	—	12	—	—
	CD	$\frac{1}{16}$	10^5 psi	—	8	—	—
Flex Str	MD	$\frac{1}{16}$	1000 psi	20.5	21	18	20
	CD	$\frac{1}{16}$	1000 psi	14.5	16.8	15	16
Compr Str		$\frac{1}{16}$	1000 psi	34	36	30	35
Edgewise Impact Str (Izod)	MD	$\frac{1}{16}$	ft-lbs/inch notch	3.0	2.2	2.3	2.1
	CD	$\frac{1}{16}$		2.9	2.0	2.1	1.9
Rockwell Hardness		$\frac{1}{16}$	—	R70	R100	R70	R82
Bond Str		$\frac{1}{16}$	1000 psi	6	19	8.5	11
Bursting Str, Mullen		$\frac{1}{64}$	1000 psi	275	—	—	175
Tear Str, Elmendorf	MD	$\frac{1}{64}$	grams	550	—	—	270
	CD	$\frac{1}{64}$	grams	700	—	—	290
ELECTRICAL							
Dielec Str (Short Time)		$\frac{1}{64}$	v/mil	400	—	230	325
		$\frac{1}{16}$	v/mil	215	200	200	225
		$\frac{1}{8}$	v/mil	200	200	195	220
Arc Res		—	sec	125	100	80	107
PHYSICAL							
Density		$\frac{1}{16}$	gm/cc	1.20	1.34	1.20	1.28
Sp Vol		$\frac{1}{16}$	cu in./lb	23.0	20.6	23.0	21.7
Ther Cond (@ 149 F)		—	Btu/hr/ft²/°F/ft	0.168	0.168	0.168	0.168
Sp Heat		—	Btu/lb/°F	.403	.403	.403	.403
Heat Res (Continuous) [a]		—	F	221	221	221	221
Coef of Expansion × 10^5	MD	—	10^5/°F	1.1	1.1	1.1	1.1
	CD	—		1.7	1.7	1.7	1.7
Dimen Change per			%	1.00	1.00	1.00	1.00
Degree Change in	MD	—	%	0.10	0.10	0.10	0.10
Moisture Content	CD	—	%	0.25	0.25	0.25	0.25
Water Abs (24 hr)		$\frac{1}{16}$	%	63	52	63	54
Coef of Friction							
Fibre on Fibre		—		0.16	0.16	0.16	0.16
Fibre on Smooth Cast Iron		—		0.21	0.21	0.21	0.21
Flammability		$\frac{1}{16}$	in./min	0.5	0.5	0.5	0.5

[a] AIEE Class A insulation.

W

WASH PRIMERS

Wash primers are a special group of corrosion-inhibitive coatings designed for use on clean metal surfaces. They are also known as "wash-coat primers," "metal conditioners" and "etch primers."

The most widely utilized primers consist of a two-part system which is prepared at the point of use by simple mixing of specified proportions. The base grind portion contains a corrosion-inhibiting pigment, basic zinc chromate (also known as zinc tetroxy chromate) and a small amount of talc extender ground in an alcohol solution of polyvinyl butyral resin. The reducer portion consists of phosphoric acid, alcohol and water. When these are mixed a slow chemical reaction ensues, resulting in partial reduction of the chromate pigment. The life of the mixed primer is usually 8 to 12 hr. Single-package primers are now coming into use which are subject to certain limitations.

WATER-SOLUBLE PLASTICS

Within the plastics industry, water-soluble materials offer a variety of desirable physical properties, yet retain the advantages inherent in a water system. These advantages include ease of handling, negligible solvent costs, low toxicity, and low flammability.

There is no sharp dividing line between water-dispersible and water-soluble polymers. Many so-called water-soluble plastics form colloidal dispersions rather than true solutions. In this text, emulsions or dispersions of water-insoluble polymers are not discussed (such as acrylics, polyvinyl acetate, styrene butadiene, polyvinyl butyral, etc.).

The water-soluble plastics covered here can be roughly divided into two general classes:

Thermoplastic resins

These plastics are usually synthesized by addition polymerization techniques. That is, small units (or monomers) are joined together to develop the final molecular weight and polymer configuration. Rarely do these polymers develop into long straight chains; considerable branching often occurs. The molecular weight, chemistry of side groups and extent of branching all determine the properties which are obtained. These plastics are available as white or light-colored powders or in solution. Films, moldings, and extrusions are also available based on some of the thermoplastic resins.

Alkali soluble polyvinyl acetate copolymers. Polyvinyl acetate itself is water-insoluble. However, copolymers are available wherein vinyl acetate is copolymerized with an acidic comonomer. Such products retain the organic solubility of polyvinyl acetate but are soluble in aqueous alkali. These polymers exhibit low viscosity in solution and deposit high gloss films which are water resistant, provided a volatile alkali such as ammonia is used. The use of a fixed alkali will result in a film with permanent water sensitivity. They generally possess good adhesion to cellulose and a wide variety of other surfaces.

Major uses include loom-finish warp sizes for dope dyed yarns, repulpable adhesives or sizes for paper and board, conditioning agent for masonry prior to painting, protective coatings for metals and leveling agent and film former in self-polishing waxes.

Ethylene-maleic anhydride copolymers. High molecular weight polymers have been prepared by copolymerizing ethylene and maleic anhydride. These resins are available either in "linear" form or crosslinked with either anhydride, free acid, or amide-ammonium salt side chains.

Major applications include general thickening and suspending in adhesives, agricultural chemicals, cleaning compounds, and ceramics. This resin is used as a thickener for latex and as a warp size for acetate filament.

Polyacrylates. Commercially important polymers are prepared by polymerizing either acrylic or methacrylic acid. Usually these products are neutralized with bases to the salt form. Solution viscosity increases during neutralization. Cast films are hard, transparent, colorless and somewhat brittle.

Polyacrylic acid itself is used as a warp size for nylon. The neutralized polymers (polyacrylates) are used in various coating and binding applications (ceramics, grinding wheels, etc.). Because of interesting solution properties, the polyacrylates are used as thickeners, flocculants, and sometimes as dispersants in applications such as ore processing, drilling muds and oil recovery.

Polyethers. Two different polymer types are covered under this heading: polyoxyethylene (includes polyethylene glycols) and polyvinyl methyl ether and copolymers.

Polyoxyethylene (polyethylene glycol). These resins are available over a wide molecular weight range. Low molecular-weight members are slightly viscous liquids whereas the medium molecular weight types (1000–20,000) are waxy solids. Polymers up through this molecular-weight level are known as polyethylene glycols. Extremely high molecular-weight (several thousand to several million) homologues are also available. All types are soluble in water and in some organic solvents.

Applications for the liquid and waxy solids include lubricants (rubber molds, textile fibers, and metal), bases for cosmetic and pharmaceutical preparations and chemical intermediates for further reaction. The very high molecular-weight types are useful principally as thickeners in many application areas.

Polyvinyl methyl ether. This unique family of vinyl polymers shows inverse solubility in that the resins precipitate out above 35 C. They do redissolve upon cooling and the addition of low molecular-weight alcohols increases the solubility in water and raises the precipitation temperature. Higher homologues are available which are water-insoluble and quite tacky. The products exhibit pressure-sensitive adhesiveness coupled with good cohesive strength and high wet tack. Copolymers are available which contain maleic anhydride to modify physical properties, particularly solubility or tolerance to water and organic solvents and ease of insolubilization.

Major uses take advantage of properties such as pressure-sensitive characteristics (adhesives), tackiness (various latex systems), thickening active, heat sensitizing (latices for dip forming), and binding power (pigments).

Polyvinyl alcohol. These resins are available commercially in a wide range of types which vary in viscosity and chemical composition. Polyvinyl alcohol exhibits good water solubility, high resistance to organic solvents, oils, and greases, high tensile strength, adhesion and flexibility. In addition, the polymer is resistant to oxidation and in film form is an excellent barrier for various gases. Certain types exhibit surface activity in solution and all types are soluble in both acid and alkaline media. The resins can be crosslinked by borax and numerous organic and inorganic agents to produce thickening or even insolubilization.

In adhesives, polyvinyl alcohol contributes machinability, viscosity control, specific adhesion and in some cases remoistenability. Other major uses include paper coating and sizing (for increased strengths, ink hold-out and grease resistance), textile sizing, wrinkle-resistant finishes (wash-and-wear fabrics in conjunction with thermosetting resins), polyvinyl acetate emulsion polymerization (protective colloid), binder (for nonwoven ribbons, filters, etc.), film (release agent in polyester and epoxy molding and water-soluble packaging), cement additive (for improved strength, toughness, and adhesion) and photosensitive coating (in the graphic arts industry).

Polyvinyl pyrrolidone. Polyvinyl pyrrolidone (PVP) exhibits good solubility in both water and various organic solvents. A nontoxic material and tacky substance when wet, the polymer is a dispersant, suspending agent, and an adhesive component for bonding difficult surfaces.

Major uses include cosmetic preparations (hair sprays, etc.), tablet binding and coating, detoxifying of dyes, drugs, and chemicals, beverage clarification and specialty textile and paper applications involving sizing, dyeing and printing.

Copolymers are also available (like PVP-vinyl acetate) which have some advantages over the homopolymers in heat sealability, pressure-sensitive adhesiveness and other properties.

Polyacrylamide. These high molecular-weight polymers are soluble in both cold and hot water and in selected organic solvents. The resin is an efficient thickener and by reaction can be changed in physical and chemical properties.

In addition to general uses for water-soluble resins, these resins have shown an outstanding ability to flocculate fines and increase the filtration rate of slurries. Consequently, polyacrylamide is used in ore processing and in other such systems where dispersed materials are encountered.

Styrene-maleic anhydride. Copolymers of these two monomers are soluble in some organic solvents and alkaline water. Styrene-maleic anhydride resins produce viscous and stable aqueous solutions. This resin is a strong polyelectrolyte.

It is used as a textile warp size, paper coating and static-electricity conductor. The polymer is also used in alkaline latex systems as a protective colloid, emulsifier, pigment dispersant and filming aid.

Cellulosic derivatives. Various commercial derivatives are prepared from alpha cellulose which is obtained from several plant sources. One class of derivatives is the water-soluble ethers. These products produce viscous aqueous solutions. All have some resistance to organic solvents, are hygroscopic and are difficult to insolubilize.

Major industries which use these polymers as well as the water-soluble synthetic resins include food, pharmaceutical, cosmetic, textile, paper, petroleum (drilling muds), ceramic, paint, emulsion polymerization and leather.

Hydroxyethylcellulose. This polymer is manufactured by reacting alkali cellulose with ethylene oxide. It can be water-soluble or only alkali-soluble depending upon the extent of reaction. The alkali-soluble types possess the advantage of increased water resistance in deposited films. This polymer is somewhat intermediate in properties between methylcellulose and sodium carboxymethylcellulose. It is a protective colloid and relatively insensitive to the inclusion of multivalent ions in solution. The polymer is soluble in both hot and cold water, nonionic, but depolymerized by strong acids.

Water-soluble hydroxyethylcellulose is used in polyvinyl acetate emulsions as a stabilizer and in latex paints as a thickener and leveling agent.

Methylcellulose. Methylcellulose exhibits inverse water solubility in being more soluble at low temperatures than at high temperatures. It is nonionic in solution and is a very efficient thickener.

A major use is in latex paints (both polyvinyl acetate and acrylic types). Methylcellulose thickens the paint and contributes to good brushing characteristics as well. Other uses include bulking in laxatives, and binding and thickening in cosmetics and pharmaceuticals.

Sodium carboxymethylcellulose. This polymer is soluble in both hot and cold water. It exhibits good thickening action and suspending ability for particulates. Since solutions are ionic in character, they are somewhat sensitive to pH shifts and salt additions.

Major uses include soil suspension in synthetic detergents and viscosity control in oil-well drilling muds.

Thermosetting resins

A number of thermosetting resins are available in water solutions or in water-soluble form. These are principally the addition reaction products of formaldehyde with either urea, phenolic or melamine. Resorcinol and thiourea may also be reacted with formaldehyde to form water-soluble precondensates, although these have not attained the volume of the three main classes defined above. Certain cyclic derivatives have gained recent prominence.

These resins develop high molecular weight by a condensation reaction. The properties may change as the reaction proceeds. A-stage resins are those wherein the degree of polymerization is minor. In some cases the degree of polymerization is such that only dimers, trimers or similar small units are prepared. These products are water-soluble or at least can tolerate the addition of significant amounts of water.

Should the reaction proceed further, the polymers enter an area roughly defined as B-stage wherein they will tolerate addition of only small amounts of water but are soluble in certain organic solvents.

Manufacturers of thermosetting resins carry the polymerization reaction to the A- or B-stage. Further reaction (to the C-stage) is carried out by consumers of these resins.

As the condensation reaction proceeds and the molecular weight builds to form a rigid, three-dimensional system, the polymers reach a point where they will not dissolve in organic solvents and are then termed "cured" (or C-staged). Further heating of the resin beyond this point may establish additional crosslinks, but the physical properties do not change drastically.

Once the resin has reached the C-stage, excessive heating leads to chemical breakdown of the material.

Thermosetting water-soluble polymers are treated with heat and/or catalyst to advance the cure after deposition on a particular surface or within a particular structure. Thus, the water-solubility feature is important in that it allows easy manipulation without the cost and hazards of organic systems. However, water is an essential ingredient when crosslinking cellulose with low molecular-weight thermosetting resins. The general characteristics of a fully cured, water-soluble, thermosetting resin are similar to those obtained by curing B-stage varnishes or molding compounds.

In many areas, the water-soluble thermosetting resins compete with one another. Certain ones may be preferred in an industry or in a certain particular application because of specific properties or cost. Generally, these resins offer high-temperature stability and hardness coupled with water and solvent resistance. Resistance to either acids or bases can also be obtained.

Cyclic thermosetting resins. In this category are cyclic ethylene urea-formaldehyde resins and triazones which can be obtained by cyclization of dimethylolurea with a primary amine (usually ethylamine) and then adding two moles of formaldehyde. These cyclic thermosetting resins were developed primarily for textile applications because they do not react or polymerize with themselves (as do other water-soluble thermosetting resins) but do react with the hydroxyls in cellulose through the methylol groups. These resins are called cellulose reactants and are by far the largest class of resins used in the wash-and-wear treatment of fabrics. Such resins impart crease resistance, wrinkle recovery, stiffness, tensile strength, water repellency, and good resistance to yellowing by chlorine-containing bleaches. In addition there are increases in resiliency, dimensional stability and permanent texturizing.

Cyclic ethylene urea-formaldehyde resins have the advantage over triazone in better color stability, absence of odor, and scorch resistance. Triazone resins have gained prominence particularly since they are more resistant to the effects of chlorine and are somewhat cheaper. Triazones are used principally to develop wash-and-wear properties on white cotton.

Melamine-formaldehyde. In general, melamine-formaldehyde (MF) resins are the most expensive of the water-soluble thermosetting types. They possess good color, lack of odor, high abrasion resistance, and high resistance to alkali.

The major use is in decorating laminants for surfacing of wood, paper and other products. Other uses include binding of rock wool and glass wool for thermal insulation, finishing of nylon for stiffness and resilience, and imparting dimensional stability to wool and some cellulosics.

Phenol-formaldehyde. This very popular class of water-soluble thermosetting resins is intermediate in cost between the melamine-formaldehyde and urea-formaldehyde (UF) types. They possess good water resistance, toughness and acid resistance although they are somewhat poorer than the MF resins in color, odor and flame resistance. In most other physical and chemical properties they are superior to the UF types.

Applications include laminates (including plywood and fabrics), grinding wheels, thermal insulation, battery separators, brake linings and foundry uses.

Urea-formaldehyde. These resins are the lowest cost as a class of the three general types but still cure to hard and somewhat brittle resins which have many desirable properties. The urea-formaldehyde (UF) resins have an added plus in that they can cure at room temperature with suitable catalysts, whereas both the melamine-formaldehyde and phenol-formaldehyde types normally require temperatures in the neighborhood of 300 F to develop their full properties. The UF resins suffer somewhat in comparison to the other two types in poorer water resistance, less toughness, and poorer resistance to cyclical changes in temperature or water exposure.

Urea-formaldehyde resins are used in plywood because of ease of handling and lower temperature of cure, on paper for increased wet strength in air filters and on certain rayon fabrics for improved stabilization and water resistance. The UF resins are used as insolubilizers for hydroxyl-containing polymers and in many of the general application areas for water-soluble, thermosetting resins.

WHITE CAST IRON

White cast iron solidifies with all its carbon in the combined state, mostly as iron carbide, Fe_3C (cementite). White iron contains no free graphite as does gray iron, malleable iron, and ductile iron. White iron derives its name from the fact that it shows a bright white fracture on a freshly broken surface.

The main use for white iron is as an intermediate product in the manufacture of malleable iron. In addition to this, white iron is made as an end product to serve specific applications that require a hard, abrasion-resistant material. White iron is very hard and resistant to wear, has a very high compressive strength, but has low resistance to impact and is very difficult to machine.

By the proper balancing of chemical composition and section size, an iron casting can be made to solidify completely white throughout its entire section. By modifying the balance and adjusting the cooling rate, the casting can be made to solidify with a layer of white iron at the surface backed up by a core of gray iron. Castings with such a duplex structure are called "chilled iron" castings.

Castings of white iron and chilled iron find their main use in resistance to wear and abrasion. Typical applications include parts for crushers and grinders, grinding balls, coke and cinder chutes, shot-blasting nozzles and blades, parts for slurry pumps, car wheels, metalworking rolls, and grinding rolls.

By using a fairly low silicon content, cast iron can be made to solidify white without the use of any additional alloy. Carbon contents are kept high (about 3.6%) when high hardness (575 Bhn) is desired. Such irons have very low toughness and a strength of about 30,000 psi. For somewhat higher toughness and strength, at some sacrifice of hardness, the carbon content is lowered to about 2.8%. Unalloyed white and chilled irons have a structure composed of particles of massive iron carbide (Fe_3C) in a matrix of fine pearlite. For highest hardness, strength and toughness, the white iron is alloyed so as to produce a martensitic matrix surrounding particles of massive carbide. The composition and properties of some typical unalloyed and alloyed white irons are as follows:

TABLE 1—TYPICAL WEAR-RESISTANT WHITE AND CHILLED IRONS

	Unalloyed White Iron	Martensitic Alloyed White Iron	High-Chromium White Iron
Typical Composition, %			
Carbon	2.8–3.6	2.8–3.6	1.8–3.5
Silicon	0.5–1.3	0.4–0.7	0.5–2.5
Manganese	0.4–0.9	0.2–0.7	0.3–1.0
Nickel	—	2.5–4.7	0–5
Chromium	—	1.2–3.5	10–35
Copper	—	—	0–3
Molybdenum	—	—	0–3
Bhn	300–575	525–600	250–700
Ten Str, 1000 psi	20–50	40–75	23–90

WHITEWARE CERAMICS

Technical whitewares include clays, porcelains, china, white stoneware and steatites. The modern oxide ceramics would also be in this group, but are treated elsewhere.

For technical use, these whitewares are usually vitrified (nonporous) or very nearly so. Most commonly the pieces are glazed.

To produce white-bodied ceramics, the raw materials must be of superior quality and selection. The range of materials available is comparatively limited. Though clays are widely distributed over the earth, large, uniform deposits of white burning clays are not common. Some of the largest deposits in the United States are in Florida, North Carolina, Georgia, Tennessee and Kentucky, with small deposits in other states east of the Mississippi.

Composition and types

Most whiteware clays are basically kaolins and chemically are hydrous aluminum silicates. Ball clays are less pure than kaolins and contain free silica and small amounts of other contaminants. Kaolins are generally not highly plastic, while ball clays are very plastic. The plasticity derives from the physical form of the minute particles which are colloidal in size. Ball clays are not as white burning as kaolins and impart color to the fired body.

Feldspars are used as fluxes to provide the alkaline oxides for the glassy phase surrounding the mullite ($3\ Al_2O_3 \cdot 2\ SiO_2$) crystals forming the mass of the body.

Porcelains, china, and stoneware are composed

of clays, silica and feldspar. Steatites are composed of talc (magnesium silicate) and clay. Minor variations are made to enhance special properties.

The porcelains, china, and stoneware are composed of alumina (Al_2O_3), silica (SiO_2), sodium and potassium oxides (Na_2O and K_2O), as well as calcia (CaO), zinc oxide (ZnO), zirconia (ZrO_2), titania (TiO_2), barium oxide (BaO), magnesia (MgO), phosphoric oxide (P_2O_5). Some other oxides may be present as traces. Iron oxide is usually present in small amounts as an undesirable impurity.

The oxides are supplied in kaolin, ball clays, quartz, feldspar, whiting, magnesia and talc. Oxide ceramics which do not use clays may use only mixtures of refined oxides.

Compounding and forming

The raw materials are intimately mixed, usually by ball milling, and then prepared for forming into ware by one of the methods listed below. For jiggering and simple mechanical pressing or extruding, the ball-milled slip is dewatered, filter pressed, deaired, and extruded into convenient size billets.

For casting, the specific gravity is adjusted to give a good casting viscosity.

For dry pressing, the body is dried, shredded and powdered or spray dried into minute granules.

The pieces are formed as close to size before firing as practical, allowing for the shrinkage during firing. This shrinkage may run as high as 25% and must be very closely controlled to avoid loss in the firing and off dimension pieces.

The ware is formed in a number of ways, some of them unchanged for centuries or even thousands of years and others unknown 25 years ago. Following are the primary methods used:

1. Throwing or jiggering on mechanical potters wheel from plastic clay body.
2. Casting in plaster of Paris mold from liquid slurry or slip.
3. Pressing from plastic clay with simple mold.
4. Pressing from dry powder in metal dies.
5. Isostatic pressing from dry powder in rubber sack with hydraulic pressure.
6. Hot pressing with heated clay blank and mold.
7. Simple extrusion through die.
8. Extrusion with thermoplastic resin in the body (injection molding).

Many pieces are formed by a combination of pressing or extrusion followed by mechanical shaping in lathes by special tools or dry grinding.

After forming, pieces are bisque fired, which drives out moisture and water of crystallization. Porcelain is not vitrified during the bisque firing. The porcelain bisque is dipped or sprayed with glaze and, in the second firing, the body and glaze mature or vitrify together.

China is vitrified in the first firing. In china manufacture, the glaze is applied to the fired body and the glaze matures in the second firing which is at a lower temperature than the initial firing. In some cases the glaze can be applied to an unfired piece and only one firing is needed.

Today, most kilns are fired with natural or manufactured gas, fuel oil, or electricity. The round beehive kiln fired periodically has been largely replaced with continuous tunnel kilns which may be built with movable cars moving from one end to the other on a straight track or as a circular tunnel with a moving floor. Periodic kilns are often used in specialized work where the volume does not justify use of a tunnel kiln. They are also more versatile.

The firing of technical ceramics is performed under the most carefully controlled conditions possible. Both the temperature and the atmosphere must be known and controlled. The ware must be properly placed in the kilns or damage to the piece will result. Warpage, uneven firing and cracking can easily occur.

Maximum firing temperatures for unglazed refractory porcelain are usually about 3200 F. Laboratory chemical porcelain which is glazed is fired to 2650 F. Hotel, sanitary china, and electrical porcelain is usually fired about 2300 F. Steatite bodies mature at around 2300 to 2400 F.

WOOD

For most purposes wood may be defined as the dense fibrous substance which makes up the greater part of a tree. It is found beneath the bark, and in the roots, stems and branches of trees and shrubs. Of the three sources the stem or trunk furnishes the bulk of raw material for lumber products.

Wood is a renewable resource. It is grown in every state of the Union and can be produced in any reasonable quantity needed for our future consumption. Wood products and the management of our forested lands are changing to meet modern conditions; hence we are growing trees to meet modern production requirements for size, quality and quantity. For example, with the developments in the modern technique of gluing, the former use of extremely wide, thick and excessively long lumber is no longer necessary. Laminated lumber and plywood have generally taken the place of these large boards and timbers. Not only are the raw materials for such products easier to grow and more economical to obtain than are solid timbers of comparable size, but the products are generally improved by the use of modern methods of fabricating.

Although there are many species of wood in the United States, the commercially important types can be grouped into two categories of about twenty-five for each group.

Hardwoods and softwoods

The terminology used in the classification of trees is confusing, but since it has become general in usage, it is important for those who make or purchase products of wood to understand it.

The terms "hardwood" and "softwood" have no direct application to hardness or softness of the materials. Basswood is a softer domestic species, yet the yellow pines, which are classed as "softwood," are often much harder. Even balsa, a foreign species that everyone knows, the lightest and softest wood used in commerce, is classed as "hardwood." For practical purposes, the "hardwoods" have broad leaves, whereas the "softwoods" have needlelike leaves. Trees (hardwoods) with broad leaves usually shed them at some time during the year, while the conifers (softwoods) retain a covering of the needlelike foliage throughout the year. There are quite a few exceptions to these criteria, but as one gains familiarity with the various woods, he will soon learn by experience into which group a species falls.

Those who use wood must know something of the botanical classification because the lumber industry is also divided into two distinct groups. Their methods of doing business and manufacturing and grading for quality differ from each other.

Generally the hardwoods are used for the manufacture of factory-made products, such as tools, furniture, flooring, instrument cases, etc. The largest market for softwoods is in the home construction field or for other building purposes.

But there is no line of demarcation that is reliable, for hardwoods and softwoods are often interchangeable in use.

The more important hardwoods are:

Alder	Hackberry	Sweet or Red
Ash	Hickory and	Gum
Aspen	Pecan	Sycamore
Basswood	Holly	Tupelo or Black
Beech	Locust	Gum
Birch	Magnolia	Walnut, Black
Cherry	Maple (Hard &	Willow, Black
Chestnut	Soft)	Yellow Poplar
Cottonwood	Oak, Red	
Elm	Oak, White	

The more important softwoods are:

Cedar (several species)	Pine, Ponderosa
Cypress	Pine, Red
Douglas Fir (not a true fir)	Pine, Southern Yellow (several species)
Firs (eastern and western)	Pine, Sugar
Hemlock (eastern and western)	Pine, Virginia
Larch	Pine, Western White
Pine, Eastern White	Redwood
Pine, Jack	Spruce, Eastern
Pine, Lodgepole	Spruce, Englemann
Pine, Pitch	Spruce, Sitka
	Tamarack

Chemical composition

Wood, regardless of the species, is composed of two principal materials: cellulose, which is about 70% of the volume, and lignin, nature's glue for holding the cells and fibers together, which is from 20 to 28%. Residues in the form of minerals, waxes, tannins, oils, etc., compose the remainder. The residues, though small in volume, often provide a species with unusual properties. The oils in cypress are responsible for its renown as a decay-resistant wood. Aromatic oils provide many of the cedars with distinctive odors that make them valuable for clothing storage chests. Other chemicals provide resistance to water absorption, which is useful for constructing light, high-speed boats that are relatively free from increase in weight due to water absorption.

Many of the chemical residues of wood can be removed by neutral solvents, such as water, alcohol, acetone, benzene and ether. Some of them may be caused to migrate from one part of the wood to another. American Black Walnut can be treated with steam so that some of the dark coloring matter will travel from the heartwood to the white sapwood.

Structure of wood

The roots, stem and branches of a tree increase in size by adding a new layer each year, just as the size of one's hand is increased by putting on a glove. The layer or growth ring will vary in thickness due to the age of the tree, growing conditions, amount of foliage and other factors.

The growth ring is divided into two parts—spring wood and summer wood. The former is usually lighter in weight than the latter and is denser and stronger. Generally, a tree or portion of a tree having the most summer wood is stronger than one which has less.

The thickness of the growth ring and the relative amounts of spring and summer wood have great affect on the appearance of wood and are often a deciding factor in the choice of furniture materials. Excessively thick growth rings usually provide a rather coarse-textured material, while narrow growth rings provide fine texture. However, in hardwoods a thick ring may and usually does have more summer wood and is therefore the stronger of the two. The situation is somewhat different for the softwoods.

In the life of a tree a gradual change usually takes place in the older rings. Due to the infiltration of minerals and other chemicals residues are absorbed and retained. Sometimes these residues are very dark in color as in black locust, walnut, redwood and cypress. This part of the tree is known as heartwood, while the outer or uncolored rings are known as sapwood. Some species have a preponderance of heartwood while others have only small amounts. In some, such as firs and spruces, it is difficult to see a distinction.

To the user of wood the differentiation between heartwood and sapwood may be of considerable importance. First of all, the heartwood of many species is more decay resistant than the sapwood of the same species. Secondly, sapwood, regardless of species, has about the same decay resistance. Therefore, when a decay-resistant species is mentioned, the reference is only to the heartwood. On the other hand, if it is desirable to treat wood with artificial chemicals for the prevention of decay the sapwood is frequently found to be better for absorption of the chemicals. The heartwood of many of the decay-resistant species is extremely difficult or even impossible to impregnate, but they have natural decay resistance and require less artificial treatment to achieve the purpose.

With some species, the sapwood is preferred for many uses. Maple and hickory are good examples. For nearly every use the white or sapwood of maple is preferred because of its appearance. For tool handles and in other areas where great strength is needed, the sapwood of hickory is preferred to the heartwood. This is based on an erroneous assumption that the white wood is the stronger. There is no basis for the assumption; in fact, there is no difference.

The cells or fibers of wood and the glue or lignin that holds them together are among the structural elements that make the material strong or relatively weak. These cells, regardless of species, are made of the same material, which is heavier than water. Wood floats only because of the millions of minute air spaces within its structure. All species, if compressed sufficiently, will have the same specific gravity and will sink when placed in water.

Most of the cells in any species of wood are roughly parallel to the axis of the tree stem, limb or root, but in all species there is another cell system that runs perpendicular to the axis. Technically, these are known as "rays." In some woods, such as oak and beech, the rays are easily seen with the naked eye, but in the softwoods they are less pronounced and seldom noted by the average person. Whether seen or not, however, they are important. If lumber is cut parallel with the rays, it is known as "quarter sawn," but if the cut is at right angles to the plane of the rays it is "plain sawn." For the species such as oak where the ray is prominent the figure of the grain or appearance of the wood when quarter sawn is entirely different than for plain sawn. Also, when lumber is quartered the edges of the annual rings on the wide face are narrow; hence, with such species as yellow pine and Douglas fir, the grain appears to be finer and is better in appearance than flat-sawn lumber with large cone-shaped markings.

In addition to appearance, quarter-sawn lumber has some different properties than plain sawn. A quarter-sawn board will shrink less across the broad face than plain sawn, but it will shrink more in thickness. For this reason a quartered board tends to warp and twist to a lesser extent.

Peculiarly, many purchasers interested in appearance make a mistake of specifying that wood shall be straight grain, whereas they actually mean, edge grain, comb grain or quartered lumber. Straightness of grain has nothing to do with its fineness; in fact, cross grain may enhance appearance.

Grades of lumber

Modern grading of lumber is the result of experience. Over a period of seventy years there has been a gradual evolution to meet the changes in industry. This trend will continue as long as wood products are used.

The grading of hardwoods and softwoods is entirely different; the former are used almost entirely as raw materials for manufactured products. For them, basic grading is on the number of usable cuttings or pieces that can be cut from a board. For some grades these cuttings need only be sound, but for others they must be clear at least on one face. The cuttings must also be of a certain size.

Hardwood grading. The grades of hardwood lumber from the highest quality to the lowest are: "Firsts," the top-quality grade; the next are known as "Seconds." These two grades are usually marketed as one and called "Firsts and Seconds"; the designation for the grade is FAS. The next lowest grade is "Selects," followed by "No. 1 Common," "No. 2 Common," No. 3A Common" and "No. 3B Common." Sometimes a grade is further differentiated, such as "FAS One Face," which means that it is of a much higher grade on one face than on the other. A prefix "WHND" is also used sometimes. It means that wormholes are not to be considered as defects in evaluating cuttings nor as a reason for disqualifying them as might otherwise be done.

For the user of hardwood lumber, the important thing to remember is that the grading is based on the amount of usable lumber that can be obtained from a board. The size and number of defects, such as knots, are of no significance except that if the damaged area is too great it may reduce the number of cuttings below the requirement for a certain grade. However, boards of the higher grades must usually have greater average lengths and widths than the lower grades.

Hence, the user has a choice of grades that he can use for the same product. He will get more clear cuttings from "Selects" than he will from "No. 1 Common," but the cuttings which he obtains from either one will be of the same quality and can be used for the same purpose. The best grade to purchase, therefore, depends on a thorough analysis of the product to be made, the relative price of the various grades, the manufacturing costs of cutting up the lumber for sizes that will suit the needs and many other factors. The most advantageous grade for a given purpose will change from time to time as the economic factors change, particularly the comparative prices. Furthermore, if properly handled, the choice of the grade will have nothing to do with the quality of the finished product but it can have everything to do with the cost.

Official rules for the grading of hardwood lumber are published by the National Hardwood Lumber Association, Chicago, Ill. These rules cover practically every domestic species of hardwood lumber and some foreign species for sale in this country, such as Philippine, South American and African mahoganies.

Many users of hardwood buy rough lumber for manufacturing their products. Usually they will kiln-dry the lumber with their own facilities, cut it to length and width, and plane it to dimensions that suit their needs. The result is a product known as "hardwood dimension." This is a specialized manufacturing process, for it involves choosing the right grade of lumber, cutting the lumber to the proper size in order to obtain the maximum from the grade and working out means for disposing of too short or narrow residues that may have use in another product.

Large wood-products manufacturers with considerable resources in equipment, experience and a diversification of products can usually handle the business themselves. But small wood-products concerns or manufacturers of products where the use of hardwood is incidental, often find it is profitable to purchase rough, semi-machined or finished dimensions. Quite a large specialized industry has been built to service these needs. This branch of hardwood manufacturers constitutes the Hardwood Dimension Industry. These firms are skilled in estimating the lumber and grade requirements and can integrate utilization of one lumber requirement with another. Usually they are located convenient to raw material sources and can therefore refine the lumber so that the shipment of waste is reduced to a minimum. This industry has its own grading rules for the products it manufactures. Information can be obtained by writing the Hardwood Dimension Manufacturers' Association.

Softwood grading. The theory of softwood lumber grading is probably somewhat more difficult for the layman to understand than that used for the hardwoods. This is due to the fact that many of the species are graded under separate association rulings, which are similar, but with some important differences.

Furthermore, softwoods are divided into three general classes of products, each of which is graded under a different set of rules.

Yard lumber, which is used for building construction and other ordinary uses, is obtainable in "Finish Grades" which are called "A," "B," "C," and "D"; "Common Boards" which are called "No. 1," "No. 2," "No. 3," "No. 4" and sometimes "No. 5," and as "Common Dimension" in grades "No. 1," "No. 2" and "No. 3."

Yard lumber is less than 5 in. thick and is intended for general building purposes. The user should keep in mind that in Yard lumber, the Finish Grades are exactly as the name implies, i.e., intended for finishing purposes, and the grading is intended to supply lumber that is suitable in appearance for these purposes. Finished lumber is less than 4 in. thick and under 16 in. wide.

Common Boards are intended for general use where appearance is not a specific requirement. They are less than 2 in. thick. Common Dimension is over 2 in. in thickness and under 5 in.; it is intended for general-utility purpose such as framing.

Structural lumber is a relatively modern concept in the field of lumber grading. It is an engineered product, intended for use where definite strength requirements are specified. The allowable stresses designated for a piece of structural lumber depend upon the size, number and placement of the defects. The relative position of a defect is of great importance; therefore, if the maximum strength of the piece is to be developed it must be used in its entirety. It cannot be remanufactured for width, thickness or length.

Structural lumber is never less than 2 in. in thickness.

Factory and shop lumber is the third general category. These grades are similar to those for hardwoods in that the lumber is graded by the number of usable cuttings that can be taken from a board, but here the resemblance ceases, for both the grade descriptions and nomenclature are different. The term "Factory and Shop" is descriptive of the uses for which the product is designed. Much of it is used for general millwork products, patterns, models, etc., and wherever it is necessary to cut up softwood lumber for the production of factory-made items.

The basic principles for modern softwood lumber grading were worked out by the industry in cooperation with the U.S. Department of Commerce. The standards themselves are issued in pamphlet form as *Simplified Practice Recommendations of the United States Department of Commerce.* Copies of the publication may be obtained by writing to that organization.

While the simplified practice recommendations are the basis for establishing softwood grading rules, and have exerted a powerful influence on them, they must not be confused with the rules specific to different species. These may be obtained from: California Redwood Association, Southern Cypress Manufacturers' Association, Southern Pine Association, West Coast Lumbermen's Association, Western Pine Association, and Northern Hemlock and Hardwood Manufacturers' Association. Information can also be obtained regarding any species by writing the National Lumber Manufacturers Association.

Mechanical properties

Density, or specific gravity, is the major factor related to the strength properties of wood. As a general rule, strength increases with density. For a given density of dry wood, the other factors influencing most mechanical properties or moisture content and defects in the wood. In endwise compression, for example, strength is about twice as great for a 12% moisture content as for green wood. However, toughness and shock resistance generally are not affected, and sometimes actually decrease as moisture content decreases.

Strength—Modulus of rupture, which is a measure of the ability of a beam to support a slowly applied load for a short time, is a widely accepted criterion of wood strength. Static bending or flexure stress at proportional limit is another strength indicator. Maximum crushing strength, which is the maximum stress sustained by a compression load slowly applied parallel to the grain, is an important criterion for compression strength. In general, tensile strength of wood parallel to the grain is forty times that perpendicular to the grain. Compressive strength parallel to the grain is three to ten times greater than that perpendicular to the grain (Table 1).

At temperatures above normal, wood strength tends to decrease. Air dried wood can be exposed to temperatures of up to about 150 F for a year or more without significant loss in most strength properties, but strength at such elevated temperatures is temporarily reduced. Exposures to temperatures of 150 F or more for extended periods of time permanently weaken wood, the extent depending chiefly on moisture content, temperature and exposure time.

Stiffness or Rigidity—Stiffness of woods is expressed in terms of modulus of elasticity in bending, and thus is a measure of resistance to deflection. The modulus is high compared to most polymeric materials, ranging from around 1.2 to 2.5 $\times$ 10^6 psi. The modulus in compression parallel to the grain is usually about 10% higher than the bending modulus.

Toughness—The ability of wood to absorb shock or impact loads is chiefly dependent on strength, and therefore density, flexibility and moisture content. As a class, woods compare favorably with other polymeric materials. Because of moisture content, green wood is generally tougher than seasoned or dry wood. As a

TABLE 1—STRENGTH PROPERTIES OF SOME AMERICAN WOODS

Kind of Wood	Specific Gravity	Mod. of Rupture Bend., 1000 psi	Bend. Stress at Prop. Limit, 1000 psi	Mod. Elast. in Bend., 10^6 psi	Max. Crush. Str., par.t grain, 1000 psi
Softwoods:					
Cedar—Port Orford	0.42	11.3	7.7	1.3	6.5
Cedar—Eastern Red	0.47	8.8	3.8	0.9	6.0
Cypress—Southern Bold	0.46	10.6	7.2	1.4	6.4
Douglas Fir—Coast	0.48	12.2	8.1	2.0	7.4
Fir—Balsam	0.36	7.6	5.2	1.2	4.5
Hemlock—Eastern	0.40	8.9	6.1	1.2	5.4
Hemlock—Western	0.42	10.1	6.8	1.5	6.2
Pine—Eastern White	0.35	8.6	6.0	1.2	4.8
Pine—Longleaf	0.58	14.7	9.3	2.0	8.4
Pine—Ponderosa	0.40	9.2	6.3	1.3	5.3
Redwood—Virgin	0.40	10.0	6.9	1.3	6.2
Spruce—Litka	0.40	10.2	6.7	1.6	5.6
Hardwoods:					
Ash—White	0.60	15.4	8.9	1.8	7.4
Basswood	0.37	8.7	5.9	1.5	4.7
Beech	0.64	14.9	8.7	1.7	7.3
Beech—Yellow	0.62	16.6	10.1	2.0	8.2
Cottonwood—Black	0.35	8.3	5.3	1.3	4.4
Elm—Rock	0.63	14.8	8.0	1.5	7.1
Hickory—Shag Bark	0.72	20.2	8.9	2.2	9.2
Locust—Black	0.69	19.4	12.8	2.1	10.2
Maple—Sugar	0.63	15.8	9.5	1.8	7.8
Oak—Red	0.63	14.3	8.4	1.8	6.8
Oak—White	0.68	15.2	8.2	1.8	7.4
Poplar—Yellow	0.42	10.1	6.1	1.6	5.5
Walnut—Black	0.55	14.6	10.5	1.6	7.6

group, hardwoods are tougher than softwoods, although some species, such as basswood and sycamore are quite brittle. Among the softwoods, longleaf pine is about the toughest.

Fatigue Resistance—Wood, as a fibrous material, is less sensitive to repeated loads than are more crystalline structural materials, particularly metals. Endurance limits are usually higher in proportion to ultimate strength than is the case for some metals. For tension in fatigue parallel to the grain, the endurance load for 30 million cycles of stress can be 40% of the strength of air dry wood.

Hardness—Hardness gives a measure of resistance of wood to scratching, indentation and wear. It is measured by the load required to imbed a 0.444 in. ball to one-half of its diameter into the wood. Hardness on end grain wood surfaces are higher than on side grain surfaces, sometimes by over 50%. An exception is lignum vitae and teak, where side grain hardness is higher. As previously stated, the names softwoods and hardwoods are not indicative of the relative hardness of woods in these two major classes. Among hardwoods, basewood, poplar, aspen, cottonwood and willow are quite soft, whereas yew and some cedars have high hardness.

Thermal properties

Thermal conductivity varies with wood density. Thus, lighter weight woods are better heat insulators, and balsa being lightest is the best. For a given density, thermal conductivity can vary considerably depending on moisture content and its distribution. In high density woods, for example, conductivity can increase 20 to 25% when moisture content increases from 12 to 30%. Other influencing factors on thermal conductivity are direction of grain, proportion of springwood to summerwood and defects such as checks and knots.

Thermal expansion of wood differs in the longitudinal and in the radial and tangential directions of wood. Along the grain values are independent of density and vary from about 1.7×10^{-6} to 2.5×10^{-6} per 1°F. Therefore, coefficient of linear thermal expansion along the grain is from one-tenth to one-third that of common metals, concrete and glass. Across the grain, thermal conductivity generally varies with density, and although usually larger than in the grain direction, it is still usually less than those of other common structural materials.

Chemical and decay resistance

Wood has good resistance to common organic acids and solutions of acidic salts, but poor resistance to most alkaline solutions and inorganic acids. All oxidizing solutions of salts, acids and alkalies rapidly attack wood.

Chemicals can affect wood's strength by 1. producing swelling; 2. hydrolysis of the cellulose by acids or salts; and 3. attacking the lignin. Of the three, swelling is almost completely reversible, so that if the swelling liquid is removed from the wood, original dimensions and strength are regained.

The combination of moisture, air and microorganisms, such as fungi, is the major cause of wood decay. If kept dry or submerged in water, wood does not decay. Decay does not take place unless moisture content is above around 20%. Rate of decay varies greatly depending on local conditions and temperature. Wood in warm humid climates deteriorates more rapidly than in cool dry areas. The sapwood of most species has much lower resistance to decay than heartwood. Woods with low decay resistance can be improved by a number of different preservative treatments.

Electrical properties

Although wood is essentially a non-conductor, its electric resistance and dielectric properties vary considerably with moisture content. The specific resistance (resistance of a cubic centimeter of wood) of some common woods varies from 3×10^{17} to 3×10^{18} ohm-cm for oven-dry wood to 10^8 ohm-cm for wood at 16% moisture content. The dielectric constant of oven-dry wood is about 4.2, but increases with moisture content, since the dielectric constant of water is approximately 81.

Working and forming qualities

One of the major advantages of wood is its good working qualities, particularly with hand tools. Although there is no test for quantitatively evaluating workability, a qualitative classification based on experience at the Forest Products Laboratory is given in Table 2. The machinability of various woods often can be rated according to the quality of the surface produced when machine-worked. Table 3 roughly rates various hardwoods for their machinability in planing, and shaping operations.

TABLE 2–CLASSIFICATION OF CERTAIN HARDWOOD AND
SOFTWOOD SPECIES ACCORDING TO EASE OF
WORKING WITH HANDTOOLS

Group 1–Easy to Work	Group 2–Relatively Easy to Work	Group 3–Least Easy to Work
HARDWOODS		
Alder, red	Birch, paper	Ash, commercial white
Basswood	Cottonwood	Beech
Butternut	Magnolia	Birch
Chestnut	Sweetgum	Cherry
Yellow-poplar	Sycamore	Elm
	Tupelo:	Hackberry
	Black	Hickory, true and pecan
	Water	Honeylocust
	Walnut, black	Locust, black
		Maple
		Oak:
		Commercial red
		Commercial white
SOFTWOODS		
Cedar:	Baldcypress	Douglas-fir
Atlantic white	Fir:	Larch, western
Incense	Balsam	Pine, southern yellow
Northern white	White	
Port-Orford	Hemlock:	
Western redcedar	Eastern	
	Western	
Pine:		
Eastern white	Pine, lodgepole	
Ponderosa	Redcedar, eastern	
Sugar	Redwood	
Western white	Spruce:	
	Eastern	
	Sitka	

The groupings in the table indicate the approximate order of ease of working based on the experience of the Forest Products Laboratory and the general reputation of the wood. Direct comparison of species within a group and comparison of hardwoods and softwoods is not intended.

TABLE 3–RELATIVE MACHINABILITY AND
BENDABILITY OF HARDWOODS

Kind of Wood	Planing	Shaping	Steam Bending
Ash	B	A	B
Basswood	C	D	D
Beech	–	C	B
Birch	C	A	B
Buckeye	–	D	D
Chestnut	B	C	C
Cottonwood	D	D	C
Elm	D	D	B
Hackberry	B	D	A
Hickory		C	B
Magnolia	C	C	A
Mahogany	B	A	C
Maple:			
Hard	C	A	C
Soft	C	C	–
Oak:			
Chestnut	–	C	A
Red	A	C	A
White	A	C	A
Pecan	A	B	B
Sweetgum	C	C	B
Sycamore	D	D	C
Tupelo:			
Black	C	C	C
Water	–	A	C
Walnut, black	C	B	B
Willow	C	D	B
Yellow-poplar	B	D	C

Note: A represents highest rating.

Although wood cannot be hot or cold-formed in the same way as metals or other polymers, it can be bent or shaped to a considerable extent by soaking in hot water or steam and then shaping and clamping in position until the wood dries. Table 3 roughly rates various hardwoods for their ability to be steam bent.

Altering properties of wood

There are several methods by which the characteristics of wood and lumber can be altered and controlled.

Laminated wood. The laminating of wood is a process of building up assemblies by laying and fastening two or more pieces of wood or lumber together so that the grain of each layer of wood is parallel to the grain of all others. (Not to be confused with plywood where each layer is positioned so that its grain is perpendicular to that of the preceding ply.)

The laminae may be fastened to each other by nailing, gluing or other means, but the modern practice is to use only glue for this purpose. Laminated products compete with heavy timbers. They can be made to any required length, width or thickness and can be curved to nearly any shape.

The glues used depend on the service requirements: for marine products, fully waterproof glues are used, for exposure requirements which are not quite as severe, water-resistant glues may be used and where exposure is not a problem, non-water-resistant glues may be used. Most of the laminated products are custom-built, except for a few stock items, therefore the customer should be explicit about the type of adhesive needed. He should also keep in mind that generally the greater the exposure resistance required the more expensive will be the product.

In laminating a large timber of great length it is customary to build the strips of laminae of any convenient length of lumber. The shorter pieces are end jointed. For example, a fifty-foot strip may have only a few or it may have many end joints. When packaged for assembly the end joints in the laminae are staggered. The thickness of the laminae depends to a considerable extent on the curvature of the piece; hence, for assemblies having severe curvature the laminae may be very thin, while for straight pieces they are frequently an inch or more thick. A curved timber built in accordance with good practice retains all the strength of the wood, since it is not stressed in bending as are steam- or dry-bent solid timbers.

In laminating, any quality can be built into the timber. For exceptional requirements all of the laminae can be free of defects, but for most uses this extreme is not necessary and provides no advantage. Excessive requirements will cost a great deal for unnecessary strength and durability properties.

Laminated timbers have many other advantages over natural timbers. A laminated timber, no matter how thick, will have a fairly uniform moisture content throughout. It will be relatively free of checking and warping, even under severe use. If desired, each thin lamina can be treated with preservative chemicals prior to gluing, or the laminae can be treated with chemicals to provide fire resistance, hardness or stability. In fact, if one is willing to pay the price, laminated wood can be made almost indestructable.

Laminated products, although most frequently thought of as being of large size, are not confined to heavy timbers. Smaller products such as laminated and impregnated knife handles have now become familiar on the American market.

Some of the uses for laminated products are trusses, arches, barge spuds, beams and stringers, elevator guides for mine shafts, hammer boards for the steel industry, springboards for swimming pools, truck-body parts, boat framing and planking decking for ships, heavy-duty trucks, knife handles, hoops and staves, furniture, wallpaper print rolls, bowling pins, tennis rackets, skis, meat chopping blocks, print blocks, hat blocks, shoe lasts and shoe trees, pulleys and blocks, and machine parts.

Laminated products are truly engineered, with each separate type of product designed for a specific use. The potentials for this type of product are practically without limit, and the industries making them are widely scattered throughout the United States. For the larger assemblies the American Institute of Timber Construction is the best source of information. For smaller items many firms manufacturing Hardwood Dimension, produce laminated parts and may be contacted by writing the Hardwood Dimension Manufacturers' Association. The National Lumber Manufacturers Association may be contacted for advice on any of the products.

End and edge-jointed lumber. The development of modern glues that will perform in any type of service has not only made the plywood and laminating industries possible, but has had a considerable effect in modernizing and creating entirely new processes and products.

Very wide panels made up of a multiple of wood boards or strips which have been glued together edgewise are available for many purposes. Large counter or bar tops are examples. Similar panels are used for sink tops, cutting boards, drafting boards and other places where large, flat working spaces are required.

For pattern lumber, where wide lumber may be necessary, this type of edge-glued product is excellent. The furniture industry annually uses millions of feet of edge-glued panels for furniture tops and other panel requirements. For the latter purpose the panels are usually covered on both faces with two or more sheets of thin veneer in which form the product is known as "furniture plywood," not to be confused with the type of plywood made throughout its thickness from thin veneers. The type of glue used again depends on the exposure requirements.

End gluing of lumber is coming into widespread use. With modern glues this operation can be performed efficiently and the joints may have all the water resistance and most of the strength that can be obtained in a solid timber. While there is no difficulty in end gluing lumber or other wood products to provide sufficient strength and durability, it is usually difficult to hide the joint, mainly because it is at right angles to the grain of two pieces of wood. This has the tendency to exaggerate the differences in the grain and color of each piece. It of course makes no difference for pieces covered with paint, veneers or other facings. Proper design of a product will sometimes offset the undesirable appearance of a horizontal glue line. For room paneling this is often handled by a V-groove, but the places where this can be done are limited.

End- and edge-jointing operations are conducted with a variety of equipment. Usually the presses are steam-heated and conventional, but in recent years some special types of production equipment permit high-frequency heating of the glue line.

The most important thing for the purchaser to know about edge- and end-jointed lumber and lumber products is that the material has all the advantages of natural solid wood, plus availability in any reasonable length or width. Also, a certain amount of additional dimensional stability is inherent in the edge-glued panel over the solid board panel. This is caused by breaking up the grain pattern; the narrower the strips, the greater the stability.

Timber connectors. A special type of connector system has been responsible for construction that was entirely unknown a few years ago. This is used mostly for buildings, bridges, etc., but it has also found limited use in other fields. Connectors are used in ship construction, appliances for mine shafts, oil drilling and other places where heavy loads must be supported.

There are several types of connectors. However, the principal one is known as the split ring. This is a ring of metal, obtainable in various sizes. A typical ring is 2½ in. in diameter, with a depth of about ¾ in. The purpose is to fasten two or more timbers together. To use the rings, the timbers are first grooved on the faces to be joined with a circular groove, the depth of which is one-half the width of the ring. The rings, when inserted in these grooves, provide a very large bearing area, much greater than can be obtained by bolts or other mechanical fasteners. To hold the joint together so that the rings do not fall out, the two timbers are also through-bored for a small bolt in the center of the circle where each connector is placed. This bolt is relatively small and not intended to carry load, except that which is necessary to hold the two timbers together and keep the connectors in place. Two or more connectors are used at each joint. The shallow grooves and the through-bolt hole are made with a special tool in one operation.

Machining for timber connector joints may be accomplished in a factory with permanently installed equipment, or in the field with simple portable equipment. Assembly can take place wherever desired.

There are many variations of the split-ring connector (the word split is used because the ring is cut through at one point similar to a piston ring of an engine, so that it can expand or contract slightly to form a perfect fit in the groove). There are also other connectors of different types.

Information may be obtained regarding these devices from the Timber Engineering Company, Washington, D. C.

Wood preservation. There are several wood-preservation methods. The system used depends upon the service requirements. For example, a railroad tie, once in place, must withstand continuous exposure to the elements and almost every conceivable condition that promotes decay, until it is worn out some twenty to thirty years later. It is therefore necessary to provide this and similar items with maximum protection.

The two general methods of treating are surface applications where the wood is exposed to the chemicals by dipping, soaking or brushing, and treatments wherein the chemicals are forced into the wood through pressure. These latter treatments are used for the most critical applications. Pressure-treating operations must be conducted in plants with considerable equipment. It is a specialized business, and a large industry has sprung up to take care of the many needs for highly treated products.

Some of the American species are easy to pressure-treat, others are difficult to penetrate with any chemical. For these latter species incising methods are sometimes used. This consists of puncturing the wood surface with tiny holes so that the chemicals will have entrance avenues below the surface of the wood. Railroad ties, telephone poles and similar timbers are often handled in this way. Fortunately, it is not necessary to impregnate thick sections of wood for their entire depth to obtain resistance to decay. A moderate penetration below the surface is usually sufficient. However, when using heavy timbers where complete penetration is not practicable it should be remembered that bolt holes or other machining will expose untreated fibers.

The most effective way to prevent damage through exposure of untreated wood is to perform all major machining operations prior to treating. If this is impossible the exposed parts should be given a surface treatment, by brushing or dipping. Unfortunately, neither of these treatments is as effective as the original pressure treating. All domestic species can be pressure-treated, but some of them require it less than others. The inherently decay-resistant species, such as cypress, redwood, some of the cedars and white oak, need no treatment for most uses. But this applies only to the heartwood of any specie.

Of the various chemicals used for wood-preservative treatments with the pressure system, creosote is one of the oldest and best. However, it has an odor which is sometimes offensive and is therefore not generally suitable for manufactured items that may be used in contact with the body, or adjacent to food or enclosed places where the fumes may become objectionable. It can also be somewhat of a fire hazard and is sometimes not used for this reason. A third objection to creosote is the difficulty of painting over it. However, creosote is very effective against decay fungi and insect damage.

There are several effective forms of copper salts. These can be painted over; they do not contribute to fire hazard and are odorless. They are extensively used in the manufacture of boats and other marine appliances.

A relatively new chemical for decay and insect prevention is pentachloraphenol. This has most of the advantages of creosote, except that it is probably not as effective against termite damage and is more expensive, but it lacks most of the disadvantages and can be used in enclosed places. The oils which are sometimes used for carriers of the chemicals are flammable, but soon disappear.

For food containers preservative treatments are not generally recommended unless the conditions are closely examined and there is assurance that the chemicals are approved under the existing laws and are in no way harmful to health.

The second method of treatment is by one of the surface systems. Generally the chemicals used for this purpose are almost the same as for pressure treating. However, they are often specially prepared and their carriers may be different in order to obtain greater natural penetration.

Of the various methods of application dipping is the most efficient, for in this way all the surface is exposed to the chemicals and a maximum amount of usable chemical is deposited, which may not be the case with brush or spray treatments. Often, heating the chemicals and cooling the pieces during dipping will increase penetration. However, long periods of soaking are not usually of much advantage. It is better to dip the wood parts for several minutes (sufficient time to assure that surface exposure is complete) and then pile the pieces closely together soon after they are removed from the chemical bath. Several days of natural absorption under these conditions will often provide a surprising amount of penetration. The success of this method depends to a considerable extent on the type of chemical used and particularly the carrier.

In the millwork industry the surface method of application is of tremendous importance and is widely used for nearly all its products. Windows, storm sash, exterior trim and most other products exposed to weather are effectively treated. In addition, the preservative chemicals are often combined with waxes or oils to provide a reasonable amount of dimensional stability.

Treatments for dimensional stabilization. All wood and wood parts will eventually reach equilibrium with average atmospheric conditions which surround them. If the air we breathe and live in did not change in moisture content there would be no swelling or shrinking of wood, except on those occasions when the wood products actually came in contact with free water. But changes in relative humidity do take place and in one way or another they affect practically every product; wood is no exception. Fortunately for the manufacturers and users of wood products these changes come frequently, so that continued exposure to one set of unbalanced conditions is unusual. One day the humidity will be high and the next day it will be lower. With each change a piece of wood makes effort to come into balance, but usually the changes in the atmosphere come so rapidly that the wood has little chance to reach more than a mean average level.

The most practical answer to this condition for 95% of our wood products is to treat them so that the product arrives in the consumer's hands with a moisture content that has reached the mean average. This is done by kiln drying the lumber for the use intended. If the product is for exterior use a moisture content of 15 to 20% is about right. If it is for interior use in an American home the moisture content should be from 6 to 9%.

Kiln drying is inexpensive and adequate for articles such as flooring, furniture, sporting equipment and the hundreds of other things commonly found around the home, office or factory. There are a few products, however, that need somewhat closer tolerances than can be obtained by usual means. Shoe lasts are an example. In the manufacture of shoes the lasts are not only subjected to high humidity conditions, but the leather is wet and changes in dimension can occur quite rapidly. For this type of product it is necessary to provide more than ordinary protection.

Aside from kiln drying lumber there are two other systems of providing dimensional stability:

1. Surface coatings: The purpose of a surface coating is to prevent atmospheric contact with the surfaces. To prevent all contact between them we would have to put the wood product in a tightly sealed bottle. There is no known surface coating that will do this. About the best that can be expected is to reduce the intimacy of contact to a minimum. Ordinary paints, lacquers, varnishes or waxes have very little efficiency for this purpose. There are some specially prepared products such as high-grade phenolic varnishes that will slow down the transfusion of moisture between the two media to such an extent, that such treatment for many uses provides a safety factor above that which is required.

However, in considering this system for use it should be kept in mind that if the object to be protected is submitted to an unbalanced condition of constant magnitude for a sufficient length of time, the coated wood will arrive at the same state as uncoated wood.

2. Impregnation treatments: The second method of dimensional stabilization is to impregnate the wood with a chemical that can be polymerized after it is deposited (producing so-called "impreg wood"). This chemical probably becomes a structural component inside the wood that resists movement of the elements. The most efficient, but expensive and hard-to-perform treatments can provide a high percent of permanent stability. Most of these treatments for ultimate stability have limited practical application. For some easily treated species of wood a chemical treatment with urea, phenol, or melamine resins, followed by heating to obtain polymerization, will provide permanent dimensional stabilization ranging from 30 to 60% of ultimate efficiency. This amount of efficiency is sufficient for more exacting purposes.

When using these treatments, the thinner the wood sections the easier it is to obtain impregnation; therefore, relatively thin veneers are treated and then made into plywood or laminated wood to obtain the desired thickness.

Hardening of wood. For certain products, the hardening of wood is desirable. Good examples are desk legs, tool handles, some furniture tops, etc. There are also many potential uses for this type of product, such as sporting goods, piano components and implements of various kinds. The treatments and chemicals used are similar to those for dimensional stabilization. However, the hardening takes place by applying high pressure at the time of polymerization of the chemicals (producing so-called "compreg wood"). The amount of hardness depends on the thoroughness of impregnation and the amount of pressure. Since impregnation is easier to obtain with thin pieces of wood, veneers are usually treated and built up to the required thickness. This type of product is known in the trade as compregnated wood.

The process used for desk legs is somewhat unusual. The solid wood is first machined to almost proper dimension, but slightly oversize for the finished width and thickness. Impregnation of slight depth is then obtained by pressure treatments. The legs are then put in a mold and pressed to final size. The hardening is confined to a fraction of an inch below the surface. The two exposed faces of the corner legs are not machined afterwards, but the hardened surfaces on the two faces that will not be exposed are surfaced off. This procedure provides extreme resistance to indentation for the exposed faces, but does not change the interior so as to alter the ease of working the natural wood. The process has come into widespread use in the production of wood office furniture and could be used in many other products.

Heat-resistant furniture tops. Many years ago it was found that a piece of aluminum foil placed under a thin piece of veneer would conduct the heat from the wood so rapidly that it would not burn from lighted cigarettes, cigars, etc. However, the wood surface must also be protected from stains due to the infiltration of residual tars from burning tobacco. Also, it is advisable to provide furniture tops with protection from common household chemicals such as alcohol, acetone and other solvents. Certain high-grade phenolic varnishes will prevent this damage and also add somewhat to the heat resistance. The combination of aluminum foil to conduct heat away and a surface finish to prevent stains is in use in the furniture industry and has many potential uses elsewhere.

The above treatment does not afford increased protection from abrasion or indentation, therefore much hotel furniture is not only protected by aluminum foil, but has face veneers of compreg wood.

Other engineered wood products

The list of processes and potential processes that are available for wood construction is almost endless. The laminating of wood can be extended to using species of one kind for the surfaces and another for the core. Also, other species having altogether different properties may be used in other places of the assembly. Decking for aircraft carriers is a good example: the surface of the decking must have high abrasion resistance, but great strength in bending is also required. The combination of several species in the same timber is therefore most effective.

Barrels and other cylindrical containers are made of laminated wood staves, plywood or veneer. Hogsheads for tobacco export are of solid staves or plywood.

Wood can be treated to provide considerable fire resistance; it then can be destroyed by heat, but will not support combustion.

Natural wood is not adversely affected by extremes of cold such as are encountered in high latitudes and is therefore much in use for shelters, tools, sporting equipment and other products where these conditions exist.

Wood is widely used for structures and products that must be nonmagnetic; hence, it is used for minesweepers and similar craft. Laminated wood has no peer for this type of product.

Special treatments have been developed for use in power-line construction. These treatment materials must act as nonconductors as well as preservatives.

Wood has excellent thermal insulating properties. It is used for refrigerated spaces, refrigerated delivery trucks and in a great many other places where light weight and thermal insulation, combined with strength are necessary. For example, specially treated milk containers are in widespread use in the dairy industry.

Special forms of timbers are manufactured for boxcar decking. Decking of this type provides a medium for fastening cargo as well as furnishing other needed functions.

In the textile industry, wood, both treated and in the natural form, is used for shuttles and bobbins, pulleys and many other types of machinery.

Wood, either in a natural form, treated or laminated, is used for water and other liquid conduits, cooling towers and chemical containers.

WOOD-BASE FIBER AND PARTICLE MATERIALS

Flat-formed board products can be classified into two groups: (1) Those primarily from fiber interfelted during manufacture with a predominantly natural bond, although extraneous material may be added to improve some property like bond (or other strength property), and water resistance, and (2) those made from distinct fractions of wood with the primary bond produced by an added bonding material.

Moldings, based on a composition of wood-base fiber or particle and binding material similar to those used for boards, have found increasing use. These products will be described also and typical present-day uses discussed.

Production statistics for molded units of fiber and wood particle are difficult to obtain because the units are used in such widely different commodities as toilet seats, croquet balls, school desk tops and seats, frames for luggage, and armrests, door panels, and other molded components for automobiles. The number of uses for these moldings is increasing because they combine shape with adequate structural strength and durability for many uses.

Classification

Following are the classifications for such materials as developed to date.

Boards. The various types of wood-base composition boards are classified by ASTM (American Society for Testing and Materials) into two major groups, depending on whether they are primarily interfelted fiber or resin-bonded particle in origin.

(1) The first type is classed as "fibrous-

felted," manufactured from refined or partially refined lignocellulosic fibers that are interfelted. The resulting board is characterized by a natural bond. Fibrous-felted boards are divided into three groups of different densities:

Group 1: Structural insulating board with a density range of approximately 10 to 26 lb per cu ft.

Group 2: Medium-density building fiberboard with a density range of approximately 26 to 50 lb per cu ft.

Group 3: Hardboard with a density range of aproximately 50 to 80 lb per cu ft and manufactured under carefully controlled conditions of pressure, heat and moisture, so that a softening of lignin occurs and the board has a characteristic natural ligneous bond.

Fibrous-felted boards may have binders or other materials added during manufacture to improve or obtain certain properties, but the bond is primarily produced by interfelting of fiber. There are fibrous-felted products of lower density than structural insulating board (less than 10 lb per cu ft). These products may be classified as semirigid insulation though their usual use is in industrial cushioning and they are usually classified with those products.

(2) The second type is "particle board," composed of small discrete pieces of wood that are bonded together with a synthetic resin adhesive in the presence of heat and pressure. Particles range in shape and size from fine elements, approaching fibers, to large flakes. Particle boards are divided into three groups by density, as are the fibrous-felted boards. Furthermore, because of method of manufacture, particle boards must be classified as either flat-platen or extrusion pressed.

About 80% of the particle board is manufactured in platen presses with the pressure applied perpendicularly to the face of the board. The remainder is made by forcing the blend of particles and resin through a long, heated die. In that instance, the applied pressure is parallel to the faces of the finished board and in the extruded direction. At the present time, extruded boards are manufactured only in the medium-density range although, in thicknesses greater than 1 in., boards may be extruded with openings in the center of the thickness to achieve an average density approximately that of insulating type particle board.

Classification of particle boards by density is:

Group 1: Insulating type particle boards with a density range of approximately 10 to 26 lb per cu ft.

Group 2: Medium-density particle boards with a density range of approximately 26 to 50 lb per cu ft.

Group 3: Hard-pressed particle boards with a density range of approximately 50 to 80 lb per cu ft.

Moldings. No clear-cut classification of moldings has been developed beyond those based on type of raw material, that is, fiber or particle. With the exception of special low-density pulp moldings like oil filter cartridges, where little or no pressure is required during manufacture, products and commodities molded to shape are usually in the density range of 50 to 80 lb per cu ft because of pressures required in forming the items to the contour of the mold.

Hardboards may be postformed to single curvatures or moderate double curvatures by heat and pressure after manufacture as a flat sheet. These materials are generally classed as postformed hardboard.

Manufacture

Methods of producing both board and moldings are little standardized. Following is a discussion of methods used.

Boards. The methods by which the various wood-base fiber and particle panel materials are manufactured vary greatly, and the products produced likewise have a wide range in properties. Fibrous-felted boards are usually manufactured by modified paper-making techniques. The fiber is prepared from wood or other lignocellulosic material like bagasse. It is handled as a pulp, and formed into mats on a deckle box, Fourdrinier or cylinder former. If the mat is dried an insulating board is produced; if the mat is dried and then hot-pressed or hot-pressed and dried simultaneously, a medium-density building fiberboard or hardboard is produced. One notable exception is in the manufacture of hardboard where, in some processes, the fiber is conveyed by air rather than water, and the mats are formed from an air suspension rather than a water suspension. Wet felting and air felting are the terms used to differentiate the two methods of manufacture.

Particle-board manufacture is more complex

than that used for fibrous-felted boards, not only because of the two types of pressing, but because of the wide range in types of particles used, the available processes and variations in process, and the kinds of resin used for binders. Particles are produced either by hammer milling, fiber-making techniques like grinders or attrition mills, or by special cutters. They are then screen or air classified, blended with resin, and handled by air or mechanical conveyors. The blend of particles and resin is either extruded and the resin cured, or it is formed into mats by continuous or batch methods and hot pressed and the resin cured in flat-platen presses. Extruded boards have the same type of particle throughout; flat-platen-pressed boards may be homogeneous (the same kind and size of particle throughout) or one type of particle for the core portion and another for the faces. Urea formaldehyde and phenol formaldehyde resins are the principal binders, although other materials are being used alone or in combination with the conventional binding materials.

Moldings. Most commodities formed from wood fiber (with or without resin binders as the principal source of bond in the final product), or wood particle and resin, are consolidated and pressed in a heated compression mold consisting of a male and female die. Since flow in the mold is limited with the amounts of resin commonly used, there are limitations on the complexity of curvature and shape of the moldings. Preforms may be required for proper distribution of material and pressure in the final molding operation.

In the instance of the fibrous-felted products, now mainly limited to the higher density products commonly called molded hardboard, these preforms are formed from both wet and air suspensions of fiber. When wet suspensions are used, the preforms are made on either male or female screen forms where the water flows through the screen and leaves the fiber on the screen in the same manner as when flat sheets are formed. Amounts of fiber deposited at different points on the form are controlled by baffling the flow. When an air suspension of fiber is used, the fibers are either blown against a screen or pulled from the air suspension by a suction fan. Again, baffling of the air currents controls the amount of fiber deposited on the different locations of the preform. In some in-

stances, the preform can be made from a previously formed flat mat of fiber by cutting it to shape and adding layers in places that are to be thickest in the final molding.

Resin-particle moldings are pressed in the same way as fiber-felted ones. With most commodities, the mixture of particle and resin is carefully apportioned to different areas of the mold itself rather than in the preformer. Larger amounts of resin are usually required in wood particle moldings than in boards because of the amount of flow required in the mold. Urea-formaldehyde, phenol-formaldehyde, melamine-formaldehyde, and blends of the melamine and urea types are the commonly used resins for binding the particles together. Because of the high unit value of moldings as compared to particle board, melamine resin may be preferred for many items because it combines a light color comparable to urea and high moisture resistance comparable to phenolic.

Basically, the difference between flat boards and moldings is one of technique of manufacture rather than difference between the two products. The manufacture of these moldings is so new that it is difficult to accurately describe the different steps in manufacture. They can be considered the same as for flat board except for the special equipment needed for forming and pressing to curved shape and different cross section.

Physical characteristics

The primary benefit of wood composition materials is low cost. But physical properties can be surprisingly high.

Boards. Board materials are manufactured products, and the characteristics and properties of the final products are influenced more by method of manufacture, additives used, and control of process than by species of wood used for raw material. It has been shown that in most instances a given kind and quality of board can be manufactured from nearly any species of wood. The advantage of kind or form is mainly one of convenience and economics. Since the board products are manufactured—often tailored to a particular use requirement—the physical characteristics and strength properties are dependent on variables of manufacture.

The insulation-board manufacturers have developed their production mainly around different

TABLE 1—STRENGTH AND MECHANICAL PROPERTIES OF WOOD-BASE FIBER AND PARTICLE PANEL MATERIALS [a]

Material	Density, lb/cu ft	Specific Gravity	Modulus of Rupture, psi	Modulus of Elasticity (bending), 1000 psi	Tensile Strength (parallel to surface), psi	Tensile Strength (perpendicular to surface), psi	Compression Strength (parallel to surface), psi	24-Hr Water Absorption, % weight	Thickness Swelling (24-hr soak), %	Maximum Linear Expansion,[b] %	Thermal Conductivity, Btu/sq ft/hr/°F/in.
FIBROUS-FELTED BOARDS											
1. Structural insulating board	10–26	0.16–0.42	200–800	25–125	200–500	10–25	—	—	—	0.5	0.27–0.45
2. Medium-density building fiberboard	26–50	0.42–0.80	400–4,000	90–700	800–2,000	—	500–3,400	6–150	—	[c]0.2–1.30	0.50–0.60
3. Hardboard											
a. Untempered	50–80	0.80–1.28	3,000–7,000	400–800	3,000–6,000	—	1,800–6,000	3–30	10–25	0.6	0.80–1.40
b. Tempered	60–80	0.96–1.28	6,500–10,000	800–1,000	4,000–7,800	—	4,200–6,000	3–20	8–15	0.4	1.10–1.50
4. Superhardboard	85–90	1.36–1.44	10,000–12,500	1,250	7,800	500	26,500	0.3–1.2	—	—	1.85
PARTICLE BOARDS											
1. Insulating type	10–26	0.16–0.42	700	—	—	—	—	—	—	—	0.36
2. Medium-density type	26–50	0.42–0.80									
a. Extrusion	Values not presented because extruded boards are always used and tested with facing applied										—
b. Flat-platen pressed			1,500–8,000	150–700	500–4,000	40–400	1,400–2,800	20–75	20–75	0.6	0.40–1.00
3. Hard-pressed type	50–80	0.80–1.28	3,000–7,500	400–1,000	1,000–5,000	275–400	3,500–4,000	15–40	15–40	0.85	1.10–1.50

[a] The data presented are general round-figure values, accumulated from numerous sources; for more exact figures on a specific product, individual manufacturers should be consulted or actual tests made. Values are for general laboratory conditions of temperature and relative humidity.

[b] Expansion resulting from a change in moisture content from equilibrium at 50 per cent relative humidity to equilibrium at 90 per cent relative humidity.

[c] For homogeneous and laminated boards respectively.

uses for the products. The hardboard industry has developed production around different classes and types of materials. For a given use it is necessary to select the suitable class and type of hardboard by comparing the requirements of that use with other similar places where a hardboard has been used successfully, or by comparing important properties of the board with performance requirements. To date, the particleboard industry has not determined whether its approach will be by class and type or by use.

Physical characteristics and other information pertinent to general properties of the wood-base fiber and particle panel materials is summarized in Table 1. Data are only indicative rather than specific. For any actual application, specific information from the manufacturer may be required. It may be obtained from their bulletins or from test programs designed to evaluate a specific property. Special products may be available for certain uses that require qualities beyond those of general manufacture, and the information in Table 1 cannot cover those special properties.

The practice of treating boards to make them resistant to insect attack and decay has been limited. While it is possible to add arsenical, pentachlorophenol, and pentachlorophenate compounds during manufacture or by dipping the boards after manufacture to give them the necessary toxicity, this practice has been limited because boards have not been used where high resistance to such attack has been required. A few of the manufacturers of insulation board and particle board have treated boards destined for use where flying termites are a problem, such as the extreme southern parts of the United States or for board exported to South America and the Caribbean area.

Treatment to give boards added resistance to fire likewise has been limited. Fire-resistant salts can be added to reduce the flammability of the board, but this has been done only in isolated instances. The only concentrated effort has been by the insulation-board industry, where they have adopted the practice of using intumescent paint coatings for all factory-finished interior type boards to reduce the rate of flame spread in case of fire.

Moldings. The physical and mechanical characteristics of moldings can be considered similar to those for boards of the same density and from the same kind of particle or fiber and resin or other additive. The molding technique used and such other factors as shape, complexity of curvature, and distribution of pressure and resulting orientation of fiber and particle may influence a property like dimensional stability or strength more than the basic components from which the molding is made. In products of this type, the strength and stability of the moldling is principally from the particle or fiber, and the resin or binder mainly serves to hold them together so that they act as a unit.

WOOD-METAL LAMINATES

Wood-metal laminates constitute a composite panel construction in which the core material is made up of a wood or wood-derivative slab to which metal facing sheets are adhesively bonded.

The core material is usually plywood or a composition board of wood fibers or chips compressed and bonded together to form a flat core slab. Balsa wood and insulation boards are frequently used as core materials where thermal insulation is a requirement.

The metal facing sheets may be steel, aluminum, stainless steel, porcelain enamelled metals, rigidized metals or metals with decorative finishes. Light-gauge metals are usually used, ranging in thickness from 0.010 to 0.060 in. depending upon the specific strength, stiffness and service requirements.

Adhesives used in bonding the metal to the core material are selected to meet the desired service requirements. For most standard laminates the adhesive is water resistant and fungus proof so that the panels may be subjected to exterior as well as interior exposure. Continuous service temperatures may range from a minimum of −60 F to a maximum of 170 F, although specially prepared laminates are available for continuous use as high as 350 F.

Greater stiffness. Wood-metal laminates are designed to utilize the best properties of each of the component parts providing a panel which is not only light in weight but has good structural strength and high flexural rigidity. Since the metal facing sheets are supported by a core material of substantial thickness, smooth flat panels free from waves and buckles are obtained with light-gauge metals. For panels subjected to flexure, the facing sheets may be considered analogous to the flanges of a structural I-beam carrying the tension and compression stresses. Similarly, the core material is analogous to the web of an I-beam carrying the shear stresses. For a given thickness of facing sheets the rigidity of such a panel varies approximately as the square of the panel thickness. Thus a ½-in. panel will have approximately four times the rigidity of a ¼-in. panel. Typical examples of the comparison of wood-metal laminates and flat metal sheets are noted below:

1. At a given weight of 2.5 lb per sq ft a ⅜-in. laminate with steel facings on both sides of plywood provides 75 times the resistance to bending of a comparable weight (16 gauge, or 0.063 in.) of steel.
2. A ⅜-in. laminate with aluminum facings on both sides of plywood weighs 1.43 lb per sq ft, but has 100 times the stiffness of 20-gauge (0.038-in.) steel which weighs 1.5 lb per sq ft.
3. A ¼-in. laminate faced on both sides with zinc-coated steel has about the same stiffness as 6-gauge (0.20 in.) steel, yet weighs only one-fourth as much.

Available sizes. Wood-metal laminates are limited in size only by the size of the press equipment and commercial sizes of the metal sheets and core material available. The most common panel sizes are 4 ft wide by 8, 10 and 12 ft long. The most common thicknesses are ¼, ⅜, ½, ¾ and 1 in. The panels may be sawed to exact size from stock size sheets. Frequently, however, the facing sheets are fabricated prior to bonding to the core material to provide special edge details.

Applications

Wood-metal laminates are used extensively in the architectural building and transportation fields. Curtain wall panels, column enclosures, partition panels, facia and soffit panels are typical applications. Truck and trailer bodies, shipping containers, railroad car partition panels and doors are common uses of the laminate. In general, wherever light weight, combined with high rigidity and structural strength are prerequisites, wood-metal laminates may be used to good advantage.

WOODS, IMPORTED

There are between 90 and 100 different species of foreign woods currently being imported into the United States. By far, the greatest number of these are used for decorative or nonengineering purposes—some are used for both.

Imported woods to be considered here are all defined as imported species, except those which are also native to the United States. Thus, we exclude all Canadian kinds because the United States has every species of wood that grows in Canada. Mexico, on the other hand, supplies some tropical woods not found in the United States. There are pines from the highlands of Central America which are imported in fairly sizable volume at times, but the engineering aspects of their utilization coincide directly with our own Southern pine; consequently, no attempt will be made to describe these. No other softwoods (conifers) are imported for use as engineering materials. Thus, all types discussed here are categorized as hardwood (broadleaf) species. Even balsa is a hardwood, because the expressions "hardwood" and "softwood" in lumber-industry parlance allude to the botanical classification rather than the actual hardness or softness of the wood.

There are 13 kinds of tropical hardwoods known to be used for engineering products or purposes. (Furniture is not considered here because furniture designers generally place more emphasis on the artistic qualities or design than on the utilization of the wood to the best engineering advantage). In certain parts of the world there are vast expanses of untouched tropical forests which, no doubt, contain several other hardwoods of potential engineering value. However, the woods described in this chapter are all in current use. Following are typical imported woods and applications:

1. Refrigeration: Balsa
2. Wharves & Docks: Greenheart, Ekki, Jarrah, Ironbark, Apitong, Angelique.
3. Boat Construction: Philippine Mahogany, Central and South American Mahogany, African Mahogany, Balsa, Apitong, Teak, Iroko, Ironbark, Jarrah, Lignum Vitae.
4. Tanks and Vats: Philippine Mahogany, Apitong.
5. Building Construction: Mahogany (all types), Apitong, Balsa, Greenheart.
6. Poles, Piling: Greenheart, Ekki.
7. Machinery: Lignum Vitae.
8. Aircraft and Missiles: Balsa.
9. Vehicles: Apitong.

Apitong (Dipterocarpus species, principally grandiflorus)

Apitong is a Philippine hardwood second in importance to Philippine mahogany in availability and in volume of export sales. It is a very tall, straight timber, having open unclogged pores in the wood. This characteristic is extremely rare among all species. American red oak is one of the few others in this category. The chief advantage of such a structure is the ease and thoroughness with which it can be creosoted or impregnated with liquids under pressure. Engineers faced with the problem of rendering a hardwood practically immune to decay or insect attack should bear this in mind.

Apitong has been used successfully by the U. S. Navy in Guam for its retractable fender construction program. Unlike ordinary dock fenders, the retractable type is engineered to move upward and inward as a complete unit under the high lateral stresses imposed by large warships in docking. The development of retractable fenders is credited to Capt. P. W. Roberts, Public Works Officer, New York Naval Shipyard and Mr. Virgil Blancato, manager of its structural branch.

Apitong is an excellent wood for any wharf or dock construction purpose. Data on comparative cost and mechanical properties of this wood, as well as others, are contained in Table 1.

Due to its remarkable stiffness and abrasion resistance, apitong is frequently employed as decking or floor planking in railroad freight cars, motor trucks, and other vehicles subjected to repeated loading abuse.

It is a rare occurrence to find knots in apitong. Most of the exported grades are straight-grained and practically free of this defect; consequently, the clear wood strength data may be used with a much lower safety factor than we find necessary in the case of determining working stresses for ordinary coniferous species.

Electric power companies have found apitong very efficient for crossarms and braces, as well as for more specialized uses. For such purposes, a three-minute dip in an open tank of 5% pentachlorophenol solution is adequate to protect the wood from decay for the expected service life of the installation.

Mahogany, tropical American and African

The original mahogany first discovered by the early explorers from Europe was found in the West Indies. It was found as far north as the Florida Keys, and even today there are many small native mahogany trees thriving on private estates on the Keys. This was the *Swietenia mahagoni*. It was a prized wood not only in America, but in Europe. Unfortunately, the original timber is no longer available to the export trade. Cuba, the last source of supply, discontinued her mahogany export business by government order in 1946.

This old West Indian mahogany was considerably harder and heavier than its nearest relative, *Swietenia macrophylla*, the principal species of mahogany now obtained from Mexico, Central America and South America. It weighed from 45 to 50 lb per cu ft after seasoning. The Mexican and Central American wood averages about 34 lb, but the mahogany coming from Peru and through the headwaters of the Amazon runs just slightly higher.

As evidence of the durability of West Indian mahogany, the author has in his possession a lovely bowl turned from a house beam that was in service over 450 years. It was taken from the Don Francisco Clavijo family home, built between the years 1509 and 1515 in Santo Domingo, and razed in 1925 to make room for a modern highway. A photostatic copy of an old Spanish Government record describes the history of the Clavijo hacienda. Most of the mahogany beams and columns from this historic house were found to be in such excellent state of preservation that they were purchased by a New York lumber importing firm and later sold to a prominent custom furniture manufacturer. It is reported that a dining room suite was made of this wood, each chair of which was valued at $1400.

The mahogany from Mexico, Central America and South America is used today primarily for expensive furniture. Its relatively high cost, compared with other mahoganies, greatly limits its use in construction, even though its mechanical strength properties are practically the same as those of African and Philippine mahogany.

African mahogany, (*Khaya*, principally *ivorensis*) is exported from West Africa to many parts of the world, mainly the United States and Europe. It has a narrower and more prominent ribbon-stripe grain on the quartered surface and the freshly cut logs reveal a decidedly

TABLE 1—MECHANICAL STRENGTHS OF IMPORTED WOODS

Species	Source of Test Data	MC, % [a]	Spec. Grav., Ov. dry, %	Static Bending			Comp. Par. to Grain		Compression Perpendicular prop. limit, psi	Hardness		Shear Par. to Grain, Max, psi	Relative Cost
				Prop. Limit, psi	Mod of Rupt, psi	Mod of Elast, 1000 psi	Prop. Limit, psi	Maximum Crushing Str, psi		Side lbs.	End lbs.		
Apitong	Phil. Forest Prod Research Institute	12	0.72	10,400	17,050	2,590	5,020	8,980	1,160	1530	1720	1,780	Low
Angelique	Report #1787, Forest Prod Lab.	12	0.62	12,980	18,590	2,240	7,160	9,020	1,420	1510	2010	1,780	Med
Balsa	ASTM 1937	5	0.144	—	2,160	456	—	1,045	—	—	—	—	Low
Ekki	Yale Univ, Survey Afr Woods, 1955	17.3	0.93	13,850	22,870	2,732	8,470	10,450	2,420	3810	4310	2,490	Med
Greenheart	Bul #7, Univ Michigan	14.8	1.06	16,200	25,500	3,700	10,000	12,920	—	2630	2140	1,830	Med
Ironbark	Bul #279, Melbourne, Aus.	12	0.69	15,100	27,100	3,540	8,750	14,000	2,600	3090	2720	2,590	Med
Jarrah	Bul #279, Melbourne, Aus.	12	0.54	10,200	16,200	1,880	4,120	8,870	1,600	1910	2070	2,100	Med
Karri	Bul #279, Melbourne, Aus.	12	0.57	11,600	19,200	2,760	7,260	10,400	1,280	2030	1980	1,810	Med
Lignum Vitae	Report #1940, Forest Prod Lab.	12	1.09	—	—	—	—	11,400	—	4500	3600	—	High
Mahogany, American	Report #1139, Forest Prod Lab.	12	0.50	8,810	11,140	1,430	—	6,430	—	760	880	1,050	High
Mahogany, African	Report #1139, Forest Prod Lab.	12	0.47	7,890	10,700	1,480	—	5,680	—	790	1080	1,340	Med
Mahogany, Philippine	Report #1139, Forest Prod Lab.	12	0.57	8,590	12,270	1,690	—	6,830	—	840	780	1,570	Low
Teak	N. Borneo Forest Records #3	52	0.60	7,090	11,440	1,670	4,075	5,870	—	1038	915	1,107	High

[a] MC refers to Moisture Content of wood at time of test.

pinkish color, as opposed to the more brownish tone of American mahogany. Both woods darken with age. Furniture manufacturers use practically all the lumber and veneers made of African mahogany. Its use in construction is restricted not because of suitability, but because of short supply.

The color of a finished piece of wood is the least reliable feature of all in identifying it. Modern wood finishers can apply practically any color tone to any wood. Blond mahogany can be made by bleaching, red by staining, and any so-called "wheat," "fruitwood," and "eggshell" finishes can be imparted to mahogany, birch, gum, or practically any other wood. It is the grain, the minute macroscopic features, luster, hardness, weight, and wood application which, in combination, reveal with more certainty the true identity.

Teak (Tectona grandis)

Commercial teak is currently imported in the form of logs or sawn lumber mainly from Thailand and Burma. Its botanical range includes Java, India, Malaya and Indochina but exports from these countries are inconsequential.

Teak is tobacco-brown in color and the planed wood surface has a waxy feeling. It is coarse-grained, of medium weight, and coincides closely in strength to American black walnut.

From an engineering standpoint, teak is of interest mainly as a ship decking material. For this purpose, it is unsurpassed by any other wood known and except for its high cost, would probably find more extensive use in wooden boat construction. Ships built in all the great maritime countries invariably contain some of this famous wood, if not in railings, doors, and other exterior woodwork, then in decking. Decay is practically unknown to teak. It is likewise very resistant, though not immune, to marine borers.

Teak has been used in ship construction for several centuries. It is almost a daily occurrence for fishing boats somewhere on our coastal waters to ensnare their nets on old sunken ships. In attempting to free their nets, pieces of wood are sometimes broken off and raised. The author has examined many such specimens, and found most of them to be either teak, mahogany, oak, or pine. Their state of preservation generally

ranks in that order, also. But the interesting thing is that whenever a piece of teak is recovered, it will be found that the iron nails or spikes originally driven into the wood are not the least bit rusty inside the wood. They may be completely eaten away where they protrude into the salt water but when the wood is split away with an ax, the embedded iron is found to be in its original condition. It is evidently the natural oil content of the teak which protects the metal, since most of the iron found in other species of wood from the same wreck are corroded both inside of the wood and out.

Australian woods

There are three Australian timbers used to some extent for heavy marine structures on our West Coast—jarrah, karri and ironbark. All three are members of the great *Eucalyptus* genus, in which there are several hundred species.

Jarrah resembles karri so closely that it is difficult to distinguish one from the other. Both are dark red in color, similar in weight and appearance. A fairly reliable and quick method of identifying them is to burn a thin match-size splinter. The jarrah will turn to a black charcoal when it burns out while the karri leaves a white ash. Ironbark is heavier than either of these and is more gray in color. Also, it is nearly always severely surface checked, a characteristic which does not detract significantly from its strength. It is very strong, having a modulus of rupture in bending of 27,100 psi, whereas jarrah and karri run about 16,000 and 19,000 psi respectively.

Ironbark and jarrah are rated "very durable" by the British Forest Products Research Laboratory. Karri is rated only "moderately durable."

An indication of the uses made of these Australian woods may be gained from the huge sizes of the timbers generally carried in the West Coast hardwood yards. They are common in dimensions as large as 12×24 in. and 30 to 40 ft long. Boat stems and keels are primary uses for these large, strong timbers. They are also purchased for dock, wharf and bridge construction.

All of these Australian timbers are difficult to machine. In planing, a 15-deg angle of the knife to wood surface is recommended. Metal fasteners can only be applied after boring the appropriate size holes in the wood. In general, such timbers are useful where great strength and durability are more important than appearance.

Philippine mahogany

The Philippine Islands are richly endowed with hardwood species numbering into the hundreds, yet only about a half dozen of these are included in exports of Philippine mahogany. These half-dozen species mentioned dominate the dense forest growth to such an extent that it took one large lumber mill 50 years to cut the timber off a 100,000-acre tract. During this period, the average annual cut was 40,000,000 board feet and the total volume removed was a staggering 2 billion board feet. Approximately 90% of this was Philippine mahogany, the balance consisting mainly of apitong.

It is a practice among the better mills of the Philippines to segregate their lumber into two types, based mainly on color—light red and dark red. The former consists mainly of white lauan and almon and the latter of tangile and red lauan. There is another light-colored wood called bagtikan which is now marketed under its own name, since it is heavier, denser, and thus better suited for flooring, planking, etc.

A large portion of the Philippine mahogany entering the United States finds its way into boat and ship construction, a use for which there is no better wood. It has strength, durability, and beauty of grain and color. It does not soak up water; it holds paint unusually well, yet is beautiful when sanded and varnished. Boards are available in wider and longer sizes than in any other species, and the cost is very modest.

Wood in a boat is subjected to virtually every type of stress during its service life, e.g., bending, torsion, compression, tension, and abrasion. Engineers are not inclined to think of Philippine mahogany as a load-bearing timber, although it is most certainly well adapted to such uses. The fact that it is nearly always available in clear form, i.e., no knots, gum pockets, wane, and other defects common to domestic soft woods, should be of particular interest. The clear wood mechanical strength data of domestic structural timbers is normally reduced quite substantially because of knots and other defects in calculating the safe working stresses. If we compare the clear wood strength data for one of the principal Philippine mahogany species (tangile) with that of the leading domestic softwoods,

there is not much difference. In computing the safe working stresses for beams, posts, columns, etc., it is customary to deduct a certain percentage from these values for knots alone, depending on the grade. If we have no knots in a timber, this particular reduction is not necessary and the allowable working stresses are consequently reduced only for the other factors, such as slope of grain, natural wood variability, factor of safety, etc.

The Bureau of Public Works in Manila has assigned an allowable unit working stress of 1,500,000 psi for modulus of elasticity in bending for tangile. This is the species most prevalent in shipments of dark red Philippine mahogany. The other, red lauan, is rated at 1,300,000 psi. Almon and white lauan, the "light red" species, are also rated at 1,300,000 psi each.

Philippine mahogany can be glue-laminated and formed into large trusses, arches, columns, etc. The cost is somewhat higher than for domestic softwoods, but for certain types of construction it is often desirable from an appearance standpoint. Arched ceilings in churches are good examples. Here the Philippine mahogany arches, purlins, etc. make a complementary match for pews and woodwork of the same wood.

At the time of this writing, there is one large laminating firm on the West Coast which is prepared to build any engineered arch truss or other member from Philippine mahogany.

Greenheart (Ocotea rodioei)

Greenheart is produced in British Guiana and has been imported into the United States in fairly substantial quantities since the days of the Panama Canal construction. Over the past 40 years its utilization has spread from piling and lock gate construction to transmission poles, warehouse floor planking, shipbuilding, highway bridges, jetties, fenders, derricks and other purposes where a timber of unusual strength and durability is required.

Greenheart has approximately twice the strength of our domestic timbers used for similar purposes. *Engineering Review*, May 15, 1920 (British), says of greenheart's durability in sea water: "Reports from widely different localities are practically unanimous in dividing timber into greenheart and others."

The outer planking on Capt. Bob Bartlett's arctic vessel "Morrissey" was greenheart. After two years of punishment in the hard arctic ice, the captain said the planking was "as good as ever." This appears to substantiate the high abrasion resistance and crushing strength properties attributed by other sources. A large warehouse at 30th Street and 10th Avenue, New York City, contains greenheart floor planking that has seen many years of service. The dead load on this floor is 3000 pounds per square foot and the live load, 2500 pounds per caster.

Another characteristic of greenheart round timbers, such as piling, is the absence of taper. Most piles are practically the same diameter at each end. Tests have been made on greenheart piling which indicate that they will sustain loads equal to those on reinforced concrete piles. An 18-in. diameter pile, 55 ft long, was driven through 34 ft of mud, sand, gravel, stiff clay, and iron ore, taking loads up to 105 tons, at eight blows per inch in the final stages. After being pulled out, the steel clad point was found to be undamaged and no splits or other injuries were noted.

Practically all leading railroads which maintain wharves, car ferries, piers, etc. at seaports have used greenheart extensively.

Balsa (Ochroma lagopus)

Balsa has the distinction of being the lightest commercial wood in the world. As an engineering material it occupies a position uncontested by any other wood. Practically the entire commercal supply comes from Ecuador, and this largely from two very sizable logging and mill operations, the competition between which has kept the price of balsa at a surprisingly low level compared with other foreign woods.

The properties of balsa should be of great interest to any engineer, even in this age of remarkable achievements in plastics, since balsa combines the advantages of light weight, strength and economy to a degree not attainable with any man-made material. Comparing a simple block of styrene foam one cubic foot in size, having a weight of 6 lb, with a cubic foot of balsa also weighing 6 lb, we find that the balsa exceeds the styrene approximately five times in both crushing strength and tensile strength, simply by using the balsa in such a way that its grain is parallel to the direction of the stresses applied. To incorporate any sort of supplemental reinforcement to the styrene designed to equal the strength of the balsa would bring its cost so far out of line that it would be ridiculous to pursue the comparison. It reminds the writer, who was engaged for several years in railroad-crosstie research, of the secret of why wood ties are still in universal use (and probably always will be as long as we have railroads). What other piece of material 7 × 9 in. × 8½ ft in size could one place under the rails that would last up to 20 years or more, take the merciless pounding of repeated rolling loads, in all kinds of weather exposure, and cost only about $2.50 anywhere in the country?

Balsa has been subjected to criticism by the promoters of competitive materials for its susceptibility to decay. While it is true that balsa, in common with other organic materials, is vulnerable to certain bacterial attack, this danger can be eliminated when it is used properly. This can be done either by chemical treatment, or simply by keeping the moisture content of the wood *under 20%*. In spite of its soft texture, balsa is impervious to the complete passage of liquid under pressure. The surface layers of the wood can be penetrated but thorough impregnation is not possible under any presently known method. Fortunately, however, our supplies of balsa are available in kiln-dried form at a moisture content well below the critical point mentioned.

At some stage of practically every boy's life, he encounters balsa in building airplane or boat models. Here he forms impressions of the wood from firsthand experience. As a result, most men, not only engineers, know something about balsa. There is no other wood in America more likely to perform the function of stimulating the latent engineering talents of a boy.

Some designers of guided missiles use balsa in primary structural panels. The bulkheads of many commercial aircraft have balsa cores and sandwich panels for all sorts of vehicles are being made lighter and stronger with balsa. There has not been a single instance of fungus attack in balsa panels faced either with metal or plastic in the past 18 years. Used structurally, balsa is covered with metal, plastics or a heavier wood veneer and therefore does not create an additional fire hazard. In modern-day vehicular fire disasters, whether on a ship, plane, or other vehicle, it is the fuel which is the hazard, not the structural parts of the vehicle.

Insulating properties. With a K-factor of 0.30 and a V-factor of 0.109, balsa is unique among all woods and other materials of equal weight and strength. Moreover, it is peculiar and unusual in that the thermal conductivity decreases as a straight-line function with reduction in temperature, e.g., the K-factor of 0.30 at 85 F decreases uniformly to 0.09 at −300 F.

Engineers designing the huge tanker vessels, which now deliver liquefied gas from Texas to Europe, use balsa to insulate the steel hull from the extreme subzero temperatures of their unusual cargo. It is carefully layered in huge blocks a foot thick between the outer hull and the tanks. No other material known could withstand the embrittling effect of such low temperatures.

More information of a technical nature is available on balsa than on any of the other foreign woods. This is generally available from the suppliers.

Angelique (Dicorynia guianensis)

Angelique is a relatively "new" wood in the United States trade. Imports from British Guiana have increased since the Yale test results were published in 1954. Their findings showed angelique to be exceptionally promising, especially for boat construction.

Angelique is a heavy, reddish-colored hardwood which may or may not contain silica. This precipitates from the sap in the vessels of the wood over a period of time, the amount varying with different localities. Its effect is to dull cutting edges of knives and saws, thus serving as a deterrent to any extensive use of the wood by millwork firms. The shipbuilding industry, accustomed to the same problem in teak, should not find it objectionable. Angelique has so many admirable properties, they outweigh this one objectionable feature. Its natural durability, strength, wearing qualities and economy, compared with other tropical woods, place it among the finest of our engineering timbers.

Wharf and dock construction appears to be a particularly good place for angelique. It seasons from the green condition with a minimum of shrinkage and warping and can be relied upon to sustain severe impact loads without splitting. An installation in the Brooklyn Navy Yard has evoked praise from Naval engineers throughout the service.

No specific instances of angelique utilization

in shipbuilding have come to our attention so far, but the Navy's Bureau of Ships advised recently that its interest has by no means abated. The construction of commercial wood boats, such as fishing vessels, ferries, tugs, and thousands of smaller fishing and pleasure craft, requires millions of board feet of timber annually. The keels, stems and frames of all such boats could be made of angelique.

In all of its strength properties, except tension perpendicular to grain, angelique exceeds both teak and white oak. It is also very resistant to marine borers, having been exposed to teredo and pholad attack for as long as 15 years without appreciable damage to piling at Balboa, Canal Zone.

Ekki (Lophira alata procera)

Ekki, sometimes called "African ironwood" or "azobe," comes from West Africa. It is an excellent timber for piling, wharf and dock construction, bridge planking, ties, and all heavy timber structures.

The wood is a chocolate-brown color, sometimes verging on dark red, and has a speckled surface caused by yellowish deposits in the pores. It is so heavy that the timbers will not float in water. Machining is difficult but not unmanageable.

Experience with ekki in the United States is quite limited, although it has been used for many years in Europe. One New York lumber importer reports having sold it to the New York City Transit Authority for subway ties and platform planking, also to the Navy for cribbing, blocking, skids, etc., for ship launching. Other orders have been filled for truck body parts and for bridge and dock timbers.

Lignum Vitae (Guajacum species)

An extremely heavy, hard, oily wood, lignum vitae is obtained commercially from the West Indies and Central America and is best known for its use in stern tube bearings in propeller-driven vessels. Its density of 69 to 83 lb per cu ft places it among the heaviest known woods on earth. There are a few less widely known woods that are consistently heavier but these are not used to any extent, since most of them are commercially unavailable or lacking in some essential property.

Lignum-vitae wood is dark olive green with pale yellow sapwood. About 36% of its weight is an oil called "guaiacum." This oil is still used as an ingredient in certain pharmaceutical preparations. Early-day apothecaries erroneously believed guaiacum to be capable of extending the life span of man; in fact the wood's name, lignum vitae, means "wood of life."

It is surprising how many uses this wood finds in modern industry. Many are allied with marine equipment. Mallet heads, saw guides, spuds, motor mountings and pulley sheaves are a few items where lignum vitae is used, in addition to the thousands of wedge-shaped sections comprising stern tube bearings.

According to the U. S. Bureau of Ships, Navy Department, lignum-vitae bearings will last about seven years. This exceeds the wearing ability of synthetic bearings even in surface vessels which are in daily service.

There is no other wood which can equal lignum vitae for resistance to splitting. Its dense grain is so finely interlocked that a log or block the size of an ordinary short fireplace log cannot be split by hand with an ax, sledge hammer, wedges, or anything else, but must be sawn.

Lignum vitae may possibly be the most durable wood in the world. Logs have been found by the author on the Florida Keys which have been exposed to the elements for many hundreds of years with no sign of decay. In recently completed decay tests by the U. S. Government, lignum vitae was found to be immune to all the common wood-destroying fungi used in the tests.

Unfortunately, lignum-vitae trees do not grow tall and straight. The logs are short and invariably crooked. This means that specifications are limited in length to around 8 ft and in width to not over 12 in. It is sold on a weight basis rather than by the board foot, as is the usual case with timbers.

For a wood of such hardness and weight, lignum vitae is surprisingly easy to saw or machine. This is due to its natural oil, which serves as a cutting lubricant.

WROUGHT IRONS

Wrought iron is a ferrous material, aggregated from a solidifying mass of pasty particles of highly refined metallic iron with which a minutely and uniformly distributed quantity of slag is incorporated without subsequent fusion.

Chemical composition

Present-day quality wrought iron is distinguished by its low carbon and manganese contents. Carbon in well-made wrought iron seldom exceeds 0.035%. Due to specification, manganese content is held at 0.06% max. Phosphorus in wrought iron usually ranges from 0.10% to 0.15% depending upon property requirements. Sulfur content is normally low, ranging from 0.006% to below 0.015%. Silicon content ranges from 0.075% to 0.15% depending upon the siliceousness of the entrapped iron silicate; silicon content of base metal is 0.015% or less. Residuals such as Cr, Ni, Co, Cu and Mo are generally low, totaling less than 0.05%.

Structural characteristics

Structurally, wrought iron is a composite material; the base metal and the slag are in physical association, in contrast to the chemical or alloy relationship that generally exists between the constituents of other metals.

The form and distribution of the iron-silicate particles may be stringerlike, ribbonlike, or platelets. Practically, the physical effects of the incorporated iron-silicate slag must be taken into consideration in bending and forming wrought iron pipe, plate, bars, and shapes, but when properly handled—cold or hot—fabrication is accomplished without difficulty.

Mechanical properties

Tensile. The tensile properties of wrought iron are largely those of ferrite plus the strengthening effect of any phosphorus content which adds approximately 1000 psi for each 0.01% above 0.10% of contained phosphorus. Strength, elasticity and ductility are affected to some degree by small variations in the metalloid content and in even greater degree by the amount of the incorporated slag and the character of its distribution.

Nickel, molybdenum, copper and phosphorus are added to wrought iron to increase yield and ultimate strengths without materially detracting from toughness as measured by elongation and reduction in area.

Average tensile properties of plain and alloyed wrought iron for different product forms are furnished in Table 1. Note that the yield point of wrought iron, plain or alloyed, is the same whether tested longitudinally or transversely.

Shear strength. The ratio of the shearing strength across the thickness of a wrought-iron plate, either with or across the grain, is about 80% of tensile strength. If the shearing forces are applied on the planes perpendicular to the plane of the plate, the shearing strength is about the same as tensile strength. Shearing resistance on a plane parallel to the plane of the plate is

TABLE 1—TENSILE PROPERTIES

	Ten Str, 1000 psi	Yld Pt, 1000 psi	Elong, in 8 in., %	Red of Area, %
Plain Wrought Iron				
Bars	50	30	32	55
(⅞ in. round)				
Pipe	48	28	25	—
(1¼ in. std.)				
⅜ in. Plate	48 [a]	30	20	—
(standard)	42 [b]	30	4	—
⅜ in. Plate	45 [a]	30	10	—
(special forming)	45 [b]	30	10	—
Alloyed Wrought Iron				
3.5% Ni	60	45	25	50
(1 in. round)				
0.30% P, 0.30% Cu	60	40	25	—
(3½ in. o.d. tubing)				
1.30% Cu	60	45	25	40
(⅞ in. round)				
1% Mn, 0.10% P	60	40	25	—
(6 in. pipe)				

[a] Longitudinal.
[b] Transverse.

about half the shearing strength across the thickness of the plate.

Torsion strength. Shafts of fibrous materials such as wrought iron, with the fibers parallel to the axis and along which fibers the shearing strength is relatively low, fail by shearing longitudinally. A hollow shaft, such as a thin-walled tube or pipe made of wrought iron and subjected to torsional failure, first flattens and then fails at a transverse section similar to a low carbon-steel pipe, which also has a shearing strength less than its tensile strength.

Impact strength. Impact strength, in ft-lb,

for wrought iron at 68 F, using various types of impact specimens, is listed as follows:

Standard Charpy (keyhole notch) ..... 24 to 28
Standard Izod (Izod V-notch) .... 50 to 60
Modified Charpy (Izod V-notch)[a] ..... 70 to 85
Modified Charpy (Izod V-notch)[b] ..... 40 to 44

[a] Specimens machined from double refined wrought iron rounds.
[b] Longitudinal specimens machined from wrought iron plates—notch in the plane of the plate, transverse to fiber direction.

Physical properties

Weight, lb per cu ft 480
Sp Gr, lb per cu in. 0.278
Melt Pt .... 2750 F (approx)
Mean Coef of Lin Exp,
  68–200 F, in./in./°F, × 10⁻⁶.. 7.41
Sp Ht, 68F 0.11
Elec Resis, 70 F,
  microhm/cm/sq cm 11.97
Tension Mod, 80 F, 10⁶ psi 29.5
Shear Mod, 80 F, 10⁶ psi 11.8
Poisson's Ratio 0.30
Ther Cond, K, Btu/hr/sq ft/
  in./°F
    64 F .... −417.89
    212 F .... −414.99

Fabrication

Forging. Wrought iron is an easy material to forge using any of the common methods. The temperature at which the best results are obtained lies in the range of 2100 to 2400 F. Ordinarily "flat and edge" working is essential for good results. Limited upsetting must be accomplished at "sweating to welding" temperatures.

Bending. Wrought-iron plates, bars, pipe and structurals may be bent either hot or cold, depending upon the severity of the operation, keeping in mind that bending involves the directional ductility of the material. Hot bending ordinarily is accomplished at a dull red heat (1300 to 1400 F) below the critical "red-short" range of wrought iron (1600 to 1700 F). The ductility available for hot bending is about twice that available for cold bending. Forming of flanged and dished heads is accomplished hot from special-forming, equal-property plate.

Welding. Wrought iron can be welded easily by any of the commonly used processes, such as forge welding, electric resistance welding, electric metallic-arc welding, electric carbon-arc welding, and gas or oxyacetylene welding. The iron-silicate or slag included in wrought iron melts at a temperature below the fusion point of the iron base metal, so that the melting of the slag gives the metal surface a greasy appearance. This should not be mistaken for actual fusion of the base metal; heating should be continued until the iron reaches the state of fusion. The siliceous slag content provides a self-fluxing action to the material during the welding operation.

Threading. The machinability or free-cutting characteristics of most ferrous metals are adversely influenced by either excessive hardness or softness. Wrought iron displays almost ideal hardness for good machinability, and the entrained silicate produces chips that crumble and clear the dies. Standard threading equipment which incorporates minor variations in lip angle, lead and clearance is usually satisfactory with wrought iron.

Protective coatings. Wrought iron lends itself readily to such cleaning operations as pickling and sand-blasting for the application of protective coatings. Where protective coatings such as paint or hot-dipped metallic coatings are to be applied, the coatings are found to adhere more firmly to wrought iron and a thicker coat will be attained compared with other wrought ferrous metals. This is because the natural surface of wrought iron is microscopically rougher than other metals after cleaning, thus providing a better anchorage for coatings. Weight of zinc taken on by wrought iron in hot dip galvanizing processes averages 2.35 oz or more per sq ft and shows excellent adherence.

Corrosion resistance. The resistance of wrought iron to corrosion has been demonstrated by long years of service life in many applications. Some have attributed successful performance to the purity of the iron base, the presence of a considerable quantity of phosphorus or copper, freedom from segregation, to the presence of the inert slag fibers disseminated throughout the metal, or to combinations of such attributes.

Laboratory corrosion testing has shown that wrought iron has very definite directional corrosion properties; that is, transverse and longitudinal sectional faces show significantly higher corrosion rates than rolled surfaces or faces.

In actual service the corrosion resistance of wrought iron has shown superior performance in such applications as radiant heating and snow-melting coils, skating-rink piping, condenser and heat-exchanger equipment, and other industrial and building piping services. Wrought iron has long been specified for steam condensate piping where dissolved oxygen and carbon dioxide present severe corrosion problems. Cooling water cycles of the once-through and open-recirculating variety are solved by the use of wrought-iron pipe.

Applications and forms available

Some of the many applications for which wrought iron is recommended and used are listed below.

Building construction. Hot and cold potable water, soil, waste, vent and downspout piping; radiant heating, snow melting, air-conditioning cooling and chilled-water lines; gas, fire protection and soap lines; condensate and steam returns, ice-rink and swimming-pool piping; underground service lines and electrical conduit.

Industrial. Unfired heat exchangers, brine coils, condenser tubes, caustic soda, concentrated sulfuric acid, ammonia, and miscellaneous process lines; sprinkler systems, boiler feed and blowoff lines, condenser water piping, runner buckets, skimmer bars, smokestacks and standpipes, salt and water well pipe and casing.

Public works. Bridge railings, fenders, blast plates, drainage lines and troughs, traffic signal conduit, sludge digestor heating coils, aeration tank piping, sewer outfall lines, large o.d. intake and discharge lines, trash racks, weir plates, dam gates, pier-protection plates, sludge tanks and lines, dredge pipe.

Railroad and marine. Tie spacer bars, diesel

exhaust- and air-brake piping, ballast and brine protection plates, brine, cargo and washdown lines on ships, hull and deck plating, rudders, fire screens, breechings, tanker heating coils, car retarder and yard piping, spring bands, car charging lines, nipples, pontoons, car and switch deicers.

Other. Gas collection hoods, staybolts, flue gas conductors, sulfur mining gut, air and trans- port lines, coal-handling equipment, chlorine, compressed air lines, distributor arms, cooling tower and spray pond piping.

Wrought iron is available in the form of plates, sheets, bars, structurals, forging blooms and billets, rivets, chain, and a wide range of tubular products including pipe, tubing and casing, electrical conduit, cold-drawn tubing, nipples and welding fittings.

Y

Z

ZINC AND ALLOYS

Zinc (Zn) is a blue to gray metallic element which is found broadly throughout the world and which now ranks only behind aluminum and copper in order of consumption among the nonferrous metals. Its basic characteristics of (a) relatively low melting point (die casting), (b) good resistance to atmospheric corrosion combined with its high place in the galvanic series of metals (galvanizing), (c) solubility in copper (brass) and (d) inherent ductility and malleability (wrought zinc and zinc base alloys) have been responsible for its high place among the materials for construction in today's civilization.

Basic properties

Zinc as freshly cast has a bright silver-blue surface, which on exposure to air forms an impervious, tenacious and protective grayish oxide film. Its atomic number is 30, atomic weight 65.38. Major properties are given in Table 1.

TABLE 1

Crystal structure and orientation:
Hexagonal close-packed,
a = 2.664 Å, c = 4.9469 Å, c/a = 1.856
Glide plane (0001); glide direction [1120]
Twinning plane (1012)
Density:
Solid at 25 C, 7.133 gm/cu cm
at 419.5 C, 6.83 gm/cu cm
Liquid at 419.5 C, 6.62 gm/cu cm
at 800 C, 6.25 gm/cu cm
Melting Pt: 419.5 C
Boiling Pt (1 atm): 907 C
Heat Capacity:

Solid—$Cp = 5.35 + 2.40 \times 10^{-3}T$ (298–692.7 K) cal/mol
Liquid—$Cp = 7.50$ cal/mol
Gas (monatomic)—$Cp = 4.969$ cal/mol
Heat of Fusion: 1765 cal/mol at 419.5 C
Heat of Vaporization: 27.430 cal/mol at 907 C
Lin Coef of Ther Exp:
Polycrystalline (20–250 C), 39.7×10^{-6} per °C
a axis, (20–100 C), 14.3×10^{-6} per °C
c axis, (20–100 C), 60.8×10^{-6} per °C
Volume Coef of Ther Exp, 20–400 C
8.9×10^{-5} per °C
Ther Cond:
Solid (18 C) 0.27 cal/sec/cm/°C
Solid (419.5 C) 0.23 cal/sec/cm/°C
Liquid (419.5 C) 0.145 cal/sec/cm/°C
Liquid (750 C), 0.135 cal/sec/cm/°C
Elec Res:
Along a axis (20 C) 5.83 microhm-cm
Along c axis (20 C) 6.16 microhm-cm
Liquid (423 C) 36.955 microhm-cm

Production of metal

Zinc in the United States is produced in the following grades (ASTM Designation B6-58):

Grade		Pb Max. %	Fe Max. %	Cd Max. %	Sum of Pb, Fe, Cd, Max. %
(1a)	Special High Grade	0.006	0.005	0.004	0.010
(1)	High Grade	0.07	0.02	0.07	0.10
(2)	Intermediate	0.20	0.03	0.50	0.50
(3)	Brass Special	0.60	0.03	0.50	1.0
(4)	Selected	0.80	0.04	0.75	1.25
(5)	Prime Western	1.60	0.08	—	—

The final stage in the production of slab zinc from ore may be thermal or electrolytic but all

541

processes are similar in their preliminary steps.

After mining and such preliminary mechanical separation of valuable minerals from gangue as is practical, the minerals are finely crushed and concentrated in flotation circuits. Consequently, when finely crushed ore is strongly agitated with suitable reagents in a suspension of small air bubbles in water, the mineral portions cling to the bubbles while the earthy materials sink. Every effort is made during processing to separate lead, copper and any other values as separate concentrates.

Selective concentration by flotation is followed by roasting—conversion of sulphide to oxide (with the production of large quantities of sulphuric acid as a major by-product)—largely on moving grates. The use of moving grates causes an agglomeration of an otherwise extremely fine calcine and a properly sintered product ready for final refining is obtained directly.

Refining by thermal processes. The metallurgy of zinc is dominated by the fact that its oxide is not reduced by carbon below the boiling point of the metal.

A large fraction of the world's zinc is still produced from relatively small horizontal retorts with one furnace (or bank) containing hundreds of such units. Each unit consists of a cylindrical refractory clay pipe closed at one end, charged with a mixture of sinter and carbon, and fitted on the open end with a clay condenser. The charged retort and condenser units are assembled and then placed horizontally in the furnaces with heat externally applied to the retorts, and the condenser ends so protruding into cooler areas as to permit the condensation and accumulation of metal after reduction by carbon. While the full cycle may occupy about 24 hr or more, "draws" of metal are made at regular intervals with the "first production" of higher purity and the final metal being least pure. The proper segregation of such materials by "draw" is used commercially in meeting specification by grade.

Since 1929, large, continuously operated vertical retorts have operated, with top charging of briquets of zinc oxide and bituminous coal, and metal tapping from an outside condenser.

Another continuous method used since 1931 involves electrothermic reduction, using a novel condenser in which the retort vapors are sucked through molten zinc.

Most recently the problem of condensing zinc from low tenor gases has been solved in England and a blast furnace is now employed for the production of metallic zinc.

Neither horizontal or vertical retorts, electrothermic units, nor blast furnaces normally produce zinc of the extreme high purity required by much of zinc's total market. Since 1935 a redistillation process has been used as the thermal means of meeting this demand. The principles of fractional distillation are utilized and zinc of 99.99+% purity is made.

Electrolytic production. An electrolytic method of producing metallic zinc was developed in the United States and Canada simultaneously about 1915. As the selective flotation process made additional quantities of zinc concentrates available in localities where electric power is cheap the production of electrolytic zinc increased and now is responsible for about 40% of U.S. production. The proportion is not greatly different throughout the rest of the world.

In the electrolytic process, the zinc content of the roasted ore is leached out with dilute sulphuric acid. The zinc-bearing solution is filtered and purified and the zinc content recovered from the solution by electrolysis, using lead alloy anodes and sheet aluminum cathodes. Current passing through the electrolytic cell, from anode to cathode, deposits the metallic zinc on the cathodes from which it is stripped at regular intervals, melted and cast into slabs. Zinc so produced is 99.9+% or 99.99+% pure depending on need and the process control exercised.

Applications

To protect steel. For many years the greatest use of zinc has been to protect iron and steel against atmospheric corrosion. Because of zinc's relatively high electropotential it is anodic to iron. If zinc and iron or steel are electrically connected and are jointly exposed in most corrosive media, the steel will be protected while the zinc will be attacked preferentially and sacrificially. This, along with the fact that zinc corrodes far less rapidly than iron in most environments, forms the basis for one of zinc's great fields of use—in galvanizing (by hot dipping or electrolytically), metallizing, sherardizing, in zinc pigmented paint systems, and as anodes in systems for cathodic protection. The six techniques are described below.

Hot dip galvanizing. Zinc alloys readily with iron. Therefore, steel articles, suitably cleaned, will be wet by molten zinc and will acquire uniform coatings of zinc the thickness of which will vary with time, temperature and rate of withdrawal. Such coats are continuous and reasonably ductile. Ductility is improved considerably by the restriction of immersion time and by the addition of small amounts of aluminum to the galvanizing bath. These latter facts are utilized to major advantage in the continuous line galvanizing of steel sheet which now accounts for about 4% of all U.S. steel sales.

Millions of tons of steel products are protected by zinc annually.

The time before first rusting of the iron or steel base is proportional to the thickness of zinc coat which in turn is subject to control—depending on product and processing—within a range from thin wiped coats on some products to as much as .008 in. on certain low alloy steels allowed to acquire a full natural coat.

Zinc's usefulness as a coating material comes from its dual ability to protect, first as a long lasting sheath, and then sacrificially when the sheath finally is perforated.

Electrogalvanizing. Zinc may be electrolytically deposited on essentially all iron and steel products. Wire and strip are commonly so treated as are many fabricated parts. Electrodeposited coats are ductile and uniform but normally are thinner and therefore find application in less rigorous service. An exception is Bethanized wire which normally has heavy coats and is used in all appropriate services.

Metallizing. Zinc wire or powder is melted and sprayed on suitably grit-blasted steel surfaces—a growing use. Its virtues are flexibility. in application and substantial thicknesses which may be applied. The method is particularly useful for renewal of heavy coatings on areas exposed to particularly critical corrosive conditions and the coating of parts too large for hot dipping. While metallized coats may be somewhat porous, zinc's sacrificial nature nevertheless makes them protective. Suitable pore sealants may be used as a part of a metallizing system.

Sherardizing. Zinc powder is packed loosely around clean parts to be sherardized in an airtight container. When sealed, heated to temperature near but below zinc's melting point, then slowly rotated, the zinc alloys with the steel

forming a thin, abrasion-resistant and uniform protective coating. Sherardizing is used commonly to coat small items such as nuts, bolts, and screws; an exception is tubular electrical conduit. Sherardized coats receive varnish, paints, and lacquers particularly well.

Zinc-pigmented paints. Evidence has accumulated to demonstrate that paints heavily pigmented with zinc dust serve similarly to zinc coats otherwise applied. Electrical contact must exist between the steel and the zinc-dust particles; consequently, special vehicles must be used and the steel surface must be clean.

Zinc anodes. High-purity zinc, normally alloyed with small additions of aluminum, with or without cadmium, is cast or rolled into anodes which, when electrically bonded to bare or painted steel, will protect large areas from the corrosive attack of such environments as sea water. The advantages of zinc in this application include self-regulation (no more current is generated than is required), a minimum generation of hydrogen, and long life. This is a growing application for the protection of ship hulls, cargo tanks in ballast, piers, pilings, etc.

General comment. Reference has been made to the importance of coating thickness—the heavier the coat the longer the time before first rusting. All evidence at hand indicates that the amount of zinc in a coat is the controlling factor and the method of application is of secondary importance. Uniformity of coat and adhesion must be good. No data are known to demonstrate that common zinc impurities normally present in amounts to or slightly above specification limits have any significantly deleterious or beneficial influence on the ability of zinc to protect iron or steel against atmospheric corrosion. While any grade of zinc may be used for galvanizing, Prime Western is the one most commonly employed.

Die casting. This is a market for zinc which may soon become its largest. In the United States the tonnage of zinc die cast is nearly twice that of its nearest competitor, aluminum. Zinc for die casting is presently supplied as either of two alloys as in Table 2 with the first of greater commercial importance.

These alloys melt readily, are highly fluid, and do not attack steel dies or equipment. When used under good temperature control and with good die design practices casting surfaces are excellent and easily finished. Physical properties

TABLE 2—ZINC-BASE DIE-CASTING ALLOYS
(ASTM B-86)

Component	% in Alloy AG40A (Alloy 3)	% in Alloy AC41A (Alloy 5)	% in Alloy 7 [a]
Cu	0.25 max	0.75–1.25	0.25 max.
Al	3.5–4.3	3.5–4.3	3.5–4.3
Mg	0.03–0.08	0.03–0.08	0.005–0.02
Fe	0.100 max.	0.100 max.	0.075 max.
Pb	0.007 max.	0.007 max.	0.003 max.
Cd	0.005 max.	0.005 max.	0.002 max.
Sn	0.005 max.	0.005 max.	0.001 max.
Ni	—	—	0.005–0.02
Zn	Remainder	Remainder	Remainder

[a] No ASTM designation available.

are good and dimensional stability excellent.

Alloy control within the specified limits insures long life. Low aluminum results in decreased casting and mechanical properties and adversely affects the performance of plated coatings. High aluminum can lead to brittleness (an alloy eutectic forms at 5% Al). High copper content decreases dimensional stability. Iron as commonly encountered is not critical. Lead, tin and cadmium, if present above specification limits, can lead to intercrystalline corrosion with objectionable growth and serious cracking or brittleness as a result. Magnesium minimizes the deleterious influence of lead, tin or cadmium but at or near specification maximum decreases ductility and castability and can lead to objectionable hot shortness. Other impurities such as chromium and nickel, which may be encountered, are not critical.

A large-scale testing and market-development program is under way with an alloy (common designation, Alloy 7) offered by several alloy-makers which is most promising. This alloy differs from the standard alloys above by virtue of a closer control of impurities, lower magnesium content, and the addition of a small amount of nickel. This material is said to be more fluid and to permit the casting of thinner and more complex sections while at the same time more faithfully reproducing smooth die surfaces and simplifying finishing.

About 60% of U. S. zinc die castings are used by the automotive, truck, and bus industry for functional, decorative-functional, or decorative purposes. A majority is plated with copper-nickel-chromium in a variety of plating systems especially adapted to withstand severe service conditions. Major improvements in plating were developed and proven during the late 1950's.

Other major outlets for zinc die casting include household appliances, business machines, machine tools, air-brake systems and communication equipment.

Only Special High Grade Zinc is used for die-casting alloys.

Zinc in brass. Zinc, when added to copper, forms a series of single phase alloys up to about 35% zinc. These alpha brasses, available in a variety of colors and tempers, as well as in all mill-product forms, have excellent corrosion resistance and good mechanical properties. They are adaptable to essentially all forms of fabrication and continue as a major metal of industry. When zinc is used in excess of about 35%, a second or beta phase appears, producing a series of alloys which are readily hot-worked from rough cast form and with proper processing can be finished and hardened by cold working.

Special High Grade Zinc is used for brasses containing less than 35% zinc which are to be hot-rolled. Less pure grades are used elsewhere, and particularly where lead is desired to facilitate machining.

Rolled zinc. Rolled zinc is produced as sheet, strip, plate, rod, and wire in numerous alloys depending on the end use intended. Historically, the commercial grades of rolled zinc contain varying amounts of the natural impurities, lead, cadmium and iron. In recent years the trend has been toward a general use of alloys starting with High Grade or Special High Grade slab and using small to very small amounts of alloying elements such as copper, magnesium, manganese, aluminum, chromium and titanium.

Casting practices are normally conventional and involve either open or closed book type molds.

Early rolling operations from the slab form are at temperatures of 150 to 260 C depending on the particular product being processed. Later rolling may be finished at somewhat lower temperatures from 50 to 175 C, again varying with alloy and form. Process annealing seldom is used since the properties of rolled zinc are established during the rolling operation. Zinc can be and is extruded.

Most promising among the alloys of zinc to be rolled is that variety which contains something less than 1% copper and on the order of 0.15 to 0.20% titanium. Alloys in the composition range show great promise, having superior resistance to creep and grain growth as well as a substantially lower coefficient of thermal expansion.

Major end uses for rolled zinc include shells for dry batteries, building materials (as for flashings, roofing, gutters, leaders, downspouts, terrazzo strips, termite shields, weather stripping), engravers' plates, lithographers' sheets, rod for screw-machine products, wire for metallizing, strip for eyelet machine products.

Drawing, spinning, impact extrusion are used. Soldering is simple and straightforward.

ZIRCONIUM AND ALLOYS

Zirconium is distributed widely in nature and, of the elements in the crust of the earth, is eleventh in abundance. This ninth most abundant metal is more plentiful than common metals such as copper, lead, nickel and zinc. The principal ore of zirconium is the orthosilicate zircon ($ZrSiO_4$), which occurs in great abundance throughout the world as beach sands. In these sands, zircon occurs with the titanium minerals ilmenite and rutile and is normally derived as a co-product in the beneficiation of these ores. Zirconium is always associated in nature with hafnium, its sister element, in amounts varying from 0.7 to 6% (usually 2% for most ores). The properties of the two elements are so similar that the presence of hafnium in zirconium was not recognized for over a hundred years.

Zirconium is primarily used as a cladding and structural material in nuclear reactors because of excellent nuclear properties and stability in severe corrosive environments. Although zirconium has a very low thermal-neutron-capture cross section, hafnium has a very high thermal-neutron-capture cross section. Reactor-grade zirconium must therefore be free of hafnium, whereas hafnium is not removed from zirconium for nonreactor purposes. Recently zirconium has been commercially developed at greatly accelerated rates because of its application in the Naval Nuclear Reactor Program, and this accelerated development has made zirconium available as a material for construction in the chemical industry.

Preparation

The element zirconium was discovered in 1789 by Klaproth but elemental metallic zirconium was not made until 1824 by Berzelius. The first practical method of producing ductile zirconium was developed in 1925 by Van Arkel, DeBoer, and Fast at the University of Leyden and at the Philips Lamp Works at Eindhoven, Holland. Their method, commonly known as the iodide process, depends on the decomposition of zirconium iodide by a hot wire or filament. In 1938, ductile zirconium was first produced in the United States by the iodide process at the Foote Mineral Company. Then, W. J. Kroll and co-workers of the U. S. Bureau of Mines investigated less expensive methods and adopted a process based on reduction of the tetrachloride of zirconium by molten magnesium in an inert atmosphere, a method similar to the familiar Kroll process for titanium. In this operation, zircon sand is first converted to tetrachloride, which can be reduced directly to elemental zirconium by a reactive metal such as sodium or magnesium. When magnesium is used as the reductant, the reaction proceeds according to the following equation: $ZrCl_{4(g)} + 2Mg_{(l)} \longrightarrow Zr + 2MgCl_2$. By-product magnesium chloride ($MgCl_2$) is removed from the zirconium by a high-temperature vacuum distillation; the resulting product is a porous, granular material generally referred to as sponge. (When reactor-grade zirconium is required, hafnium is removed by a complex liquid-liquid extraction process that results eventually in the pure, but separated, zirconium and hafnium tetrachlorides for Kroll process reduction.) The zirconium sponge resulting from the reduction step is consolidated into an ingot by consumable-electrode, vacuum arc melting. The resulting ingot can be converted to wrought forms on conventional steel-working equipment.

Because molten zirconium has a high affinity for atmospheric gases and attacks practically every common crucible material known to man, the sponge metal is melted by vacuum arc and thereby converted to ingot form without contamination. In this process the furnace is evacuated and a consumable electrode of pressed zirconium sponge is melted in a water-cooled copper crucible to form an ingot. Zirconium is available in all of the following forms: ingot, billet, sheet, strip, bar, tubing, pipe, and rod. Castings can also be made.

Properties

Zirconium is a ductile, silvery metal that appears to be much like stainless steel. It is characterized by a very low thermal-neutron-capture cross section, excellent corrosion resistance in a wide variety of corrosive environments, and high strength; it is consequently very useful in the construction of nuclear reactors and chemical-process equipment.

TABLE 1—TYPICAL PROPERTIES OF ZIRCONIUM (HAFNIUM-FREE, REACTOR GRADE, UNALLOYED)

Density, lb/cu in.	0.235
Melting Point, C	1852
Crystal Structure	
close-packed hexagonal	Below 865 C
body-centered cubic	Above 865 C
Coef of Ther Exp, 25 C, in./in./°C	5.8×10^{-6}
Ther Cond, cal/sec/cm/°C	
25	0.050
100	0.049
300	0.045
Sp Ht, 25 C, cal/gm/°C	0.067 ± 0.001
Thermal Neutron Cross Section,	
barns	0.18
Hardness, Rockwell B	76
Ten Str, 1000 psi	65
Yld Str, 1000 psi	35
Elong, in 2 in., %	19
Red in Area, %	30
Elastic Mod, 10^6 psi	13.8

Engineers may be confused by the discordance in the published properties of zirconium prepared by different reduction methods or resulting from advancements made in standard methods that have led to continual improvements in metal purity. Discrepancies in the data are caused principally by variations in the oxygen and nitrogen contents of individual samples tested. Hardness and strength increase with oxygen content, and this fact is reflected in the variations found in the data. The properties listed above are considered average for the bulk of zirconium currently being produced in the United States.

Zirconium alloys

Several zirconium alloys have been tailor-made for specific applications in the nuclear field. These alloys are all of the alpha-plus-compound type that contain only small amounts of the alloying agents.

Pressurized water-cooled reactors in the United States use a family of alloys known as "Zircaloy" for fuel cladding. These alloys contain tin as the major alloying element and minor amounts of chromium, iron and sometimes nickel. Zircaloy resists corrosion very well in high-temperature water and steam (up to about 315 C and 5000 psi) and possesses better mechanical properties than unalloyed zirconium.

In the U.S.S.R. a zirconium alloy containing $2\frac{1}{2}\%$ columbium is used in similar service.

A zirconium alloy containing $\frac{1}{2}\%$ copper and $\frac{1}{2}\%$ molybdenum has been developed for the gas-cooled reactors of the United Kingdom. This alloy has very good resistance to corrosion by the carbon dioxide coolant gas up to about 705 C. This alloy, called ATR, has better elevated-temperature mechanical properties than Zircaloy. New alloys of zirconium needed for specific applications in both the nuclear and corrosion fields are being developed at a rapid pace.

Fabrication

Zirconium and most of its alloys are ductile and easily worked. They can be fabricated on standard shop equipment with but a few modifications and special techniques. Of primary concern are the tendencies of zirconium to react with air at elevated temperatures and to gall and seize under sliding contact with other metals.

Nuclear reactor cores and chemical process equipment of a variety of designs are made of zirconium or one of its alloys. Sheets and plates are cut on the shears and punches used for stainless steels. Bars and shapes can be cut by hacksaws with 4 to 10 teeth per in. A light feed and medium stroke rate works best. On band sawing, a blade having a 16 to 20 pitch moving at 75 ft per min will cut satisfactorily.

Forming is more difficult than for stainless steel because of the lower ductility and the tendency of zirconium to gall. Press and brake forming with 5T bend radius is quite common. Springback is great and usually variable. With some experience, most metal shops have little difficulty in forming operations.

Arc welding is the most common method of joining pieces of zirconium or its alloys. Spot welding, brazing, and soldering methods, though possible, are seldom used. Because of the great affinity of zirconium for atmospheric gases at elevated temperatures, the welding zone must be shielded by an inert gas such as helium or argon. Large glove boxes filled with the inert gas are sometimes used to protect the metal during welding. When it is impractical to weld in a glove box, a flowing gas shield (such as that of the Heliarc system) is used. Direct current, straight polarity is commonly used for welding with either a consumable zirconium electrode or a nonconsumable tungsten electrode.

Zirconium is easily machined. The best metal-removal rates are obtained when slow speeds, heavy feeds and a flood of coolant are used. Tools should be kept sharp and have large clearance angles.

Zirconium is usually ground on wet-grinding machines that use carefully selected abrasives and grinding speeds. A grinding specialist will supply the correct conditions for the specific job.

Fine turnings, chips, and cuttings of zirconium will burn quite readily. Machines should be kept clean; the scrap zirconium resulting from the machining operation should be stored in small containers. In most cases it is advisable to burn fine zirconium scrap particles out-of-doors. Only a pound or so should be ignited at one time, for intense heat develops.

Corrosion resistance

The most outstanding property of zirconium, accounting for its important commercial interest, is its resistance to corrosion in a wide variety of chemical environments. The resistance of zirconium to attack by alkalies is excellent. It is completely resistant to hydroxide solutions in all concentrations up to their boiling points; even fused caustic has no effect. In acid solutions, although the corrosion resistance is not so sharply defined, zirconium resists corrosion excellently at all temperatures to the boiling point in sulfuric acid (up to 55% concentration by weight), in nitric acid at all concentrations, in hydrochloric acid up to 20% concentration by weight, and in phosphoric acid up to 60% concentration by weight. Zirconium fully resists attack by metal chloride solutions with the exceptions of ferric and cupric chlorides.

The resistance to corrosion by acids and bases makes zirconium a good substitute for tantalum in orthopedic surgery. No known compound of zirconium has a toxic effect on the body. Bone and tissue will adhere to this metal. Its mechanical strength, wear resistance, and low density make it a very useful metal for special surgical applications.

Also available are special alloys of zirconium that exhibit excellent corrosion resistance in boiling and pressurized water and high temperature-high pressure steam, carbon dioxide and liquid sodium.

In general, the optimum corrosion-resistant property of zirconium for either nuclear or chemical applications depends on the specific properties of a primordial protective zirconium oxide film. This film minimizes or eliminates the migration of oxygen and hydrogen to the metal and thereby effectively prevents continued corrosion.

There are only a few areas where zirconium is not recommended. These include fluorides, boiling sulfuric acid (over 65% concentration), wet chlorine, aqua regia, and oxygen, nitrogen or hydrogen containing gases above 540 C.

Applications

More than 90% of all zirconium being produced is used in nuclear reactors. Without this market incentive, zirconium would not be available today for the many other uses that are developing. Zirconium is finding expanding use in the chemical process industry in towers, tanks, heat exchangers, pumps, valves, and pipe. By virtue of its long service life, zirconium is able to replace equipment made of less costly materials that corrode more rapidly, especially in continuous chemical processes where shutdowns for repairs or replacement of parts are extremely costly.

The potential significant commercial uses of zirconium will be in equipment handling the following:

1. Hot sulfuric acid (0-55%, to the boiling point).

2. Hot hydrochloric acid (0-30%, to the boiling point).

3. Hot phosphoric acid (0-60%, to the boiling point).

4. Hydrogen peroxide at all temperatures and concentrations.

5. Mixed acids.

6. Strong alkalies to the fusion point.

7. High-temperature and -pressure urea processing.

8. Applications involving both acids and caustics.

Contributors

D. B. AIKEN, CERAMIC FIBERS
H. J. ALBERT, GOLD AND ALLOYS
STEPHEN H. ALEXANDER, BITUMINOUS COATINGS
F. H. ANCKER, CALENDERED SHEET
E. A. ANDERSON, ZINC AND ALLOYS
J. ANDERSON, SYNTHETIC NATURAL RUBBER
KENNETH ARATA, DECORATIVE THERMOSETTING LAMINATES
HOWARD S. AVERY, HARD FACINGS
A. S. BACKUS, MICA
F. E. BACON, CHROMIUM AND ALLOYS; SILICON
H. BAKER, MAGNESIUM AND ALLOYS
J. H. BALDRIDGE, SILICONE RUBBER
W. L. BATTEN, METAL POWDERS
M. DOUGLAS BEALS, TITANATES
E. P. BEST, WROUGHT IRONS
C. G. BIEBER, MARAGING STEELS; SUPERALLOYS
MARCEL K. BLACK, CELLULOSE PLASTICS
JOHN M. BLOCHER, JR., NITRIDES
G. C. BODINE, JR., TANTALUM AND ALLOYS; TUNGSTEN AND ALLOYS
R. G. BOURDEAU, PYROLYTIC MATERIALS
S. W. BRADSTREET, PORCELAIN ENAMELS AND CERAMIC COATINGS
J. Z. BRIGGS, MOLYBDENUM AND ALLOYS
ROBERT P. BRINGER, FLUOROCARBON PLASTICS (CFE)
M. C. BROCKWAY, CARBON AND GRAPHITE
ROY W. BROWN, REFRACTORIES, SPECIALTY
O. F. BURBANK, JR., FORGINGS
S. C. CARAPELLA, JR., ANTIMONY; ARSENIC
JAMES F. CARLEY, EXTRUDED PLASTICS
R. A. CHEGWIDDEN, MAGNETIC MATERIALS (HARD)
WILLIAM F. CHRISTOPHER, POLYCARBONATE PLASTICS
PHILIP J. CLOUGH, VACUUM COATINGS
C. M. COSMAN, MANGANESE; VANADIUM
G. A. DAUM, NITRILE RUBBER
R. F. DECKER, MARAGING STEELS
J. T. EASH, MARAGING STEELS
L. T. EBY, ETHYLENE-PROPYLENE ELASTOMER
LEON R. EGG, COLD-MOLDED PLASTICS
E. EPREMIAN, COLUMBIUM AND ALLOYS
HAROLD V. ETTORE, PLYWOOD
JOHN L. EVERHART, CENTRIFUGAL CASTINGS; SCREW-MACHINE PARTS
ROBERT J. FABIAN, DIFFUSION COATINGS; KNITTED AND WOVEN METALS; HOT-DIP COATINGS;
    IMMERSION AND CHEMICAL COATINGS; PLASTER MOLD CASTINGS; PREFINISHED METALS
B. FADER, LEAD AND ALLOYS
EDWARD A. FARRELL, ASBESTOS
ROY B. FEHR, FLUOROCARBON PLASTICS (TFE, FEP)
ELDON FENDER, INDUSTRIAL THERMOSETTING LAMINATES
EDWARD FERRARI, CAST PLASTICS
SEYMOUR S. FEUER, PREMIX MOLDINGS

Russell W. Fitch, mechanical finishes
Neil T. Flathers, vinyl plastics
S. G. Fletcher, high-speed steels
L. M. Foster, gallium
Jacob K. Frederick, Jr., textile fibers
R. R. Freeman, molybdenum and alloys
J. V. Fusco, ethylene-propylene elastomer
D. Geiselman, scandium
Robert C. Gibson, phosphate conversion coatings
Thomas J. Gillick, Jr., felts
W. E. Gloor, polypropylene plastics
Mark V. Goodyear, epoxy plastics
Bernard Gould, adhesives
A. Kenneth Graham, electroplated coatings
Robert E. Grandpre, water-soluble plastics
Charles H. Green, silica
Warren R. Griffin, fluoroelastomers
Richard Hardesty, acetal plastics
William E. Harris, Jr., rust preventives
Harold T. Harrison, powder metallurgy
P. T. Hart, polyester film
J. R. Haws, polybutadiene rubber
H. J. Heine, malleable irons
C. C. Higgins, sheet-metal parts
G. F. Hodgson, die castings
W. C. Hosford, organosol coatings; plastisol coatings
H. E. Howe, bismuth, cadmium, thallium
Louis H. Howland, natural rubber
E. H. Hull, diamond
J. B. Hunter, platinum metals
John M. Jewell, epoxy plastics
D. A. Jones, blow moldings
T. A. Kauppi, silicone resins
L. K. Keay, clad metals
Charles L. Keller, preimpregnated decorative foil
R. L. Kenyon, iron
A. S. Kidwell, silicone rubber
T. E. Kihlgren, superalloys
J. L. Kimberly, zinc and alloys
G. E. Kloote, wood-metal laminates
W. C. Klunkler, Jr., forgings
Walter J. Laird, Jr., polyvinyl fluoride film
Edwin F. Lewis, Jr., plastisols
Wayne C. Lewis, wood-base fiber and particle materials
H. S. Link, low-alloy, high-strength steels
John Y. Lomax, chlorinated polyether
John T. Long, alkyd molding compounds
Walter Lovett, foamed glass
Frederick A. Lowenheim, electroplated coatings
H. W. Lownie, Jr., alloy cast iron; cast iron; gray irons; white cast iron
R. A. Lula, stainless steel (wrought)
R. M. MacIntosh, tin and alloys
J. A. Manfre, titanium and titanium alloys

A. C. MARSHALL, SANDWICH MATERIALS
CARL F. MASSOPUST, COMPRESSION AND TRANSFER MOLDED PARTS
R. A. MATASICK, TITANIUM AND TITANIUM ALLOYS
CHARLES E. McCORMACK, NEOPRENE RUBBER
A. G. METCALFE, SOLAR, IMPREGNATING MATERIALS FOR CASTINGS
M. R. MEYERSON, NICKEL AND ALLOYS
A. J. MITCHELL, OXIDE COATINGS
EUGENE W. MOFFETT, POLYESTER PLASTICS
J. B. MOHLER, CHEMICAL MILLED PARTS
A. E. MOREDOCK, COPPER AND ALLOYS
DAVID MORELAND, POLYETHYLENE PLASTICS
F. R. MORRAL, COBALT AND ALLOYS
JOHN A. MUELLER, MECHANICAL FINISHES
T. W. MULLEN, BLOW MOLDINGS
ROBERT G. NELB, ABS PLASTICS
H. W. NORTHRUP, DUCTILE IRONS
ROBERT D. OLT, SAPPHIRE
CHARLES W. OSTRANDER, CHROMATE CONVERSION COATINGS
LOUIS PAGGI, INJECTION MOLDED PARTS
WALTER R. PASCOE, FLUIDIZED-BED COATINGS
MARLOW H. PAULSON, JR., BUTADIENE-STYRENE THERMOSETTING RESINS
DONALD PECKNER, CAST STEELS; CONTINUOUS CASTINGS; HIGH ENERGY RATE FORMED PARTS;
    INDIUM; LITHIUM; PERMANENT MOLD CASTINGS; SAND CASTINGS; SHELL MOLD CASTINGS;
    SPUN PARTS
N. PINTO, BERYLLIUM AND ALLOYS
J. H. PIPER, POWDER PLASTICS MOLDINGS
MILTON PLATT, FABRICS, WOVEN
NORBERT PLATZER, THERMOFORMED PLASTIC SHEET
J. H. PORT, RHENIUM
CARROLL F. POWELL, VAPOR DEPOSITED COATINGS
JOHN J. PRENDERGAST, POLYSULFIDE RUBBER
ROBERT R. RADCLIFF, CHLOROSULFONATED POLYETHYLENE RUBBER
R. L. RANDOLPH, JR., CERAMIC PARTS AND FORMS
JOHN C. REDMOND, CERMETS
KURT F. RICHARDS, BUTYL RUBBER
LLOYD D. RICHARDSON, SR., ALUMINIDES
M. W. RILEY, CARBON AND GRAPHITE; FABRICS, WOVEN; FIBROUS GLASS; PLASTIC FOAMS;
    MELAMINE FORMALDEHYDE PLASTICS; PAPER
CARL A. RISHELL, WOOD
LESTER E. ROBB, ELUOROCARBON PLASTICS, PVF
ROBERT B. ROHRER, CORK
MURRAY H. ROTH, WASH PRIMERS
D. A. ROTHROCK, ACRYLIC PLASTICS
ROBERT G. RUDNESS, SAPPHIRE
CLAYTON F. RUEBENSAAL, SBR RUBBER
H. J. RUTHERFORD, CALENDERED SHEET
C. S. RYLAND, WHITEWARE CERAMICS
A. M. SABROFF, EXTRUDED METALS
J. H. SCHEMEL, HAFNIUM AND ALLOYS; ZIRCONIUM AND ALLOYS
J. R. SCHETTIG, TOOL AND DIE STEELS
E. A. SCHOEFER, HEAT RESISTANT ALLOYS (CAST); STAINLESS STEEL (CAST)
C. J. SCHOLL, FORGINGS
WARREN K. SEWARD, MECHANICAL FINISHES

PETER T. B. SHAFFER, CARBIDES
HOWARD E. SHEARER, FABRICS, NONWOVEN BONDED
S. SHEEHEY, SILVER AND ALLOYS
A. P. SHEPARD, FLAME-SPRAYED COATINGS
L. T. SHERWOOD, JR., NYLON PLASTICS
PAUL SILVERSTONE, ELECTROFORMED PARTS
A. A. SMITH, JR., SELENIUM; TELLURIUM
GAIL P. SMITH, GLASS
LOTHAR A. SONTAG, PHENOLIC PLASTICS
CEDRIC C. SOVIA, FLUOROCARBON PLASTICS CFE
JOHN SPECHT, UREA FORMALDEHYDE PLASTICS
D. R. SPINK, HAFNIUM AND ALLOYS; ZIRCONIUM AND ALLOYS
J. L. STEARNS, WOODS, IMPORTED
SAMUEL STEINGISER, URETHANE MATERIALS
JOHN L STIEF, JR., PREIMPREGNATED MATERIALS FOR REINFORCED PLASTICS
S. DONALD STOOKEY, GLASS-CERAMICS
E. C. SVENDSEN, FOAM RUBBER
GORDON B. THAYER, POLYSTYRENE PLASTICS
J. L. THOMAS, ALLYLICS (DIALLYL PHTHALATE PLASTICS)
RICHARD J. THOMPSON, POLYCARBONATE PLASTICS
HANS THURNAUER, ELECTRICAL CERAMICS
K. F. TUPPER, INVESTMENT CASTINGS
R. V. VANDEN BERG, ANODIC COATINGS
R. J. VAN THYNE, RARE EARTH METALS
E. E. WACHSMUTH, IMPACT EXTRUSIONS
L. F. WAGNER, STRIPPABLE COATINGS
THOMAS E. WALLIS, HARD RUBBER
RALPH WEHRMANN, SILICIDES
J. A. WHITEHEAD, NITRIDING STEELS
R. W. WIEHELM, VULCANIZED FIBRE
G. W. WIENER, MAGNETIC IRONS AND STEELS
ROBERT K. WILLARDSON, SEMICONDUCTING MATERIALS
F. WILLS, GERMANIUM
L. A. WOERNER, ELASTOMERIC LININGS
R. B. G. YEO, MARAGING STEELS
RICHARD E. YOUNG, FILAMENT WOUND REINFORCED PLASTICS
BENJAMIN M. G. ZWICKER, POLYACRYLIC RUBBER

Index

ABS plastics, **1**, 334
Acetal Plastics, **2**
Acrylate rubber. *See* Polyacrylic rubber
Acrylic plastics, **6**
    high impact types, 8
    plastic alloys, 334
    standard types, 6
    thermoformed sheet, 471
Acrylonitrile butadiene styrene plastics. *See*
    ABS plastics
Adhesives, **8**
    alloys, 9
    elastomers, 8
    epoxies, 10
    natural, 8
    neoprene-phenolic, 10
    phenolics, 326
    structural, 9
    thermoplastics, 8
    thermosets, 8
    vinyl formal-phenolic, 10
Alkyd molding compounds, **10**
    glass-reinforced types, 11
    granular types, 11
    putty types, 11
Alligator leather, 246
Allomers, 311
Alloy cast irons, **12**
Alloy electroplates, 143
Alloy steels. *See* specific type
Alloy steel powders, 160
Allylic plastics, **13**
    laminates, 13
    molding compounds, 13
Alumina, 76, 315. *See also* Electrical ceramics;
    Whiteware ceramics; Oxide ceramics
    specialty refractories, 394
Alumina silica fibers, 66
Aluminides, **14**
Aluminized coatings, 124, 170, 495
Aluminosilicate gloss, 202
Aluminum and alloys, **15**
    anodic coatings for, 20
    cast alloys, 19, 236
    coatings 124, 170, 226, 319, 495
    extrusions, 150
    forgings, 196
    heat treatable alloys, 18
    non-heat treatable alloys, 17
    permanent mold castings, 325
    plaster mold castings, 333
    powder, 20
    solder alloys, 440
Aluminum brass, 111
Aluminum bronze, 113

Aluminum nitride, 299
Amosite, 31
Anodic coatings, **20**
    for aluminum, 20
    for beryllium, 25
    hard coatings, 21
    for magnesium, 23
    for thorium, 25
    for titanium, 25
    for zinc, 25
    for zirconium, 25
Anthophyllite, 31
Antimony, **26**
Apitong wood, 531
Aramid fibers, 470
Aromatic polyester, 219
Arsenic, **27**
Asbestos, **29**
    amosite, 31
    anthophlyllite, 31
    chrysotile, 29
    crocidolite, 30
    industrial thermosetting laminates, 233
    paper, 321
    prepregs, 386
    premix moldings, 387
Angelique wood, 536
Austenitic stainless steels, 447

Balsa wood, 535
Barium titanate, 138, 477
Beryllia, 315
Beryllium and alloys, **32**
    anodic coatings for, 25
    beryllium copper, 36, 114
    beryllium oxide, 36
    extrusions, 34
    forgings, 33, 197
    powder parts, 32
Bismuth and alloys, **36**
    solder alloys, 440
Bitumens (Bituminous), **37**
    coatings, 37
Black oxide coatings, 316
Blow moldings, **41**
    cellulosics, 42
    acetals, 42
    polyamides (nylons), 42
    polycarbonates, 42
    polyethylene, 42
    polypropylene, 42
    polyvinyl chloride, 42
Borazon, 299
    pyrolytic type, 389

Kant's Life and Thought

KANT'S LIFE
AND THOUGHT
ERNST CASSIRER

Translated by JAMES HADEN

Introduction by STEPHAN KÖRNER

NEW HAVEN AND LONDON : YALE UNIVERSITY PRESS

Published with assistance from the foundation
established in memory of James Wesley Cooper
of the Class of 1865, Yale College.

Designed by James J. Johnson
and set in Palatino Roman type.
Printed in the United States of America by
Edwards Brothers Inc., Ann Arbor, Michigan.

Library of Congress Cataloging in Publication Data

Cassirer, Ernst, 1874–1945.
  Kant's life and thought.

  Translation of: Kants Leben und Lehre.
  Includes index.
  1. Kant, Immanuel, 1724–1804.   I. Title.
B2797.C313   193   81-3354
ISBN 0-300-02358-8   AACR2

10 9 8 7 6 5 4 3 2 1

CONTENTS

121784

INTRODUCTION TO THE
ENGLISH EDITION

Ever since its first publication in 1918 Ernst Cassirer's book on Kant's life and teaching has been widely and, I believe, rightly regarded as a classic of its kind. The aim of this introduction to the English translation is not so much to justify this estimate as to indicate the relevance of Cassirer's Kant interpretation to contemporary philosophy in its renewed endeavor to understand Kant's position and in its varied reactions to it. If this aim is to be achieved, it is necessary, however briefly (1) to consider some recent or still influential interpretations and modifications of Kant's central theses; (2) to characterize Cassirer's version of transcendental idealism; and (3) to draw attention to some of Cassirer's distinctive exegetic points in the light of his historical and philosophical ideas.

1

Kant's Copernican revolution in philosophy consists in asking and answering two kinds of questions which, borrowing terms from the Roman jurists, he calls "questions of fact" (*quid facti*) and "questions of legality" (*quid iuris*).[1] The former concern factual claims to the effect that all rational beings in their thinking—theoretical, practical, aesthetic, or teleological—accept certain judgments or employ certain concepts. The latter questions concern the justification of the factual

1. See, for example, *Critique of Pure Reason*, B 116.

claims. The Copernican revolution is based on an entirely new conception of philosophy and philosophical method which Kant describes as critical or transcendental.

Within the theoretical sphere, that is to say, mathematics, natural science, and commonsense thinking about what is the case, Kant's factual claims include three theses about synthetic judgments a priori—a judgment being synthetic if, and only if, its negation is not self-contradictory and a priori if, and only if, it is logically independent of any judgment describing a sense experience. He claims (1) that there are synthetic judgments a priori; (2) that there is one and only one internally consistent set of them; and (3) that it has been completely exhibited in the *Critique of Pure Reason*. (It comprises all axioms and theorems of Euclidean geometry, all true arithmetical propositions, and certain assumptions of Newtonian physics, such as the principles of causality, of the conservation of substance, and of continuity.)

Kant's attempted justification of these three claims—of existence, uniqueness, and completeness—is a characteristic and important instance of a transcendental justification. It presupposes that we have objective experience or experience of objects and that within it we can distinguish what is given to the senses (its sensory or a posteriori content) from what, although not so given, is yet ascribed by us to the objects (its nonsensory or a priori form). And it consists in producing, or trying to produce, a cogent argument to the effect that the a priori features of objective experience are necessary conditions of its objective character. Transcendental arguments are thus based on the twofold conviction that an experience without—or deprived of—these features is not "a possible objective experience";[2] and that, in another succinct, if rather metaphorical, Kantian phrase "we can have a priori knowledge only of those features of the things which we ourselves put into them."[3]

The fundamental tasks of the *Critique of Pure Reason*, namely the factual exposition and transcendental justification of the system of

2. See, for example, ibid., B 810.
3. Ibid., B xviii.

(theoretical) a priori judgments in its uniqueness and completeness, involve the subsidiary tasks of expounding and transcendentally justifying the system of the Categories, which Kant also regards as unique and complete. The Categories, for example, causality and substance, are concepts which occur in nonmathematical, synthetic a priori judgments; which are a priori in the sense of being applicable to, but not abstracted from, sense experience; and which in being applied to what is given to the senses confer objectivity upon it. Kant calls the transcendental justification of the Categories their transcendental "deduction." In doing so he again adopts and adapts a technical legal term from those jurists who mean by "deduction" the demonstration that what has been established as being *de facto* the case is also appropriate *de jure*.[4] In this sense of the term a deduction may, but need not, coincide with a logical deduction.

The theses of Kant's First Critique which have just been outlined have all been subjected to various—in some cases very different—interpretations, criticisms, and modifications. In considering them it is advisable to observe the Kantian distinction between questions of fact and factual claims on the one hand and questions of legality and transcendental justification on the other. As regards the interpretation of Kant's factual claims to have expounded the unique and complete system of synthetic a priori judgments and Categories, there is hardly any disagreement among the commentators. There is, on the other hand, sharp disagreement about the correctness of these factual claims—even among philosophers who accept Kant's distinctions between synthetic a posteriori and synthetic a priori judgments and between a posteriori and a priori concepts.

The main reason for their disagreement lies in their different reactions to post-Kantian developments in mathematics and physics. These are on the one hand the discovery of non-Euclidean geometries, together with the incorporation of one of them into the general theory of relativity, and on the other hand the discovery of quantum mechanics, which—at least in its dominant interpretation—is incompatible with the Kantian a priori principles of causality and con-

4. See, for example, B 116, above.

tinuity. According to these reactions we may distinguish between Kantian absolutists who accept Kant's factual claims in their original form (e.g., Leonard Nelson and the so-called Göttingen school); neo-Kantian absolutists who accept Kant's uniqueness claim, but replace his system of synthetic a priori judgments and Categories by a different one (e.g., some Anglo-American analytical philosophers); and neo-Kantian pluralists who reject his claims of uniqueness and, hence, of completeness (e.g., Hermann Cohen and the so-called Marburg school, of which Cassirer was a prominent member).

Turning to Kant's transcendental justification of his factual claims, especially his transcendental deduction of the Categories, one is immediately struck by exegetic divergences which are radical and irreconcilable. For the pluralists, who deny Kant's uniqueness claim, a transcendental deduction of a unique system of Categories and, hence, of universally and necessarily true synthetic a priori judgments is impossible—whatever the alleged nature of such a "deduction" may be. To express the matter in accordance with Kant's quasi-legal terminology, the pluralists hold that since the *quaestio facti* has to be answered in the negative, that is to say, since the factual claim has to be rejected, the *quaestio iuris* does not arise (or, in a phrase of the Roman jurists, *cadit quaestio*).

In the most common contemporary interpretation, Kant's transcendental deduction of the Categories is seen as an attempted logical inference from 'x is capable of objective experience' to 'x is capable of objectifying (transforming, unifying, organizing, into an objective phenomenon) a spatiotemporally ordered, subjectively given manifold by applying *the* Categories to it.' Of Kantian and neo-Kantian absolutists, who accept the uniqueness claim for Kant's or some other system of Categories, some consider this logical inference to be valid, others consider it invalid but remediably so, while still others consider it irremediably invalid. Thus many analytical philosophers with Kantian sympathies argue that the logical inference can be validly reconstructed if the Kantian set of Categories is replaced by another.[5]

---

5. See, for example, J. Bennett, "Analytic, Transcendental Arguments," Bieri et al., eds., *Transcendental Arguments and Science* (Dordrecht, 1979), pp. 45–64.

An exception is C. D. Broad, who suggests that the invalid logical inference can at best be replaced by a more modest probabilistic argument.[6] Lastly, Leonard Nelson and his followers, though accepting Kant's factual claims, regard his transcendental deduction as an instance of the irremediably invalid logical fallacy of a vicious infinite regress. According to Nelson, Kant's correct factual claims can be justified only by showing that the Categories are applicable to an originally obscure, nonpropositional cognition, in the same way that the a priori concepts of arithmetic and geometry are applicable to the nonpropositional intuition of time and space.[7]

In order to understand Kant's thought and influence it is important to separate his fundamental conviction that "we can have a priori knowledge only of those features of the things which we ourselves put into them" from his absolutist claim that what we so put into them is determined for all times and for all rational beings. For, although the fundamental conviction, which is common to all versions of transcendental idealism, may well have inspired the absolutist claim, it is compatible with a pluralistic assumption of alternative and even changeable systems of a priori concepts and judgments.

To the distinctions drawn by the *Critique of Pure Reason* within the sphere of theoretical thinking between the a posteriori and merely subjectively given features, which are its matter, and the a priori, objective features, which are its form, there correspond analogous distinctions drawn by the *Critique of Practical Reason* within the sphere of practical thinking and by the *Critique of Judgment* within the spheres of aesthetic and teleological thinking. Within the practical sphere the difference between a posteriori matter and a priori imposed and organizing form manifests itself in the contrast—and conflict—between desires and inclinations on the one hand and the moral will and moral duties on the other. In the sphere of aesthetic experience it manifests itself in the contrast between the pleasant and the beautiful. Like the factual claims and justifications of the First Critique, Kant's factual

6. See *Kant—An Introduction* (Cambridge, 1978), pp. 183 ff.

7. See "Die Kritische Methode und das Verhältnis der Psychologie zur Philosophie" (1904), reprinted in *Ges. Schriften*, ed. Paul Bernays et al., vol. 1 (Hamburg, 1970), pp. 4–78.

claims about the a priori features of morals, aesthetics, and teleology have been subjected to very different interpretations and criticisms. And these have again suggested proposals for various reforms, ranging from versions of neo-Kantian absolutism, over versions of a pluralistic neo-Kantian pluralism, to outright rejections of the transcendental approach in these fields.

Kant's transcendental or critical method is his most distinctive contribution to philosophy, a discipline which has always been very conscious of its dependence on "the correct method." But many of his other original ideas have proved hardly less influential. Examples of Kantian themes in recent philosophy that stem from the First Critique are the antilogicist analysis of arithmetic and the identification of physics with Newtonian physics. The former has not only become a central topic in the philosophy of mathematics but has, through its acceptance by Hilbert and Brouwer, influenced the course of mathematics itself. The latter has become important in the philosophy of science, in particular in the discussion of quantum physics, for example, Einstein's rejection of its indeterministic implications. An example of a Kantian theme stemming from the Second Critique is the discussion and rejection of utilitarianism in ethics, and hence in political philosophy, where—possibly under the influence of new developments in welfare economics—the conflict between utilitarian and antiutilitarian analyses of justice has again become central.

Kant's Third Critique has been comparatively neglected by contemporary philosophers, even by those whose main concern is with aesthetics. A probable reason for this neglect is the widely accepted view that Kant's *Critique of Judgment* owes its existence less to his philosophical insight and originality than to his idiosyncratic devotion to philosophical architectonics. It is one of the great merits of Cassirer's commentary to have shown this view to be mistaken, or at least extremely doubtful.

2

Although Cassirer's own philosophical position and approach to the history of ideas developed under the continuous influence of Kant's work, they in turn, as is to be expected, influenced his judgment of Kant's achievement and life. Cassirer very early on accepted—and never abandoned—the fundamental guiding principle of Kant's transcendental philosophy, namely that the objectivity of phenomena is not given to the senses or otherwise passively received by the mind, but is the result of the mind's imposing an a priori form on the manifold given to it. An example, indeed the prototype, of such imposition of form and objectivity on a given manifold is the mind's creation of the a priori structure of Newtonian physics by the application to sense experience of the (Kantian) Categories. This application is, as Cassirer sometimes puts it, the "making of a world"—the world of Newtonian physics in which all physical objects exist and are related to each other.

Yet Cassirer rejects the identification of physics with Newtonian physics, which he regards as constituting one physical world rather than the only possible one. He holds, moreover, that man is the maker of not only one type of world, but of worlds of different types, including, apart from the world or worlds of physics, the world or worlds of language, myth, religion, and art. In a lecture given in 1921 and in the introduction to the *Philosophy of Symbolic Forms*, he quotes with approval a passage by Heinrich Hertz which sees mathematical physics as resulting from a symbolic representation of experience.[8] In commenting on it he argues that Hertz's theory of the role of symbolic representation is merely a special case of his own general theory of the function of symbols in the constitution of a world.

According to Hertz it is the main task of science to "derive the future from the past"—a task which we perform by making for ourselves "inner, apparent pictures *(Scheinbilder)* or symbols" in such a manner that "the necessarily thought *(denknotwendigen)* consequences of the pictures are also pictures of the naturally necessary *(naturnotwendigen)* consequences of the represented objects." This account of the function of symbolic representation in the service of

---

8. See "Der Begriff der symbolischen Form im Aufbau der Geisteswissenschaften," in *Wesen und Wirkung des Symbolbegriffs* (Oxford, 1956), p. 186, and *Philosophie der Symbolischen Formen*, vol. 1 (Berlin, 1923-25), p. 5; Hertz, *Prinzipien der Mechanik* (Leipzig, 1894), p. 1.

scientific prediction was in the first place intended to explain and justify Hertz's own reconstruction of classical mechanics by eliminating the concept of force from it. But, as Cassirer points out, it equally explains and justifies much more radical modifications which break through the constraints of classical physics accepted by Newton and exhibited by Kant. It moreover confirms that, "as we must recognize ever more clearly," the building of science "does not progress by rising on a foundation which is firm and fixed forever."[9]

Lastly, as has been mentioned already, Hertz's account may be taken as admitting or suggesting that since man does not live by scientific prediction alone, the symbolic formation of worlds extends beyond the world or worlds of science. It is this suggestion which for Cassirer is the most important. For in pursuing it one comes to see that science, language, and the mystical-religious world are all symbolic forms, each being "an energy of the mind by which a mental content of meaning *(ein geistiger Bedeutungsgehalt)* is tied to a concrete sensible sign and is internally incorporated into it *(ihm innerlich zugeeignet wird)*."[10]

Although this definition of the concept of a symbolic form and the necessarily brief characterization of Cassirer's idea of a general theory or "grammar" of symbolic forms may help to throw some light on his understanding and interpretation of Kant's philosophy, they cannot do justice to his own thought unless they are supplemented by detailed applications of his theory. In fairness it should, moreover, be noted that he regarded his philosophy of symbolic forms not as a finished theory but as a theory *in statu nascendi.*[11]

Cassirer's approach to the history of ideas is closely related to his philosophy; here the philosophy of history plays an important role, since "the synopsis of the mental cannot realize itself anywhere except in its history."[12] For our purposes it is not necessary to explain this statement in detail. Instead, it may be useful to mention two

9. "Zur Logik des Symbolbegriffs," *Theoria*, vol. 2 (1938), reprinted in *Wesen und Wirkung des Symbolbegriffs*, p. 230.
10. *Wesen und Wirkung des Symbolbegriffs*, p. 175.
11. Ibid., p. 229.
12. Ibid., p. 171.

guiding ideas which characterize Cassirer not so much as a philosopher of history but as a historian of philosophy. They are, first, a general view about the direction and the progress of philosophy from Descartes to Kant and beyond; second, a general view about the relation between a philosopher's personality and his philosophy.

Cassirer agrees with those who regard as progressive the tendency of modern philosophy from Descartes to Kant to replace ontology by epistemology. He approves of Descartes's program "to determine the whole domain and limits of the mind," but rejects as impossible the Cartesian and, more generally, the rationalist attempt at "deducing... the concrete totality of the mind from a single logical principle."[13] He sees in Kant's epistemology, which conceives the object of experience as "the correlate of the synthetic unity of the understanding," a decisive step in the right direction.[14] But he also holds that further progress was made—by Herder and von Humboldt among others—in rejecting the limits of epistemology as too narrow and in trying to replace it by a philosophy of culture on the lines of his own philosophy of symbolic forms.

In explaining his view of the relation between a philosopher's work and his life Cassirer frequently appeals to Goethe. Thus in his essay "Goethe and the Kantian Philosophy" he quotes with approval a passage from the Xenien which implies that the philosopher cannot, any more than the poet, be separated from his work, since "alle Wahrheit zuletzt wird nur gebildet, geschaut" (all truth is ultimately shaped, intuited), that is to say, not just passively apprehended.[15] And at the very beginning of Kant's Life and Thought he expresses his full agreement with Goethe's conviction that "the philosophers cannot present us with anything but forms of life" (Lebensformen).[16] For Cassirer the essential task in interpreting a philosopher's work is to understand and present the interaction of his form of life and his form of doctrine (Lehrform). A person's form of life is his manner of dealing

13. Philosophie der Symbolischen Formen, vol. 1, pp. 14 ff.
14. Ibid., pp. 9–10.
15. See Rousseau, Kant and Goethe (Princeton, 1945), p. 84.
16. Kant's Life and Thought, p. 5. Unless otherwise indicated, all subsequent page references in the introduction refer to this work.

with the world in which he finds himself, not the sum of his mannerisms and trivial habits. There is for example, as Cassirer shows, a deep similarity between the form of life of a Kant, by whose daily habits the citizens of Königsberg were able to set their watches, and the form of life of a Rousseau, who threw away his watch so that he would "no longer find it necessary to know what time of day it is."[17]

3

In trying to determine what is central and still alive in Kant's philosophy and what is peripheral and now of merely historical interest, Cassirer is naturally influenced both by his own philosophy and by his judgment of Kant's personality and form of life. As a neo-Kantian pluralist Cassirer sees Kant's main achievement in the discovery of the transcendental method or point of view, conceived as "the expression of the enduring and continuing tasks of philosophy" rather than as "a complete historical whole" (p. 3). As an interpreter of Kant's form of life Cassirer distinguishes between Kant's devotion to "the objective necessity of his subject matter" and his delight in a "surveyable architectonic" of thought structures (p. 171).

Kant's absolutism and his love of architectonic perfection reinforce each other. This is particularly so in the theoretical sphere, where the inconceivability to Kant of a non-Newtonian physics and his aversion from unstable foundations combine to support his firm belief that there is—and can be—only one system of synthetic a priori judgments expressing the condition of any possible objective experience and only one set of Categories the application of which to the sensory manifold confers objectivity upon it. In distinguishing between Kant's fundamental insight into the nature of objective experience and his often highly artificial elaboration of a unique categorical framework, Cassirer is, of course, not alone among the commentators. But, unlike many of them, he not only objects to the alleged finality of the Kantian categorical framework, but to any uniqueness claim for such a framework.

Another important exegetic point, which conforms to Cassirer's

17. *Rousseau, Kant and Goethe*, p. 57.

thesis that objectivity is always the result of a symbolic making, concerns the Kantian distinction between phenomena and noumena (or things in themselves). According to Cassirer this distinction must not be understood as a distinction between knowable objects and unknowable objects which, though not subject to the conditions of possible objective experience, nevertheless exist independently of it. The concept 'x is a thing in itself' is merely the negative concept 'x is not an object of possible experience.' Kant, as Cassirer argues, does not make it clear enough that the positive meaning which the concept of a noumenon acquires in the practical sphere is not that of an objective *thing*, but that of an objective *value* which belongs not to the "range of empirical existence" but to a wholly different domain of being (p. 215).

Kant's exploration of this domain—the domain of human actions rather than natural events—finds an early expression in two essays, published in 1784 and 1785 (p. 223). Cassirer regards them not only as marking an important stage in the development of Kant's thought, but also as being crucial to the development of German idealism and, consequently, of European thought. History, according to Kant, differs from the natural sciences in that it reflects not on "a sequence of mere events, but on a sequence of actions." And since "the thought of action includes the thought of freedom," Kant's essays on the philosophy of history anticipate his ethical theory (p. 227).

For Cassirer these essays represent the culmination of the philosophy of the Enlightenment by clearly expressing its fundamental conviction that the historical progress of mankind coincides with "the ever-more-exact apprehension and the ever-deepening understanding of the thought of freedom" (p. 227). He illuminates the historical role of the essays by comparing the ideas contained in them with the ideas of Rousseau and Herder. Although Kant admired Rousseau as the Newton of moral and political thinking, and although in the essays he still speaks—as Cassirer points out—the language of Rousseau, he no longer accepts Rousseau's actual or metaphysical description of the presocial state of human happiness and innocence. For Kant had by then arrived at the conclusion that social organization is a possible means of educating man toward freedom (pp. 223 ff.).

Herder was a critical disciple who tried to develop his teacher's thought in ways which Kant found wholly uncongenial. Cassirer, on the other hand, regards some of Herder's ideas as a legitimate extension of Kant's doctrine. His philosophical and human sympathies with both thinkers enable him to give an account of their intellectual and personal relations which throws much light on their work, their personalities, and their times. He shows in particular how deeply Kant and Herder differed in their form of life, as expressed in the style and substance of their works; and how, in spite of their respect for each other, this difference tended to blind them to each other's principal merits. Thus Kant can feel little sympathy for a writer who, in Cassirer's words, "is a poet as a philosopher and a philosopher as a poet" (p. 230). And Herder is likely to be repelled by a philosophical style which Kant likens to "engraving on copper for which an etching needle is required" and opposes to "working on a piece of timber for which chisel and mallet serve well enough" (p. 221). To the difference in style there corresponds a profound difference of purpose. Just as Kant's style is appropriate to, and required by, the task of criticism and abstraction, so Herder's style is rooted in his aim of showing that "whereas the philosopher must abandon one thread of perception (*Empfindung*) in order to pursue another, in nature all these threads are *one web*."[18]

In commenting on Kant's *Critique of Practical Reason* Cassirer emphasizes the crucial role in it of the transcendental approach which, as in the *Critique of Pure Reason*, distinguishes between what is passively received by the mind and what is spontaneously formed and thereby made objective. What is passively received is in the theoretical sphere an aggregate of impressions given to the faculty of sense, in the practical sphere an aggregate of desires given to the "lower faculty of desire." What spontaneously transforms these aggregates so as to confer objectivity upon them is in the theoretical sphere the faculty of the understanding, in the practical sphere the spontaneous and autonomous will. The result of this transformation is in the theoretical sphere a world of objects of experience, in the practical sphere a world of objective values (p. 239 ff.).

18. Herder, *Über den Ursprung der Sprache*, quoted in *Philosophie der Symbolischen Formen*, vol. 3, p. 39.

By comparing the transcendental approach in the First and Second Critiques Cassirer reveals some misinterpretations of Kant's aim and method in ethics, in particular the charge of an empty formalism—a charge which is particularly inappropriate when linked with praise for the transcendental approach to theoretical thinking. Another charge, for the rejection of which Cassirer gives good reasons, is the accusation that the Kantian concept of freedom (as opposed, for example, to the empiricist concept, which means no more than the absence of specific constraints in a wholly deterministic universe) is internally inconsistent or in some other sense absurd. He shows in particular that there is no inconsistency between Kant's thesis that the will is free to conform to or to violate the categorical imperative and his claim that the nature of this freedom cannot be understood or that, in Kant's words, even though we do not comprehend the practical and unconditioned necessity of the categorical imperative, "we nevertheless comprehend its incomprehensibility" (p. 263).

An important, perhaps the most important, part of Cassirer's commentary is his treatment of Kant's *Critique of Judgment.* It is there that Cassirer's own philosophy and approach to the history of ideas are most in evidence. He argues first of all that, as has been mentioned earlier, this work is not just the fairly obvious concluding bit in the mapping of the a priori features of the world or worlds which have been investigated in the First and Second Critiques. He next shows convincingly that "the domains of art and of the organic forms of nature" which are the subject matter of the Third Critique constitute for Kant "a different world from those of mechanical causality and of norms" and that this world is subject to "its own characteristic form of laws" (p. 285).

He lastly argues that Kant's conception of these laws and their objectifying function implies "nothing less than a change in the mutual systematic position of all the fundamental, critical concepts so far acquired and determined" (p. 287). There arises, therefore, the task of showing in detail the extent to which the earlier foundations are left unaltered and the extent to which they have been replaced. In undertaking it Cassirer is well served by his historical and scientific knowledge and by his ability to consider philosophical problems as a.ı independent thinker rather than as a dogmatic follower of Kant or any other philosopher.

At the very outset of his discussion of the Third Critique Cassirer warns against giving way to the temptation of seeing it as anticipating Darwin's theory of evolution or of modern biology in general. Such an interpretation would among other things obscure the connection between Kant's analysis of "being harmonious" (*zweckmässig* as used by eighteenth-century German writers) and of "having a purpose" (*zweckhaft*). According to Cassirer, in the Third Critique Kant is not investigating the conditions for the *existence* of harmonious structures in nature and art but "the peculiar direction taken by our cognitive faculty when it *judges* an entity as harmonious, as expressing an inner form" (p. 284). He advances weighty textual evidence to show that the Third Critique has indeed widened and deepened Kant's original concept of the a priori by including "pleasure"—which so far he had considered as absolutely and irradicably empirical—in the "domain of what is determinable a priori and a priori knowable" (p. 303).

Cassirer's interpretation of the Third Critique as constituting an important advance beyond Kant's earlier works is supported by drawing attention to its effect on Kant's contemporaries, notably Goethe. To show the deep impression it made on Goethe one can hardly do better than quote the following passage from Goethe's *Einwirkung der neueren Philosophie,* in which he says of the *Critique of Judgment* that he owes to it "one of the happiest periods of his life" because in it he saw his "most diverse thoughts brought together, artistic and natural production handled the same way; the powers of aesthetic and teleological judgment mutually illuminating each other."[19]

Throughout his book Cassirer tries to see Kant's work in the context of his life and his life and works in the context of his time. He draws attention to often surprising affinities and contrasts between Kant and other creators of Western culture, including—apart from Newton, Rousseau, Goethe, and Herder—Leibniz, Schiller, Lessing, and many others. And he corrects a popular picture of Kant's personality, based on a one-sided view of his regular habits and the alleged

---

19. Weimar Edition, vol. 11, p. 50, quoted in *Rousseau, Kant and Goethe*, p. 64.

rigorous formalism of his moral philosophy. He quotes Charlotte von Schiller as saying that Kant would have been one of the greatest human beings if he had been capable of feeling love, but he finds in Kant's character a "courtesy of the heart" which is close to love as well as to the true spirit of Kant's ethics (pp. 413 ff.).

The commentary contains many illuminating anecdotes which reveal Kant as the very opposite of a pedantic, provincial, and professorial professor. A characteristic example is a remark made in the year 1764 by Hamann, who said about the young Kant that although he carried quite a number of lesser and major works in his head he was so involved in the "whirl of social diversions" that he was "quite unlikely ever to finish any of them" (p. 52).

In concluding these remarks on Cassirer's book, it seems proper to emphasize once again that they are meant to hint at its exegetic and philosophical value and to indicate its relevance to contemporary philosophy and scholarship, and at best can do no more. If they induce some potential readers to study the book, without deterring others from doing so, they will have fulfilled their purpose.

New Haven   STEPHAN KÖRNER
1981

NOTE ON THE
TRANSLATION

Kants Leben und Lehre was first published in 1918, by Bruno Cassirer in Berlin, as a supplementary volume to the edition of Kant's works of which Ernst Cassirer was both general editor and also sole or coeditor of four individual volumes. The edition, entitled *Immanuel Kants Werke*, began appearing in 1912 and was completed with the eleventh, supplementary volume in 1918. An essentially unchanged second edition of *Kants Leben und Lehre* was published in 1921, also by Bruno Cassirer. The present translation has been made from the second edition.

Cassirer's notes have been translated and notes have been added where clarification was called for or a citation was missing. Any mistakes in citations that were found in Cassirer's notes have been silently corrected and missing publication information supplied. Notes added in their entirety in the English edition appear in square brackets. References in the notes consisting of a roman and an arabic numeral only are to volume and page of *Immanuel Kants Werke*, edited by Ernst Cassirer et al. References to the edition of the Prussian Academy of Sciences have in most cases been added in the English edition; following standard practice, these are indicated by the abbreviation *"Ak.,"* followed by volume and page number.

The following English translations of Kant's works have been used for quotations in the text, with modifications: *Critique of Pure Reason*, tr. Norman Kemp Smith (London, 1929); *Critique of Practical Reason and Other Writings in Moral Philosophy*, tr. Lewis White Beck (Chicago,

1949); *Kant's Critique of Aesthetic Judgment*, tr. James Creed Meredith (Oxford, 1911); *Kant's Critique of Teleological Judgment*, tr. James Creed Meredith (Oxford, 1928); *First Introduction to the Critique of Judgment*, tr. James Haden (Indianapolis, 1965); *Kant's Political Writings*, ed. Hans Reiss, tr. H. B. Nisbet (Cambridge, 1970); *Kant's Inaugural Dissertation and Early Writings on Space*, tr. John Handyside (Chicago, 1929); *Philosophical Correspondence, 1759–99*, ed. and tr. Arnulf Zweig (Chicago, 1967); *Prolegomena to Any Future Metaphysics*, tr. Lewis White Beck (Indianapolis, 1950); *Dreams of a Spirit-Seer, Illustrated by Dreams of Metaphysics*, ed. Frank Sewall, tr. E. F. Goerwitz (London, 1900); *Kant's Cosmogony as in His Essay on the Retardation of the Rotation of the Earth and His Natural History and Theory of the Heavens*, ed. and tr. W. Hastie (Glasgow, 1900).

I would like to mention here with great appreciation and respect Charles Hendel, by whose suggestion and encouragement I originally undertook this translation. Hendel has been to a large extent the moving force behind the Yale University Press's publication of Cassirer's works in English; a friend of Cassirer and a colleague during the latter's years at Yale University, Hendel shares to a rare degree Cassirer's breadth of philosophical outlook and interests. His services in support of knowledge and study of Cassirer's thought in English-speaking countries should be neither underestimated nor forgotten.

Wooster, Ohio   JAMES HADEN
1981

Kant's Life and Thought

FROM THE FOREWORD TO
THE FIRST EDITION

The work I publish here is intended to serve as a commentary on and supplement to the complete edition of Kant's works, of which it forms the conclusion. Therefore it is not addressed to those readers who consider themselves in any sense "finished" with Kant and his philosophy; rather it is for readers still in the midst of studying Kant's works. This book aims to show them a path leading from the periphery of the critical system to its center, from the host of particular questions to an open and comprehensive view of the entirety of Kant's thought. As a result, I have tried from the very beginning not to become lost in the multitude of special problems which Kant's doctrine continually presents, but, by rigorous concentration, to bring out only the plan of his system and the major, essential outlines of the Kantian intellectual edifice. The value of the detailed work done by the Kant-Philologie of the last few decades is not to be underestimated, and the results it has led to in the historical and systematic sense naturally demanded careful consideration in the present book. Yet it seems to me as though this trend toward detailed research has frequently hindered rather than furthered a living insight into the meaning of Kant's philosophy in its unity and totality. Confronted by a school of research and activity which seems to preen itself above all on the detection of Kant's "contradictions," and which in the end threatens to reduce the entire critical system to a mass of such contradictions, we may and must strive to regain the kind of synoptic view of Kant and his doctrine possessed by Schiller or Wilhelm von

Humboldt. With this aim in mind, the constant concern of the following study is to turn back from the multiplicity and almost immeasurably complex involution of particular problems to the natural and self-subsistent character, the noble simplicity and universality, of the basic formative ideas of the Kantian system. Given the limitation put on this exposition by the overall plan of the edition, this goal could, of course, only be reached by dispensing with a complete presentation of the sheer bulk of Kant's thought and its detailed explication for the reader. And for the biographical portion of the book I had to impose the same restriction on myself as for the systematic portion. Here, too, I have consciously ignored the wealth of details and anecdotal embroidery which has been handed down by the earliest biographers of Kant, and which since then has entered into all the accounts of his life. I have tried to bring out only those major and enduring characteristics of Kant's career that in the course of his human and philosophical development emerged more and more clearly as this career's consistent "meaning." Knowledge of Kant as a person has not, I hope, suffered accordingly. For Kant's characteristic and genuine individuality can be sought for only in the same basic traits of his mode of thought and character on which his objective and philosophically creative originality rests. This originality does not consist in any peculiarities and quirks of personality and the externalities of his life, but in the orientation and tendency toward the universal revealed in the shaping of both his life and his philosophy. I have tried to show how both aspects condition and complement each other, how they point to an identical origin and finally join in a single outcome, and thus how Kant's personality and work are in fact from one and the same mold. On the other hand, as far as the external circumstances of Kant's life are concerned, the intention here was to present them only insofar as they reveal and express the peculiarly distinctive content of Kant's actual existence—the essence and the growth of his fundamental ideas.

The manuscript of this book was ready for the press in the spring of 1916; only the delay inflicted by the war on the progress of the edition of the complete works is to blame for the fact that it is appearing now, more than two years after its completion. I regret this postponement of printing all the more deeply because I can no longer place the book in the hands of the man who followed it from its inception with the warmest and most helpful interest. Hermann Cohen died on April 4, 1918. The significance of his works for the renewal and the development of Kantian philosophy in Germany I have tried to state elsewhere, and I do not want to return to that here.[1] But I must mention with deep gratitude the personal impact which I myself, more than twenty years ago, received from Cohen's books on Kant. I am conscious that I was first introduced to the full earnestness and the depth of Kant's teaching through these books. Since that time, I have returned again and again to the problems of Kantian philosophy in ever-renewed studies of my own and in the context of a variety of concrete tasks, and my conception of these problems has in many respects diverged from that of Cohen. But all along I have found the underlying methodological idea which guided Cohen, and on which he founded his interpretation of the Kantian system, to be fruitful, productive, and helpful. For Cohen himself this idea—the unavoidable necessity of the "transcendental method"— became the essence of scientific philosophy. And because he construed the Kantian teaching in this way, not as a closed historical entity but as the expression of the permanent *tasks* of philosophy itself, it became for him not only an influence in history but a force which directly affects life. He experienced it and taught it as that sort of power, and in this same sense he understood the close connection between the Kantian philosophy and the general problems of the German spirit. He had suggested this connection in many of his writings, but its complete and comprehensive presentation was the project he had set himself for the present edition of Kant's complete works. Now, however, this long-planned book on Kant's significance for German culture, whose outline and structure he unfolded to me a few days before his death, can no longer be written. But even though it was not granted to us to have Cohen himself as one of the collaborators on this edition, his name is ever to be linked with it. For

---

1. "Hermann Cohen und die Erneuerung der Kantischen Philosophie," *Festheft der "Kant-Studien" zu Cohens 70. Geburtstag. Kantstudien* 17 (1912): 253 ff.

just as he himself remained until the end close to each individual collaborator on the edition, as friend and as teacher, so his mode of thinking shaped a unity of ideals and indicated the basic conviction concerning materials and method which they shared and which has continued to define and guide their labors.

Schierke i. Harz   Ernst Cassirer
August 14, 1918

INTRODUCTION

Goethe once uttered the dictum with respect to Kant that all philosophy must be both loved and lived if it hoped to attain significance for life. "The Stoic, the Platonist, the Epicurean, each must come to terms with the world in his own fashion; indeed, precisely that is the task of life from which no one is exempted, to whatever school he may belong. The philosophers, for their part, can offer us nothing but patterns of life. The strict moderation of Kant, for example, required a philosophy in accordance with his innate inclinations. Read his biography and you will soon discover how neatly he blunted the edge of his stoicism, which in fact constituted a striking obstacle to social relationships, adjusted it and brought it into balance with the world. Each individual, by virtue of his inclinations, has a right to principles which do not destroy his individuality. Probably the origin of all philosophy is to be sought for here or nowhere. Every system succeeds in coming to terms with the world in that moment when its true champion appears. Only the acquired part of human nature ordinarily founders on a contradiction; what is inborn in it finds its way anywhere and not infrequently even overcomes its contrary with the greatest success. We must first be in harmony with ourselves, and then we are in a position, if not to eliminate, at least in some way to counterbalance the discords pressing in on us from outside."[1]

1. Goethe's conversation with J. D. Falk. *Goethes Gespräche*, newly edited by F. Frhr. v. Biedermann, Leipzig, 1909–11, vol. 4, p. 468.

These words epitomize one of the essential goals toward which exact investigation and presentation of Kant's life must be directed. This aim rules out mere narration of external vicissitudes and events; the peculiar fascination and the peculiar difficulty of the task consist rather in discovering and illuminating the *Lebensform*, the form of life, corresponding to his form of thinking. As regards the latter, it has its own history, transcending all personal boundaries, for the problems of the Kantian philosophy, if one traces their origin and development, cannot be confined within the sphere of his personality. On the contrary, in those problems an independent logic of facts emerges; there dwells in them a theoretical content which, detached from all temporal and subjectively personal bonds, possesses an objective existence grounded in itself alone.

And yet, on the other hand, with Kant the relation between form of thought and form of life cannot be understood to mean that the latter came to be only the basis and passive receptacle for the former. In Kant's actual existence, as Goethe—doubtless correctly—says, thought, in its objective content and its objective "truth," not only rules life but also receives in return the characteristic stamp of the life to which it imparts its form. Here that peculiar reciprocal relationship prevails in which each of the two moments that influence each other appears simultaneously as determining and determined. What Kant is and means, not in the context of the whole history of philosophy, but as an individual intellectual personality, is manifest only in this twofold relation. How this relationship is forged, and how the unity it creates is then outwardly revealed ever more lucidly and purely, forms the basic spiritual theme of his life and therefore the focal point of his biography. For in the last analysis the essential task of every biography of a great thinker is to trace how his individuality blends ever more closely with his work and seemingly vanishes entirely, and how its spiritual outlines yet remain embedded in the work and only thus become clear and apparent.

At the threshold of modern philosophy stands a work which affords a classic exposition of this connection. The intent of Descartes's *Discourse on Method* is to develop a radical technique by means of which all the special sciences are to be deduced and demonstrated from their primary and universal "grounds"; but how clearly this objective explanation, by an inner necessity, fuses with our information on Descartes's own development, from the initial universal doubt to the unshakable certainty given him by the idea of a "universal mathematics" and by the axioms and basic propositions of his metaphysics! A rigorous deduction of objective propositions and truths is the aim of the treatise, but at the same time, unintentionally and as it were accidentally, the modern type of philosophical personality is here achieved and clearly delineated. It is as if the new unity of the "subjective" and the "objective" which composes the systematic underlying idea of the Cartesian theory were to be presented again from a completely different aspect and in a different sense.

The second masterpiece of Descartes, his *Meditations on First Philosophy*, also reveals this characteristic style. In these meditations we encounter the highest abstractions of Cartesian metaphysics, but we see them, as it were, growing out of a particular, concrete situation which is maintained in its full detail and specific coloring. The ego, the "cogito," is distilled out as the universal principle of philosophy; at the same time, however, in bold relief against this objective background, there stands out the image of the new life Descartes created for himself in his seclusion in Holland, consciously turning away from tradition and all social constraints and conventions. The literary form of the soliloquy may here point back to older prototypes, especially to Augustine's *Soliloquies* and Petrarch's philosophical confessions; the inner content, however, is nonetheless new and unique. For the confession here is not wrung from a moral or a religious emotion but springs from the pure and indomitable energy of thinking itself. Thought exhibits itself in its objective structure, as a systematic linkage of concepts and truths, or premises and consequents—but in the process the total act of judging and reasoning comes alive for us at the same time. And in this sense the personal *Lebensform* is explicated simultaneously with the form of his system. It can scarcely be asked any longer in this connection whether the former is dependent on the latter, or the latter on the former; ideal and real, world view and process of individual life, have become moments of one and the same indivisible spiritual growth.

If one tries to maintain a similar standpoint in the consideration of Kant's life and philosophy, one in fact immediately finds oneself faced with a special difficulty. For even in a superficial sense, the biographical material we possess seems totally insufficient to arrive at such a comprehensive view. The eighteenth century, more than almost any other, is characterized by its compulsion toward introspection and confession. This drive constantly draws new sustenance from the most varied sources: the trend toward psychological empiricism, toward "experiential study of the soul," joins with the religious motivations that stem from Pietism and the new cult of feeling that originated with Rousseau. Kant is intimately affected by all these spiritual and intellectual currents. His rearing as a child bears the stamp of Pietism; in his youth and manhood he turns to psychological analysis, in order to discover in it a new basis for metaphysics; and for him Rousseau is the Newton of the moral world who has uncovered its most secret laws and animating forces. But in spite of all this, what we possess in the way of personal testimony by Kant is just as meager in extent as it is thin in content. We know next to nothing of actual diary entries, unless one were to include in this category the remarks and observations he was wont to add to the text of books on which he lectured. In an age that sought and valued, above all, sentimental outpourings of the heart in correspondence with friends, he takes a coolly sceptical attitude toward all such spasms of emotion. His letters merely augment and extend the thoughts he set down in his scientific and philosophical treatises. As such, they are of extraordinary importance for knowledge of his system and its development, but they give way only occasionally and, as it were, grudgingly to a personal mood and a personal interest.

The older Kant grows, the more inveterate this basic trait becomes. His first essay, the *Thoughts on the True Estimation of Living Forces (Gedanken von der wahren Schätzung der lebendigen Kräfte)*, does indeed begin with a number of purely personal observations, in which he appears to be seeking an initial definition of the standpoint from which he proposes to judge the material at hand. Here, on a theme that belongs purely to abstract mathematics and mechanics, not only do we hear the scientific researcher, but in addition in his youthful self-confidence the thinker and writer ventures beyond the strict confines of the particular task toward greater subjective vividness of treatment and exposition. And this tone is echoed even in the writings of his mature years; in the objective critique of metaphysics contained in the *Dreams of a Spirit-Seer (Träume eines Geistersehers)* one senses everywhere expressions of the personal liberations Kant is experiencing at the time. But from the moment the foundation of the critical system has been definitively laid, Kant's style also undergoes an inner change. The phrase "De nobis ipsis silemus" [of ourselves, we say nothing], which he takes from Bacon to serve as a motto for the *Critique of Pure Reason (Kritik der reinen Vernunft)*, gathers more and more force. The more definitely and clearly Kant conceives his great objective, the more laconic he becomes about everything concerning his own person. For the biographer of Kant, the source for the systematic investigation and exposition of his work seems to have run dry at the very point where it really begins to broaden.

Yet in itself this difficulty cannot and should not constitute any crucial hindrance, for the part of Kant's life that goes on outside his work cannot in any case be of decisive importance for the more profound task that confronts philosophical biography. Whatever the work itself does not reveal to us at this stage cannot be compensated for by any knowledge, however extensive, of the inner and outer life of its author. This lack, then, is not what we feel to be the actual barrier to our understanding of Kant's nature here, but rather—however paradoxical this may sound—it is the very reverse that interferes at this point with the freedom and breadth of our view. An adequate grasp of Kant's personality suffers not from too few but from too many facts and stories handed down to us about him. The earliest biographers of Kant, from whom we derive all our knowledge of his life, have no object other than the most exact reproduction of all those petty details which constitute his external life. They thought they had grasped Kant as a human being when they described him exhaustively and faithfully in all the particulars of his actions, in the division and organization of his daily life, in his most minute inclinations and habits. They extended this description to what he wore, ate, and drank. With the help of their reports, we can calculate, watch in hand, Kant's daily activity down to the hour and minute; we know about every detail of his household furnishings and economy; we are instructed in the minutest way about all the maxims of his physical and moral regimen. And Kant's image has passed over into tradition and popular memory exactly as they sketched it. Who could ever think of him without remembering one of the peculiarities and oddities, one of the thousand anecdotes born of his habits, so zealously assembled by this tradition?

But, on the other hand, whoever labors to depict Kant's spiritual and intellectual integrity purely on the basis of a knowledge of his philosophy must at once sense an inherent contradiction. For how can one comprehend the fact that the further this philosophy develops, the more thoroughly it is permeated with a tendency toward the purely general, toward the objectively necessary and universally valid—while simultaneously the individual in shaping his life seems to fall prey more and more to sheer particularity, idiosyncrasy, and crotchetiness? Are we faced here with a truly ineradicable contradiction between the form of the critical system and Kant's own form of life, or does this contradiction perhaps disappear as soon as we indicate a different basis for our biographical considerations and deliberately redirect them?

This is the first question the biographer of Kant faces. His task could only be deemed fulfilled if he were to succeed in organizing and interpreting the chaotic mass of notes and information we possess about Kant's person and his way of life, in such a way that this conglomeration of individual details is recast into a truly unified spiritual whole, not merely into the unity of a characteristic type of behavior. The first biographers of Kant, however charming at times their naive and literal portrayals, never attained this goal; indeed, they had virtually no systematic awareness of it. Their mode of observation remained in the true sense "eccentric"; they were satisfied with selecting and assembling discrete peripheral characteristics without seeking or even suspecting the true vital and intellectual center from which they emanated, directly or indirectly. If much of what we know or think we know about Kant's personality seems to us queer and paradoxical today, we should ask ourselves whether this oddness is founded objectively in Kant's life as such or in the subjective observation to which this life was from the beginning frequently exposed; whether, in other words, the appearance of eccentricity in Kant cannot be blamed primarily on eccentricity of understanding and interpretation.

But even so it is not solely the fault of the superficial standpoint from which observation has customarily proceeded if we believe we detect one final unresolved dualism between Kant's inward and outward life, however simple that life may appear. This contradiction is not just an illusion, but is rooted in the very conditions to which this life was subjected, and from which even its steady upward course failed to release it. The complete and symmetrical unfolding of life and of creative work granted to the most fortunate of great men was not allotted to Kant. He molded his whole life with the strength and purity of an indomitable will and infused it with a single ruling idea; but this will, which in the formation of his philosophy proved itself to be a maximally positive and creative element, affects his personal life with a restrictive and negative cast. All the stirrings of subjective feeling and subjective emotion comprise for him only the material which he strives with ever-growing determination to subject to the authority of "reason" and of the objective dictates of duty.

If Kant's life lost something of its richness and harmony in this struggle, on the other hand it was through this alone that it gained its genuinely heroic nature. Nevertheless, even this process of inner self-development can be disclosed only by conceiving Kant's life history and the systematic evolution of his philosophy as one. The characteristic integrity and wholeness expressed in Kant's being cannot be made manifest if one attempts to assemble it from its separate "parts"; one must think of it as something primary and fundamental underlying both his work and his life. How this originally indeterminate substratum unfolds and becomes equally manifest in his sheer intellectual energy and in the energy with which he molded his personal life forms the essential content of the story of Kant's development.

I YOUTH AND EDUCATION

1

The story of Kant's childhood and schooling can be briefly told. Immanuel Kant was born on April 22, 1724, in the cramped circumstances of a German workingman's house, as the fourth child of the master saddler Johann Georg Kant. Concerning the origins of his family Kant says, in a letter written in his old age, that his grandfather, whose final residence was Tilsit, came from Scotland; that he had been one of the great number of those who emigrated from there toward the close of the seventeenth and at the beginning of the eighteenth centuries, some settling in Sweden and some in East Prussia.[1] Objective scrutiny has not substantiated this testimony, at least in the form in which Kant gives it; it has since been established that Kant's great-grandfather was already living as an innkeeper in Werden, near Heydekrug.[2] The statement by Borowski, his first biographer, that the family name originally read "Cant," and that Kant himself first introduced the now customary spelling of the name, has also proved incorrect; as far as the name can be traced through documentary evidence, we encounter it in the version "Kant" or

1. To Bishop Lindblom, October 13, 1797 (X, 326 f.) (*Ak.* XII, 204).

2. On this point, cf. Johann Sembritzki, *Altpreussische Monatsschrift* 36:469 ff. and 37:139 ff. In addition, see Emil Arnoldt, "Kantz Jugend und die fünf ersten Jahre seiner Privatdozentur im Umriss dargestellt," *Gesammelte Schriften*, ed. Otto Schöndörffer, vol. 3, pp. 105 ff.

"Kandt." It is possible, therefore, that the statement concerning Scottish descent, which Kant must have received from an old family tradition, is wholly without foundation. In any case, no one has so far been able to prove it with any adequate degree of certainty.

About Kant's parents we know hardly more than what little their son subsequently related from his own scanty childhood recollections. His mother's image seems to have impressed him more strongly than did his father's. Even as an old man he spoke of her with deeply felt love and emotion, although he lost her when he was only thirteen. He was conscious of having experienced through her the spiritual influences that remained decisive for his entire concept and conduct of life. "I shall never forget my mother," he once expressed himself to Jachmann, "for she implanted and nurtured the first seed of the good in me; she opened my heart to the influence of Nature; she awakened and broadened my ideas, and her teachings have had an enduring, beneficent effect on my life."[3] His mother also seems to have been the first to recognize the boy's intellectual gifts, and she decided, on the advice of her spiritual counselor, the theology professor and preacher Franz Albert Schultz, to guide him toward an academic education.

With Schultz there entered into Kant's life a man who became decisively important for the entire formative period of his youth. In his fundamental religious orientation he was a Pietist, as were Kant's parents. But as a former pupil of Wolff, one whom the latter is said to have particularly esteemed, he was at the same time thoroughly familiar with the substance of contemporary German philosophy, and hence with the tendencies of secular culture in general. In the autumn of 1732, as a boy of eight, Kant entered the Collegium Fridericianum, whose direction Schultz took over in the following year. What this school offered him was solely information, and even in this respect it remained narrowly restricted. The ideal of the old Latin and academic school still reigned, especially in Prussia, and the aim of its instruction was almost exclusively directed toward the

3. Reinhold Bernhard Jachmann, *Immanuel Kant geschildert in Briefen an einen Freund* (Königsberg, 1804), Eighth Letter, pp. 99 ff.

knowledge and skillful use of Latin. Even in Pomerania in 1690, an old ecclesiastical order dating from 1535 that expressly forbade the use of the German language during classes had been reinvoked: "The preceptors shall on all occasions address the pupils in Latin and not in German, as that is frivolous and in children is scandalous and disgraceful."[4]

If one disregards the specifically theological orientation, the condition and the internal organization of the Fridericianum in the period when Kant attended it recalls in many respects the Latin school in Stendal, in which Winckelmann, who was about seven years Kant's senior, grew up. In both schools grammatical and philological teaching composed the framework of the instruction, and while mathematics and logic were indeed included in the curriculum, they were presented in only the sketchiest way. All of natural science, history, and geography was as good as totally excluded.[5] It can be calculated how little significance the instruction imparted to him at the Fridericianum had for his deeper intellectual orientation if one considers that it is precisely these fields to which Kant later feels drawn almost exclusively during the entire initial period of his creative work and to which he dedicates himself with his first youthful zest for knowledge as soon as freedom of choice is granted to him. Kant retained a friendly memory only of the Latin teacher in the *Prima*, the philologist Heydenreich, since through him he discovered a method of elucidating classical authors that depended not merely on the grammatical and formal elements, but also on the content, and that insisted on clarity and precision of concepts. He later expressly said of the other teachers, however, that they were probably incapable of fanning into flame the spark of philosophical or scientific study that lay in him. Thus his most individual, fundamental aptitude at this time remained completely shrouded in darkness; even those of Kant's boyhood

4. See Karl Biedermann, *Deutschland im achtzehnten Jahrhundert*, vol. 2, *Deutschlands geistige, sittliche und gesellige Zustände im achtzehnten Jahrhundert*, 2d ed. (Leipzig, 1880), pt. 1, p. 480.

5. Concerning Winckelmann's schooldays, cf. Karl Justi, *Winckelmann. Sein Leben, seine Werke und seine Zeitgenossen* (Leipzig, 1866–72), vol. 1, pp. 223 ff.

friends who thought they perceived in him the earmarks of future greatness saw then only the eminent philologist-to-be. What the school did give him as a genuine contribution to his later intellectual development is confined to a respect for and an exact acquaintance with the Latin classics, which he retained into his old age. He seems to have been affected hardly at all by the spirit of Greek, which was taught exclusively by use of the New Testament.

From the earliest childhood and youthful memories of most great men there radiates a peculiar glow, illuminating them from inside as it were, even in cases where their youth was oppressed by need and harsh external circumstances. This magic is, as a rule, especially characteristic of great artists. To Kant, on the contrary, his youth appears in retrospect neither in the light of phantasy nor idealized by memory; rather, with the judgment of mature understanding, he sees in it merely the period of his intellectual tutelage and lack of moral freedom. As thoroughly as he afterwards became imbued with Rousseau's theoretical principles, he was never able to arouse in himself the sentiment for childhood and youth that is alive in Rousseau. Rink relates one of Kant's sayings, that he who as a man yearns for the time of his childhood must himself have remained a child,[6] and it is even more indicative and moving when Hippel recounts that this man, who was so reticent in all expressions of emotion, was accustomed to say that terror and apprehension overwhelmed him as soon as he reflected on his former "child slavery."[7] It can be seen in these bitter words that Kant's upbringing left on him a mark he could never fully efface from his life. The decisive factor here was not the external pressure of his social status and the exertions and privations it imposed on him, for all this he bore throughout his life with such composure that it seemed to him almost incomprehensible and offensive when others later spoke of it. The value of life, when it is reckoned according to the sum of pleasure, is "less than nothing":[8] this is no isolated theorem of Kant's philosophy, but precisely the pervasive

6. Cf. F. T. Rink, *Ansichten aus Immanuel Kants Leben* (Königsberg, 1805), pp. 22 ff.
7. Theodor Gottlieb von Hippel the Elder's biography (Gotha, 1801), pp. 22 ff.
8. *Critique of Teleological Judgment*, §83 (V, 514) (*Ak.* V, 434).

motto of his outlook on the world and of his conduct of life. From the very beginning, the goal of his life was not "happiness" but self-sufficiency in thinking and independence of will.

At this very point, however, there was interference from the spiritual discipline to which Kant was subjected in his youth. It did not end with the concrete fulfillment of definite prescriptions and duties, but strove for possession of the *whole* human being, of his opinions and convictions, of his feeling and his will. This scrutiny of the "heart," in the pietistic sense, was practiced incessantly. There was no inner stirring, be it ever so hidden, that could escape or elude this examination, and that perpetual supervision did not attempt to control. Even after thirty years, David Ruhnken, at that time a famous teacher of philology at the University of Leiden, who had attended the Fridericianum with Kant, speaks of the "pedantic and gloomy discipline of fanatics" to which their life at the school was subjected.[9] A mere glance at the curriculum of the institution, filled with uninterrupted prayers and devotional exercises, with periods of edification, sermons, and catechizations, confirms this judgment. All this gave the instruction not only its moral but also its intellectual imprint, for even the theoretical classes were expressly designed to remind one of their relation to religious and theological questions.

If we wish to form a clear picture of the spirit of this instruction, we must supplement the scanty reports we possess as to the teaching activity of the Fridericianum with the many characteristic testimonials that inform us about the growth and development of the pietistic spirit in Germany. In this context, the individual differences are of little weight, for it was the fate of Pietism that, whereas it originally aimed purely at the revivification of an inward, personal religion, as it evolved it hardened almost entirely into a common stereotype. What individuals tell of their conversion takes on little by little the marks of a fixed pattern repeated with only slight variations. And this pattern was more and more explicitly made into a sine qua non for the attainment of salvation. One of Susanna von Klettenberg's feminine correspondents detects even in the former's truly profound religious

9. Ruhnken to Kant, March 10, 1771 (IX, 94) (*Ak.* X, 112).

nature no "formal penitential struggle," without which, she claims, the inner transformation forever remains uncertain and dubious.[10] A definite religio-psychological *technique* now emerged more and more consciously and ostentatiously, in contrast to the original religious content of Pietism. One can scarcely open one of the biographies of this period without meeting its traces everywhere. Not only was the general theological education of young people at that time influenced by this technique—as has been vividly and impressively shown by Semler in his autobiography, for example—even men like Albrecht von Haller, who represent the entire scope and substance of contemporary German culture, sought vainly throughout their lives for inward liberation from it.

In Kant's critical mind, however, this emancipation seems to have been completed quite early. Even in his boyhood and youth, the dissociation that later comprises one of the essential and basic moments of his system is in preparation—the divorce of the ethical meaning of religion from all those surface manifestations that take the shape of dogma and ritual. As yet this dissociation did not involve any abstract conceptual knowledge; it was rather a feeling that grew ever firmer in him when he compared and weighed against each other the two religious attitudes he saw before him in his parental house and in the academic life of the Fridericianum. If juxtaposed purely superficially, the judgments Kant passed on Pietism in his later years sound at first remarkably discordant and contradictory, but their meaning becomes completely unambiguous when one reflects that Kant is thinking of entirely disparate forms of pietistic thought and behavior. The first, which he found embodied in his parents' home, he continued to value and praise even after he had dissociated his personal viewpoint from it. "Even though," as he once said to Rink, "the religious ideas of that time and the concepts of what they called virtue and piety were anything but clear and adequate, still they really got hold of the basic thing. You can say

10. General information on the history of Pietism can be found in A. B. Ritschl, *Geschichte des Pietismus*, 2 vols. (Bonn, 1880–86); Julian Schmidt, *Geschichte des geistigen Lebens in Deutschland von Leibniz bis auf Lessings Tod 1681–1781*, (Leipzig, 1862–64); Biedermann, *Deutschland im achtzehnten Jahrhundert*, vol. 2, part 1.

what you want about Pietism—the people who were serious about it were outstanding in a praiseworthy respect. They possessed the highest thing men can possess, that calm, that serenity, that inner peace, undisturbed by any passion. No trouble, no persecution put them in a bad humor, no dispute was able to incite them to anger and enmity. In short, even the mere observer was involuntarily compelled to respect them. I still remember how there once broke out between the harness and saddler trades quarrels over their respective privileges, from which my father suffered rather directly; regardless of that, this rift was treated by my parents, even in household conversation, with such consideration and love toward their opponents... that, although I was a boy at the time, still the thought of it will never leave me."[11]

Even more intense was the aversion Kant always felt toward the regulation and mechanization of religious life, for which Pietism likewise served him as the prototype. Not only did he condemn every self-tormenting dissection of one's own inner life (with express reference to Haller), because to him this was the direct path to "the mental disorder of alleged supernatural inspirations... illuminatism or terrorism,"[12] but in later years he also rejected and branded as hypocritical all public displays of religious feelings of any kind. His opinion of the worthlessness of prayer, which he revealed both in personal conversation and in his writings, is well known, and in all his expressions of this opinion a suppressed emotion can be sensed in which a recollection of the "fanatical discipline" of his youth still seems to echo.[13] Here for the first time we see how a fundamental tenet of Kantian philosophy, the contrast it makes between the religion of morality and the religion of "ingratiation," has its roots in one of the thinker's earliest and deepest life experiences.[14]

11. Rink, *Ansichten aus Immanuel Kants Leben*, pp. 13 ff. Cf. a similar remark to Kraus in Rudolf Reicke, *Kantiana* (Königsberg, 1860), p. 5.

12. *Anthropologie*, §4 (VIII, 17–18) (*Ak.* VII, 133).

13. See the biography of Hippel, p. 34. Cf. especially the essay "Vom Gebet" (IV, 525 ff.).

14. There is no doubt that Kant's own ideal of the religious education of youth was developed as it were *per antiphrasin* from the experiences of his childhood. He writes to Wolke, the director of the Philanthropin at Dessau, when he recommends his friend

When Kant's *Anthropology (Anthropologie in pragmatischen Hinsicht)* appeared, Schiller complained in a letter to Goethe that even this "serene and Jove-like spirit" had not quite been able to rid his wings of the "contamination of life," and that certain dark impressions from his youth had remained indelibly stamped on him. This judgment rests on a correct intuition, but it is one-sided because it retains only the negative aspect of this situation. The conflict into which Kant was thrown signifies both the first and the formative discipline of his character and will. By resolving it in harmony with his own disposition and view of life, he crystallized a basic trait of both his own nature and of his future development.

Kant's initial years at the university, to judge by the slight information about them that has been preserved, are also significant more for this education of the will than for the knowledge furnished him in the regular course of lectures. In Prussia at this time, school and university supervision were still barely distinct from each other. As late as 1778, under the reign of Frederick the Great, a ministerial edict was promulgated to the professors of the University of Königsberg expressly forbidding the free organization of academic instruction and demanding the closest adherence to prescribed textbooks, on the grounds that the worst compendium was better than none at all. The professors might, if they possessed sufficient wisdom, emend the

---

Motherby's son to him as a pupil: "In respect to religion, the spirit of the Philanthropin is actually quite in harmony with the thinking of the boy's father—so much so that he wishes that even the natural knowledge of God, to the degree that the boy gradually achieves it with increase of years and understanding, should not immediately be directed to devotional activities, save only after he has learned to realize that all of these have value only as a means of quickening a daily fear of God and a conscientiousness in the pursuit of one's duties, as divine commands. For the idea that religion is nothing but a kind of ingratiation and fawning before the Highest Being—arts in which men differ only by the variety of their opinions as to the way which might be the most pleasing to Him—is an illusion which, whether based on dogmas or free from dogmas, makes all moral thinking uncertain and ambiguous. For this illusion assumes, in addition to a good life, something else as a means of obtaining—surreptitiously as it were—the favor of the Most High and occasionally exempting oneself thereby from the most scrupulous care in respect of the good life, and yet having a sure refuge in readiness for an emergency." To Wolke, March 28, 1776 (IX, 149) (*Ak.* X, 178). [Cf. below, chap. 7.—Tr.]

author, but the reading of their own *dictata* was totally abolished. Moreover, the syllabus for each subject was laid down in detail, and particular value was put on the institution of regular examinations by the lecturers, "partly to learn how their *auditores* have grasped this and that, partly to arouse their zeal and attention and thus to become acquainted with the able and industrious."[15] Hence the area in which academic study on the part of teachers and students was constrained to move was narrowly circumscribed, to say the least.

Kant, who by a basic trait of his nature habitually adapted himself to the existing order of life and was content with it, at first seems hardly ever to have intentionally overstepped these narrow confines. It is all the more significant, therefore, that from the start he nevertheless transgresses them, involuntarily as it were. Just as he later, as a *dozent*, expands the prescribed pattern of instruction (the previously mentioned ministerial order expressly excepts Professor Kant and his collegium on physical geography, since in this field no entirely suitable textbook was yet available), so as a student less than seventeen years old he already shows all the evidences of a precocious intellectual independence in selecting and organizing his studies.

"Choose one of the faculties!" was still the universal formula of direction and guidance for the organization of the universities of the time, and in Prussia, for example, this formula had just recently been enjoined once again by a decree of Friedrich Wilhelm I dated October 25, 1735. "And henceforth," this decree runs, "shall the objections be of no avail that many young persons, when they come to the academy, do not yet know whether they should settle on theology, law, or medicine, especially since *studiosi* must already know this, and little is to be hoped for from them if they conduct their affairs so badly that, when they go to the academy, they have not yet decided on what they wish to work on there. Also, the pretext that they wish to apply themselves only to philosophy or a part thereof is under no circumstances to be accepted; but each shall declare himself in addi-

15. Concerning the condition and regulations of the University of Königsberg, cf. D. H. Arnoldt, *Ausführliche und mit urkunden versehene Historie der Königsberger Universität* (Königsberg, 1746).

tion for one of the higher faculties, and make it his business to derive at least some profit from these."[16]

In contrast to this conception of Friedrich Wilhelm I, which regards the university only as a training ground for the future civil servant, who is to be made useful and competent for some particular branch of service, Kant was convinced from the very beginning, as far as we know, of a different basic view, which he held fast to and brought to fruition undisturbed by all external pressures. When he matriculated at the University of Königsberg on September 24, 1740, he was burdened by the most straitened and needy circumstances. According to an entry in the Königsberg church register, his mother had been interred three years before "poorly and quietly," that is, without the attendance of the clergy and with remission of fees, and the same notation is found for the burial of his father on March 24, 1746. But with the certainty and unconcern of genius, Kant seems even then to have spurned any thought of mere training for a profession. For a long time tradition stamped him, on uncertain evidence, as a student of theology, but since the exhaustive investigation of this question by Emil Arnoldt, it has been established that Kant did not, in any case, belong to the theological faculty, and hence probably did not intend to educate himself for a theological calling. The statement on this score which was found in Borowski was struck out by Kant himself in the course of the scrutiny to which he subjected Borowski's biographical sketch. Particularly crucial in this regard is the report of one of the most intimate of Kant's youthful friends, Heilsberg, later the councillor for war and crown lands in Königsberg, who explicitly testifes that Kant had never been an "advanced *studiosus theologiae.*" If he attended lectures on theology, he did so only because of a conviction he also continually impressed on his fellow students: one must seek knowledge from all fields of study and therefore dare neglect none of them, not even theology, "even if one did not thereby seek his daily bread." In connection with this, Heilsberg depicts how

---

16. See D. Arnoldt's history of the University of Königsberg. Cf. with this and with the following in particular, Emil Arnoldt, "Kants Jugend," pp. 115 ff.

Kant and he, together with a third young friend, Wlömer, had attended a lecture by Franz Albert Schultz, Kant's former teacher at the Fridericianum, and had so distinguished themselves through their interest and understanding that Schultz called them to him at the close of the last class and inquired about their personal situations and plans. When Kant replied that he wished to be a doctor,[17] while Wlömer confessed himself a jurist, Schultz further demanded to know why in that case they listened to theological lectures, a question Kant answered with the simple phrase: "out of intellectual curiosity." This answer has a genuinely unsophisticated force and pregnancy. It already contains the first consciousness of an intellectual orientation that could neither be expressed by nor satisfied by a single outward goal of study. An involuntary recognition of this state of affairs is signified by Jachmann's later acknowledgment in his biography of Kant that he had fruitlessly inquired about the "syllabus" Kant had followed at the university; even the one friend and intimate of Kant known to him, Doctor Trummer in Königsberg, had been unable to give him any information on this matter. Just this much is certain, that Kant principally studied *humaniora* at the university and devoted himself to no "positive" science.[18]

The predicament in which this biographer of Kant and his friends found themselves contains an element of unconscious irony: concealed in it is the complete opposition that exists between the tangible goals of ordinary mankind and that purposiveness without purpose which governs the life of even the most reflective and self-conscious genius. Kant's turning away from the traditional academic work and subject matter of the university of his day toward the *humaniora*, looked at from the standpoint of his life history, indicates one of the earliest seeds of just that freer, "humane" form of education to whose later acceptance and realization in Germany Kant's philosophy contributed so decisively. In the evolution of this new ideal of humanity

17. Whether this answer of Kant's—as Arnoldt hints—contained an "admixture of sly humor" is uncertain; it is safer to assume that in the then-existing pattern of the division of the faculties, this was the only answer by which Kant could express his ruling interest in natural science.

18. Jachmann, *Immanuel Kant*, Second Letter, pp. 10 ff.

the most individual and the most universal, the personal and the ideal, intertwine directly. In Kant's lectures the young Herder, who had just liberated himself from the constricting intellectual coercion of his childhood and school years, first discovered that new demand for the "education of mankind" that henceforth constituted the foundation and the impetus for his own creative work.

Moreover, for Kant himself the fruits of these years of study lay less in what theoretical knowledge and insights they afforded him than in the intellectual and moral discipline to which they educated him from the beginning. From all that we know about this period of his life, the privations in even the least things, which had to be overcome daily with the most unrelenting tenacity, never disturbed his inner equanimity; they merely deepened his innate disposition to stoicism. And precisely because this stoicism was never imposed on him from without, but stemmed from a fundamental orientation of his own nature, this phase of his life was marked by a certain innocent vigor and unconcern. Throughout the sketches by Kant's comrades of that period, especially in the memoirs which Heilsberg set down at eighty as material for Wald's memorial address on Kant, this trait emerges conspicuously. One sees how an intimate and common bond of a personal and intellectual kind springs up between Kant and the fellow students with whom he lives, a bond which at the same time takes on the external form of a primitive community of goods— as Kant assists the others with his advice and tutoring, while he receives help from them in the minor tribulations of his material affairs.[19] Thus within this circle there reigns a spirit of true comradeship, a free give-and-take, in which none becomes the debtor of the other.[20] On this point Kant was extremely stern with himself from his early youth. One of the basic maxims he had laid down from the start was to maintain his economic independence, because he saw in it a condition for the self-sufficiency of his mind and character. But although with advancing age Kant's uncompromising sense of independence gradually brought something rigid and negative into his

19. See Heilsberg's account in Reicke, *Kantiana*, pp. 48 ff.
20. See in this regard the description by Emil Arnoldt, "Kants Jugend," pp. 146 ff.

life, there is still visible in his youth a freer and unaffected flexibility in this regard, which was natural to his convivial character and sociable gifts. The harmony of these two tendencies, the impulse to companionship and living communication, and at the same time the positive assertion of inward and outward freedom, is what gave Kant's student life its balance.

The biographer of Winckelmann, whose school years bear a striking resemblance to Kant's in many details of his intellectual development and of the molding of his outer life,[21] has said that there was nothing juvenile in Winckelmann's character save the strength to endure quantities of work.[22] This description is applicable to Kant as well. Even the comradeship with his associates of his own age, about which many humorous details are told, basically developed out of a community of study and work wherein there are already recognizable in Kant, who always emerges as the preeminent intellectual leader, many traits foreshadowing the future university teacher. Just as Kant himself, as Heilsberg recounts, loved "no frivolities, and still fewer revelries," so he gradually converted his audience—a significant expression—to a similar attitude; the sole recreation he permitted himself and them consisted in playing billiards and ombre, which, with the great skill they acquired, sometimes furnished them with a welcome source of income.

Yet in reconstructing the spiritual and intellectual elements of this period we should be even less content to stop with the outward pattern of Kant's life than we generally are. Everything we are told about it is totally subordinate in its significance to the new domain of the mind that must have been first revealed to Kant at that time. In this period the concept of *science*, both in its abstract generality and in its specific embodiments, became truly vivid for him. What secondary

---

21. In this respect one should especially compare the account that Paalzow gives of Winckelmann's student years (in Justi, *Winckelmann*, vol. 1, pp. 46 ff.) with what Heilsberg (in Reicke, *Kantiana*, pp. 48 ff.) relates concerning Kant; it is especially typical that Winckelmann also opposed the requirement of enrollment in one of the three "higher faculties."

22. Justi, *Winckelmann*, vol. 1, p. 44.

school had proffered him as knowledge was at bottom no more than crude material for memorization, whereas now he encounters philosophy and mathematics as essentially interrelated and interacting. The professor who introduced him to them achieved in so doing a decisive influence on the entire future course of his studies. What we know about this teacher, Martin Knutzen, and of his activity as instructor and author, does not immediately account for the depth of this influence. To be sure, Knutzen does reveal himself in his writings as a serious and keen thinker, but the problems that concern him do not essentially transcend the horizons of the current academic philosophy. Within these limits he does not fully commit himself to any particular faction, striving for originality of judgment and independence of decision; however, even the closest scrutiny to which he has been subjected as Kant's teacher can scarcely discover any truly unusual ideas and definitely novel suggestions.[23] Although Christian Jacob Kraus—of all Kant's friends and pupils the one with the most profound understanding of the significance and the content of his philosophy—nonetheless did say of Knutzen that he was the sole person in Königsberg at that time able to affect Kant's genius, this relates less to the content of his teaching than to the spirit in which it was presented. Of the teachers at the University of Königsberg, Knutzen alone represented the European concept of universal science. He alone looked beyond the limitations of the conventional textbook learning; he stood in the midst of the universal discussions that were being carried on about the foundations of rational and empirical knowledge, and he devoted equal interest to Wolff's writings and to Newton's.

Through this teacher's lectures and exercises, Kant entered into a new intellectual atmosphere. The significance of the single fact that it was Knutzen who first lent him Newton's works can hardly be overestimated, since for Kant Newton was the lifelong personification of the concept of science. A sense that he had at last taken a first and permanent step into the world of the intellect must have been alive in

23. Cf. Benno Erdmann, *Martin Knutzen und seine Zeit* (Berlin, 1878).

Kant from the beginning. Borowski tells us that from now on he "attended Knutzen's lectures in philosophy and mathematics."[24] These covered logic as well as natural philosophy, practical philosophy as well as natural law, algebra and infinitesimal analysis as well as general astronomy. A new cognitive horizon was thus disclosed to Kant, one which for his mind, oriented from the start toward systematization and methodology as it was, was bound also to transform the substance and the meaning of knowing.

This trend in his inner development emerged with full clarity in Kant's first published paper, which marks the close of his student years. He must have written it while still a student; the proceedings of the philosophical faculty of the University of Königsberg for the summer semester of 1747 contain the notice that the *Thoughts on the True Estimation of Living Forces (Gedanken von der wahren Schätzung der lebendigen Kräfte)* by the "Studiosus Immanuel Kandt" has been submitted to the censorship of the dean. The printing of the treatise was long delayed; begun in 1746, it was only completed three years later. No detailed biographical data on the intellectual motives that led Kant to choose this theme can be discovered, but simply from the content we can hazard a guess at the path by which the young Kant arrived at the problem of the measurement of force. A survey of the literature on natural philosophy and physics of the early decades of the eighteenth century yields recognition of a general question underlying the controversy over measurement of force, as it was zealously waged in Germany in particular. The defenders of the Leibnizian *measure* of force attempted at the same time to uphold the Leibnizian *concept* of force. This concept was threatened from both sides, for on the one hand it was opposed by the Cartesian geometric outlook in which matter and motion are nothing but modifications of sheer extension, while on the other hand basic Newtonian mechanics, which totally rejects any conclusion as to the essence of force and sees the description and calculation of phenomena as the sole task of empirical

24. Ludwig Ernst Borowski, *Darstellung des Lebens und Charakters Immanuel Kants* (Königsberg, 1804), pp. 28 ff.

science, was asserted ever more strongly and uncompromisingly.[25]

As this controversy progressed, the roles of the individual opponents had gradually become oddly interchanged and confused. For the "metaphysicians" no longer stood clearly and distinctly apart from the "mathematicians"—as had seemed to be the case when the discussion began—but both factions bring "metaphysics" into play, then hurl recriminations at each other for using it. Newton and Clarke see in Leibniz's concept of the monad a revival of the Aristotelian-medieval concept of substance, which conflicts with the basic principles of the modern, mathematic-scientific mode of knowledge. Leibniz, on the other hand, never misses an opportunity for indignation over the concept of forces acting at a distance, claiming that it resurrects the old "barbarism" of scholastic physics, with its substantial forms and occult qualities. The issue thus began to shift more and more from the purely physical realm to that of universal method. It was precisely in virtue of this aspect of the problem that Kant felt himself attracted to it.

Here the question was no longer the discovery and confirmation of individual, definite facts, but rather a fundamental conflict in the *interpretation* of the recognized and accepted phenomena of motion in general; here not just isolated observations and data but the *principles* on which the examination of nature is founded and their diverse areas of jurisdiction had to be weighed against one another. Kant always formulated his particular question in the light of this general task. What is noteworthy in this maiden paper is that the first step Kant takes into the realm of natural philosophy immediately turns into an inquiry into its method. His entire critique of the Leibnizian conception is subordinated to this point of view; at one point he expressly explains that he is not so much combating Leibniz's result as its foundation and derivations, "not actually the facts themselves, but the *modus cognoscendi.*"[26] This confident and deliberate focusing on the

25. For a fuller discussion of this, see my book *Das Erkenntnisproblem in der Philosophie und Wissenschaft der neueren Zeit*, 2d ed. (Berlin, 1911), vol. 2, pp. 400 ff.

26. *Thoughts on the True Estimation of Living Forces* [*Gedanken von der wahren Schätzung der lebendigen Kräfte*, 1746], chap. 2, §50 (I, 60) (Ak.I, 60).

modus cognoscendi is what gives Kant's treatment of the complicated issue its characteristic stamp. "One must have a method by which he can invariably, through a general consideration of the basic principles on which a certain belief is constructed and through the comparison of them with their implications, infer whether the nature of the premises comprises everything requisite for the propositions deduced from them. This occurs when one notes clearly the qualifications involved in the nature of the conclusion and is careful in constructing the proof to select such basic propositions as are confined to the specific qualifications contained in the conclusion. If this is not found to be the case, then we may be certain that these conclusions, thus defective, prove nothing. . . . Briefly, this entire treatise is to be regarded solely and simply as a consequence of this method."[27] Kant called his first work in the philosophy of physics a "treatise on method," as he later, at the zenith of his creative life, termed the *Critique of Pure Reason* (*Kritik der reinen Vernunft*) a treatise on method; the change which the meaning of this designation had undergone for him comprises his whole philosophy and its development.

For Kant at this point is still far away from a "critical" view in the sense of his later doctrine, and it would be arbitrary to read it into this treatise. He had already begun to doubt whether academic metaphysics was firm and solid, but this doubt had its roots more in a general feeling than in conceptual precision and clarity. "Our metaphysics," he states in this paper, "is, like many other sciences, in fact only at the threshold of truly well-founded knowledge; God knows when we shall see it crossed. To see its weakness in many of its undertakings is not difficult. Nothing is more to blame for this than the prevailing inclination among those who seek after extension of human knowledge. They would like to have a great fund of wisdom concerning the world, but it would be desirable for it to be sound as well. Almost the sole return to a philosopher for his labors is, after a painstaking inquiry, to rest at last in the possession of really well-founded, exact knowledge. Hence, it is a great deal to require of him that he but seldom trust his own approval, that he not conceal

27. Ibid., §88 (I, 95 ff.) (*Ak*,I, 93 ff.).

the imperfections of his own discoveries when it is in his power to
rectify them. The understanding is greatly inclined to self-
approbation, and indeed it is quite difficult to restrain it for long; but
one should keep oneself in check, so as to sacrifice for the sake of
well-founded knowledge everything that has a diffuse seductiveness."[28] But within Kant's own essay this considered and precocious
renunciation collides continually with his élan and youthful speculative daring. Not only is the distinction between "living" and "dead"
force, on which the whole treatise rests, itself far more "metaphysical" than "physical" in nature, but also the essay is dominated
throughout by the effort to rise from sheer description of the particular and the actual to direct insight into the most universal "possibilities" of thinking. An especially characteristic example is the
speculation that the given three-dimensional space of our empirical
world is perhaps but a special case of a system of spatial forms, which
may be diverse in their structure and their metrics. "A science of all
these possible types of space would be," the treatise adds, "unquestionably the highest geometry which a finite mind could undertake."
It would at the same time imply the idea that the various forms of
space may correspond to an actual number of different worlds, which
however are not related by any dynamic connection and interaction.[29]
In general, the paper attempts to reconcile and unify mathematics
and metaphysics. Kant himself is of course aware that this is contrary
to the dominant scientific taste of the period, yet it was an indispensable endeavor for him because it was evident that the "primary
sources of the events in Nature" must certainly constitute "a proper
subject of metaphysics."[30]

From the standpoint of Kant's life history, however, the peculiar
interest of the *Thoughts on the True Estimation of Living Forces* lies less
in the content of the paper than in the tone in which it is written. Its
content doubtless looks rather thin in purely scientific respects, especially if one compares it with earlier and with contemporary works on

28. Ibid., chap. 1, §19 (I, 29 ff.) (*Ak.* I, 30 ff.).
29. Ibid., §§8–11 (I, 20 ff.) (*Ak.* I, 22 ff.).
30. Ibid., chap. 2, §51 (I, 61) (*Ak.* I, 61).

classical mechanics, with Euler's *Mechanica sive motus scientia* of 1736 and d'Alembert's *Essai de Dynamique* of 1743. One can see that the twenty-two-year-old student, although he has absorbed an astonishing amount of knowledge from the literature of mathematics and physics, has not yet fully mastered the most fundamental content of the mathematical education of that age. Kant's mode of inquiry rests throughout on the distinctions between dead and living force, between the relations of "dead pressure" and "active motion," distinctions already undermined by the demand of modern mechanics for unambiguous definition of all basic concepts and for the exact mensurability of all relations. In this regard, Lessing's well-known caustic epigram to the effect that Kant in his estimation of living forces neglected the estimation of his own forces was not wrong. Yet even today, when almost all its conclusions are obsolete, the work radiates a peculiar charm, a charm that lies not in what it explicitly contains and offers us, but in what it aspires to and promises us. This is our first encounter with the idiosyncratic temper of Kant's thought in its full strength and clarity. This thinking is oriented exclusively toward *what is the case*, with respect to which every opinion is devoid of weight, though it seem thoroughly authenticated by tradition and by the luster of a famous name. "There was a time when there was much to fear in such an undertaking; but I fancy that this time is now past and the human understanding has happily thrown off the fetters which ignorance and awe had formerly imposed on it. Now one can venture boldly to disregard the prestige of *Newton* and *Leibniz*, if it blocks the discovery of truth, and to yield to no persuasion other than the force of the understanding." Thus viewed, the assessment of the doctrine of living forces takes on a new meaning. Its youthful critic no longer stands forth as advocate of the opinion of a particular faction, but as the partisan of the "understanding." The dignity of human reason is to be vindicated by reconciling its internal conflicts as personified by such astute men.[31] But this vindication does not remain on the purely eclectic plane; when Kant turns his special attention to a certain "mediating proposition," intended to reconcile the claims of

31. Foreword, §1, and chap. 3, §125 (I, 5, 152) (*Ak.* I, 7, 149).

both opponents,[32] the mediation thus enjoined is not supposed to represent a mere compromise between the substance of the opposing views, but is to be reached through exact testing and analysis of the conditions which govern both assertion and counterassertion and which bestow on each its basic validity.

Thus we can already sense here how the general style of Kant's mode of thought is, as it were, taking on shape and definition in every sentence, although this style still lacks a theme truly worthy of it. And consciousness of this individuality and originality is so strong in him that it compels direct self-declaration. "I imagine," he says in the foreword to the paper, "that it is sometimes useful to place a certain magnanimous reliance on one's own powers. Confidence of this sort quickens all our efforts and imparts to them a certain buoyancy which greatly assists the search for truth. If one's state of mind has the conviction of his ability to persuade himself, one is permitted the conviction that he may to some extent trust his own perceptiveness and that it is possible to detect a Herr von Leibniz in errors, thus one bends all his efforts toward verifying his suspicions. After one has gone astray a thousand times in such an undertaking, the gain accruing to our knowledge of truth will still be much more considerable than when one has only kept to the beaten path. Here I take my stand. I have already marked out the road ahead which I intend to follow. I shall embark on my course, and nothing shall hinder me from pursuing it."[33]

With such simplicity and vigor does the note of promise resound in the opening sentences of Kant's first paper. At the moment of his debut as a philosophical writer all constraint and poverty in his outward life is as though obliterated, and there emerges in almost abstract clarity only that decisive law governing his being and his mode of thought. From now on, that wonderful trait of consistency appears in his life and compensates for its lack of fullness and outer variety. He has discovered the form, not of a specific dogma, but rather of his own thinking and willing. With the limitless self-

32. Ibid., chap. 2, §20 (I, 31) (*Ak.* I, 32).
33. Ibid., foreword, §7 (I, 8) (*Ak.* I, 10).

confidence of genius, even the young man in his twenties is conscious that this form will be preserved and fulfilled. He sets at the head of the *Thoughts on the True Estimation of Living Forces* the motto from Seneca: "Nihil magis praestandum est quam ne pecorum ritu sequamur antecedentium gregem, pergentes non qua eundum est, sed qua itur."[34] It remained to be demonstrated that this motto, which Kant chooses as the maxim for his thinking, could also constitute the maxim for his life. Kant was able to achieve and ensure future freedom in the exercise of his profession as writer only by first renouncing it for a long while. Even before his first work is printed, he leaves Königsberg, "forced by the state of his circumstances," as Borowski tells us, to take a position as private tutor in the house of a country preacher.[35] This exile in the role of *Hofmeister* lasted at least seven years (if not nine); but during it Kant won the independence from society and the free self-determination which comprised all that he ever sought or expected for himself in the way of a happy life.[36]

2

In the years that follow, Kant's life recedes almost entirely into shadow—so much so that even its superficial contours can no longer be traced with certainty, and even the data concerning the places and dates of the several phases of this period are dubious and flickering. Most biographers agree that Kant initially took up residence as tutor in the household of the reform minister Andersch in Judschen, and from there removed to the von Hülsen estate in Gross-Arnsdorf near Saalfeld. But the further information that he was also active as private tutor in the house of Count Johann Gebhardt von Keyserling in

34. ["There is naught more important than that we should not follow like sheep the herd that has gone before, going not where we should but where the herd goes."]

35. See Borowski, *Darstellung des Lebens und Charakters Immanuel Kants*, pp. 30 ff.

36. "Even as a youth the master desired to make himself self-sufficient and independent of everyone, so that he might live not for men but for himself and his duty. In his old age he declared that this freedom and independence was the basis of all happiness in life, and asserted that it had always made him much happier to do without than to allow indulgence to make him the debtor of another." (Jachmann, *Immanuel Kant*, Eighth Letter, pp. 65 ff.).

Rautenburg near Tilsit is uncertain and ambiguous. Christian Jacob Kraus, at any rate, asserts flatly that Kant never entered into any relation of the sort; his testimony in this matter has special weight since it was Kraus who took over the post of private tutor and master in the Keyserling home in Königsberg following the marriage of Countess Keyserling to her second cousin Heinrich Christian Keyserling. In any event, judging by the ages of the Keyserling sons, any tutorial activity on Kant's part could hardly have taken place before 1753, and in the following year Kant must already have resumed residence in Königsberg, since a letter of that period is dated from there. Whatever the exact circumstances may have been,[37] it is obvious that such vague and uncertain data are no basis for a judgment that might shed the slightest degree of light for us on Kant's inner development in this period. Only Borowski has preserved a few scanty bits of information on this subject. "The placid rustic environment," he says, "served to foster his industry. His head already held the outlines of many undertakings, a large number of them already almost completely worked out, which he... in 1754 and the subsequent years, to the surprise of many... produced all at once in rapid succession. It was then that he assembled from all fields of learning in his commonplace books what seemed to him important in any way for human knowledge, and he still thinks back today to those years of his rural sojourn and labor with great contentment."[38]

If this account, as seems certain, rests on Kant's own statements—Kant at the very least indirectly corroborated it, since he left it unaltered in looking over Borowski's biographical sketch—the conclusion is that the new set of influences to which Kant was subjected by the pressures of his external situation had no power to destroy the calm continuity of his mental growth. Though in the recollections of Kant's old age this period appears as one totally devoid of struggle, still it was not a time that fostered harmony between his inner and outer life. The years as private tutor, to be sure a typical

---

37. All the materials bearing on the resolution of this question are compiled by Emil Arnoldt ("Kants Jugend," pp. 168 ff.); cf. also E. Fromm, "Das Kantbildnis der Gräfin K. Ch. A. von Keyserling," *Kantstudien* 2 (1898):145 ff.

38. Borowski, *Darstellung des Lebens und Charakters Immanuel Kants*, pp. 30 ff.

part of an intellectual's destiny in that age, invariably meant for any sensitive nature a stern school of spiritual deprivation. The social status of the *Hofmeister* was in every respect oppressive and troublesome. "They don't want to spend more than forty thalers for a tutor"; it is said in the letters of Frau Gottsched, "for that he is supposed to take care of the steward's accounts as well."[39] One can get a lively impression of what the situation was like, especially in East Prussia, if one pictures the conditions portrayed by Lenz, twenty-five years later, in his comedy *Der Hofmeister*, set at an estate at Insterburg. "Plague take it, Pastor!" says the Privy Councillor to the pastor who wants his son to become a tutor. "You didn't raise him to be a servant, and what is he except a servant when he sells his freedom as a private person for a handful of ducats? He's a slave, over whom his master has unlimited power, only he has to have learned enough at the academy to anticipate their heedless notions in advance so as to gloss over his servitude. . . . You complain so much about the nobility and its arrogance; those people consider a tutor one of the domestic servants. . . . But who makes you feed their arrogance? Who makes you turn servant when you have learned something and become the vassal of some numskull of a nobleman who all his life has been used to nothing but slavish obsequiousness from his household help?" The noblest and strongest characters, Fichte for example, always felt profoundly bitter about this serfdom of the private tutor. Kant, so far as we know, was completely spared experiences of this kind. He did sense mutual incompatibility between him and his occupation, and later he protested with a smile that he was perhaps the worst private tutor the world had ever known.[40]

Nonetheless, everything known to us of his relations with the families where he worked shows the high personal esteem in which he was held. Here too, within the circle in which he dwelt, intellectual leadership and a sort of moral ascendancy seem to have swiftly accrued to him. From his youth there flowed from his person, unpretentious though it was, a strength which flourished in every situation

---

39. *Letters of Frau Gottsched*, vol. 2, p. 97 (quoted in Biedermann, *Deutschlands geistige, sittliche und gesellige Zustände*, vol. 2, pt. 1, p. 522).

40. Cf. Jachmann, *Immanuel Kant*, Second Letter, pp. 11 ff.

of life in which he was put and which exacted respect from everyone. His self naturally shaped his environment and his relations with others. For a long time after he had departed, Kant remained in extremely friendly communication with the family of Count von Hülsen. The letters they sent him contained, by Rink's testimony, "the heartfelt expression of thanks, respect, and love, which thus reveals that they see to it that he participates in every interesting family event." "It is perhaps not entirely superfluous to note," Rink adds, "that the von Hüllesen family liberated their serfs during the reign of the then King of Prussia [Friedrich Wilhelm III], and, as is stated in the official notices, they were graciously favored by the humanitarian monarch by elevation to the rank of count."[41] When the Countess von Keyserling moved to Königsberg after her second marriage, Kant continued in an intimate personal and intellectual connection with the Keyserling household; Kraus has told us that since Kant customarily sat in the place of honor at table directly beside the Countess, "it was necessary for one to be a total stranger for this place to be yielded to him as a courtesy."[42]

If we put all this information together, one fact emerges, namely, that even these years as a household tutor, foreign and ill-suited to his true nature as that role appears to be, brought a deep and lasting effect for Kant himself and for others. Kant had originally been compelled to take a position as tutor, but this did not destroy his feeling of inner freedom, since his goal, for which he had risked this intrusion into the prime of his youth, stood fixed and unshakable. In the universal sweep of its view, in profundity and acuteness of conception, in the vigor and power of its language the *Universal Natural History and Theory of the Heavens (Allgemeine Naturgeschichte und Theorie des Himmels)*, which must have been largely written or drafted during Kant's days as tutor, is surpassed by but few of Kant's later works.[43]

41. Rink, *Ansichten aus Immanuel Kants Leben*, pp. 28 ff.
42. Cf. Kraus's account in Reicke, *Kantiana*, p. 60; see also the account of Elisabeth von der Recke, the daughter of Countess von Keyserling (excerpts from *Über C. F. Neanders Leben und Schriften* [Berlin, 1804], pp. 108 ff.). For further information concerning Countess Keyserling and her circle, see E. Fromm, *Kantstudien* 2 (1898):150 ff.
43. Arthur Warda (*Altpreussische Monatsschrift* 38:404) makes it plausible that Kant stayed in Judschen as tutor until 1750, and from 1750 until Easter of 1754 worked on the

It was more than mere scrapbooks of learning that Kant brought to fruition in these years; what he achieved was a free intellectual outlook and mature judgment on the whole realm of scientific problems, both of which the *Thoughts on the True Estimation of Living Forces* had lacked. Secure in both inward and outward matters, he could now return to the university. He succeeded in "assembling the means of advancing toward his future vocation less encumbered by cares,"[44] and he now also possessed a compass of knowledge that enabled him in his beginning years as instructor to lecture on logic and metaphysics, on physical geography and general natural history, and on problems of theoretical and practical mathematics and mechanics. On June 12, 1755, Kant became a doctor of philosophy on the strength of a treatise *De igne (On Fire)*; on September 27 of the same year, after the public defense of his work *Principiorum primorum cognitionis Metaphysicae nova dilucidatio (A New Explanation of the First Principles of Metaphysical Knowledge)*, he was granted permission to hold a course of lectures. Thus Kant began his new career with both a physical and a metaphysical theme. But his mind, which invariably pressed forward to organization and critical structure, was unable to rest content with a simple juxtaposition of diverse sciences. From yet another standpoint he was set the problem of putting physics and metaphysics on firm principles and delimiting their respective modes of ordering problems and of acquiring knowledge. Only when this distinction was completed would it be possible to construct that bridge between philosophy and natural science, between "experience" and "thinking," on which the new concept of knowledge—itself inaugurated and confirmed by the critical philosophy—rests.

However, before we pass on to this development, as we look back at the whole course of Kant's youth yet once more a general observation forces itself upon us. The life of the great individual, seemingly running its course in complete detachment from the grand historical

---

estate of the von Hülsen family in Gross-Arnsdorf. Since the dedication of the *Universal Natural History and Theory of the Heavens* is dated March 14, 1755, and the work unquestionably required several years of preparation, the inference is that its conception and working out largely fall within Kant's years as tutor.

44. Rink, *Ansichten aus Immanuel Kants Leben*, p. 27.

movements of the age, also stands united at its heart with the collective life of the nation and of the age. The fundamental spiritual forces in Prussia in the eighteenth century are summed up in three names: Winckelmann, Kant, and Herder. The formative years of all three display a common direction despite the diversity that stems from the uniqueness of their basic outlooks and the detailed conditions of their development, a path reflecting the general spiritual and material situation of the Prussia of that day. Prussia's accomplishments under Friedrich Wilhelm I had been achieved through an iron discipline, through the power of self-restraint and renunciation. The forces from which the new political shape of the country was to be forged were brought together under a regime of harshest compulsion and extreme penury. As this compulsion permeated all the strictures of private life, it determined, through the institutions of child-rearing and education, the view of life which was to put its stamp thereon. The life of the great individual had first of all to be liberated from the emptiness, narrowness, and lack of freedom of the political and intellectual milieu. Winckelmann and Herder waged this battle with mounting bitterness. Winckelmann, after he discovered himself in Rome, looked back upon the serfdom of his youth and on "barbaric" Prussia with intense anger; and Herder too felt that his intellectual powers fully unfolded only at the very moment when he was on the point of leaving his homeland forever. His true nature first flowered in its entirety in contact with the breadth of the world and of life; his "Travel Diary" offered the first rounded picture of his personal and literary originality. He was no longer bound to his native country by sentiment; "the States of the King of Prussia," he coolly decrees, "will never be happy until they are split into brotherhoods."

If one compares Kant's way of thinking to that of Winckelmann and Herder, the fact that Kant dedicated the first work revealing him as a mature and universal thinker, the *Universal Natural History and Theory of the Heavens*, to Friedrich II assumes a general symbolic significance. It is—considering this trend throughout the whole of Kant's future life—as if he thereby had forever sworn allegiance to his homeland, in all its narrowness and limitation. Whatever his spiritual development, in comparison with that of Winckelmann and Herder, may have lost in so doing cannot be measured, but on the other hand the gain which from then on became an integral part of the evolution of his character and his will was infinitely significant. Kant remained on the soil where birth and the circumstances of life had placed him; but with the strength and self-restraint that form a specific characteristic of his intellectual and moral genius he wrested from this soil what mental fruits it contained. Having already learned in youth and adolescence to fulfill the duty of a man, he remained faithful to this duty to the end, and from the energy of this moral will there grew the critical philosophy's new theoretical perspective on the world and on life.

II THE EARLY TEACHING YEARS AND THE BEGINNINGS OF KANTIAN PHILOSOPHY

1. The Natural Scientific World-Picture—Cosmology and the Physics of the Universe

Kant held his first lecture in the autumn of 1755, in the house of Professor Kypke, where he was then living. The spacious lecture room this house possessed, the entrance hall, and even the steps were "packed with an almost incredible crowd of students." Kant was extremely nervous at this unexpected wealth of listeners. He lost almost all his composure, spoke even more softly than usual, and had to correct himself frequently. However, even these numerous errors in his presentation did not detract from the effect of the lecture on its sizable audience; rather they gave "only an even more lively warmth" to the admiration for this unassuming thinker. The "presumption of the most comprehensive erudition" on Kant's part had by now been permanently formed, and his expositions were followed patiently and expectantly. In the next lecture the picture had changed: Kant's presentation was not only thorough but also frank and winning, and it remained so from then on.

This portrait is drawn from the biography by Borowski,[1] who was himself present in the audience at Kant's first lecture. It is a typical testimony to the strong personal impression that the young Kant made on everyone. That "presumption of the most comprehensive

1. Ludwig Ernst Borowski, *Darstellung des Lebens und Charakters Immanuel Kants* (Königsberg, 1804), pp. 185 f.

erudition" Kant encountered in his auditors can hardly have been founded on his literary reputation, since the very work that could have been the foundation of his literary fame once and for all at this period, the *Universal Natural History and Theory of the Heavens*, had by a curious mischance remained completely unknown to the public. The publisher had gone bankrupt while the work was in press; his entire warehouse was sealed up, and therefore this book never came onto the market.[2] What was known of Kant's scientific labors at the time he began his lectures was therefore restricted—aside from his initial publication in natural philosophy—to a few brief essays he had published in the *Wochentliche Königsbergische Frag- und Anzeigungs-Nachrichten* in 1754.[3] It could not have been from these few pages, which treat specific questions of physical geography, that the expectations of the audience for the young instructor in logic and metaphysics had been brought to such a pitch. Even when Kant received his master's degree on June 12, 1755, a throng of learned and highly regarded men of the city was in attendance, which "revealed its respect by the exceptional hush and attentiveness" with which he was obliged.[4] It must have been the effect of Kant's conversation and personal relationships that earned him this respect, just as later, when all his major philosophical works had finally appeared, his most intimate friends and pupils were adamant in their assertion that in personal intercourse and in his lectures Kant "was far more genial than in his books," that he "threw off ingenious ideas by the thousands," and had squandered "an immeasurable wealth of ideas." The special mark of his originality they found just here, for

2. Ibid., pp. 194 f.

3. "Untersuchung der Frage, ob die Erde in ihrer Umdrehung um die Achse, wodurch sie die Abwechselung des Tages und der Nacht hervorbringt, einige Veränderung seit den ersten Zeiten ihres Ursprungs erlitten habe" ["Inquiry into the Question Whether the Axial Rotation of the Earth, Which Produces the Alternation of Day and Night, Has Undergone Any Change since Its Earliest Period"], *Wochentliche Königsbergische Frag- und Anzeigungs-Nachrichten*, June 8 and 15, 1724; "Die Frage ob die Erde veralte physikalisch erwogen" ["Consideration of the Question Whether the Earth Has Physically Aged"], ibid., August 10 and September 14. (See I, 189 ff., 199 ff.) (*Ak.* I, 183 ff., 193 ff.).

4. Borowski, *Darstellung des Lebens und Charakters Immanuel Kants*, p. 32.

with the run-of-the-mill academician, the book is commonly more learned than its author, whereas the depth and special quality of the true "independent thinker" manifests itself precisely in that his writings do not rank above their author but remain subordinate to him.[5]

In any event, if anything could have destroyed the freshness and immediacy of Kant's mind, it would have been the life into which he now entered in the first years of teaching activity. Over and over he had to struggle with the uncertainty of his livelihood and often with worry about the immediate future. He had laid by twenty Friedrichsdor, as insurance against total destitution in the event of an illness. So as not to dip into this "hoard," he had, by Jachmann's account, "bit by bit to sell off his originally extensive and imposing library, because for several years he could not meet the costs of his most pressing needs from his wages."[6] Even some decades later, Kraus said to Poerschke that anyone who decided to attach himself to the University of Königsberg had taken a vow of poverty.[7] But the external privations, which Kant had long been accustomed to, were not the only pressures he faced. There was also the monstrous academic workload he now assumed under the compulsion of his situation, one which would have slain any other nature but his at the very outset. In the first semester, the winter of 1755–56, he lectured on logic, mathematics, and metaphysics; the next term added, along with the repetition of his previous lectures, a course on physical geography and on the foundations of general natural science. And from now on the scope of his academic activity grew wider and wider; the winter of 1756–57, which introduced ethics into the cycle of his lectures, shows twenty hours a week as compared with twelve and sixteen the preceding semesters. If we move some years further along, we find announced—for instance in the summer semester of 1761—as well as logic and metaphysics, mechanics and theoretical

5. See Karl Ludwig Poerschke's description and judgment, in his lecture on Kant's birthday celebration, April 22, 1812.
6. Reinhold Bernhard Jachmann, *Immanuel Kant geschildert in Briefen an einen Freund* (Königsberg, 1804), Second Letter, p. 13.
7. See Johannes Voigt, *Das Leben des Professors Christian Jacob Kraus* (Königsberg, 1819), p. 437.

physics; besides physical geography, arithmetic, geometry, and
trigonometry, a "disputation" on these latter every Wednesday and
Saturday morning, with the remaining class hours on both days "devoted partly to review and partly to solution of problems." In all, this
announcement comprises no less than thirty-four to thirty-six hours a
week, so one may question whether the whole compass of the program was ever carried through.[8] Is it any wonder that Kant often
complained that he felt this activity, which he performed most con-
scientiously and scrupulously and without the least interruption, to
be only laborious mental serfdom? "I sit daily," he writes to Lindner
in October of 1759, "at the anvil of my lectern and keep the heavy
hammer of repetitious lectures going in some sort of rhythm. Now
and then an impulse of a nobler sort, from out of nowhere, tempts me
to break out of this cramping sphere, but ever-present need leaps on
me with its blustering voice and perpetually drives me back forthwith
to hard labor by its threats—*intentat angues atque intonat ore* [he beholds the serpents and his mouth thunders forth]."[9]

This admission is truly unnerving, and yet one is almost inclined
to forget it when one looks at Kant's writings from this period. For
scanty as they are—his literary production from 1756 to 1763 comprises only a few pages—each of them displays a superior intellectual
mastery of his theme and a fresh and original point of view in its
treatment. In the *Monadologia Physica* [*Physical Monadology*], he posits
a theory of the "simple" atom and of forces acting at a distance that
probes the fundamental problems of the natural philosophy of that
era, especially as they were being taken up and systematically presented at that very time by Boscovich; in the *New Notes on the Explanation of the Theory of the Winds* (*Neue Anmerkungen zur Erläuterung der
Theorie der Winde*) he anticipates the explanation of Mariotte's law of

8. A list of all the lectures announced by Kant in the years 1755–96 has been
assembled by Emil Arnoldt and supplemented by Otto Schöndörffer, the editor of
Arnoldt's *Gesammelte Schriften*, through important research. On the foregoing, cf.
Gesammelte Schriften, vol. 5, pt. 2, pp. 177 ff., 193 ff.

9. To Lindner, October 28, 1759 (IX, 17 ff.). At the time to which this letter pertains,
Kant had—after finishing his "Versuch einiger Betrachtungen über den Optimismus"
["Some Experimental Reflections about Optimism"]—announced a lecture on logic
(using Meier's textbook), on physical geography (using his own manuscript), and on
pure mathematics and mechanics (using Wolff). (See II, 37) (*Ak.* II, 35).

rotation of the winds that Dove later gave in 1835; in the *New Theory of Motion and Rest (Neuer Lehrbegriff der Bewegung und Ruhe)* of 1758, he develops an insight into the relativity of motion completely opposed to the ruling conception, which stood under the aegis of Newton's name and authority. From all of this shines forth an intellectual power undimmed by daily academic drudgery, a universal active energy that allows itself to be only fleetingly confined within the narrow limits imposed on it by the conventional form of university work.

One should not look to this period for fundamental and ultimate philosophical judgments, for everything it contains shows it to belong to the process of intellectual *orientation* which Kant had first to work through for himself. In the later essay, "What Is Orientation in Thinking?" ("Was heisst: sich im Denken orientieren?") (1786), Kant, in analyzing the meaning of the words of the title, brought out three different fundamental meanings of the concept of orientation. The first, in which the sensory root of the word is still clearly recognizable, concerns orientation in *space*; it refers to the determination of the regions of the heavens, which we make by reference to the place where the sun rises. This geographical concept is then joined by the extended mathematical meaning, in which the question is to determine directions in a specific space as such, without requiring any given object and its locus (such as the place where the sun rises) as points of reference. In this sense we "orient" ourselves in a familiar dark room, if we are simply given the position of some object (any one at all, because with its place fixed, all the others can be ascertained by the known relationship of right and left). In both cases we make use of experience with a purely sensuous basis, since the opposition of the directions right and left itself rests on a felt distinction in the subject himself, namely between the right and the left hands. The last and highest stage is reached when we progress from geographical and mathematical orientation to *logical* orientation in the most general sense of that word, in which it is no longer a matter of the locus of a thing in space, but of fixing the place of a judgment or a cognition in the universal system of *reason*.[10]

The distinct stages and their sequence as Kant gives them here can

10. "Was heisst: sich im Denken orientieren?" ["What Is Orientation in Thinking?"] (IV, 351 ff.) (*Ak.* VIII, 134 ff.).

be applied to his own intellectual development. There too he starts with a physical, geographical orientation: it is the plurality and origin of the earth's formation, and equally its place in the cosmos, which are the initial objects of his natural scientific interests. The "Inquiry into the Question Whether the Axial Rotation of the Earth, Which Produces the Alternation of Day and Night, Has Undergone Any Change since Its Earliest Period" ("Untersuchung der Frage, ob die Erde in ihrer Umdrehung um ihre Achse einige Veränderung seit den ersten Zeiten ihres Ursprungs erlitten habe") and the solution of the problem of whether we can speak of an aging of the earth in the physical sense, constitute, in 1754, the beginning of his work as a writer on natural science; it is extended by special studies on the theory of the winds, as well as on the causes of earthquakes and on volcanic phenomena. But all these individual questions are conceived in relation to the one great basic theme of that period: the universal problem of cosmogony, which receives its exhaustive exposition in the *Universal Natural History and Theory of the Heavens*. Yet it appears that even this attempt at a completely universal explanation of the *phenomena* of nature remains insufficient so long as the *principles* and the ultimate empirical and theoretical grounds of the processes of nature are not clearly understood. The concern for orientation brings these increasingly into prominence. Kant sees himself ever more decisively forced out of the realm of description of nature and of natural history into that of natural philosophy. The *Monadologia physica* sets up and defends a new form of atomism, while the *New Theory of Motion and Rest* endeavors to remove an obscurity that had lodged in the foundation of physics itself, in the definition of the basic concepts of mechanics. And once again the analysis is broadened and deepened, as it turns from the elements of physics to those of mathematics. Full light on the relations and laws of magnitudes, which natural science deals with, can be expected only when the presuppositions of mathematical definition and measurement are completely transparent. In this respect the "Attempt to Introduce the Concept of Negative Magnitudes into Philosophy" ("Versuch den Begriff der negativen Grössen in die Weltweisheit einzuführen") of 1763 achieves a first important result; in it the concepts of "direction" and "opposite direction" are defined and used in a new and fruitful sense.

At the same time the conflict between syllogistic and mathematical thinking, between the logic of the schools and the logic of arithmetic, geometry, and natural science is rendered sharp and clear. The old question about the "boundaries" between mathematics and metaphysics thus is given a new substance. All the works of the next few years are related, directly or indirectly, to this central problem, which in the treatise *On the Form and Principles of the Sensible and the Intelligible World (De mundi sensibilis atque intelligibilis forma et principiis)* (1770) is finally given its complete systematic formulation. Once again it is shown that what is put forward here as a conclusive solution immediately dissolves into a complex of the knottiest questions, but the new general path is marked out once and for all, and will be confidently held to from now on. Determination of the spatial cosmos is replaced by determination of the "intellectual" cosmos; the empirical geographer is transformed into a "geographer of reason," who undertakes to map the circuit of its entire content under the guidance of definite principles.[11]

If we turn back from this preview of the general evolution of Kant's thinking to the particular tasks that are the mark and fulfillment of his work during his first decade as a teacher, a consideration of the extent of the world he had to conquer by thought is vital here. No other period in Kant's life is so highly defined and characterized by pure passion for substance. Now he begins a powerful labor aimed at mastering the material of intuition and studying what will provide the foundation for his new total conception of the world. To do this, secondary sources of all sorts have to make up for what Kant lacks in the way of firsthand impressions and experiences: geographical and scientific works, travel descriptions and reports of researches. Even the minutest detail in all this material does not escape his intense and lively notice. This way of assimilating material seems to bear with it all the dangers involved in passive reception of others' observations, but the lack of immediate sense perception is outweighed here by that

11. Cf. *Critique of Pure Reason,* Discipline of Pure Reason, second sect. A 759 = B 787 (III, 513).

gift for precise sensory imagination which is peculiar to Kant. By its means those individual strokes he gleaned from a wealth of scattered reports were composed into a unified, focused picture.

In this regard, what Jachmann has reported about his "astounding inner powers of intuition and imagination" is especially well known. "One day, for example, he described, in the presence of a born Londoner, Westminister Bridge, in its shape and orientation, length, breadth, and height and the specific masses of every particular part so precisely that the Englishman asked him how many years he had lived in London, and whether he was especially absorbed in architecture; whereupon he was assured that Kant had never gone outside Prussia and was not an architect by profession. He conversed in an equally detailed way with Brydone, so that the latter inquired how long he had stayed in Italy."[12] By virtue of this capacity of the mind he builds up—stroke by stroke, piece by piece—the whole of the visible cosmos; his inner powers of representation and thinking enlarge the scant data of the immediately given into a picture of the world that combines richness and systematic completeness. In this period, Kant's power of synthesis far outweighs that of analysis and criticism, contrary to the common notion about him. This urge toward wholeness is so strong in Kant's mind that his constructive imagination almost always outruns the patient study of particular data. The saying "Give me matter and I will build a world," which the preface to the *Universal Natural History and Theory of the Heavens* illustrates and works variations on, in this sense designates not only the special theme of Kantian cosmogony, but also the most general task under his consideration in this period. The astronomical cosmic construction is just the outcome and tangible expression of a specific fundamental power of his thinking. In two separate directions, with respect to space and time, this thinking inquires into the limits of what is empirically known and given. The seventh chapter of the *Universal Natural History and Theory of the Heavens*, which treats "Of the Creation in the Whole Extent of Its Infinitude in Space as Well as in Time," begins: "The universe, by its immeasurable greatness and the infinite variety

12. Jachmann, *Immanuel Kant*, Third Letter, pp. 18 f.

and beauty that shine from it on all sides, fills us with silent wonder. If the presentation of all this perfection moves the imagination, the understanding is seized by another kind of rapture when, from another point of view, it considers how such magnificence and such greatness can flow from a single law, with an eternal and perfect order. The planetary world in which the sun, acting with its powerful attraction from the center of all the orbits, makes the moving spheres of its system revolve in eternal circles, has been wholly formed . . . out of the originally diffused primitive stuff that constituted all the matter of the world. All the fixed stars which the eye discovers in the hollow depths of the heavens, and which seem to display a sort of prodigality, are suns and centers of similar systems. . . .

"If, then, all the worlds and systems acknowledge the same kind of origin, if attraction is unlimited and universal, while the repulsion of the elements is likewise everywhere active; if, in presence of the infinite, the great and small are small alike; have not all the universes received a relative constitution and systematic connection similar to what the heavenly bodies of our solar world have on the small scale—such as Saturn, Jupiter, and the Earth, which are particular systems by themselves, and yet are connected with each other as members of a still greater system? . . .

"But what is at last the end of these systematic arrangements? Where shall creation itself cease? It is evident that in order to think of it as in proportion to the power of the Infinite Being, it must have no limits at all. We come no nearer the infinitude of the creative power of God, if we enclose the space of its revelation within a sphere described with the radius of the Milky Way, than if we were to limit it to a ball an inch in diameter."[13]

And corresponding to this immeasurability in the duration of the world is the infinitude of its becoming. Creation is not the work of an instant; rather, after it has made a start by producing an infinity of substances and matter, it is active throughout the whole succession of eternity in ever-increasing degrees of fruitfulness. The formative principle can never cease working, and it will continuously be oc-

13. *Universal Natural History and Theory of the Heavens* (I, 309 ff.) (*Ak.* I, 306 ff.).

cupied with producing more natural events, new things, and new worlds. If thought, directed toward the past and the origin of things, must at last stop with formless matter, a "chaos," which is shaped progressively into a "world," namely a unified spatial composition and mechanical interrelation of the whole through the constructive forces of attraction and repulsion, the prospect into the future of becoming is unhampered for us, for "the remaining part of the succession of eternity is always infinite and that which has flowed is finite, the sphere of developed nature is always but an infinitely small part of that totality which has the seed of future worlds in itself, and which strives to evolve itself out of the crude state of chaos through longer or shorter periods."[14]

It is unnecessary to discuss here the significance of this theory, the so-called Kant-Laplace hypothesis, in natural science as a whole. So far as Kant's intellectual evolution is concerned, this work, which more than any other delves into the detail of empirical natural science, is significant less for its content than for its method. To reveal the essence of this method, one has at the outset to renounce labeling it by certain philosophical battle cries, such as the sectarian titles of "rationalism" and "empiricism." Whenever anyone has tried to use this schematic opposition as a plumb line for expounding Kant's intellectual development, it has confused the picture far more than it has clarified it. The original, fundamental orientation of Kant's research and thought is precisely that he has in view from the outset a deeper unity of the empirical and the rational than had heretofore been accomplished or recognized in the struggle between philosophical schools.

In this sense the *Universal Natural History and Theory of the Heavens* also asserts, as its title indicates, a thoroughgoing interrelation between the empirical and the theoretical, between experience and speculation. This work takes up the question of cosmogony at exactly the point where Newton had dropped it. Six planets, with their ten satellites, move jointly in the same direction around the sun as the central point and in fact in the selfsame direction in which the sun

14. [Ibid. (I, 317) (*Ak.* I, 314).]

itself rotates; and their orbits are so arranged that as a group they lie almost in one and the same plane, namely the equatorial plane of the sun as extended. If one takes this phenomenon as premise, one is led on to demand a cause of this complete agreement and to trace back the "unanimity of the direction and position of the planetary orbits" to it. Newton saw this problem but was unable to solve it, since he regarded (correctly, judged from the standpoint of the state of knowledge at that time) the space in which the planets moved as completely empty; thus there was no material cause discoverable which by its distribution throughout the space of the planetary bodies could have maintained the similarity of motion. Accordingly, Newton had to say that the hand of God executed this ordering directly without recourse to the forces of nature. He would have been unable to stop with this "conclusion grievous to a philosopher" if instead of seeking the physical bases of the system of astronomical phenomena exclusively in its present state he had turned his gaze backwards to the past of the system, if he had pushed forward from the consideration of the systematic *state* of the universe to its systematic becoming. The law of becoming is what first really accounts for the state of being and makes it thoroughly intelligible according to natural laws.

Thus while in Newton there is a unique blend of empiricism and metaphysics, because with him empirical causality reaches a point where it turns directly into and becomes metaphysical causality, Kant on the contrary returns to that demand for unity of method with which Descartes founded modern philosophy. This foundation itself is not alien to the astronomical problem in cosmology: the outline for an explanation of the world contained in Descartes's unpublished work *Le Monde* explicitly lays down the proposition that we can only comprehend the world in its actually given structure if we first cause it to come into being for ourselves. The *Universal Natural History and Theory of the Heavens* gives this thought the value of a general principle of the "philosophical" explanation of nature. That which for the physicist, for Newton, was the ultimate "given" in nature, must be unfolded before the mind's eye by a philosophical view of the cosmos and derived genetically. Here, hypothesis, even speculation itself, not only may but must go beyond the content of the given, under the assumption that it nonetheless submits to control by this content in that the theoretical results obtained must agree with the data of experience and observation.

In this connection it is clear that Kant, despite all his regard for the pursuit of empirical research, by no means exclusively acknowledged and applied himself to it. This becomes clearer still in the general tendency that wholly governs his own inquiries in this period. Not only the *Universal Natural History and Theory of the Heavens* but also the entire natural scientific orientation of the next decade is guided by an overall ethical and intellectual interest: it seeks nature in order to find man in it. "As I saw at the very beginning of my academic teaching," Kant wrote in his announcement of the schedule of his lectures for the year 1765–66, "that a great neglect among young people who are studying lies particularly in the fact that they learn to rationalize early, without possessing enough historical knowledge which can substitute for *experiences*, I therefore undertook the project of composing a pleasant and easy compendium of the history of the present state of the earth or geography in its broadest sense, which might prepare the way for *practical reason*, and kindle the desire to extend more and more the knowledge thus begun."[15] "Practical reason" is taken here in the widest sense of the term; it comprises the general moral vocation of man, like that totality of "knowledge of the world and of man" which plays so significant a role in every pedagogical program of the Enlightenment. In order to fulfill properly his place in creation, man must above all open his eyes to it; he must conceive himself to be part of nature and yet, by his final purpose, raised beyond it. Thus causal and teleological considerations are directly intertwined here. The way in which Kant, in the preface to the *Universal Natural History and Theory of the Heavens*, tries to reconcile the two with each other, striving to discover in the universal mechanical lawfulness of the cosmos itself the proof of its divine origin, does not as yet contain any original tendencies in comparison with the general outlook of the eighteenth century. The basic ideas of Leibniz's philos-

15. (II, 326) (*Ak.* II, 312).

ophy are merely repeated, namely, that the seamless causal order of everything is itself the highest and fully valid proof of its inner "harmony" and of its intellectual and moral "purposiveness." The world is full of miracles, but "miracles of reason": for the proof and seal of the divinity of being lies not in exceptions from the rules of nature, but in the universality and the inviolable validity of these laws themselves. Wherever the natural science of that time is philosophically oriented and grounded, it clings to this conception, which recurs not only in the scholastic doctrine of the Wolffians, but also in French philosophy with d'Alembert and Maupertuis. Since Kant unselfconsciously assumes this form of the teleological proof, all his intellectual and spiritual endeavors cohere in an unbroken unity. There is no talk of a dualism between the world of the is and the world of the ought, between physics and ethics, but his reflection moves back and forth between the two realms, without any feelings on Kant's part of any sort of shift or methodological leap.

This reflective stance is characteristically expressed also in his mood and outlook on life. Kant described this period of his *Magisterjahre* as the most peaceful of his life, when he later looked back at it.[16] Of course he still labored under the pressure of financial need and under the excess of academic work that was imposed on him, but the marvelous mental elasticity of these youthful years easily and completely overcame all constraints of this kind. Although in the later period of Kant's life, especially in the time when he was constructing and expounding the critical philosophy, concentration of every power of thought and life on a single point is characteristic of him, here, instead, there still reigns a free surrender to life and to experience in all its breadth. Just as Kant worked experiential material of the most diverse kinds and origins into his studies and lectures, so he seeks in this period the manifold stimulation of social intercourse. "Thus," Rink says, "Kant in his early years spent almost every midday and evening outside his house in social activities, frequently taking part also in a card party and only getting home around midnight. If he was

---

16. See the letter to Lagarde dated March 25, 1790 (X, 16) (*Ak.* XI, 142).

not busy at meals, he ate in the inn at a table sought out by a number of cultured people."[17] Kant gave himself to this mode of life in such an easy and relaxed way that even the most meticulous psychological observer among his intimates was occasionally puzzled about him; in 1764 Hamann says that Kant carries in his head a host of greater and lesser works, which he however probably will never finish in the "whirl of social distraction" in which he is now tossed.[18]

Kant's teaching at this time was also marked by this cosmopolitan urbanity, appropriate to the standards he had set for himself. His treatment of physical geography—"not with that completeness and philosophical exactitude in each part which is a matter for physics and natural history, but with the rational curiosity of a traveler who everywhere seeks out what is noteworthy, peculiar, and beautiful, collates his collection of observations, and reflects on its design"[19]—is not surprising because of the popular, encyclopedic character he gave to this discipline: he himself even declares about the teaching of the abstract scholarly disciplines that they ought to form in the hearer "first the man of *understanding*, then the man of *reason*," and only in the end the *learned* man. This inversion of the customary manner of instruction seems to him unavoidable for philosophy in particular, for one cannot learn "philosophy" but only "how to philosophize." Logic itself, prior to its emergence as "critique of and preface to true learnedness," must be employed as critique of and preface to "sound understanding," "just as this latter on the one hand touches crude concepts and ignorance, on the other science and learning." Ethics, too, may not start with abstract and formal prescripts of obligation, but must always reflect historically and philosophically on what *does* happen before it points out what *should* happen.[20] Thus it is in general an ideal of comprehensive practical human wisdom at which Kant aims in his own growth as well as in his teaching. Like the

17. F. T. Rink, *Ansichten aus Immanuel Kants Leben* (Königsberg, 1805), pp. 80 f.

18. *Hamanns Schriften*, ed. F. Roth (Augsburg, 1821–43), vol. 3, p. 213.

19. "Proposal for and Announcement of a College of Physical Geography" (1757) (II, 3) (*Ak.* II, 1).

20. See the announcement of the arrangement of his lectures during the winter term, 1765–66 (II, 319–28) (*Ak.* II, 303–13).

lectures on physical geography in the beginning, the later lectures on anthropology pursued this goal. The special, deeper basis for the congenial facility Kant's philosophy achieved in this period lies in the general relationship set up here between "experience" and "thinking," between "knowledge" and "life." No inner tension and contrast yet exists between these two poles. Thinking itself and its systematization, as it is here understood, is nothing but experience refined, freed of superstition and prejudice, and rounded out and extended through the power of analogy. It does not strive beyond this form.

Nowhere does Kant stand closer to the ruling eighteenth-century ideal of "philosophy," to the ideal of "popular philosophy," than at this point. Even if he does express and present this line of thought in a more clever, lively, and vital way than its other champions do, still by and large he gave it no perceptibly novel turn. He also seems to still expect the solution to basic philosophical problems from the sifting and refining of the concepts of common sense. In this sense, perhaps, his essay "Some Experimental Reflections about Optimism" ("Versuch einiger Betrachtungen über den Optimismus,") from the year 1759, aims to achieve a solution to the problem of the "best world," which however rather resembles a complete *petitio principii*. "If someone makes bold to assert," he says there, "that the Supreme Wisdom has preferred the worse to the best or that the highest Good has let itself love a lesser good rather than a greater one equally within its reach, I restrain myself no longer. One serves philosophy very ill if one uses it to overturn the principles of sound understanding and one does it little honor if one finds it necessary, in order to vanquish such efforts, to borrow their own weapons."[21]

Real radicalism is absent from his thinking and his life alike. This explains why Kant, even at a time when a complete change in his form of life and thought had been setting in for a long time, was still taken by those not close to him as the "worldly philosopher" whom they preferred to consult for decisions in questions of taste and style of life. Borowski tells us that his students were wont to ask of him,

---

21. "Versuch einiger Betrachtungen über den Optimismus" (II, 35 f.) (*Ak.* II, 27 ff.).

"straight from the shoulder," what they needed for life and for learn-ing: they not only asked him, in 1759, for a course in "eloquence and German style," which Kant turned over to Borowski instead of doing himself, but they turned to him in 1764 also, at the funeral of a Königsberg professor, for help in "setting up the ceremonies."[22] Cul-tivated Königsberg society tried increasingly to draw him into its circle; "those who didn't even understand how to estimate his superiority," Rink remarks naively, "at least sought, each for himself, the honor of seeing so highly esteemed a man in his own circle of acquaintance."[23] Kant had a close personal relation with the officers of the Königsberg garrison, and for a long time ate almost every day with them; General von Meyer, a "clear mind," in particular liked it when the officers of his regiment were instructed by Kant in mathe-matics, physical geography, and fortification.[24] His connection with distinguished merchant families is well known, especially with the eccentric Green, the model for Hippel's *Clockwork Man (Der Mann nach der Uhr)*, and Green's crony Motherby. The most amiable traits of Kant's nature emerged in this friendship, which Kant's contem-poraries loved to illustrate with a wealth of amusing anecdotes.[25] A noteworthy demonstration of the direction in which esteem for Kant was moving during his teaching years was eventually given even by the Prussian government, when, after the death of Professor Bock in 1764, it offered him the post of Professor of—Poetry, a post along with which went that of censor of all poems for official occasions and the obligation of composing German and Latin *carmina* for all academic celebrations.[26] If Kant had not, despite the hardships of his external circumstances (he shortly thereafter, when applying for the position

22. Borowski, *Darstellung des Lebens und Charakters Immanuel Kants*, pp. 189 f.; Hamann to Lindner, Easter Monday, 1764.

23. Rink, *Ansichten aus Immanuel Kants Leben*, p. 80.

24. Ibid., p. 32; Hamann to Lindner, February 1, 1764; Rudolf Reicke, *Kantiana* (Königsberg, 1860), p. 11.

25. On the friendship with Green and Motherby, cf. Jachmann, *Immanuel Kant*, Eighth Letter, pp. 75 ff.

26. The official acts in this connection are published in Friedrich Wilhelm Schubert's biography of Kant, *Sämmtliche Werke*, ed. K. Rosenkranz and F. W. Schubert (Leipzig, 1842), pp. 49 ff.

of a sublibrarian paying sixty-two dollars a year, spoke of his "very precarious subsistence at the local academy"),[27] possessed the resolution to resist this way of obtaining a livelihood, he would not have been spared the fate of acting in Königsberg as the successor to Johann Valentin Pietsch, Gottsched's renowned teacher.

Nonetheless it was just at this time that Kant's intellectual evolution took the path that in the end reversed his whole style of thought and life. The Berlin Academy of Sciences had proposed, for the year 1763, a topic that immediately attracted the attention of the entire German philosophical world. "Are the metaphysical sciences," it asked, "amenable to the same certainty as the mathematical?" Almost all the leading German thinkers—Lambert, Tetens, and Mendelssohn in particular, besides Kant—tried their hand at solving this problem. For the others it afforded them at most the chance to publicize and argue for the settled view they had already formed on the theme, through established opinion or by their own inquiries. For Kant, on the contrary, working out this task was the starting point for a movement of thought that continually advanced and gathered strength. The problem did not arise in the reply to the question he sent to the Academy but only really took hold of him after he had finished his answer. Outwardly, the circle of his interests and efforts seems hardly altered by this. Questions of natural science, psychology, and anthropology keep their grip on his thoughts,[28] and if the center of gravity of those reflections shifts gradually over from outer experience toward inner experience, only their object, not their principle, has changed. The essential novelty lies in the fact that now whenever Kant attends to a given subject, he is never occupied with it alone, but requires a justification of the essence of the *type of cognition* through which we are aware of it and which makes it knowable.

The *Universal Natural History and Theory of the Heavens* was far

27. To Friedrich II, October 24, 1765 (IX, 40) (*Ak.* X, 46).
28. Cf. "Versuch über die Krankheiten des Kopfes" ["Essay on Diseases of the Brain"], written in 1764 (II, 301 ff.) (*Ak.* II, 257 ff.); the review of Pietro Moscati's book *Von dem körperlichen wesentlichen Unterschiede zwischen der Struktur der Thiere und der Menschen* [*The Essential Physical Difference between the Structure of Animals and Men*] (Göttingen, 1771), (II, 437 ff.) (*Ak.* II, 421 ff.).

removed from this kind of analysis of the cognitive modes. It applied indiscriminately the procedures of scientific induction, of mathematical measurement and computation, and, finally, those of metaphysical thinking. The structure of the material world and the universal laws of motion that hold in it are made into the basis for a proof of God's existence, and Kant's mind leaps straight from a calculation of the different densities of the planets to speculation on the physical and mental differences of their inhabitants and to the prospects for immortality.[29] Since causal and teleological insights are so completely merged here, intuition of nature leads straight to a doctrine of the moral vocation of man, which then finds its conclusive expression in certain metaphysical propositions and requirements. "If one has satisfied his mind with such reflections," Kant concludes the *Universal Natural History and Theory of the Heavens*, "the contemplation of a starry heaven on a pleasant night affords a kind of enjoyment which is felt only by noble souls. Out of the universal stillness of Nature and the repose of the senses, the immortal soul's secret capacity for knowledge speaks an unnamed language and gives us implicit concepts which can be felt but not described. If there are among the thinking creatures of this planet base beings who, heedless of all the charms whereby so vast an object can allure them, are nevertheless able to linger firmly in the service of vanity, how unfortunate is this globe that it can produce such miserable beings! How fortunate it is, on the other hand, that amid all the constraints we must accept a way is opened to a happiness and sublimity which is exalted infinitely beyond the excellences attainable by the most advantageous course of Nature in every body in the universe."[30]

But the mind of a Kant could not dally with concepts that let themselves be "felt but not described." Where he set and acknowledged limits to conceptualizing, he demanded the proof and foundation of this "inconceivability." The need to translate the unnameable language of feeling into the precise and clear tongue of the understanding, and to make the "secret capacity for knowledge" itself man-

29. See the appendix to the *Universal Natural History and Theory of the Heavens* (I, 353 ff.) (*Ak.* I, 349 ff.).
30. Ibid. (I, 369 f.) (*Ak.* I, 367 f.).

ifest and lucid, became ever more imperative. Is the method of metaphysics—this is how the question must now be posed—interchangeable with that of mathematics and empirical science, or is there a fundamental opposition between them? And if the latter should be the case, have we in general any guarantee that thinking, purely logical concepts and logical deduction, is able to express fully the structure of "reality"? The final solution to this question still lies in the distant future for Kant, but having now been posed, it signifies a whole new orientation for the further evolution of his system.

2. The Problem of Metaphysical Method

The first step toward the gradual crumbling of the foundations on which the edifice of the *Universal Natural History and Theory of the Heavens* is raised lay in the direction of the problem of teleology. The basic intuitions that governed Kant as he worked out his thoughts on cosmology are through and through optimistic in nature. It is the Leibnizian system of "harmony" that Kant believes he recognizes in the form of Newtonian physics and mechanics. A secret plan underlies the mechanistic rise and fall of worlds, a plan we are unable to follow in detail, to be sure, but of which we are nonetheless certain that it will always lead the whole universe ever closer to its supreme goal: steadily increasing perfection. Even where this conviction is decked out in the traditional form of the teleological proof of God's existence, Kant makes no opposition. "I recognize the great value," he expressly remarks in the preface to the *Universal Natural History and Theory of the Heavens*, "of those proofs which are drawn from the beauty and perfect arrangement of the universe to establish the existence of a Supremely Wise Creator; and I hold that whoever does not obstinately resist all conviction must be won by those irrefutable reasons. But I assert that the defenders of religion, by using these proofs in a bad way, perpetuate the conflict with the advocates of Naturalism by presenting them unnecessarily with a weak side of their position."[31]

31. [Ibid. (I, 224) (*Ak.* I, 222).]

This weak side lies in confusing "material" and "formal" teleology, inner "purposiveness" and outward "intention." Not in every case where we observe the harmony of the parts within the whole and their cooperation toward a common end do we have the right to assume that that sort of agreement is only brought about through the artfulness of a mind standing outside and above the parts. For it might very well be that the nature of the object itself necessarily leads to such a harmony, that the original unity of a formative principle which unfolds itself little by little in a manifold of effects determines unaided such an internal organization of the details. We find a composition of this latter sort not only in all organic structures, but even in the pure forms through which the logical and geometrical lawfulness of space is known by us: for here a wealth of novel and surprising consequences flows from some kind of individual basic determination or relation, held together as though through a supreme "plan" and adapted to the solution of a wide variety of tasks.

Chiefly by dint of this distinction of formal and material, external and internal purposiveness, Kant is enabled to keep the idea of an end clear of any confusion with the trivial conception of utility. The *Universal Natural History and Theory of the Heavens* has already denounced this confusion and fought it with all the weapons of satire and mockery. Voltaire's *Candide*, which Kant later makes reference to,[32] could not in this regard teach him anything new. In the basic plan of nature and "providence" every creature, however insignificant, is on a par with man. For the infinity of creation embraces in itself as equally necessary all creatures which its superabundant riches bring forth: "From the most sublime sort of thinking beings to the humblest insect, no member is indifferent to her; and none can be taken away without rupturing the beauty of the whole, which consists in this interconnection."[33]

Yet it is more a personal reaction than a strict logical and systematic examination that Kant applies to the popular philosophical way of

---

32. See "The Only Possible Basis of Proof for a Demonstration of God's Existence," sect. II, Sixth Reflection, §4 (II, 138) (*Ak.* II, 131), and *Dreams of a Spirit-Seer*, pt. 2, chap. 3 (II, 390) (*Ak.* II, 373).

33. *Universal Natural History*, pt. 3 (I, 355 f.) (*Ak.* I, 354).

regarding teleology. Only gradually is keener critical analysis of the concepts and demonstrations brought to bear, probably receiving in this instance its first decisive impulse from outside. Much as Goethe, when a seven-year-old boy, was gripped by the "extraordinary event" of the Lisbon earthquake and for the first time felt moved to deeper spiritual reflection, and as the conflict between Rousseau and Voltaire over the "best of all possible worlds" was set ablaze by this same event, Kant likewise saw himself here summoned to the renunciation of intellectual justification. He tried to fulfill his obligation to inform and illuminate the public in three essays which he published in 1756, partly in the *Wochentliche Königsbergische Frag- und Anzeigungs-Nachrichten,* and partly separately,[34] but this did not silence the problem so far as he himself was concerned. "Some Experimental Reflections about Optimism" of 1759, which is no more than a hastily composed, academic occasional piece,[35] was also insufficient to settle it.

He took up the question yet again, four years later, in "The Only Possible Basis of Proof for a Demonstration of God's Existence" (Einzig möglichen Beweisgrund zu einer Demonstration des Daseins Gottes"), in order to present his view of teleology, in both the positive and the negative senses, systematically and exhaustively, and to give it a foundation. Here he finds the proof for the existence of the divine being, customarily drawn from the purposive arrangement of the world, largely proportioned "alike to the worth as to the weakness of human understanding." But this latter point he raises more acutely than before, and points out the fundamental defect clinging to the whole methodology of physico-teleology. The conviction that flows from it may be "exceedingly sensory and hence very lively and gripping and both accessible and comprehensible to the most ordinary understanding," but at no point can it stand up to the strict requirements of conceptual knowledge. For even supposing it were proved that order arose from disorder, a "cosmos" from "chaos," by specific divine actions, that primordial being which ought to be thought as infinite and all-sufficient will precisely thereby labor under a basic

34. See I, 427 ff., 439 ff., 475 ff. (*Ak.* I, 417 ff., 429, ff., 463 ff.).

35. There is information about the origin of this piece in a letter of Kant's to Lindner dated October 28, 1759 (IX, 16).

limitation laid on it from outside. If crude matter is the opponent which this being has to overcome and it displays its goodness and wisdom only in that victory, then if the proof is not to lose all its meaning and effectiveness this matter has to be recognized as something in itself, as a given stuff with which the purposeful power must occupy itself. Hence this procedure can only serve "to prove an originator of the connections and artful composition of the world, but not of matter itself and the creation of the elements of the universe." God will by this route always be shown only as master craftsman, not as creator of the world; the order and formation of matter appears as the work attributable to Him, but not its generation.

In this way that very idea of purposiveness of the wo: d which is supposed to be established is put in extreme jeopardy. For there now enters into the world a basic dualism which, no matter how hard one may try to conceal it, is ultimately ineradicable. The shaping of the sheer stuff of being by intentional will is never absolute, but always something relative and conditioned: there is, in this mode of intuition, at least a definite substrate of being which as such does not carry the form of reason in itself but rather is opposed to it. The gap in the physico-theological proof is at this point clearly visible; it can be plugged only if we succeed in showing that what we have assumed to be the real and independent "essence" of matter and from which we can deduce its universal laws of motion is not alien to reason's regulation but rather is an expression and a particular manifestation of these very rules.[36]

This conception of the task now, however, transforms for Kant the whole aim and form of the proof of the existence of God. For now we no longer work from the configuration of the actual to discover in it testimony to a supreme will, which formed it according to its own wishes, but we take our stand on the validity of the highest truths and seek to win from them a passage to certitude concerning an absolute being. It is not in the realm of empirical, contingent things but in the realm of necessary laws, not in the territory of existence but in that of

36. On the whole of this, cf. "The Only Possible Basis of Proof," sect. II, Fifth and Sixth Reflections (II, 122–44) (*Ak.* II, 116–37).

sheer "possibilities," that we shall henceforth have to choose our starting point. In putting the problem in this way, Kant is indeed aware that he has overstepped the bounds of the popular mode of expounding philosophical ideas that he had followed in his writings up to this time. "I might also be fearful," he remarks, "of offending the sensibilities of those who complain most of all about dryness. But without being hampered by this charge, I must this time ask their indulgence on this score. For although I find as distasteful as anyone else the oversubtle wisdom of those who inflate, distill, and refine definite and useful concepts in their logical smelting shops until they burn away in vapors and volatile salts, yet the object of consideration before me is of a kind which one must either abandon totally the hope of demonstrating with certainty or else endure the analysis of its concepts into their atoms."[37] The process of abstraction cannot stop before it has pressed on to the pure and simple concept of "existence" on the one hand and the pure and simple concept of logical "possibility" on the other.

With this formulation of the opposition, Kant at the same time points back to the historical origin of the problem, which underlies it here. "The Only Possible Basis of Proof" uses the language of Leibnizian philosophy throughout. But in it the distinction between the actual and the possible goes back to the more profound methodological distinction between "contingent" and "necessary" knowledge, between "truths of fact" and "truths of reason." The latter, to which belong all propositions of logic and mathematics, are independent of the state of transient existing things, for they do not express the particular existent, occurring once, here and now, in a specific locus in space and at a determinate point of time, but rather they signify relations that are valid completely universally and are binding on any given content. That $7 + 5 = 12$, that the angle inscribed in a semicircle is a right angle, are "eternal truths," which do not depend on the nature of spatiotemporal, individual *things*, and which thus remain true even if there were no things of those sorts, even if there were no

37. "The Only Possible Basis of Proof," sect. I, First Reflection, §2 (II, 79) (*Ak.* II, 74-75).

matter and no physical world. In logic, in pure geometry and number science, and moreover in the principles of the pure theory of motion, it is thus a matter of cognitions that express a purely ideal dependence between substances in general, not of a connection between determinate empirical, actual objects or events. If we translate this logical insight into the terminology of Leibnizian metaphysics, it can be said that propositions of the first class, the pure truths of reason, are valid for all possible worlds that are comprehended in the divine understanding, while the mere truths of fact pertain only to specification of the one actual world that has been lifted out of this sphere of general possibilities by an act of the divine will and "permitted" actual existence.

From this point on, the particular form that Kant gives to the problem of the proof of God's existence is fully comprehensible. In the place of the "moral" dependence of things on God, which is the customary relation in this proof, he wishes to put "nonmoral" (or better, "extramoral") dependency, that is to say, he does not wish to seem to be drawing his arguments from the realm of particular phenomena, which the reference to a specific divine act of will seems to involve, but to take his stand on universal and necessary relations, which as such are irrefragable norms for every finite and infinite understanding alike.[38] He does not want to proceed from "things" as an already given order, but rather to go back to the universal possibilities that are the presupposition for the state of all ideal truths and hence mediately for the state of everything real as well. Therefore the proof that Kant attempts bears a thoroughly aprioristic character, for it follows not from the contingent, merely factual existence of a particular thing or even from the whole array of particular empirical things, but from an interconnection of concepts that, like the concepts of geometry and arithmetic, compose an unchanging, systematic structure, free of all arbitrariness.[39] Is it possible, Kant's question is now stated, to arrive at certainty concerning an absolute existent— that is, as will appear, at certainty about God—when on our part

---

38. Cf. ibid., sect. II, Second Reflection (II, 106) (*Ak.* II, 100).
39. Ibid., sect. I, conclusion (II, 96) (*Ak.* II, 90).

nothing but the certainty of ideal truths or "universal possibilities" is presupposed? Is God certain, not insofar as another kind of thing is certain or a specific contingent sequence of events is actual, but only insofar as the true and the false are differentiated, insofar as there are rules of any kind under which a correspondence between specific concepts holds apodictically while it is denied between others, equally evidently and necessarily?

Kant now believes that he can in fact give an affirmative answer to this latter question. For, he infers, if there were no absolute existent whatsoever, there could be no ideal relations, no agreement or contradiction between pure concepts. It was generally considered that such relations are in no way adequately grounded and certified by the purely formal unity expressed in the logical principle of identity and contradiction, but that they necessarily presuppose certain material conditions of thought. A rectangle is not a circle: of that I am certain by virtue of the principle of contradiction; but that there exist in general figures such as rectangle and circle, and that some kind of qualitative differentiation between substances can be made, I am taught not by the wholly general and formal logical principle, but by that specific lawful order which I designate by the name "space." If there were no such determinate things like space and the shapes in it, number and its differences, motion and its diversities of magnitude and direction—in other words, if these could not be distinguished from one another and contrasted simply as conceptual substances— then matter would also dwindle to "potentiality," and then it would be impossible not only to assert any empirical entity but even to assert any true proposition. Thought would thus be annihilated, not because its foundations are formally contradictory but because no data would be given to it any longer and hence in general nothing more would be posited to which it could be opposed. For possibility as such drops away, "not only if an internal contradiction is encountered as the logic of impossibility, but also if there is nothing material, no datum to think. For then nothing thinkable is given; but everything possible is something which can be thought and to which logical relationship according to the principle of contradiction applies"—not immediately but by way of this principle. And herein lies the nerve of

Kant's proof: it must be shown that in fact by the cancellation not only this or that existent, but all existents whatsoever and all "matter" of thought, in the sense just specified, would be destroyed. "If all existence is canceled out, nothing whatsoever is posited, in general nothing at all is given, no material for anything whatsoever thinkable, and all possibility falls away completely. There is, to be sure, no inner contradiction in the denial of all existence. For that would require something to be simultaneously asserted and canceled; here, however, nothing at all is posited, so one can not in fact say that this cancellation contains a self-contradiction. But that some sort of possibility exists and yet nothing actual, is self-contradictory, because if nothing exists, nothing is given which would be thinkable, and we are at odds with ourselves if we still want something to be possible."[40]

Yet it seems that in fact that Kantian proof does not end with this, for even if the foregoing argument is regarded as conclusive, it has in any event only shown that "something," some sort of substance in general, must exist absolutely and necessarily, but not that this substance is "God." But this portion of Kant's conclusion is given relatively briefly. If we are sure of an absolutely necessary existence in general, it can be demonstrated that this existence must be unique and simple, unchanging and eternal, that it comprises all reality in itself and that it must be of a purely spiritual nature—in short, that we must attribute to it all those characteristics which we normally combine in the name and concept of God.[41] Accordingly, the movement here does not *proceed from* the concept of God in order to exhibit in it the predicate of existence together with other predicates, since "existence" does not designate a conceptual predicate that might also belong to another thing, but comprises the simple and not further analyzable "absolute positing" of a thing.[42] The direction of proof is rather the reverse: when absolute being is attained and guaranteed, the effort is then to derive its determinations, its essential "what," more closely, and it is thus discovered and demonstrated that its nature exhibits all the characteristics that comprise the distinctive

40. Ibid., sect. I, Second Reflection, §2 (II, 82 f.) (*Ak.* II, 78).
41. Ibid., sect. I, Third Reflection (II, 86–95) (*Ak.* II, 81–87).
42. Ibid., sect. I, First Reflection (II, 76 ff.) (*Ak.* II, 72 ff.).

content of our concept of God. Thus the ontological argument is firmly adhered to, and the cosmological and the physico-theological proofs are referred back to it.

However, an alteration in ontological thinking occurs, which promises its complete supersession in the future. While the ontological proof, in the form given it by Anselm of Canterbury and revived by Descartes, starts with the concept of the most complete being so as to deduce its existence, while it infers "existence" synthetically from "essence," Kant begins instead with pure ideal possibilities, with the system of eternal truths as such, in order then to show by progressive analysis that an absolute being must be required as the condition of the possibility of this system. We have before us essentially a prelude to the transcendental method to come, since the ultimate justification for positing existence in an absolute sense resides in the fact that without this assertion the possibility of knowledge is inconceivable. Of course, however, judged from the standpoint of the later critical system all "positings" achieved by this route are relative, not absolute; they are restricted, both as to their validity and as to their application, to experience, which they make possible.

We can, though, for the time being abandon the more exact and detailed evaluation of the fundamental problem of "The Only Possible Basis of Proof," especially since Kant's own ongoing development of this problem will of itself bring ever greater clarity and definiteness. If we pause at the point to which this development has led us, the difference between "The Only Possible Basis of Proof" and all of Kant's previous writings is revealed primarily in the fact that it belongs to a higher stage of reflection and critical self-consciousness. It now no longer suffices for Kant to produce observations and proofs for the specific object he is considering, but at the same time he questions their logical origin and the specific sort of truth that belongs to them. Kant was girded and armed like no other thinker of this era to answer the question set by the Berlin Academy the year before. In fact, he did not seem directly stimulated to undertake the task by the announcement of the prize competition itself, but rather to have felt moved to it only after completing "The Only Possible Basis of Proof," by reason of the essential link he discovered between the problem of this essay and the Academy's question.[43] "It is desired to know," the question went, "whether metaphysical truths in general and the first principles of *Theologia naturalis* and of morals in particular are susceptible of clear and evident proofs like those of geometrical truths, and if they are not susceptible of the aforesaid proofs, what the peculiar nature of their certainty is, to what degree their stated certainty can be brought, and whether this degree is sufficient for complete conviction."

The decision on the essays submitted was reached in the session of the Academy in May, 1763. First prize was awarded to the treatise by Moses Mendelssohn, but it was expressly declared that Kant's essay "had come as closely as possible" to being the prize work "and merited the highest praise." Both papers, Kant's and Mendelssohn's, appeared together in the proceedings of the Academy.[44] A special historical irony was that Formey, as the permanent secretary of the Berlin Academy, was the first to congratulate Kant on his success in a letter dated July, 1763. This scientific eclectic owed his philosophical prestige to popularization of the Wolffian system, which he had attempted in a multivolumed, monotonous, and verbose work.[45] Had he been capable of appreciating the substance of Kant's treatise, he would necessarily have had a premonition that the paper which he printed on behalf of the Academy contained the seed of a revolution in philosophy, by which the "inflated pretentiousness of whole volumes of insights" of dogmatic metaphysics[46] would one day be destroyed.

So far as Kant was concerned, he was conscious from the start of

43. The Academy's announcement was published in June, 1761, while Kant set to work on the topic only at the end of 1762, shortly before the deadline for submission. He himself calls his treatise a "hastily composed work" (II, 322 [*Ak.* II, 308]; cf. also II, 202 [*Ak.* II, 301]). "The Only Possible Basis of Proof" appeared at the end of December, 1762; it was in Hamann's hands on December 21, as can be inferred from the latter's letter to Nicolai dated that day. The manuscript of the essay was therefore probably completed in the autumn of 1762 at the latest. Cf. the comments of Kurd Lasswitz and Paul Menzer in the Academy edition of Kant's works, vol. 2, pp. 470, 492 ff. See also Adolf von Harnack, *Geschichte der Königlich preussischen Akademie der Wissenschaften* (Berlin, 1901), p. 315.

44. Cf. II, 475 (*Ak.* II, 494).

45. Jean Henri Samuel Formey, *La belle Wolffienne*, (The Hague, 1741–53).

46. Cf. Kant's letter to Mendelssohn, April 8, 1766 (IX, 55) (*Ak.* X, 66).

what was at stake: "The question posed," his exposition begins, "is of the sort that, if it is properly solved higher philosophy will have to take on a definite form. If the method by which the utmost certainty in this sort of knowledge can be attained is established, and the nature of this conviction is well understood, an unchangeable methodological rule will necessarily unite thoughtful minds in similar endeavors, in the place of the everlasting instability of opinions and schools; just as Newton's method in the natural sciences transformed the confusion of physical hypotheses into a sure procedure guided by experience and geometry."

But what was the crucial idea by which Newton effected this revolution? What differentiates the physical hypotheses current before him from the rules and laws which he established? If we ask this question, we see that the manner in which the universal is related to the particular and united with it in modern mathematical physics has turned into something quite other than what it was in the speculative physics of Aristotle and the Middle Ages. Galileo and Newton do not begin with the general "concept" of gravity so as to "explain" the phenomena of weight; they do not infer from the essence and the nature of matter and motion what must occur in freely falling bodies; they occupy themselves first of all in ascertaining the data of the problem, as presented by experience. Fall toward the earth's center, projectile motion, the motion of the moon around the earth, ultimately the revolving of the planets around the sun in elliptical orbits: all these are phenomena which are examined at the outset and defined purely quantitatively. Only then do they ask the question whether this whole complex of facts which has been ascertained cannot be brought under a common concept, that is, whether there is not a mathematical relation, an analytic function, which contains and expresses all those particular relations. In other words, here one does not proceed from a "force" which is conceived or imagined, deducing specific motions from it (as, for instance, in the Aristotelian system the physics of falling bodies is "explained" by a natural striving that draws each part of matter to its "natural place"), but what we call "weight" is here but another way of expressing and unifying known and measurable relations of magnitude.

If we now apply what this relation tells us to metaphysics, we see that metaphysics is concerned with a different realm of facts from that of mathematical physics. For its object is not outer but inner experience; not bodies and their motions, but knowledge, acts of will, feelings, and inclinations make up its basic theme. The type of knowledge, however, is neither determined nor altered by this difference in the object. Here too it is solely a matter of analyzing given complexes of experience into simple basic relations, stopping with these as the ultimate data that cannot be traced further back. Here it is equally true that determinations enter into these data which, because they are unanalyzable into simpler parts, are not further susceptible of any scholastic definition (by *genus proximum* and *differentia specifica*). For there is a kind of determinacy and evidence—and here it occurs in basic concepts and relations—which cannot be increased by a logical definition in this sense, but only muddled. "Augustine says: I know very well what time is, but when someone asks me, I do not know." And thus in philosophy one can often recognize an object clearly and certainly, and derive sure conclusions from it, before possessing its definition, indeed even if one makes no effort to provide it. "I can be immediately certain of various predicates of any thing, even though I do not know it well enough to give the explicitly determined concept of it, that is to say, the definition. If I never explained what a desire is, I would still be able to say with surety that every desire presupposes a representation of what is desired, that this representation is an anticipation of the future, that the feeling of pleasure is connected with it, etc. Everyone is always aware of all this in the immediate consciousness of desire. From similar comparative observations one could probably at last arrive at the definition of desire. But as long as what is sought can be inferred from some immediately certain qualities of the thing itself, without a definition, it is needless to attempt so delicate an undertaking."[47] Thus in the natural sciences we no longer begin with the explication of the essence of force, but what we call "force" is at most the final analytic expression for known, measured relations of motions; so also the logical essence about which metaphysics inquires can only constitute the terminus of the inquiry, not its start.

47. "Inquiry into the Distinctness of the Principles of Natural Theology and Morals," Second Observation (II, 184) (*Ak.* II, 284).

Yet every compendium of metaphysics whatsoever reveals how strongly the conventional course of inquiry, hallowed by usage and tradition, contradicts this prescription. The account of what is most universal—here Alexander Baumgarten's *Metaphysica*, on which Kant customarily based his lectures, is particularly typical—the definitions of being, of essence, of substance, of cause or effect, and of appetite are here placed at the head, and the attempt is made to derive the particular by combining these definitions. But if one looks more closely at this supposed deduction, one recognizes that in truth it tacitly presupposes the knowledge of the particular which it claims to deduce, and makes use of it, so that the ostensible philosophical grounding is merely circular. If we want to achieve actual clarity as to what metaphysics is or is not suited to, only a return to the humbler but more honest experience of physics can be of assistance. It follows from this that in both instances we do not try to expand the content of our knowledge at any price, but that we strictly observe the boundaries of what is known and unknown, what is given and what is sought, and that neither we nor others transgress them. We arrive at "being" in both cases alike only through the painstaking and continuous analysis of appearances; in this we have to resign ourselves to the fact that, since—at least in the present state of metaphysics—we can never claim with certainty the completion of this analysis, all our determination of being in this realm is nothing absolute, but rather is relative and preliminary. "The genuine method of metaphysics," as the Prize Essay concisely and expressively summarizes these observations, "is fundamentally the same as that which Newton introduced into natural science and which had so many fruitful consequences. There it is said that one should seek out the rules by which certain phenomena proceed in Nature, by means of indubitable experiences, and if need be, with the aid of geometry. Even if one has no insight into their ultimate foundation in bodies, it is still certain that they act according to this law, and the complex data of Nature are explained when it is clearly shown how they are contained under these well-demonstrated rules. Similarly in metaphysics: seek out by secure inner experience, that is, immediately evident consciousness, those properties which unquestionably lie in the concept of any sort of universal state, and if you do not know the whole essence of the matter, you can still securely make use of them to infer much about the thing."[48]

There is one preeminent respect in which Kant now parts company both with conventional metaphysics and with the procedure he himself had initially employed. Metaphysics can discover nothing; it can only make plain the pure fundamental interconnections in experience. It brings clarity and intelligibility into what is given to us as an obscure and complex totality, makes its structure transparent to us. But of its own authority it adds no substantive factor whatsoever. Kant's thought in that previous period, wherein the *Universal Natural History and Theory of the Heavens* found expression, held that metaphysics stood squarely on the soil of experience, but where experiential data were insufficient, it did not hesitate to round out and go beyond what is empirically given by the synthetic power of imagination and inference. It started with the world, with the cosmos of the natural scientist, but it was led on, in a continuous and imperceptible line, to hypotheses about the First Being, the teleology of the world, and the survival and immortality of the human soul.

Now Kant becomes aware of just how problematical this whole mode of thinking is. Can metaphysics, he asks, proceed synthetically and constructively? And the instant the question is posed this clearly, it is equally clearly answered in the negative. For synthesis has a place where the concerns are the self-created products of the understanding, which therefore are subordinate to the law of the understanding purely and exclusively.

In this sense mathematics can and must above all pursue pure geometry synthetically, for the figures it treats only arise in and together with the act of construction. They are not abstractions from something given physically and they would retain their significance and truth even if nothing physical, nothing actually material, existed. What a circle or a triangle "is" exists only through the power of the intellectual and intuitive act in which we bring them into being by a composition of separate spatial elements, and there is not a single attribute of these forms, no determinant added from elsewhere, that

<hr/>

48. Ibid. (II, 186) (*Ak.* II, 286). For more on the historical connection of these statements with the methodology of Newton and his school, see *Das Erkenntnisproblem,* II, 402 ff., 590.

is not contained in this basic act and completely deducible from it. "A cone may elsewhere mean what it will; in mathematics it arises from the arbitrary representation of a right triangle which is rotated on one of its sides. The definition obviously originates, both here and in all other cases, through the synthesis."

It is clearly different with the concepts and definitions of philosophy. In mathematics, as has been shown, the specific object that is to be defined, such as an ellipse or a parabola, does not precede the genetic construction of the figure but instead arises from it; in contrast, metaphysics is confined from the outset to a definite, fixed material that is given to it. For it is not purely ideal determinations that it proposes to unfold to our minds, but the properties and relations of the "real." It has to create its object, therefore, no more than physics does, but it grasps only the actual nature of the object. It does not describe its object in the sense in which the geometer describes a certain figure, that is, by showing its construction, but rather it can only circumscribe it in the sense that it selects from it some distinctive characteristic and comprehends that in abstraction. A metaphysical concept obtains its relative validity only by the completeness of this relation to the "given" of inner and outer experience. Metaphysical thinking is not in the least entitled to be an invention; it is not prospective as is geometry, in which new conclusions are successively formed from an original definition, but rather retrospective, so that given a state of affairs it seeks out the conditions from which that state results; for a total phenomenon it seeks the possible "grounds of explanation."[49] These explanatory grounds are only hypothetical, but they become certain in proportion to the possibility of their embracing the totality of known appearances, and through them exhibiting it as a unity that is lawful and determinate. There is no doubt in Kant's mind that in the conception and execution of metaphysics to date this task has not been performed at all: "Metaphysics is without doubt the most difficult of all human inquiries; but no metaphysics has even been written."[50]

And in fact none could be written, so long as the tool available to

49. For the whole discussion, see the Prize Essay, First Observation, §§1 and 3 (II, 176 ff.) (*Ak.* II, 276 ff.).

50. Ibid., First Observation, §4 (II, 183) (*Ak.* II, 283).

thought for the task was the conventional method of logical deduction customary in school philosophy. For the means this procedure
essentially, and moreover exclusively, uses is the syllogism: the world
as known and conceived is validated when it is analyzed into a chain
of rational conclusions. In this sense Wolff, in textbooks regarded as
classics in his day, had developed his "rational thoughts" concerning
God, the world and the soul, justice and the state and society, the
activities of nature and the coherence of the life of the mind, in short,
"concerning everything universally." Kant appreciated the methodical strictness and sobriety that imbues these works, and even at the
height of his critical system he defended them against the objections
by eclectic popular and fashionable philosophy. In the preface to the
second edition of the *Critique of Pure Reason* Wolff is extolled as the
"awakener of the spirit of thoroughness which is not extinct in Germany," because by the orderly establishment of principles, the clear
definition of concepts, and by his avoidance of daring leaps in his
inferences, he first tried to lead metaphysics into the sure path of a
science.[51] Nonetheless, in Kant's whole philosophical development
no indication can be found that he ever was intellectually dependent
on the Wolffian system, as was the case with Mendelssohn and
Sulzer. The artful technique of syllogistic proof never dazzled him,
and in one of his own writings he attempted in 1762 to expose the
"false sophistry" latent in it.[52]

More profound than this formal discussion is the charge that Kant
now draws from his new conception of the tasks of metaphysics. The
syllogistic procedure is "synthetic" in the specific sense given this
term in the Prize Essay "Inquiry into the Distinctness of the Principles
of Natural Theology and Morals" ("Über die Deutlichkeit der
Grundsätze der naturlichen Theologie und Moral"). It moves from
premises to conclusions, from general concepts and definitions laid
down at the beginning to particular determinations. Yet, does such a
cognitive procedure correspond to the one which, as we have seen, is

51. B xxxvi (III, 28 f.).
52. "Die falsche Spitzfindigkeit der vier syllogistischen Figuren erwiesen" ["The
False Subtlety of the Four Syllogistic Figures Demonstrated"]. (See II, 49 ff.) (*Ak.* II, 45
ff.).

prescribed for us in every inquiry into what is real? Further, the principles on which all logical deductive procedure rests are the principle of identity and that of contradiction—the former, as Kant shows in a logical paper of 1755, the *Nova dilucidatio*, is the highest principle for all affirmative judgments, the latter for all negative ones.[53] Every inference aims at nothing but the identity of two terms, *A* and *B*; where this is not immediately obvious, it is shown mediately by inserting a series of concepts. The system of things and events, according to the basic principles of rationalism, is to be presented thus as an ever more exact and precise system of premises and conclusions. In this view of the task of philosophy, Wolff unmistakably goes back to Leibniz, but in the further elaboration of his system he erased the delicate methodological boundary that existed for the latter between the principle of contradiction and the principle of sufficient reason. According to Leibniz, the first of these is the principle of the necessary, the second that of contingent truths; the former gives rise to the propositions of logic and mathematics, while the latter is responsible specifically for the propositions of physics. Within the Wolffian scholastic system, however, the uniformity of the schema of proof constantly pushes toward uniformity among the principles themselves. Thus, the effort to overcome the separation between the material content of knowledge and the principles of knowledge dominates throughout, so that the attempt is to reduce them to the logical principle of identity and prove them from it. In this sense Wolff essayed a proof of the "principle of a ground," which was in fact circular: if there were something without a ground, he reasoned, then nothing must be the ground of something, which is self-contradictory. He even tried to deduce the necessity of the spatial order of appearances in this way, purely from the validity of the supreme logical principle: what we think as different from us, the inference ran, we must think as existing outside us, thus as spatially separated from us. The "other than us," *praeter nos*, was here directly translated into an "outside us," *extra nos*, the abstract concept of diversity into the concrete, intuitive externality of space.

53. *Principiorum primorum cognitionis metaphysicae nova dilucidatio* [*A New Explanation of the First Principles of Metaphysical Knowledge*], sect. I, Proposition II (I, 393) (*Ak.* I, 389).

The flaw in this manner of thinking had not gone unnoticed in German academic philosophy. In his criticism Crusius, the most important of Wolff's opponents, lays the utmost emphasis on the fact that the principle of contradiction, as a purely formal principle, can of itself alone yield no specific and concrete knowledge, but that a set of original and underivable, but nevertheless certain, "material principles" is unconditionally necessary for that.[54] Kant took his final, decisive step in this direction in the treatise that was probably finished immediately after the composition of the Prize Essay,[55] the "Attempt to Introduce the Concept of Negative Magnitudes into Philosophy." Here the source of the sharp distinction between logical and real opposition is directly expressed. The former occurs where two predicates are related to each other as A and non-A, thus where the logical affirmation of the one implies the logical denial of the other. The result of this opposition is hence pure nothingness; if I try to think a man as learned and as unlearned at the same time and in the same respect, or a body as simultaneously in motion and at rest, this thought is shown to be completely empty and impossible.

Matters stand otherwise in all cases of real opposition, in those cases where, popularly speaking, it is a matter not of an opposition of conceptual characteristics but of an opposition of forces. The velocity that a freely falling body, unhindered by any external factors, possesses can be canceled by another one that is equal but opposite; the result is not, as in the first case, a logical contradiction, but that quite definite and characteristic physical state which we designate by the expression "rest" or "equilibrium." If in the first case, the attempt to unite A and non-A conceptually, the outcome was an absurdity, here it is a determinate and completely unambiguous magnitude, since the magnitude "zero" is no less definite than any other quantity signified by a positive or negative number. Thus the way in which diverse real

54. See *Das Erkenntnisproblem in der Philosophie und Wissenschaft der neueren Zeit*, 2d ed. (Berlin, 1911), vol. 2, pp. 527–34, 587 ff.; cf. what Kant says about Crusius in the Prize Essay, Third Observation, §3 (II, 194 ff.) (*Ak.* II, 293 ff.).

55. The presentation of the treatise is recorded in the proceedings of the Königsberg faculty of philosophy on June 3, 1763, while the Prize Essay was finished at the end of 1762.

causes determine each other and combine into a unified fact, a relation best seen in the parallelogram of motions or forces, is not at all equivalent to the relation holding between merely logical predicates and judgments. The "real ground" is an independent, qualitatively distinct relation, which is not only inexhaustible by the logical relation of ground to consequent, of *antecedens* to *consequens*, but is never expressible through it. Hence metaphysical method is in the last analysis different from syllogistic, for metaphysics, in the sense Kant has given it, is the doctrine of real grounds. In metaphysics the analysis of complex events leads, as in natural science, to definite, ultimately simple basic relations, which can be grasped only in their pure factuality, but which, however, cannot be made comprehensible from concepts alone.

This is above all true of the causal relation, which we cannot doubt but which is nonetheless logically indemonstrable; indeed, the formal conceptual system of logic affords no means whatsoever to grasp and to think it determinately. It is easy to see how an inference is established through its conceptual ground, or a conclusion by the rule of identity, for in such cases we need only analyze the two concepts that are here related to each other to discover in them the selfsame property. But how something arises from something different, not according to the rule of identity, is a completely different question, about which Kant avows that no "real philosopher" has so far been able to make plain. The words "cause" and "effect," "force" and "action" are no solution, but merely restate the problem. All of them assert that because something is, another, different thing must exist, not that, in conformity with purely logical proof, because something is thought, something else must be thought as fundamentally identical with it.[56]

Here the first sharp dualism in the Kantian system emerges. The view that logic in its traditional form, as syllogistic, could suffice to "construct" the system of actuality crumbles once and for all, since it and its supreme principle, the principle of contradiction, are in-

---

56. See "Attempt to Introduce the Concept of Negative Magnitudes into Philosophy," third sect., General Remark (II, 240 ff.) (*Ak.* II, 201 ff.).

adequate to express the peculiarity of even the simplest real relation, that of cause and effect. But is thought to renounce the understanding of the composition and structure of being? Are we to abandon ourselves to an empiricism that is content to array impression with impression, fact with fact? Surely this cannot be Kant's intent, nor was it in any period of the evolution of his thought. The renunciation of syllogistic and its method, which imitates the synthetic proofs of geometry, in no way implies for him the renunciation of a rational foundation of philosophy as such, for the analysis of experience itself, which he now sees as the essential task of all metaphysics, is still for him through and through the work of reason.

If we survey Kant's view at this period of the capabilities of reason with respect to reality, a double relation is revealed. On the one hand, reason has to analyze the data of experience, until it has uncovered the ultimate simple fundamental relations of which experience is composed, relations which can then be shown purely as they exist but cannot be further deduced. But on the other hand, reason can ground and give evidence for the necessity of an absolute being, which is its characteristic task and prerogative, for from the pure, ideal possibilities that comprise its particular realm there follow, as "The Only Possible Basis of Proof" demonstrated, the existence and specification of the highest, most encompassing reality, which we designate by the concept of God. If these two functions are compared, we discover that they belong to two quite different orientations of thought. It is especially discordant for Kant on the one hand to consign reason in its determination of actuality completely to the data of experience, and on the other to entrust to it the power of bringing us to unconditional certainty regarding an infinite being lying beyond all possibility of experience. The analyst of inner experience, who tries to mold himself on the model of the Newtonian method, and the speculative philosopher, who clings to the central element of rational metaphysics, the ontological proof of God's existence, though in an altered form, here have not yet been clearly and sharply separated. In this opposition lay the seed and the conditioning factor of Kant's future philosophical development; once it was clearly grasped, it demanded a definite decision, which forced Kant further and further from the system of academic philosophy.

3. The Critique of Dogmatic Metaphysics:
Dreams of a Spirit-Seer

Kant established his reputation in the German literary and philosophical world by his writings of 1763. "The Only Possible Basis of Proof for a Demonstration of God's Existence" was reviewed in the *Literaturbriefe* by Mendelssohn, who was not wholly just to the essay's idiosyncratic ideas and method of proof, but who ungrudgingly and unstintingly acknowledged Kant as an "independent thinker," even where he could not follow him. Kant later said that this review first introduced him to the public. Further, the judgment made by the Berlin Academy on the "Inquiry into the Distinctness of the Principles of Natural Theology and Morals," and the fact that this essay appeared alongside Mendelssohn's prizewinning one in the proceedings of the Academy, made Kant's name known even beyond the borders of Germany. From now on literary acclaim numbers him among the leading minds of Germany, although his place in contemporary philosophy is not by any means clearly determined and staked out in the common judgment. Men like Lambert—who undoubtedly belongs among the most original minds of that epoch and whom Kant himself esteemed as the "foremost genius in Germany" in the field of metaphysics—enter into scientific correspondence with him and submit rough drafts of their philosophy to his judgment. Kant is now commonly seen as the future creator of a new system, which Mendelssohn urged him to work out in 1763 in the above-mentioned review in the *Literaturbriefe*, eighteen years before the appearance of the *Critique of Pure Reason*.

The phase that Kant's evolution as thinker and author entered after the writings of 1763 dashed the hopes of the world and of his friends most strikingly. What was expected and hoped for from him was the project of a new, deeper, and more tenable metaphysics—an abstract, analytic dissection of its presuppositions and a careful theoretical examination of its most general conclusions; what was received is a work which in its literary form and in its stylistic dress alike upset all the traditions of the literature of scientific philosophy. *Dreams of a Spirit-Seer, Illustrated by Dreams of Metaphysics (Träume eines Geistersehers erläutert durch Träume der Metaphysik)* is the title of this work, which appeared anonymously in Königsberg in 1766. Was the learned Magister Kant, was the author of the Academy's Prize Essay the author of this work? Inevitably there was doubt on that score, so jarring must have been the strange and unfamiliar tone in which it was written. For here it is no longer a matter of the theoretical scrutiny of metaphysics and its main propositions; rather, a reflective humor sports playfully with all its concepts and divisions, with its definitions and distinctions, with its categories and its logical chains of conclusions.

Yet, for all the exuberance of the satire, there conversely runs through the book a serious vein, which can be perceived clearly through all its mockery and self-irony. It is concerned with the doubts and reflections connected with the highest spiritual and religious problems of mankind, questions such as immortality and the endurance of the self, in which Kant had a crucial interest at every period of his thinking, whatever form his theoretical answers to them might take. "It will be said," we read at one place in the book, "that this is a very serious subject for so noncommittal an exercise as our discussion is, which deserves to be called a trifle rather than an earnest undertaking, and such a judgment would not be wrong. But although one need not make a great to-do over a trifle, one can do so given the opportunity. . . . I do not find that I have any sort of partisanship or that any unexamined bias has crept in to deprive my mind of its flexibility on all grounds pro or con, with one sole exception. The scales of the understanding are not quite impartial, and one arm of them, which bears the inscription: Hope of the future, has a mechanical advantage. . . . This is the sole error which I cannot set aside, and which in fact I never want to."[57]

But in this paradoxical mixture of jest and earnestness, which was

57. *Dreams of a Spirit-Seer*, pt. 1, chap. 4 (II, 365) (*Ak*. II, 349–50).

the decisive factor? Which was the author's true face and which the mask he had assumed? Was the book just a passing by-blow of free humor, or was there concealed behind this satyr play of the mind something resembling a tragedy of metaphysics? None of Kant's friends and critics was ever able to answer this question with certainty. The most sympathetic critics, such as Mendelssohn, were unrestrainedly amazed at this ambiguity. But Kant's reply to them was very like a riddle. "The unfavorable impression you express concerning the tone of my little book," he writes to Mendelssohn, "proves to me that you have formed a good opinion of the sincerity of my character, and your very reluctance to see that character ambiguously expressed is both valuable and pleasing to me. In fact, you shall never have cause to change this opinion. For though there may be flaws that even the most steadfast determination cannot eradicate completely, I shall certainly never become a fickle or fraudulent person, after having devoted the largest part of my life to studying how to despise those things that tend to corrupt one's honesty. Losing the self-respect that stems from a sense of honesty would therefore be the greatest evil that could, but most certainly shall not, befall me. Although I am absolutely convinced of many things that I shall never have the courage to say, I shall never say anything I do not believe."[58]

If one tries to approach the problem of intellect and life lurking mysteriously behind this work of Kant's, the outward story of the origin of the *Dreams of a Spirit-Seer* affords little help. Kant himself has brilliantly depicted, in a famous letter to Charlotte von Knobloch, how he first became aware of the marvelous tales surrounding the "visionary" Swedenborg, which led him to immerse himself deeper into Swedenborg's chief work, the *Arcana coelestia*. We use this account here not to repeat it, but are content to make reference to it.[59] Who will seriously believe that because he had bought the eight quarto volumes of Swedenborg's works, at a considerable outlay of trouble and expense, Kant would have decided to perform a literary analysis on the book? Or ought we to take the humorous preface to

58. To Mendelssohn, April 8, 1766 (IX, 55) (*Ak.* X, 66).
59. To Charlotte von Knobloch (1763) (see IX, 34) (*Ak.* X, 40).

the *Dreams of a Spirit-Seer* at face value in this regard? "The author," it says, "confesses with a certain humility that he was so simpleminded as to track down the truth of some tales of the sort mentioned. He found—as usual, where one has nothing to look for—he found nothing. Now this is in itself reason enough to write a book, but in addition there was that which has more than once wrung books out of reticent authors: the impetuous perseverance of known and unknown friends."

All this would hardly have influenced Kant, who was not easily led astray by any "author's itch,"[60] to occupy himself so intensely with the "arch-phantasist" Swedenborg, the "worst visionary of them all," if it were not that what he discovered in Swedenborg had a queer, indirect link with the crucial questions that his own inner development had led him to. Swedenborg is for Kant the caricature of all supersensible metaphysics, but precisely because of this distortion and exaggeration of all its distinctive features, he set himself to hold up a mirror to this metaphysics. If it failed to recognize itself in the gentle and objective analysis of the Prize Essay, it should now see itself in this caricature of it. For what in fact does distinguish the fantastic eccentricities of the visionary from the "architects of sundry airy thought-worlds" who were wont to call their creations "systems of philosophy"? Where is the line between the visionary's imaginings and that ordering of things "hewn less out of the stuff of experience than out of fraudulent concepts" by Wolff, or produced "by Crusius by the magical powers of some oracular utterances on the thinkable and the unthinkable" as out of nothing?[61] If the philosopher proposes to conjure up "experiences," the enthusiast will not lack for all sorts of instances of positively certified supersensible data and "facts," aside from the fact that scrutiny of this claim will very often lead one to vital gaps in its justification.

Or should the form of the system, the "rational connection" of concepts and conclusions, be crucial here? But his thorough study of

60. Compare his letter to Marcus Herz of the year 1773 (IX, 114) (*Ak.* X, 136).

61. *Dreams of a Spirit-Seer*, pt. 1, chap. 3 (II, 357) (*Ak.* II, 342); a passage in the Prize Essay, Third Observation, §3 (II, 196 f.) (*Ak.* II, 293 f.), explains what is meant by Crusius's "oracular utterances on the thinkable and the unthinkable."

the *Arcana coelestia* taught Kant once again just how far this systematization can be pushed, even in patent absurdities. Just as the upshot of the writings of 1763 is that no syllogistic is sufficient to give us knowledge of a single efficient cause, conversely the absence of true realities is no obstacle to verbalizing what seems to be a valid and continuous schema of deductions. The "dreams of reason" are in this respect no better than the "dreams of feeling"; the most cunning architectonic in the structure cannot overcome the lack of building materials. Even for the systematic philosopher there is no other criterion for the reality of his conclusions than the most scrupulous and patient testing of the data at his disposal for every particular question. But what an aspect traditional metaphysics takes on if we apply this yardstick to it! Everywhere we encounter problems that are revealed not only as uncomprehended but on closer look as incomprehensible, because the form in which the problem is put is infected by an ambiguous concept or a surreptitious assumption.[62] One talks about the "presence" of the soul in the body, one studies how the "spiritual" can act on the "material" or this on that, but fails to notice that the whole idea of the spiritual is due to habit and prejudice, rather than to exact scientific analysis. This is gross self-deception, but on the other hand it is understandable enough: "For whatever one knows a lot about as a child, he is certain later, as an adult, to be ignorant of, and the profound man in the end becomes the sophist of his youthful extravagance."

Kant had, however, at the conclusion of his treatise on negative magnitudes ironically referred to the "weakness of his insight," whereby he "customarily conceived least well what everybody believed easy to understand." Thanks to this weakness, the deeper he

62. Cf. Kant's letter to Mendelssohn of April 8, 1766: "My analogy between a spiritual substance's actual moral influx and the force of universal gravitation is not intended seriously; but it is an example of how far one can go in philosophical fabrications, completely unhindered, when there are no *data*, and it illustrates how important it is, in such exercises, first to decide what is required for a solution of the problem and whether the necessary *data* for a solution may be lacking. . . . Here we must decide whether there really are not limitations established by the bounds of our reason, or rather, the bounds of the experience that contains the *data* for our reason" (IX, 55) (*Ak.* X, 66).

goes into contemporary metaphysics "with its cursed fertility," the more it seems a book with seven seals. It envelops him in a web of opinions which, like Swedenborg's accounts of the spirit world, may be mastered historically, but which cannot be understood from first principles and brought to genuine conviction. There is but one firm standpoint left for him here: honest and open confession of ignorance. The whole problem of the spiritual realm, along with all other business relating to objects beyond all experience, is no longer a matter of theoretical speculation for him. What philosophy can accomplish at this point seems, impartially considered, trivial, but it is methodologically decisive for the whole conduct of cognition and life. It transforms grudging scepticism into something free and voluntary. "When science has made the circuit of its domain, it arrives in a natural way at the point of a modest distrust, and says to itself with irritation: How many things there are which I do not comprehend! But reason which has ripened through experience, and become wise, speaks joyfully through the mouth of Socrates, in the midst of the wares of the marketplace: How many things there are which I do not need! In this wise two highly diverse currents flow together, though they originated in quite different quarters, since the first is futile and malcontent but the second firm and temperate. For to choose rationally between them one must first know what is dispensable, indeed, impossible; but in the end science determines the limits set it by the nature of human reason. All baseless projects, however, not unworthy in themselves perhaps save that they lie outside the human sphere, empty into the limbo of futility. Then metaphysics becomes what at present it is rather far from being and what one ought least to presume of it: the handmaiden of wisdom."[63]

These sentences have a twofold interest for consideration of Kant's development as a whole. On the one hand, they show him to be still extremely closely connected with the substantive tendencies of the Enlightenment; on the other, however, they indicate that the spirit of this substance has been given a new form through a novel foundation. If the philosophy of the Enlightenment was naive in the

63. *Dreams of a Spirit-Seer*, pt. 2, chap. 3 (II, 385 f.) (*Ak.* II, 369).

way it rejected the supersensible and limited reason to the "here-and-now" and to what can be apprehended empirically, the same result appears in Kant as the product of a thought process that has gone through all the stages of critical reflection. He no longer stands on the soil of "experience" because he is wary or lazy, but he has consciously stationed himself there. Thus metaphysics is still science for him; however, it is no longer a science of things in a supersensible world, but of the limits of human reason. [64]

It directs man back to his proper and allotted sphere, because that is all that is necessary to man for his ethical vocation, for regulation of his action.

The whole moral voice of the Enlightenment, as it lived in the purest and greatest spirits, has here received its theoretical justification by Kant. "No; the time of fulfillment will come, will surely come," Lessing cries out at the end of the *Education of Mankind*, "since Man, the more convinced his understanding feels itself with respect to an ever better future, will have less need to borrow motivations for his actions from this future, for he will do the Good because it is good, not because arbitrary rewards are set up to bind and to strengthen his fickle gaze so as to recognize the better, inner rewards." Out of the same ethical insight and the same intellectual pathos Kant, a decade and a half prior to the *Education of Mankind*, had rendered his decision for and against metaphysics. "What? Is it then good to be virtuous only because there is a world beyond, or are actions no longer to be praised because they are in themselves good and virtuous?" [65] Whoever still needs the perspectives of metaphysics as a basis for ethics knows he is not yet in that pure autarchy and self-sufficiency which constitutes his genuine state. In this sense of ethical immanence the *Dreams of a Spirit-Seer* concludes with a reference to the words of the "honest Candide": "Let us take heed for our happiness, and go into the garden and work."

The new theoretical ideal at this point is directly transformed into a new ideal of life. We have in a well-known description by Herder, which cannot be omitted from any description of Kant's life, classic,

64. Ibid., pt. 2, chap. 2 (II, 384) (*Ak.* II, 368).
65. Ibid., pt. 2, chap. 3 (II, 389) (*Ak.* II, 372).

definitive witness as to how both ideals were stamped on Kant's entire mental stance and on the effect he had on others. "I have enjoyed the good fortune to know a philosopher, who was my teacher. In the prime of life he had the happy cheerfulness of a youth, which, so I believe, accompanied him even in grey old age. His forehead, formed for thinking, was the seat of indestructible serenity and peace, the most thought-filled speech flowed from his lips, merriment and wit and humor were at his command, and his lecturing was discourse at its most entertaining. In precisely the spirit with which he examined Leibniz, Wolff, Baumgarten, and Hume and pursued the natural laws of the physicists Kepler and Newton, he took up those works of Rousseau which were then appearing, *Émile* and *Héloïse*, just as he did every natural discovery known to him, evaluated them and always came back to unprejudiced knowledge of Nature and the moral worth of mankind. The history of nations and peoples, natural science, mathematics, and experience were the sources from which he enlivened his lecture and converse; nothing worth knowing was indifferent to him; no cabal, no sect, no prejudice, no ambition for fame had the least seductiveness for him in comparison with furthering and elucidating truth. He encouraged and engagingly fostered thinking for oneself; despotism was foreign to his mind. This man, whom I name with the utmost thankfulness and respect, was Immanuel Kant; his image stands before me to my delight."[66]

In his "Travel Diary" as well, Herder contrasts Kant's "living instruction" and pure "humane philosophy" with the dry, abstract, fragmented style of teaching he looks back on in his years as a child and youth. When he again and again stresses freedom and joyousness of soul as the foundation of Kant's nature, he does not seem fully aware that this harmonious balance was not for Kant a direct gift of nature and fate, but that it was won instead by hard intellectual struggles. These battles appear to have come to an end with the period of the *Dreams of a Spirit-Seer*. Kant has oriented himself toward pure "this worldiness" in the theoretical and ethical sense, in

66. Herder, *Letters on the Advancement of Humanity*, letter 79.

understanding as well as action. Now he believes himself planted ever more surely and firmly in the "human condition," and to be securely opposed to every deceptive enticement which might dislodge him from that stance.[67]

This tendency emerges in him so decisively that it is immediately communicated to everyone who came into close contact with him at that time. "He has given light to many an eye, simplicity of thought and naturalness of life," it is said in a poem composed in 1770 by a young man, Jakob Michael Reinhold Lenz, "in the name of all those from Courland and Livonia studying at Königsberg, presented to Professor Kant on his entry into his new post."[68] At this period there was realized in Kant the ideal of a life equally contemplative and active, which confined itself to the most immediate round of daily obligation and was capable of the widest prospects, which pronounced on the most universal mental and spiritual relationships and yet at every moment was conscious of the limits of human knowledge. Kant himself portrayed a life of that sort in a letter sent to Herder in Riga, in 1768: "In the early unfolding of your talents I foresee with divers pleasures the time when your fruitful spirit, no longer so sorely driven by the warm impulse of youthful feeling, attains that serenity which is peaceful yet full of feeling and is the contemplative life of the philosopher, just the opposite of that dreamed of by the mystics. From what I know of you, I confidently look forward to this epoch of your genius, of all states of mind the most advantageous to its possessor and to the world, one in which Montaigne occupies the lowest place and Hume so far as I know the highest."[69]

Among all the mental and spiritual influences on Kant in this period, the contribution of this state of mind is crucial—or more accurately, he views philosophic literature from the perspective of this psychic state and takes his stance toward it on that basis. Between Kant and Montaigne, between the "critical thinkers" and the "scep-

67. See the "Fragmente aus Kants Nachlass," *Immanuel Kants sämmtliche Werke,* ed. G. Hartenstein (Leipzig, 1867–68), vol. 8, p. 625.

68. The poem is printed in the "Stürmer und Dränger" collection edited by A. Sauer, vol. 2, pp. 215 f. (Kürschners Deutsche Nat.-Lit., vol. 80).

69. IX, 60 f. (*Ak.* X, 70).

tics," between the most strictly systematic thinker and the most unsystematic thinker of all, there seems at first glance to be an unbridgeable chasm. And yet in this spiritual phase we are now considering, there is a link between them, rooted in their common position with respect to learnedness. Just as Montaigne warned over and over that we enfeeble our power of comprehension when we demand that it grasp too much, that we may become learned through others' knowledge but wise only through our own, similarly Kant's *Dreams of a Spirit-Seer* is shot through with the belief that true wisdom is the handmaiden of simplicity and that since in it the heart prescribes to the understanding, it normally renders dispensable the immense apparatus of erudition and the whole clamorous estate of learning.[70] Just as Montaigne elevates "Que sçais je?" to be the motto for his philosophy of life, Kant sees in the "methodological prattle of the toplofty schools often just an agreement to shirk a difficult question by means of shifting word patterns, because the easy, and as a rule reasonable: 'I do not know' is not readily heard in the academies."[71] Just as Montaigne, as one of the first modern thinkers, wanted to sever morality from all connections with religion, and demanded a morality uncompelled by legal or religious prescriptions, one which rather had grown "from its own root, from the seed of universal reason," Kant indignantly asks whether the human heart does not contain immediate ethical prescripts, and whether the machinery of another world is needed to arouse one's vocation in this world.[72]

When he adds that the true and essential human goals ought not to be thought dependent on such means, which forever lie beyond human power, we encounter a different realm of thought and life; we are transported into the mood of the confession of faith of the Savoyard Vicar. To express what Rousseau's work meant to him from the beginning, we do not need the famous anecdote which tells us how, on reading in 1762 the newly published *Émile*, for the first time Kant was unfaithful to his customary daily schedule, and to the amazement of his fellow citizens did not take his afternoon walk. The

70. *Dreams of a Spirit-Seer*, pt. 2, chap. 3 (II, 389 ff.) (*Ak.* II, 368 ff.).
71. Ibid., pt. 1, chap. 1, (II, 333) (*Ak.* II, 319).
72. Ibid., pt. 2, chap. 3 (II, 389) (*Ak.* II, 372); cf. Montaigne, *Essays*, vol. 3, p. 12.

historical novelty of the phenomenon of Rousseau perhaps is seen most clearly in the total inadequacy of all the established standards decreed by that age when they are applied to him. As is the case with minds which have the special stamp of individuality, he exerted a completely opposite effect on those standards. As far as the characteristic philosophy of the Enlightenment is concerned, Rousseau remains a fundamentally incommensurable magnitude, though joined with it by many threads. Although the German Enlightenment did not unhesitatingly adopt a Voltairean tone, although the thoughtful and sober Mendelssohn strove to reach an equable and just evaluation, all vision of Rousseau's true originality was denied him. Mendelssohn's review of the *Nouvelle Héloïse* in the *Literaturbriefe*, in which he ranked Rousseau far below Richardson as regards "knowledge of the human heart," is indicative of a cross-section of the literary taste of that day; Hamann alone, in his *Chimärische Einfälle*, opposed it with his whole temperament and all the force of his bilious humor, and ridiculed it most effectively. Only the next generation, that of the young "geniuses," understood the artist in Rousseau. It was willingly captivated by the force of Rousseau's feeling and language; it thought that it heard in every word the voice of life and of "nature." But in this cult of feeling ignited by Rousseau, all the sharp distinctions, all the conceptual and dialectical problems which are no less essential to the whole of his personality and his historical mission were submerged.

In contrast to these two typical conceptions and estimations of Rousseau, Kant clings to a completely independent standpoint in his own judgment. If the Enlightenment upholds against Rousseau the rights of an intellectual culture grown old and rigid, sees him with the eyes of old age, if the "geniuses" see him with the eyes of youth, Kant confronts him from the start with the open-mindedness and the ripe judgment of a mature man. Only Lessing resembles him in this respect.[73] At the peak of his powers as an author in the years when he became acquainted with Rousseau—the *Observations on the Feeling*

73. See Lessing's notice on Rousseau's Dijon prize essay ("Das Neueste aus dem Reiche des Witzes," April, 1751) *Werke*, ed. K. Lachmann and F. Muncker (Stuttgart, 1886–1924), vol. 2, p. 388.

of the Beautiful and the Sublime[74] which appeared in 1764, is that work which next to the *Dreams of a Spirit-Seer* most clearly shows what Kant was capable of as a stylist—he has the liveliest sympathy for and interest in the new personal style Rousseau introduced into philosophic literature. But he is no prisoner of this charm. "I must," he commands himself, "read Rousseau until the beauty of expression no longer moves me, and then I can look at him rationally." But the aesthetic charm of Rousseau's writing is not the only thing that hampers reflective and temperate scrutiny; beyond that lies the no less dangerous charm of his dialectic. "The first impression which the reader who is not merely reading idly and passing time gets from the writings of J. J. Rousseau is that he is in the presence of an uncommonly acute mind, a noble sweep of genius, and a soul filled with a degree of feeling so high as to have been possessed all together by no other author, of whatever age or nationality. The succeeding impression is astonishment at the peculiar and nonsensical notions, so opposed to what is generally current, so that it readily occurs to one that the author, by his extraordinary talents and the sorcery of his eloquence, wanted to display himself as an eccentric fellow, who surpasses all his intellectual competitors through bewitching and startling novelty."

Kant does not stop with these two impressions, but seeks out the philosopher Rousseau behind the "sorcerer" Rousseau. Paradox in the man's manner of expression and in his very being does not blind him and lead him astray; he is convinced that this strange phenomenon, subject to no convention or mold, nonetheless must have its own inner law, which he endeavors to uncover. And here he arrives at a quite new and special view of it. If there was anything that the judgment of his contemporaries agreed on, it was that they saw in Rousseau the champion against the tyranny of "rules." As such, he was on the one hand attacked from the standpoint of popular "reason" and bourgeois morality, on the other hailed enthusiastically as a liberator. The return to "nature" seemed to be the return to freedom of personal, inner life, to the unchaining of subjective feeling and emotion.

74. *Beobachtungen über das Gefühl des Schonen und Erhabenen* (see II, 243 ff.) (*Ak.* II, 205 ff.).

For Kant, though, coming as he did from the side of Newton, the concept of nature had quite a different ring. He sees in it the expression of the highest objectivity, the expression of order and lawfulness. It is in this sense that he interprets Rousseau's tendency of thought as well. Just as Newton had done with the objective rules of the paths of the heavenly bodies, Rousseau sought for and laid down the objective ethical norm of human inclinations and actions. "Newton first saw order and lawfulness going hand in hand with great simplicity, where prior to him disorder and its troublesome partner, multiplicity, were encountered, and ever since the comets run in geometrical paths; Rousseau first discovered amid the manifold human forms the deeply hidden nature of man, and the secret law by which Providence is justified through his observations." This sturdy "nature," firmly at one with itself, is independent both of subjective inclination and of changes in theoretical notions. It is the autonomous moral law in its pure, unchangeable validity and obligation. All differences must vanish in the face of the simplicity and the grand unity and uniformity of this law, differences by which the individual believes himself to be exceptional through preeminence of birth and estate or through gifts of mind and learnedness. Kant himself asserts that he is "by inclination an inquirer," and says he traces in himself the thirst for knowledge and eager impatience to increase it. But he refuses to seek the essential moral worth and the "honor of mankind" in man's purely intellectual capacities and intellectual progress: Rousseau has "set him straight." "This delusive superiority disappears; I learn to respect mankind, and would find myself much more dispensable than a common laborer, if I did not believe that this reflection could give to everyone else a worth which restores to mankind its rights."[75]

Now it is understandable how in the very same letter to Mendelssohn in which he says that he views with disgust the inflated arrogance of entire volumes of metaphysical ideas as they currently

75. What Kant says here about Rousseau can be found in the remarks (first published by Friedrich Wilhelm Schubert) which he inserted into the manuscript of *Observations on the Feeling of the Beautiful and the Sublime* (1764). In most editions of Kant these are included under the title "Fragmente aus Kants Nachlass." Cf. Hartenstein's edition, vol. 8, pp. 618, 624, 630.

stand, indeed even with hatred, Kant also declares he is so far from taking metaphysics itself, objectively considered, to be unimportant or dispensable that he is convinced that the genuine and lasting well-being of the human race depends on it.[76] For the orientation and goal of metaphysics have now totally changed. One categorical demand for a new foundation of ethics has replaced the diverse problems treated in the schools under the headings of ontology, rational psychology, and theology. It is here, not in scholastic, logical concepts, that the true key to the signficance of the spiritual world is to be sought.

Did Kant derive this idea from Rousseau, or did he read it into him? This is a pressing question, for in just such subtle intellectual and ideal relationships the proposition Kant expressed concerning a priori theoretical knowledge holds: that we only know in things what "we ourselves put into them." Even as Schiller later could fathom the fully developed fabric of Kantian philosophy on brief acquaintance because he conceived it from the standpoint of his own center, the idea of freedom, which was the fundamental idea of his own life, similarly Kant has here, under the guidance of this idea so essential to him, read and understood Rousseau accordingly. Newton helped him to clarify the phenomenon of the world; Rousseau shows him the way to the deeper meaning of the noumenon of freedom. But significantly, in this very distinction there lies the germ of a new fundamental problem. Now it is necessary to show how it is possible to hold to the standpoint of pure "immanence" and yet to preserve the unconditionality of ethical norms, how we can keep the "intelligible" in ethics and still, or even for that very reason, reject the supersensible world of mystical twaddle and speculative metaphysics.

Given this question, which henceforth becomes more and more central and which governs the whole intellectual advance from the *Dreams of a Spirit-Seer* to *On the Form and Principles of the Sensible and the Intelligible World*, the position Kant adopts, both positively and negatively, toward Hume's theory becomes clear.[77] He says in his

76. Letter to Mendelssohn, April 8, 1766 (IX, 55) (*Ak.* X, 66).

77. I will not go into further detail here as to the vexed question of the direction, the extent, and the period of Hume's influence on Kant; to avoid repetition, reference is made to an earlier discussion of it (*Das Erkenntnisproblem*, 2d ed., vol. 2, 606 ff.).

letter to Herder dating from 1768 that he now feels closer to Hume in his entire intellectual orientation; Hume occupies the highest place among the teachers and masters of the true philosophical "state of mind."[78] And on the purely theoretical side, Kant assumes frankly and without reservations at least one crucial result of Hume's doctrine. That a mere conceptual analysis, carried out according to the principles of identity and contradiction, cannot afford the slightest knowledge of any sort of "basic reality"—this truth, already enunciated in the essay on negative magnitudes, is now for Kant further confirmed and deepened. "How something can be a cause or possess a power," he is convinced, can never be known through "reason," that is, by comparison of concepts according to the criterion of identity and contradiction; the knowledge of this basic relation can "be drawn solely from experience."[79]

But this agreement with Hume has limits in two respects. A completely different theoretical and ethical concern speaks in Kant and in Hume. Hume's scepticism is the full and adequate expression of his whole outlook and attitude. It is delight in doubting for the sake of doubting, delight in the unfettered activity of his superior analytical intelligence which totally rules and engrosses him. To be sure, the popular moral tendencies of the philosophy of the Enlightenment have some effect on him, especially in Hume's *Dialogues on Natural Religion*, but on the whole he turns a cool, half-ironic face of superiority toward ethical questions. Kant, in contrast, has become ever more sceptical toward all dogmatic religion and theology, to the extent that it presents itself as the foundation of ethics, but his position regarding the substance of ethics itself and his recognition of its unconditionally valid claims remained unchanged throughout his lifetime. In this regard, even the *Dreams of a Spirit-Seer* admits that the "scales of understanding" Kant uses are never wholly impartial, that they cannot do away with all ethical interest. The battle against metaphysics and its conception of God and a supersensible world means to him also the battle for a new positive foundation for autonomous morality.

Within the realm of pure logic as well, there is an analogous rela-

---

78. See above, pp. 85 f.
79. *Dreams of a Spirit-Seer*, pt. 2, chap. 3 (II, 387) (*Ak.* II, 370).

tion. Kant's conception of experience has a positive substantiality which no scepticism attains, for every genuine experiential cognition includes the application of mathematics. The experience Hume speaks of dissolves into the sheer play of representations, held together by the subjective rules of the imagination and the psychological mechanism of association. For Kant, the experience in which all our knowledge of basic reality must be rooted is rather the method of physical induction, as Newton built it up with the aid of a precise and specified experimental method and continuous application of mathematical analysis and calculation. Thus while Kant is aroused by Hume to struggle against metaphysics and to make war on all transcendence, his thinking immediately takes on a new and independent direction relative to Hume; for the more purely he now strives to cleave to the "fertile lowland of experience," the more clear it becomes to him that this depth of experience is itself founded in a factor which is rooted not in sensation as such but in mathematical concepts. Thus his keener grasp of the concept of experience itself leads him to distinguish more accurately the different conditions on which it rests, and to define their specific validity relative to each other.

4. THE SEPARATION OF THE SENSIBLE AND THE INTELLIGIBLE WORLDS

One who has told the story of Kant's youth has noted that the conventional view that Kant's life unfolded in an extremely simple and disciplined way is less and less corroborated the more deeply familiar one is with the details of that life. Rather, in a perpetually surprising fashion it is shown that even the outward course of Kant's life is not measurable by everyday standards and rules. "Kant trod no ordinary way. . . . From the beginning of his independent development until his old age he never did what an ordinary man would have done in his place. Looked at closely, his life did not progress at all 'in perfect regularity,' but moved in a very irregular way toward its goals. It always ran counter to the common view of men and mystified the expectations of those observers around him. For what they had a right to expect of him, either he did not undertake, or else undertook it after they had given up hope, and then accomplished it so grandly and completely that his performance evoked amazement and thus again contradicted every expectation all the more."[80]

If this view has an air of paradox so far as it concerns the external course of Kant's life, it is entirely accurate as regards the intellectual labor that went into the formation of Kant's system. As methodical as this labor is in its deepest themes, it pursues its conclusions with little simplicity, regularity, and "linearity." Everywhere one comes upon places at which his thought, after it is just on the point of arriving at a definite solution, suddenly steps backwards. A problem is taken up, thought through, and its solution reached—but suddenly it is shown that the conditions under which it was first worked out were not appropriate and complete enough, and hence not one step of the solution is valid, but instead the whole way in which the question is put has to be framed anew. Reticent as they normally are about questions of his inner development, Kant's letters tell us again and again of reversals of this kind. A conceptual whole is not constructed bit by bit in a steady, unbroken progression, but new threads seem continually to be spun, only to be immediately severed. If Kant adhered to and defended every essential basic proposition of his critical doctrine once it had been constructed, it is characteristic of this preparatory period that he has a certain indifference toward everything that is a mere "result"; he feared premature termination of his intellectual process more than he sought it. "As for me," he writes to Herder in 1768, "since I depend on nothing and, with a profound impartiality toward the opinions of others and my own, often upend the whole structure and inspect it from a variety of points of view so as eventually to hit upon that in it which I can hope to subscribe to as true, I have, since we parted, inserted different ideas in many places."[81] And a later remark confirms the maxim of Kantian thinking even more definitely. "I am not of the view of a superior man who, once he has convinced himself of something, henceforward feels no doubt about it. That does not do in pure philosophy. The understanding

80. Emil Arnoldt, "Kants Jugend und die fünf ersten Jahre seiner Privatdozentur im Umriss dargestellt," *Gesammelte Schriften,* vol. 3, p. 205.
81. To Herder, May 9, 1768 (see IX, 59) (*Ak.* X, 70).

itself is naturally opposed to it. One must consider the propositions in all sorts of applications and even try to assume their contradictories, if they lack a special proof, and delay long enough so that the truth is illuminated from all sides."[82]

The moment one realizes this general procedure of Kant's, the primary, as it were subjective, basis of the totally surprising turn his doctrine took once again in the years between 1766 and 1770, the period between the *Dreams of a Spirit-Seer* and *On the Form and Principles of the Sensible and the Intelligible World*, is comprehensible. Yet once more the expectation the world attached to Kant's ongoing development was foiled most remarkably. We recall that in 1763, after the composition of "The Only Possible Basis of Proof" and the Prize Essay, knowledgeable philosophers saw in Kant the future creator of a new, more thorough metaphysics, a metaphysics with critically tested and examined foundations, but one which was to be constructed in general on the old "rational" model. Now, however, to their astonishment their experience was that Kant, whom they had numbered among their own, struck out on a path which seemed to cut him off from metaphysics forever. To be sure, he still confessed an old affection and weakness for it, but he did so with such an ironic air of superiority that the subjective liberation he believed he had now at last achieved could be felt all the more strongly. "Metaphysics, whom it is my lot to love though I can seldom boast of any favors from her,

---

82. *Reflexionen Kants zur kritischen Philosophie*, vol. 2: *Reflexionen zur Kritik der reinen Vernunft*, ed. Benno Erdmann (Leipzig, 1884), no. 5. I have thoroughly examined these reflections (which are marginal notations Kant made in the textbooks he read, especially Baumgarten's *Metaphysica*) and used them in an earlier exposition I did of the evolution of the critical philosophy; here they are deliberately introduced only when the date of their composition can be established with certainty, either because they contain definite evidence of a date, or when it is immediately and unambiguously apparent from their content. Where the dating is questionable, or where it can only be done indirectly by complex factual inferences, I have preferred to leave these documents aside rather than to burden a biographical exposition, which is dependent above all on accurate and unambiguous temporal evidence, with material which is not indispensable and is in many respects problematical. Presumably more exact tools for dating the *Reflexionen* will be supplied by the publication of Kant's entire *Handschriftliche Nachlass*, which Erich Adickes has begun for the Academy edition of Kant's writings.

provides two benefits. The first is to satisfy those tasks which arise in the inquiring mind when it delves rationally into the hidden properties of things. But here the outcome all too often plays our hopes false and evades our eager hands. . . . The other benefit is more proportioned to the nature of the human understanding; it consists in investigating whether the task can be decided from what man can know, and what relation the question may have to those empirical concepts on which all our judgments must always rest. In this respect metaphysics is a science of the limits of human reason. . . . I have not determined these limits with any precision here, but have indicated that the reader will find on further reflection that he can excuse himself from all vain inquiries with regard to a question the data for which are to be found in a world other than the one in which he perceives himself to be. I have thus wasted my time so that I might regain it. I have imposed on my reader, so that I might be of service to him, and if I did not afford him any new knowledge, I uprooted the folly and idle knowledge which bloats the understanding, and in its cramped space cleared a place which wise doctrines and useful instruction can occupy." Metaphysics conceived as a theoretical question and task seem thereby to be dismissed. Kant expressly declares that he lays aside the whole matter of spirits as settled and done with. It is of no future concern for him, since the foregoing considerations cut off all philosophical understanding of such beings, and henceforth there may be opinions about them but never knowledge. This assertion, he adds, may seem boastful, of course, but it is not, for the termination spoken of here is such only in a negative sense, which does not determine an object but only securely fixes the limits of our knowledge. On this basis the whole pneumatology of the human soul might be called a scientific system of our necessary ignorance regarding a supposititious species of being, and as such is easily equal to the task.[83]

After this confession, it must have been totally unforeseen that Kant, on assuming his new position as Professor of Logic and

---

83. *Dreams of a Spirit-Seer*, pt. 1, chap. 4 (II, 367 f.) (*Ak.* II, 348 f.), and pt. 2, chap. 2 (II, 384 f.) (*Ak.* II, 357 f.).

Metaphysics on August 20, 1770, would defend a treatise the very title of which promised to determine the form of the intelligible world and to distinguish its essential features from those of the sensible world. For what the comprehensive concept of the intelligible world here includes is in truth nothing other than that realm of immaterial substances, entry to which seemed to have been forbidden us. And in this instance it was not a matter of a literary pamphlet, sprung from a momentary mood; instead, a strict, systematic thinker here unfolded stroke by stroke, with the precision of a balance sheet, the entire program of his future work as teacher and scholar. We are now presented with a profound theory of the intelligible, founded on an inquiry into its principles and presuppositions and pursued through all the main areas of the familiar metaphysics. Kant never doubts for a moment that this whole investigation is prompted by questions the data for which lie in a world different from that in which we feel ourselves to be, but now he is far from scorning this inquiry as an "idle search." He strides forward, secure and undistracted, and if, as is natural in a preparatory work, he does not provide a picture of the intelligible world worked out to the last detail, still he is convinced that he has defined and clearly delineated its general outline. And nothing in this sketch points back to the earlier drafts and experiments; it is as though this new picture of the sensible and the intelligible worlds had sprung out of nothing before our eyes.

Yet we must seek out the connecting link for this work too, if not to the earlier answers, at least to the earlier problems of Kant's thought. What relation is there between the denial of the *Dreams of a Spirit-Seer* and the affirmation of *On the Form and Principles of the Sensible and the Intelligible World?* Are these both concerned with the same object, or has perhaps the theme of metaphysics undergone a change? And if the latter, what new tasks have quickened in Kant's mind in the interim and now occupy the center of his theoretical interest? There is no direct, or at least no complete, answer to all these questions in the testimony we have concerning Kant's development in the years from 1766 to 1770. But the substance of the Inaugural Dissertation itself fills in this gap, for it clearly and unmistakably points to the new region of thought that Kant has now entered. For

the first time, Leibnizian philosophy is shown to be a force that determines him inwardly. This assertion seems paradoxical, for did not
Kant's first work on the estimation of living forces treat a theme from
Leibniz's natural philosophy, and had not the totality of Leibniz's
doctrines—at least in the form they received at the hands of Wolff and
academic philosophy—from then on gone with him at every step? In
truth, however, Kant's frequent reference to the substance of these
doctrines shows that their most essential philosophic spirit remained
closed to him for the present. Even the *Monadologia Physica,* which
seems to hew most closely to Leibniz, is no exception, since as physical monadology it tries to reach the ultimate elements in the realm of
the corporeal. The monads are here conceived as centers of force,
from the mutual interaction, attraction, and repulsion of which matter, as extended mass, is constituted. This dynamic construction thus
consistently applies concepts which (like the concept of corporeal
atoms, the concept of action at a distance and physical influence)
were unqualifiedly designated as fictions in Leibniz's sense.

The concept of monads in its characteristic metaphysical meaning,
however, functions—in the "Inquiry into the Distinctness of the Principles of Natural Theology and Morals"—as a paradigm of that synthetic procedure of metaphysics Kant fought against, where the basic
concepts are not so much deduced through analysis of appearances
into their elements as rather arbitrarily "imagined."[84] This judgment
as well shows that Kant was at that time still completely unable to
survey and evaluate the monstrous analytical intellectual labor
through which Leibniz had, by contemplating phenomena, gained
his concept of substance as their "principle" and "foundation."[85] One
must have vividly in mind Kant's attitude toward Leibniz's doctrine
up until now to judge what a decisive change Leibniz's *Nouveaux
essais sur l'entendement humain* must have produced in Kant's overall
outlook when he first became acquainted with it. The manuscript of
this book had been buried in the library at Hanover for sixty years,

84. "Inquiry into the Distinctness of the Principles of Natural Theology and
Morals," First Reflection, §1 (II, 177) (*Ak.* II, 277).
85. More on this in my book *Leibniz' System in seinen wissenschaftliches Grundlagen*
(Marburg, 1902), esp. chap. 6.

until it was brought into print in 1765 by Raspe in his edition of the
Oeuvres philosophiques. Now, however, it inevitably affected the age
with all the force of a totally new impression. Leibniz once again
stood among them as a contemporary, as though raised from the
dead. Only now did the whole breadth and originality of his thought,
hitherto clouded by academic tradition, emerge. In this book, it was
universally felt, it was not a matter of an isolated learned artifact, but
of an event which was a decisive irruption into universal intellectual
history and all its problems and interests. This is the way in which
Herder and Lessing, who projected and made a start on a German
translation of the *Nouveaux essais*, understood and welcomed the
book.[86] And moreover it was just these years from 1765 to 1770 that
did the most for general knowledge and deeper understanding of the
Leibnizian philosophy in Germany, for with the appearance of Dutens's great edition in 1768, the whole of Leibniz's philosophical and
scientific labors, which till now had been scattered or unknown,
could be surveyed with some accuracy and completeness.

For Kant, too, an entire new source was opened up. His notes
from this period give abundant and unambiguous evidence that he
was intensively occupied with the *Nouveaux essais* in particular.[87] For
the first time he encountered Leibniz not only as philosopher of nature or as speculative metaphysician, but as epistemological critic.
Now he understood in what sense the doctrine of innate ideas and
truths cohered with the system of monadology, how it was on one
side the foundation of that system and on the other found only in it
full, concrete confirmation. Once more Kant sees himself face to face
with the great question about the relation between the methodology
of scientific knowledge and that of metaphysics. Leibniz leads him
back to his own fundamental problem, which now is freed of all ties
with particular, concrete questions and achieves its fully universal
expression.

86. See Lessing's *Werke*, ed. Lachmann and Muncker, vol. 15, pp. 521 f.

87. Cf. Kant's *Reflexionen*, nos. 513, 273–78; for the dating of these reflections, see
Erich Adickes, *Kant-Studien*, (Kiel and Leipzig, 1895) pp. 164 ff., and *Das Erkenntnisproblem*, 2d ed., vol. 2, pp. 622 f.

If we wish to visualize this process, we should not start with the actual historical meaning of Leibniz's system, but instead with how it presented itself to Kant's mind. Kant's interpretation of individual Leibnizian concepts and propositions is not free of misunderstandings and hardly could be, since, despite Dutens's collected edition, the most important sources for Leibniz's philosophy which are available to us—especially the major part of the philosophical and mathematical correspondence—were still undiscovered in the eighteenth century. But this is of little importance for the history of Kant's intellectual evolution, since it is not a matter of what Leibniz was, but of how Kant understood and saw him. When Kant later comprehensively surveyed Leibniz's system in the *Metaphysical Foundations of Natural Science (Metaphysische Anfangsgründe der Naturwissenschaften)* he laid all his emphasis on the point that the monadology ought not to be judged as an attempt to explain nature, but as a "Platonic concept of the world, correct in itself, insofar as the world is not an object of the senses but is regarded as a thing in itself, purely an object of the understanding, but one which is the foundation of sensory appearances."[88] In fact, this was the perspective from which he judged Leibniz's doctrine from the beginning. The monads are the "atoms" of things, but this atomicity in no way denotes that of a physical component which is an ingredient in the composition of bodies, but rather that ultimate, unanalyzable unity of which we are aware as the spiritual subject in the idea of the ego.[89] In the act of self-consciousness a unity is given us that is not derivable from something else, but that instead is the principle of all derivation. This unity is not the consequent of a more ultimate, deeper lying plurality, but forms the necessary presupposition for representing any plurality. For to think a plurality or to represent it to oneself, its diverse moments must be mutually interrelated and thought as an interconnected whole; this inclusive grasp, however, can be achieved only

88. *Metaphysical Foundations of Natural Science,* chap. 2, Proposition 4, note 2 (IV, 413) (*Ak.* IV, 507).

89. Cf. *Critique of Pure Reason,* note to the Second Antinomy, A 442 = B 470 (III, 318).

when we base it on that universal possibility of seeing the "one in the many" which we customarily signify by the name of "perception" or "consciousness."

Hence there are two views of the world opposed to one another in their principle and their origin, although they are united in the concrete whole of our experience. According to the one, we comprehend ourselves as spiritual substances, as a whole of psychic phenomena the entire manifold of which refers back to the same identical ego and only thus constitutes a single series of personal experiences, an integral "substance." According to the other, we regard ourselves, like the world around us, as a coherent corporeal whole governed by mechanical laws, those of pressure and impact. In the first form, the conception of what we call the "world" is that of a whole of purely inner states, an aggregate of representations and impulses; in the second, we contemplate the states as they might present themselves to an outside observer. For this the intensive manifold must be changed into an extensive one; the dependence of inner phenomena on one another, and their qualitative relationship of similarity must appear as an external order as we think it in the concepts of space and time.

But if we ask which of these two views of actuality possesses the higher truth, the answer cannot be in doubt. For in the former we comprehend ourselves as we are purely and simply for ourselves; the latter represents the viewpoint under which our being falls when seen from outside. In the one case a purely spiritual being is expressed and evidenced through purely spiritual concepts, such as that of the dynamic limitation of one state by another; in the other case we must transform what is in truth an inward relationship into the externality of space and time in order to make it publicly knowable. So on the one hand we have the picture of a world purely of the understanding: a community of diverse spiritual substances; on the other hand, the picture of a sense world, that is, a nexus of *appearances*, the coexistence and succession of which can be empirically observed and described. In this conception, as his comparison of Leibniz and Plato shows, Kant found the old opposition of the "phenomenon" and the "noumenon" pointed out and understood from a fresh perspective. He now views the system of monadology in relation to this universal intellectual history, which Leibniz himself had emphatically called attention to.[90] The classical division between the intelligible and the sensible world[91] seems here to be deduced from the fundamental laws of knowledge and only thus conceived as necessary.

This also altered Kant's own stand regarding the question. In his critique of metaphysics, from the Prize Essay to the *Dreams of a Spirit-Seer*, he constantly asked for the data on which knowledge of a supersensible world might rest, and he had not been able to unearth these data in the received definition of academic metaphysics, not to mention the theories and tales of a Swedenborg. Now, however, he had discovered a new point of departure: the crucial datum, as Kant only became fully aware on studying Leibniz, lies in the differing origin and the differing type of validity of the principles of our cognition. It is here, if anywhere, that metaphysical reflection is to be rooted. That which is truly spiritual is not the infinite, transcending every form of our knowledge, but it is comprised in this very form of knowledge. The distinction between universally valid and particular, between necessary and contingent truth is "given," is indubitably certain; one may investigate whether it is possible to define the boundaries of the sensible and the intelligible worlds without assuming anything besides this distinction.

In the quarrel between Leibniz and Locke, Kant had come down on the side of the former, and apparently without hesitation. The Lockean derivation of the pure concept of the understanding from experience always seemed to him to be a sort of *generatio aequivoca;* at no point in his thinking was he content with this kind of "birth certificate."[92] If Kant was an empiricist, that only meant for him the demand that the validity of concepts be shown to be grounded in the analysis of the *objective* contents of experience, but he never regarded

90. See the *Epistola ad Hanschium de Philosophia Platonica sive de Enthusiasmo Platonico; Opera,* ed. L. Dutens (Geneva, 1768) vol. 2, p. 1.
91. Cf. *De mundi sensibilis atque intelligibilis forma et principiis,* §7 (II, 411) (*Ak.* II, 395); also the *Critique of Pure Reason* A 235 ff. = B 294 ff. (III, 212 ff.).
92. Cf. *Critique of Pure Reason,* A 86 = B 119 and B 167 (III, 106 and 135).

the evidence of the subjective psychological origin of a concept and its being traceable back to simple sensations as the sufficient or necessary condition of its truth. That special concepts such as possibility, existence, necessity, substance, cause, and so on, together with everything related to them and following from them, are never to be obtained and inferred in this way is something about which he has now become totally clear. For since the relations which they express are not of a sensory nature, they can never be abstracted from the stuff of perceptions by a mere summation of particular sensations.[93]

If one wishes to say that these pure relational concepts are gotten "by abstraction" from the particular sensations of sight, hearing, and so forth, the ambiguity attaching to the concept of abstraction must first be put aside. A true logical or mathematical concept is not *abstracted* from sensory appearances (for then it would contain nothing which was not some kind of concrete component momentarily present in them), but rather it has an *abstractive relationship* to them, that is, it posits a universal relation regardless of whether such a relation is exemplified and presented in any particular sensory instances. Therefore it would be more accurately called a *conceptus abstrahens* than a *conceptus abstractus*.[94] In this sense Kant also called geometrical principles "ideas of pure reason" for some time, before he hit in the Inaugural Dissertation on the essential methodological designation of "pure intuitions" for space and time. For the latter, too, express relations which we need not have experienced particular instances of in order to know in general. In a certain sense we arrive at them "by abstraction," but the material from which the abstraction is taken is not sensations, but the activity of the mind itself, which we grasp in its immanent lawfulness and hence necessity.

"Some concepts," as it is said in a note from this period, "are abstracted from sensations, others purely from the law of the understanding, which compares abstract concepts so as to connect or to distinguish them. The origin of the latter is in the understanding; that of the former is in the senses. All concepts of such a kind are called

93. *De mundi sensibilis,* §8 (II, 411) (*Ak.* II, 395).
94. Ibid., §6 (II, 410) (*Ak.* II, 394).

pure concepts of the understanding: *conceptus intellectus puri*. To be sure, on occasion sensations can evoke these activities of the understanding, and we become conscious of certain concepts of universal relationships of abstract ideas according to laws of the understanding; thus Locke's rule that without sensation we have no clear ideas holds in this case. But although the *notiones rationales* arise through sensations and are thought only as applied to the ideas abstracted from them, they are not contained in them nor abstracted from them, just as in geometry we do not borrow the idea of space from the sensation of extended things, though we can only make this concept (similarly) clear when given the sensation of corporeal things. Hence the idea of space is a *notio intellectus puri*, which can be applied to the abstracted idea of mountains and casks. The philosophy of the concepts of *intellectus purus* is metaphysics; it is related to the rest of philosophy as *mathesis pura* to *mathesis applicata*. The concepts of existence (reality), possibility, necessity, ground, unity and plurality, part, whole, nothing, composite and simple, space, time, change, motion, substance and accident, force and action, and everything else belonging to ontology proper are related to the rest of metaphysics as common arithmetic is to *mathesis pura*."[95]

The work *On the Form and Principles of the Sensible and the Intelligible*

95. *Reflexionen* on the *Critique of Pure Reason*, no. 513 (see above, n. 86, on the dating). To see their historical connection with Leibniz vividly, one should put these sentences side by side with the following from the preface to the *Nouveaux essais*: "Perhaps our clever author (Locke) will not depart entirely from my feeling. For after having occupied his whole first book in rejecting the inner light, taken in a certain sense, he nonetheless affirms at the outset of the second, and in what follows, that ideas which do not originate in sensation come from reflection. Now, reflection is nothing but attending to what is within us, and the senses do not give us what we already carry with us. That being so, can it be denied that there is already much innate in our mind, since we are, so to speak, innate to ourselves? Or that there is in us: Being, Unity, Substance, Duration, Change, Action, Perception, Pleasure, and a thousand other objects of our intellectual ideas? And these objects being immediately and always present to our understanding (although they are not always perceived by reason of our distractions and desires), why be surprised that we say that these ideas, together with all that depends on them, are innate in us?" For space and time, see especially *Nouveaux essais*, II, 5: "Ideas... such as that of space, figure, motion, rest are... of the mind itself, for they are ideas of the pure understanding, but which have a relation to the external and to what the senses cause us to perceive."

World attaches to these thoughts the decisive terminological specification by which the ambiguous concept of "innate ideas" is bypassed. As regards the basic categories of the understanding, it is not a matter of innate ideas (*conceptus connati*), but of inherent laws of the mind (*leges menti insitae*), which we become conscious of by attending to their actions and also when experience occurs.[96] Here, too, Kant makes no material strides beyond Leibniz, but he has coined a new and significant expression for the fundamental ideas which the latter advocated, terminology which, in its pregnancy and incisiveness, leads on to a sharpening and deepening of the problem of the a priori.

But it was important first to hit upon yet another critical distinction, which necessarily led Kant into far more complex questions than the opposition between Leibniz and Locke. It was unthinkable to him to judge in Locke's favor, for he had always distinguished very definitely between "empiricism" and "empirism." But in building up pure intellectual knowledge, as he now undertook to do, by abandoning Locke did he have to abandon Newton also? And were there not between the latter and Leibniz the most severe unreconciled and seemingly irreconcilable contradictions? Since these contradictions had been given their most acute form in the polemical exchange of correspondence between Leibniz and Clarke, they had not been laid to rest. The whole philosophical and scientific literature of the eighteenth century is still full of them. Everywhere the metaphysicians' and ontologists' conceptions of the world are starkly and uncompromisingly contrasted. This division becomes a universal watchword, under which the intellectual battles of that time were fought. The greatest scientific genius of Germany, Leonhard Euler, discussed this conflict again very extensively in 1768, in a popular work, the *Letters to a German Princess*. While the metaphysician, he said, analyzes the world into ultimate, simple parts in order to understand it, the mathematician conversely must insist that the divisibility of matter, like that of space, is infinite and that therefore an unanalyzable simple entity is never reached. If the former resolves the actual into a sum of point substances which taken in their totality

96. *De mundi sensibilis*, §8 (II, 411) (*Ak*. II, 395).

manifest the phenomenon (or rather the appearance) of extension, the latter knows that a more complex spatial or temporal relation is reducible to another, simpler relation, but that extension can never be produced from points, the extensive from the intensive. Further, while according to established metaphysical theories pure space and time are nothing in themselves but are to be regarded only as qualities or "accidents" of the bodies, which alone are actual, and their motions, the mathematician and physicist for their part are unconcerned with establishing the *sort* of reality that space and time possess. That some sort of reality is to be attributed to them, however, and that extension and duration, even apart from what is extended and enduring, have a substantial being are held to unconditionally, because without these assumptions it would be impossible to achieve a clear and determinate sense of the supreme laws of motion. The law of inertia, for example, cannot be definitely and precisely formulated if one does not differentiate pure or, as Newton called it, absolute space from all that it contains and recognize it as an independent whole, in relation to which the rest or motion of a material system can be spoken of.[97]

The keenest and most decisive objection against any encroachment of metaphysics in questions of the theory of nature was raised by a thinker for whom Kant had always felt the deepest respect, and whom he was accustomed to regard as the proper arbiter of all questions concerning the exact and empirical sciences. In the preface to the "Attempt to Introduce the Concept of Negative Magnitudes into Philosophy" Kant had already referred to Euler's procedure of taking the certain results of mathematics as the necessary touchstone of the truth or falsity of universal philosophical propositions; he relied on him in the treatise "On the First Ground of the Distinction of Regions

---

97. Leonhard Euler, *Lettres à une princesse d'Allemagne sur divers sujets de physiques et de philosophie* (Petersburg, 1768); *Theoria motus corporum solidorum seu rigidorum* (Rostock and Greifswald, 1765); "Reflexions sur l'espace et le temps," *Histoire de l'Académie des Sciences et Belles Lettres* (Berlin, 1748); *Mechanica sive motus scientia analytica exposita*, 2 vols. (Petersburg, 1736–42). See *Das Erkenntnisproblem*, 2d ed., vol. 2, pp. 472 ff., 501 ff., for more about Euler and his battle against the "metaphysical" theory of space and time.

in Space" ("Von dem ersten Grunde des Unterschiedes der Gegen-
den im Raume"), dating from 1768, which is explicitly advanced as a
continuation of Euler's "Reflections on Space and Time," and in the
book *On the Form and Principles of the Sensible and the Intelligible World*
he is once more celebrated as *phaenomenorum magnus indagator et arbiter* ["great investigator and judge of phenomena"].[98]

Accordingly, one thing stood firm and indubitable for Kant, now
that he was commencing a transformation of his doctrine which
seemed to bring him closer to metaphysics once again, namely, that
whatever validity might be attributed to metaphysical principles,
mathematics, as pure and applied knowledge, had to be confirmed as
unconditionally valid and guarded against all metaphysical "chicanery." How was this aim to be accomplished, however, if one held fast
to the opposition between the sensible and the intelligible world, as
Kant did from now on? Was it possible for mathematics to be completely applicable to the physical, unless both were declared to be of
the same kind in their nature and essence? Here thought runs into a
peculiar dilemma. If it decides to assert full correspondence between
the mathematical and the physical, so that there is no proposition of
pure mathematics which is not also completely valid in applied mathematics, then it seems that the origin and cognitive value of
mathematical concepts are no different from those of empirical ones.
On the other hand, if mathematical truths are regarded as truths of
pure understanding, which derive not from things but from the laws
and activities of the intellect itself, what guarantees do we have that
things fully conform to pure concepts, that the sensible conforms to
the intelligible? If we were to fall back on a "preestablished harmony"
between the two realms, we would have mere verbiage, not a solution to the problem.[99]

And in fact the Leibnizian system of metaphysics runs aground on
just this point. The basic flaw in this system, in Kant's judgment, is
precisely that the sole form of the rational advanced and recognized
in it can be affirmed only through its applicability to empirical being,

98. II, 206, 394, 431 (*Ak.* II, 167, 378, 414).
99. Cf. Kant's later verdict in his letter to Marcus Herz dated February 21, 1772 (IX,
102) (*Ak.* X, 123), and *De mundi sensibilis*, §22, Scholium (II, 426) (*Ak.* II, 409).

and that it introduces a false concept of this latter. For the form under which the empirically actual stands is space and time; these two, however, are not acknowledged in Leibniz's system to be specifically essential and pure means of cognition, but are only treated as "confused ideas." Strict, literal truth in this system belongs exclusively to the dynamic relations between substances, the relationships of the simple monads, while nothing that we express in the language of space and time ever gives us this truth, but always merely an indirect and clouded image of it.

If this view is valid, however, the doctrine of Leibniz and Wolff has foundered on this point. For if substances are primary and space and time secondary and derivative (indeed something derived that never fully matches its archetype), all the content of mathematics depends on the actuality of things. Thus, if we wished to think through the consequences and not turn arbitrarily aside, we would be brought back to the standpoint of an empirical foundation for mathematics, and it makes no material difference that the result is reached from premises quite different from Locke's. For generally speaking, where things determine concepts, not the reverse, only contingent knowledge is attainable, not knowledge which is universally valid and necessary. Hence, if the assumptions of the Leibnizian-Wolffian systems hold, and space and time express the structure of actuality, though not adequately, but in an obscure and confused fashion, that is the end of the exactitude and unconditional necessity of all mathematics. Mathematical propositions would then be entitled to claim merely relative and comparative, not absolute universality and truth, and the idea that geometrical axioms and propositions might be changed or contradicted by further experience would no longer be absurd.[100] Only one way is open to us to avoid all these difficulties: to grant mathematics its full freedom and independence from the empirical and actual, and conversely to guarantee its complete agreement with the latter. It would have to remain a part of the realm of pure

---

100. See esp. *De mundi sensibilis*, §15 (II, 420) (*Ak.* II, 403). It might be reiterated that here it naturally is not a matter of the actual view of space and time and the cognitive value of mathematics which Leibniz and Locke historically held, but of hypothetical inferences which Kant claims to be grounded in the premises of Leibniz's system.

intellectual forms, and yet be related to the sensible realm in a special and specific way true of no other mere "concept of the understanding." It would have to rest on a principle of knowledge which would be at once rational and sensory, general and individual, universal and concrete.

It is plain that we are not dealing here merely with an arbitrary and paradoxical demand, but that, if we now move on to an exact critical analysis of the forms of space and time, a genuine datum of knowledge corresponds to what is desired. For in these forms everything that was just previously asserted as a mere postulate finds its complete and precise fulfillment. Space and time are "general"; they are what all possibility of figure and place in general rest on, and hence they must be assumed in every statement about a determinate and particular form of being, and about an individual empirical structure. But at the same time they are "concrete," for we are not dealing with generic concepts in them, which might be exemplified in a number of particular instances, but if we wish to grasp them in their essential determinacy we must think them as both particular and "single." A generic concept contains its different species under it, as the concept of tree includes pines, lindens, oaks, etc. Here, though, with regard to space and time, there are no comparable subordinate classes. Analyze the whole of space and time as we will, we are led to nothing simpler in thought, no concept with a less complex content, but to conceive every foot and every yard, every minute and every second, we must also think the totality of spatial coexistence and temporal succession. The yard would not be thought "in" space nor the second "in" time were this requirement not met, for they must be marked off from all the other parts of space and time; therefore these latter must be thought together with them.

There now appears a new psychological and epistemological terminology for this peculiar way of relating the individual to the general and vice versa, and of conceiving the whole in every part and with every part. Wherever this kind of conception is needed and is possible, we have to do not with the form of a sheer concept, but with the form of intuition.[101] Now Kant has discovered the decisive idea that

101. See *De mundi sensibilis*, §§13–15 (II, 414–22) (*Ak.* II, 398–406).

contains the solution to all his former doubts. The intuition of space and time, which must be acknowledged as independent and peculiar givens for knowledge, in fact forms the true answer to the demands which till now must have seemed incompatible. In intuition, the moment of purity is combined with the moment of sensibility. Space and time are sensible, because coexistence and succession cannot be resolved into mere conceptual determinations by any analysis, however far it is pushed; both are "pure" because even without undertaking any sort of analysis into conceptual elements, we can comprehend the function they, as wholes, have of bringing us to full clarity, and we can grasp them in their unconditional validity, divorced from all that is merely factual and empirical. Only after we have progressed to this point do we have a science of the sensible, a strict and exact application of mathematics and its necessary determinations to phenomena and their change and cessation. We have differentiated two fundamental kinds of pure knowledge: that by which we determine the relations of the intelligible, and that by which we determine order in what is sensible. Only the first kind instructs us about things as they are, while the second, intuitive knowledge of space and time, makes the world of appearances accessible and meaningful to us, but within this its domain, full universality and necessity, unlimited precision and certainty, are conserved.[102]

In this way, Kant hit upon a final resolution of the opposition between Leibniz and Newton, though it could not be expressed as simply as in the case of the conflict between Leibniz and Locke. In the latter case, Kant could in every essential respect espouse Leibniz's judgment; if he rejected the label "innate" and substituted the affirmation of fundamental laws of the mind, which, however, are only recognized when they are exercised, this was more an improvement in terminology than a completely new substantial turn he gave to Leibnizian thought. But in the battle between Leibniz and Newton, it was no longer possible simply to declare himself for one side or the other, for in posing his problem this way he went as far beyond the one as the other. When Euler, in standing up for Newton, had defended the interests of empirical research, which were to be protected

102. See ibid., esp. §§11 and 12 (II, 413 ff.) (*Ak.* II, 397 ff.); §4 (II, 408 f.) (*Ak.* II, 392 f.).

against any encroachments from elsewhere, at this point a difficult and complex problem resulted for Kant's philosophical critique. It had to substitute a positive judgment for a negative one; it had not only to secure and confirm science within its own boundaries, but at the same time to determine with precision, as the proper domain of metaphysics, what lay beyond those boundaries.

Only thus could both the infringement of metaphysics on natural science and also the meddlings of the latter in the former successfully be warded off. The development of mathematical physics in the eighteenth century afforded many cautionary examples of the second kind of meddling. Kant had willingly allowed the geometer and the physicist the use of the concept of absolute space to derive their propositions, for in fact this use was exhausted in the assertion that the meaning of "space" in geometry and mechanics is not identical with what we call the whole of the material world, but stands over against it as something unmistakably all its own. Kant's own view agreed completely with this thesis, and he sought support for it from an examination of purely geometrical relations in his treatise of 1768, "On the First Ground of the Distinction of Regions in Space."[103] What he had no right to do, on the other hand, was this: from the nature of this pure mathematical space to draw conclusions on all sides concerning the basic problems of speculative cosmology and theology, touching the relation of God to the world, of creation to eternity. Newton had led the way here as well, subjoining to the calculations and experiments of the *Principia Mathematica* and the *Opticks* sections in which he had given his theory of space as God's "sensorium" and the organ of divine omnipresence, cautious and tentative in its form, to be sure, but nevertheless quite positive and dogmatic as regards its content.[104] And in the correspondence between Leibniz and Clarke questions of this kind had in the end overridden and pushed aside almost all others.

The dialectical contradictions into which one thus falls had, how-

103. See II, 391 ff. (*Ak.* II, 375 ff.).

104. Isaac Newton, *Philosophiae naturalis principia mathematica*, bk. 3, ed. Thomae Le Seur and Francisci Jacquier (Geneva, 1739), vol. 3, pp. 673 ff.; *Optice*, translated into Latin by Samuel Clarke (Lausanne, 1740), pp. 297 f.

ever, already been pointed out keenly and clearly by Leibniz. If one assumes, he had reasoned, that space and time are predicates which apply indifferently to all that is, thus identically to the mental and the physical, to God and the world, the Creation seems necessarily to be an act taking place in absolute space and absolute time. Thus its "where" and "when" are determinate, that is, there is a fixed moment of its inception and a fixed place, a delimited portion of infinite space, which serves as the original receptacle for matter.

But if one goes on to determine rationally, somehow or other, what this place and this time were, one is soon entangled in a net of antinomies. Since in empty space and empty time no place is to be preferred to any other nor manifests any essential difference from any other, any point that we might assume hypothetically as the "beginning" or the spatial locus of the Creation is arbitrarily interchangeable with any other. Therefore in this whole way of looking at the question it is impossible to posit any "here" without its immediately turning into a "there," or any "now" without its turning under our very hands into its opposite, an "earlier" or a "later."[105] Kant took the most active interest in all these problems—the Leibniz-Clarke correspondence had been freshly brought to his attention by its publication in Dutens's edition of Leibniz's works in 1768, and the notes he made in his personal copy of Baumgarten's *Metaphysica* show how deeply engrossed he was with it over a period of time—and he understood the question propounded here, but he gave it a far more universal meaning. The contradiction Leibniz has discovered here is not an isolated one; rather, it appears everywhere and anywhere sensory predicates are applied to intelligible objects or intelligible predicates to sensible objects. Whenever this is done, any proposition which we can assert is immediately confronted by its antithesis, and both can be demonstrated with apparently equal validity and necessity.

Kant himself tells us that in the period prior to the Inaugural Dissertation he worked out such antithetical proofs, and in doing so first became fully conscious of the distinctive features of the new

105 See the correspondence between Leibniz and Clarke (in my edition of Leibniz's main writings on the foundations of philosophy [Philosophische Bibliothek, vol. 107/8], vol. 1, pp. 134 f., 137 f., 188, 190).

theory, the separation of the substantive content of the sense world from that of the intelligible world with regard to principles and method. "At first I saw this theory only in shadow. I tried very seriously to prove propositions and their opposites, not in order to justify a dubious theory but because I suspected I might discover an illusion of the understanding with which it was involved. The year 1769 brought great light for me."[106] The illusion was destroyed as soon as it was recognized that in order for the object of any judgment to be completely determined, a specific mark stating the cognitive conditions under which it stands for us was required. If this is neglected, laws which are rooted in our subjective "aptitude" (indoles), and in fact are necessarily grounded in it, are erroneously taken to be conditions of things in general, which thus must pertain to them however we regard them; there then results a characteristic subreption of consciousness. Since the boundary lines of the mode of cognition are erased, all clarity and unambiguity of objects disappears; we no longer have any fixed subject of judgment, but wander back and forth between differing interpretations and meanings of our judgments with no sure guide. The human mind becomes a magic lantern which strangely alters and distorts the outlines of things by the semblance it projects onto them. The only protection against such a "deception of the mind" is the secure delimitation of the two spheres in which all our judgment moves. If this division is made, we can no longer blunder into the attempt to apply the predicate of "where" and "when" to objects in the pure world of the understanding, for example, God and immaterial substances, just as conversely we can no longer conceive sensible objects except under the specific conditions of sensibility, space and time as the pure forms of intuition.[107]

And two things are accomplished by this. The "infection," the contagium, of the intelligible by the sensible, which emerges so clearly in Newton's theory concerning God,[108] is avoided; on the other side, unconditional certainty and total applicability is guaranteed to the

106. Reflexionen on the Critique of Pure Reason, no. 4.
107. See De mundi sensibilis, sect. V: "On the Method of Dealing with the Sensitive and the Intellectual in Metaphysics" (II, 427 ff.) (Ak. II, 410 ff.).
108. See ibid., §§22 and 23 (II, 426, 428) (Ak. II, 409, 410–11).

forms of sensibility within their own field, thus for the whole sphere of objects of experience. Metaphysics as well as mathematics is satisfied in the same way; each has found in itself its center of gravity and its essential principle of certainty. It is here that the main theme and the proper center of the Inaugural Dissertation lie for Kant himself. On September 2, 1770, he writes to Lambert, to whom he is sending the book: "The first and fourth sections can be scanned without careful consideration; but in the second, third, and fifth, though my indisposition prevented me from working them out to my satisfaction, there seems to me to be material deserving more careful and extensive exposition. The most universal laws of sensibility play an unjustifiably large role in metaphysics, where, after all, it is merely concepts and principles of pure reason that are at issue. A quite special, though purely negative science, general phenomenology (*phenomenologia generalis*), seems to me to be presupposed by metaphysics. In it the principles of sensibility, their validity and their limitations, would be determined, so that these principles could not be confusedly applied to objects of pure reason, as has heretofore almost always happened. For space and time, and the axioms for considering all things under these conditions, are, with respect to empirical knowledge and all objects of sense, very real; they are actually the *conditions* of all appearances and of all empirical judgments. But extremely mistaken conclusions emerge if we apply the basic concepts of sensibility to something that is not at all an object of sense, that is, something thought through a universal or a pure concept of the understanding as a thing or substance in general, and so on. It seems to me, too (and perhaps I shall be fortunate enough to win your agreement here by means of my very inadequate essay), that such a *propaedeutic discipline*, which would preserve metaphysics proper from any admixture of the sensible, could be made usefully explicit and evident without great strain."[109] What Kant here regards as the object of an easy effort was to engross his most profound and taxing intellectual labor for a decade. Only the appearance of the *Critique of Pure Reason* almost eleven years after this letter to Lambert brought that propaedeutic to

109. IX, 73 (*Ak.* X, 92).

metaphysics which Kant has in mind to its true "explicitness and evidence."

But before we embark on this new path leading out of the Inaugural Dissertation, let us review again the development from which the conclusions of this book arose. There are relatively few external facts that can be established as to the period between the *Dreams of a Spirit-Seer* and the Dissertation, but if they are assembled, a clear picture of the philosophical advance of these years emerges. We know that Kant became acquainted with Leibniz's *Nouveaux essais* at this time, that in agreement with it he sketched out a theory of pure intellectual concepts in which space and time stand immediately alongside pure concepts of reason such as substance, cause, possibility and necessity, etc., and that his pathway to the sharp separating-off of the elementary concepts of sensibility, the pure concepts of intuition, was only gradually cleared. We can trace how he tried to resolve the conflict between the "mathematicians" and the "metaphysicians" on the problem of space and time for himself, bolstered in particular by Euler's writings and with reference to the discussion between Leibniz and Clarke; and how he becomes ever more deeply tangled in dialectical contradictions until at last, in 1769, he comes face to face with the decisive significance of the general problem of the antinomies.[110] He is given the new solution to the question along with this precise formulation of it. The thesis and the antithesis of the antinomies can only be reconciled if we understand that the two refer to different worlds. To establish the division between these two worlds, and thus to ground and secure each in itself, from now on comprises the proper task of metaphysics. Thus it is not valid in metaphysics for "practice to dictate method," for us to start, as in the other sciences, with particular inquiries and mental steps, and only later, when a certain quantity of knowledge has been accumulated, to seek the principles by which our thinking has been guided. Rather, here the question of method is the essential and the only justifiable starting point of all cognition: *methodus antevertit omnem scientiam.*[111]

110. For more on the significance of the problem of the antinomies for Kant's development, see Benno Erdmann's preface to his edition of the *Reflexionen*, pp. xxiv ff.

111. *De mundi sensibilis*, §23 (II, 427) (*Ak.* II, 411).

Any dogmatic judgment regarding which this basic preliminary question cannot be answered is to be rejected as an empty intellectual bagatelle.

At this point it is especially clear how Kant, while arriving at a new intellectual perspective, does not break with the previous evolution of his thought. Philosophy is still for him a "science of the boundaries of human thought," but a new datum, neither the whole scope nor bearing of which he has grasped so far, is now reached as a foundation for the determination of these boundaries. The system of a priori cognitions is the basis on which any division of the sensible and the intelligible worlds rests. Leibniz was the first to sketch out this system, but he did not see and make known its finer ramifications and convolutions, for beyond the common principle, rationality, which is equally proper to all its components, to logical and ontological concepts and to mathematical concepts alike, he overlooked the specific difference in validity that nonetheless holds. The Inaugural Dissertation took the first step toward illuminating this difference; now it was a matter of not stopping there, but of drawing the detailed boundary lines more and more sharply and precisely, until reason emerged as perfectly unified and at the same time as particularized and organized in all its individual moments.

5. The Discovery of the Critical Problem

When Kant, at the age of forty-six, entered into his new academic position with his work *On the Form and Principles of the Sensible and the Intelligible World*, it might have seemed as though his philosophical development had reached its true zenith and were on the verge of coming to an end. He had by now confronted all the major intellectual powers of the age and had won his own independent place among them. It seemed that nothing more was needed than to make the intellectual domain that he had struggled to achieve secure and to extend it on all sides. Kant himself believed that all his ensuing work would be devoted to this aim, merely more detailed fleshing-out and confirmation of the insights he had won. But just at this point occurs the decisive turn that gives his life and thought their true profundity.

What would have constituted the end for others, even for major philosophical talents, was for Kant's philosophical genius only the first step down a completely new road. Later, Kant himself located the beginning of his original achievements as thinker and writer in the year 1770, and in fact everything before this time, rich as its specific content was, seems of minor significance when measured by those standards set up by his development from the Inaugural Dissertation to the First Critique.

Before we launch on the study of this most important period of Kant's inner growth, though, some of the external facts about Kant's life and progress in his academic calling should be briefly recalled. His appointment as professor *ordinarius* of logic and metaphysics formed a significant stage in this, for it was that which first gave Kant leisure to carry out his philosophical work. Although he never complained in the slightest, the letters he addressed to the minister of education and to the king when applying for the professorship are instructive as to how heavily concern about the security of his future weighed on him. "This spring," he writes, "I enter the 47th year of my life, the course of which makes my apprehensions over future poverty ever more disquieting. . . . My years, and the scarcity of opportunities to make a living in the academy if one also has the scrupulousness to apply only for those posts which one can honorably hold, would, in the event that my humble request fails of its aim, necessarily extinguish and abolish any further hope I might have of remaining in my fatherland in the future."[112] Indeed, all the earlier steps Kant had made in this direction had remained fruitless. In his earliest years as teacher, the post of instructor in the Kneiphöfische Domschule in Königsberg, for which he applied, had been denied him; as Wald tells us in his memoir, it was taken by a "notorious ignoramus" by the name of Kahnert.[113] Kant's attempt to obtain the post of professor *extraor-*

112. To Minister von Fürst, March 16, 1770; to Friedrich II, March 19, 1770 (IX, 68, 70) (*Ak.* X, 86, 88).

113. See Reicke, *Kantiana*, p. 7; Borowski, *Darstellung des Lebens und Charakters Immanuel Kants*, p. 31. There are no remaining grounds to doubt these accounts since Arthur Warda ("Zur Frage nach Kants Bewerbung um eine Lehrerstelle an der Kneiphöfischen Schule," *Altpreussische Monatsschrift* 35:578 ff.) has shown from the

dinarius of logic and metaphysics, several years after Martin Knutzen's death, also came to grief; when Kant submitted his application in April, 1756, war was just about to break out again, and the Prussian regime left the post vacant for reasons of economy.[114] The next application, which Kant submitted two years later for the professorship of logic and metaphysics, was made under even less favorable auspices. The position became vacant with the death of Professor Kypke in 1758, at a time when all of East Prussia was occupied by the Russians and under the control of their military government. So the application had to be addressed not only to the philosophical faculty in Königsberg, but also to the "Most Serene and Mighty Czarina and Ruler of All Russia," the Czarina Elizabeth. Her representative, the Russian governor of Königsberg, decided against Kant, however; instead of him, his colleague Buck received the appointment; he was the primary candidate of the Senate of the university on the grounds that he had more than twelve years' seniority over Kant.[115] But even after Königsberg had reverted to Prussian control, and when at the end of the Seven Years' War the affairs of higher education could be more vigorously attended to, the Ministry of Justice, which at that time had jurisdiction over higher education, had almost no opportunity to promote Kant. A rescript dated August 5, 1764, addressed to the East Prussian administration in Königsberg, expressly notes that a "certain instructor, Immanuel Kant by name," had come to their attention "by reason of several of his writings, which display very thorough scholarship," but the only post which could be offered to him at that time was a professorship of poetry in Königsberg. Though Kant turned down this position, he at least had assurance that as soon as another occasion arose he "would be placed," and a memorandum

---

records of the Kneiphöfische Domschule that Kahnert was a teacher there from 1757. No positive evidence that Kant applied for the position has been discovered in the records.

114. Kant's letter of application to King Friedrich II, April 8, 1756 (see IX, 2) (*Ak.* X, 3).

115. Kant's letters to the Rector and Senate, the philosophical faculty at Königsberg, and to the Czarina Elisabeth of Russia were a formality explicitly required by the Russian administration and had been enjoined by a specific ordinance. More information is given by Arthur Warda in the *Altpreussische Monatsschrift* 36:498.

expressly issued to the Senate of the University of Königsberg decreed "that the well-known and universally acclaimed teacher, Magister Kant, is to be promoted at the first opportunity."[116] But six more years elapsed before this chance presented itself.

Meanwhile, Kant had to be content with the fact that upon his application, the position of sublibrarian in the royal castle library was given to him, with an annual salary of sixty-two dollars, an amount which, as he said in his application, as modest as it was "would serve as some assistance to his highly uncertain academic subsistence."[117] Thanks to the incompetence of his superior, the senior librarian Bock, all the work to be done in the library fell almost entirely on his shoulders, but he discharged the duties of the office for some years with the care and accuracy he showed in all things, large and small. It was only in April, 1772, two years after he had become a professor *ordinarius*, that he resigned from his position as sublibrarian, since this division of his time was not compatible with his new academic obligations.[118] Further, Kant's concern over the financial security of his old age during his last years as an instructor is most clearly shown by the fact that when, in 1769, the prospect of a call to Erlangen presented itself, he did not want to reject this "chance for a small but certain prosperity" out of hand. But he took fright when the university, wishing some sort of declaration, decreed his immediate nomination and invited him, through the professor of mathematics and physics, Simon Gabriel Suckow, to take up his duties shortly.

Only now did he feel the whole force of what the change in his surroundings and his accustomed pattern of life would have meant to him. "Renewed and very strong assurances," he wrote to Suckow, "the growing likelihood of a perhaps imminent removal from this place, attachment to my natal city and to a rather extensive circle of

116. On the plan to give Kant the professorship of poetry and the related rescripts and decrees, see Schubert's biography of Kant, pp. 49 ff.

117. For the application to King Friedrich II and the Minister Freiherr von Fürst, dated October 24 and 29, 1765, respectively, see IX, 40, 41 (*Ak.* X, 46, 47). Cf. also Arthur Warda, *Altpreussische Monatsschrift* 35:477 ff.

118. To King Friedrich II, April 14, 1772 (IX, 109) (*Ak.* X, 130). For more information on Kant's position and activities as sublibrarian, see Karl Vorländer, *Kants Leben* (Leipzig, 1911), pp. 79 ff.

friends and acquaintances, but most of all my enfeebled physical constitution are suddenly so powerfully opposed in my mind and heart to this plan, so that I look forward to peace of mind only there where I have so far always found it, though in burdensome circumstances.... I am very much afraid, dear sir... that I have brought your displeasure upon myself through a vain expectation to which I gave rise. But you, my dear sir, know the weaknesses of the human character too well not to forbearingly class a mind which is unreceptive to changes that seem trifling to others with those impediments over which one is as little master as he is of his fate, although their consequences are often detrimental."[119]

This way of thinking was confirmed still more in Kant during the succeeding years, when after becoming professor of logic and metaphysics he was no longer oppressed by financial worries. When Minister of Culture and Education von Zedlitz, who not only valued him as an academic instructor but revered him as a philosopher, attempted to convince him to accept the professorship in Halle, and when, at Kant's refusal, he not only reckoned up for him "the mathematically exact improvement" but also reminded him that for a man such as he it was his duty not to forgo the greater sphere of influence open to him, Kant nonetheless stood fast in his resolution. "I wish that persons with your knowledge and gifts were not so rare in your profession," Zedlitz wrote on that occasion; "I would not trouble you so. But I would like you to be not unmindful of your duty to be as useful as you could be, in the opportunity which is offered to you, and to consider that the 1,000 to 1,200 students at Halle have a right to further their education, which I do not want to be responsible for neglecting, through you."[120] Halle, where Wolff had worked for fourteen years after Frederick the Great called him back, enjoyed the reputation of being the first-ranking university in Germany in philosophy, and in the other faculties as well Zedlitz, who labored diligently to improve the university, could hold up to Kant some great names. Voltaire had said that to see the crown of German

119. To Suckow, December 15, 1769 (IX, 66) (*Ak.* X, 78).
120. Zedlitz to Kant, March 28, 1778 (IX, 171) (*Ak.* X, 212).

scholarship, one must go to Halle. Still, Kant withstood not only the seduction of vanity—Zedlitz had offered him the title of *Hofrat* (Councillor), in case "minor circumstances, from which even the philosopher cannot stand aloof," might be able to make this title attractive to him—but also what undoubtedly meant more to him, all the representations Zedlitz had made on the basis of his duties toward the world and toward young people in school. "Monetary gain and the excitement of a grand stage are, as you know, not much of an incentive for me," he wrote to Marcus Herz at this time. "A peaceful situation, nicely fitted to my needs, occupied in turn with work, speculation, and my circle of friends, where my mind, which is easily touched but otherwise free of cares, and my body, which is cranky but never ill, are kept busy in a leisurely way without strain, is all that I have wished for and had. Any change makes me apprehensive, even if it gives the greatest promise of improving my condition, and I am persuaded by this natural instinct of mine that I must take heed if I wish the threads which the Fates spin so thin and weak in my case to be spun to any length. My great thanks, then, to my well-wishers and friends, who think so kindly of me as to undertake my welfare, but at the same time a most humble request to protect me in my present situation from any disturbance."[121]

This decision has often been deplored; fun has also been poked at the philosopher's excessively tender sensibility and his anxiety regarding every question touching his external circumstances of life; but in both cases the judgment rests more on abstract and general grounds than on a weighing of the concrete life situation from which Kant came to his decision. At that time he was squarely facing the completion of his work, which in both the intellectual and the literary respect imposed on him a labor greater than which perhaps no other thinker had ever had to accomplish. From the moment Kant conceived this work, his life no longer had any independent and separate meaning; it was but the underpinning for that intellectual task which it was vital to accomplish. All the powers of his person were solely and simply applied to the process of thinking and put at its disposal.

121. To Marcus Herz, April 1778 (IX, 174) (*Ak.* X, 214).

During this time he continuously bemoaned his frail, "incessantly fitful" health, but his body withstood the intense strain, unprecedented even for Kant, thanks to a careful, scrupulously calculated diet. It is understandable how Kant, at this period, felt any change as nothing but perilous and upsetting, however much it might outwardly appear to be an improvement in his situation. Kant's letter to Marcus Herz recalls in many details, especially in its whole tenor, the correspondence Descartes carried on with Chanut, the French ambassador in Stockholm, when the latter invited him to the court of Christina of Sweden. Descartes, too, strongly resisted this invitation, which required him to abandon the pattern of life he had methodically chosen and so far adhered to strictly systematically—a resistance that in the end he gave up less through conviction than for external reasons. Kant, on the contrary, remained true to his inner law without hesitation, and we may believe that the "natural instinct" which he appeals to was the *daimon* of the great man who orders the external course of his life clearly and positively in accordance with the pure and essential demands of his work.

In his correspondence with Marcus Herz in the decade from 1770 to 1780, we have testimony of incomparable value as to how this work took shape through the steady progression of his thinking, despite all the inner difficulties and hindrances, testimony which must speak for itself, since other accounts of this period are almost completely lacking. For if one tries to infer from the accounts that we have of Kant's lectures on metaphysics a picture of his whole philosophical view at this time, the procedure is questionable in more than one respect. Aside from the fact that the dating of these reports cannot be established with sufficient certainty, so many extraneous things have gotten into them—partly the fault of the writer, partly from the textbook on which Kant based his lectures, as was customary—that their worth as sources for Kantian philosophy is problematical in the extreme. By contrast, the letters to Herz not only reflect the objective progress of Kant's thinking, but are also an accurate mirror of the shifting personal and intellectual moods that accompanied it. Marcus Herz participated as respondent in the public defense of Kant's book *On the Form and Principles of the Sensible and the Intelligible World* and under

Kant's personal tutelage was introduced to all the details of that work. Kant could expect from him, if from anyone, an understanding of the further intellectual development connected with that work. The exchanges of letters on this topic are quite spasmodic and seem to have ceased entirely for a while, but Kant, who was rendering in them an account to himself of how his thinking was progressing, seems always to have felt the need to begin them anew.

Even the personal relation between teacher and pupil took on an increasingly intimate and cordial form in this exchange of letters. "Chosen and inestimable friend," "Most worthy and prized friend," is the way Kant, who was always chary with the title "friend," saluted Herz in his letters. Feeling this way, he allowed Herz a deeper look into the workshop of his mind than anyone before. Even the first letter, of June, 1771, not only sketches out the new results he had achieved meanwhile, but also at the same time throws a clear light on the personal method of thinking he uses from now on. "You know that I examine reasonable criticisms," Kant writes to Herz when he is apologizing for the delay in his reply to the objections of Lambert and Mendelssohn to the Dissertation, "not merely as to how they might be refuted, but also upon reflection I always weave them into my judgments and allow them to overthrow all preconceived opinions that I have previously cherished. In this way I always hope to look at my judgments impartially, from the standpoint of someone else, so as to derive a third view which is better than the one I had. Moreover, simple absence of persuasion among men of such intelligence always proves to me that my theories must lack clarity, evidence, or even something essential. Long experience has taught me that insight into the matters we have in view cannot be forced and sped up by straining, but it takes a rather long while, since one must, with respites, look at divers concepts in as many different relationships and as broad a context as possible: above all, in doing this the sceptical mind should rouse itself and test whether what has been thought is proof against the keenest doubt. From this point of view, I think I have made good use of the time which I took at the risk of incurring a charge of discourtesy, but in fact out of respect for the judgment of both scholars. You know what a great influence sure and certain

knowledge of the distinction between what rests on subjective principles of the human soul, not only of sensibility but also of understanding, and that which concerns objects directly has on the whole field of
philosophy, indeed on the most important aims of mankind in general. If one is not bewitched by the passion for systems, inquiries into
this selfsame fundamental law in its widest application are mutually
confirmatory. I am therefore presently occupied with a work which,
under the title 'The Boundaries of Sensibility and Reason,' is to cover
basic, definite concepts and laws regarding the sense world as related
to a sketch of what constitutes the nature of the theory of taste,
metaphysics, and morals, something very detailed to carry out. During the winter I went through all the pertinent materials, sifted them
all, weighed them, fitted them together, but I finished the outline
only very recently."[122]

What was the new factor distinguishing this project from the
sketch given in the Inaugural Dissertation? It appears indubitable
from Kant's further remarks in the same letter to Herz that the Dissertation was also to form the basis for the forthcoming work Kant was
now contemplating, although he had already recognized its defects in
particular details. We must assume on his part an attitude both positive and negative, an insight that affirms the fundamental procedure
of *On the Form and Principles of the Sensible and the Intelligible World* and
that nonetheless denies the result with which it had concluded. We
obtain a clear indication of what this insight consisted in if we keep
clearly in mind those objections of Lambert and Mendelssohn that
were the starting point for Kant's further reflections and that served
to arouse his "sceptical mind." The objections of both men agree in
opposing the way in which the doctrine of the "ideality of space and
time" was expressed in the Dissertation. This doctrine, simply in
itself, contained nothing surprising or paradoxical for either of them,
for it was an established proposition of Leibnizian metaphysics that
space and time were only the orderings of "phenomena," a proposition that repeatedly was given the most diverse new twists in
eighteenth-century philosophical literature. Lambert and Men-

122. To Marcus Herz, June 7, 1771 (IX, 96) (*Ak.* X, 116).

delssohn took exception only to the fact that in the Dissertation this ideality of both space and time seemed to be interpreted once again into a mere subjectivity. "Time," Mendelssohn wrote, "is according to Leibniz a phenomenon, and has, like all appearances, something objective and something subjective." And Lambert, also, emphasized that he had not so far been able to convince himself of the assertion that time was "nothing real," for if change is real (as even an idealist must admit, since he is immediately aware of it in the alteration of his inner representations), time must be also, since all change is connected with time and not "thinkable" apart from it.[123]

Both objections failed to touch the essential, deeper sense of Kant's doctrine; they interchange "transcendental" idealism with "psychological" idealism, to put it in the language of the system to come. This is easy for us to see today, and Kant himself pointed it out in a well-known place in the *Critique of Pure Reason*.[124] But was not this misunderstanding excusable? Was it not almost inevitable, given the form in which the theory of space and time was presented? Must not the subjectivity of the forms of intuition, even though they were the basis of the certainty of mathematics and natural science, have nevertheless appeared to be a blot separating them from the pure concepts of the understanding, to their disadvantage? For it was the business of these latter to enable us to recognize things not only as they appear to us but also as they are in and for themselves. Though it might be insisted over and over that although space and time are not objects in the absolute sense, their concept is nonetheless "supremely true,"[125] still this truth always remained a second-order truth as long

123. See Mendelssohn's letter to Kant dated December 25, 1770 (IX, 90 ff.) (*Ak.* X, 108 ff.); Lambert's letter to him of October 13, 1770 (IX, 80 ff.) (*Ak.* X, 98 ff.).

124. See the Transcendental Aesthetic, §7, A 36–41 = B 53–58 (III, 67 ff.).

125. Cf. the Inaugural Dissertation, §14, no. 6 (II, 418) (*Ak.* II, 401): "Quanquam autem Tempus in se et absolute positum sit ens imaginarium, tamen, quatenus ad immutabilem legem sensibilium, qua talium pertinet, est *conceptus verissimus* et per omnia possibilia sensum objecta in infinitum patens intuitivae repraesentionis conditio." ["But though time, posited in itself and absolutely, is an imaginary being, yet so far as it is related to an immutable law of sensibles as such, it is a quite genuine concept, and a condition of intuitive representation, extending *in infinitum* through all possible objects of the senses."] See also the analogous statement concerning space, Dissertation, §15 E (II, 420) (*Ak.* II, 404).

as there were other concepts which could claim to relate "directly to things," not simply to appearances and their relations.

Kant's letter to Herz shows us how his reflections as they progress implant themselves at precisely this extremely difficult point. He adheres firmly to the separation of sensible from intellectual concepts as irreversibly certain, but at the same time he now extends the distinction between what rests on subjective principles and what pertains immediately to objects to the sphere thus far unaffected by critique. He now begins to see subjectivity, "not only of sensibility but of understanding as well," ever more definitely and distinctly, but instead of his being thereby enmeshed in a universal theory of doubt, the reverse is the case: the concepts of the understanding take on the same stamp of truth as the forms of pure intuition. Also, it is now the case that they are not true because they depict for us absolute objects, but because they are unavoidable conditions within the system of cognition, in the construction of experiential actuality, and hence are universally and necessarily valid. That this is so had already been recognized and said in the Dissertation, but only a relatively minor significance was granted to this purely logical use of the concepts of the understanding in contrast to the "real" use, which is directed toward knowledge of supersensible objects.[126]

Now, however, the center of gravity of the problem begins to shift: the division between objects, the dualism of the sensible and the intelligible world, is displaced by a division between cognitive functions which are the basis of any sort of objectivity or which claim objectivity for themselves. The boundary is no longer drawn between the *mundus intelligibilis* and the *mundus sensibilis*, but between sensibility and reason. And the latter is here taken in its broadest, most comprehensive sense. Just as we can ask what the essential form of objectivity belonging to space and time is, and just as we discover this form when we clarify the structure and the mode of knowledge of pure mathematics and pure mechanics, we can and must also inquire into the principle on which the necessity of pure knowledge of the understanding or the justness and validity of our basic ethical or

126. For the contrast between *usus logicus* and *usus realis* of the concepts of the understanding, see the Dissertation, §5 (II, 409 f.) (*Ak.* II, 393 f.).

aesthetic judgments rests. The first outline of a work that is to answer all these questions, that is to fix and define relative to one another the differing claims to validity within theoretical cognition, as well as in the realm of ethics and aesthetics, now stands before Kant's eyes; all that seems required to bring it to completion is the detailed execution of a design trenchantly conceived in all its fundamental features.

But after we have gotten to this point, the crucial question looms once more. Assuming that we had specified the boundaries between sensibility and understanding, and in addition the boundaries between theoretical, ethical, and aesthetic judgment, would we actually have then reached a "system" of reason, or perhaps nothing more than an "aggregate"? Is it enough to put this multiplicity and diversity simply on the same level and to handle it that way, or must we not look for a common point of view underlying all these diverse queries? Every dividing line we draw presupposes in the very division it creates an original unity of that which is divided; every analysis presupposes a synthesis. What does this connecting link consist in, if we are now going to search for it, according to the result we have now obtained, in the structure and lawfulness of "pure reason" and never in the world of things?

Kant's letter to Marcus Herz dated February 21, 1772, gives the answer to all these questions, an answer that at one stroke clarifies all the developments which precede and follow, and illuminates them from within, as it were. It has not unjustly been said of this letter that it marks the true hour of birth of the *Critique of Pure Reason*. "You do me no injustice if you become indignant at the total absence of my replies," Kant begins, and we must let his letter speak for itself in full, if all the delicate nuances of the course of his thinking are to be grasped, "but lest you draw any disagreeable conclusions from it, let me appeal to your understanding of the way I think. Instead of excuses, I shall give you a brief account of the kind of things that have occupied my thoughts and that cause me to put off letter-writing in my idle hours. After your departure from Königsberg I examined once more, in the intervals between my professional duties and my sorely needed relaxation, the project we have debated, in order to adapt it to the whole of philosophy and other knowledge and in order to understand its extent and limits. I had already previously made considerable progress in the effort to distinguish the sensible from the intellectual in the field of morals and the principles that spring therefrom. I had also long ago outlined, to my tolerable satisfaction, the principles of feeling, taste, and power of judgment, with their effects—the pleasant, the beautiful, and the good—and was then making plans for a work that might perhaps have the title, 'The Limits of Sense and Reason.' I planned to have it consist of two parts, a theoretical and a practical. The first part would have two sections, (1) general phenomenology and (2) metaphysics, but this only with regard to its nature and method. The second part likewise would have two sections, (1) the universal principles of feeling, taste, and desire and (2) the basic principles of morality. As I thought through the theoretical part, considering its whole scope and the reciprocal relations of all its parts, I noticed that I still lacked something essential, something that in my long metaphysical studies I, as well as others, had failed to pay attention to and that, in fact, constitutes the key to the whole secret of hitherto still obscure metaphysics. I asked myself: What is the ground of the relation of that in us which we call 'representation' to the object?"[127]

This relation, the exposition continues, is easily seen in two cases: when the object produces the representation, and conversely when the latter produces the former. We then understand where the conformity between the two arises, since we believe we see that every effect is proportional to its cause and must "copy" it in the precise sense of that term. Thus the problem seems solved, when we look at it from the standpoint of sensory perception as well as when we adopt the viewpoint of an understanding that itself produces the object that it apprehends. For in the first case, that of pure passivity, there arise no difference and tension, so to speak, between what is given externally and what is caused in us; the object impresses its whole state on us and leaves a sensory imprint which tells us about it. In the second case, however, that of the "divine understanding," the agreement between knowledge and object is again easy to see, for

127. [IX, 102 f. (*Ak.* X, 123).]

here one and the same original identity of the divine essence is exhibited and explained in knowing and forming, in contemplating and in creating. Accordingly, the possibility of a pure creative understanding, an *intellectus archetypus*, is at least in general comprehensible, as well as the possibility of a purely receptive understanding, an *intellectus ectypus*. But our understanding falls under neither the one nor the other of these categories, since it neither of itself generates objects connected with its cognition nor does it simply accept its effects as they are immediately presented in sensory impressions. The Dissertation had already exhaustively ruled out the second alternative. "The pure concepts of the understanding," Kant now goes on, "must not be abstracted from sense perceptions, nor must they express the reception of representations through the senses; but though they must have their origin in the nature of the soul, they are neither caused by the object nor bring the object itself into being. In my dissertation I was content to explain the nature of intellectual representations in a merely negative way, namely, to state that they were not modifications of the soul brought about by the object. However, I silently passed over the further question of how a representation that refers to an object without being in any way affected by it can be possible. I had said: The sensuous representations present them as they are. But by what means are these things given to us, if not by the way in which they affect us? And if such intellectual representations depend on our inner activity, whence comes the agreement that they are supposed to have with objects... ?"[128]

In mathematics this may of course happen, since here the object in fact arises in an intuitive and conceptual context. The Prize Essay of 1763 had already shown what a circle or a cone "is"; in those cases I need only inquire as to the act of construction by which this figure is produced. But what counsels us if we want to grant a similar "construction" for "metaphysical" concepts as well, and if we wish to construct them in *this* sense "independently of experience"? Concepts of magnitudes may be spontaneous, because the magnitudes as wholes are built up for us in the synthesis of the manifold "by taking

128. [IX, 103 f. (*Ak.* X, 123).]

numerical units a given number of times," and accordingly the principles of the pure theory of magnitude may hold a priori and with
unconditional necessity. "But in the case of relationships involving
qualities—as to how my understanding may form for itself concepts
of things completely a priori, with which concepts the things must
necessarily agree, and as to how my understanding may formulate
real principles concerning the possibility of such concepts, with which
principles experience must be in exact agreement and which nevertheless are independent of experience—this question, of how the faculty of the understanding achieves this conformity with the things
themselves, is still left in a state of obscurity."[129] The whole of previous metaphysics leaves us in the lurch regarding this question. For
what good is it if one thinks one has solved the riddle by pushing it
back into the ultimate origin of things, into that mysterious unity
where "being" and "thought" have not yet been separated? What
advantage is it if Plato makes a prenatal intellectual intuition of what
is divine the origin of the pure concepts of the understanding, if
Malebranche postulates a continuing, present connection between
the human and the divine mind that is verified and revealed in every
cognition of a pure rational principle, if Leibniz or Crusius bases the
agreement between the order of things and the order of the laws of
the understanding in a "preestablished harmony"? In all these seeming "explanations," rather, is not something absolutely unknown
used to explain something relatively unknown, something inconceivable and unintelligible in our concepts used as the explanation of
something merely problematical? "But the *deus ex machina*," Kant
protests against all attempts of this kind, "is the greatest absurdity
one could hit upon in the determination of the origin and validity of
our knowledge. It has—besides its deceptive circle in the conclusion
concerning our cognitions—also this additional disadvantage: it en-
courages all sorts of wild notions and every pious and speculative
brainstorm."[130] The fundamental question raised by knowledge, the
question of what ensures its objective validity, its relation to the ob-

129. [IX, 104 (*Ak.* X, 123).]
130. [IX, 105 (*Ak.* X, 123).]

ject, must be answered in the clear light of reason and with the recognition of its essential conditions and limits.

The door to the *Critique of Pure Reason* was open, since this form of the question was now immutable. Kant himself says, later on in the letter to Herz, that he has projected an entire system of "transcendental philosophy," since he has reduced "all concepts of completely pure reason" to a certain number of categories—not like Aristotle who assembled his categories merely at random, but rather as they are divided into classes by a few basic laws of the understanding itself. "Without going into details here," he continues, "about the whole series of investigations that has continued right down to this last goal, I can say that, so far as my essential purpose is concerned, I have succeeded and that now I am in a position to bring out a 'Critique of Pure Reason' that will deal with the nature of theoretical as well as practical knowledge—insofar as the latter is purely intellectual. Of this, I will first work out the first part, which will deal with the sources of metaphysics, its methods and limits. After that I will work out the pure principles of morality. With respect to the first part, I should be in a position to publish it within three months."[131]

Strange as it may seem at first glance, Kant's illusion in believing himself able to finish in three months a book which was to occupy him exclusively for eight or nine years more is nonetheless understandable: having conceived this new task so positively and clearly, he might hope to have in that fact alone all the essential conditions of the solution. For all the fundamental insights from which the *Critique of Pure Reason* was wrought are actually achieved. What Kant later called his "revolution in thinking," the "Copernican" turn to the problem of knowledge,[132] is here complete. Reflection no longer begins with objects as things known and given, in order to show how an object migrates into our cognitive faculty and is pictured in it,[133] but it inquires about the meaning and stuff of the very concept of an object, about what the claim to objectivity universally means, whether in

131. [IX, 104 (*Ak.* X, 123).]
132. See the *Critique of Pure Reason*, preface to the 2d ed., B x ff. (III, 15 ff.).
133. Cf. *Prolegomena to Any Future Metaphysics*, §9 (IV, 31) (*Ak.* II, 282).

mathematics, natural science, metaphysics, or in morals and aesthetics. In this question is found the link that directly unifies all concepts and problems of "pure reason" into a system. Whereas all previous metaphysics had begun with the "what" of the object, Kant begins with the "how" of judgment about objects. While the earlier metaphysics knew how to give an account of the general quality of things, Kant examines and analyzes simply the assertion of knowledge of objects, to establish what is posited and meant by it, by the relation it expresses.

In this transformation of the question, "metaphysics" became "transcendental philosophy" in the strict sense in which the *Critique of Pure Reason* later defined the new term: "I call all knowledge transcendental which is in general concerned not with objects, but with our mode of knowledge of objects, insofar as this is to be possible a priori."[134] A whole not of things but of "modes of knowledge," to which the essential features of our moral, teleological, and aesthetic faculties of judgment belong, faces us and demands unification and division, association in a common task and recognition of their specific work. And similarly, the idea, if not the expression, of the other great question of the First Critique is arrived at: "How are synthetic judgments a priori possible?" For this is precisely the problem posed in Kant's letter to Herz: by what right can we speak of a priori knowledge which goes beyond all that is given in the passive elements of perceptions and sensibility, just as it goes beyond any sheer conceptual analysis; knowledge which as a declaration concerning "real" connection and real opposition is necessarily related to experience, but which on the other hand, because it wants to be valid for "all experience in general," is grounded in no *particular* experience? It is the universally valid and necessary—what is found not only in knowledge of quantities but also in that of qualities, that comes to light not only in the unfolding of the relations of coexistence in space or succession in time but also in "dynamic unity," in assertions about things and properties, causes and effects—which has become the

---

134. *Critique of Pure Reason*, introduction, VII, A 11–12 = B 25 (III, 49).

problem, a problem which can be unlocked only with that same new view of the "concept of the object," in which in general "the key to the whole secret of hitherto still obscure metaphysics" is to be sought.

The closer Kant comes to mastery of the details, though, the more clearly the whole complexity of the task he has undertaken confronts him. Behind every solution new questions arise; behind every categorization of the concepts of reason into fixed classes and faculties arise further subdivisions, each of which leads to a fresh and subtle inquiry. The plan of his labor had already become known, and Herz in particular, with understandable impatience, presses him to finish the work. But Kant does not allow himself to be diverted from the clear requirements of the subject and from his progress by any expectation which he himself cherishes or which he had aroused in others. "Since I have come this far in my projected reworking of a science that has been so long cultivated in vain by half the philosophical world," he writes in his next letter to Herz, separated from the earlier one by almost two years, "since I see myself in possession of a principle that will completely solve what has hitherto been a riddle and that will bring the procedure of reason isolating itself under certain and easily applied rules, I therefore remain obstinate in my resolve not to let myself be seduced by any author's itch into seeking fame in easier, more popular fields, until I shall have freed my thorny and hard ground for general cultivation."[135] Still, Kant hopes to have the book ready to deliver "at Easter" of 1774, or to be able to promise it "almost certainly" shortly after Easter; but at the same time he emphasizes how much time and effort the "planning and complete execution of a whole new conceptual science" has cost him in the matters of method, divisions, and terminology that is exactly fitting. He intends to complete the transcendental philosophy first, then he wants to move on to metaphysics, which he will carry out in two parts: the "metaphysics of Nature" and the "metaphysics of morals." He adds that he contemplates publishing the latter first, and that he is already anticipating this with pleasure.

135. [IX, 114 (Ak. X, 136).]

It is of particular interest as regards his system that here ethical questions are treated on the same presuppositions and on the same plan as the questions of pure theoretical knowledge. The period in which Kant seemed to assimilate himself to the psychological method of ethics as practiced by the English, and in which he prized the procedure of a Shaftesbury, Hutcheson, and Hume as a "beautiful discovery of our age,"[136] now lies far behind him. The Inaugural Dissertation had ranged the problem of morality on the side of the "intelligible," and divested it of all sensuous determination on grounds of pleasure and pain, in express opposition to Shaftesbury.[137] Kant saw in his transformation of the foundations of ethics, as he wrote to Lambert when sending him the Dissertation, one of the most important goals of the now altered form of metaphysics.[138] Ethics, like the doctrine of space and time and like that of the pure concepts of reason, has become an a priori discipline; the characteristic objectivity of the ought on the one side is distinguished from the objectivity of being as on the other side it illuminates and is reciprocally illuminated by it.

This correspondence between Kant and Herz will not be gone into in further detail, however, since in it the same overall picture is repeated constantly. To an outside observer it might have seemed at times as if the plan Kant was contemplating were but a will-o'-the-wisp luring him blindly into unknown reaches of thought. Time and again he believes he is at the end, but the further he goes, the longer is the path he has yet to traverse. After he thinks, toward the end of 1773, that he can promise the termination of his work "almost certainly" at a time shortly after Easter of 1774, three more years pass in which, under the continuous influx of ever-new questions, he obviously has not even begun the systematic composition and writing. The expectations and queries directed at him from the literary and scholarly circles of Germany grow increasingly impatient and pressing. "Say something to me even in a couple of lines," Lavater writes

136. See the announcement of his course of lectures in 1765–66 (II, 326) (*Ak.* II, 312).
137. *De mundi sensibilis,* §9 (II, 412) (*Ak.* II, 396).
138. To Lambert, September 2, 1770 (IX, 73) (*Ak.* X, 92).

him in February, 1774; "are you dead? why do so many write who cannot—and you, who can write so exquisitely, write nothing? why are you silent—in this, this *modern* age—make not a sound? Asleep? Kant—no, I will not praise you—but tell me why you are silent? or rather: tell me that you will speak."[139] When Lavater wrote these words, he did not in the least suspect that it was precisely the advent of the "modern age" that this silence foretold.

"I am rebuked from all sides," Kant writes to Herz on November 24, 1776, "on account of the inactive state in which I seem to have been for so long, and yet I have never been more systematically and engrossingly busy than during the years you have not seen me. Topics which I might hope to earn fleeting acclaim by treating, pile up under my hands, as is usual when one has been seized by a fruitful principle. But they are all held in check by one main thing, as by a dam, a thing by which I hope to earn an enduring gain, which I actually think I now possess and which is less necessary to think through than to carry through. . . . It is the part of persistence, if I may say so, to follow a plan undistracted, and I am often tempted by difficulties to devote myself to different, pleasanter matters, an infidelity from which I have from time to time been restrained by overcoming some obstacles, partly by the importance of the business itself. You know that it must be possible to survey the field of judgment independent of all empirical principles, that is, of pure reason, because it lies in us a priori and needs await no revelations from experience. To specify the whole scope of it, the divisions, the boundaries, the entire substance of it by sure principles and to erect boundary stones so that in the future one can know confidently whether he finds himself on the terrain of reason or of sophistry, it takes: a critique, a discipline, a canon, and an architectonic of pure reason, hence a formal science, which can use nothing of what lies ready to hand and which requires for its foundation quite special technical expression."

Not only the systematic, but also the technical, outline of the First Critique is now clear to Kant's eyes, and above all the distinction

---

139. Lavater to Kant, February 8, 1774 (IX, 117) (*Ak.* X, 141).

between "analytic" and "dialectic," between the realm of reason and that of sophistry, has been vouchsafed him. But he still could not estimate the task of composition in its entirety, for there follows again the assurance, which is already rather dubious, that he hopes to be finished, to be sure not before Easter, but probably in the following summer. Nevertheless, he begs Herz not to have any expectations "which are wont occasionally to be troublesome and detrimental."[140] Three-quarters of a year later, in August, 1777, Kant informs him that the *Critique of Pure Reason* is still a "stumbling block" to all the other plans and labors he has in mind, nevertheless he is busy clearing it away; he now believes that he will have it done "this winter." What is holding him up now is nothing more than the labor of making his thoughts as clear as possible for others, because experience shows that what one is thoroughly familiar with, and which is clarified to the highest degree for oneself, is customarily misunderstood even by experts if it lies wholly off the beaten path.[141]

In April, 1778, however, he must once again counter the rumor that several pages from his "work in hand" are ready to be printed. But if one were to conclude from this expression that at least the first outline of the book and the literary form it was to take were firm for Kant, the following sentences, which speak explicitly of a writing "not very many pages of which have been ushered into the world," inform us otherwise.[142] In August of the same year we hear of the work as a "Handbook of Philosophy," which he is still working on tirelessly; and again a year later its completion is projected for Christmas of 1779.[143] The composition must in any case have begun by then, for in May, 1779, Hamann told Herder that Kant was working briskly away on his "Ethics of Pure Reason"; in June, 1780, it was further said that he prides himself on the delay, because that very thing will contribute to the perfection of his project.[144] The actual

140. To Herz, November 24, 1776 (IX, 151) (*Ak.* X, 184).
141. To Herz, August 20, 1777 (IX, 158) (*Ak.* X, 195).
142. To Herz, April 1778 (IX, 174) (*Ak.* X, 214).
143. To Engel, July 4, 1779 (IX, 191) (*Ak.* X, 238).
144. Hamann to Herder, May 17, 1779, and June 26, 1780, *Schriften* (ed. Roth), vol. 6, pp. 83, 146.

136 BEGINNINGS OF KANTIAN PHILOSOPHY

writing, aside from preparatory sketches and drafts, can have consumed only a very short time; Kant confirms this in telling Garve and
Mendelssohn that he accomplished his exposition of the subjects
which he had carefully pondered more than twelve years running "in
some four or five months, on the wing, as it were." After a decade of
the deepest meditation, after repeated postponement, the completion
of the book is achieved only by a sudden resolution that energetically
interrupts the spinning-out of his thoughts. Only the fear that death
or the enfeeblement of old age might surprise him while still at work
gave Kant the strength at last to put an outward conclusion to his
thinking, one which he himself felt to be only preliminary and inadequate.[145] But in this as well the *Critique of Pure Reason* is a classic
book, for the works of the great thinkers, unlike great works of literary art, appear in their truest form when the seal of perfection is not
set on them, but when they still reflect the incessant movement and
the inner restlessness of thought itself.

In the particular preparatory studies for the First Critique we still
possess, this process comes to light with maximum clarity and vividness. The papers Rudolf Reicke has published under the title *Loose
Papers from Kant's Literary Remains (Lose Blätter aus Kants Nachlass)* as
well as the *Reflections (Reflexionen)* edited by Bruno Erdmann, contain
notes that unmistakably belong to this stage of his preparations; one
of the loose papers published by Reicke can be dated with fair precision, since Kant did his jotting in the empty space of a letter sheet
bearing the date May 20, 1775. If one starts with this sheet and groups
with it the other notes that are of a piece with its contents, the composite thus obtained sheds light from quite diverse angles on the
point that Kant's thinking had reached at this period.[146] We cannot go
further into the substance of these notes here; it is comprehensible
only if we presuppose the way in which the problem is put in the
Critique of Pure Reason and the basic concepts of the latter.

145. To Garve, August 7, 1783; to Mendelssohn, August 16, 1783 (IX, 223, 230) (*Ak.*
X, 315, 322).
146. Further on this in Theodor Haering (who has edited and commented on the
loose papers concerned), *Der Duisburgsche Nachlass und Kants Kritizismus um 1775*
(Tübingen, 1910).

But almost as meaningful as the purely substantial content of these notes is the glimpse they afford us of Kant's manner of working. "Kant," Borowski tells us on this score, "first made general outlines in his head; then he worked these out in more detail; he wrote what was to be inserted here or there, or was to be explained more fully, on little scraps of paper which he then attached to that first, hastily jotted-down manuscript. After some time had elapsed, he worked the whole over again, and copied it out neatly and clearly, as he always wrote, for the printer."[147] The notes we have from 1775 still belong entirely to that first stage of preparation, in which Kant tries to get the ideas established purely for himself, without regard to the reader and the literary form of the book, and tries to vary his manner of expression in the most diverse ways. No definite, strictly maintained scheme of exposition reigns here, no attachment to a fixed arrangement or terminology. Greatly varying statements and ventures cut across and crowd each other out, without any of them achieving ultimate supremacy and a fixed and final form. Anyone who might picture Kant's thought as a steel-clad structure of definitions, scholastic distinctions, and analyses of concepts must be astonished by the freedom and flexibility found here. In particular, Kant maintains a truly sovereign neutrality toward all questions of terminology. He coins designations and distinctions according to the substantive demands of the problem in hand, only to drop them again as soon as a new turn his thought has taken demands it. Nowhere is his progress with the topic at hand hindered by a previously adopted stereotype, but the content always generates its appropriate form.

Thus there results, as though casually and by chance, a wealth of ideas, which even in comparison with the later, final expression of ideas in the *Critique of Pure Reason* have their own special and independent worth. In fact, for anyone who pursues Kant's statements with that pedantry which many seem to regard as the hallmark of genuine and "exact" Kant-Philologie, in order to show the variations and "contradictions" in the particular concepts and expressions he

147. Borowski, *Darstellung des Lebens und Charakters Immanuel Kants*, p. 191 f.

uses, these loose papers can only mean a chaos of heterogeneous instances. If, however, they are read, as they must be, as various attempts to pin down thought that is in motion and to give it a first, preliminary outline, one gains from them a picture of the peculiarity and style of Kant's thinking perhaps more vivid than from many a finished and eloquent book. Moreover, one understands what powerful internal and external difficulties had to be overcome before ideas of such a kind could take on the final form we encounter in the First Critique. So Kant was perhaps not in fact wrong when he made the difficulties of exposition ultimately responsible for the slow progress of the book.

The general outlines of the critical system were laid down as early as 1775, as nearly as we can tell from the notes dating from that time, but it seems that the printing of the *Critique of Pure Reason* began only in December, 1780, according to the allusions contained in Hamann's correspondence with Hartknoch. On the first of May, 1781, Kant is able to inform Herz in a letter of the imminent appearance of the book. "In the current Easter book fair there will appear a book of mine, entitled *Critique of Pure Reason*. . . . This book contains the result of all those varied investigations, which have their origin in the concepts we discussed under the heading 'The *mundus sensibilis* and *intelligibilis*,' and it is very important to me to deliver the summation of my efforts to the same perceptive man who deigned to cultivate my ideas, a man so discerning that he penetrated those ideas more deeply than anyone else."[148] Thus Kant retrospectively couples his book with his philosophical past. But while the man who was now fifty-seven years old may have looked on the book, born of twelve years' reflection, as the terminus of his life's work, he did himself an injustice by this judgment. For this book became, for Kant himself as well as for the history of philosophy, the beginning of a wholly new development.

148. [IX, 194 (*Ak.* X, 249).]

III THE CONSTRUCTION AND CENTRAL PROBLEMS OF THE *CRITIQUE OF PURE REASON*

1

If what is said about great thinkers is true, that the style is the man, this aspect of the *Critique of Pure Reason* poses a difficult problem for the biographer of Kant. For nowhere else in the history of literature and philosophy do we find so profound and involved an alteration of style as took place with Kant in the decade between 1770 and 1780—not even in Plato, the style of whose old age, in the *Philebus*, the *Sophist*, or the *Parmenides*, differs so markedly from the manner in which the early dialogues were written. Only with difficulty can one recognize in the author of the *Critique of Pure Reason* the man who wrote *Observations on the Feeling of the Beautiful and the Sublime* or the *Dreams of a Spirit-Seer*. Strictness in abstract analysis of concepts replaces the free play of humor and imagination; a kind of academic ponderousness supplants the reflective grace and cheerfulness of those other books. To be sure, anyone who knows how to read the *Critique of Pure Reason* rightly also finds in it, along with the acuteness and depth of thought, an extraordinary strength of intuition and an exceptional power of linguistic imagery. Goethe said that when he read a page in Kant, he always felt as though he were entering a lighted room. Alongside his skill at exhaustively analyzing the most difficult and knotty complexes of ideas stands Kant's gift for expressing and focusing the comprehensive result of a long deduction and conceptual analysis at one stroke, as it were, in striking images and

epigrammatic, unforgettable turns of phrase. On the whole, however, most readers have the overpowering impression that the expository form Kant chose fetters his thought and does not foster its
adequate and limpid expression. In his concern for the stability and
definiteness of terminology, for exactness in the definition and division of concepts, and for agreement and parallelism of schemata,
Kant's natural, lively personal and intellectual form of expression
seems struck dumb. He felt this himself, and said so. "The method of
my discourse," he remarks in a diary note, "has a prejudicial counte-
nance; it appears scholastic, hence pettifogging and arid, indeed
crabbed and a far cry from the note of genius." But it is conscious
intent that holds him back from every approximation, from every
concession to the note of "genius." He says elsewhere: "I have
adopted the scholastic method and preferred it to the free play of
mind and wit, although I indeed found, since I wanted every
thoughtful mind to share in this investigation, that the aridity of this
method would frighten off readers of the kind who seek a direct
connection with the practical. Even if I had the utmost command of
wit and literary charm, I would want to exclude them from this, for it
is very important to me not to leave the slightest suspicion that I
wanted to beguile the reader and gain his assent that way, but rather I
had to anticipate no concurrence whatsoever from him except
through the sheer force of my insight. The method actually was the
result of deliberation on my part."[1] His sole ideal now, in the face of
which all other claims retreat, is to advance strict conceptual deduction and systematization.

But Kant did not renounce those claims lightheartedly. In the
years immediately preceding the composition of the *Critique of Pure
Reason*, he was continually occupied with weighing whether and how
far it might be possible to give philosophical ideas a popular form,
without loss of profundity. "For some time," he had written to Herz
in January, 1779, "I have been thinking in certain idle moments about
the principles of popularity in the sciences (such as are capable of it,
of course, since mathematics is not), especially in philosophy, and I

1. *Reflexionen* on the *Critique of Pure Reason*, nos. 9 and 14.

think that I have been able to define from this aspect not just another alternative but an entirely different arrangement from that demanded by the scholastic method, which is still the foundation."[2] In point of fact, the early drafts of the First Critique were dominated by this outlook. Along with "discursive (logical) clarity through concepts," they strove for "intuitive (aesthetic) clarity through intuitions," and concrete examples. On this point we find in the preface to the finished book what it was that finally moved Kant to abandon this plan. "For the aids to clearness, though they may be of assistance in regard to details, often interfere with our grasp of the whole. The reader is not allowed to arrive sufficiently quickly at a conspectus of the whole; the bright coloring of the illustrative material intervenes to cover over and conceal the articulation and organization of the system, which, if we are to be able to judge of its unity and solidity, are what chiefly concern us."[3] In the place of the early attempts at an intuitive, generally comprehensible exposition, a deliberate renunciation has taken place: there is no more a royal road to transcendental philosophy, Kant sees, than there is to mathematics.

The deeper reason for this stylistic change, however, lies in the fact that Kant is presenting a completely novel *type of thinking*, one in opposition to his own past and to the philosophy of the Age of Enlightenment—to Hume and Mendelssohn, whom he envies for their way of writing, which is as elegant as it is profound. In the decades of withdrawn, lonely meditation in which Kant forged for himself his special method and questions, he had gradually moved away from the common presuppositions on which the philosophical and scientific thought of his age rested, as if by a silent consensus. He still often speaks the language of this age; he still uses the concepts it coined and the scholastic classifications it enforced in its textbooks on ontology, rational psychology, cosmology, and theology, but the whole bulk of these materials of expression and thought is now put into the service of a completely different goal. For Kant this goal is immutable, but he does not disdain terminological and expository

2. To Herz (IX, 188) (*Ak.* X, 230).
3. *Critique of Pure Reason,* preface to the 1st ed., A xix.

expedients, even though their precision is no longer strictly fitted to his own thinking. In fact, he often prefers to go back to these expedients, hoping to find in them the quickest route to a direct link with the reader's habitual conceptual realm. This very flexibility, however, becomes the source of a multitude of difficulties; precisely where Kant has descended to the standpoint of his age, he has failed to raise the age to his level.

Another factor has to be taken into consideration, one that made an entry into Kant's basic vision difficult for his contemporaries and which has continued to be a source of numerous errors and misunderstandings since then. If one considers only the external form Kant gave his writings, nothing seems to be clearer than that what unfolds before us is a finished, doctrinal system, complete as a whole and in every detail. The materials for its structure lie ready to hand in their totality; the basic outline is sketched out clearly and precisely in all particulars: all that needs be done is to put the pieces together according to the plan. But only when this endeavor is actually undertaken does the full magnitude of the task emerge. Fresh doubts and questions are encountered on every side; it is shown everywhere that the particular concepts we thought we could use as assumptions themselves need definition. The concepts become more and more altered, according to the place they occupy in the ongoing systematic composition of the whole. They are not a stable foundation for the movement of thought from its beginning to its end, but they evolve and are stabilized in the course of this very movement. Anyone who does not keep this tension in mind, anyone who believes that the meaning of a specific fundamental concept is exhausted with its initial definition and who tries to hold it to this meaning as something unchangeable, unaffected by the progress of thought, will go astray in his understanding of it.

Kant's distinctive style as a writer harmonizes with what was observed to be characteristic of him as a teacher. "His lecture," Jachmann says, "was always perfectly fitted to its material, yet it was not something memorized, but rather a freshly thought-out outpouring of his mind. . . . Even his course on metaphysics, with allowances for the difficulty of the material for the beginning thinker, was luminous and interesting. Kant displayed special artistry in expounding and defining metaphysical concepts, in that he experimented in front of his audience just as if he were starting to think about the subject, gradually adding on new defining concepts, trying out continually improved explanations, and finally passing on to a full conclusion with the concepts perfectly exhausted and illuminated from all sides; he thus not only acquainted the listener who paid strict attention with the matter under discussion but also introduced him to methodical thinking. Whoever failed to grasp this course of his lecture by observing him closely, and took his first explanation as the correct and completely exhaustive one, not making the effort to follow him further, gathered only half truths, as several reports from his auditors have convinced me."[4]

This fate of Kant's auditors has been the fate of many of his commentators as well. If one approaches the definitions of analytic and synthetic judgments, the concepts of experience and of the a priori, the concepts of the transcendental and of transcendental philosophy, as they appear in the beginning of the *Critique of Pure Reason,* with the idea that one is hitting upon ready-minted coins whose value is settled once and for all, then one must inevitably be perplexed by the further progress of the book. It repeatedly becomes obvious that an inquiry which was apparently concluded is taken up again, that an earlier explanation is supplemented, broadened, even entirely transformed, that problems which had just been treated separately abruptly enter into a totally new relationship in which their original meaning is altered. At bottom, however, the only natural and necessary situation is precisely this mutability, since it is a testimony to the fact that we stand here in the midst of a living process and a steady advance of thinking itself. Much that in isolation appears contradictory is illuminated only when it is reintegrated with this flow and interpreted in its whole context. Whereas Kant, on the strength of the synthetic method he uses in the First Critique, gradually proceeds step by step from the particular to the whole, the free reproduction of

---

4. Reinhold Bernhard Jachmann, *Immanuel Kant geschildert in Briefen an einen Freund* (Königsberg, 1804), Eighth Letter, pp. 28 ff.

his system must begin with the idea of the whole and specify the meaning of the particular relative to it, in a way analogous to the one he himself pointed out in the *Prolegomena*. As more and more threads are spun together in this process, at last the ingenious web of concepts stands before us; a retrospective analysis is the converse, disentangling only the major aspects from the numerous complexes of concepts and laying down the broadest principles that guide the idea in all its ramifications and developments. The totality of the particular questions comprised by the system of critical philosophy is never exhausted by this procedure; it must suffice if the general articulation seen by Kant himself as the essential moment and the decisive criterion for judging the unity and the solidity of his doctrine becomes visible and lucid.

2

The First Critique begins with a consideration of the idea of metaphysics and its shifting fortunes through the ages. Running through the entire history of metaphysics is the inner contradiction that, while metaphysics claims to be the highest court of appeal for the problem of "being" and "truth," it has not provided for itself any norm of certainty whatsoever. The succession of systems seems to mock any attempt to bring it into the "sure path of a science." But although metaphysics is impossible as a science, judging by its historical experiences, as a "natural disposition" it nevertheless remains necessary. For every attempt to renounce its fundamental questions is soon shown to be illusory. No voluntary resolve and no logical demonstration, however acute, can enable us to forgo the tasks that are set us. Dogmatism, which teaches us nothing, and scepticism, which promises us nothing, prove to be equally unsuccessful solutions to the problem of metaphysics. After centuries of intellectual effort, we have arrived at a point from which it seems we can move neither forward nor backward—a point at which it is as impossible to fulfill the demands implicit in the concept and the name of metaphysics as it is to give them up. "The mathematician, the sophisticated man, the natural philosopher: what do they accomplish when they make wanton mock of metaphysics? In their heart lies the call which bids them constantly to make trial in that selfsame field. If, as human beings, they do not seek their ultimate ends in the satisfaction of the goals of this life, they have no choice but to ask: From whence am I? What is the origin of all that is? The astronomer is even more compelled to these questions. He cannot avoid seeking something which will satisfy him on these matters. With the first judgment he makes on them, he is in the realm of metaphysics. Does he want simply to abandon himself, without any guidance, to the opinions which spring up in him if he has no map of the territory through which he wishes to roam? The critique of pure reason thrusts a torch into this gloom, but it illuminates the dark spaces of our own understanding, not the things unknown to us beyond the sense world."[5] Thus we see that it is not the *object* of metaphysics which is to undergo fresh considera-tion and clarification in the *Critique of Pure Reason;* it is the *question* of metaphysics itself which we are to understand more profoundly than before and gain insight into through examining its source in our understanding.

This expresses the first essential difference separating Kant's doc-trine from the systems of the past. The old metaphysics was ontology: it began with definite general convictions about being as such, and attempted to press on from that basis to knowledge of the particular determinations of things. This is fundamentally just as true of those systems which call themselves "empirical" as it is of those which profess "rationalism." Empiricism and rationalism are distinguished by their intuitions about the specific cognitive means by which we assimilate being, but the fundamental view that there is such being, that there actually is a reality of things which the mind has to take into itself and copy, is common to both. Thus, although taken separately they may be thought in contrast to each other, their unity remains: both start with a specific assertion about reality—about the nature of things or of the soul—and derive as consequences from there all further propositions. Kant's initial reflection and his initial demand have their source at this point. The proud name of an ontology, which

5. *Reflexionen*, no. 128.

claims to give a systematic doctrine of universally valid and necessary cognitions of "things in general" must be replaced by the modest title of a mere analytic of pure understanding.[6] An ontology asks what being is, in order to show how it comes to be understood, that is, how it is presented and expressed in concepts and cognitions; here, in contrast, the first thing is to establish *what the question concerning being in general means*. While ontology takes being as the starting point, here being is taken as a problem or a postulate. Whereas heretofore some sort of definite structure of the world of objects was assumed as a secure beginning, and the task consisted simply in showing how this form of objectivity passes over into the form of subjectivity, as in cognition and representation, the demand here is for an explanation as to what in general the concept of reality and the claim to objectivity assert, before any theory of this transition is propounded. For objectivity, it is now recognized, is not a primordial, fixed state that is not further analyzable but a basic question of reason. It is a question which perhaps cannot be fully answered but concerning whose *meaning* a complete and exhaustive account must be given.

This might still seem obscure, but it is clarified if we go back to the first seed of the First Critique that is found in Kant's letter to Herz in 1772. In that letter Kant had stated that the "key to the whole secret of hitherto still obscure metaphysics" is the problem of what to call a representation in us that is founded on the object. He found no enlightenment in the theories of that relation up to his own time; they led either to a mere receptivity of the mind, which did not explain its capacity for universal and necessary cognitions, or else, since they did attribute this capacity to the mind, they traced it ultimately back to some kind of deus ex machina which had implanted them originally to agree with the "nature of things."[7] This mystical solution, however, is basically as unnecessary as it is unsatisfying, once we understand that the general question as to the object of knowledge is less a question of metaphysics than of logic. For in the contrast we make between representation and object it is not a matter of two fundamen-

---

6. *Critique of Pure Reason*, A 247 = B 303 (III, 217).
7. See above, pp. 129 ff.

tally different orders of absolute being, but of a definite quality and orientation of *judgment*. We attribute objectivity to a determinate connection of the contents of consciousness, we regard it as the expression of being, when we have grounds to assume that the form of this connection is not merely contingent and arbitrary but is something necessary and universally valid. For the moment, it is not yet certain what gives us the right to this assumption; nevertheless, this assumption is not only what our whole consciousness of truth and objective validity rests on, but is also what this consciousness essentially consists of. In other words, "things" are not given to us, of which certain and necessary knowledge can then be gained, but it is the certainty of these cognitions which is expressed otherwise in the assertions of "being," a "world," and "nature."

In the letter to Herz, the statement and solution of the problem had not yet progressed to this precision. In the *Critique of Pure Reason* it was reached only in the decisive chapters on the transcendental deduction of the categories. "At this point we must make clear to ourselves what we mean by the expression 'an object of representations,'" this passage states with especially impressive pregnancy. "What, then, is to be understood when we speak of an object corresponding to, and consequently also distinct from, our knowledge? It is easily seen that this object must be thought only as something in general = x, since outside our knowledge we have nothing which we could set over against this knowledge as corresponding to it. Now we find that our thought of the relation of all knowledge to its object carries with it an element of necessity; the object is viewed as that which prevents our modes of knowledge from being haphazard or arbitrary, and which determines them a priori in some definite fashion. For insofar as they are to relate to an object, they must necessarily agree with one another, that is, must possess that unity which constitutes the concept of an object.... It is only when we have thus produced synthetic unity in the manifold of intuition that we are in a position to say that we know the object.... Thus we think a triangle as an object, in that we are conscious of the combination of three straight lines according to a rule by which such an intuition can always be represented. This *unity of rule* determines all the manifold, and limits it to conditions which make unity of apperception possible. The concept of this unity is the representation of the object = x, which I think through the predicates, above mentioned, of a triangle."[8] Thus the necessity of the judgment does not stem from the unity of an object behind and beyond the cognition, but this necessity is what constitutes for us the only conceivable sense of the thought of an object. He who understands what this necessity rests on and the constitutive conditions in which it is grounded will have penetrated and solved the problem of being insofar as it is soluble from the cognitive standpoint. For it is not because there is a world of objects that there is for us, as their impression and image, a world of cognitions and truths; rather, because there are unconditionally certain judgments—judgments whose validity is dependent neither on the individual empirical subject from which they are formed nor on the particular empirical and temporal conditions under which they are formed—there is for us an order which is designated not only as an order of impressions and representation, but also as an order of objects.

The origin of the Kantian doctrine and of the disparity between it and every earlier conception of the metaphysical problem is thus defined once and for all. In order to express this contrast, Kant himself, in the preface to the second edition of the First Critique, coined that famous image in which he compares his "intellectual revolution" to what Copernicus did. "Hitherto it has been assumed that all our knowledge must conform to objects. But all attempts to extend our knowledge of objects by establishing something in regard to them a priori, by means of concepts, have, on this assumption, ended in failure. We must therefore make trial whether we may not have more success in the tasks of metaphysics, if we suppose that objects must conform to our knowledge. This would agree better with what is desired, namely, that it should be possible to have knowledge of objects a priori, determining something in regard to them prior to their being given. We should then be proceeding precisely on the lines of Copernicus's primary hypothesis. Failing of satisfactory pro-

8. *Critique of Pure Reason*, A 103 f (III, 615 ff.).

gress in explaining the movements of the heavenly bodies on the supposition that they all revolved around the spectator, he tried whether he might not have better success if he made the spectator to revolve and the stars to remain at rest."[9] The analogy here to the "revolving of the spectator" consists in our reviewing all the cognitive functions at the disposal of reason in general, and examining each one individually, both as to its necessary mode of validity and as to its characteristically determinate and limited mode of validity.

Even in the cosmos of rational knowledge we are not allowed to stop, rigid and immobile, at any particular point, but we must successively traverse the whole range of positions that we can give to truth in relation to the object. There is a definite form of objectivity for us, which we call the *spatial order* of things: we must try to understand and determine it, not because we proceed from the existence of an absolute space but because we examine and analyze the laws of geometrical construction, those laws by which points, lines, planes, and bodies arise for us by way of continuous construction. There is for us a connection and a systematic interrelation between *numerical forms,* such that each individual number has its fixed place within the number system and its relation to all other members of this system. We must conceive this interconnection as necessary, since we can base it on no other datum than the universal procedure by which, beginning with the unit, we construct the whole domain of numbers out of its first elements, according to one constant principle. And finally, there is that whole of physical bodies and physical forces which we are accustomed to designate, in its narrower sense, as the world of nature. But here, too, in order to understand this whole we must take as our point of departure not the empirical existence of objects but the special mode of the cognitive functions, that "reason" embedded in our experience itself and in each of its judgments.

The path along which the critical revolution leads us is far from agreed upon. Metaphysics as a theory of being, as general ontology, recognizes but one mode of objectivity—only material and immaterial substances which in some form or other exist and endure. For the

9. Ibid., preface to the 2d ed., p. xvi (III, 18).

system of reason, however, there are pure immanent necessities; thus there are claims to objective validity which as such are no longer expressible in the form of existence but belong to a new and totally different species. The necessity expressed in ethical or aesthetic judgment is of this kind. Both the realm of ends, the image of which is sketched by ethics, and the realm of pure shapes and forms disclosed to us in art "exist" in some sense, for they have a fixed status, divorced from any individual arbitrariness. But this status is neither the same as the empirical, spatiotemporal existence of things, nor basically comparable in any way, since it rests on special formative principles. From this essential difference in principles it follows that for us the world of duty and the world of artistic form must be other than that of concrete existence. It can be seen that it is the diversity found in reason itself, in its basic orientations and ways of questioning, that at bottom mediates and interprets the diversity of objects. Nonetheless, a universal and exhaustive systematic knowledge of it must be achieved, because the concept of reason consists in the fact that "we should be able to give an account of all our concepts, opinions, and assertions, either upon objective or, in the case of mere illusion, upon subjective grounds."[10] The revolution in thought consists in beginning with the reflection of reason on itself, on its presuppositions and principles, its problems and tasks; reflection on objects will follow if this starting point is made secure.

At the same time, this beginning indicates the peculiarity of two important basic concepts decisively significant for the question of the critique of reason. If we hold fast to what is essential in the "Copernican revolution," we have attained the full and exhaustive meaning of the Kantian concept of "subjectivity," and of the Kantian concept of the "transcendental" as well. From this we first fully understand that both are only determined jointly and by one another, because it is precisely the new *relation* they undergo toward each other that constitutes what is essential and special obtained by way of the First Critique. Let us start with the concept of the transcendental: Kant explains that he calls that knowledge "transcendental" which is con-

10. Ibid., A 614 = B 642 (III, 423).

cerned not so much with objects as with the mode of our knowledge of objects in general, insofar as this is possible a priori. "Neither space nor any a priori geometrical determination of it is a transcendental representation; what can alone be entitled transcendental is the knowledge that these representations are not of empirical origin, and the possibility that they can yet relate a priori to objects of experience."[11] If we trace this idea further, we see that the concepts of magnitude and number, of permanence or causality, can equally little be designated as transcendental concepts in the strict sense; this designation, on the contrary, properly belongs only to that theory showing us how the possibility of all knowledge of nature rests on these concepts as necessary conditions. Even the idea of freedom, taken in itself, cannot be called transcendental. This title must be reserved for the knowledge that the special quality of the consciousness of duty, and therefore the entire structure of the realm of the ought, is founded on the datum of freedom, and for the knowledge of how it is so founded.

In this way we understand only now in what sense, from the standpoint of strictly transcendental reflection, the factor of subjectivity can and must be ascribed to space and time, magnitude and number, substantiality and causality, etc. This "subjectivity" means nothing but what Kant's "Copernican revolution" implies: it signifies the result not of the object but of a specific lawfulness of cognition, to which a determinate form of objectivity (be it theoretical, ethical, or aesthetic in kind) is to be traced back. Once this is grasped, that secondary sense of "subjective," which infects it with the appearance of individuality and arbitrariness, immediately vanishes. In the relation we are establishing here, the concept of the subjective expresses a foundation in a necessary procedure and a universal law of reason. Thus, for instance, the subjective turn Kant gives to the theory of space does not imply that the essence of space is to be determined by an analysis of the spatial representation and by exhibiting the particular psychological moments which attach to it, but rather that insight into this essence follows from insight into the nature of geometrical

11. Ibid., A 56 = B 81; also A 11 = B 25 (III, 83 and 49).

knowledge and remains dependent on it. What must space be—the transcendental exposition asks—so that knowledge of it, hence cognition, may be possible such that, like the content of geometrical axioms, it is at once universal and concrete, unconditionally certain and purely intuitive. [12] To start with the special nature of the cognitive function so as to determine from it the special nature of the object of knowledge is thus the sole subjectivity in question here. Just as the ensemble of numbers is derived from the principle of numeration, so the order of objects in space and of events in time is derived from the principles and conditions of knowledge of experience, from the categories of causality and reciprocity. Thus also, in another realm of questions, the form of the ethical imperative, on which for us all obligation rests, is rendered comprehensible by the fundamental certainty disclosed to us in the idea of freedom. It is no longer possible to confuse this subjectivity of reason with the subjectivity of arbitrariness or of psychophysical organization, since the former must be assumed and implied even to dispose of the latter.

This relation emerges, even more clearly than in the First Critique itself, in several reflections and notes, in which the reconstruction of the new meaning and relation of the central concepts can be traced out in detail. Certain of these observations seem to belong to the period preceding the final completion of the *Critique of Pure Reason*, and they seem more to indicate the stage of thought in process than thought already finalized; but even where such a temporal relation is indemonstrable, the growth of the particular concepts is exhibited more keenly and vividly in these vacillating remarks and observations than in the presentation of the finished results. "Can something be discovered through metaphysics?" runs one of these reflections. "Yes; regarding the subject, but not the object." [13] But this assertion obviously only imperfectly describes the new turn, since if one took it in isolation, one would have to anticipate a metaphysics that promises us new insights concerning the "soul" but not about things, a

12. Ibid., B 40 f. (III, 59 f.).
13. *Reflexionen*, no. 102.

metaphysics fundamentally indistinguishable from earlier dogmatic systems of spiritualism.

Therefore, the concise and pregnant statement, made elsewhere, that metaphysics deals not with objects but with cognitions[14] is an essentially sharper formulation of the basic distinction. In this formulation the "subjectivity" relating to metaphysics is first rounded out and defined more exactly: it is not that of "human nature," as understood by Locke and Hume, but that which bears the stamp of the sciences, in the method of geometrical construction or arithmetic numbering procedure, in empirical observation and measurement or in the performance of physical experiments. "In all philosophy," another note explains, "that which is genuinely philosophical is the metaphysics of science. All sciences which employ reason have their metaphysic."[15] This definitely shows the sense in which the earlier, dogmatic, objective path of the old ontology is abandoned while the concept of metaphysics is held onto and deepened in the direction of the subjective.[16] What is objective in the sciences, it can now be said in Kant's sense, is their theorems; what is subjective is their principles. We view geometry objectively, for instance, when we look at it purely from the standpoint of its theoretical content as a sum of propositions concerning spatial forms and spatial relations. We view it subjectively when, instead of inquiring about its results, we ask about the principles of its construction, its basic axioms, which are valid not for this or that spatial figure but for every spatial construct as

14. *Reflexionen*, no. 91. That this *Reflexion* dates from the period which Benno Erdmann calls that of "critical empiricism," thus from the 1760s, is extremely improbable. The passage in the Prize Essay of 1763, which Erdmann invokes in support of his view, is no proof at all in this regard, since in that passage metaphysics (in the sense of πρώτη φιλοσοφία traditional since Aristotle) is described as a philosophy of the first grounds of our knowledge, but Kant could no more say here that it does not concern itself "with objects" than he could say this at any time prior to the decisive turn in his letter to Marcus Herz of 1772.

15. *Reflexionen*, no. 129.

16. Cf. *Reflexionen*, no. 215; "Progress in metaphysics has been fruitless up till now; no one has discovered anything by it. Equally, it cannot be given up. Subjective instead of objective?"

such. It is this thrust of the question which is henceforth unwaveringly adhered to. "Metaphysics is the science of the principles of all knowledge a priori and of all knowledge which follows from these principles. Mathematics contains such principles, but is not a science of the possibility of these principles."[17]

Herein, however, lies a new factor, peculiar to the Kantian definition of concepts. Even the transcendental philosophy intends to treat the various forms of objectivity, and must do so; but each objective form is conceivable by it and accessible to it only as mediated by a specific form of cognition. The material it is concerned with and relates to is hence always a material preformed in some manner. What transcendental analysis aims to discover and explain is how reality, seen through the medium of geometry or mathematical physics, is thought, or what it signifies in the light of artistic intuition or from the standpoint of ethical obligation. It has no answer to the question of what this reality may be in itself and apart from every relation to the mind's specific ways of understanding, because with that question philosophy would feel itself gone astray again in the empty space of abstraction; all firm footing would be lost. Metaphysics must be the metaphysic of the sciences, the theory of principles of mathematics and natural knowledge, or, if it claims specific content for itself, it must be the metaphysic of morality, of right, of religion, or of history. It integrates these multiple objective mental directions and activities as a single problem, not so as to make them vanish in this unity but so as to illuminate the essential individuality and proper limitation of each of them.

In this way philosophy is shown to be the necessary starting point of the entirety of intellectual and spiritual culture as it is given to us. Philosophy, however, no longer wishes to accept that culture as given, but rather to make its origin and the universally valid norms governing and guiding it comprehensible. Only now do we fully understand Kant's statement, that the torch of the critique of reason does not light up the objects unknown to us beyond the sense world, but rather the shadowy space of our own understanding. The

17. *Reflexionen*, no. 140.

"understanding" here is not to be taken in the empirical sense, as the psychological power of human thought, but rather in the purely transcendental sense, as the whole of intellectual and spiritual culture. It stands directly for that entity which we designate by the name "science" and for its axiomatic presuppositions, but further, in an extended sense, for all those orders of an intellectual, ethical, or aesthetic kind demonstrable in reason and perfectible by it. What appears in the experiential, historical life of mankind as detached and divided and laden with contingencies, is, by means of the transcendental critique, to be grasped as necessitated by its ultimate grounds, and conceived and exhibited as a system. Just as each individual figure in space is connected with the general law, which is itself grounded in the pure form of "coexistence," in the form of intuition, every "what" of the works of reason in the end goes back to a characteristic "how" of reason, to something fundamental and distinctive which all its products manifest and confirm. Philosophy now no longer has any special domain, no particular sphere of entities and objects that belongs solely and exclusively to it, distinct from the other sciences; rather, it conceives the relation of the basic functions of the mind only in their true universality and depth, a depth inaccessible to any one of them alone. The world is given over to the individual theoretical disciplines and to the particular productive powers of the mind, but the cosmos of these powers themselves, its multiplicity and its articulation, comprises the new "object" which philosophy has gained in its place.

If we begin with the structure of mathematics in order to make this clear in detail, it is less a matter of developing the stuff of mathematical principles in particular than of showing the universal procedure which is the sole means whereby there can be principles for us, that is, by means of which we can understand how every special spatial thing or every particular operation of numbering and measuring is tied to original universal conditions, from which it cannot be freed. Every geometrical proposition or proof is based on a concrete and to that extent individual intuition. But no such proof is about an individual; rather it passes from the individual to a judgment concerning an infinite totality of forms. A certain property is not asserted of this or that triangle, or of a specific circle, but of "the" triangle or "the" circle in general.

What justifies our passing in this case from the individual thing, which is all that can be given us in sensory representation, to the totality of possible cases, which as infinite is inconceivable in any empirical representation? How do we succeed in making limited, partial content the bearer of an assertion which as such is valid not of it but of an infinite content, which to us it "stands for"? To answer these questions it is sufficient, according to Kant, if we get clear in our minds the peculiar nature of the procedure of scientific geometry as it is actually practiced and as it has evolved historically. The elevation of geometry from its earliest, rudimentary state, in which it was nothing more than a practical art of measuring, to the rank of fundamental theoretical knowledge, is ascribable only to an intellectual revolution that is fully analogous to the one we previously contemplated in discussing the transcendental philosophy. "The history of this intellectual revolution—far more important than the discovery of the passage round the celebrated Cape of Good Hope—and of its fortunate author, has not been preserved. But the fact that Diogenes Laërtius, in handing down an account of these matters, names the reputed author of even the least important among the geometrical demonstrations, even of those which, for ordinary consciousness, stand in need of no such proof, does at least show that the memory of the revolution, brought about by the first glimpse of this new path, must have seemed to mathematicians of such outstanding importance as to cause it to survive the tide of oblivion. A new light flashed upon the mind of the first man (be he Thales or some other) who demonstrated the properties of the isosceles triangle. The true method, so he found, was not to inspect what he discerned either in the figure, or in the bare concept of it, and from this, as it were, to read off it properties; but to bring out what was necessarily implied in the concepts that he had himself formed a priori, and had put into the figure in the construction by which he presented it to himself."[18] If in order to carry out the geometrical proof we had to draw the figure, if it lay before us

18. *Critique of Pure Reason,* preface to the 2d ed., B xi f. (III, 15).

as something given, the specific properties of which we had to learn through observation, geometrical judgment could never go beyond the objective individual content of that particular shape. For what right would we have to infer from what is given to what is not given, from the special case as perceived to the whole sum of possible cases? In truth, however, such an inference is neither possible nor required here, for the totality of individual geometrical instances does not exist prior to and apart from construction but arises for us only in the act of construction itself. Since I not only think the parabola or the ellipse universally *in abstracto* but cause both to exist by construction, through a specific rule (as perhaps, by their definition as conic sections), I thus create the condition under which alone the particular parabola or ellipse can be thought. We now understand how the geometrical concept as constructive does not follow on the specific cases, but precedes them. Thus to that extent it is valid as a true a priori relative to them.

Seen in this connection, this designation obviously is in no way related to an empirical psychological subject and to temporal succession, to the before or after of its individual representations and cognitions, but it expresses purely and exclusively a relation within what is known, a relation of the thing itself. Geometrical construction is "prior" to the individual geometrical form, because the meaning of the individual form is established only via the construction, not the other way around—the meaning of the construction through the individual form. All the necessity belonging to geometrical judgments rests on this fact. In geometry the cases do not exist apart from the law, as things detached and independent; they basically issue from consciousness of the law. In geometry the particular does not constitute the presupposition of the universal; rather it is thought only by means of determining and specifying more exactly the universal in general. No particular shape and no particular number can contradict what is embodied in the general procedure of spatialization or of the synthesis of numeration, because only in this procedure does all that which participates in the concept of the spatial and the concept of number come into being. In this sense geometry and arithmetic furnish the immediate confirmation of a principle that Kant now puts

forward as the universal norm and the touchstone of the "new
method of thought," "namely, that we can know a priori of things
only what we ourselves put into them."[19]

The third central and crucial concept of the First Critique, that of
"synthesis a priori," emerged hand in hand with the concepts of the
subjective and the transcendental. The significance of this synthesis
becomes clear as soon as we contrast the procedure of geometry and
arithmetic, as heretofore established, with that of ordinary empirical
concept formation, as well as with the procedure of formal logic. In
empirical concept formation (especially as practiced in the purely descriptive and classificatory sciences) we are content to add case to
case, fact to fact, and inspect the resulting sum to see whether it
exhibits any "common" characteristic unifying all the particulars. A
decision as to whether there is a connection of this kind obviously can
be made only after we have run through and examined the particulars
that are relevant to our question, for since we know the determination
here affirmed only as the observed property of a given thing, it is clear
that before the thing as such is actually given, that is, established in
experience, no more precise identifying characteristic can be assigned
to it. Knowledge thus seems in this case to derive from a collection, a
mere aggregate of elements that even outside of this connection and
prior to it possess independent being and meaning.[20]

The situation seems to be totally different with those universal
propositions an examination of formal logic supplies us with. In a
genuine universal proposition of this logic, universality is not derived
from the contemplation of particulars, but precedes and determines

19. Ibid., B xviii (III, 19).
20. It must, of course, be emphasized that in this exposition of empirical knowledge
("synthesis a posteriori"), it is less a matter of describing an actual state of affairs of
cognition than of constructing a limiting case, which we use in order to show the
special nature of a priori judgment more accurately, through contrast and opposition.
Kant himself used this construction in his distinction between judgments of perception
and judgments of experience, and in his emphasis on the purely subjective nature of
the former; see the *Prolegomena*, §18 (IV, 48) (*Ak.* IV, 297 f.). In itself, though, there is
for him no "particular judgment" which does not claim some kind of "universal" form;
no "empirical" proposition which does not contain in itself some "a priori" assertion,
since the form of judgment as such contains the requirement of "objective universal
validity."

them. From the fact that all men are mortal, and from the certainty contained in this major premise, the mortality of Caius is proven as a necessary consequence. However, logic is satisfied to explicate the forms and the formulas of this proof, without reflecting on the content of cognition and its origin and justification. Hence it assumes as given the universal premise that is the starting point of a specific deduction, without further inquiry into the basis of its validity. It shows that if all A's are b's, this must also hold for a specific, particular A, while the question of whether and why the hypothetical major premise is valid lies outside its sphere of interest. Hence, general logic basically does nothing but analyze back into their parts specific complexes of concepts, which it has previously formed by composition. It defines a concept by asserting specific qualities as its content, and it then abstracts from the logical totality thus created an individual factor which it separates off from the others so as to predicate it of the whole. This predication does not create any new knowledge, but only reiterates separately what we already had before, in order to explicate and eludicate it; it serves as an "analysis of concepts which we already have of objects," in which there is no further inquiry into the cognitive source from which these concepts are derived.[21]

Now we recognize in the resulting twofold opposition the characteristic peculiarity that synthesis a priori displays. In mere judgment of experience, or a posteriori connection, the whole we are trying to achieve was composed of purely individual elements, which necessarily were given independently beforehand. In formal logical judgment, where a given logical whole was simply analyzed and divided into its parts, a priori synthesis reveals a completely different structure. Here we begin with a specific constructive connection, in and through which simultaneously a profusion of particular elements, which are conditioned by the universal form of the connection, arises for us. We think the diverse possibilities jointly in a single, comprehensive, and constructive rule: sections of a cone; and we have in that way simultaneously produced the totality of those geometric

21. Cf. *Critique of Pure Reason*, introduction, sect. III, A 5 = B 9 (III, 39). See also the *Prolegomena*, §2b (IV, 15) (*Ak.* IV, 267).

forms which we call second-order curves: circles, ellipses, parabolas, and hyperbolas. We think the construction of the system of natural numbers, according to one basic principle, and we have included in it from the first, under definite conditions, all the possible relations between the members of this set. Kant's Inaugural Dissertation had already introduced the essential expression "pure intuition" for this form of relation between part and whole. Thus the result is that all synthesis a priori is inseparably linked with the form of pure intuition, that it either is itself pure intuition, or else is mediately related to and rests on some such intuition.

When Eberhard later, in his polemic against Kant, deplored the absence in the *Critique of Pure Reason* of one clearly defined principle of synthetic judgments, Kant referred him to this relation. "All synthetic judgments of theoretical cognition," as Kant now formulates this principle, "are possible only through the relation of the given concept to an intuition."[22] Space and time hence remain the true models and archetypes, exhibiting purely and fully the characteristic relation holding between the infinite and the finite, between the universal and the particular and individual, in every a priori synthetic cognition. The infinity of space and of time assert nothing more than that all determinate individual spatial and temporal magnitudes are possible only through limitations of the one all-inclusive space or the unitary, unrestricted representation of time.[23] Space does not arise because we construct it out of points, nor time because we construct it out of instants, as though they were substantial elements; rather, points and instants (and hence indirectly all figures in space and time) can be posited only through a synthesis in which the form of coexistence or succession in general originates. Thus we do not locate these forms in space and time as given, but only produce them by means of "space" and "time," if both are understood as basic constructive acts of intuition. "Mathematics must first exhibit all its concepts in intuition, and pure mathematics in pure intuition, i.e., construct them, without which (because it cannot proceed analytically, namely

22. Cf. Kant's letter to Reinhold, dated May 12, 1789 (IX, 402) (*Ak.* XI, 33); see also the essay in reply to Eberhard (VI, 59 ff.) (*Ak.* VIII, 239 ff.).

23. Cf. Transcendental Aesthetic, §4, A 31 f. = B 47 f. (III, 64).

through analysis of concepts, but only synthetically) it is impossible for it to take one step. . . . Geometry is founded on the pure intuition of space. Arithmetic achieves its number concepts by successive additions of units in time; in particular, however, pure mechanics can only achieve its concepts of motion by means of the representation of time."[24] Because the subject matters of geometry, arithmetic, and mechanics are arrived at in this fashion, because they are not physical *things* whose properties we must discover a posteriori, but rather limitations we place on the ideal wholes of extension and duration, all propositions implicit in these fundamental forms are necessarily and universally valid for them.

But if this consideration seems to explain to us the employment and the validity of a priori synthesis in mathematics, it seems at the same time to close off every path to the claim of a similar validity for the realm of the actual, for the domain of empirical science. In fact, it was precisely this touchstone that Kant pointed out to us: "that we know a priori only what we ourselves put into things." Such a "putting into" was understandable in ideal mathematical constructions, but what would tell us if we may do it in some fashion with empirical objects as well? Is not the decisive trait which basically marks the objects as real, as "actual," precisely that they exist in all their particularity prior to all mental developments and positings, that they thus fundamentally determine our representation and thinking, not that they are determined by the latter? And would not the ground necessarily cave in under our feet the moment we try to reverse this relationship? Space and time may be conceivable for us in universal principles, because they can be construed by these principles; it is the actuality of things in space and time, the existence of bodies and their motions, that seems to constitute the insuperable limit to all constructions of that sort. Here, it seems, there is no other course than to await the influence of things and to observe them in sensory perception. We call objects "actual" to the extent that they are made known to us in this form of actuality, and by this we are thus made acquainted with their individual qualities. Hence a general assertion

24. [*Prolegomena*, §10 (IV, 32) (*Ak.* IV, 283)]

about physical existences may be possible; in no case is its possibility understandable save through amassing particular instances, through the collection and comparison of the many impressions of the things which we have experienced.

And in fact Kant's transcendental idealism does not aim to eliminate the special nature of empirical knowledge; indeed, its essential merit is to be sought in its affirmation of this nature. Kant's saying that his field is the "fertile lowland of experience" is well known. But his general counsel holds also for the new critical determination of the concept of experience itself: that here as well we have to begin not with the observation of objects, but with the analysis of knowledge. The question of what an empirical object, an individual natural thing, may be, and whether it is accessible to us in any way other than through direct perception of its particular properties, must thus be left open for the moment. For before it can be meaningfully decided as a general matter, we must have succeeded in being completely clear as to what the cognitive mode of the natural sciences means, what the structure and systematic of physics is. Here we quickly see a fundamental difficulty in the traditional way of looking at the matter. Let us pursue this approach to the point where we assume that the object of mathematics rests in fact on the pure proposition of thought and has to that extent ideal validity, while the physical object is given to us and is conceivable by us exclusively by means of the various types of sensory perception. The possibility of a pure mathematical theory on the one side and a pure empiricism on the other is then understandable, that is, how on the one hand there can be a complex of propositions which, independently of all experience, treat only of such matters as we can produce by free construction, and how on the other hand a descriptive science can be built up which consists of sheer individual, factual observations of given things. What remains completely inexplicable under this assumption is the essential mutual involvement of both moments which we encounter in the actual structure of mathematical natural science. For in this latter measurement does not simply go hand in hand with observation, nor are experiment and theory simply opposed to each other or interchangeable, but they mutually condition each other. Theory leads to experiment and decides the character of the experiment, just as experiment determines the content of theory.

Once again the preface to the second edition of the *Critique of Pure Reason* has, in its broad survey of the whole realm of knowledge, displayed this relationship with masterful, unsurpassable clarity. "When Galileo caused balls, the weights of which he had himself previously determined, to roll down an inclined plane; when Torricelli made the air carry a weight which he had calculated beforehand to be equal to that of a definite column of water; or in more recent times, when Stahl changed metal into calx, and calx back into metal, by withdrawing something and then restoring it, a light broke upon all students of nature. They learned that reason has insight only into that which it produces after a plan of its own, and that it must not allow itself to be kept, as it were, in nature's leading strings, but must itself show the way with principles of judgment based upon fixed laws, constraining nature to give answer to questions of reason's own determining. Accidental observations, made in obedience to no previously thought-out plan, can never be made to yield a necessary law, which alone reason is concerned to discover. Reason, holding in one hand its principles, according to which alone concordant appearances can be admitted as equivalent to laws, and in the other hand the experiment which it has devised in conformity with these principles, must approach nature in order to be taught by it. It must not, however, do so in the character of a pupil who listens to everything that the teacher chooses to say, but of an appointed judge who compels the witnesses to answer questions which he has himself formulated. Even physics, therefore, owes the beneficent revolution in its point of view entirely to the happy thought, that while reason must seek in nature, not fictitiously ascribe to it, whatever as not being knowable through reason's own resources has to be learned, if learned at all, only from nature, in so seeking it must adopt as its guide that which it has itself put into nature. It is thus that the study of nature has entered on the secure path of a science, after having for so many centuries been nothing but a process of merely random groping."[25]

25. *Critique of Pure Reason*, preface to the 2d ed., B xii f. (III, 16 f.).

Thus while a lone sensory perception or a mere collection of such perceptions may be able to get along without the guidance of a plan of reason, it is still the latter that first makes experiment precise and possible, "experience" in the sense of physical knowledge. By use of that rational method, isolated sense impressions can become physical observations and facts; for that to occur it is of primary importance to transform the originally purely qualitative manifold and diversity of perceptions into a quantitative manifold; the aggregate of sensations must be changed into a system of measurable magnitudes. The idea of such a system is basic to every single experiment. Before Galileo could measure the magnitude of acceleration in free fall, the conception of acceleration itself, as well as measuring apparatus, had to exist, and it was this mathematical conception which once and for all differentiated his unadorned way of putting the question from that of the medieval scholastic physics. The outcome of the experiment determined not only what magnitudes are true of free fall, but also that in general such magnitudes must be sought for and insisted upon. What Galileo laid down in advance, according to that plan of reason, is what initially made it possible for the experiment to be conceived and directed.

From this point on, the structure of mathematical physics becomes truly transparent. The scientific theory of nature is not a logical hybrid; it does not spring from the eclectic combination of epistemologically heterogeneous elements but forms a self-contained and integral method. To understand this unity and to explain it by a universal principle, analogously to the unity of pure mathematics, is the task the transcendental critique sets itself. In its conception of this task, it has overcome the onesidedness of rationalism and of empiricism alike. Neither an appeal to concepts nor an appeal to perception and experience, as is now plain to see, has anything to do with the essence of natural scientific theory, for both single out but one moment, in isolation, instead of defining the peculiar relation between the moments, on which this whole question depends.

However, this does not solve the problem, but only states it in its most comprehensive form. For what synthesis a priori in pure mathematics explained and made comprehensible was this: the "whole" of the form of intuition—pure space and pure time as a whole—was prior to and underlay all particular spatial and temporal forms. Can the same relation or one of a similar kind be asserted of the realm of nature as well? Can a statement be made about nature as a whole that does not just follow on collection of individual observations, but is rather one that first makes observation of the individual itself possible? Is there here too something particular which can be arrived at and established in no way except by the limitation of an original totality? As long as we think of nature in the usual sense, as the assemblage of material physical things, we will have to answer all these questions in the negative, since how can anything be said about a totality of things without our having run through them and examined them one by one? But the concept of nature already contains a determination that points our reflection in another direction. For we do not call every complex of things "nature," but understand by that term a whole of elements and events ordered and determined by universal laws. "Nature," as Kant defines it, "is the existence of things so far as it is determined according to universal laws." Thus although in the material sense it signifies the set of all objects of experience, in the formal sense it signifies the conformity to law of all these objects.

The general task is thus reformulated: instead of asking what the conformity to law of things as objects of experience rests on, we ask how it is possible to recognize the necessary conformity to law of experience itself in respect to its objects in general. "Accordingly," it is said in the *Prolegomena to Any Future Metaphysics,* "we shall here be concerned with experience only and the universal conditions of its possibility, which are given a priori. Thence we shall define Nature as the whole object of all possible experience. I think it will be understood that I here do not mean the rules of the observation of a Nature that is already given ... but how the a priori conditions of the possibility of experience are at the same time the sources from which all universal laws of nature must be derived."[26] Thus the question is redirected from the contents of experience, from empirical objects, to

26. *Prolegomena,* §14, §17 (IV, 44, 46 ff.) (*Ak.* IV, 294, 296 f.).

the function of experience. This function has a basic definiteness comparable to that disclosed to us in the pure form of space and time. It cannot be fulfilled without specific concepts coming into play, just as when every scientific experiment is performed, there is contained in the question which, in that act, we put to nature itself the presupposition that nature is determinately quantified, the presupposition that specific elements in it are unchanging and conserved, and the presupposition that events follow one another according to a rule. Without the idea of an equality which determines the relation of distance and time of fall, without the idea of the conservation of the quantity of motion, without the universal concept and the universal procedure of measurement and quantification, not a single experiment of Galileo's would have been possible, because without these preconditions Galileo's whole problem remains incomprehensible.

Accordingly, experience itself is a "mode of cognitions which requires understanding," that is, a process of inference and judgment which rests on specific logical preconditions.[27] And, in fact, with this we are again shown a whole which is not put together from separate parts but which is the basis for the possibility of first asserting parts and specific content. Nature, too, must be thought as a system before its details can be observed. Just as a particular spatial form previously appeared as a limitation of the one space and as a specific span of time appeared as a limitation of infinite duration, now all particular laws of nature, looked at in this connection, appear as "specifications" of universal principles of the understanding. For there are many laws which we can know only by means of experience, "but conformity to law in the connection of appearances, that is, in Nature in general, we cannot discover by any experience, because experience itself requires laws which are a priori at the basis of its possibility."[28] Extravagant and paradoxical as it sounds to say that the understanding itself is the source of the laws of nature and hence of the formal unity of nature, such an assertion is nonetheless correct and conforms with its object, namely experience. "Certainly, empirical laws as such can never de-

27. See the *Critique of Pure Reason*, preface to the 2d ed., B xviii (III, 18).
28. *Prolegomena*, §36 (IV, 71) (*Ak.* IV, 318–19).

THE *CRITIQUE OF PURE REASON*

rive their origin from pure understanding. That is as little possible as to understand completely the inexhaustible multiplicity of appearances merely by reference to the pure form of sensible intuition. But all empirical laws are only special determinations of the pure laws of understanding, under which, and according to the norm of which, they first become possible. Through them appearances take on an orderly character, just as these same appearances, despite the differences of their empirical form, must nonetheless always be in harmony with the pure form of sensibility."[29] We can establish the specific numerical constants that are characteristic of a particular realm of nature only through empirical measurement, and discover individual causal connections only by observation, but that we universally search for such constants, that we demand and presuppose universal causal lawfulness in the succession of events, issues from that plan of reason which we do not derive from nature but which we put into it. What is comprised in this plan is what alone yields a priori knowledge.

The second basic line of synthesis a priori, the synthesis of the pure concepts of the understanding, or the categories, is thus established, and it is justified by the same principle as that of pure intuition. For the pure concept does not display its true and essential action in a mere description of what is given in experience, but in construction of the pure form of experience; not where it collects and classifies the contents of experience, but where it is the foundation of the systematic unity of our way of cognition. It is in no way sufficient, as is commonly imagined, for experience to compare perceptions and to unite them by means of judgment in one consciousness, for by this alone we would never get beyond the specific validity of the perceptual consciousness, and would never reach the universal validity and necessity of a scientific principle. "Quite another judgment therefore is required before perception can become experience. The given intuition must be subsumed under a concept which determines the form of judging in general relatively to the intuition, connects empirical consciousness of intuition in consciousness in general, and thereby

29. *Critique of Pure Reason*, A 127 ff. (III, 627 ff.).

procures universal validity for empirical judgments. A concept of this nature is a pure a priori concept of the understanding, which does nothing but determine for an intuition the general way in which it can be used for judgments."

Even the judgments of pure mathematics are not exempted from this condition: the proposition, for instance, that a straight line is the shortest distance between two points presupposes as a general matter that the line is conceived from the standpoint and under the concept of magnitude, a concept "which certainly is no mere intuition, but has its seat in the understanding alone and serves to determine the intuition (of the line) with regard to the judgments which may be made about it, in respect to their quantity, that is, plurality.... For under them it is understood that in a given intuition there is contained a plurality of homogeneous parts."[30] This connection emerges still more clearly where it is not a matter of a simple mathematical determination of the object, but of a dynamic one, that is, where not only is an individual spatiotemporal form produced as a quantity by successive synthesis of similar parts,[31] but in addition its relation to another object is to be established. For it will be shown that every such relational determination—the order we give to individual bodies in space and individual events in time—is founded on a form of the actual assumed to hold between them; the idea of actuality, however, presupposes functional dependence, and hence a pure concept of the understanding.

If, however, the cooperation and the reciprocal relation of the two basic forms of a priori synthesis shed light on these simple examples, at the moment we still lack any further principle by which to develop the systematic of the second form completely. We can indeed point out and give names to individual applications of the pure concepts of the understanding, but at this point we have no criterion whatsoever

---

30. *Prolegomena*, §20 (IV, 51 f.) (*Ak.* IV, 301–02).

31. Cf. *Critique of Pure Reason*, Transcendental Doctrine of Method, chap. 1, sect. 1, "The Discipline of Pure Reason in its Dogmatic Employment": "[Thus] we can determine our concepts in a priori intuition, inasmuch as we create for ourselves, in space and time, through a homogeneous synthesis, the objects themselves—these objects being viewed simply as *quanta*." A 723 = B 751 (III, 491).

to assure us of the formal unity and completeness of our knowledge. Kant was led to precisely this latter requirement, as we recall, by the train of thought he entered upon immediately after the Inaugural Dissertation. In the letter to Marcus Herz of 1772 he posed as the task of the newly discovered science of "transcendental philosophy," "to reduce all concepts of completely pure reason to a certain number of categories, but not like Aristotle, who, in his ten predicaments, placed them side by side as he found them, in a purely chance disposition, but as they are of themselves divided into classes according to a few basic laws of the understanding."[32]

A new *fundamentum divisionis* for this long-standing demand is now reached. "The *possibility of experience*," as the basis for this division is called in the section "The Highest Principle of All Synthetic Judgments," "is, then, what gives objective reality to all our a priori modes of knowledge. Experience, however, rests on the synthetic unity of appearances, that is, on a synthesis according to concepts of an object of appearances in general. Apart from such synthesis it would not be knowledge, but a rhapsody of perceptions that would not fit into any context according to rules of a completely interconnected (possible) consciousness, and so would not conform to the transcendental and necessary unity of apperception. Experience depends, therefore, upon a priori principles of its form, that is, upon universal rules of unity in the synthesis of appearances. Their objective reality, as necessary conditions of experience, and indeed of its very possibility, can always be shown in experience. Apart from this relation synthetic a priori principles are completely impossible. For they have then no third something, that is, no object, in which the synthetic unity can exhibit the objective reality of its concepts.... Accordingly, since experience, as empirical synthesis, is, insofar as such experience is possible, the one species of knowledge which is capable of imparting reality to any nonempirical synthesis, this latter [type of synthesis], as knowledge a priori, can possess truth, that is, agreement with the object, only insofar as it contains nothing save what is necessary to synthetic unity of experience in general.... Synthetic a

32. See above, p. 130.

priori judgments are thus possible when we . . . assert that the conditions of the *possibility of experience* in general are likewise conditions of
the *possibility of the objects of experience,* and that for this reason they
have objective validity in a synthetic a priori judgment."[33]

In these sentences the entire internal structure of the *Critique of
Pure Reason* is revealed to us. Experience is the starting point—but not
as a sum of ready-made things with determinate, equally ready-made
properties, nor as a mere rhapsody of perceptions; the concept of
experience is, rather, characterized and determined by the necessity
of interconnection, the rule of objective laws. Up to this point, the
transcendental method has only established what had been valid in
mathematical physics for a long time and was recognized in it,
whether consciously or unconsciously. Kant's assertion that every
genuine experiential judgment must contain necessity in the synthesis of perceptions in fact only brings the demand already stated by
Galileo to its most concise and striking expression. In it, the sensualist
concept of experience is simply replaced by that of mathematical empiricism.[34]

At this point, however, the essential intellectual revolution begins. While until now necessity was held to be founded in objects and
only indirectly carried over from them into knowledge, now it is
understood that the reverse is true, that every idea of the object arises
from an original necessity in knowledge itself: "for this object is no
more than that something, the concept of which expresses such a
necessity of synthesis."[35] In the flow of our sensations and representations it is not arbitrariness that rules, but a strict lawfulness
which excludes every subjective whim; for that reason and that reason alone phenomena are for us objectively coherent. That which
characterizes and constitutes experience as a mode of knowledge
conditions and thus renders possible the assertion of empirical objects. Whether any other objects might be given to us apart from this

33. *Critique of Pure Reason,* A 156 ff. = B 195 ff. (III, 152 ff.).
34. *Prolegomena,* §22 (IV, 32) (*Ak.* IV, 304–05). Cf. *Critique of Pure Reason,* B 218 (III,
166): "Experience is possible only through the representation of a necessary connection
of perceptions."
35. *Critique of Pure Reason,* A 106 (III, 616).

relation is a question that is completely superfluous for us, and so it must be, according to the transcendental principles, as long as no other mode of knowledge whose structure is essentially distinguished from that of experience is demonstrated for this presumed different type of object. Here, however, where the demand for such a mode of knowledge is incomprehensible, or where at the very least its fulfillment remains completely problematical, no conclusion is possible other than the one which the highest principle draws. The conditions on which experience as a function rests are at the same time the conditions of everything it yields us, for every determination of an object rests on the interpenetration of the pure forms of intuition and the pure concepts of the understanding, through which the manifold of mere sensations is first woven into a system of rules, and thereby constituted as an object.

3

We have expressed in the foregoing reflections the great classical principles of the *Critique of Pure Reason*. Now the question of the classification and systematic division of the pure concepts of the understanding introduces us for the first time into its detailed workings. Immediately it seems as though we stand on different ground, as though here it is no longer the objective necessity of things that purely and exclusively holds sway, but instead a manner of explication and exposition that ultimately is only fully understood and evaluated by tracing it back to certain personal peculiarities of Kant's mind. Delight in comprehensive architectonic structure, in the parallelism of the art form of systematization, in the monolithic schematism of concepts seems to play a greater part in the detailed working out of the doctrine of the categories than is proper. In fact, one of the essential objections that has always been leveled against the overall form of the First Critique is that the table of the pure concepts of the understanding it draws up copies the logical table of judgments with great analytical artistry but with equal artificiality. Since judgments, according to the view of traditional logic as Kant found it, are divided into the four classes of quantity, quality, relation, and modality, the concepts of the understanding should exhibit the same structure; just as in every main class of logical judgments a triad of particular forms is assumed, where the third is the result of synthesizing the first and second, so in the structure of the concepts of the understanding this standpoint is adhered to and rigorously carried out. Under quantity, the subcategories of unity, plurality, and totality are the result; in the domain of quality, the concepts of reality, negation, and limitation; while relation is analyzed into substance, causality, and community, and modality into possibility, existence, and necessity.

Whatever complaints may be raised against this form of the deduction, however, in general all the polemics directed against the systematic relation between category and judgment fall short. For they ignore the true sense of the central and fundamental transcendental question; they overlook the fact that the significant and preeminent place that Kant allots to judgment is of necessity already rooted in the initial presuppositions of his way of putting the problem. Judgment is the natural, factually demanded correlate of the object, since it expresses in the most general sense the consummation of and demand for that combination to which the concept of the object has been reduced for us. "Consequently, we say that we recognize the object when we have effected synthetic unity in the manifold of intuition"; however, when expressed in exact logical notation, the types and forms of synthetic unity are precisely what yield the forms of judgment. Only one objection could still be validly made here, namely, that if one grants this connection, the system of formal logic might not be the court before which the forms of objective interconnection have to be defended, for is not the essence of this logic, and its basic operation, analysis rather than synthesis? Doesn't it abstract from that relation, from that content of knowledge that must be decisive and essential for us? In reply, it is important to keep clearly in mind that for Kant there is indeed such an abstraction, but that it is always to be understood only in a relative, not in an absolute sense. An analysis that is nothing but analysis, that does not in any way relate indirectly to and rest on an underlying synthesis is impossible, "for where the understanding has not previously combined, it cannot dissolve, since only as having been combined *by the under-standing* can anything that allows of analysis be given to the faculty of representation."[36] Thus general logic is concerned with "analysis of the concepts which we already have of objects,"[37] and explicates the judgments which result from presupposing such objects as a ready-made substrate, so to speak, of a proposition.

But as soon as we reflect on the origin of this substrate itself, and inquire as to the possibility of the state assumed by logic—which of course lies outside its province—we are entering the sphere of a different consideration, which demands a deeper explanation and a fundamental deduction of judgment itself. Now it appears that the function which unites diverse representations in one judgment is one and the same as that which also combines the manifold of sensory elements so that they achieve objective validity. "The same understanding, through the same operations by which in concepts, by means of analytical unity, it produced the logical form of a judgment, also introduces a transcendental content into its representations, by means of the synthetic unity of the manifold in intuition in general. On this account we are entitled to call these representations pure concepts of the understanding, and to regard them as applying a priori to objects—a conclusion general logic is not in a position to establish."[38] While general logic can similarly be employed as the "clue to the discovery of all the pure concepts of the understanding," this is not done with the aim of basing the transcendental concepts on the formal ones, but, conversely, with the aim of basing the latter on the former, and in that way yielding a more profound understanding of the ultimate ground of their validity.

"Aristotle," as Kant himself summarizes this whole development in the *Prolegomena*, "collected ten pure elementary concepts under the name of categories. To these, which are also called 'predicaments,' he found himself obliged afterwards to add five post-predicaments, some of which however (*prius, simul,* and *motus*) are contained in the former; but this rhapsody must be considered (and commended) as a

36. Ibid., B 130 (III, 113).
37. Ibid., A 5 = B 9 (III, 39).
38. Ibid., A 79 = B 105 (III, 98).

mere hint for future inquirers, not as an idea developed according to rule. . . . After long reflection on the pure elements of human knowledge (those which contain nothing empirical), I at last succeeded in distinguishing with certainty and in separating the pure elementary notions of the sensibility (space and time) from those of the understanding. Thus the seventh, eighth, and ninth categories had to be excluded from the old list. And the others were of no service to me because there was no principle on which the understanding could be exhaustively investigated, and all the functions, whence its pure concepts arise, determined exhaustively and precisely. But in order to discover such a principle, I looked about for an act of the understanding which comprises all the rest and is distinguished only by various modifications or phases, in reducing the multiplicity of representation to the unity of thinking in general. I found this act of the understanding to consist in judging. Here, then, the labors of the logicians were ready at hand, though not yet quite free from defects; and with this help I was enabled to exhibit a complete table of the pure functions of the understanding, which are however undetermined with respect to any object. I finally referred these functions of judging to objects in general, or rather to the condition of determining judgments as objectively valid; and so there arose the pure concepts of the understanding, concerning which I could make certain that these, and this exact number only, constitute our whole knowledge of things by pure understanding."[39] The course of the deduction Kant describes in this passage fully conforms with his general basic tendency. Aristotle had determined the elements of knowledge, while Kant wishes to discover the principle of these elements; Aristotle's starting point was the fundamental properties of being, while Kant's was judgment as the unity of the logical act,[40] in which we achieve permanence and necessity of the content of representation, and thus objective validity.

The essential meaning of each individual category cannot, of course, be fully gauged if we simply relate it back in this way to the form of logical judgment corresponding to it; we must also look for-

39. *Prolegomena,* §39 (IV, 75 f.) (*Ak.* IV, 323–24).

40. Cf. in particular the *Critique of Pure Reason,* A 68 f. = B 93 f. (III, 90 f.) and B 140 ff. (III, 120 ff.).

ward to the work it is responsible for in the structure of objective experience. This work, though, does not belong to the abstract categories as such, but appears only in that concrete form which is given to the concepts of pure understanding, thus transforming them into principles of pure understanding. One of the fundamental merits of Cohen's books on Kant is that they fully and clearly defined this relationship for the first time. It is urged again and again in these books[41] that the system of synthetic principles forms the true touchstone for the validity and the truth of the system of the categories. For a synthetic principle arises because the function that characterizes a specific category relates to the form of pure intuition and permeates it in a systematic unity. Empirical objects—this is settled from the first sentences of the "Transcendental Aesthetic"—can be given to us in no way except through the mediation of intuition, by the mediation of the forms of space and time. This necessary condition is not, however, sufficient. Intuition as such contains only the pure manifold of coexistence and succession; for definite forms defined relative to one another to stand out in this manifold, its elements have to be run through and combined from a definite point of view and in accordance with a fixed rule, and thus composed into relatively substantial unities. This is precisely the work of the understanding, which therefore does not discover the connection of the manifold as already existing in some way in space and time, but which itself produces it originally, since it affects both.[42]

While a synthesis of that kind is needed to produce concrete geometrical figures,[43] it proves to be completely indispensable in matters of specifying physical objects. For to determine a physical object I must indicate its "where" and "when," I must assign it a definite place in the whole of space and duration. This is, in turn, only possible if I produce a definite rule, or rather a total structure and system of rules, by which this particular thing to be established is seen as thoroughly interconnected with and functionally dependent on other

41. See esp. Hermann Cohen, *Kants Theorie der Erfahrung,* 2d ed. (Berlin, 1885), pp. 242 ff.

42. *Critique of Pure Reason,* B 155 (III, 128 f.); cf. esp. B 160, note (III, 132).

43. Cf. esp. *Prolegomena,* §38 (IV, 73) (*Ak.* IV, 320 ff.).

things. Places in space, moments in time are, in the physical sense, determinable only on the basis of forces and relations of forces; the order of coexistence and succession can be established in a lawful way only if we assume certain universally valid dynamic relationships between the individual elements of experience. To lay down the form of these presuppositions and thus to indicate the conditions of the universal possibility of a mutual interconnection of objects in space and in time is the general task the system of synthetic principles sets for itself. If one holds fast to this aim, the principle that orders this system and by which it advances from the simple to the composite immediately becomes evident.

The first step will doubtless have to be that the object, insofar as it is to be intuited in space and time, participates in the fundamental character of both orders, that is, that it is determined as an *extensive* magnitude. But if the concrete physical thing, in the customary way of looking at it, "has" magnitude, here, in conformity with the critical, transcendental view, this proposition is reversed. The predicate of magnitude does not attach to things as their most general and essential property; rather, the synthesis in which the concept of quantity arises for us is the same one by which the manifold of mere perceptions becomes a manifold organized and governed by rules through which it first becomes an order of objects. Magnitude is not a basic ontological property, which we passively receive from objects and can isolate by comparison and abstraction; nor is it some simple sensation given to us, like that of color or sound. It is, rather, an instrument of thinking itself, a pure means of knowledge, with which we originally construct for ourselves "nature" as a universal, lawful order of appearances. For "appearances cannot be apprehended, that is, taken up into empirical consciousness, except through that synthesis of the manifold whereby the representations of a determinate space or time are generated, that is, through combination of the homogeneous manifold and consciousness of its synthetic unity." The concept of the quantum in general is, however, just this consciousness of the homogeneous manifold, insofar as the representation of an object only becomes possible through it. "Thus even the perception of an object, as appearance, is only possible through the same synthetic unity of the manifold of the given sensible intuition as that whereby the unity of the combination of the manifold [and] homogeneous is thought in the concept of a *magnitude*. In other words, appearances are all without exception magnitudes, indeed *extensive magnitudes*. As intuitions in space or time, they must be represented through the same synthesis whereby space and time in general are determined."[44]

The question of the possibility of applying exact mathematical concepts to the appearances of nature—a question that had continuously engrossed not only all prior philosophy, but also Kant himself in his precritical period—is answered at one stroke. For it is now plain that the question is falsely put: it is a matter not of the application of given concepts to a world equally given, to things heterogeneous with the concepts and contrasting with them, but of a special way of ordering to which we subject the "simple" sensations and by which we transform them into objective intuitions. "It will always remain a remarkable phenomenon in the history of philosophy," Kant remarks in the *Prolegomena*, "that there was a time when even mathematicians who at the same time were philosophers began to doubt, not of the accuracy of their geometrical propositions so far as they concerned space, but of their objective validity and the applicability of this concept itself, and of all its corollaries, to nature. They showed much concern whether a line in Nature might not consist of physical points, and consequently that true space in the object might consist of simple parts, while the space which the geometer has in his mind cannot be such."[45] They failed to see that it is precisely this mental space which makes physical space possible, that is to say, that makes the extension of matter itself possible, that this is the same procedure by which we sketch out the image of ideal space in pure geometry to posit a quantitative connection and relation between the sensory empirical elements. All objections to this are but "the chicanery of a falsely instructed reason" which is unable to discover the true ground of its own cognitions because it erroneously searches for it in a world of transcendent things, instead of among its own principles. As long as we regard pure mathematical

---

44. [*Critique of Pure Reason*, A 162 = B 202 f. (III, 157).]
45. *Prolegomena*, §13, Remark I (IV, 37) (*Ak.* IV, 287–88).

determinations as *data* of experience, we cannot achieve complete certainty as to the precision of these determinations, since all empirical measurement is necessarily inexact; this becomes irrelevant, however, the moment we learn to understand magnitude as a principle instead of as a property. The only thing that joins the representation of the possibility of such a *thing* with this *concept* is that space is a formal a priori condition of external experiences, that the productive synthesis by which we construct a triangle in imagination is completely identical with the synthesis we employ when apprehending an appearance so as to make a concept of experience out of it.[46]

The deduction of the second synthetic principle, which Kant calls the principle of the "Anticipations of Perception," seems more difficult, for here it is a matter of how this designation is to indicate in advance, in a general proposition, not merely the form of perception but its content as well. Since, however, perception is simply "empirical consciousness," any such requirement must seem paradoxical; how can something be anticipated in what can only be given to us a posteriori? Quantity may be susceptible to universally valid propositions, but how such propositions are possible regarding quality, which is given us by the agency of sensation, is at first impossible to see at all. And yet there is a definite moment which we assert of all qualities in nature and which, taken strictly, cannot be sensed in any way. We differentiate extensive magnitudes according to their extent in space and in time when we ascribe to them differing extension and duration, but this means of measurement and comparison deserts us in regard to qualities. For if we think a certain quality (as, say, the speed of a body or its temperature, its electrical or magnetic potential, etc.), it is not bound up with the form of externality that is essential to space and time. We can think the speed of a body in nonuniform motion as changing from place to place, from moment to moment, without stopping to conceive it as a magnitude at each indivisible point of space and time and attributing a definite value to it with respect to other velocities. And similarly, what we call the tempera-

46. See the *Critique of Pure Reason*, A 162–66 = B 202–07 (III, 157–59), A 224 = B 271 (III, 198).

ture or the electrical energy of a body can be regarded as determinate at one particular point and as different from point to point. This qualitative magnitude specified at a point thus is not composed, like extensive magnitudes, of individual, separate parts, but rather is present in the point wholly and indivisibly all at once, so that it displays a definite "more" or "less" in relation to other magnitudes of the same sort, and hence permits an exact comparison. Extensive magnitude here contrasts with intensive magnitude; extensional or durational magnitude is opposed to magnitude of degree, which also has a definite, assignable value for the spatiotemporal differential when differentiated with respect to space and time. That this value, that the particular qualities of particular bodies cannot be determined except through empirical measurement, is of course immediately evident.

And yet it appears, if we analyze the whole of our knowledge of nature, that it is not the determinateness of individual qualities and degrees, but what is demonstrable in them that is a universal basic relation, a universal requirement which they collectively satisfy. We presuppose that the transition from one degree to another proceeds continuously, not by leaps; that a specific value a is not replaced directly by another one, larger or smaller, but that in a change of that kind all intermediate values thinkable between a and b are traversed and in fact passed through and actually assumed one after the other. This proposition also rests on empirical observation; can it be proved or disproved by sensation? Obviously not, since however one may determine the relation of a sensation to an objective quality, one thing is clear in every case: that the evidence of sensation is always related to the immediately given individual case, and therefore, however many data we collect about it, it never goes beyond a specific, finite circle of conditions. The principle of the continuity of all physical changes, however, is an assertion which does not concern a sum of finite elements, although it is essential to infinitely many elements. Between any two points in time which we think of as the starting point and the end point of a certain process, however close they may be to one another, infinitely many instants can be interpolated, on the basis of the unlimited divisibility of time, and there corresponds to each of these moments, as the assertion of continuity of change says, a definite, unambiguous quantitative value of the variable quality, a value which is actually assumed once in the course of the whole process.

Regardless of how many of these values may be empirically demonstrated or demonstrable, there always remains an infinity of values for which this demonstration is not made. Nonetheless we may assert that they fall under the same universal rule. For if we thought that the continuity of change was interrupted at any point, we would have no further way of connecting the change with a unitary, identical subject. Assume that a body exhibits a state x at moment a and at moment b a state x', without having run through the values intermediate between the two: we would conclude from this that it was no longer a question of the "same" body; we would assert that at moment a a body in state x disappeared and at moment b a *different* body in state x' appeared. From this it is seen that the assumption of continuity in physical changes is not a matter of a particular result of observation, but of a presupposition of natural knowledge in general, that it is a question not of a *theorem* but of a *principle*. As the first synthetic principle, that of the "Axioms of Intuition," subjects the physical object to the conditions of geometric and arithmetic magnitude, the second principle subjects the object in nature to conditions expressed and scientifically worked out in the analysis of the infinite. This analysis is the true *mathesis intensorum*, the mathematics of intensive magnitudes.[47] While previously appearances were determined as quanta in space and time, now their quality, which has its subjective, psychological expression in sensation, is grasped by a pure concept and thus the "real" in appearance achieves its first scientific designation and objectification.

"In all appearance," as Kant formulates the principle of the "Anticipations of Perception," "the real that is an object of sensation has intensive magnitude, that is, a degree." Empty regions of space and time would be completely identical with one another because of the thoroughgoing homogeneity of pure space and pure time, and hence

47. See esp. Cohen, *Kants Theorie der Erfahrung*, 2d ed., p. 422; *Das Prinzip der Infinitesimalmethode und seine Geschichte* (Berlin, 1883), pp. 105 ff.

indistinguishable. A way of distinguishing between them is obtained only when we put a definite content into them and conceive a difference of "greater" and "lesser," of "more" or "less" in this content. A purely sensory apprehension, though, strictly speaking, occupies only an instant; to an indivisible "now" there corresponds an indivisible sensory content, which one can think as varying from moment to moment. "As sensation is that element in the [field of] appearance the apprehension of which does not involve a successive synthesis proceeding from parts to the whole representation, it has no extensive magnitude. The absence of sensation at that instant would involve the representation of the instant as empty, therefore as = o. Now what corresponds in empirical intuition to sensation is reality (*realitas phaenomenon*); what corresponds to its absence is negation = o. Every sensation, however, is capable of diminution, so that it can decrease and gradually vanish. Between reality in the [field of] appearance and negation there is therefore a continum of many possible intermediate sensations, the difference between any two of which is always smaller than the difference between the given sensation and zero or complete negation. In other words, the real in the [field of] appearance has always a magnitude... but not extensive magnitude."[48] All sensations as such are therefore given a posteriori, but their property of having a degree, and further that this degree, so far as it undergoes change, must change continuously, can be understood a priori as necessary. In this sense the quality of the empirical, the essential determinateness of perceptions themselves, can be anticipated. "It is remarkable that of magnitudes in general we can know a priori only a single *quality*, namely, that of continuity, and that in all quality (the real in appearances) we can know a priori nothing save [in regard to] their intensive *quantity*, namely that they have degree. Everything else has to be left to experience."[49] What was true earlier of the concept of spatial and temporal magnitudes now holds for the concept of degree: it too affords less the recognition of a universal property of the thing as it is than a constitutive condition which first makes

48. [*Critique of Pure Reason*, A 167 f. = B 209 f. (III, 160 f.).]
49. Ibid., A 176 = B 218 (III, 166).

it possible to establish and differentiate empirical objects themselves.

But if the individual object is taken only in its particularity, the essential concept of "nature" is not yet fulfilled, for the system of nature claims to be a system of laws and therefore is not concerned with the isolated object as such, but instead with the thoroughgoing interconnection of appearances and the way in which they are related to one another by a form of mutual dependence. This idea leads us to a new set of principles, intended to express the principal presupposition less for the purpose of establishing individual things than for establishing relations. When Kant calls these principles the "Analogies of Experience," he is following the way of speaking of the mathematics of his time, in which the term "analogy" was used as the universal expression for any kind of proportion. The fundamental proportion to be established here, however, is the reciprocal place occupied by the particular appearances in space and time, thus the objective relation of their community and succession. In order that such a relation can be expressed, it seems necessary first of all to introduce the individual things, each separately, into space and time alike, that is, to assign them a definite point in the given manifold of space and time in general that marks their individual "here" and "now."

In this, however, we run directly into a peculiar difficulty. In order to use space and especially time in this way as the basis of determination, we would initially have to possess both as absolute and established orders. A fixed structure of space and a fixed succession of time would have to be given us, to which we could relate all motion in space and all qualitative change, just as to a permanent basic scale. But even if it is assumed that such a scale exists, is it recognizable by us in any way? Newton speaks of "absolute, true, and mathematical time," which flows in itself and by reason of its nature uniformly and without relation to any sort of external object. But if we grant him the right to this explanation, can the instants of this uniform time be differentiated independently of every relation to physical objects? Do we know temporal instants and their series directly, or is not rather all knowledge of them which we think we have mediated by our knowledge of the contents of space and time and by the dynamic interconnection which we assume between them? It is not from the absolute "where" and "when" of things that we can draw a conclusion about how they act; on the contrary, what allows us to assign to them a definite order in space and in time is the form of the action that we assume between them, on the basis of experience or inference. On the basis of the law of gravitation, thus on the basis of an assertion about the distribution and relation of forces, we mentally sketch out the picture of the cosmos as it exists in space and as it has unfolded in time. In this theoretical structure much of what is encountered in initial sensory perception, in the sheer coexistence and in the succession of impressions, is spatially and temporally separated (as, for example, we mentally relate the light of extinct fixed stars, the perception of which reaches us simultaneously with the perception of some sort of nearby body, to an object that is centuries or millennia distant in time). On the other hand, many things that are distinct from one another in sensation are combined and transformed into a unity by objective scientific judgment.

But although the *particular* order which we ascribe to the contents of space and time seems always to rest on certain particular laws of action that we assume to hold between them, it is now important, from the transcendental point of view, to convert this knowledge into something universal. There are three different basic determinations, three modes which we distinguish in time and in which the idea of time itself is fulfilled: duration, succession, and simultaneity. We must understand that these three determinations themselves are not immediately given, that they are not simply to be read off impressions, but for each of them to be comprehensible by us, a definite synthesis of the understanding is needed, which for its part is a universal presupposition of the form of experience itself. "There will, therefore, be three rules of all relations of appearances in time, and these rules will be prior to all experience, and indeed make it possible. By means of these rules the existence of every appearance can be determined in respect of the unity of all time."[50] It is these three fundamental rules which Kant formulates in his three "Analogies of

50. Ibid., B 219 (III, 167).

Experience." They are the presupposition for our success in determining objective temporal relations in general, that is, that we are not
simply abandoned to the chance sequence of impressions in ourselves
according to the mere play of association, which is different for each
individual and indeed governed by his private circumstances, but
rather that we can pronounce universally valid judgments about temporal relations. For instance, to establish in an objective sense the
occurrence of change, it is not enough that we posit diverse substances and attribute them directly to different instants of time—for
time and instant are not, as such, objects of possible experience; we
must point to something enduring and unchanging in the appearances themselves, relative to which the change can be ascertained by
certain other determinations. This idea of something relatively constant and something relatively changeable in phenomena, this category of substance and accident, is thus the necessary condition for the
emergence of the concept of the unity of time, of duration in change,
out of the totality of our representations in general. The permanent is
the "substratum of the empirical representation of time itself; in it
alone is any determination of time possible."[51] *What* quantum in
nature we have to regard as constant always remains a question, the
answer to which we must leave to factual observation, but the assumption of *some* quantum in general that remains constant in this
fashion is a fundamental presupposition without which the concept
of "nature" and of natural knowledge itself would be invalid for us.

The same consideration holds for the relation of causality and
reciprocity, which are defined in the second and third analogies of
experience. Hume's sensualist critique of the causal concept began by
attacking the objective and necessary validity of this concept, since
the criticism tries to reduce its whole content to a statement about the
more or less regular succession of representations. The coupling of
phenomena which we believe we understand in the idea of causality
signifies, according to this view, actually nothing other than that they
frequently follow one another and hence are joined by our imagination into a relatively firm psychological association of representations.

51. [Ibid., A 183 = B 226 (III, 171).]

If this view is to be refuted in principle and at its root, it can once again happen only through that reversal of the question characteristic of the basic transcendental interpretation: it has to be shown that regularity in the succession of our sensations and perceptions does not produce the concept of causality, but that, conversely, this concept, the idea of and demand for a rule which we bring to perceptions, is what first makes it possible for us to extract determinate forms and definite factually necessary connections from the uniform, flowing series of perceptions, and hence to give our representations an object.

For in fact, when we inquire as to what new property the relation to an object adds to our representations, and what the dignity, the special logical validity, may be that is bestowed on it thereby, we find "that it does nothing more than make the connection of representations necessary in a certain manner and subject them to a rule; that conversely, only a secure order of temporal relations is needed for our perceptions, to impart objective meaning to them."[52] But the causal relation does precisely this, for if I put two phenomena a and b in the relation of cause and effect, this means nothing but the assertion that the passage from one to the other cannot be performed at will (as perhaps in a dream or in fantasizing subjectively we can arbitrarily shift the individual elements around at will, like the colored bits of glass in a kaleidoscope, and combine them this way or that way), but that it obeys a fixed law by which b must always and necessarily follow a, but not also precede a. In thus subjecting a given empirical relation to the concept of causality, we have thereby first truly fixed and unambiguously determined the temporal order in the succession of its members. "Let us suppose that there is nothing antecedent to an event, upon which it must follow according to rule. All succession of perception would then be only in the apprehension, that is, would be merely subjective, and would never enable us to determine objectively which perceptions are those that really precede and which are those that follow. We should then have only a play of representations, relating to no object; that is to say, it would not be

52. [Ibid., A 197 = B 242 f. (III, 181).]

possible through our perception to distinguish one appearance from another as regards relations of time. . . . I could not then assert that two states follow upon one another in the [field of] appearance, but only that one apprehension follows upon the other. That is something subjective, determining no object; and may not, therefore, be regarded as knowledge of any object. . . . We have, then, to show . . . that we never, even in experience, ascribe succession . . . to the object, and so distinguish it from subjective sequence in our apprehension, except when there is an underlying rule which compels us to observe this order of perceptions rather than any other; nay, *that this compulsion is really what first makes possible the representation of a succession in the object.*"[53]

And in this way Hume's problem is solved—indeed, very much "contrary to the expectation of its author." In his entire psychological analysis, Hume made one uncritical presupposition: that in general certain impressions are given in an objective and regular succession. For were this not the case, it would be purely at our option that the thing *a* might now precede the thing *b*, now be altogether dissociated from it but included in a different sequence—thus it would be impossible to establish a customary association between *a* and *b*, which is the condition for repeated encounter with the same items of experience connected in the same way.[54] In this one assumption of an objective sequence of elements of experience, however—as Kant charges—what is essential to the causal concept under attack is already granted, so that all subsequent sceptical criticism which is attempted is defective. Only from the perspective of cause and effect, only by the idea of a rule whereby the appearances are understood for themselves independently of the subjective consciousness of the individual observer, can one speak of a sequence in "nature" or of "things" in contrast to the sheer mosaic of representations "in us." "It is with these," Kant notes, "as with other pure a priori representations—for instance, space and time. We can extract clear concepts of them from experience only because we have put them

53. [Ibid., A 194–97 = B 239–42, italics added (III, 179 f.).]
54. Cf. esp. *Critique of Pure Reason*, A 74 ff. = B 100 ff. (III, 613 ff.).

into experience, and because experience is thus itself brought about only by their means."[55]

For dogmatic metaphysics, causality is valid as an objective power, a sort of fate, rooted in things themselves or in the ultimate ground of things. Sceptical psychological criticism demolishes this view, but, considered more closely, it posits in place of the compulsion of things only the compulsion lying in the mechanism of representations and their connection. In contrast, the critical method bases the necessity that we think in the cause-and-effect relation on nothing other than a necessary synthesis of the understanding, which shapes disparate and isolated impressions into experience. The method cannot yield any more secure and fixed objectivity, but neither is that necessary, since its highest principle says that objects are given us in experience by means of its conditions. The causal concept is not obtained through perception and comparison of similar sequences from experience, that is, from sensory impression; rather, the fundamental principle of causality reveals "how in regard to that which happens we are in a position to obtain in experience any concept whatsoever that is really determinate."[56]

The third "Analogy of Experience" rests on the same idea, in principle; Kant expresses it as the "principle of coexistence, in accordance with the law of *reciprocity* or *community.*" "All substances, insofar as they can be perceived to coexist in space, are in thoroughgoing reciprocity." For as it was only possible to objectify succession by linking the elements by a causal rule, in a sequence regarded as necessary, so the objectivity of coexistence can only be assured when the two factors we say are so related stand in a dynamic relation, by means of which each seems to be as much the cause of the other as its effect. As long as we abandon ourselves simply to the stream of sensations and impressions, there is for us no simultaneity in the strict sense, because what we apprehend is just something flowing and successive, within which a single item can only exist by displacing and excluding its predecessor. "The synthesis of imagination in

55. [*Critique of Pure Reason*, A 196 = B 241 (III, 180).]

56. Ibid., A 301 = B 357 (III, 249); for the whole discussion, see A 189 ff. = B 232 ff. (III, 175 ff.).

apprehension would only reveal that the one perception is in the subject when the other is not there, and vice versa, but not that the objects are coexistent, that is, that if the one exists the other exists at the same time.... Consequently, in the case of things which coexist externally to one another, a pure concept of the reciprocal sequence of their determinations is required, if we are to be able to say that the reciprocal sequence of the perceptions is grounded in the object and so to represent the coexistence as objective."[57] The general nature of this concept of the understanding is established through the preceding principle: the form of action or functional dependence is that which affords us the ground for assuming a definite temporal connection in the object itself. Here, however, the elements do not, as in the case of causality, stand in a unilateral dependence, so that one element, *a*, presupposes the other, *b*, in the temporal as well as the concrete sense; rather, inasmuch as they are to be simultaneous, it must be possible to make the transition between the two equally well from *a* to *b* as from *b* to *a*. Thus we arrive at a causal system which involves both members in such a way that there can be passage from one to the other just as readily as from the other to the one. A system of this kind is presented in the set of mathematico-physical equations derived from Newton's law of gravitation. Through it, the spatial position and motion of every member of the cosmos is explained as a function of all the rest, and these in turn as a function of it, and this total reciprocity, which is perfect from mass to mass, constitutes for us the objective whole of physical space itself and the ordering and structuring of its individual parts.[58]

This last great example, however, which from early on always signified for Kant himself the essential archetype of all natural knowledge, is at the same time an indication that, with the principle we have before us here, the task of determining what an object in nature is has reached its conclusion. The principles that now follow, and that are assembled by Kant under the name of the "Postulates of Empirical Thought," add no further novelty to this determination. For, as their

57. [Ibid., B 257 (III, 189 f.).]
58. Ibid., A 211 ff. = B 256 ff. (III, 189 ff.).

name indicates, they concern the content of objective appearance itself less than the place which we ourselves give it in empirical thinking. Whether we regard a substance simply as possible, whether we regard it as empirically actual or even as necessary, changes nothing in its nature as such, and does not add a single new characteristic to its concept; but it comprises a different status in the whole of our knowledge, which we give it. Thus the categories of modality, which express this threefold status, have the peculiarity "that, in determining an object, they do not in the least enlarge the concept to which they are attached as predicates. They only express the relation of the concept to the faculty of knowledge. Even when the concept of a thing is quite complete, I can still inquire whether this object is merely possible or is also actual, or if actual, whether it is not also necessary. No additional determinations are thereby thought in the object itself; the question is only how the object, together with all its determinations, is related to understanding and its empirical employment, to empirical judgment, and to reason in its application to experience."[59] This relation to the understanding signifies, accordingly, when considered more closely and designated more precisely, the relation to the system of experience, in which alone objects can be known as given, and hence also as "actual," "possible," or "necessary." What agrees with the formal conditions of experience (intuition and concepts)—according to the three modal postulates—is *possible*; what is bound up with the material conditions of experience (sensation) is *actual*; that which in its connection with the actual is determined according to universal conditions of experience is (or exists) *necessarily*. We see that here it is not a matter of defining the purely formal logical concepts of the possible, actual, and necessary; the distinction between the three stages is, rather, the result of a completely specific epistemological concern. Something would be called possible, in the sense of general logic, which included no contradictory characteristics and hence no internal contradiction, but by the criterion we are examining here, the assurance that this is not the case is far from sufficient. For even without bothering about a formal ab-

59. Ibid., A 219 = B 266 (III, 195).

sence of contradiction, a definite concept can nevertheless be so completely empty for us that it does not unambiguously determine any object of knowledge at all. Thus there is nothing contradictory in the concept of a figure enclosed by two straight lines, since the concepts of two straight lines and their intersection do not include the negation of a figure, and yet this concept can refer to no particular spatial form that is essentially distinguishable from other forms. To get such a form, we would have to pass from the analytic rules of logic to the synthetic conditions of construction in pure intuition. But even the addition of these latter conditions is not enough to yield the full, concrete sense of the possible, which is what is to be defined here. We accomplish this only when we come to know that the pure synthesis of space as such is necessarily ingredient in every empirical synthesis of perceptions, through which alone there arises the idea of a physical, sensory "thing." Thus, for example, the act of construction by which we trace in our imagination the form of a triangle would be entirely homogeneous with that act we perform in apprehending an appearance so as to make it into a concept of experience.[60] It is not the fulfillment of this or that particular condition, but the fulfillment of all conditions essential for the object of experience which thus comprises the true conception of the "possible."

The first modal principle, however, only asserts the validity of the formal conditions of experience—pure intuition and the pure concepts of the understanding. If, on the other hand, we progress from claiming possibility to claiming actuality, we see we are directed to a totally different cognitive factor. Something real *in concreto*, a definite individual thing, is given to us neither through a pure concept nor through pure intuition. For in the bare concept of a thing absolutely no mark of its concrete existence can be formed; and as to the constructive synthesis by which geometric forms arise for us, this also never goes as far as the individual determinations that are what we mean when we speak of the existence of a particular object. We construct "the" triangle or "the" circle as a schema and general model which can be actualized in infinitely many separate examples that are

60. Ibid., A 223 f. = B 271 f. (III, 198); cf. above, p. 177.

individualized and different one from the other; but as soon as we wish to extract one actual specimen from this ensemble of possible examples, as soon as we conceive a form as particularized in every part, as in, say, the length of its sides and the size of its angles, or in the specificity of its "here," its position in absolute space, then we have gone beyond the mathematical way of stating the problem and the foundations of mathematical knowledge in general. Only sensation includes reference to such details of the individual thing. "The postulate bearing on the knowledge of things as *actual* does not, indeed, demand immediate *perception* (and, therefore, sensation of which we are conscious) of the object whose existence is to be known. What we do, however, require is the connection of the objects with some actual perception, in accordance with the analogies of experience, which define all real connection in an experience in general."[61] Thus a specific substance need not in any way be capable of being sensed in order to be designated as actual, as existing, but it must at least exhibit that link with some sort of given perceptions that we call the system and the order of empirical causality (in the broadest sense). The existence of a magnetic material which penetrates all bodies, for instance, cannot be shown through immediate sense perception, but it is enough if it can be disclosed on the basis of observable data (as, say, the attraction of iron filings) through causal laws.

Thus, the relation of perceptions to laws of this sort, and conversely, the relation of the laws to perception, is what constitutes for us the essential nature of empirical actuality. "That there may be inhabitants in the moon, although no one has ever perceived them, must certainly be admitted. This, however, only means that in the possible advance of experience we may encounter them. For everything is real which stands in connection with a perception in accordance with the laws of empirical advance."[62] Neither do we have any criterion for the difference between dreaming and waking that is other than and better than the one laid down by this proposition. For this distinction can never be shown in the bare character of the con-

---

61. Ibid., A 225 ff. = B 272 ff. (III, 198 ff.).
62. Ibid., A 493 = B 521 (III, 350).

tents of consciousness as such, in the peculiar nature of the individual representations which are given to us in the one and in the other state, since these data are alike in both cases. The only thing that makes the decisive difference is that in one case we are able to integrate the totality of these data into one whole, which agrees with itself and is governed by laws, while in the other case they remain for us merely a disjointed conglomerate of individual impressions which displace one another.[63]

In this definition, the postulate of actuality at the same time makes direct contact with the postulate of necessity. For the meaning of necessity, as here understood, is not the formal and logical necessity of conceptual connections, but refers rather to a cognitive value that has its roots in empirical thinking, hence in physics. In this way of thought we call a determinate fact necessary not by claiming it to be a fact on the basis of observation, but by regarding and demonstrating this factuality to be the consequence of a universal law. In this sense, for example, the rules of planetary motion as stated by Kepler signify a merely factual determination, but they were raised to the rank of empirical necessity when Newton succeeded in finding a universal formulation of the law of gravitation, in which those rules are contained as special cases and are deducible from it mathematically. It is obvious that this necessity is nothing absolute, but merely hypothetical. It holds only under the presupposition that the major premise of the deduction—in our case the Newtonian law of attraction as in direct proportion to the masses and in inverse proportion to the square of the distance—is regarded as established and valid. Thus the existence of sensory objects can never be known completely a priori, "but only comparatively a priori, relatively to some other previously given existence. . . ."[64] The relation of perceptions to laws is therefore treated in the postulate of necessity just as it was in the postulate of actuality, but the orientation of this relation is different in the two cases. The one instance moves from the particular to the universal, while the other leads from the universal to the particular; the former

---

63. Cf. *Prolegomena*, §13, Remark III (IV, 40) (*Ak.* IV, 290).
64. *Critique of Pure Reason*, A 226 = B 279 (III, 203).

is tied to the individual instance that appears in sensation and perception, whereas in the latter the movement of thought is from the law to the individual case. The principle of actuality thus refers to the form of physical induction, that of necessity to the form of physical deduction. In this way it is shown that neither is an independent mode of procedure, but that they are reciprocally related and that only as thus correlated do they determine the overall form of experience in general. In this context we recognize once again the special place the modal postulates occupy within the system of synthetic principles: they no longer directly relate to the interconnection of empirical objects, but rather to the coherence of empirical methods; they are aimed at determining the relative justification of each method, and defining its significance in the whole of experiential knowledge.

4

The "subjectivity" that was the starting point of transcendental reflection has until now been presented in a precisely defined, terminologically restricted sense. It meant going in no way beyond the bounds of the individual knower, nor beyond the psychological processes through which the world of sensations, of ideas and their connection, is generated for an individual. Rather, it held fast only to this: that determination of the pure *form* of knowledge must precede determination of the object of knowledge. In conceiving space as a unitary synthetic procedure, the lawfulness of geometric and physical geometric forms is revealed to us. When we analyze the method of experiment, and point out the pure concepts of magnitude and mass, and the universal presuppositions of permanence and causal dependence in it, we have thereby accounted for the universality and the objective validity of experiential judgments through their true origin. The "subject" spoken of here is hence none other than reason itself, in its universal and its particular functions. In this sense alone could we style Kant's system "idealism"; the ideality to which it is related and on which it rests is that of the highest rational principles, in which all special and derivative results are prefigured in some sense and by which they are necessarily "determined a priori."

But is there not a completely different sense of subjectivity, which, although it is not the starting point for the *Critique of Pure Reason,* still at least merits its consideration? And are there not other forms of idealism well recognized in the history of philosophy, in contradistinction to which this new doctrine must be sharply and surely delimited if it is not to foster continual misunderstandings? No problem of exposition occupied Kant's thoughts so deeply and lastingly as this one. Again and again he sought to distinguish the special nature of his "critical" idealism from Descartes's "sceptical" or "problematic" idealism; against the "dogmatic" idealism of Berkeley he tried to guard his own fundamental ideas, which concern simply the determination of the form of experience, from confusion with common and material psychological idealism. But although Kant could explain all confusion of this sort simply as due to an "almost deliberate misapprehension," it appears in a different light to purely historical judgment. For an essential basic element of the First Critique is precisely that it contains as much a novel doctrine of consciousness as it does a new theory of the object. If his contemporaries singled out the former component above all from the whole of the critical system and tried to interpret the entire *Critique* by it, the primary reason for this was their discovery of a language of philosophical concepts that seemed to afford the quickest link with accepted modes of thought. For while in the objective deduction of the categories, in the proof that the conditions of the possibility of experience are at the same time conditions of the possibility of the objects of experience, Kant had to create singlehandedly not only the concepts themselves but also their logical expression, in the subjective deduction he concurred throughout with the common psychological nomenclature of his time. Hamann says in a letter to Herder that Tetens's major work, the *Philosophische Versuche über die menschliche Natur,* lay open on Kant's table during the writing of the First Critique.[65] Thus the impression might arise that what was created here is a novel transcendental substructure for empirical psychology, that concrete psychological facts were translated into a different, metaphysical language.

65. See *Hamanns Schriften,* ed. F. Roth (Augsburg, 1821–43), vol. 6, p. 83.

In truth, however, the First Critique is aimed against psychological "idealism" as much as it is against dogmatic "realism," since it is intended no less as a critique of the concept of ego than as a critique of the concept of the object. Psychological metaphysics, which found its typical expression in Berkeley's system, is characterized through and through by its assertion of the certainty of the ego as an original datum, and the certainty of external things as merely inferred datum. In the existence of the ego we possess an immediate and indubitable concrete existent, while everything else that we designate by the name of "reality," as in particular the being of things in space, is dependent on the fundamental fact of the ego. Thus "souls" (and the infinite spirit of God which stands over them) comprise the sole truly "substantial" actuality. The totality of what we call "existence" can be expressed and understood, then, as nothing but psychic substance, as "perceiver" or "perceived." Kant divorced himself from this view chiefly in that for him the ego, the psychological unity of self-consciousness, formed a terminus, not a starting point, of the deduction. If one does not judge from the standpoint of an absolute metaphysics but from the standpoint of experience and its possibilities, it is obvious that the fact of the ego has no preeminence and no prerogative over and above other facts attested to by perception and empirical thought. For even the self is not given to us originally as a simple substance: the idea of it only arises in us on the basis of the same synthesis, the identical functions of unification of the manifold by which sense-content becomes the content of experience, impression becomes object. Empirical self-consciousness does not precede the empirical consciousness of objects temporally and concretely; rather, in one and the same process of objectification and determination the whole of experience is divided for us into the field of the inner and of the outer, the self and the world.[66]

In the "Transcendental Aesthetic" time had already been called the "form of inner sense, that is, of the intuition of ourselves and of our inner state."[67] In this first condition immediately all others are

66. See the Refutation of Idealism, *Critique of Pure Reason*, B 274 ff. (III, 200 ff.).
67. Transcendental Aesthetic, §6 (III, 65).

ultimately contained, for now it is only a matter of analyzing the consciousness of time itself, so as to elicit in detail all the determining
factors that make it up. The presence of a problem here appears most
clearly if we ask what is the basis for the possibility of comprehending
a temporal *whole* in thought and showing that it is a definite unity.
This possibility may be understandable in the case of space, for since
by its essential concept its parts are taken to be simultaneous, nothing
further seems to be needed than to pull together that which exists
simultaneously also in thought, so as to arrive at the intuition of a
definite spatial extension. A single instant of time is, on the other
hand, characterized by its being given as the fleeting, pointlike boundary between past and future, by its thus basically existing only as
something singular, excluding all other moments. Only the indivisible, present "now" actually exists, while every other point in time
must be regarded as something which is not yet or which is no longer.
Thus it is obvious that no aggregate, no sum of the individual elements in the ordinary sense, is possible in this case, for how can a
sum be formed when the first member vanishes as I move on to the
second one? But if a whole, the totality of an entire series, is to be
posited in time—and this is precisely what constitutes the necessary
presupposition   for   that   unity   we   call   the   unity   of   self-
consciousness—it must be at least indirectly possible to hold on to
the moment without losing thereby the general nature of time as a
continuous progress and transition. Temporal moments may not be
simply posited and apprehended, but they must be created anew repeatedly; the "synthesis of apprehension" must operate simultaneously and within one and the same indivisible fundamental act as a
"synthesis of reproduction."[68] In this way the present can be added
to the past, the past preserved in the present, and both thought
jointly. Basically, however, the temporal process would not yet be
grasped as a unity even so if reproduction, when complete, were not
at the same time also known to be reproduction: that is, if what is
posited severally and at diverse points in time were not nonetheless
determined by thought to be one, as identical. To all the diversities in

68. See the *Critique of Pure Reason*, A 100 ff. (III, 613).

the qualitative contents of sensation and all the multiplicity of temporal locations, however essential they are to pure intuition, the unity of the synthesis of the understanding must be superadded. "If we were not conscious that what we think is the same as what we thought a moment before, all reproduction in the series of representations would be useless. For it would in its present state be a new representation which would not in any way belong to the act whereby it was to be gradually generated. The manifold of the representation would never, therefore, form a whole, since it would lack that unity which only consciousness can impart to it. If, in counting, I forget that the units, which now hover before me, have been added to one another in succession, I should never know that a total is being produced through this successive addition of unit to unit, and so would remain ignorant of the number. For the concept of the number is nothing but the consciousness of this unity of synthesis. The word 'concept' might of itself suggest this remark. For this unitary consciousness is what combines the manifold, successively intuited, and thereupon also reproduced, into one representation. This consciousness may often be only faint, so that we do not connect it with the act itself, that is, not in any direct manner with the *generation* of the representation, but only with the outcome [that which is thereby represented]. But notwithstanding these variations, such consciousness, however indistinct, must always be present; without it, concepts, and therewith knowledge of objects, are altogether impossible."[69]

Only in this last stage of the synthesis, this "recognition in a concept," does that substance arise for us which we oppose to sheer flux and alteration of sensory impressions and representations as the "abiding and unchanging 'I'." Although it appeared that sensualism had provided an adequate explanation of the concept of the self by calling the self a loose structure of separate psychic entities, a mere "bundle of perceptions," that explanation rested on an extremely crude and incomplete analysis, as is now demonstrated. For aside from the fact that even the loosest and most superficial form of that connection would already involve a critical epistemological problem,

69. *Critique of Pure Reason*, A 103 f. (III, 614 f.).

the transcendental inversion holds here also. The self, so far from being the product of individual perceptions, actually constitutes the fundamental presupposition for something that can in general be called "perception." The "self" as identical reference point imparts to what is particular and diverse its qualitative meaning as content of consciousness. In this sense the ego of pure apperception is the "correlate of all our representations," with regard to the pure possibility of our becoming conscious of them, and "all consciousness as truly belongs to an all-comprehensive pure apperception, as all sensible intuition, as representation, does to a pure inner intuition, namely, to time."[70] The unity of time, in and through which alone there is for us a unity of empirical consciousness, is thus here traced back to universal conditions, and on closer analysis these conditions, together with the principles flowing from them, are demonstrated to be the same ones on which all assertion of objectively valid interconnections, and hence all knowledge of the object, rest.

Only now is the relation between inner and outer experience, between self-consciousness and consciousness of the object, clarified. These two do not comprise "halves" of experience as a whole, which subsist independently of each other, but they are conjoined in the same ensemble of universally valid and necessary logical presuppositions, and inseparably related to each other through this ensemble. We now no longer ask how the "I" makes contact with things in themselves, nor how things in themselves begin to participate in the "I." Now the expression for both "self" and "object" is one and the same: the lawfulness of "experience in general" signified in the concept of transcendental apperception. This is the sole mediator and agency for us of any entities whatsoever, be they of inner or of outer sense.

The moment we fail to grant this meaning and source of the concept of the ego, we are straightway entangled in all the insoluble problems that crop up in every metaphysical psychology. If we cease to think the "transcendental unity of apperception" in the form of a pure condition, if we try to intuit it as something given and existing

70. Ibid., A 123 f. (III, 625).

and to make it imaginable, we blunder onto the path of a dialectic which, step by step, consequence by consequence, becomes more and more difficult and complicated. This dialectic arises whenever we try to convert a definite relation, which is valid within experience and for the purpose of uniting its separate members, into an independent substance prior to all experience. In this conversion of a pure relation into an absolute substantiality there lurks no mere chance or personal deception for which the individual empirical subject would be responsible: we have to do here rather with a sophistry of reason itself, which is unavoidable before it is fully revealed by the transcendental critique and the reasons for it completely fathomed. A new field of questions and tasks is thus presented to the First Critique. The "Transcendental Aesthetic" and the "Transcendental Analytic" were aimed at showing the conditions of *genuine* objectification, which takes place in experience and by means of its principles, whereas the "Transcendental Dialectic" is oriented negatively, toward guarding against the false objects generated for us by transgression of these conditions; the "logic of truth" was occupied with the former, the "logic of appearance"[71] with the latter.

If we first apply this conceptual distinction simply to the psychological problem, it will be a matter of rendering comprehensible the illusion that results from the hypostatization of the universal unifying function of consciousness into a particular simple soul substance. All the paralogisms of rational psychology, all the fallacies of the pure metaphysical theory of the soul, are rooted in this hypostatization. For the whole previous conception of the soul rests on the fact that we abstract a unity that can be shown in the series of phenomena of consciousness itself and whose necessity is demonstrable within this realm from the totality of this very series, and ascribe this unity to an original, self-subsistent substrate, of which the particular phenomena of consciousness are supposed to be but an indirect consequence. Thus instead of simply thinking the phenomena themselves as interconnected, we now superadd to them in thought a nonempirical ground, from which we attempt to explain and derive

71. See ibid., A 60 f. = B 85 f., A 293 ff. = B 349 ff. (III, 86, 244 ff.).

the multiplicity of phenomena. A simple, indivisible, and unchangeable something is now posited, and, although its general form as a thing is analogous to things in space and comparable with them, nevertheless it is essentially distinguished from them by its specific structure, and hence it supposedly never enters into any relation with them except a purely contingent and dissoluble one. But this assertion, and hence all assertions about immaterial nature and about the permanence of the soul, is the basis of the same unresolved contradiction. The sole text of rational psychology is the proposition "I think," which must be able to accompany all our representations in that they are to be explained only by its means as belonging to one and the same self-consciousness, whether it be expressly attached to them or only latent in them.

But this reference of all psychic contents to one common central point neither says anything whatsoever about any sort of enduring concrete existent to which it points, nor does it determine a single actual predicate belonging to this concrete existent. It is indubitably certain that the concept of the "I" as a constant unity, identical with itself in all particular representing and thinking, is met with again and again, but that in no way gets us to the intuition of a self-existent object corresponding to this concept. Every inference from the logical unity of intellectual functioning to the real and metaphysical unity of the substantial soul means, instead, a μετάβασις εἰς ἄλλο γένος, an illicit transition to a completely different field of problems. "It follows, therefore, that the first syllogism of transcendental psychology, when it puts forward the constant logical subject of thought as being knowledge of the real subject in which the thought inheres, is palming off upon us what is a mere pretence of new insight. We do not have, and cannot have, any knowledge whatsoever of any such subject. Consciousness is, indeed, that which alone makes all representations to be thoughts, and in it, therefore, as the transcendental subject, all our perceptions must be found; but beyond this logical meaning of the 'I,' we have no knowledge of the subject in itself, which as substratum underlies this 'I,' as it does all thoughts. The proposition, '*The soul is substance*,' may, however, quite well be allowed to stand, if only it be recognized that this concept [of the soul as substance] does not carry us a single step further, and so cannot yield us any of the usual deductions of the pseudo-rational doctrine of the soul, as, for instance, the everlasting duration of the human soul in all changes and even in death—if, that is to say, we recognize that this concept signifies *a substance only in idea,* not in reality."[72] And it is precisely in this that the intellectual labor the transcendental dialectic has to perform at this point consists: to completely transform the earlier metaphysical definitions of the soul as substance into epistemological definitions of the soul as Idea. The "I," the "transcendental apperception," is permanent and unchangeable, but it is only an invariable relation *between* the contents of consciousness, not the unvarying substratum *from* which they arise. It is simple and undivided, but this is only so relative to the synthetic act of unification of the manifold, which as such can either be thought only totally and completely or else not thought at all. No bridge leads from the indivisibility of this act to the assertion of an indivisible thing that stands behind it and as its foundation. Hence my own simplicity (as soul) is not deduced from the proposition "I think," but is already involved in that very idea. "The proposition '*I am simple,*' must be regarded as an immediate expression of apperception, just as what is referred to as the Cartesian inference, *cogito, ergo sum,* is really a tautology, since the *cogito* (*sum cogitans*) asserts my existence immediately. '*I am simple*' means nothing more than that this representation, 'I,' does not contain in itself the least manifoldness and that it is absolute (although merely logical) unity."[73]

The terms in which the problem is put in the "Transcendental Dialectic" and its basic tendency are even more sharply evident in the critique of the concept of the cosmos than in that of the concept of the soul. First of all, it seems as though the "Transcendental Analytic" had arrived at a finally valid answer to the question, for what does the concept of the cosmos assert except the concept of "nature," and what is nature, according to the highest principle of all synthetic judgments, except the whole of possible experience, whose structure

72. Ibid., A 350 f. (III, 637).
73. Ibid., A 354 f. (III, 639).

and limits have been precisely established by the system of the pure principles of the understanding? But in speaking of the whole of experience we have already suggested the new problem that goes beyond the limitations of the Analytic. The experience, the possibility of which we inquired into, is not a particular sort of thing, but a specific "mode of knowledge." It signifies an ensemble of modes of understanding, which science employs less to construct a given actuality than to accomplish the universal and necessary connection of phenomena, which we call their "reality." Still, seen from this perspective, it is not a finished product but instead a process that progressively shapes itself. We are able to determine the conditions of this process, but not its outcome. In that way, to be sure, an unambiguous direction is prescribed to our cognition of experience, since its progress is furthered by universal and constant fundamental methods; however, in so doing its sum and its outcome are not equivalently indicated and fixed. What is available to us here is an ensemble of various ways of determining objects, but the goal to which these ways all point is in fact never reached by any of them.

Thus we are in command of the fundamental forms of pure space and pure time, by which we combine appearances into the orders of coexistence and succession; thus with the help of the causal concept we abstract from the manifold of events specific causal series and sets of causal series. The process of determination, however, never reaches an ultimate terminus this way, for not only does an individual member in every particular series always point to another that precedes it, without our ever succeeding to a last member, but also, when we grasp each series as a unity, the moment we wish to indicate how it coordinates with other series and depends on them, the result is a nexus of ever-new functional connections, which, when we try to follow it out and express it, leads us straight into the indefinite distance. What we call "experience" consists in such a set of progressive relations, not in a whole of absolute data. The demands posed at this point not only by dogmatic metaphysics but also by naive realism are in no way satisfied by this. For the mark of this view is not only that it wishes to think the object as progressively determined through experiential knowledge, but that it assumes the world, as a totality, prior to the process of this determination. Though we may *conceive* the world in our empirical knowledge only piecemeal and fragmentarily, it is nonetheless present as a whole, finished and perfect in every respect. But now the transcendental critique asks what this "being present" means. It is clear that it cannot be demonstrability in immediate sensation and perception that is meant, since what must be stressed here is precisely that the portion of being given us from time to time in actual perception constitutes but a vanishingly small fragment of the whole. Hence it is once again a definite hallmark of the *judgment* of objectivity that we have before us in this assertion of a present, closed world. At the very least, it is important to understand this judgment and evaluate its peculiar logic—even if we have to deny the absolute existence of the object to which it refers.

So from the standpoint of transcendental reflection, we must first of all begin by admitting that the comparison between experience and the object, as it has been maintained and understood up until now, does not contain any ultimate and unambiguous solution to our question. For the necessity for thought to reach out beyond what is empirically known and given is undeniable. If, in the critical sense, we regard experience as a product of intuition and understanding, if we distinguish within it the individual conditions of space, time, magnitude, substantiality and causality, etc., it appears that, when we single out any one of these functions, it is never exhausted in any specific result. As, for instance, according to a proposition of the "Transcendental Aesthetic," the infinity of time signifies nothing more than that all definite temporal magnitudes are possible only through limitations of one single fundamental time, an analogous infinity is attributable to each particular form of pure synthesis. Every determinate quantum is thinkable only on the basis of the universal procedure of assigning quantity and defining it; every individual case of causal connection is thinkable only as the specification of the causal principle in general. By means of this infinity, which is already incorporated in its pure logical form, each of the constitutive factors in experiential cognition insists on its exhaustive application, which exceeds every actually attained limit. Each cause that we can point to in experience has only a limited and relative being, since we can always posit it as something individual by relating it to another, more remote cause. But the principle and the idea of causality is unrestrictedly valid. To carry out this principle in a systematically complete way throughout the whole field of phenomena, so that no one phenomenon as ostensibly "ultimate," and hence not ever traceable back to something further, stands in our path and attempts to block our progress—that is a demand raised by reason itself and founded in reason. "Reason" in the specific sense this concept has, through the "Transcendental Dialectic," means nothing but this very demand.

"Understanding may be regarded as a faculty which secures the unity of appearances by means of rules, and reason as being the faculty which secures the unity of the rules of understanding under principles. Accordingly, reason never applies itself directly to experience or to any object, but to understanding, in order to give to the manifold knowledge of the latter an a priori unity by means of concepts, a unity which may be called the unity of reason, and which is quite different in kind from any unity that can be accomplished by the understanding."[74] The categories of the understanding, taken together, are only means of leading us from one conditioned thing to another, while the transcendental concept of reason invariably proceeds to the absolute totality in the synthesis of conditions, and hence never terminates in what is absolutely (i.e., in every relation) unconditioned. "Reason accordingly occupies itself solely with the employment of understanding, not indeed insofar as the latter contains the ground of possible experience (for the concept of the absolute totality of conditions is not applicable in any experience, since no experience is unconditioned), but solely in order to prescribe to the understanding its direction toward a certain unity of which it has itself no concept, and in such manner as to unite all the acts of the understanding, in respect of every object, into an *absolute whole*."[75] But the justified transcendental claim contained herein immediately becomes transcendent if one attempts to exhibit it under the image of an absolute thing; if one makes the totality of beings into an existing,

74. Ibid., A 302 = B 359 (III, 250).
75. Ibid., A 326 = B 383 (III, 264).

given object—which constitutes the endless task of experiential knowledge. That which was not only permissible but necessary, which was regarded as a maxim and as a guide for empirical inquiry, now appears as a substance that on closer analysis disintegrates into obviously contradictory factors and individual attributes. Thus, concerning the world as a given whole, we can successively demonstrate, with equal logical justification, that it has a beginning in time and a limit in space, as well as that in respect of time as well as space, it is infinite; thus with the same validity it can be shown that the world is composed of absolutely simple substances, as well as that division, in the physical realm and also in pure space, is never complete and hence that absolute simplicity is an unrealized idea.

The real basis of all these antinomies in the concept of the world, the content and systematic significance of which had already emerged in the historical evolution of Kant's thinking,[76] can now be indicated in its full import and simplicity from the general presuppositions of the critical system. That two diametrically opposite determinations and conclusions can be deduced from a single concept is possible only if the latter itself already contains an internal contradiction in its construction and in the original systhesis on which it rests. In our case, though, this contradiction, inspected more closely, lies in the fact that the content of the world concept is bound up with the definite article, that "the" world is used as a substantive. For experience as a whole is never given to us as such, as a rigid, closed entity; it is not a result lying behind us, but a goal lying before us. The state we ascribe to it is therefore ultimately grounded in nothing other than the rule of progress itself, in which, starting from what is particular, we ascend to the concept of the world as the total nexus of empirical being. This rule, for its part, has its definite objective validity also, but it cannot be thought in the form of a "thinglike" whole, which would be given together with all its parts. It cannot determine what the object may be, only how the empirical regress is to be carried on so as to arrive at the complete concept of the object.[77] "It [the rule] cannot be regarded

76. See above, pp. 110 ff.
77. *Critique of Pure Reason*, A 510 = B 438 (III, 360 f.).

as maintaining that the series of conditions for a given conditioned is in itself either finite or infinite. That would be to treat a mere idea of absolute totality, which is only produced in the idea, as equivalent to thinking an object that cannot be given in any experience. For in terms of it we should be ascribing to a series of appearances an objective reality that is independent of empirical synthesis. This idea of reason can therefore do no more than prescribe a rule to the regressive synthesis in the series of conditions; and in accordance with this rule the synthesis must proceed from the conditioned, through all subordinate conditions, up to the unconditioned. Yet it can never reach this goal, for the absolutely unconditioned is not to be met with in experience."[78]

In this sense, the idea of totality is "regulative," not "constitutive," because it contains only a prescription as to what we are to do in the regress, but does not determine and anticipate what is given in the object prior to any regress. The distinction set up herein concerns only transcendental reflection on the source of the principle, but not its actual empirical employment. In regard to this latter, it is "a matter of indifference whether I say that in the empirical advance in space I can meet with stars a hundred times farther removed than the outermost now perceptible to me, or whether I say that they are perhaps to be met with in cosmical space even though no human being has ever perceived or ever will perceive them."[79] For the presence of an empirical object, viewed more exactly, means and can mean nothing more than its determinability, direct or indirect, by way of the empirical method: by sensation or pure intuition, by the "analogies of experience" or the "postulates of empirical thought," by the synthetic principles or the regulative Ideas of reason. Accordingly, if I think to myself all existing objects of sense in the whole of time and space, I do not place them in both prior to experience; rather, this thought is nothing but the idea of a possible experience in its absolute perfection.[80] This idea is as such unavoidable, but it is entangled in contradictions the moment we elect to isolate and hypostatize its content, which is to say, the moment we fabricate a thing unrelated to it,

78. [Ibid.]
79. [Ibid., A 496 = B 524 (III, 352).]
80. Ibid., A 495 = B 523 f. (III, 352).

intended to correspond to it, instead of using it and holding fast to it as a guideline within empirical inquiry.

This insight yields the principal solution to those problems covered by the *Critique of Pure Reason* in its third and final part. The critique of rational psychology and cosmology moves on to the critique of rational theology: the analysis of the Idea of the soul and that of the cosmos terminates in the analysis of the Idea of God. Here, too, consistent with the general methodological tendency, the point is to show that in the Idea of God it is not so much that a definite absolute entity is thought, as rather that a special principle of possible experience is posited and hence a mediate relation to the general tasks of empirical inquiry is set up. But this turn contains a paradox. For does not the whole meaning of the concept of God lie in His "transcendence"? Does it not lie precisely in the fact that we assert the certainty of an ultimate Being, which exists apart from all contingency and conditions of finite empirical being? This is the sense in which the concept seems to have been assumed by all previous metaphysics from the time of Aristotle. From time immemorial it had been concluded that if there is no entity which is purely "in itself" and "through itself"—then neither is the being of any secondary and dependent thing thinkable; thus all actuality as a whole dissolves into insubstantial illusion. Even Kant's own precritical essay, "The Only Possible Basis of Proof for a Demonstration of God's Existence," stood by and large within this basic outlook; indeed, it strengthened and confirmed it, by seeking to demonstrate the absolutely necessary Being as the ground, not only of everything actual, but also of all possible being, or of all truth of conceptual and ideal relations.[81] From the critical viewpoint, however, this reflection too must be reversed. Instead of moving from a universal concept of the logically possible to the special concept of the possibility of experience, now it is instead "possible experience" which is viewed as the foundation that can confer on all concepts, as cognitive, their value and their objective validity.

And with that, the whole ontological way of reasoning, the foun-

81. Cf. above pp. 62 ff.

208 THE *CRITIQUE OF PURE REASON*

dation of all previous rational theology, has become untenable. For the core of all ontology is that inference can be made from the concept of the most perfect Being to its existence, because existence itself is a perfection which thus cannot be excluded from the predicates of this concept without contradiction. From the transcendental standpoint, though, it is finally realized that existence in general is not a specific conceptual predicate, standing on an equal footing with others, but that it is a problem of knowledge, which must be progressively defined and mastered with every epistemological means. Only the sum total of these techniques can define for us what it means in general to exist empirically. Here neither the mere analytical logical concept nor the pure intuition of space and time, nor even sensation and sense perception suffices; it is only the mutual relation of all these factors on which experience, and in and through that the object, is founded for us. Within the system of synthetic principles, it was above all the "Postulates of Empirical Thought," and of them especially the "Postulate of Actuality," that established this connection. In these we learned how sensation, intuition, and concept must cooperate to result in any valid statement about a "concrete existent." Ontology, however, not only arbitrarily and onesidedly abstracts the function of thinking from this whole complex, but it takes thinking itself to be the merely analytic dissection of a given conceptual content, instead of the synthetic function of combining in relation with the manifold of intuition.

But regarded thus, thinking is blocked from any access to being and from any foothold therein. It can now only infer, by a *petitio principii*, from the possible to the actual; for the simple reason that, purely from its own resources, it neither knows nor understands the full difference between possibility and actuality. A hundred actual dollars contain, if I simply reflect on the concept and on the predicates that can be abstracted analytically from it, not the least bit more than a hundred possible dollars. "By whatever and by however many predicates we may think a thing . . . we do not make the least addition to the thing when we further declare that this thing *is*. Otherwise, it would not be exactly the same thing that exists, but something more than we had thought in the concept; and we could not, therefore, say

THE *CRITIQUE OF PURE REASON*

that the exact object of my concept exists. . . . When, therefore, I think a being as the supreme reality, without any defect, the question still remains whether it exists or not. For though, in my concept, nothing may be lacking of the possible real content of a thing in general, something is still lacking in its relation to my whole state of thought, namely, [insofar as I am unable to assert] that knowledge of this object is also possible a posteriori. . . . For through the concept the object is thought only as conforming to the *universal conditions* of possible empirical knowledge in general, whereas through its existence it is thought as belonging to the context of experience as a whole. In being thus connected with the *content* of experience as a whole, the concept of the object is not, however, in the least enlarged; all that has happened is that our thought has thereby obtained an additional possible perception."[82]

Hence only connection with the content of experience and the context of its assertions and judgments has the power to justify every statement about actuality. If the existence of God is to be shown demonstratively at all, we seem to be guided from the a priori proof of ontology to the a posteriori forms of proof, to the cosmological and the physico-theological proofs. The former follows from the circumstance that, within the series of causes in the world, we always pass from one conditioned and dependent existence to another, so that by this route the absolute ground of the whole series is never evident; thus, it must be sought outside the series in the existence of a being which exists as *causa sui,* not through some other being but through itself. The second proof concludes from the rational and purposive order that is visible in individual parts of the cosmos and in its overall structure to a highest intelligence from which it has its origin and by which it is conserved in its continuing state. But aside from the internal logical flaws in these proofs, which Kant had already recognized and revealed early on,[83] they are invalid because they are only seemingly independent and self-sufficient. They are proposed in previous metaphysics as a support and supplement to the ontological

---

82. *Critique of Pure Reason,* A 600 f. = B 628 f. (III, 414 f.).
83. See above, pp. 58 f.

proof, but in truth they completely presuppose it in its entirety. For even if it were assumed that the path of the cosmological proof could be followed to a supreme cause of the world, or that inference can be made from purposiveness internal to appearances to a rational ground of the world, it would not be thereby demonstrated that this cause and this ground are identical with what we are accustomed to designate by the concept and the name of God. In order to make this identification, involving not only the existence of an ultimate ground but also its more precise description and its specific predicates, we are forced back into the use of the ontological proof. We must try to show that the absolutely substantial and necessary Being is at the same time the most real Being, that all reality and perfection is included in it and derivable from it. The circularity of the proof therefore becomes blatant, for what is produced here to confirm the ontological proof is devoid of any precise and unambiguous definition, so long as it itself is not presumed to be valid by presupposition.[84]

In general, the critique of the proofs of God again uncovers the basic flaw for which Kant reproaches all previous metaphysics: that in it the true relation between experience and thought is not recognized accurately and surely, and expressed with clear consciousness. The kind of thought that shuts its eyes to everything else in order to spin actuality out of itself is shown to be compelled at last to submit to this very actuality, because it smuggles into its presuppositions certain fundamental empirical determinations. When it does so, however, it muddies the character of pure thinking on the one hand, and equally, on the other, it falls short of the pure concept of experience.

In place of this, the "Transcendental Dialectic" now seeks, here as elsewhere, to transform the negative outcome of the critique of the proofs for God into a positive insight, by bringing into relief a factor in the earlier understanding of the concept of God which, when translated from the language of metaphysics into that of transcendental philosophy, has essential significance for the nature of experience itself and its ongoing process. In metaphysics, God is thought as the

---

84. *Critique of Pure Reason,* A 606 f. = B 634 f. (III, 418 f.). For the whole discussion, see A 603–30 = B 631–59 (III, 416–33).

most real Being, that is to say, as that which unites in itself all pure affirmations and perfections, while excluding all negations and defects. In this, nothing but absolute being, devoid of all nonbeing, is asserted, for that a thing is something and not something else, that a definite predicate *a* belongs to it while other predicates *b, c, d,* . . . are denied of it, is simply the expression for the fact that something is thought as limited and finite. The proposition *omnis determinatio est negatio* accurately designates the nature and method of that definition which is the only one possible here, in the realm of empirical, finite existence, since in positing such an existent, we have at the same time severed it from the totality of reality and assigned to it only a limited sphere of reality. In God, on the contrary, we do not think a determinate individual as distinguished from others; we think the perfect Ideal of total determination. Here we conceive the idea of a "sum of reality," which is not only "a concept which, as regards its transcendental content, comprehends all predicates *under itself;* it also contains them *within itself;* and the complete determination of any and every thing rests on the limitation of this *total* reality, inasmuch as part of it is ascribed to the thing, and the rest is excluded. . . ."[85]

For its own purpose, however, reason does not need the existence of such a being in accord with the ideal, but only the *Idea* of it. "The ideal is, therefore, the archetype (*prototypon*) of all things, which one and all, as imperfect copies (*ectypa*), derive from it the material of their possibility, and while approximating to it in varying degrees, yet always fall very far short of actually attaining it. All possibility of things . . . must therefore be regarded as derivative, with only one exception, namely, the possibility of that which includes in itself all reality. This latter possibility must be regarded as original. . . . All manifoldness of things is only a correspondingly varied mode of limiting the concept of the highest reality which forms their common substratum, just as all figures are only possible as so many different modes of limiting infinite space. The object of the ideal of reason, an object which is present to us only in and through reason, is therefore entitled the *primordial being (ens originarium).* As it has nothing above

85. [Ibid., A 577 = B 605 (III, 401).]

it, it is also entitled the *highest being (ens summum)*; and as everything
that is conditioned is subject to it, the *being of all beings (ens entium)*."[86]
But just as space, which underlies all particular shapes, is not to be
thought as a separate, absolute thing, this "thing of all things" that is
asserted in the concept of God is still to be understood, in the transcendental sense, as "form," although as form belonging to a realm of
validity completely separate from that of the forms of sensibility and
the concepts of pure understanding. Its essential value lies in its
regulative significance, as with all the Ideas of reason. Experience as
one and all-encompassing, and its coherence under laws, is that
wherein alone everything that is real for us in all particular appearances can be given. That this "whole" of experience precedes all
individual empirical experiences was, in point of fact, the insight on
which the *Critique of Pure Reason*'s solution to the riddle of synthetic a
priori judgments rested. This whole was to be thought primarily as an
ensemble of principles and fundamental propositions, but it is determined in and through these principles as an ensemble of objects as
well. We cannot fix a particular empirical object in any way except by
assigning it a place within this system relative to all other elements of
this ensemble, actual or even only possible.

Hence we have arrived at the transcendental analogue to the
metaphysical concept of God as the "most real being." But at the
same time we see that the totality we find ourselves referred to here is
not the totality of absolute existence but is only the expression of a
definite epistemological postulate. For the qualitative whole of the
objects of possible experience is like the quantitative whole, to which
we are accustomed to give the name "world"; it is never a whole that
is given, but always one set us as a task. The dialectical illusion of
transcendental theology is generated as soon as we hypostatize this
Idea of the ensemble of all reality. We are betrayed into doing this by
a natural illusion of the understanding, since we "substitute dialecti-
cally for the *distributive* unity of the empirical employment of the
understanding, the *collective* unity of experience as a whole; and then
think this whole [realm] of appearance as one single thing that con-

86. Ibid., A 578 = B 606 (III, 401 f.).

tains all empirical reality in itself; and then... substitute for it the concept of a thing which stands at the source of the possibility of all things, and supplies the real conditions for their complete determination."[87] Three stages of this false dialectical reification can be distinguished: the ideal of the most real being is first *realized*, that is, combined generally into the concept of an object; next it is *hypostatized*; and last of all *personified*, in that we bestow intelligence and self-consciousness on it. But from the standpoint of pure theoretical reflection, the whole idea of the divine essence and self-sufficiency formed in this way is resolved into a mere "transcendental subreption," into an intellectual fraud by which we attribute objective reality to an Idea that functions solely as a rule.[88]

With this insight we come to the end of the "Transcendental Dialectic," and hence of the entire structure of the critique of pure theoretical reason. This critique has discovered the universal and necessary conditions of all objective judgments, and therefore of all objective assertions possible within experience. Since it refers the empirical object back to these conditions and confines it within them, it has thus defined the empirical object as the object of appearance. For "appearance," understood in the purely transcendental sense, signifies nothing other than the object of a possible experience; thus it does not denote the object thought "in itself" and apart from all cognitive functions, but the object which is mediated precisely through these functions, through the forms of pure intuition and of pure thinking, and is "given through their efficacy alone." If we now wanted to inquire what the object might be if we abstract from all these constitutive moments of it, if we no longer think it in space and time, no longer as extensive or intensive magnitude, no longer in the relations of substantiality, causality, reciprocity, etc., we must acknowledge this question as such not to be self-contradictory. A contradiction arises only where I combine two positive predicates that are antithetical into a single concept and hence posit them jointly. Here, though, I have not posited anything in general; I have merely can-

87. Ibid., A 582 f. = B 610 f. (III, 404).
88. Ibid., A 509 = B 537 (III, 360).

celed out the conditions known to me under which I can posit something. The result is thus not a contradiction but pure nothing, insofar
as not the slightest further basis can be shown for the idea of such an
object existing in itself, apart from any relation to laws of the form of
cognition. The idea is, of course, possible in an analytic sense, under
the rule of formal logic, but not valid in a synthetic sense as any real
content of knowledge. And even when we do not abstract from the
conditions of knowledge to such a degree, as is conceptually possible,
if we think an absolute object not in the sense that it abstracts from all
formal principles of cognition, but only in that we assume between
these principles a different relation from that holding in any given
experiential knowledge, the same objection obtains. For what we
know as experience rests on the essential cooperation of those two
basic factors which the *Critique* has called sensibility and understanding, pure intuition and pure thinking.

On the other hand, we have no sort of positive concept of what
form an experience might have, in which one of these factors were to
be eliminated, or if a radically different relation to the other were
defined; indeed, we have no idea whether under this assumption any
form whatsoever, any definite, lawful structure of experience, would
remain. For we truly know only the relation between understanding
and intuition, not either separately as absolute element and substratum. If we detach pure thinking from its connection with pure
and empirical sensibility, its objective reference falls away for us; thus
it forfeits its specific "sense," as language characteristically expresses
it.[89] The functional unity resident in the categories is what basically
yields us positive cognitive content, so that it is schematized in the
form of space and time.

Thus the concept of magnitude cannot be explained except by
including in this explanation the "how many times" a basic unity is
iterated, but what this "how many times" means is comprehensible
only if one goes back to successive repetition, hence to time and the
synthesis of similar elements. In just the same way, if in the idea of
substance I omit the factor of permanence in time, the logical repre-

89. Ibid., A 240 = B 299 (III, 214 f.).

sentation of a subject would remain, which can never be the predicate of anything. The merely formal account has no way of determining, however, whether such a thing could be given as an object, be it of outer or inner experience. The same is true of the concepts of causality and community, which we also only deduced, that is, their validity could be shown for every determination of the empirical object, in that we recognized them as related to spatiotemporal intuition and as presuppositions of the order therein. "In a word, if all sensible intuition, the only kind of intuition which we possess, is removed, not one of these concepts can in any fashion *verify* itself, so as to show its *real* possibility. Only *logical* possibility then remains, that is, that the concept or thought is possible. That, however, is not what we are discussing, but whether the concept relates to an object and so signifies something."[90]

Thus the pure categories, shorn of the formal conditions of sensibility, have merely transcendental significance, but they have no transcendental (that is, exceeding the possibility of experience and its objects) use. Although their origin is a priori, still the application that we can make of them is invariably empirical, in the sense that they are limited to the bounds of experience "and that the principles of pure understanding can apply only to objects of the senses under the universal conditions of a possible experience, never to things in general without regard to the mode in which we are able to intuit them."[91] The concept of a "noumenon," that is, of a thing which is to be thought in no way as an object of sense, but as a thing in itself, simply through pure understanding, hence remains in every case a purely problematical concept, even if we concede the logical possibility. The object thus conceived is hence not a particular intelligible object of our understanding, but "the [sort of] understanding to which it might belong is itself a problem";[92] it is a mode of knowledge as to the possibility of which we can form not the slightest representation. Such a concept can serve as a limiting concept, to constrain sensibility (since it impresses on us that the field of sensible objects does not

90. [Ibid., B 302, note (III, 216 f.).]
91. [Ibid., A 246 = B 303 (III, 217).]
92. [Ibid., A 256 = B 311 (III, 222).]

coincide completely with that of thinkable objects in general), but it can never establish anything positive beyond the perimeter of its domain.[93]

The *Critique of Pure Reason,* taken strictly, is incapable of taking us beyond this insight into the doctrine of the noumenon "in the negative sense"; its structure stops at this point, and there is a fundamental sense in which we have to forgo a clear view into the problem thus defined, namely, giving a new positive meaning to the problematic concept. Kant himself did not shy away from this prospect, and he proclaimed ever more decisively and powerfully the new direction the question was taking, despite all the obstacles and fetters imposed by the threefold division of his system into the realms of theoretical reason, practical reason, and judgment. This new direction is no longer related to being but to obligation, as that which is essentially and truly "unconditioned." An essential defect of Kant's exposition in the First Critique was that it was unable to illuminate this relation fully, and only hinted at it through a number of vague suggestions. Thus from the beginning Kant's doctrine of the "noumenon" and of the "thing in itself," in the form in which it initially appeared in the *Critique of Pure Reason,* continued to labor under an obscurity which was to prove fateful for its reception and its historical development. We do not need here, however, to make the effort to foresee the new form and the new solution of the problem of the thing in itself, which is achieved in Kant's doctrine of freedom, for it does not affect the theory of appearance as such, the systematic analysis of pure experiential knowledge. It comprises a self-contained whole, resting on independent presuppositions, which can and must be understood by purely immanent reflection. Whether outside of this sphere of concrete, empirical existence, which has thus far been shown us as all that is determinable, there is yet another sphere that yields not so much objects as, rather, objective value judgments, and whether our whole transcendental concept of objectivity itself thereby undergoes an enrichment and deepening of its content, is a question

93. Ibid., A 248 ff. = B 305 ff. (III, 218 ff.). For the whole discussion, cf. esp. the chapter "The Ground of the Distinction of All Objects into Phenomena and Noumena," A 235 ff. = B 294 ff. (III, 212 ff.).

to which only critical ethics and aesthetics can give the final answer. It is only there that we will discover the true, positive meaning of the noumenon, the underlying, ultimate "datum" on which the separation of the sensible and the intelligible, the appearance and the thing in itself, in the last analysis rests.

IV FIRST FRUITS OF THE CRITICAL PHILOSOPHY. THE *PROLEGOMENA*. HERDER'S *IDEAS* AND THE FOUNDATION OF THE PHILOSOPHY OF HISTORY

Strengthened by the firmness of his will, shortly before the end of his fifty-seventh year Kant was deep in the continuously revised and perpetually expanding intellectual tasks implied in the Inaugural Dissertation of 1770. The *Critique of Pure Reason* was finished within the span of a few months, an accomplishment that is scarcely rivaled, even as a purely literary feat, in the entire history of thought. During the period of its execution, in supreme concentration of mind and will on the goal of completing the book, Kant had to keep in the background every question as to the consequences it might entail. As he had in the years of solitary meditation, he abandoned himself to the prosecution of the matter in hand, not asking how the book might gain the readiest reception from contemporary readers and the schools of philosophy. Indeed, the motto from Bacon that Kant later made the epigraph to the second edition of the *Critique of Pure Reason* accurately expressed his thought in this regard: "Of ourselves, we say nothing; but as concerns the matter here treated, we ask that men regard it not as opinion, but as a necessary work, and be assured that we do not here undertake to found a sect or any arbitrarily spun-out system, but to uphold the greatness and utility of mankind."

Yet Kant was immediately wrenched out of this mood, in which he had carried out the labor on the First Critique, by the initial critical

218

evaluations the *Critique* received. Whatever these judgments might be, they were unanimous on one point: where Kant had believed he was posing an absolutely necessary and universally valid problem, they saw only the expression of a personal view and dogma. Whether someone was attracted to or repelled by the First Critique depended on whether this view seemed consistent with or opposed to his own; but nowhere was there the slightest recognition of the fact that Kant's whole way of putting the question was not in any way a graft onto the traditionally prescribed branches of the schools of philosophy. The interpreters' only concern for a long time was whether the system should be called or thought of as "idealism" or "realism," "empiricism" or "rationalism." Mendelssohn leveled the strongest criticism against it when, in a well-known expression, he called Kant the "all-destroyer," at least demonstrating a correct sense of the gap between it and traditional philosophy.

To Kant himself, this type of outlook and evaluation was made crystal clear in the penetrating review that appeared in the *Göttingische Gelehrte Anzeigen* of January 19, 1782. The story of the genesis of this review is famous.[1] On a trip that took him to Göttingen, Christian Garve—a writer of popular philosophy universally esteemed in the eighteenth century—had undertaken to do a major critical piece for the *Gelehrte Anzeigen* of Göttingen, as thanks for the "many demonstrations of courtesy and friendship" he had received. He requested for this purpose the *Critique of Pure Reason,* which he had not read until then, yet which, he wrote in his letter to Kant dated July 13, 1783, "promised a very great pleasure" to him because he had "already gotten so much from Kant's previous little writings." The first few pages he read in the book must have convinced him of his error. A wealth of difficulties confronted him from the start; it was a sort of reading for which he was totally unprepared by his previous studies, which had been mainly in the areas of aesthetics and moral psychology; he was in addition suffering from the after-effects of a severe

1. It receives the best treatment from Emil Arnoldt: "Vergleichung der Garveschen und der Federschen Rezension über die Kritik der reinen Vernunft" (Arnoldt, *Gesammelte Schriften,* ed. Otto Schöndörffer, vol. 4, p. 1 ff.); see also Albert Stern, *Über die Beziehungen Chr. Garves zu Kant* (Leipzig, 1884).

illness. Only his respect for his given word motivated him to carry through his labors and compose an exceptional review, which, after he had more than once rewritten and shortened it, he finally sent to the editor of the journal.

The office of editor was held by a man who felt none of the scruples and doubts that Garve had experienced during his reading of the First Critique. Johann Georg Feder belonged to that circle of Göttingen professors who allowed themselves complete certainty in their judgment about Kant. When Christian Jacob Kraus, shortly before the First Critique appeared, asserted in this circle that there was a work lying on Kant's desk which would certainly put philosophers into yet another cold sweat of anxiety, they laughingly replied that it would be hard to expect anything of that sort from a "dilettante in philosophy."[2] Feder added to this unshatterable complacency of the members of a learned profession the facileness of the editor, who having few concrete ideas of his own, knows how to tailor the scope and content of each contribution adroitly to the momentary needs of his journal. Powerful strokes of his pen reduced Garve's review of the *Critique* to almost one-third of its original length, altering it in many stylistic respects. Further, there were many long insertions by Feder himself intended to lay out for the reader a specific standpoint for studying and understanding Kant's book. The systematic devices he added for this purpose were the most petty imaginable: they consisted of nothing but the application of the familiar rubrics of the history of philosophy laid down in every manual and hallowed by custom. "This work," Feder's version of the Göttingen review began, "which continuously exercises the reader's understanding though not always informing it, often taxes one's attention to the point of exhaustion, occasionally coming to its aid with felicitous images or rewarding it with unforeseen, widely useful consequences, is a system of the higher, or as the author puts it, of transcendental idealism. This idealism extends impartially to mind and matter, translating the world and ourselves into representations and thus making all objects

2. See Johannes Voigt, *Das Leben des Professors Christian Jacob Kraus* (Königsberg, 1819), p. 87.

arise from appearances, so that the understanding connects them into a *single* experiential series and so that reason tries inevitably, yet fruitlessly, to expand and unify them into *one* whole and complete world-system."

In these opening sentences we can imagine the impression Kant must have gotten from this review. Nothing he said about it—and he expressed himself very harshly—is excessive, taken simply at face value; his one mistake was to see a personal intent to distort and misrepresent where in fact there was only the naive and open expression of pettiness and conceit. But, provoked by the criticism from Göttingen, he proceeded to develop the basic ideas of his theory all over again, with trenchant brevity, and this book that seemed accidentally extorted from his pen quickly took on under his hands a universal and systematic significance. Out of a mere rejoinder to the Garve-Feder review there grew the *Prolegomena to Any Future Metaphysics Which Can Come Forth as Science* (*Prolegomena zu einer jeden künftigen Metaphysik, die als Wissenschaft auftreten können*).

From the standpoint of literary history, we are present here at the decisive crisis of the philosophy of the German Enlightenment. The old kind of popular philosophy, as upheld by Garve in an honorable and straightforward way, is annihilated at a stroke by the *Prolegomena*. "Hammer and chisel," says the preface, "serve quite adequately to work a piece of timber, but for engraving on copper, one needs to use the etching needle." Kant himself never practiced this subtle art of rendering visible the most delicate distinctions and nuances in the basic concepts of cognition, together with their universal interrelations, more superbly than here. Now it was he who was in the position of being reader and critic of the completed book; now he could once again fully expound the complex web, picking out the main threads with a sure hand, and showing how it held together as a whole. Though Kant had been thinking for a long time, as he writes in a letter to Marcus Herz dated January, 1779, "about the basic premises of popularity in the sciences in general, especially in philosophy," the problem he had set himself now received both a theoretical and a practical solution. For the *Prolegomena* inaugurates a new form of truly philosophical popularity, unrivaled for clarity and keenness.

We shall not explicate the detailed content of this work again here; we have already included it in the exposition of the basic ideas of the First Critique, since it contains the most certain and authentic interpretation of that book. But along with this detailed content, the *Prolegomena* has a personal meaning in Kant's development. Through his free survey of what he had achieved so far, he now felt himself spurred on to new, comprehensive productivity. The work on the *Critique* was not yet over, but he already began to lay the foundation for the future systematic working-out which was to encompass all three *Critiques*.

In 1786 the *Metaphysical Foundations of Natural Science* brought the new sketch of the Kantian natural philosophy. It gives a definition of the concept of matter, which is taken in the transcendental spirit, in that the reality of matter appears here not as something posited as ultimate, but as derived, since the existence of matter is seen only as another expression for the reality and lawfulness of forces. A defined dynamic relation, a balance between attraction and repulsion, is what our pure experiential conception of matter rests on. Our analysis does not normally go beyond this, and it cannot penetrate any deeper into the fact. For the so-called metaphysical essence of matter, the "absolutely intrinsic," which is perhaps still taken for granted in it, is an empty notion; it is "a mere something, which we could in no wise understand, even if somebody should be able to tell us what it is." In actual fact, what we can empirically grasp of it is a mathematically determinable proportion in the effect itself, thus something only relatively intrinsic which itself in turn consists of external relations.[3] How these relations are governed, how they are subordinated and fitted to universal conceptual laws, had already been shown by the *Critique of Pure Reason* in the chapter on the analogies of experience. The *Metaphysical Foundations of Natural Science* is the concrete execution of the basic ideas elaborated there. It puts forward the three *Leges motus* from which Newton had worked: the law of inertia, the law of proportionality of cause and effect, and the law of the equality of action

3. See *Critique of Pure Reason*, A 277 = B 333; for further material on Kant's dynamic construction of matter, see August Stadler, *Kants Theorie der Materie* (Leipzig, 1883).

and reaction, as specific expressions of the universal synthetic principles of relation.

As a companion to this work on the metaphysics of natural science stands Kant's novel outlook on the metaphysics of history. The two treatises "Idea for a Universal History from a Cosmopolitan Standpoint" ("Idee zu einer allgemeinen Geschichte im weltbürgerliche Absicht") and "What Is Enlightenment?" (Beantwortung der Frage: Was ist Aufklarung?") appeared in the issues of the *Berlinische Monatsschrift* for November and December, 1784; these were augmented by his review in 1785 of the first and second parts of Herder's *Ideas for a Philosophy of the History of Mankind,* in the *Allgemeine Literaturzeitung* of Jena. We seem to have before us in these essays only brief, casual works, tossed off quickly, and yet they present the whole foundation of the new conception of the essence of the state and of history that Kant had developed. These writings were hardly less momentous for the internal development of German idealism than the *Critique of Pure Reason* was within its own field. The first-mentioned essay especially, the "Idea for a Universal History from a Cosmopolitan Standpoint," reminds us of something that was significant in universal intellectual history: it was the first of Kant's writings that Schiller read, and the one that awakened in him the decision to study Kant's philosophy more deeply.[4]

But in yet another sense this document constitutes a potent watershed in the movement of intellectual development. On the one hand it is still of a piece with the political and historical ideas of the earlier eighteenth century, while on the other it clearly foreshadows the new insights of the nineteenth century. Kant still uses the language of Rousseau here, but he has gone beyond Rousseau in the systematic and methodological foundations of his ideas. While Rousseau sees all of man's history as a fall from the condition of innocence and happiness in which man lived before he entered into society and before he banded into social groups, to Kant the idea of such an original state appears utopian if taken as a fact, and ambiguous and unclear if regarded as a moral ideal. His ethics orients him toward the

4. See Schiller's letter to Körner, August 29, 1787.

individual and toward the basic concept of the moral personality and its autonomy; but his view of history and its philosophy leads to the conviction that it is only through the medium of society that the ideal task of moral self-consciousness can find its actual empirical fulfillment. The value of society may seem negative when measured by the happiness of the individual, but this shows only that this point of view for evaluating and the standard of evaluation itself have been falsely chosen. The true criterion of this value lies not in what the social and political community accomplishes for the needs of the individual, for the security of his empirical existence, but in what it signifies as an instrument in his education into freedom.

In this regard the fundamental antithesis that is the substance of Kant's whole view of history now arises. Theodicy, the inherent ethical justification of history, is the result here if one thinks that the way to true, ideal unity of mankind leads only through struggle and opposition, that the only route to autonomy is through compulsion. Because nature, because providence, has decreed that man must produce everything beyond the routine ordering of his animal existence entirely out of himself, and that he participate in no happiness or perfection other than what he himself has created through his own reason, unaided by instinct—he therefore had to be put in a position in which, physically speaking, he was inferior to all other creatures. He was made more needy and defenseless than other beings, so that this very insufficiency would be a stimulus for him to escape from his natural limitations and his natural isolation. It was not a drive toward society originally implanted in man but rather need that founded the first societal groupings, and need further formed one of the essential conditions for erecting and consolidating a social structure.

What the *Metaphysical Foundations of Natural Science* asserts in regard to physical bodies is also valid, if rightly understood, for social bodies. Society is not simply held together through an original intrinsic harmony of individual wills, on which the optimism of Shaftesbury and Rousseau had relied; but its existence, like that of matter, is rooted in attraction and repulsion, in an antagonism of forces. This opposition forms the heart and the presupposition of any social order. "Thus are taken the first true steps from barbarism to culture, which consists in the social worth of man; thence gradually develop all talents, and taste is refined; through continued enlightenment the foundations are laid for a way of thought which can in time convert the coarse, natural disposition for moral discrimination into definite practical principles, and thereby change a society of men driven together by their natural feelings into a moral whole. Without those in themselves unamiable characteristics of unsociability from whence opposition springs—characteristics each man must find in his own selfish pretensions—all talents would remain hidden, unborn in an Arcadian shepherd's life, with all its concord, contentment, and mutual affection. Men, good-natured as the sheep they herd, would hardly reach a higher worth than their beasts; they would not fill the empty place in creation by achieving their end, which is rational nature. Thanks be to Nature, then, for the incompatibility, for the heartless competitive vanity, for the insatiable desire to possess and to rule! Without them, all the excellent natural capacities of humanity would forever sleep, undeveloped."[5] Thus evil itself becomes the source of good in the course and progress of history; thus out of discord alone can true, self-confident moral harmony emerge.

The essential idea of social order consists not in bringing individual wills to a common level by force, but in preserving their individuality and hence their opposition, at the same time, however, defining the freedom of the individual in such a way that it discovers its own limits in other people. To assimilate this determination, initially enforceable only by external power, into the will itself and to acknowledge it as the realization of the will's own form and fundamental demand is the ethical goal proposed for all historical development. Herein resides the most difficult problem mankind has to master, and the end to which all external politico-social institutions, the very order of the state itself in all forms of its historical existence, are but means. A philosophical attempt to survey universal world history from this standpoint and to see in it the progressive actualization of a plan of nature, which aims at the complete unification of the

5. "Idea for a Universal History from a Cosmopolitan Standpoint," Fourth Thesis (IV, 151–166) (*Ak.* VIII, 2–31).

human race in society, is hence not only possible but must itself be regarded as conducing to this intent of nature. "Such a justification of Nature—or, better, of providence—" as Kant concludes this discussion, "is no unimportant reason for choosing a standpoint toward world history. For what is the good of esteeming the majesty and wisdom of Creation in the realm of brute nature and of recommending that we contemplate it, if that part of the great stage of supreme wisdom which contains the purpose of all the others—the history of mankind—must remain an unceasing reproach to it? If we are forced to turn our eyes from it in disgust, doubting that we can ever find a perfectly rational purpose in it and hoping for that only in another world?"[6]

Again, adopting the standpoint of the transcendental inquiry, it is the essential method of this view of history, not its content, that has a primary claim on our interest. A new perspective for contemplating the world, an alteration in the stance our historical knowledge takes relative to the flux of empirical historical existence, is the basic object of this search. Kant explicitly stresses at the end of his treatise that this stance is in no way intended to harm or displace the customary historical outlook, which tries to grasp phenomena in their pure factuality and to give a narrative account of them.[7] But hand in hand with this manner of proceeding there must be another, through which their significance is revealed totally differently from the way it is in the empirical, sequential arranging of facts. At this point, however, the basic character of this new procedure cannot be fully surveyed and defined with the exactitude of principles, for Kant's philosophy of history constitutes only one component of his universal system of teleology. Not until this system has been completely explicated in his fundamental ethical works and in the *Critique of Judgment* (*Kritik der Urteilskraft*) will the final critical verdict on the questions of historical teleology be rendered.

But in these first stirrings of the Kantian philosophy of history, we encounter a decisive turn that is completely clear. In the opening

6. Ibid., Ninth Thesis (IV, 165) (*Ak.* VIII, 30).
7. [Ibid. (IV, 165 ff.) (*Ak.* VIII, 29 ff.).]

sentences of the Kantian doctrine we are transported from the realm of being, where the critical task has been pursued until now, to the realm of obligation. According to Kant, the concept of "history" in the strict sense exists for us only where we contemplate a certain series of events in such a way that we do not look at the temporal sequence of its individual moments or its causal connectedness but relate the series to the ideal unity of an immanent end. Only when we apply this idea, this novel way of judging, and persevere, is historical process in its unique character and independence made manifest in the homogeneous stream of becoming, the complex of sheer natural causes and effects. In this connection, it is immediately evident that the question of the goal of history has quite a different ring for Kant, with his transcendental point of view, than it does for those who contemplate the world in the usual way and for traditional metaphysicians. Just as full insight into the validity of the laws of nature was only attained when we saw that nature as given does not "have" laws, but that the concept of a law is what creates and constitutes nature—so history as a well-established set of facts and events equally little has a "meaning" and a special *telos*. Rather, its own "possibility," its special significance, originates in the assumption of a meaning of this sort. "History" first truly exists where we as contemplators no longer stand in the series of sheer events, but in the series of actions; the idea of an action, however, includes the idea of freedom. Thus the principle of Kant's philosophy of history foreshadows the principle of Kantian ethics, where it will find its resting place and its full explication.

Since this correlation comprises the original *form* of Kant's conception of history—and therefore cannot be eliminated in the methodological sense—it is also decisive for its substance. The evolution of mankind's spiritual history coincides with the progress, the ever keener comprehension, and the progressive deepening of the idea of freedom. The philosophy of the Enlightenment has here reached its supreme goal, and in Kant's "An Answer to the Question: 'What Is Enlightenment?'" it now finds its lucid, programmatic conclusion. "Enlightenment is man's release from his self-incurred tutelage. Tutelage is man's inability to make use of his understanding without direction from another. Self-incurred is this tutelage when its cause lies not in lack of reason but in lack of resolution and courage to use it without direction from another. *Sapere aude!* Have the courage to use your own reason!—that is the motto of enlightenment."[8] But this is at the same time the motto of all human history, for the process of self-liberation, the progress from natural bondage toward the spirit's autonomous consciousness of itself and of its task, constitutes the only thing that can be called genuine "becoming" in the spiritual sense.

In this mood and with this conviction Kant comes to Herder's *Ideas for a Philosophy of the History of Mankind,* and the inevitable opening up of antagonism between him and Herder is understandable from now on. In the conception of this, his basic work, Herder remained the pupil of Kant, who in Herder's student years in Königsberg had first shown him the way to that "humanistic" philosophy which henceforth hovered before him as an enduring ideal. But Hamann's world view affected the whole of Herder's view of history more profoundly than Kant's did; with Hamann he felt himself truly and essentially congenial. What Herder sought in history was a vision of the infinitely manifold, infinitely diverse expressions of the life of mankind, a vision that unveils and reveals itself in them all as one and the same. The deeper he plunges into this whole, not to reduce it to concepts and rules but rather to feel it and to adapt to it, the more clearly it is impressed on him that no single abstract standard of measure, no uniform ethical norm and ideal could create its content. Every age and every period, every epoch and every nation contains the measure of its fulfillment and its perfection in itself. Here there is no valid comparison between what they are and what they want to be, no selection of common traits in which the essence binding the particular into a living unity is effaced and destroyed. The stuff of the child's life is incommensurable with that of the middle-aged or elderly man, but possesses within itself the focus of its own being and worth, and the same is true for the historical life of peoples. The idea of the perpetually ongoing intellectual and moral perfectibility of the

8. [IV, 169 (*Ak.* VIII, 35).]

human race is nothing but an audacious fiction, on the strength of which every age thinks itself superior to all its predecessors, as evolutionary stages that have been abandoned and surpassed. We only grasp the true image of history when we let it work on us with all its brilliance, all its color, and hence with all the irreducible multiplicity of its individual elements.

To this extent Herder's work makes no pretension to be history but rather philosophy of history; to this extent it lays down specific teleological guidelines through the endless manifold of becoming. A plan of providence is unveiled in the ongoing progress of history, but this plan signifies no external ultimate purpose set for becoming and no universal goal into which all particular goals are absorbed. It is rather the thoroughly individual thing itself in which the form of totality is finally achieved, in which the idea of man finds its concrete fulfillment. In the play of events and scenes, of ethnic individualities and destinies, of the rise and fall of specific historical forms of existence there ultimately stands before us a whole which is to be conceived not as the detached product of all these moments, but only as their living totality. Herder inquires no deeper into the vision of this totality. To him who has it, history has revealed its meaning; he requires no further norm to elucidate and explain it to him. While Kant needs the abstract unity of an ethical postulate to comprehend the meaning of history, while he sees in it the ever more complete solution of an endless task, Herder lingers in its pure givenness. If to Kant the stream of occurrences must be projected against an intelligible "ought" to make it intrinsically comprehensible, Herder remains equally immovable on the plane of pure becoming. Ethical insight into the world, resting on the dualism of being and obligation, of nature and freedom, is in sharp contrast to organic and dynamic insight into nature, which tries to conceive both as aspects of one and the same development.

Only when one reflects on them from the perspective of this basic contradiction over the history of the spirit can one do justice to the two reviews of Herder's *Ideas* that Kant wrote. It was Herder's tragic fate that, unable to follow the evolution of Kant and the critical philosophy since the sixties, he failed to rise to this perspective, and that consequently his polemic with Kant slipped more and more into the petty and personal. For his part, insofar as one can talk of guilt and innocence in this sort of intellectual combat, Kant cannot be said to be completely free of blame, since in the superiority afforded him by his critical analysis of the basic concepts he shut his eyes to the grand vision of the whole that was vivid everywhere in Herder, despite all the conceptual defects of his historico-philosophical deductions. He who fixed his eye above all on strictness of proofs, precise inference to principles, and sharp distinction of spheres of validity, could see in Herder's methodology nothing but "an adroitness in unearthing analogies, in wielding which he shows a bold imagination. This is combined with cleverness in soliciting sympathy for his subject—kept in increasingly hazy remoteness—by means of sentiment and sensation. Further suspicion is aroused as to whether these emotions are effects of a prodigious system of thought or only equivocal hints which cool, critical examination would uncover in them."[9] The philosophical critic and analyst here inexorably forces the renunciation of every form of methodological syncretism[10]—a renunciation which would have also meant dispensing with the most characteristic personal merits of Herder's way of looking at things.[11] For this style of contemplation consists precisely in its incessant direct passage from intuition to conception and from conception to intuition; Herder as a poet is a philosopher, and as a philosopher, a poet. The irritation with which he now took up the cudgels against Kant, and the growing bitterness with which he waged the battle, is thus explicable: he sensed and knew that it was not just an isolated question that was at issue here, but that his essence, his most intimate gift, was jeopardized by Kant's fundamental theoretical demands.

In Kant's two reviews of Herder's *Ideas*, the conflict has not yet come to a head. For as long as Kant had not completed the foun-

9. ["Review of Herder's *Ideas for a Philosophy of the History of Mankind*," Sect. I (IV, 179) (*Ak.* VIII, 45).]

10. Cf. Kant's letter to Friedrich Heinrich Jacobi, August 30, 1789 (IX, 433) (*Ak.* XI, 72).

11. For further details on Herder's battle against Kant, see Eugen Kühnemann's exceptional account: *Herder*, 2d ed. (Munich, 1912), pp. 383 ff.

dations of ethics, as long as his conception of freedom had not yet reached its ultimate clarity, one of the essential presuppositions of this conflict was missing. To be sure, as early as the *Critique of Pure Reason* the concept of freedom had been put forward and the antinomy between freedom and causality discussed, but the matter stopped at that, with what was on the whole a purely negative definition of the idea of freedom. Only with the *Foundations of the Metaphysics of Morals (Grundlegung zur Metaphysik der Sitten)* in 1785 does his progress take a new, positive turn, one destined to shake to its foundations the whole previous contrast between determinism and indeterminism, which the First Critique seemed still attached to. Only from this point on does the significance of the essays on the philosophy of history belonging to the years 1784 and 1785 emerge within the whole context of Kant's activity as a philosophical writer. They are the link with a whole new circle of problems, on which his systematic interest is ever more strongly concentrated. The Kantian conception of history only poses a single concrete example of a complex of questions, focused as a whole in the concept of "practical reason"; Kant now moves on to the more precise definition of that concept.

V THE GROWTH AND STRUCTURE OF CRITICAL ETHICS

Upon the completion of the First Critique, Kant did not tack the *Critique of Practical Reason* (*Kritik der praktischen Vernunft*) onto the theoretical portion as a second component of his system. He had conceived his philosophy from the very first as a self-contained whole, and ethical problems formed an essential, integrating constituent of it. We grasp the special and most profound concept of "reason" itself, as Kant understands it, only through this relation. When, in the Prize Essay of 1763, Kant examined the universal method of metaphysics and put it on a new foundation, he was above all concerned to include in his scrutiny, in accordance with the Berlin Academy's formulation of the question that was set, the basic concepts of morals as well. Their value and utility being beyond question, it is their distinctness that is inquired into here, and they too are to be conceived on the grounds of their universal validity. Although even an empiricist like Locke had put the type of relation dominant in moral truths on the same plane with the interconnection of geometric judgments and theorems, and had attributed to morality the selfsame demonstrative certainty as in metaphysics, Kant finds the first principles of morality in their contemporary state entirely insusceptible of the degree of evidence required. For the basic concept of obligation itself (which was the cornerstone of the deduction of natural rights and duties in Wolff's philosophy of natural law) is infected with obscurity. "One ought to do this or that and leave something else undone; this is the formula under which every obligation is enunciated. Now that 'ought' expresses a necessity of action and is capable of two meanings. That is, either I ought to do something (as a means) if I wish something else (as an end), or I ought directly to do something else and make it real (as an end). The former we can call the necessity of means (*necessitas problematica*), and the latter the necessity of ends (*necessitas legalis*). No obligation is present in necessity of the first kind; it only prescribes the solution of a problem, saying what are the means I must use if I wish to reach a particular end. When anyone prescribes to another the actions which he should do or refrain from doing if he wishes to promote his happiness, perhaps all the teachings of morals could be brought under the precepts; but they are then no longer obligations but only like what might be called an obligation to make two arcs if I wish to bisect a line. That is, they are not obligations at all but only counsels to suitable actions if one wishes to attain a particular end. Since the use of means has no other necessity than that which pertains to the end, it follows that all actions which morals prescribes under the condition of particular ends are contingent and cannot be called obligations so long as they are not subordinated to an end necessary in itself. I ought, for example, to promote the greatest possible total perfection, or I ought to act according to the will of God; to whichever of these propositions all practical philosophy were subordinated, that proposition, if it is said to be a rule and principle of obligation, must command the action as directly necessary, not commanding it merely under the condition of some particular end. And here we find that such an immediate supreme rule of all obligation would have to be absolutely indemonstrable. For from no consideration of a thing or concept, whatever it is, is it possible to know and infer what we should do, unless what is presupposed is an end and the action a means. But this it must not be, because it would then be a formula not of obligation but only of problematic skill."[1]

When Kant wrote these words, none of his contemporary readers and critics could have foreseen that in these few and simple sentences every moral system produced by the eighteenth century was essen-

1. *Inquiry into the Distinctness of the Principles of Natural Theology and Morals,* Fourth Observation, §2 (II, 199 f.) (*Ak.* II, 298-99).

tially felled. In fact, this passage contains the fundamental concept of his ethics yet to come: the strict distinction between the categorical imperative of the moral law and the hypothetical imperatives of merely mediate ends is discussed here with full precision and clarity. Nothing further can be deduced or established concerning the content of the unconditional moral law, as Kant emphasizes, for every deduction of this sort, having made the validity of the command dependent on something else—be it the existence of a thing or the presumed necessity of a concept—would locate the moral law once again in the sphere of the conditioned, from which it had just had to be liberated. Thus the formal nature of the first basic ethical certainty already immediately includes the moment of its indemonstrability. That it must bestow absolute moral worth, a good in itself not given through something else, cannot be deduced and understood by way of mere concepts; we can presuppose this assertion for the purpose of constructing a pure ethics only in the same sense as we must posit materially certain but indemonstrable propositions in the construction of logic and mathematics, together with the purely formal principles of identity and contradiction. For this special mode of knowledge and certainty will here, as regards ethical problems, trace back to the psychological faculty of feeling. "In these times we have first begun to realize that the faculty of conceiving truth is intellection, while that of sensing the good is feeling, and that they must not be interchanged. Just as there are unanalyzable concepts of the true, that is, what is met with in the objects of intellection considered by themselves, there is also an unanalyzable feeling for the good.... It is a task of the understanding to resolve the compounded and confused concept of the good and to make it distinct by showing how it arises from simpler sensations of the good. But if the sensation of the good is simple, the judgment, 'This is good,' is completely indemonstrable and a direct effect of the consciousness of the feeling of pleasure associated with the conception of the object. And since many simple sensations of the good are certainly in us, there are many simple unanalyzable conceptions of the good."[2]

2. Ibid. (II, 201) (*Ak.* II, 299).

This link with the psychological language of the eighteenth century, which refers back in particular to the theory of "moral sentiment" developed by Adam Smith and his school, carries the danger for Kant that through it the distinctiveness of the novel direction he had won for the foundation of ethics is already once more being gradually effaced. In fact, in his subsequent writings the analysis of the pure concept of "obligation," which Kant had made the locus of the special task of moral philosophy, retreats more and more into the background. His interest seems ever more energetically concentrated on being and becoming, on the viewpoint of genetic development instead of the ought; the ethical way of putting questions is crowded out by that of psychology and anthropology.

In the information on his course of lectures for the winter term, 1765–66, Kant says explicitly that he proposes to make use of the method of moral inquiry which Shaftesbury, Hutcheson, and Hume had founded, as a "beautiful discovery of our age": that method which, before pointing out what ought to happen, always examines historically and philosophically what does happen, and hence proceeds not from abstract premises but from the actual nature of man.[3] When we take a closer look at these propositions and consider the connection between them, we recognize that Kant is not tempted to subscribe to the technique of English moral psychology with no critical reservations. That human nature on which he wishes to take his stand is, as he instantly adds, to be understood not as a variable but as a constant magnitude. Man is not to be comprehended and presented in the shifting form his momentary contingent state impresses on him, but rather his eternal essence is to be sought out and revealed as the foundation for moral laws. What Kant here understands by nature and by human nature is drawn less from the influence of English psychology than from that of Rousseau. It is he who essentially determines the substance of Kant's ethics during this period. Rousseau is the one who "set him right," who freed him from an intellectual overvaluation of pure thinking and reoriented his philosophy toward the act. The delusory superiority, the false luster of pure

3. (II, 326) (*Ak.* II, 311).

knowing vanishes: "I learn to honor men, and would find myself more useless than the humblest laborer if I did not believe that this perspective could bestow worth on all others, and restore the rights of man."[4] But this opens the way, in the purely methodological sense, to yet another path of reflection, for Rousseau's concept of nature is an existential concept only in its expression, while in its pure content it is unmistakably an ideal and normative concept. In Rousseau's work both meanings indeed exist side by side yet completely intertwined: nature is that original state from which man emerged, as it is the goal and end to which he is to return. But this mixture could not long withstand Kant's analytical mind. He distinguished between the is and the ought even where he seemed to be basing the latter on the former. The more keenly and clearly this distinction took shape for him, the more progress he made in his critical analysis of the pure concept of knowledge and the more definitively he separated the question of the parentage and birth of cognitions from that of their value and their objective validity.

Since this separation receives its first full systematic expression in Kant's dissertation *On the Form and Principles of the Sensible and the Intelligible World,* the problem of ethics is hence also given a completely new foundation. Just as there is a pure cognitive a priori, there is now a moral a priori as well. In the same way as the former is not deducible from mere sensory perceptions, but has its roots in an original spontaneity of the understanding, an *actus animi,* the latter also, conceived with respect to its content and its validity, is loosed from any dependence on the sensory feelings of pleasure and pain, and cleansed of any contamination by them. Thus it was as early as this that Kant broke with all morality based on eudaemonism. He turns away so brusquely that from now on among those who make happiness the principle of ethics he even numbers Shaftesbury, who uses pleasure as a moral criterion not at all in the sense of an immediate sensory feeling but in its maximum aesthetic refinement and sublimation. Such an equation could not help evoking astonishment

---

4. *Fragmente aus Kants Nachlass;* concerning the relation between Kant and Rousseau, see above, pp. 86 ff.

in his contemporaries, and Mendelssohn was unable to contain his surprise at seeing Shaftesbury ranked beside Epicurus.[5]

But Kant now saw that between himself and the whole of previous ethics there was a difference not merely of content but of significance and basic outlook. From this point on he must have felt increasingly pressed to go beyond the meager hints about his ethical system contained in the Dissertation. But every time he resolved on a closer exposition and crystallization of his new standpoint—and his correspondence from 1772 to 1781 contains indubitable evidence that he put his hand to this at various times during this period—this labor was "blocked as though by a dam"[6] by the "chief subject" occupying his thoughts at this time. Repeatedly, Kant seemed on the verge of overcoming this hesitation by a precipitate decision to lay aside the *Critique of Pure Reason* for a while, its completion being postponed again and again, and to apply himself to working out his ethics as a desirable rest from the difficulties of his epistemological inquiry. "I have made up my mind," he writes in September, 1770, on sending the Dissertation to Lambert, "to rid myself of a long indisposition which seized on me this summer, and so as not to be without occupation in my spare time this winter, to set in order and to ready my researches into pure moral wisdom, in which no empirical principles are to be found, and also the metaphysics of morals. In various ways it will smooth the pathway for the most important points of the altered form of metaphysics, and it seems to me in addition to be equally necessary for the principles of the practical sciences which are still so ill-judged at present."[7] But however frequently this inquiry might entice him in the course of the next decade, which was filled with the most abstract speculation, nevertheless his systematic mind always stood in opposition. He demanded of himself, as the unavoidable methodological foundation, that the pure transcendental philosophy be outlined and carried out, so as to apply himself to the "metaphysics of Nature and of morals" only when both were com-

5. See *De mundi sensibilis*, §9 (II, 412) (*Ak.* II, 396); cf. Mendelssohn's letter to Kant, December 25, 1770 (IX, 90) (*Ak.* X, 108).

6. See his letter to Marcus Herz, November 24, 1776 (IX, 151) (*Ak.* X, 184).

7. To Lambert, September 2, 1770 (IX, 73) (*Ak.* X, 92).

plete. Regarding the latter, his intention was to bring it out first, and a letter to Herz in 1773 reports that he was already "rejoicing in advance" about it.[8]

Thus the *Foundations of the Metaphysics of Morals,* on its appearance in 1785, was, like the *Critique of Pure Reason,* the product of more than a dozen years' reflection. The vivacity, the suppleness, and the drive of his exposition did not suffer in the slightest from this, however. In none of his major critical works is Kant's personality so immediately evident as here; in none is the rigor of the deduction united in the same perfect way with such a free movement of thought, ethical power and stature with the sense of psychological detail, and acuteness of conceptual discrimination with the noble concreteness of a popular way of speaking rich in felicitous images and examples. Here for the first time, the subjective ethos that forms the inmost core of Kant's nature could unfold and express itself in its purity. This ethos is not something which "comes into being"; it appears already full-blown in his youthful writings, in the *Universal Natural History and Theory of the Heavens* and in the *Dreams of a Spirit-Seer.* Only here, though, does it achieve full self-consciousness and forge its adequate philosophical expression in deliberate contrast to the philosophy of the Enlightenment.

If we try to say what the most universal content of critical ethics is—here we are looking ahead to the *Critique of Practical Reason* that appeared three years later, so as not to separate what in fact belongs together—we must not be confused and led astray by the handy catchwords that have played so large a role in characterizing Kant's doctrines. Again and again people have talked about the "formalistic" character of Kantian ethics; they have charged that the principle from which it proceeds yields but one universal and hence empty formula for moral conduct, inadequate for deciding concrete individual cases and choices. Kant himself, since he granted this reproach and in a certain sense recognized it, had a counter to objections of this kind. "A critic who wished to say something against that work," he remarks, "really did better than he intended when he said that there

8. To Herz, the end of 1773 (IX, 114) (*Ak.* X, 136).

GROWTH OF CRITICAL ETHICS 239

was no new principle of morality in it but only a new formula. Who would want to introduce a new principle of morality and, as it were, be its inventor, as if the world had hitherto been ignorant of what duty is or had been thoroughly wrong about it? Those who know what a formula means to a mathematician, in determining what is to be done in solving a problem without letting him go astray, will not regard a formula which will do this for all duties as something insignificant and unnecessary."[9]

The special foundation for Kant's "formalism" is to be sought in a still deeper vein of his thought, for it lies in that universal transcendental concept of form that precedes and underlies mathematics as well. The *Critique of Pure Reason* has established that the objectivity of knowledge cannot be founded on material, sensory data nor on the "what" of individual sensations. Sensation is just the expression of the individual subject's state, varying from moment to moment; it constitutes that which is wholly contingent, different from case to case, from subject to subject, and hence is not determinable by any unambiguous rule. If judgments with universally valid truth content are to be constructed from such infinitely variable circumstances, if the appearances that originally are totally unclear are to be legible as experiences, it is necessary that there be definite basic types of relations, which as such are invariant and which produce the objective unity of cognition and only thus make possible its object and are its foundation. It was these fundamental syntheses that the critical theory discovered and raised into prominence as the "forms" of pure intuition, the "forms" of pure understanding, and so on. There is for Kant the most intimate analogy between introduction into the problem of ethics and this basic idea. As was the case earlier for mere "representation," now it is necessary to discover the factor that leads to the quality of objective validity for the realm of practice, of desire and act. Only if such a factor is demonstrable can we use it to pass from the sphere of the arbitrary into that of the voluntary. Will and cognition are alike in this respect: they exist only insofar as a permanent and stable rule constituting their unity and identity is estab-

9. *Critique of Practical Reason,* preface (V, 8) (*Ak.* V, 8).

lished. Just as this cognitive rule was not abstracted from the object, but posited through the analytic of the understanding; just as it was shown that the conditions of the possibility of experience, as an ensemble of cognitive functions, are at the same time the conditions under which definite individual objects can alone be said to be for us, so we now try to translate this way of posing the problem into the realm of the ethical. Is there here also a lawfulness that is rooted not in the concrete substance and the concrete differentia of what is willed, but in the peculiar basic orientation of willing itself and that, on the strength of this origin, has the power to form the basis of ethical objectivity in the transcendental sense of that term, that is to say, the necessity and the universal validity of moral worth?

Starting from this form of the question, it is immediately comprehensible on what grounds pleasure and pain, of whatever form and coloration, are for Kant untenable as ethical principles. For pleasure, however it may be conceived, stands on the same plane of validity as sensory perception insofar as it signifies the sheer passivity of impression. It changes in accordance with the state of the individual subject and the external attraction that influences him, and is as infinitely variable as the diversity of these two factors. To be sure, naturalistic metaphysics, accustomed to founding ethics on the pleasure principle, tries to conceal this situation, since it appeals to the psychological universality of this principle. But although it may be true that it is innate in all subjects to strive for pleasure, this biological fact is totally worthless for setting up an identical standard in which individual wills might find unity and harmony. For since everyone strives not so much for pleasure as for *his* pleasure or what he thinks is his, the sum of these strivings disintegrates into a chaotic mass, a confusion of the most diverse intertwined and entangled tendencies, each of which is qualitatively completely opposed to the others even where they are seemingly directed toward the same goal. "It is therefore astonishing," Kant remarks, "how intelligent men have thought of proclaiming as a universal practical law the desire for happiness, and therewith to make this desire the determining ground of the will merely because this desire is universal. Though elsewhere natural laws make everything harmonious, if one here attributed the universality of law to this maxim, there would be the extreme opposite of harmony, the most arrant conflict, and the complete annihilation of the maxim itself and its purpose. For the wills of all do not have one and the same object, but each person has his own (his own welfare), which, to be sure, can accidentally agree with the purposes of others who are pursuing their own, though this agreement is far from sufficing for a law because the occasional exceptions which one is permitted to make are endless and cannot be definitely comprehended in a universal rule. In this way a harmony may result resembling that depicted in a certain satirical poem as existing between a married couple bent on going to ruin, 'Oh, marvelous harmony, what he wants is what she wants'; or like the pledge which is said to have been given by Francis I to the Emperor Charles V, 'What my brother wants (Milan), that I want too.' "[10] The harmonizing of different individual acts of will thus cannot be attained by directing them toward the same concrete object, toward one and the same material goal of the will, for that would instead result in their total conflict; it is attained only through the subjecting of each to the guidance of a universal and overriding ground of determination. In the unity of such a ground that which is ethically objective, a truly self-sufficient and unconditional moral value, could be founded, just as it was the unity and indestructible necessity of the basic logical principles of cognition that enabled us to posit an object of our representations.

Hence it is not any particular *state* of pleasure, but rather its essential nature, that unfits it as the foundation of ethics. In the analysis of the problem of knowledge, the particular nature of individual sense perceptions could remain outside Kant's consideration, since for him the proposition that the "coarseness or subtlety of sensation has nothing to do with the form of possible experience" was valid; the same is true for the analysis of will. Whether one takes pleasure in its "coarse" meaning, or whether one is concerned to purify and sublimate it through all the stages of refinement on up to the most elevated intellectual pleasure, may perhaps make a difference in the content of ethical principles, but not in how they are deduced and justified.

10. Ibid., §4, Theorem III (V, 31 f.) (*Ak.* V, 27 f.).

Similarly, as every sensation, regardless of its clarity and distinctness, has a certain cognitive character that sets it off from pure intuition and from the pure concept of the understanding, in the realm of practice as well the character of subjective desire must be distinguished from that of pure will. As long as the individual in his striving is not oriented and committed to any goal other than the satisfaction of his subjective drive, he remains enclosed and fettered within his particularity, whatever the particular form of this impulse may be. In this respect, all material practical principles—all those which seat the value of willing in what is willed—are "of one and the same kind and belong under the general principle of self-love or one's own happiness." "It is astonishing," Kant says in support of this proposition, "how otherwise acute men believe they can find a difference between the lower and the higher faculty of desire by noting whether the conceptions which are associated with pleasure have their origin in the senses or in the understanding. When one inquires into the determining grounds of desire and finds them in an expected agreeableness resulting from something or other, it is not a question of where the conception of this enjoyable object comes from, but merely of how much it can be enjoyed. If a conception, even though it has its origin and status in the understanding, can determine choice only by presupposing a feeling of pleasure in the subject, then its becoming a determining ground of choice is wholly dependent on the nature of the inner sense, that is, it depends on whether the latter can be agreeably affected by that conception. However dissimilar the conceptions of the objects, be they proper to the understanding or even to the reason instead of to the senses, the feeling of pleasure, by virtue of which they constitute the determining ground of the will (since it is the agreeableness and enjoyment which one expects from the object which impels the activity toward producing it) is always the same. This sameness lies not merely in the fact that all feelings of pleasure can be known only empirically, but even more in the fact that the feeling of pleasure always affects one and the same life-force which is manifested in the faculty of desire, and in this respect one determining ground can differ from any other only in degree.... As the man who wants money to spend does not care whether the gold

in it was mined in the mountains or washed from the sand, provided
it is accepted everywhere as having the same value, so also no man
asks, when he is concerned only with the agreeableness of life,
whether the ideas are from the sense or understanding; he asks only
how much and how great is the pleasure which they will afford him
over the longest time."[11]

The common character of all types and qualities of pleasure is thus
thrown into sharp relief: it consists in consciousness keeping itself
purely passive with regard to all material attractions, so that it is
affected and determined by their influence. But such an "affect" is not
enough to serve as a basis for the concept of truth and the objective
validity of knowledge; equally little can an objective norm of what is
moral be gotten from it. What is needed is the selfsame complement
we have already encountered in its full significance in the theoretical
structure of the First Critique. To affection there must be opposed
function, to the receptivity of impressions, the spontaneity of the
concepts of reason. It is necessary to exhibit a relation between the
will and its object in which the object, the particular "matter" of
desire, determines the will less than will determines this object. If we
keep in mind the critical result of the analytic of the understanding,
no longer can any paradox be discovered in this demand, for even
the matter of sensation acquired its objective cognitive worth only in
that transcendental apperception demonstrated the fundamental lawfulness on which all synthesis of the manifold and hence all its objective significance rests.

Now we need only transfer this result from the theoretical into the
practical sphere to arrive at the basic concept of Kantian ethics: the
concept of autonomy. Autonomy signifies that binding together of
theoretical and practical reason alike, in which the latter is conscious
of itself as the bonding agent. In it, the will submits to no other rule
than that which it has itself set up as a universal norm and proposed
to itself. Wherever this form is achieved, wherever individual desire
and wish know themselves to be participants in and subject to a law
valid for all ethical subjects without exception, and where on the

11. Ibid., §3, Theorem II, Remark I (V, 25 f.) (*Ak.* V, 22 f.).

other hand they understand and affirm that this law is their own, then and only then are we in the realm of ethical questions. Popular moral consciousness, with the analysis of which the *Foundations of the Metaphysics of Morals* begins, already leads to this insight. For the conception of duty, by which it is ruled and guided, includes in itself all the essential determinations we have met so far. An action is said to be in accordance with duty only when every thought of advantage to be expected from it, every calculation of present or future pleasure likely to result from it, indeed every material aim of any other kind, is eliminated and only adherence to the universality of the law, which reins in all contingent and particular impulses, remains as the sole ground of determination. "An action done from duty does not have its moral worth in the purpose which is to be achieved through it, but in the maxim by which it is determined. Its moral value, therefore, does not depend on the reality of the object of the action but merely on the principle of volition by which the action is done without any regard to the objects of the faculty of desire....It is clear that the purposes we may have for our actions and their effects as ends and incentives of the will cannot give the actions any unconditional moral worth. Wherein, then, can this worth lie, if it is not in the will in relation to its hoped-for effect? It can lie nowhere else than in the principle of the will irrespective of the ends which can be realized by such action. For the will stands, as it were, at the crossroads halfway between its a priori principle which is formal and its a posteriori incentive which is material. Since it must be determined by some-thing, if it is done from duty, it must be determined by the formal principle of volition as such, since every material principle has been withdrawn from it."[12]

In the same fashion, the truth of a representation, according to Kant, does not consist in its likeness to an external transcendent thing, as an image to its original, but in the fact that the content of the representation is wholly and necessarily connected with other similar elements that we designate by the name of experiential knowledge; thus the predicate of good belongs to that act of will that is directed

12. *Foundations of the Metaphysics of Morals*, first sect. (IV, 256) (*Ak.* IV, 399–400).

not by a contingent and private impulse but rather by regard for the entirety of possible determinations of will and for their inner harmony. The "good" will is the will to law, and hence to agreement—an accord that concerns the relation between diverse individuals as well as the inner consequence of the manifold volitions and actions of one and the same subject, insofar as they display, above and beyond any fluctuations of particular motives and impulses, that essential self-containedness that we are accustomed to call by the name of "character." In this sense—and only in this sense—it is "form" that is the foundation of the value of truth as well as of the value of good, since it renders possible and consists in, in the one case, the interconnection of empirical perceptions into a system of necessary and a priori knowledge, in the other the unification of particular ends into the unity of a single goal and an enveloping purposiveness.

Thus we are standing directly in the presence of the ultimate statement of the basic principle of critical ethics: the formula of the "categorical imperative." An imperative is called hypothetical when it indicates which means must be willed or employed in order that something further, which is presupposed as the end, may be realized. It is called categorical when it manifests itself as an unconditional demand that has no need to borrow its validity from some further end, but instead possesses its own validity in that it presents an ultimate, self-evident value. But since this fundamental value is not to be sought in any particular content of willing, but only in its universal lawfulness, both the content and the object of the only possible categorical imperative are fully articulated herein. "Act only according to that maxim," states the fundamental rule, "by which you can at the same time will that it should become a universal law."[13] The methodical advance to the achievement of this proposition, by the power of pure analysis of the concept of duty, constitutes the clearest and most definitive exposition of its substance as well. Were any particular determinations whatsoever assimilated into this substance, were one single concrete good asserted by it to be the supreme good, we could not dismiss the question of the ground for the privilege of

13. See *Foundations*, second sect. (IV, 279 ff.) (*Ak.* IV, 421).

this value, unless we wanted merely to accept this assertion as a dogma. Every attempt to answer that question, however, would lead us straight on to discover, in this ground itself, another and higher value from which the value first posited would be derived. The categorical imperative would thus have been converted into a hypothetical one, and the unconditional value into a conditioned one.

Only by the idea of universal lawfulness in general as the substance of the supreme principle of value are we delivered from this dilemma. For here we have reached a point at which every question as to a further "why?" must fall silent, where indeed it loses its meaning and its significance. In the theoretical realm we progress synthetically from bare perceptions to judgments and complexes of judgments, from individual appearances to increasingly comprehensive associations, until we have at last discovered in the a priori principles of pure understanding the archetype and model of all theoretical lawfulness, on which we must take our stand as the ultimate justification of experience, without being able to deduce this lawfulness itself in turn from something deeper, from some concrete transcendent entity. The same relation obtains here also. We measure the singular against unity, a particular concrete psychological motive against the totality of possible determinations of the will, and we evaluate it by its relation to this totality, but we have no confirmation of this measure as such save that which is inherent in it. Critical ethics affords us no answer as to why order takes precedence over chaos, free subordination to the universality of a self-given law over arbitrariness of individual desires.[14] In the critique of reason, theoretical as well as practical, the idea of reason, the idea of a final and supreme union of knowledge and will is taken for granted. Whoever fails to acknowledge this idea thus excludes himself from the orbit of its manner of posing problems, and from its conceptions of "true" and "false," "good" and "evil," which it alone can substantiate, empowered by its method.[15]

14. Cf. in particular the opening part of the section "Of the Interest Attaching to the Ideas of Morality" (IV, 308 ff.) (*Ak.* IV, 448 ff.).

15. Cf. the preface to the *Critique of Practical Reason:* "Nothing worse could happen to all these labors, however, than that someone should make the unexpected discovery

Thus it is only here that a premise underlying all the developments hitherto finds its true substantial consummation. It is only in self-determination of the will that reason knows and comprehends itself, and it is this knowledge that comprises its peculiar, most profound essence. We encountered the pure "spontaneity" of thought also in the realm of theoretical cognition, but this spontaneity was knowable only through its image and reflection. What the unity of apperception and the individual concepts and principles founded thereon are, only becomes apparent in the growth of the objective world, which these concepts served to complete. A world of things, ordered in space and time, determined in accordance with the analogies of experience, the relations of substantiality, causality, and reciprocity, was the outcome in which the composition of the understanding and its special structure first became clearly visible to us. Consciousness of the ego, pure transcendental apperception, is given us only in and with consciousness of the object as an objective appearance. Now, however, we confront a problem in which even this last limitation vanishes. For we are constrained to think the pure will as something bound by law and hence "objective," but this objectivity belongs to a sphere totally distinct from that which is expressed in the spatiotemporal phenomenon. It is not a world of things we are assured of here, but one of free personalities; not a set of causally related objects, but a republic of self-sufficient subjects purposively united.

From this perspective, what was indicated earlier by the general theoretical expression of the appearance of the object of experience now dwindles to the value of a mere fact in comparison with the person as its own self-assured unity. Only in a person is the idea of the end in itself and the ultimate end fulfilled. Only with respect to a natural thing, embedded in a determinate web of causes and effects, can we inquire as to its "whence" and "whither." By contrast, this question becomes superfluous regarding the person who by virtue of

---

that there is and can be no a priori knowledge at all. But there is no danger of this. It would be like proving by reason that there is no such thing as reason" (V, 12) (*Ak.* V, 12).

his original legislation gives to himself the unified maxims of his volition, and hence his "intelligible character." The relativity, the reciprocal conditioning of the mean, has here uncovered its limit in an absolute value. "The ends which a rational being arbitrarily proposes to himself as consequences of his action are material ends and are without exception only relative, for only their relation to a particularly constituted faculty of desire in the subject gives them their worth. And this worth cannot, therefore, afford any universal principles for all rational beings or valid and necessary principles for every volition. That is, it cannot give rise to any practical laws. All these relative ends, therefore, are grounds for hypothetical imperatives only.... Therefore the worth of any objects to be obtained by our actions is at all times conditional. Beings whose existence does not depend on our will but on Nature, if they are not rational beings, have only a relative worth as means and are therefore called 'things'; on the other hand, rational beings are designated 'persons,' because their nature indicates that they are ends in themselves, i.e., things which may not be used merely as means. Such a being is thus an object of respect and, so far, restricts all [arbitrary] choice....Thus if there is to be a supreme practical principle and a categorical imperative for the human will, it must be one that forms an objective principle of the will from the conception of that which is necessarily an end for everyone because it is an end in itself. Hence this objective principle can serve as a universal practical law. The ground of this principle is: rational nature exists as an end in itself....The practical imperative, therefore, is the following: Act so that you treat humanity, whether in your own person or in that of another, always as an end and never as a means only."[16]

Thus the order of means coincides with the order of natural things, while the order of ends is equated with that of pure, self-determined intelligences. The concept of such a rational being, which must be regarded as legislating universally by all maxims of its will so as to judge itself and its actions from this perspective, leads directly to the correlative conception of a community of rational beings in a

16. *Foundations*, second sect. (IV, 286–87) (*Ak*. IV, 427–29).

"realm of ends." If all rational beings stand under the law so that, in constituting their personhood, they are in relation with the moral individuality of all others, and so that they also demand the fundamental worth which they thus grant themselves from every other subject and acknowledge it in all other subjects, from this there springs "a systematic union of rational beings through common objective laws. This is a realm which may be called a realm of ends (certainly only as an ideal), because what these laws have in view is just the relation of these beings to each other as ends and means."[17] In this realm there is no longer any price for things that serve as means to the attainment of a further end, but there is a worth, which each subject bestows on himself in conceiving himself—as the source of his voluntary decision—as simultaneously individual and universal.[18]

With this we certainly seem, since we are oriented toward an order totally different from that of empirical phenomenal things, to be standing once more in the precincts of metaphysics, but this metaphysics is not rooted in a new conception of things that contrasts and competes with the concept of the object of experience, but purely and exclusively in that basic certainty that we receive in our consciousness of the ethical law as the consciousness of freedom. Every other access to the world of the intelligible and to the unconditioned is closed off to us. The new standpoint, which we give to ourselves in the ought, is our sole guarantee of a sphere of validity superordinate to the purely phenomenal flux. Indeed, the antinomy between freedom and causality is once again posed for us in all its poignancy. For in the selfsame event and the selfsame action in which the idea of causality claims necessity and the impossibility of being otherwise, the idea of the pure will and the ethical law says that they might have occurred otherwise than they did. The whole sequence of causes interconnected and dependent on one another is here annulled as though by a decree; the very fundamental principle of the logic of pure natural knowledge is dissolved.

17. [Ibid. (IV, 285 f.) (*Ak.* IV, 426 ff.).]
18. [Cf. ibid.]

But putting the question in this fashion, it is then valid to consider whether here it may be a matter of opposition between two types of determinism, but not in any way of opposition between determinism and indeterminism. It is in this sense that freedom is introduced by Kant himself—to be sure, expressed imprecisely and ambiguously— as a "special mode of causality." "Since the concept of a causality entails that of laws according to which something, i.e., the effect, must be established through something else which we call cause, it follows that freedom is by no means lawless even though it is not a property of the will according to laws of Nature. Rather, it must be a causality according to immutable laws, but of a peculiar kind. Otherwise a free will would be an absurdity. Natural necessity is, as we have seen, a heteronomy of efficient causes, for every effect is possible only according to the law that something else determines the efficient cause as to its causality. What else, then, can the freedom of the will be but autonomy, i.e., the property of the will to be a law to itself? The proposition that the will is a law to itself in all its actions, however, only expresses the principle that we should not act according to any other maxim than that which can also have itself as a universal law for its object. And this is just the formula of the categorical imperative and the principle of morality. Therefore a free will and a will under moral laws are identical."[19] The will and its act are thus unfree when they are determined by an individual, given object of desire, by a particular material incentive. They are free when we allow them to be determined by the idea of the totality of determining ends and the requirement of their unity. For in the first case the essential character of a merely mechanical occurrence, as we ascribe it to the physical world of things, is not yet overcome. In the same way that the properties of and changes in a corporeal substance succeed one another and proceed from one another, and as the later state is already fully latent in the preceding one and is derivable from it by a quantitative conservation rule, there unfolds here the procession of inner stirrings and strivings. A given objective incentive releases a corresponding urge, and the latter sets off a specific action with the

19. *Foundations*, third sect. (IV, 305 f.) (*Ak.* IV, 446–47).

same necessity we see in the impact of bodies. But where the action comes under the idea of autonomy, under the requirement of obligation, limits are placed on this sort of analogy. For here the series of temporal moments and the particular empirical content located in them does not simply unroll; in this instance what was present in an earlier instant is not carried over into the succeeding instant, but instead we take our stand in a nontemporal contemplation, in which we bind past and present into one even as we anticipate the future.

According to Kant, we encounter this basic feature in every elemental moral judgment. In every one, "pure reason is practical of itself alone"; that is, it judges what has happened and thus what had to happen in accordance with the empirical causal order as something which, viewed from the standpoint of the certainty of its norms, reason has the freedom either to accept or to reject.[20] This relation to a supreme, self-evident criterion of value adds a new dimension to any contemplation of the factual order. In place of the flux of events, ever-similar to itself, the succession of which we are able to trace simply as it is and which we can shape into an objective temporal order through the understanding's basic principle of causality, in which each component is unambiguously determined in its before and after, here there is introduced the conception and the anticipation of a teleological system in which one element exists "for" another, and in which all particular material ends are ultimately comprehended under the form of one lawfulness, one unconditional value. That will which can grasp this value and subordinate itself to it is the truly free will, for it no longer submits to accidental, shifting, and momentary determinations, but instead opposes them in its pure spontaneity. And hence the order of "experience," which we were firmly restricted to in the *Critique of Pure Reason* and especially in the deduction of the categories, is transcended, although it is still the case that the transcendence does not result from any theoretical datum, and hence does not furnish us with any new theoretical datum to construct and enlarge this new "intelligible" world. Liberation from experience, from empirical objects in space and time, is not the work

20. *Critique of Practical Reason*, §7 (V, 36) (*Ak.* V, 31).

of the understanding, as if the latter had discovered another realm of knowledge divorced from the conditions of sensuous intuition, but instead comes about through the will, which beholds opportunity for its application independently of all incentives of sense and empirical, material motivations. It is the will that vaults over concrete actuality and the mere existence of things in each of its truly authentic acts, for it is not bound by the given but is purely and exclusively committed to the moral task, which lifts it above and impels it beyond all that is given. It pursues this task with all its might and in all its purity, unhampered by the opposition seemingly offered to it by the whole actual state of existing being and the entire previous empirical course of things. Anyone who tries to confine this impulse of the will and of the moral idea by pointing to the limits of experience and of feasibility, is answered by the basic conception of idealism and by the new link it sets up between idea and actuality. It is no accident that Kant invokes Plato in precisely this connection, here feeling and speaking as a complete Platonist.

"Plato," it has already been said in the *Critique of Pure Reason,* "very well realized that our faculty of knowledge feels a much higher need than merely to spell out appearances according to a synthetic unity, in order to be able to read them as experience. He knew that our reason naturally exalts itself to modes of knowledge which so far transcend the bounds of experience that no given empirical object can ever coincide with them, but which must nonetheless be recognized as having their own reality, and which are by no means mere fictions of the brain.

"Plato found the chief instances of his ideas in the field of the practical, that is, in what rests on freedom, which in its turn rests upon modes of knowledge that are a peculiar product of reason. Whoever would derive the concepts of virtue from experience and make (as many have actually done) what at best can only serve as an example in an imperfect kind of exposition, into a pattern from which to derive knowledge, would make of virtue something which changes according to time and circumstances, an ambiguous monstrosity not admitting of the formation of any rule. . . . That no one of us will ever act in a way which is adequate to what is contained in the pure idea of virtue is far from proving this thought to be in any respect chimerical. For it is only by means of this data that any judgment as to moral worth or its opposite is possible; and it therefore serves as an indispensable foundation for every approach to moral perfection—however the obstacles in human nature, to the degree of which there are no assignable limits, may keep us far removed from its complete achievement.

"The *Republic* of Plato has become proverbial as a striking example of a supposedly visionary perfection, such as can exist only in the brain of the idle thinker; and Brucker has ridiculed the philosopher for asserting that a prince can rule well only insofar as he participates in the ideas. We should, however, be better advised to follow up this thought, and, where the great philosopher leaves us without help, to place it, through fresh efforts, in a proper light, rather than to set it aside as useless on the very sorry and harmful pretext of impracticability. . . . Nothing, indeed, can be more injurious, or more unworthy of a philosophy, than the vulgar appeal to so-called adverse experience. Such experience would never have existed at all, if at the proper time these institutions had been established in accordance with ideas, and if ideas had not been displaced by crude conceptions which, just because they have been derived from experience, have nullified all good intention. . . . If we set aside the exaggerations in Plato's methods of expression, the philosopher's spiritual flight from the ectypal mode of reflecting upon the physical world-order to the architectonic ordering of it according to ends, that is, according to ideas is an enterprise which calls for respect and imitation. It is, however, in regard to the principles of morality, legislation, and religion, where the experience, in this case of the good, is itself made possible only by the ideas—incomplete as their empirical expression must always remain—that Plato's teaching exhibits its quite peculiar merits. When it fails to obtain recognition, this is due to its having been judged in accordance with precisely these empirical rules, the invalidity of which, regarded as principles, it has itself demonstrated. For whereas, so far as nature is concerned, experience supplies the rules and is the source of truth, in respect of the moral laws it is, alas, the mother of illusion! Nothing is more reprehensible than to derive the

GROWTH OF CRITICAL ETHICS

laws prescribing what *ought to be done* from what *is done,* or to impose upon them the limits by which the latter is circumscribed."[21]

The basic difference between the causality of being and the causality of obligation, on which the idea of freedom rests, is thus stated as pointedly as it can be. The causality of obligation is not confined to the actual, but is oriented toward what is not actual, indeed to what is empirically impossible. The pure content and the pure validity of the categorical imperative thus hold, even when experience affords us no proof that any actual subject has ever acted in accordance with it; in fact, no such proof may ever be provided, strictly speaking, since it is not given to us to see into the heart of the agent and determine what sort of guiding maxim he has. Nonetheless, the moral law stands as given, "as an apodictically certain fact, as it were, of pure reason, a fact of which we are a priori conscious, even if it be granted that no example could be found in which it has been followed exactly."[22] Here nothing can protect us against totally discarding our ideas of duty save the clear conviction that, even if there never have been acts which sprang from such pure sources, yet the question here is not at all whether this or that did happen, "but that reason of itself and independently of all appearances commands what ought to be done. Our concern is with actions of which perhaps the world has never had an example, with actions whose feasibility might be seriously doubted by those who base everything on experience, and yet with actions inexorably commanded by reason."[23] The essential and specific reality of the idea of freedom is precisely that, uncowed by the demand for what seems impossible, it only thus discloses the true realm of the possible, which the empiricist thought to be limited to what is already actual. Thus the concept of freedom, as the preface to the *Critique of Practical Reason* puts it, becomes "the stumbling block of all empiricists but the key to the most sublime practical principles to all critical moralists, who see, through it, that they must necessarily proceed rationally."[24] This sublimity stands out the more purely

---

21. *Critique of Pure Reason,* A 314 ff. = B 370 ff. (III, 257 ff.).
22. *Critique of Practical Reason,* "Of the Deduction of the Principles of Pure Practical Reason" (V, 53) (*Ak.* V, 47).
23. *Foundations,* second sect. (IV, 264 f.) (*Ak.* IV, 408).
24. [*Critique of Practical Reason,* preface (V, 8) (*Ak.* V, 7–8).]

where the law to which the willing subject submits himself negates and cancels the empirical existence of this subject himself; where life, regarded as physical existence, is sacrificed to the idea. It is only in a determination of this kind, by motives of action which are above and beyond the sensory, that we truly are in contact with being that is supersensory: the world of the "intelligible" in the critical sense. This being cannot be laid hold of otherwise than through the medium of the pure will. If we set that aside, the world of the intelligible is lost to our view, just as there is no longer any world of empirical forms for us if we abstract from the pure intuition of space, as there is no "nature" made up of physical things except through the understanding's fundamental principle of causality.

Even at this point, where we stand in contemplation of the sole "Absolute" to which the critical viewpoint can lead us, the characteristic nature of the basic transcendental stance is preserved. It consists in the correlation of every assertion about something objective with a basic form of consciousness, the necessity to search for the basis and justification of each assertion about an existent in a fundamental function of reason. This relation is completely preserved here. The concept of a rational world, as Kant explains clearly and definitively, is but a standpoint outside appearances which reason sees itself forced to adopt in order to think itself as practical: "If the influences of sensibility were determining for man, this would not be possible; but it is necessary unless he is to be denied the consciousness of himself as an intelligence, and thus as a rational and rationally active cause, that is, a cause acting in freedom."[25] The possibility of such a supersensuous nature thus does not call for an a priori intuition of an intelligible world, which in this case, as supersensuous must needs be impossible for us; rather, in the end it becomes a matter of the will's determining ground in its own maxims: "Is the determining ground empirical, or is it a concept of pure reason (a concept of its lawfulness in general)? And how can it be the latter?"[26]

The explanation of the Kantian proposition so well known and so widely misunderstood, that we are to take the intelligible as justified

25. *Foundations*, third sect. (IV, 318) (*Ak.* IV, 458).
26. *Critique of Practical Reason*, "Of the Deduction of the Principles of Pure Practical Reason" (V, 52) (*Ak.* V, 45).

only "in the practical respect," is thus given in full. The *causa noume-nonon* continues to be an empty concept in respect to the theoretical employment of reason, though possible and thinkable. In now using this concept as the foundation of ethics, however, we do not obtain any theoretical knowledge of the constitution of a being that has a pure will; it is enough for us merely to designate it thereby as such a being, hence simply to connect the concept of causality with that of freedom (and with what is inseparable therefrom, with the moral law as its determining ground).[27] Anyone who goes beyond this, or even tries to, who attempts to depict the intelligible world instead of think-ing of it as the norm and the task of his activity, who sees in it a state of objects instead of an order of ends and a purposive communion of free intelligences as moral persons—that man has abandoned the solid ground of critical philosophy. A sphere of the "in itself" is indeed pointed to and defined by freedom in contrast to the world of appearances, the objective reality of which is manifested in the moral law "just as through a fact," but we can approach it only in action, not in intuition and thought; we grasp it only in the form of a goal and a task, not in the form of a "thing."

Many a difficulty and finespun speculation about Kant's doctrine of the thing in itself—which is admittedly paradoxical and ambiguous as he expresses it—would have been obviated if scholars had always kept this connection in mind with complete clarity. The "in itself," construed practically, does not in the slightest define the tran-scendental cause of the world of appearances, but it leads adequately to the intelligible ground thereof, since we only become fully cogniz-ant of its meaning and import thereby, and also are shown the ultimate end of all empirical willing and acting. Thus what is accomplished here is not an extension of our knowledge of given supersensuous objects, but a broadening of theoretical reason and its knowledge regarding the supersensuous in general. The Ideas lose their character of transcendence here; they become "immanent and constitutive, since they are the grounds of the possibility of realizing the necessary object of pure practical reason (the highest good)."[28]

27. *Critique of Practical Reason*, Analytic of Pure Practical Reason, chap. I, §8, II (V, 63) (*Ak.* V, 56).
28. Ibid., Dialectic, chap. II, sect. VII (V, 146 f.) (*Ak.* V, 135 f.). Cf. introduction,

Kant's doctrine of the opposition between our empirical and our intelligible nature takes on its full significance only within this all-embracing complex of problems. If we think our intelligible nature, as Schopenhauer did, in such a way that the willing subject has given to itself its determinate essence once for all in a primitive act underlying its empirical existence, so that it now remains inexorably bound to this essence in the experiential world, we are precipitated into an absolutely insoluble labyrinth of metaphysical questions. For we have not a single category that might enable us to clarify and explain that sort of relation between the "in itself" and appearance, between what is absolutely atemporal and extratemporal and the field of the temporal. But all these hesitations vanish immediately if at this point we transplant Kant's theory once more from the soil of metaphysics and mysticism to that of pure ethics, if we take it in the sense in which Schiller and Fichte understood it. Only then does it become apparent that the significance of the intelligible essence does not orient us backward, into a mythical past, but forward into the ethical future. The givenness which our intelligible nature leads us to and which its concept truly certifies to us is nothing more than the givenness of our endless practical task. One and the same act stands on the one hand under the compulsion of causes that are past and gone, while on the other hand it is seen from the point of view of future ends and their systematic unity. It receives its empirical, concrete significance from the first consideration, its nature as value from the second; in the former sense it belongs to the series of events, in the latter to the intelligible order of obligation and free, ideal determination.

Here Kant can once again refer to the way the ordinary popular mind expresses this dual form of judging. Common human reason's legal claim to freedom of the will, he explains, is founded on consciousness and the admitted presupposition of the independence of reason from sensuous causes and motives that determine merely subjectively. The man who believes himself endowed with an autonomous will thus places himself in another order of things and relates himself to determining grounds of an entirely different sort from

---

"Of the Idea of a Critique of Practical Reason" (V, 17) (*Ak.* V, 15), and also the Deduction (V, 54) (*Ak.* V, 43).

when he perceives himself as a phenomenon in the sense-world and subordinates his causality to external determination under natural laws. The fact that he has to represent and think everything in this twofold way is not at all contradictory, for it rests in the first place on his consciousness of himself as an object affected by the senses, in the second on the consciousness of himself as intelligence, that is, as an active subject who, in using reason, is freed from any passive attachment to sensory impressions.[29] Thus, in line with the basic orientation of the transcendental method, here too determination of the object is the result of the mediation of the analysis of judgment. If I judge that I ought not to have done this or that act which I did do, such an assertion would be meaningless if the "I" in it were taken in a simple sense. For the self as a sensory, empirical phenomenon, as this determinate will amid these determinate conditions, had to carry out the act; if the empirical nature of a man were fully known to us, we could predict everything he does and all his behavior as precisely as we can calculate in advance an eclipse of the sun or moon. But the truth is that in this judgment quite another connection is postulated and intended. The act is reprehensible insofar as it is determined only by particular and contingent motives, corresponding to the passing moment, which overrode respect for teleological grounds of determination in their entirety. The self has denied its true, its intelligible "essence" when it permits this momentary contingency of a particular situation and a particular impulse to become its master; it strengthens its essence whenever it examines and sits in judgment on the particular mode of action out of the postulated integrity of its character. Thus the intelligible, the unity in thought of normative determinations, appears as the continuing standard of measure to which we submit everything empirical. The phenomenon is related to the noumenon as to its own ground, not in the sense that it is known as a given supersensory substrate, but in the sense that its own worth, its place in the realm of ends, is assured in that way.

The fact that the idea of the *mundus intelligibilis*, in the form it has had ever since the Dissertation, nonetheless retains its power, that

29. Cf. *Foundations*, third sect. (IV, 317) (*Ak.* IV, 452).

the idea of obligation in a general sense crystallizes in the shape of a "world," has its profound methodological basis. For wherever critical analysis reveals and makes known to us a specifically characteristic mode of judgment, a particular form of object is coordinated with this form of judgment as well. This objectification is a basic function of pure theoretical reason itself, which we cannot divorce ourselves from, but it is important to distinguish in each particular case precisely which sphere of validity the cognition and the judgment belong to, and what the corresponding mode of being grounded in it is. For the domain of practical reason, Kant carried out this investigation in that important section he titles "The Typic of Pure Practical Judgment." Here the contrast between sensuous and supersensuous objectification is made comprehensible through the contrast between "type" and "schema." The world of experience, that of physics and natural science in general, arises for us by the understanding relating its universal principles to the pure intuitions of space and time, inscribing them in these fundamental pure Forms. Empirical concepts of the "thing" and its physical qualities and changes come about in that we flesh out the pure categories of substance and accident, of cause and effect, with concrete intuitive content, and think of substance as not simply the bearer and the purely logical subject of individual qualities but also in terms of conservation and duration, of causality as not just the relation of the ground to the grounded and dependent but also as the determination of objective temporal relation in an empirical series of appearances. When it comes to the nature of the intelligible, all such forms of asserting things are denied us. There is indeed here an *analogon* of the law of nature: one of the best-known formulations of the categorical imperative bids the will so to act as if the maxim of its action were through it to become a "universal law of Nature."[30] But the "nature" meant here is not the sensuous existence of objects, but the systematic interrelation of individual ends and their harmonious composition in a "final end." It is a model, a type, against which we measure every particular determination of the will, not an objectively existing archetype that permits

---

30. Ibid., second sect. (IV, 279) (*Ak.* IV, 421).

itself to be intuited apart from this relationship. What it has in common with the sensuous physical world is the factor of stability, of an immovable order, which we think equally in both. But in the one case it is a matter of an order that we intuit immediately as external to ourselves; in the other, one that we actively produce by the power of the autonomy of the moral law.

So it is permissible to use the nature of the sense world as the type of an intelligible nature, "so long as we do not carry over to the latter intuitions and what depends on them but only apply to it the form of lawfulness in general. . . ."[31] Should this carrying-over occur, however, we let the boundaries of the sensuous and the supersensuous blur together inadvertently; the inevitable upshot is once again that species of mysticism which Kant has been combating tirelessly ever since the *Dreams of a Spirit-Seer*. Since obligation is transformed into an image, it loses its productive, regulative force. This path leads us to a "mysticism of practical reason," which converts into a schema what served only as a symbol, that is, it bases the application of moral concepts on real and yet nonsensuous intuitions (an invisible kingdom of God) and meanders off into balderdash. And it is importantly and methodologically significant in all this that it is not the doctrine of the pure a priori that betrays us most readily into such mystical ecstasies, but, on the contrary, the purely empirical foundation of ethics, the view of morality as a doctrine of happiness. Because this point of view recognizes nothing but sensuous motives, it can never get truly clear of sensuous descriptions in all its illusory transcendence of experience and in all its depiction of what is "beyond" sense. To the degree that practical reason is pathologically determined, that is, with the interest of the inclinations in sole command, under the sensuous principle of happiness, "Mohammed's paradise or the fusion with the deity of the theosophists and mystics, according to the taste of each, would press their monstrosities on reason, and it would be as well to have no reason at all as to surrender it in such a manner to all sorts of dreams."[32] And we should not be afraid

31. *Critique of Practical Reason*, "Of the Typic of Pure Practical Judgment" (V, 78) (*Ak.* V, 70).

32. Ibid., "On the Primacy of the Pure Practical Reason" (V, 131) (*Ak.* V, 120–21).

that, if we renounce such sensuous props and aids, the pure ethical imperative would remain abstract and formal, and hence ineffective. "The fear that, if we divest this representation of everything that can commend itself to the sense," the *Critique of Judgment* emphasizes— and in words like these we are in touch with Kant whole and entire—"it will thereupon be attended only with a cold and lifeless approbation and not with any moving force or emotion, is wholly unwarranted. The very reverse is the truth. For when nothing any longer meets the eye of sense, and the unmistakable and ineffaceable idea of morality is left in possession of the field, there would be need rather of tempering the ardor of an unbounded imagination to prevent it rising to enthusiasm, than of seeking to lend these ideas the aid of images and childish devices for fear of their being wanting in potency.... This pure, elevating, merely negative presentation of morality involves... no fear of *fanaticism*, which is a delusion that would *will some vision beyond all the bounds of sensibility*; i.e., would dream according to principles (rational raving). The safeguard is the purely negative character of the presentation. For *the inscrutability of the idea of freedom* precludes all positive presentation. The moral law, however, is a sufficient and original source of determination within us; so it does not for a moment permit us to cast about for a ground of determination external to itself."[33]

Thus here too Kant's doctrine terminates in something inscrutable, yet it is a completely different relation from the one we met in his critique of mere theoretical reason. When we speak of the "thing in itself," when we claim it has a form of being but on the other hand challenge its knowability, seemingly insoluble difficulties lurk therein. For even to assert its mere presence, aside from any closer determination of it, is impossible save by those forms of cognition any transcendent employment of which the *Critique of Pure Reason* wants to excise. In the domain of the Kantian doctrine of freedom, however, we are absolved of this conflict. Freedom and the moral—which are put forward in the categorical imperative—indeed do have to be rec-

33. *Critique of Aesthetic Judgment.* "Analytic of the Sublime," §29, General Remark (V, 347) (*Ak.* V, 274–75).

ognized as inscrutable in Kant's sense. They signify for us the ultimate "Why" of all being and becoming, since they relate becoming to its ultimate end and anchor it in one supreme value, but no further "why" can be demanded of them themselves.

Thus, in the purely logical sense, it is true that we are trapped in a sort of circle, from which it seems there is no escape. We take ourselves to be free in the order of active causes, so as to think ourselves under moral laws in the order of ends, and then we think ourselves in submission to these laws because we have attributed freedom of the will to ourselves. "Freedom and self-legislation of the will are both autonomy and thus are reciprocal concepts, and for that reason one of them cannot be used to explain the other and to furnish a ground for it. At most they can be used for the logical purpose of bringing apparently different conceptions of the same object under a single concept (as we reduce different fractions of the same value to the lowest common terms)."[34] But this logical dilemma cannot and should not confuse us in our willing and acting. We need no further explanation here for the fact of freedom, because what is indescribable is done for us. The limits of knowledge are no limitation on certainty, since there can be no higher certainty for us than that which assures us of our moral self, or our own autonomous personality. Reason would be utterly out of bounds if it ventured to explain how pure reason could be practical, which would be identical with the task of explaining how freedom is possible. For how a law might be, immediately and of itself, the basis of determination of the will, how we have to represent this sort of causality theoretically and positively to ourselves, cannot be known by any further sort of datum that theory can show us; we can and must simply assume that there is such a causality by the moral law and for its service.[35] But nevertheless we are now face to face with the inscrutable, no longer as something abstract, not as an unknown substantial being; rather, it has unveiled itself to us in the ultimate law of our intelligence as free personality, and therefore has become inwardly comprehensible to us, even though it is not further

34. *Foundations*, third sect. (IV, 310) (*Ak.* IV, 450).

35. Ibid. (IV, 319 ff.) (*Ak.* IV, 459 f.); *Critique of Practical Reason* (V, 80, 145) (*Ak.* V, 72, 134).

explicable. Thus we have no grasp of the practical unconditional necessity of the moral imperative, "yet we do comprehend its incomprehensibility, which is all that can be fairly demanded of a philosophy which in its principles strives to reach the limit of human reason."[36] But it is imperative to press on to this point, so that reason does not, on the one hand, stumble about in the sense world, in a fashion that is ethically shameful, seeking the supreme ground of action and a conceptual, though empirical interest, "and so that it will not, on the other hand, impotently flap its wings in the space (for it, an empty space) of transcendent concepts which we call the intelligible world, without being able to move from its starting point and losing itself amid phantoms."[37] The shadow that lies over theoretical knowledge as to this point is illuminated for us in acting, but this light is imparted to us only so long as we actually continue in the midst of action and do not try to analyze and interpret it by mere abstract speculations.

Thus, where knowledge ends, "rational moral faith" enters, proceeding from freedom as a basic fact, not to infer the certainty of God and immortality, but to demand it. The nature of this postulate with which Kant brings to a close the development of his ethics indeed seems in the first instance not to be defined without some purely methodological question. For, strictly speaking, it offers the idea of freedom as little further supplementation as it does additional substantiation. By it, as supreme principle, the realm of obligation is delimited and fully exhausted, but it is applicable to the realm of being only by a complete μετάβασις εἰς ἄλλο γένος. Of course, there was not the slightest doubt remaining in Kant's mind that the concept of God did not afford any newer and firmer basis for the idea of freedom than was already contained in the consciousness and the validity of the moral law itself. This concept was not intended for the deduction of the validity of the idea of self-legislation from a supreme metaphysical reality; rather, it was intended only to express and guarantee the application of this idea to empirical, phenomenal actu-

36. [*Foundations*, third sect. (IV, 324) (*Ak.* IV, 463).]
37. Ibid. (IV, 322 f.) (*Ak.* IV, 462).

ality. The decisions of pure will are determined neither by considerations of feasibility nor by foresight as to the empirical consequences of action; what characterizes pure will is precisely that it receives its worth not through what it effects or accomplishes, not through its utility in achieving any sort of predetermined end, but solely from the form of willing itself, from its disposition and the maxims from which it flows. Its fruitfulness or fruitlessness can neither add to nor subtract from this worth.[38] However, little as the will in its decisions depends on consideration of results, on the other hand, in our practical thought and action we are just as unable to determine whether in general the given empirical actuality of things lends itself to the progressive actualization of the goal of the pure will. If the is and the ought are totally disparate spheres, it is at the very least not logically contradictory to think that the two might be forever mutually exclusive, that there might be in the realm of existence insuperable obstacles to carrying out the command of the ought, the unconditional validity of which cannot be watered down. The ultimate convergence of both series, the claim that the order of nature in its empirical course will and in fact must lead to a state of the world that conforms to the order of ends, is not demonstrated thereby but only postulated. And it is the content of this demand that, according to Kant, constitutes the practical meaning of the concept of God. God is thought here not as Creator, not as the explanation of the genesis of the world, but as the guarantee of its moral goal and end. The highest good in the world, the final harmony between happiness and being worthy of happiness, is possible only insofar as we assume a supreme cause of nature, which has a causality that accords with the moral disposition. Consequently, the postulate of the possibility of the highest derivative good (the best possible world) is at the same time the postulate of the actuality of a highest original good, namely the existence of God.[39] This assumption is in no way necessary *for* morality, but rather is necessitated *by* it. We must assume a moral cause of the world in order to set before ourselves a final purpose in accordance

38. Ibid., first sect. (IV, 250) (*Ak.* IV, 394).

39. *Critique of Practical Reason*, "The Existence of God as a Postulate of Pure Practical Reason" (V, 136) (*Ak.* V, 124).

with the moral law, and insofar as the purpose is necessary, to that extent (that is, in the same degree and on the same basis) it is necessary to assume the existence of God.[40] Thus here as elsewhere, the aim is definitely not to comprehend God in the metaphysical sense as the infinite substance with attributes and properties, but to try to determine ourselves and our wills appropriately.[41] The concept of God is the concrete form under which we think our intelligible moral task and its progressive empirical fulfillment.

The idea of immortality takes on an analogous significance, according to Kant, since this idea too arises in us when we clothe the thought of the infinitude of our vocation, of the unending task set for rational beings, in the temporal form of duration and eternity. Total conformity of the will to the moral law is a perfection of which a rational being is never capable while he exists in the world of sense, "but since it is required as practically necessary, it can be found only in an endless progress to that complete fitness; on principles of pure practical reason, it is necessary to assume such a practical progress as the real object of our will."[42] Here more than at any other place in his philosophy, Kant is in continuity with the philosophical world view of the eighteenth century. Like Lessing in his *Education of Mankind*, in the idea of immortality Kant maintains the requirement of an endless potentiality for development in the ethical subject, and, like Lessing, he disdains to make this idea into the determining ground of the ethical will, which must rather pursue the immanent, self-given ground unhampered by hope of the future.[43] The power of ethical action must itself be sufficient witness on this score. Every foreign and external impulse joined to it would necessarily enfeeble it and introduce confusion into it and its peculiar energy. Even if it were assumed that there were some way to demonstrate the personal continuance of the individual, by the most compelling of arguments, so

---

40. *Critique of Teleological Judgment*, §87 (V, 531 f.) (*Ak.* V, 447 f.); see especially, V, 553, note (*Ak.* V, 471).

41. *Critique of Teleological Judgment*, §88 (V, 538) (*Ak.* V, 457).

42. *Critique of Practical Reason*, "The Immortality of the Soul as a Postulate of Pure Practical Reason" (V, 132) (*Ak.* V, 122).

43. Cf. above, pp. 82 f.

that we might have it before our very eyes as an indubitably settled fact, from the standpoint of behavior more would be lost than gained. Transgression of the moral law would then be avoided, in the certainty of a future punishment, and that which is commanded be done, "but because the disposition from which actions should be done cannot be instilled by any command . . . most actions conforming to the law would be done from fear, few would be done from hope, none from duty. The moral worth of actions, on which alone the worth of the person and even of the world depends in the eyes of supreme wisdom, would not exist at all. The conduct of man, so long as his nature remained as it now is, would be changed into mere mechanism, where, as in a puppet show, everything would gesticulate well but no life would be found in the figures."[44] Thus the factor of uncertainty, which attaches to the idea of immortality taken in the purely abstract sense, liberates our life from the rigidity of merely abstract knowledge, and gives it the dye of decision and deed. "Rational practical faith" conducts us to this point more surely than any logical deduction could because, proceeding directly from the focal point of action, it straightway reenters the domain of action and determines its course.

The critical ethical system culminates in the doctrine of the postulates, and at this point we can make a retrospective survey of the major phases in the evolution of Kant's ethical life-view. The problem of immortality can serve as our guide in this connection, for it runs through all the periods of Kant's speculation. It is evident from the very first, the period essentially oriented toward natural science and the philosophy of nature; the world-picture of modern astronomy and Newtonian cosmology and cosmophysics serves as a backdrop for metaphysical reflections on the duration of the individual soul and its capacity for development. There is no gulf here between the world of the is and the world of the ought, but rather the eye roves directly from one to the other. The conflicts between the two are resolved in

44. *Critique of Practical Reason*, "Of the Wise Man's Adaptation of Man's Cognitive Faculties to his Practical Vocation" (V, 159) (*Ak.* V, 147).

the unity of the aesthetic disposition underlying this world view. "Should the immortal soul," Kant concludes the *Universal Natural History and Theory of the Heavens*, "in the whole infinity of its future duration... remain fixed forever at the point of the universe, our earth?... Who can say that it is not its destiny to come some day to know close at hand those distant globes of the cosmos and the perfection of their economies, which from afar already entrance its curiosity? Possibly some orbs of the planetary system are being formed to prepare new dwelling places for us under other skies, on completion of the temporal course prescribed for our sojourn here. Who knows if those satellites of Jupiter do not make their rounds so as to light us perpetually?... In fact, if one has filled his mind... with such meditations, the sight of a starry heaven on a serene night yields a sort of pleasure felt by none but noble souls. In the universal stillness of Nature and the calm of the senses, the hidden cognitive faculties of the immortal spirit speak a nameless tongue, and yield muffled concepts, which can be felt but not described."[45]

Thus this passage already sets forth that penetrating analogy the *Critique of Practical Reason* later expressed and elaborated on in its familiar and famous concluding sentences. The "starry heavens above me and the moral law within me" reciprocally point to each other and interpret each other. "I do not merely conjecture them and seek them as though obscured in darkness or in the transcendent region beyond my horizon: I see them before me, and I associate them directly with the consciousness of my own existence. The former begins from the place I occupy in the external world of sense, and it broadens the connection in which I stand into an unbounded magnitude of worlds beyond worlds and systems of systems and into the limitless times of their periodic motion, their beginning and continuance. The latter begins from my invisible self, my personality, and exhibits me in a world which has true infinity but which is comprehensible only to the understanding—a world with which I recognize myself as existing in a universal and necessary (and not only, as

45. [*Universal Natural History*, part three, conclusion (I, 368 f.) (*Ak*. I, 366 f.).]

in the first case, contingent) connection, and thereby also in connection with all those visible worlds."[46] If we put these words side by side with the final remarks of the *Universal Natural History and Theory of the Heavens*, the decisive advance made by the *Critique of Pure Reason* will be plain, for all the profound affinity of basic intellectual outlook. Consideration of nature and consideration of purpose are now as much united as they are separated, as much interrelated as in opposition. We must hold fast to this twofold distinction if, on the one hand, science is to be protected in its own preserve from all foreign encroachments and if, on the other, morality is to be upheld as regards the power of its pure and characteristic motive. We are just as little entitled to seek out the absolutely unconditioned, spiritual "inner being of nature," which is rather a mere phantom and will remain so,[47] as we are permitted to seek the realm of freedom and of the ought on any other basis than that resident in the content of the highest moral law itself. In the course of the empirical history of culture, both demands have been violated. "The observation of the world began from the noblest spectacle that was ever placed before the human sense and that our understanding can bear to follow in its vast expanse, and it ended in—astrology. Morals began with the noblest attribute of human nature, the development and cultivation of which promised infinite utility, and it ended in—fanaticism or superstition."[48] Only the critique of theoretical and practical reason alike can safeguard against both of these false paths, can prevent us from explaining the orbits of the heavenly bodies by spiritual powers and guiding intelligences instead of mathematically and mechanically, and, conversely, keep us from trying to describe in terms of sensuous images the pure laws of obligation and the intelligible order it opens to us. To inculcate this distinction, this "dualism" between idea and experience, between the is and the ought, and to assert the unity of reason in and through this distinction: this can now be described as the most comprehensive task set by the critical system for itself.

Coordinate with this objective unity of his philosophy, there now

46. [*Critique of Practical Reason*, conclusion (V, 174) (*Ak.* V, 161–62).]
47. See the *Critique of Pure Reason*, A 277 = B 333 (III, 235).
48. [*Critique of Practical Reason*, conclusion (V, 175) (*Ak.* V, 162).]

stands before us in full clarity the integrity of Kant's personality, the nature of the man himself, with his incorruptible critical sense of truth and his unshakable moral conviction, immune to the confusions of doubt, with the sober strength of his thought and the fire and enthusiasm of his will. This dual pattern of his nature has been minted with growing explicitness in the course of Kant's evolution as thinker and writer. In his youthful works, in which the full power of synthesizing phantasy still rules by the side of acuteness and clarity of analytical thinking, Kant's thought is often carried away with an almost lyrical and enthusiastic excess, and many a trait of the *Universal Natural History and Theory of the Heavens* shows us that we are here in the age of sensibility. But the further Kant goes, the more he rids himself in this respect of the prevailing sentimentality. In the struggle against the moral and aesthetic ideals of the Age of Sensibility, he now stands shoulder to shoulder with Lessing. It is especially characteristic how, in his lectures on anthropology, he takes up and endorses the well-known judgment that Lessing passed on Klopstock in his *Literaturbriefe*. For Kant, Klopstock is "no real poet at all," because the essential creative power is denied him; he moves only "per sympathy," since he is himself speaking as one who is moved. Kant's literary and ethical judgment is aimed even more sharply and inexorably against the whole tribe of "novel writers," who, like Richardson, depict in their characters a chimerical idealistic perfection, hoping thereby to spur the will into emulation. All these "masters of the feeling-and-affect-laden style of writing" are for him merely "mystics of taste and sentiment."[49] Feelings naturally evoke tears, but nothing in the world dries sooner than tears; basic principles of action, in contrast, must be built on concepts. "On any other foundation only passing moods can be achieved which give the person no moral worth and not even confidence in himself, without which the consciousness of his moral disposition and character, the highest good in man, cannot arise."[50] Only in this context does full light fall on the

49. On Kant's opinions about Klopstock and Richardson, see O. Schlapp, *Kants Lehre vom Genie und die Entstehung der Kritik der Urteilskraft* (Göttingen, 1901), pp. 170, 175, and 299.

50. *Critique of Practical Reason*, Methodology (V, 166 ff.) (*Ak.* V, 157).

much-hailed and much-deplored rigorism of Kantian ethics. It is the reaction of Kant's completely virile way of thinking to the effeminacy and over-softness that he saw in control all around him. It is in this sense, in fact, that he came to be understood by those who had experienced in themselves the value and power of the Kantian act of liberation. Not only Schiller, who explicitly lamented in a letter to Kant that he had momentarily taken on the "aspect of an opponent,"[51] but Wilhelm von Humboldt, Goethe, and Hölderlin also concur in this judgment. Goethe extols as Kant's "immortal service" that he released morality from the feeble and servile estate into which it had fallen, through the crude calculus of happiness, and thus "brought us all back from that effeminacy in which we were wallowing."[52] Thus it was exactly the formalistic nature of Kantian ethics that proved historically to be the peculiarly fruitful and effective moment; by the very fact that it conceived the moral law in its maximum purity and abstraction, Kantian ethics immediately and tangibly invaded the life of Kant's nation and his age, imparting to them a new direction.

51. Schiller to Kant, June 13, 1794 (X, 242) (Ak. XI, 487).
52. Goethe to Chancellor von Müller, April 29, 1818.

VI THE *CRITIQUE OF JUDGMENT*

1

In a letter to Schütz dated June 25, 1787, informing him that the manuscript of the *Critique of Pure Reason* was finished, Kant declined to review the second part of Herder's *Ideas for a Philosophy of the History of Mankind* in the Jena *Literaturzeitung*, on the grounds that he must shun any collateral work in order to make progress on the "Foundations of the Critique of Taste." One can thus see that momentous literary and philosophical tasks were crowding in on him in this period, the most productive and fruitful of his life. No relaxation or response was afforded by tasks accomplished; instead the implications of his unfolding thought pressed on relentlessly toward fresh problems. During the decade of his life from sixty to seventy, Kant experienced, in the fullest and deepest sense, that continuous surpassing of self ordinarily granted to even the greatest men only in the happy time of youth or in the period of their maturity. His works from this time show the creative power of youth united with the ripeness and consummation of age. They build upward and outward at the same time; they extend simultaneously to the disclosure of novel realms of questions and to the more and more precise architectonic ordering of the intellectual material already assimilated. In the *Critique of Judgment*, at first glance the latter tendency seems to have overpowered the former. The conception of the work seems decided more by an external, systematic analysis of the most important basic concepts of the *Critique* than by the discovery of an essential, specifically new lawfulness of consciousness. The power of judgment itself is presented, in its initial conceptual definition, as a mediating element, an insertion between theoretical and practical reason, with the object of welding the two into a new unity. According to the fundamental ideas of the critical theory, nature and freedom, the is and the ought, are permanently separated. Nonetheless the search is for a standpoint from which we may survey them both, less for their differences than in their mutual relation, less in their conceptual separation than in their harmonious interconnection. Even in the preface, the *Critique of Judgment* is treated as a "means of uniting the two parts of philosophy into a whole." "The concepts of Nature, which contain the basis of all theoretical knowledge, rest on the legislation of the understanding. The concept of Freedom, which contains a priori the basis of all practical precepts unconditioned by sense experience, rests on the legislation of the reason. . . .But in the family of the higher faculties of knowledge there is in addition a middle term between the understanding and the reason. This is the power of *judgment*, of which one, by analogy, has reason to suspect that it, too, may contain if not a special power of legislating, still a principle peculiar to it of operating under laws, although in any case a purely subjective a priori. If no field of objects corresponds to it as its domain, still it can have some sort of grounds with a certain character for which this principle alone can be valid."[1]

It has become a standing and generally accepted opinion in the literature that the analogy Kant refers to here was precisely what guided him to the discovery of the problem of the *Critique of Judgment*. It was not out of an immediate interest in the problems of art and artistic creation—so it is said—that Kant's aesthetic grew, nor does it have an integral connection with the problem of natural purposiveness that is by a necessity rooted in the subject matter itself. In both instances, Kant's predilection for the subtle and artistic architectonic of his systems, for divisions and subdivisions of concepts, and for the coordination of the faculties of knowledge into particular families is

---

1. *Critique of Judgment*, introduction, III (V, 245) (*Ak.* V, 176–77).

impressively evinced. To the adherent of this opinion about the historical origin of the *Critique of Judgment,* its historical effect must appear almost miraculous. For a strange thing came to pass, that with this work, which seems to have grown out of the special demands of his system and to be designed only to fill a gap in it, Kant touched the nerve of the entire spiritual and intellectual culture of his time more than with any other of his works. Both Goethe and Schiller—each by his own route—discovered and confirmed his own essential relation to Kant through the *Critique of Judgment;* and it, more than any other work of Kant's, launched a whole new movement of thought, which determined the direction of the entire post-Kantian philosophy. That "happy dispensation," by which what was only a consequence of the elaboration of the transcendental schematism could grow into the expression of what were in fact the deepest intellectual and cultural problems in the eighteenth and the early nineteenth centuries, is often a source of wonder, but it has hardly been explained with complete satisfaction. It remains a most noteworthy paradox that upon simply completing the scholastic framework of his theory and working it out in detail, Kant was led to a point that can be called the crucial one of all living intellectual interests in his epoch, and that in particular he succeeded "in constructing the concept of Goethe's poetry."[2]

Something more is added to this, heightening the paradox. It was not actually the content of the *Critique of Judgment* that captivated Goethe but rather its detailed arrangement and the way it was constructed. By the special mode of composition he recognized that the work could be attributed to "one of the happiest periods of his life." "Here I saw my most diverse thoughts brought together, artistic and natural production handled the same way; the powers of aesthetic and teleological judgment mutually illuminating each other.... I rejoiced that poetic art and comparative natural knowledge are so closely related, since both are subjected to the power of judgment." It was precisely this fundamental tendency of the work, which had

---

2. Cf. Wilhelm Windelband, *Die Geschichte der neueren Philosophie,* 3d. ed. (Leipzig, 1904), vol. 2, p. 173.

attracted Goethe, that constituted a stumbling block for the technical philosophical critics who evaluated this Critique. What unlocked the gates of understanding for Goethe was regarded on the whole, and particularly for the contemporary way of thinking, as one of the oddest manifestations of Kant's views and mode of presentation. Even Stadler, although he followed the development of the *Critique of Teleological Judgment* with acute understanding, expressed his astonishment at it. He found the linking of the aesthetic problem with the problem of natural teleology almost pointless, because it leads to the imputation of entirely too much value to a moment of purely formal significance and hence results in error as to the deeper worth of the book.[3] So we face a peculiar dilemma; precisely what seems, by analysis of only the philosophical content of the *Critique of Judgment*, to be a relatively accidental and dispensable element appears to have been the essential ingredient of the immediate impression it made in its own time and in its far-reaching effect. Must we acquiesce in this conclusion—or is there perhaps a deeper connection between the formal division of the *Critique of Judgment* and its actual problem, a connection that has gradually become obscured for us, although it was still immediate for and accessible to the intellectual culture of the eighteenth century?

When the question is put this way, it points to a general difficulty standing in the way of a historical and systematic comprehension of the *Critique of Judgment*. It is fundamental to Kant's transcendental method that it is always related to a specific "fact" on which philosophical criticism is performed. Difficult and involved as the progress of this criticism itself may be, nevertheless the object to which it is directed is unmistakably clear from the outset. In the *Critique of Pure Reason* this is found in the form and structure of mathematics and mathematical physics; in the *Critique of Practical Reason* the conduct of "common human reason" and the criterion it employs in all moral judgments constitute the requisite starting point. But for the questions Kant grouped under the single concept of "judgment"

---

3. August Stadler, *Kants Teleologie und ihre erkenntnistheoretische Bedeutung* (Leipzig, 1874), p. 25.

any foundation like this for the inquiry seems to be lacking. Every special scientific discipline that might be named, and every specific psychologically characterizable aspect of consciousness sought in support, proves on closer inspection to be insufficient. No path leads straight from the problems of descriptive and classificatory natural science to the problems of aesthetic form, and, conversely, no bridge can be found from aesthetic consciousness to the concept of teleology as a particular method of observing nature. Thus the parts indeed seem susceptible of being transcendentally anchored in a unitary "datum," but not the *whole*, which, however, ought to present the intellectual connection between them. On this point we must assume an actual unity to which the philosophical question can be related and on which it is founded, if the *Critique of Judgment* is to be seen not as a leap into the void but rather as a development and deduction with methodical continuity and power from previous problems. We shall try, before taking another step and before launching into the analysis of the individual questions of the *Critique of Judgment*, to specify this underlying unity more precisely—an attempt which compels us to leave the discussion of the critical system for a moment and to go back to the concrete historical origins of metaphysics.

2

The wording of Kant's first definition of judgment, as a faculty of giving laws a priori, points more to a problem of general formal logic than to a basic question belonging to the sphere of transcendental philosophy. "Judgment in general," Kant explains, "is the faculty of thinking the particular as subsumed under the universal. If the universal (the rule, the principle, the law) is given, the judgment which subsumes the particular under it (even if it prescribes as transcendental judgment) *a priori* the conditions under which alone it can be subsumed under that universal) is *determinative*. But if only the particular is given, for which the universal is to be found, the judgment is merely reflective."[4] According to this explanation, the prob-

4. *Critique of Judgment*, introduction, IV (V, 248) (*Ak.* V, 179).

lem of judgment would be joined with the problem of concept formation, for it is precisely the concept that groups particular cases into a
higher genus, thinking them as contained under its generality. But
even cursory historical consideration quickly shows a wealth of problems lying hidden in this seemingly narrow logical question, problems relating to the theory of being that are decisive for this theory.
Aristotle called Socrates the discoverer of the concept, because he first
recognized the relation between the particular and the general—
which is expressed through the concept—as worthy of examination.
In the question of the τί ἔστι, which he addressed to the concept, he
saw the germ of a new meaning of the general question concerning
being. This meaning emerged in its full purity when the Socratic *eidos*
went on to unfold into the Platonic "Idea." In this latter conception
the problem of the relation between the general and the particular
was raised to a new stage of contemplation. For now the universal no
longer appears, as was still possible in the Socratic meaning, as mere
grouping, which the particulars undergo in and through the genus,
but it is considered the archetype of all individual form. Particular
things "are" by imitation of the universal and through participation in
it, insofar as any sort of being is to be attributed to them.

A new development for the whole history of philosophy begins
with this fundamental idea. It would doubtless be entirely too simple
a formula to label this development as transposing the question of the
connection of the universal and the particular from the sphere of logic
into that of metaphysics. For such a label would presuppose logic and
metaphysics as previously known elements, while the special interest
of the intellectual development before us lies rather in the knowledge
of how both fields are gradually shaped and have their boundaries
defined under their reciprocal influence. Aristotle's achievement of
such a sharp delimitation was only apparent. Also, Aristotle is, of
course, no empiricist; even for him what is central to his consideration
is not the individual and its basis but the understanding of essence.
But where Socrates and Plato had raised the question of the concept,
Aristotle sees a concrete ontological question confronting him. The
Socratic τί ἔστι is displaced by the τὸ τί ἦν εἶναι: the problem of
the concept is transformed into the problem of teleology. The end

itself, however, does not remain confined, as with Socrates, to the technical goals and functions of men: Aristotle tries to demonstrate it as the ultimate ground of all events in nature. The universality of the end contains the key to knowledge of the universality of being (essence). Amid all the multiplicity and particularization of empirical becoming, there emerges something universal and typical, which gives this becoming its direction. The world of "Forms" does not stand beyond phenomena as something prior to and separate from them, but it is immanent in phenomena themselves as a whole of teleological forces, which rule and guide the consummation of purely material events. Hence within the Aristotelian system it is the concept of development that is designed to reconcile the opposition of matter and form, of the particular and the universal. The individual "is" not the universal; but it strives to become the universal as it runs the course of its possible forms. In this transition from the possible to the actual, from potency to act, resides what Aristotle designates in the most general sense by the concept of motion. Natural motion thus is organic motion, according to its pure concept. The Aristotelian entelechy thus signifies the fulfillment sought earlier in the Socratic *eidos* and the Platonic Idea. The question of how the particular stands in relation to the universal, how it differs from it and how it is identical, is answered for Aristotle in the idea of the end; for by this idea we immediately grasp how every individual event is joined to the whole and is conditioned and brought forth by a comprehensive whole. Being and becoming, form and matter, the intelligible world and the sensible world appear as united in the end; the truly concrete actuality seems given, comprising in itself all these oppositions as individual determinations.

The Neoplatonic system, which in general is intended to be a union of fundamental Aristotelian and Platonic ideas, assumes this definition, but in it the concept of development receives a different stamp than it did in Aristotle. Although in Neoplatonism development is connected above all with the phenomenon of organic life, Plotinus tries to restore it to its broadest and most abstract meaning, since he understands by it not so much natural becoming itself, but rather that transition from the absolutely One and First to mediated and derivative being which constitutes the fundamental conception of his system. Development here appears in the metaphysical guise of emanation; it is that primordial process through which the descent from the intelligible ground to the sense world is accomplished by determinate stages and phases. In this conception of the question, however, there emerges distinctly for the first time in the history of philosophy the relation and intellectual parallelism between biological and aesthetic problems, between the idea of organism and the idea of the beautiful. Both, according to Plotinus, are rooted in the problem of form and express, although in differing senses, the relation of the pure world of forms to the world of appearances. As in the case of animals we see that not just material and mechanical causes are at work, but that the formative *logos* is active inwardly as the special motive force, transmitting the generic structure to the newborn individual; the creative process in the artist, as well, shows the same relation seen from another perspective. For here too the Idea, which originally is encountered only as something mental and thus as an indivisible unity, is extended into the material world; the mental archetype carried by the artist within himself commands matter and molds it into a reflection of the unity of the Form. The more perfectly this is carried out, the more purely the appearance of the Beautiful is actualized. The essential outcome of the idealistic aesthetic, insofar as it had received a strictly systematic form prior to Kant, was basically contained in this one idea. The speculative aesthetics that grew out of the circle of the Florentine Academy, thence to have its effects from Michelangelo and Giordano Bruno down to Shaftesbury and Winckelmann, is an extension and development of the fundamental motif sounded by Plotinus and Neoplatonism. In this view, the work of art is only one particularly notable specimen of that "inner Form" on which the cohesiveness of the universe as a whole rests. Its composition and articulation are the immediately intuitive isolated expressions of what the world *as a whole* is. It displays, as in a specimen of being, the all-pervasive law; it demonstrates that thoroughgoing interrelation of all individual moments, whose highest and most perfect example we behold before ourselves in the starry heavens. Where empirical observation perceives but things separated by space and time, where for it the world fragments into a manifold of unrelated parts, aesthetic intuition discerns that interpenetration of formative forces on which the possibility of the beautiful and the possibility of life equally rest; for the phenomenon of beauty and that of life both are comprised and enfolded in the single underlying phenomenon of creation.

From this point, however, speculative metaphysics presses on to a further result, which seems necessitated and adumbrated by its very way of putting the question. From the standpoint of this metaphysics, the structuralization which actuality displays as a whole as well as in its individual parts, in general and in particular, is only comprehensible if its cause can be shown to lie in a supreme absolute understanding. The abstract doctrine of the *logos* in this way receives its specific theological stamp. The actual is form and has form, because behind it stand a formative Intelligence and a supreme Will-to-form. The *logos* is the principle of explanation of the world because, and insofar as, it is the principle of the creation of the world. This thought henceforward determines not only the ontology, but along with it the entire epistemology as well. For now it is valid to distinguish two fundamentally opposed modes of knowledge, one of which corresponds to the standpoint of the finite and dependent intellect and the other to the standpoint of the unconditioned and creative intellect. For the empirical mode of observation, which proceeds from particular things and remains the prisoner of comparison and collection of particulars, there is no other way to progress to the laws of the actual than to note the likenesses and differences of particulars and to unite them in this way in classes and types, in empirical "concepts." But how would this empirical form of concept, as a union of particulars in space and time into logical species, be possible, if the actual were not in fact so ordered that it is adapted and fitted to the form of a conceptual system? Everywhere that we seem merely to array particular with particular, to pass from the special case to the genus and to divide this once more into species, a prior, implicit, deeper assumption holds sway. Without the assumption that the world as a totality possesses a pervasive, all-embracing logical structure, so that one can find no element in it which is totally unconnected with all else, sheer empirical classification and comparison would lose all force. But once this is recognized, at that moment the right is granted to reverse the whole former viewpoint. Truth in its essential and full sense will only be disclosed to us when we no longer begin with the particular, as the given and actual, but end with it, when we return to the primordial principles of formation itself instead of stationing ourselves in the midst of being already possessed of form. For these principles are what is first by nature, by which the individual form of everything particular is determinate and controlled.

For this way of thinking, which in the universality of a supreme principle of being includes and possesses the fullness of all derivative elements of existence, Plotinus coined the concept and term "intuitive understanding." The infinite, divine intellect, which does not take up into itself something lying outside it but itself produces the object of its knowledge, consists not in the mere intuition of a particular thing, out of which it derives, by the rules of empirical connection or according to logical rules of deduction, another individual—and so forth in an endless sequence—but in the totality of the actual and the possible which is enclosed in and given to it in a single glance. It has no need to link any concept to other concepts, proposition to proposition, achieving in this way an apparent whole of knowledge which still must remain but an aggregate and fragment; for this intellect, the individual is the same as the All, the nearest the same as the farthest, premises and consequences are comprised in one and the same mental act. Under this notion of the divine and archetypal understanding, temporal distinctions become as accidental as the distinction in the gradations of the Universal, with which logical classification and logical rules of validity are concerned. This understanding sees the total form of the actual, because it actively produces it each moment and because it is immanent in the formative law which underlies all existence.[5]

This basic conception runs through the whole philosophy of the Middle Ages, and even modern philosophy from Descartes on contains it almost unaltered, although it impresses on it the characteristic

---

5. On the concept of the *intellectus archetypus*, cf. the statements by Kant in his letter to Marcus Herz, above pp. 126 ff.

stamp of its own problems. Thus one finds, for example, in Spinoza's work *De intellectus emendatione* and in the form of the ontological proof of God's existence it supports, the notion of the archetypal and creative intellect still in its full force; but the entire philosophical point of view into which this idea has been woven and the consequences it leads to have changed. The world-picture within which Spinoza stands is not the organic, teleological one of Aristotle and Neoplatonism; it is the mechanical cosmos of Descartes and modern science. But this newly won content, too, now shows itself capable of adaptation to the old metaphysical form of concept, remarkable as this seems at first glance. For it is precisely mathematical thinking, which ordinarily had been construed as an illustration of a syllogistic and therefore discursive procedure, which for Spinoza becomes the token and offspring of the possibility of a different, purely intuitive sort of knowledge. All true mathematical knowledge proceeds genetically; it determines the properties and characteristics of the object, since it produces this object itself. From the adequate idea of the sphere, understood not as a mute, inert image on a blackboard but as the constructive law out of which the sphere arises, all its specific determinations can be deduced with irrefragable certainty and completeness. If one transfers the requirement contained in this geometric ideal of knowledge to the whole of the world and its contents, here too it will be a question of grasping an idea of the whole in which all its particular properties and modes are included. The thought of the one substance with infinitely many attributes presents the solution that Spinoza's system gives to this task; he attributes, as it were, the realistic copy to the thought of the archetypal and creative intellect. A universal concept of being is here conceived, in which, according to the claim of the Spinozistic system, all particular manifestations and laws of being are contained as necessary, just as it is essential to the nature of the triangle that the sum of its angles should be equal to two right angles. The true order and connection of things thus reveals itself as identical with the order and connection of ideas. But in contrast to the connection of ideas stands the merely accidental sequence of our subjective perceptions; opposed to insight into the structure of the cosmos stands bare knowledge of the empirical, temporal course of events and the empirical, spatial togetherness of bodies within a limited portion of being. If, with Spinoza, we designate the knowledge of these spatiotemporal connections of appearances as the epistemological form of imagination, then the form of pure intuition, which constitutes the only truly adequate stage of knowledge, is thus once more differentiated from it in strict, thorough opposition.

And as it becomes clear here, so it becomes clear in the history of philosophy generally that the idea of the "intuitive understanding" in its most universal sense, which has been constant since Plotinus and Neoplatonism, at the same time has a variable meaning through which it serves the expression of the concrete world view of a period, to which it adapts itself. Hence the whole development of the modern speculative systems in general can be traced in the progressive transformation this idea undergoes in modern thought. Kepler's conception of the notion of the "creative understanding," for example, contains, together with the basic mathematical motif, the basic aesthetic motif of this theory: since the Creator of the universe, the "Demiurge," bore within himself aesthetic proportions and "harmonies," besides mathematical numbers and forms, we encounter their reflected glow and splendor everywhere, even within conditioned empirical existence. Next, with Shaftesbury, this idealism directly rejoins its ancient origins, when it unites with the problem of life and the Aristotelian-Neoplatonic conception of the concept of organism. The concept of "inner form" again occupies the center of interest, to be demonstrated to be as meaningful and fruitful for the progress of speculation as for the artistic view of the world and of life. All living things owe the individuality of their particular being to the specific form actual in them; the unity of the universe, however, rests on the ultimate inclusion of all particular forms in a "Form of forms" and hence the cohesiveness of nature appears as the expression of one and the same life-bestowing and purpose-giving "Genius" of the All. The eighteenth century, especially in Germany, still holds to this fundamental view,[6] and it comprises one of the latent presuppositions at which the *Critique of Judgment* hints.

6. More detailed information on this can be found in my book *Freiheit und Form: Studien zur deutschen Geistesgeschichte*, 2d ed. (Berlin, 1918), esp. pp. 206 ff.

It is necessary to bring ourselves up to date on this general historical background of the Kantian statement of the problem, even to come to a full understanding of merely the external structure of the *Critique of Judgment*. The individual underlying concepts we have met in the metaphysical and speculative unfolding of the problem of form as the principal phases of a historical line of development simultaneously constitute, within the working out of the *Critique of Judgment*, the specific milestones of the systematic thought process. The relation of the universal and the particular is forced into the center of the inquiry by the definition of judgment itself. The relation and inner connection to be assumed between the aesthetic and the teleological problem, between the idea of the beautiful and the idea of organism, is expressed in the arrangement by which the main parts of the work, both correlative to each other and mutually supplementary, are juxtaposed. The train of thought proceeds from this point; the connection between the problem of empirical concepts and the problem of ends emerges, the meaning of the unfolding thoughts is determined more precisely, until at last the whole Kantian question is involved in that profound discussion of the possibility of an archetypal understanding, in which Fichte and Schelling deemed that philosophic reason had attained its supreme height, beyond which no further step could be made. For the time being we shall not inquire into the precise content of all these special problems but shall next fix our eye simply on the general disposition of the work, the joining of the partial questions into an overall question. Modern *Kant-Philologie* and *Kant-Kritik* have overlooked this overall question primarily because it restricted itself, in systematically judging Kant's thought, too onesidedly to that narrow concept of development that had become important in the scientific biology of the second half of the nineteenth century. Even Stadler's outstanding inquiry into Kant's teleology is limited exclusively to a comparison between Kant and Darwin. If some thought to honor Goethe's view of nature most highly by stamping him as "a Darwinian before Darwin," the attempt was made to establish the same claim for Kant too—his well-known saying, that it is "preposterous for mankind" to conceive the plan of a mechanical explanation of organic existence and to hope for a "Newton of the grass blade" ought to have been an especial admonition to prudence here. In truth, however, the historical place of the *Critique of Judgment* can be made fully clear only if one resists the attempt to project the work onto the standpoint of modern biology and regards it always within its proper context. Metaphysical teleology, as it evolved in the most varied transformations and ramifications from antiquity to the eighteenth century, constitutes the matter for Kant's critical question. This does not mean that he takes the decisive directions of his thought from it, but only that it delimits the totality of objects of inquiry to which his solution intends to do justice. Actually, perhaps nowhere else does the opposition of this solution to the traditional categories of metaphysical thought emerge so sharply and clearly as at this point; nowhere is the critical revolution in thinking revealed as so decisive as it is here, where metaphysics is tracked down in a realm that for ages has been its exclusive domain and its especial dominion.

Here too, Kant once again begins with that inversion of the question which represents his universal methodological scheme. It is not the special characteristics of the concrete things which arrest his attention; for him the question is not the conditions for the existence of purposive structures in nature and art. What he wants to establish is the peculiar orientation of our knowledge when it judges some existing thing as purposive, as the coinage of an inner form. The justification and the objective validity of such judgment is all that is in question. The assignment of the teleological and the aesthetic problem to a unitary critique of judgment finds its deeper explanation and foundation only here. The term "faculty of judgment," which had been first introduced by Baumgarten's pupil Meier, was in common use before Kant; but it is only through the whole of the basic transcendental point of view that the new and special meaning it now receives attaches to it. If one takes the point of view of naive or metaphysical realism, the treatment of the question, whose starting point is the analysis of the judgment, must always and in every way appear subjective; to proceed from the judgment seems the opposite of proceeding from the object. A completely different picture of the matter is seen, however, when one reflects that under the general conviction the *Critique of Pure Reason* had established, judgment and object are strictly correlative concepts, so that in the critical sense, the truth of the object is always to be grasped and substantiated only through the truth of the judgment. When we inquire what is meant by the relation of a representation to its object, and, accordingly, what it means to assume "things" as contents of experience in general, we find that the datum which is our ultimate support is the distinction of validity existing between those different forms of judgment which the *Prolegomena* contrasts as judgments of perception and judgments of experience. The necessity and universal validity we ascribe to the latter originally constitute the object of empirical knowledge; the a priori synthesis underlying the form and the unity of the object is also the ground of the unity of the object, insofar as it is thought as an object of possible experience. Thus theoretically considered, what we call being and empirical actuality is revealed to be founded in the specific validity and the particular nature of determinate judgments. We were presented with an analogous form of inquiry in the construction of ethics. Since one and the same action was brought sometimes under the viewpoint of empirical causality and sometimes under that of moral obligation, the realm of nature and the realm of freedom were in opposition as sharply differentiated areas.

From the presuppositions one can further see that, if the aesthetic sphere is to be put forward as self-sufficient and independent, and if, moreover, in addition to the causal and mechanical explanation of natural events the teleological view of things as natural ends is to be justified, these two results can be attained only by discovery of a new realm of judgments, to be distinguished in their structure and in their objective validity from theoretical as well as practical judgments. Only in this way do the realms of art and of organic natural forms present a different world from that of mechanical causality and ethical norms, because the connection between the individual forms that we assume in both is governed by a characteristic form of law, which is expressible neither through the theoretical analogies of experience, through the relations of substance, causality, and reciprocity, nor through the ethical imperative. What is this form of law, and on what is the necessity we attribute to it founded? Is it a subjective or an objective necessity? Does it rest on a connection residing only in our human thinking and falsely transferred to objects, or is it grounded in the essence of these objects? Is the idea of an end, as Spinoza wishes to maintain, always an *asylum ignorantiae* or, as Aristotle and Leibniz assert, does it form the objective foundation of all profounder explanations of nature? Or, if we transfer all these questions from the realm of nature to that of art, is art a sign of natural truth or a sign of illusion; is it the imitation of something existing or a free creation of fantasy, which rules the given according to its own pleasure and free choice? These problems can be traced through the entire history of the theory of organic nature just as through that of aesthetics; but now it is necessary to assign them to a firmly systematic place, so that they are already half solved.

This task does not add any wholly novel moments to the development of the critical philosophy, for, since the classic letter from Kant to Marcus Herz, in which a new foundation for judgments of taste is demanded and promised, the general transcendental question is so conceived that it subsumes under it all the various modes, by means of which in general any sort of objective validity can be grounded.[7] This objectivity may arise from the necessity of thinking or of intuition, from the necessity of being or of obligation; thus it always constitutes a determinate, unified problem. The *Critique of Judgment* brings a new differentiation of this problem; it uncovers a new type of general claim to validity, but even here it remains completely within the framework established by the first comprehensive draft of the critical philosophy. The true mediation between the world of freedom and that of nature cannot consist in our inserting between the realms of being and of willing any sort of middle realm of essence, but consists instead in our discovery of a type of contemplation that participates equally in the principle of empirical explanation of nature and in the principle of ethical judgment. The question is whether nature cannot be so thought of "that the conformity to law of its form may at least agree with the possibility in it of ends acting in accordance with laws of freedom."[8] If this question is posed, an entirely

7. See above, pp. 130 ff.
8. *Critique of Judgment,* introduction, II (V, 244) (*Ak.* V, 176).

fresh perspective is immediately opened to us—thus it comprises nothing less than a change in the mutual systematic arrangement of all the basic critical concepts previously acquired and established. The task arises of seeing in detail how far this transformation confirms the earlier foundations and how far it extends and adjusts them.

3

The problem of the individual structuring of the actual existent, which is central to the *Critique of Judgment,* has its intellectual and terminological focus in the concept of purposiveness, which is Kant's starting point. According to the modern outlook on language, this initial expression of the basic question is not entirely adequate to its true content, for we are used to attaching to the purposiveness of a specific structure the idea of conscious adaptation to an end, of deliberate creation, which we must completely lay aside here if we want to grasp the true universality of the question. The linguistic usage of the eighteenth century construes "purposiveness" in a wider sense: it sees in the term the general expression for every harmonious unification of the parts of a manifold, regardless of the grounds on which this agreement may rest and the sources from which it may stem. In this sense the word represents merely the transcription and German rendering of that concept which Leibniz, in his system, signified by the expression "harmony." A totality is called "purposive" when in it there exists a structure such that every part not only stands adjacent to the next but its special import is dependent on the other. Only in a relationship of this kind is the totality converted from a mere aggregate into a closed system, in which each member possesses its characteristic function; but all these functions accord with one another so that altogether they have a unified, concerted action and a single overall significance. For Leibniz, the exemplar of such a cohesiveness was the universe itself, in which each monad is self-existent, and, cut off from all external physical influence, follows solely its own law, yet all these individual laws are so regulated that the most precise correspondence holds between them, and their results accordingly are in complete mutual agreement.

In comparison with the metaphysical concept of a whole, the critical standpoint seems to pose what is essentially a less pretentious and simpler task. Following its fundamental tendency, it works not so much toward the form of actuality itself but toward the form of our concepts of the actual; the system of these concepts, not the system of the world, constitutes its starting point. For wherever we have before us a whole, not of things but of knowledge and truths, the same question is posed. Every such logical whole is at the same time a logical construction, in which each member conditions the community of all the rest, just as it is simultaneously conditioned by them. The elements are not arranged adjacent to one another but exist only because of one another; within the complex the relation in which they stand necessarily and essentially belongs to their own logical existence. This mode of interconnection emerges clearly in the system of pure mathematical knowledge. If one considers such a system, if one surveys, for example, the content of the theorems we ordinarily combine in the concept of Euclidean geometry, it is seen as a progressive sequence from relatively simple beginnings, according to a fixed form of intuitive connection and deductive inference, to ever-richer and more advanced results. The manner of this progression guarantees that no member can be obtained that is not perfectly definable in terms of what precedes, although on the other hand each fresh step expands the previous content of knowledge and adds synthetically to it a specific new modification. Thus there reigns here a unity of principle that maintains itself continuously and enduringly in a manifold of consequences, a simple, intuitive seed that is unfolded conceptually for us and divides itself into a series of new forms, which is in itself unlimited but fully controllable and surveyable. This yields precisely that cohesiveness and correlation of parts which constitutes the essential factor in Kant's concept of purposiveness. Purposiveness thus can be found not only in the accidental formations of nature but also in the strictly necessary formation of pure intuition and pure concept.

Before we seek it out in the realm of natural forms, it is worthwhile to discover and grasp it in the realm of geometric forms. "In such a simple figure as the circle lies the key to the solution of a host of problems every one of which would separately require elaborate materials, and this solution follows, we might say, directly as one of the infinite number of excellent properties of that figure.... All conic sections, taken separately or compared with one another, are, however simple their definition, fruitful in principles for solving a host of possible problems.—It is a real joy to see the ardour with which the older geometricians investigated these properties of such lines, without allowing themselves to be troubled by the question which shallow minds raise, as to the supposed use of such knowledge. Thus they investigated the properties of the parabola in ignorance of the law of terrestrial gravitation which would have shown them its application to the trajectory of heavy bodies....While in all these labors they were working unwittingly for those who were to come after them, they delighted themselves with a finality which, although belonging to the nature of the things, they were able to present completely a priori as necessary. Plato, himself a master of this science, was fired with the idea of an original constitution of things, for the discovery of which we could dispense with all experience, and of a power of the mind enabling it to derive the harmony of real things from their supersensible principle.... Thus inspired he transcended the conceptions of experience and rose to ideas that seemed only explicable to him on the assumption of a community of intellect with the original source of all things real. No wonder that he banished from his school the man that was ignorant of geometry, since he thought that from the pure intuition residing in the depths of the human soul he could derive all that Anaxagoras inferred from the objects of experience and their purposive combination. For it is the necessity of that which, while appearing to be an original attribute belonging to the essential nature of things regardless of service to us, is yet final, and formed as if purposely designed for our use, that is the source of our great admiration of nature—a source not so much external to ourselves as seated in our reason. Surely we may pardon this admiration if, as the result of a misapprehension, it is inclined to rise by degrees to fanatical heights."[9]

If one has fully digested the fundamental results of the "Tran-

9. *Critique of Teleological Judgment*, §62 (V, 440 f.) (*Ak.* V, 362–64).

scendental Aesthetic," however, one sees that this enthusiastic flight of the spirit, born of wonderment at the internally harmonized structure of geometric forms, weakens the calm critical transcendental insight. For here it is shown that the order and regularity that we think we perceive in spatial forms lies rather in ourselves. The unity of the manifold in the field of geometry becomes comprehensible as soon as one is convinced that the geometric manifold is not something given, but something constructed. The law governing every element by its original formation is shown to be the a priori ground of that unity and flawless consistency we admire in the deduced consequences. A completely different state of affairs, and hence a totally new problem, is presented as soon as we deal with an empirical manifold instead of a mathematical manifold (such as pure space). This is precisely the assumption that we make in any empirical inquiry: that not only the whole domain of pure intuitions but also the domain of sensations and perceptions itself can be unified into a system analogous and comparable to that of geometry. Kepler not only speculates on the interconnection of conic sections as arbitrarily produced geometrical forms, but he maintains that in these forms he possesses the model of and key to the understanding and exposition of the movements of astronomical bodies. Whence comes this confidence that not only the purely artificially constructed but the given itself must be conceivable in this sense, that is to say, that we can regard its elements as if they were not completely alien to one another, but as if they stood in a fundamental intellectual affinity which needs only to be discovered and specified more exactly?

It might seem as though this question—insofar as it may be posed in general—is already answered by the principal results of the *Critique of Pure Reason*. For the *Critique of Pure Reason* is the critique of experience; its intent is to demonstrate that the lawful order which the understanding only appears to discover in experience is something grounded in the categories and rules of this understanding itself, and to this extent necessary. That appearances are joined to the synthetic unities of thinking, that not chaos but the solidity and determinateness of a causal order reigns amid them, that out of the flux of "accidents" a permanent and constant something is raised: we comprehend all this once we have seen that the ideas of causality and substantiality belong to that class of concepts with which we "spell out appearances so as to be able to read them as experiences." The lawfulness of appearances *in general* thereby ceases to be a riddle, for it is presented merely as another expression for the lawfulness of the understanding. The concrete structure of empirical science, however, confronts us at the same time with another task, which has not been solved and overcome along with the first one. For here we find not only a lawfulness of events as such, but a connection and interpenetration of particular laws of such a type that the whole of a determinate complex of appearances is progressively combined and dissected for our thought in a fixed sequence, in a progression from the simple to the complex, from the easier to the more difficult.

If we consider the classical example of modern mechanics, it is shown in the *Critique of Pure Reason* and in the *Metaphysical Foundations of Natural Science*, which is an appendix to the former, that three general laws of the understanding correspond to and underlie the three basic laws laid down by Newton: the law of inertia, the law of the proportionality of cause and effect, and the law of equality of action and reaction. But the structure and the historical development of mechanics is not thereby adequately circumscribed and comprehended. If we trace its progress from Galileo to Descartes and Kepler, from these men to Huyghens and Newton, yet another connection than the one stipulated by the three analogies of experience is revealed. Galileo begins with observations of the free fall of bodies and motion on an inclined plane, as well as the determination of the parabolic trajectory of a projectile; Kepler adds empirical determinations of the orbit of Mars, Huyghens the laws of centrifugal motion and the oscillations of a pendulum; finally all these particular moments are combined by Newton and are demonstrated to be capable, as thus integrated, of encompassing the whole system of the universe. Thus in a steady advance from minor, relatively simple primary elements and primary phenomena the entire picture of the actual is sketched, as we encounter it in cosmic mechanics. We reach in this way not just any old order of events, but an order that our understanding can survey and comprehend. Such comprehensibility cannot be demonstrated and seen as a priori necessary through the pure laws of the understanding alone, however. According to these laws, it could be thought that empirical reality indeed obeyed the general premise of causality, but that the various causal sequences which interpenetrate to form it ultimately determine in it a complexity such that it would be impossible for us to isolate and trace out individually the individual threads in the whole sprawling tangle of the actual.

In this case, too, it would be impossible for us to grasp the given in that characteristic order which is the foundation of the essential nature of our empirical science. For this order is required more as a sheer opposition between the empirical and particular and the abstract and universal, more as a mere stuff underlying the pure forms of thought as given by transcendental logic in some fashion not subject to further determination in detail. The empirical concept must determine the given by progressively mediating between it and the universal, since it relates the data to the universal through a continuous series of intermediate conceptual stages. The highest laws themselves, since they are mutually interrelated, must be specified to the particularities of the individual laws and cases—just as conversely the latter, purely because they are juxtaposed and illuminate one another, must permit the exposition of the universal connections holding between them. Only then do we possess that concrete unification and presentation of the factual our thinking seeks and insists on.

How this task is carried out in the growth of physics was already patent in its history, but it emerges still more clearly and definitely in biology and in all descriptive natural sciences. Here we seem to confront a totally unassessable mass of individual facts, which we first of all must take one by one and simply record. The idea that this material can be analyzed according to definite points of view and that it can be divided into species and subspecies signifies only a requirement laid on experience, but fulfillment of which the latter seems in no wise to guarantee. Nonetheless, scientific thinking, undisturbed by any consideration of a philosophical and epistemological nature, does not hesitate in the slightest to pose this requirement and to carry it right through into the realm of the given. In things which are absolutely individual it seeks similarities, common qualities, and properties and no apparent lack of success is allowed to divert it from this original line. If a certain class concept has not proved correct, if it is overthrown by fresh observations, it is replaced by another one; collection into genera and division into species as such, however, remain undisturbed by these fatalities among the individual concepts. Thus there is here revealed an inviolable function of our concepts, which to be sure prescribes a priori no particular content for them, but which is decisive for the total form of the descriptive and classificatory sciences.

And in this way we have also achieved a new transcendental insight of essential significance, for the term "transcendental" must be applicable to any characteristic which does not directly concern objects themselves but which concerns the mode of our knowledge of objects. We *discover* in nature what we call the affinity of species and of natural forms only because we are constrained by a principle of our power of judgment to *seek* it in nature. This shows, of course, that the relation between principle of knowledge and object has altered if we compare this example with that established by the analytic of pure understanding. Whereas the pure understanding was revealed to be "legislator for nature" because of the demonstration that it contains the conditions of the possibility of its object, here reason approaches empirical material not as if commanding but as if questioning and inquiring; thus the relation is not constitutive, but regulative, not determinative but reflective. For in this case the particular is not deduced from the universal so as to specify its nature, but the attempt is to discover in the particular itself, by successive considerations of the relations it bears within itself, and the similarities and differences which its individual parts show with respect to one another, a connection that can be expressed in ever more comprehensive concepts and rules. However, the fact that an empirical science does exist and progressively unfolds ensures that this attempt is not undertaken in vain. The manifold of facts seems, as it were, to accommodate itself to our knowledge, to meet it halfway and to prove tractable to it. Precisely because such a harmony of the content, on which our empirical knowledge rests, with the will to form by which it is guided, is not self-evident, because it is not deduced as being necessary from universally logical premises but can only be considered as something accidental, we have no choice but to see herein a certain purposiveness: namely, an appropriateness of appearances to the conditions of our judgment. This purposiveness is "formal," since it does not relate directly to things and their inner nature, but rather to concepts and their connections in one mind; but at the same time it is thoroughly objective in the sense that it undergirds nothing less than the status of empirical science and the orientation of empirical research.

Until now we have tried to develop the problem strictly according to its purely objective content, without going into detail on the particular formulation in which we encounter it in Kant. Only this way of contemplating the problem can show clearly that it is the immanent development of the actual tasks of the critique of reason, and not merely the extension and the elaboration of the Kantian architectonic of concepts, which leads to the critique of judgment as a particular portion of the system. Once these tasks are clearly understood, the expression of them Kant chose and the synthesis to which he attaches them by content and by terminology no longer offers any essential difficulty. In the first draft of the introduction to the *Critique of Judgment*, Kant gave that exposition of the fundamental question which is both the most profound and the most comprehensive; because of its great length, however, he replaced it with a shorter version in the final process of editing the work. Only much later did he recall this first draft, when Johann Sigismund Beck asked him for contributions for the commentary to the critical works he planned; yet Beck, to whom Kant gave the draft to be used as he pleased, published it with severe and arbitrary cuts and under a misleading title. To make the full content of Kant's presentation lucid, one must go to the original manuscript of the introduction.[10] Kant here sets out to reconcile the opposition of the theoretical and the practical—which is an apparent result of this whole theory—by introducing a new concept. As a step toward the systematic reconciliation he seeks, however, he deems it advisable first to reject a different, popular reconciliation, which at first glance seems to present itself. It is sometimes thought that a

10. The first appearance in print of this manuscript is in the present edition of Kant's works: see V, 177–231. For further information on its composition and its subsequent fate, see the notes (V, 581 ff.).

unification of the practical and the theoretical sphere has been produced when a given theoretical proposition is considered not only with regard to its purely conceptual grounds and its conceptual consequences, but also when the applications it permits are taken into consideration. Insofar as we think to reckon, say, statecraft and political economy as practical science, we think it possible to call hygiene and dietetics practical medicine, pedagogy practical psychology, because in all these disciplines the problem is not so much the achievement of theoretical propositions as it is the use of certain cognitions which have their foundations elsewhere. But practical propositions of this kind are not truly and in principle differentiable from theoretical ones; this separation, in its true precision, is only present where it is a question of the opposition between motivation by natural causality and by freedom. All the rest of the so-called practical propositions are nothing but the theory of the nature of things, merely applied to the manner in which we can produce them according to a principle. Thus the solution of any problem in practical mechanics (e.g., the solution of the task of finding for a given force which is to be in equilibrium with a given weight, the ratio of the respective lever arms) in fact contains nothing else and requires no other assumptions than those which are already expressed simply in the formula for the law of the lever; and it merely indicates a different bent of temporary, subjective interest, not a difference in the content of the problem itself, whether I clothe it on one occasion in the form of a pure judgment of knowledge or on another in that of a precept for the production of a particular set of conditions. Such propositions ought to be called technical rather than practical, where technic means less something opposed to theory than its execution with respect to a given particular case. Its rules belong to the art of bringing about the realization of one's desires, "which is always merely an extension of a complete theory, and never an independent part of any species of precepts."

But now Kant's treatment of the question presses on beyond technic, the middle term thus established, and achieves a new broadening and deepening of the theoretical field. For besides technic as a particular artistic human institution which perpetually clings to the illusion of free choice, there is also, as Kant notes, a technic of nature itself, namely, so far as we regard the nature of things as if their possibility rested on art, or in other words as if they were the expression of a creative will. To be sure, such a mode of conception is not given us by the object itself—for regarded as an object of experience, "nature" is nothing but the totality of appearances, insofar as it is governed by universal and therefore mathematico-physical laws; it is a standpoint which we adopt in reflection. It therefore arises neither from the mere awareness of the given, nor from its arrangement in causal connections, but the interpretation which we attach to it is a special and independent one. It can in a certain sense, of course, be quite generally asserted, from the standpoint of the critical view of the world, that it is the form of knowledge which determines the form of objectivity. Here, however, this proportion is valid in a more restricted and specific sense, for it is a second-stage creative process, as it were, that we have before us here. A whole, which as such is contained directly under the pure concepts of the understanding and experiences its objectification through them, now embodies a new meaning, in that the interrelations and mutual dependence of its parts are subjected to a new principle of contemplation. The idea of a technic of nature, in contrast to that of the purely mechanical and causal succession of appearances, is one which "determines neither the nature of objects nor the manner of producing them; rather, nature is judged by means of them, but only by analogy with an art and, more particularly, in a subjective relation to our faculty of knowledge and not in an objective relation to the objects." Only one question can and must now be asked: whether this judgment is possible—that is, whether it is compatible with the prior judgment by which the manifold is grasped under the unifying form of the pure understanding. We cannot as yet anticipate the answer which Kant gives to this question; nevertheless, it may be expected that such consistency between the principle of knowledge via the understanding and that via reflective judgment will be able to be effected only if the new principle does not trespass upon the domain of the old one, but advances an entirely separate claim to validity, which needs to be clarified and delimited with respect to the earlier one.

The idea of a technic of nature, and what sets it off from the idea of a deliberate contrivance for attaining some external goal, emerges most clearly if one first abstracts from all relation to the will at this point and holds firmly to the relation with the understanding, thus expressing the form with which nature invests it strictly in analogy with the logical interconnection of forms. That such an analogy exists is clear the moment one reflects that nature in the critical sense means for us nothing more than the totality of the objects of possible experience; and that moreover experience consists as little of a mere sum of separate observations patched together as it does of a sheer abstract set of universal rules and principles. It is only the conjunction of the moments of individuality and universality in the concept of "experience as a system of empirical laws" that constitutes the concrete whole of the connectedness of experience. "For although experience forms a system under *transcendental* laws, which comprise the condition of the possibility of experience in general, there might still occur such an *infinite multiplicity of empirical laws* and so *great a heterogeneity of natural forms* in particular experience that the concept of a system in accordance with these empirical laws would necessarily be alien to the understanding, and neither the possibility nor still less the necessity of such a unified whole is conceivable. Yet particular experience, which is thoroughly coherent under invariable principles, demands this systematic connection of empirical laws as well, whereby it becomes possible for judgment to subsume the particular under the universal, remaining always within the empirical sphere, proceeding to the highest empirical laws and their appropriate natural forms. Hence the *aggregate* of particular experiences has to be regarded as a *system*, for without this assumption total coherence under laws, that is, the empirical unity of them, cannot come about."[11] Were the multiplicity and dissimilarity of the empirical laws so great that it would be possible to organize individual ones under a general class concept but never to comprehend the totality of them in a unitary series ordered by degrees of generality, we would have in nature, even if we thought of it as subjected to the law of causality, just a "crude chaotic

11. [*Critique of Judgment*, first introduction, II (V, 185 f.) (*Ak.* XX, 202 f.). Cf. *Critique of Judgment*, introduction, V.]

aggregate." But now the judgment confronts the idea of such formlessness, not with an absolute logical decree but with the maxim that acts as its incentive and guidepost in all its inquiries. It posits a progressive lawfulness of nature, which is contingent by the concepts of the understanding alone, but which it "assumes for its own benefit." Of course, in the midst of all this it has to remain aware that in this formal purposiveness of nature, that is to say, in its quality of comprising for us a permanently interconnected whole of particular laws and particular forms, it does not posit and establish either theoretical knowledge or a practical principle of freedom, but rather provides a firm rule for our judging and inquiry. This does not contribute any new division to philosophy as the doctrinal system of the knowledge of nature and freedom; on the contrary, our concept of a technic of nature belongs to the critique of our faculty of knowledge, as a heuristic principle for judging nature. The "aphorisms of metaphysical wisdom" with which descriptive natural science in particular is accustomed to work, and which the *Critique of Pure Reason* had censured in the section on the regulative principles of reason, are only here seen in their true light. All those maxims—that nature always chooses the shortest path, that she does nothing in vain, that she suffers no leap in the manifold of forms, and though rich in varieties is poor in species—now appear less as absolute determinations of the essence of nature than as "transcendental utterances of judgment."

"All comparisons of empirical representations, so as to perceive in natural things empirical laws and the corresponding *specific* forms, and yet through comparison of these with others to detect *generically harmonious* forms, presupposes that Nature has observed in its empirical laws a certain economy, proportional to our judgment, and a similarity among forms which we can comprehend, and this presupposition must precede all comparison, being an a priori principle of the judgment."[12] For here, too, it is a matter of an a priori principle, since this hierarchy and this formal simplicity of natural laws cannot be deduced from individual experiences, but are the presuppositions

12. [First introduction, V (V, 194). Cf. *Critique of Judgment*, introduction, V.]

that are the only basis on which we are able to systematize experiences.[13]

Here at last one can see fully the evolution through which the critical philosophy and metaphysics come to diverge at this point. Whenever the problem of individual forms of actual things had been debated in pre-Kantian metaphysics, it was united with the idea of an absolute teleological understanding, which had inserted a primordial formative act into the heart of being, of which the purposiveness confirmed by our empirical concepts is but a reflection and image. We have seen how the doctrine of the *logos* held fast to this idea from its origins with Plotinus, and how it was expressed in the most diverse ways. Here too Kant carries through the characteristic transformation that is the hallmark of the whole course of his idealism: the Idea changes from an objective and active power in things into the principle and basic premise of the knowability of things as objects of experience. To be sure, to relate the order of appearances in general, which is teleological for our understanding and conforms to its requirements, to a higher level of purposiveness and to a creative and "archetypal" Intelligence, seems to him to be a step necessarily demanded by reason itself, but deception enters in as soon as we change the idea of a relation of this kind into the idea of an actually existing primal Being. We thereupon transmute, by the power of the selfsame innate sophistry of reason which the "Transcendental Dialectic" had disclosed, a goal which experiential knowledge looks forward to and from which it cannot divorce itself, into a transcendent Being which lies behind us; we conceive as a finished and factual condition an order which is posited for us in the process of knowledge itself and is grounded more deeply and solidly with each new step. It is a sufficient critique of this position to recall the transcendental point of view that the "Absolute" is not so much given as proposed to us. Even the thoroughgoing unity of the particular forms of actuality and the particular laws of experience may thus be regarded as if an understanding (though not our own) had produced them for the benefit of our

13. On the whole of this, see *Critique of Judgment,* first introduction, I, II, IV, V (V, 179 ff.) (*Ak.* XX, 195 ff.); cf. *Critique of Judgment,* introduction, I, IV, V (V, 239 ff., 248 ff.) (*Ak.* V, 171 ff., 179 ff.).

faculties of cognition, so as to render possible a system of experience under particular laws of nature. But we do not thereby assert that this forces us really to postulate such an understanding; rather, the judgment in this case legislates only for itself, not for nature, since it is mapping a course for its own reflections. One cannot ascribe to the products of nature themselves anything like the relation to ends (even to the ends of thoroughly systematic understandability); this concept is only useful in reflecting on that relation in respect to the interconnection of appearances given under empirical laws. The judgment thus has in itself an a priori principle for the possibility of nature, but only in a subjective respect, whereby it does not legislate for nature as autonomous but for itself as heautonomous. "So when it is said that Nature specifies its universal laws on a principle of finality for our cognitive faculties, i.e., of suitability for the human understanding and its necessary function of finding the universal for the particular presented to it by perception, and again for varieties ... connexion in the unity of principle, we do not thereby either prescribe a law to Nature, or learn one from it by observation—although the principle in question may be confirmed by this means. For it is not a principle of the determinant but merely of the reflective judgment. All that is intended is that, no matter what is the order and disposition of Nature in respect of its universal laws, we must investigate its empirical laws throughout on that principle and the maxims founded thereon, because only so far as that principle applies can we make any headway in the employment of our understanding in experience, or gain knowledge."[14]

The contrast in the two methods is now sharply and unmistakably indicated. Speculative metaphysics tries to account for the individual formation of nature as arising from something universal that progressively specifies itself; the critical perspective is unable to say anything about any such self-unfolding of the Absolute as a real process, but rather, where metaphysics discovers a final solution, it sees only a question that we must needs put to nature, with the ongoing reply necessarily left to experience. There can be whole ranges of experi-

14. *Critique of Judgment*, introduction, V (V, 250–55) (*Ak.* V, 181–86).

ence (and there undoubtedly are such in each of its incompleted phases) within which this requirement is not yet fulfilled, thus where the given particular is not yet truly fused with the conceptual universal but the two confront each other as still relatively disparate. In a case like this, the judgment cannot simply impress its principle on experience, it cannot force order on the empirical material and interpret it at its own whim. It can and will assert just one thing: that simply because the problem is unsolved, it should not be taken as insoluble. Its effort at the continuous reconciliation of individual things with the particular and the universal is never-ending and is not dependent on occasional success, because this is not an effort undertaken arbitrarily but one undeniably based on an essential function of reason itself.

And here the logical technic of nature which we have discovered now points to the deeper and more comprehensive question that completes the overall orientation of the *Critique of Judgment.* If we regard nature in reflective judgment as if it specified its general basic laws so that they combine in a thoroughly comprehensible hierarchy of empirical concepts, it is regarded as *art.* The idea of the "nomothetic by transcendental laws of the understanding" that constitutes the special key for the deduction of the categories no longer suffices here, because the new standpoint that now emerges validates its right not as law but only as presupposition.[15] But now, how is the state of affairs which hereby appears from the substantive, objective side to be presented subjectively; how is the grasp of that specific "artistic" peculiarity of the laws of nature to be consciously expressed and mirrored forth? We must needs pose this question, for by the basic methodical ideas of the critical doctrine it is firmly established that each of its problems is susceptible of such a double aspect and indeed cries out for it. As the unity of space and time can equally well be called unity of pure intuition, and as the unity of the object of experience is simultaneously that of transcendental apperception, here too we may expect that for the new substantive determination the idea of the technic of nature has simultaneously revealed a corresponding

15. See *Critique of Judgment,* first introduction, V (V, 196) (*Ak.* XX, 215).

new function of consciousness. But the answer Kant gives to this question is very surprising and striking. For the psychological content to which he now points is precisely that which in all the preceding consideration—in the *Critique of Pure Reason* and even more sharply and energetically in the *Critique of Practical Reason*—had been designated as the sole example of a content not determinable according to laws and hence in no way objectifiable. The subjective expression of every purposiveness that we encounter in the order of appearance is the feeling of pleasure that is connected with it. Wherever we detect an agreement for which no sufficient reason can be found in the general laws of the understanding, but which proves to be necessary for the whole of our cognitive powers and their coordinated use, there we accompany this demand, which falls to our lot equally as a free benefit, with a sensation of pleasure. We feel ourselves—just as if in this sort of structural consonance of the substance of experience it were a matter of a lucky accident which was favorable to our way of looking at things—rejoicing in it and "relieved of a want." The universal laws of nature, of which the basic laws of mechanics can serve as a model, do not carry such an agreement with themselves. For the same is true of them as of purely mathematical relationships: wonder over them ceases the moment we have grasped their exceptionless, strictly deducible necessity.

"But it is contingent, so far as we can see, that the order of Nature in its particular laws, with their wealth of at least possible variety and heterogeneity transcending all our powers of comprehension, should still in actual fact be commensurate with these powers. To find out this order is an undertaking on the part of our understanding, which pursues it with a regard to a necessary end of its own, that, namely, of introducing into Nature unity of principle. . . .The attainment of every aim is coupled with a feeling of pleasure. Now where such attainment has for its condition a representation a priori—as here a principle for the reflective judgment in general—the feeling of pleasure also is determined by a ground which is a priori and valid for all men. . . .As a matter of fact, we do not, and cannot, find in ourselves the slightest effect on the feeling of pleasure from the coincidence of perceptions with the laws in accordance with the universal concepts of Nature (the Categories), since in their case understanding necessarily follows the bent of its own nature without ulterior aim. But while this is so, the discovery, on the other hand, that two or more empirical heterogeneous laws of Nature are allied under one principle that embraces them both, is the ground of a very appreciable pleasure. . . . It is true that we no longer notice any decided pleasure in the comprehensibility of Nature, or in the unity of its divisions into genera and species, without which the empirical concepts, that afford us our knowledge of Nature in its particular laws, would not be possible. Still it is certain that the pleasure appeared in due course, and only by reason of the most ordinary experience being impossible without it, has it become gradually fused with simple cognition, and no longer arrests particular attention. . . . As against this a representation of Nature would be altogether displeasing to us, were we to be forewarned by it that, on the least investigation carried beyond the commonest experience, we should come into contact with such a heterogeneity of its laws as would make the union of its particular laws under universal empirical laws impossible for our understanding. For this would conflict with the principle of the subjectively final specification of Nature in its genera and with our own reflective judgment in respect thereof."[16]

In these Kantian sentences we hew first and foremost to that path which makes them meaningful and striking in the methodical sense. Pleasure, which heretofore had been reckoned as totally empirical, is now included in the domain of that which can be determined and known a priori; previously regarded as completely private and arbitrary, something wherein each subject differs from the other, it now contains—at least in one of its fundamental moments—a universal significance for everyone. The principle of transcendental critique is in this way applied to a realm which up to now it seemed to exclude. The first edition of the *Critique of Pure Reason* had stigmatized the hope of the "admirable analyst Baumgarten" to achieve a scientifically based "critique of taste" as abortive, because the elements of aesthetic liking and disliking consist in pleasure and pain; these stem

16. *Critique of Judgment*, introduction, V and VI (V, 253–57) (*Ak.* V, 184–88).

from purely empirical sources, however, and hence could never furnish a priori laws.[17] This outlook is now revised: the peculiar thing
about this revision, however, is that it is not the direct consideration
of the phenomenon of art and artistic creation that leads to it, but a
step forward in the critique of theoretical knowledge. An extension
and deepening of the concept of the a priori in theory first makes
possible the a priori in aesthetics and paves the way for its determination and perfection. Because it has been shown that the condition of
the universal laws of the understanding is necessary but not sufficient
for the complete form of experience; because a singular form and a
singular teleological connection of the particular was discovered,
which in its turn first completed the systematic concept of experience,
a moment of consciousness is sought on which the lawfulness of the
particular and contingent is stamped. If this moment is found, however, the limits of the inquiry thus far have been pushed back. It no
longer stops short at the question of the individual, since it treats the
individual as that which changes from case to case and hence is determinable by nothing except immediate particular experience and by
the material factor of sensation—but it seeks to discover even in this
formerly blocked-off realm the basic moments of a priori creation.

By this route Kant transcends the purely logical theory of empirical concept formation and the question of the critical conditions of a
system and classification of natural forms, and arrives at the
threshold of critical aesthetics.[18] Here the concept of a technic of

---

17. *Critique of Pure Reason*, Transcendental Aesthetic, §1, A 21 = B 35 (III, 56 f.).

18. Kant's well-known letter to Reinhold is to be taken in this sense, which sheds
light on the origin of the *Critique of Judgment*. Here Kant writes, on December 28, 1787:
"I may without being guilty of conceit, assert that the longer I persevere on my road the
less concerned I become that a contradiction or even an alliance (which is nothing out of
the ordinary nowadays) might be capable of doing my system some grave injury. This
is an inward conviction which grows on me since in my progress to other endeavors I
not only find myself always consistent, but also, when I occasionally do not know just
how to apply my method of investigation to something, I need only look back over that
general list of the elements of knowledge and the mental powers belonging to them, to
arrive at conclusions I had not foreseen. I am now busy with the critique of taste, which
has been the opportunity to discover another sort of a priori principles from the previous ones. For there are three capacities of the mind: the faculty of knowledge, the
feeling of pleasure and pain, and the faculty of desire. For the first, I found the a priori

nature is mediated from the objective side, the transcendental psychological analysis of the feeling of pleasure and pain from the subjective side. We saw already that nature, insofar as it is thought as specified in its types and varieties according to a principle that can be grasped by our judgment, is here regarded as art, but this ingenious analysis, taken by itself, at the same time appears "artistic."[19] This is true so far as it does not disclose itself immediately to the ordinary consciousness and has to be elicited by a special application of the epistemological consideration. The average human understanding takes the stability and the systematic subordination and superordination of the laws of nature as given facts that call for no explanation. But precisely for that reason, because it sees no problem here, it overlooks the solution of the problem and the specific feeling of pleasure that is connected with it. Thus if nature revealed nothing but this logical purposiveness, this would be grounds for amazement, "but hardly anyone but a transcendental philosopher would be capable of this astonishment, and even he could cite no determinate case where

---

principles in the critique of pure (theoretical) reason, for the third in the critique of practical reason. I sought also for those of the second, and although I used to think it impossible to find them, the *systematization* which permitted me to discover the analysis of the above faculties and which yields me enough material to marvel at and, if possible, to investigate for the rest of my life, set me on this road, so that I now recognize three parts of philosophy, each of which has its a priori principles, which can be enumerated and the range of knowledge possible in each mode determined: theoretical philosophy, teleology, and practical philosophy. Of these, the middle one is found to be the poorest in a priori grounds of determination" (IX, 343) (*Ak.* IV, 487). If these declarations by Kant are taken not only in a superficial, literal sense, but also if one combines them with what the *Critique of Judgment* itself tells us about the actual connection of the problems in Kant's mind, no doubt can remain as to which role "systematization" played in the discovery of critical aesthetics. Kant has not contrived a third thing in addition to the two already existing a priori principles for the sake of symmetry; it was an extension and a keener comprehension of the concept of apriority itself that came to him on what were basically theoretical grounds—in the idea of the logical "adequation" of Nature to our cognitive faculties. But in this the consideration of ends in general—or, to put it from the transcendental psychological point of view, the realm of pleasure and pain—had been shown to him to be a possible object of a priori determination, and the trail led on further from this point, ultimately to the winning of the a priori foundation of *aesthetics* as a part of a system of universal teleology.

19. Cf. *Critique of Judgment*, first introduction, V (V, 196) (*Ak.* XX, 215).

this purposiveness is manifested *in concreto*, but would have to think it only in the universal."[20] In this limitation of the previous results there is also clearly indicated the direction wherein their systematic development and growth must be looked for. Is there, we must ask, a purposive form of appearances which is not disclosed to us only through the mediation of the concept and of transcendental reflection, but instead speaks directly to us in the feeling of pleasure and pain? Is there an individual configuration of being, a union of phenomena, which presents to the world of pure and empirical thinking an unknowable singularity and hence is in no way to be grasped by the methods of classification and systematization in scientific laws—and yet which displays an independent and fundamental lawfulness of its own? When we pose these two questions we are led directly to the point at which the metaphorical sense of art, as we encounter it in the concept of a technic of nature, goes beyond the special sense, and at which the system of universal teleology assimilates into itself the critique of aesthetic judgment as its most important member.

4

As the question of individual formation brought about the transition from the universe of the pure laws of the understanding to the world of particular laws, so the same question can also serve as an intimate and direct introduction into the basic problems of critical aesthetics. For the realm of art is a realm of pure forms, each of which is complete in itself and possesses its own individual center, while it simultaneously belongs together with other things in a peculiar unity of natures and effects. How can this interconnection of essences be shown, and how can it be expressed and characterized so as not to lose the independent individual quality and the life of the particular form? In the domain of pure theory and in that of practical, moral reason we have no truly appropriate and distinctive model for such a fundamental relation. The "individual" of theory is always but the special case of a

20. Ibid., V (V, 197).

general law, from which it derives its meaning and its truth value, just as the individual as a moral subject, from the basic viewpoint of Kantian ethics, is regarded only as the bearer of the universally valid precepts of practical reason. The free personality becomes what it is only in the full sacrifice of its contingent impulses and inclinations and in unconditional subordination to the universally governing and universally binding rules of obligation. In both cases the individual seems to find its true basis and justification only by being taken up into the universal. Only in artistic intuition does a completely new relationship in this matter emerge. The work of art is something singular and apart, which is its own basis and has its goal purely within itself, and yet at the same time in it we are presented with a new whole, and a new image of reality and of the mental cosmos itself. Here the individual does not point to an abstract universal that stands behind it, but it is this universal itself, because it comprises its substance symbolically in itself.

We saw how, in the theoretical scientific consideration of the concept of a whole of experience, the further critical insight progresses the more clearly it manifests itself as an unfulfillable demand. Insistence on comprehending the entire universe in thought led us into the midst of the dialectical antinomies of the concept of infinity. We are to conceive this whole not as given but only as proposed; it is not put before us as an object having a fixed form and delimitation, but dissolves into an unlimited process, of which we can determine the direction but not the goal. In this sense every theoretical judgment of experience necessarily remains a fragment; it recognizes its fragmentary nature as soon as it has achieved critical clarity about itself. Each member of the experiential series, in order to be scientifically conceived, requires yet another, which as its "cause" determines its fixed spatiotemporal location; but this other one in its turn succumbs to the same dependence, so that it has to seek its ground once more outside of itself. As element is joined to element, series to series, in this way, the object of experience which itself is nothing but a nexus of relations, is built up for us. A totally different sort of connection of the individual to the whole, of the manifold to unity is presented to us, however, when we proceed from the datum of art and artistic creation. The datum itself we presuppose in this case—as everywhere in the transcendental inquiry. We do not ask whether it is, but how it may be; we do not move toward its historical or psychological origin, but seek to understand its pure existence and the conditions of this existence. In so doing we necessarily see ourselves guided to a new form of judgment; for every connection between contents of consciousness, objectively grasped, expresses itself as a judgment. But the judgment itself has here outgrown the confines of its heretofore purely logical definition. It is no longer a matter of the subordination of the particular to the universal or the mere application of a universal cognition to the particular, as was taught in the *Critique of Pure Reason* (principally in the chapter on the schematism of the concepts of the understanding) as the feature of the determining judgment; instead a completely different type of relation is presented. This type must be positively described and differentiated from all other syntheses of consciousness if the special character of this new range of problems is to emerge distinctly.

But before this differentiation is carried out in detail, it is worthwhile keeping vividly in mind that we are not erasing the unity of the function of judgment and the essential critical insights we have achieved concerning it. Every judgment is for Kant an act not of receptivity but of pure spontaneity; insofar as it possesses true a priori validity it does not present a mere relation to given objects, but it is the positing of objects themselves. In this sense there exists then an essential opposition between Kant's "aesthetic judgment" and what German aesthetics of the eighteenth century called "critical power" and had tried to analyze. This critical power begins with given works of "taste" and intends to show how to move from them, by analysis and comparison, to general rules and criteria of taste. Kant's viewpoint, on the contrary, achieves its end in the opposite movement: it does not wish to abstract the rules from any sort of given objects—in this case from given examples and models—but it inquires about the basic lawfulness of consciousness, on which every aesthetic perception and every designation of an item of nature or art as beautiful or ugly rests. Thus for this view that which is already formed is only the standpoint from which it strives to reach the conditions of the possibility of creation itself. These conditions can in the main be designated only negatively, since in that way we determine less what they are than what they are not. The fact that the unity of aesthetic harmony and aesthetic form rests on a principle different from the one authorizing us to combine particular elements in common and scientific experience into complex wholes and under integrated rules has already been shown. This latter unification is in the last analysis always a matter of a relation of causal superordination and subordination, of the exposition of an unbroken coherence of conditions, which can be conceived as the analogue of a connection between premises and conclusions. One experience is joined to another in a kind of dependency relation, in which both relate to each other as ground and consequent. The aesthetic grasp of a whole and its individual partial moments, on the other hand, excludes this kind of view. Here the appearance is not dissolved into its conditions, but it is affirmed as it is immediately given to us; here we do not become swamped in conceptual grounds or consequences, but we stay with the thing itself, surrendering ourselves to the impression that pure contemplation of it arouses. Instead of analysis into parts, and their superordination and subordination for the purpose of a conceptual classification, here it is proper to grasp them all together and unify them in an overall perspective for our imagination; in place of the effects, through which they link into the causal chain of appearances and are prolonged therein, we focus on the value of their sheer presence as it is disclosed to intuition itself.

This indicates the difference dividing aesthetic consciousness from practical consciousness, the world of pure form from that of will and act. As the theoretical point of view dissolves the existent into a nexus of causes and effects, of conditions and limitations, so the practical point of view dissolves it into a web of ends and means. The given manifold of content is thereby determined and structured so that in the first case an element is there through the other, in the second for the sake of the other. In pure aesthetic contemplation, on the other hand, all this kind of decomposition of the content into correlative parts and contrasts falls away. The content here appears in that qualitative perfection which requires no external completion, no ground or goal lying outside itself, and it brooks no such addition. The aesthetic consciousness possesses in itself that form of concrete realization through which, wholly abandoned to its temporary passivity, it grasps in this very fleeting passivity a factor of purely timeless meaning. The "before" and "after" that we objectify conceptually in the idea of the causal relation and shape into the empirical time sequence and time order, are here blotted out and brought to a standstill equally with that foresight and aiming at a goal which characterizes our desire and willing. And thus we have in hand the essential and decisive moments that are fused in Kant's definition of the beautiful. If we style as "pleasant" what is attractive and pleasing to the senses in the act of sensation, if we call "good" what pleases on the basis of a rule of obligation, thus by means of reason through the concept alone, so we designate as beautiful what pleases in "mere contemplation." In this expression, "mere contemplation," is indirectly included everything making up the special nature of the aesthetic perception in general, and from it all further determinations which the aesthetic judgment may reveal are deducible.

Here we have thrust upon us a question which, with regard to method, is the counterpart and the necessary completion of the previous result. If up to this point the task was to designate the special quality of the aesthetic perception, now it is, conversely, a matter of establishing unambiguously the mode of objectivity of the aesthetic object. For each function of consciousness, however it may be constituted in detail, reveals an orientation toward the object belonging to it alone and giving it a special stamp. Once again in this connection there emerges a negative determination: the objectivity of the aesthetic content is totally divorced from actuality, as the latter is posited in empirical judgment or pursued in empirical desiring. The satisfaction that determines the judgment of taste is devoid of all interest, interest being understood as interest in the existence of the thing, in the production or existence of the object contemplated.

"If anyone asks me whether I consider that the palace I see before me is beautiful, I may, perhaps, reply that I do not care for things of that sort that are merely made to be gaped at. Or I may reply in the same strain as that Iroquois sachem who said that nothing in Paris pleased him better than the eating-houses. I may even go a step further and inveigh with the vigor of a Rousseau against the vanity of the great who spend the sweat of the people on such superfluous things. Or, in fine, I may quite easily persuade myself that if I found myself on an uninhabited island, without hope of ever again coming among men, and could conjure such a palace into existence by a mere wish, I should still not trouble to do so, so long as I had a hut there that was confortable enough for me. All this may be admitted and approved; only it is not the point now at issue. All one wants to know is whether the mere representation of the object is to my liking, no matter how indifferent I may be to the real existence of the object of this representation. It is quite plain that in order to say that the object *is beautiful,* and to show that I have taste, everything turns on the meaning which I can give to this representation, and not on any factor which makes me dependent on the real existence of the object. Everyone must allow that a judgment on the beautiful which is tinged with the slightest interest, is very partial and not a pure judgment of taste. One must not be in the least prepossessed in favor of the real existence of the thing, but must preserve complete indifference in this respect, in order to play the part of judge in matters of taste."[21] The peculiarity of aesthetic self-activity, and hence the special nature of aesthetic subjectivity, come forward clearly at this point. The logical spontaneity of the understanding concerns the determination of the object of appearance through universal laws; ethical autonomy issues from the spring of the free personality, but it wants nonetheless to introduce the demands thus grounded into the empirically given things and states of affairs and to actualize them therein. The aesthetic function alone does not ask what the object may be and do, but rather what I make of its representation in me. The actual retreats to its real status, and into its place steps ideal determination and ideal unity of the pure image.

In this sense—but only in this sense—the aesthetic world is a world of appearance. The concept of appearance is intended only to ward off the false notion of an actuality which would precipitate us

21. *Critique of Aesthetic Judgment,* §2 (V, 273) (Ak. V, 204–05).

once again into the machinations of the theoretical concept of nature or the practical concept of reason. It elevates the beautiful above the sphere of causality—for freedom is also a special kind of causality for Kant—so as to place it purely under the rule of inner creation. The latter also legislates for appearance—because appearance receives from it the essential connection of its separate moments. As in every situation where we apply the contrast of subjective and objective, here too it is valid to specify this opposition sharply and carefully, so as to avoid the dialectic concealed in it. Leaving out of account the existence of the thing is precisely the characteristic and essential reality of the aesthetic representation. For it is just this way that it becomes the intuition of pure form, in leaving out of consideration all the associated conditions and consequences which unavoidably cling to the "thing." Where both are still mingled, where the interest in the structure of the form and its analysis still intersects with and is elbowed aside by interest in the actuality of it which is the image, the essential viewpoint which constitutes and is the hallmark of the aesthetic as such has not yet been attained.

The idea of purposiveness without purpose, by which Kant designates and circumscribes the whole ambit of the aesthetic, is now divested of the final paradox that clung to it. For purposiveness means, as has been shown, nothing other than individual creation, which displays a unified form in itself and in its structure, while purpose means the external determination which is allotted to it. A purposive creation has its center of gravity in itself; one that is goal-oriented has its center external to itself; the worth of the one resides in its being, that of the other in its results. The sole function of the concept of "disinterested pleasure" is to bring this state of affairs, considered subjectively, into thought. Hence the essential sense of this central concept is missed when—as has happened—Kant's aesthetic ideal is designated as "indolent repose" and therefore Herder's and Schiller's dynamic ideal of beauty, which takes beauty to be a "living form," is thought to be opposed to it.[22] The Kantian insis-

---

22. See Robert Sommer, *Grundzüge einer Geschichte der deutschen Psychologie und Aesthetik* (Würzburg, 1892), pp. 296, 337 ff., 349.

tence on disregarding all interest leaves full and unhampered room
for the activity of the imagination; only the activity of the will and the
activity of sensory desire are routed from the threshold of the aesthetic on methodological grounds. Adherence to immediate attraction
and immediate need is precisely rejected thereby, because it hems in
and stifles that immediate life of the representation, that free figura-
tion of the formative imagination which constitutes for Kant the special characteristic of the artistic. To this extent Kant in no way contradicts the "energetic" aesthetic of the eighteenth century, but as
the focus of aesthetic interest has been shifted from the actuality of the
thing to the actuality of the image, so the passive stimulation of the
emotions is translated into the excitement of their pure play. In the
freedom of this play the whole passionate inner excitement of emotion is conserved; but in it the play is separated from its purely material foundations. Hence in the last analysis it is not emotion itself, as
an isolated psychological state, that is drawn into this arousal, but the
elements of the play compose the universal basic functions of consciousness, from which each individual psychic content issues and to
which it refers back. This universality explains the universal communicability of the aesthetic state, which we presuppose, since we
ascribe to the judgment of taste a "validity for everyone," although
we are incapable of conceptualizing the grounds of the validity thus
asserted and of deducing it from concepts. The mental state of the
aesthetic representation is that of "a feeling of the free play of the
powers of representation on a given representation for a cognition in
general."

"Now a representation, whereby an object is given, involves, in
order that it may become a source of cognition at all, *imagination* for
bringing together the manifold of intuition, and *understanding* for the
unity of the concept uniting the representations. This state of *free play*
of the cognitive faculties attending a representation by which an object is given must admit of universal communication: because cognition, as a definition of the Object with which given representations (in
any Subject whatever) are to accord, is the one and only representation which is valid for everyone.

"As the subjective universal communicability of the mode of rep-

resentation in a judgment of taste is to subsist apart from the presupposition of any definite concept, it can be nothing else than the mental state present in the free play of imagination and understanding (so
far as these are in mutual accord, as is requisite for *cognition in general*): for we are conscious that this subjective relation suitable for a
cognition in general must be just as valid for every one, and consequently as universally communicable, as is any determinate cognition, which always rests upon that relation as its subjective condition."[23]

It seems indeed as though, with this explanation of the universal
communicability of the aesthetic state, we are again diverted from its
proper domain, for its separation from the sensory, private feeling of
pleasure and pain seems at bottom attainable only by once again
reentering the path of the logically objectifying way of thinking.
When the imagination and the understanding thus are unified, as is
requisite for a "cognition in general," it is the empirical use of the
productive imagination, as developed by the *Critique of Pure Reason*,
rather than its specifically aesthetic use that is thus explained. In fact,
according to a basic insight of that *Critique*, which is enlarged on in
particular in the chapter on the schematism of the pure concepts of
the understanding, even the spatiotemporal connection of the perceptions of the senses and their unification into objects of experience
rests precisely on cooperation between the understanding and the
imagination. The mutual determination of these two functions seems
to constitute no truly new relationship, such as we would expect and
demand as an explanatory ground for the new problem that appears
here. Yet we must take into consideration at this point the fact that
the earlier insight does receive a new emphasis. A specific unity of
knowledge is achieved for theoretical as well as for aesthetic representation; but if in the former the tone and emphasis lie on the factor
of knowledge, so in the latter they lie on the factor of unity. The
aesthetic relation is "purposive for the cognition of objects *in general*,"
but exactly thereby it renounces the sorting of objects into particular
classes, designating and defining them by particular differentiating

23. *Critique of Aesthetic Judgment*, §9 (V, 286 f.) (*Ak.* V, 217 f.).

characters as they are expressed in empirical concepts. The intuitive unity of the form has no need of this ongoing discursive sorting. The free process of imaging itself is not here tied down and confined by reference to the objective existence of things, as we fix it through scientific concepts and laws. On the other hand, of course, the role of the understanding is recognizable in this creative activity of the imagination if the concept of the understanding itself is taken in a sense above and beyond the exclusively logico-theoretical one. The understanding, in its most universal meaning, is the capacity of setting limits; it is what arrests the steady activity of representation itself, and facilitates its circumscription into a definite image. When this synthesis occurs, when we succeed in fixing the movement of the imagination this way, without making a detour via the conceptual abstractions of empirical thinking, so that the imagination does not get lost in vagueness but crystallizes into solid forms and configurations, then that harmonious interpenetration of both functions is achieved which Kant calls for as a basic moment of the genuine aesthetic attitude.

For now understanding and intuition are no longer in opposition as things totally dissimilar, so that they have to be brought together through the agency of a foreign mediator and conjoined through a cunning schematism, but they are truly blended and absorbed in each other. The capacity for specification acts directly in the actual course of imaging and beholding, since it articulates and vivifies the flowing, ever-constant series of images. In the empirical judgment of subsumption a determinate individual intuition is related to a determinate concept and subordinated to it, as, for instance, the curvature of the table that we see before us is related to the geometric concept of a circle and cognized through the latter.[24] Nothing like this is found in aesthetic consciousness. For here the individual concept and the individual intuition do not stand in contrast to each other; it is rather a question of harmonizing the *function* of the understanding and that of beholding. The free play that is required concerns not representation but the powers of representation; not the results in which intuition and understanding are made concrete and in which they both come to

24. See *Critique of Pure Reason*, A 137 = B 176 (III, 141).

rest, but rather the living excitation that occupies them. In this way every utterance of this sort, wherein a particular image is not compared with a particular concept but rather the totality of the powers of the mind first is disclosed in its true completeness, lays hold immediately of the "life-feeling" of the very subject. "To apprehend a regular and appropriate building," it is remarked in the opening of the *Critique of Aesthetic Judgment*, "with one's cognitive faculties, be the mode of representation clear or confused, is quite a different thing from being conscious of this representation with an accompanying sensation of delight. Here the representation is referred wholly to the Subject, and what is more to its feeling of life—under the name of the feeling of pleasure or displeasure—and this forms the basis of a quite separate faculty of discriminating and estimating, that contributes nothing to knowledge. All it does is to compare the given representation in the Subject with the entire faculty of representations of which the mind is conscious in the feeling of its state."[25] In empirical theoretical judgment the individual experience that is present to me is held up to the system of experiences (actual or possible) and its objective truth value is only determined through this comparison; in the aesthetic situation the present individual intuition or the present impression brings the whole of the perceiving and representing powers into direct resonance. If, then, the unity of experience and its object must be built up by the labor of concept formation, line by line, and element by element, the perfected work of art in one stroke presents that unity of mood which is for us the unmediated expression of the unity of our ego, of our concrete feeling of life and self.

This new relation set up between the singular and the universal holds the exact key to the solution of the problem of what form of universality is to be ascribed to aesthetic judgment. That it must contain some sort of universality has been already settled, so far as he is concerned, through the connection in which he approaches the basic question of aesthetics, for it is in consolidating and deepening his concept of the a priori that he first encounters the problem of aesthetic judgment. At the same time, however, the behavior of ordinary con-

25. *Critique of Aesthetic Judgment*, §1 (V, 272) (*Ak.* V, 204).

sciousness affords direct confirmation of the claim to universal validity made by the judgment of taste. In what concerns judgment about
the pleasant in sensation, everyone is reconciled to the fact that,
because it is founded on a private feeling, it is confined to himself.
The situation is the reverse with the beautiful. "It would... be
ridiculous if any one who plumed himself on his taste were to think of
justifying himself by saying: This object (the building we see, the
dress that person has on, the concert we hear, the poem submitted to
our criticism) is beautiful *for me*. For if it merely pleases *him*, he must
not call it *beautiful*. Many things may for him possess charm and
agreeableness—no one cares about that; but when he puts a thing on
a pedestal and calls it beautiful, he demands the same delight from
others. He judges not merely for himself, but for all men, and then
speaks of beauty as if it were a property of things. Thus he says the
thing is beautiful; and it is not as if he counted on others agreeing in
his judgment of liking owing to his having found them in such
agreement on a number of occasions, but he *demands* this agreement
of them. He blames them if they judge differently, and denies them
taste, which he still requires of them as something they ought to have;
and to this extent it is not open to men to say: Every one has his own
taste. This would be equivalent to saying that there is no such thing at
all as taste, i.e., no aesthetic judgment capable of making a rightful
claim upon the assent of all men."[26]

And yet this pure value claim of the aesthetic is not interchangeable with its demonstrability from concepts alone, as was virtually
universally assumed in the German aesthetics of the Enlightenment
(Gottsched and the Swiss, for example, agree on this point). At this
juncture the critical task consists rather in the insight as to the possibility of universality, which nonetheless spurns mediation by means
of logical concepts. Now it was shown above that in and through
aesthetic harmony an immediate relation of the contingently given
individual content of consciousness to the totality of the powers of the
mind is established. The aesthetic state concerns the subject and his
life-feeling exclusively, but it takes this feeling not in an isolated and

26. Ibid., §7 (V, 281 ff.) (*Ak.* V, 212 ff.).

to that extent contingent moment, but rather in the ensemble of its moments. Only where this resonance of the whole in the particular and singular is present are we immersed in the freedom of the play and experience this freedom. But only with this experience do we attain the full estate of subjectivity itself. In the case of sensory perception the individual ego has no way to communicate it to another ego other than by transposing it into the sphere of the objective and defining it therein. The color that I see, the tone that I hear is presented as the joint possession of the knowing subjects, since by applying the basic principles of extensive and intensive magnitude and the categories of substance and causality, which are exactly knowable and measurable, both are translated into vibrations. But with this translation into the spheres of number and measure that is a condition of scientific objectification, color and tone as such have ceased to exist; their being in the theoretical sense is absorbed into the reality and the lawfulness of motion. If this is done, however, the method of universal communication, as practiced in the theoretical concept, has really made the content to be communicated vanish, to be replaced by a mere abstract symbol. The fact that color and tone, besides what they mean as physical elements, are also experiences in a perceiving and feeling subject is completely eliminated from this way of determining them. Here the problem of aesthetic consciousness comes in. This consciousness asserts that there is a universal communicability from subject to subject, which thus does not need to detour through the conceptually objective and be swallowed up in it. In the phenomenon of the beautiful the inconceivable thing happens, that in contemplating beauty every subject remains in itself and is immersed purely in its own inner state, while at the same time it is absolved of all contingent particularity and knows itself to be the bearer of a total feeling which no longer belongs to "this" or "that."

Only now do we understand the expression "subjective universality" that Kant coins as the mark of the aesthetic judgment. "Subjective universality" is the assertion and requirement of a universality of subjectivity itself. The designation "subjective" does not act to restrict the claim to validity made by the aesthetic, but just the opposite: it designates an enlargement of the realm of validity, which is here perfected. Universality is not prevented by the individuality of the subject, for as it is true that the subjects have their life not only in passive sensory perceptions or in pathological desires, but can stimulate themselves to the free play of the representational powers, it is equally true that in such activity they employ one and the same essential basic function. In this functioning, which first properly makes the self into a self, every ego is akin to every other, and hence the function may be assumed in each other. The artistic feeling remains a feeling of self, but precisely as such it is at the same time a universal feeling of the world and life. The "self" detaches itself from its individuality when it objectifies itself in a construction of aesthetic fantasy; its individual unique stimulation is nevertheless not destroyed in this construction, but rather dwells powerfully in it and is communicated to all those who are capable of grasping it. Thus the subject is placed in a universal medium, one which however is something completely different from the medium of reification into which the natural scientific way of contemplation plunges us. What differentiates the most complete description of a landscape resulting from concepts of descriptive natural science from its artistic presentation in a painting or a lyric poem? Only that in the latter all the features of the object, the sharper and more clearly they stand out, prove even more intensely to be features of a psychic excitation, communicated to the beholder through the graphic or lyrical construction. Here the inner passion flows out into the object only to be received back from it in a stronger and purer form. As the self in a state of aesthetic contemplation does not just remain attached to its contingent representation, but in Kant's expression "holds it up against the whole faculty of representation," a new cosmos is revealed to it, which is not the system of objectivity but the whole of subjectivity. In this whole it finds itself also to be an individuality closed to all others. In this way aesthetic consciousness solves the paradoxical task of presenting a universal which is not contrary to the individual but which is its pure correlate, because it finds its fulfillment and embodiment in it alone.

And in this way the question of universal communicability—which is not universal demonstrability—is also answered. Since in the aesthetic attitude the judge feels himself to be fully free in respect to the pleasure he centers on the object, he can discover, as grounds of this pleasure, no private conditions on which its subject might depend and must therefore regard it as grounded in what can be presupposed of everyone else; consequently, he must believe that he can justifiably attribute to everyone a similar pleasure. "Accordingly he will speak of the beautiful as if beauty were a quality of the object and the judgment logical (forming a cognition of the Object by concepts of it); although it is only aesthetic, and contains merely a reference of the representation of the object to the Subject;—because it still bears this resemblance to the logical judgment, that it may be presupposed to be valid for all men. But this universality cannot spring from concepts. . . .Here, now, we may perceive that nothing is postulated in the judgment of taste but such a *universal voice* in respect of delight that is not mediated by concepts; consequently, only the *possibility* of an aesthetic judgment capable of being at the same time deemed valid for every one. The judgment of taste itself does not *postulate* the agreement of every one (for it is only competent for a logically universal judgment to do this, in that it is able to bring forward reasons); it only *imputes* this agreement to every one, as an instance of the rule in respect of which it looks for confirmation, not from concepts, but from the concurrence of others."[27]

Hence Kant has arrived at the principal question standing at the crossroads of all aesthetic discussions in the eighteenth century by a new route and in a completely different systematic connection. Is a rule to be abstracted from given works of art, from classical prototypes and models, which prescribes specific objective limits to creation—or does the freedom of the imagination, which is bound to no external norm, reign here? Is there a conceptually determinable law of artistic creation, from which one cannot depart if its goal is not to be aborted—or is everything ceded to the creative will of the gifted subject, which moves from an unknown beginning to an unknown end? These questions, which recur in the aesthetic doctrines of the eighteenth century in the most diverse forms, were brought to a

27. Ibid., §§6, 8 (V, 280, 285) (*Ak.* V, 211, 216).

sharp and clear dialectical formulation in the area of literary criticism by Lessing. The struggle between genius and rule, between imagination and reason—so run the decisive discussions of the *Hamburg Dramaturgy*—is pointless, for the creation of genius receives no rule from outside, but it is this rule itself. In it is shrouded an inner lawfulness and purposiveness, which, however, appears and leaves its imprint nowhere else than in the concrete and individual art form itself. Kant unhesitatingly adheres to this conclusion of Lessing's, but it now guides him back into the whole depth and universality of the questions that for him are comprised in the idea of the self-legislation of the spirit. "Genius"—thus he even defines it—"is the talent (natural gift) which gives the rule to art. . . .For every art presupposes rules which are laid down as the foundation which first enables a product, if it is to be called one of art, to be represented as possible. The concept of fine art, however, does not permit of the judgment upon the beauty of its product being derived from any rule that has a *concept* for its determining ground. . . .Consequently fine art cannot of its own self excogitate the rule according to which it is to effectuate its product. But since, for all that, a product can never be called art unless there is a preceding rule, it follows that nature in the individual (and by virtue of the harmony of his faculties) must give the rule to art, i.e., fine art is only possible as a product of genius."[28]

Thus the unity of the harmony precedes the objective unity of the form. Genius and its act stand at the point where supreme individuality and supreme universality, freedom and necessity, pure creation and pure lawfulness indissolubly coalesce. In every line of its activity it is thoroughly original but nonetheless thoroughly exemplary. For just where we stand in the true focus of personality, where the latter gives itself purely without any external consideration and expresses itself in the individually necessary law of its creating, all the accidental limitations clinging to the individual in his particular empirical existence and his particular empirical interests fall away. In its immersion in this unadornedly personal sphere genius finds the secret and the power of universal communicability, and each great work of art

28. Ibid., §46 (V, 382) (*Ak.* V, 307).

presents nothing but the objectification of this basic power. As a temporally unique psychic event, never recurring in the same way, the work of genius testifies straightforwardly and unambiguously how the most intimate subjective feeling at the same time reaches down into the deepest sphere of pure validity and timeless necessity. And this highest form of communication is also the only one which is at the disposal of genius. Were genius to try to speak to us elsewhere than in the immediate creation of its work, it would have precisely in that act cut itself off from the soil in which it is rooted. Hence what it is and what it signifies, as a "natural gift," cannot be expressed in a general formula and thus put forward as a prescription; the rule must, so far as it does exist, be abstracted from the act, that is, from the product, which serves as an example not for imitation but for comparable creation. Herein Lessing's saying that a genius can only be kindled by a genius is also taken up by Kant. "The artist's ideas arouse like ideas on the part of his pupil, presuming nature to have visited him with a like proportion of mental powers." It is this "proportion" which is the characteristic generative motive in the creation of genius.

And from this aspect artistic productivity can also be differentiated from scientific productivity. Kant's assertion that there can be no genius in the sciences[29] can only be rightly evaluated if one keeps in mind that in this discussion it is for him always a matter of the systematic difference of meaning of these two cultural realms, not of the psychological difference of individuals. Whether the scientific discoverer may not also make "one case stand for a thousand," whether along with the discursive comparison of individual things an intuitive anticipation of the whole may not also be possible and actual: these are questions about which nothing can be decided at this point. The decisive difference lies solely in that everything which pretends to be scientific insight, as soon as it is to be communicated and established, possesses no form for this save that of the objective concept and objective deduction. The personality of the creator must be expunged if the accuracy of the result is to be protected. Only in the great artist

29. See ibid., §47 (V, 383 f.) (*Ak.* V, 308 f.).

is this division nonexistent, for everything he gives is endowed with its peculiar and supreme value only through what he is. He does not alienate himself in any work which then continues to exist as an isolated thing of value in itself, but in each particular work he creates a new symbolic expression of that univocal basic relation given in his "nature," in the "proportion of his mental powers."

Considered historically, this Kantian doctrine of genius signifies the achievement of a reconciliation between two diverse spiritual worlds, for it shares a crucial motive with the fundamental outlook of the Enlightenment, while on the other hand it shatters the conceptual schema of the philosophy of the Enlightenment from within. Kant's theory of genius became the historical point of departure for all those romantic, speculative developments of the concept of genius that attributed significance to the productive aesthetic imagination as begetter of the world and reality. Schelling's theory of intellectual intuition as the basic transcendental faculty, Friedrich Schlegel's theory of the ego and of "irony" were developed along this line. However, what distinguishes Kant's own view once and for all from these attempts is the form and the direction of his concept of the a priori. That his a priorism is a critical one is further manifested in that the a priori is not traced back to one single basic metaphysical power of consciousness but is firmly kept within the strict particularity of its specific applications. Thus the concept of "reason," as it was evolved by the eighteenth century, is for Kant expanded into the deeper concept of "spontaneity" of consciousness, but he does not regard the latter as exhausted in any finished work and activity of consciousness. It is impossible for the aesthetic spontaneity of fantasy here to become, as it did in romanticism, the final founding and unifying principle, since the essential intention aims at differentiating it strictly and decisively from the logical spontaneity of judgment and from the ethical spontaneity of will. The whole scale of degrees of subjectivity and objectivity which Kant sets forth, and which receives its most important completion and its essential conclusion only in the *Critique of Judgment*, subserves this task above all. The being of the laws of nature, the ought of the moral law, should not be abandoned in favor of the

play of the imagination, but on the other hand this play is in possession of its own autonomous realm into which no conceptual demand
and no moral imperative may intrude.

The essential meaning of the restriction of the concept of genius to
art lies in assisting this thought to its clear expression. The concept of
the "sciences of the beautiful" had won a dangerous importance and
degree of dissemination in the second half of the eighteenth century.
Sterner, more profound minds, like Lambert—who expressed himself
on the topic in a letter to Kant in 1765[30]—never tired of opposing
this by demanding exact conceptual definition as the foundation
of all scientific knowledge, but the muddling together of the realms
nonetheless remains the characteristic mark of popular philosophy.
Lessing as a young man once remarked, in opposition to the fashionable current of the time, that the true *beaux esprits* were generally the
really shallow minds. At this Kant's theory of genius draws a sharp
line. Whatever the great scientific mind may discover, still he is not to
be called a genius for that reason: "For what is accomplished in this
way is something that *could* have been learned. Hence it all lies in the
natural path of investigation and reflection according to rules, and so
is not specifically distinguishable from what may be acquired as the
result of industry backed up by imitation. So all that *Newton* has set
forth in his immortal work on the Principles of Natural Philosophy
may well be learned, however great a mind it took to find it all out,
but we cannot learn to write in a true poetic vein, no matter how
complete all the precepts of the poetic art may be, or however excellent its models. The reason is that all the steps that Newton had to
take from the first elements of geometry to his greatest and most
profound discoveries were such as he could make intuitively evident
and plain to follow, not only for himself but for every one else. On the
other hand no *Homer* or *Wieland* can show how his ideas, so rich at
once in fancy and in thought, enter and assemble themselves in his
brain, for the good reason that he does not himself know, and so
cannot teach others. In matters of science, therefore, the greatest
inventor differs only in degree from the most laborious imitator and

30. See Lambert's letter to Kant, dated November 13, 1765 (IX, 42) (*Ak.* X, 48).

apprentice, whereas he differs specifically from one endowed by nature for fine art."[31] This insight as to the "unconscious" creativity of artistic genius becomes yet more meaningful where it comprises less the opposite to theoretical grounding than the opposite to the intent of desire and action. In this direction, too, Kant's theory transcends philosophical systematic and joins with the essential cultural problems of the age. In Baumgarten's doctrine, which contains the first elevation of aesthetics to the rank of an independent science, the concept of the beautiful is subordinated to that of perfection. All beauty is perfection, however of a kind such that it is not known in pure concept but can be grasped only mediately in a sensory, intuitive image. The whole of German academic philosophy is dominated by this view, which is further developed by Mendelssohn and put on a universal metaphysical foundation, and from this vantage point it works itself out into the circle of artistic creativity. Even Schiller's "artists" present little more than a poetic circumscription and spinning out of Baumgarten's idea.

Kant's critique constituted on this point also a clear historical boundary line. "Purposiveness without purpose," which he finds actualized in the work of art, excludes equally the mundane concept of need and the idealistic concept of perfection. For any concept of perfection presupposes an objective measure, to which the art work can be related and with which it can be compared; and to propose a formal objective purposiveness without purpose, that is, the mere form of a perfection (without content and concept of what is harmonized in it) would be a genuine contradiction.[32] Thus it was Kant the ethical rigorist who in his foundation of aesthetics was the first to break with the ruling moral rationalism. This constitutes no paradox, but is rather the necessary completion and the exact confirmation of his basic ethical view. As he founded obligation on the pure concept of reason and tried to repel all appeals to "moral feeling," to subjective perception and inclination, so on the other side the aesthetic aspect of feeling is to be held onto firmly and is not to be abandoned

---

31. *Critique of Aesthetic Judgment*, §47 (V, 383 f.) (*Ak.* V, 308–09).
32. Ibid., §15 (V, 296 f.) (*Ak.* V, 226–28).

in favor of the logical and moral concept. The exclusion of pleasure and pain from the basis of ethics does not mean, as now appears, an unconditional rejection, but it opens the way to a new objectification and makes possible another specific form of universality of which they are susceptible. Thus only the overcoming of ethical utilitarianism and hedonism paves the way for the idea of the autonomy and self-purpose of art. The concept of disinterested delight in the beauty of nature and art, regarded purely substantively, presents no radically new tendency in the evolution of aesthetics. It was already laid down by Plotinus, and was independently carried further in the modern era by Shaftesbury, by Mendelssohn, and Karl Philipp Moritz in his work "On the Creative Imitation of the Beautiful."[33] But only through the systematic exposition that it received in Kant's theory could it unfold its essential meaning, could it, in opposition to the philosophy and poetics of the Enlightenment, lay down nothing less than a new concept of the nature and the origin of the spiritual itself.

Kant, however, reached the highest synthesis between his ethical and his aesthetic basic principles only in the second part of the *Critique of Aesthetic Judgment,* in the "Analytic of the Sublime." In the concept of the sublime itself aesthetic and ethical interest undergo a new fusion, and here the critical separation of the two points of view is even more compellingly demonstrated. In the discussions he directs to this point, Kant moves once again on terrain that is personally and genuinely his. In the "Analytic of the Beautiful" one still detects, behind all the precision and acuity of the conceptual development, a certain foreign quality as soon as the inquiry leaves the region of pure principles and turns to concrete applications, for the fullness of individual artistic intuition is denied to Kant. The "Analytic of the Sublime," on the contrary, displays all the moments of the Kantian spirit and all those properties indicative of the man as well as of the writer in genuine fulfillment and in the most felicitous mutual interpenetration. Here the trenchancy of the analysis of pure concepts is found united with the moral sensitivity that forms the core of Kant's person-

---

33. [*Über die bildende Nachahmung des Schönen* (Braunschweig, 1788).]

ality; here the eye for psychological detail that Kant had already evidenced in the precritical *Observations on the Feeling of the Beautiful and the Sublime* is allied with the encompassing transcendental perspectives, which he had achieved since that time over the whole domain of consciousness.

The place occupied by the problem of the sublime within the total system of critical aesthetics can be seen most clearly if one looks back to the peculiar relation between the basic faculties of consciousness presented in the phenomenon of the beautiful. This phenomenon ought to emerge from a free play of the power of imagination and the understanding; in this case, however, "understanding" does not mean the capacity for logically conceiving and judging, but the capacity for simply delimiting. It is this that invades the movement of the imagination and extracts from it a closed form.[34] But from this there results a new question. Does limitation constitute an essential moment of the aesthetic—or is it not rather the boundless that presents a true aesthetic value? Does not precisely the thought of the unlimited, indeed of the illimitable, also in its way contain a factor with fundamental aesthetic meaning? The concept of the sublime provides the answer to this question. For the impression of sublimity in fact arises wherever we confront an object that surpasses any and all means by which we may conceive it, and which we hence are unable to bring together into a bounded whole either intuitively or conceptually. We call that "sublime" which is absolutely great—here it may be a matter of the magnitude of sheer extension or of power: of the "mathematically" or "dynamically" sublime. A relation of this sort cannot be given in objects as such, for all objective measure and estimation of size is nothing but comparison of magnitudes, in which, according to the basic standard of measure applied, the content can be called now small, now large, and thus magnitude itself is always to be taken as just a pure expression of a mental relation, never as an absolute quality and as an equally inalterable aesthetic essence. This latter determination enters in, however, when the standard of measure is transferred from the object to the subject, if it is no longer sought in

34. See above, pp. 313 f.

an individual, spatially given thing, but in the totality of the functions of consciousness. If now this totality encounters something unmeasurable, we then no longer face the sheer infinity of number, which at bottom means nothing but the power of numerical repeatability and thus an indeterminate progression, but the cancellation of all limits has yielded us a new positive determination of consciousness.

Thus the infinite, which as soon as theoretical consideration tried to grasp it as a given whole evaporated into a dialectical Idea, attains a felt totality and truth. "That is sublime [is Kant's own explanation] in comparison with which all else is small. Here we readily see that nothing can be given in Nature, no matter how great we may judge it to be, which, regarded in some other relation, may not be degraded to the level of the infinitely little, and nothing so small which in comparison with some still smaller standard may not for our imagination be enlarged to the greatness of a world. Telescopes have put within our reach an abundance of material to go upon in making the first observation, and microscopes the same in making the second. Nothing, therefore, which can be an object of the senses is to be termed sublime when treated on this footing. But precisely because there is a striving in our imagination towards progress *ad infinitum*, while reason demands absolute totality, as a real idea, that same inability on the part of our faculty for the estimation of the magnitude of things of the world of sense to attain to this idea, is the awakening of a feeling of a supersensible faculty within us; and it is the use to which judgment naturally puts particular objects on behalf of this latter feeling, and not the object of sense, that is absolutely great, and every other contrasted employment small. Consequently it is the disposition of soul evoked by a particular representation engaging the attention of the reflective judgment, and not the Object, that is to be called sublime. ...*The sublime is that, the mere capacity of thinking which evidences a faculty of mind transcending every standard of sense.*"[35]

Since in this fashion the basis of the sublime is shifted from objects to the "harmony of the spirit," since it is discovered to be not a quality

35. *Critique of Aesthetic Judgment*, §25 (V, 321 f.) (*Ak*. V, 250).

of being but a quality of contemplation, it is truly lifted up into the sphere of aesthetic reflection. But this sphere here no longer touches the region of the understanding and intuition, as it did in the contemplation of the beautiful, but rather that of the Ideas of reason and their supersensory meaning. While in judging the beautiful the imagination was interwoven in a free play with the understanding, in judging a thing to be sublime it is related to reason, so as to evoke a harmony of the mind "conformable to that which the influence of definite . . . ideas would produce upon feeling, and in common accord with it."[36] For Kant, however, all concord of reason passes ultimately into the one idea of freedom, and it is this which also everywhere underlies our use of the category of the sublime. What properly belongs to the feeling of ourself and our intelligible task here is transformed into a predicate of the given things of nature only through a peculiar subreption. Under deeper analysis and self-awareness this illusion also vanishes. "Who would apply the term 'sublime' even to shapeless mountain masses towering one above the other in wild disorder, with the pyramids of ice, or to the dark tempestuous ocean, or such like things? But in the contemplation of them, without any regard to their form, the mind abandons itself to the imagination and to a reason placed, though quite apart from any definite end, in conjunction therewith, and merely broadening its view, and it feels itself elevated in its own estimate of itself on finding all the might of imagination still unequal to its ideas. . . .In this way external Nature is not estimated in our aesthetic judgment as sublime so far as exciting fear, but rather because it challenges our power (one not of Nature) to regard as small those things of which we are wont to be solicitous (worldly goods, health, life), and hence to regard its might (to which in these matters we are no doubt subject) as exercising over us and our personality no such rude dominion that we should bow down before it, once the question becomes one of our highest principles and of our asserting or forsaking them. Therefore Nature is here called sublime merely because it raises the imagination to a presentation of those cases in which the mind can make itself sensible of the appro-

36. Ibid., §26 (V, 327) (*Ak.* V, 256).

priate sublimity of the sphere of its own being, even above Nature."[37]

To be sure, this critical solution of the problem of the sublime, when looked at more closely, carries a new critical question within itself. For through the relation of the sublime to the idea of self-legislation and the free personality the sublime seems, since it is cut loose from nature, to fall wholly into the realm of the ethical. Its special aesthetic character and its independent aesthetic value, however, would be erased equally thoroughly in either case. In fact, the execution of Kant's analysis reveals how near we are to this peril. For the psychology of the sublime leads us back to that basic emotion of awe, which we have already recognized as the universal form in which the consciousness of the moral law presents itself to us. In the phenomenon of the sublime we again recognize that mingling of pleasure and pain, of resistance and freely willed submission which constitutes the peculiar character of the feeling of awe. In it we feel ourselves at once overwhelmed, as physically finite subjects, by the grandeur of the object, while at the same time we feel exalted above all finite and conditioned being through the discovery that this grandeur is rooted in the consciousness of our intelligible task and in our faculty of Ideas. But since the sublime is founded on the same feeling as the moral in general, we thereby seem to have overstepped the boundaries of disinterested delight and to have passed into the domain of the will. The difficulty lying herein can only be removed when one sees that the subreption through which in the sublime we think a determination of ourselves as a determination of the natural object does not vanish when it is recognized as such. Our intuition remains aesthetic only when it views the self-determination of our mental faculties not in and for themselves but equally through the medium of the intuition of nature; when it reflects the inner in the outer and vice versa. Such a mutual mirroring of the ego and the world, of feeling of self and feeling of nature, comprises for us both the essence of aesthetic contemplation in general and also the essence of that contemplation which finds its expression in the sublime. Here a new form of the investment of nature with soul is advanced, ulti-

37. Ibid., §§26, 28 (V, 327 f., 333 f.) (*Ak.* V, 256, 262).

mately leading on beyond the shape of nature as it is symbolically copied in the appearance of the beautiful—and yet which on the other hand perennially leads back to nature because it can be grasped only in this very opposition. Only therein does the infinity of nature, which previously was a mere thought, receive its concrete felt truth, because it is seen in light reflected from the infinity of the self.

The proposition from the introduction to the *Critique of Judgment*, which says that in it the "ground of the unity of the supersensible that lies at the basis of Nature, with what the concept of freedom contains in a practical way" is to be demonstrated, is now made completely determinate. And henceforward we can also understand the reason for attaching the limitation that the concept indicating this unity itself affords neither any theoretical nor any practical knowledge of this unity; hence it has no proper domain, but only makes feasible transition from the mode of thinking governed by the principles of the one to that according to the principles of the other.[38] As to how the unity of the "supersensory ground" is able to differentiate itself so that it is presented to us now in the guise of nature and again under the image of freedom and the moral law, we are not once granted even a hint, much less a theoretical explanation. But even if we refuse to speculate on this topic, there still remains an undeniable phenomenon in which the contemplation of nature and that of freedom undergo a totally novel relation to one another. This phenomenon is that of artistic perception. Every genuine work of art is completely determined in the sensory respect and seems to desire nothing more than to remain in the circle of the sensory, and each nonetheless necessarily extends beyond this circle. It contains a portion of a purely concrete and personal life, and still it reaches back into a depth where the ego feeling turns out to be the feeling of the whole as well. Looked at conceptually, that might be called a miracle, but in all supreme creations of art (one need think only of the highest examples of Goethe's lyrics) this miracle is truly accomplished, so that the question of its possibility is silenced. In this respect—but only in this respect—the actual existence of art, if we do not shatter it by abstract hairsplitting,

38. See *Critique of Judgment*, introduction, II (V, 244) (*Ak.* V, 176).

points to a new unity of the sensible and the intelligible, of nature and freedom; indeed, it is itself the expression and the immediate guarantee of this unity. Thus the route by which we here arrive at the thought of the supersensory in every respect fits the general critical orientation, for we do not begin with the essence of the supersensory, so as then to dissect it into its individual expressions, but rather its idea arises in us, since we bring together the basic directions given in consciousness itself and make them intersect in an imaginary perspective, a point beyond possible experience.

Accordingly, the doctrine of the "supersensible substrate" of nature and of freedom is not about a primal *thing*, but about the primal *function* of the *spiritual*, which is disclosed for us with novel meaning and profundity in the aesthetic. For the universal communicability that every real aesthetic judgment claims for itself points us to a basic agreement, to which the subjects as such belong, independently of their contingent individual differences, and in which therefore not so much the intelligible ground of objects as rather the intelligible ground of humanity is presented. Kant concludes this discussion thus: "This is that intelligible to which taste . . . extends its view. It is, that is to say, what brings even our higher cognitive faculties into common accord, and is that apart from which sheer contradiction would arise between their nature and the claims put forward by taste. In this faculty judgment does not find itself subjected to a heteronomy of laws of experience as it does in the empirical estimate of things—in respect of the objects of such a pure delight it gives the law to itself, just as reason does in respect of the faculty of desire. Here, too, both on account of this inner possibility in the Subject, and on account of the external possibility of a nature harmonizing therewith, it finds a reference in itself to something in the Subject itself and outside it, and which is not Nature, nor yet freedom, but still is connected with the ground of the latter, i.e., the supersensible—a something in which the theoretical faculty gets bound up into unity with the practical in an intimate and obscure manner.[39]

This "obscure manner" is known at least to the extent we can

39. [*Critique of Aesthetic Judgment*, §59 (V, 430) (*Ak.* V, 353).]

precisely designate the general higher concept on which the connection rests. Once again it is the concept of autonomy, of the self-legislation of the spirit, that manifests itself as the center of gravity of the Kantian system. Because this concept receives fresh confirmation and illumination in the aesthetic, it leads us to a deeper level of the intelligible. From the autonomy of the pure understanding and its universal laws came nature as the object of scientific experience—from the autonomy of the ethical proceeded the idea of freedom and the self-determination of reason. These two, however, do not stand in isolation but are necessarily related to each other, for the world of freedom ought to have an influence on the world of nature, ought to execute its demands in the empirical world of men and things. Nature must hence at the very least be thinkable "so that the lawfulness of its form may at least agree with the possibility of ends working within it according to the laws of freedom." But every attempt actually to think it in this way perpetually collides, in the purely theoretical area, with the antinomy between causality and freedom. No matter how much progress we may make, we finally confront the tremendous gulf between the realm of the concept of nature as sensible and the realm of the concept of freedom as supersensible.[40] Only artistic insight discloses a new path to us. Even if the objective agreement of nature and freedom remains a never-completed task, even if the paths of the two intersect only at infinity, their full subjective unity is actualized within the sphere of concrete consciousness itself, in the feeling of art and the creating of art. Here, in the free play of the powers of the mind, nature appears to us as if it were a work of freedom, as if it were shaped in accordance with an indwelling finality and were formed from the inside out—while on the other hand the free creation, the work of artistic genius, delights us as something necessary and therefore as a creation of nature. Thus here we wed what simply, as existent, is distinct and must remain so with a new manner of contemplation, the special content of which continues to exist for us only if we resist the attempt to interpret it as an independent mode of theoretical cognition of the actual. The supersensible substrate the judgment of

40. See *Critique of Judgment*, introduction, II (V, 244) (*Ak.* V, 175–76).

taste points us to is hence impossible to display conceptually from appearances as objective phenomena of nature, but it is immediately confirmed in a peculiar relation of consciousness itself, which is as sharply and characteristically distinct from all relation to knowledge through concepts and laws as from any relation to the pure determination of the will. Once this relationship is clearly and unambiguously seated in the subject, this consequence has a reverse effect on the image of objective reality. The harmonious play of the mind's powers is what endows nature itself with the content of life: aesthetic judgment passes over into teleological judgment.

5

The result of the *Critique of Judgment* thus far can be summed up by saying that the concept of an end has now undergone that transformation corresponding to the Kantian revolution in thinking. The end is not an objectively acting power of nature in and behind things but is a mental principle of union that our judgment applies to the totality of experiences. As a principle of this sort, it manifests itself to us in the idea of formal purposiveness as well as in that of aesthetic purposiveness. We encounter formal purposiveness when we analyze nature into a system of particular laws and particular natural forms, but for the critical enterprise it constitutes less a new factor in appearances themselves than a concurrence of the appearances with the demands of our understanding. Aesthetic creation was also introduced directly into reality itself, but the more deeply and purely it was grasped, the more clearly it could be seen that the unity of being presented to us neither wants to be nor can be anything other than a reflection of the unity of the mood and the feeling which we experience in ourselves. But now the question arises whether these alterations in the idea of an end also exhaust its sphere of application in its entirety. Is there not some perspective in which the end not only expresses a relation of the given appearance to the beholder, but in which it is to be seen as an objectively necessary moment of the appearance itself? And to the extent that there is such a point of view, what is it and how can it be critically established and justified?

The idea of finality differs from all other categories in that through it, wherever it appears, a new type of unity of the manifold is asserted, a novel relation between a formed whole and its individual partial moments and conditions. Thus in the concept of formal purposiveness the substance of the particular laws of nature was thought in a way such that it presented not a mere aggregate, but a system which "specifies" itself according to a definite rule. Thus a totality of consciousness and its powers was disclosed in the aesthetic feeling, preceding and underlying all dissection of consciousness into individual faculties contrasted with each other. From each of these two standpoints the whole here under consideration is regarded not as if it were made up of its parts but as if it were itself the origin of the parts and the basis of their concrete determinateness. But this whole itself was purely ideal nature: it was a presupposition and requirement which our reflection saw itself compelled to apply to objects, although without entering directly into the formation of these objects and combining indissolubly with them. There is, nevertheless, one area of facts and problems in which this peculiar transition is actually made, where purpose seems to confront us not as a mere principle of subjective contemplation but as the very creature and substance of nature itself. Wherever we do not conceive nature as an aggregate of mechanical causal laws arranged in a hierarchy from the universal to the particular, but rather as a whole of life forms, this step is taken. For life is conceptually distinguished precisely by the assumption of a type of actuality proceeding not from plurality to unity but from unity to plurality, not from the parts to the whole but from the whole to the parts. An event in nature becomes a life process for us when we think it not as a mere flux of miscellaneous individual things, one following the other, but when all these particular entities are for us expressions of *one* occurrence and one essence, which reveals itself in them only in manifold structures.

A movement toward this sort of unity of being, as distinguished from a mere stream of events all of equal importance, is what constitutes for us the character of "development." Where true development is present, a whole is not formed *out of* parts, but it is contained *in* them, as a guiding principle. Instead of a uniform passage of temporal before and after, where every previous moment is swallowed up by the present and its existence is lost, in the phenomenon of life we think a mutual interpenetration of the individual moments, such that the past is conserved in the present and in both the tendency toward future formation is active and knowable. We conventionally signify this sort of connection by the concept of organism. In an organism, according to the explanation early given to it by Aristotle, the whole precedes the parts, because the former is not possible through the latter, but rather the latter only through the former. A particular stage of life receives its meaning only from the totality of the expressions of life to which it belongs; we conceive it not by divorcing it from the event as a causal condition, but by regarding it as a means, which is a means "to" that totality. "In such a natural product as this every part is thought as *owing* its presence to the *agency* of all the remaining parts, and also as existing *for the sake of the others* and of the whole, that is as an instrument, or organ. But this is not enough—for it might be an instrument of art. . . .On the contrary the part must be an organ *producing* the other parts—each, consequently, reciprocally producing the others. No instrument of art can answer to this description, but only the instrument of that nature from whose resources the materials of every instrument are drawn—even the materials for instruments of art. Only under these conditions and upon these terms can such a product be an *organized* and *self-organizing being*, and, as such, be called a *natural end*."[41] Since the idea of the end now is not referred to the relation between our cognitive powers and other powers of the mind but is immediately intuited concretely and objectively, there arises the idea of organism: "things as ends of Nature are organized beings."

However, this purely objective view ought not to seduce us into a misunderstanding. We are not here involved in a metaphysic of nature, but in a critique of judgment. The question, therefore, is not whether nature acts purposively in some of its products, whether its creative activity might be guided by a conscious or unconscious intention, but rather whether our judging is compelled to posit and assume

41. *Critique of Teleological Judgment*, §65 (V, 451 f.) (*Ak.* V, 373–74).

a special "thing-form" distinct from that of the bodies of abstract mechanics and going beyond them. And it must first of all be established, in accordance with the transcendental method, that this postulation, whatever the ultimate decision as to its justification may be, is undeniably a simple fact. We are as unable to blot out of our conception of nature the thought of organic life as to dispense with the fact of will or of aesthetic intuition and creation in our view of spiritual being. The distinction between the two ways of operating—mechanical-causal and inner-purposive—belongs to the image of nature itself, which we have to sketch according to the conditions of our knowledge; however we may answer the metaphysical question, it presents a state of cognitive consciousness demanding recognition and explanation. The contrast between the sort of events we find in a clockwork and those presented to us in a living body is immediately demonstrable in the phenomenon and as a phenomenon. "In a watch one part is the instrument by which the movement of the others is effected, but one wheel is not the efficient cause of the production of the other. One part is certainly present for the sake of another, but it does not owe its presence to the agency of that other. . . .Hence one wheel in the watch does not produce the other, and, still less, does one watch produce other watches, by utilizing, or organizing, foreign material; it does not of itself replace parts of which it has been deprived, nor; if these are absent in the original construction, does it make good the deficiency by the subvention of the rest; nor does it, so to speak, repair its own causal disorders. But these are all things which we are justified in expecting from organized Nature. And organized being is, therefore, not a mere machine. For a machine has solely *motive power,* whereas an organized being possesses inherent *formative* power, and such, moreover, as it can impart to material devoid of it. . . . This, therefore, is a self-propagating formative power, which cannot be explained by the capacity of movement alone, that is to say, by mechanism."[42]

Thus a tree produces another tree by a known law of nature and hence reproduces itself in accordance with its species; secondly, how-

---

42. [Ibid., §65 (V, 452) (*Ak.* V, 374). Cf. also §64 (V, 447 ff.) (*Ak.* V, 369 ff.).]

ever, it produces itself as an individual, insofar as it enlarges and
renews its individual parts in an orderly way. Although customarily
we just call this latter activity "growth," we must nonetheless not lose
sight of the fact that it is totally different from every other increase in
size governed by merely mechanical laws, for the matter added in the
growth process is used in a specifically characteristic creativity, and
thus constitutes both a recreation of its species and a further development, not a mere enlargement of its mass and its quantity.[43] The
natural object, which was determined as a magnitude through the
basic principles of pure understanding, substantiality, causality, and
reciprocity, only here receives a quality peculiar to it and which distinguishes it from all other formations; this quality, however, is not so
much a property of its being as a property of its becoming, and it
designates the individual direction of this becoming.

Thus the individual appearances of nature here achieve a new
meaningfulness, enriching and deepening their own substance, but
unrelated to an alien end lying outside them. For as was done previously in the founding aesthetics, once again the idea of purposiveness without a purpose is strictly and thoroughly worked out. This
task is all the more pressing since Kant at this point once more sets
himself consciously in opposition to his era. The whole teleology of
the Age of Enlightenment is characterized by the thoroughgoing confusion of the idea of finality with that of general utility. The profounder elements of the Leibnizian concept of purpose were degraded by
Wolff into an insipid utilitarian outlook and calculation. The universal
metaphysical ideas of theodicy had here become lost in a narrow and
pendantic pettiness, which sought to detect in every single feature of
the course of the universe the advantage of mankind and hence the
wisdom and goodness of the Creator. Wolff even bestows on sunlight
a teleological justification of his kind: "The light of day," he once
remarked, "is very useful for us, for with it we can conveniently carry
on our duties, which cannot be done in the evening at all, or at least
not so handily and with difficulties."[44] In German literature Brockes

43. Cf. ibid., §§64 and 65 (V, 448 ff.) (*Ak.* V, 369 ff.).

44. Cf. Josef Kremer, *Das Problem der Theodizee in der Philosophie und Literatur des 18.
Jahrhunderts mit besonderer Rücksicht auf Kant und Schiller* (Berlin, 1909), p. 95.

became the poet of this outlook and orientation. Even when a young man, however, Kant combated it with serene and abundant irony, having been attracted by the problem of natural teleology and occupied with it since the *Universal Natural History and Theory of the Heavens,* and he had a predilection for referring to Voltaire's sarcastic saying that God surely had given us noses so that we—might put spectacles on them.[45]

The *Critique of Judgment* recurs to this authority (without naming him), but no less clearly and definitively it vanquishes the basic positive intuition of Voltairean deism. The world no longer is a clockwork mechanism finding its ultimate explanation in the hidden, divine "watchmaker," for the metaphysical form of the cosmological proof of God's existence is seen to be as fallacious as that of the teleological proof. From now on if the finality of nature is to be discussed, this cannot mean a signpost pointing to an external transcendant ground on which nature depends, but only a reference to its own immanent structure. This structure is purposive—so long as the *relative* finality for mankind or any other created being is kept clearly separate from *inner* finality, which possesses no point of comparison other than the appearance itself and the structure of its parts. As to the former, relative finality, it is clear without further ado that its demonstration remains dubious in every case. For even if we assumed that we had proved an individual phenomenon of nature or nature as a whole to be necessarily for the sake of another and teleologically constrained, what is our guarantee of the necessity of this other? If we wanted to designate it as its own end we would introduce a completely new yardstick, inadmissible and futile. The concept of something which is its own end belongs, as the establishment of the Kantian ethics has shown, not to the realm of nature but to that of freedom. If we remain within the bounds of nature, there is no escape from the circle of relativity. "We can easily see that the only condition on which extrinsic finality, that is, the adaptability of a thing for other things, can be looked on as an extrinsic physical end, is that the existence of the

45. Cf. "The Only Possible Basis of Proof for God's Existence," pt. 2, Observation 6, §4 (II, 138) (*Ak.* II, 131).

thing for which it is proximately or remotely adapted is itself, and in its own right, an end of Nature. But this is a matter that can never be decided by any mere study of Nature. Hence it follows that relative finality, although, on a certain supposition, it points to natural finality, does not warrant any absolute teleological judgment."[46] Strictly construed, the idea of self-purpose, like that of self-value, is restricted to the sphere of the ethical, to the idea of the subject of willing; but in the domain of objective existence it possesses a symbolic counterpart in the phenomenon of the organism (as previously in that of the work of art). For all the parts of an organism are oriented as if to a single center; this center, however, lies in itself and is related only to itself. The existence of the organism and its individual form interpenetrate each other: the one seems to be there for the sake of the other.

Here, however, there begins a new question, in contrast to the whole of aesthetic contemplation. No conflict can occur between the concept of natural beauty and that of natural lawfulness, for the validity each claim is of a totally different sort. Aesthetic consciousness creates its own world and elevates it beyond all intercourse and clash with empirical actuality, since it constructs it as a world of play and of semblance. But this way out is denied to teleological judgment, which we apply to nature and its products, for its object is one and the same as that of judgment of experience and cognitive judgment. But can nature in general mean for the critical philosopher anything but the object of experience, presented under the form of space and time as well as the categories of magnitude and reality, of causality and reciprocity, and which is exhausted in the totality of these forms? It is impossible—so it seems—either to haggle away any of this determination of the object of experience or to add anything to it. What does it thus signify if now the idea of purpose comes forward claiming to justify or to fulfill the idea of causality? We recall that the basic premise of causality in the critical sense means nothing but the unavoidable means of objectifying the temporal succession of appearances. The causal connection of phenomena is not inferred from their temporal order, but the reverse: only by applying to a given sequence of

46. *Critique of Teleological Judgment*, §63 (V, 446) (*Ak.* V, 368–69).

perceptions the concept of cause and effect, of condition and the conditioned, can the objective time order determine its elements unambiguously.[47] If we hold fast to the result, we see immediately that there is no possibility of exempting any particular realm of nature whatsoever from the all-encompassing validity of the causal principle. For that would instantly banish it from the one objective temporal order as well; it would no longer be an "event," in the empirical sense of this word. Hence, since the development which we attribute to the organism is truly and permanently such an event, it must also be thought as unqualifiedly subject to the fundamental law of causal connection. Every particular emergent formation in a developmental series must be explicable from what precedes and from the conditions of the environment. All determination of what is now given by a future something not yet given must remain excluded; what is earlier conditions and posits what succeeds it, because generally only in this form of conditioning is the objective phenomenon of an unambiguous temporal order constituted. In this view of nature there is no place for the assumption of a special class of purposive forces, because no gap exists into which the new idea could be inserted.

The result of this connection is that for Kant purpose cannot enter into the picture as a special principle of the explanation of natural phenomena, be they inorganic or organic. There is only one principle and one ideal of natural scientific explanation, and this is defined by the form of mathematical physics. A phenomenon is "explained" when it is known and determined in all its individual moments as magnitude and when its existence can be deduced from universal quantitative laws, and similarly from knowledge of certain constants that characterize the particular instance. That this deduction can never really be done completely, that every individual case and every individual form comprises in itself an unlimited complexity, is equally true. For where the analysis of mathematical physics has not yet been actually completed it must nevertheless be regarded as completable—if the object in question is not to fall outside the realm of nature as bounded by the universal law of conservation and its

47. See above, pp. 184 ff.

corollaries. The reduction of all events to comparisons of magnitude, and the transformation of "organism" into "mechanism" is thus set up as an unconditional demand at the very least, even in the face of all the limitations of our present knowledge. The *Critique of Teleological Judgment* leaves no doubt whatsoever concerning this result. It begins with the premise that in the "general idea of Nature," as the ensemble of sense objects, there is no basis at all for the assumption that things of nature serve each other as means to ends and that this sort of causality renders their possibility sufficiently comprehensible. For this can neither be demanded nor predicted a priori, nor can experience ever show us such a form of causality "save on the assumption of an antecedent process of mental jugglery that only reads the conception of an end into the nature of the things, and that, not deriving this conception from the objects and what it knows of them from experience, makes use of it more for the purpose of rendering Nature intelligible to us by an analogy to a subjective ground upon which our representations are brought into inner connexion, than for that of cognizing Nature from objective grounds."[48]

But were this the final result, the inquiry would have gone in a circle. For this was the very question thrust upon us after the analysis of the aesthetic finality of the powers of the mind and after the discussion of the formal finality among our concepts; namely, whether the idea of an end did not at least mediately participate in the building up of our experiential world and its objects, and to this extent possess some sort of objective validity. If the latter is denied to it, then the teleology of nature poses no new problem in the critical sense. There would be only one way to make compatible the seemingly irreconcilable demands of the purposive principle and the causal principle. If the causal principle is to remain the sole constitutive basic concept of nature and experience—and if on the other hand the idea of purpose is nonetheless to possess an independent relation to experience, this is conceivable only if this relation is itself effected and established through the mediation of the causal concept. Then and only then would a new field of activity for the concept of an end be found, if this

48. *Critique of Teleological Judgment*, §61 (V, 438) (*Ak.* V, 359–60).

concept is not to oppose causal explanation but is to foster and guide it. It is in fact here that its true and legitimate use lies. The final principle does not have a constitutive but rather a regulative meaning; it serves not for the conquest of the causal interpretation of phenomena, but rather the reverse, for its deepening and its universal application. It does not resist this interpretation but paves the way for it, since it points out the appearances and the problems to which the causal principle should address itself.

That such a preparation is fruitful, indeed indispensable, in the phenomena of organic nature is easy to show. For here the direct application of the causal principle and the universal causal laws, far from being thought of as conflicting with purpose, finds no content whatsoever on which it might be exercised. The laws of mechanics and physics do not concern the "things" of nature as they are presented to direct observation; they speak instead of "masses" and "mass-points." The object must be stripped of all its erstwhile concrete determinateness, reduced to the pure abstractions of analytical mechanics, if there is to be a possibility of subjecting it to those same laws. Where we are concerned, as in the appearances of organic nature, with matter not as mass in motion but as the substrate of the phenomena of life, where the natural form in its full complexity is our particular interest—there, before the causal deduction of the singular can be undertaken, the whole toward which the inquiry is directed must as a rule first be designated and brought out purely descriptively. Out of the general stuff of spatiotemporal being, in which everything can be basically related to everything else, some sort of specifically determined individual series must be extracted, in which the members show a particular form of affinity with one another. This is the function fulfilled by the concept of an end. Finality, unlike the fundamental concepts of mathematical physics, does not assist deduction but induction, not analysis but synthesis, for it initially sets up the relative unities which we subsequently can dissect into their individual causal elements and causal conditions. The visual process in all its particularities must be explained causally, but the structure of the eye is studied from the point of view and under the assumption that the eye is "determined for seeing," though not intentionally so

constructed. Thus teleological judging is justifiably connected, prob-
lematically at least, with research into nature, for the concept of teleological relations and forms in nature is at any rate *"one more principle*
for reducing its phenomena to rules in cases where the laws of its
purely mechanical causality do not carry us sufficiently far. . . .But this
is a different thing from crediting Nature with causes acting *de-
signedly,* to which it may be regarded as subjected in following its
particular laws. The latter would mean that teleology is based, not
merely on a *regulative* principle, directed to the simple *estimate* of
phenomena, but is actually based on a *constitutive* principle available
for *deriving* natural products from their causes: with the result that the
conception of a physical end no longer exists for the reflective, but for
the determinant, judgment. But in that case the conception would not
really be specially connected with the power of judgment. . . . It
would, on the contrary be a conception of reason, and would introduce a new causality into science—one which we are borrowing all
the time solely from ourselves and attributing to other beings, although we do not mean to assume that they and we are similarly
constituted."[49]

That is the critical distinction that Kant wields in the old battle for
and against finality. The received metaphysical interpretation of the
concept of purpose is in fact the *asylum ignorantiae* that Spinoza said it
was, but in its purely empirical use it is more the means to an ever-
richer and more precise knowledge of the connections and the structural relations of organic nature. As a "maxim of the reflective judgment" assisting knowledge of natural laws in experience it does not
serve to "institute the intrinsic possibility of natural forms, but to
become acquainted with Nature in accordance with its empirical laws."[50]
At this point the guide to research and the principle of the explanation of particular natural phenomena diverge. One must keep firmly
in mind that no trace of that mystical aura surrounding the insistence
on and the longing to penetrate "into the heart of nature" clings any
longer to the concept of the explanation of nature as Kant conceives it,

49. Ibid.
50. See ibid., §69 (*Ak.* V, 385 f.).

but rather it designates an inevitable and decisive but nonetheless individual, logical function of knowledge. All causal explanation of one phenomenon by another ultimately reduces to the determination of one's spatiotemporal location by the other. In this the "how" of the transition from one to the other is not conceived; only the fact of the necessary conjunction of the elements in the empirical sequence is indispensable.

The principle of finality, when it is used in the critical sense, also renounces the task of unriddling the secret of this transition, but it orders the phenomena around a new center and thereby offers a different type of mutually interrelating form. As far as causal derivation can penetrate and as much leeway as we give it, it can never shoulder this form aside and render it dispensable. For within the phenomena of life it can of course be shown purely causally how one member of the development arises from its predecessor, but we eventually arrive, however far we may trace it back, at an initial condition of organization which we must accept as a presupposition. The causal point of view tells us the rules by which one structure is transformed into another; however that such individual "seeds" exist, that there are primitive formations specifically different from one another which are the basis of the development, cannot be made further intelligible but must be assumed as a fact. The antinomy between the concepts of finality and causality thus disappears, as soon as we think of the two of them as different *modes of ordering*, through which we try to unify the manifold of phenomena. Opposition between two basic metaphysical factors of events is supplanted by agreement between two maxims and demands of reason which complement each other. "If I say: I must *estimate* the possibility of all events in material Nature, and, consequently, also all forms considered as its products, on mere mechanical laws, I do not thereby assert that they *are solely possible in this way*, that is, to the exclusion of every other kind of causality. On the contrary this assertion is only intended to indicate that I *ought* at all times to *reflect* upon these things *according to the principle* of the simple mechanism of Nature, and, consequently push my investigation with it as far as I can, because unless I make it the basis of research there can be no knowledge of Nature in the true sense of the term at all. Now this does not stand in the way of the second maxim . . . that is to say, in the case of some natural forms (and, at their instance, in the case of entire Nature), we may, in our reflection upon them, follow the trail of a principle which is radically different from explanation by the mechanism of Nature, namely the principle of final causes. For reflection according to the first maxim is not in this way superseded. On the contrary, we are directed to pursue it as far as we can. Further it is not asserted that those forms were not possible in the mechanism of Nature. It is only maintained that *human reason*, adhering to this maxim and proceeding on these lines, could never discover a particle of foundation for what constitutes the specific character of a physical end, whatever additions it might make in this way to its knowledge of natural laws."[51]

The consequent reconciliation between the principles of finality and mechanism binds both to the condition that they aspire only to be different and specific ways of ordering natural phenomena, and that they renounce any dogmatic unfurling of a theory of the ultimate origin of nature itself and the individual forms in it. In such an undertaking, both the concept of purpose and that of causality would be shipwrecked. For the concept of a being which, by virtue of its purposive reason and will, is the primal ground of nature is possible in the formal, analytical sense, to be sure, but indemonstrable in the transcendental sense, for the reason that since it cannot be abstracted from experience nor is it required for the possibility of experience, its objective reality can in no way be secured. To this extent, where research into nature is concerned, the concept of an end remains "a stranger in natural science" threatening to erase the steady progress of its methodology and to detach the very concept of a cause, which designates a *relation internal to appearance,* from this its basic meaning.[52] On the other side, however, the causal idea also, if it remains aware of its essential task of "spelling out appearances so that we can read them as experiences," must renounce the claim to be the means to a true insight into the first and absolute grounds of organized life.

51. Ibid., §70 (V, 465 f.) (*Ak.* V, 387–88).
52. Cf. ibid., §§72 and 74 (V, 467 ff., 474 ff.) (*Ak.* V, 389 ff., 395 ff.).

For within phenomena themselves, the infinite complexity which every organic natural form possesses for us points also to the limits of its powers. "It is, I mean, quite certain that we can never get a sufficient knowledge of organized beings and their inner possibility, much less get an explanation of them, by looking merely to mechanical principles of Nature. Indeed, so certain is it, that we may confidently assert that it is absurd for men even to entertain any thought of so doing or to hope that maybe another Newton may some day arise, to make intelligible to us even the genesis of but a blade of grass from natural laws that no design has ordered. Such insight we must absolutely deny to mankind. But, then, are we to think that a source of the possibility of organized beings amply sufficient to explain their origin without having recourse to a design, *could* never be found buried among the secrets even of Nature, were we able to penetrate to the principle upon which it specifies its familiar universal laws? This, in its turn, would be a presumptuous judgment on our part. For how do we expect to get any knowledge on the point? Probabilities drop entirely out of account in a case like this, where the question turns on judgments of pure reason."[53] We can try, of course, here as well, to make the lines which diverge for us intersect in the supersensible; we can assume that the transcendent ground on which the world of appearance rests is so constituted that a thoroughly purposive order of the universe must proceed from it, according to universal laws and hence without the intrusion of any sort of willed intent. In this direction—for example, in Leibniz's metaphysics of preestablished harmony—lies the attempt to reconcile the realm of final causes with that of efficient causes, the concept of God with the concept of nature.

For Kant, however, the "supersensible" here means less the substrate and the ultimate explanatory ground of things than the projection beyond the bounds of experience of a goal unattainable in experience. No theoretical certainty as to the absolute genesis of being is asserted thereby; there is merely the indication of a direction we have to keep to when applying our basic cognitive methods. The possibility of reconciling mechanism and teleology in the supersensible asserts

53. Ibid., §75 (V, 478 ff.) (*Ak.* V, 400).

one thing above all: that we ought unswervingly to utilize both modes of procedure for experience itself and for investigation into the connection of its phenomena, since each is necessary and, within its own area of validity, irreplaceable. To explain the purposiveness of nature, metaphysics had involved now inanimate matter or a lifeless God, now living matter or a living God; from the standpoint of transcendental philosophy, however, the only thing left for all these systems is "to break away from all these *objective assertions,* and weigh our judgment *critically* in its mere relation to our cognitive faculties. By so doing we may procure for their principle a validity which, if not dogmatic, is yet that of a maxim, and ample for the reliable employment of our reason."[54] In this sense it is the case here also that the union of the principles of finality and causality "cannot rest on one basis of *explanation* setting out in so many terms how a product is possible on given laws so as to satisfy the *determinant* judgment, but can only rest on a single basis of *exposition* elucidating this possibility for the *reflective* judgment."[55] Nothing is said as to whence nature, regarded as a thing in itself, comes and whither it is going, but we establish in this way the concepts and cognitions indispensable for comprehension of the totality of phenomena as a self-contained and systematically articulated unity.

Thus that very principle which above all seemed destined to reach into the primal transcendent ground and the origin of all experience only probes more deeply into the structure of experience, and illuminates, instead of this primal ground, only the richness and the content of appearance itself. The reality which, under the idea of causality and mechanism, appears as a product of universal laws, is integrated into a whole of life forms for and through the principle of finality. In this consists both the connection and the contrast holding between the idea of purpose as coined in aesthetics and in natural teleology. Aesthetic judgment implied a complete reversal from the reality of the pure understanding and its universal laws; through it a new form of being was revealed and grounded in a new function of

54. Ibid., §72, note (V, 470) (*Ak.* V, 392).
55. Ibid., §78, (V, 491) (*Ak.* V, 412).

consciousness. But the domain thus granted its independence conserved this independence and insularity of character; it was separated
off from the world of empirical realities and empirical ends as a self-
sufficient world of play, centered only on itself. In the teleological
view of organic nature, however, this sort of separation does not
occur; there a stable reciprocity holds between the concept of nature
advanced by the understanding and that put forward by the teleological judgment. The principle of finality itself calls in the causal principle and instructs it in its tasks. We cannot regard a structure as purposive without becoming involved in research into the grounds of its
origin; for the assertion that it owes its genesis to intention on the part
of nature or of providence is meaningless, since it is purely tautologi-
cal and only restates the question.[56] So the attempt, at least, must be
made to hold to the idea of mechanism and to follow it as far as
possible, although we are sure, on the other hand, that we will never
thus arrive at any ultimately valid answer to the question. For knowledge means just this continuous extensibility of its own fruitfulness.
Thanks to this procedure, the secret of organic life is never solved in
an abstract and purely conceptual fashion, but the knowledge and the
intuition of the individual forms of nature are steadily broadened and
deepened by it. More than this, however, the "maxim of the reflective
judgment" may not do, nor does it desire anything more, for its goal
does not consist in a solution to the "riddle of the universe" in the
sense of a metaphysical monism, but in continuously sharpening
one's eyes for the wealth of the phenomena of organic nature and in
penetrating ever further into the particularities and the individualities
of the phenomenon of life and its conditions.

Having arrived at this point, Kant may then once again, with
the utmost methodological acuity and sensitivity, contrast his philosophical principle with the principle of the received metaphysics.
The opposition between discursive and intuitive understanding, to
which the *Critique of Pure Reason* had already called attention, is
here given a new and even more comprehensive meaning. For an
absolutely infinite and absolutely creative understanding—such as

56. See ibid., §78 (V, 489 ff.) (*Ak.* V, 410 ff.).

that from which metaphysics derives the purposiveness of the forms and order of nature—the contrast between the possible and the actual, which binds us in all our cognition, would drop away, since for such a mind the mere positing of an object in thought and in will would imply its existence. The distinction between being that is thought and being that is actual, between contingent and necessary being, would be nonsensical for such an intellect; since in the first member of the sequence of being it contemplates there would be comprised for it the sequence as a totality as well as the whole of its structure, both ideal and actual.[57] For human understanding, on the contrary, the notion of such a survey signifies a completely unattainable Idea. For it is not given to human understanding to grasp the whole and to raise it up before itself except through a progressive composition of parts. Its proper locus is not the cognition of the primal, original grounds of being, but the comparison of individual perceptions and their subordination to universal rules and laws. And there too, where it pursues the path of pure deduction, seeming to move from the universal to the particular, it invariably achieves no more than the analytical universality proper to concepts as such. "It is, in fact, a distinctive characteristic of our understanding that in its cognition—as, for instance, of the cause of a product—it moves from the *analytic universal* to the particular, or, in other words, from conceptions to given empirical intuitions. In this process, therefore, it determines nothing in respect of the multiplicity of the particular. On the contrary, understanding must wait for the subsumption of the empirical intuition—supposing that the object is a natural product—under the conception, to furnish this determination for the faculty of judgment. But now we are also able to form a notion of an understanding which, not being discursive like ours, but intuitive, moves from the *synthetic universal*, or intuition of a whole as a whole, to the particular—that is to say, from the whole to the parts. To render possible a definite form of the whole, a *contingency* in the synthesis of the parts, is not implied by such an understanding or its representation of the whole. But that is what our understanding requires. It

57. Cf. above, pp. 278 ff.

must advance from the parts as the universally conceived principles to different possible forms to be subsumed thereunder as consequences. . . . How then may we avoid having to represent the possibility of the whole as dependent upon the parts in a manner conformable to our discursive understanding? May we follow what the standard of the intuitive or archetypal understanding prescribes, and represent the possibility of the parts as both in their form and synthesis dependent upon the whole? The very peculiarity of our understanding in question prevents this being done in such a way that the whole contains the source of the possibility of the nexus of the parts. This would be self-contradictory in knowledge of the discursive type. But the *representation* of a whole may contain the source of the possibility of the form of that whole and of the nexus of the parts which that form involves. This is our only road. But, now, the whole would in that case be an effect or product the *representation* of which is looked on as the *cause* of its possibility. But the product of a cause whose determining ground is merely the representation of its effect is termed an end. Hence it follows that it is simply a consequence flowing from the particular character of our understanding that we should figure to our minds products of Nature as possible according to a different type of causality from that of the physical laws of matter, that is, as only possible according to ends and final causes. In the same way we explain the fact that this principle does not touch the question of how such things themselves, even considered as phenomena, are possible on this mode of production, but only concerns the estimate of them possible to our understanding. . . .Here it is also quite unnecessary to prove that an *intellectus archetypus* like this is possible. It is sufficient to show that we are led to this idea of an *intellectus archetypus* by contrasting with it our discursive understanding that has need of images (*intellectus ectypus*) and noting the contingent character of a faculty of this form, and that this idea involves nothing self-contradictory."[58]

All the lines previously established by the critique of reason here converge in a single point; all its concepts and presuppositions unite

58. *Critique of Teleological Judgment,* §77 (V, 486 f.) (*Ak.* V, 407–08).

352 THE *CRITIQUE OF JUDGMENT*

to determine unambiguously and precisely the position occupied by the idea of an end in the whole of our cognition. The inquiry here passes into depths which are genuine and ultimate, into the very foundations of the Kantian conceptual structure. Schelling said of these propositions from the *Critique of Judgment* that perhaps never before had so many profound thoughts been crammed into so few pages as here. At the same time, however, all the difficulties surrounding Kant's doctrine of the "thing in itself" and his conception of the "intelligible" are presented afresh. The principal conclusion to be drawn from this whole consideration is that the *methodological* orientation points out the distinction between the *intellectus archetypus* and the *intellectus ectypus,* between the understanding which is primal and that which is secondary and craves images. The two members of this contrast are not opposed to each other in their existence, nor do we look in them to a difference in actual things. But through them two systematic orientations are to be created, lending themselves to a supportive relation with the nature of our specific means of cognition, and its meaning and its validity. This task can be facilitated by ranging the systematic orientation beside the historical. In the history of metaphysics, the concept of an end encounters two opposing basic viewpoints and evaluations. On the one hand there is the doctrine of Aristotle, on the other that of Spinoza: in the former case teleology is the highest form of adequate cognition of being and insight into it; in the latter it is a special "human" way of knowing, which is put into things themselves and their formation only through a deception of the imagination. For Aristotle the end means the τὸ τί ἦν εἶναι: the ultimate intelligible ground of all being and change; for Spinoza it is merely an imaginative frippery that smirches and obscures the pure image of being, the image of substance that produces the totality of its modifications with geometrical necessity. Between these two extremes the entire evolution of metaphysics moves. The inner freedom Kant achieved with respect to the results of this evolution is shown anew by the way he eschews in equal measure both typical solutions that had been offered for the problem of finality. For him the end is neither the basic concept of the *intellectus archetypus,* as it was for Aristotle, nor, as for Spinoza, a creature of the *intellectus ectypus* that falls short of the true vision of being. The teleological way of thinking arises instead through a new *relationship* that ensues when our conditioned and finite understanding responds to the demand of the unconditioned; thus its basis is an opposition which is possible only from the standpoint of our mode of cognition, but which on the other hand is shown to be unavoidable and necessary under its previously established presuppositions.

An end is accordingly no more a product of absolute thinking than a purely anthropomorphic type of representation, which in the highest knowledge we leave behind as a mere subjective deception. Rather, its "subjectivity" itself has a universal nature: the conditioned state of human reason itself finds its expression herein. The concept of finality issues from the mirroring of experience in idea, from the comparison of the form of our categorical thinking with that other type of understanding that the demand or reason for systematic unity and perfection in use of the understanding shows us. Its nature and the particularities of its methodology are here similarly misconstrued if we relax our hold on either of the members of this correlation. If we take our stance with the absolute and archetypal understanding, the ground is cut away from under every application of the concept of an end. For purposiveness is, according to Kant's definition, the "lawfulness of the contingent"; for such an understanding, however, the concept of the contingent would be empty. That which comprises the part and the whole, the particular and the universal, in one indivisible mental gaze, the contrast between possibility and actuality, to which we are bound thanks to the basic laws of our mode of cognition, would be obliterated; there would exist for it only the absolutely unitary series of being, which would tolerate the thought of nothing in addition to and outside of itself. The survey of a set of possible cases, which is the presupposition of every judgment of finality, would also drop away here; where the insight that the whole of reality can be nothing other than what it in fact is holds sway, the assertion of some particular preferential end of this specific being loses its meaning and validity.[59] On the other hand, to say this is

59. At this point, therefore, Kant indirectly completes his criticism of the Leibnizian conception of the concept of purpose and the metaphysic Leibniz founded on it. In Leibniz's theodicy, it is God's understanding which chooses among the infinite number of "possible worlds" and "permits the actualizing" of the best of them. The

absolutely not to say that the concept of *empirical* actuality, that our thinking about *phenomena,* either must or even could be renounced in applying the concept of finality. For this thinking is stimulated precisely by that dualism of logical and intuitive conditions that is the basis for the application of this concept and cannot escape this duality without surrendering itself. Its locus is the antithesis of the universal and the particular and yet it feels itself compelled to progressively conquer that antithesis. The form of this conquest, a never-ceasing, never-completed pursuit which yet is feasible to the end, is the concept of purpose. Hence it is unavoidable for us; in no way does it go beyond the ensemble of our methods of cognition, but rather it is

---

basic flaw in this conception, according to Kant, consists in the false hypostatization of a "subjective" antithesis which attaches to the form of our cognition, and attribution of it to the Absolute itself. The basis for the fact that for *us* the possibility of things does not coincide with their actuality is that, in our mode of cognition, the sphere of the understanding and that of intuition, the realm of what is thought and what is given, are not identical in their scope, so that here something can be *thought* as possible which finds no correlate in intuition and no instance of actualization. For the "intuitive understanding," however, whose thinking is seeing and whose seeing is thinking—even if we only admit the idea of such an understanding—the distinction between the potential and the actual must be regarded as canceled out. "This means that if our understanding were intuitive it would have no objects but such as are actual. Conceptions, which are merely directed to the possibility of an object, and sensuous intuitions, which give us something and yet do not thereby let us cognize it as an object, would both cease to exist.... To say, therefore, that things may be possible without being actual ... is to state propositions that hold true for human reason, without such validity proving that this distinction lies in the things themselves.... An understanding into whose mode of cognition this distinction did not enter would express itself by saying: All Objects that I know *are,* that is, exist; and the possibility of some that did not exist, in other words, their contingency supposing them to exist ... would never enter into the imagination of such a being. But what makes it so hard for our understanding with its conceptions to rival reason is simply this, that the very thing that reason regards as constitutive of the Object and adopts as principle is for understanding, in its human form, transcendent, that is, impossible under the subjective conditions of its knowledge" (*Critique of Teleological Judgment,* §76). Here, as can be seen, the Leibnizian theodicy is vanquished, since the critical attack is directed not so much against its result as against the very basis of its way of posing the question. The Leibnizian application of the concept of purpose, in the idea of the "best of all possible worlds," is taxed by Kant for being an "anthropomorphism"; however, it is not an anthropomorphism of a psychological but of a "transcendental" kind that he discovers in it, and which he therefore thinks can at last be unseated through the whole of his transcendental analysis and its conclusions.

valid just precisely for this ensemble itself, though not for that "absolute" being which metaphysics in its received form treats of. The idea of an end and that of organic life are what initially give to our experience and our knowledge of nature the immanent infinity proper to them; it converts conditioned and isolated experiences into a totality, into the intuition of a living *whole,* but it simultaneously points to the limits of this whole since it comes to know it as a whole of phenomena. "If at last I rest in the ultimate phenomenon," Goethe once said, "this is but resignation; but still there is a great difference whether I submit to the bounds of human nature or to a hypothetical restriction of my limited individuality." For Kant the appearance of organic life, and also the idea of purpose in which it is expressed for our cognition, is such an ultimate phenomenon. It is neither the expression of the absolute itself nor of a merely contingent and dispensable subjective restriction of judgment, but it leads to the limits of human nature itself, so as to grasp them as such and to accommodate itself to them.

From the totality of these abstract reflections, however, we are plunged straight into the midst of the realm of intuitive contemplation, as soon as Kant goes on to make secure the basic insight he achieved in his critique of the concept of purpose in the facts of nature and in their detailed interpretation. The synthesis of the principle of finality and the principle of mechanism, and the reciprocal conditioning between the two that is to be assumed within experience, present themselves with concrete immediacy and clarity in Kant's concept of evolution. Evolution is itself a purposive concept, for it posits an "imprinted form," a unitary "subject" of the phenomena of life, which conserves itself in all changes, while it is transformed as well. But it must at the same time be explained purely causally in all its individual phases, so these truly compose a temporally ordered whole. The requirement stands inviolable for Kant from the beginning, because it was in the world of *cosmic* phenomena, the world of mechanism itself, that the full meaning of the idea of evolution first came upon him. In his initial youthful attempt to crystallize the whole of his natural scientific view of the world, the universal *theory* of the heavens had been transformed for him into the universal *natural his-*

tory of the heavens. This standpoint bore fruit not only in a fullness of
new, detailed results, but, what is decisive in the philosophical sense,
in a new ideal of knowledge clearly and consciously opposed to the
ruling technique of systematic classification of existing natural forms,
as carried out, for instance, in Linnaeus's doctrine. "Natural history,
which is almost nonexistent for us," as Kant characterized this ideal
in a later work on the diversity of the races of men, "would teach us
about the changes in the figure of the earth, and as well the earthly
creatures (plants and animals) which have changed through natural
migrations and the deviations from the archetype of the basic species
which have arisen in them therefrom. It would probably trace back a
host of apparently diverse types of races to one and the same genus,
and change the scholastic system of the description of Nature which
is presently too widespread into a physical system for the understanding."[60] The basic notion is here already being advanced that
nature initially comprises for the understanding a clear and surveyable unity, when we do not grasp it as a rigid entity of juxtaposed
forms, but pursue it in its continuous becoming. The *Critique of Judgment* gives this thought new breadth and depth, since in the principle
of formal purposiveness it erects its universal critical foundation.
Here it is shown that we only understand any particular manifold
insofar as we think it as proceeding from a principle which "specifies"
itself, and that such a *judging* of the manifold from the standpoint of
our faculties of cognition constitutes the inevitable means of making
its structure conceivable and transparent. If we apply this logical result to the consideration of *physical* existents, we then immediately
arrive at a new concept of nature, which, unlike that of Linnaeus,
does not just dispose species and genera in ranks, separated one from
the other by fixed, unchangeable characteristics, but rather tries to
make the coherence of nature comprehensible through the transformation of species.

Now we can understand that it is not at all an *aperçu* of genius but
a necessary consequence of his methodological presuppositions when
Kant assumes this postulate in the *Critique of Judgment,* and when he

60. See "On the Various Races of Men," §3, note (II, 451) (*Ak.* II, 434).

attempts to carry it out over the entire realm of natural forms. He begins with the universal requirement for every "natural explanation," which for him is already posited through the concept and the form of scientific experience itself. "It is of endless importance for reason to keep in view the mechanism which Nature employs in its productions, and to take due account of it in explaining them, since no insight into the nature of things can be attained apart from that principle. Even the concession that a supreme Architect has directly created the forms of Nature in the way they have existed from all time, or has predetermined those which in their course of evolution regularly conform to the same type, does not further our knowledge of Nature one whit. The reason is that we are wholly ignorant of the manner in which the supreme Being acts and of His ideas, in which the principles of the possibility of the natural beings are supposed to be contained, and so cannot explain Nature from Him by moving from above downwards, that is, a priori."[61]

On the other hand, the preceding discussions have established the equally necessary maxim of reason that the principle of ends in the products of nature is not to be ignored, because, although it makes it no more comprehensible how these products have arisen, yet it is a heuristic principle with which to investigate the particular laws of nature. Even if the two principles are mutually exclusive as basic premises of explanation and deduction regarding a given thing in nature, they are nonetheless thoroughly compatible as basic premises for discussion. Our cognition has the authority to explain all the products and events in nature, even those which are purposive, as mechanical, so far as it is in our power, and indeed that is its vocation. But it must be resigned to arriving ultimately at a primordial "organization" of them for which no mechanical "why" can be seen, but only a teleological "why." Since, however, prior to this point no impediment to the question is permissible, it is praiseworthy to go through the great creation of organic nature with the aid of a "comparative anatomy" to see whether something like a system, and indeed one governed by the genetic principle, can be found. "When we consider

61. *Critique of Teleological Judgment,* §78 (*Ak.* V, 410).

the agreement of so many genera of animals in a certain common schema, which apparently underlies not only the structure of their bones, but also the disposition of their remaining parts, and when we find here the wonderful simplicity of the original plan, which has been able to produce such an immense variety of species by the shortening of one member and the lengthening of another, by the involution of this part and the evolution of that, there gleams upon the mind a ray of hope, however faint, that the principle of the mechanism of Nature, apart from which there can be no natural science at all, may yet enable us to arrive at some explanation in the case of organic life. This analogy of forms, which in all their differences seem to be produced in accordance with a common type, strengthens the suspicion that they have an actual kinship due to descent from a common parent. This we might trace in the gradual approximation of one animal species to another, from that in which the principle of ends seems best authenticated, namely from man, back to the polyp, and from this back even to mosses and lichens, and finally to the lowest perceivable stage of Nature. Here we come to crude matter; and from this, and the forces which it exerts in accordance with mechanical laws (laws resembling those by which it acts in the formation of crystals), seems to be developed the whole technic of Nature which, in the case of organized beings, is so incomprehensible to us that we feel obliged to imagine a different principle for its explanation.

"Here the *archaeologist* of Nature is at liberty to go back to the traces that remain of Nature's earliest revolutions, and, appealing to all he knows of or can conjecture about its mechanism, to trace the genesis of that great family of living things. . . .He can suppose that the womb of mother earth as it first emerged, like a huge animal, from its chaotic state, gave birth to creatures whose form displayed less finality, and that these again bore others which adapted themselves more perfectly to their native surrounding and their relations to each other, until this womb, becoming rigid and ossified, restricted its birth to definite species incapable of further modification. . . .Yet, for all that, he is obliged eventually to attribute to this universal mother an organization suitably constituted with a view to all these forms of life, for unless he does so, the possibility of the final form of the products of the animal and plant kingdoms is quite unthinkable. But when he does attribute all this to Nature he has only pushed the explanation a stage farther back. He cannot pretend to have made the genesis of those two kingdoms intelligible independently of the condition of final causes."[62]

We must follow these Kantian propositions—as widely known and renowned as they are—just as far as they will lead us, for aside from the fundamental natural scientific insights anticipated in them, they express once more the whole essence of Kantian thinking. Kant's keen eye for detail and his synthetic power of imagination, his acuteness of intuition and his critical wariness of judgment, all these appear here as though brought to a focus. The idea of a unified derivative and evolutionary series of organisms appeared to Kant as an "adventure of reason"; but he was, like Goethe, barred from entering boldly on this adventure so long as in so doing he had to commit himself to the compass of the critical philosophy. He conceived the bounds set to the journey even before he began it; he saw the Pillars of Hercules, that betokened *nihil ulterius*,[63] clearly and steadily before him from the start. For Kant evolution is no metaphysical concept, which pushes back into the transcendent source of being and enfolds in it the secret of life; it is the principle by means of which the whole fullness and coherence of the phenomena of life might be presented for our cognition. We are not accustomed to ask whence life stems when we see before us in intuitive clarity and conceptual order only the totality of its forms and its arrangement in stages. In this conclusion one of the most profound factors of the Kantian doctrine speaks out once again from a new direction. The *Critique of Judgment* holds fast to the dualism of the "thing in itself" and "appearance"; but again this dualism is mediated by the idea that the "thing in itself," *regarded as an idea*, is what first brings the reality of experience to true completion. For only the idea is what ensures the systematic perfection of the use of the understanding, in which the objects are given to us not as disparate singular things, and hence as fragments of being,

62. Ibid., §80 (V, 497–99) (*Ak.* V, 418–20).
63. Cf. *Critique of Pure Reason*, A 395 (III, 661).

but in their concrete totality and in their thoroughgoing, unbroken interconnection.

Thus the *Critique of Judgment* adheres to the basic presuppositions of Kantian thought, while on the other hand it far transcends their previous sphere of application. The trial of precritical metaphysics launched by Kant comes to its close here: the *Critique of Judgment* confirms the verdict that the *Critique of Pure Reason* and the *Critique of Practical Reason* had pronounced on dogmatic metaphysics. And yet the critical philosophy now enters into another relationship with metaphysics. For the former has pursued the latter in its most central domain and has taken its measure by deciding and solving precisely those fundamental problems which from ancient times have seemed to be the peculiar property of metaphysics. In doing all this Kant's doctrine has not exceeded the ramifications of "transcendental philosophy": the general task of analyzing the contents and means of knowledge. As the content of the ethical could only be established by exhibiting the necessary and universally valid principles of all moral judgment, so this analysis could approach the problem of art, indeed that of life itself, only through the mediation of a critique of aesthetic and teleological judgment. But now it can be seen yet more clearly that this course, which is rooted in the essence of Kantian methodology, does not abort the wealth of intuitive actuality and water it down into a system of flimsy abstractions, but that, on the contrary, Kant's original concept of knowledge has undergone an extension and deepening that only now makes it feasible to survey the *whole* of natural and spiritual life and to conceive it as intrinsically a single organism of "reason."

VII LAST WORKS, LAST BATTLES. *RELIGION WITHIN THE LIMITS OF REASON ALONE,* AND THE CONFLICT WITH THE PRUSSIAN GOVERNMENT

If we turn aside from the structure and development of the Kantian system and consider Kant's outward life after the completion of the *Critique of Judgment,* we find it precisely at the point where we left it a decade before. Nothing in his way of life or in his relation to the world and his surroundings had changed during this epoch, which was so fertile and inwardly active. It is as if every occurrence and all progress were devoted purely and exclusively to his labors and withdrawn from him as a person. Since he had consciously and methodically set the style of his outward existence, he adhered to it with scrupulous exactness and regularity down to the smallest detail. In 1783 he changed his dwelling place for the last time: he moved to the house on Schlossgraben, where he lived until his death. Kant's first biographers depicted the arrangement of this house. It had eight rooms, of which Kant kept for himself only two, a study and a bedroom. "When one entered the house," Hasse tells us, "a serene calm reigned. ... When one climbed the steps, ... one went left through the entirely plain, unadorned, slightly dingy entrance hall into a larger room, which led to the parlor, but where nothing showy was on display. A sofa, some chairs with linen covers, a glass-fronted cupboard with some porcelain, a bureau which held his silver and ready cash, a nearby thermometer and a pier table ... were all the furniture, which took up one portion of the white walls. And thus one pushed through a quite shabby door into the equally shabby 'Sans-Souci,' which one was invited to enter, upon knocking, by a cheerful 'Come in!'. . . . The entire room breathed simplicity and quiet detachment from the hurly-burly of the city and the world." Two tables, which usually were covered with books, a plain sofa, some chairs, and a chest of drawers comprised the total furnishings of the space, its sole decoration consisting of a portrait of Rousseau, which hung on the wall.[1] Kant was more than ever confined to his house, since in 1787 he had decided to give up the lunch table in the inn, which had been almost the only diversion of his years as a young man and *privatdozent*, and to establish a group of his own. He had not renounced his pleasure in companionship in so doing; almost every day he had at table several of his friends, with whom he spent the luncheon hour in lively and stimulating conversation. This intellectually diverting round table remained unforgettable to the younger members of Kant's circle in particular. Poerschke, Kant's student and later his colleague at the University of Königsberg, says about it that Kant here lavished an immeasurable wealth of ideas, that he uttered a myriad of genial thoughts, of which he was scarcely conscious afterward. "In him," Poerschke adds, "one saw how childlike innocence and brilliance interacted with each other; his mind bore, along with the most marvelous fruits, numberless flowers, which often amused and served but for a moment."[2] A profusion of the richest personal ideas and hints were thus confined to a very narrow circle, for Kant was very particular that, in accordance with a maxim of sociability, the number of his companions at table amount to no less than three, but no more than nine. Although he felt at that time no brooding, melancholy inclination toward solitude, by conscious intent he strongly protected himself from the press of the outside world. He himself set the limits of his own involvement in and consideration of it, since this was an area in which he put to the test his basic rule of autonomy in the smallest and most intimate things.

1. See Johann Gottfried Hasse, *Letzte Äusserungen Kants von einem seiner Tischgenossen* (Königsberg, 1804), pp. 6 ff.

2. Cf. ibid., pp. 39 f.

This tendency comes out most conspicuously in Kant's mode of existence in relation to the new element that had come into his life since the middle of the 1780s. Only now had literary fame in its full extent fallen to Kant's lot, along with all the demands and burdens. Since Reinhold's "Letters on the Kantian Philosophy" ("Briefe über die Kantische Philosophie"), which appeared in 1786 and 1787 in Wieland's *Deutscher Merkur*,[3] and since the founding of the *Jenaische Allgemeine Literaturzeitung* by Schütz and Hufeland, which soon evolved into the special organ of the critical philosophy, the victory of Kantian philosophy in Germany was settled. A long struggle against misunderstandings and attacks from opponents of all kinds was still to follow, but these battles would only undergird and confirm anew the place it henceforth assumed in the entire intellectual life of Germany. All the forces of tradition were now summoned up once more against it. Almost no manner and almost no degree of polemic were not met with here. From Nicolai's flat jokes to the objections (at least intended to be profound) of the Wolffian philosophical school, which had created for itself a special literary organ in the *Philosophisches Magazin* in Halle, founded by Eberhard and Maas, every variety of criticism was to be found. The Berlin Academy of Sciences' orientation toward popular philosophy and popular science, in its fight against the Kantian doctrine, was at one with the "adepts" and fanatical minds with new metaphysical revelations; "sound human understanding" and the outlook of philosophical "intuition" closed ranks to ward off the "presumptions" of the transcendental philosophy. Even as the Kantian doctrine spread and even with its greater and greater influence, this countermovement ran unaltered through it all. The Kantian philosophy prevailed, although it itself rapidly disintegrated into various warring parties, each of which claimed for itself the sole correct and valid interpretation of the fundamental ideas of the First Critique.

With this development, however, more and more demands from outside were laid on Kant, which tended to force him out of his self-chosen circle of life, away from his plans for philosophical writ-

3. [These appeared in book form in two volumes in 1790-92.—Tr.]

ing, and declare himself definitely in the battle going on around him. In general, Kant remained chilly toward all these efforts: he saw too clearly before him the road he had to traverse and the positive task he still had to accomplish to let himself be held back by mere repetition and interpretation of his old books. Where, as in the case of Feder's and Eberhard's criticism, he believed he saw a conscious distortion of the root intention of his philosophy, he pursued it with a ruthless and unembittered keenness. On the whole, though, he held firmly to the conviction that once the discussion was guided to the right point, the sense of the main critical problems would grow increasingly more clear out of the welter of interpretations. Besides, he had only a very limited feeling for the struggle for personal fame in the present and the future, so unshakable was his consciousness of the content and value of his philosophy. "Author's itch," which he had avoided so perseveringly during the long gestation and ripening of the *Critique of Pure Reason,* still held no power over him. It looked almost as though he simply could not see himself in the role of celebrated writer that now fell to him. Those marks of childlike innocence, which Poerschke emphasizes in his portrait of Kant and which he found to be intimately related to the basic traits of Kant's genius, often come to the fore in a surprising way. When Schütz dealt with him about his participation in the *Jenaische Literaturzeitung,* he could not be astonished enough at Kant's modesty; not only did he profess his voluntary renunciation of the author's honorarium, but he even begged that he wanted to work out his review of Herder's *Ideas* only as a trial, and that the decision between their respective views of society, which were the basis on which the *Literaturzeitung* had been founded, be abandoned.[4] "Kant," Poerschke explains over again in a letter to

4. Cf. Schütz's letters to Kant of August 23, 1784, and February 18, 1785 (IX, 257, 260) (*Ak.* X, 372, 374). "Your review of Herder," Schütz writes in the latter, "you will by now probably already have seen in print. Everyone who is an unbiased judge regards it as a masterpiece of precision.... My God, and you could believe that a review like yours might not be acceptable! It brought involuntary tears to my eyes when I read that. Such modesty in a man like you! I cannot describe the feeling that I had. It was joy, horror, and indignation all in one, particularly the last, when I think of the immodesty of so many learned men of this *seculum,* who are not worthy to unloose the latchets of a *Kant's* shoes."

Fichte, "is a model of a modest writer, of all human souls he feels his greatness the least; I often hear him judging an opponent magnanimously, only they must not attack him personally and like monks."[5] Such a nature was not to swerve one step from its road by reason of success or failure: in the whole of Kant's career as an author no sign can be found that worry over it ever upset him and that it interfered with his intellectual development in any way whatsoever.

This is not the place for us to trace out the universal effect on history wrought by the Kantian theory and the transformation it underwent in the process. Only certain personal witnesses, who inform us as to the impact of the new philosophy on individuals, might be referred to briefly. Fichte's famous saying, that he had the Kantian philosophy to thank not only for his basic convictions but also for his character, nay, for the effort to will to have a character of that kind, is typical in this regard: it expresses most pregnantly a feeling which, especially after the appearance of Kant's ethical works, spread abroad and became more and more intense. Kant's correspondence offers the most numerous evidences for this. In a letter of May 12, 1786, the twenty-year-old physician Johann Benjamin Erhard tells how he immersed himself in Kant's writings, led in the first instance by his wish to refute the Kantian philosophy, until, on pressing further, he had been completely taken captive by it. "Six months ago, awakened by the call to do it, I began to read your Critique. No other book have I taken in hand with such bitterness; to enter the lists against you was my warmest wish and prayer. My pride was in fact to blame for my blindness, for as long as I had the idea that it is Kant who frustrates in me the hope of my own system to come, my inmost being revolted against you, but as soon as I became aware that Truth had chosen as my lot to lead me out of a stormy land where I wanted to build on unfirm ground a palace to protect myself, into a paradisiacal region where a perpetual springtime did not compel me to seek safety under a heap of stone, I pressed it to my bosom and am certain it will never leave my hand. . . . Your metaphysics of morals, however, quite made

5. Immanuel Hermann von Fichte, *Johann Gottlieb Fichtes Leben und literarischer Briefwechsel* (Seidel, 1830–31), vol. 2, p. 447.

me one with you; a sensation of bliss streams through all my limbs, as often as I recall the hour when I read it for the first time."[6] In his autobiography as well, Erhard admits that he had Kant's ethical works to thank for a "rebirth of his whole inner man."[7] For Reinhold, too, this was the moment that forever bound him to Kant. Although in his later writings he tried above all to define the highest theoretical principle of the transcendental philosophy, still it was practical and religious motives that originally led him to it. Here that "concord of head and heart" which he had hitherto sought in vain was born in him. And even a man like Jung-Stilling, who was certainly not driven to the Kantian theory by any deeper speculative need, found his way into it under the viewpoint and the influence of Reinhold's "Letters on the Kantian Philosophy"; the token of its powerful and universal effect is that even this simple and modest mind dared to say that the Kantian theory would soon have effected "a far greater, more blessed, and more universal revolution than Luther's Reformation."[8] It is obvious on all sides how Kant's philosophy, even before it was fully accepted and taken over in the theoretical sense, was immediately felt to be an inescapable new force in life. Because this foundation of the critical philosophy stood firm amid the strife of the Kantian schools, which seemed more dangerous than any attacks by opponents, its essential historical power remained unweakened. The aim of the system was put forward clearly in the transcendental doctrine of freedom: people believed they could hold fast to it, even if the path supposedly leading to it seemed ever and again to be lost in darkness and in dialectical confusion.

For Kant himself there was no such separation between his results and his method, between the critical theory and its applications. For him, within the system, every part conditioned and supported the other, and the convenient and traditional division of theory from practice, with which German popular philosophy tried to mitigate the "rigorism" of its ethics, he opposed yet again as keenly as possible in

6. See Kant's correspondence (IX, 299) (*Ak.* X, 422).

7. K. A. Varnhagen von Ense, *Denkwürdigkeiten des Philosophen and Arztes Johann Benjamin Ehrhard* (Stuttgart & Tübingen, 1830).

8. See Jung-Stilling's letter to Kant, March 1, 1789 (IX, 378) (*Ak.* XI, 7).

a treatise in reply to Garve dating from 1793.[9] But still, after the theoretical foundations of his system had been completed with the conclusion of the *Critique of Judgment,* he turned once more by preference to the immediate questions of life that were exercising his era. It is now primarily political problems that press into the center of his interest, more than before. Kant used the essay against Garve to develop a complete outline of his politics and his theory of civil law, as an appendix to the particular question with which he set out. Kant's shorter treatises, too, which in this period appeared in the *Berlinische Monatsschrift,* are advanced with an eye to the specific political relations and situations of that time. The critical philosopher, who had just completed the whole of his theoretical edifice, turns journalist. He is not content to lay down abstract doctrines and claims, but is driven to become involved in the tasks of the day and to enter directly into the shaping of concrete actuality, although only by way of providing enlightenment and theory. Looked at from this perspective, Kant's literary activity during this period, which at first glance seems as conflicting as it is multifarious, immediately takes on a fixed and integral focus. Kant allies himself with the Berlin school of enlightenment philosophy, whose main organ was the *Berlinische Monatsschrift* managed by von Biester, in order to take up in concert with it the fight against political and intellectual reaction in Prussia, the portents of which he recognized earlier and more clearly than anyone else. Whatever in his basic philosophical outlook set him off from this enlightenment movement became minor for him in the face of this new common task. As early as 1784, in "An Answer to the Question: 'What Is Enlightenment?',", he had gathered up all the threads clustering around the name of this party, and endeavored to define their one most profound integrating tendency. Here the concept of enlightenment is recast by means of the critical conception of autonomy, and grounded and secured in it. "Enlightenment is man's release from his self-incurred tutelage. Tutelage is man's inability to

---

9. "On the Common Saying: 'This May Be True in Theory, but It Does Not Apply in Practice'" ["Über den Gemeinspruch: 'Das mag in der Theorie richtig sein, taugt aber nicht für die Praxis'"], sect. 1 (VI, 355 ff.) (*Ak.* VIII, 273 ff.).

make use of his understanding without direction from another. Self-incurred is this tutelage when its cause lies not in lack of reason but in lack of resolution and courage to use it without direction from another. *Sapere aude!* Have the courage to use your own reason!—that is the motto of enlightenment."[10] On the strength of this idea and this motto, Kant opposed all efforts to put the critical philosophy in the services of an irrationalism which, in making feeling and faith an element also of all *theoretical* knowledge, threatened in the end to demolish the foundations of the theoretical concepts of truth and certainty themselves. He turned sharply and definitively against Friedrich Heinrich Jacobi's philosophy of faith. And here too he directly combined his conceptual analysis, in which he disclosed the difference between Jacobi's concept of faith and his own doctrine of rational faith, with a political view and a political admonition. The epistemological exposition ends with a personal warning and apostrophe. "Men of intellectual abilities and broad sentiments! I respect your talents and love your feeling for mankind. But have you also considered well what you are doing, and where your attacks on reason are leading? Doubtless you want the *idea of freedom* to be preserved in sound health; for without that there would soon be an end to your free flights of genius.... Friends of the human race and of what is most holy to it! Assume what seems to you most worthy of belief, upon careful and sincere examination, be it facts, be it the grounds of reason; only do not contest reason in what it makes the highest good on earth, namely, the privilege of being the ultimate touchstone of truth! Otherwise, unworthy of this freedom, you will certainly forfeit it too, and drag down this misfortune on the head of those who are innocent, who otherwise would be fully minded to serve freedom in accordance with law, and thereby also for the purpose of the best of all worlds!"[11] Kant's style rose but seldom to that kind of urgent personal feeling: we sense in these words, written in the year in which Frederick the Great died, how lucidly Kant saw the coming of the new regime, which soon thereafter found its voice in

10. "An Answer to the Question: 'What Is Enlightenment?'" (1784) (IV, 169) (*Ak.* VIII, 35).

11. "What Is Orientation in Thinking?" (1786) (IV, 363 ff.) (*Ak.* VIII, 144–47).

naming Wöllner as minister and in the promulgation of the Prussian edict concerning religion.

Thus for the man of almost seventy, after a decade of the most comprehensive and most profound productivity, there was not a moment's respite; he saw himself straightway embroiled all over again in fresh battles, which he had to wage on diverse fronts. On the one side, it was important to ward off misunderstandings and distortions of his philosophy, which threatened its essential content and specific worth. While the reigning academic philosophy had seen in Kant primarily the "all-destroyer," as Mendelssohn had honestly felt and said, this opinion gradually gave way to another feeling and a different tactic. The initial impression of sheer negativity made by the critical theory had to lessen in the degree that its positive content emerged more and more distinctly, at least indirectly, in its *effect*. Now the attempt had to be made to conceive this content, little as people might relevantly and truly assimilate it, at least by given *historical* categories and models. In the same way that on its first appearance the First Critique was compared with Berkeley, just as Hamann hailed Kant as "the Prussian Hume," now the voices that drew attention to the relationship between Kantian and Leibnizian idealism grew louder. But Leibniz's idealism was not understood in its genuine universality and depth; rather it was seen through the medium of Wolffian philosophy and in the light of the recognized handbooks of metaphysics stemming from the Wolffian school of thought. When the Kantian results were translated back into the language of these handbooks, they seemed at first to shed their strangeness and to be incorporated into the circle of accepted ideas. But surprise was growing at what strange forms and formulas the transcendental philosophy, taken as a result already known in its essential points, had been tricked out in. All the basic methodological distinctions of the First Critique: the contrast between sensibility and understanding, the difference between analytic and synthetic judgments, the opposition between a priori and a posteriori, were affected by this view of them. Since as individual moments they were disengaged from all the systematic relations and connections to which they belong and only in which do they have their peculiar foothold and their meaning, they were thus stamped with the character of particular bits of doctrine, for which an analogue and counterpart could easily be pointed out in an alien world of ideas. The critical studies undertaken by Eberhard and Maas regarding the basic question of the First Critique, in the *Philosophisches Magazin* for 1788/89, are entirely slanted this way, despite the semblance of scientific strictness and thoroughness given to them.

Kant set has face against this procedure with a sharpness and bitterness that recalls his polemic against Feder. He, in whose mind the critical philosophy was conceivable as a living, methodical whole and only as such a whole, could see in this willful and disjointed way of treating it nothing but its "almost deliberate" falsification and misconstruction. In this, regarded simply from the psychological aspect, he doubtless did his opponents injustice; he was so little able to transport himself into the scholastic and disciplinary limitation of their way of thinking that he was inclined to attribute this fault to their will rather than to their intellect. But now he felt all the more forced, in his declaration of war against Eberhard, to set before the reader all the essential leading ideas of his system yet once more in a comprehensive survey, and to illuminate them reciprocally each by the other. In this respect, the treatise "On a Discovery by Which All New Critique of Pure Reason Is to Be Made Unnecessary by an Older One" presents an outline which in its clarity and significance stands directly alongside the *Prolegomena*. The specific nature assigned to sensibility in distinction from the understanding, the methodological peculiarity of the pure forms of space and time, the meaning of the a priori and its contrast with the innate, all this comes out once again with maximum specificity, and this yields, as if spontaneously, the proof of that singularly decisive originality of his system, an originality measurable not by the sum of its results but by the power and the unity of its creative conceptual motifs. [12]

Though in the essay he wrote in reply to Eberhard the total energy of Kant's polemical style is once more displayed, his defense against

12. "Über eine Entdeckung nach der alle neue Kritik der reinen Vernunft durch eine ältere entbehrlich gemacht werden soll." (See VI, 3–71) (*Ak.* VIII, 185–251).

an attack of Garve's, which followed shortly, is pitched in a milder tone. It was the fate of this man, noble and lovable though a very mediocre thinker, to cross Kant's trail at every point. Kant had forgiven him the part he played in the notorious review of the First Critique in the *Göttingische Gelehrte Anzeigen* the moment he explained it openly and candidly. But the *Foundations of the Metaphysics of Morals* necessarily aroused Garve's opposition afresh: the austerity of Kantian ethics ran counter to his reconciling nature, averse to any mordancies and oppositions, just as much as it affronted the commonplace ideas of his popular philosophy. Hence he turned, not so much against the principle of critical ethics directly, as against its unrestricted realization. He conceded the rule willy-nilly, so as at once to demand and to plead exceptions to it. For Kant, however, there was no weakening and no compromise on this question—and even silence would seem a compromise to him. Goethe wrote to Schiller on a later occasion, "I like the fact that the old man was always willing to keep reiterating his principles and to hammer at the same mistake at every opportunity. The younger practical man does well to take no notice of his opponents; the older theoretician must not let slip an untoward word to anyone. We will adhere to that in the years to come, too."[13] Kant took up the platitude about the difference between theory and practice as such an "untoward word." In the supposed relativity of the empirical possibilities for applying the moral law there is no deliverance from the unconditional nature of the moral claim raised by the categorical imperative. "In a theory founded on the *concept of duty,* any worries about the empty ideality of the concept completely disappear. For it would not be a duty to strive after a certain effect of our will if this effect were impossible to experience (whether we envisage the experience as complete or as progressively approximating to completion). And it is with theory of this kind that the present essay is exclusively concerned. For to the shame of philosophy, it is not uncommonly alleged of such theory that whatever may be correct in it is in fact invalid in practice. We usually hear this said in an arrogant, disdainful tone, which comes of presum-

13. Goethe to Schiller, July 27, 1798.

ing to use experience to reform reason itself in the very attributes which do it most credit. Such illusory wisdom imagines it can see further and more clearly with its molelike gaze fixed on experience than with the eyes which were bestowed on a being designed to stand upright and to scan the heavens. This maxim, so very common in our sententious, inactive times, does very great harm if applied to matters of morality. . . . For in such cases, the canon of reason is related to practice in such a way that the value of the practice depends entirely upon its appropriateness to the theory it is based on; all is lost if the empirical (hence contingent) conditions governing the execution of the law are made into conditions of the law itself, so that a practice calculated to produce a result which *previous* experice makes probable is given the right to dominate a theory which is in fact self-sufficient."[14]

This unshatterable claim of pure theory over against all particular conditions stemming from the concrete empirical material in which it is applied is shown in three directions: in relation to subjective ethical reflection, which in fact is directed to establishing valid maxims for the individual's moral behavior; in relation to the imperative of obligation toward political life and the political constitution; and finally in the cosmopolitan sense, which extends the idea of legal and moral organization to the totality of nations and states, and thus broadens it into the ideal of a universally valid law of nations. In the first regard, the exposition only needs to repeat the specifications given in the *Foundations of the Metaphysics of Morals* and the *Critique of Practical Reason* as to the relation between the "matter" of desire and the pure "form" of the will. But now it makes a further stride into the area of concrete and individual psychological problems, insofar as it takes into consideration not only the pure validity of the moral law as such, but also the factual effectiveness of its application to individual cases. And here the distinction entirely in favor of form as against matter, in favor of pure idea over against the empirical feeling of pleasure and striving for happiness, is also shown to vanish. To the concept of duty

14. "On the Common Saying," introduction (VI, 359) (*Ak.* VIII, 276–77).

belongs not only the sole truly normative meaning, but also the only effective motivating force. It is, "in its full purity," not only incomparably simpler, clearer, more comprehensible and more natural for everyone's practical use, and more natural than every motive derived from happiness and mingled with consideration for it, but also in the judgment of the most ordinary human reason it is far stronger, more urgent, and more promising of result than any grounds of motivation borrowed from the latter principle.[15] But if Kant's basic ethical ideas are extended into the area of pedagogy, an actual broadening of his general theoretical horizon can also be seen at the point where Kant turns his consideration to political life. Here he confronts a new decision in principle: the question of the relation of "theory" to "practice" is transformed into the particular question of the relation between ethics and politics.

In his basic political outlook Kant's feet are planted firmly on the soil of those ideas which found their theoretical expression in Rousseau, and their visible practical efficacy in the French Revolution. Kant sees in the French Revolution the promise of the actualization of the pure law of reason. For to him, the peculiar problem of every political theory consists in the question of the possibility of unifying diverse individual wills into one total will: nonetheless, this does not nullify the autonomy of the particular wills, but its validation and acknowledgement is achieved in a new sense. The intent of every theory of right and of the state, philosophically considered, can be nothing other than the solution of the task of how the freedom of each individual has to limit itself, under the necessity of a recognized law of reason, in such a way that it permits and confirms the freedom of everyone else in so doing. Thus Kant's theory of right and of the state consistently adheres to the universal presuppositions of the eighteenth century: the idea of the inalienable fundamental rights of man and the idea of the social contract. Friedrich Gentz said, not unjustly, about Kant's essay in reply to Garve that it contained "the complete theory of the rights of man, so copiously praised and so

15. Ibid., sect. I (VI, 369) (*Ak.* VIII, 286).

little understood, . . . which issue in a modest but complete form from the calm and incisive reasoning of the German philosopher."[16] Indeed, Kant has no doubt that if he succeeds in unifying the theory of civil law and political practice, if he achieves the conformation of actual political life with the idea of the social contract, the methodological dualism between being and obligation is not erased. The theory itself is here a pure theory of obligation, capable of seeing always but a conditioned and relative prominence in empirical being, perfect as we may think it. Only the *claim* to actualization is unconditioned and unfettered by any temporal and contingent limitations, while its fulfillment remains forever incomplete.

Hence even the concept of the social contract does not signify something factual, done in any sort of past time or to be done in any sort of future time, but rather in fact a task, which still is to be used and held onto as a yardstick for every judgment of what is factual. A coalition of individual wills, as is conceptually assumed here, need never have occurred in such a way that to consider a civil constitution like this as binding it would have to be proven from previous history that a people did once actually perform an act of that kind, and an indubitable spoken or written record of it left for us. "It is in fact merely an *idea* of reason, which nonetheless has undoubted practical reality; for it can oblige every legislator to frame his laws in such a way that they could have been produced by the united will of a whole nation, and to regard each subject, insofar as he can claim citizenship, as if he had consented within the general will. This is the test of the rightfulness of every public law."[17] Where, on the contrary, this rule is not fulfilled, where the sovereign arrogates rights to himself, rights which are incompatible with the rule, then the individual possesses as little right to opposition by force as does the people as an empirical totality. For to concede such a right is to destroy the factual basis on which every political order as such rests. The autonomy of the head of state must continue unimpugned in its concrete existence; pure

---

16. Friedrich Gentz, "Nachtrag zu dem Räsonnement des Herrn Prof. Kant über das Verhältnis zwischen Theorie und Praxis," *Berlinische Monatsschrift*, December, 1793.
17. "On the Common Saying," sect. II, conclusion (VI, 381) (*Ak.* VIII, 297).

theory, however, and the universally valid ethical principles can insist that nothing stand in the way of their unhindered exposition and discussion. The opposition that is justified against the power of the state, but against it under certain circumstances both necessary and called for, is thus of a purely intellectual kind. In every commonwealth obedience to the mechanism of the constitution under laws that compel is what must rule, but at the same time also a spirit of freedom and hence of public criticism of things as they are. The right to opposition, which many theories of civil law arrogate to the citizen, thus dissolves for Kant into mere freedom of the pen; this however must remain inviolable by the sovereign, as "the sole palladium of the rights of the people."

We see in this once more the double nature of the struggle in which Kant was involved during this whole period. He begins with defending the purity and the unrestricted validity of his conception of duty, but this defense drives him back to the general question of the relation of ethical theory to practice. Previously it was not clear and unambiguous which of the two opposing moments here is the standard of measurement and which the one that is measured; the question of whether the actual serves as norm for the idea or the idea for the actual is not forwarded by a single systematic step. The substance of this division, however, is fixed for Kant on the basis of his first critical presuppositions. Just as in the theoretical realm knowledge does not conform to the object, but rather the object to knowledge, so pure obligation provides a universal rule with respect to what is empirically present and actual. Since in fact Kant upholds in this fashion the unlimited applicability of theory as such, this at the same time definitely circumscribes the scope of its means. Theory remains within its own territory: it renounces all use of force as means for the practice of opposition and resistance, so as to make use of rational means alone. This at the same time indicates the role science has in the life of the state, in its positive as well as its negative aspect. Science, in all forms of its public existence and organization, cannot avoid the power of the state and its guardianship, but it submits to the latter only under the condition that the state for its part leaves unchallenged the right of science as the principal examiner and critic of all the institutions of the state. Thus the general task broadens for Kant: from an inquiry into the basic questions of his system and a defense of the purity of his method, he is led to question the place of philosophical theory in the whole of intellectual culture, of which science and religion, civil life and the life of justice, are but individual parts. The need to indicate the bounds of the unique faculty of consciousness all over again, and to keep watch over its exact confines, merges with the relevant particular motives provided Kant by the political situation at that time. We have here anticipated Kant's reply to Garve, which appeared in 1793, because as the culmination of a certain development of thought it indicates most plainly that whole trend. Now, however, we must turn back, to follow closely the course Kant's actions as philosopher and journalist had taken since the death of Friedrich II.

Two years after Friedrich's death, Zedlitz was removed from his post as minister of cultural and educational affairs. The post was entrusted "by particular confidence" of the new king to Johann Christoph Wöllner, a man whom Friedrich had once described, in a short note to a file copy of one document, as a "swindling, scheming parson." Wöllner launched his official activity by decreeing the famous religious edict, which was followed shortly by the issuance of a censorship edict and the institution of a special censorship commission for all printed matter appearing in Prussia. The state considered it important to prosecute the battle of orthodoxy against freethinking and enlightenment with all the means within its power. The religious edict pledged to the subjects toleration of their religious convictions "so long as each quietly fulfills his duties as a good citizen of the state, but keeps his particular opinion in every case to himself, and takes care not to propagate it or to convert others and cause them to err or falter in their faith." Two years later, on December 9, 1790, it was supplemented by a rescript which was issued to the consistories and subjected the examination of candidates in theology to a meticulously prescribed schema.[18] The personal creed of the candidate was to be

18. For more on the religious edict and Wöllner's regime, see Dilthey, "Der Streit Kants mit der Zensur über das Recht seiner Religionsforschung," *Archiv für Geschichte*

determined by rigorous questioning, and each of them was to give his oath, sealed by a handshake, not to go beyond the bounds of this creed in the conduct of his office as teacher and preacher.

To get fully and clearly in mind the impression all these measures made on Kant, one must recall the position he had adopted toward all confessions of faith and toward the essence of the church from his youth on, since he had arrived at a firm and independent conviction regarding religious matters. When Johann Caspar Lavater in 1775 besought him as to his view of Lavater's essay on faith and prayer, Kant answered him with the utmost decisiveness and candor: "Do you then know," Kant wrote him, "to whom you are applying on this score? To someone who knows no means which stands the test at the final instant of life other than the purest rectitude in regard to the most secret sentiments of the heart, and who, with Job, deems it a sin to wheedle God and to make inward confessions which may have been extorted by fear and which the mind does not assent to when believing freely. I separate the *teaching* of Christ from the *report* we have about Christ's teaching, and to discover the former in its purity, I try first of all to extract the moral teaching, in distinction from all the legalisms of the New Testament. This is surely the basic teaching of the Gospel, the rest can only be a doctrine which is added on. . . . But when the doctrine of righteous conduct and purity of heart (with faith that God will fulfill the rest . . . *without the so-called worshipful supplications which have perennially constituted the religious delusion* in a fashion which is emphatically totally unnecessary for us to know), is sufficiently widespread so that it can survive in the world, then the scaffolding must crumble when the building is finished. . . . Now I frankly confess that in the historical respect our New Testament documents will never be able to be brought to such authority that we might dare to surrender ourselves to every part of them with unbounded confidence and thus mainly weaken our attention to the one thing needful, namely, the moral faith of the Gospel, whose excel-

der Philosophie, vol. 3; E. Fromm, *I. Kant und die preussische Zensur* (Hamburg and Leipzig, 1894); and Emil Arnoldt, *Beiträge zu dem Material der Geschichte von Kants Leben und Schriftstellertätigkeit in Bezug auf seine "Religionslehre" und seinen Konflikt mit der preussischen Regierung, Gesammelte Schriften,* vol. 6, ed. Otto Schöndörffer (1898).

lence consists just in the fact that all our striving for purity of our hearts and conscientiousness in righteous conduct of our lives is bound up with it. In this way the holy law is always before our eyes and makes every falling-away from the divine will, even the least, an incessant reproach to us as if we were judged by a stern and just judge, *against whom no confessions of faith, invoking holy names, or attention to worshipful observances can help at all.* . . . Now it is very clearly seen that the apostles regarded the side-doctrine of the Gospel as its fundamental teaching, and . . . instead of extolling the practical religious doctrine of the holy Teacher as essential, they preached reverence for the Teacher himself, and a kind of ingratiation by flattering and by eulogizing him, something against which he spoke so explicitly and often."[19]

Such a "religion of ingratiation," which he had branded as the immemorial, peculiar delusion of religion, Kant now saw expressly recognized and demanded by the state, and in the given circumstances the tangible, political and practical sense of "ingratiation" threatened to be ranked alongside the transcendent sense. From now on he unweariedly lodged in every quarter the sharpest protests against this, which he felt to be both a religious and a political corruption. Almost all the brief essays which he submitted at this time to the *Berlinische Monatsschrift* are related, directly or indirectly, to this fundamental, overriding theme.[20] Reference to the Book of Job, already found in his letter to Lavater, seems to have been frequently on his lips in this connection: now he elaborates it further, since he sets in contrast to the honorable doubt and honorable despair over insight into the divinity of the world order, which is Job's hallmark, the portrait of the "wheedler of God," and gives this portrait traits that obviously are taken from those in power in Prussia at that time, and aimed at them. "Job," it is said in the essay "On the Failure of All Philosophical Attempts at Theodicy" ("Über das Misslingen aller philosophischen Versuche in der Theodizee"), "speaks the way he thinks and the way he is expected to, and speaks as probably every man in his situation would be expected to do; his friends speak in the

19. To Lavater, April 28, 1775 (see IX, 138 ff.) (*Ak.* X, 171 ff.).

20. Arnoldt in particular has drawn attention to this relation (*Beiträge zu dem Material der Geschichte von Kants Leben,* pp. 107 ff.).

opposite way, as if they were being covertly overheard by the Mighty
One, whom they are justifying and to stand in whose favor is, in their
judgment, dearer than to be truthful. This malice of theirs in saying,
for the sake of appearances, things which they still had to confess
they did not understand, and in shamming a conviction which they in
fact did not have, contrasts with Job's straightforward courage and
frankness, which is so far removed from false flattery that it almost
borders on presumption, much to Job's advantage." And the reference to contemporary relationships concealed in this antithesis is increasingly unveiled as the essay goes on. "Job would most probably
have experienced a nasty fate at the hands of any tribunal of dogmatic
theologians, a synod, an inquisition, a pack of reverends, or any
consistory of our day (with one sole exception).[21] Thus only sincerity
of heart, not superiority of insight, the sincerity to openly confess his
doubt, and disgust at counterfeiting a conviction not felt . . . : these
are the marks which distinguished the superiority of the honest man,
in the person of Job, in relation to the religious wheedler, in the
words of the divine Judge." A "Concluding Remark," clearly referring to Wöllner's examination order and the oath of orthodox belief it
specified, was then directed against the *tortura spiritualis* in things
which by their nature are never amenable to theoretical, dogmatic
conviction. In such matters he who makes an affirmation of faith
simply because it is required of him, without even having glanced
into himself, whether he in fact makes this asseveration deliberately
or even intentionally to some degree—"he *lies* not merely the most
outright falsehood in the face of his own heart's repudiation, but also
the most impious, because it undermines the foundation of every
virtuous intention: sincerity. How quickly such blind and superficial
creeds (which very readily become compatible with an equally untrue
inward creed), when they provide the basis for employment, can bit
by bit bring the commonwealth to a certain falsity in thinking, is
easily to be seen."[22] A more definite and unrestrained declaration on
Kant's part about the new direction he saw taking over the common-

21. An allusion to the Berlin Upper Consistory, which under Spalding's leadership
had raised energetic resistance to Wöllner's actions.

22. On all of this, see "Über das Misslingen aller philosophischen Versuche in der
Theodizee" (1791) (VI, 132 ff.) (*Ak.* VIII, 265 ff.).

wealth was scarcely possible: only Wöllner's name, which was immaterial, was suppressed here, while the aim and the consequences of
his politics were erected as danger signals so plain that in this regard
not the slightest doubt nor any misunderstanding could prevail.

On this ground, conflict between Kant and the ruling circles of the
Prussia of that period was inevitable, and foreseeable long before it
broke out. The government had spared Kant in the beginning; it
probably shied away from attacking the famous author, who moreover enjoyed the personal confidence of the king, and had been especially singled out by him at the coronation ceremonies in Königsberg.
Kiesewetter, who had been dispatched from Berlin to Königsberg
especially to study Kantian philosophy, acted upon his return to the
court as tutor for the king's children, and displayed a lively enthusiasm for the universal propagation of the critical doctrine, which,
as a matter of fact, he understood and lectured on only in a popular
and dilute form. But the real opposition was pressing ever more
strongly toward a clear decision. A proposal to prohibit Kant's literary
activity altogether—according to a rumor that Kiesewetter mentions
in a letter to Kant, and which he himself in fact believed to be
unreliable—was submitted to the king in June, 1791, by the *Oberkon-
sistorialrat* (High Ecclesiastical Councillor) Woltersdorf. "He is feeble
now, in body and soul," Kiesewetter writes about the king; "he sits
and weeps for hours at a time. Bischofswerder, Wöllner, and Rietz are
the ones who tyrannize the king. A new edict on religion is expected,
and the commoners are muttering that they will be compelled to go to
church and to communion; as to this, they feel for the first time that
there are things which no prince can require, and one has to be
careful not to ignite the spark."[23] However, when Kant's essay "On
the Radical Evil in Human Nature" was submitted to him, the censor
appointed by Wöllner, Gottlob Friedrich Hillmer, could not immediately decide to refuse permission to print it; he allowed it to
appear in the April number of the *Berlinische Monatsschrift*, while he
contented himself with the thought that "only deep thinkers read
Kant's writings." But the continuation of this treatise, the essay "On

23. See Kiesewetter's letter to Kant of June 14, 1791 (X, 77) (*Ak.* XI, 252).

the Struggle of the Good Principle with the Evil for Mastery over Mankind," which this time also had the theological censor Hermes as coreader, since its content was regarded as belonging to biblical theology, offended the latter, and its publication was forbidden. A complaint by the editor of the *Berlinische Monatsschrift*, Biester, to the board of censors and to the king went in vain. It was necessary for Kant to see to publication in some other way, if he did not want to give it up altogether, and since he supplemented the two essays written specifically for the *Monatsschrift* with two further pieces, he had the whole appear as an independent book: *Religion within the Limits of Reason Alone (Religion innerhalb der Grenzen der blossen Vernunft)*, at Easter of 1793. He had previously asked the theological faculty at Königsberg if it regarded the book as belonging to "biblical theology," and hence if it claimed the right of censorship;[24] as the reply proved negative, in order to obtain an expert opinion about the book from a scholarly body, he turned to the philosophical faculty of the University of Jena, whose dean at the time, Justus Christian Hennings, issued the imprimatur.[25]

If we look at the substantive content of the work, before we go into its further fortunes, it must first of all be stressed that Kant's book on religion cannot be measured by the same standards as his fundamental, principal critical works. It is not on a par with the writings on the foundation of his system, with the *Critique of Pure Reason* or of *Practical Reason*, with the *Foundations of the Metaphysics of Morals*, or the *Critique of Judgment*. For one thing, the Kantian system does not in general recognize the philosophy of religion as a fully independent member of the system, as a way of looking at things that is idiosyncratic and rests on autonomous and independent assumptions. The kind of validity which Schleiermacher later claims for the philosophy of religion is foreign to Kant, for the substance of his philosophy of religion comprises for him only a confirmation of and a corollary of the substance of his ethics. Religion "within the limits of reason

---

24. On this see Kant's letter to Stäudlin of May 4, 1793 (X, 205) (*Ak.* XI, 414).

25. It was Arnoldt (*Beiträge zu dem Material der Geschichte von Kants Leben*, pp. 31 ff.) who first proved the issuance of permission for printing by Hennings.

alone," which thus does not need to take heed of the concept of revelation and is not allowed to do so, has no essential content other than that of pure morality; it only expounds this content from a different viewpoint and in certain symbolic dress. Religion is for Kant the "knowledge of our duties as divine commands." Here, too, the concept of duty stands at the center; but contemplation of its origin and of the basis of its validity takes another direction from what was the case in the foundation of ethics. Instead of regarding the concept of duty purely as to its meaning and what it commands, we here join the substance of the demand with the idea of a supreme being, which we think as the creator of the moral law. Such a change is humanly inevitable, for every idea, even the highest such as that of freedom, can be grasped by man only in an image and by "schematization." We always require a certain analogy with nature to make the supersensible properties conceivable to us, and cannot avoid this "schematism of the analogy."[26]

In this, what governs is not only a peculiarity of our sensitive and intuitive nature, which even has to present everything spiritual in a spatiotemporal metaphor, but at the same time—and this has become fully clear to Kant only since completion of the *Critique of Judgment*—a basic tendency of our pure *aesthetic* consciousness.[27] Although the

---

26. *Religion within the Limits of Reason Alone* [Religion innerhalb der Grenzen der blossen Vernunft], pt. 2, sect. 1, note (VI, 205 f.) (*Ak.* VI, 64 f.).

27. This point of view comes out especially clearly when Kant opposes his own standpoint as to rational ethical faith to the standpoint of a mere religion of feeling, for according to him feeling has a positive and constructive significance only for the construction of the aesthetic world. The consequence of this for him was the possibility of a mediation which does not unconditionally reject the new factor—one especially fertile and contrary to the eighteenth-century enlightenment, which was contained, for example, in Jacobi's philosophy of feeling—but gives it an altogether different interpretation and application. "But why is there all this conflict," he concludes his essay "On a Condescending Tone Recently Raised in Philosophy" (1796), "between two parties which at bottom have one and the same good intent, namely, to make mankind wise and righteous? It is much ado about nothing, disunion out of misunderstanding, in which no reconciliation is required, but only mutual clarification. . . . The veiled goddess, before whom we both bend the knee, is the moral law within us, in its inviolate majesty. We hear its voice, to be sure, and also understand its command well enough, but we are in doubt, as we hearken, whether it proceeds from man, from the perfection of the power of his own reason, or whether it comes from some other being, unknown

powers conducive to natural and to positive religion are not only psychologically understood thereby, but also critically justified, a careful lookout must be kept so that they presume to no false independence. The preface to the first edition of *Religion within the Limits of Reason Alone* already says that morality, insofar as it is founded on the concept of man as a being who is free, but for that very reason binding himself by his reason to unconditional laws, neither requires the idea of another being superior to him to know his duty, nor any motive for observing the law except the law itself. At least it is man's own fault if such a need is found in him, for which there is no other remedy, because what does not spring from himself and from his freedom does not compensate for deficiency in his morality. Morality "thus in no way needs religion for its own service (objectively, as regards willing, as well as subjectively, as regards ability), but in virtue of pure practical reason it is sufficient unto itself."[28] Where this is misunderstood, where the religious way of thinking is permitted even the slightest influence on the essential basis of morality, then

---

to him, and which speaks to man through this reason of his. Fundamentally, we would do far better to desist from this inquiry of ours, since it is purely speculative and what we are obligated to do remains (objectively) always the same, whether the one or the other principle is used to support it: only the didactic process of making the moral law in us conceptually clear by a logical kind of instruction is purely and simply *philosophical*, but that process of personifying that law and making a veiled Isis out of reason as it commands us morally (whether we attribute to this any other properties than those discovered by this method) is an aesthetic mode of thinking this very same object. We can well make use of the latter, when the principles are purified by the former, so as to enliven these ideas by a sensory exposition, though it is only analogical, yet always with the danger of lapsing into muddleheaded visions, which is death to all philosophy" (VI, 494 f.) (*Ak.* VIII, 405). The major difficulty in giving religion a truly independent role in the whole of the transcendental critique is manifested quite characteristically here. By its content it ought, as rational religion, to merge with pure ethics, from which it is distinguished only by its form: the "personification" of just this content. But this form itself does not belong essentially to it; rather, it goes back—even if the universal, purely *theoretical* meaning of the transcendental "schematism" is disregarded—to the basic *aesthetic* function of consciousness. Accordingly, the religious appears under Kantian presuppositions not as a proper *domain* of consciousness with its own laws, but only as a new *relation*, in which the domains and faculties previously defined and demarcated with respect to each other come closer together.

28. *Religion within the Limits of Reason Alone,* preface to the 1st ed. (VI, 141) (*Ak.* VI, 3).

something happens not only to the pure fundamental ideas of ethics, but also to those of religion itself—then worship is subverted into idolatry.

Ever since he had expressed this thought in correspondence with Lavater, Kant held steadily to it. This book also calls it the peculiar "delusion of religion" when man supposes that he can do something else, apart from the good conduct of his life, to become acceptable to God; however, our action can only be called good when it is based purely on the principle of autonomy and when in thus recognizing the law as such the particular relation to the "legislator" is disregarded. No contortion of external behavior, whatever form it may take, helps to overcome a deficiency in this basic temper. "Once one has gone over to the maxim of worshipping supposedly to please God for his own sake, and also to placate Him, but to a worship which is not purely moral, among the ways of worshipping Him mechanically, so to speak, there is no essential difference which would give preference to one or the other way. They are all of equal worth (or rather worthlessness) among themselves, and it is sheer affectation to regard oneself as more elect by reason of a more subtle deviation from the sole intellectual principle of genuine reverence for God than by something making one guilty of a forgivable gross lapse into sensuousness. Whether the hypocrite makes his legalistic visit to church or a pilgrimage to the shrines of Loretto or Palestine, whether he brings his prayer formulas to the heavenly authorities by his lips or, like the Tibetan . . . does it by a prayer wheel, or whatever kind of surrogate for the moral service of God it may be, it is all worth just the same. It is here a matter not so much of difference in the outer form, but entirely of the acceptance or abandonment of the sole principle, of becoming pleasing to God either through moral conviction alone, exhibited in a living way in actions as its epiphany, or through pious gewgaws and passivity."[29]

The difficult methodological problem bound up with religion, and the special dialectic it raises, comes to the fore in just this connection.

29. Ibid., fourth part: "Of Worship and Idolatry under the Mastery of the Good Principle, or of Religion and Priestcraft," pt. 2, §2 (VI, 320 ff.) (*Ak.* VI, 170–73).

On one side stands the sensory "schematism" of the essence of religion, inseparable from it and at the same time unavoidable: religion would cease to be what it is if it wished to renounce that. On the other side, though, this very factor means for religion a continuing threat to its deepest and most basic content; as soon as religion surrenders itself to it uncritically, it sees itself necessarily perverted into the opposite of its fundamental tendency. We see ourselves faced with the alternative of either dissolving religion purely into ethics, and thereby allowing it to disappear as an independent form, or maintaining it by the side of ethics, but in that way also in opposition to ethics. For the deduction and substantiation of the moral law suffers any sensory support as little as it does any transcendent "supplement": every heteronomous element which we permit must necessarily unhinge this foundation. For Kant, the solution to this antinomy lies once again in the strict separation of the empirical and the intelligible, the given and the commanded. The conversion of pure rational religion into pure ethics is required, but in the world of historical appearances it is never completed; yet it is at any time capable of completion therein. The point of juncture we are looking for and to which we must hold fast lies at infinity. But it does not become in any way an imaginary point for that reason; rather, it strictly and precisely points out the direction from which religious development must not deviate, if it does not wish to miss its goal. Religion, where it appears in historical actuality, must take on the forms which are alone appropriate to this actuality. To be communicable, it must clothe itself in the sensuous signs of communication; it requires, in order to affect the life of the community, the firm outer rules and bonds of this community life. Thus in its empirical existence it necessarily becomes a church. But it submerges itself, on the other hand, in this form of existence only so as to continually transcend it and to ask what is beyond. Ever anew the idea of what religion is purely "in itself" must be contrasted with its particular and limited temporal modes of appearance; ever anew its special fundamental teaching must be affirmed over against the mere "side-doctrine" and elevated to authority. Thus the struggle between the infinite content to which it is directed and the finite modes of presentation in which alone it can be understood in fact resides in every one of its stages and phases; but it is just this struggle that gives it its historical life and its historical effectiveness. In this sense, Kant, like Lessing, regards "positive" religions as moments and transitional points in the education of mankind; in this sense he demands of them that they recognize the standard of rational ethical religion for themselves, instead of rigidifying into a narrow dogmatism, and hence indeed preparing their own overcoming and dissolution.

The general theme of the Kantian doctrine of religion is thus indicated, while the clear execution and realization of this theme, in *Religion within the Limits of Reason Alone,* is beset with many limitations. These reside mainly in the particular nature of the book, which in no wise claims to give a complete exposition of Kant's basic ideas on the philosophy of religion, but only to set forth, by the example of a specific, presupposed dogmatism, how an embodiment of purely rational basic ethical truths is obtained from a system of given articles of faith by deepening and interpreting those articles. But on the other hand, quite definite boundaries are set by reason of this link to the critical approach. Not that Kant in using this approach in general shows any desire to renounce its principle, but he now exercises this principle in a material which is accepted as given from outside. *Religion within the Limits of Reason Alone* thus has from beginning to end the character of compromise. It selects one particular dogmatic state in order to peel it away and expose in its purity the moral kernel lying hidden in the dogmatic husk. Everything that seems discordant with this central message is either eliminated from the essence of the doctrines of faith under consideration as a subsequent, falsifying accretion, or interpreted in a sense such that it harmonizes in some fashion or other with the overall method of treatment.[30] In this way, not only is an arbitrary and accidental baseline set for the treatment, but it seems, as a result of this dependence on one given set of dogmas, that a scholasticism is also suffered and reintroduced, which could be thought finally and definitively overthrown by the theoretical foun-

30. This compromise which is characteristic of *Religion within the Limits of Reason Alone* has been emphasized particularly acutely by E. Troeltsch, to whose detailed exposition I refer here: "Das Historische in Kants *Religionsphilosophie,*" *Kantstudien* 9 (1904), pp. 57 ff.

dations of the First Critique. One must, though, be wary of wanting to explain this defect, which as such is unmistakable, all too quickly by the purely accidental limitations of Kant's personality and character. It was not in the least mere intellectual timidity that restrained him here. His outward consideration for the political and ecclesiastical authorities may have led to much vagueness and many instances of camouflage in his expression, but this did not disturb the core of his thought. Kant stood in opposition to traditional religion in the main no differently from the way he opposed traditional metaphysics.

Here, however, it was a matter of a different task set him: for the "fact" of a definite religion is given in a vastly stricter sense than that of metaphysics, in which each successive system seems to negate its predecessor; it is given as a relatively enduring historical datum and as one that is stable in its main outlines. He who strives to overcome it theoretically must also reckon with this empirical factuality. Idealization joins hands with the given, not to justify it at any cost, but to indicate that point in it from which it can surpass itself, because of the unfolding of the proper rational germ presupposed in it. Kant is here only following a method used by the whole Enlightenment in full subjective sincerity. He displays that cleverness in separating the exoteric and esoteric which Lessing, in his analysis and critique of Leibniz's theology, had expressly emphasized and raised in the latter. He too sought to strike fire from the flint, but he did not conceal his fire in the flint.[31] In this sense *Religion within the Limits of Reason Alone* does not so much belong to Kant's purely philosophical works as it does to his pedagogical works. Here he was speaking as educator of the people and of the government as well, and hence he had, at the very least, to begin with the form of popular faith as much as with the form of the dominant state religion. In the process, the critical mode of thought did not directly change into a dogmatic form, but it did become "positive" in a special sense: it stopped tearing down, since it could not succeed in doing that, so as instead to build up what was

31. See Lessing, "Leibniz von den ewigen Strafen," *Werke*, ed. K. Lachmann and E. Muncker), vol. 9, pp. 461 ff.

there just as it stood, in order to gradually reform it from the inside out in such a way that it took on a new form, concordant with the requirements of pure reason. In this project, Kant was personally full of that optimism toward the historical that was true of Lessing and Leibniz also. The very preservation of Christianity through the centuries proved to him that in Christianity there must reside a factor of absolutely universally valid significance, for without the creative force of the fundamental motive of pure rational ethical religion, its endurance and existence would be inconceivable.

Thus we confront at the same time a second factor in the Kantian doctrine of religion, in which the breadth of its original plan is revealed equally with the narrowness of its execution. The religion of reason, as Kant thinks of it, in its relation to the historical and empirical from the start is in no way pointed toward or restricted to any specific form in which religion appears in history. Biblical theology, in the field of the sciences, stands opposed to a philosophical theology, which in order to confirm and elucidate its assertions makes use of the history, languages, and books of all peoples, among which the Bible is included but always as just one outstanding example.[32] Along with it, the Vedas, the Koran, and the Zendavesta can also be named without hesitation, and the same right to consideration and study is granted to them. But this is for Kant only a matter of a theoretically granted right, which in his particular practical carrying-out of the basic conception comes to nothing. For at bottom Kant values the collective religious literature outside of Christianity in the anthropological sense only, not in the ethical and religious. His stance toward it is that of the connoisseur, one who shows interest in every strange phenomenon, but he is not inwardly stirred by it. Toward Judaism in its entirety and toward the Old Testament, Kant has all along so strongly subjective a prejudice that he can see in the religion of the prophets and in the psalms nothing more than a collection of statutory laws and usages. In this, however, quite apart from the substantive right and worth of such individual judgments, an essential methodological circle, contained in Kant's view of the philosophy

---

32. *Religion within the Limits of Reason Alone,* preface to the 1st ed. (VI, 147) (*Ak.* VI, 9).

and history of religion, is revealed immediately. The ethical measuring rod is help up to the specific forms of religion as a universally valid and objective criterion, but in the way it is applied subjective feeling and outlook unmistakably play a role. Because he had been certain of the moral *effect* of the New Testament scriptures from his youth on, the question of their unique and incomparable *content* is settled for Kant from the start. Rational analysis was here only to confirm and explicate in detail what, as an overall result, was already sure for him in advance.

The power of the initial pietistic impression of his youth is shown nowhere so clearly as in Kant's book on religion. For it had been Pietism, to be precise, which had brought into undeniable currency once again that principle of "moral" written interpretation on which Kant's theory of religion is also based. As early as the Middle Ages, in fact, this form of interpretation, among others, was well known and current. Thomas Aquinas already makes a systematically acute and specific distinction between the *sensus allegoricus,* the *sensus anagogicus,* and the *sensus moralis* or *mysticus* of a scriptural passage. In Pietism, this kind of biblical interpretation had then taken on that specifically Protestant cast in which it affected Kant. Filled with the idea of the unconditional primacy of practical reason, he now sought out the exclusively ethical meaning behind every religious symbol familiar to him. The whole set of Protestant dogmas—the dogma of the Fall and redemption, of being born again and of justification through faith—is traversed with this intent. Kant has unqualified subjective confidence that the fundamental and leading idea of his rational religion must be capable of having dominion over this set of dogmas and conforming to its actuality; but for exactly this reason he does not strive beyond it, since he is sure that he can fully demonstrate in it the universal application of his principle.

In fact, the whole analysis and critique of the dogmas that runs through his book on religion concentrates from the outset on one point. Kant's theory of "radical evil" in human nature, like his conception of the doctrine of the personhood of Christ, the interpretation that he gives to original sin and to the idea of justification, his concept of the kingdom of God, and his opposition between the purely moral and the statutory laws—all this is related to one single basic philosophical question, and only in that does it find its true unity. Everywhere in this theory it is for Kant a matter of particular moments and particular interpretations of the *concept of freedom*. Freedom and the opposition between heteronomy and autonomy, between the sensory and the intelligible worlds, is the original fact to which all fundamental religious doctrines point in a veiled and symbolic form. The method of the Kantian philosophy of religion is consistently aimed at making this connection evident. Attempts have been made to draw a sharp line between Kantian philosophy of religion and Kantian moral philosophy, so that the concept of redemption can be called the specific substance of the former, but it has rightly been maintained to the contrary that the motive of redemption means for Kant's philosophy of religion nothing but a specified limitation of the problem of freedom. He knows and allows no "redemption" in the sense of a supernatural, divine interference, which takes the place of the moral subject's own act; rather, he sees in it only the expression for the intelligible act itself, in virtue of which the self-legislation of the pure will and of practical reason wins mastery over the empirical sensuous drives.[33] Thus, even for the Kantian theory of religion, freedom remains at the same time the sole mystery and the sole principle of explanation. It illuminates the essential meaning and aim of the doctrine of faith, but there is no further theoretical "explanation" of it itself—on grounds given in the critical ethics. All we can do with respect to it consists in conceiving it precisely in its inconceivability.[34] But in establishing and acknowledging the bounds of our theoretical knowledge in this way, we are not led into a mere mystical darkness, for little as any question can be raised about "why," about a further ground of freedom, yet freedom itself and its content are given in the unconditional demand of the ought as something absolutely certain and necessary. Religion and ethics both, each

33. On this, see Kuno Fischer, *Geschichte der neueren Philosophie*, 4th ed., (Heidelberg, 1914), vol. 5, pp. 289 ff., and the objections Troeltsch ("Das Historische in Kants Religionsphilosophie," pp. 80 ff.) raises to Fischer's interpretation.

34. See above, pp. 261 ff.

in its own language, express this content; but basically it remains one and the same, so truly is the moral law in its essence only one thing, however many the forms and symbols we may try again and again to express it by.

Thus, despite all its complications, the Kantian philosophy of religion shows that it is governed by a basic, integral systematic idea, while in Kant's *book* on religion this unity is presented in only a qualified and inadequate way. Hence it is understandable that the initial effect wrought by *Religion within the Limits of Reason Alone* was ambiguous through and through. The two poles between which judgment oscillated are visible as soon as we set Schiller's assessment against the impression Goethe received from the work. Goethe recoiled indignantly from the book, in which he could see merely a concession to ecclesiastical orthodoxy and dogmatism; he remarked bitterly in a letter to Herder that Kant has disgracefully "slobbered on" his philosopher's cloak "with the blot of radical evil, so that even Christ would be enticed to kiss its hem." Schiller, in contrast, whose feeling about the Kantian doctrine of radical evil was at first no less antagonistic, in the end let himself be captured by the Kantian definition of and argument for the concept, since he was forced to recognize in it, though oddly disguised, the fundamental idea of the Kantian doctrine of freedom, to which he had for a long time been inwardly devoted. He too expressed his concern, in contrast to Körner, that Kant's basic tendency would be misunderstood: while Kant's intention had only been not to throw away what had been attained, and to that end he was especially adept at relating philosophical thinking to reason still in leading strings, the dominant dogmatism would straightway seize it all and exploit it for its own ends—and thus in the long run Kant would have done nothing other than "shore up the rotting edifice of stupidity." Sceptical as his estimate of the Kantian doctrine of religion was, he believed himself clear on its essential content. He believed he could ascertain in Kant a completely independent intellectual attitude toward the stuff of dogmas: Kant bypassed them, just as the Greek philosophers and poets had treated their mythology.[35] As far as the church orthodoxy itself is concerned, it could not be deceived for an instant as to the unbridgeable gap between the Kantian persuasion and the system favored and required by the state. Yet the government tried to avoid open conflict. Even when Kant published in book form the essay banned by the censor, it did not bestir itself immediately. However, Kant's essay in reply to Garve in September, 1793, which moved in a threatening way from general ethics to encroach on the theory of the state, and which asserted not only religious freedom of conscience but also freedom of the pen as the sole palladium of the rights of the people, deducing this from the basic concepts of natural right, inevitably aroused anew the suspicion and concern of the political powers.

Kant foresaw the conflict that would necessarily ensue, and, little as he sought it, he was disdainful of the timid holding back that could still perhaps have averted it. "I hasten, esteemed friend," he wrote to Biester in May, 1794, when he sent him his essay "The End of All Things" ("Das Ende aller Dinge"), "to send you the promised treatise, rather than bring an end to correspondence between you and me. . . . I thank you for the information you shared with me, and convinced of having acted in every case scrupulously and lawfully, I look forward calmly to the end of these singular events. If new laws *command* what is not contrary to my principles, I will conform to them immediately; that will happen even if they should merely forbid that my principles be made public, as I have done heretofore (and which in no sense do I regret). Life is short, especially what is left after 70 years have passed; to bring it to an untroubled close, some corner of the earth can be found, I suppose."[36] Certainly these words are not an expression of a fighting mood, but still this man of seventy, who by all his habits and by his whole pattern of life was totally rooted in his native city, and who two decades earlier had called it an instinct of his physical and mental nature to avoid every external change, now was even ready to give up his teaching post and his right to live in Prussia, his homeland, if he could safeguard his independence in no

---

35. Goethe to Herder, June 7, 1793; Schiller to Körner, February 28, 1793.
36. To Biester, May 18, 1794 (X, 240 f.) (*Ak.* XI, 481 f.).

other way. As far as the essay Kant sent to Biester is concerned, it contains such clear references to the current situation, and such bitter disquisitions against the Prussian rulers, that these could scarcely overlook them. "Christianity," it says, "above and beyond the greatest respect which the holiness of its laws infuses, has something else in it which is worthy of love. . . . If now some type of authority is added on to Christianity to ensure this, be the intent thereof ever so well meant and the aim thereof ever so good, its lovableness has vanished. For it is a contradiction to *command* someone not only to do something but also that he should do it gladly. . . . Thus it is the *free* way of thinking—equally distant from the slave's outlook and from anarchy—from which Christianity expects its teaching to be effective, by which it has the power to win over the hearts of men to itself, men whose understanding is already illuminated by the idea of the law of their duty. The feeling of freedom in choosing one's ultimate goal is what makes the giving of the law lovable to them. . . . Should it once happen that Christianity stops being lovable (which could indeed occur were it armed with imperious authority, instead of its gentle spirit), then rejection and rebellion against it would inevitably come to be the dominant way of thought among men, because there is no neutrality in moral things (still less a combining of opposing principles). . . ; thereupon, however, since Christianity is *destined* to be the universal religion of the world, but it would not be *favored* by fate to become so, the (erroneous) *end of all things* as regards morality would occur."[37] These sentences are written in the baroque style of Kant's old age, but their essential meaning and thrust were nonetheless unmistakable. The government was forced into the decision to take action against the embarrassing remonstrator who was gradually edging more and more out of the circle of "deep-thinking scholars," where they at first thought him safely enclosed, and who now turned against them with the weapons of mockery and satire in particular. Thus there was issued to Kant, on October 1, 1794, the famous letter signed by the king personally, in which he was reproached for having

37. "The End of All Things" ["Das Ende aller Dinge"] (VI, 422–24) (*Ak.* VIII, 337–39).

"misused" his philosophy over a long period of time "for the distortion and debasing of many principal and basic teachings of Holy Scripture and of Christianity," and in which he was instructed, to avoid the royal disfavor, henceforth to be guilty of nothing similar: "otherwise you can unfailingly expect, on continued recalcitrance, unpleasant consequences."[38]

Kant's attitude toward the reproaches and threats leveled against him is well known. In writing his defense, he first rejected the accusation that he as a *teacher of the young*, that is, as he understood it, in his academic lectures, had ever mixed in any judgment on the Bible and Christianity, and he made reference on this score to the nature of Baumgarten's textbooks, which he squarely based his lectures on and which of themselves ruled out any such connection. Also, in his book he had not in any way spoken as a "teacher of the people," but exclusively intended a "discussion among the scholars of the faculties," so that it must needs be an incomprehensible and closed book to the general public. Further, his book on religion could not contain a "debasing" of Christianity and the Bible, for the reason that in it the sole theme was the evolution of pure rational religion, not the critique of definite historical forms of belief; moreover, so far as he had dealt with the specific content of Christianity, he had left no doubt that he recognized in it the fullest historical product of pure rational faith. "As concerns the second point," Kant's explanation concludes, "I will henceforth be guilty of no distortion and debasement of Christianity of the (alleged) kind: thus to prevent even the least suspicion on this score, I hold it as most certain herewith, cheerfully to declare myself Your Royal Majesty's most faithful subject: that I will refrain entirely in the future from all public discourses concerning religion, natural or revealed, in lectures and in writing alike."[39]

In his reply to the royal rescript, Kant thus gives in to the government's demands on practically all points; in the process, he tries to find a justification for this retreat only in that, by a mental reserva-

---

38. See the wording of the letter in the preface to *The Contest of Faculties* (VII, 316) (*Ak.* VII, 6).

39. See *The Contest of Faculties*, preface (VII, 317–21) (*Ak.* VII, 7–11).

tion, he limits it to the reign and the lifetime of Friedrich Wilhelm II. The addition that "as His Majesty's most faithful subject" he pledged himself henceforward to silence in religious matters, as Kant himself later explained, explicitly contained this meaning. This reservation has often been harshly censured, but these reproaches have at best not touched the essential, decisive point. If Kant, being conscious of and sensitive to the philosophical life's work still lying ahead of him—he himself never regarded this work as finished, and in his eighties still complained that important parts of it were yet incomplete—had resolved to renounce the battle against the Wöllner regime because this fight would have robbed him of the best part of the strength to live and work still remaining to him, it would be narrow-minded and petty to wish to take him to task on this score. It is the basic right of a genius to determine for himself his own path and his own tasks out of his personal necessity, which is at the same time the highest impersonal necessity; and it is always shortsighted and unproductive to want to substitute an external, abstract and doctrinaire measuring rod for this internal one.

If Kant therefore had now sacrificed his activity as publicist or had put it off to a more favorable time, in order to gain room and leisure for accomplishing the other problems that still awaited him, all complaint about this would be baseless. But in fact there is a sign in his attitude toward the government's writ of accusation that shows he now, in full inner freedom, no longer opposed settlement of the conflict he so clearly foresaw and so resolutely went forth to meet. Indeed, he thrust the idea of a merely apparent retraction from him with all his strength of mind. "Retraction and betrayal of one's inward conviction," runs one of his notes from this period, "is base; but keeping silence in a case like the present one is the duty of the subject; and even if everything one says is necessarily true, there is no duty to utter all truth publicly." Even here he thus weighed the scope and extent of individual duties carefully against each other, in his strict and methodical way; but in all this, quite apart from the personal privileges he granted himself with respect to the ruling political authorities, he at least underestimated the personal power he actually possessed against them. "When the strong men of the world," he wrote to Karl Spener about this time, "are in a state of intoxication, whether it originates from a breath of the gods or from noxious vapors,[40] it is advisable for a pygmy who is fond of his skin not to meddle in their quarrel, even by the gentlest and most respectful persuasion; mostly because, while they wouldn't listen to him, he would be misunderstood by others who are tattletales. Four weeks from today I enter on the seventieth year of my life. What particular thing can someone this old still hope to effect with men of spirit? And with the common masses? Labor thus employed would be labor lost, indeed labor harmful to him. In this half a life that is left, an old man is well advised that *non defensoribus istis tempus eget* [these defenders are not short of time] and to consider the extent of his powers, which allows for almost no wish beyond that for peace and quiet."[41] The ironic undertone in these sentences is unmistakable; but on the other hand, they reveal the full native timidity and self-consciousness of the lonely scholar and thinker, who feels a deeper and deeper aversion to every development in the "squabbles of the world." It was not fear of losing his post that was crucial for Kant in all this; he had already reckoned in advance with the possibility that he would have to resign from it, without any effect on his attitude. Even more foreign to him was any false esteem for rank and eminence as such: all reports about his personal communication with King Friedrich Wilhelm II, whom he had to welcome as rector of the university at the coronation ceremonies in Königsberg, celebrate the unaffectedness and natural frankness he displayed then. But Kant has a modest enough opinion of the role the individual might play in the polity as a whole, under an absolutist government. Here he was held back by that scepticism that caused him to renounce early on any directly practical activity of reform. As concerns the *theories* of morality, religion, and civil right,

40. Kant's word here translated as "noxious vapors" is *Mufette*, derived from the French *mofette* or *moufette*. It is hard to say precisely what Kant had in mind in using it, since in older chemistry *mofette* stood for any nonrespirable gas, while Buffon used it to apply to the firedamp or chokedamp found in mines and also to a Mexican mammal, the *ysquiepail* (perhaps the skunk), which emits a foul smell. In addition, its German adaptation can mean "bad wine." Kant would certainly have been familiar with its scientific meanings, and may well have had them in mind here.—Tr.]

41. To Spener, March 22, 1793 (X, 197 f.) (*Ak*. XI, 402 f.).

he believed them to have been taken to the point from which they could, progressing gradually and step by step, gain their increasingly extensive influence on "praxis." But he did not feel called upon to lay his own hand directly and actively to this. Objectively, he doubtless did have too low an opinion of the influence his personality could have exercised, because he was utterly incapable of surveying and assessing what his philosophy already meant as an ideal force in the whole life of the nation. In this, perhaps, lies the essential defect and error in Kant's attitude toward the rescript of the Prussian government: but to avoid it he would have had to feel himself elevated above his historical setting in a totally different degree than in fact he felt; he would have had to ascribe to his individual self an immediately influential force that he never credited it with.

Within the confines of philosophical speculation, however, Kant's thinking remains, as before, oriented toward basic political problems, which now undergo a fresh expansion and deepening. From the constitution of the individual state, the question trenches onto the idea of a "federation of nations," the indispensable empirical and historical prerequisites of which Kant tries to found and establish in his work *Perpetual Peace* (*Zum ewigen Frieden*) (1795). In the methodological sense, however, the whole series of ideas connected with this once again leads back to one indivisible foundation, which heretofore had had no independent and creative treatment in the critical system. The Kantian conception of the state rests on his conception of the idea of freedom, but the idea of freedom in itself alone does not suffice to constitute the concrete concept of the state. If the state, in its ideal task, points to the sphere of freedom, in its factual existence and its historical actualization it belongs to the sphere of coercion. This places it in a contradictory position, to mediate which is precisely one of its most essential definitions. The "Idea for a Universal History from a Cosmopolitan Standpoint" had already hinted at this connection, but a most important factor was still missing in it, through which alone the conflict between force and freedom, and the link between the two, is brought to its sharpest and most exact conceptual expression. In the concept of force lies the necessary preparation and prerequisite for the concept of right. For precisely what distinguishes moral duty from the duty of right, according to Kant, is that the former asks not only about action but at the same time and above all about its subjective maxims and motives, while the duty of right abstracts from every consideration of that sort, so as to judge action as such merely with respect to its objective circumstances and execution. It is sheer agreement or disagreement of an action with the laws, without regard to its motives, that constitutes its legality, while its morality is only assured when it is established that it proceeds from the idea of duty as its sole motivating ground. It is the latter agreement which, since it relates to something purely inward, simply is *commanded;* the former is what can at the same time be *demanded.* The external coercibility of an action is hence joined with the concept of right itself. "Right in its strict sense," in which any contribution of moral concepts is disregarded, can and must "be envisaged as the possibility of a general and reciprocal coercion consonant with the freedom of everyone in accordance with universal laws." "For just as the only object of right in general is the external aspect of actions, right in its strict sense, that is, right unmixed with any ethical consideration, requires no determinants of the will apart from purely external ones; for it will then be pure and will not be confounded with any precepts of virtue. Thus only a completely external right can be called right in the *strict* (or narrow) sense. This right is certainly based on each individual's awareness of his obligations within the law; but if it is to remain pure, it may not and cannot appeal to this awareness as a motive which might determine the will to act in accordance with it, and it therefore depends rather on the principle of the possibility of an external coercion which can coexist with the freedom of everyone in accordance with universal laws ... thus right and the authority to apply coercion mean one and the same thing. The law of reciprocal coercion, which is necessarily consonant with the freedom of everyone within the principle of universal freedom, is in a sense the *construction* of the concept of right: that is, it represents this concept in pure a priori intuition by analogy with the possibility of free movement of bodies within the law of the *equality of action and reaction.* Just as the qualities of an object of pure mathematics cannot be directly

deduced from the concept but can only be discovered from its construction, it is not so much the *concept* of right but rather a general,
reciprocal and uniform coercion, subject to universal laws and harmonizing with the concept itself, which makes any representation of
the concept possible."[42]

It is this exposition Kant tries to provide in the *Metaphysical Elements of the Theory of Right* (*Metaphysische Anfangsgründe der Rechts-
lehre*), which appeared at the beginning of 1797. It forms the final
book belonging wholly to the sphere of the great chief systematic
works and having their nature, since it sets up a universal principle
for a specific, objective and intellectual cultural field, intended to
explain the nature and the necessity of its construction. This already
is no longer the case to the same degree in the *Metaphysical Elements of
Virtue* (*Metaphysische Anfangsgründe der Tugendlehre*), which follows in
the same year. For the principle of ethics here is laid down in advance, as something already firmly based: now it is only a matter of
how to trace it through a wealth of applications, in which Kant's
discussion frequently loses itself in a laborious schematism and a
thorny casuistry. Even the development of the concept of private
right, which is laid out in the first part of the *Metaphysical Elements of
the Theory of Right*, with its division of rights into personal, real, and
real personal, is not free of this increasingly overpowering tendency
toward a schematic, by which the detailed questions are frequently
classified and to which they are subordinated. Kant's construction of
honor as a real personal right is especially typical in this regard.

His treatment only rises to a greater freedom of overview when it
applies itself to the questions of public right: political right and international right. What Kant had earlier put forward separately in his
short treatises now is substantiated by and deduced from a single
fundamental idea. The questions of the sovereignty of the ruler and
its origin in the sovereignty of the people, the division of powers
flowing from this and the delimitation of their rights with respect to
each other, are discussed with systematic completeness and together

---

42. *Metaphysical Elements of the Theory of Right*, introduction, §E (VII, 33 f.) (*Ak.* VI,
232 f.); cf. introduction, III (VII, 19) (*Ak.* VI, 218).

with their latent relation to empirical historical detail. The method Kant relies on in doing this seems at first glance not to differ at all from the natural right point of view governing the philosophy of right of the whole Enlightenment and revolutionary period. The theory of the social contract—especially in the form Rousseau gave it—is here assumed to hold throughout. But yet once more the tendency already apparent in the essay against Garve on the relation between theory and practice puts in an appearance, lending its special hallmark to Kant's overall view, as within the evolution of the conception of natural rights. The social contract is raised from the sphere of the empirical and the specifically historical into the sphere of the "Idea." "The act by which the people constitutes a state of itself, or more precisely, the mere idea of such an act (which alone enables us to consider it valid in terms of right), is the *original contract*. By this contract, all members of the people (*omnes et singuli*) give up their external freedom in order to receive it back at once as members of a commonwealth, that is, of the people regarded as a state (*universi*). And we cannot say that men within a state have sacrificed a *part* of their inborn external freedom for a specific purpose; they have in fact completely abandoned their wild and lawless freedom, in order to find again their entire and undiminished freedom in a state of lawful dependence (i.e., in a state of right), for this dependence is created by their own legislative will."[43] Thus the intelligible in the idea of freedom guarantees for Kant the intelligible in the concept of the state and of right, and guards it against being confused with something purely factual, which is founded exclusively in the actually existing relationships of power and rule.[44]

The community of the body politic, into which the individual is assimilated and to which he gives himself as an individual without reservation, however, includes by its own ideal nature a totality of ideal conditions, which can be summed up in the proposition that what the whole people cannot decide concerning themselves, no legislator can decide either.[45] This universal spirit of the original con-

43. Ibid., §47 (VII, 122) (*Ak.* VI, 315–16).
44. On all of this, cf. above, pp. 223 ff.
45. *Metaphysical Elements of the Theory of Right*, General Remarks on the Legal Consequences of the Nature of the Civil Union, §C (VII, 135) (*Ak.* VI, 327). On Kant's

tract furnishes the guiding principle and norm for all particular types and forms of government, which is "to alter the mode of government by a gradual and continuous process... until it accords *in its effects* with the only rightful constitution, that of a pure republic. The old empirical (and statutory) forms, which serve only to effect the *subjection* of the people, should accordingly resolve themselves into the original (rational) form which alone makes *freedom* the principle and indeed the condition of all *coercion*. For coercion is required for a just political constitution in the truest sense, and this will eventually be realized in letter as well as in spirit."[46]

While here it is a matter of the most universal basic questions concerning the philosophy of right and of the state, in his next work Kant returns once more to his personal experience with the existing political powers in regard to his literary and philosophical activities. This relation scarcely comes to the surface, save in the preface to the work, but it clearly forms the motive from which its basic ideas arose and which explains its whole structure. Once again it is the system of the sciences and the connection and order of its chief components that Kant undertakes to establish here: but instead of inquiring into the content and the pertinent presuppositions of the sciences, he now takes hold of them exclusively from the perspective of their relations with the state and its administration. It is not so much their logical status as their disciplinary activity that is in question here, and for which a fixed principle is required. On the strength of this turn of consideration, the quarrel between the sciences has become a contest between the faculties. For the state needs to take notice of the sciences insofar as they confront it as specific associations with settled boundary lines with respect to one another, as independent corporations based on historical right. Only as externally organized in this way does the state recognize them as members of its own organization, for which it assumes a right of supervision as well as a duty of protection. From this point of view, a whole discipline is considered and

---

theory of right, cf. esp. Erich Cassirer, *Natur- und Völkerrecht im Lichte der Geschichte und der systematischen Philosophie* (Berlin, 1919).

46. Ibid., §52 (VII, 148 f.) (*Ak.* VI, 340 f.).

evaluated only according to its place within the entire political hierarchy, so the scholar can count on a hearing only insofar as he is able to
show that he is at the same time a representative and *official* of the
state. This is the manner of framing the question which Kant adheres
to throughout *The Contest of Faculties* (*Streit der Fakultäten*); but in the
midst of the dry profundity with which he pursues it, one detects
clearly a sportive humor which once again reminds us of the style of
Kant's youthful works. And here too, as in those other books, the
humor is the expression and reflection of an inward philosophical
self-liberation. This self-liberation, as was fitting and natural to Kant,
consists in his converting his personal conflict with the administration
of the state into a conflict of method, and in his trying to settle it that
way. Since he places himself entirely at the standpoint of the political
practitioner, intentionally narrowing his political horizon, he tries to
demonstrate the right and the inalienable freedom of philosophical
theory and of science from precisely that perspective. Through the
attitude and aim of the politician, which he has assumed, the true
outlook and conviction of the critical thinker gleams at every point,
and this duality is what gives *The Contest of Faculties* that amalgam of
cheerful, reflective irony and dogged, businesslike solemnity which
comprises its peculiar character.

The ironic undertone becomes audible as early as the first section,
wherein Kant, in alliance with tradition, distinguishes the "higher"
theological, juristic, and medical faculties from the "lower"
philosophical faculty. The genesis of this received distinction is, as he
comments, readily recognizable: it stems from the government, with
which it is, indeed, never a matter of knowledge as such, but simply
of the effects on the people that the government expects from knowledge. On this basis, it sanctions certain theories, from which it promises itself a useful influence, but it does not condescend to propose
any kind of definite theory itself. "It does not teach, but only gives
orders to those who do (as for truth, that has to look out for itself),
because on taking up their posts they came to an understanding by
the agency of a contract with the government. A government which
busied itself with theories, thus also with the spread or the improvement of the sciences, and hence wished to play the scholar itself in the

shape of its highest personage, would only bring shameful attention on itself by such pedantry, and it is beneath its dignity to lower itself to the level of the vulgar (to its scholarly class), who don't understand jokes and who mold everybody involved with the sciences into the same shape."[47] In this sense it obliges the individual disciplines to specific statutes, on the strength of its authority as magistrate, since "truth" cannot exist for it except in the form of such a statute, and does not need to exist. The Bible is prescribed to the theologian, the universal law of the land to the jurist, to the physician the system of medical regulation, as rule and guide. Punctilious adherence to this rule is what ensures theology, jurisprudence, and medicine their place in public life, and what elevates them to the dignity and rank of a "higher" faculty.

Only one thing: knowledge purely for the sake of knowledge, is left empty-handed in this allotment and division, because no essential potentiality for directly practical ends is to be expected from it. If one still wants to assign a place to it, too, it must then content itself tamely with the rank of a "lower faculty." In it, reason remains free and independent of governmental orders, but for that very reason it remains ineffective, and must be meek, devoid of influence on the course of affairs. What its inalienable prerogative is, viewed objectively, allots to it the lowest place in conventional estimation. The philosophic faculty as such stands quite outside the circle of command and obedience, and it is human nature "that he who can command, though he be also a humble servant of another, still thinks himself superior to someone else who, to be sure, is free but has nobody to order around."[48]

Out of this difference in the basis of the rights of the faculties, there now results a "lawful controversy" between them: a conflict which is grounded in their very being and which hence cannot be eliminated through any sort of compromise, but must continue and be fought out. As parts and members of the political hierarchy, the higher faculties continue to be defined as much by their appetite for power as by

---

47. *The Contest of Faculties*, First sect., introduction (VII, 329) (*Ak.* VII, 19).
48. [Ibid. (VII, 330) (*Ak.* VII, 20).]

their desire to know, while the philosophical faculty, to the extent that it wants to remain true to its task, has to receive all its directives from this latter. Thus its natural role is that of opposition, but an opposition of a kind requisite and unavoidable for the prosperity and the positive progress of the whole itself. The philosophical faculty wages the eternal conflict of the "rational" against all that is merely "statutory," of scientific reason against power and against tradition. In this its basic function, it may not be hampered and confined, even by the state, insofar as it understands its own needs and its own vocation. All that the state may require of it is this: that it not trench directly on state administrative activities. The training and education of men of practical affairs, whom the state needs for its ends, is left to the higher faculties, which for this reason are subject to its legitimate oversight. But it may also be expected, conversely, of the members of the higher faculties that they not overstep the boundaries drawn for them. If the biblical theologian refers to reason for any one of his statements, "he thus leaps over (like Romulus's brother) the wall of the faith of the church, which alone confers salvation, and goes astray in the open meadow of judging for oneself and of philosophy, where he, having fled spiritual government, is exposed to all the perils of anarchy."[49] Just so the jurist, as an appointed judicial officer, has simply to apply the existing legal decrees, and it would be preposterous if instead before doing so he demanded or wished to prove them to be reasonable. Only the philosophical faculty, as guardian of pure theory, can never regard itself as exempted from this proof. It can happen that a practical doctrine is followed out of obedience, "but thereby to assume it to be true because it is commanded, is totally impossible, not only objectively (as a judgment which *ought* not to be) but also subjectively (as a judgment which *can* pass)."[50] Accordingly, if the quarrel is over truth and falsity, not about the utility or harmfulness of a theory, there is no higher principle than reason: to limit its autonomy in any way at all is nothing other than to destroy the essential concept of truth itself.

49. [Ibid., first sect., I, sect. IA (VII, 334) (*Ak.* VII, 24).]
50. Ibid., first sect., I, sect. II (VII, 337) (*Ak.* VII, 27).

What consequences issue from this for the contest between rational religion and church faith, between the pure philosophy of religion and biblical orthodoxy—this *Religion within the Limits of Reason Alone* has already expounded. What *The Contest of Faculties* does in this connection is only to supplement and confirm the earlier expositions, in which memory of the specific phases of the personal battle Kant had to wage echoes everywhere. However, reflection takes a new turn when—in the form of the exposition of the conflict between the juristic and philosophical faculties—it takes into its attack the question of the relation between the natural right and the positive right foundation of the constitution. Is all right simply the expression of actual empirical power relationships, and can it be resolved into them, as its essential basis, or does an ideal factor cooperate in it, which asserts itself slowly and steadily as a political effective factor as well? The answer to this question includes, according to Kant, nothing more trifling than the judgment as to whether human history and the human race are conceived as ascending and progressing steadily toward the better, or whether both persist at the selfsame stage of perfection, with minor oscillations, or even are as a whole exposed to decay and retrogression. If one tries to decide this from the standpoint of pure reflection on happiness, the upshot can be nothing but negative: Rousseau's pessimism toward culture holds here absolutely correctly. Eudaimonism, with its sanguine hopes, seems to be untenable and to promise little in support of a "prophetic history of mankind" which bears on continuing further progress on the road toward the good.[51]

But here immediately there intervenes the methodological consideration that the problem in general cannot be brought to clarification and solution in a purely empirical way. For to raise the question as to the moral progress of mankind is already a paradox. It is a matter herein of predicting something about the inquiry that of its nature cannot and ought not be predicted. The fate of the human race is not a fate imposed by any sort of blind "nature" or "providence," but it is the outcome and the handiwork of humanity's own free self-

51. Ibid., second sect., §3b (VII, 394) (*Ak.* VII, 82).

determination. But how is one to trace and to make visible the course and path the intelligible determination takes, in the empirical, causal running of events, in the sheer flux of appearances? The two realms are nowhere actually congruent, thus a relation of this kind is only possible in that the world of appearance, that is to say, the continuance of historical occurrences in the world, at least contains a symbolic event whose interpretation leads us back to the realm of freedom necessarily and of itself.

Is there such a historical sign, to which the hope and expectation that the human race as a whole is conceived to be continuously progressing may be joined? Kant answers this question by referring to the French Revolution, which is to be understood here not in its empirical course and outcome, but exclusively with regard to its ideal meaning and its tendency. "The occurrence in question does not involve any of those momentous deeds or misdeeds of men which make small in their eyes what was formerly great or make great what was formerly small, and which cause ancient and illustrious states to vanish as if by magic, and others to arise in their place as if from the bowels of the earth. No, it has nothing to do with all this. We are here concerned only with the attitude of the onlookers as it reveals itself *in public* while the drama of great political changes is taking place: for they openly express universal yet disinterested sympathy for one set of protagonists against their adversaries.... Their reaction (because of its universality) proves that mankind as a whole shares a certain character in common, and it also proves (because of its disinterestedness) that man has a moral character, or at least the makings of one. And this does not merely allow us to hope for improvement; it is already a form of improvement in itself, insofar as its influence is strong enough for the present. The revolution which we have seen taking place in our own times in a nation of gifted people may succeed, or it may fail. It may be so filled with misery and atrocities that no right-thinking man would ever decide to make the same experiment again at such a price, even if he could hope to carry it out successfully at the second attempt. But I maintain that this revolution has aroused in the hearts and desires of all spectators who are not themselves caught up in it a *sympathy* which borders almost on enthusiasm, although the very utterance of this sympathy was fraught with danger. It cannot therefore have been caused by anything other than a moral disposition within the human race."[52] It is the certainty of this disposition on which is founded the hope of the evolution of a condition of natural right in the relation of the individual to the state and in the relation of separate states to each other. A phenomenon such as the French Revolution was, will never be forgotten, because it has revealed a capacity for the better in human nature, the like of which no politician would have rationalized from the course of things till now, and which alone unites nature and freedom in accordance with the inner principles of right in mankind.

Now it is shown that the ideal of the state, as the great social theoreticians have regarded it—as the the ideal of a constitution concordant with the natural rights of man—is no empty chimera, but rather the standard for every civil constitution in general. And with this insight "perpetual peace" ceases to be a mere dream: for the establishment within a nation of a constitution strictly democratic and republican in spirit also offers the external guarantee—as the book *Perpetual Peace* had already put forward—that the intent of unjustly oppressing one nation by another, and likewise the means of realizing this intent, are progressively weakened, so that approximation to the "cosmopolitan" condition is also progressively fulfilled in the history of nations.[53]

With its prospect on this goal of human history, in which the idea of freedom is to find its concrete fulfillment and its empirical political actualization, Kant's philosophical activity comes to a close. The idea of freedom forms the terminal point of his philosophy, just as it had formed its beginning and middle. What Kant's activity as writer adds to those discussions is only a sparse literary gleaning, adding no further dimension to the essential substance of his philosophical system. The final section of *The Contest of Faculties*, which treats of the contest of the philosophical faculty with the medical faculty, is only

52. [Ibid., second sect., §6 (VII, 397 ff.) (*Ak.* VII, 85).]
53. For the whole of this, see VII, 391–404 (*Ak.* VII, 79–94). Cf. *Perpetual Peace* (VI, 427–74) (*Ak.* VIII, 341–81).

superficially hooked on: this treatise, "On the Power of the Mind to Control Its Feelings of Illness by Sheer Willpower," is concerned only with a number of dietetic precepts, loosely arrayed, that Kant had tested out on himself personally and with methodical self-observation. Even the *Anthropology* of 1798 cannot in any sense take its place beside the essential main systematic works by virtue of its content and structure: it compiles merely "in a pragmatic respect" the rich material on human history and anthropology that Kant had assembled over a long lifetime from his own observations and from odd sources, and had enriched over and over by the notes and studies for his lectures.

On the other hand, that work in which Kant's entire inner concern was wrapped up during this period, and which he himself saw as belonging intimately to the whole of his systematic labor, never came to maturity, as untiringly as Kant devoted himself to its continuance on into the closing years of his life and until the complete expiration of his physical and mental powers. With a perpetually renewed effort of will he applied himself to this book, "The Transition from the Metaphysical Principles of Natural Science to Physics" ("Übergang von den metaphysischen Anfangsgründen der Naturwissenschaft zur Physik"), which was to lead to a complete and concluding survey of the "system of pure philosophy in its coherence." His biographers unanimously bear witness to the affection he attached to this work, about which he was wont to speak "with veritable animation" and which he many times declared to be "his most important work."[54] He often thought that he had reached the conclusion of this *chef d'oeuvre;* he believed that only a brief editing of the manuscript was needed to be able to publish it, "his system as a complete whole."[55] Was it simply a natural self-deception on the part of the old man that misled him into this judgment? We are tempted to suppose that, when we look at the superficial form of the manuscript.[56] The same sentences and ex-

54. See Reinhold Bernhard Jachmann, *Immanuel Kant geschildert in Briefen an einen Freund*, Third Letter, pp. 17 f.; E. A. C. Wasianski, *Immanuel Kant. Ein Lebensbild nach Darstellungen der Zeitgenossen Jachmann, Borowski, Wasianski* (Halle a.S., 1902), p. 95.

55. Cf. Hasse, *Letzte Äusserungen Kants von einem seiner Tischgenossen*, pp. 21 ff.

56. Parts of the manuscript were published by Rudolf Riecke in the *Altpreussische Monatsschrift*, 1882–84, under the title, "Ein ungedrucktes Werk Kants aus seinen

pressions recur over and over in innumerable repetitions; the important and the trivial tumble over each other in a motley jumble;
nowhere is there to be found a realized systematic arrangement and
strict structure and movement of thought. And yet, the further one
reads, the more it becomes evident that the real defect is not so much
in the ideas themselves as in their exposition. It is as if his original
creative power of thought lingered longer in Kant than the lesser
powers of arrangement and division. His power of recall breaks
down; his memory is not sufficient for him to recollect the beginning
of a sentence when he writes the end of it; stylistic periods get confused; and yet there gleam forth from time to time in the midst of this
chaos individual ideas of astonishing power and depth—ideas which
are devoted to illuminating once again the whole of his system and
exposing to view its ultimate foundations. In particular, information
can be found here about the methodical meaning of the opposition of
the "thing in itself" and "appearance" the equal of which one might
seek for in vain in the earlier works. The attempt to coin in detail the
intellectual handiwork of his old age seems, in view of the state of the
manuscript, probably doomed to remain forever vain—thus the deeper one immerses oneself in the samples from the book published so
far, there well up only pangs of regret that it was not granted to Kant
himself to recover this treasure.

In 1795 Wilhelm von Humboldt could still advise Schiller, following information he had gotten from Memel, that Kant was carrying a
monstrous host of unworked-out ideas in his head, which he intended to elaborate in a certain order, so that he reckoned the life
span remaining to him more in accordance with the size of that stock
of ideas than by the usual probability.[57] Schiller himself found in the
"Proclamation of the Approaching Conclusion of a Treaty of Perpetual Peace in Philosophy" ("Verkündigung des nahen Abschlusses
eines Traktats zum ewigen Frieden in der Philosophie"), aimed at
Schlosser, which Kant published in 1797, a fresh and genuinely
youthful character, which might almost be called aesthetic, as he

---

letzten Lebensjahren." For the substance of the entire work, now cf. esp. Erich Adickes's exposition, *Kants Opus posthumum* (Berlin, 1920).
57. Humboldt to Schiller, *Briefwechsel*, ed. A. Leitzmann (Halle, 1908), p. 153.

added in a letter to Goethe, if one were not plunged into embarrass-
ment by the drab style, which might be called a philosophical of-
ficialese.[58] The young Count von Purgstall reports, from his personal
association, the deep impression he received from Kant's lectures in
April, 1795, and the brilliance and clarity they shed over the whole of
his thought; and Kant's colleague Poerschke testifies in a letter to
Fichte in 1798 that Kant's mind was not yet gone, although he no
longer possessed the capacity for continuous intellectual labor.[59]
Even in the conduct of his personal affairs and in carrying on the
business of his post, Kant repeatedly showed in this period that his
old force of will and energy had not left him. He had indeed given up
his lectures as of the summer of 1796: on July 23, 1796, he seems to
have mounted to his lecture-desk for the last time.[60] He even refused
the post of rector when it was offered to him again in 1796, referring to
his age and his physical feebleness.[61] But two years later, when the
attempt was made to restrict his functioning in the university senate
and substitute in his place an "adjunct," who was to safeguard his
rights for him and conduct his business, he rebelled against such a
request in powerful language and terse legalistic argumentation.[62]
The painful feeling of "a total end to his counting for anything in
matters concerning *all* of philosophy" hung before his eyes, and the
sense of no longer being able to reach this goal never left him from
this time on; he himself called it, in a letter to Garve, a "sorrow of
Tantalus."[63] Despite the inward inclination that drove him forcefully
again and again back to the basic and main theme of this period, the
problem of the "transition from metaphysics to physics," he now
fended off questions about his philosophical labors, for the most part
with clear awareness and lack of pretense. "Oh, what it is like *sarcinas*

58. Schiller to Goethe, September 22, 1797.

59. *Fichtes Leben und literarischer Briefwechsel,* vol. 2, p. 451.

60. On this question, see the material in Arnoldt, *Beiträge zu dem Material der Geschichte von Kants Leben und Schriftstellertätigkeit,* and also Arthur Warda, *Altpreussische
Monatsschrift* 38:75 ff.

61. Letter to the Rector, February 26, 1796 (*Ak.* XII, 461).

62. To the Rector, December 3, 1797 (X, 330 f.) (*Ak.* XII, 463).

63. To Garve, September 21, 1798 (X, 351) (*Ak.* XII, 254).

colligere [to pack for the journey]! That is all I can think about now," he often said to his friends, by Borowski's account.[64]

It is a signal literary dispensation that we are so precisely and fully informed about no other portion of Kant's life as we are about this last one. In the reports of his faithful friend and nurse-companion Pastor Wasianski, poignant in their simplicity and calm objectivity, we can trace the distinct phases of his complete decline from year to year, almost from week to week. We do not need to go into the details of these reports here, though, since they do not transcend a mere record of illness. Wasianski reports the statement of a "passing scholar," who sought Kant out some two years before his death, that he had not seen Kant, but only Kant's shell.[65] Increasingly now Kant himself felt the pressure of such visits, to which more and more people were tempted, partly through personal interest, partly through sheer curiosity. "In me," he was wont to reply to the compliments of such visitors, "you see an old, decrepit, and feeble man, who has lived out his life." In December, 1803, he could no longer write his name, nor comprehend any expression of sociability; ultimately he began to fail to recognize those around him. Only the basic traits of his character remained true to him, even as his intellectual powers crumbled, and one can give all the more unqualified credence to what Wasianski tells us about this since his account overall is pitched in the tone of plain truth, scorning all decorative rhetorical trappings. "Every day," he recounts concerning his intercourse with Kant in the last years, "I profited; for daily I discovered one more lovable facet of his good heart; daily I received new assurances of his trust. . . . Kant's greatness as a scholar and thinker is known to the world, I cannot evaluate it; but no one has had such opportunity as I to observe the finest traits of his unassuming good nature." "Ever and again there were some moments when his great mind, though it no longer shone as blindingly as before, was nonetheless visible, and when his kind heart was even more luminous. In the hours when he was less bur-

64. Ludwig Ernst Borowski, *Darstellung des Lebens und Charakters Immanuel Kants* (Königsberg, 1804), p. 184.
65. Wasianski, p. 202.

dened by his weakness, he acknowledged every measure that alleviated his fate for him with heartfelt thanks to me and active thanks
to his servant, whose extremely burdensome labors and unwearied
loyalty he rewarded with considerable gifts."

There is a particular incident from the final days of Kant's life,
preserved by Wasianski, that makes the retention of the fine human
traits in Kant's personality more clearly visible than any merely indirect reference might. "On February 3," roughly a week before Kant's
death, "all urge to live seemed to be entirely slackened and to wane
completely, for from this day on he ate essentially nothing. His existence seemed to be only due to a kind of momentum of a motion
which had been going on for eighty years. His physician had arranged with me to visit him at a certain hour, and wished my presence close by. . . . When he arrived, Kant being almost unable to see
any longer, I said to him that his doctor had come. Kant stands up
from his chair, extends his hand to his doctor, and talks about *posts,*
repeats the word frequently in a tone as though he wants to be helped
out. The doctor calms him by saying that everything is taken care of
about the posts, because he takes this utterance for a delusion. Kant
says: 'many posts, troublesome posts,' then quickly: 'great kindness,'
shortly thereafter: 'thankfulness,' all this disjointed, yet with increasing warmth and a certain degree of consciousness of himself. I guessed his meaning quite well, however. He wanted to say that with his
many and troublesome posts, especially that of Rector, it was very
kind of his doctor to visit him. 'Exactly right' was Kant's reply; he was
still standing, and about to collapse from weakness. The doctor requested him to sit down. Kant hesitated uncertainly and uneasily. I
was too familiar with his way of thinking to have made any mistake
about the real cause of the delay, why Kant did not change his position which was fatiguing and weakened him. I made the doctor aware
of the true cause, to wit Kant's courteous way of thinking and civil
manners, and assured him that Kant would sit down just as soon as
he, the stranger, had first taken a chair. The doctor seemed dubious
about this reason, but he was quickly convinced of the truth of my
statement and moved almost to tears when Kant, having collected his
powers with main force, said: 'The sense of humanity has not yet

abandoned me.' That is a noble, refined, and good man! we cried to each other as one."

It is a chance utterance, arising from a particular situation, we are told of here, but it has a universal and symbolic value, seen in the context of Kant's personality. Kant's biographers say that at a time when it was very difficult for him to follow ordinary, everyday speech, his grasp of general ideas was retained undiminished: one needed only to turn the conversation to a general philosophical or scientific topic for him to be moved immediately to lively participation. Just as this trait testifies to the force and durability of fundamental theoretical ideas in Kant's mind, the uninterrupted controlling guidance of his will is again mirrored in what is told us about the expressions of his character in the last years. He was and he remained, as Wasianski puts it, "the man of determination, whose feeble foot often tottered, whose stern soul never." As difficult as it often was for him to grasp a simple decision relating to a present concrete situation, he persevered in his resolution, even under what were for him the most difficult circumstances, as soon as he had once laid hold of it and safeguarded it by a consciously formulated maxim.

Together with this energy and consistency of will, the essential tenderness of his personal nature came to light more and more as well. Charlotte von Schiller said about Kant that he would have been one of the greatest phenomena of mankind in general if he had been able to feel love; but since this was not the case there was something defective in his nature.[66] In fact, even in Kant's relation to people in his immediate surroundings, in all the sympathy and all the selfless devotion of which he was capable, a certain limit set by reason was never crossed, and this rational control, where one thinks himself justified in expecting and demanding a direct expression of emotion, can easily arouse the illusion of an impersonal coolness in reflection on human things and relationships. Actually, all emotions of the "softhearted kind," as he himself called them, were alien to Kant's disposition and nature. But all the more richly and delicately de-

---

66. On this and the following, cf. Otto Schöndörffer, "Kants Briefwechsel," *Altpreussische Monatsschrift* 38:120 ff.

veloped in him was the emotion which he himself regarded as the fundamental ethical emotion, and in which he thought he recognized the motive force for all concrete ethical behavior. His relationship to individuals was guided and ruled by universal respect for the freedom of the moral person and his right of self-determination. And this respect was no abstract demand, but it acted on him as an immediately living motive, determining every particular utterance. By this disposition Kant acquired that "courtesy of the heart" which, if not precisely the same as love, is nonetheless related to love. His "sense of humanity," which he held fast to and guarded until the last days of his life, was divorced from every merely sentimental subsoil. It was precisely in this respect that his particular special quality prevailed against his times and his environment, against the Age of Sensibility. Kant's attitude toward mankind was defined by the pure and abstract medium of the moral law; but even in this law itself he recognized and at the same time honored the highest force of human personality. Therefore the idea of humanity and of freedom was not for him a politico-social and pedagogic ideal, but it became the lever by which he displaced the entire intellectual and spiritual world, and lifted it from its hinges. The idea of the "primacy of practical reason" implied a transformation of the basic conception of theoretical reason itself: the new feeling and the new consciousness of humanity led to a universal "intellectual revolution," only in which did it find its final and decisive footing.

On the morning of February 12, 1804, Kant died. His funeral turned into a great public ceremony, in which the whole city and the inhabitants of all quarters of it took part. His body had been laid out in his home previously, and a great host of people "of the highest and lowest condition" streamed in to see it. "Everyone hastened to seize the last opportunity . . . for many days the pilgrimage went on, every day. . . . Many came back two and even three times, and many days later the public had not yet fully satisfied its desire to see him." The obsequies were organized by the university and by the students, who were intent on showing special honor to Kant. Amid the tolling of every bell in Königsberg, young students came to Kant's house to take up his body, from whence the innumerable procession, accompanied by thousands, wound to the university cathedral. Here it was laid in the so-called professor's vault; later a special hall, the Stoa Kantiana, was erected on this spot.

But splendid as its external forms were and great as was the participation in Kant's funeral, and however much "the clearest marks of universal reverence, ceremonial pomp, and taste," in Wasianski's phrase, were united in it, Kant himself had become almost a stranger to his environment and his native city when he died. In 1798—six years before his death—Poerschke had already written to Fichte that Kant, since he did not lecture any more and had withdrawn from all social intercourse save that in Motherby's house, was gradually becoming unknown even in Königsberg.[67] His name shone with the old luster; but his person had begun to be increasingly forgotten. The historical effect of his philosophy waxed greater and greater and the most distinctive part of his philosophy was being spread abroad by entire hosts; but in the final years of his life his personality already seemed to belong more to memory and to legend than to the actual historical present. And this, too, reveals a typical trait essential to Kant's life and significant for it. For the greatness and power of this life did not consist in all the personal and individual factors of Kant's mind and will achieving an ever-richer unfolding, but in its putting itself always more definitely and exclusively in the service of relevant demands, ideal problems, and tasks. The personal forms of life and of existence here had no independent worth purely as such; their whole significance is merged into their becoming the stuff and means of the life of abstract thinking, which moves according to its own law and by the force of its immanent necessity. On this relation of person and thing is founded the entire form and structure of Kant's life, is founded that which constitutes its profundity and that which might appear to be its peculiar limitation and narrowness. Full devotion to purely impersonal goals seems eventually to have as its inevitable consequence impoverishment of

---

67. Poerschke to Fichte, July 2, 1798; see *Fichtes Leben und literarischer Briefwechsel*, vol. 2, p. 451.

the concrete substance and the individual fullness of life, but on the other hand, it is only here that the full, compelling power of the universal emerges—that universal expressed equally as a theoretical and as a practical idea in the world of Kant's thought and of his will.

We recall with what power and freshness, with what immediate subjective vivacity Kant's fundamental orientation speaks even at the outset of his activity as philosopher and writer, in *Thoughts on the True Estimation of Living Forces.* "I have already marked out the road ahead," the young man of twenty-two wrote, "which I intend to follow. I shall embark on my course, and nothing shall hinder me from pursuing it."[68] Kant's thought had traversed this road in a far more comprehensive sense than his youthful enthusiasm could have foreseen. The path from detail and the particular to the whole, from the individual to the universal, had been trodden in the most diverse directions. His reflection had begun with the problem of cosmology and cosmogony, with the questions of how the world arose and how it is ordered. What was most important in that was to establish a new standpoint for judgment. It was not only necessary to go beyond direct sensory perception, which remains bound to spatiotemporal particularity, to the respective here and now, but also to supplement and deepen the mathematical scientific world-picture of Newton, since it took up the question of the temporal origin of the cosmos and at the same time created a new dimension of reflection. Only then did the empirical terrestrial horizon expand into the truly comprehensive and universal horizon of astronomical conception and judgment. An analogous broadening of the concept of human history then took place, in Kant's researches into the foundation of a physical geography and an empirical anthropology, since history was classified as a special case of the general problem of organic evolution and subordinated to it.

Kant's critical period retains this basic tendency, but it shifts the center of gravity from the "natural" to the "mental," from physics and biology to the realm of logic and ethics. Here too, by the indica-

68. See above, p. 31.

tion of their universally valid a priori foundations, the full force and depth of the authority of judgment and action are first brought into clear consciousness, but at the same time the boundaries that the application of these principles cannot overstep without becoming lost in the void are fixed. Both moments, that of establishing and that of limiting, are for Kant directly implied in each other, for only in binding the understanding and the will by a universal and necessary law is the objective order of the worlds of the understanding and of the will, on which their essential content rests, produced.

In the famous parallel he draws between Plato and Aristotle in his history of the theory of colors, Goethe compared two basic types of philosophic reflection with each other. "Plato comports himself in the world like a blessed spirit, whose will it is to sojourn in it a while. For him it is less a matter of learning to know the world, because he already assumes it, than of sharing with it as a friend what he brings with him and what it needs so sorely. He presses into the depths more to fill them with his own being than to explore them. He moves longingly to the heights, to participate once again in his source. Everything that he utters is related to an eternal One, Good, True, Beautiful, whose demands he strives to enliven in his bosom. . . . Aristotle, on the contrary, stands in the world like a man, an architect. He is just here, and is going to work and to produce here. He studies the earth, but no farther than until he strikes hard ground. From there to the center of the earth the rest is all the same to him. He traces round a monstrous circle for his foundation, gathers materials from all sides, sorts them, piles them up, and thus ascends to the heights like a pyramid, in a geometric form, while Plato reaches for the heavens like an obelisk, indeed like a pointed flame. When two such men, who apportion human nature between them to a certain extent, appeared as distinct representatives of glorious qualities which are not easy to combine, when they had the fortune to educate themselves fully, to utter their education completely, not in short laconic sentences like oracular sayings but in exceptional, extensive, numerous works; when these works for the best part remain to mankind and are more or less continuously studied and reflected on: it naturally follows that the world, insofar as it is regarded as feeling and thinking, was obliged to devote itself to one or the other, to acknowledge one or the other as master, teacher, leader."

It is indicative of the scope and depth of Kant's philosophical genius that, as regards the fundamental orientation of his mind, he is an exception to the universal contrast in intellectual history that Goethe expresses typically here. The alternatives posed here held no force and validity for him. In place of this struggle between philosophy's intellectual motivations in world history up to the present, in him there enters a unification that is novel in the world. If Plato and Aristotle seem to be divided into representatives of separate qualities in mankind, Kant, in his philosophical achievement, erects a new total conception of what man can do and attain in conceiving and performing, in thinking and doing. Perhaps in this lies the peculiar secret of the historical effect his doctrine has exercised. For in Kant the basic tendencies Goethe contrasted in his portrait of Aristotle and Plato join and fuse, and both are in such perfect equilibrium that one can hardly speak any longer of a contrary preeminence of one over the other. Kant felt himself a Platonist specifically in the foundations of his ethics, and in the *Critique of Pure Reason* he declared himself forcefully and decisively for the correctness of the Platonic "Idea" and against all objections to it stemming from the "vulgar appeal to so-called adverse experience."[69] But then a current of passing fashion tried to substitute for Plato the dialectician and moralist the mystical theologue, lauding Plato as a technician in this sense, as the philosopher of the supersensuous and of "intellectual intuition"; whereas Kant no less vigorously attached himself to the "worker" Aristotle, on whom that "philosophy with a condescending tone" thought it could look down. "It can occur to no other than the philosopher of *intuition*, who does not prove himself by the herculean labor of self-knowledge from below, but rather from above, soaring on high, by an apotheosis which costs him nothing, to give himself airs: because he speaks from his own insight and therefore is obligated to be called to account by no one." In contrast, the philosophy

69. See above, p. 253.

of Aristotle is hard work, for Aristotle's aim as metaphysician is directed in every instance toward division of a priori knowledge into its elements and toward its reconquest and reassembling from these elements, no matter whether and by what means he accomplishes it.[70]

The dual orientation of Kant's conception of philosophy is here indicated in a few words. The critical philosophy also strives from the empirical and sensory to the "intelligible," and it finds its fulfillment and its true conclusion only in the intelligible of the idea of freedom. But the path to this goal leads "through the herculean labor of self-knowledge." Accordingly, no "flights of genius" and no appeal to any sort of intuitive flashes have any weight here, but strict conceptual demands and necessities rule; here no immediate feeling of evidence, psychological or mystical, decides, but methodically performed scientific analysis and the "transcendental deduction" of the basic forms of knowledge. The genuine intelligible, which underlies experience, is only attained in the strengthening and securing, in the full critical understanding, of precisely this experience itself. Even this endeavor which leads on beyond experience to the supersensuous and to the Idea pulls back all the more deeply into the "fertile lowland of experience" to do this. It is even a proof of the power of the Idea and of idealism that both, in raising themselves beyond experience, only achieve complete understanding of the form of experience and the law of its structure in thus elevating themselves. The Idea strives into the absolute and unconditioned, but the critical attitude finds that the true unconditioned is never given, but is always imposed, and that in this sense it is one with the demand for the totality of conditions. Hence, to step into the infinite it suffices to penetrate the finite in all its aspects. Fully developed, the empirical itself guides us to "metaphysics"—as metaphysics, in the transcendental sense, should be presented and expressed as nothing else than the whole stuff of the empirical. The endeavor toward the unconditioned is innate and native to reason, but the total system of

70. See "On a Condescending Tone Recently Raised in Philosophy" (1796) (VI, 478, 482) (*Ak.* VIII, 390, 393).

conditions of theoretical and practical reason itself is shown to be the ultimate unconditioned to which we can advance. In this sense, the concepts of "what can be investigated" and "what cannot be investigated" are demarcated and defined in Kant's doctrine. Something that cannot be investigated remains unknown; it is no longer a mere negation, however, but rather it becomes the rule of knowledge and action. It is no longer the expression of an impotent and hopeless scepticism, but it aims to point out the path and the direction in which inquiry has to move and by which it has to unfold itself comprehensively. Thus, in the truly intelligible, in the intelligible of reason's task, the world of being is transformed into a world of deed. In this new relation between the conditioned and the unconditioned, between the finite and the infinite, between experience and speculation, Kant has wrought a new type of philosophical thinking in contrast to that of Plato and Aristotle: in him, the specifically modern conception of idealism, inaugurated by Descartes and Leibniz, achieves its systematic perfection and fulfillment.

INDEX